A Note from the Authors

You are about to embark on an amazing journey of discovery. The study of life spans from the inner workings of cells to the complex interactions of entire ecosystems, through the information stored in DNA to the ways genetic information evolves over time. At the same time that our understanding of biology is growing in leaps and bounds, so too are great insights into how learners acquire new knowledge and skills. We are thrilled to join Scott Freeman on *Biological Science*, a book dedicated to active, research-based learning and to exploring the experimental evidence that informs what we know about biology. The next few pages highlight the features in this book and in MasteringBiology® that will help you succeed.

From left to right: Michael Black, Emily Taylor, Jon Monroe, Lizabeth Allison, Greg Podgorski, Kim Quillin

To the Student: How to Use This Book

New chapter-opening Roadmaps visually group and organize information to help you anticipate key ideas as well as recognize meaningful relationships and connections between them.

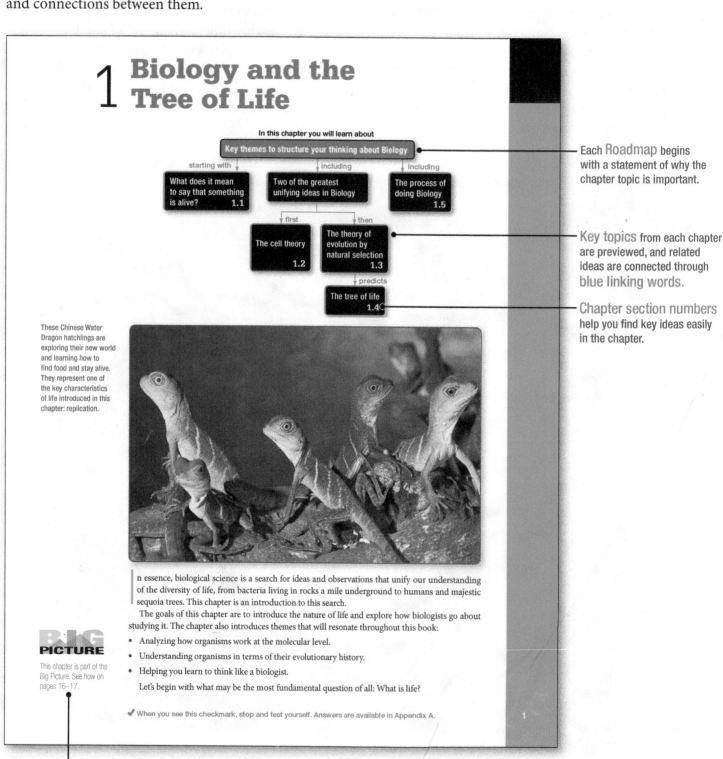

1 Biology and the Tree of Life

In this chapter you will learn about

Key themes to structure your thinking about Biology

starting with

What does it mean to say that something is alive? **1.1**

including

Two of the greatest unifying ideas in Biology

including

The process of doing Biology **1.5**

first

The cell theory **1.2**

then

The theory of evolution by natural selection **1.3**

predicts

The tree of life **1.4**

Each Roadmap begins with a statement of why the chapter topic is important.

Key topics from each chapter are previewed, and related ideas are connected through blue linking words.

Chapter section numbers help you find key ideas easily in the chapter.

These Chinese Water Dragon hatchlings are exploring their new world and learning how to find food and stay alive. They represent one of the key characteristics of life introduced in this chapter: replication.

In essence, biological science is a search for ideas and observations that unify our understanding of the diversity of life, from bacteria living in rocks a mile underground to humans and majestic sequoia trees. This chapter is an introduction to this search.

The goals of this chapter are to introduce the nature of life and explore how biologists go about studying it. The chapter also introduces themes that will resonate throughout this book:

- Analyzing how organisms work at the molecular level.
- Understanding organisms in terms of their evolutionary history.
- Helping you learn to think like a biologist.

Let's begin with what may be the most fundamental question of all: What is life?

BIG PICTURE

This chapter is part of the Big Picture. See how on pages 16–17.

✔ When you see this checkmark, stop and test yourself. Answers are available in Appendix A.

1

Big Picture Concept Maps are referenced on the opening page of related chapters, pointing you to summary pages that help you synthesize challenging topics.

Big Picture Concept Maps integrate visuals and words to help you synthesize information about challenging topics in biology that span multiple chapters and units.

Three New Big Picture topics
have been added to the Fifth Edition:

- NEW! Doing Biology
- NEW! The Chemistry of Life
- Energy for Life
- Genetic Information
- Evolution
- NEW! Plant and Animal Form and Function
- Ecology

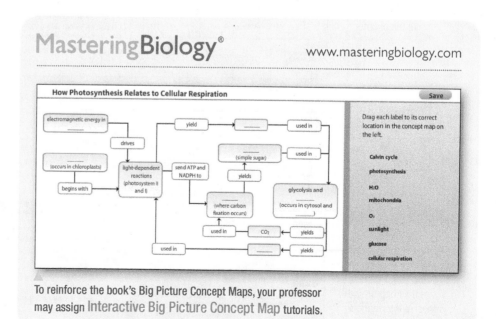

To reinforce the book's Big Picture Concept Maps, your professor may assign Interactive Big Picture Concept Map tutorials.

Practice for success
on tests and exams

Intertwined color-coded "active learning threads" are embedded in the text. The gold thread helps you to identify important ideas, and the blue thread helps you to test your understanding.

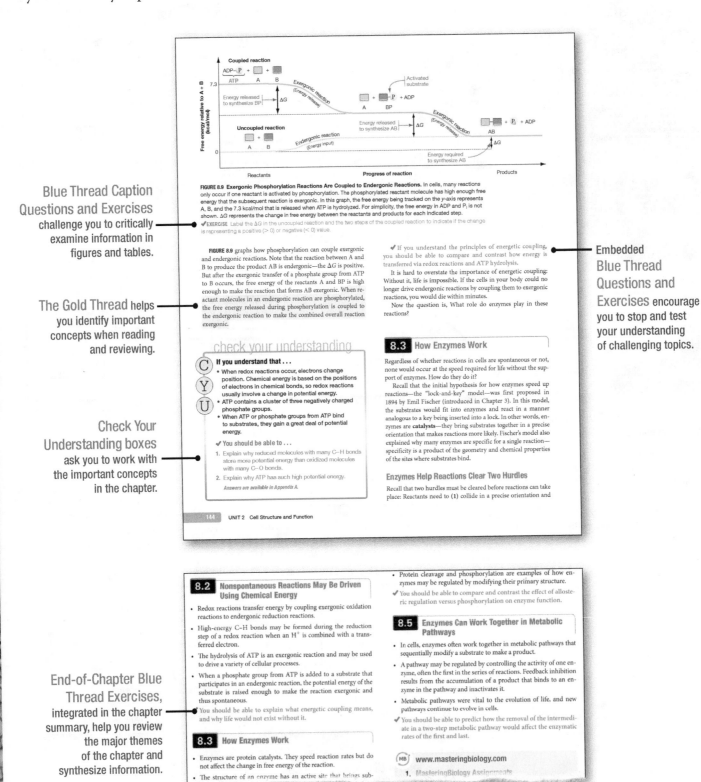

FIGURE 8.9 Exergonic Phosphorylation Reactions Are Coupled to Endergonic Reactions. In cells, many reactions only occur if one reactant is activated by phosphorylation. The phosphorylated reactant molecule has high enough free energy that the subsequent reaction is exergonic. In this graph, the free energy being tracked on the y-axis represents A, B, and the 7.3 kcal/mol that is released when ATP is hydrolyzed. For simplicity, the free energy in ADP and P_i is not shown. ΔG represents the change in free energy between the reactants and products for each indicated step.

✔**EXERCISE** Label the ΔG in the uncoupled reaction and the two steps of the coupled reaction to indicate if the change is representing a positive (> 0) or negative (< 0) value.

Blue Thread Caption Questions and Exercises challenge you to critically examine information in figures and tables.

FIGURE 8.9 graphs how phosphorylation can couple exergonic and endergonic reactions. Note that the reaction between A and B to produce the product AB is endergonic—the ΔG is positive. But after the exergonic transfer of a phosphate group from ATP to B occurs, the free energy of the reactants A and BP is high enough to make the reaction that forms AB exergonic. When reactant molecules in an endergonic reaction are phosphorylated, the free energy released during phosphorylation is coupled to the endergonic reaction to make the combined overall reaction exergonic.

The Gold Thread helps you identify important concepts when reading and reviewing.

✔ If you understand the principles of energetic coupling, you should be able to compare and contrast how energy is transferred via redox reactions and ATP hydrolysis.

It is hard to overstate the importance of energetic coupling: Without it, life is impossible. If the cells in your body could no longer drive endergonic reactions by coupling them to exergonic reactions, you would die within minutes.

Now the question is, What role do enzymes play in these reactions?

Embedded Blue Thread Questions and Exercises encourage you to stop and test your understanding of challenging topics.

check your understanding

C Y U

If you understand that . . .
- When redox reactions occur, electrons change position. Chemical energy is based on the positions of electrons in chemical bonds, so redox reactions usually involve a change in potential energy.
- ATP contains a cluster of three negatively charged phosphate groups.
- When ATP or phosphate groups from ATP bind to substrates, they gain a great deal of potential energy.

✔ **You should be able to . . .**
1. Explain why reduced molecules with many C–H bonds store more potential energy than oxidized molecules with many C–O bonds.
2. Explain why ATP has such high potential energy.

Answers are available in Appendix A.

Check Your Understanding boxes ask you to work with the important concepts in the chapter.

8.3 How Enzymes Work

Regardless of whether reactions in cells are spontaneous or not, none would occur at the speed required for life without the support of enzymes. How do they do it?

Recall that the initial hypothesis for how enzymes speed up reactions—the "lock-and-key" model—was first proposed in 1894 by Emil Fischer (introduced in Chapter 3). In this model, the substrates would fit into enzymes and react in a manner analogous to a key being inserted into a lock. In other words, enzymes are **catalysts**—they bring substrates together in a precise orientation that makes reactions more likely. Fischer's model also explained why many enzymes are specific for a single reaction—specificity is a product of the geometry and chemical properties of the sites where substrates bind.

Enzymes Help Reactions Clear Two Hurdles

Recall that two hurdles must be cleared before reactions can take place: Reactants need to **(1)** collide in a precise orientation and

8.2 Nonspontaneous Reactions May Be Driven Using Chemical Energy

- Redox reactions transfer energy by coupling exergonic oxidation reactions to endergonic reduction reactions.
- High-energy C–H bonds may be formed during the reduction step of a redox reaction when an H^+ is combined with a transferred electron.
- The hydrolysis of ATP is an exergonic reaction and may be used to drive a variety of cellular processes.
- When a phosphate group from ATP is added to a substrate that participates in an endergonic reaction, the potential energy of the substrate is raised enough to make the reaction exergonic and thus spontaneous.

✔ You should be able to explain what energetic coupling means, and why life would not exist without it.

End-of-Chapter Blue Thread Exercises, integrated in the chapter summary, help you review the major themes of the chapter and synthesize information.

8.3 How Enzymes Work

- Enzymes are protein catalysts. They speed reaction rates but do not affect the change in free energy of the reaction.
- The structure of an enzyme has an active site that brings sub-

- Protein cleavage and phosphorylation are examples of how enzymes may be regulated by modifying their primary structure.

✔ You should be able to compare and contrast the effect of allosteric regulation versus phosphorylation on enzyme function.

8.5 Enzymes Can Work Together in Metabolic Pathways

- In cells, enzymes often work together in metabolic pathways that sequentially modify a substrate to make a product.
- A pathway may be regulated by controlling the activity of one enzyme, often the first in the series of reactions. Feedback inhibition results from the accumulation of a product that binds to an enzyme in the pathway and inactivates it.
- Metabolic pathways were vital to the evolution of life, and new pathways continue to evolve in cells.

✔ You should be able to predict how the removal of the intermediate in a two-step metabolic pathway would affect the enzymatic rates of the first and last.

MB www.masteringbiology.com

1. MasteringBiology Assignments

Identify gaps in your understanding, then fill them

The Fifth Edition provides many opportunities for you to test yourself and offers helpful learning strategies.

◀ **Bloom's Taxonomy** describes six learning levels: Remember, Understand, Apply, Analyze, Evaluate, and Create. Questions in the book span all levels, including self-testing at the higher levels to help you develop higher-order thinking skills that will prepare you for exams.

Analyze: Can I recognize underlying patterns and structure?

Evaluate: Can I make judgments on the relative value of ideas and information?

Create: Can I put ideas and information together to generate something new?

Apply: Can I use these ideas in the same way or in a new situation?

Understand: Can I explain this concept in my own words?

Remember: Can I recall the key terms and ideas?

Steps to Building Understanding
Each chapter ends with three groups of questions that build in difficulty:

✔ TEST YOUR KNOWLEDGE

Begin by testing your basic knowledge of new information.

✔ TEST YOUR UNDERSTANDING

Once you're confident with the basics, demonstrate your deeper understanding of the material.

✔ TEST YOUR PROBLEM-SOLVING SKILLS

Work towards mastery of the content by answering questions that challenge you at the highest level of competency.

BIOSKILL 16   using Bloom's taxonomy

Most students have at one time or another wondered why a particular question on an exam seemed so hard, while others seemed easy. The explanation lies in the type of cognitive skills required to answer the question. Let's take a closer look.

▲
NEW! BioSkill Covering Bloom's Taxonomy helps you to recognize question types using the Bloom's cognitive hierarchy, and it provides specific strategies to help you study for questions at all six levels.

Answer Appendix Includes Bloom's Taxonomy Information
Answers to all questions in the text now include the Bloom's level being tested. You can simultaneously practice assessing your understanding of content and recognizing Bloom's levels. Combining this information with the guidance in the BioSkill on Bloom's Taxonomy will help you form a plan to improve your study skills.

▶
✔ Test Your Problem-Solving Skills

13. analyze A scientific theory is not a guess—it is an idea whose validity can be tested with data. Both the cell theory and the theory of evolution have been validated by large bodies of observational and experimental data.
14. apply If all eukaryotes living today have a nucleus, then it is logical to conclude that the nucleus arose in a common ancestor of all eukaryotes, indicated by the arrow you should have added to the figure. See **FIGURE A1.2**. If it had arisen in a common ancestor of Bacteria or Archaea, then species in those groups would have had to lose the trait—an unlikely event.
15. evaluate The data set was so large and diverse that it was no longer reasonable to argue that noncellular life-forms would be discovered. **16.** apply b

---

MasteringBiology®
www.masteringbiology.com

NEW! End-of-chapter questions from the book are now available for your professor to assign as homework in MasteringBiology.

Practice scientific thinking and scientific skills

A unique emphasis on the process of scientific discovery and experimental design teaches you how to think like a scientist as you learn fundamental biology concepts.

RESEARCH

QUESTION: Do horses minimize the cost of locomotion?

HYPOTHESIS: Horses choose gaits that minimize energy use at different speeds.

NULL HYPOTHESIS: Horses do not choose gaits based on cost of locomotion.

EXPERIMENTAL SETUP:

1. Measure oxygen consumption of horses trained to walk, trot, and gallop at a range of speeds on a treadmill. Calculate energy used per distance travelled at different speeds.

2. Videotape the same horses locomoting freely in the paddock, and measure the gaits and speeds they choose to use naturally.

PREDICTION: For each gait, there is a range of speeds where energy use is minimized. Horses will favor these gaits and speeds.

PREDICTION OF NULL HYPOTHESIS: There will be no correlation between chosen gaits and energy consumption.

RESULTS:

CONCLUSION: Horses choose gaits that minimize energy use at different speeds and avoid speeds with high energy consumption.

FIGURE 48.16 Horses Minimize the Cost of Locomotion by Choosing Appropriate Gaits.
SOURCE: Hoyt, D. F., and C. R. Taylor. 1981. Gait and the energetics of locomotion in horses. Nature 292: 239–240.

✔ **QUANTITATIVE** Use the graph to estimate the relative energy expense of galloping rather than trotting at 3.5 meters/second (m/s).

All of the Research Boxes cite the original research paper and include a question that asks you to analyze the design of the experiment or study.

Research Boxes explain how research studies are designed and give you additional practice interpreting data. Each Research Box consistently models the scientific method, presenting the research question, hypotheses, experimental setup, predictions, results, and conclusion. 15 Research Boxes are new to the Fifth Edition.

MasteringBiology®
www.masteringbiology.com

NEW! Solve It Tutorials are available for homework assignments in MasteringBiology and give you an opportunity to work like a scientist through a simulated investigation that requires you to analyze and interpret data.

Experimental Inquiry Tutorials based on some of biology's most seminal experiments give you a chance to analyze data and the reasoning that led scientists from the data to their conclusions.

Experimental Inquiry tutorial topics include:

- What Can You Learn About the Process of Science from Investigating a Cricket's Chirp?
- Which Wavelengths of Light Drive Photosynthesis?
- What Is the Inheritance Pattern of Sex-Linked Traits?
- Does DNA Replication Follow the Conservative, Semiconservative, or Dispersive Model?
- How Do Calcium Ions Help to Prevent Polyspermy During Egg Fertilization?

- Did Natural Selection of Ground Finches Occur When the Environment Changed?
- What Effect Does Auxin Have on Coleoptile Growth?
- What Role Do Genes Play in Appetite Regulation?
- Can a Species' Niche Be Influenced by Interspecific Competition?
- What Factors Influence the Loss of Nutrients from a Forest Ecosystem?

Build important skills scientists use to perform, evaluate, and communicate scientific research.

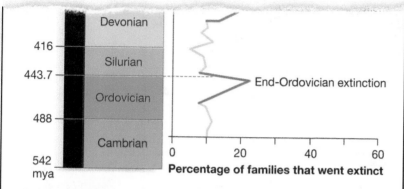

FIGURE 28.14 The Big Five Mass Extinction Events. This graph shows the percentage of lineages called families that went extinct over each interval in the fossil record since the Cambrian explosion. Over 50 percent of families and 90 percent of species went extinct during the end-Permian extinction.

DATA: Benton, M. J., 1995. *Science* 268: 52–58.

✔**QUANTITATIVE** Which extinction event ended the era of the dinosaurs 65 million years ago? About what percentage of families went extinct?

NEW! Graphs and tables now include their data sources, emphasizing the research process that leads to our understanding of biological ideas.

NEW! Quantitative questions are identified throughout the text, helping you practice computational problem solving and data analysis.

Expanded BioSkills Appendix helps you build skills that will be important to your success in biology. At relevant points in the text, you'll find references to the BioSkills appendix that will help you learn and practice foundational skills.

BioSkills Topics include:

- The Metric System and Significant Figures
- Some Common Latin and Greek Roots Used in Biology
- Reading Graphs
- Using Statistical Tests and Interpreting Standard Error Bars
- Combining Probabilities

- Using Logarithms
- Reading a Phylogenetic Tree
- Reading Chemical Structures
- Separating and Visualizing Molecules
- Separating Cell Components by Centrifugation
- Biological Imaging: Microscopy and X-ray Crystallography

- Cell and Tissue Culture Methods
- Model Organisms
- NEW! Primary Literature and Peer Review
- Making Concept Maps
- NEW! Using Bloom's Taxonomy

MasteringBiology®
www.masteringbiology.com

You can access self-paced BioSkills activities in the Study Area, and your instructor can assign additional activities in MasteringBiology.

Visualize biology processes and structures

A carefully crafted visual program helps you gain a better understanding of biology through accurate, appropriately detailed figures.

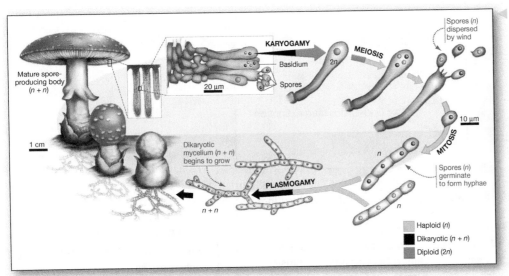

◀ NEW! Redesigned Life Cycle diagrams in Unit 6 and 7 help you compare and contrast processes among different organisms.

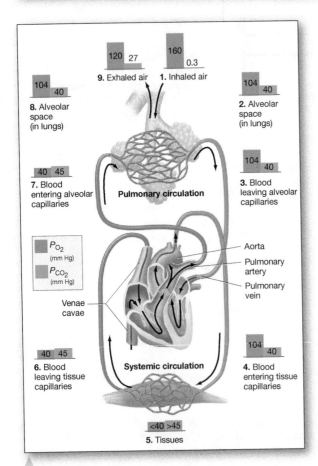

▲ Informative figures help you think through complex biological processes in manageable steps.

SUMMARY TABLE 28.3 **Branch Lengths in Phylogenetic Trees**

Graphical Symbol	Meaning of Branch Lengths
	Branch lengths are *arbitrary*. Emphasis is on the branching *pattern*, which estimates evolutionary relationships among populations. This is the type of tree used in this book.
0.1 Number of base substitutions per site	Horizontal branch lengths show the extent of *genetic difference* among populations. A scale bar is included.
4 3 2 1 0 Millions of years ago	Horizontal branch lengths show the extent of *evolutionary time* between nodes. A scale bar is included.

▲ Visual Summary Tables pull together important information in a format that allows for easy comparison and review.

Instructor and Student Resources

For Instructors

Instructor Resource DVD-ROM
978-0-321-86112-2 • 0-321-86112-4

Everything you need for lectures in one place, including video segments that demonstrate how to incorporate active-learning techniques into your own classroom. Enhanced menus make locating and assessing the digital resources for each chapter easy. The Instructor Resource CD/DVD-ROM includes PowerPoint® Lecture Outlines that integrate figures and animations for classroom presentations. All textbook figures, art, and photos are in JPEG format, and all PowerPoint slides and JPEGs have editable labels. Over 300 Instructor Animations accurately depict complex topics and dynamic processes described in the book.

Instructor Guide (Download only)

Available in the instructor resource area of MasteringBiology.®

TestGen® (Download only)

All of the exam questions in the Test Bank have been peer reviewed and student tested, providing questions that set the standard for quality and accuracy. To improve the Test Bank, Metadata from MasteringBiology users has been incorporated directly into the software. Test questions that are ranked according to Bloom's taxonomy and improved TestGen® software makes assembling tests that much easier.

For Students

Study Guide
978-0-321-85832-0 • 0-321-85832-8

The Study Guide presents a breakdown of key biological concepts, difficult topics, and quizzes to help students prepare for exams. Unique to this study guide are four introductory, stand-alone chapters that introduce students to foundational ideas and skills necessary for classroom success: Introduction to Experimentation and Research in the Biological Sciences, Presenting Biological Data, Understanding Patterns in Biology and Improving Study Techniques, and Reading and Writing to Understand Biology. "Looking Forward" and "Looking Back" sections help students make connections across the chapters instead of viewing them as discrete entities.

Practicing Biology: A Student Workbook
978-0-321-88647-7 • 0-321-88647-X

This workbook focuses on key ideas, principles, and concepts that are fundamental to understanding biology. A variety of hands-on activities such as mapping and modeling suit different learning styles and help students discover which topics they need more help on. Students learn biology by doing biology. An instructors guide can be down-loaded from the Instructor Area of MasteringBiology.

MasteringBiology®

www.masteringbiology.com

MasteringBiology is an online homework, tutorial, and assessment system that delivers self-paced tutorials that provide individualized coaching, focus on your course objectives, and respond to each student's progress. The Mastering system helps instructors maximize class time with customizable, easy-to-assign, and automatically graded assessments that motivate students to learn outside of class and arrive prepared for lecture. MasteringBiology is also available with a complete Pearson eText edition of *Biological Science*.

Highlights of the Fifth Edition Item Library include:

* NEW! **assignment options** include Solve It activities, end-of-chapter problems, and questions that accompany new BioSkills and new Big Picture Interactive Concept Maps.
* NEW! **"best of" homework pre-built assignments** help professors assign popular, key content quickly, including a blend of tutorials, end-of-chapter problems, and test bank questions.

* *Get Ready for Biology* **and Chemistry Review assignment options** help students get up to speed with activities that review chemistry, mathematics, and basic biology.

MasteringBiology® Virtual Labs

978-0-321-88644-6 • 0-321-88644-5

MasteringBiology: Virtual Labs is an online environment that promotes critical-thinking skills using virtual experiments and explorations that might be difficult to perform in a wet-lab environment due to time, cost, or safety concerns. MasteringBiology: Virtual Labs offers unique learning experiences in the areas of microscopy, molecular biology, genetics, ecology, and systematics.

For more information, please visit
www.pearsonhighered.com/virtualbiologylabs

BIOLOGICAL SCIENCE

VOLUME ❸ How Plants & Animals Work

After you explore this book . . . you should be able to . . .

• Pose an evolutionary hypothesis to explain why meter-long male water dragons are larger and have more colorful throats than the females.

• Propose how DNA sequences could be used to determine the relationships among populations of these lizards in China, India, and Southeast Asia.

• Design an experiment to study the relative importance of swimming, tree climbing, and running to the ability of these semi-aquatic lizards to find food and escape from predators.

• Create questions of your own and suggest methods for finding the answers!

Chinese Water Dragon,
Physignathus cocincinus

BIOLOGICAL SCIENCE

FIFTH EDITION

SCOTT FREEMAN
University of Washington

LIZABETH ALLISON
College of William & Mary

MICHAEL BLACK
California Polytechnic State University in San Luis Obispo

GREG PODGORSKI
Utah State University

KIM QUILLIN
Salisbury University

JON MONROE
James Madison University

EMILY TAYLOR
California Polytechnic State University in San Luis Obispo

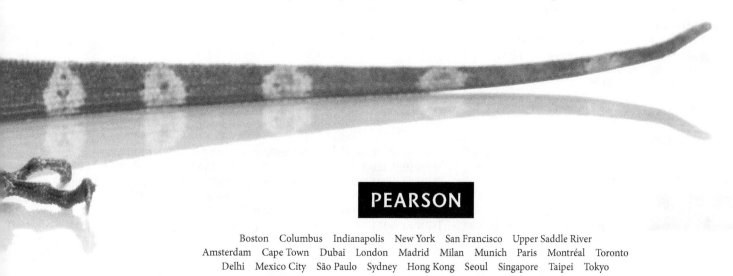

PEARSON

Boston   Columbus   Indianapolis   New York   San Francisco   Upper Saddle River
Amsterdam   Cape Town   Dubai   London   Madrid   Milan   Munich   Paris   Montréal   Toronto
Delhi   Mexico City   São Paulo   Sydney   Hong Kong   Seoul   Singapore   Taipei   Tokyo

Editor-in-Chief: Beth Wilbur
Senior Acquisitions Editor: Michael Gillespie
Executive Director of Development: Deborah Gale
Senior Development Editor: Sonia DiVittorio
Project Editor: Anna Amato
Development Editors: Mary Catherine Hager, Moira Lerner Nelson, Bill O'Neal
Art Development Editors: Fernanda Oyarzun, Adam Steinberg
Associate Editor: Brady Golden
Assistant Editor: Leslie Allen
Editorial Assistant: Eddie Lee
Executive Media Producer: Laura Tommasi
Media Producer: Joseph Mochnick
Associate Media Producer: Daniel Ross
Associate Media Project Manager: David Chavez
Text Permissions Project Manager: Joseph Croscup
Text Permissions Specialist: Sheri Gilbert
Director of Production: Erin Gregg
Managing Editor: Michael Early

Production Project Manager: Lori Newman
Production Management: S4Carlisle Publishing Services
Copyeditor: Christianne Thillen
Compositor: S4Carlisle Publishing Services
Design Manager: Mark Ong
Interior Designer: Integra Software Services
Cover Designer: tt eye
Illustrators: Imagineering Media Services
Photo Permissions Management: Phutu
Photo Researchers: Kristin Piljay, Eric Schrader, Maureen Spuhler
Senior Photo Editor: Travis Amos
Manufacturing Buyer: Michael Penne
Director of Marketing: Christy Lesko
Executive Marketing Manager: Lauren Harp
Sales Director for Key Markets: David Theisen

Cover Photo Credit: *Physignathus cocincinus*
Eric Isselée/Fotolia

Library of Congress Cataloging-in-Publication Data
Freeman, Scott, 1955- Biological science / Scott Freeman.—Fifth edition.
    pages cm
  ISBN-13: 978-0-321-74367-1 (student edition)
  ISBN-10: 0-321-74367-9 (student edition)
  ISBN-13: 978-0-321-84159-9 (instructors review copy)
  ISBN-10: 0-321-84159-X (instructors review copy) [etc.]
  1. Biology—Textbooks.   I. Title.
  QH308.2.F73 2014
  570—dc23

3 4 5 6 7 8 9 10—V0UD—16 15 14

ISBN 10: 0-321-74367-9; ISBN 13: 978-0-321-74367-1 (Student Edition)
ISBN 10: 0-321-84159-X; ISBN 13: 978-0-321-84159-9 (Instructor's Review Copy)
ISBN 10: 0-321-86216-3; ISBN 13: 978-0-321-86216-7 (Books a la Carte Edition)
ISBN 10: 0-321-84180-8; ISBN 13: 978-0-321-84180-3 (Volume 1)
ISBN 10: 0-321-84181-6; ISBN 13: 978-0-321-84181-0 (Volume 2)
ISBN 10: 0-321-84182-4; ISBN 13: 978-0-321-84182-7 (Volume 3)

www.pearsonhighered.com

Detailed Contents

5 An Introduction to Carbohydrates 72

6 Lipids, Membranes, and the First Cells 84

UNIT 2 CELL STRUCTURE AND FUNCTION 106

7 Inside the Cell 106

8 Energy and Enzymes: An Introduction to Metabolic Pathways 136

30 Protists 552

31 Green Algae and Land Plants 577

About the Authors

A Letter from Scott:

I started working on *Biological Science* in 1997 with a simple goal: To help change the way biology is taught. After just shy of 20,000 hours of work on four editions of this text, that goal still gets me out of bed in the morning. But instead of focusing my energies on textbook writing, I've decided to devote myself full-time to research on student learning and developing new courses for undergraduate and graduate students at the University of Washington.

So with this edition I am passing the torch—to an all-star cast of leading scientists and educators who have enthusiastically taught from, and contributed to, previous editions of *Biological Science*. Working with them, I have seen the new team bring their passion, talent, and creativity to the book, with expertise that spans the breadth of the life sciences. Just as important, they work beautifully together because they think alike. They are driven by a shared concern for student learning, a commitment to the craft of writing, and a background in evidence-based teaching.

These pages provide a brief introduction to Liz Allison, Michael Black, Greg Podgorski, Kim Quillin, Jon Monroe, and Emily Taylor. As a group, they've built on the book's existing strengths and infused this edition with fresh energy, perspective, and ideas. I'm full of admiration for what they have accomplished, and excited about the impact this edition will have on biology students from all over the world.—*Scott Freeman*

Scott Freeman received a Ph.D. in Zoology from the University of Washington and was subsequently awarded an Alfred P. Sloan Postdoctoral Fellowship in Molecular Evolution at Princeton University. He has done research in evolutionary biology on topics ranging from nest parasitism to the molecular systematics of the blackbird family and is coauthor, with Jon Herron, of the standard-setting undergraduate text *Evolutionary Analysis*. Scott is the recipient of a Distinguished Teaching Award from the University of Washington and is currently a Senior Lecturer in the UW Department of Biology, where he teaches introductory biology for majors, a writing-intensive course for majors called The Tree of Life, and a graduate seminar in college science teaching. Scott's current research focuses on how active learning affects student learning and academic performance.

Lizabeth A. Allison is professor and chair of the Biology Department at the College of William & Mary. She received her Ph.D. in Zoology from the University of Washington, specializing in molecular and cellular biology. Before coming to William & Mary, she spent eight years as a faculty member at the University of Canterbury in New Zealand. Liz teaches introductory biology for majors and upper-division molecular biology courses. She has mentored graduate students and more than 80 undergraduate research students, many of them coauthoring papers with her on intracellular trafficking of the thyroid hormone receptor in normal and cancer cells. The recipient of numerous awards, including a State Council for Higher Education in Virginia (SCHEV) Outstanding Faculty Award in 2009, Liz received one of the three inaugural Arts & Sciences Faculty Awards for Teaching Excellence in 2011, and a Plumeri Award for Faculty Excellence in 2012. In addition to her work on this text, she is author of *Fundamental Molecular Biology*, now in its second edition.
Lead Author; Chapter 1 and BioSkills
laalli@wm.edu

Michael Black received his Ph.D. in Microbiology & Immunology from Stanford University School of Medicine as a Howard Hughes Predoctoral Fellow. After graduation, he studied cell biology as a Burroughs Wellcome Postdoctoral Fellow at the MRC Laboratory of Molecular Biology in Cambridge, England. His current research focuses on the use of molecules to identify and track the transmission of microbes in the environment. Michael is a professor of Cell & Molecular Biology at California Polytechnic State University in San Luis Obispo, where he teaches introductory and advanced classes for majors in cell biology and microbiology. In addition to his teaching and research activities, Michael serves as the director of the Undergraduate Biotechnology Lab, where he works alongside undergraduate technicians to integrate research projects and inquiry-based activities into undergraduate classes.
Chapters 2–12, 36, and 51
mblack@calpoly.edu

Greg Podgorski received his Ph.D. in Molecular and Cellular Biology from Penn State University and has been a postdoctoral fellow at the Max Plank Institute for Biochemistry and Columbia University. His research interests are in biology education, developmental genetics, and computational biology. Greg's most recent work has been in mathematical modeling of how patterns of different cell types emerge during development and how tumors recruit new blood vessels in cancer. Greg has been teaching at Utah State University for more than 20 years in courses that include introductory biology for majors and for nonmajors, genetics, cell biology, developmental biology, and microbiology, and he has offered courses in nonmajors biology in Beijing and Hong Kong. He's won teaching awards at Utah State University and has been recognized by the National Academies as a Teaching Fellow and a Teaching Mentor.

Chapters 13–24
greg.podgorski@usu.edu

Jon Monroe is professor of Biology at James Madison University in Harrisonburg, Virginia. Jon completed his undergraduate work in Botany at the University of Michigan and his graduate work in Plant Physiology at Cornell University. He began his current position after a postdoc in biochemistry at Michigan State University. He currently teaches Plant Biology, and Cell and Molecular Biology. Jon's interest in plants is broad, ranging from systematics and taxonomy to physiology and biochemistry. His research, mostly with undergraduates, uses Arabidopsis thaliana to study the functions of a family of β-amylase genes in starch metabolism. Jon has been active in promoting undergraduate research through his work with the American Society of Plant Biologists (ASPB) and the Council on Undergraduate Research. He has received ASPB's Excellence in Teaching award and James Madison University Alumni Association's Distinguished Faculty Award.

Chapters 29–32; 37–41
monroejd@jmu.edu

Kim Quillin received her B.A. in Biology at Oberlin College *summa cum laude* and her Ph.D. in Integrative Biology from the University of California, Berkeley (as a National Science Foundation Graduate Fellow). Kim has worked in the trenches with Scott Freeman on every edition of *Biological Science*, starting with the ground-up development of the illustrations in the first edition in 1999 and expanding her role in each edition, always with the focus of helping students to think like biologists. Kim currently teaches introductory biology at Salisbury University, a member of the University System of Maryland, where she is actively involved in the ongoing student-centered reform of the concepts-and-methods course for biology majors. Her current research focuses on the scholarship of teaching and learning with an emphasis on measuring science process skills and the advantages and pitfalls of active multimedia learning.

Chapters 25–28; 33–35; 48; 52–57
kxquillin@salisbury.edu

Emily Taylor earned a B.A. in English at the University of California, Berkeley followed by a Ph.D. in Biological Sciences from Arizona State University, where she conducted research in the field of environmental physiology as a National Science Foundation Graduate Research Fellow. She is currently an associate professor of Biological Sciences at the California Polytechnic State University in San Luis Obispo, California. Her student-centered research program focuses on the endocrine and reproductive physiology of free-ranging reptiles, especially rattlesnakes. She teaches numerous undergraduate and graduate courses, including introductory biology, anatomy and physiology, and herpetology, and received the California Faculty Association's Distinguished Educator Award in 2010 and Cal Poly's Distinguished Teaching Award in 2012. Her revision of Unit 8 is her first foray into textbook writing.

Chapters 42–50
etaylor@calpoly.edu

Preface to Instructors

The first edition of *Biological Science* was visionary in its unique emphasis on the process of scientific discovery and experimental design—teaching how we know what we know. The goal was for students not only to learn the language of biology and understand fundamental concepts but also to begin to apply those concepts in new situations, analyze experimental design, synthesize results, and evaluate hypotheses and data—to learn how to think like biologists. Each edition since has proudly expanded on this vision. The Fifth Edition is no exception.

A team of six dedicated teacher-scholars has joined Scott to build on and refine the original vision, and by so doing, make the book an even better teaching and learning tool. The pace of biological discovery is rapid, and with each novel breakthrough it becomes even more challenging to decide what is essential to include in an introductory biology text. Pulling together an author team with firsthand expertise from molecules to ecosystems has ensured that the content of the Fifth Edition reflects cutting-edge biology that is pitched at the right level for introductory students and is as accurate and as exciting as ever for instructors and students alike.

New findings from education research continue to inform and inspire the team's thinking about *Biological Science*—we know more today than ever before about how students learn. These findings demand that we constantly look for new ways to increase student engagement in the learning process, and to help instructors align course activities and learning goals with testing strategies.

The New Coauthors

The new coauthor team brings a broad set of talents and interests to the project, motivated by a deep commitment to undergraduate teaching, whether at a small liberal arts college or a large university. Kim Quillin has been a partner in this textbook in every edition. For the Fifth Edition, she revised chapters across three units in addition to spearheading the continued effort to enhance the visual-teaching program. Michael Black, Greg Podgorski, Jon Monroe, and Emily Taylor, who served as unit advisors on the Fourth Edition, were already familiar with the book. And most of the authorial team have been avid users of previous editions for many years.

Core Values

Together, the coauthor team has worked to extend the vision and maintain the core values of *Biological Science*—to provide a book for instructors who embrace the challenge of boosting students to higher levels of learning, and to provide a book for

students that helps them each step of the way in learning to think like scientists. Dedicated instructors have high expectations of their students—the Fifth Edition provides scaffolding to help students learn at the level called for by the National Academy of Sciences, the Howard Hughes Medical Institute, the American Association of Medical Academies, and the National Science Foundation.

What's New in This Edition

The Fifth Edition contains many new or expanded features, all of them targeted at ways to help students learn to construct their own knowledge and think like biologists.

- **Road Maps** The new Road Maps at the beginning of each chapter pair with the Big Picture concept maps introduced in the Fourth Edition. Together they help students navigate chapter content and see the forest for the trees. Each Road Map starts with a purpose statement that tells students what they can expect to learn from each chapter. It then goes on to visually group and organize information to help students anticipate key ideas as well as recognize meaningful relationships and connections between the ideas.

- **The Big Picture** Introduced in the Fourth Edition, Big Picture concept maps integrate words and visuals to help students synthesize information about challenging topics that span multiple chapters and units. In response to requests from instructors and students, three new Big Pictures focused on additional tough topics have been added: Doing Biology, The Chemistry of Life, and Plant and Animal Form and Function. In addition, the Ecology Big Picture is completely revised to reflect changes to that unit.

- **New Chapters** Two new chapters are added to better serve instructors and students. Unit 2 now contains a new Chapter 8, Energy and Enzymes: An Introduction to Metabolic Pathways. This chapter consolidates these critical topics in a place where students and instructors need it most—right before the chapters on cellular respiration and photosynthesis. In the Fourth Edition, animal movement was discussed in a chapter largely focused on animal sensory systems. In the Fifth Edition, this important topic is treated in depth in a new Chapter 48, Animal Movement, that explores how muscle and skeletal systems work together to produce locomotion.

- **New BioSkills** Instructors recognize that biology students need to develop foundational science skills in addition to content knowledge. While these skills are emphasized throughout the book, *Biological Science,* beginning with the Third

Edition, has provided a robust set of materials and activities to guide students who need extra help. To promote even fuller use of this resource, the BioSkills are now updated, expanded, and reorganized. New in this edition are a discussion of significant figures within the BioSkills on the Metric System, and two new BioSkills on Primary Literature and Peer Review and Using Bloom's Taxonomy. BioSkills are located in Appendix B, and practice activities can be assigned online in MasteringBiology®.

- **Promotion of Quantitative Skills** Reports like *Biology 2010, Scientific Foundations for Future Physicians*, and *Vision and Change* all place a premium on quantitative skills. To infuse a quantitative component throughout the text, new and existing quantitative questions are flagged in each chapter to encourage students to work on developing their ability to read or create a graph, perform or interpret a calculation, or use other forms of quantitative reasoning.

- **Bloom's Taxonomy** In the Fifth Edition, all questions in the text are assigned a Bloom's Taxonomy level to help both students and instructors understand whether a question requires higher-order or lower-order cognitive skills. Questions span all six Bloom's levels. (Bloom's levels are identified in Appendix A: Answers.) The coauthors were trained by experts Mary Pat Wenderoth and Clarissa Dirks[1] to ensure we followed a process that would result in high inter-rater reliability—or agreement among raters—in assigning Bloom's levels to questions. The new BioSkill, Using Bloom's Taxonomy, explains the six Bloom's levels to students and offers a practical guide to the kinds of study activities best suited for answering questions at each level.

- **Expanded Emphasis on "Doing Biology"** A constant hallmark of this text is its emphasis on experimental evidence—on teaching how we know what we know. To reflect the progress of science, in the Fifth Edition, the coauthor team replaced many experiments with fresh examples and added new Research Boxes. And as noted earlier, they added a new Big Picture on Doing Biology, focusing on the process of science and the organizational levels of biology. Data sources are now cited for all graphs and data tables to model the importance of citing data sources to students. Updated Research Box questions continue to encourage students to analyze some aspect of experimental design. Also new to this edition is a BioSkill on Primary Literature and Peer Review.

- **Art Program** The art program is further enhanced in this edition by the addition of more illustrated summary tables. These tables make subject areas more accessible to visual learners and reinforce key concepts of the chapter. Many of the life-cycle figures in Unit 6 are significantly overhauled.

[1] Crowe, A., C. Dirks, and M. P. Wenderoth. 2008. Biology in Bloom: Implementing Bloom's Taxonomy to enhance student learning in biology. *CBE–Life Sciences Education* 7: 368–381.

Updated Blue Thread Scaffolding

In the Third and Fourth editions of *Biological Science*, a metacognitive tool was formulated as the now popular feature known as "Blue Thread"—sets of questions designed to help students identify what they do and don't understand. The fundamental idea is that if students really understand a piece of information or a concept, they should be able to do something with it.

In the Fifth Edition, the Blue Thread is revised to reflect changes in chapter content, and to incorporate user feedback. Blue-Thread questions appear in the following locations:

- **In-text "You should be able to's"** offer exercises on topics that professors and students have identified as the most difficult concepts in each chapter.

- **Caption questions and exercises** challenge students to examine the information in a figure or table critically—not just absorb it.

- **Check Your Understanding boxes** present two to three tasks that students should be able to complete in order to demonstrate a mastery of summarized key ideas.

- **Chapter summaries** include "You should be able to" problems or exercises related to each key concept.

- **End-of-chapter** questions are organized in three levels of increasing difficulty so students can build from lower to higher-order cognitive questions.

Integration of Media

The textbook continues to be supported by MasteringBiology®, the most powerful online homework, tutorial, and assessment system available. Tutorials follow the Socratic method, coaching students to the correct answer by offering feedback specific to a student's misconceptions as well as providing hints students can access if they get stuck. Instructors can associate content with publisher-provided learning outcomes or create their own. Content highlights include the following:

- **NEW! Solve It Tutorials** These activities allow students to act like scientists in simulated investigations. Each tutorial presents an interesting, real-world question that students will answer by analyzing and interpreting data.

- **Experimental Inquiry Tutorials** The call to teach students about the process of science has never been louder. To support such teaching, there are 10 interactive tutorials on classic scientific experiments—ranging from Meselson–Stahl on DNA replication to the Grants' work on Galápagos finches and Connell's work on competition. Students who use these tutorials should be better prepared to think critically about experimental design and evaluate the wider implications of the data—preparing them to do the work of real scientists in the future.

- **BioFlix® Animations and Tutorials** BioFlix are movie-quality, 3-D animations that focus on the most difficult core topics and are accompanied by in-depth, online tutorials that provide hints and feedback to guide student learning. Eighteen BioFlix animations and tutorials tackle topics such as meiosis, mitosis, DNA replication, photosynthesis, homeostasis, and the carbon cycle.

- **NEW! End-of-Chapter Questions** Multiple choice end-of-chapter questions are now available to assign in MasteringBiology.

- **Blue-Thread Questions** Over 500 questions based on the Blue-Thread Questions in the textbook are assignable in MasteringBiology.

- **Big Picture Tutorials** Interactive concept map activities based on the Big Picture figures in the textbook are assignable in MasteringBiology, including tutorials to support the three new Big Pictures: Doing Biology, The Chemistry of Life, and Plant and Animal Form and Function.

- **BioSkills Activities** Activities based on the BioSkills content in the textbook are assignable in MasteringBiology, including activities to support the new BioSkills on Primary Literature and Peer Review and Using Bloom's Taxonomy.

- **Reading Quiz Questions** Every chapter includes reading quiz questions you can assign to ensure students read the textbook and understand the basics. These quizzes are perfect as a pre-lecture assignment to get students into the content before class, allowing you to use class time more effectively.

Serving a Community of Teachers

All of us on the coauthor team are deeply committed to students and to supporting the efforts of dedicated teachers. Doing biology is what we love. At various points along our diverse paths, we have been inspired by our own teachers when we were students, and now are inspired by our colleagues as we strive to become even better teacher-scholars. In the tradition of all previous editions of *Biological Science*, we have tried to infuse this textbook with the spirit and practice of evidence-based teaching. We welcome your comments, suggestions, and questions.

Thank you for your work on behalf of your students.

Content Highlights
of the Fifth Edition

As discussed in the preface, a major focus of this revision is to enhance the pedagogical utility of *Biological Science*. Another major goal is to ensure that the content reflects the current state of science and is accurate. The expanded author team has scrutinized every chapter to add new, relevant content, update descriptions when appropriate, and adjust the approach to certain topics to enhance student comprehension. In this section, some of the key content improvements to the textbook are highlighted.

Chapter 1 Biology and the Tree of Life   A concept map summarizing the defining characteristics of life is added. The process of doing biology coverage is expanded to include discussion of both experimental and descriptive studies, and more rigorous definitions of the terms hypothesis and theory.

Chapter 2 Water and Carbon: The Chemical Basis of Life   A stronger emphasis on chemical evolution is threaded throughout the chapter to bring chemistry to life for the student reader. Two prominent models for chemical evolution are introduced; the historic Miller prebiotic soup experiment was moved here. Advanced discussion of energy and chemical reactions was moved to a new chapter (see Chapter 8).

Chapter 3 Protein Structure and Function   The chapter is reorganized to emphasize the link between structure and function, from amino acids to folded proteins. Updated content illustrates that protein shapes are flexible and dynamic, and may remain incompletely folded until the protein interacts with other molecules or ions. Details of how enzymes work were moved to Chapter 8.

Chapter 4 Nucleic Acids and the RNA World   New experimental results concerning the synthesis of nucleotides and nucleic acids in a prebiotic environment are discussed. The section on the RNA world is expanded to include the artificial evolution of a novel ribozyme involved in nucleotide synthesis.

Chapter 5 An Introduction to Carbohydrates   The molecular basis for resistance of structural polymers, such as cellulose, to degradation is clarified. A new research box illustrates the role of carbohydrates in cellular recognition and attachment using the egg and sperm of mice as a model system.

Chapter 6 Lipids, Membranes, and the First Cells   New content on lipid and membrane evolution and the proposed characteristics of the first protocell is introduced. The aquaporin and potassium channel figures are updated; how key amino acids serve as selectivity filters is now highlighted.

Chapter 7 Inside the Cell   Several new electron micrographs were selected to more clearly illustrate cell component structure and function. A new figure is added to better depict the pulse–chase assay used to identify the secretory pathway. Coverage of nuclear transport is expanded to differentiate between passive diffusion and active nuclear import. Updated content emphasizes the role of the cytoskeleton in localizing organelles, and how polarity of microtubules and microfilaments influences their growth rate.

Chapter 8 Energy and Enzymes: An Introduction to Pathways   This new chapter pulls together concepts in energy, chemical reactions, and enzymes that previously were covered in three different chapters. Oxidation and reduction reactions are emphasized to prepare students for Chapters 9 and 10. The energetics behind ATP hydrolysis and its role in driving endergonic reactions is discussed, and figures are revised to better illustrate the process. Updated content on enzyme regulation and a new process figure show a model for how metabolic pathways may have evolved.

Chapter 9 Cellular Respiration and Fermentation   Two new summary tables for glycolysis and the citric acid cycle are added that provide the names of the enzymes and the reaction each catalyzes. New content is introduced to propose a connection between the universal nature of the proton motive force and the story of the chemical evolution of life.

Chapter 10 Photosynthesis   More extensive comparison between the chemical reactions in mitochondria and chloroplasts is added. A new figure is introduced to illustrate noncyclic electron flow in the context of the thylakoid membrane. Greater emphasis is placed on the number of ATPs and NADPHs required for each cycle of carbon fixation and reduction.

Chapter 11 Cell–Cell Interactions   Coverage of extracellular matrix structure and function is expanded, including its role in intercellular adhesions and cell signaling. The plant apoplast and symplast are now introduced as key terms in the text and illustrated in a new figure. New content and a new figure on unicellular models for intercellular communication via pheromone sensing (yeast) and quorum sensing (slime mold) are added.

Chapter 12 The Cell Cycle   A new figure helps explain the pulse–chase assay for identifying phases of the cell cycle. Content is added to the text and to a figure that illustrates the similarities between chromosome segregation in eukaryotes and prokaryotes. A revised description of anaphase emphasizes how microtubule fraying at the kinetochore can drive chromosome movement. The explanation of how phosphorylation and dephosphorylation turns on MPF activity is updated to reflect current research.

Chapter 13 Meiosis   To improve the flow of the chapter, the section on advantages of sexual reproduction was moved to before mistakes in meiosis. The discussion of the role and timing of crossing over during meiosis I is updated. A new study that supports the hypothesis that sex evolved in response to the selective pressure of pathogens is introduced.

Chapter 14 Mendel and the Gene Material on gene linkage is revised to emphasize the importance of genetic mapping. A new matched set of figures on pedigree analysis brings together the various modes of transmission that were previously shown in four individual figures. A new summary table on characteristics of different patterns of inheritance is added.

Chapter 15 DNA and the Gene: Synthesis and Repair A new research figure is added that focuses on the relationship between telomere length and senescence in cultured somatic cells.

Chapter 16 How Genes Work Coverage of the evolving concept of the gene and of different types of RNA is expanded. A figure showing the karyotype of a cancer cell is revised to improve clarity.

Chapter 17 Transcription, RNA Processing, and Translation The sections on transcription in bacteria and eukaryotes are now separated, and content on charging tRNAs was moved to a new section. The discussion of translation is reorganized, first to emphasize the process in bacteria and then to highlight differences in eukaryotes.

Chapter 18 Control of Gene Expression in Bacteria Coverage of *lac* operon positive regulation is updated to reflect current research. A new section and new process figure on global gene regulation are added, using the *lexA* regulon as an example.

Chapter 19 Control of Gene Expression in Eukaryotes Extensive updates to the discussion of epigenetics include a new research box and a section on DNA methylation. Coverage of transcription initiation is updated to reflect current science. A new figure illustrates the role of p53 in the cell cycle in normal and cancerous cells.

Chapter 20 Analyzing and Engineering Genes The material on sequencing the Neanderthal genome is updated, including evidence of limited Neanderthal genetic material in some modern human populations. New information on current generation sequencing technologies and massive parallelism is added. Recent advances in gene therapy are highlighted.

Chapter 21 Genomics and Beyond Extensive updates throughout reflect recent advances in genomics. Changes include sequence database statistics, genomes that have been sequenced to study evolutionary relationships, and new figures illustrating gene count versus genome size in prokaryotes and eukaryotes and functional classes of human DNA sequences. A new section on systems biology is added. Also included are notes on the discovery of widespread transcription of eukaryotic genomes, deep sequencing, and the spectrum of mutations in human tumors.

Chapter 22 Principles of Development New information is added on dedifferentiation in induced pluripotent stem cells, maternal genes in *Drosophila* development, how morphogens work, and tool-kit genes. The order of topics in the discussion

of developmental principles is reorganized. The figure on *Hox* genes in *Drosophila* and the mouse is updated.

Chapter 23 An Introduction to Animal Development The chapter is streamlined by focusing on principles of animal development. The discussion of gametogenesis was moved to Chapter 50 (Animal Reproduction). The presentation of fertilization is simplified, and a new figure summarizing steps of fertilization is added. The figure on gastrulation is modified to better depict the arrangement of the germ layers and their movement.

Chapter 24 An Introduction to Plant Development The chapter is modified to impart an evolutionary perspective on the similarities and differences in plant and animal development. The chapter also was streamlined by removing material such as details of gametogenesis, which now appears in Chapter 41 (Plant Reproduction).

Chapter 25 Evolution by Natural Selection Several new key passages are included, among them the use of the Grand Canyon as a context for understanding relative dating of fossils, Darwin's artificial selection experiments with fancy pigeons, and Malthus's concept of struggle for existence. A new example of people living at high altitude in Tibet clarifies the difference between acclimatization and adaptation. An illustrated summary table of common misconceptions is added.

Chapter 26 Evolutionary Processes Discussion of sexual selection now falls within the section on natural selection, and the terms intersexual and intrasexual selection are added. Several new examples replace those in the Fourth Edition, including inbreeding depression in Florida panthers, gene flow in Oregon steelhead trout, and lateral gene transfer in aphids. An illustrated summary table is added on modes of selection.

Chapter 27 Speciation Several points are clarified, such as the gradient-like (rather than all-or-nothing) nature of reproductive isolation. The section on polyploidy is reorganized and the figures revised, including a side-by-side comparison of autopolyploidy and allopolyploidy.

Chapter 28 Phylogenies and the History of Life The phylogenetics section is reorganized and expanded to include three illustrated summary tables and updated life-history timelines. New content is added, including the concept of the Anthropocene, the calendar analogy to the history of the Earth, the Chengjiang fossils, and a Life-in-the-Cambrian illustration. Evidence for the impact hypothesis is combined into an illustrated summary table.

Chapter 29 Bacteria and Archaea The chapter is updated to include a description of metagenomic experiments with an emphasis on the role of gut bacteria in digestion. A newly recognized phylum of Archaea, the Thaumarchaeota, is included, and the table comparing key characteristics of the Bacteria, Archaea, and Eukarya is streamlined.

Chapter 30 Protists For simplicity, protist lineages are now referred to throughout the chapter by their more familiar common names. Also, some key lineage boxes were consolidated to trim the number to one box per major lineage. Discussion of the origin of the nuclear envelope and mitochondria is expanded to reflect new thinking on the evolution of eukaryotic cells. Protist life cycle figures are significantly overhauled.

Chapter 31 Green Algae and Land Plants Coverage of the evolution of land plants is expanded to include the importance of UV light and UV-absorbing molecules on the colonization of land. The "Redwood group" is now referred to as the Cupresophyta. Updates emphasize the role(s) of each stage of a life cycle in dispersal and in increasing genetic variation and individual numbers. A new research box is added, showing the importance of flower color to pollinator preference. Plant life-cycle figures are significantly overhauled.

Chapter 32 Fungi Coenocytic fungal hyphae are illustrated with a new image showing GFP-labeled nuclei in *Neurospora crassa*. The chapter now points out the similarity between fungal and animal modes of nutrition, in terms of extracellular digestion and absorption of small molecules. The discussion of lignin degradation is updated, and new descriptions of mutualisms of fungi with animals are included. Fungal life cycle figures are significantly overhauled.

Chapter 33 An Introduction to Animals The chapter is extensively revised to streamline and modernize the presentation, including emphasis on genetic tool kits and symmetry in the phylogeny of animals. The "Themes of Diversification" section is reorganized around five illustrated summary tables. The discussion of life cycles is revised to be more general. Insect metamorphosis has moved to Chapter 34, and a sea urchin life cycle replaces *Obelia*.

Chapter 34 Protostome Animals Two themes are threaded throughout the chapter: the water-to-land transition and modular body plans. The section on lophotrochozoans emphasizes spiral cleavage, indirect versus direct development, hemocoels, and radulas. The section on ecdysozoans highlights segmentation and *Hox* genes, including discussion of the origin of the wing and metamorphosis. Key lineage boxes include new phylogenies for annelids, crustaceans, and chelicerates.

Chapter 35 Deuterostome Animals Updates to reflect current research include revised phylogenies, evolution of flight and feathers, *Australopithecus sediba*, human migration out of Africa, and genetic evidence for interbreeding of *Homo neanderthalensis* and *Homo sapiens*.

Chapter 36 Viruses New content focuses on how viruses contribute to evolution via lateral gene transfer and direct addition of genes to cellular genomes. Content is updated and expanded on viral structure and function, and on lytic and latent infections. Three new figures are added, including a comparison of replication of viruses and cells, how pandemic strains of influenza arise via reassortment, and the devastating impact of the 1918 influenza pandemic.

Chapter 37 Plant Form and Function The use of terminology is streamlined for consistency and clarity. For example,

"lateral meristem" is replaced with "vascular cambium" to avoid confusion with lateral buds, and the description of bark is clarified to avoid using the term phelloderm. Several complex figures were converted to illustrated summary tables.

Chapter 38 Water and Sugar Transport in Plants The chapter is revised to improve accuracy, and points out that water loss is also a means for transporting minerals from roots to shoots. Updated content clarifies the role of energy expenditure in moving water across roots, and the Casparian strip as a barrier to the back diffusion of ions and water out of the root. A new research figure shows the importance of the sucrose proton symporter in long-distance transport in *Arabidopsis*.

Chapter 39 Plant Nutrition In this chapter the coverage of mycorrhizae is modified to emphasize their overall role in nutrient acquisition. In the section on nitrogen fixation, a description of the worldwide practice of crop rotation involving legumes and grains is added.

Chapter 40 Plant Sensory Systems, Signals, and Responses A new research box is added that reveals the essential role of PHOT1 phosphorylation in phototropism. A section on the effect of day length on flowering was moved here from Chapter 41 and is integrated with the discussion of phytochromes. New content on the role of plasmodesmata in plant action potentials is added. The section on how plants respond to pathogens is simplified and updated with an example of control of stomata during a bacterial infection.

Chapter 41 Plant Reproduction To provide a clearer example of a gametophyte-dominant life cycle, the liverwort life cycle has been replaced with a moss life cycle. The term "pollination syndrome" is clarified. Also, a research box on how capsaicin prevents seed predation and facilitates dispersal was reinstated from an earlier edition of *Biological Science*.

Chapter 42 Animal Form and Function The chapter includes a new experiment illustrating physiological trade-offs, along with improved examples of thermoregulatory strategies in animals. Several complex figures were converted to illustrated summary tables.

Chapter 43 Water and Electrolyte Balance in Animals The chapter is reorganized to better integrate the relationship between excretion and water and electrolyte balance. Osmoregulatory strategies are now organized according to the challenges presented by marine, freshwater, and terrestrial habitats. Coverage of osmoregulation in bony fishes versus cartilaginous fishes, mammalian kidney function, and how nonmammalian vertebrates concentrate their urine is expanded and clarified.

Chapter 44 Animal Nutrition This chapter contains new information on nutritional imbalances, including diabetes and obesity.

Chapter 45 Gas Exchange and Circulation Discussion of the insect tracheal system is expanded, including new content on how respiration restricts upper limits of body size of insects.

Details regarding the lymphatic system and heart anatomy in vertebrates are updated, and a new section on cardiovascular disease is added.

Chapter 46 Electrical Signals in Animals This chapter is greatly expanded to reflect recent research and growing interest in neuroscience. New information includes comparative anatomy of vertebrate brains, more case studies of brain injuries or dysfunctions that have led to major discoveries in neuroscience, and the concept of neuroplasticity—especially neurogenesis.

Chapter 47 Animal Sensory Systems Content from the Fourth Edition has been split into two chapters (47, Animal Sensory Systems; 48, Animal Movement). The chapter on sensory systems is now organized by type of sensory reception: mechanoreception (with new coverage on the lateral line system of fishes), photoreception, chemoreception (with new coverage of pheromones), and a new section introducing thermoreception, electroreception, and magnetoreception.

Chapter 48 Animal Movement This new chapter introduces the importance of movement in animals, building from small to large scale. The mechanism of muscle contraction (with revised figures) is covered, followed by discussions of types of muscle tissue (with new content on skeletal-muscle fiber types and parallel- versus pennate-muscle fiber orientation), and skeletal systems (hydrostatic skeletons, exoskeletons, endoskeletons). A completely new final section discusses how biologists study locomotion on land, in the air, and in the water.

Chapter 49 Chemical Signals in Animals Figures and content are updated for clarity. The chapter includes a new section on endocrine disruptors.

Chapter 50 Animal Reproduction New content includes details of sperm and egg structure and function, reproduction in the spotted hyena, and human contraceptive methods.

Chapter 51 The Immune System in Animals Coverage of the innate immune response is expanded to include more detail on Toll-like receptors and how they transmit signals. The section on adaptive immunity is reorganized to improve flow. Updated content on inappropriate immune responses (autoimmunity and allergies) and inadequate responses (immunodeficiency) is grouped together in one section. The hygiene hypothesis is introduced to explain the growing trend of inappropriate immune responses in populations that have reduced exposure to common pathogens and parasites.

Chapter 52 An Introduction to Ecology The first section on levels of ecological study is expanded to include global ecology. The rest of the chapter is reorganized, beginning with the factors that determine the distribution and abundance of organisms (including a new Argentine ant case study) and ending with biomes. The biome section is streamlined with a new emphasis on human impacts, including an introduction to anthropogenic biomes and the Anthropocene.

Chapter 53 Behavioral Ecology The introduction is revised to provide a clearer framework for types of behavior. Sections are now organized around proximate versus ultimate causation. These new examples replace those in the Fourth Edition: Argentine ant territorial behavior (replacing spiny lobsters), optimal foraging in gerbils (replacing white-fronted bee-eaters), sexual selection in *Anolis* lizards (replacing barn swallows), and map orientation in green sea turtles.

Chapter 54 Population Ecology A new introductory section focuses on the distribution of organisms in populations, including dispersion patterns and a consolidated discussion of measurement methods. The human population content is updated and separated into a new section. The quantitative methods boxes and life table are now more student friendly.

Chapter 55 Community Ecology Several changes to content are made, such as a clarification of competitive exclusion versus niche differentiation. New content includes a summary table on constitutive defenses, a discovery story on mimicry (including Bates and Müller), an introduction to food webs, and the process of soil formation in primary succession.

Chapter 56 Ecosystems The title and scope of the chapter are updated to include global ecology. Extensive revisions include many content updates and new figures on the relationship between GPP and NPP, the one-way flow of energy and cycling of nutrients, the food web, open versus closed aquifers, the High Plains Aquifer, the biomagnification of DDT, and the greenhouse effect. The climate change section is expanded and updated, including an illustrated summary table.

Chapter 57 Biodiversity and Conservation Biology Throughout the chapter, there is more emphasis on conserving ecosystem function. Many new examples are added, including Smits's restoration project in Borneo, the Census of Marine Life, the IUCN Red List, the Sinervo lizard extinction experiment, and Florida panther genetic restoration. New summary tables highlight ecosystem services and threats to biodiversity.

Acknowledgments

Reviewers

The peer review system is the key to quality and clarity in science publishing. In addition to providing a filter, the investment that respected individuals make in vetting the material—catching errors or inconsistencies and making suggestions to improve the presentation—gives authors, editors, and readers confidence that the text meets rigorous professional standards.

Peer review plays the same role in textbook publishing. The time and care that this book's reviewers have invested is a tribute to their professional integrity, their scholarship, and their concern for the quality of teaching. Virtually every paragraph in this edition has been revised and improved based on insights from the following individuals.

Tamarah Adair, *Baylor University*
Sandra D. Adams, *Montclair State University*
Marc Albrecht, *University of Nebraska at Kearney*
Larry Alice, *Western Kentucky University*
Leo M. Alves, *Manhattan College*
David R. Angelini, *American University*
Dan Ardia, *Franklin & Marshall College*
Paul Arriola, *Elmhurst College*
Davinderjit K. Bagga, *University of Montevallo*
Susan Barrett, *Wheaton College*
Donald Baud, *University of Memphis*
Vernon W. Bauer, *Francis Marion University*
Robert Bauman, *Amarillo College*
Christopher Beck, *Emory University*
Vagner Benedito, *West Virginia University*
Scott Bingham, *Arizona State University*
Stephanie Bingham, *Barry University*
Wendy Birky, *California State University, Northridge*
Jason Blank, *California Polytechnic State University*
Kristopher A. Blee, *California State University, Chico*
Margaret Bloch-Qazi, *Gustavus Adolphus College*
Lanh Bloodworth, *Florida State College at Jacksonville*
Catherine H. Borer, *Berry College*
James Bottesch, *Brevard Community College*
Jacqueline K. Bowman, *Arkansas Tech University*
John Bowman, *University of California, Davis*
Chris Brochu, *University of Iowa*
Matthew Brown, *Dalhousie University*
Mark Browning, *Purdue University*
Carolyn J. W. Bunde, *Idaho State University*
David Byres, *Florida State College at Jacksonville*
Michael Campbell, *Penn State Erie*
Manel Camps, *University of California, Santa Cruz*
Geralyn M. Caplan, *Owensboro Community and Technical College*
Richard Cardullo, *University of California, Riverside*

David Carlini, *American University*
Dale Casamatta, *University of North Florida*
Deborah Chapman, *University of Pittsburgh*
Joe Coelho, *Quincy University*
Allen Collins, *Smithsonian Museum of Natural History*
Robert A. Colvin, *Ohio University*
Kimberly L. Conner, *Florida State College at Jacksonville*
Karen Curto, *University of Pittsburgh*
Clarissa Dirks, *Evergreen State College*
Peter Ducey, *SUNY Cortland*
Erastus Dudley, *Huntingdon College*
Jeffrey P. Duguay, *Delta State University*
Tod Duncan, *University of Colorado, Denver*
Joseph Esdin, *University of California, Los Angeles*
Brent Ewers, *University of Wyoming*
Amy Farris, *Ivy Tech Community College*
Bruce Fisher, *Roane State Community College*
Ryan Fisher, *Salem State University*
David Fitch, *New York University*
Elizabeth Fitch, *Motlow State Community College*
Michael P. Franklin, *California State University, Northridge*
Susannah French, *Utah State University*
Caitlin Gabor, *Texas State University*
Matthew Gilg, *University of North Florida*
Kendra Greenlee, *North Dakota State University*
Patricia A. Grove, *College of Mount Saint Vincent*
Nancy Guild, *University of Colorado, Boulder*
Cynthia Hemenway, *North Carolina State University*
Christopher R. Herlihy, *Middle Tennessee State University*
Kendra Hill, *South Dakota State University*
Sara Hoot, *University of Wisconsin, Milwaukee*
Kelly Howe, *University of New Mexico*
Robin Hulbert, *California Polytechnic State University*
Rick Jellen, *Brigham Young University*
Russell Johnson, *Colby College*
William Jira Katembe, *Delta State University*
Elena K. Keeling, *California Polytechnic State University*
Jill B. Keeney, *Juniata College*
Greg Kelly, *University of Western Ontario*
Scott L. Kight, *Montclair State University*
Charles Knight, *California Polytechnic State University*
Jenny Knight, *University of Colorado, Boulder*
William Kroll, *Loyola University Chicago*
Dominic Lannutti, *El Paso Community College*
Brenda Leady, *University of Toledo*
David Lindberg, *University of California, Berkeley*
Barbara Lom, *Davidson College*
Robert Maxwell, *Georgia State University*
Marshall D. McCue, *St. Mary's University*
Kurt A. McKean, *SUNY Albany*
Michael Meighan, *University of California, Berkeley*
John Merrill, *Michigan State University*

Richard Merritt, *Houston Community College*
Alan Molumby, *University of Illinois at Chicago*
Jeremy Montague, *Barry University*
Chad E. Montgomery, *Truman State University*
Kimberly D. Moore, *Lone Star College System, North Harris*
Michael Morgan, *Berry College*
James Mulrooney, *Central Connecticut State University*
John D. Nagy, *Scottsdale Community College*
Margaret Olney, *St. Martin's University*
Nathan Okia, *Auburn University at Montgomery*
Robert Osuna, *SUNY Albany*
Daniel Panaccione, *West Virginia University*
Stephanie Pandolfi, *Michigan State University*
Michael Rockwell Parker, *Monell Chemical Senses Center*
Lisa Parks, *North Carolina State University*
Nancy Pelaez, *Purdue University*
Shelley W. Penrod, *Lone Star College System, North Harris*
Andrea Pesce, *James Madison University*
Raymond Pierotti, *University of Kansas*
Melissa Ann Pilgrim, *University of South Carolina Upstate*
Paul Pillitteri, *Southern Utah University*
Debra Pires, *University of California, Los Angeles*
P. David Polly, *Indiana University, Bloomington*
Vanessa Quinn, *Purdue University North Central*
Stacey L. Raimondi, *Elmhurst College*
Stephanie Randell, *McLennan Community College*
Marceau Ratard, *Delgado Community College*
Flona Redway, *Barry University*
Srebrenka Robic, *Agnes Scott College*
Dave Robinson, *Bellarmine University*
George Robinson, *SUNY Albany*
Adam W. Rollins, *Lincoln Memorial University*
Amanda Rosenzweig, *Delgado Community College*
Leonard C. Salvatori, *Indian River State College*
Dee Ann Sato, *Cypress College*
Leena Sawant, *Houston Community College*
Jon Scales, *Midwestern State University*
Oswald Schmitz, *Yale University*
Joan Sharp, *Simon Fraser University*
Julie Schroer, *North Dakota State University*
Timothy E. Shannon, *Francis Marion University*
Lynnette Sievert, *Emporia State University*
Susan Skambis, *Valencia College*
Ann E. Stapleton, *University of North Carolina, Wilmington*
Mary-Pat Stein, *California State University, Northridge*
Christine Strand, *California Polytechnic State University*
Denise Strickland, *Midlands Technical College*
Jackie Swanik, *Wake Technical Community College*
Billie J. Swalla, *University of Washington*
Zuzana Swigonova, *University of Pittsburgh*
Briana Timmerman, *University of South Carolina*
Catherine Ueckert, *Northern Arizona University*
Sara Via, *University of Maryland, College Park*
Thomas J. Volk, *University of Wisconsin–La Crosse*
Jeffrey Walck, *Middle Tennessee State University*
Andrea Weeks, *George Mason University*
Margaret S. White, *Scottsdale Community College*
Steven D. Wilt, *Bellarmine University*
Candace Winstead, *California Polytechnic State University*
James A. Wise, *Hampton University*

Correspondents

One of the most enjoyable interactions we have as textbook authors is correspondence or conversations with researchers and teachers who take the time and trouble to contact us to discuss an issue with the book, or who respond to our queries about a particular data set or study. We are always amazed and heartened by the generosity of these individuals. They care, deeply.

Lawrence Alice, *Western Kentucky University*
David Baum, *University of Wisconsin–Madison*
Meredith Blackwell, *Louisiana State University*
Nancy Burley, *University of California, Irvine*
Thomas Breithaupt, *University of Hull*
Philip Cantino, *Ohio University*
Allen Collins, *Smithsonian Museum of Natural History*
Robert Full, *University of California, Berkeley*
Arundhati Ghosh, *University of Pittsburgh*
Jennifer Gottwald, *University of Wisconsin–Madison*
Jon Harrison, *Arizona State University*
David Hawksworth, *Natural History Museum, London*
Jim Herrick, *James Madison University*
John Hunt, *University of Exeter*
Doug Jensen, *Converse College*
Scott Kight, *Montclair State University*
Scott Kirkton, *Union College*
Mimi Koehl, *University of California, Berkeley*
Rodger Kram, *University of Colorado*
Matthew McHenry, *University of California, Irvine*
Alison Miyamoto, *California State University, Fullerton*
Sean Menke, *Lake Forest College*
Rich Mooi, *California Academy of Sciences*
Michael Oliver, *MalawiCichlids.com*
M. Rockwell Parker, *Monell Chemical Senses Center*
Andrea Pesce, *James Madison University*
Chris Preston, *Monterey Bay Aquarium Research Institute*
Scott Sakaluk, *Illinois State University*
Kyle Seifert, *James Madison University*
Jos Snoeks, *Royal Museum for Central Africa*
Jeffrey Spring, *University of Louisiana*
Christy Strand, *California Polytechnic State University, San Luis Obispo*
Torsten Struck, *University of Osnabrueck, Germany*
Oswald Schmitz, *Yale University*
Ian Tattersal, *American Museum of Natural History*
Robert Turgeon, *Cornell University*
Tom Volk, *University of Wisconsin–La Crosse*
Naomi Wernick, *University of Massachusetts, Lowell*

Supplements Contributors

Instructors depend on an impressive array of support materials—in print and online—to design and deliver their courses. The student experience would be much weaker without the study guide, test bank, activities, animations, quizzes, and tutorials written by the following individuals.

Brian Bagatto, *University of Akron*
Scott Bingham, *Arizona State University*
Jay L. Brewster, *Pepperdine University*

Mirjana Brockett, *Georgia Institute of Technology*
Warren Burggren, *University of North Texas*
Jeff Carmichael, *University of North Dakota*
Tim Christensen, *East Carolina University*
Erica Cline, *University of Washington—Tacoma*
Patricia Colberg, *University of Wyoming*
Elia Crisucci, *University of Pittsburgh*
Elizabeth Cowles, *Eastern Connecticut State University*
Clarissa Dirks, *Evergreen State College*
Lisa Elfring, *University of Arizona, Tucson*
Brent Ewers, *University of Wyoming*
Rebecca Ferrell, *Metropolitan State University of Denver*
Miriam Ferzli, *North Carolina State University*
Cheryl Frederick, *University of Washington*
Cindee Giffen, *University of Wisconsin–Madison*
Kathy M. Gillen, *Kenyon College*
Linda Green, *Georgia Institute of Technology*
Christopher Harendza, *Montgomery County Community College*
Cynthia Hemenway, *North Carolina State University*
Laurel Hester, *University of South Carolina*
Jean Heitz, *University of Wisconsin–Madison*
Tracey Hickox, *University of Illinois, Urbana–Champaign*
Jacob Kerby, *University of South Dakota*
David Kooyman, *Brigham Young University*
Barbara Lom, *Davidson College*
Cindy Malone, *California State University, Northridge*
Jim Manser, retired, *Harvey Mudd College*
Jeanette McGuire, *Michigan State University*
Mark Music, *Indian River State College*
Jennifer Nauen, *University of Delaware*
Chris Pagliarulo, *University of California, Davis*
Stephanie Scher Pandolfi, *Michigan State University*
Lisa Parks, *North Carolina State University*
Debra Pires, *University of California, Los Angeles*
Carol Pollock, *University of British Columbia*
Jessica Poulin, *University at Buffalo, the State University of New York*
Vanessa Quinn, *Purdue University North Central*
Eric Ribbens, *Western Illinois University*
Christina T. Russin, *Northwestern University*
Leonard Salvatori, *Indian River State College*
Joan Sharp, *Simon Fraser University*
Chrissy Spencer, *Georgia Institute of Technology*
Mary-Pat Stein, *California State University, Northridge*
Suzanne Simon-Westendorf, *Ohio University*
Fred Wasserman, *Boston University*
Cindy White, *University of Northern Colorado*
Edward Zalisko, *Blackburn College*

Book Team

Anyone who has been involved in producing a textbook knows that many people work behind the scenes to make it all happen. The coauthor team is indebted to the many talented individuals who have made this book possible.

Development editors Mary Catherine Hager, Moira Lerner-Nelson, and Bill O'Neal provided incisive comments on the revised manuscript. Fernanda Oyarzun and Adam Steinberg used their artistic sense, science skills, and love of teaching to hone the figures for many chapters.

The final version of the text was copyedited by Chris Thillen and expertly proofread by Pete Shanks. The final figure designs were rendered by Imagineering Media Services and carefully proofread by Frank Purcell. Maureen Spuhler, Eric Schrader, and Kristen Piljay researched images for the Fifth Edition.

The book's clean, innovative design was developed by Mark Ong and Emily Friel. Text and art were skillfully set in the design by S4Carlisle Publishing Services. The book's production was supervised by Lori Newman and Mike Early.

The extensive supplements program was managed by Brady Golden and Katie Cook. All of the individuals mentioned—and more—were supported with cheerful, dedicated efficiency by Editorial Assistant Leslie Allen for the first half of the project; Eddie Lee has since stepped in to skillfully fill this role.

Creating MasteringBiology® tutorials and activities also requires a team. Media content development was overseen by Tania Mlawer and Sarah Jensen, who benefited from the program expertise of Caroline Power and Caroline Ross. Joseph Mochnick and Daniel Ross worked together as media producers. Lauren Fogel (VP, Director, Media Development), Stacy Treco (VP, Director, Media Product Strategy), and Laura ensured that the complete media program that accompanies the Fifth Edition, including MasteringBiology, will meet the needs of the students and professors who use our offerings.

Pearson's talented sales reps, who listen to professors, advise the editorial staff, and get the book in students' hands, are supported by tireless Executive Marketing Manager Lauren Harp and Director of Marketing Christy Lesko. The marketing materials that support the outreach effort were produced by Lillian Carr and her colleagues in Pearson's Marketing Comunications group. David Theisen, national director for Key Markets, tirelessly visits countless professors each year, enthusiastically discussing their courses and providing us with meaningful editorial guidance.

The vision and resources required to run this entire enterprise are the responsibility of Vice President and Editor-in-Chief Beth Wilbur, who provided inspirational and focused leadership, and President of Pearson Science Paul Corey, who displays unwavering commitment to high-quality science publishing.

Becky Ruden recruited the coauthor team, drawing us to the project with her energy and belief in this book. The editorial team was skillfully directed by Executive Director of Development Deborah Gale. Finally, we are deeply grateful for three key drivers of the Fifth Edition. Project Editor Anna Amato's superb organizational skills and calm demeanor assured that all the wheels and cogs of the process ran smoothly to keep the mammoth project steadily rolling forward. Supervising Development Editor Sonia DiVittorio's deep expertise, creative vision, keen attention to detail, level, and clarity, and inspiring insistence on excellence kept the bar high for everyone on every aspect of the project. Lastly, Senior Acquisitions Editor Michael Gillespie's boundless energy and enthusiasm, positive attitude, and sharp intellect have fueled and united the team and also guided the book through the hurdles to existence. The coauthor team thanks these exceptional people for making the art and science of book writing a productive and exhilarating process.

1 Biology and the Tree of Life

In this chapter you will learn about

Key themes to structure your thinking about Biology

starting with → **What does it mean to say that something is alive?** **1.1**

including → **Two of the greatest unifying ideas in Biology**

including → **The process of doing Biology** **1.5**

first → **The cell theory** **1.2**

then → **The theory of evolution by natural selection** **1.3**

predicts → **The tree of life** **1.4**

These Chinese water dragon hatchlings are exploring their new world and learning how to find food and stay alive. They represent one of the key characteristics of life introduced in this chapter—replication.

BIG PICTURE

This chapter is part of the Big Picture. See how on pages 16–17.

In essence, biological science is a search for ideas and observations that unify our understanding of the diversity of life, from bacteria living in rocks a mile underground to humans and majestic sequoia trees. This chapter is an introduction to this search.

The goals of this chapter are to introduce the nature of life and explore how biologists go about studying it. The chapter also introduces themes that will resonate throughout this book:

- Analyzing how organisms work at the molecular level.

- Understanding organisms in terms of their evolutionary history.

- Helping you learn to think like a biologist.

Let's begin with what may be the most fundamental question of all: What is life?

✔ When you see this checkmark, stop and test yourself. Answers are available in Appendix A.

1.1 What Does It Mean to Say That Something Is Alive?

An **organism** is a life-form—a living entity made up of one or more cells. Although there is no simple definition of life that is endorsed by all biologists, most agree that organisms share a suite of five fundamental characteristics.

- *Energy* To stay alive and reproduce, organisms have to acquire and use energy. To give just two examples: plants absorb sunlight; animals ingest food.

- *Cells* Organisms are made up of membrane-bound units called cells. A cell's membrane regulates the passage of materials between exterior and interior spaces.

- *Information* Organisms process hereditary, or genetic, information encoded in units called genes. Organisms also respond to information from the environment and adjust to maintain stable internal conditions. Right now, cells throughout your body are using information to make the molecules that keep you alive; your eyes and brain are decoding information on this page that will help you learn some biology, and if your room is too hot you might be sweating to cool off.

- *Replication* One of the great biologists of the twentieth century, François Jacob, said that the "dream of a bacterium is to become two bacteria." Almost everything an organism does contributes to one goal: replicating itself.

- *Evolution* Organisms are the product of evolution, and their populations continue to evolve.

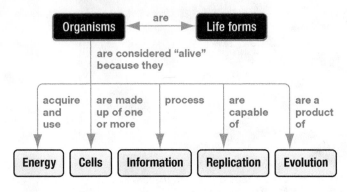

You can think of this text as one long exploration of these five traits. Here's to life!

1.2 The Cell Theory

Two of the greatest unifying ideas in all of science laid the groundwork for modern biology: the cell theory and the theory of evolution by natural selection. Formally, scientists define a **theory** as an explanation for a very general class of phenomena or observations that are supported by a wide body of evidence. The cell theory and theory of evolution address fundamental questions: What are organisms made of? Where do they come from?

When these concepts emerged in the mid-1800s, they revolutionized the way biologists think about the world. They established

two of the five attributes of life: Organisms are cellular, and their populations change over time.

Neither insight came easily, however. The cell theory, for example, emerged after some 200 years of work. In 1665 the Englishman Robert Hooke devised a crude microscope to examine the structure of cork (a bark tissue) from an oak tree. The instrument magnified objects to just 30× (30 times) their normal size, but it allowed Hooke to see something extraordinary. In the cork he observed small, pore-like compartments that were invisible to the naked eye. Hooke coined the term "cells" for these structures because of their resemblance to the cells inhabited by monks in a monastery.

Soon after Hooke published his results, a Dutch scientist named Anton van Leeuwenhoek succeeded in developing much more powerful microscopes, some capable of magnifications up to 300×. With these instruments, van Leeuwenhoek inspected samples of pond water and made the first observations of a dazzling collection of single-celled organisms that he called "animalcules." He also observed and described human blood cells and sperm cells, shown in **FIGURE 1.1**.

In the 1670s an Italian researcher who was studying the leaves and stems of plants with a microscope concluded that plant tissues were composed of many individual cells. By the early 1800s, enough data had accumulated for a German biologist to claim that *all* organisms consist of cells. Did this claim hold up?

All Organisms Are Made of Cells

Advances in microscopy have made it possible to examine the amazing diversity and complexity of cells at higher and higher magnifications. Biologists have developed microscopes that are tens of thousands of times more powerful than van Leeuwenhoek's and have described over a million new species. The basic conclusion made in the 1800s remains intact, however: All organisms are made of cells.

The smallest organisms known today are bacteria that are barely 200 nanometers wide, or 200 *billionths* of a meter. (See **BioSkills 1** in Appendix B to review the metric system and its prefixes.[1]) It would take 5000 of these organisms lined up side by side to span a millimeter. This is the distance between the smallest hash marks on a metric ruler. In contrast, sequoia trees can be over 100 meters tall. This is the equivalent of a 20-story building. Bacteria and sequoias are composed of the same fundamental building block, however—the cell. Bacteria consist of a single cell; sequoias are made up of many cells.

Today a **cell** is defined as a highly organized compartment that is bounded by a thin, flexible structure called a plasma membrane and that contains concentrated chemicals in an aqueous (watery) solution. The chemical reactions that sustain life take place inside cells. Most cells are also capable of reproducing by dividing—in effect, by making a copy of themselves.

The realization that all organisms are made of cells was fundamentally important, but it formed only the first part of the cell

---

[1]BioSkills are located in the second appendix at the back of the book. They focus on general skills that you'll use throughout this course. More than a few students have found them to be a life-saver. Please use them!

(a) van Leeuwenhoek built his own microscopes—which, while small, were powerful. They allowed him to see, for example . . .

Lens

(b) . . . human blood cells (this modern photo was shot through one of van Leeuwenhoek's original microscopes) . . .

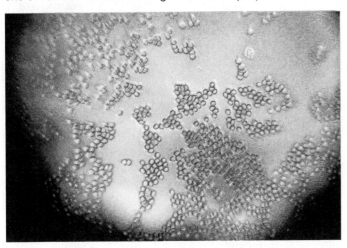

(c) . . . and animal sperm (drawing by van Leeuwenhoek of canine sperm cells on left, human on right).

FIGURE 1.1 Van Leeuwenhoek's Microscope Made Cells Visible.

theory. In addition to understanding what organisms are made of, scientists wanted to understand how cells come to be.

Where Do Cells Come From?

Most scientific theories have two components: The first describes a pattern in the natural world; the second identifies a mechanism or process that is responsible for creating that pattern. Hooke and his fellow scientists articulated the pattern component of the cell theory. In 1858, a German scientist named Rudolph Virchow added the process component by stating that all cells arise from preexisting cells.

The complete **cell theory** can be stated as follows: All organisms are made of cells, and all cells come from preexisting cells.

Two Hypotheses   The cell theory was a direct challenge to the prevailing explanation of where cells come from, called spontaneous generation. In the mid-1800s, most biologists believed that organisms could arise spontaneously under certain conditions. For example, the bacteria and fungi that spoil foods such as milk and wine were thought to appear in these nutrient-rich media of their own accord—springing to life from nonliving materials. In contrast, the cell theory maintained that cells do not spring to life spontaneously but are produced only when preexisting cells grow and divide. The all-cells-from-cells explanation was a **hypothesis:** a testable statement to explain a phenomenon or a set of observations.

Biologists usually use the word theory to refer to proposed explanations for broad patterns in nature and prefer hypothesis to refer to explanations for more tightly focused questions. A theory serves as a framework for the development of new hypotheses.

An Experiment to Settle the Question   Soon after Virchow's all-cells-from-cells hypothesis appeared in print, a French scientist named Louis Pasteur set out to test its predictions experimentally. An experimental **prediction** describes a measurable or observable result that must be correct if a hypothesis is valid.

Pasteur wanted to determine whether microorganisms could arise spontaneously in a nutrient broth or whether they appear only when a broth is exposed to a source of preexisting cells. To address the question, he created two treatment groups: a broth that was not exposed to a source of preexisting cells and a broth that was.

The spontaneous generation hypothesis predicted that cells would appear in both treatment groups. The all-cells-from-cells hypothesis predicted that cells would appear only in the treatment exposed to a source of preexisting cells.

FIGURE 1.2 (on page 4) shows Pasteur's experimental setup. Note that the two treatments are identical in every respect but one. Both used glass flasks filled with the same amount of the same nutrient broth. Both were boiled for the same amount of time to kill any existing organisms such as bacteria or fungi. But because the flask pictured in Figure 1.2a had a straight neck, it was exposed to preexisting cells after sterilization by the heat treatment. These preexisting cells are the bacteria and fungi that cling to dust particles in the air. They could drop into the nutrient broth because the neck of the flask was straight.

In contrast, the flask drawn in Figure 1.2b had a long swan neck. Pasteur knew that water would condense in the crook of the swan neck after the boiling treatment and that this pool of water

QUESTION: Do cells arise spontaneously or from other cells?

SPONTANEOUS GENERATION HYPOTHESIS: Cells arise spontaneously from nonliving materials.

ALL-CELLS-FROM-CELLS HYPOTHESIS: Cells are produced only when preexisting cells grow and divide.

(a) Pasteur experiment with straight-necked flask:

1. Place nutrient broth in straight-necked flask.

Cells

2. Boil to sterilize the flask (killing any living cells that were in the broth).

No cells

Cells

3. Preexisting cells enter flask from air.

(b) Pasteur experiment with swan-necked flask:

1. Place nutrient broth in swan-necked flask.

Cells

2. Boil to sterilize the flask (killing any living cells that were in the broth).

No cells

Condensation settles in neck

Cells

3. Preexisting cells from air are trapped in swan neck.

PREDICTION OF SPONTANEOUS GENERATION HYPOTHESIS: Cells will appear in broth.

PREDICTION OF ALL-CELLS-FROM-CELLS HYPOTHESIS: Cells will appear in broth.

PREDICTION OF SPONTANEOUS GENERATION HYPOTHESIS: Cells will appear in broth.

PREDICTION OF ALL-CELLS-FROM-CELLS HYPOTHESIS: Cells will not appear in broth.

RESULTS:

Cells

Both hypotheses supported

No cells

Spontaneous generation hypothesis rejected

CONCLUSION: Cells arise from preexisting cells, not spontaneously from nonliving material.

FIGURE 1.2 The Spontaneous Generation and All-Cells-from-Cells Hypotheses Were Tested Experimentally.

✔**QUESTION** What problem would arise in interpreting the results of this experiment if Pasteur had (1) put different types of broth in the two treatments, (2) heated them for different lengths of time, or (3) used a ceramic flask for one treatment and a glass flask for the other?

would trap any bacteria or fungi that entered on dust particles. Thus, the contents of the swan-necked flask were isolated from any source of preexisting cells even though still open to the air.

Pasteur's experimental setup was effective because there was only one difference between the two treatments and because that difference was the factor being tested—in this case, a broth's exposure to preexisting cells.

One Hypothesis Supported And Pasteur's results? As Figure 1.2 shows, the treatment exposed to preexisting cells quickly filled with bacteria and fungi. This observation was important because it showed that the heat sterilization step had not altered the nutrient broth's capacity to support growth.

The broth in the swan-necked flask remained sterile, however. Even when the flask was left standing for months, no organisms appeared in it. This result was inconsistent with the hypothesis of spontaneous generation.

Because Pasteur's data were so conclusive—meaning that there was no other reasonable explanation for them—the results persuaded most biologists that the all-cells-from-cells hypothesis was correct. However, you will see that biologists now have evidence that life did arise from nonlife early in Earth's history, through a process called chemical evolution (Chapters 2–6).

The success of the cell theory's process component had an important implication: If all cells come from preexisting cells, it follows that all individuals in an isolated population of single-celled organisms are related by common ancestry. Similarly, in you and most other multicellular individuals, all the cells present are descended from preexisting cells, tracing back to a fertilized egg. A fertilized egg is a cell created by the fusion of sperm and egg—cells that formed in individuals of the previous generation. In this way, all the cells in a multicellular organism are connected by common ancestry.

The second great founding idea in biology is similar, in spirit, to the cell theory. It also happened to be published the same year as the all-cells-from-cells hypothesis. This was the realization, made independently by the English scientists Charles Darwin and Alfred Russel Wallace, that all species—all distinct, identifiable types of organisms—are connected by common ancestry.

1.3 The Theory of Evolution by Natural Selection

In 1858 short papers written separately by Darwin and Wallace were read to a small group of scientists attending a meeting of the Linnean Society of London. A year later, Darwin published a book that expanded on the idea summarized in those brief papers. The book was called *The Origin of Species*. The first edition sold out in a day.

What Is Evolution?

Like the cell theory, the theory of evolution by natural selection has a pattern and a process component. Darwin and Wallace's theory made two important claims concerning patterns that exist in the natural world.

1. Species are related by common ancestry. This contrasted with the prevailing view in science at the time, which was that species represent independent entities created separately by a divine being.

2. In contrast to the accepted view that species remain unchanged through time, Darwin and Wallace proposed that the characteristics of species can be modified from generation to generation. Darwin called this process descent with modification.

Evolution is a change in the characteristics of a population over time. It means that species are not independent and unchanging entities, but are related to one another and can change through time.

What Is Natural Selection?

This pattern component of the theory of evolution was actually not original to Darwin and Wallace. Several scientists had already come to the same conclusions about the relationships between species. The great insight by Darwin and Wallace was in proposing a process, called **natural selection,** that explains *how* evolution occurs.

Two Conditions of Natural Selection Natural selection occurs whenever two conditions are met.

1. Individuals within a population vary in characteristics that are **heritable**—meaning, traits that can be passed on to offspring.

A **population** is defined as a group of individuals of the same species living in the same area at the same time.

2. In a particular environment, certain versions of these heritable traits help individuals survive better or reproduce more than do other versions.

If certain heritable traits lead to increased success in producing offspring, then those traits become more common in the population over time. In this way, the population's characteristics change as a result of natural selection acting on individuals. This is a key insight: Natural selection acts on individuals, but evolutionary change occurs in populations.

Selection on Maize as an Example To clarify how selection works, consider an example of **artificial selection**—changes in populations that occur when *humans* select certain individuals to produce the most offspring. Beginning in 1896, researchers began a long-term selection experiment on maize (corn).

1. In the original population, the percentage of protein in maize kernels was variable among individuals. Kernel protein content is a heritable trait—parents tend to pass the trait on to their offspring.

2. Each year for many years, researchers chose individuals with the highest kernel protein content to be the parents of the next generation. In this environment, individuals with high kernel protein content produced more offspring than individuals with low kernel protein content.

FIGURE 1.3 shows the results. Note that this graph plots generation number on the *x*-axis, starting from the first generation (0 on the graph) and continuing for 100 generations. The average percentage of protein in a kernel among individuals in this population is plotted on the *y*-axis.

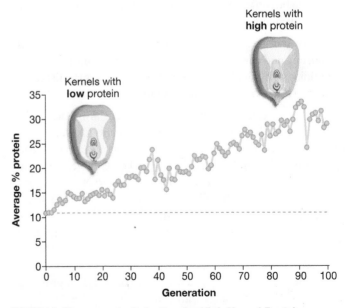

FIGURE 1.3 Response to Selection for High Kernel Protein Content in Maize.

DATA: Moose, S. P., J. W. Dudley, and T. R. Rocheford. 2004. *Trends in Plant Sciences* 9: 358–364; and the Illinois long-term selection experiment for oil and protein in corn (University of Illinois at Urbana–Champaign).

To read this graph, put your finger on the x-axis at generation 0. Then read up the y-axis, and note that kernels averaged about 11 percent protein at the start of the experiment. Now read the graph to the right. Each dot is a data point, representing the average kernel protein concentration in a particular generation. (A generation in maize is one year.) The lines on this graph simply connect the dots, to make the pattern in the data easier to see. During a few years the average protein content goes down, because of poor growing conditions or chance changes in how the many genes responsible for this trait interact. However, at the end of the graph, after 100 generations of selection, average kernel protein content is about 29 percent. (For more help with reading graphs, see **BioSkills 3** in Appendix B.)

This sort of change in the characteristics of a population, over time, is evolution. Humans have been practicing artificial selection for thousands of years, and biologists have now documented evolution by *natural* selection—where humans don't do the selecting—occurring in thousands of different populations, including humans. Evolution occurs when heritable variation leads to differential success in reproduction.

✔ QUANTITATIVE If you understand the concepts of selection and evolution, you should be able to describe how protein content in maize kernels changed over time, using the same x-axis and y-axis as in Figure 1.3, when researchers selected individuals with the *lowest* kernel protein content to be the parents of the next generation. (This experiment was actually done, starting with the same population at the same time as selection for high protein content.)

Fitness and Adaptation   Darwin also introduced some new terminology to identify what is happening during natural selection.

- In everyday English, fitness means health and well-being. But in biology, **fitness** means the ability of an individual to produce viable offspring. Individuals with high fitness produce many surviving offspring.

- In everyday English, adaptation means that an individual is adjusting and changing to function in new circumstances. But in biology, an **adaptation** is a trait that increases the fitness of an individual in a particular environment.

Once again, consider kernel protein content in maize: In the environment of the experiment graphed in Figure 1.3, individuals with high kernel protein content produced more offspring and had higher fitness than individuals with lower kernel protein content. In this population and this environment, high kernel protein content was an adaptation that allowed certain individuals to thrive.

Note that during this process, the amount of protein in the kernels of any individual maize plant did not change within its lifetime—the change occurred in the characteristics of the population over time.

Together, the cell theory and the theory of evolution provided the young science of biology with two central, unifying ideas:

1. The cell is the fundamental structural unit in all organisms.

2. All species are related by common ancestry and have changed over time in response to natural selection.

1.4   The Tree of Life

Section 1.3 focuses on how individual populations change through time in response to natural selection. But over the past several decades, biologists have also documented dozens of cases in which natural selection has caused populations of one species to diverge and form new species. This divergence process is called **speciation.**

Research on speciation has two important implications: All species come from preexisting species, and all species, past and present, trace their ancestry back to a single common ancestor.

The theory of evolution by natural selection predicts that biologists should be able to construct a **tree of life**—a family tree of organisms. If life on Earth arose just once, then such a diagram would describe the genealogical relationships between species with a single, ancestral species at its base.

Has this task been accomplished? If the tree of life exists, what does it look like?

Using Molecules to Understand the Tree of Life

One of the great breakthroughs in research on the tree of life occurred when American biologist Carl Woese (pronounced *woze*) and colleagues began analyzing the chemical components of organisms as a way to understand their evolutionary relationships. Their goal was to understand the **phylogeny** of all organisms—their actual genealogical relationships. Translated literally, phylogeny means "tribe-source."

To understand which organisms are closely versus distantly related, Woese and co-workers needed to study a molecule that is found in all organisms. The molecule they selected is called small subunit ribosomal RNA (rRNA). It is an essential part of the machinery that all cells use to grow and reproduce.

Although rRNA is a large and complex molecule, its underlying structure is simple. The rRNA molecule is made up of sequences of four smaller chemical components called ribonucleotides. These ribonucleotides are symbolized by the letters A, U, C, and G. In rRNA, ribonucleotides are connected to one another linearly, like the boxcars of a freight train.

Analyzing rRNA Why might rRNA be useful for understanding the relationships between organisms? The answer is that the ribonucleotide sequence in rRNA is a trait that can change during the course of evolution. Although rRNA performs the same function in all organisms, the sequence of ribonucleotide building blocks in this molecule is not identical among species.

In land plants, for example, the molecule might start with the sequence A-U-A-U-C-G-A-G (**FIGURE 1.4**). In green algae, which are closely related to land plants, the same section of the molecule might contain A-U-A-U-G-G-A-G. But in brown algae, which are not closely related to green algae or to land plants, the same part of the molecule might consist of A-A-A-U-G-G-A-C.

The research that Woese and co-workers pursued was based on a simple premise: If the theory of evolution is correct, then rRNA sequences should be very similar in closely related organisms but less similar in organisms that are less closely related. Species that are part of the same evolutionary lineage, like the plants, should share certain changes in rRNA that no other species have.

To test this premise, the researchers determined the sequence of ribonucleotides in the rRNA of a wide array of species. Then they considered what the similarities and differences in the sequences implied about relationships between the species. The goal was to produce a diagram that described the phylogeny of the organisms in the study.

A diagram that depicts evolutionary history in this way is called a phylogenetic tree. Just as a family tree shows relationships between individuals, a phylogenetic tree shows relationships between species. On a phylogenetic tree, branches that share a recent common ancestor represent species that are closely related; branches that don't share recent common ancestors represent species that are more distantly related.

The Tree of Life Estimated from Genetic Data To construct a phylogenetic tree, researchers use a computer to find the arrangement of branches that is most consistent with the similarities and differences observed in the data.

Although the initial work was based only on the sequences of ribonucleotides observed in rRNA, biologists now use data sets that include sequences from a wide array of genetic material. **FIGURE 1.5** shows a recent tree produced by comparing these sequences. Because this tree includes such a diverse array of

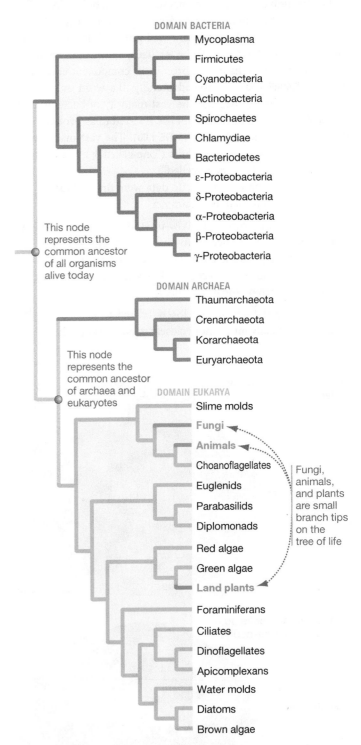

FIGURE 1.5 The Tree of Life. A phylogenetic tree estimated from a large amount of genetic sequence data. The three domains of life revealed by the analysis are labeled. Common names are given for lineages in the domains Bacteria and Eukarya. Phyla names are given for members of the domain Archaea, because most of these organisms have no common names.

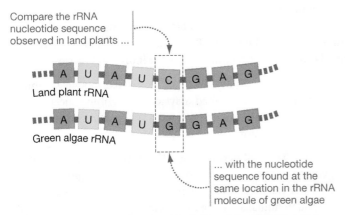

FIGURE 1.4 RNA Molecules Are Made Up of Smaller Molecules. A complete small subunit rRNA molecule contains about 2000 ribonucleotides; just 8 are shown in this comparison.

✔**QUESTION** Suppose that in the same section of rRNA, molds and other fungi have the sequence A-U-A-U-G-G-A-C. Are fungi more closely related to green algae or to land plants? Explain your logic.

species, it is often called the universal tree, or the tree of life. (For help in learning how to read a phylogenetic tree, see **BioSkills 7** in Appendix B.) Notice that the tree's main node is the common ancestor of all living organisms. Researchers who study the origin of life propose that the tree's root extends even further back to the "*last universal common ancestor*" of cells, or **LUCA.**

The tree of life implied by rRNA and other genetic data established that there are three fundamental groups or lineages of organisms: (**1**) the Bacteria, (**2**) the Archaea, and (**3**) the Eukarya. In all **eukaryotes,** cells have a prominent component called the nucleus (**FIGURE 1.6a**). Translated literally, the word eukaryotes means "true kernel." Because the vast majority of bacterial and archaeal cells lack a nucleus, they are referred to as **prokaryotes** (literally, "before kernel"; see **FIGURE 1.6b**). The vast majority of bacteria and archaea are unicellular ("one-celled"); many eukaryotes are multicellular ("many-celled").

When results based on genetic data were first published, biologists were astonished. For example:

- Prior to Woese's work and follow-up studies, biologists thought that the most fundamental division among organisms was between prokaryotes and eukaryotes. The Archaea were virtually unknown—much less recognized as a major and highly distinctive branch on the tree of life.

- Fungi were thought to be closely related to plants. Instead, they are actually much more closely related to animals.

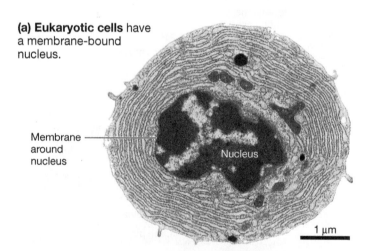

(a) Eukaryotic cells have a membrane-bound nucleus.

Membrane around nucleus

Nucleus

1 μm

(b) Prokaryotic cells do *not* have a membrane-bound nucleus.

No nucleus

0.1 μm

FIGURE 1.6 Eukaryotes and Prokaryotes.

✔**QUANTITATIVE** How many times larger is the eukaryotic cell in this figure than the prokaryotic cell? (Hint: Study the scale bars.)

- Traditional approaches for classifying organisms—including the system of five kingdoms divided into various classes, orders, and families that you may have learned in high school—are inaccurate in many cases, because they do not reflect the actual evolutionary history of the organisms involved.

The Tree of Life Is a Work in Progress   Just as researching your family tree can help you understand who you are and where you came from, so the tree of life helps biologists understand the relationships between organisms and the history of species. The discovery of the Archaea and the accurate placement of lineages such as the fungi qualify as exciting breakthroughs in our understanding of evolutionary history and life's diversity.

Work on the tree of life continues at a furious pace, however, and the location of certain branches on the tree is hotly debated. As databases expand and as techniques for analyzing data improve, the shape of the tree of life presented in Figure 1.5 will undoubtedly change. Our understanding of the tree of life, like our understanding of every other topic in biological science, is dynamic.

How Should We Name Branches on the Tree of Life?

In science, the effort to name and classify organisms is called **taxonomy.** Any named group is called a **taxon** (plural: **taxa**). Currently, biologists are working to create a taxonomy, or naming system, that accurately reflects the phylogeny of organisms.

Based on the tree of life implied by genetic data, Woese proposed a new taxonomic category called the **domain.** The three domains of life are the Bacteria, Archaea, and Eukarya.

Biologists often use the term **phylum** (plural: **phyla**) to refer to major lineages within each domain. Although the designation is somewhat arbitrary, each phylum is considered a major branch on the tree of life. Within the lineage called animals, biologists currently name 30–35 phyla—each of which is distinguished by distinctive aspects of its body structure as well as by distinctive gene sequences. For example, the mollusks (clams, squid, octopuses) constitute a phylum, as do chordates (the vertebrates and their close relatives).

Because the tree of life is so new, though, naming systems are still being worked out. One thing that hasn't changed for centuries, however, is the naming system for individual species.

Scientific (Latin) Names   In 1735, a Swedish botanist named Carolus Linnaeus established a system for naming species that is still in use today. Linnaeus created a two-part name unique to each type of organism.

- *Genus*   The first part indicates the organism's **genus** (plural: **genera**). A genus is made up of a closely related group of species. For example, Linnaeus put humans in the genus *Homo*. Although humans are the only living species in this genus, at least six extinct organisms, all of which walked upright and made extensive use of tools, were later also assigned to *Homo*.

- *Species*   The second term in the two-part name identifies the organism's species. Linnaeus gave humans the species name *sapiens.*

An organism's genus and species designation is called its **scientific name** or Latin name. Scientific names are always italicized. Genus names are always capitalized, but species names are not—as in *Homo sapiens*.

Scientific names are based on Latin or Greek word roots or on words "Latinized" from other languages. Linnaeus gave a scientific name to every species then known, and also Latinized his own name—from Karl von Linné to Carolus Linnaeus.

Linnaeus maintained that different types of organisms should not be given the same genus and species names. Other species may be assigned to the genus *Homo*, and members of other genera may be named *sapiens*, but only humans are named *Homo sapiens*. Each scientific name is unique.

Scientific Names Are Often Descriptive Scientific names and terms are often based on Latin or Greek word roots that are descriptive. For example, *Homo sapiens* is derived from the Latin *homo* for "man" and *sapiens* for "wise" or "knowing." The yeast that bakers use to produce bread and that brewers use to brew beer is called *Saccharomyces cerevisiae*. The Greek root *saccharo* means "sugar," and *myces* refers to a fungus. *Saccharomyces* is aptly named "sugar fungus" because yeast is a fungus and because the domesticated strains of yeast used in commercial baking and brewing are often fed sugar. The species name of this organism, *cerevisiae*, is Latin for "beer." Loosely translated, then, the scientific name of brewer's yeast means "sugar-fungus for beer."

Scientific names and terms often seem daunting at first glance. So, most biologists find it extremely helpful to memorize some of the common Latin and Greek roots. To aid you in this process, new terms in this text are often accompanied by a translation of their Latin or Greek word roots in parentheses. (A glossary of common root words with translations and examples is also provided in **BioSkills 2** in Appendix B.)

check your understanding

If you understand that . . .

- A phylogenetic tree shows the evolutionary relationships between species.
- To infer where species belong on a phylogenetic tree, biologists examine genetic and other characteristics of the species involved. Closely related species should have similar characteristics, while less closely related species should be less similar.

✔ **You should be able to . . .**

Examine the following rRNA ribonucleotide sequences and draw a phylogenetic tree showing the relationships between species A, B, and C that these data imply:

Species A: A A C T A G C G C G A T

Species B: A A C T A G C G C C A T

Species C: T T C T A G C G G T A T

Answers are available in Appendix A.

1.5 Doing Biology

This chapter has introduced some of the great ideas in biology. The development of the cell theory and the theory of evolution by natural selection provided cornerstones when the science was young; the tree of life is a relatively recent insight that has revolutionized our understanding of life's diversity.

These theories are considered great because they explain fundamental aspects of nature, and because they have consistently been shown to be correct. They are considered correct because they have withstood extensive testing.

How do biologists go about testing their ideas? Before answering this question, let's step back a bit and consider the types of questions that researchers can and cannot ask.

The Nature of Science

Biologists ask questions about organisms, just as physicists and chemists ask questions about the physical world or geologists ask questions about Earth's history and the ongoing processes that shape landforms.

No matter what their field, all scientists ask questions that can be answered by observing or measuring things—by collecting data. Conversely, scientists cannot address questions that can't be answered by observing or measuring things.

This distinction is important. It is at the root of continuing controversies about teaching evolution in publicly funded schools. In the United States and in Turkey, in particular, some Christian and Islamic leaders have been particularly successful in pushing their claim that evolution and religious faith are in conflict. Even though the theory of evolution is considered one of the most successful and best-substantiated ideas in the history of science, they object to teaching it.

The vast majority of biologists and many religious leaders reject this claim; they see no conflict between evolution and religious faith. Their view is that science and religion are compatible because they address different types of questions.

- Science is about formulating hypotheses and finding evidence that supports or conflicts with those hypotheses.

- Religious faith addresses questions that cannot be answered by data. The questions addressed by the world's great religions focus on why we exist and how we should live.

Both types of questions are seen as legitimate and important.

So how do biologists go about answering questions? After formulating hypotheses, biologists perform experimental studies, or studies that yield descriptive data, such as observing a behavior, characterizing a structure within a cell by microscopy, or sequencing rRNA. Let's consider two recent examples of this process.

Why Do Giraffes Have Long Necks? An Introduction to Hypothesis Testing

If you were asked why giraffes have long necks, you might say based on your observations that long necks enable giraffes to reach food that is unavailable to other mammals. This hypothesis is expressed in African folktales and has traditionally been accepted by many biologists. The food competition hypothesis is so plausible, in fact, that for decades no one thought to test it.

In the mid-1990s, however, Robert Simmons and Lue Scheepers assembled data suggesting that the food competition hypothesis is only part of the story. Their analysis supports an alternative hypothesis—that long necks allow giraffes to use their heads as effective weapons for battering their opponents, and that longer-necked giraffes would have a competitive advantage in fights.

Before exploring these alternative explanations, it's important to recognize that hypothesis testing is a two-step process:

Step 1 State the hypothesis as precisely as possible and list the predictions it makes.

Step 2 Design an observational or experimental study that is capable of testing those predictions.

If the predictions are accurate, the hypothesis is supported. If the predictions are not met, then researchers do further tests, modify the original hypothesis, or search for alternative explanations. But the process does not end here. Biologists also talk to other researchers. Over coffee, at scientific meetings, or through publications, biologists communicate their results to the scientific community and beyond. (You can see the Big Picture of the process of doing biology on pages 16–17.)

Now that you understand more about hypothesis testing, let's return to the giraffes. How did biologists test the food competition hypothesis? What data support their alternative explanation?

The Food Competition Hypothesis: Predictions and Tests The food competition hypothesis claims that giraffes compete for food with other species of mammals. When food is scarce, as it is during the dry season, giraffes with longer necks can reach food that is unavailable to other species and to giraffes with shorter necks. As a result, the longest-necked individuals in a giraffe population survive better and produce more young than do shorter-necked individuals, and average neck length of the population increases with each generation.

To use the terms introduced earlier, long necks are adaptations that increase the fitness of individual giraffes during competition for food. This type of natural selection has gone on so long that the population has become extremely long necked.

The food competition hypothesis makes several explicit predictions. For example, the food competition hypothesis predicts that:

* neck length is variable among giraffes;
* neck length in giraffes is heritable; and
* giraffes feed high in trees, especially during the dry season, when food is scarce and the threat of starvation is high.

The first prediction is correct. Studies in zoos and natural populations confirm that neck length is variable among individuals.

The researchers were unable to test the second prediction, however, because they studied giraffes in a natural population and were unable to do breeding experiments. As a result, they simply had to accept this prediction as an assumption. In

(a) Most feeding is done at about shoulder height.

(b) Typical feeding posture in giraffes

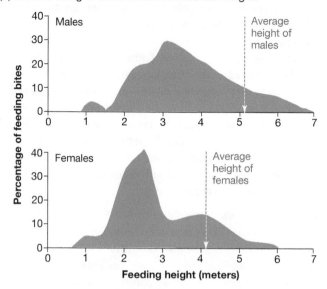

FIGURE 1.7 Giraffes Do Not Usually Extend Their Necks Upward to Feed.
DATA: Young, T. P., L. A. Isbell. 1991. *Ethology* 87: 79–89.

general, though, biologists prefer to test every assumption behind a hypothesis.

What about the prediction regarding feeding high in trees? According to Simmons and Scheepers, this is where the food competition hypothesis breaks down.

Consider, for example, data collected by a different research team on the amount of time that giraffes spend feeding in vegetation of different heights. **FIGURE 1.7a** plots the height of vegetation versus the percentage of bites taken by a giraffe, for males and for females from the same population in Kenya. The dashed line on each graph indicates the average height of a male or female in this population.

Note that the average height of a giraffe in this population is much greater than the height where most feeding takes place. In this population, both male and female giraffes spend most of their feeding time eating vegetation that averages just 60 percent of their full height. Studies on other populations of giraffes, during both the wet and dry seasons, are consistent with these data. Giraffes usually feed with their necks bent (**FIGURE 1.7b**).

These data cast doubt on the food competition hypothesis, because one of its predictions does not appear to hold. Biologists have not abandoned this hypothesis completely, though, because feeding high in trees may be particularly valuable during extreme droughts, when a giraffe's ability to reach leaves far above the ground could mean the difference between life and death. Still, Simmons and Scheepers have offered an alternative explanation for why giraffes have long necks. The new hypothesis is based on the mating system of giraffes.

The Sexual Competition Hypothesis: Predictions and Tests Giraffes have an unusual mating system. Breeding occurs year round rather than seasonally. To determine when females are coming into estrus or "heat" and are thus receptive to mating, the males nuzzle the rumps of females. In response, the females urinate into the males' mouths. The males then tip their heads back and pull their lips to and fro, as if tasting the liquid. Biologists who have witnessed this behavior have proposed that the males taste the females' urine to detect whether estrus has begun.

Once a female giraffe enters estrus, males fight among themselves for the opportunity to mate. Combat is spectacular. The bulls stand next to one another, swing their necks, and strike thunderous blows with their heads. Researchers have seen males knocked unconscious for 20 minutes after being hit and have cataloged numerous instances in which the loser died. Giraffes are not the only animals known to fight in this way—male giraffe weevils also use enormously long necks to fight for mating rights.

These observations inspired a new explanation for why giraffes have long necks. The sexual competition hypothesis is based on the idea that longer-necked giraffes are able to strike harder blows during combat than can shorter-necked giraffes. In engineering terms, longer necks provide a longer "moment arm." A long moment arm increases the force of an impact. (Think about the type of sledgehammer you'd use to bash down a concrete wall—one with a short handle or one with a long handle?)

The idea here is that longer-necked males should win more fights and, as a result, father more offspring than shorter-necked males do. If neck length in giraffes is inherited, then the average neck length in the population should increase over time. Under the sexual competition hypothesis, long necks are adaptations that increase the fitness of males during competition for females.

Although several studies have shown that long-necked males are more successful in fighting and that the winners of fights gain access to estrous females, the question of why giraffes have long necks is not closed. With the data collected to date, most biologists would probably concede that the food competition hypothesis needs further testing and refinement and that the sexual competition hypothesis appears promising. It could also be true that both hypotheses are correct. For our purposes, the important take-home message is that all hypotheses must be tested rigorously.

In many cases in biological science, testing hypotheses rigorously involves experimentation. Experimenting on giraffes is difficult. But in the case study considered next, biologists were able to test an interesting hypothesis experimentally.

How Do Ants Navigate? An Introduction to Experimental Design

Experiments are a powerful scientific tool because they allow researchers to test the effect of a single, well-defined factor on a particular phenomenon. Because experiments testing the effect of neck length on food and sexual competition in giraffes haven't been done yet, let's consider a different question: When ants leave their nest to search for food, how do they find their way back?

The Saharan desert ant lives in colonies and makes a living by scavenging the dead carcasses of insects. Individuals leave the burrow and wander about searching for food at midday, when temperatures at the surface can reach 60°C (140°F) and predators are hiding from the heat.

Foraging trips can take the ants hundreds of meters—an impressive distance when you consider that these animals are only about a centimeter long. But when an ant returns, it doesn't follow the same long, wandering route it took on its way away from the nest. Instead, individuals return in a straight line (**FIGURE 1.8**).

Once individuals are close to the nest, they engage in a characteristic set of back-and-forth U-turns until they find their nest hole. How do they do know how far they are from the nest?

The Pedometer Hypothesis Early work on navigation in desert ants showed that they use the Sun's position as a compass—meaning that they always know the approximate direction of the nest relative to the Sun. But how do they know how far to go?

After experiments had shown that the ants do not use landmarks to navigate, Matthias Wittlinger and co-workers set out to test a novel idea. The biologists proposed that Saharan desert ants know how far they are from the nest by integrating information from leg movements.

According to this pedometer hypothesis, the ants always know how far they are from the nest because they track the number

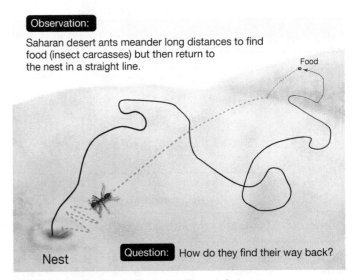

Observation:

Saharan desert ants meander long distances to find food (insect carcasses) but then return to the nest in a straight line.

Food

Question: How do they find their way back?

Nest

FIGURE 1.8 Navigation in Foraging Desert Ants.

of steps they have taken and their stride length. The idea is that they can make a beeline back toward the burrow because they integrate information on the angles they have traveled *and* the distance they have gone—based on step number and stride length.

If the pedometer hypothesis is wrong, however, then stride length and step number should have no effect on the ability of an ant to get back to its nest. This latter possibility is called a **null hypothesis.** A null hypothesis specifies what should be observed when the hypothesis being tested isn't correct.

Testing the Hypothesis To test their idea, Wittlinger's group allowed ants to walk from a nest to a feeder through a channel—a distance of 10 m. Then they caught ants at the feeder and created three test groups, each with 25 individuals (**FIGURES 1.9** and **1.10**):

- *Stumps* By cutting the lower legs of some individuals off, they created ants with shorter-than-normal legs.

- *Normal* Some individuals were left alone, meaning that they had normal leg length.

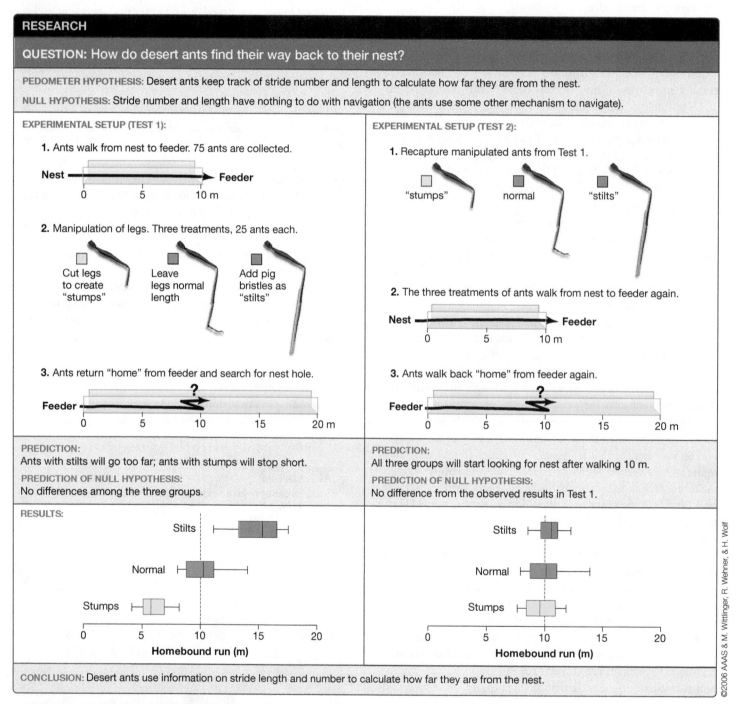

RESEARCH

QUESTION: How do desert ants find their way back to their nest?

PEDOMETER HYPOTHESIS: Desert ants keep track of stride number and length to calculate how far they are from the nest.

NULL HYPOTHESIS: Stride number and length have nothing to do with navigation (the ants use some other mechanism to navigate).

EXPERIMENTAL SETUP (TEST 1):

1. Ants walk from nest to feeder. 75 ants are collected.

Nest ————————————→ Feeder
0   5   10 m

2. Manipulation of legs. Three treatments, 25 ants each.

Cut legs to create "stumps"   Leave legs normal length   Add pig bristles as "stilts"

3. Ants return "home" from feeder and search for nest hole.

Feeder
0   5   10   15   20 m

EXPERIMENTAL SETUP (TEST 2):

1. Recapture manipulated ants from Test 1.

"stumps"   normal   "stilts"

2. The three treatments of ants walk from nest to feeder again.

Nest ————————————→ Feeder
0   5   10 m

3. Ants walk back "home" from feeder again.

Feeder
0   5   10   15   20 m

PREDICTION: Ants with stilts will go too far; ants with stumps will stop short.

PREDICTION OF NULL HYPOTHESIS: No differences among the three groups.

PREDICTION: All three groups will start looking for nest after walking 10 m.

PREDICTION OF NULL HYPOTHESIS: No difference from the observed results in Test 1.

RESULTS:

Stilts / Normal / Stumps
Homebound run (m)

Stilts / Normal / Stumps
Homebound run (m)

CONCLUSION: Desert ants use information on stride length and number to calculate how far they are from the nest.

©2006 AAAS & M. Wittlinger, R. Wehner, & H. Wolf

FIGURE 1.9 An Experimental Test: Do Desert Ants Use a "Pedometer"?

SOURCE: Wittlinger, M., R. Wehner, and H. Wolf. 2006. The ant odometer: Stepping on stilts and stumps. *Science* 312: 1965–1967.

✔**QUESTION** What is the advantage of using 25 ants in each group instead of just one?

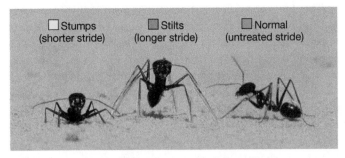

□ Stumps
(shorter stride)   ■ Stilts
(longer stride)   □ Normal
(untreated stride)

FIGURE 1.10 Manipulation of Desert Ant Stride Length.

- *Stilts* By gluing pig bristles onto each leg, the biologists created ants with longer-than-normal legs.

Next they put the ants in a different channel and recorded how far they traveled in a direct line before starting their nest-searching behavior. To see the data they collected, look at the graph on the left side of the "Results" section in Figure 1.9.

- *Stumps* The ants with stumps stopped short, by about 5 m, before starting to search for the nest opening.

- *Normal* The normal ants walked the correct distance—about 10 m.

- *Stilts* The ants with stilts walked about 5 m too far before starting to search for the nest opening.

To check the validity of this result, the researchers put the test ants back in the nest and recaptured them one to several days later, when they had walked to the feeder on their stumps, normal legs, or stilts. Now when the ants were put into the other channel to "walk back," they all traveled the correct distance—10 m—before starting to search for their nest (see the graph on the right side of the "Results" section in Figure 1.9).

The graphs in the "Results" display "box-and-whisker" plots that allow you to easily see where most of the data fall. Each box indicates the range of distances where 50 percent of the ants stopped to search for the nest. The whiskers indicate the lower extreme (stopping short of the nest location) and the upper extreme (going too far) of where the ants stopped to search. The vertical line inside each box indicates the median—meaning that half the ants stopped above this distance and half below. (For more details on how biologists report medians and indicate the variability and uncertainty in data, see **BioSkills 4** in Appendix B.)

Interpreting the Results The pedometer hypothesis predicts that an ant's ability to walk home depends on the number and length of steps taken on its outbound trip. Recall that a prediction specifies what we should observe if a hypothesis is correct. Good scientific hypotheses make testable predictions—predictions that can be supported or rejected by collecting and analyzing data. In this case, the researchers tested the prediction by altering stride length and recording the distance traveled on the return trip. Under the null hypothesis in this experiment, all the ants—altered and unaltered—should have walked 10 m in the first test before they started looking for their nest.

Important Characteristics of Good Experimental Design In relation to designing effective experiments, this study illustrates several important points:

- It is critical to include **control** groups. A control checks for factors, other than the one being tested, that might influence the experiment's outcome. In this case, there were two controls. Including a normal, unmanipulated individual controlled for the possibility that switching the individuals to a new channel altered their behavior. In addition, the researchers had to control for the possibility that the manipulation itself—and not the change in leg length—affected the behavior of the stilts and stumps ants. This is why they did the second test, where the outbound and return runs were done with the same legs.

- The experimental conditions must be as constant or equivalent as possible. The investigators used ants of the same species, from the same nest, at the same time of day, under the same humidity and temperature conditions, at the same feeders, in the same channels. Controlling all the variables except one—leg length in this case—is crucial because it eliminates alternative explanations for the results.

- Repeating the test is essential. It is almost universally true that larger sample sizes in experiments are better. By testing many individuals, the amount of distortion or "noise" in the data caused by unusual individuals or circumstances is reduced.

✔ If you understand these points, you should be able to explain: (1) What you would conclude if in the first test, the normal individual had not walked 10 m on the return trip before looking for the nest; and (2) What you would conclude if the stilts and stumps ants had not navigated normally during the second test.

From the outcomes of these experiments, the researchers concluded that desert ants use stride length and number to measure how far they are from the nest. They interpreted their results as strong support for the pedometer hypothesis.

The giraffe and ant studies demonstrate a vital point: Biologists practice evidence-based decision making. They ask questions about how organisms work, pose hypotheses to answer those questions, and use experimental or observational evidence to decide which hypotheses are correct.

The data on giraffes and ants are a taste of things to come. In this text you will encounter hypotheses and research on questions ranging from how water gets to the top of 100-meter-tall sequoia trees to how the bacterium that causes tuberculosis has become resistant to antibiotics. As you work through this book, you'll get lots of practice thinking about hypotheses and predictions, analyzing the nature of control treatments, and interpreting graphs.

A commitment to tough-minded hypothesis testing and sound experimental design is a hallmark of biological science. Understanding their value is an important first step in becoming a biologist.

CHAPTER 1 | REVIEW

For media, go to MasteringBiology

If you understand . . .

1.1 What Does It Mean to Say That Something Is Alive?

- There is no single, well-accepted definition of life. Instead, biologists point to five characteristics that organisms share.

✓ You should be able to explain why the cells in a dead organism are different from the cells in a live organism.

1.2 The Cell Theory

- The cell theory identified the fundamental structural unit common to all life.

✓ You should be able to describe the evidence that supported the pattern and the process components of the cell theory.

1.3 The Theory of Evolution by Natural Selection

- The theory of evolution states that all organisms are related by common ancestry.

- Natural selection is a well-tested explanation for why species change through time and why they are so well adapted to their habitats.

✓ You should be able to explain why the average protein content of seeds in a natural population of a grass species would increase over time, if seeds with higher protein content survive better and grow into individuals that produce many seeds with high protein content when they mature.

1.4 The Tree of Life

- The theory of evolution predicts that all organisms are part of a genealogy of species, and that all species trace their ancestry back to a single common ancestor.

- To construct this phylogeny, biologists have analyzed the sequences in rRNA and in an array of genetic material found in all cells.

- A tree of life, based on similarities and differences in these molecules, has three fundamental lineages, or domains: the Bacteria, the Archaea, and the Eukarya.

✓ You should be able to explain how biologists can determine which of the three domains a newly discovered species belongs to by analyzing its rRNA.

1.5 Doing Biology

- Biology is a hypothesis-driven, experimental science.

✓ You should be able to explain (1) the relationship between a hypothesis and a prediction and (2) why experiments are convincing ways to test predictions.

(MB) MasteringBiology

1. **MasteringBiology Assignments**

   Tutorials and Activities An Introduction to Graphing; Experimental Inquiry: What Can You Learn about the Process of Science from Investigating a Cricket's Chirp?; Introduction to Experimental Design; Levels of Life Card Game; Metric System Review; The Scientific Method

   Questions Reading Quizzes, Blue-Thread Questions, Test Bank

2. **eText** Read your book online, search, take notes, highlight text, and more.

3. **The Study Area** Practice Test, Cumulative Test, BioFlix® 3-D Animations, Videos, Activities, Audio Glossary, Word Study Tools, Art

You should be able to . . .

1. Anton van Leeuwenhoek made an important contribution to the development of the cell theory. How?
   a. He articulated the pattern component of the theory—that all organisms are made of cells.
   b. He articulated the process component of the theory—that all cells come from preexisting cells.
   c. He invented the first microscope and saw the first cell.
   d. He invented more powerful microscopes and was the first to describe the diversity of cells.

2. What does it mean to say that experimental conditions are controlled?
   a. The test groups consist of the same individuals.
   b. The null hypothesis is correct.
   c. There is no difference in outcome between the control and experimental treatment.
   d. All physical conditions except for one are identical for all groups tested.

3. The term *evolution* means that _____ change through time.

4. What does it mean to say that a characteristic of an organism is heritable?
   a. The characteristic evolves.
   b. The characteristic can be passed on to offspring.
   c. The characteristic is advantageous to the organism.
   d. The characteristic does not vary in the population.

5. In biology, to what does the term *fitness* refer?

6. Could *both* the food competition hypothesis and the sexual competition hypothesis explain why giraffes have long necks? Why or why not?
   a. No. In science, only one hypothesis can be correct.
   b. No. Observations have shown that the food competition hypothesis cannot be correct.
   c. Yes. Long necks could be advantageous for more than one reason.
   d. Yes. All giraffes have been shown to feed at the highest possible height and fight for mates.

7. What would researchers have to demonstrate to convince you that they had discovered life on another planet?

8. What did Linnaeus's system of naming organisms ensure?
   a. Two different organisms never end up with the same genus and species name.
   b. Two different organisms have the same genus and species name if they are closely related.
   c. The genus name is different for closely related species.
   d. The species name is the same for each organism in a genus.

9. What does it mean to say that a species is adapted to a particular habitat?

10. Explain how selection occurs during natural selection. What is selected, and why?

11. The following two statements explain the logic behind the use of molecular sequence data to estimate evolutionary relationships:

    "If the theory of evolution is true, then rRNA sequences should be very similar in closely related organisms but less similar in organisms that are less closely related."

    "On a phylogenetic tree, branches that share a recent common ancestor represent species that are closely related; branches that don't share recent common ancestors represent species that are more distantly related."

    Is the logic of these statements sound? Why or why not?

12. Explain why researchers formulate a null hypothesis in addition to a hypothesis when designing an experimental study.

13. A scientific theory is a set of propositions that defines and explains some aspect of the world. This definition contrasts sharply with the everyday usage of the word theory, which often carries meanings such as "speculation" or "guess." Explain the difference between the two definitions, using the cell theory and the theory of evolution by natural selection as examples.

14. Turn back to the tree of life shown in Figure 1.5. Note that Bacteria and Archaea are prokaryotes, while Eukarya are eukaryotes. On the simplified tree below, draw an arrow that points to the branch where the structure called the nucleus originated. Explain your reasoning.

    BACTERIA

    ARCHAEA

    EUKARYA

15. The proponents of the cell theory could not "prove" that it was correct in the sense of providing incontrovertible evidence that all organisms are made up of cells. They could state only that all organisms examined to date were made of cells. Why was it reasonable for them to conclude that the theory was valid?

16. Some humans have heritable traits that make them resistant to infection by HIV. In areas of the world where HIV infection rates are high, are human populations evolving? Explain your logic.
    a. No. HIV infection rates would not affect human evolution.
    b. Yes. The heritable traits that confer resistance to HIV should increase over time.
    c. No. The heritable traits that confer resistance to HIV should decrease over time.
    d. Yes. The heritable traits that confer resistance to HIV should decrease over time.

![BIG PICTURE]

The Big Picture

Biologists study the characteristics of life. The cell theory, the theory of evolution by natural selection, and the tree of life are some of the great ideas in biology that came about by biologists asking questions that can be answered by observing or measuring things—that is, by collecting data.

Notice that the study of life is not a series of linear steps with a beginning and an end. Instead, the process of doing biology is dynamic and ongoing. The answer to one question may lay the foundation for twenty more questions. Working together, biologists from different disciplines integrate data across many levels, from atoms to the biosphere.

Note that the gray numbers in boxes tell you where to go for more information. Also, be sure to do the blue exercises in the Check Your Understanding box.

Characteristics of living things
- Energy
- Cells
- Information
- Replication
- Evolution

1.1

Text section where you can find more information

focuses on

Scientists regularly integrate across many of these levels

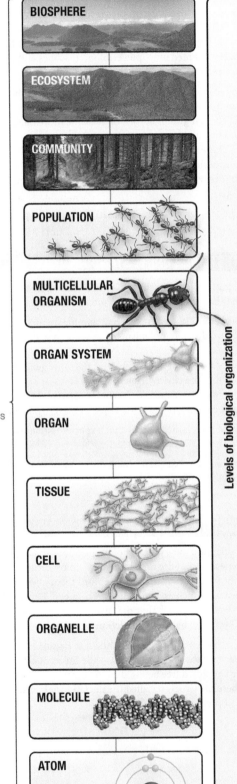

check your understanding

Ⓒ Ⓨ Ⓤ

If you understand the big picture . . .

✔ **You should be able to . . .**

1. Describe how biologists go about testing their ideas.
2. Provide an example of how an experimental study could span more than one level of biological organization.
3. Compare and contrast a hypothesis with a theory.
4. Propose the next step to take if data support the hypothesis you are testing.

Answers are available in Appendix A.

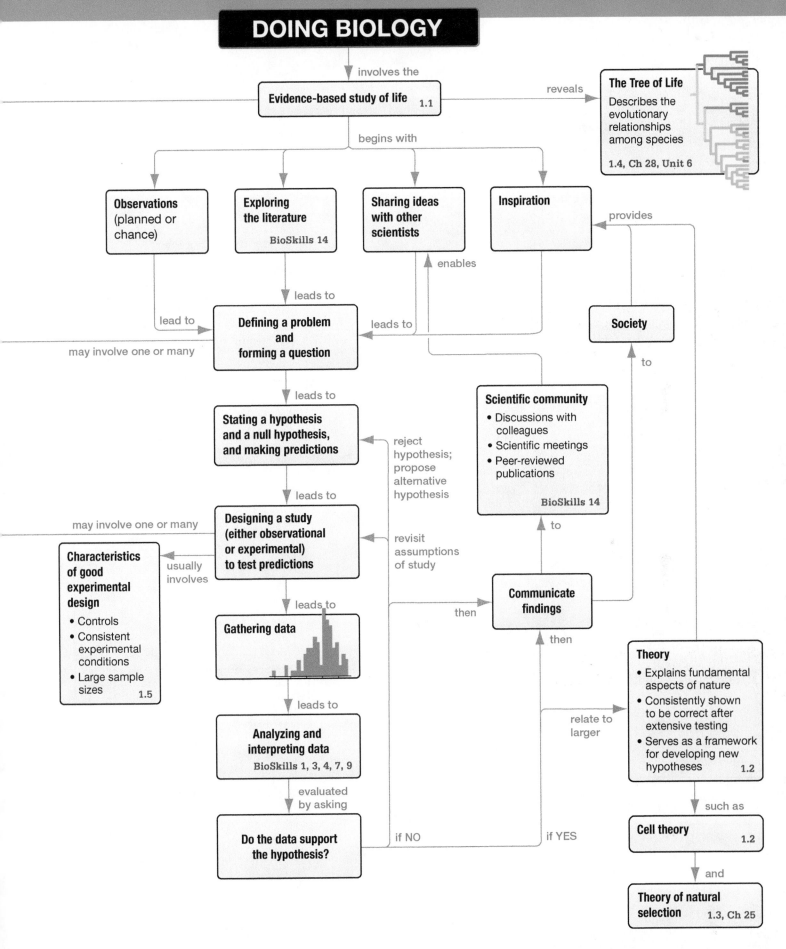

DOING BIOLOGY

Evidence-based study of life 1.1

involves the

reveals → **The Tree of Life**
Describes the evolutionary relationships among species
1.4, Ch 28, Unit 6

begins with

Observations (planned or chance)

Exploring the literature
BioSkills 14

Sharing ideas with other scientists

Inspiration

provides

lead to

leads to

enables

leads to

Defining a problem and forming a question

may involve one or many

Society

leads to

to

Stating a hypothesis and a null hypothesis, and making predictions

reject hypothesis; propose alternative hypothesis

Scientific community
- Discussions with colleagues
- Scientific meetings
- Peer-reviewed publications

BioSkills 14

leads to

to

Designing a study (either observational or experimental) to test predictions

may involve one or many

usually involves

revisit assumptions of study

Characteristics of good experimental design
- Controls
- Consistent experimental conditions
- Large sample sizes
1.5

leads to

Communicate findings

then

Gathering data

then

leads to

relate to larger

Theory
- Explains fundamental aspects of nature
- Consistently shown to be correct after extensive testing
- Serves as a framework for developing new hypotheses
1.2

Analyzing and interpreting data
BioSkills 1, 3, 4, 7, 9

evaluated by asking

such as

Do the data support the hypothesis?

if NO

if YES

Cell theory 1.2

and

Theory of natural selection 1.3, Ch 25

37 Plant Form and Function

In this chapter you will learn how

Plant structure lays the foundation for plant function

by examining

Themes and variations in plant form **37.1**

looking closer at

Roots   Shoots   Leaves

which develop from

Primary growth (extends the plant body) **37.2**

and

Secondary growth (widens shoots and roots) **37.4**

and

Plant tissue systems (dermal, ground, vascular) and their cells **37.3**

All plants are able to harvest diffuse resources and concentrate them in cells and tissues, but their forms and strategies are diverse. These baobab trees in Madagascar may live to be hundreds of years old despite drought conditions, in part by storing water in their enormous trunks.

This chapter is part of the Big Picture. See how on pages 840–841.

Photosynthetic plants carry out the most remarkable biochemistry of any terrestrial organisms. Using the energy in sunlight and the simplest of starting materials—carbon dioxide, water, and ions containing nitrogen, phosphorus, potassium, and other key atoms—plants synthesize thousands of different carbohydrates, proteins, nucleic acids, and lipids. They use these compounds to build bodies that may live for thousands of years.

This feat is even more impressive when you consider that the simple starting materials that plants need to grow are tiny and diffuse—carbon dioxide molecules, water molecules, nitrate ions, and other resources are usually found at low concentrations over a large area. To gather the raw materials required for their sophisticated biosynthetic machinery, a plant's roots and shoots grow outward, extending the individual into the soil and atmosphere.

✔ When you see this checkmark, stop and test yourself. Answers are available in Appendix A.

In essence, a plant's body harvests diffuse resources and concentrates them in cells and tissues. The structure of its body is dynamic, because most plants exhibit **indeterminate growth**; that is, they grow throughout their lives. A 4750-year-old bristlecone pine has roots and shoots that are still growing. In response to favorable conditions, a plant sends shoots and roots in the most promising directions, seeking light and the simple compounds it requires.

The contrast between the plant and animal way of life is striking. Most animals move around, eat concentrated sources of food, and avoid stressful conditions. But plants stay in one place, extend their roots and shoots to harvest diffuse resources, make their own food, and cope with stress where they stand.

This chapter focuses on three fundamental questions:

1. How is the plant body organized?

2. Why are plants so diverse in size and shape?

3. How do plants grow throughout their lives?

Instead of surveying the entire catalog of land plants, though, the focus here is on the flowering plants, or angiosperms. Recall that angiosperms are the most recent major group of plants to appear in the fossil record (see Chapter 31). With about 250,000 species described, they are by far the most abundant, species rich, and geographically widespread plant group today. It is also difficult to overstate their economic and medical importance to humans. Most of the food we eat and many of the medicines we use are derived from angiosperms.

If you look outside or down the produce aisle of a grocery store, you will see a wide diversity of plants and plant products. By the time you finish this chapter, you'll understand how these plant bodies are put together and how they grow. Exploring questions about the anatomy of flowering plants is vital to understanding the world at large as well as the other chapters in this unit. (You can review the importance of plant form and function in the Big Picture of plant and animal physiology on pages 840–841.)

37.1 Plant Form: Themes with Many Variations

Plants—along with algae, cyanobacteria, and a variety of protists—obtain energy and carbon to grow and reproduce (see Chapter 10). Plants use light energy to synthesize carbohydrates using carbon dioxide from the air and water from the soil.

For photosynthesis to occur efficiently, plants need large amounts of light and carbon dioxide, along with water as an electron source. Plants also need large amounts of water to fill their cells and maintain them at normal volume and pressure.

To synthesize nucleic acids, enzymes, phospholipids, and the other molecules needed to build and maintain cells, plants must obtain nitrogen (N), phosphorus (P), potassium (K), magnesium (Mg), and a host of other nutrients. Most of these key elements exist in nature as ions that dissolve in water found in soil.

FIGURE 37.1 labels the major structures in the two basic systems that plants use to acquire the resources they need for photosynthesis. A belowground portion called the **root system** anchors

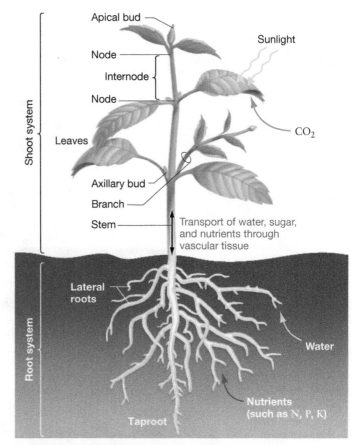

FIGURE 37.1 Typical Root and Shoot Systems Acquire and Transport Resources. Shoot systems are specialized for harvesting light and CO$_2$. Root systems absorb water and key nutrients such as nitrogen (N), phosphorus (P), and potassium (K).

✔**QUESTION** Suppose that this plant's growth was limited by access to light and nutrients, and that the plant had a new deposit of nutrient-rich soil to the left and much more sunlight suddenly available to the right. Explain what you would expect this individual to look like in one month.

the plant and takes in water and nutrients from the soil; an aboveground portion called the **shoot system** harvests light and carbon dioxide from the atmosphere to produce sugars. Both systems grow throughout the life of the individual, allowing the plant to increase in size, acquire resources, and reproduce.

In most plants, vascular tissue connects the root and shoot systems. Through vascular tissue, water and nutrients are transported from roots to shoots; sugars can be transported in both directions. As you learn about how the plant form has evolved to serve necessary functions, compare and contrast what you learn with how animals have adapted to new environmental challenges. In many ways the two groups are remarkably similar.

The Importance of Surface Area/Volume Relationships

Before exploring the nature of root and shoot systems in more detail, it's important to recognize a key structural relationship that is critical to their function.

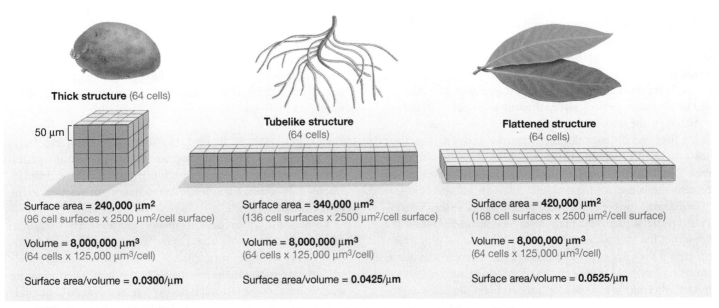

Thick structure (64 cells)

50 μm

Tubelike structure
(64 cells)

Flattened structure
(64 cells)

Surface area = **240,000 μm²**
(96 cell surfaces x 2500 μm²/cell surface)

Surface area = **340,000 μm²**
(136 cell surfaces x 2500 μm²/cell surface)

Surface area = **420,000 μm²**
(168 cell surfaces x 2500 μm²/cell surface)

Volume = **8,000,000 μm³**
(64 cells x 125,000 μm³/cell)

Volume = **8,000,000 μm³**
(64 cells x 125,000 μm³/cell)

Volume = **8,000,000 μm³**
(64 cells x 125,000 μm³/cell)

Surface area/volume = **0.0300/μm**

Surface area/volume = **0.0425/μm**

Surface area/volume = **0.0525/μm**

FIGURE 37.2 The Morphology of Roots and Leaves Gives Them a High Surface-Area-to-Volume Ratio. In this example, the "thick structure" represents a tree trunk or potato-like storage organ; the "tubelike structure" represents a root; the "flattened structure" represents a leaf. Note that each schematic structure has the same number of cells and the same total volume—but a very different surface area.

Root and shoot systems both function in absorption—of water and key ions, or of light, respectively. Absorption takes place across a surface. But the cells that use the absorbed molecules and light occupy a volume. Thus, a plant body is more efficient as an absorption-and-synthesis machine when it has a large surface area relative to its volume.

FIGURE 37.2 illustrates this point. In this example, the cells in a plant are represented by cubes; the side of each cell is 50 μm long. Thus, each face of a cell has a surface area of $50 \times 50 = 2500$ μm²; each cell has a volume of $50 \times 50 \times 50 = 125,000$ μm³. Follow the calculations in the figure and note that:

- If 64 cells are arranged in a cube, the surface area/volume relationship is 0.0300/μm.

- If 64 cells are arranged in a long tube, the surface area/volume relationship is 0.0425/μm.

- If 64 cells are arranged in a flat sheet, the surface area/volume relationship is 0.0525/μm.

This simple exercise has an important punch line: Tubes and sheets have much more surface area relative to their volume than cubes. It's no surprise, then, that the absorptive regions of a root system are tubelike, and the absorptive regions of a shoot system are the flattened structures called leaves. Storage tissues such as tubers and seeds have a low surface-area-to-volume ratio because they are not involved in absorption.

The Root System

Many root systems have a vertical **taproot,** as well as numerous **lateral roots** that run more or less horizontally. The root system anchors the plant in soil, absorbs water and ions from the soil, conducts water and selected ions to the shoot, obtains energy in

the form of sugar from the shoot, and stores material produced in the shoot for later use.

Root systems can be impressive in extent. For example, in 1937 botanist Howard Dittmer grew a winter rye plant in a container full of soil for four months. He then unearthed the plant and meticulously measured its roots. The root system of this single individual contained more than 14 billion identifiable structures with a combined length of over 11,000 km—over one-fourth of Earth's circumference! The total surface area of the root system was almost 640 m², or about the size of 1.5 basketball courts. A root system like this is clearly adapted to absorbing diffuse resources located underground.

Other studies have shown that (**1**) the roots of trees routinely extend beyond the width of their aboveground canopy, and (**2**) it is not unusual for a plant's root system to represent over 80 percent of its total mass. Many plants devote a great deal of energy and resources to the growth of their root systems.

Although most root systems contain the same general structures, the root systems are diverse. This diversity can be analyzed on three levels:

1. morphological diversity among species;

2. phenotypic plasticity, or changes in the structure of an individual's root system in response to the environment; and

3. modified roots that are specialized for unusual functions.

Let's consider each level in turn.

Morphological Diversity in Root Systems As an example of the range of morphological diversity observed in the root systems of angiosperms, consider prairie plants.

Prairies are grassland ecosystems found in areas of the world such as central North America, the Serengeti Plain of East Africa, the Pampas region of Argentina, and the steppes of central Asia (see Chapter 52). Rain is abundant enough in these areas to support a lush growth of **herbaceous plants**—seed plants that lack woody tissue. Rain is scarce enough to exclude trees and most shrubs, however. The growth of woody species is also discouraged by fires that regularly sweep through these ecosystems.

Although the aboveground portions of prairie plants burn during fires and die back during the winter or dry season, their root systems are **perennial,** meaning that they live for many years. The root system sends up a new shoot system each spring and also after a fire.

To examine the root systems of prairie plants, researchers excavate around a particular plant to expose the roots. **FIGURE 37.3** shows that the root systems of prairie plants can be very different, even if they live next to each other. For example, the dense, fibrous root systems of junegrass and switchgrass do not have a taproot, whereas the taproot of a compass plant can reach depths of over 4.5 m.

Diversity in root system structure has important consequences for competition between species. The tips of the smallest roots are where most water and nutrient absorption take place, but in prairies, the small root tips of different species are often found at different depths in the soil.

To explain the diversity of root systems observed among species that grow in the same habitat, biologists suggest that natural selection has favored structures that minimize competition for water and nutrients.

Phenotypic Plasticity in Root Systems Morphological diversity in roots occurs within species as well as among species. Some of the within-species variation is due to genetic diversity among individuals, but some is due to how roots respond to the environment.

Roots show a great deal of **phenotypic plasticity**—meaning that their form is changeable, depending on environmental conditions. For example, spruce trees growing in waterlogged soils tend to have root systems less than a meter deep. Their roots are shallow because the wet soil lacks oxygen, and root cells, like animal cells, suffocate in the anoxic conditions. The same tree growing in drier soil would develop a root system several meters deep.

The key point is that even genetically identical individuals will have very different root systems if they grow in different environments.

Phenotypic plasticity is particularly important in plants because they grow throughout their lives, and sometimes their environment changes. Root systems that grow into nutrient-rich septic fields or sewer pipes that leak human waste are a prime example. Roots actively grow into areas of soil where resources are abundant; roots stop growing or die back in areas where resources are used up or lacking.

Modified Roots The taproots and fibrous roots illustrated earlier do not begin to exhaust the types of roots found among plants (**TABLE 37.1**). For example, some roots are **adventitious**—meaning they develop from an unusual source, the shoot system instead of the root system.

- In ivy, anchor roots—adventitious roots that grow from nodes in the shoot system—help anchor individuals to brick walls or other structures.

- The prop roots of corn are adventitious roots that help brace individuals in windy weather.

Even roots that are part of the root system can have specialized functions—meaning that they do things other than absorbing water and nutrients and anchoring the shoot system.

The pneumatophores of mangroves in the genus *Avicennia* are specialized lateral roots that function in gas exchange. These mangroves grow in submerged habitats where fine silt is deposited, cutting off the oxygen supply to their roots. Their root cells do not suffocate, however, because oxygen from the atmosphere can diffuse into the root system through the pneumatophores. These roots grow upward—not downward—in response to gravity.

The thick taproot of some biennial plants, such as carrots, stores starch during the first of the plants' two growing seasons. Carrot roots, the ones we eat, are normally harvested at this point. If carrots were not harvested, the plants would resume

FIGURE 37.3 Plants Have Diverse Root Systems. The roots of prairie plants that live side by side can be very different.

✔ QUESTION Why are lawns not drought tolerant?

Form and Function of Modified Roots

Modified Root	Example
Anchor roots These adventitious roots anchor stems to walls and to other plants	
Prop roots These adventitious roots stabilize the stem	
Pneumatophores Pneumatophores allow gas exchange between roots and atmosphere	

✓**QUESTION** Why do root cells need oxygen?

growth after the winter and use that stored energy during a second summer to "go to seed," producing a large flowering shoot. Starch in the taproot makes the energy-intensive reproductive process possible.

The Shoot System

As Figure 37.1 indicates, the shoot system has an array of important anatomical features.

- The shoot system consists of one or more **stems,** which are vertical aboveground structures.
- A stem consists of **nodes,** where leaves are attached, and **internodes,** or segments between nodes.

- A **leaf** is an appendage that projects from a stem laterally. Leaves usually function as photosynthetic organs.
- The nodes where leaves attach to the stem are also the site of **axillary** (or **lateral**) **buds,** which form just above the site of leaf attachment.
- If conditions are appropriate, an axillary bud may grow into a **branch**—a lateral extension of the shoot system.
- The tip of each stem and branch contains an **apical bud,** where growth occurs that extends the length of the stem or branch.
- If conditions are appropriate, apical or axillary buds may develop into flowers or other reproductive structures.

In essence, the shoot system is a repeating series of nodes, internodes, leaves, and apical and axillary buds. As plants grow, the number of nodes, internodes, and leaves increases. After an initial period of growth, however, an internode does not increase much in length over time.

As with root systems, diversity in shoots can be analyzed on three levels: morphological diversity among species, phenotypic plasticity within individuals, and modified shoots with specialized functions.

Morphological Diversity in Shoot Systems The shoot systems of plants, essentially the visible part of the plant, range in size from species like the tiny (<5 mm diameter) duckweed that you may have seen growing on the surface of stagnant ponds to redwood trees that reach heights of over 100 m (300 ft) and giant sequoias with trunks that weigh 2.6 million kg (over 5.7 million lb)—about the same as 10 diesel locomotives.

The shape of the shoot system also varies a great deal among species. For example, the manner in which new branches are added as the shoot system grows affects the shape of the individual and its ability to compete for light. As **FIGURE 37.4** shows, a plant growing with wide branching angles and short internodes has a very different shape from that of a plant with narrow branching angles and long internodes.

FIGURE 37.4 Plant Form Can Vary as a Function of Branch Angle and Internode Length.

Variation in the size and shape of the shoot system is important: It allows plants of different species to harvest light at different locations and thus minimize competition. It also allows plants to thrive in a wide array of habitats.

As an example of how the shape of a shoot system varies among species in different environments, consider the silversword plants native to Hawaii. You might recall that all of the silverswords are believed to be descended from the same ancestor—a species of tarweed that arrived in Hawaii from the west coast of North America about 5 million years ago (see Chapter 28).

Silverswords represent an adaptive radiation: a lineage that rapidly split into many species occupying a wide array of habitats. Their shoot systems are particularly diverse in size, shape, and growth habit (see Figure 28.9). Some silverswords grow low to the ground in dense mats; some form bunched rosettes of leaves; some are vines; others are woody shrubs or even small- to medium-sized trees.

Biologists interpret this diversity of shoot systems as a suite of adaptations for harvesting light and carbon dioxide in different environments.

- In lush habitats, where competition for light is intense, woody individuals that grow tall are favored by natural selection.

- In dry, windblown habitats, individuals with short stems or rosettes thrive because they require less water than taller individuals, and they don't blow over.

The adaptive radiation of silverswords has been based in part on diversification in their shoot systems.

Phenotypic Plasticity in Shoot Systems The size and shape of an individual's shoot system can vary dramatically based on variation in growing conditions: temperature, exposure to wind, and availability of water, nutrients, and light.

This conclusion was driven home in an experiment conducted by Jens Clausen and colleagues in the late 1930s. These biologists transplanted several species of herbaceous plants between sites along an elevational gradient: from sea level to alpine habitats. In each case, the transplanted individuals were propagated from cuttings—meaning that they were genetically identical to individuals growing at the other locations. As the "Results" section in **FIGURE 37.5** shows, the overall size and shape of the shoot system varied markedly among locations.

Because a plant's shoot system continues to grow over the course of its lifetime, it can respond to changes in environmental conditions just like the root system can. Experiments on phototropism, for example, established that shoot systems can bend toward light if an individual is shaded on one side (see Chapter 40). Plants also undergo differential growth, producing more branches and leaves in regions of the body that are exposed to the highest light levels. A plant's shoot system grows in directions that maximize its chances of capturing light.

Modified Stems As in root systems, the stems of many plants have modified structures to serve various functions (**TABLE 37.2**).

- Many desert cacti have highly modified stems. Instead of functioning primarily to support leaves, cactus stems often

---

QUESTION: How much does a plant's growth form depend on its environment?

HYPOTHESIS: (No explicit hypothesis—the goal of this experiment was to explore the interaction between genetic makeup versus environmental influence on size and shape.)

EXPERIMENTAL SETUP:

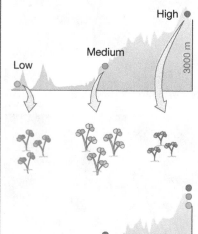

1. **Take cuttings** from individuals of *Potentilla glandulosa* growing in low-, medium-, and high-elevation habitats in the Sierra Nevada mountains.

2. **Propagate** genetically identical individuals.

3. **Transplant** individuals from each source population into each habitat (low, medium, and high elevation). Allow to grow and observe mature plants.

PREDICTION: (No explicit predictions.)

RESULTS:

Examples of mature plants observed:

| "High" plant **grown at** low elevation | "High" plant **grown at** medium elevation | "High" plant **grown at** high elevation |

CONCLUSION: Environmental conditions have a profound influence on body size and shape (genetically identical plants look different at each site). BUT, genetic makeup also has a large influence on plant morphology. (Though not pictured here, plants from each source population look different, even when grown in the same habitat.)

FIGURE 37.5 Experimental Evidence for Phenotypic Plasticity in Shoot Systems.

SOURCE: Clausen, J., D. D. Keck, and W. M. Hiesey. 1945. Experimental studies on the nature of species. II. Plant evolution through amphiploidy and autoploidy, with examples from the Madiinae. Publication No. 564. Washington, DC: Carnegie Institute of Washington.

✔ **QUESTION** Why was it important for the researchers to propagate the individuals from cuttings?

Modified Stem	Example
Water-storage structures	
Cactus stems (shown in cross section) store water	
Stolons	
Strawberry stolons produce new individuals at nodes aboveground	
Rhizomes	
Rhizomes produce new individuals at nodes belowground, and store carbohydrates	
Tubers	
Tubers such as potatoes store carbohydrates such as starch	
Thorns	
Thorns provide protection from herbivores	

enlarge into water-storage organs. Water accounts for up to 98 percent of the weight of a cactus stem. These cactus stems also contain the plant's photosynthetic tissue.

- **Stolons** are modified stems that grow horizontally along the soil surface, producing adventitious roots and leaves at each node. Because new plants form at these nodes, stolons function in asexual reproduction (see Chapter 13).

- Like stolons, **rhizomes** are stems that grow horizontally instead of vertically. They produce new plants at nodes and thus participate in asexual reproduction. But while stolons grow aboveground, rhizomes spread belowground. One quaking aspen plant in Utah was found to have 47,000 stems rising from a network of rhizomes, making it one of the largest organisms known. What we see as individual aspen trees are actually stems that are connected underground.

- **Tubers** are underground, swollen rhizomes that function as carbohydrate-storage organs. The eyes of a potato—a typical tuber—are nodes in the stem where new branches may arise.

- **Thorns** are modified stems that help protect the plant from attacks by large **herbivores,** or plant eaters, such as deer, giraffe, or cattle.

The Leaf

In most plant species, the vast majority of photosynthesis occurs in the part of the shoot system we call leaves. The total area of leaf produced by a single plant can be enormous—a single tree can have hundreds of thousands of leaves with a total surface area equivalent to that of a football field. All of this area is available for absorbing photons and supporting photosynthesis.

A simple leaf (**FIGURE 37.6a** on page 738) is composed of just two major structures: an expanded portion called the **blade** and a stalk called the **petiole.** But leaves exhibit many variations on the central theme of a flattened structure specialized for performing photosynthesis.

Morphological Diversity in Leaves   Glance outside or stroll through a garden, and you'll find many types of simple leaves with an easily recognizable blade and petiole. (Grass leaves will stump you, though, because they lack petioles entirely.) You will also find compound leaves that have blades divided into a series of leaflets (**FIGURE 37.6b**). You may even encounter doubly compound leaves, which have leaflets that are again divided (**FIGURE 37.6c**).

Not all leaf blades are thin with a large surface area, however. For example, plants that thrive in deserts and in cold, dry habitats tend to have needle-shaped leaves (**FIGURE 37.6d**). The leading hypothesis to explain this pattern is based on two observations: (1) Water is often in short supply in these environments because it is scarce in deserts, or frozen and thus unavailable in cold habitats; and (2) leaves with large surface areas lose large amounts of water through an evaporative process called **transpiration** (discussed in Chapter 38).

Thus, needlelike leaves are interpreted as adaptations that minimize transpiration in water-scarce habitats. Small, narrow

(a) Simple leaves have a petiole and a single blade.

(b) Compound leaves have blades divided into leaflets.

(c) Doubly compound leaves are large yet rarely damaged by wind or rain.

(d) Needlelike leaves are characteristic of species adapted to very cold or hot climates.

Blade

Petiole

FIGURE 37.6 Leaves Vary in Size and Shape. The structures in parts **(a)** through **(c)** represent single leaves; part **(d)** shows two leaves.

✔**QUESTION** For capturing photons, what is the advantage of having a leaf with a large surface area? In terms of wind damage and water loss, what is the disadvantage of having a leaf with a large surface area?

leaves are also much less susceptible to wind damage than are large, broad leaves.

The arrangement of leaves on a stem can vary as much as leaf shape. For example, leaves can be:

- arranged to alternate on either side of the stem (**FIGURE 37.7a**);
- paired opposite each other on the stem (**FIGURE 37.7b**);
- arranged in a whorl (**FIGURE 37.7c**);
- found in a compact basal arrangement where internodes are extremely short—leading to the rosette growth form (**FIGURE 37.7d**).

Leaf shape and the arrangement of leaves on a stem are usually determined genetically, so these characters are often used for plant identification.

Phenotypic Plasticity in Leaves Even though leaves do not grow continuously, they exhibit phenotypic plasticity just like root systems and stems do.

Leaves from the same individual that grow in sun versus shade serve as a prominent example of phenotypic plasticity in leaf morphology. As the oak tree leaves in **FIGURE 37.8** show:

- *Sun leaves* are thicker and have a relatively small surface area, which reduces water loss in areas of the body where light is abundant.
- *Shade leaves* are relatively thin and broad, providing a high surface area that maximizes absorption of rare photons.

Water loss is less of a problem for shade leaves because temperatures are cooler in shade than in bright sun.

(a) Alternate leaves

(b) Opposite leaves

(c) Whorled leaves

(d) Rosette

FIGURE 37.7 The Arrangement of Leaves on Stems Varies.

Grown in shade
Grown in sun

FIGURE 37.8 Phenotypic Plasticity in Leaves. These leaves came from the same tree.

Modified Leaves   Not all leaves function primarily in photosynthesis; some perform other roles (**TABLE 37.3**).

- Onion bulbs consist of thickened leaf bases that store nutrients, separated by highly condensed internodes.

- The thick leaves of plants called succulents, such as aloe vera, store water.

- The tendrils that enable garden peas and other vines to climb are modified leaflets or leaves.

- The bright red leaves of poinsettias attract pollinators to the tiny yellow flowers that they surround.

check your understanding

If you understand that . . .

- The plant body is organized into a root system and a shoot system.
- Roots and shoots explore the environment via continuous growth and efficiently absorb diffuse resources like water, ions, carbon dioxide, and sunlight.
- Roots and stems may also function to anchor the plant, store water, produce offspring asexually, provide protection, or store carbohydrates.
- Leaves vary among species and within individuals, and they may be modified to store food or water, capture insects, make climbing possible, attract pollinators, or protect the stem.

✔ You should be able to . . .

1. Diagram a generalized version of the angiosperm body, labeling each major part.

2. Provide two examples each of root systems, stems, and leaves that differ in structure and/or function from the generalized body shown in Figure 37.1.

Answers are available in Appendix A.

SUMMARY TABLE 37.3   **Form and Function of Modified Leaves**

Modified Leaves	Example
Bulbs Onion leaves store food	 Stem — Leaves
Succulent leaves Aloe vera leaves store water	
Tendrils Pea tendrils aid in climbing	
Floral mimics Red poinsettia leaves attract pollinators	Flowers Leaves
Traps Pitcher plant leaves trap entering insects, which are discouraged from flying out by the "hood," then are digested	 Hood Digestive enzymes or bacteria

- The tubelike leaves of the pitcher plant trap insects. When insects enter, they feed on the plant's nectar and appear to become dizzy. Eventually they fall into the bottom of the tube and drown in water that has accumulated. The insects are then digested by bacteria or enzymes secreted by the plant, and the released nutrients are taken up by epidermal cells.

- Cactus **spines** are modified leaves that protect the stem (see top image in Table 37.2).

The variability of plant roots, stems, and leaves is impressive. Diversity, plasticity, and dynamism are recurring themes in the study of plant anatomy.

37.2 Primary Growth Extends the Plant Body

Plants grow continuously because they have many **meristems**—populations of undifferentiated cells that retain the ability to undergo mitosis and produce new cells. When meristematic cells divide, some of the daughter cells remain in the meristem, allowing the meristem to persist. Other cells, though, undergo differentiation. You might recall that differentiation is a developmental process that produces a specialized cell—one that expresses only certain genes and has a distinctive structure and function (see Chapter 22).

Apical meristems are located at the tip of each root and shoot. As cells in apical meristems divide, enlarge, and differentiate, root and shoot tips extend the plant body outward, allowing it to explore new space.

This process is **primary growth,** which is common to all plants. The major consequence of primary growth is to increase the length of the root and shoot systems. Cells that are derived from apical meristems form the primary plant body.

To understand how primary growth occurs, let's look at the overall organization of the primary plant body and then delve into a detailed look at its tissues and cells.

How Do Apical Meristems Produce the Primary Plant Body?

Whether located in the root or the shoot, apical meristems give rise to three distinct populations of **primary meristem** cells: protoderm, ground meristem, and procambium. These cells are partially differentiated but retain the character of meristematic cells because they keep dividing.

The three types of primary meristematic cells are important because they give rise to three major tissue systems that extend throughout the plant body. A **tissue** is a group of cells that functions as a unit.

FIGURE 37.9 indicates where the apical meristems and the primary meristems—protoderm, ground meristem, and procambium—are found in shoots and roots.

- **Protoderm** gives rise to the **dermal** (literally, "skin") **tissue system.** The dermal tissue system, or **epidermis,** is a single layer of cells that covers the plant body and protects it.

- **Ground meristem** gives rise to the **ground tissue system,** which makes up the bulk of the plant body and is responsible for photosynthesis and storage.

- **Procambium** gives rise to the **vascular tissue system,** which provides support and transports water, nutrients, and photosynthetic products between the root system and shoot system.

(a) Apical and primary meristems in a shoot

Newly forming leaves

Apical meristem at tip of shoot

Apical meristem in lateral bud

100 μm

(b) Apical and primary meristems in a root

Primary meristems:

Procambium

Protoderm

Ground meristem

Apical meristem at tip of root

300 μm

FIGURE 37.9 The Structure of Apical and Primary Meristems in the Shoot and Root. Apical meristems consist of small, similar-looking cells that divide when conditions are favorable. Three types of primary meristem cells—protoderm, ground meristem, and procambium—are derived from the apical meristem and consist of partially differentiated cells that can still divide.

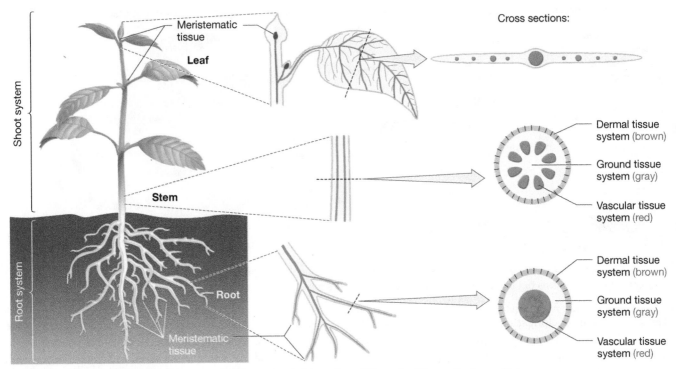

FIGURE 37.10 The Primary Plant Body Comprises the Dermal, Ground, and Vascular Tissue Systems. The dermal, ground, and vascular tissue systems arise from the protoderm, ground meristem, and procambium, respectively, as illustrated in Figure 37.9.

Vascular tissue runs through ground tissue, so the cells that make up ground tissue are usually close to cells that conduct the water and nutrients they need. **FIGURE 37.10** shows how the dermal, ground, and vascular tissues are distributed in the plant body. In contrast to these tissues, apical meristematic cells are localized at the tips of shoots and roots.

The key point to remember is that the dermal, ground, and vascular tissue systems are originally derived from cells in primary meristems, which originated from apical meristems. Thus, they represent the primary plant body.

How Is the Primary Root System Organized?

Roots have several features that allow them to grow into new regions of the soil, so they can furnish cells throughout the plant body with water and key nutrients.

As **FIGURE 37.11** shows, a group of cells called the **root cap** protects the root apical meristem. Cells produced by the meristem constantly replenish the cap, which regularly loses cells.

In addition to protecting the root tip, root cap cells are important in sensing gravity and determining the direction of growth. They also synthesize and secrete a slimy, polysaccharide-rich substance that helps lubricate the root tip, reducing friction and protecting the meristem as it is pushed through the soil.

Three distinct populations of cells exist behind the root cap:

1. The **zone of cellular division** contains the apical meristem, where cells are actively dividing, along with the protoderm, ground meristem, and procambium, where additional cell division occurs.

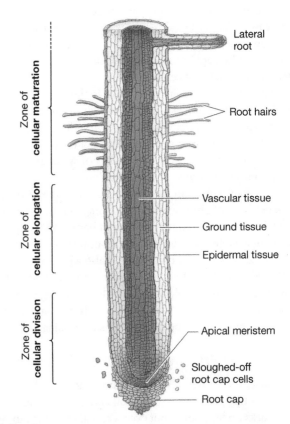

FIGURE 37.11 Roots Extend into the Soil via Growth of Apical Meristems and Cell Elongation. This is a longitudinal (lengthwise) section. The zone of cellular maturation is actually much larger than can be shown here. Most absorption of water and nutrients occurs at root hairs.

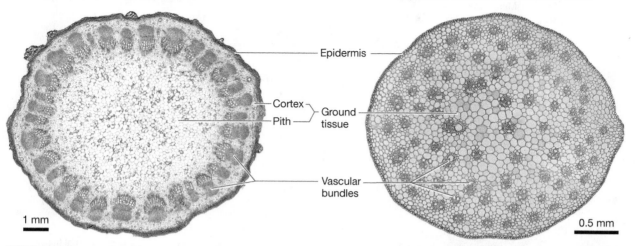

(a) Cross section of a eudicot stem   **(b)** Cross section of a monocot stem

Epidermis

Cortex — Ground
Pith — tissue

Vascular
bundles

1 mm

0.5 mm

FIGURE 37.12 Stems Contain a Variety of Cell and Tissue Types. Vascular bundles are **(a)** arranged in a ring near the perimeter of eudicot stems but **(b)** scattered throughout the ground tissue in monocots.

2. The **zone of cellular elongation** is made up of cells that are recently derived from the primary meristematic tissues and that increase in length.

3. The **zone of cellular maturation** is where older cells complete their differentiation into dermal, vascular, and ground tissues.

The zone of cellular elongation is the region most responsible for the growth of roots through the soil. The cells in this region increase in length by taking up water. Their expansion provides the force that pushes the root cap and apical meristem through the soil. When environmental conditions are good, roots can extend by as much as 4 centimeters per day.

The zone of cellular maturation is the most important root segment in terms of water and nutrient absorption. In this region, epidermal cells produce outgrowths called **root hairs,** which greatly increase the surface area of the dermal tissue.

Root hairs furnish the actual sites of water and nutrient absorption. The rest of the root system provides structural support for the root hairs, conducts water and ions to the shoot, stores the products of photosynthesis, and anchors the plant in the soil. Uptake of water and nutrients by root hairs is vital to plants (portions of Chapters 38 and 39 focus on how these processes occur).

The zone of cellular maturation is also where lateral roots begin to grow. In contrast to lateral branches in the shoot, which arise from meristems in axillary, or lateral buds (see Figure 37.1), lateral roots arise from cells within a ring of cells surrounding the vascular tissue and then erupt through the surrounding ground tissue.

How Is the Primary Shoot System Organized?

If you visit a garden regularly, you can imagine how the growing root tips push through the soil and expand to form complex networks underground. But even a casual observer can watch the growth of shoot systems over time, as the tips of stems extend and branch and as new leaves form and expand.

Just behind each shoot apical meristem, the primary meristematic cells give rise to dermal, ground, and vascular tissues.

FIGURE 37.12a shows how these tissues are arranged in the stem of a sunflower plant when it matures. Note that the vascular tissues are grouped into **vascular bundles,** which form strands running the length of the stem.

In sunflowers and other eudicots, the vascular bundles are arranged in a ring near the stem's perimeter (see Chapter 31). The ground tissue that the vascular tissue runs through is divided into two major regions: **pith,** the ground tissue that is inside the vascular bundles, and **cortex,** the ground tissue that is outside the vascular bundles.

The arrangement of the vascular bundles and ground tissue is dramatically different in the stems of monocots such as grasses, lilies, and orchids, however. As **FIGURE 37.12b** shows, vascular bundles in monocot stems are normally scattered throughout the ground tissue. (Review other comparisons of monocots and eudicots in Chapter 31.)

Now let's drill down a bit deeper, and look at the composition of the dermal, ground, and vascular tissue systems. Each of these tissue systems is made up of an array of distinct cell and tissue types. What do they look like, and what are their functions?

37.3 Cells and Tissues of the Primary Plant Body

Recall that all eukaryotic cells, including those of plants and animals, share most of their key characteristics: chromosomes enclosed in a nuclear envelope, a plasma membrane studded with proteins that regulate the passage of materials in and out, mitochondria that produce ATP by oxidizing sugars, and an array of other organelles that synthesize or degrade key molecules (see Chapter 7).

In addition, plant cells have several features that are absent in animal cells (**FIGURE 37.13a**):

1. All plant cells are surrounded by a stiff, cellulose-rich **cell wall** that supports the cell and defines its shape.

(a) Plant cells have cell walls, vacuoles, and chloroplasts.

Cell wall

Plasma membrane

Vacuole

Chloroplast

Mitochondrion

Rough endoplasmic reticulum

Smooth endoplasmic reticulum

Golgi apparatus

(b) Adjacent plant cells are connected by plasmodesmata.

Cell 2

Smooth ER

Plasmodesma

Plasma membrane

Cell wall

Cell wall

Plasma membrane

Cell 1

FIGURE 37.13 A Generalized Plant Cell. Plant vacuoles are similar to animal lysosomes; however, the cell wall, chloroplasts, and plasmodesmata are unique to plants. Unlike the extracellular matrix of animals, the plant cell wall is rigid.

2. The cytoplasm of adjacent plant cells is often connected via **plasmodesmata** (singular: **plasmodesma**) (see Chapter 11). Plasmodesmata consist of cytoplasm and segments of smooth endoplasmic reticulum (smooth ER) that run through tiny, membrane-lined gaps in the cell wall (**FIGURE 37.13b**).

3. Plant cells often contain several types of organelles that are not found in animals—specifically chloroplasts and a large, membrane-bound organelle called a vacuole, which fills most of the cell's volume.

    Chloroplasts are the site of photosynthesis (see Chapter 10). Non-photosynthetic cells found in roots, seeds, flower petals, and other locations may have organelles that are related to chloroplasts but are specialized for storing pigments, starch, oils, or proteins.

    Vacuoles, which contain an aqueous solution called **cell sap,** store wastes and in some cases also digest wastes, as do animal lysosomes. In addition, plant vacuoles store water and nutrients. They may also hold pigments that provide color, or poisons that deter herbivores.

    Another important distinction between plant cells and animal cells is that plant cells do not change position once they form. Some animal cells migrate within the body either early in the development of an individual or as mature (differentiated) cells.

    Let's examine the cells and tissues of the three primary plant tissue systems in turn.

The Dermal Tissue System

Dermal tissue is the interface between the organism and the external environment. Its primary function in shoots is to protect the plant body—from water loss, disease-causing agents, and herbivores. In roots the dermal tissue includes root hairs, and it functions primarily in water and nutrient absorption.

Most tissues are made up of several different cell types, each with a distinct structure and function. Let's consider the cell types found in dermal tissue.

Epidermal Cells Protect the Surface Most of the cells in the dermal tissue system are epidermal cells, which in shoots are flattened and usually lack chloroplasts.

Epidermal cells in the shoot system fulfill their protective role in part by secreting the **cuticle**—a waxy layer that forms a continuous sheet on the surface (see Chapter 31). Waxes are lipids and are therefore highly hydrophobic. As a result, the presence of cuticle on stems and leaves drastically reduces the amount of water that these structures lose by evaporation.

From a human perspective, the water-repellent properties of cuticle make it a valuable ingredient in polishes and lipsticks. The carnauba wax used in car and floor polishes, for example, is secreted by epidermal cells in the leaves of carnauba palms native to Brazil.

Besides minimizing water loss, cuticle forms a barrier to protect the plant from viruses, bacteria, and the spores or growing hyphae of parasitic fungi. In this way, the plant epidermis forms the first line of defense against disease-causing agents, or **pathogens.**

Waxes found in the cuticle can also be detrimental to the plant, however, by reducing gas exchange. This can be a serious problem because photosynthesis depends on the free flow of carbon dioxide to photosynthetic cells. The problem is solved by specialized structures in dermal tissue, called stomata.

Stomata Regulate Gas Exchange and Water Loss Most land plants have structures called **stomata** (singular: **stoma**) that allow carbon dioxide to enter photosynthetically active tissues.

Guard cells

Pore

Stoma closed

Stoma open

10 μm

FIGURE 37.14 A Stoma's Guard Cells Regulate the Opening of a Pore.

200 μm

FIGURE 37.15 Epidermal Cells Produce Trichomes That Provide Protection. Some trichomes on this leaf are hairlike extensions of a single cell, while others are multicellular structures. (The structures that are colored orange here hold toxins.)

A stoma consists of two specialized **guard cells,** which change shape to open or close an opening in the epidermis known as a **pore (FIGURE 37.14).**

When stomata are open, CO_2, O_2, water vapor, and other gases can move between the atmosphere and the interior of the plant by diffusion. Stomata open when CO_2 is needed inside the leaf. When stomata are open, water vapor diffuses from the moist interior of the leaf to the atmosphere. Stomata close when CO_2 is not needed. This prevents large amounts of water from being lost by transpiration. In a later chapter, we explore the molecular mechanisms responsible for stomatal opening and closing (see Chapter 40).

Trichomes Perform an Array of Functions   In addition to minimizing water loss and regulating gas exchange in shoots, cells in dermal tissue may protect the individual from the damaging effects of intense sunlight and attacks by herbivores.

Trichomes are hairlike appendages made up of specialized epidermal cells. They are found in shoot systems and come in a wide variety of shapes, sizes, and abundances.

Depending on the species, trichomes may **(1)** keep the leaf surface cool by reflecting sunlight; **(2)** reduce water loss by forming a dense mat that limits transpiration; **(3)** provide barbs, or store toxic compounds that thwart herbivores **(FIGURE 37.15);** or **(4)** trap and digest insects.

The Ground Tissue System

Most photosynthesis, as well as most carbohydrate storage, takes place in ground tissue. Cells in ground tissue are also responsible for most of the synthesis and storage of specialized products such as colorful pigments, the chemical signals called hormones, and toxins required for defense.

If the primary function of the dermal tissue system is protection, the ground tissue is all about producing and storing valuable molecules.

Ground tissue is made up of three distinct tissue types: parenchyma (pronounced *pa-REN-ki-ma*), collenchyma (*ko-LEN-ki-ma*), and sclerenchyma (*skle-REN-ki-ma*).

Parenchyma Are "Workhorse" Cells   **Parenchyma cells** have relatively thin cell walls and are the most abundant and versatile plant cells. Groups of parenchyma cells form parenchyma tissue.

The parenchyma tissue in leaves consists of parenchyma cells filled with chloroplasts, and it is the primary site of photosynthesis **(FIGURE 37.16a).** But in other organs, parenchyma cells store starch granules **(FIGURE 37.16b).** When you eat a potato or an apple, you are ingesting primarily parenchyma cells in ground tissue.

Many parenchyma cells are **totipotent,** meaning they retain the capacity to divide and develop into a complete, mature plant. The totipotency of parenchyma cells is important in healing wounds and in reproducing asexually via stolons or rhizomes. In each case, parenchyma cells may begin to divide, grow, and differentiate to form new roots and shoots.

The totipotency of parenchyma cells also allows gardeners to clone plants by making cuttings. For example, if you cut a piece of stem from a coleus plant and place it in water, parenchyma cells will divide to produce a mass of undifferentiated cells called a **callus.** Roots then develop from the callus **(FIGURE 37.17),** and the new individual can be planted in soil.

Bananas, seedless grapes, and several other commercially important plants cannot undergo sexual reproduction to produce seeds. Instead, they are propagated entirely by cuttings.

Collenchyma Cells Function Primarily in Shoot Support   **Collenchyma cells** have cell walls that are thicker in some areas than others, and their overall shape is longer and thinner than that of parenchyma cells. Groups of collenchyma cells form collenchyma tissue that supports the plant body.

Even when collenchyma cells are mature, their cell walls retain the ability to stretch and elongate. As a result, collenchyma cells can continue to lengthen as they provide structural support to the growing regions of shoots.

Collenchyma tissue is often found just under the epidermis of stems, especially outside vascular bundles. The ability of collenchyma cells to stretch allows stems to flex in the wind without tearing or breaking. The "strings" you may have peeled from a

(a) In leaves: photosynthesis and gas exchange

Chloroplasts

25 µm

(b) In roots: carbohydrate storage

Starch granules
(stained purple)

25 µm

FIGURE 37.16 Parenchyma Cells Perform a Wide Array of Tasks.

✔**EXERCISE** Give an example of a gene that is likely to be expressed in these leaf cells, but not in the root cells.

FIGURE 37.17 Parenchyma Cells in Cut Stems Can Form Adventitious Roots. Parenchyma cells in a cut coleus stem (left) divide to form a mass of undifferentiated cells called a callus, which then sprouts roots (right).

stalk of celery or rhubarb—which is actually the petiole of a celery or rhubarb leaf—include many strands of collenchyma cells (**FIGURE 37.18**).

Sclerenchyma: Two Types of Specialized Support Cells The cells that are classified as **sclerenchyma** produce a thick **secondary cell wall** in addition to the relatively thin **primary cell wall** found in all cells. Unlike a primary cell wall, the secondary cell wall contains the tough, rigid compound **lignin** in addition to **cellulose** (see Chapter 31).

Collenchyma cells can support actively growing parts of the plant because they have an expandable primary cell wall. In contrast, the nonexpandable secondary cell wall of sclerenchyma cells specializes them for supporting stems and other structures after growth has ceased. Another key difference between collenchyma and sclerenchyma is that sclerenchyma cells are usually dead at maturity—meaning they contain no cytoplasm.

Ground tissue typically contains two types of sclerenchyma cells: fibers and sclereids.

- **Fibers** are extremely elongated. The fiber cells from ramie plants, a species of Asian nettle, can be over half a meter long.

(a) Cross section of celery stalk

(b) Close-up of "string," in cross section

(c) Collenchyma cells, in cross section

100 µm

One cell
Cytoplasm
Cell wall

10 µm

FIGURE 37.18 Collenchyma Cells Support Growing Tissues. A celery stalk is actually a petiole; the strands you can peel from it are columns of collenchyma cells.

(a) Fibers

(b) Sclereids

20 μm

50 μm

FIGURE 37.19 Sclerenchyma Cells Support Mature Tissues.
(a) Fibers and **(b)** sclereids (stained pink) have thickened secondary cell walls. These cells provide support or protection for tissues that are no longer growing.

Fiber cells are important in the manufacture of paper, hemp or jute ropes, or linen and other fabrics (**FIGURE 37.19a**).

- **Sclereids** are relatively short, have variable shapes, and often function in protection. The tough coats of seeds and the thick shells of nuts are composed of sclereids; these cells are also responsible for the gritty texture of pears (**FIGURE 37.19b**).

The Vascular Tissue System

The vascular tissue system functions in support and in long-distance transport of water and dissolved nutrients. It also moves the products of photosynthesis that are made and stored in ground tissue.

Plant tissues that consist of a single cell type are called simple tissues; tissues that contain several types of cells are termed complex tissues. The vascular tissue system is made up of two complex tissues, xylem and phloem.

- **Xylem** (pronounced *ZYE-lem*) conducts water and dissolved ions in one direction: from the root system to the shoot system.

- **Phloem** (*FLO-em*) conducts sugar, amino acids, chemical signals, and other substances in two directions: from roots to shoots and from shoots to roots.

Xylem Structure   Two important cell types in xylem tissue are tracheids and vessel elements.

- In all vascular plants, xylem contains water-conducting cells called **tracheids** (*TRAY-kee-ids*).

- In angiosperms and species in the group Gnetophyta, xylem also contains conducting cells called **vessel elements** (see Chapter 31).

Tracheids and vessel elements have thick, lignin-containing secondary cell walls that are often deposited in ringlike or spiral patterns. Both tracheids and vessel elements are dead at maturity. As a result, they have no membranes and are filled with the fluids that they conduct instead of with cytoplasm.

Tracheids are long, slender cells with tapered ends (**FIGURE 37.20a**). The sides and ends of tracheids have **pits**, which are gaps in the secondary cell wall where only the primary cell wall is present. Because the cell is dead, pits have no plasma membrane spanning the opening. When water is moving up a plant through tracheids, it moves from cell to cell both vertically and laterally through pits, because that is where resistance to flow is lowest.

Vessel elements, in contrast, are shorter and wider than tracheids (**FIGURE 37.20b**). In addition to having pits, vessel elements have **perforations**—openings that lack both primary and secondary cell walls. In some species, the ends of vessel elements

(a) Tracheids are long, tapered, and have pits.

Pits

Pits

25 μm

(b) Vessel elements are short and wide and have perforations as well as pits.

Perforations

Pits

50 μm

(c) Tracheids and vessel elements are found together in vascular tissue.

100 μm

Tracheids

Vessel elements

FIGURE 37.20 Xylem May Contain Two Types of Water-Conducting Cells. (a) Tracheids are long and thin compared to **(b)** vessel elements, which are much shorter and wider. **(c)** Both types of water-conducting cells are found in the vascular tissue of angiosperms. The image in (a) is a light micrograph of stained tissue; the images in (b) and (c) are colorized scanning electron micrographs.

lack any cell wall at all, and stacked cells form open pipes called vessels. Vessel elements conduct water more efficiently than do tracheids, because their width and perforations offer less resistance to flow.

In angiosperms, tracheids and vessel elements are found adjacent to each other (**FIGURE 37.20c**). Xylem also contains sclerenchyma cells (fibers) and some parenchyma cells that transport materials laterally in the stem—not vertically.

Phloem Structure   In most vascular plants, phloem is made up primarily of two specialized types of parenchyma cells: sieve-tube elements and companion cells. Both are alive at maturity, lack lignified secondary cell walls, and arise from division of a common precursor cell.

- **Sieve-tube elements** are long, thin cells that have perforated ends called **sieve plates** (**FIGURE 37.21**). They are responsible for transporting sugars and other nutrients.

- **Companion cells** are not conducting cells, but instead provide materials to maintain the cytoplasm and plasma membrane of sieve-tube elements.

Interestingly, sieve-tube elements lack nuclei and most other organelles, but they are directly connected to adjacent companion cells by means of numerous plasmodesmata (see Figure 37.13b). Companion cells contain most of the organelles normally found in a plant cell and support the function of sieve cells.

Companion cells are also involved in loading and unloading carbohydrates and other nutrients into and out of sieve-tube elements (see Chapter 38 for more detail). Phloem also contains sclerenchyma tissue (fibers and sclereids), which provides mechanical support.

TABLE 37.4 (page 748) summarizes the major tissues and cell types found in the dermal, ground, and vascular tissue systems. Once you've mastered the structure and function of the primary plant body, you're ready to consider the next level of complexity: secondary growth.

If you understand that . . .

- Primary growth results from cell division in apical meristems. Its function is to extend the shoot system into the air and the root system into the soil.

- Apical meristems produce three types of primary meristematic cells: protoderm, ground meristem, and procambium. The dermal, ground, and vascular tissue systems that arise from these meristematic cells extend throughout the individual and make up the primary plant body.

- The dermal system protects the individual and controls the exchange of gases and nutrients with the environment; the ground system produces and stores the carbohydrates that make life possible; the vascular system moves those molecules from place to place and holds the plant up. Each of these systems consists of an array of distinctive cell and tissue types.

✔ **You should be able to . . .**

1. Explain the relationship between an apical meristem and the three primary meristematic tissues.

2. Describe the structure and function of epidermal cells, parenchyma cells, tracheids, and sieve-tube elements.

Answers are available in Appendix A.

Sieve-tube element (few organelles)   Companion cell (many organelles)

Sieve plate

Longitudinal section

Sieve plate

50 µm

Cross section

Companion cell

Sieve plate

Companion cell

25 µm

FIGURE 37.21 Phloem Consists of Sieve-Tube Elements and Companion Cells. Sieve-tube elements conduct sucrose throughout the body; companion cells support sieve-tube elements.

Tissues Present	Description of Tissue	Function
Dermal Tissue System (arises from protoderm)		
Epidermal	Complex tissue consisting of epidermal cells, guard cells, trichome cells, and root hair cells	Shoots: Protection, gas exchange Roots: Protection, water and nutrient absorption
Ground Tissue System (arises from ground meristem)		
Parenchyma	Simple tissue consisting of parenchyma cells	Synthesis and storage of sugars and other compounds
Collenchyma	Simple tissue consisting of collenchyma cells	Support (flexible cell walls)
Sclerenchyma	Simple tissues consisting of sclerenchyma cells: sclereids or fibers	Support and protection (rigid cell walls; dead at maturity)
Vascular Tissue System (arises from procambium)		
Xylem	Complex tissue consisting of tracheids, vessels, and parenchyma cells and sclerenchyma cells (fibers)	Transport of water and ions; structural support (tracheids, vessels and fibers are dead at maturity)
Phloem	Complex tissue consisting of parenchyma cells (sieve-tube elements, companion cells) and sclerenchyma cells (fibers)	Transport of sugars, amino acids, hormones, etc.; support (fibers are dead at maturity)

37.4 Secondary Growth Widens Shoots and Roots

Primary growth increases the length of roots and shoots; its major function is to extend the reach of the root and shoot system and thus increase a plant's ability to absorb photons and acquire carbon dioxide, water, and ions.

In some plants, **secondary growth** increases the width of the plant body. Its major function is to increase the amount of conducting tissue available and provide the structural support required for extensive primary growth. Recall that the evolution of conducting tissues allowed the vascular plants to grow taller to compete for light (see Chapter 31). Without the support provided by secondary growth, however, roots would not be massive enough to anchor large shoot systems, and long stems would fall over or break. Only with secondary growth could vascular plants become the giants seen today.

What Is a Cambium?

Secondary growth produces **wood** and occurs in species that have a cambium in addition to apical meristems. A **cambium** (plural: cambia) is a special type of meristem (also called a lateral meristem) that differs from an apical meristem in two ways:

1. A cambium forms a cylinder that runs the length of a root or stem and is made up of a single layer of meristematic cells. In contrast, apical meristems are localized at root tips and shoot tips and are dome shaped.

2. In a cambium, cells divide in a way that increases the width of roots and shoots (**FIGURE 37.22a**). Cells in an apical meristem divide in a way that extends the root and shoot tips.

As **FIGURE 37.22b** shows, there are two distinct types of cambium in plants that undergo secondary growth.

- A cylinder of meristematic cells called the **vascular cambium** is located between the secondary xylem and secondary phloem, inside the stem and root.

- Another cylinder of meristematic cells called the **cork cambium** is located near the perimeter of the stem and root.

One other observation is critical to understanding how cambia work: The cork cambium produces new cells primarily to the outside. Vascular cambium, in contrast, generates new layers of cells both to the inside and outside. The new cells formed to the inside push all of the other cells toward the outside, causing an increase in girth.

TABLE 37.5 summarizes the major mature tissue types and cell types that arise from each type of meristem. Now let's consider how these meristems work.

What Does Vascular Cambium Produce?

Vascular cambium in roots and stems produces both phloem and xylem (see Figure 37.22b). New cells that are produced to the outside of the vascular cambium differentiate into secondary phloem; new cells produced to the inside differentiate into secondary xylem. (In contrast, the procambium at each apical meristem produces *primary* phloem and *primary* xylem.)

(a) Cross section of a 3-year-old *Tilia* stem

(b) Vascular cambium and cork cambium produce wood and bark.

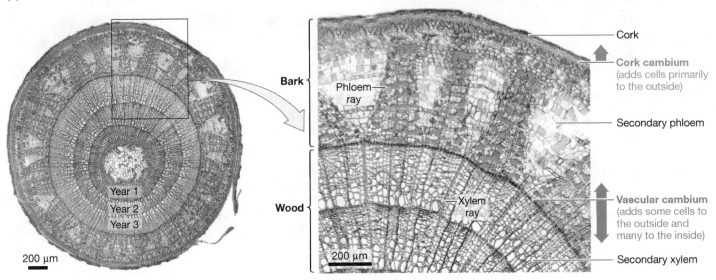

FIGURE 37.22 **Vascular Cambium Is Responsible for Secondary Growth.** Tree trunks contain two types of cylindrical cambia: cork cambium and vascular cambium. Bark consists primarily of the cork cells produced by the cork cambium and secondary phloem. Wood consists of secondary xylem and parenchyma cells produced by vascular cambium.

Secondary phloem and secondary xylem cells are not always produced simultaneously. In most cases, the vascular cambium produces many more secondary xylem cells than secondary phloem cells.

Primary phloem and xylem are produced by all vascular plants, but secondary phloem and xylem are found only in some lycophytes, all gymnosperms, and certain angiosperms.

Structurally, primary and secondary phloem and primary and secondary xylem are complex tissues, made up of more than one cell type. Functionally, primary and secondary phloem are similar, as are primary and secondary xylem.

- Primary and secondary phloem function in sugar transport.

- Primary and secondary xylem function in water transport and structural support.

Secondary xylem makes up the structural material called wood, while secondary phloem makes up the inner part of a tree's bark, which we describe later.

Besides producing conducting cells such as sieve-tube elements, tracheids, and vessel elements, the vascular cambium also produces sclerenchyma cells (fibers) for additional strength, along with some parenchyma cells. The parenchyma cells are formed in rows that radiate laterally across the xylem and phloem. These rows of cells are called **rays** (see Figure 37.22b), and they form a living conduit through which water and nutrients are transported laterally across the stem.

It's important to realize, though, that the results of cell division in vascular cambium are highly asymmetrical. As the

SUMMARY TABLE 37.5 **Components of Secondary Growth**

Meristematic Tissue	Mature Tissue	Direction of Growth	Mature Cell Composition	Mature Tissue Function
Cork Cambium	**Cork**	Produced to the outside	Cork cells	Protection
Vascular Cambium	**Secondary phloem**	Produced to the outside	Parenchyma cells (sieve-tube elements, companion cells) and sclerenchyma cells (fibers)	Transport of sugars, amino acids, hormones, etc.
	Secondary xylem*	Produced to the inside	Tracheids, vessels, parenchyma cells (arranged in rays) and sclerenchyma cells (fibers)	Transport of water and ions; structural support

*Secondary xylem is also called wood.

FIGURE 37.23 Lenticels in the Bark of Paperbark Cherry (*Prunus serrula*). Lenticels are natural breaks in cork tissue through which oxygen can enter a stem.

(a) Heartwood and sapwood have different functions.

Heartwood   Sapwood   Bark

Provides structural support but no longer transports water

Includes active water-conducting xylem tissue

(b) Growth rings result from seasonal variation in cell size.

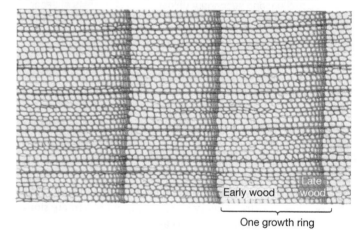

Early wood   Late wood

One growth ring

(c) Patterns in growth rings can tell a tree's history.

Thick growth rings before onset of acid rain   Thin growth rings after onset of acid rain   Bark

FIGURE 37.24 Anatomy of a Tree Trunk. (a) Unstained section of wood. **(b)** Section of wood, stained to show individual cells. **(c)** Unstained section through a fir tree from Germany's Black Forest.

✔**EXERCISE** In part (c), pick a thick growth ring and label the early wood and late wood.

---

vascular cambium grows, all of the secondary xylem is retained and accumulates, but the primary xylem in the center of the stem eventually clogs and may rot away. In addition, the outermost secondary phloem becomes fragmented and compressed as the stem increases in diameter. As a result, mature woody roots and stems are dominated by secondary xylem, or wood.

✔ If you understand this concept, you should be able to explain why a nail hammered into a tree at eye level doesn't move up as the tree grows.

What Does Cork Cambium Produce?

The cork cambium's primary role is to produce **cork cells** to the outside (Figure 37.22b). Together with the secondary phloem, the cork cambium and cork cells make up the tissue called **bark.**

Bark provides a particularly tough barrier in species whose cork cells secrete a strong secondary cell wall containing lignin. Bark also helps prevent water loss because cork cells produce a layer of wax and other molecules inside their cell walls, making them impermeable to water and gases. Gas exchange can still occur between the atmosphere and living tissues inside the stem, though—through small, spongy openings in the bark called **lenticels** (**FIGURE 37.23**).

Cork cells die when they mature. As a stem continues to widen, the cork layer often cracks and flakes, and the outer layers might even slough off the tree.

Bark is important because it protects the woody stem from damage and pathogens as it increases in girth. As a woody stem or root matures, the epidermal tissue produced by the apical meristem during primary growth is replaced by the bark, which takes over the role of preventing water loss and protecting the stem and root from pathogens and herbivores. In some species, exceptionally thick bark can even protect the shoot system from fire damage. Redwood trees, for example, are adapted to fire-prone habitats and can have bark that is 20 cm (12 in.) thick.

The Structure of a Tree Trunk

Trees are perennial plants that live for many years. As a tree matures and grows in width, the innermost xylem layers stop transporting water—only the xylem from the most recent years actually transports fluid.

Heartwood and Sapwood Xylem that no longer transports begins accumulating protective compounds secreted by other tissues. These compounds form resins, gums, and other complex mixtures. The deposition of these molecules causes the oldest portions of secondary xylem to become darker than the younger portions.

The darker-colored, inner xylem region is called **heart-wood,** while the lighter-colored, outer xylem is called **sapwood** (**FIGURE 37.24a**). If you look closely at wood furniture or flooring, you may see a color difference between sapwood and heartwood in some boards.

Annual Growth Rings Another important phenomenon occurs in environments where the vascular cambium stops growing for a portion of each year. This period of **dormancy** occurs during the winter in cold climates and during the dry season in tropical habitats.

When the vascular cambium resumes growth in the spring or at the start of the rainy season, it produces large, relatively thin-walled cells, called early wood. As the growing season nears its end, conditions tend to dry out or become cooler; the secondary xylem cells that are produced at this time tend to be smaller, thicker walled, and darker and are called late wood. Thus, when growth is seasonal, regions of large, thin-walled cells alternate with layers of small, thick-walled cells, resulting in annual growth rings in the secondary xylem (**FIGURE 37.24b**).

Analyzing patterns in tree growth rings is an important field of study in biology. Because trees grow faster when moisture and nutrients are plentiful, wide tree rings are reliable indicators of wet years. In contrast, narrow rings signal drought years—or in the case of the fir tree shown in **FIGURE 37.24c**, years when abundant acid rain, due to air pollution, reduced growth.

By studying the growth rings in fossil trees and extremely old living trees, biologists can often assemble a continuous record that dates back thousands of years. In doing so, they gain a better understanding of climate changes that occurred in the past. With continued research, researchers also hope to predict how forests might respond to the rising temperatures associated with the global climate change that is currently under way.

check your understanding

If you understand that . . .

- Secondary growth occurs in species with a vascular cambium and cork cambium and results in a broadening of the shoot and root systems.
- Vascular cambium gives rise to secondary xylem and secondary phloem tissues. Cork cambium gives rise to the protective tissue called bark.

✔ **You should be able to . . .**

1. Describe what the rings in a cross section of a tree trunk would look like if the individual is heavily shaded on one side but exposed to sunlight on the other side.

2. Draw from memory a three-year-old woody stem in cross section, labeling secondary xylem and phloem, vascular and cork cambium, and cork. Add arrows showing the direction of growth in each meristem.

Answers are available in Appendix A.

CHAPTER 37 REVIEW

For media, go to MasteringBiology

If you understand . . .

37.1 Plant Form: Themes with Many Variations

- The root and shoot systems of plants are specialized for harvesting the light, water, and nutrients required for performing photosynthesis. Structures involved in absorption have a high surface-area-to-volume ratio.

- Roots extract water and nutrients such as nitrogen, phosphorus, and potassium from the soil.

- The shoot system consists of all aboveground portions of the plant, including stems and leaves.

- Leaves carry out photosynthesis by capturing light and carbon dioxide from the atmosphere and usually consist of a flattened blade that extends from a petiole.

- The overall morphology of root and shoot systems varies widely among plant species, allowing individuals to reduce competition for resources.

- Roots and shoots may be modified to perform a variety of other functions, however, including nutrient storage, water storage, protection, and asexual reproduction.

- Because roots and shoots grow throughout life, a plant is able to respond appropriately to changes in environmental conditions.

✔ You should be able to explain why phenotypic plasticity in roots and shoots is expected to be more important (1) in environments where conditions are variable versus stable, and (2) in long-lived versus short-lived species.

37.2 Primary Growth Extends the Plant Body

- Each apical meristem gives rise to three primary meristematic tissues: protoderm, ground meristem, and procambium.

- The primary meristematic tissues give rise to the dermal, ground, and vascular tissue systems, which extend throughout the plant body.

- Behind the root apical meristem, or the zone of cellular division, cells become longer in the zone of cellular elongation and acquire specialized functions in the zone of cellular maturation.

✓ You should be able to explain why most of the water and nutrient absorption by roots takes place in the zone of cellular maturation.

37.3 Cells and Tissues of the Primary Plant Body

- The dermal tissue system is usually one cell layer thick and plays a role in protection and water absorption by roots. In shoots the epidermis synthesizes the cuticle, which aids in water conservation.

- Stomata are formed by pairs of epidermal guard cells that open and close to control CO_2 uptake and water loss by leaves.

- The ground tissue system performs photosynthesis and stores carbohydrates and other compounds.

- The vascular tissue system transports materials throughout the plant. Within the vascular system, xylem tissue transports water and dissolved ions up the plant; phloem tissue transports sugars up and down.

- Ground tissue contains (1) parenchyma cells, which function in material synthesis, transport and storage; (2) collenchyma cells, which provide structural support for growing regions; and (3) sclerenchyma cells, fibers and sclereid cells, that strengthen regions of the body that have stopped growing.

✓ You should be able to predict the results of an experiment where a drug was used to (1) poison an apical meristem in the shoot and an apical meristem in the root, and (2) selectively poison protoderm cells in a shoot apical meristem.

37.4 Secondary Growth Widens Shoots and Roots

- In some plant species, shoots and roots are widened by cylindrical meristems called vascular cambium that produce secondary xylem and secondary phloem, and cork cambium that produces cork.

- Cork cells produce waxes that protect stems from water loss.

- Wood consists of secondary xylem, while bark consists of all tissue outside of the vascular cambium.

✓ You should be able to predict the results of an experiment where a drug was used to (1) slow the growth of the vascular cambium but not the cork cambium, and (2) slow the growth of both the vascular and cork cambia on one side of a tree trunk.

(MB) MasteringBiology

1. **MasteringBiology Assignments**

   Tutorials and Activities Plant Growth; Primary and Secondary Growth; Primary and Secondary Growth in Plants; Root, Stem, and Leaf Sections

   Questions Reading Quizzes, Blue-Thread Questions, Test Bank

2. **eText** Read your book online, search, take notes, highlight text, and more.

3. **The Study Area** Practice Test, Cumulative Test, BioFlix® 3-D Animations, Videos, Activities, Audio Glossary, Word Study Tools, Art

You should be able to . . .

✓TEST YOUR KNOWLEDGE

1. Which of the following functions is not performed by parenchyma cells?
   a. synthesizing key molecules
   b. providing mechanical support
   c. performing photosynthesis
   d. storing nutrients

2. What is a sieve-tube element?
   a. the sugar-conducting cell found in phloem
   b. the widened, perforation-containing, water-conducting cell found only in angiosperms
   c. the nutrient- and water-absorbing cell found in root hairs
   d. the nucleated and organelle-rich support cell found in phloem

3. Which statement best characterizes primary growth?
   a. It does not occur in roots, only in shoots.
   b. It leads to the development of cork.
   c. It produces the dermal, ground, and vascular tissues.
   d. It produces rings of xylem and phloem tissue as well as rings of cork tissue.

4. Which statement best characterizes secondary growth?
   a. It results from divisions of the vascular and cork cambium cells.
   b. It increases the length of the plant stem.
   c. It results from divisions in the apical meristem cells.
   d. It often produces phloem cells to the inside and xylem cells to the outside of the vascular cambium.

5. To illustrate the concept that a plant structure can be modified for many different functions, give an example of a leaf and a stem that protect individuals against large herbivores.

6. Compare and contrast the structure and function of collenchyma and sclerenchyma cells.

7. How do tracheids differ from vessel elements, in addition to their overall shape?
   a. Tracheids are stacked end to end to form continuous, open columns.
   b. In tracheids, water flows from cell to cell primarily through gaps in the secondary cell wall called pits.
   c. Tracheids are dead at maturity.
   d. Tracheids have secondary cell walls reinforced with lignin.

8. Describe the general function of the shoot system and the general function of the root system. Which tissues are continuous throughout these two systems? Suggest a hypothesis to explain why the shoot and root systems of different species are so variable in size and shape.

9. Explain why continuous growth enhances the phenomenon known as phenotypic plasticity.

10. What does cuticle do? What do stomata do? Predict how the thickness of cuticle and the number of stomata differ in plants from wet habitats versus dry habitats.

11. Compare and contrast the roles of parenchyma cells in the ground tissue system versus the vascular tissue system.

12. Describe how the vascular cambium produces secondary xylem and secondary phloem.

13. Which one of the following tissues is most likely to be the source of nutrition for insects that kill trees?
    a. primary xylem
    b. secondary xylem
    c. vascular cambium
    d. cork

14. Identify the structure you are consuming when you eat each of the following vegetables: asparagus, Brussels sprouts, celery, spinach, carrot, potato.

15. Why do mahogany trees, and other trees that grow in tropical rain forests, lack growth rings?

16. Trees can be killed by girdling—meaning the removal of bark and vascular cambium in a ring all the way around the tree. Explain why.

38 Water and Sugar Transport in Plants

In this chapter you will learn how

Plants use evaporation and active transport to move materials long distances

by first exploring ↓

Water potential and the factors that affect it
38.1

which explains →

Translocation of sugars
38.4

then asking ↓

How does water move from roots to shoots?
38.2

then analyzing ↓

Water absorption and water loss
38.3

looking closer at

Sources and sinks

The pressure-flow hypothesis

Phloem loading and unloading

This chapter explores how plants move water from their roots to their leaves and how they transport sugars to all of their tissues—sometimes over great distances.

BIG PICTURE

This chapter is part of the Big Picture. See how on pages 840–841.

On a hot summer day, a large deciduous tree can lose enough water to fill three 55-gallon drums. To understand why, recall that the surfaces of leaves are dotted with stomata (see Chapter 37), which in most plants open during the day so gas exchange can occur between the atmosphere and the cells inside the leaf. This exchange is crucial. For photosynthesis to take place, leaf cells must acquire carbon dioxide (CO_2).

There's a catch, however. While stomata are open, the moist interior of the leaf is exposed to the dry atmosphere. As a result, large quantities of water evaporate from the leaf. If the lost water is not replaced with water absorbed by roots, plant cells will dry out and die.

Evaporation from leaves is a challenge for plants, but it is also beneficial. Because water enters a plant through its roots and exits through its leaves, there is a regular flow of water from roots to shoots. Minerals that are absorbed by roots are carried to the leaves by this flow. Without this

✔ When you see this checkmark, stop and test yourself. Answers are available in Appendix A.

movement of water, shoots would be unable to receive the minerals they need to grow.

In addition, under hot conditions, evaporation cools the plant, just like sweating cools your body. Heavy rates of evaporation can lower leaf temperatures by as much as $10-15°C$.

But the source of water can be a long distance from where it is needed. In the case of certain redwood trees, the leaves that lose water may be 100 meters (m) from the root hairs that absorb water. How do plants transport water against the force of gravity—in some cases, the length of a football field? And how do plants move the sugar they produce from active photosynthetic sites to sites where sugar is stored or needed for growth? These questions are the heart and soul of this chapter. Answering them is a fundamental part of understanding how plants work. (You can review plant form and function in the Big Picture of Plant and Animal Form and Function, on pages 840–841.)

38.1 Water Potential and Water Movement

Loss of water via evaporation from the aerial parts of a plant is called **transpiration.** Transpiration occurs whenever two conditions are met: **(1)** Stomata are open, and **(2)** the air surrounding leaves is drier than the air inside leaves.

The first condition is usually met during the day, when photosynthesis takes place. The second condition occurs whenever atmospheric humidity is less than 100 percent.

One of the most astonishing observations in biology is that water moves from roots to leaves passively—that is, with no expenditure of ATP. Plants do not need a heart muscle to pump water from roots to shoots. Even in 100-m-tall redwood trees, water flows passively from the root system to the shoot system. This movement occurs because of differences in the potential energy of water.

What Is Water Potential?

Recall that potential energy is stored energy (see Chapter 2). Changes in potential energy are associated with changes in position, such as the position of a molecule or an electron.

Biologists use the term **water potential** to indicate the potential energy that water has in a particular environment compared with the potential energy of pure water at room temperature and atmospheric pressure. Under these conditions, pure water has a water potential of 0.

Water potential is symbolized by the Greek letter ψ (psi, pronounced sigh). Differences in water potential determine the direction that water moves. Water always flows from areas of higher water potential to areas of lower water potential. Several factors can contribute to differences in water potential.

What Factors Affect Water Potential?

To understand how water moves from cell to cell in a plant, first consider the cell in the beaker on the left in **FIGURE 38.1a**. Notice

that it is suspended in a **solution**—a homogenous, liquid mixture containing several substances. In this case, the solution consists of water and dissolved substances, or **solutes.**

In the beaker on the left, the solute concentrations in the cell and in the surrounding solution are the same. Such a solution is said to be **isotonic** to the cell (see Chapter 6). When two solutions are isotonic, there is no net movement of water between them.

The Role of Solute Potential   What happens when the cell is transferred to the beaker on the right in Figure 38.1a? This beaker contains pure water, which has no solutes. As a result, the solution surrounding the cell is strongly **hypotonic** relative to the cell.

(a) Solute potential is the tendency of water to move by osmosis.

Solute potential inside cell and in surrounding solution is the same. No net movement of water.

Cell is placed in pure water. The cell's solute potential is low relative to its surroundings. Water moves into cell via osmosis.

(b) Pressure potential is the tendency of water to move in response to pressure.

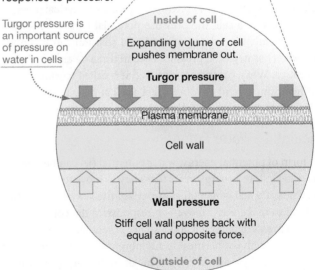

Turgor pressure is an important source of pressure on water in cells

FIGURE 38.1 Water Potential Has Two Major Components: Solute Potential and Pressure Potential.

The concentration of the solution is important because when solutions are separated by a selectively permeable membrane, such as the plasma membrane of a cell, water passes through the membrane from regions of low solute concentration to regions of high solute concentration. This movement of water across membranes, in response to differences in solute concentration, is called **osmosis** (see Chapter 6).

The tendency for water to move in response to differences in solute concentrations is called the **solute potential (ψ_S)**. The presence of solutes can only decrease a solution's solute potential below that of pure water.

The solute potential of a solution is defined by its total solute concentration relative to pure water. If water contains a high concentration of solutes, then it has a low solute potential compared with pure water. Water moves down its potential gradient toward the region of higher solute concentration.

The Role of Pressure Potential

When an animal cell is placed in a hypotonic solution and water enters the cell via osmosis, the volume of the cell increases until the cell bursts. This does not happen to plant cells, however.

If a plant cell swells in response to incoming water, its plasma membrane pushes against the relatively rigid cell wall. The cell wall resists expansion of the cell volume by pushing back, much as the walls of a basketball push back when the ball is inflated.

The force exerted by the wall is called **wall pressure** (**FIGURE 38.1b**). As water moves into the cell, the pressure inside the cell, known as **turgor pressure,** increases until wall pressure is induced. Cells that are firm and that experience wall pressure are said to be **turgid.**

Turgor pressure is important because it counteracts the movement of water due to osmosis. In the example on the right in Figure 38.1a, the solute potential favors water moving into the cell. However, the rigid cell wall limits the amount of water that can enter the cell.

Pressure potential (ψ_P) refers to any kind of physical pressure on water. Inside a cell, the pressure potential consists of turgor pressure. While the solute potential of a solution can be only 0 (pure water) or negative, pressure potential can be either positive or negative. When pressure is negative, it is called tension.

When osmosis and pressure affect a cell at the same time, how do biologists determine the direction of water movement?

Calculating Water Potential

As a form of potential energy, water potential (ψ) can be thought of as water's stored energy or its tendency to move to a new position. Thus, a water potential summarizes the stored energy that will tend to make water move—in response to the combined effects of a pressure potential and a solute potential.

When selectively permeable membranes are present, water tends to move by osmosis from areas of high solute potential to areas of low solute potential. When no membranes are present to stop it, water moves from areas of high pressure potential to areas of low pressure potential.

If we ignore the effects of gravity, water potential is defined by the following equation:

$$\psi = \psi_P + \psi_S$$

In words, the potential energy of water in a particular location is the sum of the pressure potential and the solute potential that it experiences.

Assigning Units of Pressure and Signs   Water potential is measured in units called **megapascals (MPa, 10^6 Pa)**. A **pascal (Pa)** is a unit of measurement commonly applied to pressures—force per unit area. A car tire is inflated to about 0.2 MPa, and the water pressure in home plumbing is usually 0.2 to 0.3 MPa.

Solute potentials (ψ_S) are always negative because they are measured relative to the solute potential of pure water. Because pure water contains no dissolved substances, it has a solute potential of 0 megapascal (MPa). And because there are always some solutes inside a cell, the water inside always has a solute potential lower than that of pure water. Therefore pure water will always tend to move *into* the cell. Increasing the concentration of solutes in a cell lowers its solute potential even more—making it more negative.

In contrast to the solute potential in cells, the pressure potential (ψ_P) from turgor pressure is positive inside *living* cells. But as you will see later, the pressure potential in dead cells can sometimes be negative. Let's first consider how solute and pressure potentials combine to drive water movement into and out of living cells.

Water Movement in the Absence of Pressure   In the U-shaped tube on the left side of **FIGURE 38.2a**, two solutions are separated by a selectively permeable membrane. The system is open to the atmosphere and thus is not under additional pressure, meaning $\psi_P = 0$ MPa.

Note that the left side of the tube contains pure water, which has a ψ_S of 0 MPa. The ψ_S for the solution on the right side of the membrane is −1.0 MPa. Because water potential is higher on the left side of the tube than on the right side, water moves from left to right.

The right side of Figure 38.2a models the same situation with a cell that is initially **flaccid**—meaning it has no turgor pressure and thus a pressure potential of 0. Note that the cell has been placed in a solution of pure water. Because the cell has low solute potential (−1.0 MPa) and the pure water has a higher water potential than the cell, water enters the cell via osmosis.

✔ QUANTITATIVE If you understand the concept of solute potential, you should be able to explain (1) how you would change the solute potential of the pure water on the left of Figure 38.2a to make it lower than the solute potential in the solution to the right, and (2) what the consequences for water movement in the tube would be.

Water Movement in the Presence of a Solute Potential and Pressure Potential   On the left side of **FIGURE 38.2b**, the concentrations in the U-shaped tube are the same as in Figure 38.2a, but the solution on the right side of the tube experiences pressure exerted by a plunger. If the force on the plunger produces a

(a) Solute potentials differ.

Pure water	Solution
$\psi = 0$ MPa	$\psi_P = 0$ MPa
	$\psi_S = -1.0$ MPa
	$\psi = -1.0$ MPa

Water moves left to right—from area with higher water potential to area with lower water potential

Pure water	Flaccid cell
$\psi = 0$ MPa	$\psi_P = 0$ MPa
	$\psi_S = -1.0$ MPa
	$\psi = -1.0$ MPa

Water moves into cell—from area with higher water potential to area with lower water potential

(b) Solute and pressure potentials differ.

Pure water	Solution
$\psi = 0$ MPa	$\psi_P = +1.0$ MPa
	$\psi_S = -1.0$ MPa
	$\psi = 0.0$ MPa

Water potentials are equal— no net movement

Pure water	Turgid cell
$\psi = 0$ MPa	$\psi_P = +1.0$ MPa
	$\psi_S = -1.0$ MPa
	$\psi = 0.0$ MPa

Water potentials are equal— no net movement

FIGURE 38.2 Solute Potential and Pressure Potential Interact.

✔ QUANTITATIVE In the left side of part (a), does the solute potential on the right side of the tube increase, decrease, or stay the same when water flows from left to right by osmosis?

pressure potential of 1.0 MPa on the right side, and if the ψ_S for the solution on the right side of the membrane is still -1.0 MPa, then the water potential of the right side is -1.0 MPa $+ 1.0$ MPa $= 0$.

In this case, the water potential on both sides of the membrane is equal and there will be no net movement of water. If the force on the plunger is greater than 1.0 MPa, the solution on the right side would have a *higher* water potential and water would flow from right to left.

The right side of Figure 38.2b models this situation in a living cell. In this case, the incoming water creates turgor pressure. When the positive turgor pressure ($+1.0$ MPa) plus the cell's negative solute potential (-1.0 MPa) equals 0 MPa—the water potential of pure water—the system reaches equilibrium. At equilibrium, there is no additional net movement of water. In this way, turgor pressure acts like the plunger in Figure 38.2b.

✔ QUANTITATIVE If you understand the concept of pressure potential, you should be able to explain how you would use the plunger in Figure 38.2b to create a negative pressure instead of a positive pressure.

Water Potentials in Soils, Plants, and the Atmosphere

The water contained within leaf, stem, or root tissues has a pressure potential and a solute potential, just like the water inside a cell does. Likewise, both the soil surrounding the root system and the air around the shoot system have a water potential.

Water Potential in Soils In moist soil, the water that fills crevices between soil particles usually contains relatively few solutes and normally is under little pressure. As a result, its water potential tends to be high relative to the water potential found in a plant's roots, which is higher in solutes.

There are important exceptions to this rule, however.

- *Salty soils* The soils near ocean coastlines have water potentials as low as -4MPa or less due to high solute concentrations. This is much lower than the water potential typically found inside plant roots.

- *Dry soils* When soils dry, water no longer flows freely in the spaces between soil particles. All of the remaining water adheres tightly to soil particles, creating a tension that lowers the water potential of soil water.

When the water potential in soil drops, water is less likely to move from soil into plants. If soil water potential is low enough, water may even move from plants to the soil. This situation would be deadly to plants.

This is an enormously important issue for world agriculture. When soils are irrigated to boost crop yields, much of the water evaporates. The solutes in the irrigation water are left behind and tend to collect in the first few inches below the surface. Over time, then, irrigated soils tend to become salty. In some parts of the world, formerly productive soils have become so salty that they are now abandoned as cropland.

How Do Plants That Are Adapted to Salty or Dry Habitats Cope? Salt-adapted species often respond to low water potentials in soil by accumulating solutes in their root cells, which lowers their solute potential. These plants have enzymes that increase the concentration of certain organic molecules in the cytoplasm. As a result, they can keep the water potential of their tissues even lower than the water potential of salty soils.

Species that are adapted to dry sites cope by tolerating low solute potentials. For example, let's look at the changes in solute

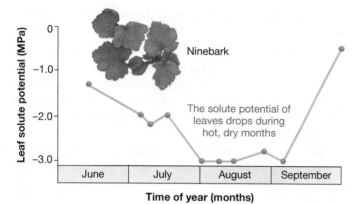

FIGURE 38.3 **Plants with Low Solute Potentials Can Grow in Dry Soils.** As soils dry, their water potential declines. But if a plant's solute potential also drops, it can maintain a water-potential gradient that continues to bring water into the plant.

DATA: Cline, R. G., and W. S. Campbell. 1976. *Ecology* 57: 367–373.

potential that were recorded in tissue from a shrub species called ninebark. In **FIGURE 38.3** each data point represents the solute potential recorded on a particular day; the lines between the data points are drawn simply to make the trend clear.

Ninebarks thrive on dry sites in the Rocky Mountains of western North America. Notice that the solute potential of tissue is relatively high in June, at the start of the growing season. In the Rockies, June tends to be rainy and cool. July and August are progressively hotter and drier, however. As the graph indicates, the solute potential in tissue drops dramatically during this period.

This is a key observation. As the summer progresses and water potential in soil drops, ninebark shrubs are able to keep acquiring water and grow because the solute potentials of their tissues can drop to stay below the soil water potential.

Plants that are adapted to wetter sites cannot tolerate such low solute potentials in their tissues. When conditions get hot and dry, they have to close their stomata and stop photosynthesis—meaning that their growth will stop or slow down dramatically compared to dry-adapted plants.

If plants keep their stomata open and leaf cells lose water faster than the water is replaced, the positive turgor pressure normally inside living cells drops to 0. If the cells do not regain turgor quickly, they are at risk of dehydration and death as cells shrivel like grapes drying into raisins. When an entire tissue loses turgor pressure, it will **wilt** (**FIGURE 38.4**).

Turgor pressure is required for growth to occur; without it, cells cannot expand once cell division is complete. In addition, turgor pressure provides structural support—which is why turgid plants do not wilt. Unless corrected, extensive wilting may lead to the death of the tissue and, eventually, the plant.

Water Potential in Air   In the atmosphere, water exists as a vapor with a solute potential of 0 MPa. The pressure exerted by water vapor in the atmosphere depends on temperature and humidity. The lower the pressure potential, the faster the rate of evaporation from liquid water to the atmosphere.

FIGURE 38.4 **Wilting Occurs When Water Loss Leads to Loss of Turgor Pressure.** Wilting is a life-threatening condition in plants, analogous to severe dehydration in humans.

- When air is dry, there are few water molecules present and the pressure they exert is low, increasing the rate of evaporation.
- When air is warm, water molecules move farther apart and also exert lower pressure.

Warm, dry air has an extremely low water potential, often approaching −100 MPa. When the weather is cool and rainy or foggy, however, the water potential of the atmosphere may be equal to the water potential inside a leaf. But normally, the water potential of the atmosphere is much lower than the water potential inside a leaf, so water evaporates quickly.

check your understanding

If you understand that . . .

- Water moves along a water-potential gradient, from areas of high water potential to areas of low water potential.
- In areas separated by a selectively permeable membrane, part of water's potential energy is made up of its solute potential—its tendency to move via osmosis.
- Water also has a pressure potential. In living plant cells, for example, the cell wall can exert pressure on water and affect its pressure potential.
- Soils, plant tissues, and the atmosphere all have a water potential made up of a solute potential and a pressure potential.

✔ **You should be able to . . .**

1. Explain why the water potential of a cell in equilibrium with pure water can be zero even though it contains solutes.
2. Compare and contrast the solute potential of an irrigated agricultural field before and after a long, hot day of evaporation.

Answers are available in Appendix A.

Low water potential

Atmosphere ψ: –90 MPa
(Changes with humidity; usually very low)

Leaf ψ: –0.8 MPa
(Depends on transpiration rate; low when stomata are open)

Root ψ: –0.6 MPa
(Medium–high)

Soil ψ: –0.3 MPa
(High if moist; low if extremely dry)

High water potential

FIGURE 38.5 A Water-Potential Gradient Exists between Soil, Plants, and Atmosphere. Water moves from regions of high water potential to regions of low water potential.

In most cases, water potential is highest in soil, lower in roots, lower yet in leaves, and lowest in the atmosphere. This situation sets up a **water-potential gradient** that causes water to move up through the plant. To move *up* a plant, water moves *down* the water-potential gradient that exists between the soil, its tissues, and the atmosphere. When it does so, it replaces the water lost through transpiration (**FIGURE 38.5**).

38.2 How Does Water Move from Roots to Shoots?

Suppose that you are caring for the wilted plant in Figure 38.4. If you add water to the soil, the water potential of the soil increases and water will move into the plant along a water-potential gradient.

Water moves into root cells by osmosis, but how does it move up to the shoots, against gravity? Biologists have tested three major hypotheses for how the water could be transported to shoots:

1. Root pressure—a pressure potential that develops in roots—could drive water up against the force of gravity.

2. Capillary action could draw water up the cells of xylem.

3. Cohesion-tension, a force generated in leaves by transpiration, could pull water up from roots.

Note that all three hypotheses could be correct—they are not mutually exclusive.

Let's begin by considering how water and solutes move from the soil across the root and into the **vascular tissue,** which contains xylem and phloem. Then we can analyze each of the three hypotheses for how water and solutes finish the trip up the xylem to the shoot system.

Movement of Water and Solutes into the Root

To understand how water enters a root, consider the cross section through a young buttercup root shown in **FIGURE 38.6** (see also Chapter 37). Starting at the outside of the root and working inward, notice that several distinct tissues are present:

- The **epidermis** (literally, "outside skin") is a single layer of cells. In addition to protecting the root, some epidermal cells produce **root hairs,** which greatly increase the total surface area of the root.

- The **cortex** consists of ground tissue—usually parenchyma cells—and stores carbohydrates.

FIGURE 38.6 In Roots, Water Has to Travel through Several Tissue Layers to Reach Vascular Tissue. This cross section through a buttercup root shows the anatomy that is typical of roots in eudicots.

- The **endodermis** ("inside skin") is a cylindrical layer of cells that forms a boundary between the cortex and the vascular tissue. The function of the endodermis is to control ion uptake and prevent ion leakage from the vascular tissue.

- The **pericycle** ("around-circle") is a layer of cells that forms the outer boundary of the vascular tissue. The pericycle can become meristematic and produce lateral roots.

- Conducting cells of the vascular tissue function in transport between roots and shoots, and are located in the center of roots in buttercups and other eudicots. Notice that, in these plants, phloem is situated between each of four arms formed by xylem, which is arranged in a cross-shaped pattern.

Three Routes through Root Cortex to Xylem   When water enters a root along a water-potential gradient, it does so through root hairs. As water is absorbed, it moves through the root cortex toward the xylem through three distinct routes (**FIGURE 38.7**).

1. The *transmembrane route* is based on flow through aquaporin proteins—water channels located in the plasma membranes of root cells (see Chapter 6). Some water may also diffuse directly across plasma membranes.

2. The *apoplastic route* is outside the plasma membranes. The **apoplast** consists of cell walls, which are porous, and the spaces that exist between cells. The name apoplastic is apt because movement takes place outside the plasma membrane and cell interior. Water moving along the apoplastic route *must* eventually pass into the cytoplasm of endodermal cells before entering xylem.

3. The *symplastic route* is inside the plasma membranes. The **symplast** consists of the cytosol and the continuous connections through cells that exist via plasmodesmata (see Chapter 11).

In essence, water can flow through tissues by crossing membranes via aquaporins, or by passing around cells via the apoplast, or by passing through the symplast via plasmodesmata.

The Role of the Casparian Strip   The situation changes when water reaches the endodermis. Endodermal cells are tightly packed and secrete a narrow band of wax called the **Casparian strip.** This layer is composed primarily of a compound called **suberin,** which forms a water-repellent cylinder at the endodermis.

The Casparian strip blocks the apoplastic route by preventing water from moving through the walls of endodermal cells and

FIGURE 38.7 How Water Travels from Root Hair to Xylem.

✔**EXERCISE** Add dots representing aquaporins in the transmembrane route.

proceeding to the vascular tissue (Figure 38.7). The Casparian strip does not affect water and ions that move through the symplastic route.

The Casparian strip is important because it means that for water and solutes to reach vascular tissue, they have to move into the cytoplasm of an endodermal cell. Endodermal cells, in turn, act as filters.

By forcing water and ions to cross at least two membranes on their way from the soil to the xylem, one entering the symplast outside the Casparian strip and one leaving the symplast inside the Casparian strip, plants can use specific channel and carrier proteins (see Chapter 6) to control what moves to the shoots. Endodermal cells allow ions such as potassium (K^+) that are needed by the plant to pass through to the vascular tissue. In contrast, these cells can prevent the passage of ions such as sodium (Na^+) or heavy metals that are not needed or may be harmful.

✔ If you understand this concept, you should be able to predict what would happen if a plant had a mutation that prevented synthesis of suberin and formation of the Casparian strip.

Water Movement via Root Pressure

Have you ever walked barefoot through grass early in the morning and gotten your feet wet? Where does that water come from? Let's explore this phenomenon.

Movement of ions and water into the root xylem is responsible for the process known as **root pressure.** Recall that root pressure is one of three hypothesized mechanisms for moving water up xylem, from root to shoot.

The Casparian strip in endodermal cells is essential for root pressure to develop. Without an apoplastic barrier between the xylem and the environment, ions and water would simply leak out of roots.

Stomata normally close during the night, when photosynthesis is not occurring and CO_2 is not needed. Their closure minimizes water loss and slows the movement of water through plants. But roots often continue to accumulate ions that their epidermal cells acquire from the soil as nutrients, and these nutrients are actively pumped into the xylem. The influx of ions lowers the water potential of xylem below the water potential in the surrounding cells.

As water flows into xylem from other root cells in response to the solute gradient, a positive pressure is generated at night that forces fluid up the xylem. More water moves up xylem and into leaves than is being transpired from the leaves.

In low-growing plants, enough water can move to force water droplets out of the leaves, a phenomenon known as **guttation.** Some of the water that gets your feet wet early in the morning spent part of the night moving through plants (**FIGURE 38.8**).

At one time, positive root pressure formed the basis of a leading hypothesis to explain how water moves from roots to leaves in trees. However, research showed that over long distances, such as the height of a tree, the force of root pressure is not enough to overcome the force of gravity on the water inside xylem.

FIGURE 38.8 Root Pressure Causes Guttation. When ions accumulate in the xylem of roots at night, enough water enters the xylem via osmosis to force water up and out of low-growing leaves.

In addition, researchers demonstrated that cut stems, which have no contact with the root system, are still able to transport water to leaves. Biologists concluded that there must be some other mechanism involved in the long-distance transport of water. What is it?

Water Movement via Capillary Action

Researchers have also evaluated a hypothesis based on the phenomenon of **capillarity** (also called capillary action), or movement of water up a narrow tube. When a thin glass tube is placed upright in a pan of water, water creeps up the tube (**FIGURE 38.9**). The movement occurs in response to three forces: (**1**) adhesion, (**2**) cohesion, and (**3**) surface tension. Let's consider each force in turn.

Adhesion is a molecular attraction among unlike molecules. In this case, water interacts with a solid substrate—such as the glass walls of a capillary tube or the cell walls of tracheids or vessel elements—through hydrogen bonding. As water molecules

Surface tension: Pulls water column up to minimize air–water interface

Adhesion: Water attracted to glass is pulled up against the force of gravity

Cohesion: Holds water molecules in the water column together

Meniscus

Thin glass tube placed in water

FIGURE 38.9 Water Can Rise in Xylem via Capillarity. Capillary action occurs through a combination of three forces, all of which are generated by hydrogen bonding and together are acting against gravity.

bond to each other and adhere to the side of the tube, they are pulled upward.

Cohesion is a molecular attraction among like molecules, such as the hydrogen bonding that occurs among molecules in water. Because water molecules cohere, the upward pull by adhesion is transmitted to the rest of the water column. The water column rises against the pull of gravity.

The effects of adhesion, cohesion, and gravity are responsible for the formation of a concave surface boundary called a **meniscus** (plural: **menisci**). A meniscus forms at most air–water interfaces—including those found in narrow tubes. Menisci form because adhesion and cohesion pull water molecules up along the sides of the tube, while gravity pulls the water surface down in the middle.

Surface tension is a force that exists among water molecules at an air–water interface. In the body of a water column, all the water molecules present are surrounded by other water molecules and form hydrogen bonds in all directions. Water molecules at the surface, however, can form hydrogen bonds only with the water molecules beside and below them. As a result, surface molecules share stronger attractive forces. This enhanced attraction results in tension that minimizes the total surface area. When a meniscus forms, surface tension exerts a pull against gravity.

Capillary action results when adhesion creates an upward pull at the water–container interface, surface tension creates an upward pull all across the surface, and cohesion transmits both forces to the water below. All three forces counteract the effect of gravity, and the result is capillarity.

Like root pressure, capillarity can transport water only a limited distance. Capillary action moves water along the surfaces of mosses and other low-growing, non-vascular plants; but it can raise the water in the xylem of a vertical stem only about 1 m.

Thus, root pressure and capillary action cannot explain how water moves from soil to the top of a redwood tree—an organism that can grow 5 to 6 stories higher than the Statue of Liberty. How then does it actually happen?

The Cohesion-Tension Theory

The leading hypothesis to explain long-distance water movement in vascular plants is the **cohesion-tension theory,** which states that water is pulled to the tops of trees along a water-potential gradient, via forces generated by transpiration at leaf surfaces.

To understand how the cohesion-tension force works, start with step 1 in **FIGURE 38.10**. Notice that spaces inside the leaf are filled with moist air as a result of evaporation from the surfaces of surrounding cells. When a stoma opens, this humid air is exposed to the atmosphere, which in most cases is much drier. This creates a steep water-potential gradient between the leaf interior and its surroundings. The steeper the gradient, the faster water vapor diffuses out through the stomata.

Step 2 shows that as water is lost from the leaf to the atmosphere, the humidity of the gas-filled space inside the leaf drops. In response, more water evaporates from the walls of the parenchyma cells. At the microscopic level, cell walls are not perfectly smooth, so menisci form at the air–water interface. As more water evaporates, the menisci become steeper and the total area of the air–water interface increases.

In 1894 Henry Dixon and John Joly hypothesized that the formation of steep menisci produces a force capable of pulling water up from the roots, dozens or hundreds of meters into the air. Is this really possible?

The Role of Surface Tension in Water Transport   The key concept in the cohesion-tension theory is that the negative force or pull (tension) generated at the air–water interface is transmitted through the water outside of leaf cells (step 3 in Figure 38.10), to the water in xylem (step 4), to the water in the vascular tissue of roots (step 5), and finally to the water in the soil (step 6).

The transmission of pulling force from the leaf surface to the root is possible because (**1**) there are continuous columns of water throughout the plant, and (**2**) all of the water molecules present hydrogen bonds to one another (cohesion).

Note that the plant does not expend energy to create the pulling force. The force is generated by energy from the Sun, which drives evaporation from the leaf surface. Water transport is solar powered.

In effect, the cohesion-tension theory of water movement states that, because of the hydrogen bonding between water molecules, water is pulled up through xylem in continuous columns.

Creating a Water Potential Gradient   You can also think about the cohesion-tension theory in terms of water potentials. Note that because tracheids and vessels are dead at maturity, the water in xylem does not cross plasma membranes. As a result, water does not move between cells by osmosis. In xylem, water movement is driven entirely by differences in pressure potential. The water in a column of xylem cells moves by **bulk flow**—a mass movement of molecules along a pressure gradient.

The pulling force generated at menisci in leaf cell walls lowers the pressure potential of water in leaves. Even though the tension created at each meniscus is relatively small, there are so many menisci in the leaves of an entire plant that the tension created by summing many small pulling forces is remarkable. It creates a water-potential gradient between leaves and roots that is steep enough to overcome the force of gravity and pull water up long distances.

To appreciate just how great the forces involved are, think of the vessel elements or tracheids in xylem as groups of drinking straws. When you use a straw, the vacuum that you create causes liquid to rise. Sucking on a drinking straw creates a pressure difference of about -0.1 MPa, which can draw water up a maximum of about 10 m. In contrast, the negative pressure exerted by the menisci in leaves can be as low as -2.0 MPa. The tension in xylem tissue may be 10 times the amount of pressure on a fully inflated car tire. (But note that the pressure on a car tire is positive, not negative.) The force is enough to draw water up 100 m.

The Importance of Secondary Cell Walls   If you suck on a drinking straw hard enough, the pressure gradient between the

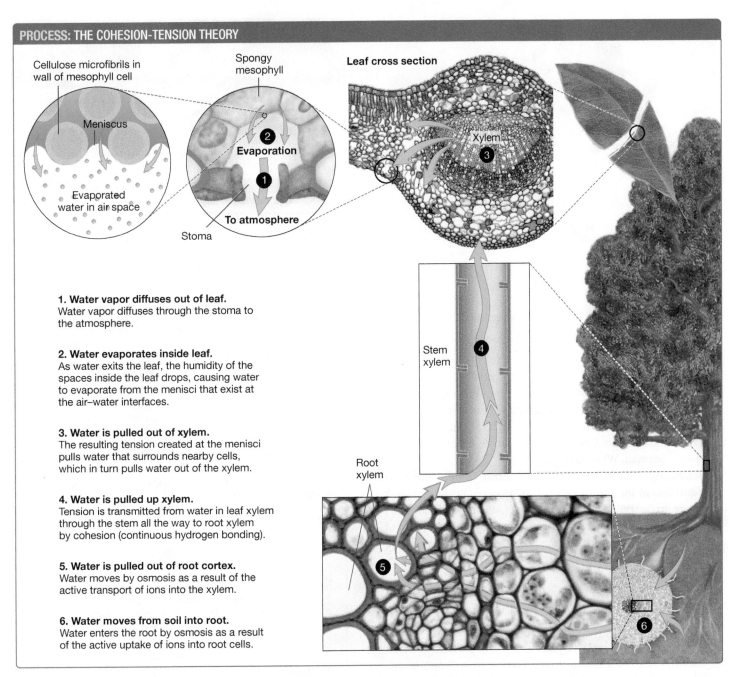

PROCESS: THE COHESION-TENSION THEORY

Cellulose microfibrils in wall of mesophyll cell

Meniscus

Evaporated water in air space

Spongy mesophyll

2 Evaporation

1

To atmosphere

Stoma

Leaf cross section

Xylem

3

Stem xylem

4

Root xylem

5

6

1. Water vapor diffuses out of leaf.
Water vapor diffuses through the stoma to the atmosphere.

2. Water evaporates inside leaf.
As water exits the leaf, the humidity of the spaces inside the leaf drops, causing water to evaporate from the menisci that exist at the air–water interfaces.

3. Water is pulled out of xylem.
The resulting tension created at the menisci pulls water that surrounds nearby cells, which in turn pulls water out of the xylem.

4. Water is pulled up xylem.
Tension is transmitted from water in leaf xylem through the stem all the way to root xylem by cohesion (continuous hydrogen bonding).

5. Water is pulled out of root cortex.
Water moves by osmosis as a result of the active transport of ions into the xylem.

6. Water moves from soil into root.
Water enters the root by osmosis as a result of the active uptake of ions into root cells.

FIGURE 38.10 Transpiration Creates Tension That Is Transmitted from Leaves to Roots.

inside of the straw and the atmosphere can overcome the stiffness of the straw and cause it to collapse. How can vascular tissue withstand negative pressures as low as −2.0 MPa without collapsing?

The answer is found in a key adaptation: the secondary thickenings characteristic of the cell walls in tracheids and vessel elements. The cells in vascular tissue have walls that are reinforced with tough lignin molecules (see also Chapter 31).

The evolution of lignified secondary cell walls was an important event in the evolution of land plants because it allowed vascular tissue to withstand extremely negative pressures. The result? Tall trees.

What Evidence Do Biologists Have for the Cohesion-Tension Theory? If the cohesion-tension theory is correct, the water present in xylem should experience a strong pulling force. A simple experiment supports this prediction.

If you find a leaf that is actively transpiring and cut its petiole, the watery fluid in the xylem, or xylem sap, withdraws from the edge toward the inside of the leaf (**FIGURE 38.11**, see page 764). According to the cohesion-tension theory, this observation is due to a transpirational pull at the air–water interface in parenchyma cells of leaves.

Although the observation that xylem sap is under tension was important, the cohesion-tension theory remained controversial.

FIGURE 38.11 Xylem Sap in Cut Stems "Snaps Back."

Petiole

Pull

Pull

When you cut a petiole, watery fluid (xylem sap) withdraws from the cut surface

For example, several studies failed to document rapid changes in xylem pressure when temperature and humidity—and thus the severity of the water-potential gradient—were changed experimentally. Advocates of the theory blamed the negative results on instruments that could not document small, rapid changes in pressure potential. Who was right?

Chunfang Wei, Melvin Tyree, and Ernst Steudle answered the question in the late 1990s with an instrument called a xylem pressure probe (**FIGURE 38.12**). A xylem pressure probe includes an oil-filled glass tube that can be inserted directly into the xylem of a leaf. The oil transmits changes in pressure within the xylem to a gauge, allowing researchers to record changes in xylem pressure instantly and directly.

To test the cohesion-tension theory, the researchers altered xylem pressure by raising light levels to alter transpiration rates in the leaves of corn plants. Their results?

- The graph in Figure 38.12 plots how xylem pressure changed as the researchers changed light levels over a 7-minute period. Note that as light intensity increased, the xylem pressure probe documented increased tension, or pull—negative pressure.

- In addition, higher light levels reduced the weight of the entire plant—suggesting that higher transpiration rates caused water loss.

Both observations are consistent with predictions that follow from the cohesion-tension theory.

These experiments convinced most biologists that increased transpiration leads to increased tension on xylem sap. Rising tension, in turn, lowers the water potential of leaves and exerts a pull on water in the roots, where the water potential is high. On the basis of these and other results, most biologists now accept the cohesion-tension theory.

RESEARCH

QUESTION: Do direct measurements of pressure in xylem tissue support the cohesion-tension theory?

HYPOTHESIS: Increasing transpiration by raising light intensity will lower xylem pressure in leaves.

NULL HYPOTHESIS: Increasing light intensity will not affect xylem pressure in leaves.

EXPERIMENTAL SETUP:

Xylem pressure probe

Pressure gauge

Variable-intensity light source

Oil-filled tube (transmits pressure in xylem to gauge)

Corn

PREDICTION: Xylem pressure will decrease as transpiration increases with higher light levels.

PREDICTION OF NULL HYPOTHESIS: Xylem pressure will not be affected by light intensity.

RESULTS:

Medium light intensity

Medium–high light intensity

High light intensity

Insertion of pressure probe into the xylem vessel

Xylem pressure (MPa)

Time (s)

CONCLUSION: Xylem pressure decreases when light intensity increases. The data support the cohesion-tension theory.

FIGURE 38.12 Measuring Changes in Pressure inside Xylem.

SOURCE: Wei, C., M. T. Tyree, and E. Steudle. 1999. Direct measurement of xylem pressure in leaves of intact maize plants. A test of the cohesion-tension theory taking hydraulic architecture into consideration. *Plant Physiology* 121: 1191–1205.

✓**QUANTITATIVE** Suppose the researchers had chosen to plot changes in the water-potential gradient between the xylem in roots and leaves on the *y*-axis, during the same experiment. What would the graph look like?

If you understand that . . .

- Water can move a short distance in xylem via root pressure or capillarity.
- Long-distance transport of water depends on movement along a steep water-potential gradient. This gradient is created primarily by the negative pressure potential of water in leaves, due to surface tension that develops in response to transpiration.

✔ **You should be able to . . .**

1. Explain how transpiration at the surface of a meniscus near a stoma creates a pull on the water in leaves and xylem.
2. Predict what happens to the meniscus when each of the following occurs: A nearby stoma closes, a rain shower starts, and weather changes as dry air blows in.

Answers are available in Appendix A.

38.3 Water Absorption and Water Loss

One of the most important features of the cohesion-tension theory is that it does not require plants to expend energy to lift the water that the roots take in by osmosis. To absorb water, roots must expend energy to accumulate the ions that maintain a solute potential that is lower than the soil. Once the water with dissolved nutrients is in the root xylem, the Sun furnishes the energy required to pull the solution from roots to shoots—not ATP supplied by the plant.

1. Energy from the Sun heats water molecules at the air–water interface inside leaves enough to break the hydrogen bonds between them and cause transpiration.

2. Rapid transpiration creates deep menisci in the walls of leaf cells, causing tension that lowers the water potential of leaves.

3. Hydrogen bonding between water molecules transmits this tension down to water molecules in the roots.

Xylem acts as a passive conduit—a set of narrow pipes that allows water to move from a region of high water potential (the roots) to a region of low water potential (the leaves). Water flows from roots to shoots as long as the water-potential gradient—from root to leaf to atmosphere—is intact.

When soils begin to dry, however, it becomes difficult for plants to replace water being lost via transpiration. If water is not replaced fast enough, the solute potentials of leaves drop and leaves and branches begin to wilt. In response, stomata may close down partially or completely to reduce transpiration rates and conserve water. However, closing stomata affects the ability of plants to carry on photosynthesis, because CO_2 acquisition slows or stops. The balance between conserving water and maximizing photosynthesis is termed the photosynthesis-transpiration compromise.

This compromise is particularly delicate for species that grow on dry sites. How do they cope?

Limiting Water Loss

Plants that thrive in dry sites have several adaptations that help them slow transpiration and limit water loss. Consider the oleander plant, which is native to the dry shrub-grassland habitats of southern Eurasia. The micrograph in **FIGURE 38.13** shows a cross section through an oleander leaf, which has the following special features:

- A particularly thick cuticle covers the upper surface of oleander leaves. This waxy layer minimizes water loss from cells that are directly exposed to sunlight. In general, species that are adapted to dry soils have much thicker cuticles than do species adapted to wet soils.

- The stomata of oleanders are located on the undersides of their leaves, inside deep pits in the epidermis. Hairlike extensions of epidermal cells called **trichomes** (see Section 37.3) shield these pits from the atmosphere. The leading hypothesis to explain these traits is that they slow the loss of water vapor from stomata to the dry air surrounding the leaf by creating a layer of still air surrounding the stomata.

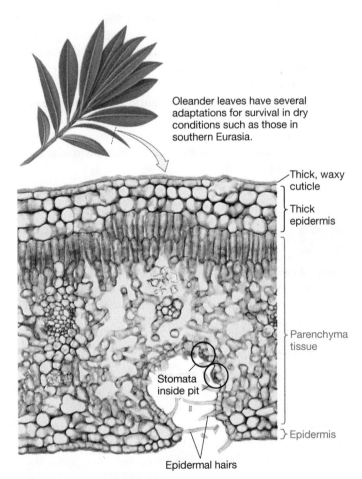

Oleander leaves have several adaptations for survival in dry conditions such as those in southern Eurasia.

Thick, waxy cuticle

Thick epidermis

Parenchyma tissue

Stomata inside pit

Epidermis

Epidermal hairs

FIGURE 38.13 Species That Are Adapted to Dry Habitats Have Modified Leaf Structures.

Other species have other adaptations for limiting water loss. For example, recall that many species adapted to water-limited habitats—either cold environments where water is often frozen or deserts where rainfall is rare—have needlelike leaves (see Chapter 37). Long, thin leaf shapes minimize the surface area exposed to sunlight and thus minimize transpiration. Plants adapted to wetter habitats, in contrast, tend to have broad leaves with a large surface area.

Obtaining Carbon Dioxide under Water Stress

Many of the species that thrive in deserts and other hot, dry habitats can continue photosynthesizing even when soil moisture content is low. Recall that two novel biochemical pathways, **crassulacean acid metabolism (CAM)** and **C_4 photosynthesis,** allow plants to increase CO_2 concentrations in their leaves and conserve water (see Chapter 10).

CAM plants open their stomata at night and store the CO_2 that diffuses into their tissues by adding the carbon dioxide molecules to organic acids. When sunlight is available during the day and photosynthesis begins, the CO_2 molecules are released from the organic acids and converted to sugar by **rubisco**—the enzyme, found in all green plants, that initiates the Calvin cycle. In this way, CAM plants can photosynthesize and grow even with their stomata closed during the day.

C_4 plants minimize the extent to which their stomata open because they use CO_2 so efficiently. Mesophyll cells in C_4 plants take up CO_2 and add it to organic acids. The CO_2 is then transferred to specialized cells called **bundle-sheath cells** (see Chapter 10), where the Calvin cycle operates. In effect, the C_4 cycle is a mechanism for concentrating carbon dioxide in cells deep inside the leaf, so stomata do not have to be wide open continuously.

Like the cuticle and stomata-containing pits of oleanders, CAM and C_4 plants have adaptations that help them conserve water by limiting transpiration.

38.4 Translocation of Sugars

Translocation is the movement of sugars by bulk flow throughout a plant—specifically, from sources to sinks. In vascular plants, a **source** is a tissue where sugar enters the phloem; a **sink** is a tissue where sugar exits the phloem.

Where do sources and sinks occur in a plant? The answer often depends on the time of year.

- *During the growing season* Mature leaves and stems that are actively photosynthesizing produce sugar in excess of their own needs. These tissues act as sources. Sugar moves from leaves and stems to a variety of sinks, where sugar use is high and production is low. Apical meristems, lateral meristems, developing leaves, flowers, developing seeds and fruits, and storage cells in roots all act as sinks (**FIGURE 38.14**).

- *Early in the growing season* When a plant resumes growth after the winter or the dry season, sugars move from storage

FIGURE 38.14 Sugars Move from Sources to Sinks.

areas to growing areas. Storage cells in roots and seeds act as sources; developing leaves act as sinks.

Tracing Connections between Sources and Sinks

To explore the relationship between sources and sinks in more detail, consider research on sugar-beet plants that were exposed to carbon dioxide molecules containing the radioactive isotope ^{14}C. The goal was to track where carbon atoms moved after they were incorporated into sugars via photosynthesis.

The location of ^{14}C atoms inside a plant can be documented in two ways: (**1**) by measuring the amount of radioactive CO_2 emanating from different tissues, or (**2**) by laying plant parts on X-ray film and allowing the radioactivity to expose and blacken the film.

Many researchers have enclosed individual leaves of intact plants in a bag and introduced a known amount of radioactive CO_2 for a short period of time. One typical experiment documented that mature leaves retained just over 9 percent of the labeled carbon. In contrast, growing leaves retained 67 percent. These data are consistent with the prediction that fully expanded leaves act as sources of sugar, while actively growing leaves and roots act as sinks.

In similar experiments, researchers have exposed all the leaves on a growing plant to labeled carbon. One such experiment found that over 16 percent of the total carbon was translocated to root tissue within 3 hours. This result is consistent with the prediction that, during the growing season, roots also act as sinks.

Similar experiments support two generalizations.

1. Sugars can be translocated rapidly—typically 50–100 centimeters per hour (cm/hr).

2. There is a strong correspondence between the physical locations of sources and sinks.

The second point is particularly interesting. For example, mature leaves that act as sources send sugar to tissues on the same side of the plant (**FIGURE 38.15a**). In addition, experiments with tall herbaceous plants show that leaves on the upper part of the stem send sugar to apical meristems, but leaves on the lower part of the plant send sugar to the roots (**FIGURE 38.15b**).

Why would leaves send sugar to tissues on a certain side or part of the body? The answer hinges on understanding the structure of phloem.

The Anatomy of Phloem

Two specialized parenchyma cell types make up phloem: **sieve-tube elements** and **companion cells** (see Chapter 37). Unlike the tracheids and vessel elements that make up most of the xylem, sieve-tube elements and companion cells are alive at maturity.

Recall that, in most plants, sieve-tube elements lack nuclei and most other organelles. They are connected to one another, end to end, by perforated **sieve plates** (**FIGURE 38.16**). The pores create a direct connection between the cytoplasms of adjacent cells.

(a) Source leaves send sugar to the same side of the plant.

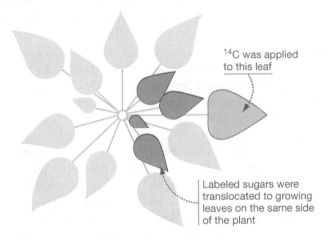

^{14}C was applied to this leaf

Labeled sugars were translocated to growing leaves on the same side of the plant

(b) Source leaves send sugar to tissues on the same end of the plant.

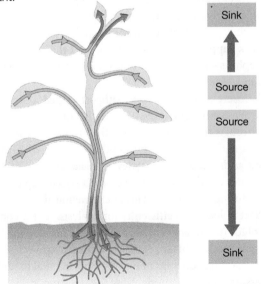

Sink

Source

Source

Sink

FIGURE 38.15 Sources Supply Sinks on the Same Side and Same End of the Body.

Companion cells, in contrast, have nuclei and a rich assortment of ribosomes, mitochondria, and other organelles. Companion cells function as "support staff" for sieve-tube elements.

You might also recall that the phloem in primary and secondary vascular tissue is continuous throughout the plant—meaning that there is a direct anatomical connection between the phloem from the tips of shoots to the tips of roots (see Chapter 37). The sieve-tube elements in phloem represent a continuous system for transporting sugar throughout the plant body.

Each vascular bundle runs the length of stems and roots, and certain bundles extend into specific branches, leaves, and lateral roots. Phloem sap does not move from one vascular bundle to another—instead, each bundle is independent.

Based on these results, the physical relationships observed between sources and sinks in herbaceous plants are logical. For example, the phloem in the leaves on one side of a plant connects directly with the phloem of branches, stems, and roots on the same side of the individual, through a specific set of vascular bundles.

The phloem sap that flows through vascular tissue is often dominated by the disaccharide **sucrose**—table sugar. Phloem sap can contain small amounts of minerals, amino acids, mRNAs, hormones, and other compounds as well. How does this solution move? What mechanism is responsible for translocating sugars from sources to sinks?

Sieve-tube element (few organelles)

Companion cell (many organelles)

Phloem sap (high concentration of sucrose) passes vertically through pores in the wall between sieve-tube elements

Phloem sap

The cytoplasm of sieve-tube elements and companion cells is connected by plasmodesmata

Sieve plate

FIGURE 38.16 Sieve-Tube Elements Are Connected by Pores.

The Pressure-Flow Hypothesis

In 1926 Ernst Münch proposed the **pressure-flow hypothesis,** which states that events at source tissues and sink tissues create a pressure potential gradient in phloem (**FIGURE 38.17**). The water in phloem sap moves down this pressure gradient, and sugar molecules are carried along by bulk flow.

Like the cohesion-tension theory for water transport in xylem, the pressure-flow hypothesis is based on movement along a water-potential gradient created by changes in pressure potential. Unlike the cohesion-tension model, however, transpiration does not provide the driving force to move phloem sap. Instead, differences between turgor pressure in the phloem near source tissues and turgor pressure in the phloem near sink tissues generate

Phloem
Xylem
Sieve-tube element
High turgor pressure
Water
Sucrose movement
Sugar

SOURCE (leaf cell)
Companion cell

Water movement in phloem and xylem via bulk flow

Low turgor pressure

Sucrose movement

Water

SINK (root cell)
Companion cell
Phloem sap movement via pressure flow
Xylem sap movement via transpirational pull

FIGURE 38.17 The Pressure-Flow Hypothesis: High Turgor Pressure Near Sources Causes Phloem Sap to Flow to Sinks. The pressure-flow hypothesis predicts that water cycles between xylem and phloem, and that water movement in phloem, is a response to a gradient in pressure potential.

✔**EXERCISE** This diagram shows how water moves between xylem and phloem in the middle of the growing season, when leaves are sources and roots are sinks. Add new arrows, in new colors, to indicate the direction of water and phloem sap flow in spring, when roots act as sources and leaves act as sinks.

the necessary force. Creating these differences in turgor pressure usually requires an expenditure of ATP.

Creating High Pressure Near Sources and Low Pressure Near Sinks To understand how Münch's model works, start with the source cell at the upper left in Figure 38.17. The small red arrows reflect Münch's proposal that sucrose moves from source cells into companion cells and from there into sieve-tube elements.

Because of this phloem loading, the phloem sap in the source tissue has a high concentration of sucrose. Compared to the water in the adjacent xylem cells in a vascular bundle, the phloem sap has a very low solute potential.

As the blue arrows in the upper part of the diagram show, water moves along a water-potential gradient—flowing passively from xylem across the selectively permeable plasma membrane of sieve-tube elements. In response, turgor pressure begins to build in the sieve-tube elements in the source region.

What is happening at the sink? Münch proposed that cells in the sink (bottom left in Figure 38.17) remove sucrose from the phloem sap by passive or active transport. As a result of this phloem unloading—a loss of solutes—the water potential in sieve-tube elements increases until it is higher than the water potential in adjacent xylem cells. As the blue arrows at the bottom of the figure show, water flows across the selectively permeable membranes of sieve-tube elements into xylem along a water-potential gradient. In response, turgor pressure in the sieve-tube elements in the sink drops.

The net result of these events is high turgor pressure in phloem at a source tissue and low turgor pressure in phloem at the sink, created by the loading and unloading of sugars, respectively. This difference in pressure potential drives phloem sap from source to sink via bulk flow. There is a one-way flow of sucrose and a continuous loop of water movement. Water returns to the source tissue via the xylem.

Testing the Pressure Flow Model The pressure-flow hypothesis is logical, given the anatomy of vascular tissue and the principles that govern water movement. But has any experimental work supported the theory? Some of the best tests have relied on aphids—small insects that make their living ingesting phloem sap.

Aphids insert a syringe-like mouthpart, called a stylet, into sieve-tube elements. The pressure on the fluid in these cells forces it through the stylet and into the aphid's digestive tract. Excess water and sucrose that the aphid does not need is excreted out its anus as droplets of "honeydew" (**FIGURE 38.18**).

If the aphids are then severed from their stylets, sap continues to flow out through the stylets. This phenomenon allows researchers to collect phloem sap efficiently for analysis. It also confirms that the aphids do not actively suck the fluid. As predicted by the pressure-flow model, phloem is indeed under pressure.

This observation supports one of the fundamental predictions of the pressure-flow hypothesis. Now the question is, How does sucrose enter and leave phloem in a way that sets up the water-potential gradient?

FIGURE 38.18 Aphids Feed on Phloem Sap. The tip of this aphid's mouthpart (the stylet) is in a sieve-tube element within the plant stem. The droplet emerging from the aphid's anus is honeydew, which consists of sugary phloem sap.

Phloem Loading

In contrast to the cohesion-tension model of water movement in xylem, pressure flow often requires that plants expend energy to set up a water-potential gradient in phloem. To establish a high-pressure potential in sieve-tube elements near source cells, large amounts of sugar have to be transported into the phloem sap—enough to raise the solute concentration of sieve-tube elements. This requirement is illustrated in Figure 38.17, top left.

In some cases, loading is active—it requires an expenditure of ATP and some sort of membrane transport system. But when sucrose concentrations in source cells are extremely high, movement of sucrose into sieve-tube elements can also occur via passive diffusion through plasmodesmata.

Phloem unloading at sinks can also be active or passive, depending on the tissue. When sugar is unloaded against its concentration gradient, an expenditure of ATP and a second membrane transport mechanism is required. How do phloem loading and unloading occur? What specific membrane proteins are involved?

To answer these questions, let's start by reviewing how transport proteins make it possible for sugars and other large or charged substances to cross a phospholipid bilayer.

How Are Sucrose and Other Solutes Transported across Membranes? **Passive transport** (**FIGURE 38.19**) occurs when ions or molecules move across a plasma membrane by diffusion—that is, along their electrochemical gradient. The adjective passive is appropriate because no expenditure of energy is required for the movement to occur.

Recall that small, uncharged molecules diffuse across phospholipid bilayers rapidly (Figure 38.19, left). But ions and many large molecules diffuse across phospholipid bilayers slowly if at all, even when their movement is favored by a strong electrochemical gradient. To diffuse rapidly, they must avoid direct contact with the phospholipid bilayer by passing through a membrane protein.

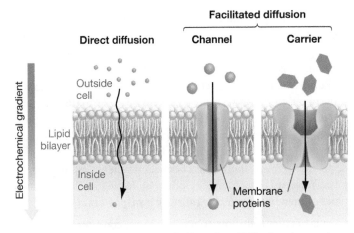

FIGURE 38.19 Passive Transport Is Based on Diffusion. In passive transport, ions or molecules diffuse across membranes—meaning they follow their electrochemical gradient. The movement can occur directly through the phospholipid bilayer or be facilitated by a channel or carrier protein.

Two types of membrane protein—channels and carriers—facilitate the passive diffusion of specific ions or molecules (see Chapter 6).

- **Channel proteins** form pores that selectively admit certain ions (Figure 38.19, middle).
- **Carrier proteins** work like enzymes, undergoing a conformational change that transports a bound molecule across the lipid bilayer (Figure 38.19, right).

Channels and carriers are responsible for **facilitated diffusion.**

Active transport (**FIGURE 38.20**) occurs when ions or molecules move across a plasma membrane against their electrochemical gradient. The adjective active is appropriate because cells must expend energy in the form of ATP to move solutes in an energetically unfavorable direction.

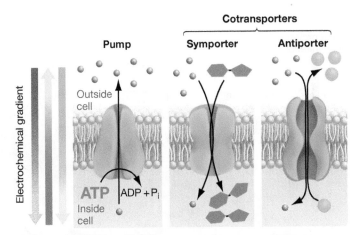

FIGURE 38.20 Active Transport Moves Ions or Molecules against an Electrochemical Gradient. All forms of active transport require an expenditure of ATP. Pumps use ATP directly; cotransport proteins use ATP indirectly. Cotransport depends on a previous expenditure of ATP by a pump.

Active transport always involves membrane proteins. **Pumps** are proteins that change shape when they bind ATP or a phosphate group from ATP. As they move, pumps transport ions or molecules against an electrochemical gradient.

Pumps establish an electrochemical gradient that favors the movement of an ion or molecule across the plasma membrane. For example, the pump on the left side of Figure 38.20 has established an electrochemical gradient for bringing the ions or molecules symbolized by the light gray balls into the cell.

In many cases, the electrochemical gradients established by pumps are used to transport other molecules or ions by two types of membrane proteins called **cotransporters.**

- **Symporters** transport solutes *against* a concentration gradient, using the energy released when a different solute moves in the same direction *along* its electrochemical gradient. The red molecules in the middle of Figure 38.20 are moving through a symporter.

- **Antiporters** work in a similar way, except that the solute being transported against its concentration gradient moves in the direction *opposite* that of the solute moving down its concentration gradient. The orange ions in Figure 38.20, right, are moving through an antiporter.

When solutes move through cotransport proteins, **secondary active transport** occurs.

Active transport, secondary active transport, and passive transport are all involved in moving sugars around plants. Let's look first at events at source tissues, where the active transport of sucrose into sieve-tube elements results in a high-pressure potential. The chapter concludes with a look at how the same molecules are unloaded to maintain low-pressure potentials at sinks.

How Are Sugars Concentrated in Sieve-Tube Elements at Sources?

Because sucrose may be more highly concentrated in companion cells than in photosynthetic cells where it is produced,

researchers hypothesized that sucrose transport from source cells into companion cells may be active. Another key observation—that strong pH differences exist between the interior and exterior of phloem cells—suggested that sucrose might enter companion cells with protons, by secondary active transport.

FIGURE 38.21 explains the logic behind this hypothesis. Note this key claim: A membrane protein in companion cells hydrolyzes ATP and uses the energy that is released to transport protons (H^+) across the membrane to the exterior of the cell. Proteins like these are called **proton pumps,** or more formally, H^+-**ATPases.**

Proton pumps establish a large difference in charge and in hydrogen ion concentration on the two sides of the membrane. The resulting electrochemical gradient favors the entry of protons into the cell.

The right side of the figure shows the second key claim: A symporter acts as a conduit for protons and sucrose to enter the cell together. With this symporter, protons move along their electrochemical gradient; sucrose moves against its concentration gradient.

If phloem loading depends on the activity of a proton–sucrose symporter, a mutation in the gene encoding this protein should result in a plant that cannot export sucrose from leaves to various sink tissues. What would this mutant plant look like?

Michael Sussman's lab used the model plant *Arabidopsis thaliana* (see **BioSkills 13** in Appendix B) to isolate a mutant lacking a gene called *SUC2*. This gene was known to be expressed in vascular tissue and to encode a proton–sucrose symporter.

Mutant plants grew much more slowly than wild-type plants, indicating that something was indeed wrong with the mutant (**FIGURE 38.22a**). When the researchers looked at leaves under a microscope, they found that chloroplasts from mutant plants accumulated huge amounts of starch, an indication that they could not efficiently export sugar to the rest of the plant (**FIGURE 38.22b**). To test this hypothesis, the researchers then applied radioactive sucrose to a single source leaf on mutant and wild-type plants and measured where it went 5 hours later.

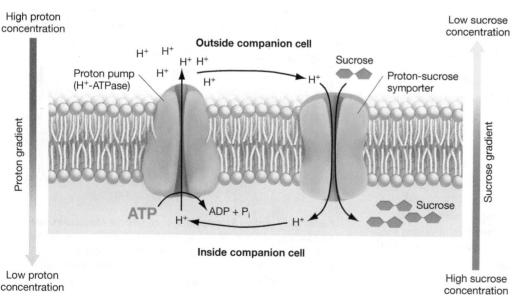

FIGURE 38.21 A Model for Cotransport of Protons and Sucrose. According to the model of cotransport, a proton pump hydrolyzes ATP to move hydrogen ions to the exterior of the cell. The resulting high concentration of H^+ outside the cell establishes an electrochemical gradient that allows the transport of sucrose into the cell against its concentration gradient.

(a) 21-day-old wild-type versus mutant plants

(b) Starch accumulation in wild type versus mutant

Normal chloroplasts   Starch-filled chloroplasts

(c) Sucrose movement in wild type versus mutant

FIGURE 38.22 The Role of a Proton–Sucrose Symporter in Long-Distance Transport in Plants.

DATA: Gottwald, J. R., P. J. Krysan, J. C. Young, et al. 2000. *Proceedings of the National Academy of Sciences* 97: 13979–13984.

✔ **QUESTION** Why did the researchers measure radioactivity after only 5 hours, and not wait a longer period such as 24 hours?

As **FIGURE 38.22c** shows, mutant leaves that were supplied with radioactive sucrose retained much more of the sucrose than wild-type leaves. Also, sink tissues in the mutant plants received much less radioactive sucrose than wild-type sink tissues. Without the proton–sucrose symporter in companion cells, mutant plants were impaired in their ability to load sucrose into phloem cells for long-distance transport and the excess sugar accumulated in leaves as starch.

There is now considerable genetic and biochemical evidence to support the following model for phloem loading:

1. Proton pumps in the membranes of companion cells create a strong electrochemical gradient that favors a flow of protons into companion cells.

2. A symporter in the membranes of companion cells uses the proton gradient to bring sucrose into companion cells from the surrounding cell walls and intercellular spaces around source cells.

3. Once inside companion cells, sucrose travels into sieve-tube elements via plasmodesmata.

Although work on the mechanism of phloem loading in *Arabidopsis* and other species continues, most researchers are convinced that proton pumps and proton–sucrose symporters play a key role. Now, once sucrose has been loaded into sieve-tube elements near sources and follows a water-potential gradient to sinks, how is it unloaded?

Phloem Unloading

The membrane proteins that are involved in transporting sucrose molecules out of the phloem, and the mechanism of movement, vary among different types of sinks within the same plant. Mechanisms for phloem unloading also vary among different species of plants.

To appreciate this diversity, consider how sucrose is unloaded in the phloem of sugar beets—a crop grown for the storage tissues in its root, which is a major source of the granulated and powdered sucrose sold in grocery stores.

check your understanding

If you understand that . . .

- Phloem sap moves from areas of high water potential to areas of low water potential.
- In phloem, high turgor pressure is observed in the sieve-tube elements near source cells. This pressure is created by pumps that actively load sucrose into companion cells against a concentration gradient. Water follows by osmosis, creating high turgor pressure inside sieve tubes.
- At sinks, turgor pressure is much lower than it is at sources because storage cells or growing tissues remove sucrose from phloem sap. The removal of solutes in phloem causes water to leave phloem and enter xylem.

✔ You should be able to . . .

1. Explain why flow of phloem sap in a stem often changes direction between early spring and midsummer.

2. Explain the adaptive value of sieve cells having no nuclei and few other organelles.

Answers are available in Appendix A.

(a) Phloem unloading into growing leaves of sugar beets

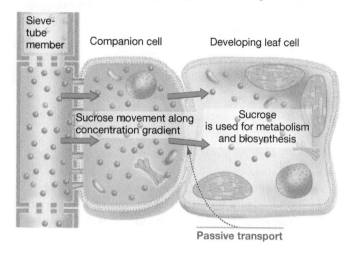

(b) Phloem unloading into roots of sugar beets

FIGURE 38.23 Phloem Unloading Occurs when Sucrose Is Moved Out of Phloem Tissue in Sinks.
The mechanism of phloem unloading can vary from sink to sink, such as in **(a)** young leaves and **(b)** roots of the same plant.

In sugar beets, sucrose is unloaded along a concentration gradient into an important sink: young, growing leaves. The passive transport occurs because sucrose is rapidly used up inside the cells to provide energy for ATP synthesis and carbon for the synthesis of cellulose, proteins, nucleic acids, and phospholipids needed by the growing cells (**FIGURE 38.23a**).

In the roots of the same plant, however, an entirely different mechanism is responsible for unloading sucrose. Root cells in this species have a large vacuole that stores sucrose. The membrane surrounding this organelle is called the **tonoplast.** It contains two proteins that work together to accumulate sucrose in the vacuole, much like the phloem loading process described earlier. A different proton pump hydrolyzes ATP and uses the energy released to transport protons into the vacuole, against its concentration gradient (**FIGURE 38.23b**). A proton–sucrose cotransporter then uses the proton gradient to move sucrose up its concentration gradient. An important distinction in this case is that the cotransporter is an antiporter, while in source tissues it is a symporter.

The active transport of sucrose into the vacuole allows sucrose to move passively from phloem into the storage cells, keeping the water potential of the phloem sap near the sink low—as required by the Münch pressure-flow model.

To summarize, more than seven decades of research provide convincing evidence that the pressure-flow hypothesis is fundamentally correct.

The cohesion-tension theory for water movement and the pressure-flow model for phloem sap movement represent major advances in our understanding of how plants work.

CHAPTER 38   REVIEW

For media, go to MasteringBiology

If you understand . . .

38.1 | Water Potential and Water Movement

- Plants lose water as an inevitable consequence of exchanging gases with the atmosphere. The flow of water from soil to air via plant tissues transports minerals to the shoot and follows a water-potential gradient.

- Water potential (ψ) is a measure of the tendency of water to move down its potential energy gradient and is expressed as megapascals (MPa).

- In plants, water potential has two components: **(1)** a solute potential, formed by the concentration of solutes in a cell or tissue; and **(2)** a pressure potential, provided by the cell wall and other factors. The water potential of a cell, tissue, or plant is the sum of its solute potential and pressure potential.

- When selectively permeable membranes are present, water moves by osmosis from areas of high solute potential to areas of low solute potential.

- When no membranes are present, water moves by bulk flow from areas of high pressure to areas of low pressure, independently of differences in solute potential.

✓ You should be able to explain why water in the xylem of a root moves up through the shoot in response only to a pressure gradient—not a solute gradient.

38.2 How Does Water Move from Roots to Shoots?

- According to the cohesion-tension theory, water is pulled in one continuous column from the soil to roots to shoots, against the force of gravity, by the surface tension caused by transpiration from leaves.

- Surface tension occurs at menisci that form as water evaporates from the walls of leaf cells, and it is transmitted downward via hydrogen bonding between water molecules. In this way, the energy in sunlight is responsible for the movement of water from roots to shoots.

✓ You should be able to suggest a hypothesis for why the pressure potential of water in a leaf, and thus the rate of water movement up a stem, change when the Sun goes behind a cloud.

38.3 Water Absorption and Water Loss

- Plants that occupy dry habitats have traits that limit the amount of water they lose to transpiration. These include narrow, needle-like leaves or stomata that are located in pits on the undersides of their leaves.

- CAM and C_4 photosynthetic cycles are adaptations in some plants that limit water loss in dry habitats.

✓ You should be able to explain how CAM plants avoid losing very much water despite living in deserts.

38.4 Translocation of Sugars

- Translocation is the movement of sucrose and other products through the plant.

- According to the Münch pressure-flow model, sugars move from sources to sinks via bulk flow along a pressure gradient that develops in phloem.

- A pressure gradient is generated by the active transport of sugars into sieve-tube elements in source tissues, coupled with the transport of sucrose out of sieve-tube elements at sink tissues. Water moves osmotically from xylem into sieve-tube elements near sources and cycles back to xylem near sinks.

✓ You should be able to explain why the pressure-flow model would not work if xylem and phloem were not bundled together.

(MB) MasteringBiology

1. **MasteringBiology Assignments**

   Tutorials and Activities Solute Transport in Plants; Translocation of Phloem Sap; Transport of Xylem Sap; Water Transport in Plants: The Transpiration-Cohesion-Tension Mechanism; Water Transport in Plants: Transpiration

   Questions Reading Quizzes, Blue-Thread Questions, Test Bank

2. **eText** Read your book online, search, take notes, highlight text, and more.

3. **The Study Area** Practice Test, Cumulative Test, BioFlix® 3-D Animations, Videos, Activities, Audio Glossary, Word Study Tools, Art

You should be able to . . .

✓ TEST YOUR KNOWLEDGE

Answers are available in Appendix A

1. Under what conditions does the rate of transpiration increase?
   a. when the temperature of a leaf decreases
   b. when stomata close at night
   c. during rainstorms, when atmospheric pressure is low
   d. when the weather changes and air becomes drier

2. The cells of a certain plant species can accumulate solutes to create very low solute potentials. Which of the following statements is correct?
   a. The plant's transpiration rates will tend to be extremely low.
   b. The plant can compete for water effectively and live in dry soils.
   c. The plant will grow most effectively in soils that are saturated with water year round.
   d. The plant's leaves will wilt easily.

3. What forces are responsible for capillarity?
   a. adhesion of water molecules to the sides of xylem cells, cohesion of water molecules to each other, and surface tension
   b. surface tension created by transpiration and cohesion of water molecules in a continuous flow from leaf to root

   c. high solute potentials created by the entry of ions during the night, when transpiration rates are low, followed by an influx of water
   d. gravity and wall pressure (from the sides of xylem cells)

4. What is a proton pump?
   a. a membrane protein that transports sucrose against a concentration gradient
   b. a membrane protein that transports protons against an electrochemical gradient
   c. a membrane protein that transports protons *along* an electrochemical gradient and sucrose *against* a concentration gradient
   d. any membrane protein that acts as a channel—meaning it does not consume ATP

5. The Casparian strip is located in what type of root cells?

6. Proteins in sieve-tube elements must be encoded by nuclei in what cells?

7. Why is the transport of phloem sap considered an active process?
   a. The manufacture of sucrose via photosynthesis is driven by the energy in sunlight.
   b. Transpiration is driven by the energy in sunlight.
   c. ATP is used to transport sucrose into companion cells near sources, against a concentration gradient.
   d. In spring, phloem sap moves against the force of gravity.

8. Draw a plant cell in pure water. Add dots to indicate solutes inside the cell. Now add dots to indicate an increase in solute potential inside the cell. Add an arrow showing the direction of water movement in response. Add arrows showing the direction of wall pressure and turgor pressure in response to water movement. Repeat the same exercise; but this time, add solutes to the solution outside the cell at a concentration that is greater than the inside of the cell.

9. Compare and contrast the forces involved in transporting water in xylem via root pressure, capillarity, and transpiration. Which of these mechanisms is/are passive?

10. Why are "cohesion-tension" and "pressure-flow" sensible names for the hypotheses analyzed in this chapter?

11. How does the movement of solutes through cotransport proteins result in phloem loading?

12. A seed is a sink when it is forming inside the parent plant. When is it a source?

13. A mutant plant lacking the ability to pump protons out of leaf companion cells will be unable to do which of the following?
    a. transpire from leaves
    b. load sucrose into sieve cells
    c. carry out photosynthesis
    d. transport water through the xylem

14. Suppose that plants over 1 meter tall had to expend energy to transport water from their roots to their leaves. What would be the consequences in terms of growth rates and overall height?

15. When young trees are transplanted to a new site, it takes several weeks or months for their root systems to grow and establish a high capacity to take up water. If a heat wave occurs during this period, the trees are likely to die—but not of starvation or loss of turgor. What kills them?

16. A recent paper indicates that the aquaporins in plasma membranes of cells throughout a plant close in response to drought stress. How does the closing of these channels help plant cells maintain turgor?

39 Plant Nutrition

In this chapter you will learn how

Plants acquire mineral nutrients they need to grow and reproduce

starting with ↓

Nutritional requirements 39.1

then asking ↓

How are nutrients incorporated?

examining ↓

The living and nonliving components of soil 39.2

then looking closer at ↓

Nutrient uptake 39.3   Nitrogen fixation 39.4

exploring further ↓

Nutritional adaptations 39.5

such as →

Epiphytism

Parasitism

Carnivory

In most plants, roots obtain the water and key nutrients required for individuals to survive and thrive.

The most urgent tasks facing any organism are to acquire (1) carbon-containing molecules that will be used as cellular building blocks and (2) the chemical energy required to make ATP. Plants acquire both of these by producing sugar through the process of photosynthesis.

Yet plants cannot live on sugar alone. Besides making the carbohydrates they need, plants synthesize all of their own nucleic acids, amino acids, enzymes, chlorophylls, enzyme cofactors, and other molecules necessary to grow and reproduce. Plants do some of the world's most impressive synthetic organic chemistry.

A plant's ability to perform sophisticated reactions depends on its capacity to harvest as raw materials a wide variety of elements in the form of simple ions and molecules. In addition to carbon dioxide and water, plants have to obtain nitrogen, phosphorus, potassium, sulfur, magnesium, and other elements. Most of these nutrients exist in soil, the majority as ions that are dissolved in soil

This chapter is part of the Big Picture. See how on pages 840–841.

✔ When you see this checkmark, stop and test yourself. Answers are available in Appendix A.

water at low—sometimes extremely low—concentrations. Once these ions are inside roots, the one-way flow of water up the xylem carries the nutrients throughout the plant body.

The plant body is an efficient machine for harvesting these diffuse resources and concentrating them in cells and tissues. You have been introduced to how the organization and growth of the plant root and shoot systems make resource acquisition possible (see Chapter 37) and how water and nutrients are transported throughout the plant body (see Chapter 38). This chapter concentrates on how plants take up elements from the soil so that these nutrients can be transported to the cells that need them.

Questions about nutrition are fundamental to understanding how plants work, increasing agricultural productivity and maintaining the productivity of forests that supply lumber and fuel—as well as mitigate temperature increases associated with global climate change (see Chapter 56) by transforming atmospheric CO_2 into wood. Let's begin by analyzing the basic nutritional needs of plants—the equivalent of the minimum daily requirements in humans.

39.1 Nutritional Requirements of Plants

What do plants need to live? In the early 1600s, Jean-Baptiste van Helmont performed a classic experiment designed to answer this question. Van Helmont wanted to know where the mass of a growing plant comes from, and he used a willow tree as a study organism.

As **FIGURE 39.1** shows, he began by placing 90 kilograms (kg) of soil in a pot with a 2-kg willow sapling. He allowed the plant to grow for five years, adding only water. At the end of the experiment, he weighed the willow and the soil. The willow had gained about 74 kg while the soil had lost only 60 grams (g).

Where had the additional 74 kg of tree come from? Because he was not aware that gases have mass, van Helmont hypothesized that the new plant material came from water. He also ignored the loss of 60 g in the soil, chalking it up to measurement error.

As it turned out, van Helmont's measurements were not the problem—his conclusions were. Most of the added mass of the tree came from carbon dioxide in the atmosphere. The 60 g removed from the soil contained vital elements—the nutrients that are the focus of this chapter. What are they?

About half the elements in the periodic table—more than 60—can be found in the tissues of one or more plant species. The question that biologists, farmers, and foresters ask is, Which of these elements are essential for growth and reproduction in most species, and in what quantities?

Which Nutrients Are Essential?

Biologists define an **essential nutrient** as an element or compound that is required for normal growth and reproduction—meaning that the plant cannot complete its life cycle without this nutrient. Essential nutrients cannot be synthesized by the organism.

Researchers test whether a nutrient is essential by denying a specific element to plants and documenting what happens to the plants over time. For most vascular plants, 17 elements are

RESEARCH

QUESTION: Where does the mass of a growing plant come from?

HYPOTHESIS: The mass of a growing plant comes from soil.

NULL HYPOTHESIS: The mass of a growing plant does not come from soil.

EXPERIMENTAL SETUP:

2-kg willow sapling + 90 kg soil = Day 1

1. Measure the weight of a willow sapling and a large amount of soil.

2. Plant the sapling and allow it to grow for 5 years.

PREDICTION: After 5 years, the soil mass will decrease by the same amount that the plant mass increased.

PREDICTION OF NULL HYPOTHESIS: The soil mass will not decrease.

RESULTS:

3700% increase in weight

0.066% decrease in weight

5 years later = 76 kg + 89.94 kg

CONCLUSION: The mass of a growing plant does not come from soil.

FIGURE 39.1 An Early Experiment on the Role of Soil in Plant Nutrition.

✓**QUESTION** Many nonbiologists think that most of a plant's mass comes from soil or water. Describe an experiment that would convince someone that most of a plant's mass comes from CO_2.

essential. Just three of these—carbon, hydrogen, and oxygen—typically make up about 96 percent of the dry weight of a plant. The remaining 14 elements are sometimes called mineral nutrients, because they originate in soil.

Although different classification schemes for the essential elements have been proposed, the most common is based on distinguishing nutrients that are obtained from water or air versus soil and then dividing soil nutrients into macronutrients and micronutrients based on their abundance in plants (**TABLE 39.1**, see page 778).

Macronutrients   Certain elements in the soil are required by plants in relatively large quantities; these are called **macronutrients.** Some of these are major components of nucleic acids, proteins, and phospholipids, all of which are plentiful in plants.

Among the macronutrients, nitrogen (N), phosphorus (P), and potassium (K) are particularly important because they often act as **limiting nutrients,** meaning their availability limits plant growth. If N, P, and/or K are added in appropriate quantities to soil as fertilizer, plant growth usually increases. This observation explains why the leading ingredients in virtually every commercial fertilizer are N, P, and K.

Micronutrients   In contrast to macronutrients, **micronutrients** are required in small quantities. When plant tissues are dried and analyzed, micronutrients are typically present in only trace amounts. Instead of acting as components of macromolecules, micronutrients usually function as cofactors for specific enzymes—substances that are required for normal enzyme function (see Chapter 3).

It's important not to underestimate the importance of micronutrients, even though only tiny amounts are needed. For example, a typical plant contains just one molybdenum atom for every 60 million hydrogen atoms in its body, not including water. Yet plants die without molybdenum, because it functions as a cofactor for several enzymes involved in nitrogen processing. What happens to plants when other essential nutrients are missing?

What Happens When Key Nutrients Are in Short Supply?

In some cases, biologists can examine a plant that is growing poorly and diagnose a nutrient deficiency (**FIGURE 39.2**). For example, if older leaves are in poor condition, the problem is probably due to lack of N, P, K, or magnesium. These elements are mobile—meaning they are readily transported from older leaves to younger leaves when they are in short supply—so older leaves deteriorate first when these elements are scarce. Immobile nutrients like iron or calcium, in contrast, stay tied up in older leaves. When they are in short supply, younger leaves are the first to show deficiency symptoms. Deficiencies in the various nutrients produce symptoms specific to each nutrient (see Table 39.1).

In large part, the ability to diagnose nutrient deficiencies is based on studies involving hydroponic growth systems. **Hydroponic growth** takes place in liquid cultures, without soil, so researchers can precisely control the availability of each nutrient.

Consider an experiment on copper deficiency in tomatoes (**FIGURE 39.3**). Researchers grew seedlings in two types of treatments. One treatment consisted of flasks containing water and

RESEARCH

QUESTION: How does copper deficiency affect plants?

HYPOTHESIS: Plants denied copper will grow poorly.
NULL HYPOTHESIS: Plants denied copper will grow normally.

EXPERIMENTAL SETUP:

Air bubbled in to provide oxygen

Nutrient solution containing copper

Nutrient solution without copper

PREDICTION: The copper-deficient plant will grow less than the normal plant.

PREDICTION OF NULL HYPOTHESIS: Both plants will grow the same.

RESULTS:

Normal   Copper-deficient

CONCLUSION: Copper deficiency leads to poor growth. All tissues appear to be affected adversely.

FIGURE 39.3 Hydroponics Is Used to Study Nutrient Deficiencies.
SOURCE: Arnon, D. I., and P. R. Stout. 1939. *Plant Physiology* 14: 371–375.

✔ **QUESTION** What problem would arise if this experiment had been done in soil with and without added copper?

(a) Normal barley   **(b)** N deficiency   **(c)** P deficiency

Yellowing of leaves

Stunted growth, dead spots

FIGURE 39.2 Nutrient Deficiencies Can Have Distinctive Symptoms.

all the essential nutrients in the relative concentrations that are optimal for tomato growth. The second treatment was identical, except that the nutrient solution lacked copper.

As Figure 39.3 shows, copper-deprived individuals have stunted shoots, unnaturally light-colored foliage, and curled leaves. Given copper's role as a cofactor or component of several enzymes involved in redox reactions required for ATP production, it is understandable that all tissues in the plant were severely affected. And because copper is a micronutrient, it is reasonable to expect that a relatively small amount would cure the deficiency. In line with this prediction, the researchers found that the symptoms were prevented if the plants were cultured in a solution containing just 0.002 milligrams per liter (mg/L) of copper. Analogous studies have been done on the other essential nutrients.

For farmers, foresters, and plant ecologists, understanding which nutrients are essential, and why, is basic to understanding why certain plants thrive and others fail. Now, where do these nutrients come from? The answer—soil—is simple. But soil itself is astonishingly complex.

39.2 Soil: A Dynamic Mixture of Living and Nonliving Components

The process of soil building begins with solid rock. As **FIGURE 39.4** shows, **weathering**—the forces applied by rain, running water, temperature changes, and wind—continually breaks tiny pieces off large rocks. The weathering process is accelerated if small cracks develop in the rock. If plant roots grow into the crack, they expand as they grow, widen the crack, and break off small flakes or pebbles. A similar effect occurs in high latitudes or at high elevations when water enters the cracks, freezes in winter, expands, and breaks off pieces.

Depending on their size and composition, the particles resulting from these processes are called gravel, sand, silt, or clay. These rock fragments are the first ingredient in soil. As

organisms occupy the substrate, they add dead cells and tissues and feces. This decaying organic matter is called **humus** (pronounced *HEW-muss*).

With time, soil eventually becomes a complex and dynamic mixture of inorganic particles, organic particles, and living organisms. In a single gram of soil there are normally dozens of small animals and plant roots; hundreds of thousands of protists, fungi, and microscopic animals; and hundreds of millions of bacteria and archaea (**FIGURE 39.5**).

Both the parent rock that contributes inorganic soil components and the organisms and organic matter that occupy soils vary from one site to another. **Texture**—the proportions of gravel, sand, silt, and clay—and other soil qualities vary as well. Soil texture is important for several reasons:

- Texture of a soil affects the ability of roots to penetrate and obtain water and nutrients, as well as to anchor and support the plant body. For example, soil that is dominated by clay-sized particles tends to compact and resist root penetration.

- Texture affects a soil's ability to hold water and make it available to plants. Water tends to adhere to clay and silt particles but runs through sand and gravel.

- A soil's texture and water content dictate the availability of oxygen. Like other eukaryotes, plants have to take in oxygen to use as an electron acceptor during cellular respiration. The oxygen used by plant root cells is found in air pockets among soil particles. This explains why overwatering a plant is just as detrimental as underwatering it: Overwatering drowns a plant's roots.

The best soils for plants, called loams, contain roughly equal amounts of sand, silt, and clay, along with a high proportion of humus.

Loams that have good texture and large amounts of organic matter can take thousands of years to develop through the weathering of rocks and the continual addition of humus. Unfortunately, it can take just a few years of abuse by humans for loams to blow or wash away.

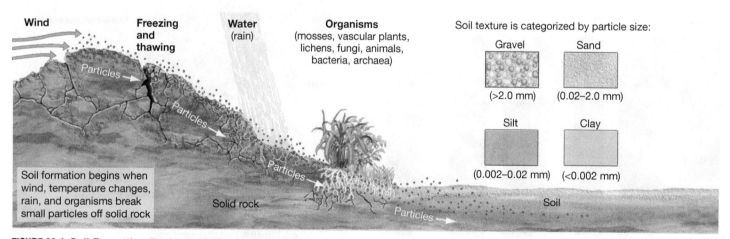

FIGURE 39.4 Soil Formation Begins with Weathering of Rock.

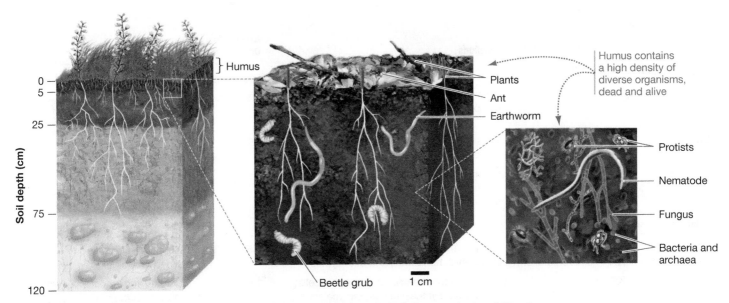

FIGURE 39.5 **Mature Soils Are a Complex Mixture of Organic and Inorganic Components.** In addition to mineral particles, soil contains humus—organic material derived from dead organisms—and a wide array of living organisms.

The Importance of Soil Conservation

Soil erosion occurs when soil is carried away from a site by wind or water. Soil erosion occurs naturally, such as when rivers cut away at their banks and carry material downstream. In most natural environments, though, the rate of soil formation exceeds the rate of soil erosion, so soils build up over time. Unfortunately, the situation can change dramatically when humans exploit an area.

When plant cover is removed for forestry, farming, or suburbanization, stems and leaves can't lessen the force of wind and rain, and plant roots no longer hold soil particles in place. The results can be devastating. At some locations in the United States,

8–10 centimeters (cm) of topsoil were blown away during the Dust Bowl of the 1930s, when drought and poor farming practices left thousands of hectares of soil unprotected (**FIGURE 39.6a**).

U.S. soil erosion rates have declined dramatically since then. Still, researchers estimate that almost 30 percent of all croplands in the United States are eroding too fast to maintain their long-term productivity. Deforestation is also exposing forest soils, contributing to disasters such as the mudslides and flooding that occurred in the Dominican Republic and Haiti in 2004, killing close to 5000 people. **FIGURE 39.6b** provides an aerial view of devastating mudslides that occurred in 2007 in the U.S. state of Washington, after recent deforestation. Worldwide, an estimated 36 billion tons of soil are lost to erosion every year.

(a) Wind erosion in the United States, 1930s

(b) Mudslides caused by deforestation

FIGURE 39.6 **Soil Erosion Can Have Devastating Consequences.**

TABLE 39.1 Essential Nutrients

Element	Form Available to Plants	Functions	Average % Dry Weight*	Deficiency Symptoms
Obtained from Water or Air: H_2O or CO_2				
Oxygen	O_2, H_2O	Electron acceptor in cellular respiration; major component of organic compounds	45	Usually affects roots: cells suffocate, leading to root rot and wilting
Carbon	CO_2	Substrate for photosynthesis; major component of organic compounds	45	Slow growth (starvation)
Hydrogen	H_2O	Major component of organic compounds; functions in electrical balance and establishment of electrochemical gradients	6	Slow growth due to cell death (desiccation)
Obtained from Soil: Macronutrients				
Nitrogen	NO_3^- (nitrate) NH_4^+ (ammonium ion)	Component of proteins, nucleic acids, ATP, chlorophyll, hormones, and coenzymes	1.5	Failure to thrive; chlorosis (yellowing of older leaves)
Potassium	K^+	Necessary for osmotic adjustment in cells; required for synthesis of organic molecules; cofactor for some enzymes	1.0	Chlorosis at margins of leaves or in mottled pattern; weak stems; short internodes
Calcium	Ca^{2+}	Regulatory functions; role in cell wall structure; stabilizes membranes; involved in signal transduction; enzyme cofactor	0.5	Necrosis (small spots of dead cells) in meristems; deformation of young leaves; stunted, highly branched root system
Magnesium	Mg^{2+}	Chlorophyll component; activates many enzymes	0.2	Chlorosis between leaf veins; premature leaf drop
Phosphorus	$H_2PO_4^-$ (dihydrogen phosphate ion) HPO_4^{2-} (hydrogen phosphate ion)	Component of ATP, nucleic acids, phospholipids, and several coenzymes	0.2	Stunted growth in young plants; dark green leaves with necrosis
Sulfur	SO_4^{2-} (sulfate ion)	Component of proteins containing methionine and cysteine; electron transport proteins and coenzymes	0.1	Stunted growth; chlorosis
Obtained from Soil: Micronutrients				
Chlorine	Cl^- (chloride ion)	Needed for water-splitting step of photosynthesis; functions in water balance and electrical balance	0.01	Wilting at leaf tips; general chlorosis and necrosis of leaves or development of bronze color
Iron	Fe^{3-} (ferric ion) Fe^{2-} (ferrous ion)	Necessary for chlorophyll synthesis; component of cytochromes and ferredoxin; enzyme cofactor	0.01	Chlorosis between veins of young leaves
Manganese	Mn^{2+}	Involved in photosynthetic O_2 evolution; enzyme activator; important in electron transfer	0.005	Chlorosis between leaf veins and small necrotic spots
Zinc	Zn^{2+}	Involved in synthesis of the plant hormone auxin, maintenance of ribosome structure, enzyme activation	0.002	Small internodes; stunted and distorted ("puckered") leaves
Boron	$H_2BO_3^-$ (borate ion)	Strengthens cell walls; required for pollen tube growth and normal membrane function	0.002	Black necrosis in young leaves and buds
Copper	Cu^+ (cuprous ion) Cu^{2+} (cupric ion)	Cofactor of some enzymes; present in lignin of xylem	0.0006	Light-green leaves with necrotic spots; twisted and malformed leaves
Nickel	Ni^{2+}	Cofactor for enzyme functioning in nitrogen metabolism	[no data]	Necrosis at leaf tips
Molybdenum	MoO_4^{2-} (molybdate ion)	Cofactor in nitrogen reduction; essential for nitrogen fixation	0.00001	Chlorosis between veins; necrosis of older leaves

*These percentages were obtained by drying vascular plants and documenting what proportion of the waterless mass consists of various elements.

If soil is managed carefully, however, it can be a renewable resource. Techniques that maintain long-term soil quality and productivity are the basis of **sustainable agriculture** and sustainable forestry. Farmers can reduce soil loss dramatically by

- planting rows of trees as windbreaks;

- using techniques that minimize the amount of plowing and tilling needed to control weeds; and

- planting crops in strips that follow the contour of hillsides.

Farmers can also maintain soil quality by adding organic material in the form of manures and, soon after a harvest, planting cover crops that are plowed in and allowed to decompose.

What Factors Affect Nutrient Availability?

The elements required for plant growth are found in the soil not as atoms, but as **ions**. The second column of Table 39.1 lists the ion forms of some essential nutrients available to plants. Notice that some of these nutrients are available as elemental ions, such as K^+ or Cl^-, while others exist as molecular ions, such as HPO_4^{2-} or NO_3^-.

Anions and Cations Behave Differently The ions present in soil tend to behave in one of two ways, depending on their charge (**FIGURE 39.7**). Anions—ions with negative charges—usually dissolve in soil water, because they interact with water molecules via hydrogen bonding. Phosphate ions, an exception to this rule, tend to form insoluble complexes with iron, aluminum, calcium, or other positively charged cations.

Because they exist as solutes, negatively charged anions are readily available to plants for absorption. They are also easily washed out of the soil by rain, however. The loss of nutrients via the movement of water through soil is called **leaching.**

Cations—ions with positive charges—dissolve in soil water but are not as immediately available as anions. In solution, cations interact with the negative charges found on two types of soil particles: (1) organic matter that is rich in negatively charged organic acids; and (2) the surfaces of the tiny, sheetlike particles called clay, which are rich in mineral anions (see Figure 39.7).

Organic soils that contain clay tend to retain nutrients, because few positively charged cations leach away and because these soils hold water (and thus anions) better than sandy soils do. The presence of clay makes cations more difficult for plants to extract and use because the cations are tightly bound to the clay.

The Role of Soil pH In addition to soil texture, other factors can influence the availability of essential elements. Perhaps the most important of these is soil pH.

Recall that the pH scale indicates the relative concentration of hydrogen ions in a solution, and that it ranges from 0 to 14 (see Chapter 2). Soils with low pH have a relatively high concentration of hydrogen ions and are considered acidic. Soils with a high pH contain relatively few hydrogen ions and are termed basic or alkaline.

FIGURE 39.7 Cations Tend to Bind to Clay Particles and Organic Matter; Anions Stay in Solution.

Acidic soils are found in regions such as conifer forests, where the decomposition of organic matter produces carbonic acid, phosphoric acid, or nitric acid. Alkaline soils, in contrast, are common in regions where limestone ($CaCO_3$) is abundant. When limestone reacts with water, the calcium ions that are released take the place of protons that cling to soil particles. The protons then react with CO_3^{2-} to form bicarbonate ions (HCO_3^-), lowering the hydrogen ion concentration of the soil and raising its pH.

Most plants thrive in soils with a relatively neutral pH—between 6 and 7. But some plants, such as blueberries, grow best in acidic soils (pH around 3 to 5); others such as lavender, rosemary, and fennel, thrive on alkaline soils (pH above 8).

Cation Exchange Soil pH affects the availability of plant nutrients in a number of ways. For example, the presence of protons in soil water can cause the release of cations that are bound to soil particles. The process responsible is called **cation exchange.** Cation exchange occurs when protons or other soluble cations bind to negative charges on soil particles and

(a) Cation exchange releases nutrients...

(b) ... which are absorbed by roots...

(c) ... or leached in heavy rains.

FIGURE 39.8 Cation Exchange Releases Nutrients Bound to Soil Particles. When cation exchange occurs, a proton binds to negative charges on clay or organic matter, releasing bound cations.

cause bound cations, such as magnesium or calcium, to be released from the soil (**FIGURE 39.8a**). Nutrients that are released by cation exchange become available for uptake by nearby plant roots (**FIGURE 39.8b**).

Plants influence cation exchange because root cells release CO_2 as a by-product of cellular respiration (see Chapter 9). The CO_2 reacts with H_2O to form carbonic acid, which releases protons.

If soil is too acidic, however, rain may leach cations away before the roots can take up nutrients (**FIGURE 39.8c**). In many tropical rain forests around the world, deforestation for agriculture has resulted in acidic, nutrient-poor soils that are not productive.

To summarize, anions stay in solution in soil water. They are readily available to plants but may wash away easily. Positive ions, in contrast, tend to bind to soil particles but can be released by cation exchange.

TABLE 39.2 details how the presence of sand, clay, and organic matter affects ion availability and other soil properties, including water availability. Based on these properties, it is not surprising that the most productive soils are loams that are a mixture of sand, clay, and organic matter.

Nutrients are found at extremely low concentrations in soil but at high concentrations in plant cells. How are plants able to bring N, P, K, and other key elements into their bodies against a concentration gradient?

39.3 Nutrient Uptake

In most species of plants, the root system is the site of nutrient uptake. You have been introduced to the general features of root anatomy (see Chapter 37) and the role of roots in water uptake (see Chapter 38). Now we need to explore the detailed anatomy of roots and analyze events that occur in the plasma membranes of their epidermal and cortex cells.

Most nutrient uptake occurs just above the growing root tip, in the region called the **zone of maturation.** Recall that epidermal cells in this part of the root have extensions called **root hairs** (**FIGURE 39.9**). Root hairs dramatically increase the surface area available for nutrient and water absorption. For example, the root system of a single annual rye plant can have a total surface area the size of a basketball court. Most of this area—60 percent—is found in an estimated 10^{10} root hairs.

Root hairs are so numerous and so efficient at absorbing nutrients from soil that, over time, they create a "zone of nutrient depletion" in soil immediately surrounding them. The creation of this mined-out region is why continued root growth is vital to a plant's health. Because roots continue to grow throughout a plant's life, the zone of maturation is continually entering new and potentially nutrient-rich areas of soil.

If an epidermal cell encounters soil where nutrients are available, what happens? How do ions enter?

TABLE 39.2 Effects of Soil Composition on Soil Properties

Composition	Water Availability	Nutrient Availability	Oxygen Availability	Root Penetration Ability
Sand	Low: water drains through	Low: poor capacity for cation exchange; anions leach out	High: many air-containing spaces	High: does not pack tight
Clay	High: water clings to charged surface	High: large capacity for cation exchange; anions remain in solution	Low: few air-containing spaces	Low: packs tight
Organic matter	High: water clings to charged surface	High: source of nutrients; large capacity for cation exchange; anions remain in solution	High: many air-containing spaces	High: does not pack tight

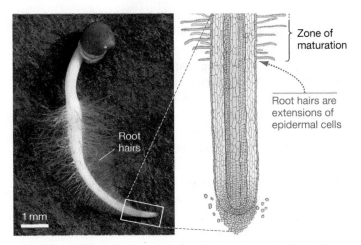

FIGURE 39.9 Root Hairs Increase the Surface Area Available for Nutrient Absorption.

Mechanisms of Nutrient Uptake

Ions, small molecules, and even large molecules can pass through the plant cell wall freely. The plasma membrane, in contrast, is highly selective. Recall that the plasma membrane is a fluid, sheetlike structure consisting of a phospholipid bilayer studded with proteins (see Chapter 6). Because the interior of the phospholipid bilayer is nonpolar and hydrophobic, it resists the passage of ions—even ions that are required by the plant. Some membrane proteins span the bilayer, however, and allow specific ions to cross the membrane.

Because root hairs have such a large surface area, they contain large numbers of membrane proteins that bring nutrients into the cytosol of root cells. These membrane proteins cannot import the ions that the plant needs completely on their own, however. They often function in tandem with another protein, a proton pump.

Establishing a Proton Gradient Plants harvest diffuse nutrients and concentrate them in their tissues. In an epidermal cell, ions such as potassium (K^+), hydrogen phosphate (HPO_4^{2-}), and nitrate (NO_3^-) are many times more concentrated than they are in the soil water outside the cell. For these and other ions to enter the cell, they have to cross the plasma membrane against a strong concentration gradient. How is this possible?

As **FIGURE 39.10a** shows, the answer hinges on **proton pumps,** or H^+-ATPases. These proteins are found in the plasma membranes of root epidermal and cortex cells, and they are similar to the pumps that make it possible for companion cells to load sucrose into phloem against a strong concentration gradient. Recall that when a phosphate group from ATP binds to the pumps, they change conformation in a way that allows them to transport protons to the exterior of the cell (see Chapter 38).

The activity of proton pumps leads to a strong excess of protons on the exterior of the plasma membrane relative to the interior. In the case of root cells, this differential results in a strong pH gradient favoring the movement of protons back into the cell. In addition, the outside of the membrane becomes positively charged relative to the inside. Stated another way, there is a separation of charge—a **voltage**—across the membrane.

Proton pumps are found in all types of plant cells, and all plant cell membranes carry a voltage. Because voltage is a form of potential energy, the charges that are separated by the membrane create a **membrane potential,** or difference in electrical charge across a cell membrane. How is the membrane potential used to transport ions?

Using a Proton Gradient to Import Cations In the membranes of root cells, the electrical gradient established by proton pumps, which favors the entry of positive ions, is strong enough to overcome the pH gradient, which opposes the entry of these cations. In essence, plant cells are batteries that are charged up to attract nutritionally necessary cations.

(a) Proton pumps establish an electrochemical gradient.

(b) Cations enter root hairs via channels.

(c) Anions enter root hairs via cotransporters.

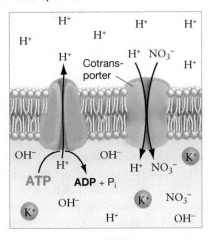

FIGURE 39.10 Ions Enter Roots along Electrochemical Gradients Created by Proton Pumps.

✔**EXERCISE** In part (b), add an arrow labeled "Proton gradient." In part (c), add an arrow labeled "Electrical gradient" and indicate its positive–negative polarity.

To drive this point home, consider potassium cations. Potassium is an essential nutrient in plants because it is required as a cofactor by over 40 enzymes. In addition, it is found at relatively high concentrations inside plant cells and plays a key role in bringing water into cells via osmosis and maintaining normal turgor pressure.

Researchers who added radioactive potassium ions to the solution outside a root cell and followed their movement found that K^+ does indeed accumulate in cells due to an electrochemical gradient (**FIGURE 39.10b**). In follow-up experiments, biologists were able to isolate the membrane protein responsible for the uptake of K^+ and sequence the gene that encodes the protein.

✔ If you understand this concept, you should be able to explain what happens to cation uptake if (1) H^+-ATPases fail, and (2) a molecule blocks a cation channel.

Cations like K^+ enter root hairs through membrane proteins along an electrochemical gradient, but how is it possible for anions to enter? The negatively charged interior of the cell should repel these ions.

Using a Proton Gradient to Import Anions **FIGURE 39.10c** shows how anions such as NO_3^- are able to enter root hairs against their electrochemical gradient. The key is that anions enter through membrane transport proteins called cotransporters (see Chapter 38), which transport two solutes at once. In this case, the cotransporter is a symporter: It brings NO_3^- into the cell along with a proton, and both solutes move in the same direction.

Here's the key observation: So much energy is released when a proton enters the cell along its electrochemical gradient that nitrate, phosphate ions, or other anions can be cotransported against their electrochemical gradients.

To summarize, the electrochemical gradient set up by proton pumps makes it possible for plant roots to absorb key cations and anions via ion channels and symporters, respectively. ✔ If you understand this concept, you should be able to explain what happens to anion uptake if (1) H^+-ATPases fail, and (2) a molecule blocks an anion symporter.

Once ions have crossed the plasma membrane of an epidermal or cortex cell in the root, they continue to move via the symplast and through the endodermis toward the central vascular tissue. Recall that the endodermis is the site of the Casparian strip, which blocks the movement of ions and water through the apoplast (see Chapter 38). Once inside the endodermal layer, ions are actively transported out of cells, where they accumulate in the xylem sap. The ions are then transported passively by bulk flow to tissues throughout the plant.

Nutrient Transfer via Mycorrhizal Fungi To synthesize proteins and nucleic acids, plants need to extract large quantities of nitrogen and phosphorus from the soil. But soils typically contain such low concentrations of these nutrients that they limit plant growth. For example, in many habitats soil nitrogen is largely unavailable to plants because it is found in proteins that are part of the organic matter. How do plants absorb enough N and P to satisfy their nutritional needs? In most cases, they rely on partnerships with fungi that live in close association with their roots.

You might recall that fungi and plant roots that live in association are called **mycorrhizae** (literally, "fungus-root") (see Chapter 32). Mycorrhizal fungi and plants are **symbiotic** ("living-together"), meaning that they live in physical contact with each other. Biologists estimate that more than 80 percent of all vascular plant species associate with mycorrhizal fungi.

In some habitats, plants receive large quantities of nitrogen from mycorrhizal fungi. The fungal symbionts in these associations are particularly efficient at digesting macromolecules (proteins and nucleic acids) in decaying organic material and absorbing the amino acids (a source of nitrogen) and phosphate ($H_2PO_4^-$) that are released. As supported by experimental evidence described earlier (see Chapter 32), the mycorrhizal fungi then transport the nutrients from soil to plant roots (**FIGURE 39.11**). In exchange for these nutrients, the plant symbionts transfer sugars and other photosynthetic products to the fungi. Because there is a reciprocal exchange of nutrients, the symbiotic relationship is considered **mutualistic,** or mutually beneficial.

Fungi are particularly efficient at acquiring the nutrients required by plants, for two reasons:

1. Networks of filamentous hyphae increase the surface area available for absorbing nutrients by up to 700 percent.

2. Fungi can acquire nutrients from macromolecules in the soil that are unavailable to non-mycorrhizal plants.

Most plant species grow slowly and are overwhelmed by competitors if denied their mycorrhizal associates.

FIGURE 39.11 Most Mycorrhizal Fungi and Plants Are Mutualists. The fungus illustrated here provides nutrients to its host plant in exchange for photosynthetic products.

Mechanisms of Ion Exclusion

Plants have sophisticated systems for absorbing nutrients using membrane transporters and energy from the proton gradients they generate. Not all ion uptake is beneficial, however.

Certain types of natural soils—as well as soils that have been contaminated by waste products from mining or smelting operations—contain enough cadmium, zinc, nickel, lead, or other metals to poison enzymes in most plants. Sodium is also detrimental at high concentrations: Too much Na^+ inside cells can disrupt enzyme function, and too much Na^+ in extracellular spaces can create a solute potential that pulls enough water out of cells to result in a loss of turgor.

Sodium poisoning is a key issue in an array of environments, including (1) ocean coastlines, (2) habitats near roads that are treated with salt to melt ice and snow, and (3) irrigated farmlands. When soils are irrigated, solutes are left behind when water evaporates from the surface—leading to salt buildup over time.

How do plants exclude ions that are detrimental?

Passive Exclusion Ions move across roots following the same routes that water follows: the apoplastic, transmembrane, and symplastic pathways (see Chapter 38). Many ions do not enter the root symplast, simply because epidermal and cortex cells lack the requisite membrane transporters. If cells lack the membrane protein required for a certain ion to enter the cell, the ion won't enter. This is a form of passive exclusion.

Ions that move across the root cortex via the apoplastic pathway may never make it to xylem. To understand why, recall that the Casparian strip forces all solutes that travel through the root apoplast to cross the plasma membrane of endodermal cells before moving into the xylem (see Chapter 38). Ions that cannot enter endodermal cells via membrane transporters are excluded from entering the rest of the plant.

By having membrane transporters only for the ions that it requires, a plant can control which ions enter the symplast. Ions that reach the endodermis via the apoplastic pathway are blocked by the presence of the Casparian strip, so endodermal cells act as a selective filter—preventing unnecessary ions from reaching the xylem. The Casparian strip also qualifies as a mechanism for the passive exclusion of metals, sodium, or other ions (**FIGURE 39.12**).

The number of ion transporters present in a root is also important in determining the toxicity to certain ions. For example, salt-tolerant strains of corn transport much less salt into their roots than salt-intolerant strains—possibly because of genetic variation in the number of sodium channels found in root cells. Among species, rice is notoriously sensitive to salt buildup. Barley, in contrast, is relatively tolerant to high salt concentrations in soil.

Active Exclusion by Metallothioneins Plants also have mechanisms for coping with toxins once they are inside their cells. This is important because even essential nutrients are toxic at high enough concentrations.

Copper channels, for example, admit the small amounts of copper ions required for normal cell function. But plants that grow on soils near copper-mining operations experience large concentration gradients that favor an influx of this nutrient, so a surplus is likely to build up inside the plant body. How do plants deal with copper and other types of excess nutrients before they poison key enzymes?

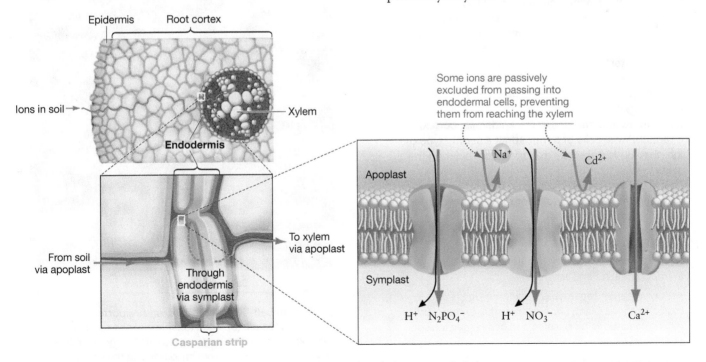

FIGURE 39.12 Passive Exclusion Occurs at the Plasma Membrane. Certain ions are excluded from cells because it is difficult for them to cross the plasma membrane.

One mechanism for coping with toxic concentrations of metals involves small proteins called **metallothioneins**. Metallothioneins bind to metal ions and prevent them from acting as a poison. Producing metallothionein proteins requires an expenditure of energy and represents a form of active exclusion.

Genes for metallothioneins have been found in a wide variety of organisms—bacteria, fungi, and animals as well as plants. Recent research on the mustard-family species *Arabidopsis thaliana* has shown that individuals from populations with a high tolerance for copper produce many more metallothionein proteins than do individuals from populations with a low tolerance for copper. (For more on the importance of *Arabidopsis thaliana* as a model organism, see **BioSkills 13**, Appendix B.)

Active Exclusion by Antiporters A second mechanism for actively neutralizing specific toxins involves transport proteins located in the **tonoplast**—the membrane surrounding the large, central vacuole. Proteins in the tonoplast membrane allow plants to actively remove toxic substances from the cytosol and store them in the vacuole.

Perhaps the best-studied example involves proteins that move sodium ions from the cytosol into vacuoles, where they cannot poison enzymes (**FIGURE 39.13a**). H^+-ATPases in the tonoplast move protons into the vacuole, creating an electrochemical

check your understanding

If you understand that . . .

- Plants absorb most of the nutrients they need via membrane proteins in root hairs.
- Proteins in the plasma membranes of root-hair, cortex, and endodermal cells pump protons out of the cells, creating an electrochemical gradient that favors the entry of selected ions via ion channels or cotransporters.
- In many species, mycorrhizal fungi are important for bringing ions that contain nitrogen or phosphorus atoms into the root.
- If no membrane proteins in the epidermal, cortex, or endodermal cells of a root admit a certain type of ion, that ion is passively excluded from entering the root.
- Plants can actively exclude toxic ions by pumping them into vacuoles or binding them to metallothionein proteins.

✔ **You should be able to . . .**

1. Explain how plants generate and use electrical power to import cations and anions.

2. Make a diagram that traces a cadmium ion (Cd^{2+}) from the soil into the plant, showing how it is excluded at different locations by passive and active mechanisms of ion exclusion.

Answers are available in Appendix A.

(a) In the tonoplast, antiporters send H^+ out of, and Na^+ into the vacuole.

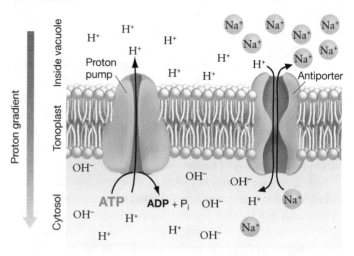

(b) Plants with additional H^+/Na^+ antiporters tolerate salt.

Transformed plants: additional antiporter gene present

Control plants: no additional antiporter gene

Salt concentration in water
Low ◄————————————————————► High

FIGURE 39.13 In Salt-Tolerant Plants, an Antiporter Concentrates Sodium in Vacuoles. (a) In the tonoplast of salt-tolerant species, proton pumps establish a proton gradient. This gradient allows antiporters to concentrate sodium ions inside the vacuole, where they cannot poison enzymes in the cytoplasm. **(b)** *Arabidopsis* plants that were transformed with the gene for the antiporter protein grow well even when watered with salty water.

gradient that favors the movement of H^+ out of the vacuole. A transport protein that functions as an **antiporter** then uses this gradient to conduct protons *out* of the vacuole, move them down their electrochemical gradient, and bring sodium ions *into* the vacuole—against the sodium concentration gradient.

Recall that an antiporter is a cotransporter involved in moving the solutes in opposite directions instead of the same direction—as solutes do through symporters (see Chapter 38).

Knowledge of this H^+/Na^+ antiporter has created a great deal of excitement among biologists, because it offers a way to alter crop plants genetically so that they can grow well in salty soils created by poor irrigation practices. As an example, consider an experiment in which normal *Arabidopsis* plants and individuals transformed with extra copies of the *Arabidopsis* gene for the H^+/Na^+ antiporter were exposed to different levels of NaCl. The genetically transformed plants accumulated more sodium in their vacuoles, and as **FIGURE 39.13b** shows, they were able to grow efficiently even when exposed to high concentrations of salt.

39.4 Nitrogen Fixation

Nitrogen gas (N_2) makes up 80 percent of the atmosphere. Unfortunately, plants and other eukaryotes cannot use nitrogen in this form. Nitrogen gas is unreactive because it takes a great deal of energy to break the triple bond between the two nitrogen atoms.

To synthesize amino acids, nucleic acids, and other nitrogen-containing compounds, plants normally absorb nitrogen in forms such as ammonium (NH_4^+) or nitrate ions. But these ions are in short supply in many soils, meaning that plant growth is often limited by the availability of usable nitrogen. Most crop plants grow much faster when they receive a nitrogen-containing fertilizer.

Nitrogen-based fertilizers have drawbacks, however. Fertilizer production is extremely energy intensive, and in many parts of the world, ammonia and other nitrogen-based fertilizers are too expensive for farmers. In more affluent regions, these fertilizers are used so extensively that they are causing serious pollution problems (see Chapter 29). For these and other reasons, there is intense interest in understanding the molecular basis of a phenomenon called biological nitrogen fixation.

The Role of Symbiotic Bacteria

Among all the organisms on the tree of life, only a few species of bacteria and archaea are able to absorb N_2 from the atmosphere and convert it to ammonia (NH_3), nitrites (NO_2), or nitrates. This process is called **nitrogen fixation.**

Nitrogen fixation requires a series of specialized enzymes and cofactors, including a large multi-enzyme complex called nitrogenase. The process is extremely energy demanding. An expenditure of 8 high-energy electrons and 16 ATP molecules is required for nitrogenase to reduce one molecule of N_2 to two molecules of NH_3. The process can be summarized as follows:

$$N_2 + 8\,e^- + 8\,H^+ + 16\,ATP \longrightarrow 2\,NH_3 + H_2 + 16\,ADP + 16\,P_i$$

In some cases, bacterial cells that are capable of nitrogen fixation take up residence *inside* plant root cells. Although several different bacteria and plant hosts can be involved, the best-studied nitrogen-fixing bacteria are members of the genus *Rhizobium* that associate with plants in the **legume** family, which includes peas, soybeans, clover, and alfalfa. Members of the bacterial genus *Rhizobium* and closely related species are often called **rhizobia.**

Because of their ability to fix nitrogen, legumes are extremely important agricultural crops. A legume crop is often planted after harvesting a grain crop such as corn or wheat, so it can add nitrogen to the soil. This type of crop rotation results in "free" nitrogen fertilizer that benefits the next grain crop. Because legumes are rich in nitrogen, they are an important part of the human diet in many parts of the world.

As **FIGURE 39.14** shows, the infected root cells of legumes form distinctive structures called **nodules,** where nitrogen-fixing rhizobia are found. The nodules are pink because they contain an iron-containing molecule called leghemoglobin (short for legume hemoglobin). **Leghemoglobin** is related to the hemoglobin that carries oxygen in your blood. Like hemoglobin, leghemoglobin binds oxygen.

Roots of pea plant

Nodules

Nodule in cross section
(red dots are infected plant cells)

Infected plant cell

Uninfected plant cell

2 μm

Close-up of rhizobia inside vesicles, inside infected plant cell

FIGURE 39.14 In Some Plants, Roots Form Nodules Where Nitrogen-Fixing Bacteria Live.

Leghemoglobin is important because nitrogenase—the enzyme complex responsible for nitrogen fixation—is poisoned by the presence of oxygen. Inside root nodules, oxygen molecules bind to leghemoglobin instead of binding to nitrogenase. In this way, leghemoglobin keeps levels of free oxygen low enough for nitrogenase to function while maintaining a source of oxygen for electron transport to occur. Intense cellular respiration and ATP synthesis are necessary to supply nitrogenase with the energy it requires.

If roots allowed bacterial pathogens to enter, the plant would possibly die from the subsequent infection. How do roots distinguish beneficial bacteria from pathogens and allow only the rhizobia to infect the roots?

PROCESS: INFECTION BY NITROGEN-FIXING BACTERIA

Root hair

1. Flavonoid released by root hairs attracts rhizobia. Rhizobia move into hairs.

Flavonoid Rhizobia

Infection thread

2. Rhizobia proliferate inside root hair and cause an infection thread to form.

3. Infection thread grows into the cortex of the root.

4. Infection thread buds off, releasing rhizobia inside cortex cells.

Nodule

5. Nodule begins to form from rapidly dividing cortex cells.

FIGURE 39.15 Infection by Nitrogen-Fixing Bacteria Is a Multistep Process. After rhizobia bind to a root hair, they enter the cytoplasm in a membrane-lined infection thread and travel into the root cortex, where they enter cortex cells.

How Do Nitrogen-Fixing Bacteria Infect Plant Roots?

Like mycorrhizal fungi and their host plants, legumes and rhizobia have a mutualistic relationship. The nitrogen-fixing bacteria provide the plant with ammonia, while the legume provides the bacteria with carbohydrates and protection. The association is costly for the host plant, however. For the bacteria to synthesize the enormous amounts of high-energy electrons and ATP required to manufacture ammonia, plants have to supply them with large quantities of sugar.

Given the cost of the plant–bacterium interaction and its importance to both species, it's not surprising that the infection process is carefully regulated.

When a legume seed germinates, its roots do not contain a population of rhizobia. Instead, the roots make contact with bacterial cells existing in the soil, the rhizobia infect root cells, and the root cells grow into a nodule.

The first event in infection is a recognition step that occurs between a legume-family plant and its symbiotic bacterium. First, young roots release compounds called flavonoids. When rhizobia contact the flavonoids, the bacteria respond by producing sugar-containing molecules called **Nod factors** (for *nod*ule formation). Nod factors, in turn, bind to signaling proteins on the membrane surface of root hairs.

The recognition step is specific to the species involved. Each legume species produces a different flavonoid that acts as a recognition signal, and each rhizobium species responds with one or more unique Nod factors. When researchers have switched recognition signals or Nod factors between species, the recognition step fails.

When Nod factors bind to the root-hair surface, they set off a chain of events that leads to dramatic morphological changes in the host legume, as **FIGURE 39.15** shows.

1. The bacterial cells contact the root-hair surface.

2. Rhizobia begin to multiply and move, through an invagination of the root-hair plasma membrane called an **infection thread**.

check your understanding

If you understand that . . .

- Certain species of bacteria infect plant root cells and fix nitrogen, which is then made available to the plant.

✔ **You should be able to . . .**

Generate hypotheses addressing two additional questions about nitrogen fixation:

1. Why don't all plants have symbiotic bacteria that fix nitrogen?

2. Why is the interaction that establishes the infection by nitrogen-fixing bacteria so complex?

Answers are available in Appendix A.

3. The infection thread extends into the root, invading the cortex.

4. Portions of the infection thread bud off, forming membrane-bound clusters within cortex cells, each filled with rhizobia.

5. The infected cortex cells divide, forming root nodules.

The interaction between rhizobia and legumes is one of the most complex and best studied of all mutualisms.

39.5 Nutritional Adaptations of Plants

Based on the data available so far, over 95 percent of vascular plants take up nutrients from soil. And over 80 percent of all plants supplement their "diet" with nutrients acquired from mycorrhizal fungi, while a small but significant fraction associate with nitrogen-fixing bacteria. Perhaps 99 percent of all living plant species make their own sugar through the process of photosynthesis.

What about the small number of plant species that don't follow these rules? Some appear to live on air, some parasitize other plants, and others catch insects and digest them.

Epiphytic Plants

Species from a diverse array of plant lineages do not absorb nutrients from soil. In fact, these species never even make contact with soil. As **FIGURE 39.16a** shows, they often grow on the trunks or branches of trees. For this reason they are called **epiphytes** ("upon-plants").

In northern forests, it is common for mosses and ferns to grow epiphytically on tree trunks. The so-called Spanish moss that hangs from oak trees in the southeastern United States is another familiar epiphyte—although this species is actually not a moss but a bromeliad, a relative of the pineapple. In the tropics and subtropics, one-third of all ferns grow as epiphytes, and there are thousands of species of epiphytic orchids, bromeliads, and lycophytes.

Epiphytes absorb most of the water and nutrients they need from rainwater, dust, and particles that collect in their tissues or in the crevices of bark. As **FIGURE 39.16b** shows, some epiphytic bromeliads have leaves that grow in rosettes and form "tanks" that collect water and organic debris. In such cases, nutrients are actually absorbed through the leaves themselves.

Parasitic Plants

Parasites are organisms that live in close physical contact with individuals from another species and that lower the fitness of those individuals, usually by obtaining water or nutrients from the host. Biologists estimate that about 3000 species of angiosperms are parasitic—less than 1 percent of the plant species that have been studied and named to date.

Some plant parasites are non-photosynthetic and obtain all of their nutrition by tapping into the vascular tissue of the host individual. But most parasitic plants make their own sugars through photosynthesis and tap the xylem of other species for water and essential nutrients.

For example, the mistletoe (**FIGURE 39.17**, see page 790) is green and photosynthetic but has structures called **haustoria** (singular: **haustorium**) that penetrate a host's xylem and extract water and ions that the mistletoe uses as nutrients. In the forests of western North America, mistletoe infection is a serious cause of economic loss.

Carnivorous Plants

Carnivorous plants trap insects and other animals, kill them, and then digest the prey to absorb its nutrients. Carnivorous species make their own carbohydrates via photosynthesis but use carnivory to supplement the nitrogen available in the environment. Most are found in bogs or other habitats where nitrogen is scarce or unavailable.

(a) Epiphytes grow on trees.

(b) Water-holding "tanks" formed by leaves of an epiphyte

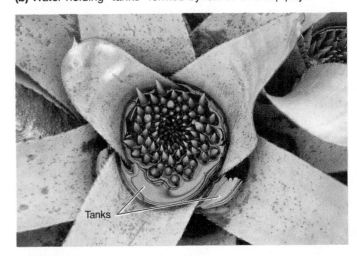

Tanks

FIGURE 39.16 Epiphytes Are Adapted to Grow in the Absence of Soil. The environments where many epiphytes grow—tree trunks and branches—are dry and nutrient poor.

FIGURE 39.17 Some Plant Parasites Tap into the Vascular Tissue of Their Hosts.

✔ **QUESTION** The bromeliads in Figure 39.16 and mistletoe both grow on other plants. How do they differ in their relationship to the trees that support them?

Haustorium

Host xylem

Mistletoe

Host tree

Mistletoe haustoria penetrate host xylem and extract water and ions

Modified Leaves Form an Array of Trapping Mechanisms Carnivorous plants use modified leaves or roots to trap insects. For example, the leaves of pitcher plants form tubes (see Table 37.3). Prey are enticed to the tube by an attractive odor and then have a difficult time climbing back out. Eventually they fall into the pool of water below, where they drown. Enzymes released by the plant digest the prey, and the plant absorbs the nutrients obtained from the insects' dead bodies.

Sundews have an alternative hunting strategy: They have modified leaves that function like flypaper. The leaf surfaces develop hairs that exude a sticky substance and trap insects (**FIGURE 39.18**). Glands near the sundew's trap release enzymes that digest the prey. The leaf then absorbs the nutrients that are released.

In the Venus flytrap, modified leaves trap insects mechanically. Sensory hairs protrude from the epidermis of each leaf. When an insect lands on the trap and bumps two or three of these hairs, the hair cells produce an electrical signal and the leaf responds by snapping shut (see Chapter 40). If the prey continues to stimulate hairs, the leaf secretes enzymes that digest the prey.

Costs and Benefits of Carnivory Initially, research on carnivorous plants focused on documenting that meat-eating was adaptive. Early experiments in natural environments supported this hypothesis by showing that, compared to individuals of the same species that are not fed fruit flies, carnivorous plants that were fed fruit flies grew faster and flowered more often.

Follow-up work explored the costs as well as the benefits of carnivory. For example, several teams of researchers showed that when carnivorous plants are provided with nitrogen-based fertilizers, they produce fewer of their specialized insect-trapping leaves and a higher proportion of leaves that function primarily in photosynthesis. These results suggest carnivory is a trait that shows phenotypic plasticity (see Chapter 37): Plants increase investment in prey-capture devices when nitrogen is rare, but they decrease investment in prey-capturing structures when nitrogen is readily available.

Currently, some research groups are doing "evo-devo" research (see Chapter 22) on carnivory. They are working to identify the novel alleles that made the evolution of carnivory possible in certain groups. Meat-eating may be relatively rare in plants, but it has inspired interesting new directions for research.

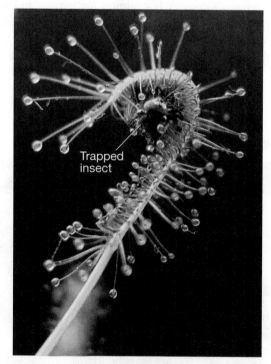

Trapped insect

FIGURE 39.18 Sundews Have Modified Leaves with Sticky Surfaces That Catch Insects.

If you understand . . .

39.1 Nutritional Requirements of Plants

- About 96 percent of the dry weight of a typical plant consists of carbon, hydrogen, and oxygen. Plants obtain these elements by absorbing carbon dioxide from the atmosphere and water from the soil.

- The other 4 percent of the plant body consists of a complex suite of elements, acquired from soil, that are required for normal growth and reproduction. These essential elements are usually absorbed in the form of ions.

- Nutrients are usually classified as macronutrients, needed in large supply, and micronutrients, needed in only trace amounts.

- When plants are lacking in an essential nutrient, they usually develop predictable deficiency symptoms.

✓ You should be able to design a field experiment that would determine whether nitrogen is a limiting nutrient in a habitat near your campus.

39.2 Soil: A Dynamic Mixture of Living and Nonliving Components

- Soil is a complex and dynamic mixture of inorganic particles such as clay and sand, organic matter, and organisms. Soil provides plants with oxygen, water, and nutrients as well as a physical substrate for anchoring and supporting the plant body.

- Plants absorb nutrients in the form of ions. Anions in soil water are normally available to roots, but cations adhere to soil particles and are made available only after cation exchange.

✓ You should be able to explain why loam is an ideal soil for most plants.

39.3 Nutrient Uptake

- Nutrients are present at low concentrations in the soil surrounding roots. Plants import nutrients against a concentration gradient by pumping protons into the extracellular space.

- A large excess of protons outside the root hair creates a strong electrochemical gradient favoring the entry of positively charged ions through membrane transporters.

- The proton gradient in root-hair membranes also allows negatively charged ions to cross the plasma membrane via symporters.

- In addition to acquiring nutrients by establishing a proton gradient, many plants obtain nitrogen or phosphorus through a mutualistic relationship with mycorrhizal fungi in exchange for carbohydrates derived from photosynthesis.

- Passive and active systems also exist for excluding certain ions. This is important because all nutrients are toxic at high concentrations.

✓ You should be able to predict the consequences of mutations that allow root cells to establish a membrane voltage that is much more negative than normal, without a large increase in energy expenditure.

39.4 Nitrogen Fixation

- Certain plants, including those in the legume family, are capable of symbiosis with nitrogen-fixing bacteria. The bacteria colonize the interior of root cells after exchanging species-specific signals.

- Nitrogen-fixing bacteria receive protection and sugar from the host plant, in exchange for producing ammonia, a useable form of nitrogen for the plant, from nitrogen gas.

✓ You should be able to design an experiment, using radioactive carbon and the heavy isotope of nitrogen ($^{15}N_2$), that would test whether the rhizobia–pea plant interaction is mutualistic.

39.5 Nutritional Adaptations of Plants

- Epiphytes grow on the trunks and branches of other plants and absorb mineral nutrients from rainwater, dust, or decaying matter trapped in the bark of their host plant.

- Some parasitic plants produce their own carbohydrates through photosynthesis but obtain water and nutrients by penetrating the xylem of host plants.

- Carnivorous plants have evolved mechanisms for trapping and digesting insects and other animals. The primary benefit of carnivory is to obtain nitrogen in low-nitrogen environments.

✓ You should be able to explain why epiphytes in tropical rain forests and desert plants often have similar adaptations to minimize water loss

(MB) **MasteringBiology**

1. **MasteringBiology Assignments**

   Tutorials and Activities Global Soil Degradation; How Plants Obtain Minerals from Soil; Nitrogen Nutrition in Plants; Soil Formation and Nutrient Uptake

   Questions Reading Quizzes, Blue-Thread Questions, Test Bank

2. **eText** Read your book online, search, take notes, highlight text, and more.

3. **The Study Area** Practice Test, Cumulative Test, BioFlix® 3-D Animations, Videos, Activities, Audio Glossary, Word Study Tools, Art

You should be able to . . .

1. Which of the following characteristics defines an element as essential for a particular species?
   a. It has to be added as fertilizer to achieve maximum seed production.
   b. If it is missing, a plant cannot grow or reproduce normally.
   c. If it is present at high concentration, plant growth increases.
   d. If it is absent, other nutrients may be substituted for it.

2. Why is the presence of clay particles important in soil?
   a. They provide macronutrients—particularly nitrogen, phosphorus, and potassium.
   b. They bind metal ions, which would be toxic if absorbed by plants.
   c. They allow water to percolate through the soil, making oxygen-rich air pockets available.
   d. The negative charges on clay bind to positively charged ions and prevent them from leaching.

3. Where does most nutrient uptake occur in roots?
   a. at the root cap, where root tissue first encounters soil away from the zone of nutrient depletion
   b. at the Casparian strip, where ions must enter the symplast before entering xylem cells
   c. in the symplastic and apoplastic pathways
   d. in root hairs, in the zone of maturation

4. Why are proton pumps in root-hair plasma membranes important?
   a. They pump protons into cells, generating a membrane potential (voltage).
   b. They allow toxins to be concentrated in vacuoles, so the toxins do not poison enzymes in the cytoplasm.
   c. They set up an electrochemical gradient that makes it possible for roots to absorb cations and anions.
   d. They set up the membrane voltage required for action potentials to occur.

5. Some plants obtain nitrogen from symbiotic fungi or bacteria. What were the original sources of N that the fungi and bacteria obtained?

6. Before the formation of a root nodule, which partner secretes a Nod factor, the plant or the bacterium?

7. Why is the relationship between most mycorrhizal fungi and their host plants considered mutualistic?
   a. They live in close physical association.
   b. Both species benefit from the association.
   c. The host plant cannot live without the mycorrhizae.
   d. The mycorrhizae cannot live without the host plant.

8. A farmer is concerned that her corn crop is suffering from iron deficiency. Design an experiment that uses hydroponic cultures to identify (1) the symptoms of iron deficiency in corn and (2) the minimum amount of iron required to support normal growth rates.

9. Explain why carnivorous and parasitic plants are most common in nutrient-poor habitats.

10. Would you expect plant productivity to be higher in sandy soils or in soils containing both sand and clay? Why?

11. Why is it important for plants to exclude certain ions? Summarize the difference between active and passive exclusion mechanisms.

12. Why is it logical for biologists to (1) distinguish nutrients that are obtained from the atmosphere or water (C, H, and O) from mineral nutrients obtained from soil, and (2) distinguish macronutrients from micronutrients?

13. **QUANTITATIVE** There is a conflict between van Helmont's data on willow tree growth and the data on essential nutrients listed in Table 39.1. According to the table, nutrients other than C, H, and O should make up about 4 percent of a willow tree's weight. Most or all of these nutrients should come from soil. But van Helmont claimed that the soil in his experiment lost just 60 g, while the tree gained 74,000 g. Calculate the percentage of added weight contributed by soil. State at least one hypothesis to explain the conflict between expected and observed results. How would you test this hypothesis?

14. Acid rain occurs when sulfur oxides and nitrous oxides released by cars or factories react with water vapor, forming sulfuric and nitric acids, respectively. In areas affected by acid rain, cations can quickly leach out of the soil. Explain why.

15. Design an experiment using the drug vanadate, which poisons proton pumps, to test the hypothesis that phosphate ions enter cells via an H^+/HPO_4^{2-} cotransporter.

16. Alder trees have nitrogen-fixing bacteria associated with their roots and have a much higher proportion of nitrogen in their tissues than do spruce trees that grow in the same habitat. Which species should decay faster after death: alder or spruce? Explain your logic.

40 Plant Sensory Systems, Signals, and Responses

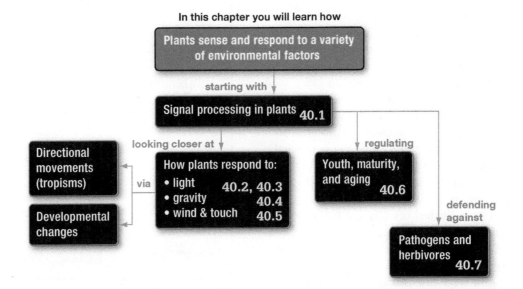

In this chapter you will learn how

Plants sense and respond to a variety of environmental factors

starting with ↓

Signal processing in plants 40.1

looking closer at ↓

Directional movements (tropisms)

Developmental changes

via

How plants respond to:
• light   40.2, 40.3
• gravity   40.4
• wind & touch   40.5

regulating ↓

Youth, maturity, and aging   40.6

defending against ↓

Pathogens and herbivores   40.7

Plants have sophisticated information processing systems. These radish seedlings sense the presence of light and are bending toward it.

This chapter is part of the Big Picture. See how on pages 840–841.

Imagine standing in place for several hundred years, like an oak tree. Each spring you produce flowers. All summer you absorb light from the Sun, carbon dioxide from the atmosphere, and water and nutrients from the soil. Water, ions, and sugars flow up and down your vascular tissue. For six months or more your body grows upward, outward, and downward. In fall, thousands of your offspring drop to the ground as acorns. Then as winter approaches, your metabolism slows and you stop growing. Like a hibernating animal, you spend the long, hard months of winter in a state of suspended animation, before resuming growth the following spring.

✔ When you see this checkmark, stop and test yourself. Answers are available in Appendix A.

To stay alive, an oak needs to gather information about its environment. It has to sense the season of the year, the time of day, the pull of gravity, the force of wind, and attacks by enemies. It needs to sense when its leaves are being shaded and which leaves are receiving the wavelengths of light that are required to support photosynthesis.

Plants may not have eyes or ears, but they can sense light, gravity, pressure, and wounds. They have the equivalent of a sense of smell, because they can perceive certain airborne molecules. It can even be argued that they have a sense of taste, because their roots sense the presence of nutrients in the soil.

In addition to gathering information about the conditions around them, plants have to respond in an appropriate way. They do not jump or swim or run, but their shoot systems grow toward light or end up shorter and stockier in response to wind. In response to gravity, shoots grow up and roots grow down. In response to touch, the modified leaves of a Venus flytrap shut fast enough to catch flying insects. If a plant is being attacked or if it senses that a neighboring individual is under attack, it may lace its tissues with toxic compounds or mobilize other defenses.

The message of this chapter is simple: Plants have sophisticated systems for collecting information about their environment and responding in ways that maximize their chances of surviving, thriving, and producing offspring. The ability to gather information and respond to it is one of the five fundamental attributes of life (see Chapter 1).

40.1 Information Processing in Plants

Every environment is full of information. But, like other organisms, plants monitor only aspects of the environment that matter to them—those that affect their ability to stay alive and produce offspring.

FIGURE 40.1 summarizes the three-step process by which plants gather, process, and respond to the information they monitor:

1. Receptor cells receive an external signal and change it to an intracellular signal.

2. Receptor cells send a signal to cells in other parts of the body that can respond to the information.

3. Responder cells receive this signal, transduce it to an intracellular signal, and change their activity in a way that produces an appropriate response.

Let's briefly consider how each of these steps works, then explore some of the details of how plants sense and respond to light, gravity, and other signals from the environment.

How Do Cells Receive and Transduce an External Signal?

When you text a friend, the signal travels from you to their receiver via airwaves. When the message arrives, the receiving cell

PROCESS: STEPS IN INFORMATION PROCESSING

External stimulus on receptor cell

Internal signal

1. Receptor cell perceives external stimulus and transduces the information to an internal signal.

Cell–cell signal

2. A cell–cell (hormone) signal released by the receptor cell travels throughout the body.

Internal signal

3. Responder cells receive the cell–cell (hormonal) signal, transduce it to an internal signal, and change activity.

FIGURE 40.1 In Many Cases, Cell–Cell Signals Link a Stimulus and a Response.

phone changes the information in the airwaves into electrical signals and then into words that your friend can understand.

Plants work in much the same way. When light strikes a sensory cell, for example, the information that it carries has to be changed into a form that is meaningful to that cell. Signals from the environment are usually detected by proteins specialized for that function. Receptor proteins change shape in response to an environmental stimulus, such as being struck by a particular wavelength of light, having pressure applied, or binding to a particular type of molecule.

When a receptor changes shape in response to a stimulus, the information changes form—from an external signal to an intracellular signal. This process is called **signal transduction** (**FIGURE 40.2**); the verb transduce means "to convert energy from one form to another." Once information has been transduced to an intracellular form, it travels down what biologists call a signal transduction pathway.

You might recall that there are two basic types of signal transduction pathways in all organisms: phosphorylation cascades and second messengers (see Chapter 11). Both begin with a receptor protein in the plasma membrane.

- **Phosphorylation cascades** are triggered when the change in the receptor protein's shape leads to the transfer of a phosphate group (PO_4^{3-}) from ATP to the receptor or an associated protein. Phosphorylation activates proteins involved in signal transduction cascades, causing them to phosphorylate and activate a different set of proteins, which in turn catalyze the phosphorylation and activation of still other proteins, and so on.

PROCESS: SIGNAL TRANSDUCTION

1. Signal arrives.

Cell wall

2. Receptor protein changes in response to signal.

Cell membrane

ATP

P

ADP

3. Receptor or associated protein catalyzes phosphorylation reaction.

ATP

ADP

4. Phosphorylated protein triggers phosphorylation cascade (left) ...

Phosphorylation cascade

...**OR** release of second messenger (right).

P

ATP

ADP

P

ATP

ADP

P

Second messenger

Ca^{2+}
Ca^{2+}
Ca^{2+}
Ca^{2+}

Vacuole

Ca^{2+}

Ca^{2+} Ca^{2+}

Ca^{2+}

P

5. Phosphorylated proteins or second messengers initiate response.

Nuclear envelope

DNA

6. Activate or repress transcription.

OR **OR**

6. Activate or repress translation.

Nucleus

6. Change ion flow through channel or pump.

FIGURE 40.2 Signal Transduction Changes an External Signal to an Internal Signal. After an environmental or cell–cell signal is transduced to an intracellular signal, it triggers a change in the cell's activity.

- **Second messengers** are produced when receptor proteins trigger the production of intracellular signals or their release from storage areas. Calcium ions (Ca^{2+}) stored in the vacuole, ER, or cell wall are one of several ions or molecules that function as the second messenger in plants.

In some cases, phosphorylation cascades and second messengers interact.

Signal transduction primes the receptor cell for action. In many cases, though, the cells that receive information from the environment are located in a part of the body that is distant from the cells that need to respond to the information. How does information from activated receptor cells get to responder cells?

In most cases, the answer involves a **hormone**—an organic compound that is produced in small amounts in one part of a plant and transported to target cells, where it causes a physiological response. Signal transduction in a receptor cell often results in the release of a hormone that carries information to responder cells.

How Are Cell–Cell Signals Transmitted?

Because plants perceive such a wide array of stimuli, they have a wide array of hormones coursing through their bodies. These signals may be transmitted from cell to cell by specialized transport proteins in cell membranes, in xylem sap or phloem sap, or by simple diffusion from the originating cell. In most cases the signals act on target tissues throughout the body.

Plant cells routinely receive information from several different hormones at the same time, so it is common for different types of hormones to interact with each other and modulate the cell's response.

How Do Cells Respond to Cell–Cell Signals?

Cells are exposed to a constant stream of hormones. Many of these signals have little to no effect on what is happening inside a given cell. The reason is simple: Hormones can elicit a response only if a cell has an appropriate receptor.

If a receptor exists on or in the cell, an arriving hormone binds to it. When the receptor changes shape in response, the effect is like a knock on the door of your room. The signal—often received at the cell's periphery—is rapidly transduced to a series of phosphorylated proteins or the release of a second messenger inside the cell. The result may be a dramatic change in the cell's activity.

To understand why these binding events can have such significant effects on cell activity, look again at Figure 40.2. Activation of a signal transduction cascade results in the production of many phosphorylated proteins, or the release of many second messengers. These events amplify the original signal many times. In this way, low concentrations of plant hormones can have a large impact on target cells. Hormones are small molecules in low concentrations, but because the information they carry is amplified during signal transduction, they produce big results.

As Figure 40.2 shows, the response to hormone binding can include

- activation of membrane transport proteins (see Chapter 6), which produces a change in the membrane's electrical potential or the cell wall's pH; or

- changes in gene expression—via alterations in transcription activators or repressors (see Chapter 19) or the translation machinery—that result in new suites of proteins or RNAs in the cell.

The response to the signal is important. When cells respond to a hormone, the change in their activity helps the plant cope with the environmental change sensed by the receptor cell.

40.2 Blue Light: The Phototropic Response

Most of the general principles of plant communication emerged from studies of how plants respond to light. For example, consider the claim that plants are highly selective about the information they process. Light is made up of a wide array of wavelengths (see Chapter 10), but plants sense and respond to only a few.

This conclusion traces back to experiments that Charles Darwin and his son Francis performed in 1881 with coleoptiles of a monocot called reed canary grass. A **coleoptile** is a modified leaf that forms a sheath protecting the emerging shoots of young grasses.

The Darwins germinated seeds in the dark, placed the young, straight coleoptiles next to a light source, and noted that they grew toward the light—a response, due to differential cell elongation, called bending (**FIGURE 40.3a**). You have probably seen the same response in houseplants that are near a window (see

(a) Shoots bend toward full-spectrum light.

(b) Shoots bend specifically toward blue light.

FIGURE 40.3 Experimental Evidence that Plants Sense Specific Wavelengths of Light. Though a candle is shown here, the Darwins used paraffin lamps or other sources of artificial light in their original experiments.

the photo at the start of the chapter). Directed movement in response to light is called **phototropism** (literally, "light-turn").

When the Darwins exposed coleoptiles to light filtered through a solution of potassium dichromate, however, the coleoptiles did not bend toward the light. Potassium dichromate solutions filter out wavelengths in the blue part of the visible spectrum. The photo in **FIGURE 40.3b** shows the results of a follow-up experiment with coleoptiles that have been exposed to blue, yellow, green, orange, and red light. Bending occurs only toward light that contains blue wavelengths.

It is important to understand the specificity of this response. Recall that **(1)** chlorophylls *a* and *b* are the primary photosynthetic pigments, and **(2)** these pigments absorb strongly in the blue and red parts of the spectrum (see Chapter 10). Plants exhibit a phototropic response if blue wavelengths are available, but show no response if blue wavelengths are not present. Plants move toward blue light, which is important in photosynthesis.

Let's look at what happens when blue light strikes a receptor cell, and follow the signal to examine the cells that undergo differential growth in response.

Phototropins as Blue-Light Receptors

Although biologists knew that the blue-light receptor must be a **pigment**—a molecule that absorbs certain wavelengths of light—it took decades to find it. A key breakthrough came in the early 1990s, when researchers found a membrane protein in the tips of emerging shoots that gains a phosphate group in response to blue light. Researchers hypothesized that the membrane protein becomes activated when it is phosphorylated in response to blue light, and that the activated protein then triggers the phototropic response.

Subsequent work succeeded in isolating the gene that codes for the membrane protein. The gene, named *PHOT1*, was found by analyzing mutant *Arabidopsis thaliana* individuals that do *not* show a phototropic response to blue light. Further experiments by Shin-ichiro Inoue and colleagues showed that *PHOT1* codes for a blue-light receptor. An important question remained, however. Is the PHOT1 protein activated by phosphorylation, and does this trigger bending of the plant?

As a first step in answering this question, Inoue and colleagues identified an amino acid in the PHOT1 protein that becomes phosphorylated after plants are exposed to blue light. During this analysis, they constructed a mutated *PHOT1* gene, called *phot1-m*, which coded for an altered protein with an amino acid that cannot be phosphorylated. This altered protein provided the researchers with a way to test whether phosphorylation of the PHOT1 protein was required for a phototropic response. **FIGURE 40.4** shows their experimental setup.

The key to the experiment was to compare bending in response to blue light in wild-type *Arabidopsis* with the response in mutant plants that lacked a PHOT1 protein. When Inoue and colleagues introduced the *phot1-m* gene into a mutant plant, the altered protein failed to rescue phototropism. However, when the normal *PHOT1* gene was introduced into a mutant plant, phototropism was restored.

The current consensus is that *PHOT1* encodes a blue-light receptor in plants. A phototropic response is initiated when this receptor protein is phosphorylated.

Recent research indicates that *Arabidopsis* has a second blue-light receptor related to PHOT1, called PHOT2. Collectively, **photoreceptors** that detect blue light and initiate phototropic responses are known as **phototropins.**

Given its importance to photosynthesis, it's not surprising that blue light triggers an array of responses in addition to bending. The phototropins, for example, trigger signal transduction cascades that result in at least two other responses.

1. Chloroplast movements inside leaf cells. These movements put chloroplasts in positions to optimize light absorption. In high light the chloroplasts move to the sides of cells to shade each other while in low light they spread out.

2. Opening of stomata. As a result, carbon dioxide can diffuse into cells as blue light triggers photosynthesis.

In addition to the phototropins, several other kinds of blue-light receptors have been identified in plants that also help control the opening of stomata, as well as stem elongation and flower production. Phototropism, however, ranks as the best studied of all blue-light responses. How is the signal from PHOT proteins transmitted to the cells that respond by bending?

RESEARCH

QUESTION: Is PHOT1 activated by phosphorylation?

HYPOTHESIS: Phototropism in *Arabidopsis* is triggered by phosphorylation of PHOT1 protein.

NULL HYPOTHESIS: Phototropism in *Arabidopsis* is not triggered by phosphorylation of PHOT1 protein.

EXPERIMENTAL SETUP:

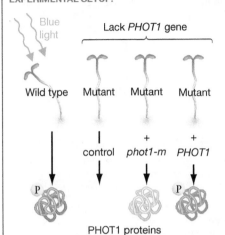

1. Mutant plants that lack the *PHOT1* gene fail to bend toward blue light.

2. Express normal PHOT1 protein and a phosphorylation-defective form of PHOT1 (phot1-m) in mutant plants to see if either rescues normal bending.

PREDICTION: Mutant plants expressing the normal *PHOT1* gene will bend toward blue light, but mutant plants expressing the defective form of *PHOT1* (*phot1-m*) will not bend toward blue light.

PREDICTION OF NULL HYPOTHESIS: Mutant plants expressing either the normal or the defective form of the *PHOT1* gene will both bend toward blue light.

RESULTS:

CONCLUSION: Phosphorylation of PHOT1 triggers phototropic bending in response to blue light.

FIGURE 40.4 The Role of PHOT1 Phosphorylation in Phototropic Bending: An Experimental Test.

SOURCE: Inoue, S., T. Kinoshita, M. Matsumoto, et al. 2008. Blue light-induced autophosphorylation of phototropin is a primary step for signaling. *Proceedings of the National Academy of Sciences* 105: 5626–5631.

✔ **QUESTION** Why did the researchers put the mutant and normal PHOT1s into PHOT1-deficient plants and not wild-type plants?

Auxin as the Phototropic Hormone

Long before the phototropins were identified, biologists knew that receptor cells responded to blue light by releasing a hormone. The Darwins established this result when they followed

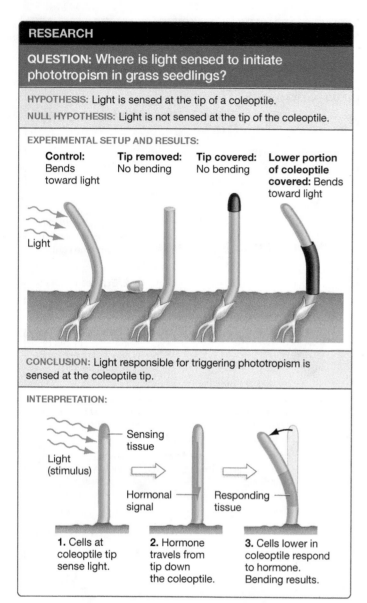

QUESTION: Where is light sensed to initiate phototropism in grass seedlings?

HYPOTHESIS: Light is sensed at the tip of a coleoptile.

NULL HYPOTHESIS: Light is not sensed at the tip of the coleoptile.

EXPERIMENTAL SETUP AND RESULTS:

Control: Bends toward light

Tip removed: No bending

Tip covered: No bending

Lower portion of coleoptile covered: Bends toward light

Light

CONCLUSION: Light responsible for triggering phototropism is sensed at the coleoptile tip.

INTERPRETATION:

Light (stimulus) — Sensing tissue

Hormonal signal — Responding tissue

1. Cells at coleoptile tip sense light.

2. Hormone travels from tip down the coleoptile.

3. Cells lower in coleoptile respond to hormone. Bending results.

FIGURE 40.5 The Sensory and Response Cells Involved in Phototropism Are Not the Same.

SOURCE: Darwin, C., and F. Darwin. 1897. *The Power of Movement in Plants.* New York: D. Appleton & Co.

✔ **QUESTION** A critic could argue that this experiment lacked appropriate controls for the treatments labeled "Tip removed" and "Tip covered." Suggest better controls for these treatments than the unmanipulated individual at the far left.

up on their initial experiments. **FIGURE 40.5** shows their experimental setup and results.

- If they removed the tips of coleoptiles, they found that the decapitated seedlings stopped bending toward the light.

- If they covered the tips of coleoptiles with opaque material, the seedlings did not bend toward light.

- If they put opaque collars below the tips, in the area where bending occurs, the seedlings bent toward light normally.

These data provided evidence that the blue-light sensors were located in the tips of the coleoptiles, and not in the part of the coleoptile that bends.

How did the sensory cells in the tip communicate with the cells that actually elongate? The Darwins proposed that phototropism depends on "some matter in the upper part which is acted on by light, and which transmits its effects to the lower part." Their hypothesis was that a substance produced at the tip of the coleoptile acts as a signal and is transported to the area of bending. This was the first explicit hypothesis stating that hormones—signaling molecules that can act at a distance—must exist.

The hormone hypothesis was not tested rigorously until 1913, when Peter Boysen-Jensen

1. cut the tips off young oat coleoptiles;

2. put a tip and a porous block of the gelatinous compound called agar on some of the decapitated coleoptiles; and

3. put a tip and a piece of mica (a nonporous mineral) on the rest of the decapitated coleoptiles.

As **FIGURE 40.6a** shows, only the coleoptiles treated with the porous agar block showed normal phototropism. Based on this observation, Boysen-Jensen concluded that the phototropic signal was indeed a chemical and that it could diffuse. He also inferred that the molecule was water soluble, because the agar that he used was a water-based gelatin.

Twelve years later, Frits Went extended these results. He placed the decapitated tips of oat coleoptiles on agar blocks, with the goal of collecting the hypothesized hormone controlling phototropism. Then he did something clever: He placed agar blocks that had been exposed to oat tips off-center on the decapitated coleoptiles of other individuals (**FIGURE 40.6b**). He also did the same with agar blocks that had *not* been exposed to oat tips, as a control (not shown).

Even though the coleoptiles were kept in the dark during the entire experiment, they responded by bending if their agar block had been exposed to oat tips. In this way, Went succeeded in producing the phototropic response without the stimulus of light.

Because it promotes cell elongation in the shoot, Went named the hormone **auxin** (from the Greek *auxein*, "to increase"). Auxin was the first plant hormone ever discovered.

Isolating and Characterizing Auxin After years of effort, researchers in two laboratories independently succeeded in isolating and characterizing auxin. The hormone turned out to be indoleacetic acid, or IAA. It was difficult to identify because it is present in such low concentrations. Depending on the species and tissue involved, IAA concentrations range from 3 to 500 nanograms per gram of tissue. (The prefix *nano–* refers to billionths.)

Like the other plant hormones introduced in this chapter, auxin is a small molecule with a relatively simple structure. It is present in quantities so small that its concentration is difficult to measure. Yet its impact is huge. Auxin can bend stems—producing, in some cases, tree trunks that are permanently bowed.

(a) The phototropic signal is a chemical.

Light

Permeable agar: Coleoptile bends toward light

Impermeable mica: No bending

Chemical diffuses through agar

(b) The hormone can cause bending in darkness.

Allow time for hormone to diffuse into agar block.

Offset blocks cause bending of coleoptiles not exposed to light

(c) The hormone causes bending by elongating cells.

Cells on the shaded side elongate in response to the hormone (red dots)

FIGURE 40.6 Experimental Evidence Supports the Hormone Hypothesis for Phototropism. (a) Coleoptiles bend in response to light if substances from the tip are allowed to move downward. **(b)** If bending can take place in darkness, then light is not directly required for the response. Only the hormone is required. **(c)** During the phototropic response, bending occurs because cells on the shaded side of the coleoptile elongate.

The Cholodny–Went Hypothesis Went's experiments were a breakthrough in research on information processing: They confirmed the hormone hypothesis and led to the discovery and characterization of IAA. But Went's experiments also inspired an important hypothesis for *how* the hormone produces the bending response. Working independently, both N. O. Cholodny and Went proposed that phototropism results from an asymmetric distribution of auxin.

The Cholodny–Went hypothesis makes three predictions:

1. Auxin produced in the tips of coleoptiles moves from the lighted side of the tip to the shaded side.

2. Auxin is then transported straight down one side of the coleoptile.

3. The asymmetric distribution of auxin causes cells on the shaded side of the coleoptile to elongate more than cells on the illuminated side (**FIGURE 40.6c**).

Bending results. In essence, bending results from differential elongation.

To test the Cholodny–Went model, Winslow Briggs used thin sheets of mica inserted into coleoptile tips to block the movement of auxin. As predicted by the model, auxin concentrations were the same on the lighted and shaded sides of divided tips. If a path for auxin movement was maintained, more of the auxin accumulated on the shaded side of the tip.

The Cholodny–Went asymmetric distribution model was correct. It explained how auxin leads to asymmetric cell elongation, and thus the bending response called phototropism. Biologists now know that the stems of eudicots and the coleoptiles of monocots respond to phototropic signals in essentially the same way.

The Cell Elongation Response How do cells in the stem respond to auxin? Experiments in corn plants and *Arabidopsis* suggest that auxin-binding proteins are found in stem and leaf cells. Researchers proposed that once auxin binds to these receptors, the signal transduction cascade that follows increases the activity of membrane H^+-ATPases, or proton pumps, in the plasma membrane.

Recall that **proton pumps** use the energy in ATP to drive protons out of the cell against an electrochemical gradient (see Chapter 38). Because the pH of the cell wall decreases when

check your understanding

If you understand that . . .

The chain of events involved in phototropism can be summarized as follows:

- When phototropins in coleoptile-tip cells absorb blue light, auxin is redistributed to the shaded side of the tip.
- Auxin is transported down the shoot and binds to receptors in target cells.
- These cells elongate when activated receptors lead to the activation of proton pumps, acidification of the cell wall, and activation of expansin proteins.

✔ **You should be able to . . .**

1. Assuming that you have a large supply of purified auxin, state how you would manipulate a large bed of roses so their stems bend toward the east.

2. Based on the information in this section, predict the phenotype of a mutant plant in which the auxin-binding protein in stems is always in the active state, even in the absence of auxin.

Answers are available in Appendix A.

1. **Proton pumps acidify cell wall** outside the plasma membrane.

2. **Wall loosens** as activated expansins "unzip" hydrogen bonds connecting cellulose microfibrils to other cell-wall polymers. Electrochemical gradient brings ions into cell.

3. **Water follows by osmosis.** Increased turgor pressure pushes loosened wall out, elongating the cell.

FIGURE 40.7 The Acid-Growth Hypothesis Requires Expansion of Cell Wall and Intake of Water. When activated proton pumps lower the pH outside the cell membrane in response to auxin; a series of events lead to elongation of the cell.

H^+-ATPases are active, the idea that these pumps are responsible for cell elongation became known as the **acid-growth hypothesis.**

To understand the rationale behind the acid-growth hypothesis, it's important to realize that two things have to happen for a plant cell to get larger:

1. The cell wall has to expand to create a larger volume.
2. Water has to enter the cell and generate turgor pressure on the cell wall to make an increase in volume possible.

As **FIGURE 40.7** shows, both processes are triggered by pumping protons into the cell wall.

When proton pumping lowers the pH of the wall to 4.5, cell-wall proteins called **expansins** are activated. Expansins "unzip" the hydrogen bonds that form between cellulose microfibrils and other polymers in the cell wall, loosening the structure.

As protons are pumped out of the cell, an electrochemical gradient is established. The inside of the membrane becomes much more negative than the outside, favoring the entry of potassium (K^+) or other positively charged ions. As the concentration of solutes increases inside the cell, water follows via osmosis.

This is a key point: Cells don't move water directly. Instead, they create an osmotic gradient that favors water movement. Pumping protons *out* of a cell is a way to bring water *into* the cell. The incoming water increases turgor pressure, which pushes out the loosened cell wall. The cell gets bigger.

The upshot? When cell walls on one side of a stem are acidified in response to a signal from auxin, bending results. Auxin's role in phototropism is an exquisite example of information processing in plants.

Given the importance of light to plants, it shouldn't be surprising that they have photoreceptors that detect other wavelengths of light. We will explore one of these photoreceptors in the next section.

40.3 Red and Far-Red Light: Germination, Stem Elongation, and Flowering

Plants are sensitive to wavelengths in the red and far-red portions of the visible spectrum, as well as to blue light. This sensitivity is interesting because red wavelengths (about 660 to 700 nm) and far-red wavelengths (over 710 nm) signal very different things to a plant.

- Red light drives photosynthesis, just as blue light does.
- Far-red wavelengths are not absorbed strongly by photosynthetic pigments, so they tend to pass through leaves. As a result, far-red wavelengths are prominent in light that is filtered through tree leaves before it reaches the forest floor. Far-red light indicates shade.

The Red/Far-Red "Switch"

The first hint that plants monitor red and far-red light emerged from studies by H. A. Borthwick and colleagues in the early 1950s on how lettuce seeds germinate. By exposing lettuce seeds to various wavelengths of light and plotting their frequency of germination, researchers discovered that germination rates peak when seeds receive red light (about 660 nm).

This observation made sense, because lettuce thrives best when it grows in bright sunlight. But the stimulatory effect of just 1 minute of red light disappeared if seeds were then exposed to 4 minutes of far-red light.

TABLE 40.1 How Do Red Light and Far-Red Light Affect the Germination of Lettuce Seeds?

Biologists exposed moistened lettuce seeds to flashes of light containing one of two wavelengths: red or far-red (FR). After exposure to light, the seeds were held in the dark for several days.

Light Exposure	Germination (%)
None (control)	9
Red	98
Red → FR	54
Red → FR → Red	100
Red → FR → Red → FR	43
Red → FR → Red → FR → Red	99
Red → FR → Red → FR → Red → FR	54
Red → FR → Red → FR → Red → FR → Red	98

SOURCE: H. A. Borthwick, S. B. Hendricks, M. W. Parker, et al. 1952. A reversible photoreaction controlling seed germination. *PNAS* 38: 662–666, Table 1.

✔**QUANTITATIVE** According to the data above, what is the average germination rate of lettuce seeds that were last exposed to red light? To far-red light? How would these values compare with the germination rate of seeds that are buried underground and receive no light at all?

This observation also made sense because far-red light indicates that the seeds are shaded. Wavelengths near 735 nm inhibit germination the most effectively.

Follow-up experiments showed that red and far-red light act like an on-off switch for lettuce seed germination (**TABLE 40.1**). Red light promotes lettuce germination; far red inhibits it.

The key observation, though, is that the last wavelength sensed by the seed determines whether germination occurs at a high rate. How could this happen?

Phytochromes as Red/Far-Red Receptors

To interpret the red/far-red switch in seed germination, biologists hypothesized that the same pigment absorbs both wavelengths. Further, they suggested that the pigment exists in two shapes, or conformations: One shape absorbs red light, and one shape absorbs far-red light.

The idea was that switching behavior, or **photoreversibility,** occurs because light absorption makes the photoreceptor pigment change shape, like a light switch moving up or down in response to touch. Each conformation would be responsible for a different response.

Biologists called the hypothesized pigment **phytochrome** ("plant-color"). Phytochrome was thought to be a specialized light receptor, different from any of the pigments involved in absorbing light during photosynthesis or phototropism.

FIGURE 40.8 illustrates the photoreversibility hypothesis. One conformation of phytochrome, called P_r (phytochrome red), absorbs red light. Another conformation of the same molecule,

FIGURE 40.8 The Photoreversibility Hypothesis for Phytochrome Behavior. According to the photoreversibility hypothesis, phytochrome switches between the P_r conformation and the P_{fr} conformation when it absorbs red or far-red light, respectively. Only the P_{fr} form promotes germination.

called P_{fr} (phytochrome far-red), absorbs far-red light. According to the photoreversibility hypothesis, each conformation switches to the other when it absorbs its specific wavelength.

How Were Phytochromes Isolated?

Young corn and bean plants lengthen their stems in response to light deprivation or exposure to excessive far-red light. If you have grown these seeds indoors, you have seen this response to far-red light firsthand. The shoots get long and spindly.

Corn and beans normally grow in open sunlight. If they are grown indoors under incandescent light, which is deficient in red wavelengths as compared with far-red wavelengths, the plants react as though they are being shaded—they attempt to grow high enough to reach full sunlight. This behavior suggests that young corn and bean plants have a receptor protein for far-red light.

To follow up on this observation, researchers purified proteins from corn coleoptiles and succeeded in isolating one that was photoreversible. Specifically, when the protein was placed in solution and exposed to alternating red and far-red light, the color of the solution switched back and forth between blue and blue-green. The color switches supported the hypothesis that the same protein absorbed red as well as far-red light. Phytochrome was found.

Follow-up work showed that a region of the phytochrome molecule changes shape in response to red and far-red light. The shape changes cause phytochrome to take on or lose a phosphate group, just as the shape changes triggered by blue light result in phosphorylation of the phototropins. Conformational changes in proteins are frequently associated with a change in phosphorylation.

The photoreversibility hypothesis explains the red/far-red switch seen in seed germination and stem elongation. But can it also explain other plant responses? Let's consider another key event in plant growth—flowering—and the signals that trigger it.

Signals That Promote Flowering

Anatomically, a flower is a modified shoot that develops from a compressed stem and highly modified leaves. In essence, flower formation begins when an apical meristem stops making energy-harvesting stems and leaves and begins to produce the modified stems and leaves that make up flowers (see Chapter 24). Instead of making more food through photosynthesis, a sporophyte commits to investing energy in sexual reproduction—by producing gametophytes that will produce gametes.

When does this happen? Early experiments on the environmental signals that promote flowering focused on the number of hours of light and dark during a day. As it turns out, the switch that triggers flowering in response to changes in day length also shows red/far-red photoreversibility. Phytochrome plays a key role in a phenomenon called photoperiodism.

Photoperiodism is any response by an organism that is based on photoperiod—the relative lengths of day and night. In plants, the ability to measure photoperiod is important because it allows individuals to respond to seasonal changes in climate—for example, to flower when pollinators are available and when resources for producing seeds are abundant. Let's take a closer look.

Responding to Changes in Photoperiod   Experiments on photoperiodism in plants have shown that, with respect to flowering, plants fall into three main categories:

1. **Long-day plants** bloom in midsummer, when days are longest and nights shortest. Radishes, lettuce, spinach, corn, irises, and other long-day plants flower only when days are longer than a certain length—usually between 10 and 16 hours, depending on the species.

2. **Short-day plants** bloom in spring, late summer, or fall. Asters, chrysanthemums, poinsettias, and other short-day plants flower only if days are shorter than a certain species-specific length.

3. **Day-neutral plants** flower without regard to photoperiod. Day length has no effect on flowering in roses, snapdragons, dandelions, tomatoes, cucumbers, and many **weeds**—plants that are adapted to grow in soils that have been disturbed enough to remove or damage existing vegetation.

How do plants sense changes in day length and initiate the chain of events that leads to flowering? Early experiments suggested that phytochrome is involved.

FIGURE 40.9 summarizes some of the key experiments. Researchers showed that interrupting the night period with a light flash changed the flowering response. In short-day plants, for example, flowering was inhibited if the dark period was interrupted by red light—wavelengths around 660 nm. But a subsequent flash of far-red light—wavelengths around 735 nm—erased the effect. In some way, the phytochrome switch linked changes in day length to flowering.

FIGURE 40.9 Flashes of Red Light and Far-Red Light Switch the Photoperiod Response On and Off. If flashes of red light (R) and far-red light (FR) are alternated during the night, the plant's flowering response correlates with the last light it experiences. A flash of red light turns a long night into a short night; a flash of far-red light restores the perception of a long night.

More recent research has shown that phytochrome's effect is closely tied to the molecular mechanisms of timekeeping in plants. Plants have a clock that is reset each morning. Clock protein levels rise during the day and trigger expression of a gene called *CONSTANS* (*CO*). The CO protein is a transcription factor that affects the production of a flowering hormone. When phytochrome is activated by light, it stabilizes CO—so that CO accumulates in cells.

- In long-day plants, high levels of CO stimulate production of the flowering hormone.

- In short-day plants, high levels of CO *inhibit* production of the flowering hormone.

Many questions remain, however, and research is continuing. In the meantime, it is clear that a complex interaction takes place between clock proteins, phytochrome activation, and the production of the signal that initiates flowering. What is that signal?

Discovery of the Flowering Hormone   Since the 1930s, biologists have known that exposing even one leaf on a plant to the conditions necessary to induce flowering may result in the whole plant flowering. This result suggests that the signal to flower comes from leaves and travels to the apical meristem.

Grafting experiments, in which an organ from one individual is physically attached to a different individual, support this result. If you provide an experimental plant with the appropriate flower-triggering night length and then cut off a leaf or stem and graft it onto a second, experimental plant that has never been exposed

QUESTION: Can signals from different plants induce flowering?

HYPOTHESIS: Signals from grafted leaves can induce flowering.

NULL HYPOTHESIS: Signals from grafted leaves cannot induce flowering.

EXPERIMENTAL SETUP:

Leaf exposed to **short-night** photoperiod

Graft

Day-neutral plant exposed to **long-night** photoperiod

PREDICTION: Plant will flower.

PREDICTION OF NULL HYPOTHESIS: Plant will not flower.

RESULTS:

Plant flowers

CONCLUSION: A signal from a grafted leaf can induce flowering.

FIGURE 40.10 Experimental Support for the Hypothesis that a Hormone for Flowering Exists.

SOURCE: Lang, A., M. K. Chailakhyan, and I. A. Frolova. 1977. Promotion and inhibition of flower formation in a day-neutral plant in grafts with a short-day plant and a long-day plant. *Proceedings of the National Academy of Sciences, USA* 74: 2412–2416.

✔ **QUESTION** For this experiment to support the claim that a flowering hormone exists, a control treatment needs to be carried out. What is this control treatment? If the hypothesis is correct, what is the predicted outcome of the control treatment?

to the correct photoperiod for flowering, the experimental plant will flower (**FIGURE 40.10**).

Like the experiments on phototropism with agar blocks (Figure 40.6), this result supported the hypothesis that some substance produced in the transplanted leaf must travel up the recipient plant. This substance would then trip a developmental switch in the apical meristem, causing the change from vegetative growth to flowering. Biologists were so convinced that flowering must be induced by a hormone that they named it **florigen**, even though the actual hormone had not been discovered.

Almost 80 years later, researchers finally found the florigen molecule. This work began by focusing on the *FLOWERING LOCUS T (FT)* gene in *Arabidopsis*, which is known to promote flowering when activated. When researchers exposed leaves to short nights, they found that the gene was expressed in the leaf vascular tissue. Subsequently, some research indicates that the protein product of the *FT* gene is transported from leaves to the shoot apical meristem, and that the protein's presence triggers the activation of genes required for converting the stem to a flower.

Details are emerging on how the information present in the P_r/P_{fr} switch is translated into action that affects germination, stem elongation, flowering, timekeeping, and other responses to red and far-red light.

Given their reliance on photosynthesis for both energy and reduced carbon compounds, it isn't surprising that plants are highly tuned to their light environment. They also need to tell which end is up. Next, let's explore how they sense gravity.

check your understanding

If you understand that . . .

- Plants respond to red and far-red light via phytochromes, which change shape and activity when they absorb red or far-red light.
- Red light acts as a sunlight or day indicator, while far-red light acts as a shade or night indicator.
- Flowering occurs at specific times of the year and is controlled by night length through phytochrome signals.

✔ **You should be able to . . .**

1. Explain why it is adaptive for red light to trigger germination in lettuce seeds while far-red light inhibits it.
2. Explain why it is adaptive for changes in night length to affect the tendency to flower in some species.

Answers are available in Appendix A.

40.4 Gravity: The Gravitropic Response

The wavelengths, quantity, and direction of light that a plant receives change with the season, weather, time of day, and shading by other plants. However, gravity is constant and unidirectional. Light means food; gravity provides information about how the plant should orient itself in space.

Shoots usually respond to gravity by growing in an upward direction; roots usually respond by growing downward or laterally. How do plants sense gravity, so that it can be used as a signal to orient the body?

In 1881 Charles and Francis Darwin published one of the first experimental results on **gravitropism** ("gravity-turn")—the ability to move in response to gravity. Recall that the ends of root tips are covered by a protective tissue called the **root cap** (see

FIGURE 40.11 Gravity Sensing Occurs in the Root Cap. The root cap is a protective structure. Gravity-sensing cells in the center of the cap contain starch granules (stained dark purple).

FIGURE 40.12 The Statolith Hypothesis States that Amyloplasts Stimulate Sensory Cells. Amyloplasts are filled with starch. They are dense, so they sink in response to gravity. The statolith hypothesis predicts that pressure receptors in the plasma membrane become activated as a result.

✔**EXERCISE** Assuming that the receptor is a protein, suggest a hypothesis to explain how signal transduction occurs in response to pressure from a statolith.

Chapter 37). The Darwins found that roots stop responding to gravity if the root caps are removed. This observation suggested that gravity sensing occurs somewhere inside the root cap.

Recently biologists demonstrated precisely which cells are involved in gravity sensing in *Arabidopsis* roots. By killing tiny blocks of cells with laser beams, researchers showed that cells located at the center of the root cap are the most important for regulating the gravitropic response. Importantly, these cells always contain starch (**FIGURE 40.11**). Cells in the root cap respond to gravity and initiate gravitropism. Does starch play a role in sensing gravity?

The Statolith Hypothesis

The leading explanation for how plants sense gravity—the **statolith hypothesis**—is based on two interconnected ideas:

1. **Amyloplasts** are organelles that contain starch granules; starch is denser than water. These organelles are pulled to the bottom of root cap cells by the force of gravity (**FIGURE 40.12**).

2. The position of the amyloplasts activates sensory proteins located in the plasma membrane. These sensory proteins initiate the gravitropic response.

The statolith hypothesis was inspired by animals that use dense particles to sense gravity. Lobsters, for example, take up grains of sand that become positioned in specialized gravity-sensing organs in their antennae. The grains of sand are called **statoliths** ("place-stones"). When the animal tilts or flips over, the statoliths move in response to gravity. Inside the organ, the sand grain ends up pushing against a sensory cell. When this cell is activated, it indicates that the animal is no longer upright.

According to the statolith hypothesis, the same thing happens in root cap cells. If the wind tips a plant over, for example, the amyloplasts settle onto the new "lower" cell walls. The weight activates receptors, which signal that the root no longer faces in the correct direction.

Although recent experiments strongly support the statolith hypothesis, the search for the gravity receptor itself continues. In contrast, the second and third steps of information processing in response to gravity—the production of a cell–cell signal and the response of target cells—are much better understood.

Auxin as the Gravitropic Signal

Root cap cells that sense changes in the direction of gravitational pull respond by changing the distribution of auxin in the root tip. **FIGURE 40.13** illustrates this chain of events.

Step 1 Under normal conditions, auxin flows down the middle of the root, then toward the perimeter, and finally away from the root cap.

Step 2 If the root is tipped, sensory receptors trigger changes in the position of auxin transport proteins that redistribute auxin.

Step 3 Auxin is redistributed: The lower portion of the root receives increased concentrations of auxin; the upper portion receives lower concentrations.

Step 4 Because high auxin concentrations inhibit growth in roots, the differences in auxin concentrations trigger differential growth. Cells in the lower portion of the root elongate more slowly compared with cells in the upper portion. The result is bending.

Note that the way root cells respond to auxin redistribution during the gravitropic response is opposite to the way that cells in the stem respond during phototropism. In stems, high concentrations of auxin lead to *increased* cell elongation and bending. In roots, high concentrations of auxin lead to *decreased* cell division and elongation. Roots bend as cells on the upper side of the zones of cellular division and elongation (see Chapter 37) continue to grow.

check your understanding

If you understand that . . .

- Cells in root caps sense gravity via pressure that amyloplasts exert on receptors.
- Changes in gravity-sensing cells result in a redistribution of auxin and changes in the growth rate of root tips.

✔ **You should be able to . . .**

1. Explain the similarities and differences that exist between auxin redistribution in shoots and roots and the responses called phototropism and gravitropism.
2. Explain why auxin could be considered a gravitropic hormone.

Answers are available in Appendix A.

40.5 How Do Plants Respond to Wind and Touch?

Light and gravity are not the only physical forces that plants sense and respond to. Plants also react to mechanical stresses such as wind and touch.

Let's analyze two of the best-studied responses to pushing or pulling forces: growth responses that lead to exceptionally stout, stiff stems, and movement responses that make a Venus flytrap snap shut.

Changes in Growth Patterns

When plants are buffeted by wind, receptor cells transduce the mechanical force into an internal signal in the form of phosphorylated proteins or second messengers. Although the exact receptor and transduction pathway are not known, studies in *Arabidopsis thaliana* have shown that transcription of a large suite of genes is turned up or down in response to touch or other mechanical stimuli that mimic the effect of wind. Some of the new protein products act to stiffen cell walls, resulting in plants that are shorter and stockier than plants that do not experience repeated vibrations or touching.

FIGURE 40.14 shows what this response looks like in *Arabidopsis*. In this experiment, plants were either touched lightly 2 times

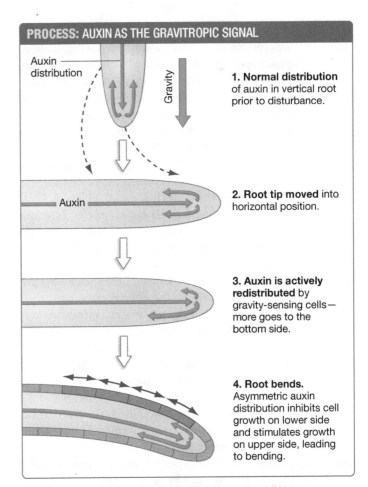

FIGURE 40.13 The Auxin Redistribution Hypothesis for Gravitropism. This sequence of events might begin when a growing root tip hits a rock and is displaced horizontally or when a plant is tipped in a windstorm and partially uprooted.

per day or left untouched. The flowing stalks of the touched plants were much shorter.

In response to wind, then, plants change their growth patterns in ways that make them more likely to withstand the force, stay upright, and live long enough to produce flowers and fruit.

FIGURE 40.14 Plant Growth Changes in Response to Wind or Touch. The *Arabidopsis* plants on the right were touched several times daily, while plants on the left were untouched.

Movement Responses

In some cases, plants respond to touch by moving. This response, **thigmotropism** ("touch-bending"), can be fast. For example, species that grow by climbing up objects or other plants may have modified leaves or stems that form long, thin structures called tendrils. When a tendril makes contact with an object, it responds by wrapping itself around the item as fast as one or more times per hour (**FIGURE 40.15**). Plants that climb in this way are able to grow taller toward light without needing to invest considerable resources in making wood.

Movement is even faster in "touch-sensitive" plants. A Venus flytrap, for example, closes fast enough to catch insects. Extremely rapid movements like this are possible when a touch-receptor cell transduces the mechanical signal to an electrical signal.

To understand how plants use electrical signaling, recall that proton pumps in the plasma membrane of most plant cells give them a negative charge relative to the exterior environment (see Chapter 38). The separation of charges creates a membrane voltage, or **membrane potential.** If a receptor protein responds to touch by allowing ions to flow across the membrane—which changes the amount of charge on either side—then the membrane potential changes. In this way, the mechanical signal (touch) can be transduced to an electrical signal.

To travel from a sensory cell to a response cell, electrical signals are propagated in a characteristic form called an **action potential.** In animals, action potentials carry signals along the plasma membrane of nerve cells (see Chapter 46). In a Venus flytrap, action potentials generated by sensory hairs inside the trap race across the leaf at a rate of about 10 cm/sec.

How do plant action potentials move across tissues, when those tissues lack nerve cells? Recall that most plant cells are connected by membrane-lined **plasmodesmata** (see Chapter 37). Action potentials move more slowly in plants than in animals, but they can still carry information long distances in a hurry.

When the action potentials in a Venus flytrap reach cells on the outer surface of the trap, the cells change shape and push the

FIGURE 40.16 Venus Flytraps Close in Response to Action Potentials Generated by Sensory Hairs.

trap shut (**FIGURE 40.16**). The response is rapid enough to resemble the way an animal's muscle contracts in response to an action potential.

Next, let's explore how various signaling molecules control a plant's growth throughout its life cycle.

40.6 Youth, Maturity, and Aging: The Growth Responses

Plants grow throughout their lives, from the time they germinate until the time they die. As the plant body grows, it matures into an efficient machine for absorbing sunlight, water, nutrients, and other diffuse resources. But growth is not constant. It speeds up when water, light, and nutrients are abundant and slows or stops when conditions are poor, or when it is time for leaves to drop or fruits to ripen.

Controlling growth in response to changes in age or environmental conditions is one of the most important aspects of information processing in plants. Hormones play key roles in regulating growth.

To explore how plants grow in response to changing conditions, let's consider six of the best-studied hormones involved in growth responses, which are appropriately referred to as plant growth regulators. We'll start with auxin—the molecule responsible for phototropism and gravitropism.

Auxin and Apical Dominance

When **apical dominance** occurs, growth is restricted to the main stems; the lateral buds in the axils of each leaf remain dormant. But if the apical bud dies or is eaten by an herbivore, the dormancy of the lateral buds is broken and lateral branches begin to grow. **FIGURE 40.17** shows this phenomenon in action.

What caused the change? Because auxin is produced in shoot tips, researchers suspected that it might have a role in apical dominance as well as phototropism. This hypothesis was

FIGURE 40.15 Thigmotropism Is Movement in Response to Touch. Portions of grape plants called tendrils wind around support structures after contacting them. The attachment provided by the tendrils allows plants to climb.

(a) Apical meristem intact

(b) Apical meristem cut off

Dormant lateral buds

Lateral shoots

FIGURE 40.17 When Apical Dominance Occurs, Growth of Lateral Buds Is Suppressed. (a) The stem of a coleus plant is shown still intact. **(b)** The same plant, several weeks after the stem was cut. The lateral shoots will orient themselves vertically.

confirmed when it was shown that apical dominance could be sustained by applying auxin to a shoot's cut surface after its tip had been removed.

Auxin's role in apical dominance suggests that tip cells send a constant stream of information down to other organs and tissues. A stop in the signal means that apical growth has been interrupted. In response, lateral branches sprout and begin to take over for the main shoot. Now the question is, How does this signal move?

Polar Transport of Auxin   Auxin transport is **polar,** or unidirectional. If radioactively labeled auxin is added to the top end of a stem segment, the hormone emerges from the basal end. But if labeled auxin is added to the basal end of the segment, it is not transported toward the top. Auxin is the only plant hormone known to be transported through individual cells in one direction only.

Studies with labeled auxin have also shown that the hormone is transported all the way down the stem through the root, via parenchyma cells in the ground tissue and vascular tissue. Auxin enters the apical end of cells via a specialized membrane protein, diffuses to the other end of the cell, and then is transported out by carrier proteins located only in the basal portion of the plasma membrane. Labeled auxin moves from cell to cell at about 10 cm/hr—approximately 10 times more slowly than substances traveling in the phloem or xylem.

Because enzymes break down some auxin molecules as they travel down the long axis of the plant, polar transport sets up a strong gradient in auxin concentration. Auxin concentrations are much higher in shoots than they are in roots.

What Is Auxin's Overall Role?   Auxin clearly plays a key role in controlling growth via apical dominance, phototropism, and gravitropism. But this chemical messenger has other important effects as well:

- Auxin produced by seeds within the fruit promotes fruit development.

- Falling auxin concentrations are involved in the **abscission,** or shedding of leaves and fruits, associated with the genetically programmed aging process called senescence.

- The presence of auxin in growing roots and shoots is essential not only for the proper differentiation of xylem and phloem cells in vascular tissue but also for the development of vascular cambium (see Chapter 37).

- Auxin stimulates the development of adventitious roots in tissue cultures and cuttings.

Auxin has so many different effects on plants that it has been difficult for biologists to understand its overall role. Recently, several investigators have proposed that auxin's overall function is to signal where cells are in the plant body. The idea is that auxin concentration identifies where a cell is located relative to the root–shoot axis; a high concentration would inform the cell that it was near the tip of the shoot.

If conditions relating to the root–shoot axis change—for instance, a windstorm tips the plant or a deer eats the shoot apex—changes in auxin concentration effectively signal how the individual's tissues should respond. Phototropism, gravitropism, apical dominance, and the production of adventitious roots are all ways of coping with changes along the root–shoot axis of the plant.

Cytokinins and Cell Division

Cytokinins are a group of plant hormones that promote cell division. (*Cyto* is the Greek root for cell; *kinin* refers to kinesis, meaning movement or division.)

The Discovery of Cytokinins   When biologists were first attempting to grow plant cells and embryos in culture, they found that coconut milk, which stores nutrients used by growing coconut embryos, promoted cell division. This was the first hint that certain molecules can promote cell division in plants. Later experiments showed that molecules derived from the nitrogenous base adenine also stimulate the growth of cells in culture.

Eventually, naturally occurring adenine derivatives that stimulate growth were discovered in corn and apples, and were named cytokinins. Zeatin, the cytokinin found in the most species, is derived from adenine.

Cytokinins are synthesized in root tips, young fruits, seeds, growing buds, and other developing organs. But most of the zeatin and other cytokinins that are active in plants are synthesized in the apical meristems of roots and transported up into the shoot system via the xylem. Biologists still add cytokinins to plant cells growing in culture, to stimulate cell division (see **BioSkills 12** in Appendix B).

How Do Cytokinins Promote Cell Division?   After years of searching, biologists have now isolated and characterized a group of closely related proteins that act as cytokinin receptors. When cytokinins bind to these receptors in the plasma membranes of target cells, the receptors activate genes that regulate cell division.

Recent research on cytokinins has explored whether they affect molecules that regulate the cell cycle, including the cyclins and the cyclin-dependent kinases (Cdks; see Chapter 12). Recall that activated cyclins and Cdks allow cells to progress through checkpoints in the cell cycle and continue dividing.

To assess whether cytokinins affect cell-cycle genes, researchers grew *Arabidopsis* cells in culture so the nutrients and other molecules available could be carefully controlled. The biologists starved the cells of cytokinins for a day and then added the hormones again to half of the cells. When they assessed the level of mRNA from a cyclin gene called *CycD3*, they documented significant increases in the cells that were treated with cytokinins compared with the level in untreated cells.

This is strong evidence that cytokinins regulate growth by activating genes that keep the cell cycle going. In the absence of cytokinins, cells arrest at the G_2 checkpoint in the cell cycle and stop dividing (**FIGURE 40.18**).

Gibberellins and ABA: Growth and Dormancy

In high latitudes and at high elevations, most seeds and mature plants start growing in spring. Conditions for growth are good at that time of year, because temperatures are warming and soil moisture levels are usually at their peak. Seedlings and mature plants continue to grow throughout the summer and early fall if moisture and nutrients are still available.

During drought conditions, however, growth stops. Growth also stops in the embryos inside seeds. Embryos begin to develop as seeds mature, but cease this initial growth and remain dormant throughout the cold winter months. **Dormancy** is a temporary state of reduced metabolic activity.

Which signals initiate growth in response to changing environmental conditions, and which signals stop it? Two hormones provide the answer. **Gibberellins,** a large family of closely related compounds, stimulate growth in plants. **Abscisic acid,** commonly abbreviated **ABA,** inhibits growth. In at least some cases, the two hormones interact like start and stop signals.

The Discovery of Gibberellins Over 100 years ago, Japanese farmers noticed that some of their rice seedlings grew exceptionally quickly but fell over before they could be harvested. Biologists found that the plants with this "foolish seedling disease" were infected with the fungus *Gibberella fujikuroi.*

Researchers confirmed a causal connection between *Gibberella* infection and rapid stem elongation when they treated rice seedlings with an extract from the fungus. As predicted, the treated seedlings produced abnormally long shoots.

The active component in the extract was eventually isolated and named gibberellic acid (GA), which is a gibberellin. Follow-up research showed that rice plants produce their own gibberellin but respond to applications of additional hormone by elongating their stems. In effect, the infected rice seedlings were suffering from a gibberellin overdose. Fungi that manipulate their host

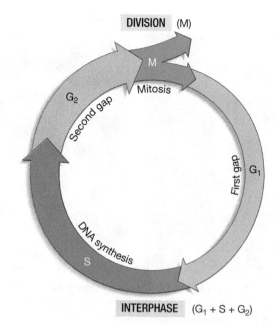

FIGURE 40.18 Cytokinins Affect the Cell Cycle.

plants in this way release their spores higher off the ground, so they are likely to blow farther away from their parent.

Gibberellins are found in a wide array of fungi and plants. Most plant species produce several different gibberellins that are active hormones.

Even though gibberellins have dramatic effects on growth, they are present in vanishingly small concentrations. In growing stems and leaves, active forms of gibberellin may be present in concentrations of about 10 nanograms per gram of tissue.

Defective Gibberellin Genes Cause Dwarfing To find the genes that are responsible for producing gibberellins, biologists analyzed mutant plants with abnormal stem length (**FIGURE 40.19**). Recall that Gregor Mendel analyzed the transmission of two alleles at a single gene that affected stem height in garden peas (see Chapter 14). One allele was associated with tall stems; the other was associated with dwarfed growth. The tall allele was dominant to the dwarf allele.

The gene responsible for the stem-length differences in garden peas is known as *Le* (for *le*ngth). Early work on dwarf mutants showed that they attain normal height if they are treated with the gibberellin called GA_1. This observation suggested that dwarf peas can respond to gibberellins normally—meaning that the problem is not with a hormone receptor.

In follow-up experiments, biologists treated dwarf peas with a radioactively labeled molecule used in the synthesis of GA_1. These plants did not produce radioactively labeled GA_1, even though plants with the normal allele did. Based on these results, researchers became convinced that the *Le* locus encodes an enzyme involved in GA synthesis.

Investigators recently identified the gene at the *Le* locus and confirmed that in dwarf plants, it contains a mutation that renders the enzyme inactive. Over 100 years after Mendel did his

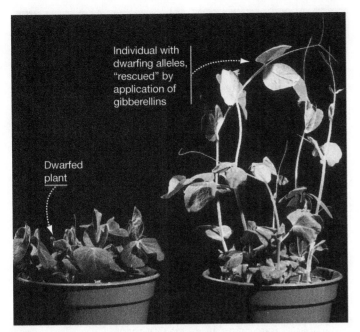

FIGURE 40.19 Dwarfed Individuals May Have Mutations That Affect Gibberellins. Dwarfed plants have much shorter stems than do normal individuals of the same age. By analyzing dwarfed individuals, biologists were able to identify genes involved in gibberellin synthesis.

✔ **QUESTION** If dwarfed individuals receive the same amount of sun and thus perform as much photosynthesis as taller individuals, they often produce more flowers and seeds. Why?

experiments, biologists finally understood the molecular basis of the dwarfing phenotype he studied.

In stems, gibberellins appear to promote both cell elongation and rates of cell division. But it is well established that auxin also promotes cell elongation, and that cytokinins also promote cell division. Research continues on how GAs, cytokinins, and auxin interact on the molecular level to control plant growth and development.

Gibberellins and ABA Interact during Seed Dormancy and Germination Many plants produce seeds that have to undergo a period of drying or a period of cold, wet conditions before they are able to germinate. A requirement for drying ensures that mature seeds will not sprout on the parent plant; an obligatory cold period prevents seeds from germinating just before the onset of winter, when young seedlings would be more likely to perish. Some seeds also have to receive a dose of red light, which indicates that they are in a sunny location.

In essence, then, seeds have an "off" setting that discourages germination and an "on" setting that initiates growth. The appropriate state for the on/off switch is determined by environmental cues such as temperature, moisture, and light.

By applying hormones to seeds, researchers learned that in many plants, ABA is the signal that inhibits seed germination and gibberellins are the signal that triggers germination. To understand how these messengers interact, researchers have studied a specific event: the production of an enzyme called α-amylase in germinating oat or barley seeds.

α-Amylase acts as a digestive enzyme that breaks the bonds between the sugar subunits of starch. (Your saliva contains an amylase that acts on the starch in food. In humans and other mammals, this enzyme initiates carbohydrate digestion in the mouth.)

FIGURE 40.20 shows that, during the germination of a barley seedling, α-amylase is released from a tissue called the aleurone layer. The enzyme diffuses into the starch-filled storage tissue in the seed called endosperm, which is dead when seeds germinate. Digestion of starch releases sugars that can be used by the growing embryo. Adding GA to the aleurone layer increases the production and release of α-amylase; adding ABA to that layer decreases α-amylase levels.

Research on the molecular interaction between GA and ABA carries several important messages:

- A cell's response to a hormone often occurs because specific genes are turned on or off.

- Hormones don't act on genes directly. Instead, a receptor on the surface of a cell or in the cytosol receives the message and

---

PROCESS: GIBBERELLINS (GAs) ACTIVATE PRODUCTION OF α-AMYLASE

1. The seed absorbs water, and germination begins.

2. Gibberellins diffuse from the embryo to the aleurone layer.

3. Cells in the aleurone layer respond by releasing digestive enzymes, such as α-amylase.

4. The enzymes digest starch, releasing sugars and other molecules to the growing plant.

FIGURE 40.20 The Molecular Mechanism of Gibberellin Action in Seed Germination.

responds by initiating a signal transduction cascade, which activates specific gene regulatory proteins—the activators and repressors of transcription (see Chapter 19).

- Different hormones interact at the molecular level because they induce different gene regulatory proteins, which increase or decrease expression of key genes.

The logic runs as follows: Different hormones trigger the production of different regulatory transcription factors. Hormone concentration affects the amount of each transcription factor produced. Changes in transcription factors are responsible for changes in gene expression.

GA's role in seed germination has important commercial applications. For example, in the production of beer, brewers routinely use gibberellins in the malting process—the conversion of starches stored in barley seeds to sugars. The sugars that are released in response to GA treatment support fermentation by yeast and provide flavor and alcohol in the finished beer.

ABA Closes Guard Cells in Stomata One of the major challenges faced by land plants is replacing water that is lost to the atmosphere when **stomata** are open (see Chapter 38). Because stomata open in response to blue light, they allow gas exchange to occur while the plant is receiving the wavelengths of light used in photosynthesis.

However, if plant roots are unable to obtain enough water to replace what is lost from the leaves, stomata close. Closing stomata is an adaptive response when roots cannot find adequate water, because continued transpiration would lead to wilting and potential tissue damage.

Early work on the mechanism of stomatal closing suggested that ABA is involved. For example, applying ABA to the exterior of stomata causes them to close within seconds.

To explore the hypothesis that a hormone regulates stomatal closing, researchers performed the experiments summarized in **FIGURE 40.21**. The fundamental idea was to grow plants whose roots had been divided. Only one side of the experimental plants was watered, while both sides of the control plants were watered. During this treatment, investigators documented that the water potential of the leaves remained the same in both control and experimental plants. Yet the stomata of experimental plants began to close. This result suggested that roots from the dry side of the pot were signaling drought stress, even though the leaves were not actually experiencing a water shortage.

Follow-up experiments have supported two important predictions: ABA concentrations in roots on the dry side of the pot are extraordinarily high relative to the watered side, and ABA concentrations in the leaves of experimental plants are much higher than in the leaves of control plants.

These results suggest that ABA from roots is transported to leaves and that it serves as an early warning of drought stress. In doing so, ABA overrides the signal from the blue-light photoreceptors introduced earlier in this chapter and causes stomata to close to conserve water.

RESEARCH

QUESTION: Can roots communicate with shoots?

HYPOTHESIS: Roots that are dry can signal shoots to close stomata.
NULL HYPOTHESIS: Roots cannot communicate with the shoot.

EXPERIMENTAL SETUP:

1. Divide roots of many plants into two sides.

2. In experimental group, water one side.

3. In control group, water both sides.

4. In both groups, measure water potential of leaves and observe stomata.

PREDICTION: Stomata in experimental plants will close; stomata in control plants will stay open.

PREDICTION OF NULL HYPOTHESIS: Stomata in both experimental and control plants will stay open.

RESULTS: No difference between experimental and control plants in water potential of leaves.

Stomata began to close.

Stomata stay open.

Experimental group   Control group

CONCLUSION: Roots can communicate with shoots. Dry roots signal the shoot and cause stomata to close, even though leaves are receiving sufficient water (from roots on the wet side of the plant).

FIGURE 40.21 Experimental Evidence that Roots Produce an "It's Too Dry" Signal.

SOURCE: Blackman, P. G., and W. J. Davies. 1985. Root to shoot communication in maize plants of the effects of soil drying. *Journal of Experimental Botany* 36: 39–48.

✓ **QUESTION** Why was it important to show that the water potential of leaves was the same in the two treatments?

The Molecular Mechanism of Guard-Cell Closure To understand how stomata open and close, recall that a stoma consists of two **guard cells** (see Chapter 37). When the vacuoles of guard cells are filled with water, the cells are turgid. The shape of turgid guard cells results in an open pore between the two cells, which allows gas exchange between the atmosphere and the interior of the leaf. But when vacuoles lose water and guard cells become flaccid and lack turgor, cell shape changes in a way that closes the pore and stops both gas exchange and loss of water via transpiration.

Based on these observations, pursuing the question of how stomata open and close is the same as asking how guard cells become turgid or flaccid—meaning, how water flows into or out of vacuoles. Activation of PHOT by blue light leads to water entry and stomatal opening; activation of ABA receptors leads to water exit and stomatal closing.

Remember that cells do not transport water directly. Instead, they change ion concentrations, creating osmotic gradients that result in water movement.

Considering this process, it shouldn't be surprising to learn that guard-cell opening and closing is based on changes in the activity of H^+-ATPases and ion transporters in the plasma membrane.

When photoreceptors are stimulated by blue light, large numbers of protons are pumped out of each guard cell. As **FIGURE 40.22a** shows, increased H^+-ATPase activity creates a strong electrochemical gradient that brings potassium and chloride ions into

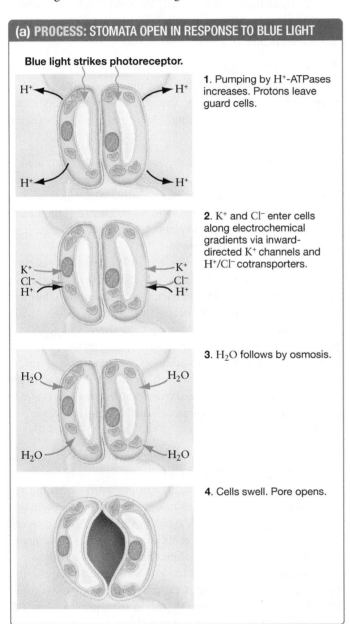

(a) PROCESS: STOMATA OPEN IN RESPONSE TO BLUE LIGHT

Blue light strikes photoreceptor.

H^+ H^+

H^+ H^+

1. Pumping by H^+-ATPases increases. Protons leave guard cells.

K^+ K^+
Cl^- Cl^-
H^+ H^+

2. K^+ and Cl^- enter cells along electrochemical gradients via inward-directed K^+ channels and H^+/Cl^- cotransporters.

H_2O H_2O

H_2O H_2O

3. H_2O follows by osmosis.

4. Cells swell. Pore opens.

(b) PROCESS: STOMATA CLOSE IN RESPONSE TO ABA

ABA binds to receptors on guard cells.

Cl^- Cl^-

Cl^- Cl^-

1. Pumping by H^+-ATPases stops. Outward-directed Cl^- channels open. Cl^- exits along electrochemical gradient.

K^+ K^+

K^+ K^+

2. Change in membrane potential opens outward-directed K^+ channels. K^+ exits along electrochemical gradient.

H_2O H_2O

H_2O H_2O

3. H_2O follows by osmosis.

4. Cells shrink. Pore closes.

FIGURE 40.22 Changes in Ion Flows Are Responsible for Opening and Closing Stomata. (a) Activated blue-light receptors trigger ion flows into guard cells. The cells swell when water follows by osmosis. **(b)** Activated ABA receptors trigger ion flows out of guard cells. The cells shrink when water follows by osmosis.

the interior of the guard cells. Water follows the incoming ions via osmosis. The upshot? The cells swell and pores open.

But as **FIGURE 40.22b** shows, guard cells respond to ABA in a very different way. When ABA reaches the guard cells, two things happen:

1. H^+-ATPases are inhibited.

2. Channels that allow chloride and other anions to leave along their electrochemical gradients are opened.

When the anions leave guard cells, the change in membrane potential causes outward-directed potassium channels to open. Large amounts of K^+ leave the cells, and water follows by osmosis. The result is a loss of turgor and closing of the pore.

Whether it acts on guard cells or seeds, ABA fulfills a general role in plants as a dormancy or "no-growth" signal. In many cases, its action depends on input from other hormones and photoreceptors. To survive and reproduce successfully, plants have to integrate information from a variety of sources.

Brassinosteroids and Body Size

When you went through puberty, you experienced a growth spurt triggered by surges in steroid hormones called testosterone and estradiol. In plants, growth spurts are triggered by surges in similar steroid hormones called **brassinosteroids.**

The name brassinosteroid was inspired by two observations:

1. The hormones were initially discovered in 1979 in *Brassica napus*—a crop plant that is the source of the canola oil you may use in cooking.

2. They are **steroids**—part of a family of lipid-soluble compounds (see Chapter 6).

Brassinosteroids promote growth and are a key regulator of overall body size in plants. In the model organism *Arabidopsis thaliana*, for example, mutant individuals that cannot synthesize brassinosteroids are extremely dwarfed (see **BioSkills 13**, see Appendix B).

Recent research has highlighted a fascinating difference between the brassinosteroids and the steroid hormones found in animals. Steroid hormones in animals act by entering cells, binding to receptors in the cytosol, and forming a hormone-receptor complex that enters the nucleus (see Chapter 49). This complex then binds to DNA and directly changes gene expression. Because steroids are lipid soluble, researchers were not surprised to find that testosterone and estradiol—animal hormones—cross the plasma membrane before binding to a receptor.

Brassinosteroids, in contrast, are more hydrophilic than animal hormones and never enter the cell. Instead, they bind to receptors on the plasma membrane and activate signal transduction events—probably phosphorylation cascades—that lead to changes in gene expression.

The genes for the synthesis of brassinosteroids appear to be homologous with the genes required for steroid hormone synthesis in animals, meaning that they are derived from a similar gene in the common ancestor of plants and animals. So how do brassinosteroids interact with auxin, cytokinins, and gibberellins to regulate growth and body size? These are questions for future research.

Ethylene and Senescence

Senescence is a regulated process of aging and eventual death of an entire organism or organs such as fruits and leaves. Like most aspects of plant growth and development, senescence is regulated by complex interactions between several hormones in response to changes in temperature, light, and other factors.

The hormone most strongly associated with senescence is **ethylene.** Like other plant hormones, ethylene is simple in structure and active at small concentrations. Unlike other plant hormones, however, ethylene is a gas at normal temperatures.

Ethylene is synthesized from the amino acid methionine and is involved in the following senescence events:

1. fruit ripening, which eventually leads to the aging and rotting of fruit;

2. flowers fading; and

3. leaf abscission—meaning their detachment and fall.

In addition, ethylene influences plant growth, and it is a stress hormone induced by drought and other conditions. Ethylene regulates a surprisingly large range of physiological responses.

The Discovery of Ethylene Ethylene was initially discovered in ancient China, when fruit growers noticed that burning incense in closed rooms made pears ripen faster. Westerners made a similar observation in the late 1800s, when gas street lamps came into wide use in cities and the plants growing near leaky gas lines dropped their leaves prematurely.

Researchers showed that ethylene in lamp gas was the molecule responsible for the leaf loss; ethylene is likewise present in incense smoke. In the 1930s ethylene was also found in the gases that are released by ripe apples.

Subsequently, biologists documented sharp spikes in ethylene production during fruit ripening in tomatoes, bananas, and certain other species in addition to apples. Follow-up research on these species showed that ethylene induces (1) the production of some of the enzymes required for the ripening process, and (2) an increase in cellular respiration, which furnishes ATP.

Ethylene and Fruit Ripening During ripening, stored starch is converted to sugar, enhancing sweetness; protective toxins are removed or destroyed; cell walls are degraded, softening the fruit; chlorophyll is broken down; and pigments and aromas that signal ripeness are produced. Biologists interpret fruit ripening as an adaptation that enhances its attractiveness to birds, mammals, and other animals that disperse seeds to new locations.

Today, fruit sellers manipulate ethylene levels to control fruit ripening. For example, they treat green bananas with ethylene after the bananas have been shipped, to stimulate ripening (**FIGURE 40.23**). Conversely, apples are stored in warehouses with high concentrations of CO_2 and low concentrations of O_2, which inhibits ethylene production in the fruit. Apples stored under these conditions can be sold long after their original harvest date, when untreated fruits have rotted.

+ Ethylene

FIGURE 40.23 Ethylene Speeds Ripening and Other Aspects of Senescence. These bananas are identical, except that the bunch on the right was exposed to the plant hormone ethylene.

Ethylene and Leaf Abscission Ethylene's effects on leaf senescence and leaf abscission involve complex interactions with auxin and cytokinins. In addition to being synthesized in apical meristems, auxin is produced in healthy leaves. It is then transported from the leaf to the stem through the petiole. In response to age or to changes in ambient temperature or day length, leaves sometimes produce less auxin (**FIGURE 40.24**). As a result, cells in a region of the leaf petiole called the **abscission zone** become more sensitive to ethylene in the tissue.

Increased ethylene sensitivity activates enzymes that weaken the walls of cells in the abscission zone. At the same time, chlorophyll in the leaf is broken down, and nutrients are withdrawn and stored in parenchyma cells in the stem. Eventually the cell walls at the base of the petiole degrade enough that the leaf falls.

Applications of cytokinins, in contrast, reverse these effects and dramatically extend the life span of leaves. As a result, ethylene and cytokinins are thought to have opposite effects on at least some of the processes involved in senescence.

An Overview of Plant Growth Regulators

Understanding how different signals interact is an exciting frontier in research on plant hormones that are growth regulators. Knowledge of the environmental signals that change hormone levels and the genes that they regulate is rapidly improving. In contrast, a great deal remains to be learned about how the information is carried by signal transduction pathways, how those pathways interact, and how plant cells respond.

check your understanding

(C)(Y)(U)

If you understand that . . .

- Auxin's primary role is to signal the position of cells along the long axis of the plant body. This is possible because auxin is produced in developing leaves and undergoes polar transport to the roots, forming a concentration gradient.
- GAs are general signals to initiate or continue growth.
- ABA is a general signal to stop growth or remain dormant.
- Ethylene is a signal that controls senescence.
- Brassinosteroids promote large body size.
- GAs and ABA interact at the molecular level to control seed germination and dormancy.
- Signals from blue light and ABA interact to control the opening and closing of stomata.

✔ **You should be able to . . .**

1. Predict the effects of watering a large number of daisy plants, each in a pot in a greenhouse, with water that contains either ABA or GAs.

2. Write a hypothesis to explain the adage "one rotten apple spoils the whole bushel."

Answers are available in Appendix A.

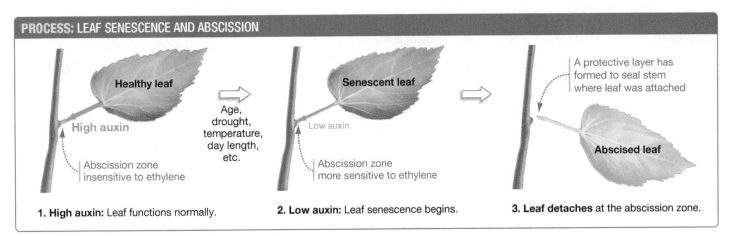

PROCESS: LEAF SENESCENCE AND ABSCISSION

Healthy leaf

High auxin

Abscission zone insensitive to ethylene

Age, drought, temperature, day length, etc.

Senescent leaf

Low auxin

Abscission zone more sensitive to ethylene

A protective layer has formed to seal stem where leaf was attached

Abscised leaf

1. High auxin: Leaf functions normally.

2. Low auxin: Leaf senescence begins.

3. Leaf detaches at the abscission zone.

FIGURE 40.24 Leaves Drop in Response to Signals from Auxin and Ethylene. Older leaves produce much less auxin than young leaves do, leading to an increase in ethylene sensitivity and leaf abscission.

TABLE 40.2 provides notes on the structure and function of hormones discussed in this section. As you study this table, two key observations should emerge:

1. It is common for a single hormone to affect many different target tissues. This means that there can be an array of responses to the same cell–cell signal. To interpret this pattern, biologists point out that hormones may carry a common message to a variety of tissues and organs. Auxin can define the root–shoot axis of the body; gibberellins trigger stem growth; cytokinins promote cell division; ABA slows or prevents growth; ethylene signals senescence; and brassinosteroids increase overall mass.

2. In most cases, several different hormones can affect the same response. Stated another way, hormones do not work independently—they interact with each other. To make sense of this pattern, biologists point out that individual hormones tend to be produced in response to an environmental cue at a certain location, such as water availability at root tips. Many environmental cues may be changing at the same time, however. For plants to respond appropriately, they need to integrate information from various environmental cues perceived at various locations in the body.

Cross-talk is the interaction between the signal transduction cascades triggered by different hormones (see Chapter 11). The

SUMMARY TABLE 40.2 **Plant Growth Regulators**

Hormone	Notes	Chemical Structure
Auxin	• Involved in cell elongation and apical dominance • Promotes differentiation of xylem and phloem • Helps to define long axis of body (phototropism and gravitropism responses) • First plant hormone isolated and characterized • Produced in shoot apical meristems and young leaves	
Cytokinins	• Promote cell division in the presence of auxin • Promote chloroplast development and break lateral bud dormancy • Delay senescence (aging) • Produced in root apical meristems, many other tissues	
Gibberellins (GAs)	• Promote stem growth via both cell elongation and division • Promote seed germination • Produced in apical meristems, immature seeds, and anthers (pollen-producing organs)	
Abscisic Acid (ABA)	• Inhibits bud growth and seed germination • Induces closure of stomata in response to water stress • Acts as a stress hormone analogous to cortisol in humans • Produced in almost all cells	
Brassinosteroids	• Promote cell elongation in stems and leaves • Structurally related to steroid hormones in animals • Act on receptors at cell surface • Produced in almost all tissues	
Ethylene	• Exists in gas form • Involved in fruit ripening • Induces senescence of fruits, flowers and leaves • Produced in all organs when plants are under stress	

key insight to remember is that signaling systems form communication networks. Cross-talk is the molecular mechanism responsible for integrating information from many sensory cells and signals.

The complex interactions among hormones involved in the growth response have a purpose: allowing individuals to survive and thrive long enough to reproduce. The same can be said for the hormones involved in protecting plants from danger.

40.7 Pathogens and Herbivores: The Defense Responses

Plants cannot run away from danger. Instead, they have to stand and fight.

Like humans and other animals, plants are constantly threatened by an array of disease-causing viruses, bacteria, and parasitic fungi. In addition, plant roots are susceptible to attacks by nematodes—the soil-dwelling roundworms (see Chapter 34).

Disease-causing agents are termed **pathogens.** If plants were not able to sense attacks by pathogens and respond to them quickly and effectively, the landscape would be littered with dead and dying vegetation.

The waxy cuticle that covers epidermal cells is an effective barrier to most viruses, bacteria, fungi, and other disease-causing agents, and the structures called thorns, spines, and trichomes help protect leaves and stems from damage by herbivores (see Chapter 37).

In addition, many plants lace their tissues with **secondary metabolites**—molecules that are closely related to compounds in key synthetic pathways but are not found in all plants. Some secondary metabolites function in plant defense by poisoning herbivores.

- The flavorful oils in peppermint, lemon, basil, and sage have insect-repellent properties.

- The pitch that oozes from pines and firs contains a molecule called pinene, which is toxic to bark beetles.

- The pyrethroids produced by *Chrysanthemum* plants are a common ingredient in commercial insecticides.

- Molecules called tannins are found in a wide array of plant species; when they are ingested by animals, tannins bind to digestive enzymes and make the herbivores sick.

- Compounds like opium, caffeine, cocaine, nicotine, and tetrahydrocannabinol (THC) disrupt the nervous systems of plant-eating insects and vertebrates.

Although these defenses are effective, they are also expensive to produce in terms of the ATP and materials invested. It is not surprising that plants often produce defenses or increase their existing defenses only in direct response to attacks by pathogens or herbivores.

Responses to attacks in both plants and animals are called **inducible defenses,** because they are induced by the presence of a threat. Let's first consider how plants sense and respond to viruses and other pathogens and then explore what they do when attacked by insects and other herbivores.

How Do Plants Sense and Respond to Pathogens?

Like animals, plants have remarkably swift and diverse defense mechanisms to resist attacks by viruses, bacteria, and fungi. Plants use a wide variety of sensory proteins to detect invading pathogens and use signaling pathways to carry out their defense. Successful pathogens, in turn, have evolved elaborate mechanisms to evade these defenses.

An Evolutionary Arms Race Most plant pathogens are very specific, infecting only one or a few host species. This is in part due to the long history of coevolution of the pathogen and its host (see Chapter 55).

If a pathogen evolves a mechanism to exploit its host more effectively, the host may evolve new abilities to detect the pathogen and defend itself from attack. In response, the pathogen may then evolve an alternate strategy to evade detection by the host.

The complex interactions that have resulted from this back-and-forth "arms race" have proved challenging for biologists to understand. The human food supply is estimated to be reduced 50 percent worldwide because of plant disease. In the mid-1800s, Ireland experienced a potato famine in which millions of people faced starvation because of a potato pathogen. Pathogens can also be devastating to native species. The once-dominant American chestnut tree was nearly wiped out by a fungal pathogen in the early 1900s.

To cause disease, plant pathogens must first get past the surface of a host plant and begin growing in the apoplast, outside the plasma membranes of cells. Most pathogens enter through stomata or wounds. Once inside the plant, some pathogens kill host cells and feed on the cell debris. More commonly, a pathogen feeds on nutrients from a host cell while it is still alive. How do the plant cells respond to an attack in progress?

The Hypersensitive Response Plant cells contain a variety of receptor proteins that can bind to molecules derived from pathogen cells. When contact is made, these 'surveillance' proteins signal to the plant cell that a pathogen is present, which triggers a cascade of signals that lead to

1. stomatal closure, preventing more pathogen cells from entering the plant;

2. production of toxin molecules targeted to the pathogen;

3. reinforcement of the neighboring plant cell walls to limit movement of the pathogen; and

4. rapid suicide of cells in the infected region.

These events take place within hours of the start of an infection and are termed the **hypersensitive response (HR).** The HR results in small brown spots in leaves—those cells dead by

FIGURE 40.25 The Hypersensitive Response Protects Plants from Pathogens. This red maple leaf is riddled with evidence of past infections.

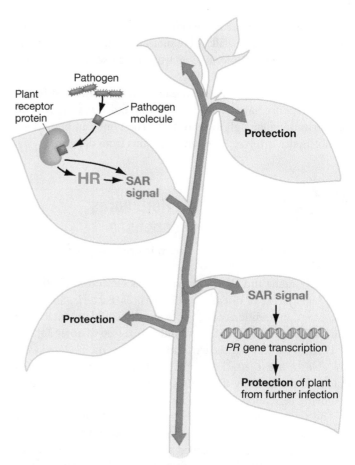

FIGURE 40.26 The Hypersensitive Response Produces a Signal That Induces Systemic Acquired Resistance (SAR). This diagram summarizes the current consensus on how the HR and SAR interact.

suicide—that accumulate over the growing season (FIGURE 40.25). By the end of the growing season, it is rare to find leaves lacking these brown spots. Plants that can detect an infection quickly and induce an HR remain healthy and show no further signs of disease. The HR is similar to the cell-mediated immune response in mammals, which leads to the death of infected cells (see Chapter 51).

✔ If you understand this concept, you should be able to explain the adaptive value of the hypersensitive response despite the death of plant cells.

In the 1990s biologists discovered that a mutant of *Pseudomonas*, a bacterial pathogen, could infect plant cells if it was artificially introduced inside leaves, but the pathogen was unable to cause disease if applied to the outside of leaves. This mutant could not produce a small molecule called coronatine, but the role coronatine played in infection was not clear.

Plants normally respond to *Pseudomonas* by closing their stomata within an hour of detecting the pathogen, effectively slamming the door on future invasions. The normal bacteria, however, are able to interfere with this defense by causing the stomata to re-open several hours later, allowing more bacteria to enter the leaf. In 2006, Maeli Melotto and coworkers discovered that coronatine is the molecule that causes the stomata to re-open. Without the ability to make coronatine, the mutant bacteria couldn't re-open the stomata and were at a clear disadvantage.

Plant defenses are based on good communication not only from cell to cell, as in the HR, but also throughout the plant and even to neighboring plants.

An Alarm Hormone Extends the HR Once the HR is under way in a localized area of infection, a hormone produced at the infection site travels throughout the body and triggers a slower and more widespread set of events called **systemic acquired resistance (SAR).** Over the course of several days, SAR primes cells throughout the root or shoot system for resistance to assault by a pathogen—even cells that have not been directly exposed to the disease-causing agent.

FIGURE 40.26 illustrates how the HR and SAR are thought to work together. In addition to triggering the HR, detection of a pathogen leads to the release of a hormone that initiates SAR. This signal acts globally as well as locally—that is, at the point of infection—and results in the expression of a large suite of genes called the pathogenesis-related (*PR*) genes.

When biologists set out to locate the hormone responsible for SAR, they found that levels of **methyl salicylate (MeSA)**—a molecule derived from salicylic acid—increase dramatically after tissues are infected with a pathogen. Follow-up work revealed these findings:

- Phloem sap leaving infected sites has elevated levels of MeSA.

- Treatments that reduce MeSA reduce or abolish SAR.

- Adding MeSA to the lower leaves of tobacco plants leads to SAR in the upper, untreated leaves.

Biologists are working to understand these complex interactions in order to genetically engineer crop plants that can resist diseases. Despite these interventions, however, pathogens continue to evolve, in a never-ending arms race.

How Do Plants Sense and Respond to Herbivore Attack?

Over a million species of insects have been discovered and named so far. Most of them make their living by eating leaves, stems, phloem sap, seeds, roots, or pollen. Plants have effective induced defenses in response to pathogens like viruses, bacteria, and fungi. But can they ramp up defenses in response to insect attack?

The Role of Proteinase Inhibitors When researchers started studying why some plant tissues are more palatable and digestible than others, biochemists discovered that many seeds and some storage organs, such as potato tubers, contain proteins called **proteinase inhibitors.**

Proteinase inhibitors block the enzymes—found in the mouths and stomachs of animals—that are responsible for digesting proteins. When an insect or a mammalian herbivore ingests a large dose of a proteinase inhibitor, the herbivore gets sick. As a result, herbivores learn to detect proteinase inhibitors by taste and avoid plant tissues containing high concentrations of these molecules.

Although many plant tissues contain proteinase inhibitors in low concentrations, biologists wanted to test the hypothesis that these proteins might also be part of an induced defense by the plant. To evaluate this idea, researchers allowed herbivorous beetles to attack one leaf on each of several potato plants.

- In liquid extracted from the other leaves on the attacked plants, proteinase inhibitor concentrations averaged 336 µg per mL of leaf juice.

- In control plants, where no insect damage had occurred, proteinase inhibitor levels averaged just 103 µg per mL.

This result supported the hypothesis that a hormone produced by wounded cells travels to undamaged tissues and induces the production of proteinase inhibitors.

The Discovery of Systemin Biologists isolated the wound-response hormone by purifying the compounds found in tomato leaves and testing them for the ability to induce the synthesis of proteinase inhibitors. The hormone that is active in tomato plants and closely related species turned out to be **systemin,** a polypeptide just 18 amino acids long. Systemin was the first peptide hormone ever described in plants.

Researchers who labeled copies of systemin with a radioactive carbon atom, injected the hormone into plants, and then monitored its location confirmed that systemin moves from damaged tissues to undamaged tissues.

Currently, work on the production of systemin and proteinase inhibitors focuses on determining each step in the signal

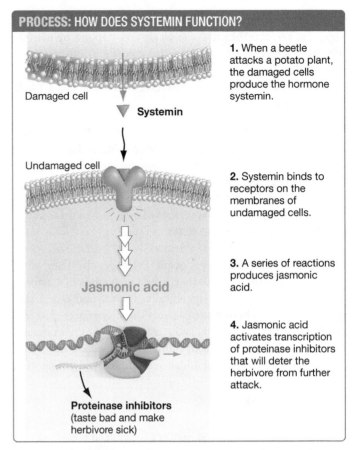

PROCESS: HOW DOES SYSTEMIN FUNCTION?

Damaged cell

Systemin

Undamaged cell

Jasmonic acid

Proteinase inhibitors (taste bad and make herbivore sick)

1. When a beetle attacks a potato plant, the damaged cells produce the hormone systemin.

2. Systemin binds to receptors on the membranes of undamaged cells.

3. A series of reactions produces jasmonic acid.

4. Jasmonic acid activates transcription of proteinase inhibitors that will deter the herbivore from further attack.

FIGURE 40.27 Signals from Insect-Damaged Cells Prepare Other Cells for Attack. Systemin, a hormone produced by herbivore-damaged cells, initiates a protective response in undamaged cells.

transduction pathway that alerts undamaged cells to danger (**FIGURE 40.27**). The data have yielded the following results:

1. Systemin is released from damaged cells.

2. Systemin travels through the body via phloem and binds to membrane receptors on target cells.

3. The activated receptor triggers a series of chemical reactions that eventually synthesize a molecule called jasmonic acid.

4. Jasmonic acid activates the production of at least 15 new gene products, including proteinase inhibitors.

In this way, plants build potent concentrations of insecticides in tissues that are in imminent danger of attack.

"Talking Trees": Responses from Nearby Plants When an insect starts munching on a leaf, volatile compounds evaporate from the leaf's surface and travel through the air. In the 1980s, researchers began to suspect that plants growing near other plants under herbivore attack "eavesdrop" on these volatiles. In response, they increase their own defenses—even though they've yet to be attacked.

The "talking trees" hypothesis has received extensive support, primarily through experiments based on exposing plants to volatiles in the absence of any insects. Even if the volatiles come from an entirely different species, research has shown that plants can sense the presence of these chemicals and respond by increasing the production of proteinase inhibitors and other defense compounds.

In some cases, plants that are under attack even call for help from other organisms.

Pheromones Released from Plant Wounds Recruit Help from Wasps Caterpillars and other herbivorous insects have enemies of their own—often wasps that lay their eggs in the insects' bodies. When wasp eggs hatch inside a caterpillar, the wasp larvae begin eating the host from the inside out. An organism that is free living as an adult but parasitic as a larva, such as these wasps, is called a **parasitoid** (FIGURE 40.28).

Biologists observed that parasitoids are particularly common when insect outbreaks occur in croplands. They wondered whether wounded plants release compounds that actively recruit parasitoids. More specifically, they hypothesized that plants produce **pheromones**—chemical messengers that are synthesized by an individual and released into the environment, and that elicit a response from a different individual. Hormones act on cells inside an individual; pheromones act on another individual.

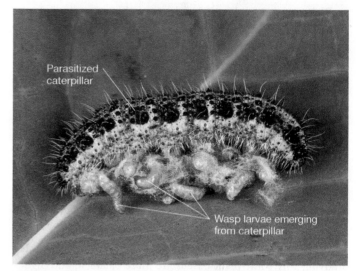

Parasitized caterpillar

Wasp larvae emerging from caterpillar

FIGURE 40.28 Parasitoids Kill Herbivores. When this plant was first attacked by this caterpillar, it produced pheromones that attracted a female wasp. The wasp laid her eggs in the caterpillar. As the wasp larvae grew and developed into pupae, they devoured the caterpillar.

SUMMARY TABLE 40.3 **Selected Sensory Systems in Plants**

Stimulus	Receptor	Signal Transduction	Response	Adaptive Significance
Blue light	PHOT1 and other phototropins in stem and leaves	PHOT1 autophosphorylates; remainder of signal transduction systems unknown	Phototropism occurs; also involved in stomatal opening	Stems grow toward light with wavelengths needed for photosynthesis
Red light	Phytochrome in seeds and elsewhere	Phytochrome changes to P_{fr} form and activates responses	Seed germinates	Sunlight triggers germination
Far-red light	Phytochrome in stem and elsewhere	P_r moves into nuclei and induces genetic responses	Stems lengthen	Species that require full sunlight attempt to escape shade from leaves
Gravity	Proteins located in plasma membrane	Details unknown	Cells on opposite side of root or shoot elongate; tissue curves	Roots grow down; shoots grow up
Touch or wind	Stretch receptors; location unknown	Details unknown, but result is transcription activation in target genes	Stems grow shorter and thicker; in some cases vining (twining) growth	Individual is more resistant to damage; plants grow toward light
Touch	Receptor hair cell in Venus flytrap	Electrical changes in receptor cell's plasma membrane trigger action potentials	Target cells change shape; trap shuts	Plant can capture prey
Pathogens	Receptor protein products	Details unknown	Hypersensitive response (HR); death of infected cells	Pathogens starve, so infection is slowed or stopped
Herbivores	Unknown; activated in response to molecule from herbivore	Details unknown	Insecticide production; signals to parasitoids	Herbivores are sickened or killed

To explore this idea, researchers analyzed compounds that were released from corn seedlings during attacks by caterpillars. The insect-damaged leaves produced 11 molecules that were not produced by undamaged leaves. These compounds were not released by leaves that had been cut with scissors or crushed with a tool; only insect damage triggered their production.

To follow up on this result, the investigators put female wasps in an arena that contained leaves damaged by insects and leaves that had suffered only mechanical damage. In more than two-thirds of the tests that were performed, the wasps preferred to fly toward the insect-damaged leaves.

These results support the hypothesis that plants produce wasp attractants in response to attack by caterpillars. Biologists are increasingly convinced that plants can produce pheromones that recruit help in the form of egg-laden wasps.

From the research that has been done on plant sensory systems, it is abundantly clear that plants don't just sit there. These organisms may be stationary, but they constantly monitor and respond to a wide array of information about their environment (**TABLE 40.3**). Gaining a better understanding of phototropism, gravitropism, response to disease, and other aspects of plant behavior forms an exciting frontier in biology.

If you understand that . . .

- The presence of receptor protein products allows plants to recognize attacks by parasites such as viruses, bacteria, and fungi.
- When a receptor protein binds to a molecule derived from a pathogen, a signal transduction sequence results in the hypersensitive response (HR).
- When plants are attacked by insect herbivores, damaged tissues release signals that trigger the production of toxins in undamaged leaves and that recruit enemies of the insects.

✔ **You should be able to . . .**

1. Explain the fitness advantage of having a wide array of receptor proteins in a single individual.
2. Explain why plants do not maintain high proteinase inhibitor concentrations in their tissues at all times.

Answers are available in Appendix A.

CHAPTER 40   REVIEW

For media, go to MasteringBiology

If you understand . . .

40.1   Information Processing in Plants

- In most cases, information processing in plants starts when a receptor protein changes shape in response to a stimulus.

- When signal transduction occurs, an external signal is changed into an intracellular signal.

- In receptor cells, signal transduction culminates in the production of hormones that are transported throughout the plant body.

- If a hormone binds to a receptor on a target cell, signal transduction occurs and culminates in changes in gene expression, altered translation rates, or changes in the activity of specific membrane pumps, channels, or ion carriers.

✔ You should be able to explain the analogy between how a plant sensory cell works and the following events: A person sees a barn on fire and calls the rural fire department; the dispatcher rings a siren; in response, members of the volunteer fire department race to the station to get the pumper truck.

40.2   Blue Light: The Phototropic Response

- Phototropins are blue-light receptors that become phosphorylated and then initiate signal transduction pathways.

- In the phototropic response, cells near the tips of coleoptiles sense changes in blue light and respond by altering the distribution of the hormone auxin. Cells on one side of the coleoptile elongate in response to auxin much more than do cells on the other side of the coleoptile. In this way, plants bend toward sunlight.

✔ You should be able to predict whether the phototropic response differs in plants that require high-light conditions versus plants that thrive best in low-light conditions.

40.3   Red and Far-Red Light: Germination, Stem Elongation, and Flowering

- Phytochrome is a red/far-red switch that allows plants to detect shade from other plants and regulates a wide variety of plant growth responses including germination, stem elongation, and flowering.

- The phytochrome protein changes shape when it absorbs red light. The same protein changes to an alternate shape when it absorbs far-red light.

✔ You should be able to state a hypothesis to explain why animals do not have phytochromes (a red/far-red light switch).

40.4 Gravity: The Gravitropic Response

- Cells in the center of root caps contain amyloplasts with starch granules that function as statoliths and cause roots to grow down in response to gravity.

- Redistribution of auxin in response to gravity results in downward growth of roots and upward growth of shoots.

✓ You should be able to explain why starch granules make good gravity-sensing bodies.

40.5 How Do Plants Respond to Wind and Touch?

- Some plants respond to touch, including wind, by growing more slowly and becoming stocky.

- Thigmotropism is a direction response to touch, such as tendril coiling and closing of Venus flytraps.

✓ You should be able to describe the adaptive significance of the response to wind.

40.6 Youth, Maturity, and Aging: The Growth Responses

- Several key hormones regulate plant growth.

- Auxin establishes and maintains the long axis of the plant body, playing a key role in phototropism and gravitropism and maintaining apical dominance. All of these responses rely on the polar transport of auxin, which establishes a gradient in auxin from the plant's shoot tips to its root tips.

- Cytokinins promote cell division by regulating the cell cycle, and they delay senescence.

- Gibberellins (GAs) signal that conditions for growth are good and promote the initiation or continuation of growth and development.

- Abscisic acid (ABA) signals that environmental conditions are bad by suppressing growth, enforcing dormancy, and closing stomata.

- Regulation of dormancy and growth by ABA and by GAs are examples of how hormones interact—allowing plants to integrate information from several different stimuli and respond appropriately.

- Brassinosteroids are steroid hormones that regulate growth.

- Ethylene is a gaseous hormone that triggers senescence and fruit ripening, and it plays a role in leaf abscission.

✓ You should be able to suggest a hypothesis explaining why some hormones are not transported in a single direction.

40.7 Pathogens and Herbivores: The Defense Responses

- Plants have receptor proteins that detect pathogen infection and trigger the hypersensitive response, which limits the spread of disease.

- Herbivores trigger plant responses that lead to the synthesis of molecules that make plants less palatable.

- Localized infection or herbivory stimulates long-distance signaling and defense responses throughout the plant.

- Volatile signals triggered by herbivory attract parasitoids that attack the herbivores.

✓ You should be able to explain why cell suicide is beneficial to a plant.

(MB) **MasteringBiology**

1. **MasteringBiology Assignments**

   Tutorials and Activities Experimental Inquiry: What Effect Does Auxin Have on Coleoptile Growth?; Flowering Lab Leaf Abscission; Plant Defenses; Plant Hormones; Plant Responses to Light; Sensing Light

   Questions Reading Quizzes, Blue-Thread Questions, Test Bank

2. **eText** Read your book online, search, take notes, highlight text, and more.

3. **The Study Area** Practice Test, Cumulative Test, BioFlix® 3-D Animations, Videos, Activities, Audio Glossary, Word Study Tools, Art

You should be able to . . .

✓ TEST YOUR KNOWLEDGE

Answers are available in Appendix A

1. Which of the following statements about phytochrome is *not* correct?
   a. It is photoreversible.
   b. Its function was understood long before the protein itself was isolated.
   c. The P_{fr} form activates the responses to red light.
   d. It is involved in guard-cell opening.

2. If a plant is touched repeatedly over the course of many days or if it experiences long-term exposure to wind, what happens?
   a. Growth of the root system is accelerated, making the plant more stable.
   b. Large-scale changes in gene expression occur, resulting in shoots that are short and stout.
   c. Electrical signals cause leaves to fold up, avoiding damage.
   d. Continued mechanical stimulation indicates that the individual is threatened with destruction, so it initiates flowering in an attempt to reproduce before it dies.

3. Which of the following statements about hormones is correct?
   a. They tend to be large molecules.
   b. They exert their effects only on the same cells that produce them.
   c. They can exert strong effects only when they are present in high concentrations.
   d. They trigger a response by binding to receptors in target cells.

4. What evidence suggests that ABA from roots can signal guard cells to close?
   a. If roots are given sufficient water, guard cells close anyway.
   b. If roots are dry, guard cells begin to close—even though leaves are not experiencing water stress.
   c. Applying ABA on guard cells directly causes them to close.
   d. If roots are dry, ABA concentrations in leaf cells drop dramatically.

5. What is the rapid death of plant cells in response to a pathogen attack called?

6. Unlike animal steroid hormone receptors, where are brassinosteroid receptors located?

Answers are available in Appendix A

✔TEST YOUR UNDERSTANDING

7. Why was it logical to predict that amyloplasts function as statoliths?
   a. They are dense and settle to the bottom of gravity-sensing cells.
   b. They are present only in gravity-sensing cells.
   c. They make a direct physical connection with membrane proteins that have been shown to be the gravity-receptor molecule.
   d. Their density changes in response to gravity.

8. Phytochromes can be considered "shade detectors," while phototropins such as PHOT1 can be considered "sunlight detectors." Explain why these characterizations are valid.

9. What does transduce mean? Compare and contrast two types of transduction processes described for a sensory or a defense process.

10. A plant's response to a given hormone depends on the cells or tissues that receive the signal, the plant's developmental stage or age, the concentration of the hormone, and the concentration of other plant hormones that are present. Provide examples that support each of these claims.

11. Discuss the general role that ethylene serves in plants, and predict how produce dealers deal with benefits or costs of the effects of ethylene.

12. Suppose that a mutant plant is unable to make methyl salicylate. Explain why it is not likely to survive in the wild.

Answers are available in Appendix A

✔TEST YOUR PROBLEM-SOLVING SKILLS

13. To explore how hormones function, researchers have begun to transform plants with particular genes. In one experiment, a gene involved in cytokinin synthesis was introduced into tobacco plants. Which one of the following results would be expected?
    a. Individuals produced more lateral branches.
    b. Stems grew extremely tall and slender.
    c. Roots were incapable of responding to gravity.
    d. Stomata were closed most of the time.

14. In general, small seeds that have few food reserves must be exposed to red light before they will germinate. (Lettuce is an example.) In contrast, large seeds that have substantial food reserves typically do not depend on red light as a stimulus to trigger germination. State a hypothesis to explain these observations.

15. In many species native to tropical wet forests, seeds do not undergo a period of dormancy. Instead, they germinate immediately. Make a prediction about the role of ABA in these seeds. How would you test your prediction?

16. Researchers have shown that transforming plants with genes encoding resistance proteins makes these plants resistant to more pathogens. How might the pathogens evolve when exposed to these genetically altered plants?

41 Plant Reproduction

In this chapter you will learn that

| Reproduction is key to the success of flowering plants |

starting with ↓

| Reproductive systems: unifying principles **41.1** |

looking closer at ↓

| Reproductive structures **41.2** | —*in*→ | Flowering plants |

involved in ↓

| Pollination and fertilization **41.3** | —*using*→ | Animals and wind to move pollen from flower to flower |

resulting in ↓

| The seed (and embryo) **41.4** | —*which uses*→ | Animals, wind, and water to be dispersed |

This chapter focuses on the structure and function of plant reproductive structures like those of this flower.

t would be difficult to overemphasize the importance of the reproductive organs and processes that are analyzed in this chapter—for plants, for biologists, and for you.

- For plants, every structure in the body and every physiological process—from water transport to photosynthesis—exist for one reason: to maximize the chances that the individual will produce offspring. Reproduction is the unconscious goal of everything that an organism does (see Chapter 1).

- For biologists, plant reproduction is not only fundamental to understanding how plants work but also the basis for major industries. Agriculture, horticulture, forestry, biotechnology, and ecological restoration draw extensively on what biologists know about plant reproduction.

This chapter is part of the Big Picture. See how on pages 840–841.

✔ When you see this checkmark, stop and test yourself. Answers are available in Appendix A.

- For you, plant reproduction for the most part means food. Human diets are based on consuming plant reproductive structures—primarily the seeds and fruits derived from flowers. A **flower** is a reproductive structure that produces gametes, attracts gametes from other individuals, nourishes embryos, and develops seeds and fruits. **Seeds** consist of an embryo and nutrient stores surrounded by a protective coat. **Fruits** develop from the flower's seed-producing organ and contain seeds.

This chapter's analysis of plant reproduction focuses on angiosperms, for three reasons: (**1**) Angiosperms represent over 85 percent of the land plants described to date; (**2**) virtually every important domesticated plant is an angiosperm; and (**3**) aspects of reproduction in other land plant lineages were described previously (see Chapter 31). By the end of this chapter, you'll appreciate the practical aspects of flowers as well as their beauty.

41.1 An Introduction to Plant Reproduction

Plant reproductive structures and processes vary among species. Consider just one aspect of reproductive organs—size. Flowers vary from microscopic to the size of a small child; seeds and fruits range from dustlike particles to coconuts.

Fortunately for students of plant biology, several basic principles unify this diversity of reproductive systems. Let's begin by defining sex.

Sexual Reproduction

Most plants reproduce sexually. **Sexual reproduction** is based on meiosis and fertilization, and it results in offspring that are genetically unlike each other and unlike their parents (see Chapter 13). In plants, meiosis and fertilization occur in alternate phases of a life cycle.

To review briefly, **meiosis** is a type of nuclear division that results in four daughter cells, each of which has half the number of chromosomes present in the parent cell. **Fertilization** is the fusion of haploid cells termed **gametes.** The result of fertilization is the production of a single, diploid cell called the **zygote,** which will develop into a multicellular individual. Male gametes, or **sperm,** are small cells that contribute genetic information in the form of DNA but few or no nutrients to the offspring. Female gametes, or **eggs,** also contain DNA. But the female parent contributes nutrient stores as well. The important point is that although both sperm and egg contribute DNA, the male and female parents contribute vastly different amounts of other resources to the offspring.

Meiosis and fertilization result in offspring that are genetically unlike the parents. Sexual reproduction produces genetically diverse offspring that can potentially be more successful than their parents at warding off attacks from viruses, bacteria, and other pathogens (see Chapter 13).

The advantages of sexual reproduction are common to all eukaryotes that undergo meiosis. When and where meiosis occurs, however, is highly variable. Let's take a closer look.

The Land Plant Life Cycle

In most animals, meiosis leads directly to the formation of gametes. In plants, the situation is much different.

Land plants are characterized by a life cycle with two distinct multicellular forms—one diploid and one haploid. An individual in the diploid phase of the life cycle is called a **sporophyte,** while an individual in the haploid phase of the life cycle is called a **gametophyte.**

This type of life cycle, called **alternation of generations,** has evolved independently in various protists and land plant groups (see Chapters 30 and 31).

What Is Alternation of Generations? Key features of alternation of generations are reviewed in **FIGURE 41.1.** When alternation of generations occurs, meiosis does not lead directly to the formation of gametes, as it does in humans and other animals. Instead, it leads to the production of haploid cells called spores. A **spore** is a cell that grows directly into an adult individual. Several structures and processes are common to all land plant life cycles:

1. Meiosis occurs in sporophytes and results in the production of haploid spores. Unlike zygotes, spores are not produced by the fusion of two cells. Unlike gametes, spores produce an adult without fusing with another cell. Meiosis and spore production occur inside structures called **sporangia.**

2. Spores undergo mitosis and divide to form multicellular, haploid gametophytes.

3. Gametophytes produce gametes using mitosis.

4. Fertilization occurs when two gametes fuse to form a diploid zygote.

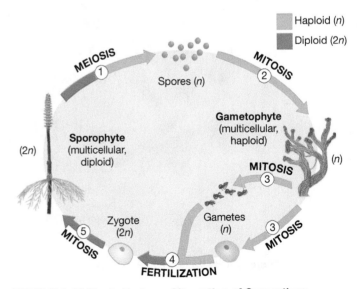

FIGURE 41.1 All Plants Undergo Alternation of Generations.

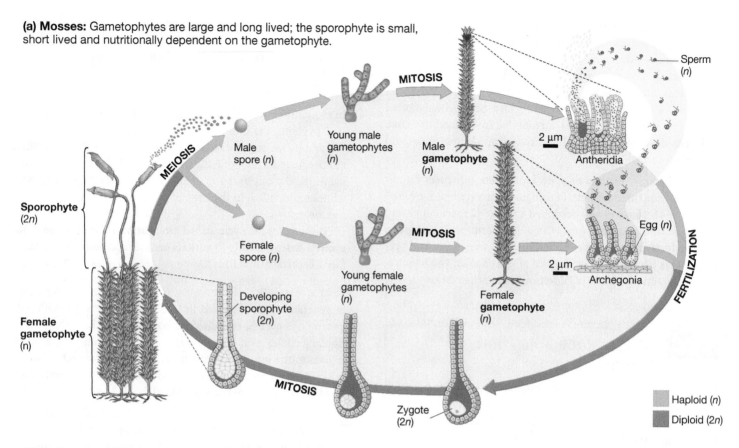

(a) Mosses: Gametophytes are large and long lived; the sporophyte is small, short lived and nutritionally dependent on the gametophyte.

MITOSIS

Sperm (n)

MEIOSIS

Male spore (n)

Young male gametophytes (n)

Male **gametophyte** (n)

2 μm

Antheridia

Sporophyte (2n)

Female spore (n)

MITOSIS

Young female gametophytes (n)

Female **gametophyte** (n)

2 μm

Archegonia

Egg (n)

FERTILIZATION

Female gametophyte (n)

Developing sporophyte (2n)

MITOSIS

Zygote (2n)

Haploid (n)

Diploid (2n)

(b) Angiosperms: Sporophyte is large and long lived; gametophytes are small (microscopic), short lived, and nutritionally dependent on the sporophyte.

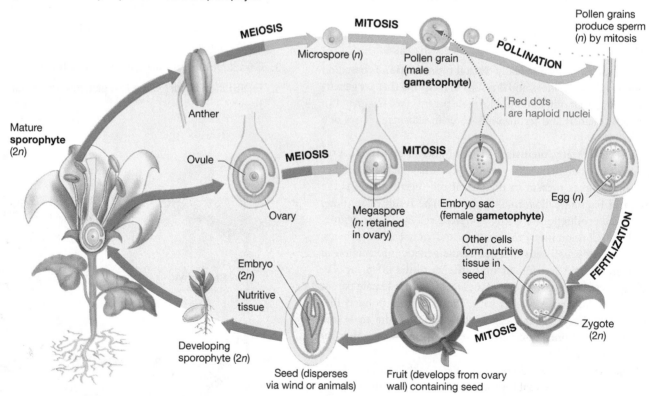

MEIOSIS

MITOSIS

Pollen grains produce sperm (n) by mitosis

Microspore (n)

Pollen grain (male **gametophyte**)

POLLINATION

Mature **sporophyte** (2n)

Anther

Ovule

MEIOSIS

MITOSIS

Red dots are haploid nuclei

Ovary

Megaspore (n: retained in ovary)

Embryo sac (female **gametophyte**)

Egg (n)

FERTILIZATION

Other cells form nutritive tissue in seed

Embryo (2n)

Nutritive tissue

Zygote (2n)

MITOSIS

Developing sporophyte (2n)

Seed (disperses via wind or animals)

Fruit (develops from ovary wall) containing seed

FIGURE 41.2 There Is Wide Variation in Plant Life Cycles. (a) Mosses, **(b)** Angiosperms.

5. The zygote undergoes mitosis and grows to form the multi-cellular sporophyte.

A good way to keep these terms straight is to remember that sporophyte means spore-plant, while gametophyte means gamete-plant. Sporophytes produce spores by meiosis. Gametophytes produce gametes by mitosis.

Variation in Plant Life Cycles Land plant species vary a great deal in how large and long lived the gametophyte and sporophyte are relative to each other (see Chapter 31). The examples in **FIGURE 41.2** represent the extremes of the variation in life cycles observed within land plants.

- Figure 41.2a diagrams the life cycle of a moss. Notice that the largest stage in the moss life cycle is the gametophyte. The sporophyte is dependent on the gametophyte for nutrition.

- Figure 41.2b diagrams the life cycle of an angiosperm. Notice that the male and female gametophytes are microscopic, physically separate, and completely dependent on the sporophyte for nutrition.

✔ If you understand the angiosperm life cycle, you should be able to (1) identify the male spore and female spore in Figure 41.2b, and (2) provide evidence to support the statement that in angiosperms, female gametophytes never leave their parent plant.

The non-vascular plants (sometimes called bryophytes), which include mosses, are the sister group to all other land plants, and they are found early in the fossil record. Angiosperms, in contrast, are the most recent lineage of land plants to appear in the fossil record and the most derived branch on the evolutionary tree. Based on these observations, biologists have concluded that over the course of land plant evolution, gametophytes became reduced while sporophytes became more conspicuous.

Why this trend occurred is still unknown. The adaptive significance of the gametophyte and sporophyte phases remains a challenge for biologists interested in plant reproduction.

Asexual Reproduction

Asexual reproduction does not involve fertilization and results in the production of **clones**—genetically identical copies of the parent plant.

Some plants extend their life indefinitely by asexual reproduction. The oldest of all known plants is a ring of creosote bushes in the Mojave Desert of California. The bushes comprise a clone that originated from a parent plant estimated to have germinated 12,000 years ago.

Although all asexual reproduction is based on mitosis, a wide array of mechanisms exists.

- **FIGURE 41.3a** shows grass shoots and roots emerging from nodes on underground horizontal stems called **rhizomes.** If the individuals emerging from the nodes become separated from the parent plant, they represent asexually produced offspring.

- The gladiolus plant in **FIGURE 41.3b** has propagated itself via modified stems called **corms,** which grow under the surface of the soil.

- The kalanchoe in **FIGURE 41.3c** produces "plantlets" (small plants) from meristematic tissue located along the margins of its leaves. When the plantlets mature, they drop off the parent plant and grow into independent individuals.

- In dandelion and certain other species, mature seeds can form without fertilization occurring. This phenomenon, known as **apomixis,** results in seeds that are genetically identical to the parent.

The key characteristic of asexual reproduction is efficiency. If an herbivore or a disease wipes out the plants surrounding a grass plant, the grass can quickly send out horizontal stems. Its asexually produced offspring are likely to fill the unoccupied space before seeds from competitors can establish themselves and grow. The parent plant can also nourish these progeny as they become established.

(a) Rhizome

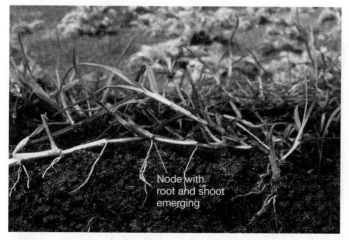

Node with root and shoot emerging

(b) Corm

Parent corm

New corms

(c) Plantlets

FIGURE 41.3 The Mechanisms of Asexual Reproduction Are Diverse. These are just three of many mechanisms of asexual reproduction in plants.

Although asexual reproduction is extremely common in plants, it does have a downside. The most important is that a fungus or other disease-causing agent that infects an individual plant will probably succeed in infecting the plant's cloned offspring as well, even if they are no longer physically connected.

This hypothesis is based on the observation that plants fight disease with a wide variety of molecules (see Chapter 40). Because sexually produced offspring are genetically unlike their parents, these offspring have unique combinations of disease-fighting molecules and may be able to resist infections that devastate their parents.

This is an important point in agriculture and horticulture because asexually propagated apples, bananas, and other crops are more susceptible to epidemics than are sexually propagated species. How does sexual reproduction occur?

41.2 Reproductive Structures

Each major group of plants, from mosses to angiosperms, has a characteristic variation on the theme of alternation of generations, as well as characteristic male and female reproductive structures (see Chapter 31). Here, though, the focus is on the flower.

The General Structure of the Flower

Structurally, all flowers are variations on a theme. They are made up of four basic organs that are essentially modified leaves: (1) sepals, (2) petals, (3) stamens, and (4) one or more carpels. These organs are attached to a compressed portion of stem called the receptacle (**FIGURE 41.4a**).

Not all four organs are present in all flowers, however; and as **FIGURE 41.4b** shows, the coloration, size, and shape of these four components are fabulously diverse. Let's consider each of the four parts, in turn.

Sepals Form an Outer, Protective Whorl **Sepals** are leaflike structures that make up the outermost parts of a flower. Sepals are usually green and photosynthetic, and they are relatively thick compared with other parts of the flower.

Because they are attached to the receptacle in a circle or whorled arrangement, sepals enclose the flower bud as it develops and grows—protecting young buds from damage by insects or disease-causing agents. The entire group of sepals in the flower is called the **calyx.**

Petals Furnish a Visual Advertisement Like sepals, **petals** are arranged around the receptacle in a whorl. Often brightly colored and scented, petals function to advertise the flower to bees, flies, hummingbirds, and other pollinators.

In some cases, the color of the petals correlates with the visual abilities of particular animals. Bees, for example, respond strongly to wavelengths in the blue and purple regions of the light spectrum, as well as yellow (they don't see red well). Flowers

(a) Basic parts of a flower

(b) Examples of flower diversity

FIGURE 41.4 The Basic Structures in Flowers Are Highly Variable.
(a) Flowers comprise sepals, petals, stamens, and carpels. **(b)** The four parts vary among species.

(a) What you and a bee see

(b) What a bee sees in addition

FIGURE 41.5 Insects See in the Ultraviolet Range. (a) The inflorescence (flower cluster) of a black-eyed Susan, seen by the unaided human eye. **(b)** The same structure, photographed with a camera that records ultraviolet wavelengths that are visible to bees but invisible to humans.

that attract bees, in turn, often have yellow, blue, or purple petals with ultraviolet patches.

The ultraviolet regions of petals in "bee flowers" frequently highlight the center of the flower (**FIGURE 41.5**), where the stamens and carpels are located. Why? In these flowers, the base of the petals contains a gland called a **nectary.** The nectary produces the sugar-rich fluid **nectar,** which is harvested by many of the animals that visit flowers, along with pollen. In the process of collecting pollen or nectar, the visiting animal usually deposits pollen from a different plant on the female parts—accomplishing pollination.

The entire group of petals in a flower is called the **corolla.** In some species, the petals within the corolla vary in size, shape, and function:

- Flattened petals may provide a landing pad for flying insects.
- Elongated, tubelike petals frequently have a nectary at their base that can be reached only by animals with a long beak or tongue-like proboscis.
- Some petals protect the reproductive organs located inside the corolla.
- Specialized cells in some petals synthesize and release molecules that provide a scent attractive to certain species of pollinating animals.

In contrast, wind-pollinated angiosperms such as oaks, birches, pecans, and grasses have flowers that have small petals or no petals at all, and they lack nectaries. These species do not invest in structures that aren't required for pollination.

Stamens Produce Pollen   Stamens are reproductive structures that produce male gametophytes—also known as pollen grains. The male gametophytes, in turn, produce sperm.

Each stamen consists of two components:

1. a slender stalk termed the **filament,** and
2. the pollen-producing organs called **anthers** (see Figure 41.4a).

The anther is the business end of the stamen—where meiosis and pollen formation take place. The filament holds the stamen in a place where wind, insects, hummingbirds, bats, or other agents can make contact with the pollen grains produced in the anther.

Carpels Produce Ovules   The fourth reproductive structure is the **carpel,** which produces female gametophytes. A carpel consists of three regions:

1. The **stigma** is a sticky tip that receives pollen.
2. The **style** is a slender stalk.
3. The **ovary** is an enlarged structure at the base of the carpel (see Figure 41.4a).

Inside the ovary, female gametophytes are produced in structures called **ovules.** An ovary may contain more than one ovule. When the female gametophytes that are produced inside ovules mature, they produce eggs.

The "Sex" of Flowers Varies   In most angiosperm species, stamens and carpels are produced on the same individual. Flowers that contain both stamens and carpels are referred to as **perfect.**

Flowers can also be **imperfect,** however, meaning they contain either stamens *or* carpels, but not both. Imperfect flowers that contain only stamens can be considered "male" flowers. Similarly, imperfect flowers that contain only carpels can be considered "female" flowers.[1]

In some cases, separate stamen- or carpel-producing flowers occur on the same individual. Species like these, including the corn plants illustrated in **FIGURE 41.6a** (see page 828), are **monoecious** (literally, "one-house"). In corn, the tassel is a collection of stamen-producing "male" flowers, and the ear contains a group of carpel-producing "female" flowers.

Species that are monoecious or that have perfect flowers are like hermaphroditic species of animals, which contain both male and female reproductive organs and produce sperm and eggs (see Chapter 50).

In contrast, some species with imperfect flowers are **dioecious** ("two-houses")—meaning that each individual plant produces either stamen-bearing flowers only, and could be considered "male," or carpel-bearing flowers only and could be considered "female." *Cannabis sativa* is a dioecious species (**FIGURE 41.6b**). Dioecious plants are like animal species with individuals that have either male or female reproductive organs—not both.

How Are Female Gametophytes Produced?

What purposes do the three parts of the carpel serve? The function of the stigma and style will become clear in Section 41.3; for now, let's concentrate on what happens inside the ovary.

---

[1]Technically, flowers are not referred to as male and female. Instead, they are staminate or carpellate. Staminate flowers produce stamens, which produce pollen grains, which produce male gametes (sperm). Carpellate flowers produce carpels, which contain ovaries. Female gametophytes develop inside ovaries and produce female gametes (eggs). For convenience, though, the text will sometimes refer to male and female flowers and reproductive structures.

(a) Corn is monoecious.

(b) *Cannabis* is dioecious.

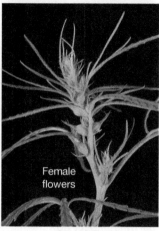

FIGURE 41.6 "Male" and "Female" Flowers Can Occur on the Same Individual or on Different Individuals. (a) The tassels of corn contain male flowers; ears contain female flowers. **(b)** In *Cannabis sativa*, male and female flowers are found on different individuals.

FIGURE 41.7 illustrates a longitudinal section showing the inside of a typical angiosperm ovary. Notice that it contains one or more ovules. Each ovule contains a structure called the megasporangium, which contains a diploid cell called the megasporocyte. (The use of "mega" is appropriate, because these structures are much larger than their counterparts in the stamen.) The megasporangium is comparable to spore-producing organs found in other plants, such as the sporangia found on the back of fern leaves.

When you study Figure 41.7, note four important points:

1. The megasporocyte divides by meiosis.

2. Four haploid nuclei called **megaspores** result from meiosis, but three degenerate. No one is sure how or why this happens.

3. The surviving megaspore divides by mitosis to produce a structure with haploid nuclei. This is the female gametophyte—usually known as the **embryo sac.**

4. The haploid nuclei segregate to different positions in the embryo sac, and cell walls form around some of them.

In the carpel, then, a diploid megasporocyte divides by meiosis to form a megaspore, which then divides by mitosis to form the female gametophyte. Female gametophytes are encased in an ovary, are retained in the flower, and produce an egg.

In many angiosperms, the embryo sac contains eight haploid nuclei and seven cells. Typically, two **polar nuclei** stay together within one central cell—the largest cell in the ovule. The number of polar nuclei varies among species, however.

The egg cell is located at one end of the female gametophyte, near an opening in the ovule called the **micropyle** ("little-gate"). The micropyle is where a sperm nucleus will enter the ovule before fertilization.

How Are Male Gametophytes Produced?

FIGURE 41.8 provides a detailed look at the stamen and the steps that occur in the production of male gametophytes. Recall that a stamen consists of two major parts: an anther and a filament.

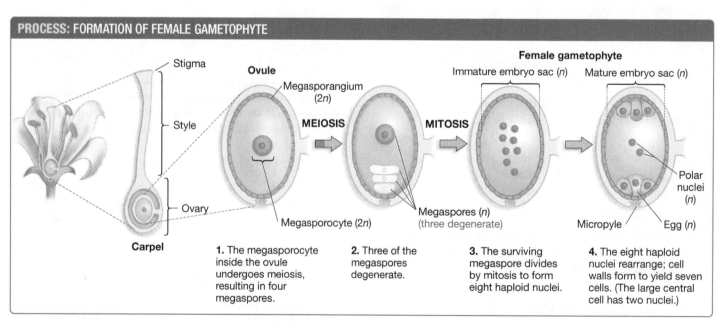

FIGURE 41.7 In Angiosperms, Megaspores Produce Female Gametophytes.

Microsporangium

Microsporocyte (2n)

Microspores (n)

Pollen grain (n)
(male gametophyte)

Generative cell   Tube cell   Tough outer coat

MEIOSIS

MITOSIS

(Each microspore)

Anther

Filament

Stamen

1. Many microsporocytes reside inside an anther. Here, the red dot represents a diploid nucleus. A microsporocyte divides by meiosis, resulting in four microspores.

2. Each microspore divides once by mitosis, forming the tube cell and generative cell.

3. The two resulting cells mature into a single pollen grain containing an immature male gametophyte. The generative cell will later divide to form sperm cells.

FIGURE 41.8 In Angiosperms, Microspores Produce Male Gametophytes.

✔ **QUESTION** Define a gametophyte. Why do pollen grains conform to the definition of a gametophyte?

Inside the anther, structures known as microsporangia contain diploid cells called microsporocytes.

When you study Figure 41.8, note three important points:

1. Microsporocytes undergo meiosis.

2. Each haploid cell that results is a **microspore.** Normally, all of the microspores survive. Microspores divide by mitosis.

3. The two nuclei that result from mitotic division in a microspore form a haploid, immature male gametophyte, also known as the **pollen grain.**

In the anther, then, a diploid microsporocyte divides by meiosis to form microspores, which then divide by mitosis to form male gametophytes. Male gametophytes are dispersed from the flower and eventually produce sperm.

At the immature stage—before it has produced sperm—the male gametophyte consists of two cells: a small generative cell enclosed within a larger tube cell. The male gametophyte is considered mature when the haploid generative cell undergoes mitosis and produces two sperm cells.

In some species, this maturation step occurs while pollen is still in the anther. In other species, maturation and sperm production don't occur until after the pollen grain lands on a stigma and begins to grow. The reasons for this difference are not understood.

The wall of a pollen grain develops a tough outer coat that includes the watertight compound called **sporopollenin** (introduced in Chapter 31). This coat protects the male gametophyte when the pollen is released from the parent plant into the environment. Depending on the species, pollen grains may be dispersed by an animal, the wind, or water currents.

And now we're finally ready to explore the moment of truth: How does a pollen grain get to the mature carpel of the same species, where an egg cell is waiting?

check your understanding

If you understand that . . .

- In angiosperm sporophytes, flowers produce megaspores and microspores that develop into female and male gametophytes, respectively.
- The female reproductive structures called carpels contain ovaries. Ovaries enclose structures called ovules. Female gametophytes are produced inside ovules.
- Formation of a female gametophyte begins when a diploid megasporocyte inside an ovule undergoes meiosis. The product of meiosis is a haploid megaspore. The megaspore divides by mitosis to form the female gametophyte—including the egg and polar nuclei.
- Male gametophytes are produced inside reproductive structures called anthers.
- Formation of the male gametophyte begins when a diploid microsporocyte undergoes meiosis to form haploid microspores. Microspores divide by mitosis to form the male gametophyte—including a tube cell and a generative cell that will divide by mitosis to form two sperm cells.

✔ **You should be able to . . .**

1. Compare and contrast a megasporocyte and a microsporocyte.

2. Compare and contrast a female gametophyte and a male gametophyte.

Answers are available in Appendix A.

41.3 Pollination and Fertilization

Pollination is the transfer of pollen grains from an anther to a stigma; fertilization occurs when a sperm and an egg actually unite to form a diploid zygote. The two events are separated in space and time.

Pollination is not restricted to angiosperms. Gymnosperms (introduced in Chapter 31) also package their male gametophytes into pollen grains. This section will focus on pollination and fertilization in flowering plants. Managing pollination and fertilization in angiosperms is a critical challenge for fruit growers and plant breeders.

In addition, angiosperms' pollination and fertilization systems are thought to be key to their evolutionary success. What aspects of pollination and fertilization allowed flowering plants to become so successful in terms of their numbers of species?

Pollination

Pollen can fall on the stigma of the same individual or the stigma of a different individual. **Self-fertilization,** or selfing, occurs when a sperm and an egg from the same individual combine to produce an offspring. In most cases, though, plants **outcross**—meaning that sperm and eggs from different individuals combine to form an offspring. Outcrossing is the result of **cross-pollination**—when pollen is carried from the anther of one individual to the stigma of a different individual.

Selfing versus Outcrossing: Costs and Benefits Selfing and outcrossing each have advantages and disadvantages. The primary advantage of selfing is that successful pollination is virtually assured—it doesn't depend on agents other than the plant itself.

Biologists have documented the benefit of pollination assurance by hand-pollinating plants that normally outcross. In most cases, the hand-pollinated individuals produce far more seed than do individuals that are pollinated naturally.

Other things being equal, self-fertilization should result in the production of many more seeds than outcrossing. Other things are not equal, however. Selfing has a distinct disadvantage: Even though it still involves meiosis, selfed offspring are usually much less diverse genetically than outcrossed offspring are (see Chapter 13). In some cases, selfed offspring may also suffer from inbreeding depression (see Chapter 26).

Although outcrossing is riskier in terms of the chances that pollination will occur, it results in genetically diverse offspring that may be much more successful at warding off attacks from viruses, bacteria, and other pathogens (see Chapter 13).

Outcrossing is much more common than selfing. In many cases, plants have elaborate mechanisms to prevent selfing:

- *Temporal avoidance* In some species that have perfect flowers, male and female gametophytes mature at different times. Thus, selfing cannot occur.

- *Spatial avoidance* Selfing isn't possible in dioecious species and may be rare in monoecious species, unless pollinators transfer pollen between different-"sexed" flowers on the same individual. And in some species with perfect flowers, the anthers and stigma are so far apart that self-pollination is extremely unlikely—if pollen falls inside the flower, there is almost no chance that it will land on the stigma.

- *Molecular matching* In species that normally outcross, it is common to find that the interaction of proteins on the surfaces of the pollen and stigma determines whether pollination will be successful. Pollination is blocked if proteins on the pollen-grain surface match proteins on the stigma—indicating that the pollen and stigma are from the same individual.

Pollination Syndromes Cross-pollination can be accomplished in various ways: Pollen can be carried from flower to flower by physical agents such as wind or water, or by organisms such as insects, birds, or bats.

Animals visit flowers to eat pollen grains, harvest nectar, or both. As an animal feeds from a flower, pollen grains adhere to its body incidentally. When the same individual visits another flower of the same species to feed, some of these grains are deposited on a stigma of the second flower.

In most cases, animal pollination is an example of **mutualism:** a mutually beneficial relationship between two species. Pollinators usually benefit by receiving food; flowering plants gain by having their male gametophytes transferred to a different individual so that outcrossing takes place.

Pollination syndromes are suites of flower characters that are associated with certain types of pollinators. For example, many bird-pollinated flowers tend to be red and unscented, and they open during the day when birds are active. In contrast, moths and bats are usually active at night. If they feed on nectar or pollen, the flowers they visit tend to be white—and thus more visible in low light, have a strong scent, and open at night.

Structures associated with pollination syndromes are thought to be adaptations: traits that increase the fitness of individuals in a particular environment. In this case, flowers and pollinators have adaptations that increase pollination frequency and feeding efficiency, respectively. To capture this point, biologists say that **coevolution** has occurred (see Chapter 55).

One of the most famous examples of coevolution in pollination involves a species called Darwin's orchid (**FIGURE 41.9**), which is native to Madagascar. When first discovered by Western scientists, the orchid attracted a great deal of attention because it has a "spur" that can be as much as 28 cm (11 in) long, and there is a nectary at its base. Charles Darwin hypothesized that it must be pollinated by a moth with a tongue-like proboscis as long as the spur. The idea seemed preposterous at the time, but 40 years later, a hawkmoth species with a proboscis that averages 24 cm in length was discovered pollinating the orchid in the wild.

Why Did Pollination Evolve? In mosses, ferns, and other groups that do not form pollen, sperm have flagella and swim to the egg through droplets of water, or are transferred on water droplets that cling to the legs of tiny soil insects, such as springtails and mites. In conifers and most other gymnosperms, wind transmits

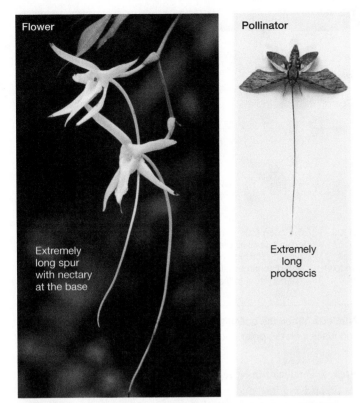

Flower

Pollinator

Extremely long spur with nectary at the base

Extremely long proboscis

FIGURE 41.9 Pollinators Can Sometimes Be Inferred from Flower Structure and Color. Charles Darwin hypothesized that this white orchid flower with an extremely long spur must be pollinated by a moth with a tongue-like proboscis that was long enough to reach the nectar at the end. He was right.

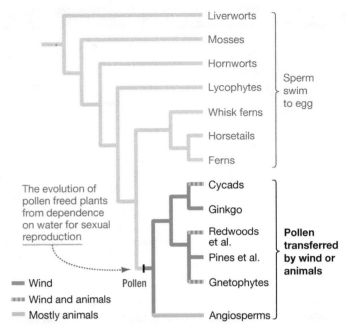

Liverworts
Mosses
Hornworts
Lycophytes

Sperm swim to egg

Whisk ferns
Horsetails
Ferns

The evolution of pollen freed plants from dependence on water for sexual reproduction

Cycads
Ginkgo
Redwoods et al.
Pines et al.

Pollen transferred by wind or animals

Gnetophytes
Angiosperms

Pollen

Wind
Wind and animals
Mostly animals

FIGURE 41.10 Pollen Is a Relatively Recent Innovation in Land Plant Evolution. Evolutionary relationships among the major groups of plants. In the lineages colored gray, sperm have flagella and swim to the egg. In the lineages colored blue or orange, sperm are produced by pollen grains.

pollen from male to female. In some of the other groups that produce pollen, such as the cycads, gnetophytes, and angiosperms, many species are pollinated by animals—particularly by insects.

When these observations are mapped onto the phylogenetic tree shown in **FIGURE 41.10**, two important patterns emerge.

1. Pollination evolved late in land plant evolution. Mosses and other groups that do not form pollen appear first in the fossil record of land plants. Conifers and other groups that are strictly or primarily wind pollinated evolved later but before angiosperms.

2. Seed plants do not need water for sexual reproduction to occur. As a result, the evolution of pollen allowed these species to be much less dependent on wet habitats. Along with the evolution of the seed—highlighted in Section 41.4—pollen paved the way for the colonization of drier environments.

In addition, it's critical to realize that pollination became a much more precise process when plants began to recruit animals to act as pollinators. Wind-borne pollen grains have a low probability of landing successfully on a flower stigma. Animal-borne pollen, in contrast, is much more likely to be successfully transferred to flowers of the same species.

Wind-pollinated species invest in making large numbers of pollen grains; animal-pollinated species make fewer pollen grains but invest in structures that attract and reward animals.

In effect, plants "pay" nectar- and pollen-eating animals to work for them. Wind is free, but animals are more precise. Animal pollination is an important adaptation because it makes sexual reproduction much more efficient.

Does Pollination by Animals Encourage Speciation?   In addition to affecting the fitness of individual plants, does pollination by animals make the formation of entirely new species more likely?

To answer this question, consider the following situation. A biologist has documented that two populations of a mountain-dwelling species called the alpine skypilot have flowers with different characteristics.

• Alpine skypilots that grow in forested habitats at or below timberline have small flowers with short stalks and an aroma described as "skunky."

• Individuals that grow in the tundra habitats above timberline have large flowers with long stalks and a sweet smell.

These differences are interesting, because different insects pollinate the two populations. Small flies are abundant at slightly lower elevations, are attracted to skunky odors, and pollinate the timberline individuals; large bumblebees are abundant at higher elevations, are attracted to sweet odors, and pollinate the tundra flowers.

Experiments have shown that bumblebees prefer to pollinate big flowers—probably because larger flowers can support their larger mass. Because flies and bumblebees prefer to visit different types of flowers, gene flow between the two skypilot populations is low and they are evolving distinct characteristics. The two populations may be on their way to becoming different species.

1. Pollen grain absorbs water and germinates on the stigma. Pollen tube begins growing down the style.

2. The tube-cell nucleus moves into pollen tube, and the generative cell nucleus divides by mitosis to form two sperm in pollen tube.

3. Pollen tube completes growth toward the egg by passing through micropyle and discharging the two sperm into a cell adjacent to egg.

4. One sperm unites with egg to form zygote. The other fuses with the two polar nuclei to form endosperm (nutrient tissue).

FIGURE 41.11 Double Fertilization Produces a Zygote and an Endosperm Nucleus. When the pollen tube reaches the female gametophyte, one sperm nucleus fertilizes the egg while the second fuses with the polar nuclei.

The message here is that evolutionary changes in the size or food-finding habits of a pollinator affect the angiosperm populations they pollinate. In return, changes in flower size and shape affect the insects pollinating that population. If a small population of Darwin's orchids evolved longer spurs, for example, the hawkmoth pollinators that lived in that area would be under intense selection that favored the evolution of a longer proboscis.

Because mutation continuously introduces variations in traits, insect and angiosperm populations frequently change, diverge, and form new species. Changes in pollination can trigger the evolution of new species. It is no surprise that insects and angiosperms are exceptionally species-rich groups.

It is clear that pollination was a crucial innovation during plant evolution. Now let's get down to mechanics. What happens once a pollen grain is deposited on a stigma?

Fertilization

FIGURE 41.11 walks you through the steps in fertilization.

Step 1 After landing on the stigma of a mature flower from the same species, a pollen grain absorbs water and germinates. **Germination** is a resumption of growth and development.

Step 2 When the male gametophyte germinates, a long tubular cell called a **pollen tube** grows through the stigma and down the length of the style. The direction of growth is affected by chemical attractants, called LUREs, which are small proteins released by synergids. Synergids are haploid cells in the female gametophyte that lie close to the egg. In the species illustrated in the figure, the tube-cell nucleus and the generative cell travel down the length of the tube, and the generative cell divides to form two sperm.

Step 3 When the pollen tube reaches the micropyle of the ovule, it grows through it and enters the interior of the female gametophyte.

Step 4 Fertilization—the fusion of sperm and egg to form a diploid zygote—occurs. In most plant groups, fertilization is straightforward—sperm and egg simply combine, and a diploid nucleus is formed. In angiosperms, however, an unusual event called **double fertilization** takes place. One sperm nucleus unites with the egg nucleus to form the zygote. The other sperm nucleus moves through the female gametophyte and fuses with the polar nuclei in the central cell. In most cases, two polar nuclei are present and a large triploid ($3n$) cell forms.

check your understanding

If you understand that . . .

- Wind, water, or animals carry pollen grains from one plant to another.
- When a pollen grain lands on a stigma, the grain germinates. A pollen tube forms and grows until it reaches the ovule.
- Sperm cells produced by the male gametophyte fertilize the egg and combine with the polar nuclei, forming a diploid zygote and in most cases a triploid endosperm, respectively.

✓ **You should be able to . . .**

1. Explain why insects increase their fitness by visiting flowers and why flowers increase their fitness by rewarding insects.

2. Describe the function of the cells produced by double fertilization.

Answers are available in Appendix A.

✔ If you understand double fertilization, you should be able to identify in a female gametophyte, immediately after fertilization occurs, which cells are haploid, diploid, and triploid.

The triploid nucleus resulting from this second fertilization undergoes mitosis and cytokinesis to form the **endosperm** ("inside-seed") tissue. In most species, endosperm is triploid and functions to store nutrients. Endosperm cells are loaded with starch or oils (lipids) plus proteins and other nutrients that the embryo will need after it germinates. But before germination can occur, seeds must develop and be dispersed.

41.4 The Seed

Fertilization triggers the development of a young, diploid sporophyte. In angiosperms, the first stage in the sporophyte's life is the maturation of the seed.

As a seed matures, the embryo and endosperm develop inside the ovule and become surrounded by a covering called a **seed coat.** At the same time, the ovary around the ovule develops into a fruit, which encloses and protects the seed (or seeds, if a single ovary contains multiple ovules). In addition to providing protection, fruits often aid in dispersing seeds away from the parent plant. The mature seed consists of an embryo, a food supply—originating with endosperm—and a seed coat. In most cases, mature seeds leave the parent plant encased in a fruit.

Along with pollen, the evolution of the seed was a crucial innovation as land plants diversified. Because seeds contain stored nutrients, they allow offspring to be much more successful in colonizing habitats that are crowded with competitors than are offspring produced from spores, which are single cells. As a young plant emerges from the seed, it can subsist on stored nutrients until it is well enough established to absorb water from the soil and feed itself via photosynthesis.

Let's analyze how seeds work, beginning with a closer look at how the embryo develops.

Embryogenesis

Recall that **embryogenesis** is the process by which a single-celled zygote becomes a multicellular embryo (see Chapter 24). As **FIGURE 41.12** shows, embryogenesis in angiosperms can be analyzed as a four-step process.

Step 1 The zygote divides to form the two daughter cells.

Step 2 The lower daughter cell, or basal cell, divides to form a cell that forms part of the root tip and a cell that produces a row of cells. This row of cells, called the suspensor, provides a route for nutrient transfer from the parent plant to the developing embryo. The upper daughter cell, or apical cell, is the parent of almost all the cells in the embryo. It divides to form a mass of cells.

Step 3 Cells within the mass differentiate into groups conforming to one of the three adult tissue types (see Chapter 37). The exterior layer of embryonic cells, the **protoderm,** is the progenitor of the adult dermal tissue, or epidermis. The **ground meristem**—the cells just within the protoderm—gives rise to the ground tissue found in adults. And the **procambium** is a group of cells in the core of the embryo that becomes the vascular tissue.

Step 4 As the embryo continues to develop, the long axis of the plant begins to emerge and several important structures take shape: (**1**) **cotyledons** ("seed-leaves"), which absorb nutrients from the endosperm and supply them to the rest of the embryo, (**2**) the **hypocotyl** ("under-cotyledon"), which is the embryonic stem, and (**3**) the **radicle,** or embryonic root. Some embryos also have an **epicotyl** ("above-cotyledon"), which is a portion of the embryonic stem that extends above the cotyledons.

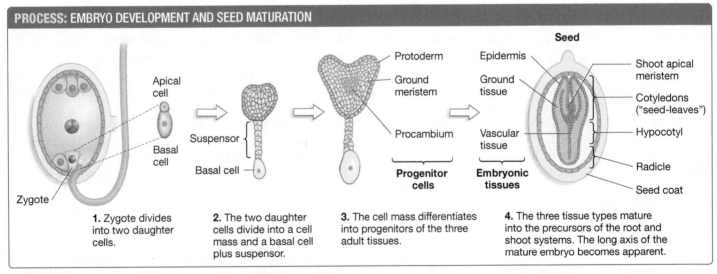

PROCESS: EMBRYO DEVELOPMENT AND SEED MATURATION

1. Zygote divides into two daughter cells.

2. The two daughter cells divide into a cell mass and a basal cell plus suspensor.

3. The cell mass differentiates into progenitors of the three adult tissues.

4. The three tissue types mature into the precursors of the root and shoot systems. The long axis of the mature embryo becomes apparent.

FIGURE 41.12 Embryonic Tissues and Structures Develop Inside Seeds. The embryo inside a seed has cotyledons and the beginnings of root and shoot systems. The embryonic epidermis and the ground and vascular tissues are organized in distinct layers.

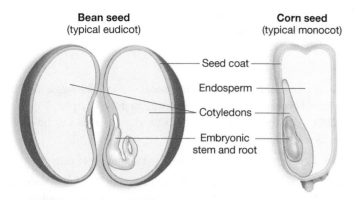

Bean seed
(typical eudicot)

Corn seed
(typical monocot)

Seed coat

Endosperm

Cotyledons

Embryonic
stem and root

FIGURE 41.13 Seeds Contain an Embryo and a Food Supply Surrounded by a Tough Coat. In beans (left), the nutrients in the endosperm have been absorbed by the cotyledons and stored. In corn (right), the endosperm is intact.

✔ If you understand the basic steps in embryonic development, you should be able to compare and contrast a cotyledon with a radicle in terms of their tissue composition and the progenitor cells they originated from.

Recall that one prominent lineage of angiosperms—the monocots—have just one cotyledon, whereas eudicots have two (see Chapter 31). In most eudicots, the cotyledons take up the nutrients that were initially stored in the endosperm and store them again. In these species, there is no endosperm left by the time the seed matures—instead, the cotyledons function as the nutrient storage organ in mature seeds. **FIGURE 41.13** compares the seed structure in beans and corn—a representative eudicot and monocot, respectively.

By the time a seed matures, then, the three major embryonic tissue types have developed. The precursors of the root and shoot systems, along with the seed leaves, have formed. Once these events are accomplished, the seed tissues dry and the embryo becomes quiescent—meaning it stops growing and waits.

The Role of Drying in Seed Maturation

The seeds of many species dry out, or desiccate, as they mature. Water makes up 90 percent of normal plant cells, but dried seeds contain just 5–20 percent water.

Loss of water is an adaptation that prevents seeds from germinating until after they are dispersed. The dry condition of seeds ensures that they will not germinate until water is available. This is adaptive in temperate species because water is crucial to the survival of germinated seedlings. Dry seeds are also less susceptible than wet seeds to damage from freezing.

How do the cellular structures in the embryo and endosperm survive the drying process? When researchers reduce the amount of water surrounding isolated membranes or isolated proteins to the levels observed in extremely dry seeds, the membranes disintegrate and the proteins denature. Clearly, something is happening at the molecular level in seeds to keep these cell components intact.

Researchers established that one of these "somethings" involves sugars. As water leaves the seed during drying, sugars

become concentrated and maintain the integrity of plasma membranes and proteins. If drying is extreme, the sugars form an extremely viscous liquid that contains little if any water. Substances such as this are considered vitrified, or glass-like (glass is a liquid solution with the viscosity of a solid). Biologists propose that this glassy state helps maintain the integrity of plasma membranes and proteins in seeds that experience extremely dry conditions. When seeds imbibe water, the glassy sugars dissolve and germination proceeds.

Drying is only one part of the seed maturation process, however. Equally important is the development of tissues surrounding the seed itself. In many cases, these tissues are required for the seed to be dispersed from the parent plant.

Fruit Development and Seed Dispersal

Fertilization in angiosperms initiates the development of the fruit as well as the seed and embryo.

Fruit Structure   Fruits come in three basic types (**FIGURE 41.14**).

- **Simple fruits** like the cherry develop from a single flower that contains a single carpel or several carpels that are fused together. This is the most common type of fruit.

- **Aggregate fruits** like the blackberry also develop from a single flower, but one that contains many separate carpels.

- **Multiple fruits** like the pineapple develop from many flowers and thus many carpels.

As a fruit matures, the walls of the ovary thicken to form the **pericarp,** the part of the fruit that surrounds and protects the seed or seeds (**FIGURE 41.15**). Fruits can be dry when they are mature, as in nuts, or fleshy, as in cherries and tomatoes. Note that fruits are formed from tissues derived from the mother—not the embryo.

Fruit Function   Fruits have two functions: They protect seeds from physical damage and seed predators, and they frequently aid in seed dispersal. Dispersal is important to the fitness of the young sporophyte. This is especially true in long-lived species, in which the offspring would compete with the parent plant for light, water, and nutrients if there were no dispersal.

Fruits sometimes split open and release seeds to be dispersed directly. In many cases, however, seeds are dispersed to new locations while they are still enclosed in the fruit.

Dry fruits may simply fall to the ground or be dispersed by wind, propulsion, or animals. Some dry fruits have hooks or barbs that adhere to passing animals, while nuts are dispersed by seed predators. Fruits that are dispersed by wind often have external structures to catch the breeze and extend the distance they travel; the fruits of dandelions and maple trees are familiar examples. Fruits that float, such as coconuts, can disperse seeds in water.

Some plants actually disperse dry fruits via propulsion. The sandbox tree, for example, produces a seed pod that shrinks as it dries. Eventually the pod splits apart violently, spraying seeds in all directions with so much force that the plant is sometimes

(a) Simple fruit (e.g., cherry):
Develops from a single flower with one carpel or fused carpels

One carpel

(b) Aggregate fruit (e.g., blackberry):
Develops from a single flower with many separate carpels

Many carpels

(c) Multiple fruit (e.g., pineapple):
Develops from many flowers with many carpels

Many flowers

FIGURE 41.14 Three Major Types of Fruits. The structure of a fruit depends on the number of carpels found in each flower and whether or not the carpels are fused.

✔ **QUESTION** When fruits ripen, their color changes in a way that makes them more conspicuous to fruit eaters. State a hypothesis to explain why this color change might increase the fitness of an individual.

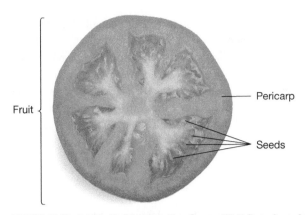

Fruit

Pericarp

Seeds

FIGURE 41.15 As Fruits Mature, the Ovary Wall Develops into a Pericarp That Surrounds the Seed or Seeds. A fruit consists of a pericarp and the enclosed seeds.

called the dynamite tree. The bursting seed pod sounds like a pistol shot, and seeds can be scattered as much as 40 m away from the parent plant. Similarly, the dwarf mistletoe fruit fills with sugars as it matures. Enough water follows via osmosis to make the fruit explode and shoot seeds as far as 5 m.

Animals are the most common dispersal agent for fleshy fruits. Just as plants have evolved to use animals for pollination, they have also evolved to use specific types of animals for seed dispersal. In cases like this, seed dispersal is an example of mutualism. The plant provides a fruit rich in sugars and other nutrients; in return, the animal carries the fruit to a new location and excretes the seeds along with a supply of fertilizer.

Mammals and birds are the most common seed dispersers. Mammals are often active at night and use their well-developed sense of smell to locate fruits. As you might imagine, fruits of mammal-dispersed seeds are usually dull colored and fragrant. In contrast, birds are usually active in the day and see well. Fruits of bird-dispersed seeds are usually brightly colored.

Some animals are seed predators that compete with dispersers. For example, birds usually swallow fruit whole and disperse seeds over long distances. Fruit-eating mice, on the other hand, chew their food, often killing the seeds inside. How can plants discourage seed predators such as mice from eating their fruits? The answer is one that might be familiar to you. Some plants such as chili peppers lace their fruits with a spicy-hot repellent called capsaicin. But does it work?

An experiment by ecologists Joshua Tewksbury and Gary Nabhan in 2001 tested the hypothesis that capsaicin promoted dispersal by the curve-billed thrasher and deterred predation by cactus mice (**FIGURE 41.16**, see page 836). Each animal was offered three kinds of fruit: hackberries, fruits from a strain of chilies that can't synthesize capsaicin (non-pungent chilies), and pungent chilies that have lots of capsaicin. All three fruits looked similar and had equivalent nutritional value. For each animal tested, the researchers recorded the percentage of each fruit eaten during a specific time interval. They then calculated the average amount of fruit that was eaten by five test individuals from each species.

Curve-billed thrashers ate all three fruits equally, but the mice ate fewer non-pungent chilies and avoided the capsaicin-laced fruits entirely. In a follow-up experiment, the researchers fed non-pungent chilies to each kind of animal. When the seeds had passed through the animal's digestive tract and were excreted, the researchers collected and planted the seeds. About 60 percent of the seeds that passed through the birds germinated, but none of those eaten by the mice germinated. Capsaicin appears to be an effective deterrent to seed predation.

Seed Dormancy

Once they have dispersed from the parent plant, seeds may not germinate for a period of time. This condition is known as **dormancy.**

Dormancy is usually a feature of seeds from species that inhabit seasonal environments, where for extended periods of time conditions may be too cold or dry for seedlings to thrive. Based on this observation, dormancy is interpreted as an adaptation that allows seeds to remain viable until conditions improve.

Consistent with this hypothesis, dormancy is rare or nonexistent in seeds produced by plants that inhabit tropical wet forests or other areas where conditions are suitable for germination year-round.

What molecular mechanisms are responsible for dormancy?

How Do Hormones Regulate Dormancy? The answer to this question is not well understood, but the hormone abscisic acid (ABA) plays important roles in seed development and dormancy (see Chapter 40). Mutants of some plants that cannot either make or respond to ABA exhibit a property called vivipary, in which the seeds germinate on the parent plant as soon as they are mature. In many species ABA triggers the accumulation of storage compounds, desiccation tolerance, and the prevention of germination.

After a seed has been dispersed from the parent plant, it may remain dormant in the soil for years or even centuries before it will germinate. The current record for longevity is a lotus seed that germinated after being submerged in an ancient lakebed in China for 1450 years.

What finally causes dormant seeds to germinate? Thanks to the wide range of habitats in which plants are adapted, germination depends on a wide variety of internal and external factors.

How Dormancy Is Broken? The coats of some seeds are thick enough to prevent water and oxygen from physically reaching the embryo. For germination to occur, these seed coats must be disrupted, or **scarified.**

Crop seeds that require scarification are placed in large, revolving drums with pieces of sandpaper that abrade and scarify the seeds. When planted, these scarified seeds germinate quickly and uniformly. In nature, impermeable seed coats protect nutritious seeds from being killed by soil bacteria and fungi, and they often prevent seeds from germinating for many years.

To germinate, most seeds must experience particular environmental conditions. Species native to high latitudes or high

---

RESEARCH

QUESTION: Does the presence of capsaicin in chilies deter some predators but not others?

HYPOTHESIS: Capsaicin deters cactus mice (seed predators) but not birds (seed dispersers).

NULL HYPOTHESIS: Cactus mice and birds respond to capsaicin in the same way.

EXPERIMENTAL SETUP:

PREDICTION: Both will eat hackberry, but only thrashers will eat pungent chilies.

PREDICTION OF NULL HYPOTHESIS: No difference between thrashers and mice in fruit consumed.

RESULTS:

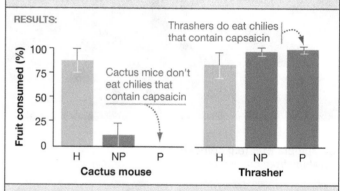

CONCLUSION: The presence of capsaicin deters cactus mice but not thrashers.

FIGURE 41.16 Experimental Support for the Hypothesis that Capsaicin Deters Seed Predators but Not Seed Dispersers.
SOURCE: Tewksbury, J. J., and G. P. Nabhan. 2001. Directed deterrence by capsaicin in chilies. *Nature* 412: 403–404.

✔ QUESTION Why were hackberries used in this experiment?

---

elevations often produce seeds that must undergo cool, wet conditions before they will germinate. Without this level of control, germination before the ensuing winter would likely result in death of the seedling. Recent studies indicate that germination is regulated by two hormones, gibberellin and ABA, and that the levels of these hormones in seeds are affected by temperature (see Chapter 40). After experiencing a winter, gibberellin levels increase and germination begins.

Because small seeds have few nutrient reserves in their cotyledons or endosperm, many small-seeded species need to germinate near the soil surface, where individuals are exposed to light and can feed themselves via photosynthesis. Lettuce seeds and other small seeds must be exposed to red light before they will break dormancy and germinate (see Chapter 40). Red light is an important environmental cue, because wavelengths in the red portion of the light spectrum are absorbed by chlorophyll. Red light and blue light indicate that sunlight is abundant and that the seedling will not be shaded by other plants.

Finally, many of the seeds produced by species native to habitats where wildfires are frequent, such as the California chaparral and South African fynbos, have an unusual chemical requirement to break dormancy: They must be exposed to fire or smoke before they will germinate. In fact, the commercial food product "liquid smoke" induces germination in these seeds as well as actual smoke does. In fire-prone habitats, it is advantageous for seeds to germinate after fire has cleared away existing vegetation.

The message here is that dormancy can be broken in response to a wide variety of environmental cues. In general, the cue that triggers germination is a reliable signal that conditions for seedling growth are favorable for a particular species in a particular environment.

Seed Germination

Even if specific environmental signals are required to break dormancy, seeds do not germinate without water. Water uptake is the first event in germination. Once the seed coat allows water penetration, water enters by moving along a steep water-potential gradient, because the seed is so dry.

Water uptake in a typical angiosperm seed has three distinct phases:

Phase 1 Germination begins with a rapid influx of water. Oxygen consumption and protein synthesis in the seed increase dramatically, but no new messenger RNAs are transcribed. Based on these observations, biologists have concluded that some of the key early events in germination are driven by mRNAs that are stored in the seed before maturation.

Phase 2 The second phase is an extended period during which water uptake stops. New mRNAs are transcribed and translated into protein products. Mitochondria also begin to multiply. In effect, seeds take up enough water in phase 1 to hydrate their existing proteins and membranes and then begin to manufacture the proteins and mitochondria needed to support growth.

Phase 3 Water uptake resumes as growth begins. This renewed phase of water uptake enables cells to develop enough turgor pressure to enlarge. Eventually, the seedling bursts from the seed coat.

FIGURE 41.17 shows what happens as eudicot and monocot embryos emerge from the seed. The radicle emerges first, and it subsequently develops into the mature root system. This is important because the seedling must have a source of water in order to grow. Initially, leaves are less important because of the nutrients stored in the seed.

In eudicots, the shoot system with its cotyledons usually emerges shortly after the radicle appears. In monocots such as corn, the emerging shoot is protected by the coleoptile. Note that in eudicots, the emerging stem has a hook shape. Like the coleoptile of monocots, this trait is thought to protect the apical meristem from damage as the shoot works its way upward through rough soil particles.

The next major event in the seedling's life occurs when either the cotyledons or the earliest leaves produced by the growing seedling commence photosynthesis. The seedling is said to be established when the young plant no longer relies on food reserves in its endosperm or cotyledons; instead, it receives all of its nourishment from its own photosynthetic products. With this, a new generation is under way.

(a) Beans are eudicots with cotyledons that emerge aboveground.

(b) Corn is a monocot with a cotyledon that remains belowground.

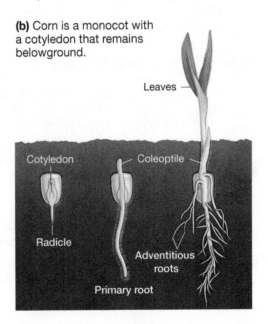

FIGURE 41.17 The Germination Sequence Varies among Species.

✔ QUESTION In which of these species are cotyledons photosynthetic?

If you understand . . .

41.1 An Introduction to Plant Reproduction

- Plants undergo alternation of generations, in which a diploid sporophyte phase alternates with a haploid gametophyte phase. Sporophytes produce spores by meiosis. Gametophytes produce gametes by mitosis.

- The relative size and life span of the gametophyte and sporophyte phases vary a great deal among plant groups.

- In the most basal groups of land plants, gametophytes are larger and longer-lived than sporophytes, and sporophytes depend on gametophytes for nutrition.

- In angiosperms, or flowering plants, sporophytes are the large and long-lived phase where photosynthesis takes place; gametophytes consist of just a few cells.

✔ You should be able to state the ploidy of a sporophyte, gametophyte, spore, gamete, and zygote.

41.2 Reproductive Structures

- In angiosperms, male and female gametophytes are microscopic and are produced inside flowers. Flowers are made up of sepals, petals, stamens, and one or more carpels.

- The lower part of the carpel, the ovary, contains one to many ovules. Within the ovule, a megasporocyte undergoes meiosis, producing a megaspore that develops into the female gametophyte.

- In the anthers of stamens, microsporocytes undergo meiosis. The resulting microspores develop into male gametophytes, which are enclosed in pollen grains.

✔ You should be able to describe what would happen if all meiotic products in megasporangia went on to form female gametophytes.

41.3 Pollination and Fertilization

- Pollination occurs when pollen grains are transported to the stigma of the carpel. In most cases, the structure of a flower correlates with the morphology and behavior of its pollinator.

- If allowed to germinate on the stigma, a pollen grain sends a long pollen tube down the style. Two sperm travel down the pollen tube and enter the female gametophyte.

- In double fertilization, one sperm fuses with the egg to form a zygote, while the other fuses with polar nuclei within the female gametophyte. The fusion of sperm and polar nuclei produces endosperm—nutritive tissue that in most species is triploid.

✔ You should be able to explain why the endosperm of a seed is called a triploid tissue and how it compares with the zygote.

41.4 The Seed

- Seeds contain an embryo and a food supply surrounded by a coat.

- The development of an angiosperm embryo includes the formation of dermal tissue (epidermis), ground tissue, and vascular tissue layers and the development of the radicle, hypocotyl, and cotyledons.

- As an embryo develops, endosperm cells divide to form a nutrient-rich tissue. In addition, cells along the outside of the ovules form a protective seed coat, and the ovary develops into a fruit.

- In many cases, the mature fruit contains structures that help disperse the mature seed via wind, water, propulsion, or animals.

- Many seeds do not germinate immediately but instead experience a period of dormancy.

- A wide variety of conditions, ranging from scarification to exposure to red light, may break seed dormancy. In many cases, the event that triggers germination ensures that the seed germinates when environmental conditions are favorable.

- Germination begins when the seed takes up water and mRNAs already present in the seed are translated. It ends when the radicle breaks the seed coat and begins to penetrate the soil.

✔ You should be able to explain why it is adaptive for many seeds from plants native to high latitudes to require a cold period before germination.

(MB) MasteringBiology

1. MasteringBiology Assignments

Tutorials and Activities Angiosperm Life Cycle; Angiosperm Life Cycle; Fruit Structure and Development; Reproduction in Flowering Plants; Seed and Fruit Development

Questions Reading Quizzes, Blue-Thread Questions, Test Bank

2. eText Read your book online, search, take notes, highlight text, and more.

3. The Study Area Practice Test, Cumulative Test, BioFlix® 3-D Animations, Videos, Activities, Audio Glossary, Word Study Tools, Art

You should be able to . . .

1. What is the major evolutionary trend in land plant life cycles?
   a. Instead of being approximately the same size and shape, gametophytes and sporophytes began to look different.
   b. Sporophytes became larger and long lived while gametophytes became drastically reduced.
   c. In lineages that evolved more recently, such as angiosperms, spores are no longer produced.
   d. Sporophytes began to rely on gametophytes for all of their nutritional needs.

2. What happens when double fertilization occurs?
   a. Two zygotes are formed, but only one survives.
   b. Two sperm fertilize the egg, forming a triploid zygote.
   c. One sperm fertilizes the egg, while another sperm fuses with the polar nuclei.
   d. One sperm fertilizes the egg, while two other sperm fuse with a polar nucleus.

3. What is a fruit?
   a. an expanded ovary that includes a seed or seeds
   b. a structure consisting of an embryo and a food supply surrounded by a tough coat

   c. a female gametophyte
   d. a male gametophyte

4. Why is the interaction between angiosperms and pollinators considered mutualistic?
   a. New species can form if mutant flowers attract new types of pollinators.
   b. Flowers may have an array of traits, including corolla shape, color, scent, and the presence of nectar, to attract a specific type of pollinator.
   c. Wind pollination is much "cheaper," but animal pollination is much more precise.
   d. Angiosperms get their pollen dispersed, while pollinators get food.

5. In the angiosperm life cycle, which cells undergo meiosis? Which cells are spores? Which structures are gametophytes?

6. What events take place during pollination and fertilization?

7. Why is the emergence of the radicle an important first step in germination?
   a. Its hook helps protect the shoot that emerges later.
   b. It carries out photosynthesis to supply the embryo with food.
   c. It is important for establishing a supply of water to the growing embryo.
   d. It is necessary to break the seed coat.

8. In terms of maximizing reproductive success, what is the advantage of asexual reproduction? What is the disadvantage?

9. How do the structure and function of sepals and petals differ? How would you expect these structures to differ in species that are pollinated by wind versus bumblebees?

10. What are the advantages and disadvantages of self-fertilization versus those of outcrossing?

11. Consider a carpel, an ovary, and an ovule. Which is responsible for producing the female gametophyte, and which produces the pericarp of a fruit? Are these structures part of the sporophyte, the gametophyte, or a combination of the two?

12. What is the relationship between the endosperm of corn and the cotyledons of beans?

13. What happens when outcrossing occurs?
    a. Inbred offspring are produced.
    b. The same flower can be visited by many different types of pollinators—not a single specialist species.
    c. Gametes from different individuals fuse to form a zygote.
    d. Gametes from the same individual fuse.

14. QUANTITATIVE Because wind is less efficient than animals at moving pollen from plant to plant, wind-pollinated plants generate far more pollen than animal-pollinated plants. How much of that pollen is wasted? Consider a typical field-corn plant that makes one ear. Ears contain about 600 kernels, each of which is fertilized by an individual pollen grain. Male flowers clustered in the tassel at the top of the plant release about 10 million pollen grains per plant. Using these numbers, what percentage of corn pollen is wasted?

15. Some flowering plants "cheat" their pollinators because they offer no food reward. Likewise, certain pollinators cheat plants by removing nectar from flowers without picking up pollen. (In some cases, they do so by chewing through the petals that hold the store of nectar.) Speculate on the types of mutations that might modify insect behavior and/or plant structure in a way that limits cheating and enforces mutualism.

16. Consider the following fruits: an acorn, a cherry, a burr, and a dandelion seed. Based on the structure of each of these fruits, predict how the seed is dispersed. Design a study that would estimate the average distance that each type of seed is dispersed from the parent plant.

The Big Picture

Plants and animals are diverse lineages of multicellular eukaryotes. They are different in important ways. Each lineage evolved independently from a different single-celled protist—plants with the ability to make their own food by photosynthesis, and animals reliant on obtaining energy from other organisms. Further, plants are sessile, while most animals are capable of complex movements and locomotion.

Yet despite these differences, plants and animals face many of the same challenges to survive and reproduce in water and on land. Use this concept map to explore some of their similarities and differences in form and function.

Note that each box in the concept map indicates the chapters where you can go for more information. Also, be sure to do the blue exercises in the Check Your Understanding box below.

Plant and Animal FORM AND FUNCTION

are the product of

Evolutionary processes
25 & 26
including

Abiotic environment
• Temperature
• Light
• Water availability
• pH, salinity, dissolved gases, nutrients
• Habitat structure
52

create → *occurring in*

Changing ecological contexts Unit 9

Biotic environment
• Parental care
• Competitors
• Predators, prey
• Parasites, hosts
• Mutualists
54, 55

because plants and animals have

Key FUNCTIONS to survive and reproduce ← *correlate with*

including

REGULATION OF WATER AND IONS

NUTRITION

PLANTS

REGULATION OF WATER AND IONS
• Water moves along a water potential gradient
• Turgor pressure provides structural support
• Terrestrial plants lose water evaporatively by transpiration, regulated by stomata and waxy surfaces
38

NUTRITION
• Autotrophic; make their own food by photosynthesis

• Take in sunlight, CO_2, and minerals
• Obtain ions from soil, symbiotic fungi, or bacteria
39

ANIMALS

• Osmotic stress varies in marine, freshwater, and terrestrial habitats
• Urinary system maintains homeostasis of water and electrolytes while managing excretion of nitrogenous wastes
• Terrestrial animals limit evaporative loss
42

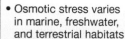

• Heterotrophic; must eat food to acquire energy
• Take in carbohydrates, fats, proteins, vitamins, and minerals
• Obtain nutrients by ingestion, digestion, and absorption
43

check your understanding

If you understand the big picture . . .
✔ You should be able to . . .

1. Propose one mechanism that both plants and animals use on land to limit the evaporative loss of water.

2. Give an example of a method that both plants and animals use to protect their eggs and sperm from drying out.

3. Explain one constraint of large body size for both plants and animals.

4. Explain where cellular respiration fits into this map.

Answers are available in Appendix A.

C Y U

Big Picture activities are available at MasteringBiology (MB)

Natural selection
25 & 26

produces →

Adaptations
• Anatomical
• Physiological
• Behavioral
25, 37, 42

limited by →

Constraints
• Genetic
• Developmental
• Historical
• Mechanical 25

Fitness tradeoffs
42

including →

Body size
• Length, mass, weight
• Surface area/volume ratio
37, 42

require ability to maintain →
Homeostasis
42

influence →
Phenotypic plasticity 37

changes →
Key STRUCTURES to survive and reproduce

can be studied at different →

Levels of organization
• Populations
• Organisms
• Organ systems
• Organs
• Tissues
• Cells
• Molecules
1, 37, 42

Plant tissues
• Dermal
• Ground
• Vascular
• Meristem 37

including →

Animal tissues
• Connective
• Nervous
• Muscle
• Epithelial 42

INTERNAL TRANSPORT

• Move water by bulk flow in xylem tissue, driven by transpiration

• Move sugars in phloem tissue along pressure gradient from sources to sinks
38

• Most move air or water through tracheae, lungs, or gills
• Circulatory systems may be open or closed
• In most, pumping heart moves blood through the body
45

SENSORY SYSTEMS, SIGNALS & RESPONSES

• Detect light, gravity, heat, touch, wind, pathogens, herbivores, chemical signals
• Hormones (plant growth regulators) carry information and induce responses
40

• Detect light, gravity, heat, touch, sound, pathogens, predators, chemical signals
• Nervous system enables rapid response
• Endocrine system enables slower, longer-term responses via hormones
46, 47, 49

MOVEMENT

• Incapable of locomotion, but gametes and seeds can disperse
• Some can move slowly via differential growth or rapidly via cell shape changes
40, 41

• Most are capable of movement using muscles

• Muscles work in combination with skeletal systems: hydrostatic skeletons, exoskeletons, and endoskeletons
48

REPRODUCTION

• Most can reproduce sexually, some asexually
• Undergo alternation of generations (haploid and diploid stages are both multicellular)
• Angiosperms have internal fertilization (within ovary of flower)
41

• Most can reproduce sexually, some asexually
• Multicellular stage is usually diploid; gametes are haploid
• Fertilization may be external or internal
50

DEFENSE

• Have both physical and chemical defenses against pathogens and herbivores
• Infection launches a hypersensitive response
40

• Have both physical and chemical defenses against pathogens and predators
• Infection launches: (1) nonspecific, innate immune response, and sometimes also (2) specific, adaptive immune response
51

42 Animal Form and Function

In this chapter you will learn that

Animal structure is fundamentally related to function

starting with

Adaptations and fitness trade-offs 42.1

then asking

| How does function correlate with structure? **42.2** | Why is body size important? **42.3** | How do animals regulate their internal environment (homeostasis)? **42.4** |

exploring

Tissues, organs, and organ systems

looking closer at

Heat exchange and thermoregulation 42.5

Oryx are adapted to desert life. They have an exceptional ability to withstand heat and can acquire most of their water from the food they eat.

This chapter is part of the Big Picture. See how on pages 840–841.

The Sahara and Arabian deserts are extreme environments. In some parts of the Sahara, several years can pass between rainfalls. In a single day, temperatures in these deserts can fluctuate between −0.5°C and 37.5°C; midday temperatures in summer can exceed 50°C. Yet few places in the Sahara or Arabian deserts are devoid of life. Even large animals, such as the oryx, thrive in both regions.

How do animals native to these areas cope? Small animals avoid the midday heat by retreating to a cool burrow deep underground or the shade of a small shrub. Large mammals have a harder time hiding from sunlight but possess traits that allow them to keep cool and conserve water.

The Arabian oryx, for example, seldom drinks. It gets 86 percent of the water it needs from the vegetation it eats and the other 14 percent from water synthesized as a by-product of cellular

✔ When you see this checkmark, stop and test yourself. Answers are available in Appendix A.

respiration—what biologists call **metabolic water.** To conserve what water they have, Arabian oryx produce extremely concentrated urine and exceptionally dry fecal pellets.

In addition, oryx do not sweat to cool off. Instead, their body temperature rises from a normal 37°C to just over 40°C as temperatures increase during the day. The excess body heat is released during the cool desert nights, when their body temperature returns to normal.

In contrast to oryx, humans may die of dehydration when denied water for just three days. And for a human, having a body temperature of 40°C is a life-threatening situation—equivalent to running a fever of 104°F. How do oryx do it? What aspects of their anatomy and physiology allow them to thrive in such an extreme environment?

Anatomy is the study of an organism's physical structure. **Physiology** is the study of how the physical structures in an organism function. The anatomy and physiology of an oryx are clearly different from those of a human—or those of a shark, frog, bird, fruit fly, or crab, for that matter. This chapter is an introduction to the study of anatomy and physiology in animals. You can review the importance of animal form and function in the Big Picture on pages 840–841.

42.1 Form, Function, and Adaptation

Biologists who study animal anatomy and physiology are studying **adaptations**—heritable traits that allow individuals to survive and reproduce in a certain environment better than individuals that lack those traits (see Chapter 25).

Recall that adaptation results from evolution by natural selection. Natural selection, in turn, occurs whenever individuals with certain alleles leave more offspring that survive to reproductive age than do individuals with different alleles. Because of this difference in reproductive success, the frequency of the selected alleles increases from one generation to the next.

Oryx with alleles that allow them to extract more water from their feces survive better and produce more offspring than do oryx with alleles that allow water to be lost in feces. The ability to produce extremely dry fecal pellets is an adaptation that helps oryx thrive in water-short environments.

The Role of Fitness Trade-Offs

Adaptations increase fitness—the ability to produce viable offspring. But no adaptation is "perfect." Instead, adaptations are limited by which alleles are present in a population and by the nature of the traits that already exist—because all adaptations derive from preexisting traits.

The human spine, for example, is a highly modified form of the vertebral column in ancestors that walked on all fours (see Chapter 35). The modifications in the human spine can be considered adaptations to support our upright posture, but they are far from perfect—most adults experience back pain at some point during their lives. The evolution of the human spine has

been constrained by the nature of the ancestral trait and by a lack of alleles that would improve its structure and function.

The most important constraint on adaptation, though, may be **trade-offs**—inescapable compromises between traits. For example, it takes a lot of energy to produce offspring through the process of reproduction and also to mount an immune response during an infection. Animals sometimes do not have enough energy to satisfy both needs. In these cases, a trade-off emerges: The animal may devote more energy to reproduction at the expense of strong immune function, or vice versa, or both traits might be negatively affected.

How do biologists study trade-offs in animal physiology? Let's consider experimental work on trade-offs in crickets. During mating, a male cricket deposits a **spermatophore** on the female's genital opening (**FIGURE 42.1a**). This spermatophore consists of a packet of sperm surrounded by a large, gelatinous mass. After mating, the female begins to eat the gelatinous mass (**FIGURE 42.1b**). The sperm packet remains behind and the sperm slowly begin to enter her reproductive tract.

The longer it takes for the female to eat the mass, the more sperm are transferred to her, increasing the number of eggs fertilized by the male. It might seem advantageous for the male to make as large a spermatophore as possible. However, there is a

(a) Decorated crickets mating

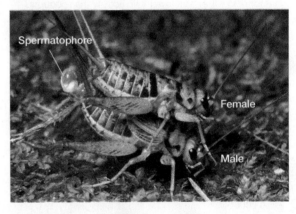

(b) The female eats the gelatinous mass.

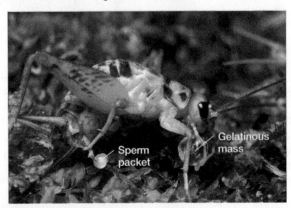

FIGURE 42.1 Mating in Crickets Involves Transfer of a Spermatophore from the Male to the Female.

QUESTION: Is there a trade-off between reproductive and immune function in male crickets?

HYPOTHESIS: Male crickets need to make an energy trade-off between reproductive function and immune function.

NULL HYPOTHESIS: No energy trade-off between reproductive function and immune function is required.

EXPERIMENTAL SETUP:

1. Remove spermatophores from male crickets:

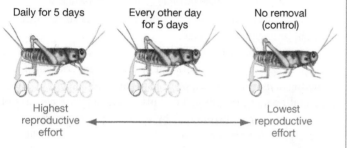

Daily for 5 days   Every other day for 5 days   No removal (control)

Highest reproductive effort ←→ Lowest reproductive effort

2. Draw hemolymph samples from the three sets of crickets.

3. Add bacteria to samples; measure lysis of bacteria.

Spectrophotometer

EXPERIMENTAL SETUP:

1. Inject male crickets:

With lipopolysaccharide (LPS) to simulate bacterial infection   With placebo (control)

2. Remove spermatophores and measure size of gelatinous mass.

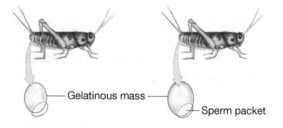

Gelatinous mass —— Sperm packet

PREDICTION: Hemolymph from males forced to produce more spermatophores will exhibit lower lytic activity than controls.

PREDICTION OF NULL HYPOTHESIS: There will be no difference in lytic activity between treated males and control males.

PREDICTION: Spermatophores from LPS-injected males will have smaller gelatinous masses than those from control males.

PREDICTION OF NULL HYPOTHESIS: There will be no difference in gelatinous mass size between treated and control males.

RESULTS:

Treated males had lower lytic activity than controls

(graph: Lytic activity (change in optical density) vs Daily, Every other day, Control)

RESULTS:

Spermatophores from treated males had smaller gelatinous masses than controls

(graph: Size of gelatinous mass (mg) vs LPS-injected, Control)

CONCLUSION: Male crickets must make an energy trade-off to support increased reproductive effort.

FIGURE 42.2 Trade-offs Between Reproduction and Immune Function in Male Crickets. This study showed that males with experimentally increased investment in reproduction had a lower ability to kill bacteria than control males, and that males with experimentally stimulated immune function had lower reproductive function than control males.

SOURCE: Kerr, A. M., S. N. Gershman, and S. K. Sakaluk. 2010. Experimentally induced spermatophore production and immune responses reveal a trade-off in crickets. *Behavioral Ecology* 21: 647–654.

✔**QUESTION** If the null hypothesis were supported by the results, what would each graph look like?

cost involved. Spermatophores weigh up to 6 percent of a male cricket's body mass, and it takes males more than three hours to make a new one. During the time males spend making new spermatophores, they lose other mating opportunities.

Losing out on new mates while making spermatophores may not be the only cost a male cricket accrues. When an insect is infected with a pathogen such as a bacterium, it mounts an immune response. Activation of the insect defense system is energetically expensive. Is there a trade-off in male crickets between producing spermatophores and mounting an immune response?

To answer this question, biologists carried out an experiment using a powerful method—a reciprocal design. They altered the energetic investment crickets made into each of the systems in question—reproduction and immune function—and observed the effect on the other system (**FIGURE 42.2**).

First, the biologists removed spermatophores from male crickets, causing them to make new ones, daily or every other day. The biologists then took a sample of the crickets' hemolymph, or blood, and added bacteria to find out whether the hemolymph was able to destroy the pathogen. As bacteria are lysed, or ruptured, the opaque hemolymph becomes clear. Lytic activity was measured using a spectrophotometer to calculate the change in the opacity, or optical density, of the hemolymph over time.

Compared to control crickets, which did not have their spermatophores removed, the experimental crickets had lower lytic activity (fewer bacteria destroyed). Even more striking, the crickets that had to make new spermatophores daily had lower lytic activity than those that had to make new spermatophores only every other day. This suggested that increased investment into making spermatophores was traded off against investment into immune function.

Next, the biologists injected crickets with a component of the cell walls of bacteria called lipopolysaccharide (LPS), which caused the crickets to mount an immune response. What was the effect on spermatophore size? The result was dramatic. Male crickets injected with LPS produced smaller spermatophores than control crickets. When the researchers looked more closely, they realized the size of the sperm packet remained largely unchanged, but the size of the gelatinous mass decreased in injected crickets. This suggests that mounting an immune response may reduce male crickets' reproductive success. How? By causing males to make smaller gelatinous masses, which decreases the time females spend eating them, reducing sperm transfer.

Taken together, these two sets of experiments provide very strong evidence that there is a trade-off between reproductive and immune function in crickets.

Trade-offs, such as the compromise between energetic investment and the competing demands of reproductive and immune function, are common in nature. Desert animals that sweat to cool off are threatened with dehydration. An eagle's beak is superbly adapted for tearing meat but not for weaving nesting materials together. In studying animal anatomy and physiology, biologists study compromise and constraint as well as adaptation.

Adaptation and Acclimatization

In biology, adaptation refers to a genetic change in a population in response to natural selection exerted by the environment. Short-term, reversible responses to environmental fluctuations are referred to as **acclimatization.** Acclimatization is a phenotypic change in an individual in response to short-term changes in the natural environment. Acclimation is similar to acclimatization, but acclimation refers to changes that occur in organisms in a laboratory setting.

If you moved to Tibet, your body would acclimatize to high elevation by making more of the oxygen-carrying pigment hemoglobin and more hemoglobin-carrying red blood cells. But populations that have lived in Tibet for many generations are adapted to this environment through genetic changes. Among native Tibetans, for example, an allele that increases the ability of hemoglobin to hold oxygen has increased to high frequency. In populations of Tibetans that do not live at high elevations, this allele is rare or nonexistent.

The ability to acclimatize is itself an adaptation. Light-skinned humans, for example, vary in the ability to tan in response to sunlight. Some individuals tan easily—they have alleles that allow them to acclimatize efficiently to environments with intense sunlight—while others do not. In this and many other cases, the ability to acclimatize is a genetically variable trait that can respond to natural selection.

42.2 Tissues, Organs, and Systems: How Does Structure Correlate with Function?

If a structure found in an animal is adaptive—meaning that it helps the individual survive and produce viable offspring—it is common to observe that the structure's size, shape, or composition correlates closely with its function.

For example, recall that biologists have documented extensive changes in beak size and shape in medium ground finches from the Galápagos Islands (see Chapter 25). Such changes are due to natural selection. Individuals with deep beaks are better able to crack the large fruits that predominate during drought years, while individuals with small beaks are better able to harvest the small seeds that predominate during wet years.

As **FIGURE 42.3** (see page 846) shows, a strong correlation between diet and beak structure is also found *among* species of Galápagos finch. Species with large, cone-shaped beaks eat large seeds; species with small, cone-shaped beaks eat small seeds. Species with long, tweezer-like beaks pick insects off tree trunks or other surfaces.

The mechanism responsible for these structure–function correlations is straightforward: If a mutant allele alters the size or shape of a structure in a way that makes it function more efficiently, individuals who have that allele will produce more offspring than will other individuals. As a result, the allele will increase in frequency in the population over time.

Species of Galápagos finch		Food source
Geospiza fuliginosa		Small seeds
Geospiza fortis		Medium seeds
Geospiza magnirostris		Large seeds
Certhidea olivacea		Insects, nectar

FIGURE 42.3 In Animal Anatomy and Physiology, Form Often Correlates with Function.

Structure–Function Relationships at the Molecular and Cellular Levels

Correlations between form and function start at the molecular level. For example, earlier chapters emphasized that the shape of proteins correlates with their role as enzymes, structural components of the cell, or transporters. The membrane proteins called channels form pores that allow specific ions or molecules to pass in or out of cells (see Chapter 6). The ends and interior of a channel are hydrophilic, which allows the protein to interact with the surrounding solution or the interior of the cell, while the perimeter is hydrophobic—allowing the protein to interact with the lipid bilayer. The protein's structure fits its function.

Similar correlations between structure and function occur at the level of the cell. In fact, it is possible to predict a cell's specialized function by examining its internal structure. Cells that manufacture and secrete hormones or digestive enzymes are packed with rough endoplasmic reticulum (ER) and Golgi apparatuses; cells that store energy are dominated by large fat droplets; cells that ingest and destroy invading bacteria have many lysosomes.

The overall shape of a cell can also correlate with its function. For example, cells that are responsible for transporting materials into or out of the body often have extremely large areas of plasma membrane. As a result, they have room to accommodate the thousands of membrane channels, transporters, and pumps required for extensive transport.

Tissues Are Groups of Cells That Function as a Unit

Animals are **multicellular,** meaning that their bodies contain distinct types of cells that are specialized for different functions. Frequently, animal cells that are similar in structure and function are physically attached to each other and form a tissue. A **tissue** is a group of cells that function as a unit.

Type	Example
Loose *soft* extracellular matrix; provides padding	Reticular tissue — Capillary — Cells — Matrix — Protein fibers — 10 μm
Dense *fibrous* extracellular matrix; provides connections	Tendon — Collagen fibers — Tendon cell nuclei — 10 μm
Supporting *firm* extracellular matrix; functions in structural support and protection	Bone — Bone cells — Matrix — 50 μm
	Cartilage — Cartilage cells — Matrix — 50 μm
Fluid *liquid* extracellular matrix; functions in transport	Blood — Red blood cell — White blood cell — Plasma — 5 μm

The embryonic tissues called ectoderm, mesoderm, and endoderm are found in most animals (see Chapter 23 and Chapter 33). As an individual develops, the embryonic tissues give rise to four adult tissue types: (**1**) connective tissue, (**2**) nervous tissue, (**3**) muscle tissue, and (**4**) epithelial tissue. In each case, the structure of the tissue correlates closely with its function. Let's consider each type in turn.

Connective Tissue **Connective tissue** consists of cells that are loosely arranged in a liquid, jellylike, or solid matrix. The matrix comprises extracellular fibers and other materials, and it is secreted by the connective tissue cells themselves (**TABLE 42.1**). Each type of connective tissue secretes a distinct type of extracellular matrix. The nature of the matrix determines the nature of the connective tissue.

- **Loose connective tissue** contains an array of fibrous proteins in a soft matrix; it serves as a packing material between organs or padding under the skin. Reticular connective tissue—in lymphoid organs such as the spleen and bone marrow—and adipose tissue, or fat tissue, are loose connective tissues made up of cells suspended in a matrix of fibers and fluid.

- **Dense connective tissue** is found in the tendons and ligaments that connect muscles, bones, and organs. As Table 42.1 shows, the matrix in tendons and ligaments is dominated by tough collagen fibers (introduced in Chapter 11).

- **Supporting connective tissue** has a firm extracellular matrix. **Bone** and **cartilage** are connective tissues that provide structural support for the vertebrate body as well as protective enclosures for the brain and other components of the nervous system.

- **Fluid connective tissue** consists of cells surrounded by a liquid extracellular matrix. **Blood,** which transports materials throughout the vertebrate body, contains a variety of cell types and has a specialized extracellular matrix called plasma (see Chapter 45).

Nervous Tissue **Nervous tissue** consists of nerve cells, which are also called **neurons,** and several types of supporting cells. Neurons transmit electrical signals, which are produced by changes in the permeability of the cell's plasma membrane to ions (see Chapter 46). Supporting cells have many functions, including regulating ion concentrations in the space surrounding neurons, supplying neurons with nutrients, or serving as scaffolding or support for neurons.

Although they vary widely in shape, all neurons have projections that contact other cells. As **FIGURE 42.4** shows, most neurons have two distinct types of projections from the cell body, where the nucleus is located: (**1**) highly branched, relatively short processes called **dendrites,** and (**2**) a relatively long structure called an **axon.** Dendrites contact other cells and transmit electrical signals from them to the cell body; the axon carries electrical signals from the cell body to other cells.

Muscle Tissue **Muscle tissue** was a key innovation in the evolution of animals—like nervous tissue, it appears in no other

FIGURE 42.4 Neurons Transmit Electrical Signals. In a neuron, information is transmitted from dendrites to the cell body to the axon.

✓**QUESTION** Based on the structure of a neuron, does it provide signals to specific cells and tissues, or does it broadcast signals widely throughout the body?

lineage on the tree of life (see Chapter 33). Some of the functions of muscle include movement of the body, pumping of the heart, and mixing of food in the gastrointestinal tract. There are three types of muscle tissue (**FIGURE 42.5**); you, along with other vertebrates, have all three.

1. **Skeletal muscle** attaches to the bones of the skeleton and exerts a force on them when it contracts. Skeletal muscle is responsible for most body movements. In addition, it encircles the openings of the digestive and urinary tracts and controls swallowing, defecation, and urination.

Long cells, voluntary movement **Branched cells,** involuntary movement **Tapered cells,** involuntary movement

FIGURE 42.5 Muscle Tissues Comprise Cells That Contract. The three types of muscle tissue have distinctive structures and functions.

Faces outer environment

Apical side of epithelium

Epithelium

Basal lamina connects epithelium to other tissues

Basolateral side of epithelium

10 μm

FIGURE 42.6 Epithelial Cells Provide Protection and Regulate which Materials Pass across Body Surfaces.

2. **Cardiac muscle** makes up the walls of the heart and is responsible for pumping blood throughout the body.

3. **Smooth muscle** cells, which are tapered at each end, form a muscle tissue that lines the walls of the digestive tract and the blood vessels.

Muscle tissue and movement are explored in detail in Chapter 48.

Epithelial Tissues **Epithelial tissues** are also called **epithelia** (singular: **epithelium**). Epithelium covers the outside of the body, lines the surfaces of organs, and forms glands. An **organ** is a structure that serves a specialized function and consists of several tissues; a **gland** is an organ that secretes specific molecules or solutions such as hormones or digestive enzymes.

Epithelia form the interface between the interior of an organ or body and the exterior. In addition to providing protection, epithelial tissues are gatekeepers. Water, nutrients, and other substances are transported, often selectively, across epithelia.

Because the primary function of epithelium is to act as a barrier and protective layer, it's not surprising to observe that epithelial cells typically form layers of closely packed cells (**FIGURE 42.6**).

(a) Tissues are organized into organs.

TISSUES:
- Epithelial tissue
- Connective tissue
- Nervous tissue
- Muscle tissue

ORGAN:
Small intestine

(b) Organs are organized into organ systems.

DIGESTIVE SYSTEM:

Salivary glands secrete enzymes that begin to digest food.

The **esophagus** is a long, muscular tube that transports food to the stomach.

The **stomach** is a thick, muscular sac whose contractions help break up food.

The **liver** and **pancreas** contain cells that secrete enzymes and other molecules that aid digestion.

The **small intestine** is a long, coiled tube where enzymes digest food and nutrients are absorbed.

The **large intestine** is a large tube where water is reabsorbed and wastes are compacted.

FIGURE 42.7 Organs Are Composed of Tissues; Organ Systems Are Made Up of Organs. (a) The human small intestine is an organ composed of all four major tissue types. **(b)** The human digestive system is essentially one long tube divided into chambers where food is processed and nutrients are absorbed. The salivary glands, liver, and pancreas are organs that secrete specific enzymes or compounds into the tube.

In many cases, adjacent epithelial cells are joined by structures that hold them tightly together, such as tight junctions and desmosomes (introduced in Chapter 11).

Epithelial tissue has polarity, or sidedness. An epithelium has an **apical** side, which faces away from other tissues and toward the environment, and a **basolateral** side, which faces the interior of the animal and connects to connective tissues. This connection is made by a layer of fibers called the **basal lamina.**

The apical and basolateral sides of an epithelium have distinct structures and functions. Epithelial cells, for example, line the surface of your trachea, or windpipe. The apical side of these cells secretes mucus and is covered with cilia that help sweep away dust, bacteria, and viruses. The basolateral side lacks these features but is cemented to the basal lamina.

Epithelial cells have short life spans. The cells that line your esophagus—the tube connecting your mouth and stomach—live for 2 to 3 days, while the cells that line your large intestine live for a maximum of 6 days. Muscle cells and neurons, in contrast, normally live as long as the individual animal does. Epithelial cells are short lived because they are exposed to harsh environments, where they are likely to be killed or scraped away.

The tissue as a whole does not wear away, however, because it includes cells that actively undergo mitosis and cytokinesis—producing new epithelial cells to replace those lost on the side that faces the environment.

Organs and Organ Systems

Cells with similar functions are organized into tissues, and tissues are organized into specialized structures called organs. Recall that an organ is a structure that serves a specialized function and consists of several types of tissues. The small intestine, for example, consists of muscle, nervous, connective, and epithelial tissues (**FIGURE 42.7a**).

An **organ system** consists of groups of tissues and organs that work together to perform one or more functions. Using the digestive system as an example, **FIGURE 42.7b** illustrates how the structure of organs correlates with their function and how the components of an organ system work together in an integrated fashion.

Because an animal's body contains molecules, cells, tissues, organs, and organ systems, biologists who study animal anatomy and physiology must work at various levels of organization to understand how that body operates.

FIGURE 42.8 illustrates these levels of organization, using the human nervous system as an example. Because the structure and function of each component in the body are integrated with those of other components, and because each level of organization is integrated with other levels of organization, the organism as a whole is greater than the sum of its parts. In other words, an organism is more than just a collection of individual systems, and each system is more than just a collection of individual cells or tissues or even organs.

Atomic and molecular levels: Membrane protein in neurons regulates flow of ions.

Cellular level: Electrical signal travels down length of neuron.

Tissue level: Electrical signals travel from cell to cell in nervous tissue.

Organ level: Nervous tissue and connective tissue in brain aid in sight, smell, memory, and thought.

Organ system level: Brain and nerves send signals throughout the body to control breathing, digestion, movement, and other functions.

Organism level: Nervous system coordinates the functions of other systems to support life.

FIGURE 42.8 Biologists Study Anatomy and Physiology at Many Levels. The levels of organization within an organism are not independent of each other. Instead, they are tightly integrated.

In effect, each subsequent chapter in this unit focuses on a different organ system found in animals, beginning with the excretory system and ending with the immune system. Each of these systems can be interpreted as a suite of adaptations and trade-offs. Each system accomplishes a specific task required for survival and reproduction, and each works in conjunction with other systems.

Before delving into the various systems, however, it's essential to examine general phenomena that affect all systems in animals. Let's start by looking at how body size affects animal physiology.

check your understanding

If you understand that . . .

- Biologists study structure and function at the molecular, cellular, tissue, organ, and organ system levels.
- Events at each level of organization in an individual interact to form an integrated whole that responds to the environment in appropriate ways.

✔ **You should be able to . . .**

Describe and compare the structure and function of the four major types of animal tissues.

Answers are available in Appendix A.

42.3 How Does Body Size Affect Animal Physiology?

Animals are living machines, made up of molecules, cells, tissues, organs, and organ systems that have changed over time in response to natural selection.

The laws of physics affect the anatomy and physiology of a living machine. The force of gravity, for example, limits how large an animal can be and still move efficiently. Or consider the forces exerted by the medium in which animals live. Because water is much denser than air, it is harder for animals to move through water. As a result, fish and aquatic mammals have much more streamlined bodies than terrestrial animals do.

Physical laws clearly affect body size. Just as clearly, body size has pervasive effects on how animals function. Large animals need more food than small animals do. Large animals also produce more waste, take longer to mature, reproduce more slowly, and tend to live longer. Conversely, small animals are more susceptible to damage from cold and dehydration than large animals are, because they lose heat and water faster. Juveniles and adults of the same species face different challenges simply because their body sizes are different.

Why is body size such an important factor in how animals work? How do biologists study the consequences of size? Let's consider each question in turn.

Surface Area/Volume Relationships: Theory

From microscopic roundworms to gigantic blue whales, animals span an incredible range of body masses—a total of 12 orders of magnitude. Many of the challenges posed by increasing size are based on this fundamental relationship between surface area and volume.

You learned previously that the relationship between the surface area and volume of the roots and shoots of plants affects water and light absorption by the plant (see Chapter 37). Similarly, surface area is important in animals because oxygen and nutrients such as glucose must diffuse into the animal's body, and waste products such as urea and carbon dioxide must diffuse out. The rate at which these and other molecules and ions diffuse depends in part on the amount of surface area available for diffusion. In contrast, the rate at which nutrients are used and heat and waste products are produced depends on the volume of the animal.

The contrast between processes that depend on surface area and those that depend on volume is important for a simple reason. As an animal gets larger, its volume increases much faster than its surface area does.

Reviewing a little basic geometry will convince you why this is so. As **FIGURE 42.9a** shows:

- The surface area of a cube increases as a function of its linear dimension *squared*. Because a cube has six sides, the surface area of a cube of length l is $6l^2$ (six times the area of any one side).

- The volume of the same structure increases as a function of its linear dimension *cubed*. Hence, the volume (or mass) of a cube of length l is l^3.

Area has two dimensions; volume has three. In general:

$$\text{Surface area} \propto (\text{length})^2$$
$$\text{Volume (or mass)} \propto (\text{length})^3$$
$$\text{Surface area} \propto (\text{volume})^{2/3}$$

(The symbol $\propto$ means "is proportional to.")

FIGURE 42.9b graphs the consequences of these relationships. The *x*-axis plots the length of a side in a cube; the *y*-axis plots the cube's volume (orange line) or surface area (yellow line). As a cube gets bigger, its surface area increases much more slowly than does its volume (or mass).

The same general relationship holds for cells, tissues, organs, and organ systems. Larger cells, for example, have lower surface area/volume ratios than smaller cells. Quantities that are based on volume, such as body mass, increase disproportionately fast with increases in linear dimensions.

✔ If you understand the relationship between surface area and volume, you should be able to predict which of the following has the higher surface area/volume ratio: a newborn or an adult human.

How does the relationship between surface area and volume affect animal form and function?

(a) What are the surface area and volume of each cube?

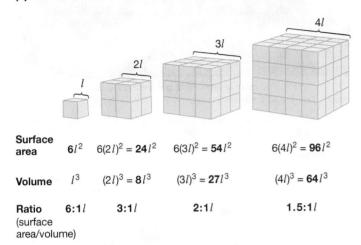

Surface area	$6l^2$	$6(2l)^2 = 24l^2$	$6(3l)^2 = 54l^2$	$6(4l)^2 = 96l^2$
Volume	l^3	$(2l)^3 = 8l^3$	$(3l)^3 = 27l^3$	$(4l)^3 = 64l^3$
Ratio (surface area/volume)	$6{:}1l$	$3{:}1l$	$2{:}1l$	$1.5{:}1l$

(b) Surface area and volume of a cube versus length of a side

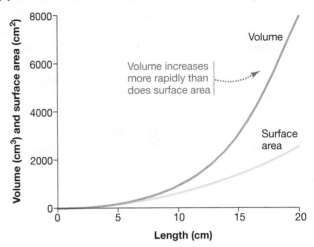

FIGURE 42.9 Surface Area and Volume Change as a Function of Overall Size. (a) The surface area of an object increases as the square of the length (l). The volume increases as the cube of the length. **(b)** Volume increases much more rapidly than does surface area as linear dimensions increase.

Surface Area/Volume Relationships: Data

As an example of how surface area/volume relationships affect an animal's physiology, consider the metabolic rate of mammals. **Metabolic rate** is the overall rate of energy consumption by an individual. Because consumption and production of energy in mammals depend largely on aerobic respiration, metabolic rate is often measured in terms of oxygen consumption, and it is typically reported in units of milliliters of O_2 consumed per hour.

Because it is so much larger, an elephant consumes a great deal more oxygen per hour than a mouse does. But what is going on at the levels of cells and tissues in these animals?

Comparing Mice and Elephants To compare metabolic rates in different species, biologists divide metabolic rate by overall mass and report a mass-specific metabolic rate in units of milliliters of oxygen per gram per hour (mL O_2/g/hr). This mass-specific metabolic rate gives the rate of oxygen consumption per gram of tissue.

Because an individual's metabolic rate varies dramatically with its activity, the accepted convention is to report the **basal metabolic rate (BMR)**—the rate at which an animal consumes oxygen while at rest, with an empty stomach, under normal temperature and moisture conditions.

FIGURE 42.10 plots per-gram or "mass-specific" BMR as a function of average body mass. Notice that the x-axis on the graph is logarithmic, to make it easier to compare very small species with very large ones. For help with logarithms, see **BioSkills 6** in Appendix B.

The graph's take-home message? On a per-gram basis, small animals have higher BMRs than do large animals. An elephant has more mass than a mouse, but a gram of elephant tissue consumes much less energy than a gram of mouse tissue does.

The leading hypothesis to explain this pattern is based on surface area/volume ratios. Many aspects of metabolism—including

oxygen consumption, food digestion, delivery of nutrients to tissues, and removal of wastes and excess heat—depend on exchange across surfaces. As an organism's size increases, its mass-specific metabolic rate must decrease. Otherwise the surface area available for exchange of materials would fail to keep up with the metabolic demands generated by the organism's enzymes.

Changes during Development A king salmon weighs a few milligrams or less at hatching but grows into an adult weighing 50 kg or more—a millionfold increase in body mass. To explore the consequences of this change, biologists have studied how gas

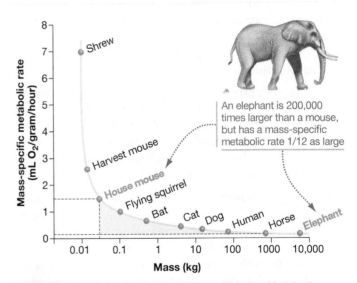

FIGURE 42.10 Small Animals Have Higher Relative Metabolic Rates than Large Animals Do. Overall body mass, plotted on a logarithmic scale, versus metabolic rate per gram of tissue.

✓ **QUESTION** Which mammal has to eat more to support each gram of its tissue: a puppy or an adult dog?

exchange—uptake of oxygen and removal of carbon dioxide—occurs in newly hatched Atlantic salmon.

Like most fish species, young salmon have rudimentary gills but also exchange gases across their skin. In aquatic animals, **gills** are organs that allow the exchange of gases and dissolved substances between the animals' blood and the surrounding water.

To document the amount of gas exchange that occurs in the gills versus the skin, researchers inserted the heads of individual larval and juvenile salmon through a pinhole in a soft rubber membrane and then recorded the rate of oxygen uptake on either side of the membrane. As the "Experimental Setup" section of **FIGURE 42.11** indicates, the gills were responsible for oxygen uptake on one side of the membrane; skin was responsible for oxygen uptake on the other side of the membrane.

The graph in the figure's "Results" section plots the percentage of total oxygen uptake that took place across the skin (green line) versus gills (purple line), as a function of body mass. Each data point represents the recordings from an individual fish.

Note that newly hatched larvae take up most of the oxygen they need by diffusion across the body surface. As an individual grows, however, its skin surface area decreases in relation to its volume. To avoid suffocation, individuals switch from skin-breathing to gill-breathing at a body mass of about 0.1 grams. What makes gills so effective in oxygen uptake?

Adaptations That Increase Surface Area

If the function of a cell or tissue depends on diffusion, it usually has a shape that increases its surface area relative to its volume. Flattening, folding, and branching are effective ways for structures to have a high surface area/volume ratio:

- *Flattening* Fish have **gill lamellae** (**FIGURE 42.12a**)—thin sheets of epithelial cells that provide the gill with an extremely high surface area relative to its volume. Because the surface available is so large, gases are able to diffuse across the gills rapidly enough to keep up with the growth in the volume of a developing fish.

- *Folding* In portions of the digestive tract where nutrients are transported into the body, the surface of the structure is folded. Extending from these folds are narrow projections called **villi** (**FIGURE 42.12b**). Together, the folds and villi make an extensive surface area available. Folded surfaces are common in diffusion-dependent organs.

- *Branching* The highly branched network shown in **FIGURE 42.12c** is a system of small, thin-walled blood vessels called **capillaries.** Capillaries have a high surface area available for gases, nutrients, and waste products to diffuse into and out of blood; branching increases the surface available in each square centimeter of tissue. In general, highly branched structures increase the surface area available for diffusion.

The amount of surface area created by flattening, folding, and branching can be impressive. The highly branched capillaries in a human have a total surface area of up to 1000 m^2; extensive folding gives a total surface area of about 140 m^2 in your lungs and

RESEARCH

QUESTION: Newly hatched salmon can breathe through their skin and through their gills. Which predominates?

HYPOTHESIS: The relative amount of gas exchange across gills and skin changes as a salmon grows.

NULL HYPOTHESIS: The relative amount of gas exchange across gills and skin does not change as a salmon grows.

EXPERIMENTAL SETUP:

One electrode measures oxygen uptake through skin

One electrode measures oxygen uptake through gills

Water flow

Water flow

Head of young salmon is inserted through a small hole in rubber membrane

PREDICTION: Juveniles will exchange a higher percentage of gas across gills and a lower percentage of gas across skin than larvae.

PREDICTION OF NULL HYPOTHESIS: Juveniles and larvae will exchange the same percentage of gas across gills and skin.

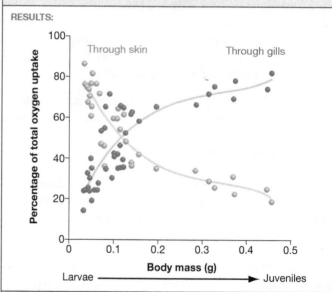

RESULTS:

Through skin

Through gills

Percentage of total oxygen uptake

Body mass (g)

Larvae ⟶ Juveniles

CONCLUSION: Breathing changes from skin to gills as larvae grow. Interpretation: Gills provide larger surface area relative to increasing volume of body.

FIGURE 42.11 How Do Young Salmon Breathe?

SOURCE: Wells, P. R., and A. W. Pinder. 1996. The respiratory development of Atlantic salmon. *Journal of Experimental Biology* 199: 2737–2744.

✔ **QUESTION** Suppose the experimenters had measured oxygen uptake on either side of the apparatus in the absence of a fish. What would the results be?

(a) Flattening: fish gill lamellae   **(b) Folding:** intestinal folds and villi   **(c) Branching:** capillaries

Villi

Lamellae

10 μm

100 μm

25 μm

FIGURE 42.12 Certain Structures Have High Surface Area/Volume Ratios. The micrograph in part (c) has been colorized to highlight capillaries, colored pink.

250 m² in your small intestine. For comparison, a doubles tennis court has a surface area of 261 m².

Surface area/volume relationships have a pervasive influence on the structure and function of animals. They will be an issue in almost every chapter in this unit.

check your understanding

If you understand that . . .

- An animal's overall size is important in part because body mass is affected by an array of physical forces.
- The amount of heat and waste that an animal produces and the amount of food and oxygen that it requires are proportional to its mass or volume.
- The amount of surface area available relative to that mass or volume is critical, because heat exchange and other important processes take place across surfaces.

✔ **You should be able to . . .**

1. Explain why large animals have a relatively low surface area/volume ratio.
2. Explain why a squirrel needs to eat more energy per unit body mass than a bear.

Answers are available in Appendix A.

42.4 Homeostasis

Adaptation and surface area/volume ratios are important themes in the analysis of animal form and function. So is homeostasis.

Homeostasis (literally, "alike-standing") is defined as stability in the chemical and physical conditions within an animal's cells, tissues, and organs. Although conditions may vary as an animal's environment changes, internal chemical and physical states are usually kept within a tolerable range.

Homeostasis: General Principles

Many of the structures and processes observed in animals can be interpreted as mechanisms for maintaining homeostasis with respect to some quantity, such as pH, temperature, or calcium ion concentration. Let's review some important general ideas about homeostasis and then analyze how homeostasis can be maintained in the face of environmental fluctuations.

Two Approaches to Achieving Homeostasis   Constancy of physiological state can be achieved by two processes: **(1)** conformation or **(2)** regulation.

1. *Conformation*   The body temperature of Antarctic rock cod closely matches that of the surrounding seawater, which is typically −1.9°C (seawater is still liquid at this temperature because of the high concentration of solutes). The rock cod does not actively regulate its body temperature to match that of seawater. Instead, its body temperature remains constant because it conforms to the temperature of its constant surroundings. If rock cod were placed into a different environment, for example warmer water, they would not maintain a constant body temperature of −1.9°C. Because these animals are adapted to such constant environmental conditions, their physiological systems cannot function properly outside of these conditions.

2. *Regulation*   Regulatory homeostasis is based on mechanisms that adjust the internal state to keep it within limits that can be tolerated, no matter what the external conditions. For example, a dog maintains a body temperature of about 38°C whether it's cold or hot outside. If the dog's body temperature rises, it might pant to cool off and maintain homeostasis. If body temperature falls, the dog might shiver to bring its temperature back up to the target value. There

are, however, limits to regulatory homeostasis. If the external temperature gets too cold or hot, animals cannot maintain homeostasis and will die.

The Role of Epithelium Because epithelium is the interface between the internal and external environments, it plays a key role in achieving homeostasis. Epithelium is responsible for forming an internal environment that can be dramatically different from the external environment such that physical and chemical conditions inside an animal can be maintained at relatively constant levels.

As subsequent chapters will show, many epithelial cells are studded with membrane proteins that regulate the transport of ions, water, nutrients, and wastes. No molecule can enter or leave the body without crossing an epithelium. Homeostasis is possible because epithelia control this exchange.

Why Is Homeostasis Important? Much of the answer to this question is based on enzyme function. Recall that enzymes are proteins that catalyze chemical reactions within cells (see Chapter 4). Temperature, pH, and other physical and chemical conditions have a dramatic effect on the structure and function of enzymes. Most enzymes function best under a fairly narrow range of conditions.

Other processes depend on homeostasis, too. Temperature changes affect membrane permeability and how quickly solutes diffuse. To take an extreme case, the expansion of water as it freezes can rip cells apart if tissues are allowed to drop much below 0°C. Conversely, extremely high temperatures can denature proteins—meaning that they lose their tertiary structure and cease to function.

When homeostasis occurs, conditions inside the body allow molecules, cells, tissues, organs, and organ systems to function at an optimal level. However, occasional departures from homeostasis can represent important adaptations. For example, a fever is a response to an infection by a pathogen. This increase in body temperature can help fight off the pathogen.

The Role of Regulation and Feedback

To achieve homeostasis, most animals have regulatory systems that monitor internal conditions such as temperature, blood pressure, blood pH, and blood glucose. If one of these variables changes, a homeostatic system acts quickly to modify it. Like the thermostat in a home heating system, each of these systems has a **set point**—a normal or target range of values for the controlled variable.

Animals have a set point for blood pH, blood oxygen and nutrient concentrations, and other parameters. In most mammals, the set point for body temperature is somewhere between 35°C and 39°C. How does an individual maintain its tissues at the set point despite changes in activity and the environment?

A homeostatic system consists of three general components: a sensor, an integrator, and an effector. **FIGURE 42.13** shows how these components interact:

1. A **sensor** is a structure that senses some aspect of the external or internal environment.

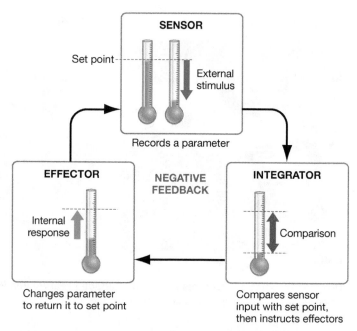

FIGURE 42.13 Animals Achieve Homeostasis through Negative Feedback. Many animals use homeostatic systems similar to this one to maintain a preferred range of hydration, blood pH, blood pressure, calcium ion concentration, body temperature, and so on.

2. An **integrator** evaluates the incoming sensory information and "decides" whether a response is necessary to achieve homeostasis. (The word decides is in quotation marks because the decision is not a conscious one.)

3. An **effector** is any structure that helps restore the desired internal condition.

Without these three elements, homeostatic control systems can't maintain a desired set point; homeostasis is impossible.

Homeostatic systems are based on negative feedback. When **negative feedback** occurs, effectors reduce or oppose the change in internal conditions. For example, a rise in blood pH triggers effectors that act to reduce that rise. Blood pH returns to the set point in response to this negative feedback.

Homeostatic systems are a key aspect of one of the five attributes of life (see Chapter 1): acquiring information from the environment and responding to it. Subsequent chapters in this unit explore how animals use sensor–integrator–effector systems to achieve homeostasis with respect to the solute concentrations of their cells and tissues, their oxygen supply, and nutrient availability. In this chapter, let's focus on how different animals achieve homeostasis with respect to body temperature.

42.5 How Do Animals Regulate Body Temperature?

All animals exchange heat with their environment. Heat flows "downhill," from regions of higher temperature to regions of lower temperature. If an individual is warmer than its surroundings, it will lose heat; if it is cooler than its environment, it will gain heat.

How does heat exchange occur?

Mechanisms of Heat Exchange

As **FIGURE 42.14** shows, animals exchange heat with the environment in four ways: conduction, convection, radiation, and evaporation.

1. **Conduction** is the direct transfer of heat between two physical bodies that are in contact with each other. For instance, when a turtle sits on a warm rock, heat is transferred from the rock to its body. The rate at which conduction occurs depends on the surface area of transfer, the steepness of the temperature difference between the two bodies, and how well each body conducts heat.

2. **Convection** is a special case of conduction. During conduction, heat is transferred between two solids; but during convection, heat is exchanged between a solid and a moving liquid or gas. For example, the heat loss that occurs when wind blows on your skin is due to convection. As the speed of the air or water flow increases, so does the rate of heat transfer.

3. **Radiation** is the transfer of heat between two bodies that are not in direct physical contact. All objects, including animals, radiate energy as a function of their temperature. The Sun radiates heat; so does your body, but to a much lesser degree.

4. **Evaporation** is the phase change that occurs when liquid water becomes a gas. Conduction, convection, and radiation can cause heat gain or loss, but evaporation leads only to heat loss. The turtle in the photograph is losing heat as water evaporates off its shell and skin. Because of the extensive hydrogen bonding in liquid water, a large amount of energy is needed

to heat water and produce evaporation (see Chapter 2). If you get overheated on a summer day, splashing water on your skin and sweating will allow you to use evaporative heat loss to cool your body. Conversely, getting wet on a cold day can be deadly. The water on your skin absorbs so much heat from your body that your temperature may drop dangerously.

Heat exchange is critical in animal physiology because individuals that get too hot or too cold may die. Overheating can cause enzymes and other proteins to denature and cease functioning. It may also lead to excessive water loss and dehydration. A sharp drop in body temperature, in contrast, can slow enzyme function and energy production. In humans, both heat stroke and hypothermia ("under-heating") are life-threatening conditions.

Although most organisms conform to the environmental temperature, some animals can regulate body temperature to keep it in an optimal range. Let's take a closer look.

Variation in Thermoregulation

The ability of animals to **thermoregulate,** or control body temperature, varies widely. Two ways to organize this variation are by examining (1) how animals obtain heat, and (2) whether body temperature is held constant.

An **endotherm** ("inner-heat") produces adequate heat to warm its own tissues, while an **ectotherm** ("outer-heat") relies principally on heat gained from the environment. Endotherms and ectotherms represent two extremes along a continuum of heat sources. Many animals fall somewhere in between.

There are also two extremes on a continuum describing whether animals hold their body temperature constant: **Homeotherms** ("alike-heat") keep their body temperature constant, while **heterotherms** ("different-heat") allow their body temperature to rise or fall depending on environmental conditions.

Humans, along with most birds and most other mammals, are strictly endothermic homeotherms. These species produce their own heat and maintain a constant body temperature. In contrast, most freshwater and terrestrial invertebrates, fishes, amphibians, and reptiles are ectothermic heterotherms whose body temperatures change throughout the day and seasonally. But many animal species lie somewhere between these extremes:

- Some desert-adapted mammals, such as the oryx featured in the introduction to this chapter, allow their body temperature to rise during the hotter part of the day—meaning they are somewhat heterothermic.

- Small mammals that inhabit cold climates lose heat rapidly because their surface area is large relative to their volume. To survive when temperatures are cold, species such as dormice reduce their metabolic rate and allow their body temperature to drop, a form of heterothermy. This condition is called **torpor.** In animals in which torpor persists for weeks or months, it is called **hibernation.**

- Naked mole rats are mammals, but lack insulation because they have no fur. They live in underground tunnels and allow their body temperature to rise and fall with burrow

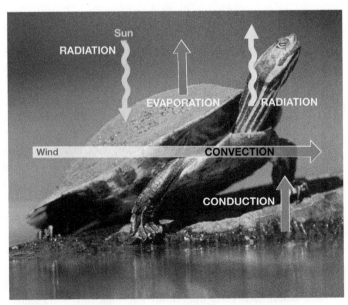

FIGURE 42.14 Four Methods of Heat Exchange. The arrows indicate the direction of heat exchange.

(a) A hornet preys on a honeybee.

1 mm

(b) A swarm of bees surrounds a hornet ...

(c) ... forming a hot defensive ball.

FIGURE 42.15 Honeybees Use Heat to Kill Predators.

temperatures. They are heterothermic and intermediate between ectotherms and endotherms.

- Japanese honeybees exhibit heterothermy when defending their hives from predatory hornets. The honeybees swarm an invading hornet and contract their flight muscles repeatedly to collectively produce heat endothermically (**FIGURE 42.15**). The temperature within the swarm rises to 47°C (117°F), killing the hornet but not the honeybees, which can tolerate temperatures up to 50°C (122°F).

Even in a homeothermic endotherm such as a mammal or bird, body temperature can vary widely in different body regions. When a Canada goose is standing on ice, its feet may be at a temperature of just 9°C, even though its body core is at 35°C. Similar variations exist in tuna and mackerel. These fish are ectotherms but generate heat to warm certain sections of their bodies, such as their eyes or swimming muscles.

Endothermy and Ectothermy: A Closer Look

Endotherms can warm themselves because their basal metabolic rates are extremely high—the heat given off by the high rate of chemical reactions is enough to warm the body. Mammals and birds retain this heat because they have elaborate insulating structures such as feathers or fur.

Ectotherms can also generate heat as a by-product of metabolism. The amount of heat they generate is small compared with the amount generated by endotherms, however, because ectotherms have relatively low metabolic rates. The most important sources of heat gain in ectotherms are radiation and conduction: Ectotherms bask in sunlight or lie on warm rocks or soil.

Endothermy and ectothermy are best understood as contrasting adaptive strategies. Because endotherms maintain a high body temperature at all times, they can be active in winter and at night. Their high metabolic rates also allow them to sustain high levels of aerobic activities, such as running or flying.

These abilities come at a cost, however: To fuel their high metabolic rates, endotherms have to obtain large quantities of energy-rich food. The energy used to produce heat is then unavailable for other energy-demanding processes, such as reproduction and growth.

In contrast, ectotherms are able to thrive with much lower intakes of food. And because they are not oxidizing food to provide heat, they can use a greater proportion of their total energy intake to support reproduction.

What's the downside of ectothermy? Chemical reaction rates are temperature dependent, so muscle activity and digestion slow dramatically as the body temperature of an ectotherm drops. As a result, ectotherms are more vulnerable to predation in cold weather and in general are less successful than endotherms at inhabiting cold environments or remaining active in cool nighttime temperatures.

In short, each suite of adaptations has advantages and disadvantages. Like all adaptations, endothermy and ectothermy involve trade-offs.

Temperature Homeostasis in Endotherms

Thermoregulation is an important aspect of homeostasis in many animals. **FIGURE 42.16** illustrates how the sensor–integrator–effector components of mammals function in thermoregulation.

Temperature receptors located throughout the body constantly monitor information about body temperature. For example, temperature receptors in the skin sense cooling or heating, and they respond by altering the pattern of electrical signals that they send to adjacent neurons. Receptors in the brain region called the **hypothalamus** respond in a similar fashion to changes in blood temperature.

The electrical signals that originate with temperature receptors are transmitted to an integrator, also located in the hypothalamus. Current evidence indicates that separate centers in the hypothalamus of the brain sense and integrate increases and decreases in body temperature.

If a mammal is cold, cells in the hypothalamus send signals to effectors that return body temperature to the set point. Signals from the hypothalamus might induce shivering to generate

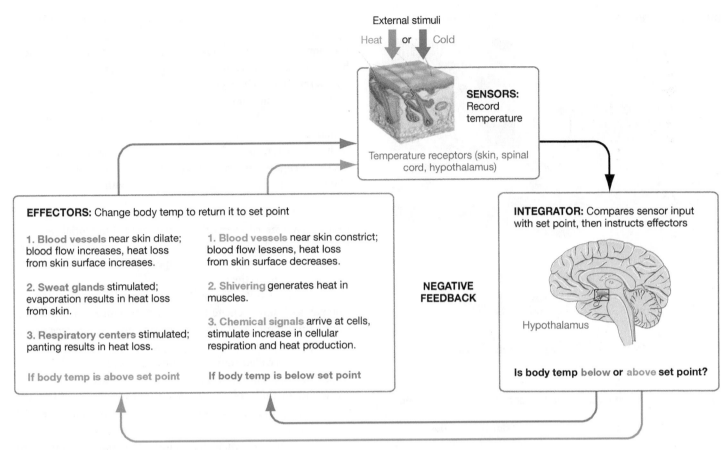

External stimuli

Heat or Cold

SENSORS: Record temperature

Temperature receptors (skin, spinal cord, hypothalamus)

EFFECTORS: Change body temp to return it to set point

1. **Blood vessels** near skin dilate; blood flow increases, heat loss from skin surface increases.

2. **Sweat glands** stimulated; evaporation results in heat loss from skin.

3. **Respiratory centers** stimulated; panting results in heat loss.

If body temp is above set point

1. **Blood vessels** near skin constrict; blood flow lessens, heat loss from skin surface decreases.

2. **Shivering** generates heat in muscles.

3. **Chemical signals** arrive at cells, stimulate increase in cellular respiration and heat production.

If body temp is below set point

NEGATIVE FEEDBACK

INTEGRATOR: Compares sensor input with set point, then instructs effectors

Hypothalamus

Is body temp below or above set point?

FIGURE 42.16 Mammals Regulate Temperature through Negative Feedback. In mammals, a set point for temperature is maintained by a complex negative feedback system that includes integrators in the anterior and posterior hypothalamus and sensors located throughout the body. The set point varies among species, from 30°C in monotremes to over 39°C in rabbits.

warmth, and fluffing of fur or feathers to improve insulation and retain heat. Signals from the same or nearby cells can also result in the release of blood-borne chemical signals that increase the rate at which cellular respiration takes place throughout the body—generating more body heat.

But if the same individual is too hot, an integrator in the hypothalamus sends signals that initiate sweating or panting—responses that cool the body. Other signals can induce behavioral changes that slow heat production, such as seeking shade or a cool burrow and then resting. Within cells, temperature spikes that are dramatic enough to denature proteins may activate **heat-shock proteins** (introduced in Chapter 3). Heat-shock proteins speed the refolding of proteins—a key step in the recovery process.

In response to either cooling or heating, behavioral and physiological responses move the body temperature back toward the set point via negative feedback. Figure 42.16 makes several points about the effectors that maintain homeostasis:

- It is common to observe redundancy in feedback systems—there are usually several ways to change a parameter.

- Feedback systems usually work in "antagonistic pairs": One set of responses increases a parameter while a corresponding set of responses decreases it.

- Input from sensors and integrators is constant, so feedback systems are constantly making fine adjustments relative to the set point.

Countercurrent Heat Exchangers

Homeothermic endotherms such as birds and mammals have sophisticated systems for thermoregulation. One of their most impressive adaptations allows them to minimize heat loss from limbs. Heat loss is a particularly important problem for mammals that live in aquatic environments. If you've ever gone swimming in cold water, you can appreciate the problem faced by seals, otters, and whales. Water is such an effective conductor of heat that aquatic organisms lose metabolic heat rapidly. To conserve heat, otters have dense, water-repellent fur that maintains a layer of trapped air next to the skin. Seals and whales are insulated by thick layers of fatty blubber. In addition, some marine mammals have specially arranged blood vessels that minimize heat loss.

An Example: Gray Whale Tongues Gray whales have a feature that minimizes heat loss from their tongue, which is exposed to cold water during feeding. The tongue contains bundles of arteries and veins. Each bundle contains an artery that carries warm, oxygenated blood from the body core. The artery is encircled by

(a) Tongue of gray whale

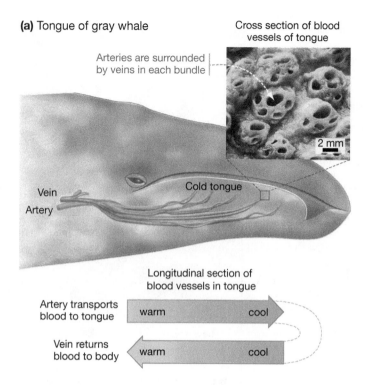

Cross section of blood vessels of tongue

Arteries are surrounded by veins in each bundle

2 mm

Vein

Artery

Cold tongue

Longitudinal section of blood vessels in tongue

Artery transports blood to tongue   warm   cool

Vein returns blood to body   warm   cool

(b) Contrasting countercurrent with "concurrent" heat exchange

Countercurrent flow:
small gradients are maintained

| 33° | 32° | 31° | 30° | 29° | 28° |

heat

| 32° | 31° | 30° | 29° | 28° | 27° |

"Concurrent" flow:
large gradients disappear quickly

| 33° | 31° | 30° | 30° | 30° | 30° |

| 27° | 29° | 30° | 30° | 30° | 30° |

FIGURE 42.17 Countercurrent Exchangers Conserve Heat.
(a) Bundles of arteries and veins in whale tongues form heat exchangers that minimize heat loss from the tongue to the cold ocean water during feeding. **(b)** Countercurrent arrangements are much more efficient than "concurrent" arrangements. The data given here are hypothetical.

smaller veins, which transport cool blood from the tongue surface back toward the body core (**FIGURE 42.17a**).

The key is that the two types of blood vessel are arranged in an antiparallel fashion. Vessels with warm blood traveling out to the tongue are in close contact with vessels containing cool blood traveling back to the body.

This type of arrangement, with fluids flowing through adjacent pipes in opposite directions, is called a **countercurrent exchanger.** The "exchanger" part of the name is apt because in a case like the whale's tongue, heat is exchanged between the warm blood in the artery and the cool blood in the veins.

To see how the countercurrent exchange system works, study the longitudinal section in **FIGURE 42.17b**. Note that the fluid that

enters the countercurrent heat exchanger is initially warm but steadily transfers heat to the adjacent, cooler fluid flowing in the opposite direction. There is a warmer-to-cooler gradient between the two currents at every point along the length of the countercurrent exchanger.

If the two solutions ran in the same direction, as in the right side of Figure 42.17b, the gradient between the two solutions would disappear quickly as the source current cooled and the recipient current heated. Countercurrent exchangers, including heat exchangers, are efficient because they maintain a gradient between the two fluids along their entire length.

A Multiplier Effect   A second key point about countercurrent heat exchangers concerns the temperature differential between the solutions. Although the differential is small at any point along the pipes, there is a large temperature differential from one end of each pipe to the other.

In effect, small differences in heat along the length of the exchanger sum up to create a large overall temperature gradient from beginning to end. The longer the system, the greater the overall differential will be. To highlight this property, the systems are sometimes called countercurrent multipliers.

Similar heat-conserving arrangements of arteries and veins are found in the flippers of whales and dolphins, and in the legs of many mammals and birds that live in cold terrestrial environments.

Countercurrent exchangers are just one of many sophisticated adaptations you'll encounter in this unit—structures and systems that allow animals to thrive in a wide array of environments.

check your understanding

C Y U

If you understand that . . .
- Two major aspects of temperature regulation vary among animal species: the amount of heat generated by the animal's own tissues and the degree to which body temperature varies over time.
- Negative feedback allows endothermic homeotherms to maintain homeostasis with respect to body temperature.
- Countercurrent heat exchangers are efficient adaptations for minimizing heat loss from extremities.

✔ **You should be able to . . .**
1. Discuss the advantages and disadvantages of endothermy and ectothermy.
2. Predict how the temperature of a whale's tongue would change if it employed concurrent rather than countercurrent flow of blood in its arteries and veins. Explain.

Answers are available in Appendix A.

If you understand . . .

42.1 Form, Function, and Adaptation

- Animal structures and their functions represent adaptations, which are heritable traits that improve survival and reproduction in a certain environment.

- Adaptations involve trade-offs, or inescapable compromises between traits.

- Acclimatization is a reversible response to the environment that improves physiological function in that environment.

✓ You should be able to provide examples of an adaptation and acclimatization of a mammal to a cold environment.

42.2 Tissues, Organs, and Systems: How Does Structure Correlate with Function?

- Cells with a common function are grouped together into four general types of tissue: connective tissue, nervous tissue, muscle tissue, and epithelial tissue.

- Epithelium forms the interface between the animal's external and internal environments.

- Organs are structures that are composed of two or more tissues that together perform specific tasks.

- Organ systems comprise organs that work together in an integrated fashion to perform one or more functions.

✓ You should be able to predict the effect of a drug that loosens the tight junctions between epithelial cells.

42.3 How Does Body Size Affect Animal Physiology?

- Large animals have smaller surface area/volume ratios than small animals. As animals grow, the volume increases more rapidly than the surface area.

- Large animals have low mass-specific metabolic rates, because they have a relatively small surface area for exchanging the oxygen and nutrients required to support metabolism, the wastes produced by metabolism, and heat.

- The relatively high surface area of small animals means that they lose heat extremely rapidly.

✓ You should be able to explain why it would be impossible for a gorilla the size of King Kong to have fur. (In answering this question, explain how the surface area/volume ratio of a normal-sized gorilla would compare to Kong's; then relate this to the role of surface area and volume in heat generation and transfer and the function of fur.)

42.4 Homeostasis

- Homeostasis refers to relatively constant physical and chemical conditions inside the body.

- Homeostasis can be achieved by conformation to a constant environment or by regulation in a fluctuating environment.

- Animals have a set point, or target value, for blood pH, tissue oxygen concentration, nutrient concentrations, temperature, and other parameters.

- When a condition in the body is not at its set point, negative feedback occurs. Responses to negative feedback return conditions to the set point and result in homeostasis.

✓ You should be able to explain how, due to acclimatization as well as adaptation, the homeostatic system for maintaining body temperature in a species of mammal might change as global temperatures rise.

42.5 How Do Animals Regulate Body Temperature?

- Animals vary from endothermic to ectothermic and from homeothermic to heterothermic.

- Most mammals have a set point for body temperature. If an individual overheats, it will pant or sweat and seek a cool environment; if an individual is cold, it will shiver, bask in sunlight, or fluff its fur.

- Countercurrent heat exchangers work by placing warm and cool fluids next to each other and running in opposite directions.

✓ The dinosaur *Apatosaurus* is one of the largest terrestrial animals that has ever lived—over 20 m in length and weighing over 20 metric tons. Do you believe that *Apatosaurus* was homeothermic or heterothermic? Explain.

MB MasteringBiology

1. **MasteringBiology Assignments**

   Tutorials and Activities Connective Tissue; Epithelial Tissue; Homeostasis; Muscle Tissue; Nervous Tissue; Overview of Animal Tissues; Surface Area/Volume Relationships; Thermoregulation

   Questions Reading Quizzes, Blue-Thread Questions, Test Bank

2. **eText** Read your book online, search, take notes, highlight text, and more.

3. **The Study Area** Practice Test, Cumulative Test, BioFlix® 3-D Animations, Videos, Activities, Audio Glossary, Word Study Tools, Art

You should be able to . . .

1. True or False: The increase in red blood cell count in tourists visiting Tibet is an example of acclimatization.

2. _____ tissues form the interface between the inside of an animal's body and the environment.

3. As an animal gets larger, which of the following occurs?
   a. Its surface area grows more rapidly than its volume.
   b. Its volume grows more rapidly than its surface area.
   c. Its volume and surface area increase in perfect proportion to each other.
   d. Its volume increases, but its total surface area decreases.

4. Which of the following best describes the set point in a homeostatic system?
   a. the cells that collect and transmit information about the state of the system
   b. the cells that receive information about the state of the system and that direct changes to the system
   c. the various components that produce appropriate changes in the system
   d. the target or "normal" value of the parameter in question

5. What does it mean to say that an animal is a heterothermic endotherm?
   a. Its body temperature varies, but most of its heat is produced by its own tissues.
   b. Its body temperature varies because it gains most of its heat from sources outside its body.
   c. Its body temperature does not vary, because most of its heat is produced by its own tissues.
   d. Its body temperature does not vary, even though it gains most of its heat from sources outside its body.

6. Which of the following is an advantage that ectotherms have over endotherms of the same size?
   a. They require much less food.
   b. They are less vulnerable to predation events during cold weather.
   c. They can remain active in cold weather or at nighttime—when temperatures cool.
   d. They have higher metabolic rates and grow much more quickly.

7. Why is epithelium a particularly important tissue in achieving homeostasis with respect to temperature?

8. The metabolic rate of a frog in summer (at 35°C) is about eight times higher than in winter (at 5°C). Compare and contrast the individual's ability to move, exchange gases, and digest food at the two temperatures. During which season will the frog require more food energy, and why?

9. Consider the following:
   - absorptive sections of digestive tract
   - capillaries
   - beaks of Galápagos finches
   - fish gills

   In each case, how does the structure relate to the function?

10. Why is the surface area/volume ratio different in a small sphere versus a large sphere? If materials diffuse into and out of each sphere, in which of the two will diffusion occur more efficiently relative to the volume? Explain your answer.

11. QUANTITATIVE Consider three cubes with lengths of 1 cm, 5 cm, and 10 cm. Calculate the surface area and the volume of each cube, and plot them on a graph with length on the x-axis and surface area and volume on the y-axis. Explain how the graph shows the relationship between size and surface area/volume ratio.

12. Which of the following is an example of negative feedback? Explain why each is or is not negative feedback.
   a. Uterine contractions during childbirth stimulate release of a hormone that stimulates more uterine contractions.
   b. A viral infection stimulates the hypothalamus to increase the set point for body temperature.
   c. High blood sugar concentrations stimulate the release of the hormone insulin, which stimulates the sugars to move from the blood into cells.
   d. The body temperature of the Arabian oryx rises during the hottest parts of the day.

13. When food is scarce, bigger *Geospiza fortis* (medium ground finches) win contests over seeds. Biologists have documented that there is strong natural selection in favor of large body size under these conditions, and that the finch population evolves in response. If so, why aren't *G. fortis* a lot bigger than they are?

14. What data would you need to collect in order to document that adaptation is occurring in a particular population of lizards, as the warming associated with global climate change intensifies?

15. An engineer has to design a system for dissipating heat from a new type of car engine that runs particularly hot. Recall that heat is gained and lost as a function of surface area. Suggest ideas to consider that are inspired by biological structures with exceptionally high surface area/volume ratios.

16. Suppose a friend of yours is trying to decide whether to buy a pet turtle or a pet mouse. In making the decision, all he cares about is how much it will cost to feed the animal. Your friend says he can't decide because the two animals weigh the same and their food costs the same per pound. As a biologist, what's your advice?
   a. Buy the turtle, because it is an ectotherm and will therefore consume less food than the mouse.
   b. Buy the turtle, because it is an endotherm and will therefore consume less food than the mouse.
   c. Buy the mouse, because it is an ectotherm and will therefore consume less food than the turtle.
   d. Buy the mouse, because it is an endotherm and will therefore consume less food than the turtle.

43 Water and Electrolyte Balance in Animals

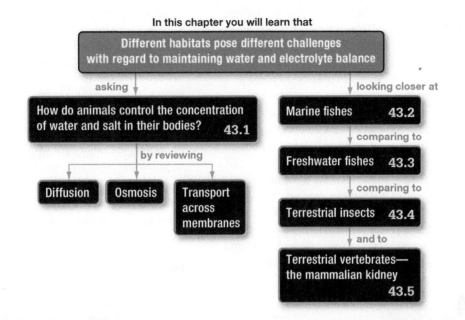

In this chapter you will learn that

Different habitats pose different challenges with regard to maintaining water and electrolyte balance

asking

How do animals control the concentration of water and salt in their bodies? **43.1**

by reviewing

Diffusion   Osmosis   Transport across membranes

looking closer at

Marine fishes   **43.2**

comparing to

Freshwater fishes   **43.3**

comparing to

Terrestrial insects   **43.4**

and to

Terrestrial vertebrates— the mammalian kidney   **43.5**

Terrestrial animals lose water every time they breathe, defecate, and urinate. For many animals, drinking is an important way to gain water and achieve homeostasis. This chapter explores how terrestrial and aquatic animals maintain water balance.

This chapter is part of the Big Picture. See how on pages 840–841.

The chemical reactions that make life possible occur in an aqueous solution. If the balance of water and dissolved substances in the solution is disturbed, those chemical reactions—and life itself—may stop. Humans can stay alive for weeks without eating but can survive just three days without drinking water. If hurricanes introduce enough freshwater to the ocean shore to disrupt normal salt concentrations, marine animals die. Maintaining water balance is a matter of life or death.

An animal achieves water balance when its intake of water equals its loss of water. Water balance is an important element in homeostasis—the ability to keep cells and tissues in constant and favorable conditions.

✔ When you see this checkmark, stop and test yourself. Answers are available in Appendix A.

Water balance is intimately associated with sustaining a balanced concentration of electrolytes throughout the body. An **electrolyte** is a compound that dissociates into ions when dissolved in water. Electrolytes got their name because they conduct electrical current.

In many animals, the most abundant electrolytes are sodium (Na^+), chloride (Cl^-), potassium (K^+), and calcium (Ca^{2+}). Cells require precise concentrations of these ions to function normally. In humans, electrolyte imbalances can lead to muscle spasms, confusion, irregular heart rhythms, fatigue, paralysis, or even death.

Water and electrolyte balance is also associated with excretion. Animals produce urine to excrete wastes. Urine contains water, so excretion of urine inevitably leads to water loss. The amount of water lost in the urine of an animal depends both on its hydration state and on the type of wastes it produces.

This chapter is focused on a single question: How do animals maintain water and electrolyte balance in marine, freshwater, and terrestrial environments? Answering it will introduce you to some of the most complex and important homeostatic systems known. You can review the importance of maintaining water and electrolyte balance in the Big Picture of Plant and Animal Form and Function, on pages 840–841.

43.1 Osmoregulation and Excretion

In organisms, electrolytes and water move by two processes—diffusion and osmosis (introduced in Chapter 6).

- **Diffusion** is the net movement of substances from regions of higher concentration to regions of lower concentration.

- **Osmosis** is a special case of diffusion. It is the net movement of water from regions of higher water concentration to regions of lower water concentration, across a selectively permeable membrane.

A membrane that exhibits **selective permeability,** such as a phospholipid bilayer, is a membrane that some solutes can cross more easily than other solutes can. When either diffusion or osmosis occurs, net movement of ions and molecules occurs along their **concentration gradient.**

FIGURE 43.1a illustrates how dissolved substances, or **solutes,** move down their concentration gradients via diffusion across a selectively permeable membrane. When the solutes are randomly distributed throughout the solutions on both sides of the membrane, equilibrium is established. Molecules continue to move back and forth across the membrane at equilibrium, but at equal rates.

As **FIGURE 43.1b** shows, water can also move down its concentration gradient (see Chapter 6). The concentration of dissolved substances in a solution, measured in osmoles[1] per liter, is the solution's **osmolarity.** Because solutes interact with water molecules, an increase in solute concentration reduces the amount

of water available for osmosis, effectively lowering water concentration. If dissolved substances are separated by a selectively permeable membrane and the solutes cannot cross that membrane, water moves from areas of lower osmolarity—that is, where solute concentrations are lower and water concentration is higher—to areas of higher osmolarity—where solute concentrations are higher and water concentration is lower.

Now let's examine how osmosis and diffusion affect water and ion balance living in different environments.

What Is Osmotic Stress?

Osmotic stress occurs when the concentration of dissolved substances in a cell or tissue is abnormal. It means that water and solute concentrations are different from their set point.

Many organisms respond to osmotic stress by osmoregulating, just as they respond to heat or cold stress by thermoregulating (see Chapter 42). **Osmoregulation** is the process by which living organisms control the concentration of water and electrolytes in their bodies.

Not all animals encounter osmotic stress. For many marine invertebrates such as sponges, jellyfish, and flatworms, achieving homeostasis with respect to water and electrolyte balance is straightforward. Seawater is a fairly constant ionic and osmotic environment, and it nearly matches the normal electrolyte concentrations found within these animals. These species are **osmoconformers.**

Seawater is **isosmotic** with respect to tissue in these species. Stated another way, solute concentrations inside and outside these animals are equal. Because the tissues of osmoconforming marine invertebrates are isosmotic to seawater, diffusion and osmosis don't alter water and electrolyte balance and induce osmotic stress.

Osmotic Stress in Seawater, in Freshwater, and on Land

In contrast to most marine invertebrates, marine bony fishes are **osmoregulators.** These fishes actively regulate osmolarity inside their bodies to achieve homeostasis.

By osmoregulating, marine bony fishes keep the osmolarity of their tissues lower than seawater. The difference in osmolarity is most important in the gills, which are organs involved in gas exchange. For gas exchange to occur with the environment, the epithelial cells on the surfaces of the gills must be in direct contact with seawater. But because there is a large difference in the concentration of solutes between the inside of each cell and the seawater outside, water tends to flow by osmosis out of the gill epithelium (**FIGURE 43.2**).

Seawater is **hyperosmotic** in comparison to the tissues of marine fishes—the solution inside the body contains fewer solutes than does the solution outside. If the water that marine fishes lose across their gills is not replaced, the cells will shrivel and die. These species face a trade-off between gas exchange and water and electrolyte balance.

Marine fishes replace the lost water by drinking large quantities of seawater. Drinking brings in excess electrolytes, however.

---

[1]The unit osmole is similar to a mole except that it takes into account molecules that dissociate in solution. For example, because NaCl dissociates into Na^+ and Cl^- in solution, adding 1 mole of NaCl is equivalent to adding 2 osmoles.

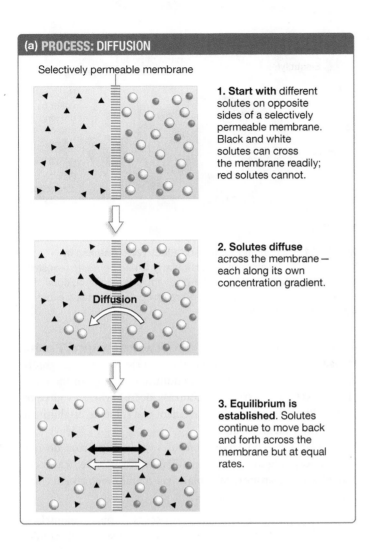

(a) PROCESS: DIFFUSION

Selectively permeable membrane

1. Start with different solutes on opposite sides of a selectively permeable membrane. Black and white solutes can cross the membrane readily; red solutes cannot.

Diffusion

2. Solutes diffuse across the membrane — each along its own concentration gradient.

3. Equilibrium is established. Solutes continue to move back and forth across the membrane but at equal rates.

(b) PROCESS: OSMOSIS

Selectively permeable membrane

Low osmolarity | High osmolarity

1. Start with more solute on one side of a selectively permeable membrane than the other. In this case, the solute cannot cross the membrane.

Osmosis

2. Water diffuses across the membrane — from the region of lower osmolarity (lower solute concentration) to the region of higher osmolarity (higher solute concentration).

FIGURE 43.1 Solutes Move Down a Concentration Gradient via Diffusion; Water Moves Down a Concentration Gradient via Osmosis. (a) Diffusion occurs any time a solute is at higher concentration in one location than another. **(b)** Osmosis is a special case of diffusion involving the movement of water across a selectively permeable membrane.

✔**QUESTION** Why doesn't the presence of red molecules on the right side of part (a) affect the movement of the black and white molecules?

Electrolyte balance is thrown even further out of whack because ions and other solutes diffuse into the gill epithelium, following a concentration gradient from seawater to tissues.

To rid themselves of these excess electrolytes, marine fishes have to actively pump ions out of their bodies and back into seawater, using membrane proteins found in the gill epithelium.

Freshwater fishes osmoregulate in an environment dramatically different from the ocean. Marine fishes are under osmotic stress because they lose water and gain solutes; freshwater fishes are under osmotic stress because they gain water and lose solutes.

Why? In the gills of freshwater fishes, epithelial cells contain more solutes than the solution outside. The freshwater is **hyposmotic** relative to the fishes' tissues. As a result, the fishes gain water via osmosis across the gill epithelium (**FIGURE 43.3**, see page 864), which puts the body under osmotic stress. Just as in marine fishes, there is a trade-off between gas exchange and osmoregulation.

If a freshwater fish does not get rid of incoming water, its cells will burst and the individual will die. To achieve homeostasis and survive, freshwater fishes excrete large amounts of water in their urine and do not drink.

In addition to gaining water, freshwater fishes undergo osmotic stress because ions and other solutes tend to diffuse out of the gill epithelium into the environment, along a concentration

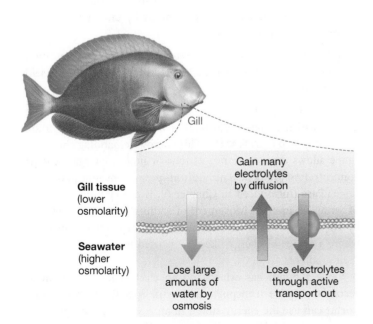

Gill tissue (lower osmolarity)

Seawater (higher osmolarity)

Gain many electrolytes by diffusion

Lose large amounts of water by osmosis

Lose electrolytes through active transport out

FIGURE 43.2 Marine Fishes Lose Water by Osmosis and Gain Electrolytes by Diffusion.

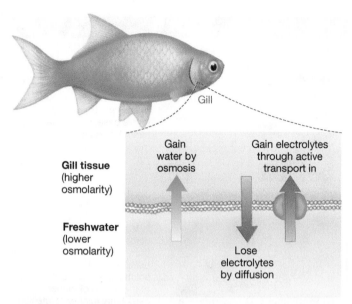

FIGURE 43.3 Freshwater Fishes Gain Water by Osmosis and Lose Electrolytes by Diffusion.

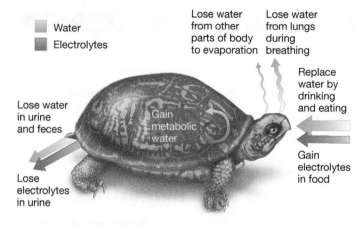

FIGURE 43.4 Maintaining Water and Electrolyte Balance in Terrestrial Environments Is Particularly Challenging. In terrestrial environments, animals lose water by evaporation from the body surface and as water vapor during breathing. Electrolytes are lost primarily in the urine and feces.

gradient. Freshwater animals must replace electrolytes that are lost by obtaining them in food or by actively transporting them from the surrounding water—usually across the gills.

What about land animals? In terms of water balance, terrestrial environments are similar to the ocean. Land animals constantly lose water to the environment, just like many marine animals do. In this case, however, the process involved is not osmosis but evaporation (see Chapter 42).

The epithelial cells that line a turtle's lung and a fruit fly's gas exchange structures have a moist surface, in order to protect the integrity of their plasma membranes and to facilitate diffusion of gases across respiratory epithelia. Because the atmosphere is almost always drier than the wet gas exchange surface, terrestrial animals lose water by evaporation. Once again, there is a trade-off between breathing and balance of water and electrolytes.

Water balance is further complicated because animals lose water in the form of urine, and some terrestrial species lose additional water when they sweat or pant to lower their body temperature (**FIGURE 43.4**). The lost water has to be replaced by drinking, ingesting water in food, or gaining metabolic water—the H_2O produced during cellular respiration (see Chapter 9). The relative importance of each of these methods of replacing water depends on the animal. For example, desert animals like the Arabian oryx (see Chapter 42) do not have access to drinking water throughout much of the year, so they rely more on ingested and metabolic water.

How Do Electrolytes and Water Move across Cell Membranes?

What molecular mechanisms allow animals to cope with the diverse challenges they face in maintaining water and electrolyte balance? You might recall from earlier chapters that solutes

move across membranes by passive or active transport. **Passive transport** is driven by diffusion along an electrochemical gradient and does not require an expenditure of energy in the form of ATP. **Active transport,** in contrast, occurs when a source of energy like ATP powers the movement of a solute to establish an electrochemical gradient.

Because ions and large molecules such as glucose do not cross phospholipid bilayers readily, most passive transport and all active transport take place via membrane proteins (see Chapter 6).

- In many cases, passive transport occurs through **channels**— proteins that form a pore, or opening, that selectively admits specific ions or molecules (**FIGURE 43.5a**).

- Passive transport also occurs via **carrier proteins,** which are transmembrane proteins that bind a specific ion or molecule and transport it across the membrane by undergoing a conformational change.

- When solutes move from an area of high to low concentration via channels or carrier proteins, **facilitated diffusion** is said to occur.

Active transport is based on membrane proteins called pumps, which change conformation when they bind ATP or are phosphorylated (**FIGURE 43.5b**). This energy-demanding change in shape allows them to transport ions or molecules against their concentration gradient. The **sodium–potassium pump,** or Na^+/K^+-ATPase (introduced in Chapter 6), is a very important type of pump in animals. Movement of sodium and potassium ions by Na^+/K^+-ATPase is an example of **primary active transport,** where a source of energy like ATP is used to move ions against their gradients.

Once a pump has established an electrochemical gradient, **secondary active transport** can occur. Specifically, a **cotransporter** can use the energy stored in an electrochemical gradient across a cell membrane to transport a different solute *against* its

(a) Passive transport is based on diffusion down a gradient.

Facilitated diffusion

Direct diffusion | Channel | Carrier

Electrochemical gradient

Outside cell

Lipid bilayer

Inside cell

Membrane proteins

(b) Active transport requires an energy source to move molecules against a gradient.

Primary active transport (driven by energy from ATP)

Secondary active transport (cotransport; driven by energy from gradient)

Pump | Symporter | Antiporter

Electrochemical gradient

ATP   ADP + P_i

FIGURE 43.5 Mechanisms of Passive and Active Transport.
(a) In passive transport, solutes diffuse along their electrochemical gradient either directly or through a membrane protein.
(b) In primary active transport, a pump uses ATP to move solutes against their electrochemical gradients. In secondary active transport (cotransport), energy stored in an electrochemical gradient for one molecule is used to move another molecule against its gradient.

electrochemical gradient. A cotransporter that moves solutes in the same direction is called a **symporter;** a cotransporter that moves solutes in opposite directions is called an **antiporter.**

How does water move? To date, there are no known mechanisms for actively transporting water across plasma membranes. Instead, cells use pumps to transport ions and set up an osmotic gradient; water then follows by osmosis—often through the specialized membrane channels called **aquaporins** (see Chapter 6). In essence, cells move water by moving solutes.

The body excretes excess solutes, along with waste products, using a urinary system. Because solutes and wastes often must be dissolved in water to be excreted, water balance is fundamentally related to excretion.

Types of Nitrogenous Wastes: Impact on Water Balance

Animal cells contain amino acids and nucleic acids that are used to synthesize proteins, RNA, and DNA. Excess amino acids and nucleic acids can be broken down in catabolic reactions that result in the production of **ammonia** (NH_3). Ammonia is toxic to cells, because at high concentrations it raises the pH of intracellular and extracellular fluids enough to poison enzymes.

Because ammonia must be dissolved in water, excretion of ammonia inevitably leads to water loss. However, the amount of water lost during excretion depends on the method of ammonia excretion.

Forms of Nitrogenous Waste Vary among Species How do animals get rid of ammonia safely and efficiently? Different species solve the problem in different ways (see **TABLE 43.1**, page 866).

- In freshwater fishes, ammonia is diluted to low concentration and excreted in watery urine.

- In freshwater and marine fishes, ammonia diffuses across the gills into the surrounding water along a concentration gradient.

- In mammals (including humans) and adult amphibians, enzyme-catalyzed reactions convert ammonia to a much less toxic compound called **urea,** which is excreted in urine.

- In terrestrial arthropods, birds, and other reptiles, reactions convert ammonia to **uric acid,** the white, paste-like substance that you have probably seen in bird feces.

Uric acid is a particularly interesting form of nitrogenous waste. Compared with urea and ammonia, uric acid is extremely insoluble in water—which explains why it is so difficult to wash bird droppings off a car. As a result, birds, snakes, lizards, and terrestrial arthropods can get rid of excess nitrogen while losing little water.

Why Do Nitrogenous Wastes Vary among Species? The type of nitrogenous waste produced by an animal correlates with its lineage—its evolutionary history. For example, mammals excrete urea while reptiles (including birds) and insects excrete uric acid (Table 43.1).

Evolutionary history is not the entire story, however. Waste production is also related to the habitat that a species occupies, and thus the amount of osmotic stress it endures.

- Terrestrial birds conserve water by excreting about 90 percent of their nitrogenous waste as uric acid and only 3–4 percent as ammonia, but aquatic birds such as ducks excrete just 50 percent of their excess nitrogen as uric acid and 30 percent as ammonia.

- Tadpoles are aquatic and excrete ammonia, but many adult frogs and toads are terrestrial and excrete urea.

- Production of urea and uric acid is particularly common in animals—such as reptiles—that live in dry habitats.

Attribute	Ammonia	Urea	Uric Acid
Solubility in water (moles/liter)	high	medium	very low
Water loss (amount required for excretion of waste)	high	medium	very low
Energy cost (amount of ATP required)	low	high	high
Toxicity	high	medium	low
Groups where it is the primary waste	bony fishes, aquatic invertebrates	mammals, amphibians, cartilaginous fishes	birds and other reptiles, most terrestrial insects and spiders (arthropods)
Method of synthesis	product of breakdown of amino acids and nucleic acids	synthesized in liver, starting with amino groups from amino acids	synthesis starts with amino acids and nucleic acids
Method of excretion	in urine, and diffuses across gills	in urine (mammals); diffuses across gills (sharks)	with feces

To make sense of these observations, biologists point out that there is a fitness trade-off between the energetic cost of excreting each type of waste and the benefit of conserving water. Ammonia excretion requires a large water loss but little energy expenditure, because the molecule isn't processed by enzymes. Uric acid excretion, in contrast, requires almost no loss of water but a sophisticated series of enzyme-catalyzed, energy-demanding reactions. Different trade-offs are favored in different environments.

Now that you have a basic understanding of the osmoregulatory and excretory challenges facing animals, let's take a closer look at how marine, freshwater, and terrestrial animals maintain water and electrolyte homeostasis.

43.2 Water and Electrolyte Balance in Marine Fishes

Marine fishes experience severe osmotic stress because they live in water with a very high osmolarity. Since their divergence over 400 million years ago, marine bony and cartilaginous fishes have evolved distinct strategies for dealing with osmotic stress.

Osmoconformation versus Osmoregulation in Marine Fishes

Marine bony fishes are osmoregulators. Recall that they maintain a lower blood osmolarity than seawater by drinking to replace water lost via osmosis, and by actively transporting electrolytes out of the body. This process comes with a significant energetic cost.

Cartilaginous fishes, such as sharks, have a different strategy. They are **osmoconformers**—their blood osmolarity nearly matches that of the sea water. However, the composition of their blood is quite different from seawater. Sharks have low

concentrations of ions in their blood, but they maintain relatively high blood concentrations of urea. This increases their blood osmolarity so that it is isosmotic with seawater. The result? Sharks lose less water by osmosis.

If sharks avoid water loss to the surrounding seawater by osmoconforming, then why don't all marine fishes osmoconform? The answer lies in a biological trade-off (see Chapter 42). Sharks must expend energy to make proteins that protect their cells from the toxic effects of high concentrations of urea. Osmoregulation and osmoconformation are two strategies for living in the ocean, and each has its own costs and benefits.

Even though they are osmoconformers, sharks still maintain relatively low concentrations of salt (NaCl) in their blood. Sharks need to excrete salt because sodium and chloride ions diffuse into their gill cells from seawater, down the concentration gradients for the ions. Research on the molecular mechanism of salt excretion in sharks revealed two key points:

1. **The mechanism is general.** The salt-excreting system discovered in sharks is found in a wide array of species, including *Homo sapiens*. It is functioning in your kidneys, right now.

2. **The mechanism represents a critically important concept in physiology.** Plant and animal cells use active transport to set up a strong electrochemical gradient for one ion—typically Na^+ in animals and H^+ in plants. The sodium or proton gradient is then used to transport a wide array of other substances, without further expenditure of energy.

This is fundamental stuff. Let's dig in.

How Do Sharks Excrete Salt?

Research on salt excretion in sharks focused on an organ called the **rectal gland,** which secretes a concentrated salt solution. To determine how this gland works, researchers studied it in vitro—meaning outside the shark's body, in a controlled laboratory environment. The basic approach was to dissect rectal glands, immerse them in a solution with a defined composition and osmolarity, and analyze the fluid that the rectal gland produced in response.

Early experiments showed that normal salt excretion occurred only if the solution in the rectal gland contained ATP. This result supported the hypothesis that salt excretion involves active transport. Ions can be concentrated only if they are actively transported against a concentration gradient. The question was, How does concentration occur?

The Role of Na^+/K^+-ATPase An energy-demanding mechanism for salt excretion implies that a protein in the plasma membrane of epithelial cells is actively pumping Na^+, Cl^-, or both. The best-characterized candidate was the Na^+/K^+-ATPase.

To test the hypothesis that the sodium–potassium pump is involved in salt excretion by sharks, biologists used a plant defense compound called **ouabain** (pronounced *WAA-bane*). This molecule is toxic to animals because it binds to Na^+/K^+-ATPase and prevents it from functioning.

Just as predicted, rectal glands that are treated with ouabain stop producing a concentrated salt solution. This was strong evidence that Na^+/K^+-ATPase is essential for salt excretion.

A Molecular Model for Salt Excretion Subsequent work has shown that salt excretion in sharks is a multistep process, summarized in **FIGURE 43.6**.

1. Na^+/K^+-ATPase pumps sodium ions out of epithelial cells across the basolateral surface, into the extracellular fluid. The pump creates an electrochemical gradient favoring the diffusion of Na^+ into the cell. This "master gradient" allows the cell to transport other ions without an additional expenditure of energy.

2. Na^+, Cl^-, and K^+ all enter the cell by secondary active transport through a cotransporter, powered by the Na^+ master gradient. Note that this master gradient causes Na^+ to diffuse into the cell *along* its electrochemical gradient, allowing Cl^- and K^+ to move into the cell *against* their electrochemical gradients. Movement of Cl^- and K^+ is possible only because these ions move through a cotransporter with Na^+.

3. As Cl^- builds up inside the cell, Cl^- diffuses down its concentration gradient into the lumen of the gland through a chloride channel located in the apical membrane.

4. Following their electrochemical gradient, Na^+ ions diffuse into the lumen of the gland through spaces between the cells.

In many animals, epithelial cells that transport sodium and chloride ions contain the same combination of membrane proteins found in the shark rectal gland. These species include

- marine birds and reptiles that drink seawater and excrete NaCl via salt glands in their nostrils;

- marine fishes that excrete salt from their gills (see Figure 43.2);

- mammals that transport salt in their kidneys.

Research on the shark rectal gland also produced an unforeseen benefit for biomedical research. Several years after the shark chloride channel was characterized, investigators identified a human protein called cystic fibrosis transmembrane regulator

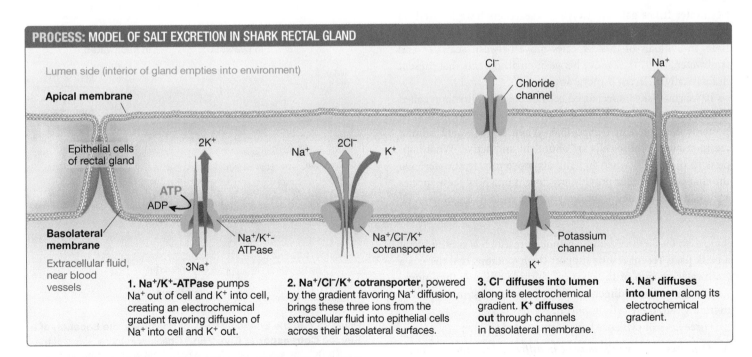

PROCESS: MODEL OF SALT EXCRETION IN SHARK RECTAL GLAND

1. **Na^+/K^+-ATPase** pumps Na^+ out of cell and K^+ into cell, creating an electrochemical gradient favoring diffusion of Na^+ into cell and K^+ out.

2. **$Na^+/Cl^-/K^+$ cotransporter**, powered by the gradient favoring Na^+ diffusion, brings these three ions from the extracellular fluid into epithelial cells across their basolateral surfaces.

3. **Cl^- diffuses into lumen** along its electrochemical gradient. **K^+ diffuses out** through channels in basolateral membrane.

4. **Na^+ diffuses into lumen** along its electrochemical gradient.

FIGURE 43.6 The Shark Rectal Gland Rids the Body of Excess Salt.

✔ **QUESTION** Which of these membrane proteins are involved in (1) primary active transport, (2) secondary active transport, or (3) passive transport?

(CFTR). Cystic fibrosis is the most common genetic disease in populations of northern European extraction (see Chapter 6). Although the disease was known to be associated with defects in the CFTR protein, no one knew what the molecule did.

When investigators realized that the amino acid sequence of CFTR is 80 percent identical to that of the shark chloride channel, it was their first hint that CFTR is involved in Cl^- transport. Subsequent studies supported the hypothesis that cystic fibrosis results from a defect in a chloride channel. With this result, studies on water and electrolyte balance in sharks shed light on an important human disease.

43.3 Water and Electrolyte Balance in Freshwater Fishes

Research on the shark rectal gland and the gills of marine bony fishes has uncovered the molecular mechanisms of salt balance in marine fishes. How do freshwater fishes achieve homeostasis with respect to electrolytes?

How Do Freshwater Fishes Osmoregulate?

Recall from Figure 43.3 that freshwater fishes have to cope with an osmotic stress that is opposite the challenge facing marine fishes. Freshwater fishes lose electrolytes across their gill epithelium by diffusion across a concentration gradient. To maintain homeostasis, they have to actively transport ions back into the body across the gill epithelium. How do they do this?

Salmon and Sea Bass as Model Systems   To understand how freshwater fishes gain electrolytes, researchers have focused on sea bass and several species of salmon. Over the course of a lifetime, individuals of these species move between seawater and freshwater. Thus, they move between environments that present dramatically different osmotic stressors.

In marine fishes, specialized cells in the gill epithelium called chloride cells move salt using the combination of membrane proteins illustrated in Figure 43.6. When sea bass and salmon are in seawater, these cells are abundant and active. What happens to these cells when individuals move into freshwater? Do the changes that occur provide any insight into how these species acclimatize to the new environment and avoid dying of osmotic stress?

A Freshwater Chloride Cell?   Although research is continuing at a brisk pace, recent results support the hypothesis that there is a freshwater version of the classical chloride cell: one that moves ions in the opposite direction of the seawater version. Instead of excreting salt, these cells import it.

Three lines of evidence have accumulated to date:

1. ***Osmoregulatory cells may be in different locations.***   Young salmon taken from freshwater versus seawater have chloride cells in different locations on the gills. Adult salmon

from freshwater versus seawater show the same pattern, and similar changes have been observed in other fish species that switch between freshwater and seawater habitats. The pattern suggests that when the nature of osmotic stress changes, the nature of the gill epithelium changes. Specifically, active pumping of ions takes place in a different population of cells in seawater versus freshwater.

2. ***Different forms of Na^+/K^+-ATPase may be activated.***   The salmon genome contains genes for several different forms of the Na^+/K^+-ATPase. There is now strong evidence that different forms are activated when individuals are in seawater versus freshwater.

3. ***The orientation of key transport proteins "flips."***   In sea bass, researchers have been able to stain epithelial cells to determine the location of the cotransporter illustrated in Figure 43.6—the one that brings Na^+, Cl^-, and K^+ into the cell. When individuals are in seawater, the protein is located in the basolateral side of chloride cells. But when individuals are in freshwater, the protein is located in the apical side (**FIGURE 43.7**).

Taken together, the data suggest that freshwater fishes have a freshwater version of the chloride cell, with different forms of Na^+/K^+-ATPase and transporters that result in the import rather than export of ions. Identifying the mechanisms of electrolyte uptake in freshwater fishes is an important challenge for researchers who want to know how aquatic organisms cope with osmotic stress.

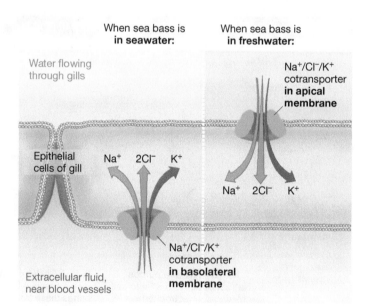

FIGURE 43.7 In the Epithelial Cells of Bass Gills, the Location of a Key Ion Cotransporter Can "Flip." Changes in the position of the $Na^+/Cl^-/K^+$ cotransporter help sea bass deal with osmotic stress in both seawater and freshwater environments. (Other membrane channels and pumps are present but not shown here.)

If you understand that . . .

- Marine fishes lose water by osmosis. To replace it, they have to drink seawater.
- Marine fishes have to rid themselves of salt. They gain salt when they drink seawater or when sodium and chloride ions diffuse into their cells along a concentration gradient.
- Freshwater fishes gain water by osmosis. They have to rid themselves of excess water by urinating.
- Freshwater fishes lose electrolytes to the surrounding water by diffusion. They gain electrolytes in their food and by active transport from the surrounding water.

✔ **You should be able to . . .**

Predict what happens to the osmolarity of a freshwater fish's tissues when epithelial cells in its gills are treated with ouabain—a molecule that poisons Na^+/K^+-ATPase.

Answers are available in Appendix A.

43.4 Water and Electrolyte Balance in Terrestrial Insects

By studying extreme situations or unusual organisms, biologists can often gain insight into how organisms cope with more moderate environments. In studies on the molecular mechanisms of water and electrolyte balance in terrestrial insects, the most valuable model organisms have been the desert locust and a common household pest called the flour beetle. (You may have seen the larvae of flour beetles, called mealworms, in bags of flour that were not shut tightly enough to keep adults from entering and breeding.)

Desert locusts and flour beetles live in environments where osmotic stress is severe. These insects rarely, if ever, drink—simply because little or no water is available in the habitats they occupy.

How do they maintain water and electrolyte balance? The answer has two parts: They minimize water loss from their body surface, and they carefully regulate the amount of water and electrolytes that they excrete in their urine and feces. Let's look at each issue in turn.

How Do Insects Minimize Water Loss from the Body Surface?

Terrestrial animals breathe by exposing an extremely thin layer of epithelium to the atmosphere (see Chapter 45). Oxygen diffuses into this epithelium, and carbon dioxide diffuses out. But water constantly leaks across the thin respiratory surface and is lost to the atmosphere via evaporation.

Evaporation from the body surface itself is another threat—a particular challenge to insects, because they are small. Small organisms have a high surface area/volume ratio (see Chapter 42). Insects have a relatively large surface area from which to lose water but a small volume in which to retain it.

How do desert locusts, flour beetles, and other insects minimize water loss during gas exchange? In these species, gas exchange occurs across the membranes of epithelial cells that line the **tracheae,** an extensive system of tubes. The insect tracheal system connects with the atmosphere at openings called **spiracles** (FIGURE 43.8a). Muscles just inside each spiracle open or close the pore, much as guard cells open or close the pores in plant leaves and stems.

When investigators manipulated insects called *Rhodnius* so that their spiracles stayed open and placed the animals in a dry environment, the insects died within three days. These data support the hypothesis that the ability to close spiracles is an important adaptation for minimizing water loss during respiration. If an insect is under osmotic stress, it may be able to close its spiracles and wait until conditions improve before resuming activity.

FIGURE 43.8b shows how insects minimize evaporation from the surface of their bodies. This diagram is a cross-sectional view of the exoskeleton of an insect, which consists of a tough,

(a) Spiracles can be closed to minimize water loss from tracheae.

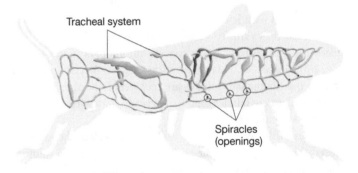

(b) Except at spiracles, the insect body is covered with wax.

FIGURE 43.8 In Desert Locusts, Adaptations Limit Water Loss during Respiration and from the Body Surface.

✔ **QUESTION** In desert grasshoppers, spiracles are located in a row along the underside of the abdomen. Why would this location help minimize water loss?

nitrogen-containing polysaccharide called chitin and layers of protein, with a layer of wax on the outside. This combination of chitin, protein, and wax is known as the **cuticle.**

Recall that waxes, a type of lipid, are highly hydrophobic and thus highly impermeable to water (see Chapter 6). Researchers who removed the wax from insect exoskeletons have confirmed that the rate of water loss from the body surface increases sharply. Based on this observation, the wax layer is interpreted as an adaptation that minimizes evaporative water loss.

For insects, minimizing water loss is only half the battle in avoiding osmotic stress. To maintain homeostasis, insects must also carefully regulate the composition of a blood-like fluid called **hemolymph.** Hemolymph is pumped by the heart and transports electrolytes, nutrients, and waste products.

To maintain water and electrolyte balance, insects rely on excretory organs called **Malpighian tubules,** and on their hindgut—the posterior portion of their digestive tract (**FIGURE 43.9a**).

Filtrate Forms in the Malpighian Tubules

As Figure 43.9a shows, Malpighian tubules have a large surface area, are in direct contact with the hemolymph, and empty into the hindgut. The Malpighian tubules are responsible for forming a **filtrate** from the hemolymph. This "pre-urine" then passes into the hindgut, where it is processed and modified before excretion.

To explore how the Malpighian tubules work, biologists collected filtrate from the lumen of Malpighian tubules in mealworms and compared its composition with that of hemolymph from the same individuals. The two solutions were roughly isosmotic, but not identical.

How do they differ? Compared to hemolymph, the solution inside the tubules has a lower concentration of Na^+ and a higher concentration of K^+. This observation suggests that the epithelial cells in the Malpighian tubules are relatively impermeable to sodium ions but contain a pump that actively transports potassium ions into the tubules.

To test this hypothesis, researchers dissected the tubules, rinsed them, and bathed their interior and exterior in a solution containing a known concentration of potassium ions. When they measured the concentration of electrolytes on either side of the membrane over time, they found that K^+ accumulated in the tubule lumen, against its concentration gradient.

This result supported the hypothesis that cells in the membranes of Malpighian tubules contain pumps that actively transport potassium ions into the lumen of the organ. Subsequent work has shown that a high concentration of potassium ions brings water into the tubules by osmosis. Other electrolytes and nitrogenous wastes then diffuse into the filtrate along their concentration gradients.

The Hindgut: Selective Reabsorption of Electrolytes and Water

The filtrate that accumulates inside the Malpighian tubules flows into the hindgut, where it joins material emerging from the digestive tract. If an insect is osmotically stressed due to a shortage of electrolytes and water, electrolytes and water from the filtrate are reabsorbed in the hindgut and returned to the hemolymph,

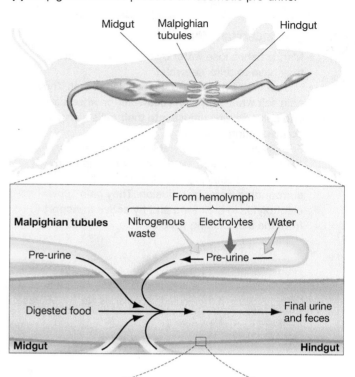

(a) Malpighian tubules produce an isosmotic pre-urine.

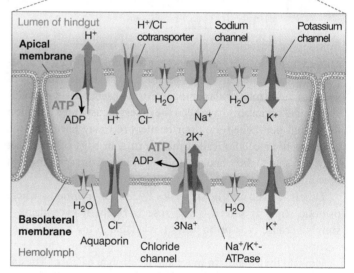

(b) Under osmotic stress, the hindgut reabsorbs electrolytes and water to form a hyperosmotic urine.

FIGURE 43.9 In Insects, Urine Forms in the Malpighian Tubules and Hindgut. (a) The isosmotic filtrate that forms in the Malpighian tubules empties into the hindgut. (b) In the hindgut, the primary driving forces for reabsorption of electrolytes and water are H^+ pumps in the apical membrane and Na^+/K^+-ATPases in the basolateral membrane.

while the uric acid remains in the hindgut. Reabsorption results in formation of hyperosmotic final urine, conservation of water, and efficient elimination of nitrogenous wastes.

In desert locusts, flour beetles, and other species in extremely dry environments, 80 to 95 percent of the water in the filtrate is reabsorbed and kept inside the body. The ability to recover this water allows these insects to live in dry habitats such as deserts and flour bins. How does reabsorption happen?

The mechanism involves a series of specific membrane pumps and channels, not unlike the system found in the chloride cells of fishes. To study it, researchers positioned the hindgut epithelium from a desert locust as a sheet dividing two solutions. They manipulated electrolyte concentrations on either side of the hindgut wall and measured changes in the solutions over time.

For example, when investigators removed K^+ and Na^+ from the solution on the lumen side of the organ, water reabsorption stopped. These data established that the hindgut's ability to recover water from urine depends on ion movement: The epithelial cells in the hindgut transport ions out of the filtrate and into the hemolymph. Water follows by osmosis, forming concentrated urine.

How do the ions move? By poisoning the experimental membranes with ouabain, biologists confirmed that Na^+/K^+-ATPase is involved in moving ions out of the lumen and into the hemolymph. Ouabain-treated membranes continued to transport Cl^-, however. Years of experiments on the locust hindgut resulted in the model in **FIGURE 43.9b**:

- H^+ is pumped into the hindgut lumen, creating an electrical gradient that favors movement of K^+ into the cells through potassium channels. Water follows via osmosis.

- H^+ diffuses along its gradient from the lumen into the cell through a H^+/Cl^- cotransporter, bringing Cl^- into the cell against its gradient.

- In the basolateral membrane, Na^+/K^+-ATPase sets up electrochemical and osmotic gradients that favor diffusion of Na^+ from the hindgut lumen into the epithelial cell, and diffusion of Cl^- and K^+ from the epithelial cell into the hemolymph. Water moves into the hemolymph by osmosis.

Regulating Water and Electrolyte Balance: An Overview Several general principles that have emerged from studies of insect excretion turn out to be relevant to vertebrate systems as well:

- Water moves only by osmosis—it is not pumped directly. Water moves between cells or body compartments via osmotic gradients that are set up by the active transport of ions.

- The formation of the filtrate is not particularly selective. Most of the molecules present in the hemolymph are also present in the Malpighian tubules.

- In contrast to filtrate formation, reabsorption is highly selective. The protein pumps and channels involved in reabsorption are highly specific for certain ions and molecules. Waste products do not pass through the hindgut membrane. Instead, they remain in the hindgut and are eliminated along with the feces. Only valuable ions and molecules are reabsorbed.

- In contrast to filtrate formation, reabsorption is tightly regulated. The membrane pumps and channels involved in reabsorption are activated and deactivated in response to osmotic stress. If an insect is dehydrated, virtually all of the water in the filtrate is reabsorbed. But if it has plenty to drink, reabsorption

does not occur and the urine is watery and hyposmotic to the individual's hemolymph. The system is dynamic and allows precise control over water and electrolyte balance.

Given the success of insects in terms of the numbers of species and individuals and the array of habitats they occupy, it is clear that their systems for maintaining water and electrolyte balance are remarkably effective.

check your understanding

If you understand that . . .

- Terrestrial insects are prone to dehydration, primarily via evaporation from their respiratory surfaces.
- Terrestrial insects have a cuticle, respiratory system, and excretory system that are adapted to conserve water.
- In terrestrial insects, urine production begins with formation of a filtrate that is isosmotic with hemolymph, followed by selective and tightly regulated reabsorption of ions, nutrients, and water.

✔ You should be able to . . .

Explain how the following traits are involved in water retention:

1. Excretion of ammonia in the form of uric acid.
2. Selective reabsorption of electrolytes in the hindgut.

Answers are available in Appendix A.

43.5 Water and Electrolyte Balance in Terrestrial Vertebrates

With respect to water loss, terrestrial vertebrates face the same hazards that terrestrial insects do. Crocodiles, turtles, lizards, frogs, birds, and mammals lose water from their body surfaces, from the surface of their lungs every time they breathe, and in their urine. Electrolytes are also lost in urine and, in some species, in sweat. To replace the water they lose, most terrestrial vertebrates drink. They also ingest electrolytes in food.

In land-dwelling vertebrates, osmoregulation occurs primarily through events that take place in the key organ of the urinary system, the **kidney.** The kidney is responsible for water and electrolyte balance, as well as the excretion of nitrogenous wastes. In terms of function, it is analogous to the Malpighian tubules and hindgut of insects. Let's take a closer look at the kidney in mammals.

The Structure of the Mammalian Kidney

Mammalian kidneys occur in pairs and tend to be bean shaped. A large blood vessel called the renal artery brings blood that contains nitrogenous wastes into the organ; the renal vein is a large blood vessel that carries "clean" blood away.

(a) Urinary system   **(b)** Kidney

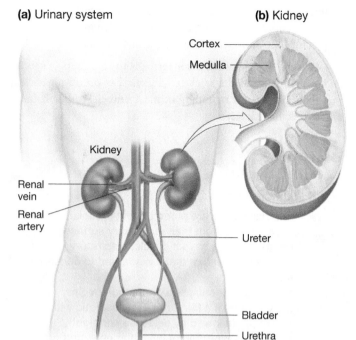

FIGURE 43.10 **Anatomy of the Human Urinary System and Kidney.** **(a)** In mammals, kidneys are paired and are located near the spinal column. **(b)** The kidney has an outer region called the cortex and an inner area called the medulla.

The urine that forms in the kidney is transported via a long tube called the **ureter** to a storage organ, the **bladder.** From the bladder, urine is transported to the body surface through the **urethra** and then excreted. In most vertebrates, the kidneys are located near the dorsal (back) side of the body. **FIGURE 43.10a** summarizes the parts of the urinary system.

Most of the kidney's mass is made up of small structures called nephrons. The **nephron** is the basic functional unit of the kidney. The work involved in maintaining water and electrolyte balance occurs in the nephron.

Most of the approximately 1 million nephrons in a human kidney are located in the outer region of the organ, or **cortex** (**FIGURE 43.10b**). But some nephrons extend from the cortex into the kidney's inner region, or **medulla.**

The Function of the Mammalian Kidney: An Overview

The nephron shares important functional characteristics with the insect excretory system:

- Water cannot be transported actively—it moves only by osmosis.
- To move water, cells in the kidney set up strong osmotic gradients.
- By regulating these gradients and specific channel proteins, kidney cells exert precise control over loss or retention of water and electrolytes.

(a) The structure of the nephron and collecting duct   **(b)** Blood vessels serve each nephron.

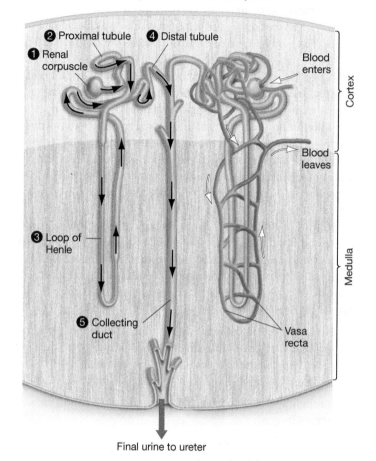

Final urine to ureter

FIGURE 43.11 **A Nephron Has Four Major Parts, Empties into a Collecting Duct, and Is Served by Blood Vessels.** Urine formation begins in the renal corpuscle and ends in the collecting duct.

FIGURE 43.11a provides a detailed view of the nephron. Note that it has four major regions and is closely associated with a tube called the collecting duct. **FIGURE 43.11b** illustrates a second key point about the anatomy of the nephron: It is served by blood vessels that wrap around each of its four regions.

The four major nephron regions and the collecting duct each have a distinct function:

1. The *renal corpuscle* filters blood, forming a filtrate or "pre-urine" consisting of ions, nutrients, wastes, and water.

2. The *proximal tubule* has epithelial cells that reabsorb nutrients, valuable ions, and water from the filtrate into the bloodstream.

3. The *loop of Henle* establishes a strong osmotic gradient in the interstitial fluid surrounding the loop, with osmolarity increasing as the loop descends.

4. The *distal tubule* reabsorbs ions and water in a regulated manner—one that helps maintain water and electrolyte balance according to the body's needs.

5. The *collecting duct* may reabsorb more water to maintain homeostasis. In addition, urea leaves the base of the collecting

duct and contributes to the osmotic gradient in the interstitial fluid set up by the loop of Henle.

The blood vessels associated with the nephron play a key role as well: They bring "dirty" blood into the nephron and then take away the molecules and ions that are reabsorbed from the initial filtrate.

Now let's delve into the details. The sections that follow trace the flow of material through each component of the nephron and out of the collecting duct.

Filtration: The Renal Corpuscle

In terrestrial vertebrates, urine formation begins in the **renal corpuscle** (literally, "kidney-little-body"). Notice in Figure 43.11a that the nephron consists of a tube that is closed at one end and open at the other. The closed end is the beginning of the nephron; the open end is the terminus of the collecting duct.

As **FIGURE 43.12a** shows, the closed end of the nephron forms a capsule that encloses a cluster of tiny blood vessels, or capillaries. These vessels bring blood to the nephron from the renal artery. Collectively, the cluster of capillaries is called the **glomerulus** ("ball of yarn"). The region of the nephron that surrounds the glomerulus is named **Bowman's capsule.** Together, the glomerulus and Bowman's capsule make up the renal corpuscle.

FIGURE 43.12b illustrates a key feature of the glomerular capillaries: They have large pores, or openings. In addition, they are surrounded by unusual cells whose membranes fold into a series of slits and ridges.

The structure of the renal corpuscle allows it to function as a **filtration** device. Water and small solutes from the blood pass through the pores and slits into the Bowman's capsule. Filtration is based on size: Proteins, cells, and other large components of blood do not fit through the pores and do not enter the nephron. They remain in the blood instead.

Stated another way, urine formation starts with a size-selective filtration step—with blood pressure supplying the force required to perform filtration. In vertebrates, blood is under higher pressure than the surrounding tissues because it is pumped by the heart through a closed system of vessels. This pressure is enough to force water and small solutes through the pores in the glomerulus, so the renal corpuscle strains large volumes of fluid without expending energy in the form of ATP.

✔ If you understand this concept, you should be able to describe the contents of blood on one side of the filter (glomerulus) and the contents of the filtrate on the other side.

It is critical to note two additional facts about the filtration step in urine formation:

1. The renal corpuscles of a human kidney are capable of producing about 180 liters of filtrate per day. This is an impressive volume—think of 180 one-liter bottles of soft drink arranged on a supermarket shelf.

2. About 99 percent of the filtrate is reabsorbed—only a tiny fraction of the original volume is actually excreted as urine.

Filtering large volumes from the blood allows wastes to be removed effectively; pairing this process with selective

(a) Anatomy of the renal corpuscle

Blood leaves glomerulus.

Bowman's capsule

Glomerulus

Pre-urine leaves Bowman's capsule.

Blood enters glomerulus.

(b) Filtration

Pores in glomerulus

Filtration slits in cells that wrap around vessel

Large molecules and cells remain in bloodstream.

Fluid and small solutes are pushed through the pores and the filtration slits into Bowman's capsule.

Direction of blood movement

FIGURE 43.12 Urine Formation Begins When Blood Is Filtered in the Renal Corpuscle. (a) The renal corpuscle consists of Bowman's capsule and the glomerulus. **(b)** The capillaries in the glomerulus have pores and are surrounded by cells that have filtration slits. Blood pressure forces water and small molecules out of the capillaries, through the slits, and into Bowman's capsule.

reabsorption allows waste excretion to occur with a minimum of water and nutrient loss.

Reabsorption: The Proximal Tubule

Where does filtrate reabsorption begin? Filtrate leaves Bowman's capsule and enters a convoluted structure called the **proximal tubule.** The filtrate inside this tubule contains water and small solutes such as urea, glucose, amino acids, vitamins, and electrolytes. Some of these molecules are waste products; others are valuable nutrients.

Active Transport Occurs in Epithelial Cells The epithelial cells of the proximal tubule have a prominent series of small projections, called **microvilli** ("little shaggy hairs"), facing the lumen. The microvilli greatly expand the surface area of this epithelium.

(a) Microvilli expand surface area of lumen of proximal tubule.

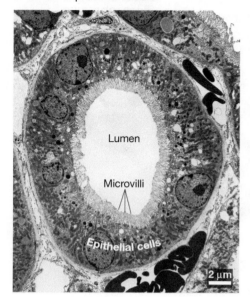

(b) Model of selective reabsorption in proximal tubules

FIGURE 43.13 Water, Electrolytes, and Nutrients Are Reabsorbed in the Proximal Tubule.

A large surface area provides space for membrane proteins that act as pumps, channels, and cotransporters (**FIGURE 43.13a**).

Epithelial cells in the proximal tubule are also packed with mitochondria, which suggests that ATP-demanding active transport is occurring. Based on these observations, biologists hypothesized that the proximal tubule functions in the active transport of selected molecules out of the filtrate.

By injecting solutions of known composition into isolated rabbit and rat proximal tubules in the presence or absence of ATP, researchers confirmed that selected electrolytes and nutrients are actively reabsorbed from the filtrate that enters the tubules.

When solutes move from the proximal tubule into epithelial cells, and then into the bloodstream, water follows along the osmotic gradient. In this way, valuable solutes and water are reabsorbed and returned to the bloodstream.

Ion and Water Movement Is Driven by a "Master Gradient"

FIGURE 43.13b summarizes the current model of the molecular mechanisms involved in selective reabsorption in the proximal tubule:

1. Na^+/K^+-ATPase in the basolateral membranes removes Na^+ from the interior of the cell. The active transport of sodium ions out of the cell creates a gradient favoring the entry of Na^+ from the lumen.

2. In the apical membrane adjacent to the lumen, Na^+-dependent cotransporters use this gradient to remove valuable ions and nutrients selectively from the filtrate. The movement of Na^+ into the cell, *along* its concentration gradient, provides the means for moving other solutes *against* a concentration gradient.

3. The solutes that move into the cell diffuse across the basolateral membrane into the interstitial fluid and then nearby blood vessels.

4. Water follows the movement of ions from the proximal tubule into the cell and then out of the cell and into blood vessels. Recall that water moves by osmosis across the membranes of these epithelial cells through membrane proteins called aquaporins.

Almost all of the nutrients, along with about two-thirds of the NaCl and water that is originally filtered by the renal corpuscle, are reabsorbed in the proximal tubule. The osmolarity of the tubular fluid is unchanged despite this huge change in volume, however, because water reabsorption is proportional to solute reabsorption.

In effect, then, the cells that line the proximal tubule act as a recycling center. The filtration step in the renal corpuscle is based on size; the reabsorption step in the proximal tubule selectively retrieves small substances that are valuable. The pumps and cotransporters in the proximal tubule reabsorb nutrients, electrolytes, and water but leave wastes. As the filtrate flows into the loop of Henle, it has a relatively high concentration of waste molecules and a relatively low concentration of nutrients.

Creating an Osmotic Gradient: The Loop of Henle

In mammals, the fluid that emerges from the proximal tubule enters the **loop of Henle**—named for Jacob Henle, who described it in the early 1860s. In most nephrons, the loop is short and barely enters the medulla. But in about 20 percent of the nephrons present in a human kidney, the loop is long and plunges from the cortex of the kidney deep into the medulla.

In 1942 Werner Kuhn offered a hypothesis, inspired by countercurrent heat exchangers, to explain what the loop of Henle does. Recall that a countercurrent heat exchanger is a system in which two adjacent fluids flow through pipes in opposite directions (see Chapter 42). Kuhn proposed that the loop of Henle is a countercurrent exchanger and multiplier. It doesn't exchange heat, however. Instead, it sets up an osmotic gradient.

Specifically, Kuhn proposed that the osmolarity of the filtrate in the loop of Henle is low in the cortex and high in the medulla. Further, Kuhn maintained that the osmolarity in the interstitial fluid surrounding the loop mirrors the gradient inside the loop. This is a key point.

Testing Kuhn's Hypothesis A series of papers published during the 1950s and early 1960s supplied important experimental support for the countercurrent exchange model. **FIGURE 43.14** reproduces two particularly important data sets, obtained by comparing the osmolarity of kidney tissue slices. In both graphs, the x-axis shows the location in the kidney, from cortex to medulla. The y-axis indicates osmolarity, measured either as the percentage of the maximum observed or as solute concentration.

- Figure 43.14a shows data on the osmolarity of fluid inside the loop of Henle. The vertical lines represent the range of values observed at a particular location. As predicted by Kuhn's model, a strong gradient in osmolarity exists from the cortex to the medulla.

- Figure 43.14b shows that the concentrations of Na^+, Cl^-, and urea also increase sharply from the cortex to the medulla in the interstitial fluid outside the loop of Henle.

These results suggested that the solutes responsible for the gradient outside the loop are Na^+, Cl^-, and urea. The change in concentration of urea turned out to be particularly important.

How Is the Osmotic Gradient Established? The loop of Henle has three distinct regions: the descending limb, the thin ascending limb, and the thick ascending limb (**FIGURE 43.15a**, see page 876). The thin and thick ascending limbs differ in the thickness of their walls. Do the three regions also differ in their permeability to water and solutes?

It took over 15 years of experiments performed in laboratories around the world to formulate a definitive answer to that question. Researchers punctured the loop of Henle of rodents with a micropipette, analyzed the composition of the fluid inside, and compared it with the nephron's final product—urine.

In the ascending limb of the loop of Henle, Na^+ and Cl^- constituted at least 60 percent of the solutes; urea constituted about 10 percent. But in the distal tubule, urea was the major solute. These data suggested that Na^+ and Cl^-, but not urea, were being removed somewhere in the ascending limb. How?

Na^+ and Cl^- were also present at high concentrations in the tissue surrounding the thick ascending limb, so researchers hypothesized that sodium might be actively pumped out of this portion of the nephron. The hypothesis was that the active transport of Na^+ out of the filtrate in the thick ascending limb would create an electrical gradient that would also favor the loss of Cl^-.

(a) Fluid inside the loop of Henle

(b) Interstitial fluid outside the loop of Henle

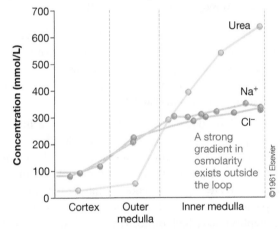

FIGURE 43.14 Data Confirm the Existence of a Strong Osmotic Gradient Both Inside and Outside the Loop of Henle. As the nephron plunges into the inner medulla, the concentration of dissolved solutes increases both inside **(a)** and outside **(b)** the loop of Henle.

DATA: K. J. Ullrich, K. Kramer, and J.W. Boyer. 1961. *Progress in Cardiovascular Diseases* 3: 395–431.

Follow-up experiments using ouabain and other poisons supported the hypothesis that sodium ions are actively transported out of the solution inside the thick ascending limb, and chloride ions follow along an electrochemical gradient. The epithelial cells responsible for salt excretion are configured almost exactly like the epithelium of the shark rectal gland (see Figure 43.6).

What is happening in the descending limb and the thin ascending limb of the loop of Henle? By injecting solutions of known concentration into the nephrons of rabbits, biologists documented that the descending limb is highly permeable to water but almost completely impermeable to solutes. The thin ascending limb of the loop, in contrast, is highly permeable to Na^+

(a) Three regions in the loop of Henle

(b) Water and ion movement differ in the three regions.

Descending limb is highly permeable to water but impermeable to solutes

Ascending limb is nearly impermeable to water but highly permeable to Na⁺ and Cl⁻

FIGURE 43.15 The Loop of Henle Maintains an Osmotic Gradient because Water Leaves the Descending Limb and Salt Leaves the Ascending Limb. The values inside the loop in part (b) represent the osmolarity of the filtrate.

and Cl^-, moderately permeable to urea, and almost completely impermeable to water.

A Comprehensive View of the Loop of Henle All of the observations just summarized came together in 1972 when two papers, published independently, proposed the same comprehensive model for how the loop of Henle works. To understand this model, follow the events in **FIGURE 43.15b**:

1. As fluid flows down the descending limb, the fluid inside the loop loses water to the interstitial fluid surrounding the nephron. This movement of water is passive—it does not require an expenditure of ATP. The water follows an osmotic gradient created by the ascending limb. At the bottom of the loop—in the inner medulla—the fluids inside and outside the nephron have high osmolarity. The filtrate does not continue to lose water, though, because the membrane in the ascending limb is nearly impermeable to water.

2. The fluid inside the nephron loses Na^+ and Cl^- in the thin ascending limb. The ions move passively, along their concentration gradients.

3. Toward the cortex, the osmolarity of the surrounding interstitial fluid is low. Additional Na^+ and Cl^- ions are actively transported out of the nephron in the thick ascending limb.

The countercurrent flow of fluid, combined with changes in permeability to water and in the types of channels and pumps that are active in the epithelium of the nephron, creates a self-reinforcing system. The presence of an osmotic gradient stimulates water and ion flows that in turn maintain an osmotic gradient.

Here's how it works: Movement of NaCl from the ascending limb into surrounding tissue increases the osmolarity of the fluid outside the descending limb, which results in a flow of water out of the water-permeable walls of the descending limb, via osmosis. This loss of water in the descending limb increases the osmolarity

in the fluid entering the ascending limb. The high concentration of salt in the fluid at the base of the ascending limb triggers a passive flow of ions out—reinforcing the osmotic gradient.

✔ If you understand this concept, you should be able to predict what happens to the osmotic gradient when the drug furosemide inhibits membrane proteins that pump sodium and chloride ions out of the thick ascending limb. Specifically, how does this drug affect (1) water reabsorption in the descending limb, (2) the osmolarity of the filtrate in the bottom of the loop of Henle, and (3) reabsorption of salt in the thin ascending limb?

The Vasa Recta Removes Water and Solutes That Leave the Loop of Henle What happens to the water and salt that move out of the loop from the filtrate into the interstitial fluid? They quickly diffuse into the **vasa recta,** a network of blood vessels that runs along the loop and eventually joins up with small veins in the kidney. As a result, the water and electrolytes that are reabsorbed are returned to the bloodstream instead of being excreted in urine (**FIGURE 43.16**).

The removal of water that leaves the descending limb is particularly important. If it were not drawn off into the bloodstream, it would dilute the concentrated fluid outside the loop of Henle and quickly destroy the osmotic gradient.

The Collecting Duct Leaks Urea Urea is the solute that is most responsible for the steep osmotic gradient in the space surrounding the nephron. Urea is at high concentration in the interstitial fluid in the inner medulla and low concentration in the outer medulla. This gradient exists because the innermost section of the collecting duct is permeable to urea.

Although the system created by the nephron, vasa recta, and collecting duct may seem complex, its outcome is simple: the creation and maintenance of a strong osmotic gradient with the minimum possible expenditure of energy.

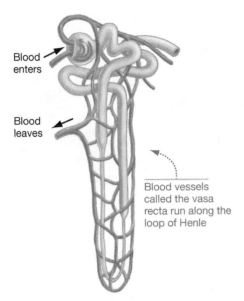

FIGURE 43.16 Blood Supply to the Loop of Henle. Water and solutes reabsorbed from the loop of Henle move into blood vessels (vasa recta).

Regulating Water and Electrolyte Balance: The Distal Tubule and Collecting Duct

The first three steps in urine formation—filtration, reabsorption, and establishment of an osmotic gradient—result in a fluid that is slightly hyposmotic to blood. Once the filtrate has passed through the loop of Henle, the major solutes that it contains are urea and other wastes along with a low concentration of ions.

The filtrate that enters the distal tubule is always dilute. In contrast, the urine that leaves the collecting duct is dilute when the individual is well hydrated but concentrated when the individual is dehydrated. How is this possible?

Urine Formation Is under Hormonal Control   The answer is based on two observations about the **distal tubule** and **collecting duct:** Their activity is (1) highly regulated, and (2) altered in response to osmotic stress. The amount of Na^+, Cl^-, and water that is reabsorbed in the distal tubule and in the collecting duct varies with the animal's hydration.

Changes in the distal tubule and collecting duct are controlled by **hormones**—signaling molecules (introduced in Chapter 11 and explored further in Chapter 49). Specifically:

- If Na^+ levels in the blood are low, the adrenal glands release the hormone **aldosterone,** which leads to activation of sodium–potassium pumps and reabsorption of Na^+ in the distal tubule. Water follows by osmosis. Aldosterone saves sodium and water.

- If an individual is dehydrated, the brain releases **antidiuretic hormone (ADH).** (The term diuresis refers to increased urine production, so antidiuresis means inhibition of urine production.) ADH is also referred to as vasopressin or arginine vasopressin. ADH saves water.

How Does ADH Work?   Epithelial cells of the collecting duct are joined by tight junctions (see Chapter 11), making them impermeable to water and solutes. ADH has two important effects on the permeability of epithelial cells in the collecting duct:

1. ADH triggers the insertion of aquaporins into the apical membrane. As a result, cells become much more permeable to water and large amounts of water are reabsorbed.

2. ADH increases permeability to urea, which is reabsorbed into the surrounding fluid. This helps create a concentration gradient favoring water reabsorption from the filtrate.

As **FIGURE 43.17a** shows, water leaves the collecting duct passively—following the concentration gradient maintained by

(a) ADH present: Collecting duct is highly permeable to water.

(b) No ADH present: Collecting duct is not permeable to water.

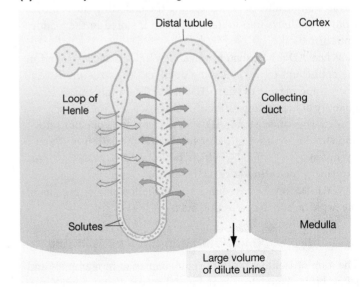

FIGURE 43.17 ADH Regulates Water Reabsorption by the Collecting Duct.

Structure	Function
Renal Corpuscle (Bowman's capsule and glomerulus)	Size-selective filtration: forms filtrate from blood (water and other small substances enter nephron)
Proximal Tubule	Reabsorbs electrolytes (active transport), nutrients, water
Loop of Henle	Maintains osmotic gradient in interstitial fluid from outer to inner medulla
• Descending limb	• Permeable to water (passive transport out of filtrate)
• Thin ascending limb	• Permeable to Na^+, Cl^- (passive transport out of filtrate)
• Thick ascending limb	• Active transport of Na^+, Cl^- out of filtrate
Distal Tubule	With aldosterone: Reabsorbs Na^+. Without aldosterone: Does not reabsorb Na^+.
Collecting Duct	Regulates water retention and loss
• Main portion	• With ADH: Water leaves filtrate; produces small volume of urine that is hyperosmotic to blood Without ADH: Water stays in filtrate; produces large volume of urine that is hyposmotic to blood
• Innermost portion	• Urea leaks out by passive transport to establish and/or maintain high osmolarity of inner medulla

FIGURE 43.18 The Cloaca Is a Cavity into Which the Urinary, Gastrointestinal, and Reproductive Tracts Empty.

the loop of Henle. When ADH is present, water is conserved and urine is strongly hyperosmotic relative to blood. The collecting duct is the final place in which the composition of the filtrate can be altered. Urine exiting the collecting ducts moves from the kidneys into ureters and then is stored in the bladder until urination.

When ADH is absent, however, few aquaporins are found in the epithelium of the collecting duct, and the structure is relatively impermeable to water (**FIGURE 43.17b**). In this case, a larger quantity of hyposmotic urine is produced.

✔ If you understand ADH's effect on the collecting duct, you should be able to predict how urine formation is affected by ingestion of ethanol, which inhibits ADH release, versus nicotine, which stimulates ADH release.

TABLE 43.2 reviews the functions of the four major regions of the nephron and the collecting duct.

Urine Formation in Nonmammalian Vertebrates

The loop of Henle is an important adaptation in mammals and some birds. Water loss is reduced because these animals can produce urine that is hyperosmotic to their blood. In contrast, fishes, amphibians, and non-avian reptiles lack loops of Henle, and their kidneys are therefore unable to produce concentrated urine.

Many fishes and amphibians do not need to produce concentrated urine. But conserving water is important in reptiles, especially those inhabiting deserts. Recall that reptiles produce nitrogenous wastes in the form of uric acid, which is excreted with very little water and is therefore hyperosmotic to their body tissues. If reptiles' kidneys produce isosmotic urine, how does it become hyperosmotic?

In most reptiles, the ureters empty isosmotic urine into the **cloaca,** a cavity into which the urinary, gastrointestinal, and reproductive tracts all empty (**FIGURE 43.18**). Reptiles are able to absorb water from urine across the wall of the cloaca into the bloodstream. Eventually, a semisolid uric acid paste is excreted along with the feces.

If you understand that . . .

• The mammalian kidney is specialized for the production of hyperosmotic urine.
• The loop of Henle is a countercurrent system that maintains a strong osmotic gradient in the medulla.
• The characteristics of urine change in response to hormonal signals. The signals trigger changes in the nephron and collecting duct that either save or eliminate water and solutes.

✔ **You should be able to . . .**

Predict how the following events would affect urine production: drinking massive amounts of water, eating large amounts of salt, and refraining from drinking water for 48 hours.

Answers are available in Appendix A.

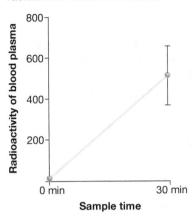

(a) Change following injection of radioactive water into bladder

(b) Change following injection of pure water into bladder or stomach

○ Bladder
● Stomach

@2007 The Company of Biologists

FIGURE 43.19 Gila Monsters Can Reabsorb Water from Their Bladders.
(a) Radioactively labeled water passes from bladder into bloodstream. **(b)** Pure water reduces blood osmolarity at a similar rate in dehydrated lizards whether added to stomach or to bladder.
DATA: Davis, J. R., and D. F. DeNardo. 2007. *The Journal of Experimental Biology* 210: 1472–1480.

✔**QUESTION** What physiological trade-offs might be imposed by carrying a large amount of water in the bladder?

Some reptiles also have a bladder that collects the isosmotic urine from the ureters and stores it before emptying it into the cloaca. When water is available, these reptiles drink a lot, and their bladders fill up with dilute urine. Researchers hypothesized that the bladder acts as a "canteen" to store water during times when water is unavailable.

To test this hypothesis, the investigators injected the bladders of dehydrated Gila monsters—a large desert lizard—with radioactively labeled water and drew blood samples over time. The radioactivity of the lizards' blood increased within a half hour, which meant that the radioactively labeled water from the bladder was being absorbed into the bloodstream (**FIGURE 43.19a**).

The researchers also examined whether water in the bladder can rehydrate lizards to the same extent as drinking water.

Working with two groups of dehydrated Gila monsters, they injected water into the bladders of one group, and—to mimic drinking—they injected water into the stomachs of the second group. They found that the osmolarity of the plasma decreased at the same rate in both groups (**FIGURE 43.19b**).

So, the bladder can indeed act as a "canteen," allowing lizards to carry a water supply that they can access when water is scarce. As water is reabsorbed from the bladder, the urine becomes more and more concentrated, but the osmolarity of the blood remains low.

Whether considering the mammalian nephron or the Gila monster bladder, the vertebrate urinary system is remarkably effective in regulating water and electrolyte balance and achieving homeostasis.

CHAPTER 43 REVIEW

For media, go to MasteringBiology

If you understand . . .

43.1 Osmoregulation and Excretion

- Solutes move across membranes via passive transport, facilitated diffusion, or active transport. Water moves across membranes by osmosis.

- In most animals, epithelial cells that selectively transport water and electrolytes are responsible for homeostasis.

- The mechanisms involved in regulating water and electrolyte balance vary widely among animal groups because different habitats present different types of osmotic stress.

- The type of nitrogenous waste excreted by an animal is affected by its phylogeny and its habitat type. Most fishes excrete ammonia; mammals and most adult amphibians excrete urea; and insects, reptiles, and birds excrete uric acid.

✔ You should be able to explain how an electrochemical gradient for sodium ions makes it possible for other ions and water to move across a membrane passively.

43.2 Water and Electrolyte Balance in Marine Fishes

- Seawater is strongly hyperosmotic in relation to the tissues of marine bony fishes, so the fishes tend to lose water by osmosis and gain electrolytes by diffusion.

- Marine bony fishes are osmoregulators, whereas cartilaginous fishes such as sharks are osmoconformers.

- Epithelial cells in the shark rectal gland and in the gills of marine bony fishes excrete excess salt using sodium–potassium pumps and sodium–potassium–chloride cotransporters located in the basolateral membrane.

- Salt-excreting cells similar to those found in the shark rectal gland occur in the gills of marine fishes, the salt glands of marine birds and other reptiles, and the kidneys of mammals.

✔ You should be able to compare and contrast the mechanisms by which cartilaginous and bony fishes achieve water and electrolyte balance in the ocean.

43.3 Water and Electrolyte Balance in Freshwater Fishes

- Freshwater is strongly hyposmotic in relation to the tissues of freshwater fishes, so the fishes tend to gain water by osmosis and lose electrolytes by diffusion.

- Epithelial cells in the gills of freshwater fishes import ions using sodium–potassium pumps (Na^+/K^+-ATPases) located in the basolateral membrane and sodium–potassium–chloride cotransporters located in the apical membrane.

✓ You should be able to describe the evidence for the hypothesis that new chloride cells form when salmon migrate from the ocean to freshwater.

43.4 Water and Electrolyte Balance in Terrestrial Insects

- A waxy coating on the insect exoskeleton limits evaporative water loss. The openings to insect respiratory organs close when osmotic stress is severe.

- The Malpighian tubules of insects form a filtrate that is isosmotic with the hemolymph. If pumps in the epithelium of the hindgut are activated, then electrolytes and water are reabsorbed from the filtrate and returned to the hemolymph.

- Insects can form hyperosmotic urine that minimizes water loss during the excretion of nitrogenous wastes.

✓ You should be able to predict what type of nitrogenous waste is excreted by aquatic insects.

43.5 Water and Electrolyte Balance in Terrestrial Vertebrates

- Nephrons in the vertebrate kidney form a filtrate in the renal corpuscle and then reabsorb valuable nutrients, electrolytes, and water in the proximal tubule.

- A solution containing urea and electrolytes flows through the loop of Henle of mammalian kidneys, where changes in the permeability of epithelial cells to water and salt—along with active transport of salt—create a steep osmotic gradient.

- If antidiuretic hormone (ADH) increases the water permeability of the collecting duct, water is reabsorbed along the osmotic gradient and hyperosmotic urine is produced.

- The nephrons of fishes, amphibians, and reptiles do not have loops of Henle and therefore cannot produce urine that is hyperosmotic to the body fluids. Some of these vertebrates are able to reabsorb water from the cloaca or bladder to produce hyperosmotic urine.

✓ You should be able to explain why desert-dwelling kangaroo rats, which are under extreme osmotic stress due to lack of drinking water, have extremely long loops of Henle relative to their body size.

(MB) MasteringBiology

1. MasteringBiology Assignments

Tutorials and Activities Control of Water Reabsorption; Kidney Structure and Function; Mammalian Kidney; Nephron Function; Structure of the Human Excretory System

Questions Reading Quizzes, Blue-Thread Questions, Test Bank

2. eText Read your book online, search, take notes, highlight text, and more.

3. The Study Area Practice Test, Cumulative Test, BioFlix® 3-D Animations, Videos, Activities, Audio Glossary, Word Study Tools, Art

You should be able to . . .

✓ TEST YOUR KNOWLEDGE
Answers are available in Appendix A

1. True or False: Sodium–potassium ATPase pumps move three sodium molecules out of a cell in exchange for two potassium ions, establishing an electrochemical gradient.

2. Which one of the following statements is true of fishes that live in freshwater?
   a. The environment is isosmotic with respect to their tissues. As a result, they do not require a specialized organ to maintain water and electrolyte balance.
   b. They lose water to their environment primarily through the gills. They replace this water by drinking.
   c. Water enters epithelial cells in their gills via osmosis. Electrolytes leave the same cells via diffusion.
   d. They have specialized cells that actively pump Na^+ and Cl^- from blood into epithelial cells, so the ions can be excreted.

3. Which of the following organisms would lose the most water by osmosis across its gills?
   a. marine bony fish
   b. shark
   c. freshwater bony fish
   d. freshwater invertebrate

4. Which of the following gives the correct sequence of the regions of the nephron with respect to fluid flow?
   a. distal tubule → ascending limb of loop of Henle → descending limb of loop of Henle → proximal tubule → collecting duct
   b. ascending limb of loop of Henle → descending limb of loop of Henle → proximal tubule → distal tubule → collecting duct
   c. proximal tubule → ascending limb of loop of Henle → descending limb of loop of Henle → distal tubule → collecting duct
   d. proximal tubule → descending limb of loop of Henle → ascending limb of loop of Henle → distal tubule → collecting duct

5. What effect does antidiuretic hormone (ADH) have on the nephron?
   a. It increases water permeability of the descending limb of the loop of Henle.
   b. It decreases water permeability of the descending limb of the loop of Henle.
   c. It increases water permeability of the collecting duct.
   d. It decreases water permeability of the collecting duct.

6. Fill in the blank: In Gila monsters, the organ in which water from urine is reabsorbed into the bloodstream is the _____.

✔ TEST YOUR UNDERSTANDING   Answers are available in Appendix A

7. Which of the following organisms would you expect to have the highest concentrations of proteins that protect cells from the toxic effects of ammonia?
   a. shark
   b. freshwater bony fish
   c. terrestrial reptile
   d. mammal

8. Compare and contrast the types of nitrogenous wastes observed in animals. Identify which compound can be excreted with a minimum of water, which is most toxic, and which waste product is found in fishes, in mammals, and in insects. Which type of nitrogenous waste would you expect to find produced by embryos inside eggs laid on land?

9. The chloride cells of fish gills are sometimes called mitochondria-rich cells due to their high density of mitochondria. How does high mitochondrial density relate to the functional role of chloride cells? Would you expect other epithelial cells involved in ion transport to contain large numbers of mitochondria? Explain.

10. This chapter introduced a number of features that help terrestrial animals reduce water loss, including the layer of wax found on insect exoskeletons, the ability of insects to close spiracles, the excretion of nitrogenous wastes as insoluble uric acid, and long loops of Henle in the mammalian kidney. Predict how each of these traits differs in animals that live in very humid versus very dry habitats. How would you test your predictions?

11. Explain why mammals would not be able to produce concentrated urine if they lacked loops of Henle.

12. Scientists have noted that marine invertebrates tend to be osmoconformers, while freshwater invertebrates tend to be osmoregulators. Suggest an explanation for this phenomenon.

✔ TEST YOUR PROBLEM-SOLVING SKILLS   Answers are available in Appendix A

13. Biologists recently have been able to produce mice that lack functioning genes for aquaporins. How does their urine compare to that of individuals with normal aquaporins?
    a. lower volume and lower osmolarity
    b. lower volume and higher osmolarity
    c. higher volume and lower osmolarity
    d. higher volume and higher osmolarity

14. Suppose a grazing mammal ingested the leaves of a plant containing the toxin ouabain. What effect would this toxin have on the mammal's urine volume? Explain.

15. You have isolated a segment of a rat nephron and introduced a solution of known composition. When you compare the fluid collected at the end of the segment with the test solution introduced at the beginning, you find that the volume has decreased by 30 percent and the Na^+ concentration has decreased by 30 percent, but the urea concentration has increased by 50 percent. What processes in this segment might account for these changes?

16. QUANTITATIVE To test the hypothesis that mussels are osmoconformers, researchers exposed mussels to water of varying osmolarities and then drew hemolymph samples from the mussels. Is the researchers' hypothesis supported by the data, shown below? Explain.

Water Osmolarity	Hemolymph Osmolarity
250	261
500	503
750	746
1000	992

44 Animal Nutrition

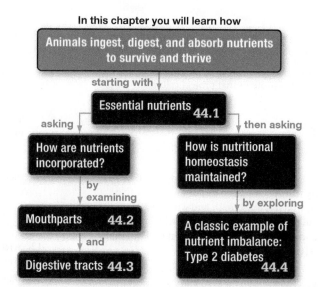

In this chapter you will learn how

Animals ingest, digest, and absorb nutrients
to survive and thrive

starting with ↓

Essential nutrients **44.1**

asking ↓   then asking ↓

How are nutrients
incorporated?

How is nutritional
homeostasis
maintained?

by
examining ↓   by exploring ↓

Mouthparts **44.2**

A classic example of
nutrient imbalance:
Type 2 diabetes
44.4

and ↓

Digestive tracts **44.3**

A crocodile has just
caught a fish. Animals
obtain nutrients by
ingesting food.

This chapter is part of the
Big Picture. See how on
pages 840–841.

Animals get the two basic requirements for life—(1) chemical energy for synthesizing ATP
and (2) carbon-containing compounds for building complex macromolecules—by ingesting
other organisms. In short, they are heterotrophs: They eat to live.

The types of food that are available to different animals vary widely, and food is often in dangerously short supply. From these observations, you might expect animals to have evolved many different means for obtaining food and to be under intense natural selection for making efficient use of
the food they have. How do animals get their food, and how do they process it? Which substances
in food are used as nutrients, and how do humans and other animals maintain appropriate levels of
key nutrients in their bodies?

✔ When you see this checkmark, stop and test yourself. Answers are available in Appendix A.

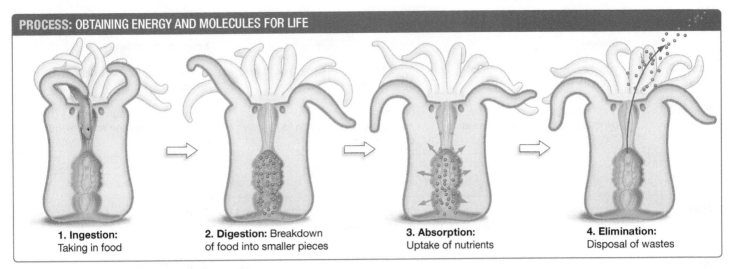

1. **Ingestion:** Taking in food
2. **Digestion:** Breakdown of food into smaller pieces
3. **Absorption:** Uptake of nutrients
4. **Elimination:** Disposal of wastes

FIGURE 44.1 Animal Nutrition Is a Four-Step Process.

If you're like most people, you've probably given a fair amount of thought to the food you eat but little thought to what happens to that food once it's inside you. As a meal moves through a digestive system, its physical characteristics change dramatically. Large packets of food that enter the mouth are reduced to monomers that can be absorbed into the bloodstream.

Research on feeding and digestion is fundamental to understanding these basic aspects of animal biology, but research on animal nutrition has important practical applications as well. For example, this chapter addresses questions about why several nutrition-related diseases, including diabetes mellitus and obesity, are on the rise in many human populations.

Just as plant nutrition is fundamental to the productivity of agricultural and natural ecosystems, nutrition is a basic component of animal health and welfare. For you or any other animal to stay alive, food must be ingested, digested, and absorbed, and then the wastes must be eliminated (**FIGURE 44.1**). Let's begin with a look at what animals must eat to live.

44.1 Nutritional Requirements

Humans and other animals get the chemical energy and carbon-containing building blocks they need from carbohydrates, proteins, and fats. All these substances are carbon compounds with high potential energy. Previous chapters analyzed the structures of carbohydrates and fats (see Chapter 5 and Chapter 6), and detailed how these compounds are used to synthesize ATP and key macromolecules (see Chapter 9).

A carbohydrate, protein, or fat is an example of a **nutrient**: a substance that an organism needs to remain alive. **Food** is any material that contains nutrients.

The amount of energy provided by foods is measured in kilocalories (on food labels, kilocalories are referred to as Calories).

Because fats and other lipids contain more C—H bonds than do carbohydrates such as starch and sugars, fats provide over twice the energy per gram. Specifically, carbohydrates and proteins provide about 4 kcal/g, and fats provide about 9 kcal/g.

Understanding which nutrients an individual needs, and in what amounts, are basic issues in research on human and animal nutrition. Let's consider what humans need to maintain good health.

In 1943 the Food and Nutrition Board of the U.S. National Academy of Sciences[1] published the first Recommended Dietary Allowances (RDAs). The goal of the RDAs was to specify the amount of each essential nutrient that an individual must ingest to meet the needs of most healthy people. **Essential nutrients** are nutrients that cannot be synthesized and must be obtained from the diet.

In addition to obtaining chemical energy and the "building blocks" for chemical synthesis from carbohydrates and other compounds in food, humans require several other essential nutrients:

- Of the 20 amino acids required to manufacture most proteins, humans can synthesize 12 from simpler building blocks (see Chapter 3). The others are called **essential amino acids** because they must be obtained from food.

- **Vitamins** are organic, or carbon-containing, compounds that are vital for health but are required in only minute amounts. They have a variety of roles; several function as coenzymes in critical reactions (see Chapter 3). **TABLE 44.1** (see page 884) lists a few of the vitamins for which RDAs have been established, notes their functions, and indicates the problems that develop if they are missing in the diet.

---

[1]The National Academy of Sciences is a group of scientists and engineers that advises the U.S. Congress on scientific and technical matters. Its Food and Nutrition Board is made up of biologists who specialize in animal nutrition.

TABLE 44.1 Some Important Vitamins Required by Humans

	Source	Function	Symptoms if Deficient
Vitamin B$_1$ (thiamine)	legumes, whole grains, potatoes, peanuts	formation of coenzyme in citric acid cycle	beriberi (fatigue, nerve disorders, anemia)
Vitamin B$_{12}$	red meat, eggs, dairy products; also synthesized by bacteria in intestine	coenzyme in synthesis of proteins and nucleic acids and in formation of red blood cells	anemia (fatigue and weakness due to low hemoglobin content in blood)
Niacin	meat, whole grains	component of coenzymes NAD$^+$ and NADP$^+$	pellagra (digestive problems, skin lesions, nerve disorders)
Folate	green vegetables, oranges, nuts, legumes, whole grains; also synthesized by bacteria in intestine	coenzyme in nucleic acid and amino acid metabolism	anemia
Vitamin C (ascorbic acid)	citrus fruits, tomatoes, broccoli, cabbage, green peppers	used in collagen synthesis, prevents oxidation of cell components, improves absorption of iron	scurvy (degeneration of teeth and gums)
Vitamin D	fortified milk, egg yolk; also synthesized in skin exposed to sunlight	aids absorption of calcium and phosphorus in small intestine	rickets (bone deformities) in children; bone softening in adults

- **Minerals** are inorganic substances used as components of enzyme cofactors or structural materials (see **TABLE 44.2**). Some, such as calcium and phosphorus, are needed in relatively large quantities. Others, such as iron and magnesium, are required in small or trace amounts.

- **Electrolytes** are mineral ions that influence osmotic balance and are required for normal membrane function. Sodium (Na$^+$), potassium (K$^+$), and chloride (Cl$^-$) are the major ions in the human body.

To obtain nutrients, animals must ingest them, usually via a mouth. The structure of animal mouthparts is therefore often highly specialized to capture specific types of food.

44.2 Capturing Food: The Structure and Function of Mouthparts

Instead of making their own food, as plants do, animals obtain the energy and nutrients they need by feeding on other organisms. Biologists assign animal feeding techniques to one of four strategies (see Chapter 33):

1. **Suspension feeders,** such as sponges and tubeworms, filter small organisms or bits of organic debris from water by means of cilia, mucus-lined "nets," or other structures.

2. **Deposit feeders,** including earthworms and sea cucumbers, swallow sediments and other types of deposited material rich in organic matter.

3. **Fluid feeders** suck or lap up fluids—like blood or nectar.

4. **Mass feeders** are the majority of animals. They seize and manipulate chunks of food by using mouthparts such as jaws and teeth, beaks, or special toxin-injecting organs.

Mouthparts as Adaptations

The types of food that animals harvest range from soupy solutions in decaying carcasses to nuts inside of hard shells. Solutions have to be lapped up; nuts have to be cracked. Given the diversity of food sources that animals exploit, it is not surprising that a wide variety of mouthpart structures have evolved to facilitate capturing and processing food.

Natural selection has closely matched the structure of animal mouthparts to their function in obtaining food. For example,

- Mammals chew their food and swallow distinct packets or boluses. The extinct mammal in **FIGURE 44.2a** illustrates one example of the many tooth shapes that evolved from the relatively simple and uniform teeth in the common ancestor of all mammals. Diversification of tooth shape has allowed mammals to exploit a wide range of foods.

- Complex skull, jawbones, and associated musculature have evolved in snakes. The highly flexible structure that results allows snakes to ingest large prey whole (**FIGURE 44.2b**).

The reason that there is such a close correlation between the structure and function of mouthparts is simple: Natural selection is particularly strong when it comes to food capture, because obtaining nutrients is so fundamental to fitness—the ability to produce offspring.

Let's pursue the correlation between mouthparts and food sources further, by analyzing the structure and function of jaws and teeth in what may be the most diverse lineage within any vertebrate family: the cichlid fishes of Africa.

A Case Study: The Cichlid Jaw

The cichlids that inhabit the Rift Lakes of East Africa are a spectacular example of **adaptive radiation**—the diversification of a

(a) The large canines of saber-toothed cats stabbed and sliced prey.

(b) A highly mobile cranium and jaws allows snakes to swallow large prey whole.

FIGURE 44.2 Mouthpart Structure Correlates with Function.

single ancestral lineage into many species, each of which lives in a different habitat or employs a distinct feeding method (see Chapter 28). Lake Victoria, for example, is home to 300 cichlids that live nowhere else.

In general, each Lake Victoria cichlid species feeds on a different specific item, but as a group they exploit almost every food source in the lake: planktonic organisms, crust-forming algae,

leaflike algae, eggs, fish scales, fish fins, whole fish, plants, insects, and snails.

How can a group of closely related species exploit so many different food sources? Many fish species have pharyngeal (throat) jaws located well behind the normal oral (mouth) jaws. Non-cichlids use their pharyngeal jaws to move food down their throats, but cichlids can also use theirs to bite. This is possible

TABLE 44.2 Major Minerals and Electrolytes Required by Humans

	Source in Diet	Function	Symptoms if Deficient
Calcium (Ca)	dairy products, green vegetables, legumes	bone and tooth formation, nerve signaling, muscle response	loss of bone mass, slow growth
Chloride (Cl⁻)	table salt or sea salt, vegetables, seafood	fluid balance in cells, protein digestion in stomach (HCl), acid–base balance	weakness, loss of muscle function
Fluorine (F)	fluoridated water, seafood	maintenance of tooth structure	higher frequency of tooth decay
Iodine (I)	iodized salt, algae, seafood	component of the thyroid hormones thyroxine and T_3	goiter (enlarged thyroid gland)
Iron (Fe)	meat, eggs, whole grains, green leafy vegetables, legumes	enzyme cofactor; synthesis of hemoglobin and electron carriers	anemia, weakness
Magnesium (Mg)	whole grains, green leafy vegetables	enzyme cofactor	nerve disorders
Phosphorus (P)	dairy products, meat, grains	bone and tooth formation; synthesis of nucleotides and ATP	weakness, loss of bone
Potassium (K⁺)	dairy products; meat; nuts; fruits; potatoes, legumes, and other vegetables	nerve signaling, muscle response, acid–base balance	weakness, muscle cramps, loss of muscle function
Sodium (Na⁺)	table salt or sea salt, seafood	nerve signaling, muscle response, blood pressure regulation	weakness, muscle cramps, loss of muscle function, nausea, confusion
Sulfur (S)	any source of protein	amino acid synthesis	swollen tissues, degeneration of liver, mental retardation

FIGURE 44.3 Rift Lake Cichlids Have Two Sets of Biting Jaws.
Oral jaws capture food; pharyngeal jaws process it.

because the upper pharyngeal jaw attaches to the skull in cichlids, and because the muscles of their lower pharyngeal jaw allow it to move against the upper jaw (**FIGURE 44.3**).

In addition to acting as a second set of biting jaws that make food processing more efficient, cichlid pharyngeal jaws provide a more specialized set of toothlike structures. These protuberances vary in size and shape among cichlids, correlating with their function, such as crushing snail shells, tearing fish scales, or compacting algae (**FIGURE 44.4**).

These observations are part of a large body of evidence supporting a general pattern in animal evolution: In response to natural selection, mouthparts have diversified to exploit a diversity of food sources. The structures of jaws, teeth, and other mouthparts correlate with their functions in harvesting and processing food.

How Are Nutrients Digested and Absorbed?

Animals ingest just about every type of food conceivable. **Ingestion** is the process of bringing food into the **digestive tract**—also known as the alimentary (literally, "nourishment") canal or gastrointestinal (GI) tract. The digestive tract is a chamber or tube where **digestion,** the breakdown of food, takes place. Interestingly, food in the digestive tract is technically *outside* the body. Various glands secrete enzymes into the digestive tract that digest food into particles small enough for efficient **absorption**—the uptake of specific ions and molecules across the membrane of the digestive tract.

Digestion is a key process in animals because they, unlike plants, unicellular organisms, and certain parasites, do not acquire nutrients as individual molecules. Instead, they take in packets of food that must be broken down into small pieces. Nutrients must be extracted from the small pieces, and waste materials must be eliminated. How and where does this processing occur?

An Introduction to the Digestive Tract

Digestive tracts come in two general designs:

1. **Incomplete digestive tracts** have a single opening that doubles as the location where food is ingested and wastes are eliminated. The mouth opens into a chamber, called a gastrovascular cavity, where digestion takes place (**FIGURE 44.5**).

2. **Complete digestive tracts** have two openings—they start at the mouth and end at the anus. The interior of this tube communicates directly with the external environment via these openings (**FIGURE 44.6**).

Trematocranus placodon

Crushes snails

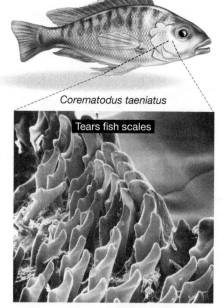

Corematodus taeniatus

Tears fish scales

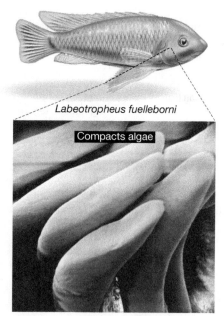

Labeotropheus fuelleborni

Compacts algae

FIGURE 44.4 In Cichlids, the Structure of the Pharyngeal Jaw Correlates with the Type of Food Ingested.

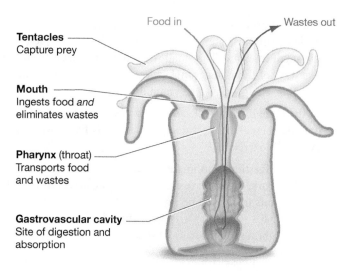

Tentacles
Capture prey

Mouth
Ingests food *and*
eliminates wastes

Pharynx (throat)
Transports food
and wastes

Gastrovascular cavity
Site of digestion and
absorption

Food in

Wastes out

FIGURE 44.5 Anemones Have an Incomplete Digestive Tract. Anemones use stinging cells located on their tentacles to capture small fishes, crustaceans, and other prey. Prey are taken into the mouth and digested in the gastrovascular cavity; then wastes are excreted through the mouth.

A tubelike digestive tract has three advantages:

1. It allows animals to feed on large pieces of food, expanding the range of food sources that can be ingested.

2. Different chemical and physical processes can be confined to different compartments within the canal, so that they occur independently of each other and in a prescribed sequence. The stomach, for example, provides an acidic environment for digestion. Ingested material proceeds from there to the small intestine, where enzymes are specialized to function in a slightly alkaline environment.

3. Because there is a one-way flow of food and wastes, material can be ingested and digested without interruption, instead of alternating with waste removal as occurs in an incomplete digestive tract.

The digestive tract is only one part of the digestive system, however. Several vital organs and glands are connected to the digestive tract. These accessory structures contribute digestive enzymes and other products to specific portions of the tract. They include the salivary glands, liver, gallbladder, and pancreas (see Figure 44.6).

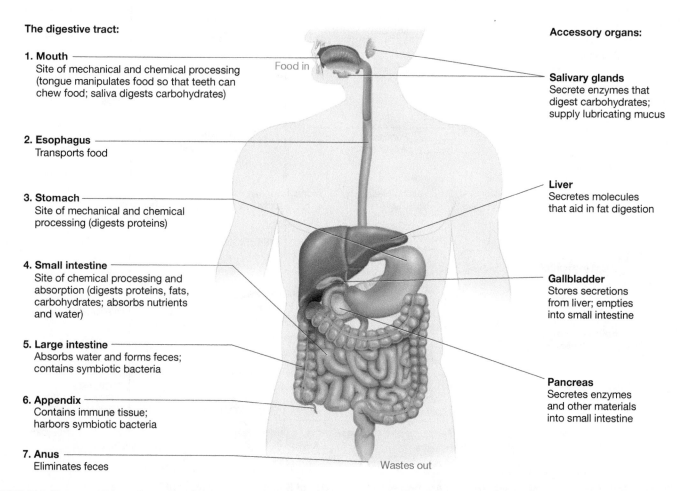

The digestive tract:

1. **Mouth**
   Site of mechanical and chemical processing
   (tongue manipulates food so that teeth can
   chew food; saliva digests carbohydrates)

2. **Esophagus**
   Transports food

3. **Stomach**
   Site of mechanical and chemical
   processing (digests proteins)

4. **Small intestine**
   Site of chemical processing and
   absorption (digests proteins, fats,
   carbohydrates; absorbs nutrients
   and water)

5. **Large intestine**
   Absorbs water and forms feces;
   contains symbiotic bacteria

6. **Appendix**
   Contains immune tissue;
   harbors symbiotic bacteria

7. **Anus**
   Eliminates feces

Food in

Accessory organs:

Salivary glands
Secrete enzymes that
digest carbohydrates;
supply lubricating mucus

Liver
Secretes molecules
that aid in fat digestion

Gallbladder
Stores secretions
from liver; empties
into small intestine

Pancreas
Secretes enzymes
and other materials
into small intestine

Wastes out

FIGURE 44.6 Humans Have a Complete Digestive Tract. In humans, the digestive tract is a tube that runs from the mouth to the anus. The salivary glands, liver, gallbladder, and pancreas are not part of the tract itself. Instead, they secrete material into the tract at specific points.

An Overview of Digestive Processes

Before analyzing the function of each component of the digestive system in detail, let's consider the general changes that happen to food on its way through the digestive tract. In this brief overview and in the detailed discussion that follows, humans will serve as a model species—simply because so much is known about human digestion.

In mammals, digestion begins with the tearing and crushing activity of teeth. Chewing reduces the size of food particles and softens them. Humans augment the mechanical breakdown of food by their use of knives and cooking. In fact, the invention of cutting tools and cooking, which make food easier to chew, is the leading hypothesis to explain why average tooth size has declined steadily over the past several million years of human evolution.

Distinct chemical changes occur as food moves through each compartment in the digestive tract (**FIGURE 44.7**):

- In the mouth, enzymes in the saliva begin the chemical breakdown of carbohydrates.

- Chemical digestion of protein begins in the acidic environment of the stomach.

- Chemical processing of the three major types of macromolecules—carbohydrates, proteins, and lipids—is completed in the small intestine. The small molecules that result from the digestion of these macromolecules are absorbed in the small intestine, along with water, vitamins, and ions.

- In the large intestine, or colon, more water is absorbed. The material remaining in the large intestine is **feces,** which are eventually excreted.

Because digestion is so important to understanding how animal bodies work, let's analyze each step in more detail. As the following sections track food on its journey from the mouth to the anus in humans—a particularly well-studied animal—watch for notes highlighting the diversity of structures found in the digestive tracts of nonhuman animals.

The Mouth and Esophagus

If you hold a cracker in your mouth long enough, it will start to taste sweet. The sensation occurs because an enzyme in your saliva hydrolyzes some of the starch molecules in the cracker to

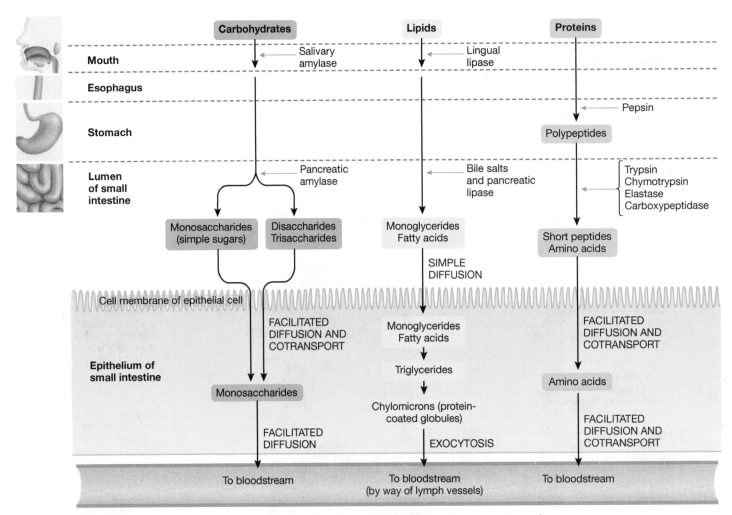

FIGURE 44.7 Carbohydrates, Lipids, and Proteins Are Processed in a Series of Steps. Three key types of macromolecules enter the digestive system (top of diagram). As they proceed through the digestive tract, they are broken apart by various enzymes. Simple sugars, fatty acids, and amino acids then enter epithelial cells in the small intestine and are transported to the bloodstream.

FIGURE 44.8 In the Esophagus, Peristalsis Transports Food to the Stomach. Peristalsis is a wave of muscle contraction that constricts the esophagus behind food and expands it in front of food. The wave of contraction begins at the oral end of the esophagus and propels the food toward the stomach.

1. Muscle constricts esophagus above food

2. Muscle contraction moves like a wave down the esophagus, pushing the food below it into the stomach

maltose. Maltose is a disaccharide that is split in the small intestine to form two glucose monomers.

Starch breakdown was actually the first enzyme-catalyzed reaction ever discovered. In the early 1800s several researchers found that a component of certain plant extracts digested starch; in 1831 the same activity was discovered in human saliva.

Digestion Starts in the Mouth **Salivary amylase,** the enzyme responsible for starch digestion in the mouth is one of the best-studied enzymes. Amylase cleaves bonds to release maltose from starch and glycogen, initiating the digestion of starch.

Cells in the tongue synthesize and secrete another important salivary enzyme, **lingual lipase** (lingual refers to the tongue), which begins the digestion of lipids by breaking triglycerides into diglycerides and fatty acids.

Salivary glands in the mouth not only produce amylase but also release water and glycoproteins called mucins. When mucins contact water, they form the slimy substance called **mucus.** The combination of water and mucus makes food soft and slippery enough to be swallowed.

Peristalsis Moves Material Down the Esophagus Once food is swallowed, it enters a muscular tube called the **esophagus,** which connects the mouth and stomach. A wave of muscular contractions called **peristalsis** propels food down the esophagus. About 6 seconds after being swallowed, food reaches the bottom of the esophagus. Because peristalsis actively moves material along the esophagus, you can swallow even when your mouth is lower than your stomach, such as when you bend over to drink from a stream.

In response to nerve signals, the muscles in the esophagus contract and relax in a coordinated fashion (**FIGURE 44.8**). The resulting wave of muscle contractions propagates down the tube, propelling the food mass ahead of it. These nerve signals are not the result of conscious choice but are a reflex—an automatic reaction to a stimulus—that is stimulated by the act of swallowing.

A Modified Esophagus: The Bird Crop In an array of bird species, the esophagus has a prominent, widened segment called the **crop** where food can be stored and, in some cases, processed.

The structure and function of the crop varies among the bird species that have one. In many groups, the crop is a simple sac that holds food and regulates its flow into the stomach. In these species, the crop is interpreted as an adaptation that allows individuals to eat a large amount in a short time; they then retreat to a safe location while digestion occurs. In addition, some birds store food in their crops and then regurgitate it into the mouths of their young.

The crop has independently evolved into a digestive organ in two leaf-eating species of bird. Leaves are difficult to digest because they contain a large amount of cellulose (see Chapter 5). In the enlarged crop of species called the hoatzin and kakapo, bacteria that are capable of breaking down cellulose perform digestion. The bacterial cells, along with the fatty acids that result from bacterial metabolism, leak out of the crop into the stomach and are used as food by the bird.

The Stomach

Although little if any digestion occurs in the esophagus of most animals, the situation changes dramatically when food reaches the stomach. The **stomach** is a tough, muscular pouch in the digestive tract, bracketed on both ends by ringlike muscles called **sphincters,** which control the passage of material (**FIGURE 44.9**).

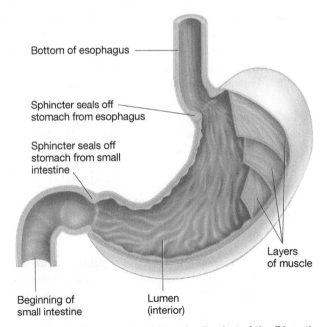

Bottom of esophagus

Sphincter seals off stomach from esophagus

Sphincter seals off stomach from small intestine

Layers of muscle

Beginning of small intestine

Lumen (interior)

FIGURE 44.9 The Stomach Is a Muscular Pocket of the Digestive Tract. The stomach provides an acidic environment for protein digestion. Its muscular contractions mix food and break it into smaller pieces.

When a meal fills the stomach, muscular contractions churn and mix the stomach contents to a uniform consistency and solute concentration. A certain amount of mechanical breakdown of food also results from this churning. The other main function of the stomach is the partial digestion of proteins.

Compared with the mouth or esophagus (or virtually any other tissue), the lumen of the stomach is highly acidic. Early researchers documented this fact by analyzing vomit or material collected by sponges that were tied to strings, swallowed, and pulled back up; chemists confirmed that the predominant acid in the stomach is hydrochloric acid (HCl).

Not long after this discovery, a physician named William Beaumont established that digestion takes place in the stomach. He reached this conclusion through an extraordinary series of experiments on a young man named Alexis St. Martin.

The Stomach as a Site of Protein Digestion In 1822, when St. Martin was 19 years old, a shotgun accidentally discharged into his abdomen, leaving a series of wounds. Despite repeated attempts, Beaumont was unable to close a hole in St. Martin's stomach. Eventually Beaumont inserted a small tube through the opening; the tube remained in St. Martin's body for the rest of his life. (Today, biologists insert tubes into various parts of the digestive tract of cows or sheep to study how these animals digest different types of feed.)

With the tube in place, Beaumont was able to tie a string onto small pieces of meat or vegetables, insert the food directly into St. Martin's stomach, and draw it out after various intervals. Beaumont also removed liquid from inside the stomach and observed how this gastric (stomach) juice acted on food in vitro. His experiments showed that gastric juice digests food—particularly meat.

Theodor Schwann later purified the enzyme that is responsible for digesting proteins in the stomach and named it **pepsin.** Because it destroys proteins, biologists hypothesized that pepsin must be synthesized and stored in cells while it is in an inactive form—otherwise it would kill the cells that make it.

In 1870 a biologist established through careful microscopy the presence of granules in specialized stomach cells called chief cells. These granules were hypothesized to be a pepsin precursor. Follow-up work confirmed this hypothesis. The precursor compound, which came to be called pepsinogen, is converted to active pepsin by contact with the acidic environment of the stomach.

Secretion of a protein-digesting enzyme in inactive form is important: It prevents destruction of proteins in the cells where the enzyme is synthesized.

Which Cells Produce Stomach Acid? The acidic environment of the human stomach denatures (unfolds) proteins so that pepsin can digest them efficiently. But where does the acid come from?

Researchers who were studying the anatomy of the stomach wall noticed clusters of distinctive **parietal cells** located in pits in the stomach lining (**FIGURE 44.10a**). An investigator also

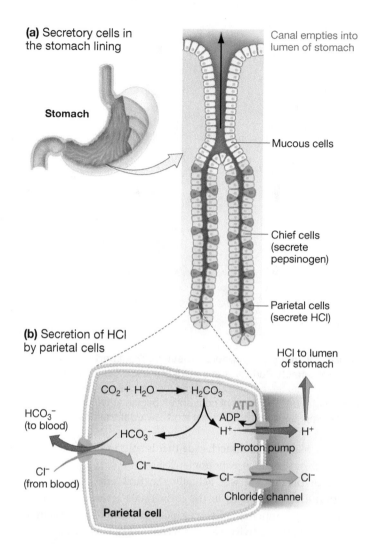

(a) Secretory cells in the stomach lining

Stomach

Canal empties into lumen of stomach

Mucous cells

Chief cells (secrete pepsinogen)

Parietal cells (secrete HCl)

(b) Secretion of HCl by parietal cells

HCl to lumen of stomach

$CO_2 + H_2O \longrightarrow H_2CO_3$

ATP

HCO_3^- (to blood)

ADP

HCO_3^-

H^+

H^+

Proton pump

Cl^- (from blood)

Cl^-

Cl^-

Cl^-

Chloride channel

Parietal cell

FIGURE 44.10 Cells in the Stomach Lining Secrete Mucus, Pepsinogen, and Hydrochloric Acid.

documented that the shape and activity of these cells appeared to vary as the digestion of a meal proceeded. Based on these observations, he inferred that parietal cells are the source of the HCl in gastric juice, which may have a pH as low as 1.5.

Earlier microscopists had shown that another type of cell, called a **mucous cell,** secretes additional mucus that is found in gastric juice. Mucus lines the gastric epithelium and protects the stomach from damage by HCl. To summarize, these anatomical studies showed that the epithelium of the stomach contains several types of secretory cells, each of which is specialized for a particular function.

How Do Parietal Cells Secrete HCl? The first clue to how parietal cells manufacture HCl emerged in the late 1930s, when a researcher found a high concentration of an enzyme called carbonic anhydrase in parietal cells.

This result was interesting because **carbonic anhydrase** catalyzes the formation of carbonic acid (H_2CO_3) from carbon dioxide and water. In solution, the carbonic acid that is formed immediately dissociates to form a proton (H^+) and the bicarbonate ion (HCO_3^-):

$$CO_2 + H_2O \rightleftharpoons H_2CO_3 \rightleftharpoons H^+ + HCO_3^-$$

A second clue to the formation of HCl came in the 1950s, when transmission electron microscopes allowed researchers to analyze parietal cells at high magnification (see **BioSkills 11** in Appendix B). The micrographs showed that parietal cells are packed with mitochondria. Because mitochondria produce ATP, the structure of parietal cells suggested that they might function in active transport.

Later work confirmed this hypothesis by showing that the protons formed by the dissociation of carbonic acid are actively pumped into the lumen of the stomach. Subsequent studies showed that chloride ions from the blood enter parietal cells in exchange for bicarbonate ions, via a cotransport protein, and then move into the lumen through a chloride channel. **FIGURE 44.10b** diagrams the current model for HCl production.

Ulcers as an Infectious Disease

An **ulcer** is an eroded area in an epithelium; it exposes the underlying tissues to damage. Ulcers in the lining of the stomach or in the duodenum—the initial section of the small intestine—can cause intense abdominal pain.

For decades, physicians thought that gastric and duodenal ulcers resulted from the production of excess acid in the stomach. They treated ulcers by prescribing alkaline compounds (bases—see Chapter 2) that neutralized hydrochloric acid in the stomach. In the 1980s, however, scientists discovered that most ulcers are associated with infections from a bacterium called *Helicobacter pylori*. Physicians now routinely prescribe antibiotics to relieve ulcers.

The Ruminant Stomach

It is common for animals to have a stomach or stomach-like organ. The structure and function of this organ can vary, however, depending on the nature of the diet. In cattle, sheep, goats, deer, antelope, giraffe, and pronghorn—species that are collectively called **ruminants**—the stomach is specialized for digesting cellulose—not proteins.

Animals do not produce the enzymes required to digest cellulose. Yet cellulose is the main carbohydrate in the leaves, stems, and twigs that ruminants ingest.

Like the hoatzin and kakapo birds described earlier in this chapter, ruminants are able to harvest energy from cellulose thanks to a combination of specialized anatomical structures and symbiotic relationships with bacteria and unicellular protists. The microbes ferment the cellulose to produce food for themselves; meanwhile, other by-products of the fermentation, as well as some of the microbes themselves, are used as food by the ruminant. This is an example of **symbiosis,** in which members of two different species live in close contact with each other.

As **FIGURE 44.11** shows, ruminant mammals have four-chambered stomachs.

1. Food initially enters the largest chamber, the rumen, which serves as a fermentation vat. The rumen is packed with symbiotic bacteria and protists. These organisms have enzymes

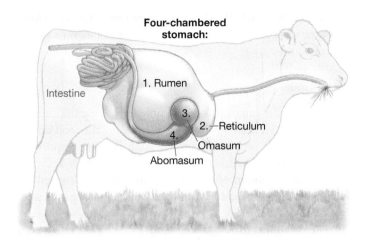

FIGURE 44.11 Ruminant Stomachs Facilitate the Digestion of Cellulose by Symbiotic Organisms. Ruminants obtain many of their nutrients from symbiotic bacteria and protists that live in the rumen and reticulum chambers of the stomach.

called **cellulase** capable of breaking apart the chemical bonds in cellulose, yielding glucose. The rumen is an oxygen-free environment, and the symbiotic organisms produce ATP from this glucose via fermentation, releasing fatty acids as a by-product (see Chapter 9). These fatty acids are absorbed by the ruminant and used as an energy source.

2. The chamber adjacent to the rumen, called the reticulum, is similar in function. After plant material has been partially digested in the rumen and the reticulum, the animal regurgitates portions of that material into its mouth, forming a cud. The ruminant chews that regurgitated material further to enhance mechanical breakdown and then re-swallows it.

3. Processed food that moves out of the rumen enters the third chamber, or omasum, where water and some minerals are absorbed.

4. The final chamber is the abomasum, which contains the ruminant's own digestive enzymes and corresponds to a true stomach.

Most of a ruminant's food consists of (1) fatty acids and other compounds produced as waste products of fermentation reactions in symbiotic organisms, and (2) the symbiotic cells themselves.

The Avian Gizzard

The avian gizzard is another prominent type of modified stomach. Birds do not have teeth and cannot chew food into small pieces. Instead, most species swallow sand and small stones that lodge in the gizzard. As this muscular sac contracts, food is pulverized by the grit.

The gizzard is particularly large and strong in bird species that eat coarse foods such as seeds and nuts. The gizzard of a wild turkey, for example, can crack large walnuts.

Like the crop, the gizzard is interpreted as an adaptation that allows birds to ingest food quickly—without needing to chew—and digest it later. Biologists invoke the same hypothesis to explain why ruminants chew cud. The ability to regurgitate material and finish chewing while hiding in a place safe from predators is thought to increase fitness.

The Small Intestine

In humans, the stomach is responsible for mixing the contents of a meal into a homogenous slurry, breaking the food up mechanically, and providing the acid and enzymes required to partially digest proteins. Peristalsis in the stomach wall then moves small amounts of material through the valve created by a sphincter muscle at the base of the stomach and into the small intestine.

The **small intestine** is a long tube that is folded into a compact space within the abdomen. In the small intestine, partially digested food mixes with secretions from the pancreas and the liver and begins a journey of about 6 m (20 ft). When passage through this structure is complete, digestion is finished, and most nutrients—along with large quantities of water—have been absorbed.

Folding and Projections Increase Surface Area   The surface area available for nutrient and water absorption in the small intestine is nothing short of remarkable (see Chapter 42). As **FIGURE 44.12** shows, the organ's epithelial tissue is folded and covered with fingerlike projections called **villi** (singular: **villus**). In turn, the cells that line the surface of villi have tiny projections on their apical surfaces called **microvilli** (singular: **microvillus**). Microvilli project into the lumen of the digestive tract.

If the small intestine lacked folds, villi, and microvilli, it would have a surface area of about 3300 cm² (3.6 ft²). Instead, the epithelium covers about 2 million cm² (over 2200 ft²)—an area about the size of a tennis court.

The enormous surface area of the small intestine increases the efficiency of nutrient absorption. And because each villus contains blood vessels and a lymphatic vessel called a **lacteal**, nutrients pass quickly from epithelial cells into the body's transport systems. (The circulatory system and lymphatic system are analyzed in Chapter 45 and Chapter 51, respectively.)

✔ If you understand the importance of surface area in the small intestine, you should be able to explain why surface area is so much higher in this structure than it is in the stomach or esophagus.

To understand how digestion is completed and absorption occurs, let's explore what happens to proteins, lipids, and carbohydrates as they move through this section of the digestive tract—again using humans as a model organism.

Protein Processing by Pancreatic Enzymes   The acidic environment of the stomach destroys the secondary and tertiary structures of proteins. In addition, pepsin cleaves the peptide bonds next to certain amino acids, reducing long polypeptides to relatively small chains of amino acids. In the small intestine, protein digestion is completed so that individual amino acids can enter the bloodstream and be transported to cells throughout the body.

How do these final stages of protein digestion occur? By the end of the nineteenth century, it was established that enzymes in the small intestine digest polypeptides to monomers. Later work showed that each of these protein-digesting enzymes, or **proteases,** is specific to certain types or configurations of amino acids in a polypeptide chain. Thus, a suite of proteases is required to completely digest polypeptides to amino acid monomers.

In addition, by 1900 biologists had determined that proteases are synthesized in an inactive form in the **pancreas,** which is connected to the small intestine by the pancreatic duct. Like the production of inactive pepsinogen by chief cells in the stomach,

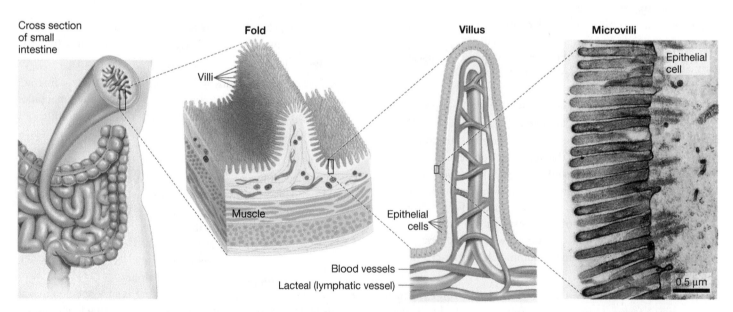

FIGURE 44.12 The Small Intestine Has an Extremely Large Surface Area. The villi that project from folds in the small intestine are covered with microvilli (colorized brown in the micrograph at the far right).

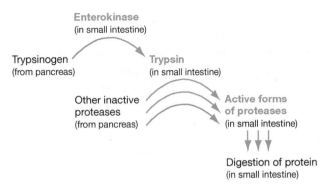

Enterokinase
(in small intestine)

Trypsinogen
(from pancreas)

Trypsin
(in small intestine)

Other inactive
proteases
(from pancreas)

Active forms
of proteases
(in small intestine)

Digestion of protein
(in small intestine)

FIGURE 44.13 Enterokinase Triggers an Enzyme-Activation Cascade in the Small Intestine.

the production of digestive enzymes in an inactive conformation prevents pancreatic cells from digesting themselves.

It took decades of work to understand how pancreatic enzymes are activated in the small intestine. In 1900 a researcher showed that contact with juice from the upper part of the small intestine activates pancreatic enzymes. Activation ceased to occur when he heated the intestinal juice, and because he knew that heat denatures enzymes (like all proteins), he hypothesized that the agent responsible for activating the pancreatic enzymes was also an enzyme. He called the unknown enzyme enterokinase.

Decades later, a researcher succeeded in purifying a pancreatic enzyme called trypsinogen and demonstrated that enterokinase activates it in vitro by phosphorylating it, resulting in the active enzyme **trypsin.** Trypsin then triggers the activation of other protein-digesting enzymes, such as chymotrypsin, elastase, and carboxypeptidase (**FIGURE 44.13**). These enzymes are also synthesized by the pancreas and secreted in an inactive form. Once these enzymes are activated in the upper reaches of the small intestine, each begins cleaving specific peptide bonds. Eventually, polypeptides are broken up into amino acid monomers.

What Regulates the Release of Pancreatic Enzymes? Digestive enzymes are needed only when food reaches the small intestine. Based on this simple observation, it was logical to predict that their release would be carefully controlled.

A classic experiment by William Bayliss and Ernest Starling, published in 1902, established how pancreatic enzymes are controlled. Bayliss and Starling began by cutting the nerves that connect to the pancreas and small intestine of a dog. Electrical signaling between the two organs via neurons was now impossible. But when the researchers introduced a weak HCl solution into the upper reaches of the animal's small intestine, to simulate the arrival of material from the stomach, its pancreas secreted enzymes in response.

This observation was startling: The small intestine had successfully signaled the pancreas that food had arrived, even though the nerves connecting the two organs had been cut.

Starling hypothesized that a chemical messenger must be involved, and that the chemical messenger must originate in the small intestine and travel to the pancreas via the blood. He tested this idea by cutting off a small piece of the small intestine,

grinding it up, and injecting the resulting solution into a vein in the animal's neck (see Figure 49.4). Minutes later, the pancreas sharply increased secretion.

Bayliss and Starling had discovered the first **hormone**—a chemical messenger that influences physiological processes at very low concentrations. The molecule they detected, which they called **secretin,** is produced by the small intestine in response to the arrival of food from the stomach.

Follow-up work showed that secretin's primary function is to induce a flow of bicarbonate ions (HCO_3^-) from the pancreas to the small intestine. The bicarbonate is important because it neutralizes the acid arriving from the stomach.

Researchers also discovered a second hormone produced in the small intestine. Called cholecystokinin (pronounced *ko-la-sis-toe-KIN-in*), it induces secretion from the liver and gallbladder as well as the pancreas. **Cholecystokinin** (literally, "bile-bag-mover") stimulates the secretion of digestive enzymes from the pancreas and the secretion of molecules from the gallbladder that aid in processing lipids.

Hormones are involved in stomach function as well. For example, after being stimulated by nerves or the arrival of food, certain stomach cells produce the hormone **gastrin.** In response, parietal cells begin secreting HCl.

How Are Carbohydrates Digested and Transported? In addition to manufacturing protein-digesting enzymes, the pancreas produces nucleases and an amylase that is similar to the salivary enzyme introduced earlier. **Nucleases** digest the RNA and DNA in food; **pancreatic amylase** continues the digestion of carbohydrates that began in the mouth. Carbohydrate digestion ends with the release of monosaccharides such as glucose.

When digestion of proteins and carbohydrates is complete in the small intestine, the resulting slurry is a mixture of nutrients, water, indigestible plant fibers from food, and bacterial cells that live symbiotically in the gut. What molecular mechanisms make it possible for epithelial cells to transport nutrients, like glucose, from the lumen of the small intestine into the bloodstream?

Two general principles apply to carbohydrate and protein absorption: (1) It is highly selective, meaning proteins in the plasma membranes of microvilli are responsible for bringing specific nutrients into the cell; and (2) it is active, meaning ATP is expended to bring nutrients into the epithelium across a concentration gradient.

Work over the past several decades has demonstrated both these principles. One of the key results grew out of a series of experiments during the 1980s, which established that glucose absorption depends on the presence of an electrochemical gradient favoring an influx of sodium ions into the epithelium. Based on this finding, biologists hypothesized that the apical membranes of epithelial cells must contain a series of cotransporters—membrane proteins that would bring a nutrient molecule into the cell along with sodium ions. To confirm that such a sodium–glucose cotransporter exists, investigators set out to find the gene that codes for the hypothesized membrane protein.

The researchers began by purifying mRNAs from rabbit intestinal cells (**FIGURE 44.14**), which presumably were transcribing the cotransporter genes. Then the team separated the mRNAs by size via gel electrophoresis (see **BioSkills 9** in Appendix B) and injected one of each type of mRNA into a series of frog eggs—cells that do not normally transport glucose. The frog cells translated the rabbit mRNAs into proteins.

When tested, one of the experimental eggs was able to import Na^+ and glucose in tandem. The biologists inferred that this egg had received the mRNA for the rabbit Na^+-glucose cotransporter. The researchers made a DNA copy of the mRNA (using techniques introduced in Chapter 20), analyzed it to determine the sequence of the gene, and from that inferred the amino acid sequence of the membrane protein.

The discovery of the Na^+-glucose cotransporter inspired a three-step model for glucose absorption:

1. Na^+/K^+-ATPase (sodium–potassium pump) in the basolateral membrane of the epithelial cells creates an electrochemical gradient that favors the entry of Na^+.

2. Glucose from digested food enters the cell along with sodium via the Na^+-glucose cotransporter in the apical membrane.

3. Glucose diffuses into nearby blood vessels through a glucose carrier in the basolateral membrane.

If this configuration of pumps, cotransporters, and carriers sounds familiar, there is a good reason: The same combination of membrane proteins occurs in the proximal tubule of the kidney, where the proteins are responsible for the reabsorption of sodium and glucose from urine (see Chapter 43).

Follow-up work showed that in the small intestine—just as in the proximal tubule—other cotransporters are responsible for the absorption of other monosaccharides and of amino acids, and specific channels and carriers in the basolateral membrane are responsible for the transport of each substance to the blood.

Digesting Lipids: Bile and Transport The pancreatic secretions include digestive enzymes that act on fats, in addition to enzymes that act on proteins and carbohydrates. Like the lingual lipase added to saliva in the mouth, the enzyme **pancreatic lipase** breaks certain bonds present in complex fats and results in the release of fatty acids and other small lipids.

Recall that fats are insoluble in water (see Chapter 6). As a result, they tend to form large globules as they are churned in the stomach. Before pancreatic lipase can act, the large fat globules that emerge from the stomach must be broken up—a process known as **emulsification.**

In the small intestine, emulsification results from the action of small molecules called bile salts. As **FIGURE 44.15** shows, bile salts function like the detergents that researchers use to break up the lipids in plasma membranes (see Chapter 6).

Bile salts are synthesized in the **liver,** an organ that performs an array of functions related to digestion, and secreted in a complex solution called **bile,** which is stored in the **gallbladder.** When bile enters the small intestine, it raises the pH and emulsifies fats. Once fats are broken into small globules, which increases their surface area, they can be attacked by pancreatic lipase and digested. **TABLE 44.3** summarizes the major digestive enzymes.

The monoglycerides and fatty acids released by lipase activity enter small intestine epithelial cells by simple diffusion and are processed into protein-coated globules called chylomicrons, which move by exocytosis into lacteals—see Figure 44.12—near

---

RESEARCH

QUESTION: How is glucose transported into epithelial cells of the small intestine?

HYPOTHESIS: Glucose enters epithelial cells along with sodium ions via a Na^+-glucose cotransporter protein.

NULL HYPOTHESIS: Glucose transport does not depend on Na^+ transport.

EXPERIMENTAL SETUP:

1. Purify mRNA from intestinal cells.

2. Separate mRNAs by size via gel electrophoresis.

3. Inject individual mRNAs into frog eggs. Test each egg—can it absorb Na^+ and glucose?

PREDICTION: An egg will be able to absorb Na^+ and glucose, because it received the mRNA that codes for the Na^+-glucose cotransporter.

PREDICTION OF NULL HYPOTHESIS: None of the eggs will be able to absorb Na^+ and glucose.

RESULTS:

Na^+, glucose in medium	Na^+, glucose in medium	Na^+, glucose in medium
Egg 1	Egg 2	Egg 3
ABSORBED	Not absorbed	Not absorbed

CONCLUSION: The egg that absorbs Na^+ and glucose received the mRNA from the Na^+-glucose cotransporter gene.

FIGURE 44.14 The Experimental Protocol for Locating the Na^+-Glucose Cotransporter Gene.

SOURCE: Wright, E. M. 1993. The intestinal Na^+/glucose cotransporter. *Annual Review of Physiology* 55: 575–589.

✔**QUESTION** Why did the researchers inject the RNAs into frog eggs instead of into rabbit epithelial cells?

Monoglycerides

Lipase

Fatty acids

1. Large fat globules are not digested efficiently by lipase.

2. Bile salts (produced in liver) act as emulsifying agents.

3. Small fat droplets result from emulsification.

4. Lipase digests the small fat droplets into monoglycerides and free fatty acids.

FIGURE 44.15 Emulsifying Agents and Lipases Digest Lipids in the Small Intestine. Once bile salts break up large fat globules, lipase can digest fats efficiently.

SUMMARY TABLE 44.3 **Digestive Enzymes in Mammals**

Digestion is accomplished by the enzymes listed here, by HCl produced in the stomach in response to the hormone gastrin, and by bile salts from the liver. Bile salts are stored in the gallbladder. They are released in response to the hormone cholecystokinin and emulsify fats in the small intestine.

	Where Synthesized	Regulation	Function
Carboxypeptidase	Pancreas	Released in inactive form in response to cholecystokinin from small intestine; activated by trypsin	In small intestine, breaks peptide bonds in polypeptides—releasing amino acids
Chymotrypsin	Pancreas	Released in inactive form in response to cholecystokinin from small intestine; activated by trypsin	In small intestine, breaks peptide bonds in polypeptides—releasing amino acids
Elastase	Pancreas	Released in inactive form in response to cholecystokinin from small intestine; activated by trypsin	In small intestine, breaks peptide bonds in polypeptides—releasing amino acids
Lingual lipase	Salivary glands	Released in response to taste and smell stimuli	In mouth and stomach, breaks bonds in fats—releasing fatty acids and monoglycerides
Nucleases	Pancreas	Released in response to cholecystokinin from small intestine	In small intestine, break apart nucleic acids—releasing nucleotides
Pancreatic amylase	Pancreas	Released in response to cholecystokinin from small intestine	In small intestine, breaks apart carbohydrates—releasing sugars
Pancreatic lipase	Pancreas	Released in response to cholecystokinin from small intestine	In small intestine, breaks bonds in fats—releasing fatty acids and monoglycerides
Pepsin	Stomach	Released in inactive form (pepsinogen); activated by low pH in stomach lumen	In stomach, breaks peptide bonds between certain amino acids in proteins—releasing polypeptides
Salivary amylase	Salivary glands	Released in response to taste and smell stimuli	In mouth, breaks apart carbohydrates—releasing sugars
Trypsin	Pancreas	Released in inactive form (trypsinogen) in response to cholecystokinin from small intestine; activated by hormone enterokinase from small intestine	In small intestine, breaks specific peptide bonds in polypeptides—releasing amino acids

the epithelial cells. The lacteals merge with larger lymph vessels, which then merge with veins. In this way, fats enter the bloodstream without clogging small blood vessels. Eventually, the products of fat digestion end up in adipose tissue and other tissues.

How Is Water Absorbed? When solutes from digested material are absorbed into the epithelium of the small intestine, water follows passively by osmosis. This is an important mechanism for (1) absorbing water that has been ingested, and (2) reclaiming liquid that was secreted into the digestive tract in the form of saliva, mucus, and pancreatic fluid.

This mechanism of water absorption inspired an important medical strategy called oral rehydration therapy. If a patient has diarrhea, clinicians frequently prescribe dilute solutions of glucose and electrolytes to be taken orally. When the glucose in the drink is absorbed in the small intestine through sodium-glucose cotransporters, enough water and sodium follow to prevent the life-threatening effects of dehydration. This simple medication saves thousands of lives every year. ✔ If you understand this concept, you should be able to predict at least two effects of a molecule that selectively blocks the sodium-glucose cotransporter.

The Large Intestine

By the time digested material reaches the large intestine of a human, a large amount of water (approximately 5 liters per day in total) and virtually all of the available nutrients have been absorbed. The primary function of the **large intestine** is to compact the wastes that remain and to absorb enough water to form feces.

These processes occur in the **colon**—the main section of the large intestine. Feces are held in the **rectum,** which is the final part of the large intestine, until they can be excreted. Although the kidneys are responsible for maintaining water balance, water absorption in the large intestine is important for keeping the body well hydrated.

In addition to compacting wastes and absorbing water, the human colon contains symbiotic microorganisms that digest cellulose. Although fermentation of cellulose is not as important to an omnivorous human as it is to an herbivore like a ruminant, bacteria in the human colon also produce several important nutrients, such as vitamin K.

Aquaporins Play a Key Role in Water Reabsorption Water is absorbed in the large intestine through aquaporins. Recall that **aquaporins** are proteins that act as water channels in plasma membranes, thus increasing the rate of water movement via osmosis (see Chapter 6). The best-studied of these proteins, called AQP1 for aquaporin 1, is common in the nephrons of the kidney. AQP1 is one of the proteins responsible for water reabsorption from urine along the osmotic gradient (described in Chapter 43). To date, four distinct aquaporins have been found in the large intestines of rats, mice, and humans. Researchers are working to unravel exactly how the activity of these aquaporins is regulated.

Variations in Structure and Function The size and function of the large intestine varies dramatically among animals. In insects,

the posteriormost portion of the digestive tract, called the hindgut, functions to reabsorb water and ions and excrete uric acid and feces (see Chapter 43). Among vertebrates, the various lineages of fish have no large intestine at all.

In some herbivorous species, the **cecum,** a blind sac at the proximal end of the large intestine, is greatly enlarged and functions in digesting cellulose. These species include rabbits, many rodents, some marsupials, horses, elephants, tapirs, and leaf-eating primates. The cecum contains symbiotic bacteria and protists that ferment cellulose, in much the same way that some birds house cellulose-digesting symbionts in their crop and ruminants house cellulose-digesting symbionts in a modified stomach.

In humans, a narrow pouch called the **appendix** emerges from the cecum. The function of the appendix has long been debated. It has often been described as vestigial—a reduced or incompletely developed trait that is a vestige of evolutionary ancestry (see Chapter 25)—partly because it does not perform any obvious vital function. Indeed, if it becomes inflamed, it can be surgically removed from a patient and cause no ill effects. However, although not vital, the appendix may have important functions: it is imbued with immune cells, and it appears to act as a haven for symbiotic microorganisms that inhabit the colon. For example, serious infections can result in diarrhea that flushes symbiotic bacteria from the colon. The appendix may provide the additional bacteria needed for recolonizing the colon after such an episode.

Rabbits and some other species with a cecum are able to excrete the cecum's contents as pellets, which the animal then reingests and passes through the digestive tract a second time. This particular example of **coprophagy,** or excrement eating, allows the animal to absorb more nutrients from the food.

check your understanding

If you understand that . . .
- In humans and many other animals, digestion begins with the mechanical breakdown of food into small pieces that are then acted on by acids, enzymes, emulsifying agents, and other chemical treatments.
- Distinct compartments within a digestive tract have distinct structures and functions.
- Nutrients are absorbed through specific membrane proteins in epithelial cells of the digestive tract.

✔ **You should be able to . . .**

1. Explain how each compartment in the human digestive tract aids the ingestion and digestion of food, absorption of nutrients, and excretion of wastes.

2. Predict the consequences of treating a person with drugs that inhibit the release of bile salts, that inactivate trypsin, or that block the action of the Na⁺-glucose cotransporter in the epithelial cells of the small intestine.

Answers are available in Appendix A.

Another striking structural variation occurs in the rectum of amphibians, reptiles, and birds. Their urine flows from the kidneys into an enlarged portion of the large intestine called the **cloaca** (see Chapter 43). Instead of having two orifices for excretion, these species have one.

44.4 Nutritional Homeostasis— Glucose as a Case Study

When digestion is complete, amino acids, fatty acids, ions, and sugars enter the bloodstream and are delivered to the cells that need them. The body uses or stores these nutrients in order to maintain homeostatic levels in the blood and avoid imbalances. Too much of a nutrient or too little can be problematic or even fatal.

The illness **diabetes mellitus** is a classic example of nutrient imbalance. People with diabetes mellitus experience abnormally high levels of glucose in their blood. Over the course of a lifetime, chronically elevated blood glucose can lead to an array of complications, including blindness, impaired circulation, and heart failure. What causes the imbalance? More important, how is glucose homeostasis normally maintained?

The Discovery of Insulin

In 1879, researchers removed the pancreas from a dog and observed that the dog's blood glucose became very high. This experiment suggested that the pancreas secretes a compound needed for removing glucose from the blood.

When other investigators cut up pancreatic tissues and injected extracts into diabetic dogs, however, they observed no response. Eventually, they decided that digestive enzymes in the pancreas were probably destroying the active agent during the extraction process.

Then, in 1921, Frederick Banting and Charles Best conducted a breakthrough experiment:

1. They began by tying off a dog's pancreatic duct—the tube through which digestive enzymes flow to the small intestine. The logic was that blocking the secretion of digestive enzymes might kill the cells that synthesize them.

2. The investigators waited several weeks for the cells near the duct to die and then removed the pancreas.

3. They froze the pancreatic tissue, ground it up, and injected an extract into a diabetic dog.

4. To their delight, the dog's blood sugar levels stabilized, and the dog became more active and healthy looking.

After Banting and Best repeated the experiment and observed the same result, they grew increasingly confident that they had located the source of the molecule responsible for lowering blood glucose levels. The molecule came to be called insulin.

Insulin's Role in Homeostasis

Insulin is a hormone that is secreted by the pancreas when blood glucose levels are high. It travels through the bloodstream and binds to receptors on cells throughout the body (see Chapter 49 for more detail on the structure and function of hormones and other chemical signals).

In response, cells that have insulin receptors increase their rate of glucose uptake and processing. Specifically, insulin stimulates cells in the liver and skeletal muscle to import glucose from the blood and synthesize glycogen from glucose monomers. As a result, glucose levels in the blood decline.

If blood glucose levels fall, as they do when food is lacking, cells in the pancreas secrete a hormone called **glucagon.** In response to glucagon, cells in the liver catabolize glycogen, and also produce glucose via **gluconeogenesis,** the synthesis of glucose from non-carbohydrate compounds. As a result, glucose levels in the blood rise (**FIGURE 44.16**).

Insulin and glucagon interact to form a negative feedback system capable of achieving homeostasis with respect to glucose concentrations in the blood.

FIGURE 44.16 Insulin and Glucagon Provide Negative Feedback in a Homeostatic System. Both insulin and glucagon are secreted by cells in the pancreas but have opposite effects on blood glucose concentrations.

✔QUESTION After reading the descriptions of type 1 and type 2 diabetes mellitus, decide which arrow is disrupted in individuals with type 2 diabetes mellitus and which arrow is disrupted in individuals with type 1 diabetes mellitus.

Diabetes Mellitus Has Two Forms

Diabetes mellitus develops in people who (1) do not synthesize sufficient insulin or (2) are resistant to insulin, meaning that insulin does not effectively activate its receptor in target cells. The first condition is called type 1 diabetes mellitus; the second condition is type 2 diabetes mellitus. In both cases, effector cells do not receive the signal that would result in a drop in blood glucose levels.

Glucose imbalance has a direct effect on urine formation. Normally, signals from insulin keep blood glucose levels low enough that all of the glucose can be reabsorbed from the filtrate formed in the kidney. But when blood glucose levels are high, so much glucose enters the nephron that it cannot all be reabsorbed. High glucose concentrations in the filtrate increase its osmolarity and decrease the amount of water reabsorbed from it. More water leaves the body, leading to high urine volume in both types of diabetes mellitus.

The word diabetes means "to run through"; water "runs through" people with diabetes. In addition, the word mellitus means "honeyed (sweet)." Before chemical methods of analyzing urine were available, physicians would taste the patient's urine. If the urine was sweet, it was highly likely that the patient was suffering from diabetes mellitus.

Why do these diseases develop? Type 1 diabetes mellitus is an autoimmune disease, meaning that the body's immune system mistakenly targets its own cells for destruction. In the case of type 1 diabetes mellitus, the insulin-producing cells of the pancreas are destroyed. Type 2 diabetes mellitus occurs when the receptors for insulin no longer function correctly or are reduced in number. The primary risk factor for developing type 2 diabetes mellitus is obesity.

Currently, type 1 diabetes mellitus is treated with insulin injections and careful attention to diet; type 2 diabetes mellitus is managed primarily through prescribed diets, exercise, and monitoring blood glucose levels, as well as taking drugs that increase cellular responsiveness to insulin. The challenge is to achieve homeostasis with respect to blood glucose in the absence of the body's normal regulatory mechanisms.

✔ If you understand the difference between type 1 and type 2 diabetes mellitus, you should be able to explain how, in individuals with each disease and without disease, insulin receptors in the plasma membrane of liver cells interact with insulin and how the liver cells respond to high glucose concentrations in the blood.

The Type 2 Diabetes Mellitus Epidemic

An epidemic of type 2 diabetes mellitus is currently under way in certain human populations. For example, as of 2011, a total of 11 percent of U.S. adults aged 20 and over were diabetic, and this figure jumps to 27 percent in U.S. adults aged 65 and over. About 90% to 95% of cases are type 2 diabetes mellitus. Minority groups such as African Americans and Hispanics are experiencing much higher rates of new cases of type 2 diabetes mellitus than white non-Hispanic Americans.

Because there is a strong association between the prevalence of diabetes mellitus in parents and in children, researchers have

long suspected that some individuals have a genetic predisposition for developing the disease. Researchers have so far identified alleles at 38 different genes that predispose individuals to type 2 diabetes mellitus (using techniques introduced in Chapter 20 and Chapter 21).

However, there is also strong evidence that environmental conditions have an important impact on the incidence of type 2 diabetes mellitus. As an example, consider the Pima Indians of North America.

The Pima consist of two main populations—one in southwest Arizona in the United States and one in a remote area of the Sierra Madre mountains of Mexico. The bars graphed in **FIGURE 44.17a** indicate the average percentage of individuals with type 2 diabetes mellitus, by gender, in Mexican people who are

(a) Variation in average incidence of type 2 diabetes mellitus

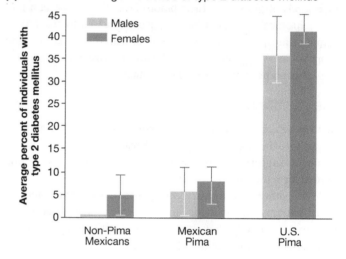

(b) Variation in average body mass index

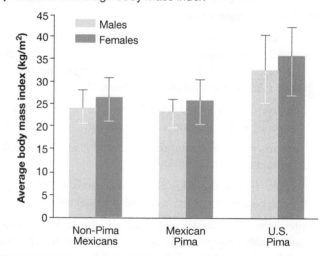

FIGURE 44.17 The Incidence of Type 2 Diabetes Mellitus Is Correlated with Obesity.

DATA: Schultz, L. O., P. H. Bennett, E. Ravussin, et al. 2006. *Diabetes Care* 29: 1866–1871.

✔**QUANTITATIVE** In the graphs above, it is apparent that U.S. Pima have higher rates of type 2 diabetes mellitus and higher body mass indices than the other groups. Does this prove that obesity causes diabetes mellitus? Why or why not?

not members of the Pima and in the two populations of Pima. Although researchers have been unable to find any significant genetic differences between the two groups of Pima, the data indicate there are dramatic differences in the incidence of type 2 diabetes mellitus.

The incidence of type 2 diabetes mellitus is correlated with obesity. As the data graphed in **FIGURE 44.17b** show, Pima from the United States are more likely to be obese than individuals in the other two populations in this study. A person who has a body mass index greater than or equal to 30 is considered obese. The **body mass index** is calculated as weight (in kilograms) divided by height squared (in meters).

The differences in body mass index among the Pima are explained by differences in physical activity and diet. Pima men in Mexico are physically active an average of 33 hours a week, while their counterparts in the United States are active just 12.1 hours per week. Pima women in Mexico average 22 hours of physical activity a week; Pima women in the United States average just 3.1. In addition, Pima in the United States get approximately 40 percent of their calories from fat, compared to 15 percent in the traditional Pima diet.

Although Pima people in the United States have the highest rates of type 2 diabetes mellitus recorded for any people to date, the same trends are occurring in populations all over the world. Nutrition-related diseases have become a major public health concern.

Type 2 diabetes mellitus used to be called adult-onset diabetes because the disease typically appeared in adults. However, that name has been abandoned as more and more young people are diagnosed with the disease. While a multitude of genes can predispose a child or adult to developing diabetes mellitus, the disease is triggered by environmental factors including diet, lack of exercise, and obesity.

In a five-year span in the past decade, obesity rates increased by 10 percent in American children overall and by over 30 percent in children from low-income families and certain racial minority groups. Scientists estimate that one in three children born nowadays will develop diabetes mellitus, and that rates are likely to continue rising. It is clear that diabetes mellitus is a modern-day epidemic.

check your understanding

If you understand that . . .

- Insulin and glucagon interact to regulate glucose concentrations in the blood.
- Diabetes mellitus results from a lack of homeostasis in the concentration of the nutrient glucose in the blood.

✔ **You should be able to . . .**

1. Identify the causes of each type of diabetes mellitus.

2. Prescribe what a person with type 1 diabetes mellitus should do when the blood glucose level is too high and when the blood glucose level is too low.

Answers are available in Appendix A.

CHAPTER 44   REVIEW

For media, go to MasteringBiology

If you understand . . .

44.1   Nutritional Requirements

- The diets of animals include fats, carbohydrates, and proteins that provide energy; vitamins that serve as coenzymes and perform other functions; ions required for water balance and for nerve and muscle function; and elements that are incorporated into molecules synthesized by cells.

- Fats contain the most energy of all nutrients: 9 kcal/g compared to 4 kcal/g for carbohydrates and proteins, making fats an efficient way to store energy in the body.

- ✔ QUANTITATIVE You should be able to calculate and compare the caloric content of skim milk and whole milk. Per serving, skim milk contains 12 g carbohydrates, 8 g protein, and no fat; whole milk contains 12 g carbohydrates, 8 g protein, and 8 g fat.

44.2   Capturing Food: The Structure and Function of Mouthparts

- Most animals are mass feeders that obtain food by seizing and manipulating it with mouthparts such as teeth, jaws, or beaks or with special toxin-injecting organs.

- Natural selection has modified mouthparts in different species to act as efficient tools for obtaining particular types of food.

- ✔ You should be able to predict the types of mouthparts and the nature of the digestive tract in a mammal that eats only nectar from flowers.

44.3 How Are Nutrients Digested and Absorbed?

- In most animals, the digestive tract begins at the mouth and ends at the anus.

- In many species, chemical digestion of food also begins in the mouth. In mammals, an enzyme in saliva called salivary amylase begins to hydrolyze the bonds linking glucose monomers in starch, glycogen, and other carbohydrates.

- Once food is swallowed, it is propelled down the esophagus by peristalsis.

- Digestion continues in the stomach. In humans, the stomach is a highly acidic environment that denatures proteins and in which the enzyme pepsin begins the cleavage of peptide bonds that link amino acids.

- Food passes from the stomach into the small intestine, where it is mixed with secretions from the pancreas and liver.

- In the small intestine, carbohydrate digestion is continued by pancreatic amylase; fats are emulsified by bile salts and digested by pancreatic lipase; protein digestion is completed by a suite of pancreatic enzymes.

- Cells that line the small intestine absorb the nutrients released by digestion. In many cases, uptake is driven by an electrochemical gradient established by Na^+/K^+-ATPase that favors a flow of Na^+ into the cell.

- As solutes leave the lumen of the small intestine and enter cells, water follows by osmosis.

- Water reabsorption is completed in the large intestine, where feces form.

- The structure of organs in the digestive tract varies widely among species, in ways that support efficient processing of the food a particular species ingests.

✔ You should be able to compare the optimal pH for amylase and pepsin.

44.4 Nutritional Homeostasis— Glucose as a Case Study

- Diabetes mellitus is a condition in which concentrations of glucose in the blood are chronically too high.

- Type 1 diabetes mellitus is caused by a defect in the production of insulin—a hormone secreted by the pancreas that promotes the uptake of glucose from the blood.

- Type 2 diabetes mellitus is characterized by a failure of cells to respond to insulin.

- The development of type 2 diabetes is correlated with obesity and is reaching epidemic proportions in some populations.

✔ You should be able to explain how each type of diabetes mellitus results in high blood glucose.

(MB) MasteringBiology

1. **MasteringBiology Assignments**

   Tutorials and Activities Digestion and Absorption of Food; Homeostasis: Regulating Blood Sugar; Understanding Diabetes Mellitus

   Questions Reading Quizzes, Blue-Thread Questions, Test Bank

2. **eText** Read your book online, search, take notes, highlight text, and more.

3. **The Study Area** Practice Test, Cumulative Test, BioFlix® 3-D animations, Videos, Activities, Audio Glossary, Word Study Tools, Art

You should be able to . . .

✔ TEST YOUR KNOWLEDGE

Answers are available in Appendix A

1. What does secretin stimulate?
   a. secretion of HCO_3^- from the pancreas
   b. secretion of HCl from the stomach epithelium
   c. secretion of digestive enzymes from the pancreas
   d. uptake of glucose from the bloodstream by cells throughout the body

2. What is an incomplete digestive tract?
   a. a digestive tract that is missing a major compartment—for example, fish lack a large intestine
   b. a ruminant-type tract, where food is regurgitated and re-chewed before digestion is completed
   c. a reduced or vestigial tract like that found in animals that absorb nutrition directly across the body wall
   d. a digestive system with a single opening for ingestion and excretion

3. In mammals, how and where are carbohydrates digested?
   a. by lipases in the small intestine
   b. by pepsin and HCl in the stomach
   c. by aquaporins in the large intestine
   d. by amylases in the mouth and small intestine

4. Cellulose is fermented in which of the following structures in rabbits?
   a. small intestine
   b. cecum
   c. abomasum
   d. rumen

5. A hormone that reduces blood glucose is _____, while a hormone that increases blood glucose is _____.

6. True or False: Type 1 diabetes mellitus is more prevalent in the United States than type 2 diabetes mellitus.

7. Explain how the structure of each of the following digestive tract compartments correlates with its function: the bird crop, cow rumen, and elephant large intestine.

8. Why are many digestive enzymes produced in an inactive form and then activated in the lumen of the digestive tract?

9. Explain why oral rehydration therapy with a solution of sodium and glucose is effective.
   a. The sodium and glucose decrease urine output.
   b. The sodium and glucose facilitate water absorption by the small intestine.
   c. The sodium and glucose help kill the intestinal bacteria.
   d. The sodium and glucose make the person thirsty.

10. How does the high type 2 diabetes mellitus rate among Pima Indians in the United States provide evidence for an environmental effect on development of the disease?

11. What do the shapes of human teeth suggest about their function in terms of the types of food humans are adapted to eat?

12. Predict what would happen if a type 1 diabetic took too much insulin.

13. Doctors sometimes prescribe proton pump inhibitors to treat patients with acid reflux. How do these drugs work?

14. An elderly man arrives at the hospital severely dehydrated. The physicians find that there is glucose in his urine and that his blood glucose does not decline much upon administration of an insulin shot. What is he likely to be suffering from?
   a. type 1 diabetes mellitus
   b. type 2 diabetes mellitus
   c. diarrhea
   d. gastric ulcer

15. Scientists who backed the hypothesis that secretion from the pancreas is under nervous control strenuously objected to the conclusions reached in the experiment that led to the discovery of hormonal control. They said the experiment was inconclusive, because of the likelihood that not all the nerves contacting the small intestine had been cut. The biologists who did the experiment replied that even if not all the nerves had been cut, the result was still valid. In your opinion, who is correct? Why?

16. Among vertebrates, the large intestine occurs only in lineages that are primarily terrestrial (amphibians, reptiles, and mammals). Propose a hypothesis to explain this observation.

45 Gas Exchange and Circulation

In this chapter you will learn that

Animals have adaptations for gas exchange across body surfaces and circulation within their bodies

via ↓

Respiratory and circulatory systems **45.1**

comparing ↓   exploring ↓

O_2 and CO_2 exchange in air vs. water **45.2** O_2 and CO_2 transport in blood **45.4**

asking ↓   asking ↓

How do different gas exchange organs work? **45.3**   How do different circulatory systems work? **45.5**

During intense exercise, animal circulatory systems deliver large amounts of oxygen to tissues and remove large amounts of carbon dioxide. This chapter explores how gas exchange occurs in animals that live in aquatic or terrestrial environments.

This chapter is part of the Big Picture. See how on pages 840–841.

Animal cells are like factories that run 24 hours a day. Inside the plasma membrane, the chemical reactions that sustain life require a steady input of raw materials. Those reactions also produce a steady stream of wastes. Earlier in this unit, you learned how solid and liquid waste materials are removed from the cells and excreted from the body (see Chapter 43). You also examined how nutrients enter the body (see Chapter 44). Let's now turn our attention to two other major questions:

1. How are two of the most important molecules in the economy of the cell—the oxygen (O_2) required for cellular respiration and the carbon dioxide (CO_2) produced by cellular respiration—exchanged with the environment?

✔ When you see this checkmark, stop and test yourself. Answers are available in Appendix A.

2. How are these gases—along with wastes, nutrients, and other types of molecules—transported throughout the body?

Understanding gas exchange and circulation is fundamental to understanding how animals work. If either process fails, the consequences are dire. Consider the maladies that can develop when gas exchange or circulation is disrupted in humans: anemia, pneumonia, tuberculosis, malaria, heart disease, and stroke are just a few examples.

Let's begin with an overview of animal respiratory and circulatory systems and then plunge into the details of how gases are exchanged and transported.

45.1 The Respiratory and Circulatory Systems

When the mitochondria inside animal cells are producing ATP via cellular respiration, they consume oxygen and produce carbon dioxide. To support continued ATP production, cells have to obtain oxygen and expel excess carbon dioxide continuously (see The Big Picture: Energy, pages 198–199).

How does this gas exchange occur between an animal's environment and its mitochondria? In most cases, gas exchange involves the four steps illustrated in **FIGURE 45.1**:

1. *Ventilation,* the movement of air or water through a specialized gas exchange organ, such as lungs or gills.

2. *Gas exchange,* the diffusion of O_2 and CO_2 between air or water and the blood at the respiratory surface.

3. *Circulation,* the transport of dissolved O_2 and CO_2 throughout the body—along with nutrients, wastes, and other types of molecules—via the circulatory system.

4. *Cellular respiration,* the cell's use of O_2 and production of CO_2. In tissues, where cellular respiration has led to low O_2 levels and high CO_2 levels, gas exchange occurs between blood and cells.

Steps 1 and 2 are accomplished by the **respiratory system,** the collection of cells, tissues, and organs responsible for gas exchange between the individual animal and its environment. In essence, a respiratory system consists of structures for conducting air or water to a surface where gas exchange takes place.

In some animals the gas exchange surface is the skin, but in most species it is located in a specialized organ like the lungs of tetrapods, the tracheae of insects, or the gills found in mollusks, arthropods, and fishes. Section 45.3 analyzes the structure and function of gills, tracheae, and lungs in detail.

Step 3 in Figure 45.1 is usually accomplished by a **circulatory system,** which moves O_2, CO_2, and other materials around the body. In many cases, a muscular heart propels a specialized, liquid transport tissue throughout the body via a system of vessels.

Keeping in mind this broad overview of gas exchange and circulatory systems, let's dive into the details by exploring how oxygen and carbon dioxide move between an animal's body and its environment.

45.2 Air and Water as Respiratory Media

Gas exchange between the environment and cells is based on diffusion. Under normal conditions, oxygen concentrations are relatively high in the environment (for example, in air that you inhale or in ocean water) and low in tissues, while carbon dioxide levels are relatively high in tissues and low in the environment. Thus oxygen tends to move from the environment into tissues, and carbon dioxide tends to move from tissues into the environment.

How much oxygen and carbon dioxide are present in the atmosphere versus the ocean? What factors in air and water influence how quickly these gases move by diffusion?

How Do Oxygen and Carbon Dioxide Behave in Air?

The atmosphere is composed primarily of nitrogen (76%) and oxygen (21%) and has trace amounts of argon (0.93%) and CO_2 (0.03%). Nitrogen (N_2) and argon are not important to animals living at sea level and are usually ignored in analyses of gas exchange.

However, the data are actually a little misleading. To understand why, consider that the *percentage* of O_2 in the atmosphere does not vary with elevation. The atmosphere at the top of Mt. Everest is

1. Ventilation

Environment

Air or water

Respiratory system

2. Gas exchange

Respiratory surface

O_2

CO_2

3. Circulation

Blood

Circulatory system

4. Cellular respiration

O_2

CO_2

Mitochondria

FIGURE 45.1 Gas Exchange Involves Ventilation, Circulation, and Respiration. In animals, oxygen and carbon dioxide are exchanged across the surface of a lung, a gill, the skin, or some other gas exchange organ. In many species these gases are then transported to and from cells—where gas exchange again takes place—in a fluid tissue such as blood.

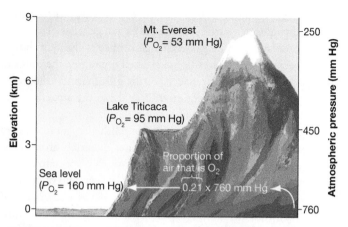

FIGURE 45.2 Oxygen Makes Up 21 Percent of the Atmosphere, but Its Partial Pressure Depends on Altitude. Atmospheric pressure, and thus oxygen partial pressure (P_{O_2}), fall with increasing elevation.

composed of 21 percent oxygen, just as it is at sea level. The key difference is that far fewer molecules of oxygen and other atmospheric gases are present at high elevations than at sea level. Air at the top of Mt. Everest is less dense than air at sea level, so less oxygen is present. To understand how gases move by diffusion, it is important to express their presence in terms of partial pressures instead of percentages. Pressure is a type of force. A **partial pressure** is the pressure of a particular gas in a mixture of gases.

To calculate the partial pressure of a particular gas, multiply the fractional composition (the fraction of air the gas comprises) by the total pressure exerted by the entire mixture (atmospheric pressure). The calculation is valid because the total pressure of a mixture of gases is the sum of the partial pressures of all the individual gases. This relationship is called Dalton's law.

For example, **FIGURE 45.2** shows that the total atmospheric pressure at sea level is 760 mm Hg (millimeters of mercury). If you multiply this value by 0.21, which is the fraction of air that is O_2, you obtain a partial pressure of oxygen, abbreviated P_{O_2}, at sea level of 160 mm Hg. Because the atmospheric pressure is only about 250 mm Hg at the top of Mt. Everest, the P_{O_2} there is only $0.21 \times 250 = 53$ mm Hg.

Oxygen and carbon dioxide diffuse between the environment and cells along their respective partial pressure gradients, just as solutes diffuse along their electrochemical gradients. In both air and water, O_2 and CO_2 move from regions of high partial pressure to regions of low partial pressure. It is hard to breathe at the top of Mt. Everest because the partial pressure of oxygen is low there—meaning that the diffusion gradient between the atmosphere and your lung tissues is small, so fewer molecules of O_2 diffuse into your tissues when you take a breath.

How Do Oxygen and Carbon Dioxide Behave in Water?

To obtain oxygen, water breathers face a much more challenging environment than air breathers do. Aquatic animals live in an environment that contains much less oxygen than the environments

inhabited by terrestrial animals. At 15°C, a liter of air can contain up to 209 milliliters (mL) of O_2, while a liter of water may contain a maximum of only 7 mL of O_2. To extract a given amount of oxygen, an aquatic animal has to process 30 times more water than the amount of air a terrestrial animal breathes.

In addition, water is about a thousand times denser than air and flows much less easily. As a result, water breathers have to expend much more energy to ventilate their respiratory surfaces than do air breathers.

What Affects the Amount of Gas in a Solution? Oxygen and carbon dioxide diffuse into water from the atmosphere, but the amount of gas that dissolves in water depends on several factors:

- *Solubility of the gas in water* Oxygen has very low solubility in water. Only 0.003 mL of oxygen dissolves in 100 mL of water for each increase of 1 mm Hg in oxygen partial pressure. Because of this low solubility, blood contains a molecule that binds to oxygen and delivers it to tissues. Without this carrier molecule, the rate of blood flow to tissues would have to increase dramatically to meet oxygen demand.

- *Temperature of the water* As the temperature of water increases, the amount of gas that dissolves in it decreases. Other things being equal, warm-water habitats have much less oxygen available than cold-water habitats do. For a fish, breathing in warm water is comparable to a land-dwelling animal breathing at high elevation.

- *Presence of other solutes* Because seawater has a much higher concentration of solutes than does freshwater, seawater can hold less dissolved gas. At 10°C, up to 8.02 mL of O_2 can be present per liter of freshwater versus only 6.35 mL of O_2 per liter of seawater. As a result, freshwater habitats tend to be more oxygen rich than marine environments.

- *Partial pressure of the gas in contact with the water* Gases move from regions of high partial pressure to regions of low partial pressure. So if the partial pressure in a liquid exceeds that in the adjacent gas, the gas will bubble up out of the liquid. This is what happens when the cap is removed from a bottle of carbonated beverage. The partial pressure of carbon dioxide in the newly opened drink is much higher than it is in the atmosphere.

What Affects the Amount of Oxygen Available in an Aquatic Habitat? The partial pressure of oxygen varies in different types of aquatic habitats, just as it varies with altitude on land. In addition to the four factors just listed, other important considerations affect oxygen's availability in water. These include the presence of photosynthetic organisms and decomposers, the amount of mixing that occurs, and the surface area of the body of water. For example,

- Habitats with large numbers of photosynthetic organisms tend to be relatively oxygen rich. In contrast, oxygen content is extremely low in bogs and other stagnant-water habitats because oxygen is quickly depleted by decomposers that use it as an electron acceptor in cellular respiration.

- Unless currents mix water almost continuously, water near the surface has much higher oxygen content than water near the bottom of the same habitat.

- Shallow ponds and streams tend to be much better oxygenated than deep bodies of water because shallower bodies have a higher ratio of surface area to volume.

- Rapids, waterfalls, and other types of whitewater are the most highly oxygenated of all aquatic environments because a large surface area is exposed to the atmosphere as water splashes over rocks and logs and because air bubbles are incorporated into the water.

Now let's consider the structure and function of ventilatory organs. How do the gills of fishes, the tracheae of insects, and the lungs of mammals cope with the differences between air and water?

check your understanding

If you understand that . . .

- O_2 and CO_2 move from regions of high partial pressure to regions of low partial pressure.
- The partial pressure of oxygen in a body of water depends on the water's temperature, surface area, and amount of mixing, and on the impacts of photosynthetic organisms and decomposers, among other factors.
- Water breathing is much more difficult than air breathing, in part because the partial pressure of oxygen in water is much lower than its partial pressure in air.

✔ You should be able to . . .

1. Explain why oxygen partial pressures are relatively high in mountain streams and relatively low at the ocean bottom.

2. For each of three aquaria, decide whether a large or a small amount of air should be bubbled in. One contains warm-water species, one contains vigorous algal growth, and one contains sedentary animals.

Answers are available in Appendix A.

45.3 Organs of Gas Exchange

Many small animals lack specialized gas exchange organs, such as gills or lungs. Instead, they obtain O_2 and eliminate CO_2 by diffusion across the body surface. This is possible because their size and shape give them an extraordinarily high ratio of surface area to volume (see Chapter 42). In sponges, jellyfish, flatworms, and other species, diffusion across the body surface is rapid enough to fulfill their requirements for taking in O_2 and expelling CO_2.

Most of these animals are restricted to living in wet environments, however. For gas exchange to take place efficiently, the surface where it occurs has to be thin, and thin tissues are prone to water loss. Living in wet or humid environments allows animals to exchange gases across their outer surface while avoiding dehydration.

In contrast, animals that are large or that live in dry habitats need some sort of specialized respiratory organ. Respiratory organs provide a greater surface area for gas exchange—enough to meet the demands of a large body filled with cells. In terrestrial animals, respiratory organs are located inside the body, which helps minimize water loss.

Biologists have long marveled at the efficiency of gills and lungs. To appreciate why, let's examine the physical factors that control diffusion rates and then look at the structure and function of these respiratory organs.

Physical Parameters: The Law of Diffusion

In 1855 Adolf Fick derived an equation regarding diffusion, based on the results of experiments he had performed on the behavior of gases. **Fick's law of diffusion** states that the rate of diffusion of a gas depends on five parameters: the solubility of the gas in the aqueous film lining the gas exchange surface; the temperature; the surface area available for diffusion; the difference in partial pressures of the gas across the gas exchange surface; and the thickness of the barrier to diffusion (**FIGURE 45.3**).

Fick's law identifies traits that allow animals to maximize the rate at which oxygen and carbon dioxide diffuse across surfaces. Specifically, Fick's law states that gases diffuse at the highest rates when three conditions are met:

1. A is large, meaning a large area is available for gas exchange. Based on Fick's law, it is not surprising that the respiratory surface in the human lungs would cover about 140 m²—about a quarter of a basketball court—if the epithelium were spread flat.

2. D is small, meaning the respiratory surface is extremely thin. In the human lung, this barrier to diffusion is only 0.2 μm thick—about 1/200th of the thickness of this page.

3. $P_2 - P_1$ is large, meaning the partial pressure gradient of the gas across the surface is large. High partial pressure gradients are maintained in part by having an efficient circulatory

$$\text{Rate of diffusion} = k \times A \times \frac{(P_2 - P_1)}{D}$$

Diffusion constant (depends on solubility of gas and temperature)

Difference in partial pressure of gas on either side of barrier to diffusion

Area for gas exchange

Distance (thickness of barrier to diffusion)

FIGURE 45.3 Fick's Law Describes the Rate of Diffusion.

system in close contact with the gas exchange surface. When blood flows close to the respiratory surface, oxygen is rapidly taken away from the area where inward diffusion is occurring, and carbon dioxide is rapidly brought into the area where outward diffusion is occurring. As a result, $P_2 - P_1$ stays high.

What other aspects of gill and lung structure affect diffusion rate? To answer this question, let's delve into the anatomy of these respiratory organs.

How Do Gills Work?

Gills are outgrowths of the body surface or throat that are used for gas exchange in aquatic animals. Gills are efficient solutions to the problems posed by water breathing, primarily because they present a large surface area for the diffusion of gases across a thin epithelium.

In some species of invertebrates, such as the red tubeworm (**FIGURE 45.4a**), gills project from the body surface and contact the surrounding water directly. In other invertebrate species, such as the crayfish (**FIGURE 45.4b**), gills are located inside the exoskeleton or body wall. If gills are internal, water must be driven over them by cilia, the limbs, or other specialized structures.

In contrast to the diversity of gills found in invertebrates, the gills of bony fishes are all similar in structure. Fish gills are located on both sides of the head and in teleosts (see Chapter 35) consist of four arches, as **FIGURE 45.5** shows.

How Do Fishes Ventilate Their Gills? To move water through their gills so gas exchange can take place, most fishes open and close their mouth and **operculum,** the stiff flap of tissue that covers the gills. The pumping action of the mouth and operculum creates a pressure gradient that moves water over the gills.

In contrast, tuna and other fishes that are particularly fast swimmers force water through their gills by swimming with their mouths open. This process is called ram ventilation.

Regardless of how fish gills are ventilated, water flows in one direction through gills, passing over long, thin structures called **gill filaments** that extend from each gill arch. Each gill filament is composed of hundreds or thousands of **gill lamellae.** Gill lamellae are sheetlike structures, shown in detail at the bottom of Figure 45.5. Note that a bed of small blood vessels called capillaries runs through each lamella.

The Fish Gill Is a Countercurrent System The one-way flow of water through gill lamellae has a profound impact on gill function, for a simple reason: The flow of blood through the capillary bed in each lamella is in the opposite direction to the flow of water. As a result, each lamella functions as a countercurrent exchanger.

Recall that countercurrent exchangers are based on two adjacent fluids flowing in opposite directions (see Chapter 42). **FIGURE 45.6** illustrates why the countercurrent flow is so critical.

Note two key points about the left side of the figure, where water and blood flow in opposite directions:

1. A slight gradient in partial pressure of oxygen (here, 10 percent) exists along the entire length of the lamella.

2. A large difference in oxygen partial pressure exists between the start and end of the system. In this example, the difference is $100\% - 15\% = 85\%$ in the water and $90\% - 5\% = 85\%$ in the blood.

The upshot? Most of the oxygen in the incoming water has diffused into the blood.

Now look at the right side of the figure, where water and blood flow in the same direction.

1. A large gradient in partial pressure of oxygen (here, 100 percent) exists at the start of the system. The gradient in oxygen partial pressures declines rapidly and eventually disappears.

(a) External gills are in direct contact with water.

(b) Internal gills must have water brought to them.

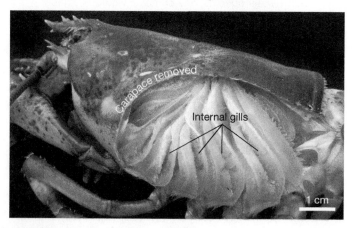

FIGURE 45.4 Gills Can Be External or Internal. (a) Red tubeworms are marine polychaetes with gills that protrude from the body. **(b)** Lobster gills are located inside the main body wall. A portion of the crayfish exoskeleton has been removed to expose the gills.

✔**QUESTION** Predict what the advantages and disadvantages of external versus internal gills might be in terms of fitness.

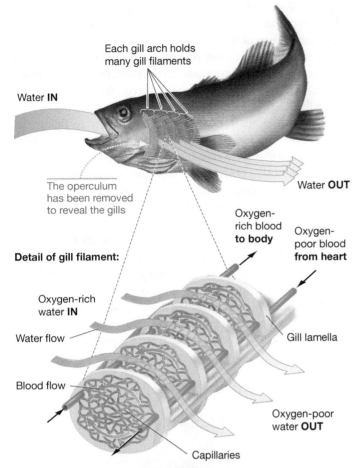

Each gill arch holds many gill filaments

Water **IN**

The operculum has been removed to reveal the gills

Water **OUT**

Detail of gill filament:

Oxygen-rich blood **to body**

Oxygen-poor blood **from heart**

Oxygen-rich water **IN**

Water flow

Gill lamella

Blood flow

Oxygen-poor water **OUT**

Capillaries

FIGURE 45.5 Fish Gills Are a Countercurrent Exchange System. In fish gills, water and blood flow in opposite directions. In the blood vessels, red represents oxygenated blood, blue represents deoxygenated blood, and purple represents mixed blood.

Countercurrent flow (seen in fish gills)

Water flow over lamellae (% oxygen)

| 100% | 70% | 40% | 15% |

O₂

| 90% | 60% | 30% | 5% |

Blood flow through lamellae (% oxygen)

Concurrent flow (not seen in fish gills)

Water flow over lamellae (% oxygen)

| 100% | 90% | 70% | 50% |

O₂

Diffusion stops

| 0% | 10% | 30% | 50% |

Blood flow through lamellae (% oxygen)

FIGURE 45.6 Countercurrent Exchange Is Much More Efficient than Concurrent Exchange. In the countercurrent system of fish gills, oxygen is transferred along the entire length of the capillaries.

✔**QUESTION** If you understand this concept, you should be able to predict what would happen to oxygen transfer from water to blood if flow were concurrent.

2. A relatively small difference in oxygen partial pressure exists between the start and end of the system. In this example, the difference is 100% − 50% = 50% in the water and 50% − 0% = 50% in the blood.

With this arrangement, only half of the oxygen in the incoming water has diffused into the blood.

Countercurrent flow makes fish gills extremely efficient at extracting oxygen from water because it ensures that a difference in the partial pressure of oxygen and carbon dioxide in water versus blood is maintained over the entire gas exchange surface.

The effect of countercurrent exchange is to maximize the $P_2 - P_1$ term in Fick's law of diffusion, averaged over the entire gill surface. Based on this observation, biologists cite countercurrent exchange as another example of how gills are optimized for efficient gas exchange.

How Do Insect Tracheae Work?

As noted earlier, air and water are dramatically different ventilatory media, because they have different densities, viscosities, and abilities to hold oxygen and carbon dioxide. In addition, the consequences of exposing the gas exchange surface to air versus water differ.

In aquatic habitats, ventilation tends to disrupt water and electrolyte balance, and homeostasis must be maintained by an active osmoregulatory system. Osmosis causes marine animals to lose water across their gas exchange surface and freshwater animals to gain water (see Chapter 43). Diffusion tends to cause marine animals to gain sodium, chloride, and other ions, and freshwater animals to lose them.

In contrast, breathing leads to a loss of water by evaporation in terrestrial environments. How do terrestrial animals minimize water loss while maximizing the efficiency of gas exchange?

To answer this question, consider the tracheal system of insects. Recall that insects have an extensive system of air-filled tubes called **tracheae** located within the body (see Chapter 43). These connect to the exterior through openings called **spiracles,** which can be closed to minimize the loss of water by evaporation. The ends of tracheae are tiny and highly branched. This design allows the tracheal system to transport air close enough to cells for gas exchange to take place directly across their plasma membranes.

Air moves from the atmosphere into the tracheae and then through the tracheae to the tissues in the insect's body. Is simple diffusion efficient enough to ventilate the system, or is some type of breathing mechanism involved?

If insects ventilate their tracheal system with simple diffusion, then according to Fick's law, the distance that the air would have to diffuse down the tracheae would be very important. In other words, the rate of diffusion of air down the tracheae would be limited in part by the length of the tracheae. In effect, only very small insects have tracheae that are short enough to be efficiently ventilated by simple diffusion.

Many insects are far too large to deliver gases to and from their tissues by simple diffusion alone. How do these large insects ventilate their tracheal system?

Consider a study on the sweet potato hawkmoth. Yutaka Komai set out to investigate how the amount of O_2 delivered to the wing muscles changes during flight. To document the partial pressure of oxygen in these muscles, he inserted a needlelike electrode into the flight muscles of a hawkmoth. The electrode was attached to an instrument that measured the partial pressure of oxygen, and the hawkmoth was tethered to a stand (**FIGURE 45.7**).

Komai recorded P_{O_2} as the insect rested and then stimulated the moth to fly by exposing it to wind. The "Results" section of

Figure 45.7 shows how P_{O_2} changed during one such experiment. (The procedure was repeated on several individuals, and the same pattern was observed.)

To read this graph, put your finger on the left-hand end of the orange line and move it toward the right, to where the insect started flying and the line changes to red. Note that P_{O_2} levels in the flight muscles dropped initially. This is not surprising, because flight is an energetically demanding activity. As flying continued, however, P_{O_2} levels recovered until they were nearly as high as when they were at rest.

To explain this observation, biologists propose that the tracheae are alternately compressed and dilated as the muscles around them contract and relax. The muscle contractions and relaxations produce pressure changes that cause the volume of space inside the tracheal system to change (**FIGURE 45.8**). This is a key point: If the volume occupied by a fixed amount of gas expands, the gas pressure decreases. But if the volume that a fixed amount of gas occupies declines, the gas pressure increases.

The volume of the tracheal system increases when muscles relax, causing pressure inside the system to go down and air from the atmosphere to rush in. What happens when muscles

RESEARCH

QUESTION: Does physical activity affect air movement through the tracheal system?

HYPOTHESIS: Air moves through the tracheal system faster during physical activity.

NULL HYPOTHESIS: Physical activity does not affect the rate of air movement.

EXPERIMENTAL SETUP:

Measure oxygen in hawkmoth flight muscle during rest and during flight.

Tether

PREDICTION: Flying will increase ventilation of tracheal system, causing increase in P_{O_2} in wing muscle.

PREDICTION OF NULL HYPOTHESIS: Flying will not increase ventilation of tracheal system; P_{O_2} in flight muscle will decline steadily during flight.

RESULTS:

Flight

Recovery of oxygen levels as flying continues

Initial drop in oxygen

Rest

Rest

Time (min)

P_{O_2} (kPa)

CONCLUSION: Muscular contractions may help ventilate the tracheal system in at least some insects.

FIGURE 45.7 Research Suggests the Tracheal System Is Ventilated during Movement. On the graph's y-axis, kPa stands for kilopascal, a unit of pressure (or force per unit area).

SOURCE: Komai, Y. 1998. *The Journal of Experimental Biology* 201: 2359–2366.

✔**QUESTION** Why was it important for the researcher to repeat this experiment with several different individuals?

Ground beetle

Tracheae

1 mm

Tracheae dilate when muscles relax

Tracheae compress when muscles contract

200 µm

FIGURE 45.8 Tracheae Are Dilated and Compressed during Ventilation. Muscle relaxation and contraction during flight or other activity alternately dilates and compresses tracheae, causing pressure changes that promote air flow in and out of the tracheae. Asterisks in the left image mark tracheal tubes that are compressed in the right image.

contract? The opposite—the volume of the tracheae decreases, pressure inside the system increases, and gas moves out of the tracheae into the atmosphere.

The action of the abdominal and flight muscles therefore stimulates air flow through the insect tracheal system, causing gases to move more quickly than they would by diffusion alone.

In large insects, the movement of gases is further promoted by larger tracheae diameters. These larger diameters increase another variable in Fick's law, the cross-sectional area for gas exchange. As evidence, researchers have shown that the diameter of the tracheae in large beetles is proportionally much larger than that of small beetles. Disproportionally large tracheae allow effective ventilation even in the largest living beetles, which can reach impressive lengths of over 15 cm (6 inches) and weigh over 50 g (3 ounces), the size of a submarine sandwich!

Why don't beetles get even larger than this? Researchers hypothesize that if they did, the size of their tracheae would be so large that there would not be enough space in their limbs for much else. The beetles would lack sufficient muscles and other tissues to support their huge bodies.

In the Paleozoic era about 300 million years ago (mya), however, giant insects flourished. While the largest dragonflies today are about 15 cm in wingspan, during the Paleozoic they grew to be as large as 70 cm (2.3 feet)! How can this be explained?

One major hypothesis posits that giant insects evolved during the Paleozoic era because the atmospheric oxygen concentration was much higher than it is today (**FIGURE 45.9**). Recall that according to Fick's law, the rate of diffusion of a gas from one point to another is affected by the difference in partial pressures of the gas at the two points. During the latter part of the Paleozoic, the value of $P_2 - P_1$ was increased in the equation shown in Figure 45.3. This may have increased the rate of movement of gases through the tracheal system, permitting efficient ventilation in extremely large insects.

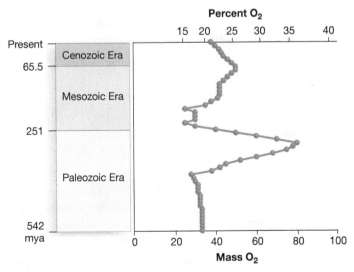

FIGURE 45.9 Atmospheric Oxygen Levels Peaked during the Paleozoic Era.

DATA: Berner, R. A. 1999. *Proceedings of the National Academy of Sciences* 96: 10955–10957.

About 50 to 100 million years later, at the start of the Mesozoic era, atmospheric oxygen concentrations declined precipitously. Giant insects went extinct, likely in part because they could no longer supply their tissues with enough oxygen.

How Do Vertebrate Lungs Work?

In most terrestrial vertebrates, air enters the body through both the nose and mouth. A tube known as the **trachea** (not to be confused with the tracheae of insects) carries the inhaled air to narrower tubes called **bronchi** (singular: **bronchus**). The bronchi branch off into yet narrower tubes, the **bronchioles.** The organs of ventilation, the lungs, enclose the bronchioles and part of the bronchi (**FIGURE 45.10a**, see page 910).

Lungs are internal organs that are used for gas exchange. In addition to being found in terrestrial vertebrates—amphibians, reptiles (including birds), and mammals—they occur in many fishes and certain invertebrates.

Lung Structure and Ventilation Vary among Species   The amount of lung surface area available for gas exchange varies a great deal among species. In frogs and other amphibians, the lung is a simple sac lined with blood vessels. The lungs of mammals, in contrast, are finely divided into tiny sacs called **alveoli** (singular: **alveolus; FIGURE 45.10b**).

Each human lung contains approximately 150 million alveoli, which give mammalian lungs about 40 times more surface area for gas exchange than an equivalent volume of frog lung tissue. As **FIGURE 45.10c** shows, an alveolus provides an interface between air and blood that consists of a thin aqueous film, a layer of epithelial cells, some extracellular matrix (ECM) material, and the wall of a capillary.

In addition to total surface area, the other major feature of lungs that varies among species is mode of ventilation. In the lungs of snails and spiders, air movement takes place by diffusion only. Vertebrates, in contrast, actively ventilate their lungs by pumping air via muscular contractions.

One mechanism for pumping air is **positive pressure ventilation,** used by frogs and related animals. A frog lowers the floor of its throat, increasing the volume there and thus drawing in air from the atmosphere through the nasal passages and into the oral cavity. The animal then closes the nasal passages and contracts its throat muscles. These actions increase the pressure on air in the oral cavity and force it into the lungs.

In effect, frogs push air into their lungs. In contrast, humans and other mammals pull air into their lungs. How does this **negative pressure ventilation** work?

Ventilation of the Human Lung   The pressure inside the human chest cavity is about 5 mm Hg less than atmospheric pressure. This negative pressure surrounding the lung is just enough to keep the lung expanded. If a wound penetrates the chest wall and the pressure differential between the chest cavity and the atmosphere disappears, the lung on the side of the injury will collapse like a deflated balloon.

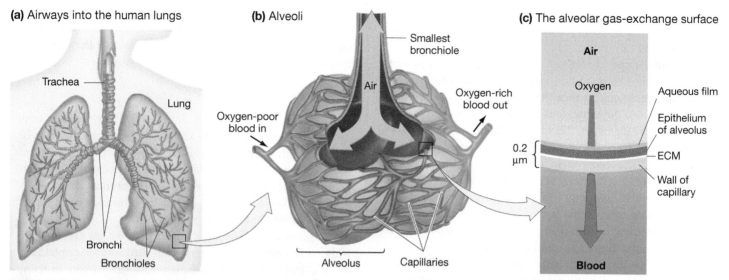

(a) Airways into the human lungs

Trachea

Lung

Bronchi

Bronchioles

(b) Alveoli

Smallest bronchiole

Air

Oxygen-poor blood in

Oxygen-rich blood out

Alveolus

Capillaries

(c) The alveolar gas-exchange surface

Air

Oxygen

Aqueous film

Epithelium of alveolus

0.2 µm

ECM

Wall of capillary

Blood

FIGURE 45.10 Lungs Offer a Thin Membrane with a Large Surface Area for Gas Exchange between Air and Blood. (a) The human respiratory tract branches repeatedly from the largest airway, the trachea, to the smallest, called bronchioles. The system of airways ends in clusters of tiny sacs called alveoli. **(b)** Alveoli are covered with capillary networks and are **(c)** the site of gas exchange.

Humans ventilate their lungs by changing the pressure within their chest cavity between about −5 mm Hg and −8 mm Hg relative to the atmosphere. As **FIGURE 45.11a** shows, inhalation is based on increasing the volume of the chest cavity and thus lowering the pressure. The change is caused by a downward motion of the thin muscular sheet called the **diaphragm** and an outward motion of the ribs. As the pressure surrounding the lungs drops, air flows into the airways along a pressure gradient.

Exhalation, in contrast, is a passive process—the volume of the chest cavity decreases as the diaphragm and rib muscles relax. Because the lung is elastic, it returns automatically to its original, collapsed shape if it is not stretched or compressed. During exercise, though, exhalation is an energy-demanding, active process.

The changes in pressure that occur during negative pressure ventilation are analogous to changing the pressure within a jar, as shown in **FIGURE 45.11b**.

About 450 mL of air moves into and out of the lungs in an average breath. Only about two-thirds of this volume actually participates in gas exchange, however, because 150 mL of the air occupies **dead space**— air passages that are not lined by a respiratory surface. The trachea and bronchi shown in Figure 45.10a, for example, represent dead space.

Breathing is much more efficient during exercise, when the chest cavity undergoes larger changes in volume. When a person is breathing hard, over 2500 mL of air can move with each inhalation–exhalation cycle, but the 150 mL of dead space stays the same.

Ventilation of the Bird Lung Flight is one of the most energy-demanding activities performed by animals. Even so, some birds fly tens of thousands of kilometers during annual migrations.

Even more impressive, geese regularly fly over the top of Mt. Everest. At this elevation, the partial pressure of oxygen is so low that most humans would immediately black out if they were flown there and dropped off. How are birds able to extract enough oxygen from the atmosphere to support long flights and to fly at high elevations?

FIGURE 45.12 provides a diagram of ventilation in birds.

1. During inhalation, air flows through the trachea and enters two large air sacs posterior to the lungs.

2. During exhalation, air leaves the posterior air sacs and enters tiny, branching airways, called parabronchi, in the posterior portion of the lungs.

3. During the next inhalation, air moves into parabronchi in the anterior part of the lungs and on to a system of air sacs anterior to the lungs.

4. During the next exhalation, air moves out of the anterior sacs, through the trachea, and out to the atmosphere. Meanwhile, air from the second inhalation is now flowing through the lungs.

The key conclusion from these observations? Airflow through the avian lung is unidirectional.

Why is the avian respiratory system so efficient?

• Dead space is restricted to the short stretch of trachea between the mouth and the opening of the anterior air sacs. As a result, birds use inhaled air much more efficiently than mammals do.

• Gas exchange occurs during both inhalation and exhalation. In contrast, no gas exchange occurs during the exhalation half of the respiratory cycle in mammals. Bird ventilation resembles the continuous ventilation of fish gills in this respect.

• Blood circulates through the bird lung in capillaries that cross the parabronchi perpendicularly. This crosscurrent pattern is less efficient than the countercurrent circulation of fish gills but far more efficient than the weblike arrangement of capillaries that surrounds mammalian alveoli.

(a) Lungs expand and contract in response to changes in pressure inside the chest cavity.

INHALATION   EXHALATION

Pressure more negative

Pressure less negative

Diaphragm

(b) Ventilatory forces can be modeled by a balloon in a jar.

Pressure more negative

Pressure less negative

When the diaphragm is pulled down, the balloon inflates.

When the diaphragm is released, the balloon deflates.

FIGURE 45.11 Changes in the Volume of the Chest Cavity Drive Negative Pressure Ventilation. (a) Inhalation: When the diaphragm and rib muscles contract, the volume of the lung cavity increases, lowering pressure. In response, air flows into the lungs. Exhalation: When the diaphragm and rib muscles relax, the volume of the lung cavity decreases, causing internal pressure to increase. In response, lung volume decreases—due to elasticity of the lungs—and air flows out. **(b)** A model of negative pressure ventilation.

Homeostatic Control of Ventilation

An animal is in trouble if its mechanisms of homeostasis fail to maintain blood oxygenation or to eliminate carbon dioxide. Adequate adenosine triphosphate (ATP) production depends on maintaining the partial pressures of oxygen and carbon dioxide within a narrow range, both during rest and vigorous exercise. How is ventilation controlled to achieve this critical homeostasis?

Lung, where gas exchange occurs

Air sacs enable one-way airflow through lung

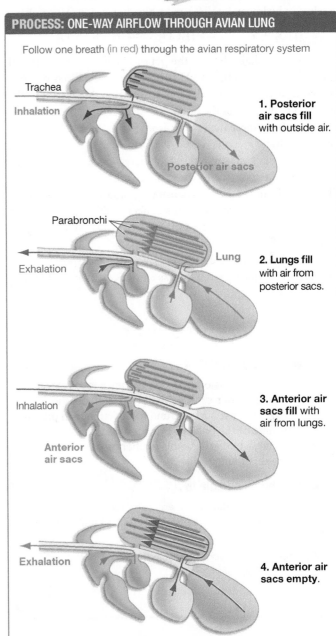

PROCESS: ONE-WAY AIRFLOW THROUGH AVIAN LUNG

Follow one breath (in red) through the avian respiratory system

Trachea

Inhalation

1. Posterior air sacs fill with outside air.

Posterior air sacs

Parabronchi

Exhalation

Lung

2. Lungs fill with air from posterior sacs.

Inhalation

3. Anterior air sacs fill with air from lungs.

Anterior air sacs

Exhalation

4. Anterior air sacs empty.

FIGURE 45.12 Air Flows in One Direction through the Bird Lung, Maximizing Gas Exchange.

When mammals are resting, the rate of breathing is established by the medullary respiratory center, an area at the base of the brain, just above the spinal cord. This center stimulates the rib and diaphragm muscles to contract about 12 to 14 times per minute in humans.

But during exercise, things change. Active muscle tissue takes up more oxygen from the blood. As a result, the partial pressure of oxygen (P_{O_2}) in blood drops. Those same muscles release larger quantities of carbon dioxide to the blood, raising its partial pressure (P_{CO_2}) in blood. When carbon dioxide reaches the brain, it rapidly diffuses from the blood into the cerebrospinal fluid that bathes the brain. In both blood and cerebrospinal fluid, CO_2 reacts with water to form carbonic acid (H_2CO_3), which then dissociates to release a hydrogen ion (H^+) and a bicarbonate ion (HCO_3^-):

$$CO_2 + H_2O \rightleftharpoons H_2CO_3 \rightleftharpoons H^+ + HCO_3^-$$

The result is a slight drop in the pH of blood and cerebrospinal fluid. The change in blood pH is sensed by specialized neurons located near the large arteries that travel from the heart into the neck and to the base of the brain.

Signals from these neurons or from pH detectors in the medullary respiratory center cause the breathing rate to increase. The resulting rise in ventilation rate increases the rate of oxygen delivery to the tissues and the rate at which carbon dioxide is eliminated from the body, restoring P_{O_2} and P_{CO_2} to their resting levels. This control system is so effective that it can maintain stable blood levels of oxygen and carbon dioxide even during intense exercise.

Now let's look more closely at how blood transports oxygen and carbon dioxide between the gas exchange surface and an animal's tissues.

check your understanding

If you understand that . . .

- The rate of diffusion of gases across a respiratory surface depends on the area of the surface, the thickness of the surface, and the difference in partial pressures of the gases across the surface.
- Most large-bodied animals exchange gases via gills, tracheae, or lungs.

✔ **You should be able to . . .**

1. Identify two features that are common to gills, tracheae, and lungs, as well as one trait that is unique to each.
2. When you hold your breath, what happens to the P_{O_2}, P_{CO_2}, and pH of your blood?

Answers are available in Appendix A.

45.4 How Are Oxygen and Carbon Dioxide Transported in Blood?

Blood is a connective tissue that consists of cells in a watery extracellular matrix. Besides carrying oxygen and carbon dioxide between cells and the lungs, blood transports nutrients from the digestive tract to other tissues in the body, moves waste products to the kidney and liver for processing, conveys hormones from glands to target tissues, delivers immune system cells to sites of infection, and distributes heat throughout the body.

Given the wide variety of functions that blood serves, it is not surprising that it is a complex tissue. In an average human, 50–65 percent of the blood volume is composed of an extracellular matrix called **plasma.** The remainder of the volume comprises cells and cell fragments that are collectively called formed elements.

The formed elements in blood include platelets, several types of white blood cells, and red blood cells.

- **Platelets** are cell fragments that act to minimize blood loss from ruptured blood vessels. They do so by releasing material that helps form the blockages known as clots.

- **White blood cells** are part of the immune system. They fight infections (as Chapter 51 will explain in detail).

- **Red blood cells** transport oxygen from the lungs to tissues throughout the body. They also play a role in transporting carbon dioxide from tissues to the lungs. In humans, red blood cells make up 99.9 percent of the formed elements.

The human body synthesizes new red blood cells at the rate of 2.5 million per second to replace old red blood cells, which die at the same rate. These red blood cells last for about 120 days. Red blood cells, white blood cells, and platelets develop from stem cells located in the tissue inside bone (bone marrow).

Vertebrates other than mammals transport oxygen in red blood cells that retain their nuclei. But in mammals, red blood cells lose their nuclei as they mature, along with their mitochondria and most other organelles. Mammalian red blood cells are essentially bags filled with approximately 280 million copies of the oxygen-carrying molecule **hemoglobin.**

Structure and Function of Hemoglobin

Even though oxygen is not highly soluble in water, it is often found in high concentrations in blood. Blood has a high oxygen-carrying capacity because O_2 readily binds to the hemoglobin molecules in red blood cells.

The evolution of hemoglobin was a key event in the diversification of animals. By increasing the oxygen-carrying capacity of blood, hemoglobin made it possible for cellular respiration rates to increase. High rates of ATP production, in turn, support high rates of growth, movement, digestion, and other activities.

Hemoglobin is a tetramer, meaning that it consists of four polypeptide chains (**FIGURE 45.13**). Each of the four polypeptide chains binds to a nonprotein group called **heme** (represented by black circles in the figure). Each heme molecule, in turn, contains an iron ion (Fe^{2+}) that can bind to an oxygen molecule. As a result, each hemoglobin molecule can bind up to four oxygen molecules. In blood, 98.5 percent of the oxygen is bound to hemoglobin; only 1.5 percent is dissolved in plasma.

What Is Cooperative Binding? Blood leaving the human lungs has a P_{O_2} of about 100 mm Hg, while at rest the muscles and other tissues have a P_{O_2} of about 40 mm Hg. This partial pressure difference creates a diffusion gradient that unloads O_2 from hemoglobin to the tissues.

Hemoglobin

Each hemoglobin molecule can bind up to four molecules of oxygen

O_2

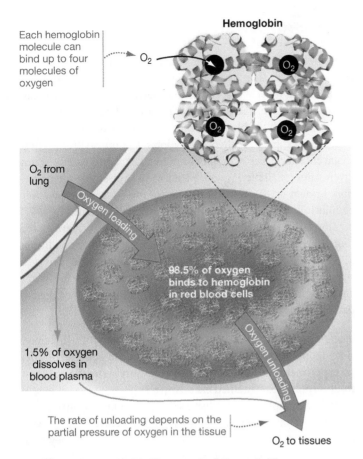

O_2 from lung

Oxygen loading

98.5% of oxygen binds to hemoglobin in red blood cells

1.5% of oxygen dissolves in blood plasma

Oxygen unloading

The rate of unloading depends on the partial pressure of oxygen in the tissue

O_2 to tissues

FIGURE 45.13 Hemoglobin Transports Oxygen to Tissues.

% O_2 changes slowly

In this part of the graph, a small change in P_{O_2} leads to a large change in how much O_2 unloads from hemoglobin

% O_2 changes slowly

P_{O_2} (mm Hg) in blood within tissue

(y-axis: O_2 saturation (%) of hemoglobin)

FIGURE 45.14 The Oxygen–Hemoglobin Equilibrium Curve Is Sigmoidal. A sigmoidal curve has three distinct regions.

Researchers who studied the dynamics of O_2 unloading in tissues found the pattern shown in **FIGURE 45.14**. This graph plots the percentage of O_2 saturation of hemoglobin in red blood cells versus the P_{O_2} in the blood within tissues. If saturation of hemoglobin is 100 percent, it means that every possible binding site in hemoglobin contains an oxygen molecule.

The graph in Figure 45.14 is called an **oxygen–hemoglobin equilibrium curve,** or an oxygen dissociation curve or hemoglobin saturation curve. Note that the *x*-axis plots the partial pressure of oxygen in tissues. In effect, this represents "demand." Oxygen-depleted tissues, where demand for oxygen is high, are on the left-hand part of the horizontal axis; oxygen-rich tissues are toward the right. The *y*-axis, in contrast, plots the percentage of hemoglobin molecules in blood that are saturated with oxygen—a measure of "supply," or how many oxygen molecules on average are bound to hemoglobin. Each 25 percent change in saturation corresponds to an average of one additional oxygen molecule bound per hemoglobin molecule or one oxygen molecule delivered to tissues.

The most remarkable feature of the oxygen–hemoglobin equilibrium curve is that it is sigmoidal, or S-shaped. The pattern occurs because the binding of each successive oxygen molecule to a subunit of the hemoglobin molecule causes a conformational change in the protein that makes the remaining subunits much more likely to bind oxygen.

This phenomenon is called **cooperative binding.** Conversely, the loss of a bound oxygen molecule changes hemoglobin's conformation in a way that makes the loss of additional oxygen molecules more likely.

Why Is Cooperative Binding Important? To understand why cooperative binding is important, use **FIGURE 45.15a** (see page 914) to figure out what happens to hemoglobin saturation when oxygen demand in tissues changes.

Let's begin with some basic observations. When blood arrives at tissues from the lungs, hemoglobin saturation is close to 100 percent. At rest, tissue P_{O_2} is typically about 40 mm Hg. But during exercise, cells are using so much oxygen in cellular respiration that tissue P_{O_2} drops to about 30 mm Hg. Now,

- Put your finger on the *x*-axis at 40 mm Hg, trace the dashed line up until it hits the equilibrium curve, and check where this point is on the *y*-axis. The answer is about 52 percent. This means that when tissues are at rest, hemoglobin unloads about half of its oxygen (about 48 percent) to the tissues.

- Now put your finger on the *x*-axis at 30 mm Hg, trace the dashed line up, and check hemoglobin saturation at this point on the curve. The answer is about 22 percent. This means that when tissues are exercising, hemoglobin unloads about 78 percent of its oxygen to tissues.

Here's the punch line: In response to a relatively small change in tissue P_{O_2}, there is a relatively large change in the percentage of saturation of hemoglobin. The large change occurs because the equilibrium curve is extremely steep in the range of P_{O_2} values commonly observed in tissues. The curve is steep because of cooperative binding. Cooperative binding is important because it makes hemoglobin exquisitely sensitive to changes in the of P_{O_2} tissues.

✔ If you understand this concept, you should be able to explain the consequences of an oxygen–hemoglobin equilibrium curve with a middle section that is even steeper than the one shown in Figure 45.15a.

(a) With cooperative binding, large amounts of O_2 are delivered to resting and exercising tissues.

(b) Without cooperative binding, smaller amounts of O_2 would be delivered to resting and exercising tissues.

FIGURE 45.15 Cooperative Binding of O_2 by Hemoglobin Results in Greater O_2 Delivery than Noncooperative Binding. The hemoglobin is almost 100 percent saturated with oxygen until it arrives at tissues.

If cooperative binding did not occur, all four subunits of hemoglobin would load or unload oxygen independently of each other. They would lose or gain oxygen in direct proportion to the partial pressure of oxygen in the blood, until the molecule was either completely saturated or unsaturated with oxygen. There would be a linear relationship between hemoglobin saturation and tissue P_{O_2}—not a sigmoidal one.

FIGURE 45.15b shows noncooperative binding. As the dashed lines on this graph indicate, a relatively small change in oxygen delivery would occur when tissue P_{O_2} changes from its resting level of about 40 mm Hg to 30 mm Hg. Specifically, hemoglobin would unload about 100% − 68% = 32% of its oxygen when tissues are at rest, and about 100% − 52% = 48% of its oxygen during exercise. This difference—about 16 percent—is much less than the 30 percent change observed with cooperative binding.

How Do Temperature and pH Affect Oxygen Unloading from Hemoglobin? Cooperative binding is only part of the story behind oxygen delivery. Hemoglobin—like other proteins—is sensitive to changes in pH and temperature.

As noted above, the temperature and partial pressure of CO_2 rise in active muscle tissue during exercise. The CO_2 produced by exercising muscle reacts with the water in blood to form carbonic acid, which dissociates and releases a hydrogen ion. As a result, the pH of the blood in exercising muscle drops.

Decreases in pH alter hemoglobin's conformation. These shape changes make hemoglobin more likely to unload O_2 at any given value of tissue P_{O_2}. As **FIGURE 45.16** shows, this phenomenon, known as the **Bohr shift,** causes the oxygen–hemoglobin equilibrium curve to shift to the right when pH declines.

The Bohr shift is important because it makes hemoglobin more likely to release oxygen during exercise or other conditions in which P_{CO_2} is high, pH is low, and tissues are under oxygen stress. Increasing temperature has the same result: It shifts the curve to the right, representing a greater unloading of oxygen to tissues at any given P_{O_2}.

Oxygen Delivery by Hemoglobin Is Extremely Efficient To appreciate cooperative binding and the Bohr shift in action, consider an experiment on how the oxygen transport system in rainbow trout responds to sustained exercise.

To begin, biologists had fish swim continuously against a current in a water tunnel. As the researchers increased the speed of the current and thus the swimming speed of the fish, they periodically sampled the O_2 content of arterial and venous blood. Arterial blood is freshly oxygenated and moving away from the gills; venous blood is returning to the gills from the rest of the body.

Not surprisingly, the biologists found that arterial O_2 levels remained fairly constant as swimming speed increased—meaning the gills continued to work efficiently enough to saturate hemoglobin with oxygen.

FIGURE 45.16 The Bohr Shift Makes Hemoglobin More Likely to Release Oxygen to Tissues with Low pH. As pH drops, oxygen becomes less likely to stay bound to hemoglobin at all values of tissue P_{O_2}. Exercising tissues have lower pH than resting tissues and thus receive more oxygen from hemoglobin.

✓**QUANTITATIVE** Estimate how much more oxygen is unloaded from hemoglobin at pH 7.2 than at pH 7.4 when the tissue has a P_{O_2} of 30 mm Hg.

In contrast, the O_2 content of venous blood, which had undergone gas exchange with the tissues, dropped steadily as swimming speed increased. When the fish had reached their maximum sustainable speed, virtually all the oxygen that had been available in the blood had been extracted.

The data show that in hard-working tissues, the combination of increased temperature, lower pH, and lower P_{O_2} caused hemoglobin to become almost completely deoxygenated. Oxygen-delivery systems based on hemoglobin are extremely efficient.

Comparing Hemoglobins Hemoglobin molecules from different individuals or species may vary in ways that affect fitness—the ability to survive and produce viable offspring. As an example, consider the oxygen–hemoglobin equilibrium curves in **FIGURE 45.17**. The curve in dark red is from a pregnant woman; the curve in light red is from a fetus she is carrying.

The hemoglobin found in fetuses is encoded by different genes than adult hemoglobin and has a distinctive structure and function. Specifically, the oxygen–hemoglobin equilibrium curve for fetal hemoglobin is shifted to the left with respect to the curve for adult hemoglobin.

Recall that the rightward shift of the curve means that hemoglobin is more likely to release oxygen. The leftward shift in Figure 45.17 means that fetal hemoglobin is less likely to give up oxygen—it binds oxygen more tightly. Stated another way, fetal hemoglobin has a higher affinity for oxygen than adult hemoglobin does at every P_{O_2}.

This shift is crucial. In the placenta, hemoglobin from the mother's blood gets close to hemoglobin from the fetus's blood. Because fetal hemoglobin has such a high affinity for oxygen, there is a transfer of oxygen from the mother's blood to the fetus's blood. The difference in hemoglobin structure and function ensures an adequate supply of oxygen as the fetus develops.

CO₂ Transport and the Buffering of Blood pH

The carbon dioxide that is produced by cellular respiration in the tissues enters the blood, where it reacts with water to form carbonic acid, which dissociates into bicarbonate and hydrogen ions. Recall that the resulting drop in blood pH stimulates an increase in breathing rate. Rapid exhalation of CO_2 then counteracts the drop in blood pH.

Homeostasis with respect to blood pH is reinforced by an elegant series of events that take place inside red blood cells. Biologists were able to work out what was happening when they discovered large amounts of the enzyme **carbonic anhydrase** in red blood cells.

The Role of Carbonic Anhydrase and Hemoglobin Recall that carbonic anhydrase catalyzes the formation of carbonic acid from carbon dioxide in water (see Chapter 44). Consequently, CO_2 that diffuses into red blood cells is quickly converted to bicarbonate ions and protons. The same reaction occurs in the plasma surrounding red blood cells, although much more slowly in the absence of the enzyme.

Why is the carbonic anhydrase activity in red blood cells so important? The answer has two parts.

1. The protons produced by the enzyme-catalyzed reaction induce the Bohr shift, which makes hemoglobin more likely to release oxygen.

2. The partial pressure of CO_2 in blood drops when carbon dioxide is converted to soluble bicarbonate ions, maintaining a strong partial pressure gradient favoring the entry of CO_2 into red blood cells.

Thus, carbonic anhydrase activity makes CO_2 uptake from tissues more efficient (**FIGURE 45.18**).

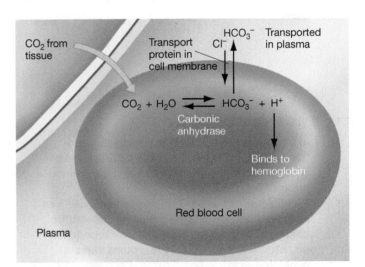

FIGURE 45.18 Carbonic Anhydrase Is Key to CO₂ Transport in Blood. When CO_2 diffuses into red blood cells, carbonic anhydrase quickly converts it to carbonic acid, which dissociates into a bicarbonate ion (HCO_3^-) and a proton (H^+). This reaction maintains the partial pressure gradient favoring the entry of CO_2 into red blood cells. The protons produced by the reaction bind to deoxygenated hemoglobin. Most CO_2 in blood is transported to the lungs in the form of HCO_3^-.

✔ QUESTION This diagram shows the sequence of events in tissues. After reading the rest of Section 45.4, explain what happens when the red blood cell in the diagram reaches the lungs.

FIGURE 45.17 Fetal Hemoglobin Binds Oxygen More Tightly than Maternal Hemoglobin.

Once bicarbonate ions form in the red blood cell, they are transported into the blood plasma. The outcome is that most CO_2 is transported in blood (specifically in plasma) in the form of HCO_3^-—the bicarbonate ion. In contrast, the protons produced by the reaction stay inside red blood cells.

What ultimately happens to these protons? When hemoglobin is carrying few oxygen molecules, it has a high affinity for protons. As a result, it takes up much of the H^+ that is produced by the dissociation of carbonic acid. The hemoglobin acts as a **buffer**—a compound that minimizes changes in pH.

What Happens When Blood Returns to the Lungs? When deoxygenated blood reaches the alveoli, its environment changes dramatically. In the lungs, a partial pressure gradient favors the diffusion of CO_2 from plasma and red blood cells to the atmosphere within the alveoli. As CO_2 diffuses from the blood into the alveoli, P_{CO_2} in blood declines.

The drop in blood P_{CO_2} reverses the chemical reactions that occurred in tissues:

1. Hydrogen ions (protons) leave their binding sites on hemoglobin.
2. Protons react with bicarbonate to form CO_2.
3. CO_2 diffuses into the alveoli and is exhaled from the lungs.

In the meantime, hemoglobin has picked up O_2. Hemoglobin's affinity for oxygen is high in the alveoli because blood pH rises as P_{CO_2} declines.

When blood leaves the lungs, it has unloaded its carbon dioxide, and its hemoglobin is saturated with oxygen. The cycle begins anew.

check your understanding

If you understand that . . .

- In blood, oxygen is bound to hemoglobin and transported inside red blood cells. Carbon dioxide is converted to bicarbonate ions and transported in plasma.
- Hemoglobin has several properties that make it an effective transport protein, including cooperative binding of oxygen, the Bohr shift response to low pH, and the ability to bind the protons that are generated when carbon dioxide is converted to bicarbonate ions.

✔ **You should be able to . . .**

Predict how the oxygen–hemoglobin equilibrium curves measured in Tibetan people, whose ancestors have lived at high elevations for thousands of years, compare to curves from people whose ancestors have lived at sea level for many generations.

Answers are available in Appendix A.

45.5 Circulation

According to Fick's law, differences in the partial pressure of gases are only part of the story when it comes to understanding diffusion rates. Surface area—A in the equation featured in Figure 45.3—also plays a key role.

Animals without circulatory systems have various ways of maximizing the surface area available for diffusion of gases and other key solutes:

- Animals that are only a few millimeters in size, like rotifers and tardigrades, have a small enough volume that diffusion over their body surface is adequate to keep them alive.
- The flattened bodies of flatworms and tapeworms give these animals a high surface-area-to-volume ratio (see Chapter 42). In these species, too, molecules are exchanged with the environment directly across the outer body surface.
- Diffusion across the body wall also occurs in roundworms, where gas exchange is facilitated by muscular contractions in the body wall. Diffusion is enhanced as roundworms circulate fluids by sloshing them back and forth.
- Jellyfish and corals have a large, highly folded gastrovascular cavity that offers a large surface area for exchange of molecules with the environment.

In larger animals, however, the problem of providing a large enough surface area for diffusion is solved by a circulatory system. A circulatory system carries transport tissues called blood or hemolymph into close contact with every cell in the body. In this case, "close contact" is a distance of about 0.1 mm between the smallest blood vessels and cells within tissues. Diffusion is efficient at this scale.

To explore how circulatory systems work, let's start by distinguishing the two most basic types—open and closed. Open circulatory systems occur in most invertebrates, while all vertebrates have closed systems.

What Is an Open Circulatory System?

In an **open circulatory system,** a fluid connective tissue called hemolymph is actively pumped throughout the body in vessels. The hemolymph is not confined exclusively to the vessels, however. Instead, hemolymph comes into direct contact with tissues. As a result, the molecules being exchanged between hemolymph and tissues do not have to diffuse across the wall of a vessel. Hemolymph transports wastes and nutrients and may also contain oxygen-carrying pigments, some cells, and clotting agents.

FIGURE 45.19 illustrates the open circulatory system in a clam. Note that a muscular organ called the **heart** pumps hemolymph into vessels that empty into an open, fluid-filled space. Hemolymph is returned to vessels and then the heart when the heart relaxes and its internal pressure drops below that in the body cavity. General body movements also move hemolymph to and from the heart.

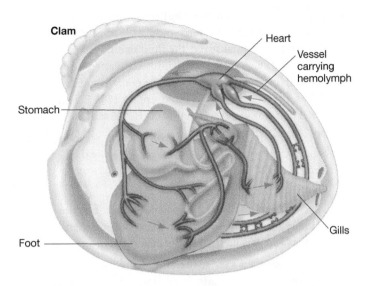

Clam

Heart

Vessel carrying hemolymph

Stomach

Foot

Gills

FIGURE 45.19 Clams Have Open Circulatory Systems. This drawing shows a clam with the top of the shell removed. Red arrows show the direction of hemolymph flow.

Because it moves throughout the volume of the body, hemolymph is under relatively low pressure in open circulatory systems. As a result, hemolymph flow rates may also be low. These features make open circulatory systems most suitable for relatively sedentary organisms, which do not have high oxygen demands.

Insects, with their rapid movements and more active lifestyles, are an exception to this rule. In the open circulatory systems of insects, the limitations imposed by low hemolymph pressure are overcome by their tracheal respiratory system, which delivers oxygen directly to the tissues.

Another characteristic of open circulatory systems, because they lack discrete, continuous vessels, is that the flow of hemolymph cannot be directed toward tissues that have a high oxygen demand and CO_2 buildup. An open circulatory system moves hemolymph throughout an animal's body in much the same way that a ceiling fan moves air throughout a room in a house.

Crustaceans are an important exception to this rule, however. Even though their circulatory system is classified as open, these species have a network of small vessels that can preferentially send hemolymph to tissues with the highest oxygen demands.

What Is a Closed Circulatory System?

In a **closed circulatory system,** blood flows in a continuous circuit through the body, under pressure generated by a heart. Because the blood is confined to vessels, a closed system can generate enough pressure to maintain a high flow rate.

In a closed circulatory system, blood flow can also be directed in a precise way in response to the tissues' needs. For example, blood can be shunted to leg muscles during exercise, to the intestines after a meal, or to regions of the brain engaged in particular mental tasks.

Which Lineages Have Closed Circulatory Systems?   Closed circulatory systems are found in vertebrates and a few other lineages where individuals tend to be active. Earthworms and other annelids, for example, have a closed circulatory system and exchange gases with the environment across their thin, moist skin, which has a dense supply of capillaries. As a result, annelids are able to obtain and circulate enough oxygen to support intense muscular activity. Most live as active burrowers and hunters.

A similar situation occurs in squid, octopuses, and other cephalopods that hunt down prey. The closed circulatory system of these mollusks generates high rates of blood flow, which oxygenates their muscles well enough to support rapid movements and a predatory lifestyle.

Closed circulatory systems contain various types of blood vessels, each having a distinct structure and function. Let's review the major types of blood vessels and then consider how the vessels of a closed circulatory system interact with the lymphatic system.

Types of Blood Vessels   An enormous amount of tubing is required to distribute blood within "diffusion distance" of every cell in the body. If all the blood vessels in a human body were laid end to end, they would stretch about 100,000 km (over 60,000 miles).

Blood vessels are classified as follows:

- **Arteries** are tough, thick-walled vessels that take blood away from the heart. Small arteries are called **arterioles.**

- **Capillaries** are vessels whose walls are just one cell thick, allowing exchange of gases and other molecules between blood and tissues. Networks of capillaries are called **capillary beds.**

- **Veins** are thin-walled vessels that return blood to the heart. Small veins are called **venules.**

The structure of arteries, capillaries, and veins correlates closely with their functions in a closed circulatory system. For example, the heart ejects blood into a large artery, usually called the **aorta.** All arteries have both muscle fibers and elastic fibers in their walls, but elastic fibers dominate the walls of the aorta. As a result, the aorta can expand when blood enters it under high pressure from the heart.

When a contraction of the heart ends, the diameter of the aorta returns to its resting state. This elastic response propels blood away from the heart and augments the force generated by the heart contraction. Similar types of secondary pumping action occur to some extent in other arteries as well. This feature helps maintain forward blood flow in the period between heart contractions.

The walls of arteries and arterioles have a thick layer composed of smooth muscle fibers. When the muscle fibers relax, the vessel diameter increases, resistance to flow is reduced, and blood flow increases in the tissues served by the vessel. But when these muscle fibers contract, the vessel diameter decreases, increasing resistance to flow and slowing the flow of blood in the vessel. In this way, blood flow to tissues can be carefully regulated by signals from the nervous system.

Capillaries are the smallest blood vessels. Their walls are only one cell layer thick, and they are just wide enough to let red blood cells through one at a time (**FIGURE 45.20a**). The extreme thinness of capillaries and the dense network they form throughout the body make them suitable sites for the exchange of gases, nutrients, and wastes between blood and the other tissues.

In some organs, such as the liver, the walls of capillaries contain many small openings that further diminish the barrier to diffusion between blood and the tissues. Despite their thinness, it is rare for capillaries to rupture, because blood pressure drops dramatically as blood passes through arterioles on its way to capillary beds.

After blood from arteries and arterioles passes through capillaries, veins carry it back to the heart. Because blood is under relatively low pressure by the time it exits the tissues, veins have thinner walls and larger interior diameters than arteries do (**FIGURE 45.20b**).

Blood flow in veins is speeded by muscle activity in the extremities, which compresses large veins. Larger veins also contain one-way **valves,** which are thin flaps of tissue that prevent any backflow of blood.

All veins contain some muscle fibers, which contract in response to signals from the nervous system, decreasing the diameter and overall volume of the vessels. Blood pressure in a closed circulatory system is regulated, in part, by actively adjusting the volume of blood contained within the veins.

What Is Interstitial Fluid?   The relatively high operating pressure of closed circulatory systems, combined with the thinness of capillaries, produces a small but steady leakage of fluid from these blood vessels into the surrounding space. The area between cells is called interstitial space; the fluid that fills it is **interstitial fluid.** Blood cells are retained within capillaries, so interstitial fluid resembles plasma in its electrolyte composition.

Why does interstitial fluid build up? In 1896 Ernest Starling proposed that two forces were at work (**FIGURE 45.21**):

1. There is an outward-directed hydrostatic force in capillaries, created by the pressure on blood generated by the heart. This force is analogous to the pressure that drives water through the wall of a leaky garden hose.

2. There is also an inward-directed osmotic force across the capillary walls, created by the higher concentration of solutes in the blood plasma than in the interstitial space.

Starling reasoned that at the end of the capillary nearest to an arteriole, the hydrostatic force (the blood pressure) would exceed the osmotic force. If so, then in that location fluid would move out of the capillary into the interstitial space. But because blood pressure drops as fluid passes through a long, thin tube, Starling proposed that at the venous end of the capillary, the inward-directed osmotic force would exceed the outward-directed hydrostatic force. Thus, the fluid that was lost at the arteriolar end of the capillary would be largely reclaimed at the venous end.

Note the adverb "largely," however—not all interstitial fluid is reabsorbed by capillaries. In Figure 45.21, not all of the fluid entering the interstitial space at the arteriolar end of the capillary

(a) Capillaries are small and extremely thin walled.

Red blood cells

10 µm

Capillary

- Nucleus
- Endothelial cells
- Basement membrane

(b) Veins and arteries differ in structure.

Red blood cells

20 µm

Vein
(Small-sized)

Artery
(Small-sized)

- Fibrous tissue
- Muscle tissue
- Elastic tissue
- Endothelium

FIGURE 45.20 The Structures of Capillaries, Veins, and Arteries Reflect Their Different Functions.
Notice the differences in relative wall thickness and overall size.

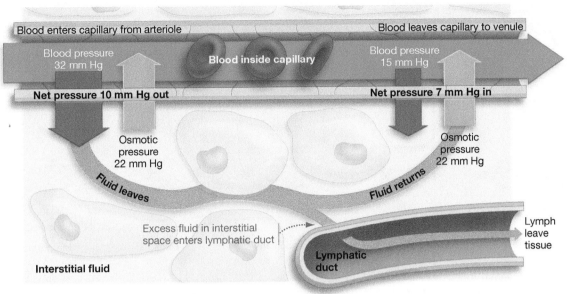

The balance of blood pressure and osmotic forces favors fluid loss from the beginning (inflow end) of capillaries and fluid recovery at the other (outflow) end. Fluid that is not recovered by the capillaries is transported out of the tissue as lymph, which eventually rejoins the blood circulation.

bed has reentered the bloodstream at the venous end—there is a net buildup of interstitial fluid. What happens to this fluid?

The Role of the Lymphatic System   Starling proposed that because interstitial fluid is continually added to the interstitial space, there must be a mechanism for draining the excess fluid. In fact, the fluid is collected in the **lymphatic system:** a collection of thin-walled, branching tubules called lymphatic vessels that permeate all tissues. Interstitial fluid that enters the lymphatic ducts is called **lymph.** Lymphatic vessels join with one another, like the tributaries of a river, to form larger vessels. The largest lymphatic vessels return excess interstitial fluid, in the form of lymph, to the major veins entering the heart.

The importance of the lymphatic system becomes evident when lymphatic vessels are damaged or blocked. For example, a disease called elephantiasis results when the lymphatic vessels in the extremities are blocked by parasitic worms that are transmitted from person to person via mosquito bites. The affected limbs swell dramatically because the lymph cannot be drained, and the skin thickens, cracks, and becomes very painful.

How Does the Heart Work?

In animals with closed circulatory systems, the heart contains at least two chambers: There is at least one thin-walled **atrium** (plural: **atria**), which receives blood, and at least one thick-walled **ventricle,** which generates the force required to propel blood out of the heart and through the circulatory system. Atria are separated from ventricles by atrioventricular valves.

The phylogenetic tree in **FIGURE 45.22** (see page 920) shows the evolutionary relationships among some major vertebrate groups and a simplified sketch of the heart and circulatory system for each lineage. Two points are particularly important to note:

1. The number of atria and ventricles in the heart increased as vertebrates diversified. Fish hearts have one atrium and one ventricle; amphibians, turtles, lizards, and snakes have two atria and one ventricle; crocodilians, birds, and mammals have two atria and two ventricles. It is common to refer to these as two-, three-, and four-chambered hearts, respectively.

2. In fishes, the circulatory system forms a single circuit—one loop services the gills and the body. In other lineages, there are separate circuits to the lungs and to the body.

Why Did Multichambered Hearts and Multiple Circulations Evolve?   To understand these evolutionary patterns, let's first consider the relatively simple heart and circulatory system found in fishes. A single atrium and ventricle and single circuit are adequate in fishes. Even though blood pressure drops as blood passes through the gills—due to the mechanical resistance to flow that occurs in the gills' capillary beds—blood pressure stays high enough to move blood throughout the body. This is largely because fishes live in the neutrally buoyant environment of water, where gravity does not have a large impact on blood flow.

In contrast, gravity has a much larger effect on circulation in land-dwelling vertebrates. Gravity has a particularly antagonistic effect on the flow of blood to elevated portions of the body. To overcome gravity in terrestrial environments, blood must be pumped at high pressure. However, the capillaries and alveoli of the lungs are too thin to withstand high pressures.

The successful solution in terrestrial vertebrates was the evolution of two separate pumping circuits:

1. The **pulmonary circulation** is a lower-pressure circuit to and from the lung.

2. The **systemic circulation** is a higher-pressure circuit to and from the rest of the body.

The pulmonary and systemic circulations are completely separated in the four-chambered hearts of crocodilians, birds, and mammals. On the other hand, the pulmonary and systemic

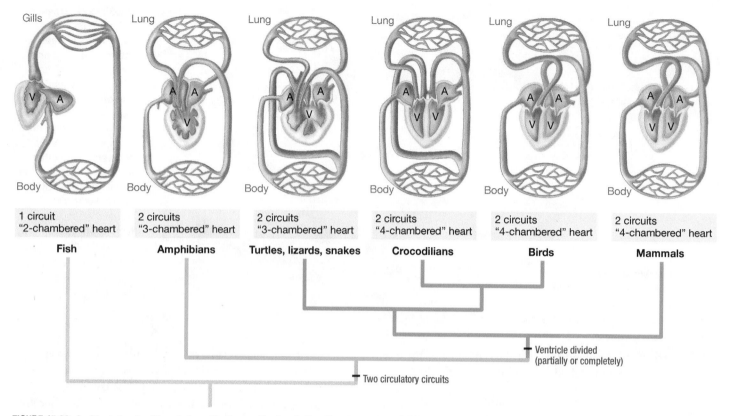

FIGURE 45.22 As Vertebrate Circulatory Systems Evolved, the Number of Atria and Ventricles Increased.
"A" denotes the atria—chambers that receive blood coming into the heart from the body and the gills or lungs.
"V" denotes the ventricles—chambers that pump blood out to the gills or lungs and the body.

circulations are only partially separated in the hearts of amphibians, turtles, lizards, and snakes. In these lineages, blood from the right and left atria may mix in the common ventricle before being expelled from the heart to the lungs or to the body. Turtles, lizards, and snakes, however, have partially divided ventricles that can limit the amount of mixing that occurs there (Figure 45.22).

In addition, turtles, lizards, and snakes have a bypass vessel running from the right side of the ventricle directly into the

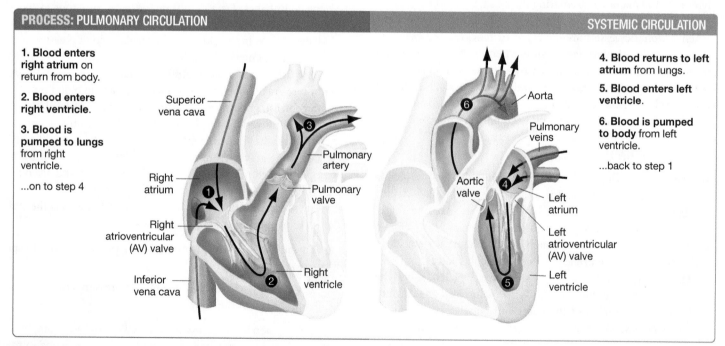

FIGURE 45.23 The Human Heart Maintains Separation of Oxygenated and Deoxygenated Blood. Blood flows through the chambers in the sequence shown.

systemic circulation. This bypass vessel is also observed in the unusual four-chambered hearts of crocodilians. The bypass vessels have an important function: They shunt blood from the pulmonary to the systemic circulation when the animal is underwater and not breathing. The result is a great reduction in blood flow to the lungs at those times.

Unlike reptiles, the hearts of birds and mammals have completely divided ventricles and lack a bypass vessel. This configuration completely separates the pulmonary and systemic circuits.

The Human Heart   Your heart is located in the chest cavity, between your lungs, and is roughly the size of your fist. As **FIGURE 45.23** shows, the human circulatory system returns blood from the body to the right atrium of the heart. This blood is low in oxygen, and it arrives via two large veins called the inferior (lower) and superior (upper) **venae cavae** (singular: **vena cava**).

When the muscles that line the right atrium contract, they send deoxygenated blood to the right ventricle. The right ventricle, in turn, contracts and sends blood out to the lungs, via the **pulmonary artery.** In this way, the right ventricle powers the movement of blood through the pulmonary circulation.

Blood flows from atrium to ventricle to artery in only one direction, because one-way valves separate the heart's chambers from each other and from the adjacent arteries. As Figure 45.23 indicates, the valves are flaps, oriented to ensure a one-way flow of blood with little or no backflow. If heart valves are damaged or defective, the resulting backflow can be heard through a stethoscope. The backflow reduces the organ's efficiency and is called a **heart murmur.**

After blood circulates through the capillary beds in the lung's alveoli and becomes oxygenated, it returns to the heart through the **pulmonary veins.** The oxygenated blood enters the left atrium.

When the left atrium contracts, it pushes blood into the left ventricle. The walls of the left ventricle are so thick with muscle cells that their contraction sends oxygenated blood at high pressure through the aorta and into the arteries and capillaries that make up the systemic circulation.

FIGURE 45.24 summarizes the flow pattern through the human circulatory system and the blood gas concentrations at various points in the pulmonary and systemic circulations. Notice that blood vessels are called arteries or veins according to the direction of blood flow relative to the heart, not because of the oxygen content of the blood in them. Thus, the pulmonary artery is called an artery because it takes blood away from the heart, even though this blood is low in oxygen.

In healthy individuals, gas exchange in the lungs and tissues is rapid relative to the rate at which blood flows through capillaries. This situation is beneficial: It ensures that the maximum amount of oxygen is taken up and the maximum amount of carbon dioxide is released.

Similarly, the slow passage of blood through capillaries means that a maximum amount of oxygen is released from blood to tissues and a maximum amount of carbon dioxide is taken up. As

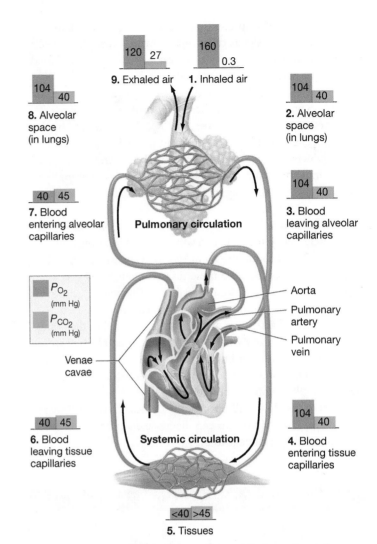

FIGURE 45.24 Partial Pressures of Gases Vary throughout the Human Circulatory System.

✔ **QUESTION** Why are the partial pressures of oxygen and carbon dioxide in exhaled air intermediate in magnitude between the partial pressures in inhaled and alveolar air?

a result, partial pressures for oxygen and carbon dioxide in the tissues and the systemic veins are equal.

Electrical Activation of the Heart   Like other muscle cells, cardiac muscle cells contract in response to electrical signals. In invertebrates, the electrical impulses that trigger contraction come directly from the nervous system. But in vertebrates, a group of cells in the heart itself is responsible for generating the initial signal.

The cells that initiate contraction in the vertebrate heart are known as **pacemaker cells.** They are located in a region of the right atrium called the **sinoatrial (SA) node.**

The SA node and the muscle cells of the heart receive input from the nervous system and from chemical messengers carried in the blood. These inputs are important for regulating both the heart rate and the strength of ventricular contraction. In this way, the amount of blood moving through the circulatory system varies in response to electrical signals and hormones. During the "fight or flight" response (introduced in Chapter 49), for example, a chemical signal called epinephrine causes both heart rate and contraction strength to increase—sending more blood through the body in preparation for rapid movement.

As carefully regulated as the heart rate and contraction strength may be, a vertebrate heart will continue to beat even if all nerves supplying it are severed. Why? An electrical impulse that stimulates contraction is generated in the SA node and rapidly conducted throughout the right and left atria. The signal spreads quickly from cell to cell because of a striking property of cardiac muscle cells: They form physical and electrical connections with each other.

All cardiac muscle cells branch to contact several other cardiac muscle cells, join end to end with these neighboring cells (see Figure 42.5), and connect to them by specialized structures called **intercalated discs.** Because these discs contain many gap junctions (cell-to-cell connections described in Chapter 11), electrical signals pass directly from one cardiac muscle cell to the next.

The electrical activation of the heart is reflected in the orange line on the bottom of **FIGURE 45.25**. This line is an **electrocardiogram,** or **EKG**—a recording of the electrical events that occur over the course of a cardiac cycle. An EKG recording is generated by amplifying the overall electrical signal conducted from the heart to the chest wall through the tissues of the body. By inspecting an EKG, physicians can diagnose disturbances of heart rhythm and detect damage to the heart muscle.

The drawings above the graph in Figure 45.25 show where the key electrical events are happening.

1. The SA node generates an electrical signal.

2. The signal from the SA node is quickly propagated to atrial muscle cells. As a result, the atria contract simultaneously and eject blood into the ventricles.

3. As the atria begin to contract, the signal is conducted to an area of the heart called the **atrioventricular (AV) node.** The AV node delays the signal slightly before passing it to the ventricles. The delay allows the ventricles to fill completely with blood from the atria before they contract.

4. After the delay, the electrical impulse is rapidly transmitted through specialized fibers in the muscular wall that separates the ventricles. The impulse spreads through both ventricles, causing them to contract as the atria relax. The ventricles empty efficiently because the signal and the resulting muscular contraction move from the bottom up to the top of each ventricle—toward the arteries that allow blood to exit.

5. The final electrical event occurs as the ventricles relax and their cells recover—restoring their electrical state before contraction.

✔ If you understand these concepts, you should be able to predict how the amount of blood ejected from the ventricles would change if there were no delay at the AV node.

PROCESS: ELECTRICAL ACTIVATION OF THE HEART

SA node

AV node

1. **Signal originates at SA node.**

2. **Signal spreads over atria;** atria contract.

3. **Signal delays at AV node.**

4. **Signal spreads** down conducting fibers to bottom of ventricles; ventricles contract.

5. **Ventricles relax.**

FIGURE 45.25 Sequential Electrical Activation Leads to Coordinated Contraction of the Heart. The rate and strength of contractions control the pressure in the chambers and arteries.

The Cardiac Cycle The electrical signals originating from the SA node ensure that the atria contract simultaneously, and the delay at the AV node ensures that the atria are relaxed by the time the ventricles contract. The contraction phases of the atria and the ventricles, called **systole,** are therefore closely coordinated with their relaxation phases, or **diastole.**

This sequence of contraction and relaxation is called the **cardiac cycle.** It consists of one diastole and one systole for both atria and ventricles.

Ventricular contraction (ventricular systole) leads to a rapid increase in pressure within both ventricles, as recorded in the dark purple line in **FIGURE 45.26.** Blood is ejected into the pulmonary artery and the aorta when ventricular pressure exceeds the pressure within each respective artery. Blood pressure measured in the systemic arterial circulation at the peak of ventricular ejection into the aorta is called the **systolic blood pressure.** Blood pressure measured just before ventricular ejection is called the **diastolic blood pressure.**

Clinicians report blood pressure measurements in fractional notation, where systolic pressure is the numerator and diastolic pressure is the denominator. People with blood pressures consistently higher than 140/90 mm Hg have high blood pressure, or **hypertension.**

Hypertension is a serious concern to physicians because it can lead to a variety of defects in the heart and circulatory system. Abnormally high blood pressure puts mechanical stress on arteries. If the walls of an artery fail, the individual may experience heart attack, stroke, kidney failure, and burst or damaged vessels.

Patterns in Blood Pressure and Blood Flow

As blood moves through capillaries, blood pressure drops dramatically—as the top graph in **FIGURE 45.27** indicates. The bottom graph on this figure shows why. Trace the line labeled "Total area" on this graph to see that as arteries branch, rebranch, and eventually form networks of capillaries, the total cross-sectional area of blood vessels in the circulatory system increases—even though the size of individual blood vessels decreases as blood moves away from the heart. Note two key points:

1. Blood pressure in the capillaries drops dramatically (on the top graph) because the amount of mechanical resistance to flow is a function of total cross-sectional area of the vessels—meaning that friction losses are high in capillary beds. The pulsing nature of the pressure observed in the larger arteries diminishes as a result of this pressure drop.

2. As the line labeled "Velocity" in the bottom graph indicates, the velocity of blood flow also decreases significantly in capillary beds relative to arteries and veins, because the same

FIGURE 45.26 Blood Pressure Changes during the Cardiac Cycle. These data show the pressures created by the left ventricle and left atrium in the course of a cardiac cycle. In this example, the blood pressure measured in the upper arm would be 120/80 mm Hg. Right ventricle and pulmonary artery pressures would produce a similar pattern except that the blood pressure in the pulmonary artery would be much lower—closer to 25/8 mm Hg.

FIGURE 45.27 Blood Pressure Drops Dramatically in the Circulatory System. The top graph shows how blood pressure changes as blood leaves the heart and travels through arteries, capillaries, and veins, as in the branching pattern of the middle diagram. In arteries near the heart, each heartbeat causes fluctuations in blood pressure. These pressure pulses disappear in the capillaries, so blood flows there at a steady speed. The bottom graph plots the total area of blood vessels shown in the diagram, as well as the velocity of blood flow through the vessels.

amount of fluid is passing through a much larger area. Recall that the slow rate of blood flow through capillaries is important: It provides sufficient time for gases, nutrients, and wastes to diffuse between tissues and blood.

Why Is Regulation of Blood Pressure and Blood Flow Important?

The general patterns of blood pressure and blood flow diagrammed in Figure 45.27 don't tell the entire story, however. Blood movement is carefully regulated at an array of points throughout the circulatory system.

Recall that the walls of arterioles are comprised partially of smooth muscle. Contraction or relaxation of this muscle can constrict or allow blood flow to specific regions of the body. This means that the nervous system, along with certain chemical messengers in the circulation, can accurately control blood flow to various tissues by contracting or relaxing the muscle. For example, arterioles in the skin may dilate during exercise, diverting blood flow to the skin to eliminate excess heat. This accounts for the flushed facial appearance induced by vigorous exercise. Regulating blood flow is important for maintaining homeostasis with respect to body temperature.

As another example of how blood pressure and blood flow are regulated, consider what happens if you sit long enough for blood to pool in your lower extremities, under the influence of gravity. If you stand up rapidly, your blood pressure can drop enough to reduce blood flow to the brain and cause dizziness or even a blackout. More serious drops in blood pressure, due to severe dehydration or blood loss, can be fatal. Fortunately, decreases in blood pressure elicit a powerful homeostatic response.

Homeostatic Control of Blood Pressure   Recall that all homeostatic responses involve (1) sensors that detect the change in condition, (2) an integrator that processes information about the change, and (3) effectors that diminish the impact of the change (see Chapter 42).

Specialized pressure-sensing receptors called **baroreceptors** detect changes in blood pressure. Baroreceptors are found in the walls of the heart and the major arteries, both as they leave the heart and as they enter the neck. This distribution is logical, because the head tends to be the most elevated part of terrestrial animals, and preserving blood flow to the brain is the highest priority.

When baroreceptors transmit nerve signals to the brain (the integrator) indicating a serious fall in blood pressure, a rapid, three-component effector response ensues:

1. Cardiac output—the volume of blood leaving the left ventricle—increases. This is due to an increase in heart rate and an increase in stroke volume, which is the amount of blood ejected from the ventricles during each cardiac cycle. (Cardiac output = heart rate × stroke volume.)

2. Arterioles serving the capillaries of certain tissues constrict to divert blood to more critical organs. (This occurs in tissues like the skin and intestines, which can endure short-term restrictions in their blood supply without damage.)

3. Veins constrict, decreasing their overall volume. Because more than half of the blood in the circulatory system is contained within the veins, constriction of these vessels shifts blood volume toward the heart and arteries to maintain blood pressure and flow to vital organs.

This coordinated response is mediated both by a portion of the nervous system called the sympathetic nervous system (Chapter 46) and by hormones (Chapter 49) produced by the adrenal glands. Sympathetic nerves and the hormones involved in blood pressure regulation deliver their messages directly to (1) the SA node to increase heart rate, (2) the ventricles of the heart to increase stroke volume, and (3) the muscular walls of the arteries and veins to modify their total volume.

Cardiovascular Disease   A healthy circulatory system is obviously critical to your well-being. Indeed, **cardiovascular disease,** which is a group of ailments collectively affecting the heart and blood vessels, is the number one cause of death in humans worldwide.

Many factors contribute to cardiovascular disease, including age, tobacco use, poor diet, obesity, inactivity, and genetics. As people age, their blood vessels harden and lose elasticity—a condition called **arteriosclerosis.** Nicotine in tobacco causes constriction of blood vessels, further reducing the diameter of blood vessels and increasing blood pressure.

High-fat diets and lack of physical activity can compound the problem by leading to the deposition of fatty plaques on the walls of blood vessels, which effectively reduces their diameter. The loss of elasticity and the decline in the diameter of the vessels combine to cause increased blood pressure, which can weaken the walls of arteries.

check your understanding

If you understand that . . .

- Animal circulatory systems may be open or closed, but both types of systems circulate blood or hemolymph via pressure generated by one or more hearts.

- In closed systems, regulated changes in the diameter of blood vessels can direct blood to specific regions, and overall blood pressure is carefully regulated through changes in cardiac output.

✔ **You should be able to . . .**

1. Explain how the mammalian lymphatic system and circulatory system interact.

2. Make a labeled diagram showing how blood circulates through the mammalian heart.

Answers are available in Appendix A.

If the arteries that deliver blood to the heart muscle become completely blocked, a **myocardial infarction,** or heart attack, can occur. In a myocardial infarction, a portion of heart tissue dies within minutes when it is deprived of oxygen. Depending on the location and extent of damage, myocardial infarction can lead to rapid death.

Over 17 million people worldwide died from cardiovascular disease in 2004, and this number is projected to reach 23 million by the year 2030. With the incidence of obesity and diabetes reaching epidemic proportions, these numbers may become even higher. Effectively combating cardiovascular disease will require improved diet, reduced tobacco use, greater amounts of physical activity, and better access to healthcare across the globe.

CHAPTER 45 REVIEW

For media, go to MasteringBiology

If you understand . . .

45.1 The Respiratory and Circulatory Systems

- Animal gas exchange involves ventilation, exchange of gases between the environment and the blood, and exchange of gases between blood and tissues.

- Animal circulation involves transportation of gases, nutrients, wastes, and other substances throughout the body.

✓ You should be able to identify the process that uses oxygen inside animal cells.

45.2 Air and Water as Respiratory Media

- As media for exchanging oxygen and carbon dioxide, air and water are dramatically different.

- Compared with water, air contains much more oxygen and is much less dense and viscous. As a result, terrestrial animals have to process a much smaller volume of air to extract the same amount of O_2, and the amount of work required to do so is less than in aquatic animals.

- Both terrestrial and aquatic animals pay a price for exchanging gases: Land-dwellers lose water to evaporation; freshwater animals lose ions and gain excess water; marine animals gain sodium and chloride and lose water.

✓ You should be able to explain why no water breathers have the extremely high rates of metabolism required for endothermy.

45.3 Organs of Gas Exchange

- The structure of gills, tracheae, lungs, and other gas exchange organs minimizes the cost of ventilation while maximizing the rate at which O_2 and CO_2 diffuse.

- Consistent with predictions made by Fick's law of diffusion, respiratory epithelia tend to be extremely thin and to be folded to increase surface area.

- In fish gills, countercurrent exchange ensures that the differences in O_2 and CO_2 partial pressures between water and blood are high over the entire length of the ventilatory surface.

- Insect tracheae carry air directly to and from tissues.

- In bird lungs, structural adaptations lead to a high ratio of useful ventilatory space to dead space.

- Breathing rate is regulated to keep the carbon dioxide content of the blood stable during both rest and exercise.

✓ You should be able to explain why no large animals have gas exchange occurring only across the skin.

45.4 How Are Oxygen and Carbon Dioxide Transported in Blood?

- The tendency of hemoglobin to give up oxygen varies as a function of the P_{O_2} in surrounding tissue in a sigmoidal fashion. As a result, a relatively small change in tissue P_{O_2} causes a large change in the amount of oxygen released from hemoglobin.

- Oxygen binds less tightly to hemoglobin when pH is low. Because CO_2 tends to react with water to form carbonic acid, the existence of high CO_2 partial pressures in exercising muscle tissues lowers their pH and makes oxygen less likely to stay bound to hemoglobin and more likely to be unloaded into tissues.

- The CO_2 that diffuses into red blood cells from tissues is rapidly converted to carbonic acid by the enzyme carbonic anhydrase. The protons that are released as carbonic acid dissociates bind to deoxygenated hemoglobin. In this way, hemoglobin acts as a buffer that takes protons out of solution and prevents large fluctuations in blood pH.

✓ You should be able to explain why exposure to large doses of carbon monoxide (CO) can lead to suffocation. (The CO found in furnace and engine exhaust and in cigarette smoke binds to the heme groups in hemoglobin 210 times more tightly than does oxygen.)

45.5 Circulation

- In many animals, blood or hemolymph moves through the body via a circulatory system consisting of a pump (heart) and vessels.

- In open circulatory systems, overall pressure is low and tissues are bathed directly in hemolymph.

- In closed circulatory systems, blood is contained in vessels that form a continuous circuit. Containment of blood allows higher pressures and flow rates, as well as the ability to direct blood flow accurately to tissues that need it the most.

- In organisms with a closed circulatory system, a lymphatic system returns excess fluid that leaks from the capillaries back to the circulation.

- In amphibians and some reptiles, blood from the pulmonary and systemic circuits may be mixed in the single ventricle.

- In mammals and birds, a four-chambered heart pumps blood into two circuits, which separately serve the lungs and the rest of the body. Crocodilians have a similar heart with a bypass vessel that can shunt blood from the pulmonary to the systemic circuit.

- The cardiac cycle is controlled by electrical signals that originate in the heart itself.

- Heart rate, cardiac output, and contraction of both arterioles and veins are regulated by chemical signals and by electrical signals from the brain.

- Cardiovascular disease is the leading cause of death in humans.

✔ You should be able to predict the characteristics of the circulatory systems of terrestrial animals with high activity rates.

MB **MasteringBiology**

1. MasteringBiology Assignments

Tutorials and Activities Gas Exchange; Gas Exchange in the Lungs and Tissues; Gas Transport in Blood; Human Heart; Human Respiratory System; Mammalian Cardiovascular System Function; Mammalian Cardiovascular System Structure; Path of Blood Flow in Mammals; Transport of Respiratory Gases

Questions Reading Quizzes, Blue-Thread Questions, Test Bank

2. eText Read your book online, search, take notes, highlight text, and more.

3. The Study Area Practice Test, Cumulative Test, BioFlix® 3-D Animations, Videos, Activities, Audio Glossary, Word Study Tools, Art

You should be able to . . .

✔ TEST YOUR KNOWLEDGE

Answers are available in Appendix A

1. O_2 will diffuse from blood to tissue faster in response to which of the following conditions?
   a. an increase in the P_{O_2} of the tissue
   b. a decrease in the P_{O_2} of the tissue
   c. an increase in the thickness of the capillary wall
   d. a decrease in the surface area of the capillary

2. Which of the following does blood *not* do?
   a. transport O_2 and CO_2
   b. distribute body heat
   c. produce new red blood cells and other formed elements
   d. buffer against pH changes

3. In insects, what is the adaptive significance of spiracles?
   a. They dilate and constrict during flight or other types of movement, functioning as a "breathing" mechanism.
   b. They open into the body cavity, allowing direct contact between hemolymph and tissues.
   c. They are thin and highly branched, offering a large surface area for gas exchange.
   d. They close off tracheae to minimize water loss.

4. Which of the following is *not* an advantage of breathing air over breathing water?
   a. Air is less dense than water, so it takes less energy to move during ventilation.
   b. Oxygen diffuses faster through air than it does through water.
   c. The oxygen content of air is greater than that of an equal volume of water.
   d. Air breathing leads to high evaporation rates from the respiratory surface.

5. Which of the following promotes oxygen release from hemoglobin?
   a. a decrease in temperature
   b. a decrease in CO_2 levels
   c. a decrease in pH
   d. a decrease in carbonic anhydrase

6. Describe the ways in which an open circulatory system is less efficient than a closed circulatory system.

✔ TEST YOUR UNDERSTANDING

Answers are available in Appendix A

7. Describe the changes in oxygen delivery that occur as a person proceeds from a resting state to intense exercise.

8. Explain how most carbon dioxide is transported in the blood. In humans, why don't intense exercise and rapid production of CO_2 lead to a rapid reduction in blood pH?

9. Why is ventilation in birds considered much more efficient than the respiratory system of humans and other mammals?

10. Explain how each parameter in Fick's law of diffusion is reflected in the structure of the mammalian lung.

11. Carp are fishes that thrive in stagnant-water habitats with low oxygen partial pressures. Compared with the hemoglobin of many other fish species, carp hemoglobin has an extremely high affinity for O_2. Is this trait adaptive? Explain your answer.

12. Frog lungs have lower surface area for gas exchange than mammalian lungs. Which of the following observations explains this?

a. Frog tissue needs more oxygen than mammalian tissue.
b. Frogs breathe more quickly than mammals.
c. Frogs also obtain oxygen via diffusion across the skin.
d. Frog lung tissue has a greater density of capillary beds than mammalian lung tissue.

✔ TEST YOUR PROBLEM-SOLVING SKILLS

Answers are available in Appendix A

13. Predict how the icefish native to the Antarctic are able to transport oxygen and carbon dioxide in their blood even though they lack hemoglobin.

14. Compare and contrast the respiratory and circulatory systems of an insect and a human.

15. Why did separate systemic and pulmonary circulations evolve in species that have the high-pressure circulatory system required for rapid movement of blood?

16. Under certain conditions, cells produce a metabolic byproduct called 2,3-diphosphoglycerate (DPG). Which of the following would support the hypothesis that DPG helps increase oxygen unloading from tissues?

a. DPG shifts the oxygen–hemoglobin equilibrium curve to the left.
b. DPG decreases percentage saturation of hemoglobin with oxygen at a P_{O_2} of 30 mm Hg.
c. DPG increases the pH of the tissues.
d. DPG is produced in greater quantity at low altitude than at high altitude.

46 Animal Nervous Systems

In this chapter you will learn how

Animals use electrical signaling along neurons for internal communication

beginning with ↓

Principles of electrical signaling   **46.1**

↓ then asking

How does the action potential work?   **46.2**

↓ finally looking at

Synapses—the interface between neurons   **46.3**

→ then exploring

The vertebrate nervous system   **46.4**

↓ including

The brain, learning, and memory

Diffusion spectrum imaging reveals the activity and trajectory of neurons in the brain. Such advances in brain imaging are allowing neurobiologists to study how neurons in the brain communicate with one another.

BIG PICTURE

This chapter is part of the Big Picture. See how on pages 840–841.

Most students and professional biologists are attracted to the study of neurobiology because they want to understand the human brain as well as higher-order processes like consciousness, intelligence, emotion, learning, and memory. But early in the history of neurobiology, researchers realized that the human brain—with its billions of cells—was much too complex to study productively.

Instead biologists did the same thing they did in the early days of studying cells, genetics, and evolution: They started simply. Early research in neurobiology focused on the function of individual nerve cells, or **neurons,** the cells mainly responsible for the working of the brain and the rest of the nervous system. Neurons conduct information in the form of electrical signals from point to point

✔ When you see this checkmark, stop and test yourself. Answers are available in Appendix A.

in the body at speeds of up to 200 m/sec (450 mph). Thus electrical signaling is a crucial aspect of information processing—one of the five attributes of life (introduced in Chapter 1).

Initial research on the electrical properties of single neurons laid a broad foundation for more recent studies of how the human brain works. This chapter proceeds in the same way. Let's begin by focusing on the neurons themselves and how they use electrical signaling to communicate within the body and with each other. By the end of the chapter, you'll be considering how the brain is organized and how phenomena like memory work.

46.1 Principles of Electrical Signaling

The evolution of neurons was a key event in the diversification of animals, along with the evolution of muscles (see Chapter 33). All animals except sponges have neurons and muscle cells. Neurons transmit electrical signals; muscles can respond to signals from neurons by contracting.

Neurons and muscle cells made rapid movement possible. These innovations in cell structure and function enabled animals to evolve into today's flatworms, squid, beetles, sharks, and birds.

Recall that neurons are organized into two basic types of nervous systems:

1. The diffuse arrangement of cells called a **nerve net,** found in cnidarians (jellyfish, hydra, anemones) and ctenophores (comb jellies).

2. A **central nervous system (CNS)** that includes large numbers of neurons aggregated into clusters called ganglia.

In most cases, animals with a CNS have a large cerebral ganglion, or brain, located in their anterior end. You also might recall that this phenomenon—the evolution of a bilaterally symmetric body with structures for information gathering and processing located at the head end—is known as cephalization (see Chapter 33).

Cephalization made animals into efficient eating and moving machines: They face the environment in one direction, and sensory appendages take in information and send it to a nearby brain for processing. After integrating information from an array of sensory cells, the brain sends electrical signals to muscles and other tissues that respond to the sensory stimuli.

Types of Neurons in the Nervous System

The sensory cells that are responsible for gathering information respond to light, sound, touch, or other stimuli. Sensory cells in the animal's skin, eyes, ears, mouth, and nose transmit information about the environment. Sensory cells inside the body monitor conditions that are important in homeostasis, such as blood pH and temperature. In this way, sensory cells monitor conditions both outside and inside the body. Many sensory cells are **sensory neurons** (FIGURE 46.1), which carry information to the brain and spinal cord.

FIGURE 46.1 **The Brain Integrates Sensory Information and Sends Signals to Effector Cells.** In most cases, sensory neurons send information to the brain, where it is integrated with information from other sources. Once integration is complete, a response is sent to effector cells through motor neurons.

Together, the brain and spinal cord form the CNS of vertebrates. One function of the CNS is to integrate information from many sensory neurons. Cells in the CNS called **interneurons** (literally, "between-neurons"), which pass signals from one neuron to another, perform this integration.

Interneurons also make connections between sensory neurons and **motor neurons,** which are nerve cells that send signals to effector cells in glands or muscles. Motor neurons and sensory neurons are bundled together into long, tough strands of nervous tissue called **nerves.** Recall that effectors are structures that bring about a physiological change in an organism (see Chapter 42).

All neurons and other components of the nervous system that are outside the CNS are considered part of the **peripheral nervous system,** or **PNS.** Section 46.4 describes the structure and function of the PNS in more detail.

Typically, sensory information from receptors in the PNS is sent to the CNS, where it is processed. Then a response is transmitted back to appropriate parts of the body via motor neurons. When you stub your toe, pain receptors in your toe relay sensory information to the brain, which then modifies your movements to avoid further injury. You might hop a bit, for example, and then limp to avoid further pressure on that toe.

The Anatomy of a Neuron

Neurons are difficult to study because they are small, transparent, and morphologically complex. So when Camillo Golgi discovered that some neurons become visible when samples of preserved tissue are treated with a solution containing silver nitrate, his finding was a major advance. The year was 1898.

(a) Information flows from dendrites to the axon.

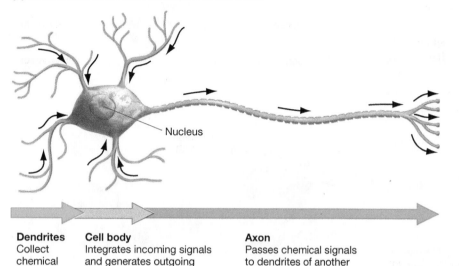

Dendrites	Cell body	Axon
Collect chemical signals	Integrates incoming signals and generates outgoing electrical signal to axon	Passes chemical signals to dendrites of another cell or to an effector cell

(b) Neurons form networks for information flow.

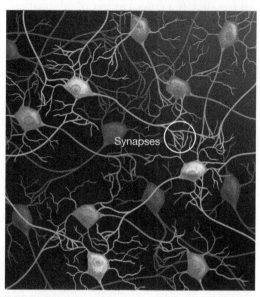

FIGURE 46.2 How Does Information Flow in a Neuron? (a) The structure of a generalized neuron. **(b)** Most neurons receive inputs from many different neurons and send projections to several different neurons.

Through the early decades of the twentieth century, the work of Golgi and Santiago Ramón y Cajal revealed several important points about the anatomy of neurons. Most neurons have the same three parts, shown in **FIGURE 46.2a:**

1. A cell body, which contains the nucleus

2. A highly branched group of relatively short projections called dendrites

3. One or more relatively long projections called axons

Dendrites are rarely more than 2 mm long, but axons can be over a meter in length. The number of dendrites and their arrangement vary greatly from cell to cell. Further, many neurons in the brain have only dendrites and lack axons.

A **dendrite** receives a signal from the axons of adjacent cells; a neuron's **axon** sends a signal to the dendrites and cell bodies of other neurons (**FIGURE 46.2b**). In short, dendrites receive signals; axons pass them on. The **cell body,** or **soma,** also receives signals from axons of other cells. Incoming signals are integrated in the dendrites and cell body, and an outgoing signal is sent to the axon.

Ramón y Cajal maintained that the plasma membrane of each neuron is separated from those of adjacent neurons. This hypothesis was confirmed in the 1950s, when images from electron microscopes showed that most neurons are separated from one another at their junctions by tiny spaces called synapses.

Given that neurons transmit electrical signals, how do they produce them?

An Introduction to Membrane Potentials

Ions carry an electric charge. In virtually all cells, the cytoplasm and extracellular fluids adjacent to the plasma membrane contain unequal distributions of ions. As a result, cells are inherently electrical in nature.

A difference of electrical charge between any two points creates an **electrical potential,** or a **voltage.** If the positive and negative charges of the ions on each side of a plasma membrane do not balance each other, the membrane will have an electrical potential.

When an electrical potential exists across a plasma membrane, the separation of charges is called a **membrane potential.** If there is a large separation of charges across the membrane, the membrane potential is large.

It is important to remember that membrane potentials refer only to a separation of charge immediately adjacent to the plasma membrane, on either side of the membrane. Even if there is a large membrane potential, there may be no charge separation slightly farther away from the membrane.

Units and Signs Membrane potentials are measured in units called millivolts. The **volt** is the standard unit of electrical potential, and a **millivolt (mV)** is 1/1000 of a volt. As a comparison, an AA battery that you buy in the store has an electrical potential of 1500 mV between its + and − terminals. In neurons, membrane potentials are typically about 65 to 80 mV.

By convention, membrane potentials are always expressed in terms of inside relative to outside (the outside value is defined as 0). Because there are usually more negatively charged ions and fewer positively charged ions on the inside surface of a membrane relative to its outside surface, membrane potentials are usually negative in sign.

Electrical Potential, Electric Currents, and Electrical Gradients Membrane potentials are a form of potential energy. Recall that potential energy is energy based on the position of matter.

To convince yourself that ions have potential energy when a membrane potential exists, consider what would happen if the membrane were removed. Ions would spontaneously move from the region of like charge to the region of unlike charge—causing a flow of charge. This flow of charge, called an **electric current,** would occur because like charges repel and unlike charges attract.

However, charge is not the only form of potential energy. Ions also have different concentrations across membranes. Therefore, a membrane potential also includes energy stored as the concentration gradients of charged ions on the two sides of the membrane. Recall that the combination of an electrical gradient and a concentration gradient is an **electrochemical gradient** (see Chapter 6).

What do all these facts have to do with neuron function? Neurons use the electrochemical gradient of ions across their membranes to power the signals that allow neurons to communicate with one another and with other tissues.

How Is the Resting Potential Maintained?

When a neuron is not communicating with other cells or tissues, the difference in charge across its membrane is called the **resting potential.** To understand why the resting potential exists, consider the distribution of the various ions and other charged molecules on the two sides of the neuron's plasma membrane, shown in **FIGURE 46.3**:

- The interior side of the membrane has relatively low concentrations of sodium (Na^+) and chloride (Cl^-) ions, a relatively high concentration of potassium ions (K^+), and some organic anions—amino acids and other organic molecules that have dropped a proton and thus carry a negative charge.

- In the extracellular fluid, sodium and chloride ions predominate.

If each type of ion diffused across the membrane in accordance with its concentration gradient, organic anions and K^+ would leave the cell, while Na^+ and Cl^- would enter.

Ions cannot cross phospholipid bilayers readily, however. They cross plasma membranes efficiently in only three ways (see Chapter 6):

1. Flowing along their electrochemical gradient through an **ion channel**—a protein that forms a pore in the membrane and allows a specific ion to diffuse along its gradient

2. Carried, via a membrane cotransporter protein, with another ion that diffuses along its electrochemical gradient

3. Pumped against an electrochemical gradient by a membrane protein that hydrolyzes adenosine triphosphate (ATP)

How does the distribution of ions illustrated in Figure 46.3 come to be? The answer hinges on two types of membrane proteins: an ion channel that allows potassium ions to diffuse along their gradient, and a pump that actively transports sodium and potassium ions against their gradients. Let's consider each in turn.

The K^+ Leak Channel At rest, the plasma membrane of neurons is relatively impermeable to most cations. However, neurons have a relatively high number of potassium channels, called K^+ **leak channels,** that allow K^+ to leak out of the cell. K^+ ions slowly

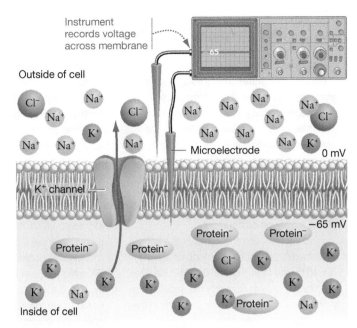

FIGURE 46.3 Neurons Have a Resting Potential. In resting neurons, the membrane is selectively permeable to K^+. As K^+ leaves the cell along its concentration gradient, the inside of the membrane becomes negatively charged relative to the outside. To measure a neuron's membrane potential, researchers insert a microelectrode into the cell and compare that reading with the reading outside the cell.

✓**QUESTION** Will K^+ continue to leave the cell indefinitely? Explain why or why not.

exit the neuron through the leak channels along their electrochemical gradient.

As K^+ moves from the interior of the cell to the exterior through leak channels, the inside of the cell becomes more and more negatively charged relative to the outside. This buildup of negative charge inside the cell begins to attract K^+ and counteract the concentration gradient that had favored the movement of K^+ out.

As a result of the counteracting influences, the membrane reaches a voltage at which equilibrium exists between the concentration gradient that favors movement of K^+ out and the electrical gradient that favors movement of K^+ in. At this voltage, there is no longer a net movement of K^+. This voltage is called the **equilibrium potential** for K^+. **Quantitative Methods 46.1** (see page 932) shows how equilibrium potentials for individual ions are calculated.

Although Cl^- and Na^+ cross the plasma membrane much less readily than does K^+, some movement of these ions also occurs through a small number of leak channels that are selective for each of them. As a result, each type of ion has an equilibrium potential. The membrane as a whole has a membrane potential that combines the effects of the individual ions.

The Role of the Na^+/K^+-ATPase What is responsible for the higher concentration of Na^+ on the outside and K^+ on the inside in the first place? It is the sodium–potassium pump, Na^+/K^+-ATPase, which actively pumps Na^+ out of the cell and K^+ into

An equilibrium potential exists whenever there is a concentration gradient for an ion across a membrane and the membrane is permeable to that ion. At the equilibrium potential, the rate at which an ion moves across the membrane down its concentration gradient is equal to the rate at which the ion moves, in the opposite direction, down its electrical gradient. It's the voltage at which the concentration and electrical gradients acting on an ion balance out.

The ion concentration gradients that contribute to equilibrium potentials are produced by the action of Na^+/K^+-ATPase. This membrane protein converts chemical energy in the form of ATP to electrical energy in the form of a membrane potential.

To calculate an equilibrium potential, biologists use the Nernst equation—a formula that converts the energy stored in a concentration gradient to the energy stored as an electrical potential. The concentration gradient for an ion is symbolized

as $[ion]_o/[ion]_i$, where $[ion]_o$ and $[ion]_i$ are the concentrations of the ion outside and inside the cell, respectively. The equilibrium potential for that ion is symbolized as E_{ion}. The Nernst equation specifies the equilibrium potential for a given ion as

$$E_{ion} = 2.3 \frac{RT}{zF} \log \frac{[ion]_o}{[ion]_i}$$

In this expression, z is the valence of the ion (for instance, +1 for potassium) and the expression RT/F is known as the thermodynamic potential. The three terms in the thermodynamic potential are the gas constant (R), which acts as a constant of proportionality; the absolute temperature (T), measured in Kelvins; and the Faraday constant (F), which specifies the amount of charge carried by 1 mol of an ion with a valence of +1 or −1.

Note that the Nernst equation is based on the base-10 logarithm of the ion concentration ratio, $[ion]_o/[ion]_i$. Thus, the thermodynamic potential specifies the voltage required to balance a tenfold concentration ratio

across the membrane. (For more on using logarithms, see **BioSkills 6** in Appendix B.)

The Nernst equation applies to only a single ion. For example, suppose that the potassium concentration inside the axon of a squid neuron has been measured as 400 mM, yet the outside concentration for this ion is only 20 mM (**Table 46.1**). Potassium has a charge of +1, so at 20°C the RT/zF part of the equation yields +25 mV. In this case, the equilibrium potential for potassium becomes

$$E_K = 2.3 \times 25 \text{ mV} \times \log \frac{20 \text{ mM}}{400 \text{ mM}}$$

$$E_K = 58 \text{ mV} \times \log 0.05 = -75 \text{ mV}$$

The minus sign indicates that the interior of the axon is negatively charged with respect to the exterior.

Repeating this process for Na^+ ions and Cl^- ions yields the following results for the squid axon:

$$E_{Na} = 58 \text{ mV} \times \log \frac{440 \text{ mM}}{50 \text{ mM}} = +54.8 \text{ mV}$$

$$E_{Cl} = -58 \text{ mV} \times \log \frac{560 \text{ mM}}{51 \text{ mM}} = -60 \text{ mV}$$

Again, notice that the equilibrium potential given by the Nernst equation is calculated independently for each ion.

✔**QUANTITATIVE** If you understand this concept, you should be able to calculate the equilibrium potential at 20°C for calcium (Ca^{2+}) in a cell where the intracellular concentration is 0.0001 mM and the extracellular concentration is 1 mM.

TABLE 46.1 **Concentration of Important Ions across a Squid Neuron's Plasma Membrane at Rest**

Ions	Cytoplasm Concentration	Extracellular Concentration	Equilibrium Potential
Na^+	50 mM	440 mM	+54.8 mV
K^+	400 mM	20 mM	−75 mV
Cl^-	51 mM	560 mM	−60 mV
Organic anions	385 mM	—	—

the cell. More specifically, the energy gained by the sodium–potassium pump when it receives a phosphate group from one ATP is used to move three Na^+ ions out of the cell and two K^+ ions into the cell (**FIGURE 46.4**).

Active transport via Na^+/K^+-ATPase ensures that eventually the concentration of K^+ is much higher on the inside of the plasma membrane than outside, while the concentration of Na^+ is lower inside than outside. In addition to setting up concentration gradients of K^+ and Na^+, the pump establishes an electrical gradient: The outward movement of three positive charges and inward movement of two positive charges each time the ATPase pumps makes the interior of the membrane less positive (more negative) than the outside.

To summarize, the neuron has a negative resting membrane potential because the Na^+/K^+-ATPase pumps three cations out for only two cations in, and because K^+ exits the neuron through

leak channels along its electrochemical gradient. The resting potential represents energy stored as concentration and electrical gradients in various ions. ✔ If you understand this concept, you should be able to predict what would happen to the membrane potential if Na^+ or K^+ were allowed to diffuse freely across the membrane.

Using Microelectrodes to Measure Membrane Potentials

During the 1930s and 1940s, A. L. Hodgkin and Andrew Huxley focused on what has become a classic model system in the study of electrical signaling: the axons of squid.

Squid live in the ocean and are preyed on by fishes and whales. When a squid is threatened, electrical signals travel down the axons to muscle cells. When these muscles contract,

Outside cell

Inside cell

ATP → ADP

1. Sodium loaded. Three sodium ions (Na⁺) enter the protein from within the cell.

2. Sodium released. ATP phosphorylates the pump, which changes shape and releases 3 Na⁺ to the outside of the cell.

3. Potassium loaded. Two potassium ions (K⁺) enter the protein from outside the cell.

4. Potassium released. The phosphate group drops off the pump. The protein changes shape and releases 2 K⁺ to the interior of the cell.

FIGURE 46.4 In Neuron Membranes, Na⁺/K⁺-ATPase Imports Potassium Ions and Exports Sodium Ions. By following radioactive Na⁺ and K⁺, biologists found that the pump transports 3 Na⁺ out of the cell for every 2 K⁺ brought in. Notice that this pump operates via a conformational change.

water is expelled from a cavity in the squid's body. As a result, the squid lurches away from danger by jet propulsion (see Chapter 34). The extremely rapid electrical signal produced in the squid is an adaptation that helps avoid predation.

Hodgkin and Huxley decided to study the squid's axon simply because it is so large. Many of the axons found in humans are a mere 2 µm in diameter, but the squid axon is about 500 µm in diameter—large enough that the researchers could record membrane potentials by inserting a wire down its length.

By measuring the voltage difference between the wire inside the cell and an electrode outside the cell, Hodgkin and Huxley could record the voltage across the plasma membrane and observe how it changed in response to stimuli. Later researchers developed glass microelectrodes that were tiny enough to be inserted into smaller neurons to record membrane voltage.

With their relatively simple early equipment, Hodgkin and Huxley were able to record the axon's resting potential. They documented that it can be disrupted by an event called the action potential when the axon is stimulated.

What Is an Action Potential?

An **action potential** is a rapid, temporary change in a membrane potential. It may qualify as the most important type of electrical signal in cells. When stimulated, neurons mount action potentials that allow them to communicate with other neurons, muscles, or glands.

Although Hodgkin and Huxley initially studied action potentials in the squid giant axon, subsequent work has shown that the action potential has the same general characteristics in all species and in all types of neurons.

A Three-Phase Signal **FIGURE 46.5** shows the form of the action potential that Hodgkin and Huxley recorded from the squid's

giant axon—the signal that allows the squid to jet away from predators. The action potential has three distinct phases:

1. **Depolarization** of the membrane. In its resting state, a membrane is said to be polarized because the charges on the two sides are different. Depolarization means that the membrane becomes less polarized than before. During the depolarization phase, the membrane potential becomes less polarized as it moves from highly negative toward zero and then is briefly positive.

2. A rapid **repolarization,** which changes the membrane potential back to negative.

3. A **hyperpolarization** phase, when the membrane is slightly more negative than the resting potential.

FIGURE 46.5 Action Potentials Have the Same General Shape. An action potential is a stereotyped change in membrane potential—meaning that it occurs the same way every time.

Together, all three phases of an action potential occur within a few milliseconds.

For an action potential to begin in a squid giant axon, the membrane potential must shift from its resting potential of −65 mV to about −55 mV. If the membrane depolarizes less than that, an action potential does not occur. But if this **threshold potential** is reached, certain channels in the axon membrane open and ions rush into the axon, following their electrochemical gradients. The inside of the membrane becomes less negative and then positive with respect to the outside.

When the membrane potential reaches about +40 mV, an abrupt change occurs and the repolarization phase begins. The change is triggered by the closing of certain ion channels and the opening of other ion channels in the membrane.

To summarize, an action potential occurs because specific ion channels in the plasma membrane open or close in response to changes in voltage. An action potential always has the same three-phase form, even though the size of the resting potential, threshold potential, and peak depolarization may vary among species or even among types of neurons in one species.

An "All-Or-None" Signal That Propagates Hodgkin and Huxley made other important observations about the action potential. Besides being fast and having three distinct phases, it is an all-or-none event.

- There is no such thing as a partial action potential.
- All action potentials for a given neuron are identical in magnitude and duration.
- Action potentials are always propagated down the entire length of the axon.

For example, when an action potential was recorded at a particular point on a squid axon, an action potential that was identical in shape and size would be observed farther down the same axon soon afterward. Neurons are said to have **excitable membranes**

because they are capable of generating action potentials that propagate rapidly along the length of the axons.

Taken together, these observations suggested a mechanism for electrical signaling. In the nervous system, information is coded in the form of action potentials that propagate along axons. The frequency of action potentials—not their size—is the meaningful signal. In the squid's giant axon, increased frequency of action potentials signals muscles to contract. As a result, the animal escapes from danger.

46.2 Dissecting the Action Potential

Of Na^+, Cl^-, and K^+, which ion or ions are most important in the currents that form the action potential? Are different ions responsible for the depolarization and repolarization phases of the event?

Hodgkin made a crucial start in answering this question when he realized that +40 mV was close to the equilibrium potential for Na^+ in the squid giant axon (see **Quantitative Methods 46.1**). If sodium channels opened early in the action potential, then Na^+ should flow into the neuron until the membrane potential was about +40 mV. How could this hypothesis be tested?

Distinct Ion Currents Are Responsible for Depolarization and Repolarization

To understand the currents responsible for the action potential, Hodgkin and Huxley recorded electrical activity in squid axons that were bathed in solutions containing different concentrations of ions.

- Removing Na^+ from the solution surrounding the axon abolished the production of action potentials.
- When axons were bathed in solutions with various concentrations of Na^+, the peak membrane potential paralleled the concentration of Na^+. If Na^+ concentration outside the cell was high, the peak was high. If Na^+ concentration outside the cell was low, the peak was low.

These experiments furnished strong support for the hypothesis that the action potential begins when Na^+ flows into the neuron. In other words, sodium ions are responsible for the depolarization phase.

What happens during the repolarization phase? Using radioactive K^+, Hodgkin and Huxley showed that there was a strong flow of potassium ions out of the cell during the repolarization phase.

The action potential consists of a strong inward flow of sodium ions followed by a strong outward flow of potassium ions. ✔ If you understand this concept, you should be able to add labels that read, "Sodium channels open—Na^+ enters" and "Potassium channels open—K^+ leaves" to Figure 46.5.

How Do Voltage-Gated Channels Work?

The action potential depends on **voltage-gated channels**—membrane proteins that open and close in response to changes

If you understand that . . .

- The plasma membranes of neurons maintain a negative resting potential because Na^+/K^+-ATPase pumps 3 Na^+ ions out of the cell and 2 K^+ ions into the cell and because the membrane is selectively permeable to K^+ ions, which leak out.
- Action potentials are three-phase, all-or-none signals that propagate down the length of a neuron.

✔ **You should be able to . . .**

1. Predict what would happen to the resting potential of a squid axon if potassium leak channels were blocked.

2. Explain why only the frequency of action potentials—not their size—contains information.

Answers are available in Appendix A.

At the resting potential, voltage-gated Na⁺ channels are closed

Conformational changes open voltage-gated channels when the membrane is depolarized

FIGURE 46.6 The Shape of a Voltage-Gated Channel Depends on the Membrane Potential. Changes in the conformation of voltage-gated channels are responsible for changes in a neuron's permeability to ions.

in membrane voltage. The shape of a voltage-gated channel, and thus its ability to admit ions, changes in response to the charges present at the inside of the membrane. **FIGURE 46.6** shows a simple model of how voltage-gated sodium channels change as a function of membrane potential.

Hodgkin and Huxley confirmed that voltage-gated channels exist using a technique called voltage clamping. **Voltage clamping** allows researchers to hold the voltage of a region of an axon at any desired level and record the electrical currents that occur at that voltage. When the researchers held the squid axon at various voltages, different currents resulted. These experiments supported the hypothesis that the behavior of the ion channels depends on voltage.

Patch Clamping and Studies of Single Channels   Studying individual ion channels became possible in the 1980s when Erwin Neher and Bert Sakmann perfected a technique known as **patch clamping.** As **FIGURE 46.7** shows, the researchers touched a membrane with a fine-tipped microelectrode and applied suction to capture a single ion channel within the electrode's tip.

Using this technique, the researchers were able to document for the first time the currents that flowed through individual

Tip of microelectrode

Axon membrane

Single Na⁺ or K⁺ channel

FIGURE 46.7 Patch Clamping Provided Insights into the Action Potential. Patch clamping makes use of extremely fine-tipped microelectrodes. The goal is to isolate one channel and record from it.

channels, and they showed that different ion channels behave differently:

- Voltage-gated channels are either open or closed.   There is no gradation in channel behavior. This conclusion is based on the shape of the recorded current: Current flow starts and stops instantly, and the size of the current is always the same.

- Sodium channels open quickly after depolarization.   They stay open for about a millisecond, close, and remain inactive for 1 to 2 msec. That explains why the cell can repolarize: Once the sodium channels close, there is a lag before they can open again.

- Potassium channels open with a delay after depolarization. They continue to flip open and closed until the membrane repolarizes. Once the membrane returns to the resting potential, these channels remain closed.

Positive Feedback Occurs during Depolarization   More detailed experiments on Na⁺ channels also explained why the action potential is an all-or-none event. The key observation was that Na⁺ channels become more likely to open as a membrane depolarizes. As a result, an initial depolarization leads to the opening of more Na⁺ channels, which depolarizes the membrane further, which leads to the opening of additional Na⁺ channels.

The opening of Na⁺ channels exemplifies **positive feedback**—meaning that the occurrence of an event makes the same event more likely to recur. When a fuse is lit, for example, the heat generated by the oxidation reaction accelerates the reaction itself, which generates still more heat and leads to additional oxidation reactions and keeps the fuse burning. Positive feedback is rare in organisms: It cannot be employed as a feedback mechanism under many circumstances because it often leads to uncontrolled events. The opening of Na⁺ channels during an action potential is one of the few examples known.

✔ If you understand how voltage-gated sodium and potassium channels work, you should be able to (1) explain how positive feedback occurs in the opening of Na⁺ channels, and (2) predict what would happen if Na⁺ and K⁺ channels were open at the same time.

Using Neurotoxins to Identify Channels and Dissect Currents   In addition to using voltage clamping and patch clamping, researchers have used neurotoxins to explore the dynamics of voltage-gated channels. **Neurotoxins** are poisons that affect neuron function—often resulting in convulsions, paralysis, or unconsciousness. They come from sources as diverse as venomous snakes and foxglove plants.

For example, when biologists treated giant axons from lobsters with tetrodotoxin from puffer fish, they found that the resting potential in treated neurons was normal, but action potentials were abolished. More specifically, the flow of K⁺ out of the cell was normal but the influx of Na⁺ was wiped out. Researchers concluded that tetrodotoxin blocks the voltage-gated Na⁺ channel, probably by binding to a specific site on the channel protein.

How Is the Action Potential Propagated?

To explain how action potentials propagate down an axon, Hodgkin and Huxley suggested the model illustrated in **FIGURE 46.8a**.

Step 1 The influx of Na^+ at the start of an action potential repels intracellular cations, causing them to spread away from the sodium channels.

Step 2 As positive charges are pushed farther from the initial sodium channels, they depolarize adjacent portions of the membrane.

Step 3 Nearby voltage-gated Na^+ channels open when the membrane reaches threshold, resulting in an action potential.

In this way, action potentials are continuously regenerated at adjacent areas of the plasma membrane (**FIGURE 46.8b**). The response is all or none because action potentials are always regenerated along the entire length of the axon.

Why don't action potentials propagate back up the axon in the direction of the cell body? To answer this question, recall that once Na^+ sodium channels have opened and closed, they are less likely to open again for a short period. This is known as the **refractory** state. Action potentials are propagated in one direction only, because "upstream" sodium channels, in the direction of the cell body, are in the refractory state.

The hyperpolarization phase, in which the membrane is more negative than the resting potential, also keeps the charge that spreads upstream from triggering an action potential in that direction, because a much stronger stimulus would be necessary to raise the membrane potential to threshold potential.

Axon Diameter Affects Speed Understanding how the action potential propagates helped researchers explain why the squid's axons are so large. When sodium ions enter the axon interior at the start of an action potential, they repulse intracellular cations, causing them to flow along the inside of the membrane. Cations moving down axons with larger diameters meet less resistance to flow than those moving down narrow axons. As a result, the charge spreads down the membrane more quickly.

The upshot is that the squid's giant axon and other large-diameter neurons transmit action potentials much faster than small axons can. The squid axon's large size is an adaptation that makes particularly rapid signaling possible.

(a) PROCESS: PROPAGATION OF ACTION POTENTIAL

Neuron — Axon

Na^+ Na^+ Na^+ Na^+ Na^+ Na^+ Na^+ Na^+

1. Na^+ enters axon.

Na^+ Na^+ Na^+ Na^+ Na^+ Na^+

2. Charge spreads; membrane "downstream" depolarizes.

Depolarization at next ion channel

Na^+ Na^+ Na^+ Na^+ Na^+ Na^+ Na^+ Na^+ Na^+

3. Downstream voltage-gated channel opens in response to depolarization.

FIGURE 46.8 Action Potentials Propagate because Charge Spreads Down the Membrane. (a) An action potential starts with an inflow of Na^+. The influx of positive charge attracts negative charges inside the cell and repels positive charges. As a result, cations spread away from the channel where the Na^+ enters and depolarize nearby regions of the neuron. Voltage-gated Na^+ channels open in response. **(b)** The action potential is propagated down the axon as a wave of depolarization, but there is no loss of signal because action potentials at one area of the membrane always stimulate action potentials at adjacent areas of the membrane.

(b) Action potential spreads as a wave of depolarization.

Myelination Affects Speed Relatively few vertebrates have giant axons. Instead, vertebrates—and some invertebrates—have specialized accessory cells whose membranes wrap around the axons of certain neurons and increase the efficiency of action potential propagation.

In the central nervous system, the specialized accessory cells are called **oligodendrocytes.** In the peripheral nervous system, described in Section 46.4, the cells are **Schwann cells** (**FIGURE 46.9a**). Oligodendrocytes and Schwann cells are two of several types of nervous system cells that support neurons. Collectively, these accessory cells are called **glia.**

When oligodendrocytes or Schwann cells wrap around an axon, they form a **myelin sheath,** which acts as a type of electrical insulation. As action potentials spread down an axon, the myelin sheath prevents charge in the form of ions from leaking back out across the plasma membrane of the neuron.

Consequently, the cations moving down the membrane are able to spread until they hit an unmyelinated section of the axon, called a **node of Ranvier** (**FIGURE 46.9b**). The node has a dense concentration of voltage-gated Na^+ and K^+ channels, so new action potentials can occur at the node.

Electrical signals "jump" from node to node down a myelinated axon much faster than they can move down an unmyelinated axon of the same diameter. In an unmyelinated axon, sodium and potassium channels are found in all locations, and action potentials occur continuously down its length. Myelination is interpreted as an adaptation that makes rapid transmission of electrical signals possible in axons that have a small diameter.

To appreciate the importance of myelination, consider what happens when it is disrupted. If myelin degenerates, the transmission of electrical signals slows considerably. The autoimmune disease **multiple sclerosis (MS)** develops when the immune system mistakenly targets the oligodendrocytes that make up the myelin sheath in the CNS. As damage to myelin increases and electrical signaling is impaired, coordination among neurons is

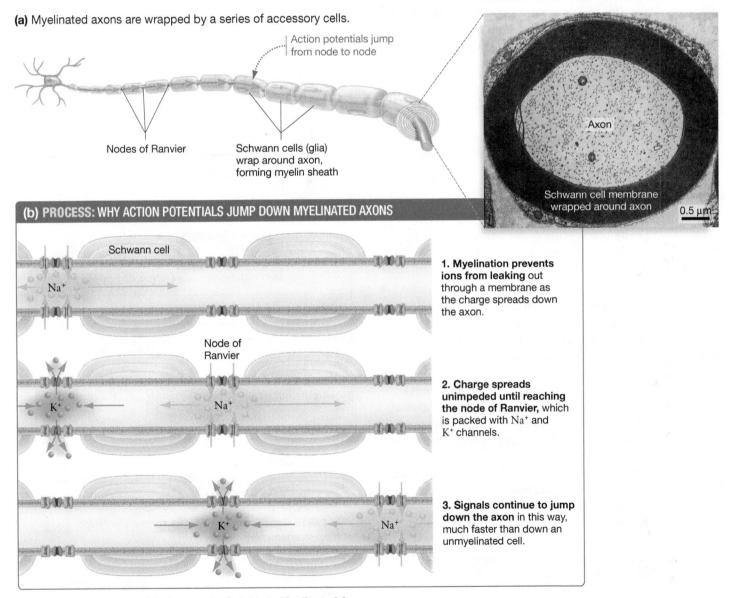

(a) Myelinated axons are wrapped by a series of accessory cells.

Action potentials jump from node to node

Nodes of Ranvier

Schwann cells (glia) wrap around axon, forming myelin sheath

Axon

Schwann cell membrane wrapped around axon

0.5 μm

(b) PROCESS: WHY ACTION POTENTIALS JUMP DOWN MYELINATED AXONS

Schwann cell

Na^+

1. Myelination prevents ions from leaking out through a membrane as the charge spreads down the axon.

Node of Ranvier

K^+ Na^+

2. Charge spreads unimpeded until reaching the node of Ranvier, which is packed with Na^+ and K^+ channels.

K^+ Na^+

3. Signals continue to jump down the axon in this way, much faster than down an unmyelinated cell.

FIGURE 46.9 Action Potentials Propagate Quickly in Myelinated Axons.

affected and muscles weaken. The symptoms of MS are highly variable; in severe cases, the disease progresses and can be crippling.

What happens once an action potential has traveled the length of the axon? In most neurons, the membrane at the end of the axon is adjacent to the membrane of another neuron's dendrite but separated from its surface by a tiny gap. Let's investigate what happens when an action potential arrives at this interface between cells.

check your understanding

If you understand that . . .

- During an action potential, membrane voltage undergoes rapid changes due to an inflow of sodium ions followed by an outflow of potassium ions.
- Action potentials propagate down an axon because inrushing sodium ions repel intracellular cations that move down the membrane and depolarize adjacent portions of the membrane.

✔ **You should be able to . . .**

1. Explain why the action potential is an all-or-none phenomenon.
2. Predict what would happen if batrachotoxin, the poison produced by some species of poison dart frogs, were applied to the membrane. (Batrachotoxin binds to voltage-gated sodium channels and maintains them in an open state.)

Answers are available in Appendix A.

46.3 The Synapse

The cytoplasm of most neurons is not directly connected to the cytoplasm of other neurons, yet neurons are able to communicate with one another. Based on this observation, there must be some indirect mechanism that transmits signals from cell to cell, across their plasma membranes.

In the 1920s, Otto Loewi showed that this indirect mechanism involves **neurotransmitters.** Neurotransmitters are chemical messengers that transmit information from one neuron to another neuron, or from a neuron to a target cell in a muscle or gland.

Loewi knew that signals from the vagus nerve slow the heart. To test the hypothesis that the signal from nerve to heart muscle is delivered by a chemical, he performed the experiment diagrammed in **FIGURE 46.10.**

First, Loewi isolated the vagus nerve and heart of a frog. As predicted, the heart rate slowed when he stimulated the vagus nerve electrically. Next, he took the solution that bathed the first heart, applied it to another heart—this time one that was not

RESEARCH

QUESTION: How is information transferred from one neuron to another?

HYPOTHESIS: Molecules called neurotransmitters carry information from one neuron to the next.

NULL HYPOTHESIS: Information is not transferred between neurons in the form of molecules.

EXPERIMENTAL SETUP:

Frog heart 1

1. **Stimulate vagus nerve** electrically.

Frog heart 1

Stimulus

2. **Heartbeat slows** after stimulus.

Frog heart 1

3. **Collect solution** surrounding nerve end.

Frog heart 2

4. **Add solution** to second heart.

PREDICTION: The heartbeat will slow.

PREDICTION OF NULL HYPOTHESIS: There will be no change in heartbeat.

RESULTS:

Frog heart 2

Stimulus

Heartbeat slows after solution is added.

CONCLUSION: The vagus nerve releases molecules that slow heartbeat. Neurotransmitters carry information.

FIGURE 46.10 Experimental Evidence for the Existence of Neurotransmitters.

SOURCE: Loewi, O. 1921. Über humorale Übertragbarkeit der Herznervenwirkung. *Pflügers Archiv European Journal of Physiology* 189: 239–242.

✔ **QUESTION** What would be an appropriate control for this experiment?

accompanied by a vagus nerve—and showed that the second heart rate slowed as well.

This result provided strong evidence for the chemical transmission of electrical signals. The vagus nerve had released a neurotransmitter into the bath.

Synapse Structure and Neurotransmitter Release

When transmission electron microscopy became available in the 1950s, biologists finally understood the physical nature of the interface, or **synapse,** between neurons. As **FIGURE 46.11** shows, (1) the membranes of axons and dendrites are separated by a tiny space, the **synaptic cleft,** and (2) the ends of axons contain numerous sac-like structures, called **synaptic vesicles.** Synaptic vesicles were hypothesized to be storage sites for neurotransmitters.

Anatomical observations such as these, combined with chemical studies of the synapse, led to the model of synaptic transmission illustrated in **FIGURE 46.12.** Notice that the "sending" cell is the **presynaptic neuron** and the "receiving" cell is the **postsynaptic neuron.**

Step 1 An action potential arrives at the end of the axon.

Step 2 The depolarization created by the action potential opens voltage-gated calcium channels located near the synapse, in the presynaptic membrane. The electrochemical gradient for Ca^{2+} results in the inflow of calcium ions through the open channels.

Step 3 In response to the increased calcium concentration inside the axon, synaptic vesicles fuse with the membrane and release neurotransmitters into the gap between the cells, the synaptic cleft. The delivery of neurotransmitters into the cleft is an example of exocytosis (see Chapter 7).

FIGURE 46.11 Synaptic Vesicles Cluster Near Synapses. A cross section of the site where an axon meets a dendrite.

Step 4 Neurotransmitters bind to receptors on the postsynaptic cell, leading to changes in the membrane potential of the postsynaptic cell. The combined effect on membrane potential of many neurotransmitters binding may trigger an action potential in the postsynaptic cell.

PROCESS: ACTION POTENTIAL TRIGGERS RELEASE OF NEUROTRANSMITTER

1. **Action potential arrives** near synaptic cleft.

2. **Voltage-gated Ca^{2+} channels open.** Ca^{2+} enters presynaptic cell.

3. **Synaptic vesicles fuse** with presynaptic membrane, then release neurotransmitter.

4. **Ion channels in the postsynaptic membrane open** when neurotransmitter binds; flow of ions causes change in postsynaptic cell potential.

5. **Ion channels in the postsynaptic membrane then close** as neurotransmitter is broken down or taken back up by presynaptic cell (not shown).

FIGURE 46.12 Neurons Meet and Transfer Information at Synapses. The sequence of events that occurs when an action potential arrives at a synapse.

Step 5 The response ends as the neurotransmitter is broken down and taken back up by the presynaptic cell.

Is the model correct? Let's begin by analyzing the role of neurotransmitters.

What Do Neurotransmitters Do?

Researchers can look for neurotransmitters by stimulating a neuron, collecting the molecules that are released, and analyzing them chemically. To find the receptor for a particular neurotransmitter, researchers can attach a radioactive atom or other type of label to the neurotransmitter and add it to neurons. Once the labeled transmitter has bound to its receptor, the receptor protein can be isolated and analyzed.

Using techniques such as these, biologists have discovered and characterized a wide array of neurotransmitters and receptors. Some of them are listed in **TABLE 46.2**.

By patch-clamping receptors, biologists confirmed that many neurotransmitters function as ligands. A **ligand** is a molecule that binds to a specific site on a receptor molecule. Many neurotransmitters are ligands that bind to receptors called **ligand-gated channels.** These are channel proteins that open in response to binding by a specific ligand—just as voltage-gated channels open in response to a change in voltage.

When a neurotransmitter binds to a ligand-gated ion channel in the postsynaptic membrane, the channel opens and admits a flow of ions along an electrochemical gradient. In this way, the neurotransmitter's chemical signal is transduced to an electrical signal—a change in the membrane potential of the postsynaptic cell. ✔ If you understand this concept, you should be able to envision a membrane with a resting potential of −65 mV and explain what happens to the membrane potential when a ligand-gated ion channel opens and allows chloride ions to leave the cell. How does this affect the likelihood that the postsynaptic cell will mount an action potential?

Not all neurotransmitters bind to ion channels, however. Some bind to receptors that activate enzymes whose action leads to the production of a second messenger in the postsynaptic cell. Recall that **second messengers** are chemical signals produced inside a cell in response to a chemical signal that arrives at the cell surface (see Chapter 11).

The second messengers induced by neurotransmitters may trigger changes in enzyme activity, gene transcription, or membrane potential. (Chapter 49 explores the cellular role of second messengers in detail.)

Postsynaptic Potentials

What happens when a neurotransmitter binds to a ligand-gated ion channel in the postsynaptic cell?

Ligand-gated sodium channels on the membranes of dendrites are in particularly high concentration near synapses. When neurotransmitters bind, these channels open to let sodium into the cell, causing depolarization (**FIGURE 46.13a**). In most cases, depolarization makes an action potential in the postsynaptic cell more likely because the membrane potential approaches threshold. Changes in the postsynaptic cell that bring the membrane potential closer to threshold are called **excitatory postsynaptic potentials (EPSPs).**

If the receptor activity at the synapse leads to an outflow of potassium ions or an inflow of chloride ions or other anions in the postsynaptic cell, the postsynaptic membrane hyperpolarizes—making action potentials less likely to occur in the postsynaptic cell (**FIGURE 46.13b**). Changes in the postsynaptic cell that make the membrane potential more negative are called **inhibitory postsynaptic potentials (IPSPs).**

If an EPSP and an IPSP occur at the same time in the same place, they cancel each other out (**FIGURE 46.13c**). Synapses also can be modulatory—meaning that their activity modifies a neuron's response to other EPSPs or IPSPs.

Postsynaptic Potentials Are Graded It is critical to realize that, unlike action potentials, EPSPs and IPSPs are not all-or-none events. Instead, they are graded in size.

The size of an EPSP or IPSP depends on the amount of neurotransmitter that is released at the synapse at a given time. A higher concentration of neurotransmitter in the synaptic cleft leads to a larger EPSP or IPSP. Both types of signal are short lived because neurotransmitters do not bind irreversibly to channels

(a) EPSPs bring the membrane potential closer to the threshold.

(b) IPSPs take the membrane potential farther from threshold.

(c) Simultaneous EPSPs and IPSPs cancel each other out.

FIGURE 46.13 Events at the Synapse May Lead to Depolarization or Hyperpolarization of the Postsynaptic Membrane. These recordings show what happens to the membrane potential of a postsynaptic neuron with the arrival of signals that cause **(a)** depolarization, **(b)** hyperpolarization, or **(c)** no change. The last result occurs because simultaneous depolarizing and hyperpolarizing signals cancel each other out.

TABLE 46.2 Categories of Neurotransmitters

Neurotransmitter	Site of Action	Action*	Drugs That Interfere†
Acetylcholine	Neuromuscular junction, some CNS pathways	Excitatory (inhibitory in some parasympathetic neurons)	• Botulism toxin blocks release • Black widow spider venom increases, then eliminates, release • Alpha-bungarotoxin (in some snake venoms) binds to and blocks receptor
Monoamines			
Norepinephrine	Sympathetic neurons, some CNS pathways	Excitatory or inhibitory	• Ritalin (used for attention deficit hyperactivity disorder) increases release • Some antidepressants prevent reuptake
Dopamine	Many CNS pathways	Excitatory or modulatory	• Cocaine prevents reuptake • Amphetamine prevents reuptake
Serotonin	Many CNS pathways	Inhibitory or modulatory	• MDMA (ecstasy) causes increased release
Amino Acids			
Glutamate	Many CNS pathways	Excitatory	• PCP (angel dust) blocks receptor
Gamma-aminobutyric acid (GABA)	Some CNS pathways	Inhibitory	• Ethanol mimics response to GABA
Peptides			
Endorphins, enkephalins, substance P	Sensory pathways (pain)	Excitatory, modulatory, or inhibitory	

*Excitatory neurotransmitters make action potentials more likely in postsynaptic cells; inhibitory neurotransmitters make action potentials less likely; modulatory neurotransmitters modify the response at other synapses.

† Drugs that prevent reuptake of neurotransmitters increase their activity.

in the postsynaptic cell. Instead, they are quickly inactivated by enzymes or taken up by the presynaptic cell and recycled.

If either the amount or life span of neurotransmitters is artificially altered, the normal functioning of neurons is altered. The drugs cocaine and amphetamine, for example, exert their effects by inhibiting the uptake and recycling of particular neurotransmitters (see Table 46.2).

Summation and Threshold How do EPSPs and IPSPs affect the postsynaptic cell? As **FIGURE 46.14a** shows, the dendrites and the cell body of a neuron typically make hundreds or thousands of synapses with other cells. At any instant, the EPSPs and IPSPs that occur at each of these synapses lead to short-lived changes in membrane potential in the dendrites and cell body of the postsynaptic cell.

(a) Most neurons receive information from many other neurons.

(b) Postsynaptic potentials sum.

FIGURE 46.14 Neurons Integrate Information from Many Synapses. (a) The dendrites and cell body of a neuron typically receive signals from hundreds or thousands of other neurons. **(b)** When action potentials arrive close together in time, the postsynaptic potentials sum. Here the first EPSP is insufficient to generate an action potential; two EPSPs arriving close together undergo summation but do not reach threshold; the summing of three EPSPs arriving close together does exceed the threshold. This example is simplified—in reality, hundreds or thousands of IPSPs and EPSPs sum to determine action potential frequency.

If an IPSP and EPSP occur close together in space or time, the changes in membrane potential tend to cancel each other out. But if several EPSPs occur close together in space or time, they sum and make the neuron more likely to reach threshold and fire an action potential (**FIGURE 46.14b**). The additive nature of post-synaptic potentials is termed **summation.**

The sodium channels that trigger action potentials in the postsynaptic cell are located near the start of the axon at a site called the **axon hillock** (see Figure 46.14a). As IPSPs and EPSPs are received and interact throughout the dendrites and cell body, charge spreads to the axon hillock. If the membrane at the axon hillock depolarizes past the threshold potential, enough voltage-gated sodium channels open to trigger positive feedback and an action potential. Once an action potential starts at the axon hill-ock, it propagates down the axon to the next synapse.

Summation is critically important. Because neurons receive input from many synapses, and because IPSPs and EPSPs sum, information in the form of electrical signals is modified at the synapse before being passed along. An action potential in a pre-synaptic neuron does not always lead to an action potential in a postsynaptic neuron—the response by the postsynaptic cell depends on the information it receives from a wide array of neurons.

check your understanding

If you understand that . . .
- At a synapse, electrical information in the form of changes in membrane voltage is transduced to chemical information in the form of released neurotransmitters.
- Binding of a neurotransmitter to its receptor causes a change in the membrane potential of the postsynaptic cell.

✔ You should be able to . . .
Predict the effect on an EPSP of an increase in the concentration of a synaptic enzyme that breaks down an excitatory neurotransmitter.

Answers are available in Appendix A.

46.4 The Vertebrate Nervous System

The first three sections of this chapter examined electrical signaling at the level of molecules, membranes, and individual cells. This section discusses electrical signaling at the levels of tissues, organs, and systems.

To begin, let's consider the overall anatomy of the vertebrate nervous system, and how researchers explore the function of the most complex organ known: the human brain. The chapter concludes by returning to the molecular level (and to studies of invertebrates) to introduce recent work on learning and memory.

What Does the Peripheral Nervous System Do?

Recall from Section 46.1 that the central nervous system (CNS) is made up of the brain and spinal cord and is concerned primarily with integrating information. The peripheral nervous system (PNS) is made up of neurons outside the CNS.

What functions do the cells of the PNS control? Anatomical and functional studies indicate that the PNS consists of two divisions with distinct functions:

1. The **afferent division** transmits sensory information to the CNS.

2. The **efferent division** carries commands from the CNS to the body.

Neurons in the afferent division monitor conditions inside and outside the body. Once information from afferent neurons has been processed in the CNS, neurons in the efferent division carry signals that allow the body to respond to changed conditions in an appropriate way.

As **FIGURE 46.15** shows, the afferent and efferent divisions are part of a hierarchy of PNS functions. The efferent division is further divided into a **somatic nervous system,** which controls voluntary movements, and an **autonomic nervous system,** which controls internal processes such as digestion and heart rate.

- The somatic nervous system carries out voluntary responses, which are under conscious control. Skeletal muscles serve as the effectors.

- The autonomic nervous system carries out involuntary responses, which are not under conscious control. Smooth muscle, cardiac muscle, and glands serve as the effectors.

Many organs and glands are served by two functionally distinct types of autonomic nerves, summarized in **FIGURE 46.16**: One type inhibits activity, whereas the other promotes it.

1. Nerves in the **parasympathetic nervous system** promote "rest and digest" functions that conserve or restore energy.

FIGURE 46.15 The Nervous System Has Several Functional Divisions.

For example, the parasympathetic nerves that synapse on the heart slow it down, while those that serve the digestive tract stimulate its activity.

2. Nerves in the **sympathetic nervous system** typically prepare organs for stressful "fight or flight" situations. Sympathetic nerves speed up the heart rate, stimulate the release of glucose from the liver, and inhibit action by digestive organs.

Functional Anatomy of the CNS

Parasympathetic nerves originate at the base of the brain or the base of the spinal cord. Most sympathetic nerves also originate in the spinal cord, but they emerge along the middle of its length. Similarly, most sensory neurons project axons to the spinal cord, and most somatic motor neurons project from the spinal cord.

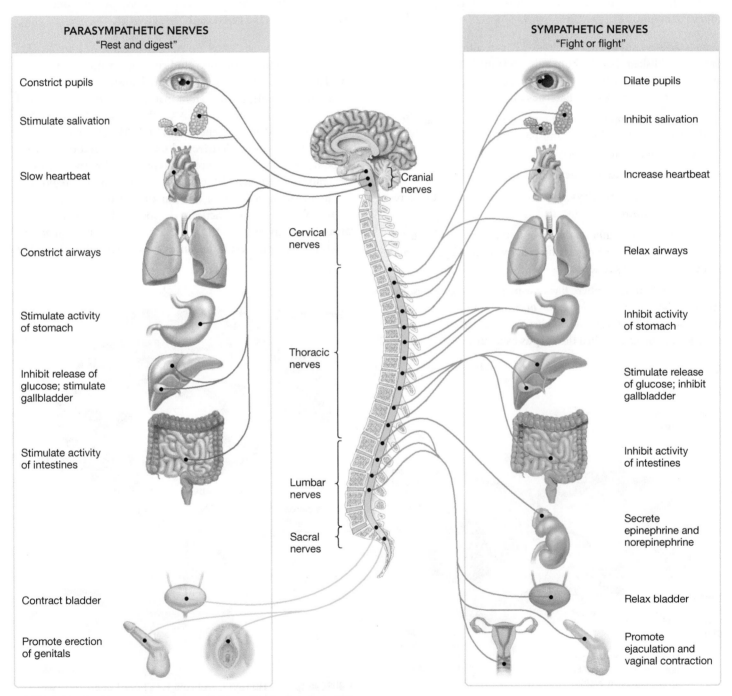

FIGURE 46.16 The Autonomic Nervous System Controls Internal Processes.

✔**QUESTION** Explain how the responses listed here for pupils, heartbeat, and liver support the rest-and-digest versus fight-or-flight functions.

In effect, then, the spinal cord serves as an information conduit. It collects and transmits information throughout the body. Virtually all the information that travels to or from the spinal cord is sent to the brain for processing. An exception is a spinal **reflex,** in which sensory neurons stimulate interneurons in the spinal cord that then directly stimulate motor neurons. This process allows you to withdraw your hand rapidly from a hot burner before your brain becomes aware of the pain.

What happens once sensory signals arrive at the brain? How are thousands of signals integrated to allow an animal to respond to stimuli? Let's begin our exploration of the brain by delving into its anatomy.

General Anatomy of the Human Brain Nineteenth-century anatomists established that the human brain is made up of the four structures labeled in **FIGURE 46.17**: the cerebrum, cerebellum, diencephalon, and brain stem. Each has a distinct function.

- The **cerebrum** makes up the bulk of the human brain. It is divided into left and right hemispheres and is the seat of conscious thought and memory.

- The **cerebellum** coordinates complex motor patterns.

- The **diencephalon** relays sensory information to the cerebrum and controls homeostasis.

- The **brain stem** connects the brain to the spinal cord. It is the autonomic center for regulating cardiovascular, digestive, and other involuntary functions.

Each cerebral hemisphere has four major areas, or lobes: the **frontal lobe,** the **parietal lobe,** the **occipital lobe,** and the **temporal lobe** (**FIGURE 46.18**). The two hemispheres are connected by a thick band of axons called the **corpus callosum.**

The relative size of the entire brain, and of its component structures, varies greatly among vertebrates (**FIGURE 46.19**). For example, compare the size of the cerebrum in fishes and mammals. In fishes, the cerebrum is quite small and is involved mainly in olfactory sensation. In mammals it is very large and contains regions specialized for memory and reasoning, in addition to the processing of multiple sensory and motor functions. Your own cerebrum is three times bigger than in other species of comparably sized mammals, reflecting its role in the higher brain functions that are so advanced in humans.

What methods do researchers use to explore the function of each area within the cerebrum?

Mapping Functional Areas in the Cerebrum I: Lesion Studies
Early work on brain function studied people with specific mental deficits caused by areas of brain damage, or lesions. Paul Broca, for example, studied an individual who could understand language but could not speak. After the person's death in 1861, Broca examined the patient's brain and discovered a damaged area, or lesion, in the left frontal lobe of the cerebrum. Broca hypothesized that this region is responsible for speech. More generally, he formulated the hypothesis that specific regions of the brain are specialized for coordinating particular functions.

Broca's claim that functions are localized to specific brain areas has been verified through extensive efforts to map the cerebrum. In some cases, advances were made by studying the

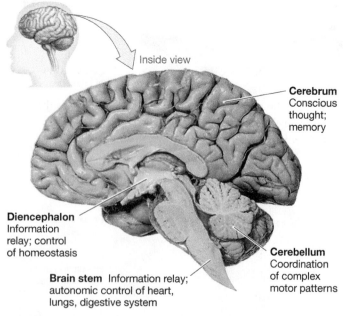

FIGURE 46.17 The Human Brain Contains Four Distinct Structures.

Cerebrum
Conscious thought; memory

Diencephalon
Information relay; control of homeostasis

Brain stem Information relay; autonomic control of heart, lungs, digestive system

Cerebellum
Coordination of complex motor patterns

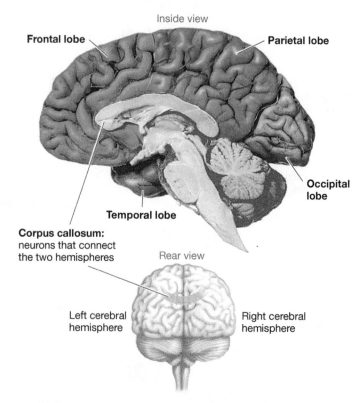

Frontal lobe

Parietal lobe

Occipital lobe

Temporal lobe

Corpus callosum: neurons that connect the two hemispheres

Left cerebral hemisphere

Right cerebral hemisphere

Inside view

Rear view

FIGURE 46.18 The Human Cerebrum Has Four Lobes and Two Hemispheres.

✔**EXERCISE** On your own head, point to each of the labeled areas.

Bony fish   Amphibian   Reptile   Bird   Mammal

FIGURE 46.19 The Relative Sizes of Brain Regions Differ among Vertebrate Lineages. For example, the cerebrum (in orange) tends to be larger in birds and mammals than in basal vertebrates.

behavior of people known to have had lesions to specific brain regions or to have had specific portions of their brains removed.

In 1848, for example, Phineas Gage was working on a railroad construction site when an iron rod over an inch in diameter was blasted through his skull. The rod entered beneath his left eye, exited through his forehead (**FIGURE 46.20**), and landed some 80 feet away. Miraculously, Gage survived this accident. However, his personality did not fare as well. After the accident, Gage's physician reported him to be "fitful," "irreverent," and "obstinate," a dramatic change from his previous personality. The iron rod had damaged Gage's frontal lobe, providing some of the first clues that this part of the brain plays a role in personality and emotion.

Another fascinating case is that of a 27-year-old man named Henry Gustav Molaison (referred to as "H.M." until his death). In 1953, surgeons treated him for life-threatening seizures by removing a small portion of his temporal lobe and about two thirds of his hippocampus, a structure at the inner edge of the temporal lobe. The man recovered (he died in December 2008, at age 83),

FIGURE 46.20 The Frontal Lobe Injury of Phineas Gage Affected His Personality. In 1848 a railroad worker named Phineas Gage recovered from a brain injury in which an iron rod was blasted through the frontal lobe of his cerebrum. The resulting change in his personality cued scientists that the frontal lobe plays a role in emotion and temperament.

had normal intelligence, and vividly remembered his childhood, but he had no short-term memory. Brenda Milner, who studied this individual for over 40 years, had to introduce herself to him every time they met; he could not even recognize a recent picture of himself. Based on case histories like Molaison's and studies of memory in laboratory animals, a consensus has emerged that several aspects of memory are governed by the hippocampus and interior sections of the temporal lobe.

Mapping Functional Areas in the Cerebrum II: Electrical Stimulation of Conscious Patients While working with severe epileptics (people suffering from seizures), Wilder Penfield pioneered a different approach to studying brain function. These individuals were scheduled to have seizure-prone areas of their brains surgically removed. While the patients were awake and under a local anesthetic, Penfield electrically stimulated portions of their cerebrums. His immediate goal was to map essential areas that should be spared from removal if possible.

When Penfield stimulated specific areas, patients reported sensations or experienced movement in particular regions of the body. From these responses, Penfield was able to map the sensory regions of the cerebrum shown in **FIGURE 46.21** (see page 946), as well as the adjacent motor regions. Brain surgeons still use this technique to map critical areas near tumors and seizure-prone areas.

Perhaps the most striking of Penfield's findings was that, on occasion, patients would respond to stimulation of their temporal lobe by having what appeared to be flashbacks. After one region was stimulated, a woman said, "I hear voices. It is late at night around the carnival somewhere—some sort of traveling circus. . . . I just saw lots of big wagons that they used to haul animals in."

Was this a memory, stored in a small set of neurons that Penfield happened to stimulate? The hypothesis that memories are somehow stored in specialized cells is controversial. However, a recent study provided support for it.

Researchers attached tiny electrodes that recorded the electrical activity of individual neurons in the temporal lobes and hippocampi of study subjects. They then showed each subject a set of images of celebrities, places, and objects. In one subject,

(a) Top view of cerebrum

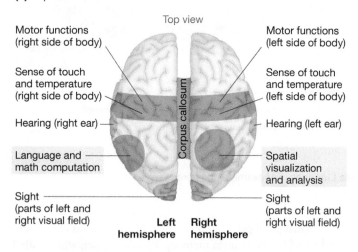

Top view

Motor functions
(right side of body)

Sense of touch
and temperature
(right side of body)

Hearing (right ear)

Language and
math computation

Sight
(parts of left and
right visual field)

Corpus callosum

Motor functions
(left side of body)

Sense of touch
and temperature
(left side of body)

Hearing (left ear)

Spatial
visualization
and analysis

Sight
(parts of left and
right visual field)

**Left
hemisphere**

**Right
hemisphere**

(b) Cross section through area responsible for sense of touch
and of temperature

Head
Trunk
Leg
Hip
Arm
Fingers
Hand
Thumb
Nose
Eye
Lips
Teeth
Jaw
Tongue
Genitals

Left
hemisphere

FIGURE 46.21 Specific Brain Areas Have Specific Functions.
(a) Map of the brain, in top view (as if the person is looking at the
top of the page), showing the functions of some major regions.
The map was compiled from studies of people with damaged brain
areas or in whom brain regions had been removed surgically. Note
that the corpus callosum is not actually as wide as shown here.
(b) Researchers mapped the area responsible for the sense of
touch and of temperature by stimulating neurons in the brains of
patients who were awake. The size of the icons corresponds to the
amount of brain area devoted to sensing those parts.

✔**QUESTION** Is there a correlation between the size of the brain area
devoted to sensing a particular body part and the size of that body
part? Explain.

a specific neuron fired when the subject viewed any of several
images of the actress Jennifer Aniston (**FIGURE 46.22**). The neuron
did not fire for other images showing spiders, buildings, other
actresses, or even Aniston with another person. It appeared that,
through experience, at least one of this patient's neurons became
singularly devoted to the concept of Jennifer Aniston.

It is important to note that the researchers were testing only a
tiny subset of the neurons present in the human brain, so there

may be more than one "Jennifer Aniston neuron" in the brain.
But the take-home message is that the formation or retrieval of
memories associated with specific concepts involves specific
neurons. Researchers are still examining exactly how these neurons interact with other parts of the brain during the processes of
learning and remembering.

Have other approaches to studying memory been productive?

How Does Memory Work?

Learning is an enduring change in behavior that results from a
specific experience in an individual's life. **Memory** is the retention of learned information. Learning and memory are thus
closely related and are often studied in tandem. As an introduction to how researchers explore these phenomena, let's first examine work that focuses on neurons and then review research at
the molecular level.

Recording from Single Neurons during Memory Tasks How do
the action potentials generated by a cell change as learning and
memory take place? Researchers have attempted to answer this
question by recording from individual neurons in the temporal
lobes of humans.

Before operating on patients who were still awake and about
to undergo surgery to remove seizure-prone areas of their brains,
physicians placed electrodes in specific brain regions. Then they
projected words or names of objects on a screen and asked the
individuals to read them silently, read them aloud, or remember them and repeat them later. The data showed that individual
neurons in the cerebrum's temporal lobe are relatively quiet while
patients identify objects but extremely active when the individual
remembers the objects and repeats their names aloud.

What do such data mean? Neurons in the temporal lobe are
most active during memory tasks. So how could action potentials from particular cells make memory possible?

Documenting Changes in Synapses Research on the molecular
basis of memory is based on the idea that learning and memory
must involve some type of short-term or long-term change in
the neurons responsible for these processes. This change could
be structural or chemical in nature. Structural changes might include modifications in the number of synapses that a particular
neuron makes, or even the destruction and creation of neurons.
Chemical changes might involve alterations in the amount of
neurotransmitter released at certain synapses or changes in the
number of receptors present in postsynaptic cells.

To explore the molecular basis of learning and memory, Eric
Kandel's group has focused on an animal much easier to study
than any vertebrate, the sea slug *Aplysia californica* (**FIGURE 46.23a**).
Much of their work has explored the reflex diagrammed in
FIGURE 46.23b: When the siphon, a structure on the animal's back,
is touched—for example, by a stream of water—the sea slug
responds by withdrawing its gill. The reflex is produced by a
sensory neuron that is activated by touch and a motor neuron
that projects to a gill muscle. Retracting the gill protects it from
predators.

| 0 action potentials/sec | 8 action potentials/sec | 0 action potentials/sec | 6 action potentials/sec | 0 action potentials/sec |

FIGURE 46.22 Single-Neuron Recording Reveals that Some Neurons in the Brain Recognize Specific Concepts. The graphs below each image show how a single neuron fires in response to images of actress Jennifer Aniston but not to other images.

DATA: Quiroga, R. Q., L. Reddy, G. Kreiman, et al. 2005. *Nature* 435: 1102–1107.

Early work established that this simple reflex is modified by learning. For example, *Aplysia* also withdraw their gills when their tails are given an electrical shock. If shocks to the tail are repeatedly combined with a very light touch to the siphon—too light to normally get a response on its own—an *Aplysia* will learn to withdraw its gills in response to a light siphon touch alone.

Follow-up studies on this reflex showed that the neurons involved in learning release the neurotransmitter **serotonin,** which causes an EPSP in the motor neuron to the gill. Repeated application of serotonin mimics what happens at the synapse during learning, when the neurons fire repeatedly. As **FIGURE 46.24a** (see page 948) shows, experimental application of serotonin leads to higher EPSPs, meaning that the motor neuron is more likely to

generate action potentials. These results suggest that in *Aplysia*, changes in the nature of the synapse form the molecular basis of learning and memory. This change in responsiveness of the synapse is termed **synaptic plasticity.**

Recently Kandel's team replicated these results with sensory and motor neurons growing in culture. **FIGURE 46.24b** shows two *Aplysia* motor neurons on a culture plate. Each motor neuron synapses with the same sensory neuron. To mimic the learning process, the investigators applied serotonin to one synapse five times over a short period. When they stimulated the sensory neuron a day later, they found a huge increase in EPSPs in the motor neuron postsynaptic to that synapse (**FIGURE 46.24c**). The experimental postsynaptic cells had also established additional synapses. The structure and behavior of the experimental

(a) Sea slug *Aplysia californica*

(b) Gill-withdrawal reflex protects the gills during an attack.

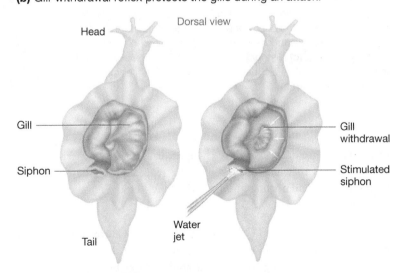

FIGURE 46.23 The Gill-Withdrawal Reflex in *Aplysia* Is a Model System in Learning and Memory.

(a) Postsynaptic potentials changing as a result of learning

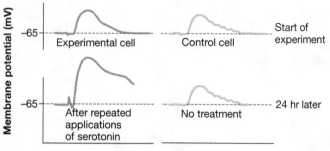

(b) Applying neurotransmitters to neurons in culture

(c) Repeating *Aplysia* experiment in culture

FIGURE 46.24 Learning and Memory Involve Changes in Synapses. (a) Experiment showing that repeated application of serotonin changes the behavior of postsynaptic neurons. **(b)** An in vitro study of interactions between *Aplysia* motor neurons and sensory neurons. **(c)** Histogram depicting the percentage increase in postsynaptic potentials that occurs after repeated application of serotonin, as in part **(a)**.

DATA: Martin, K. C., A. Casadio, H. Zhu, et al. 1997. *Cell* 91: 927–938.

postsynaptic motor neuron had changed, based on its experience. Neurons that had not received the repeated stimulation with serotonin had normal postsynaptic responses and numbers of synapses.

Documenting Changes in Neurons Synapses change over time, but what about the cells of the central nervous system themselves? When early neurobiologists dissected adult cadaver brains, they

found that none of the neurons showed signs of mitosis. From these observations they concluded that **neurogenesis**—the formation of new neurons—does not occur in adults.

In 1983, Steven Goldman and Fernando Nottebohm revisited this long-standing theory. They examined whether new neurons can be formed in the brains of songbirds. Songbirds were chosen because in their brains, specific regions that control singing behavior undergo dramatic seasonal changes in size. The researchers hypothesized that these changes result from an increased rate of neurogenesis during the reproductive season.

To test this hypothesis, the researchers injected canaries (**FIGURE 46.25a**) with radiolabeled thymidine, which becomes incorporated into newly synthesized DNA. The researchers later collected the birds' brains, cut them into very thin slices, mounted the slices onto microscope slides, and brought them into contact with photographic film. In this technique, called autoradiography (see **BioSkills 9** in Appendix B), any new cells with radiolabeled thymidine in their DNA would expose the film, thereby indicating the presence and location of cells that were produced after the bird was injected.

The researchers found that adult songbirds are indeed able to produce large numbers of new neurons. Dozens of studies since Goldman and Nottebohm's experiment have confirmed that new neurons are incorporated into the song-control regions of the brain each spring, causing them to grow dramatically in size (**FIGURE 46.25b**). The new neurons promote learning and memory in the song-control system.

Other scientists around the same time and, indeed, even earlier obtained autoradiographic evidence of neurogenesis in adult mammals. However, the sheer number of new cells was much lower in mammals than in birds, so these reports did

check your understanding

C Y U

If you understand that . . .
- The CNS and PNS work together to gather information about the external and internal environments, process that information, and signal muscles, glands, and other tissues to make appropriate responses to the information.
- Brain regions have specific functions.
- Changes in synapses and neurogenesis allow animals to learn and create memories.

✔ You should be able to . . .
1. Describe the research strategies that allowed biologists to localize particular functions to specific regions in the brain.
2. Identify the key mechanisms behind learning and memory.

Answers are available in Appendix A.

(a) Male canaries (left) sing to attract females (right).

(b) In songbirds, the size of the brain's song-control region increases in spring.

Winter

Spring

FIGURE 46.25 Neurogenesis Occurs in the Adult Songbird Brain.

not convince all researchers. Skeptics initially suggested that the labeled cells in mammals might actually be glia rather than neurons.

A breakthrough came in the early 1990s with the development of new techniques that specifically label neurons. The markers used in these techniques bind to proteins that are found only in neurons and not in glial cells. Since that time, adult neurogenesis has been definitively shown to occur in all vertebrates, including humans.

Scientists are only beginning to understand the functional role of these new neurons. Is it the same as that of neurons made during development? Can neurogenesis be used to help heal brain injuries? Some studies have even suggested that chronic stress during childhood (the stress of poverty or bullying, for example) can reduce the ability to make new neurons, thereby negatively affecting learning in adults.

Results like these reinforce a growing consensus that learning, memory, and the control of complex behaviors involve not only molecular and structural changes in synapses but also changes in the number of neurons. Further, most researchers now agree that at least some aspects of long-term memory involve changes in gene expression. Chemical messengers called hormones also cause changes in gene expression in target cells (see Chapter 49). But before investigating how hormones work, let's focus on the electrical signals involved in vision, hearing, taste, and movement—the subject of Chapters 47 and 48.

CHAPTER 46 REVIEW

For media, go to MasteringBiology

If you understand . . .

46.1 Principles of Electrical Signaling

- All neurons have a cell body and multiple short dendrites that receive signals from other cells. Most neurons also have a single axon that transmits electrical signals to other neurons or to effector cells in glands or muscles.

- Studies of the squid giant axons established that neurons have a resting potential created by the sodium–potassium pump and potassium leak channels. When Na^+/K^+-ATPase hydrolyzes ATP, it transports 3 Na^+ out of the cell and 2 K^+ in.

✔ You should be able to diagram the plasma membrane of a neuron. Add symbols to show the relative concentrations of Na^+, K^+, and Cl^-. Add labels indicating the role of K^+ leak channels and the Na^+/K^+-ATPase.

46.2 Dissecting the Action Potential

- Studies of the squid axon established that the action potential is a rapid, all-or-none change in membrane potential.

- An action potential begins with an inflow of Na^+ that depolarizes the membrane. An outflow of K^+ follows and repolarizes the membrane.

- Both Na^+ and K^+ flow through voltage-gated channels.

- As Na^+ flows in, it repulses cations that spread from the site of the action potential, causing the nearby membrane to depolarize enough to trigger an action potential in that adjacent portion of the membrane.

- Propagation takes place most rapidly in large axons or myelinated axons.

✔ You should be able to explain why every action potential in a given neuron is identical.

46.3 The Synapse

- When action potentials arrive at a synapse, synaptic vesicles fuse with the axon's membrane and deliver neurotransmitters into the synapse that bind to receptors on the membrane of a postsynaptic cell.

- One class of receptors functions as ligand-gated channels. In response to binding by a neurotransmitter, these channels open and admit ions that depolarize or hyperpolarize the postsynaptic cell's membrane.

- Postsynaptic potentials sum.

- If the membrane at the axon hillock of the postsynaptic neuron depolarizes to a threshold value, an action potential is triggered.

✔ You should be able to explain how summation in the postsynaptic cell relates to the claim that the cell body integrates information from many different synapses.

46.4 The Vertebrate Nervous System

- The CNS consists of the brain and spinal cord (in vertebrates); the PNS consists of all nervous system components outside the CNS.

- In vertebrates, the PNS contains somatic and autonomic components. The somatic PNS signals the effector cells producing movement; the autonomic PNS monitors internal conditions and signals the effector cells that change the activity of tissues and organs.

- Early efforts to map functional regions of the brain were based on analyzing deficits in individuals with brain lesions or on stimulating certain regions of the cerebrum.

- Research has established that learning and memory involve modifications in synapses. After learning takes place, certain neurons release more or less neurotransmitter, or make additional synapses, in response to stimulation.

- Recent research shows that vertebrates are capable of adult neurogenesis, which may be an important component of learning and memory.

✔ You should be able to predict the effects of a brain stem lesion in an animal.

(MB) MasteringBiology

1. MasteringBiology Assignments

Tutorials and Activities Action Potentials, How Neurons Work (1 of 3): Neuron Structure and Resting Potential; How Neurons Work (2 of 3): The Action Potential; How Neurons Work (3 of 3): Conduction of an Action Potential; How Synapses Work (1 of 2): Chemical Synapses; How Synapses Work (2 of 2): Postsynaptic Potentials; Membrane Potentials; Nerve Signals: Action Potentials; Neuron Structure; Signal Transmission at a Chemical Synapse; The Vertebrate Nervous System

Questions Reading Quizzes, Blue-Thread Questions, Test Bank

2. eText Read your book online, search, take notes, highlight text, and more.

3. The Study Area Practice Test, Cumulative Test, BioFlix® 3-D Animations, Videos, Activities, Audio Glossary, Word Study Tools, Art

You should be able to . . .

✔ TEST YOUR KNOWLEDGE

Answers are available in Appendix A

1. Which ion leaks across a neuron's membrane to help create the resting potential?
   a. Ca^{2+}
   b. K^+
   c. Na^+
   d. Cl^-

2. Why did the squid axon become a model system for studying electrical signaling in animals?
   a. Its action potentials are particularly large and frequent.
   b. It is the tissue from which researchers initially isolated Na^+/K^+-ATPase.
   c. Squids are abundant and easy to obtain.
   d. It was large enough to support intracellular recording by the first microelectrodes.

3. How does myelination affect the propagation of an action potential?
   a. It speeds propagation by increasing the density of voltage-gated channels.
   b. It speeds propagation by increasing electrochemical gradients favoring Na^+ entry.

   c. It speeds propagation because cations do not leak out of the membrane as they spread down the axon.
   d. It slows down propagation because Na^+ channels exist only at unmyelinated nodes (nodes of Ranvier).

4. In a neuron, what creates the electrochemical gradient favoring the outflow of K^+ when the cell is at rest?
   a. Na^+/K^+-ATPase
   b. voltage-gated K^+ channels
   c. voltage-gated Na^+ channels
   d. ligand-gated Na^+/K^+ channels

5. Which of the following brain regions is responsible for formation of new memories?
   a. brain stem
   b. hippocampus
   c. frontal lobe
   d. cerebellum

6. Draw a graph of an action potential and label the axes. Label the parts of the graph, and explain which ion flow or flows are responsible for each part.

7. Explain why the Na^+/K^+-ATPase is "electrogenic"—meaning that it creates a voltage across a membrane.

8. Explain the difference between a ligand-gated K^+ channel and a voltage-gated K^+ channel.

9. Describe the role of summation in postsynaptic cells.

10. Compare and contrast the somatic and autonomic components of the PNS.

11. Predict the effect on heart rate of a drug that activates sympathetic nervous system activity.

12. Why is memory thought to involve changes in particular synapses?
    a. In some systems, an increased release of neurotransmitters occurs after learning takes place.
    b. In some systems, the type of neurotransmitter released at the synapse changes after learning takes place.
    c. When researchers stimulated certain neurons electrically, individuals replayed memories.
    d. People who lack short-term memory have specific deficits in synapses within the brain regions responsible for memory.

13. Explain why drugs that prevent neurotransmitters from being taken back up by the presynaptic cell have dramatic effects on the activity of postsynaptic neurons.

14. Discuss the pros and cons of lesion studies and electrical stimulation of conscious patients in determining the functions of particular brain structures.

15. QUANTITATIVE The data points on the graph record the movement of potassium ions (in μSiemens, a unit of conductance) observed when neuron membranes were voltage-clamped at the array of voltages indicated on the x-axis. The "Neuron with venom" data were generated by adding black mamba venom to the solution bathing the neuron (black mambas are snakes). Which of these conclusions can you draw from the data?

a. The venom opens all potassium channels.
b. The venom blocks all potassium channels.
c. There is more than one type of potassium channel; at least one is affected by the venom while others are not.
d. Black mamba venom does not affect potassium ion movement.

16. Studies of the brain show that mature neurons do not undergo mitosis to give rise to new neurons, yet adult neurogenesis occurs in all vertebrates. Suggest a hypothesis as to the source of these new neurons.

47 Animal Sensory Systems

In this chapter you will learn how

Animals transform sound, smell, and other stimuli into signals the brain can understand

↓ by asking

How do sensory organs convey information to the brain? **47.1**

then looking closer at

which is part of

The nervous system **Ch. 46**

→ Mechanoreception **47.2**

→ Photoreception **47.3**

→ Chemoreception **47.4**

→ Sensing temperature and electric and magnetic fields **47.5**

In many species of moth, males have much larger antennae than females do. Receptor cells on the males' feathery antennae detect airborne chemical signals that are produced by sexually mature females. As a result, males can locate females in total darkness.

BIG PICTURE

This chapter is part of the Big Picture. See how on pages 840–841.

Many adult moths are active at night, when it is difficult or impossible to see. Instead of *looking* for a mate under these challenging circumstances, sexually mature female moths release a chemical attractant into the air. Male moths can detect even a single molecule of the attractant, due to receptor cells located on their large, feathery antennae. Guided by an airborne gradient of attractant molecules, a male moth flies unerringly toward a female.

As they patrol in search of these airborne molecules, however, male moths are hunted by bats. Like moths, bats are active almost exclusively at night. Instead of hunting by sight, like a falcon or a cheetah, bats hunt with the aid of sonar: They emit a train of high-pitched sounds as they fly and then listen for echoes that indicate the direction and shape of objects in their path. If the object is a moth, the bat flies toward it, catches the moth in its mouth, and eats it.

✔ When you see this checkmark, stop and test yourself. Answers are available in Appendix A.

But some moth species can hear bat calls. When moths detect sounds from an onrushing bat, they tumble out of the sky in chaotic escape flights.

If you were out at night as these dramas unfolded, at best you might be dimly aware that bats and moths were flying about. Humans cannot smell moth attractants or hear the sounds that most bats emit when flying. It took decades of careful experimentation for biologists to understand how moths and bats sense the world around them and respond to the information they receive.

Sensing and interpreting changes in the environment are fundamental to how animals work. Let's begin with a basic question: How are sounds, smells, and other stimuli transformed into a signal that the brain can understand?

47.1 How Do Sensory Organs Convey Information to the Brain?

As a moth flies through the night, its brain receives streams of signals from an array of sensory organs. Antennae provide information about the concentration of female attractant molecules; ears located on various parts of the body send data on the presence of high-pitched sounds; detectors for balance and gravity transmit signals about the body's orientation in space.

Each type of sensory information is detected by a sensory neuron or by a specialized receptor cell that makes a synapse with a sensory neuron. As **FIGURE 47.1** shows, the moth's nervous system integrates the sensory input—the information from sensory neurons—and responds with motor output, via electrical signals, to specific muscle groups (effectors).

The ability to sense a change in the environment depends on three processes:

1. *Transduction,* the conversion of an external stimulus to an internal signal in the form of action potentials along sensory neurons

2. *Amplification* of the signal

3. *Transmission* of the signal to the central nervous system (CNS)

The first step in the sequence, transduction, requires a sensory receptor cell specialized for converting light, sound, touch, or some other signal into an electrical signal. Sensory receptors are located throughout the body and are categorized by type of stimulus:

- **Mechanoreceptors** respond to distortion caused by pressure.
- **Photoreceptors** respond to particular wavelengths of light.
- **Chemoreceptors** perceive specific molecules.
- **Thermoreceptors** detect changes in temperature.
- **Nociceptors** sense harmful stimuli such as tissue injury.
- **Electroreceptors** detect electric fields.
- **Magnetoreceptors** detect magnetic fields.

With such a broad range of possible sensory detectors, it is no wonder that animals can monitor and respond to a wide array of changes in their environments.

Now, how do sensory cells receive information from the environment and report it to the brain, so an appropriate response can occur?

Sensory Transduction

During the resting state in most sensory cells, the inside of the plasma membrane is more negative than the exterior (see Chapter 46). When ion flows cause the interior to become more positive (less negative), the membrane is **depolarized.** When ions cause the cell interior to become more negative than the resting potential, the membrane is **hyperpolarized.**

Although sensory receptors can detect a remarkable variety of stimuli, they all transduce sensory input—such as light, sounds, touch, and odors—to a change in membrane potential. In this way, different types of information are transduced to a common type of signal—one that can be interpreted by the brain.

If a sensory stimulus induces a large change in a sensory receptor's membrane potential, there is a change in the firing rate of action potentials sent to the brain. The amount of depolarization that occurs in a sound-receptor cell, for example, is proportional to the loudness of the sound. If the depolarization passes

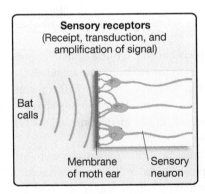
Sensory receptors
(Receipt, transduction, and amplification of signal)

Bat calls

Membrane of moth ear   Sensory neuron

Sensory input

Central nervous system
(Integration of information)
CNS

Ears   Top view

Motor output

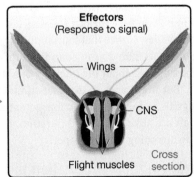
Effectors
(Response to signal)

Wings

CNS

Flight muscles   Cross section

FIGURE 47.1 Sensory Systems, the CNS, and Effectors Are Linked. Sensory neurons relay information about conditions inside and outside an animal to the central nervous system. After integrating information from many sensory neurons, the CNS sends signals to muscles.

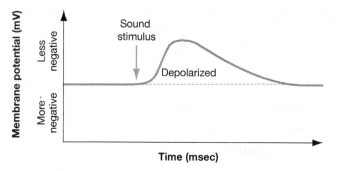

(a) Sound-receptor cells depolarize in response to sound.

Membrane potential (mV) — Less negative / More negative

Sound stimulus

Depolarized

Time (msec)

(b) Sound-receptor cells respond more strongly to louder sounds.

Number of action potentials per second

150

100

50

0

500 1000 1500 2000 2500

Louder sound

Softer sound

Frequency (Hz)

FIGURE 47.2 Sensory Inputs Change the Membrane Potential of Receptor Cells. (a) In response to sensory stimuli, ions flow across the membranes of receptor cells and either depolarize or hyperpolarize the membrane. **(b)** The rate at which action potentials occur in an auditory neuron transmits information about the nature and intensity of the sensory stimulus. (The frequency of a sound, measured in hertz, determines its pitch.)

DATA: Rose, J. E., J. E. Hind, D. J. Anderson, et al. 1971. *Journal of Neurophysiology* 34: 685–699.

threshold, enough voltage-gated sodium channels open to trigger action potentials that are relayed to the brain.

FIGURE 47.2a shows the membrane potential from a sound-receptor cell. If you put your finger on the red line and trace to the right, you'll notice that when the experimenter played a sound, the sound-receptor cell depolarized for a short time in response. Other sensory cells work in a similar way.

Recall that all action potentials from a given neuron are identical in size and shape (see Chapter 46). **FIGURE 47.2b** graphs the action potential "firing rate" recorded from a sound-receptor cell when sounds at various frequencies were played at two distinct intensities. Notice that loud sounds induce a higher rate of action potentials than do soft sounds. In this way receptor cells provide information about the intensity of a stimulus.

But if all types of external stimuli are converted to electrical signals in the form of action potentials, and if all action potentials are alike in size and duration, how does the brain interpret the incoming signals properly?

Transmitting Information to the Brain

There are two keys to understanding how the brain interprets sensory information. First, receptor cells tend to be highly specific. For example, each receptor cell in a human ear responds best

to certain pitches of sound. Some receptors are more sensitive to low-pitched sounds, and others respond best to high-pitched sounds. The pattern of action potentials from a cell contains information about the pitch of the sound that is being received, its intensity, and how long the stimulus lasts.

The second key point: Each type of sensory neuron sends its signal to a specific portion of the brain. Axons from sensory neurons in the human ear project to the temporal lobes at the sides of the brain, but axons from sensory receptors in the eye deliver action potentials to the occipital lobe at the back of the brain. Different regions of the brain are specialized for interpreting different types of stimuli.

Now that the basic principles of sensory reception and transduction have been introduced, let's delve into the details of the major sensory systems.

47.2 Mechanoreception: Sensing Pressure Changes

Animals have a variety of mechanisms for **mechanoreception**— the sensation of pressure changes. Crabs, for example, have a fluid-filled organ that helps them sense the pressure created by gravity. The organ, known as a **statocyst,** is lined with pressure-receptor cells and contains a small calcium-rich particle that normally rests on the bottom of the organ. But if the crab is tipped or flipped over, this particle presses against receptors that are *not* on the bottom of the organ. When the brain receives action potentials from these receptors, it responds by activating muscles that restore the animal to its normal posture.

It's also common for animals to have cells that are responsible for detecting direct physical pressure on skin, as well as pressure-receptor cells that monitor how far muscles or blood vessels are stretched. Animals detect sound waves, which produce pressure in air, with hearing, and some aquatic animals detect pressure waves in water via a lateral line system. These pressure-sensing systems are all based on the same mechanism.

Let's briefly examine the general design of a mechanoreceptor cell and its response to pressure, then investigate the specific structures involved in vertebrate hearing and the lateral line system.

How Do Sensory Cells Respond to Sound Waves and Other Forms of Pressure?

The mechanoreceptors responsible for sensing sound, vibrations, and other pressure waves in the environment are relatively simple. In every case, direct physical pressure on a plasma membrane or distortion of membrane structures by bending changes the conformation of ion channels in the membrane and causes the channels to open or close.

The consequent change in ion flow through the channel proteins results either in a depolarization or a hyperpolarization. This changes the pattern of action potentials in a sensory neuron.

The Structure of Hair Cells In vertebrates, ion channels that respond to pressure are often found in hair cells. **Hair cells** are

(a) Hair cells have many stereocilia and one kinocilium.

(b) PROCESS: BENDING OPENS ION CHANNELS

Pressure wave

K^+

Depolarization

K^+

Synaptic vesicle

Calcium channel

Ca^{2+} Ca^{2+}

Neurotransmitter released into synapse

Afferent neuron (to brain)

1. **Pressure wave bends stereocilia.**

2. **Potassium channels open** in response to bending.

3. **Membrane depolarizes** due to influx of K^+.

4. **Depolarization triggers inflow of calcium ions.**

5. **Synaptic vesicles fuse with plasma membrane.**

6. **Neurotransmitter is released** and diffuses to afferent neuron.

FIGURE 47.3 Hair Cells Transduce Sound Waves to Electrical Signals.

pressure-receptor cells, illustrated in **FIGURE 47.3a**, named for their stiff outgrowths called **stereocilia** (singular: **stereocilium**). The "hairy-looking" stereocilia are microvilli that are reinforced by actin filaments.

Many hair cells also have a single **kinocilium,** a true cilium that contains a 9 + 2 arrangement of microtubules (introduced in Chapter 7). Hair cells are found in the ears of land-dwelling vertebrates and the lateral line system in many species of fishes and some amphibians.

As Figure 47.3a shows, the stereocilia in hair cells are arranged in order of increasing height; if a kinocilium is present, it is the tallest of all the projections. These structures extend into a fluid-filled chamber.

Signal Transduction in Hair Cells If stereocilia are bent in the direction of the kinocilium in response to pressure (**FIGURE 47.3b**), the distortion causes potassium ion (K^+) channels in the stereocilia to open. This is the common theme connecting pressure-sensing cells: Bending opens ion channels.

Recall that the opening of K^+ channels usually causes an outflow of K^+ that hyperpolarizes neurons (see Chapter 46). Hair cell plasma membranes respond differently, however, because they are bathed by extracellular fluid with an extraordinarily high K^+ concentration. As a result, the equilibrium potential for K^+ in hair cells is 0 mV instead of the -85 mV in a typical neuron. The resting potential of the hair cell plasma membrane is -70 mV, so when the channels open, K^+ rushes in and causes a depolarization of approximately 20 mV.

In hair cells, depolarization causes an inflow of calcium ions, which triggers an increase in the amount of neurotransmitter released at the synapse between the hair cell and a sensory neuron. The end result is excitation of the postsynaptic cell, meaning that it becomes more likely to fire an action potential to the brain via an afferent sensory neuron. You might recall that afferent neurons are part of the peripheral nervous system and conduct information to the CNS (see Chapter 46).

If sound pressure waves bend stereocilia the other way, however, the K^+ channels close, and the cell hyperpolarizes by 5 mV. Hyperpolarization of the hair cell decreases the amount of neurotransmitter released at the synapse and inhibits the postsynaptic sensory neuron, making it less likely to trigger action potentials.

How can bending affect ion channels? Electron micrographs show that tiny threads connect the tips of stereocilia to each other. One hypothesis contends that when the stereocilia are bent, the threads somehow pull on the ion channels in the wall of the next tallest stereocilium and open them like tiny trapdoors (see Figure 47.3b, step 2). This hypothesis remains to be confirmed, however. Researchers still do not fully understand how the ion channels involved in pressure reception work.

Hearing: The Mammalian Ear

Hearing is the sensation produced by the wavelike changes in air pressure called sound. A sound consists of waves of pressure in air or in water. The number of pressure waves that occur in 1 second is called the **frequency** of the sound, reported in units called hertz (Hz), or cycles per second. When you hear different sound frequencies, you perceive them as different **pitches.** A high-pitched sound may have a frequency in the range of 8000 Hz, whereas a low-pitched sound frequency might be 1000 Hz.

The ear transduces sound waves into action potentials that send information to the brain. To understand how changes in the membrane potential of hair cells result in hearing, let's focus on the human ear as a case study. The human ear has three sections:

the **outer ear, middle ear,** and **inner ear** (**FIGURE 47.4**). A membrane separates each section from the next.

The path of sound through the ear is traced in Figure 47.4. The outer ear, which projects from the head, collects incoming pressure waves and funnels them into a tube known as the ear canal. At the inner end of the ear canal (see the lower part of Figure 47.4), the waves strike the **tympanic membrane,** or eardrum, which separates the outer ear from the middle ear.

The repeated cycles of air compression cause the tympanic membrane to vibrate back and forth with the same frequency as the sound wave. The vibrations are passed to three tiny bones in the middle ear called the **ear ossicles,** which vibrate against one another in response. The last ossicle, the **stapes** (pronounced *STAY-peez*), vibrates against a membrane called the **oval window,** which separates the middle ear from the inner ear. The oval window oscillates in response and generates waves in the fluid inside a chamber known as the **cochlea** (pronounced *KOK-lee-ah*). These pressure waves are sensed by hair cells in the cochlea.

In effect, the ear translates airborne waves into fluid-borne waves. The system seems extraordinarily complex, though, for

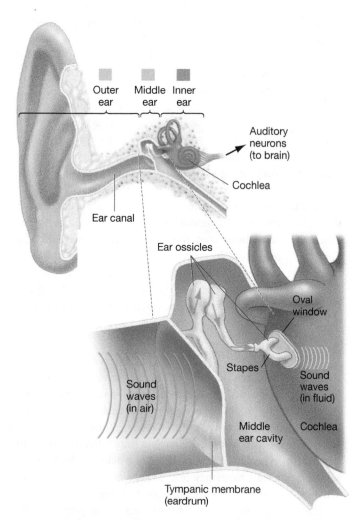

FIGURE 47.4 Mammals Have an Outer Ear, Middle Ear, and Inner Ear. The middle ear starts with the tympanic membrane and ends in the oval window of the cochlea.

such a simple result. Why doesn't the outer ear canal lead directly to the oval window? Why have a middle ear at all?

The Middle Ear Amplifies Sounds   Biologists began to understand the function of the middle ear when they recognized two key aspects of its structure. First, the size difference between the tympanic membrane and the oval window is important. The tympanic membrane is about 15 times larger than the oval window, causing the amount of vibration induced by sound waves to increase by a factor of 15 when it reaches the oval window. This phenomenon is similar to taking the same amount of force used to bang on a very large door and applying it to a very small door.

In addition, the three ossicles act as levers that further amplify vibrations from the tympanic membrane. The overall effect in mammals is to amplify sound by a factor of 22—meaning that soft sounds are amplified enough to stimulate hair cells in the cochlea. Thus, biologists interpret the mammalian middle ear as an adaptation for increasing sensitivity to sound.

To summarize, the mammalian outer ear transmits sound waves from the environment to the middle ear; the middle ear amplifies these waves enough to stimulate the hair cells within the cochlea of the inner ear.

If all hair cells responded equally to all frequencies of sound, you would be able to perceive only one pitch. Everyone's voice—indeed, every noise—would sound the same. How can hair cells distinguish different frequencies?

The Cochlea Detects the Frequency of Sounds   As **FIGURE 47.5a** shows, the cochlea is a coiled tube with a set of internal membranes that divide it into three chambers. Hair cells, forming rows in the middle chamber, are embedded in a tissue that sits atop the **basilar membrane** (**FIGURE 47.5b**). In addition, the hair cells' stereocilia touch yet another, smaller surface called the **tectorial membrane.** (The kinocilium is not present in a mature cochlear hair cell.) In effect, hair cells are sandwiched between membranes.

Researchers struggled for decades to understand how these membranes affect hair cell function. It is virtually impossible to study cochleas in living organisms, because the cochleas are tiny, complex, coiled, and buried deep inside the skull. During the 1920s and 1930s, however, Georg von Békésy pioneered work on the structure and function of these organs by performing experiments on cochleas that he had dissected from fresh human cadavers.

Von Békésy was able to vibrate the oval window and record how the cochlea's internal membranes moved in response. He found that when a pressure wave traveled down the fluid in the upper and lower chambers, the basilar membrane vibrated in response. His key finding, though, was that sounds of different frequencies caused the basilar membrane to vibrate maximally at specific points along its length (**FIGURE 47.6**). When the basilar membrane vibrated in a particular location, the stereocilia of the hair cells there were bent one way and then the other by the tectorial membrane.

Von Békésy also noted that the basilar membrane is stiff near the oval window and flexible at the other end. This is why each

(a) The middle chamber of the fluid-filled cochlea contains hair cells.

Cochlea

Auditory nerve (to brain)

Three fluid-filled chambers

Tectorial membrane

Hair cells

Neurons (of auditory nerve)

(b) Hair cells are sandwiched between membranes.

Tectorial membrane

Stereocilia

Hair cells

Basilar membrane

Axons of sensory neurons

FIGURE 47.5 The Human Cochlea Contains Fluid-Filled Chambers Separated by Membranes.

Oval window

Cochlea

Base of cochlea (near oval window)

Uncoiled cochlea

Basilar membrane

20,000 Hz   2000 Hz   200 Hz   20 Hz

Narrow part of basilar membrane is **stiff**—vibrates in response to **high frequencies**

Wide part of basilar membrane is **flexible**—vibrates in response to **low frequencies**

FIGURE 47.6 The Basilar Membrane Varies in Stiffness. Different parts of the basilar membrane respond to different sound frequencies.

segment vibrates in response to a different frequency of sound. Just as a stiff drumhead produces a high-pitched sound and a loose drumhead yields a low-pitched sound, high-frequency sounds cause the stiff part of the basilar membrane to vibrate; low-frequency sounds cause the flexible part to vibrate.

To summarize, certain portions of the basilar membrane vibrate in response to specific frequencies and result in the bending of hair cell stereocilia. In this way, hair cells in a particular place on the membrane respond to sounds of a certain frequency. When the occipital lobe of the cerebrum receives action potentials from neurons associated with specific hair cells, it interprets them as a particular pitch—meaning a specific frequency of sound. The result is the sense called hearing.

Complex sounds contain a wide variety of frequencies and trigger particular combinations of hair cells. Through experience, the brain learns which combinations of frequencies represent music, a fire alarm, or a best friend's voice.

Compared with the hearing of many mammals, human hearing is not particularly acute. Humans can hear sounds between 20 Hz and 20,000 Hz (20,000 Hz is equal to 20 kHz, or kilohertz). But some mammals can sense low-frequency infrasounds that are too low for humans to hear (*infra–* means below or under); others are aware of high-frequency ultrasounds that are above the range of human hearing (*ultra–* means beyond).

Elephants Detect Infrasound When Katherine Payne was observing elephants at a zoo in the mid-1980s, she noticed a subtle throbbing in the air. Payne knew that infrasound can produce such sensations.

To test the hypothesis that the elephants were producing infrasonic vocalizations, Payne returned to the zoo with microphones that could pick up sounds at extremely low frequency. At normal speed, the tape she made was silent. But when she raised the pitch of the sounds by speeding up the tape, she heard a chorus of cow-like noises. The elephants were calling to each other, using low-frequency sounds.

According to follow-up research, elephants have the best infrasonic hearing of any land mammal. Because infrasound can travel exceptionally long distances, biologists hypothesize that infrasonic calls allow wild elephants to coordinate their movements when they are miles apart.

Recent research suggests that in addition to detecting infrasound via their large ears, elephants use their feet to detect a by-product of infrasound: seismic vibrations traveling in the ground.

Bats Detect Ultrasound Ultrasonic hearing in bats was discovered in the late 1930s, when Donald Griffin borrowed the only ultrasonic apparatus then in existence from Robert Galambos,

FIGURE 47.7 Bats Emit Ultrasonic Sound Waves That Bounce Off Surfaces. Returning sound waves are sensed by the bat's inner ear.

a fellow graduate student. Griffin used the machine to demonstrate that flying bats constantly emit ultrasounds. In follow-up experiments, he documented that a bat with cotton in its ears, or with its mouth taped shut, crashed into walls when released in a room. Blindfolded bats, in contrast, never crashed.

Griffin and Galambos concluded that bats use sound echoes (sonar) to navigate. This concept, termed **echolocation,** was an outlandish idea at the time. When Galambos described it at a meeting in 1940, another scientist shook him by the shoulders and said, "You can't really mean that!"

Bats generate high-frequency sound waves with the larynx, or voice box. These waves "bounce" off of surfaces (including those of insects), producing echoes that the bat perceives in its inner ear (**FIGURE 47.7**). Recall that different sections of the basilar membrane are specialized for sensing sound waves of different frequencies. In bats, a huge area of the basilar membrane is specialized for sensing the high-frequency sounds of the returning

echoes. Similarly, the part of the brain used to process sound is unusually large in bats, highlighting the extremely important role of echolocation in their navigation and hunting.

More recent research has shown that dolphins, shrews, and certain other animals besides bats use sonar. In fact, it is likely that at least some of these species perceive shapes with their ears better than they do with their eyes.

The Lateral Line System in Fishes and Amphibians

Hair cells in the ear allow mammals to sense changes in air pressure that are perceived as sound. But in fishes and aquatic amphibians, hair cells in a different organ allow the perception of pressure changes in water. In most fishes and larval amphibians, groups of hair cells are embedded in gel-like domed structures called cupulae inside canals that run the length of the body (**FIGURE 47.8a**), forming a sensory organ called the **lateral line system.**

Pressure changes in the surrounding water—whether resulting from waves, an animal swimming nearby, or some other force—cause changes in the pressure of water moving through the lateral line system (**FIGURE 47.8b**). These changes cause kinocilia and stereocilia on the hair cells to bend, and the distortion leads to a change in action potentials along sensory neurons to the brain. In this way, most aquatic animals get information about pressure changes at specific points along the head and body.

What use do fishes make of the lateral line system? It seems reasonable that the lateral line system could be helpful for identifying mates, locating prey, or avoiding predators. However, fishes could also use vision, smell, or other senses for these functions. How important is the lateral line system?

To answer this question, Kirsten Pohlmann, Jelle Atema, and Thomas Breithaupt studied how nocturnal catfish locate prey. At night, the catfish cannot use vision to hunt smaller fishes, so they must use stimuli that persist in the wake of their prey after it has moved on. The researchers hypothesized that the nocturnal catfish hunt smaller fishes using the lateral line system.

(a) The lateral line system consists of a series of canals running along the head and body.

(b) Water enters the canals through pores and bends kinocilia on hair cells, activating sensory neurons.

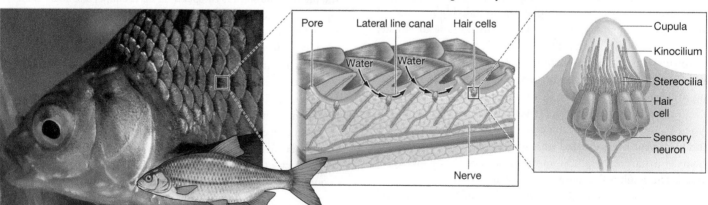

FIGURE 47.8 The Lateral Line System Detects Pressure Waves in Water. (a) Lateral line canals lie just under the fish's epidermis. **(b)** Hair cells within the canals have kinocilia and stereocilia embedded in a gel-like cupula.

To test this hypothesis, the researchers conducted an ablation experiment on the catfish (**FIGURE 47.9**). Ablation is the removal or blocking of a process. In this case, the researchers ablated the lateral line system in one group of catfish, using a chemical that blocks the hair cells from responding to pressure waves. In another group of catfish, they ablated the ability to smell or taste the water by surgically removing the lobes of the brain responsible for these senses.

While the catfish whose smell and taste were ablated took the same amount of time as the non-ablated control group to locate and capture prey, it took the catfish whose lateral line was ablated over twice as long. The researchers concluded that the lateral line system is much more important than other senses in the successful hunting of nocturnal fishes.

RESEARCH

QUESTION: What cues do predatory fish use to detect prey at night?

HYPOTHESIS: Predatory fish detect prey using the lateral line system.

ALTERNATE HYPOTHESIS: Predatory fish detect prey using smell and taste.

NULL HYPOTHESIS: Predatory fish use neither sense to detect prey.

EXPERIMENTAL SETUP:

1. Three experimental groups of catfish (*n*=16 for each group):

Lateral line ablated	Smell and taste ablated	Control fish

2. Acclimate catfish to dark tank at night.

3. Add guppies to tank and record capture success rate.

PREDICTION: Fish with their lateral line ablated will have lower capture success than control fish.

PREDICTION OF ALTERNATE HYPOTHESIS: Fish with smell and taste ablated will have lower capture success than control fish.

PREDICTION OF NULL HYPOTHESIS: There will be no capture success difference among groups.

RESULTS:

Percentage of Guppies Captured

Lateral line ablated	Smell and taste ablated	Control fish
17%	60%	65%

CONCLUSION: Nocturnal predatory catfish use the lateral line system to detect prey.

FIGURE 47.9 The Lateral Line System Is Used for Predation.

✔ **QUESTION** If you wanted to conduct a study similar to this one on a diurnal species of catfish, what other sense would you need to control for, and how would you do it?

SOURCE: Pohlmann, K., J. Atema, and T. Breithaupt. 2004. The importance of the lateral line in nocturnal predation of piscivorous catfish. *The Journal of Experimental Biology* 207: 2971–2978.

---

 check your understanding

If you understand that . . .

- Mechanoreception is a type of pressure detection that begins when the stereocilia on hair cells bend in response to changes in pressure. The bending movement opens ion channels and results in a change in membrane potential.
- The mammalian ear consists of specialized structures that function in transmitting and amplifying sound and in responding to specific frequencies as well as recognizing changes in intensity.
- The lateral line system in fishes and aquatic amphibians allows them to sense pressure waves in water caused by prey, predators, and potential mates.

✔ **You should be able to . . .**

1. Predict and explain the effect on hearing in each of the following cases: a punctured eardrum, a mutation that results in dramatically shortened stereocilia, and an age-related loss of flexibility in the basilar membrane.

2. Describe the function of the lateral line system in fishes.

Answers are available in Appendix A.

47.3 Photoreception: Sensing Light

Most animals have a way to sense light. The organs involved in **photoreception** range from simple light-sensitive eyespots in flatworms to the sophisticated, image-forming eyes of vertebrates, cephalopod mollusks, and arthropods.

Variation in the structure of light-sensing organs illustrates an important general principle about the sensory abilities of animals: In most cases, a species' sensory abilities correlate with the environment it lives in and its mode of life—how it finds food and mates. Eyes and other sensory structures are adaptations that allow individuals to thrive in a particular environment. Salamander species that live in meadows and forests have sophisticated eyes; those that live in lightless caves have no functional eyes at all.

Keep this point in mind as you delve into the details of how insects and vertebrates see.

The Insect Eye

Insects have a **compound eye,** composed of hundreds or thousands of light-sensing columns called **ommatidia.** As **FIGURE 47.10** shows, each ommatidium has a lens that focuses light onto a small number of receptor cells—usually four. The receptor cells, in turn, send axons to the brain.

Each ommatidium contributes information about one small piece of the visual field, not unlike the contribution of a single pixel on a computer monitor. Thus, the more ommatidia in a compound eye, the better the resolution—meaning the resolving power, or ability to distinguish objects.

In addition, the presence of many light-sensing columns makes species with compound eyes particularly good at detecting movement. Insects that hunt by sight, such as damselflies and dragonflies, have particularly large numbers of ommatidia.

The Vertebrate Eye

Compound eyes are found in insects, crustaceans, and certain other arthropods. Because they appear only in species that are part of the same monophyletic group (see Chapter 28), researchers conclude that this type of eye structure evolved just once—in an ancestor of today's arthropods.

In contrast, the **simple eye**—a structure with a single lens that focuses incoming light onto a layer of many receptor cells—evolved independently in several widely divergent groups, including annelids, cephalopod mollusks (squid and octopuses), and vertebrates. Let's examine the vertebrate version of the simple eye more closely.

The Structure of the Vertebrate Eye **FIGURE 47.11a** shows the major structures in a typical vertebrate eye:

- The outermost layer of the structure is a tough rind of white tissue called the sclera. This is the "white of the eye."

- The front of the sclera forms the **cornea,** a transparent sheet of connective tissue.

- The **iris** is a pigmented, round muscle just inside the cornea. The iris can contract or expand to control the amount of light entering the eye.

- The **pupil** is the hole in the center of the iris.

- Light enters the eye through the cornea and passes through the pupil and a curved, clear **lens.**

- Together, the cornea and lens focus incoming light onto the retina in the back of the eye. The **retina** contains a thin layer of cells with photoreceptors and several layers of neurons.

FIGURE 47.11b provides a closer look at the retina, which is attached to the rest of the eye by a single layer of pigmented epithelial cells. From back to front, the retina comprises three distinct cell layers:

1. The photoreceptors, sensory cells that respond to light, are held in place by the pigmented epithelium.

2. Photoreceptors synapse with an intermediate layer of connecting neurons called **bipolar cells.**

3. Bipolar cells in the intermediate neuron layer connect with one another and with neurons called **ganglion cells,** which form the innermost layer of the retina. The axons of the ganglion cells project to the brain via the **optic nerve.**

Note that vertebrate eyes—including yours—have a blind spot because there are no photoreceptor cells where the optic nerve leaves the retina. If light falls in this area, there are no sensory cells available to respond. No signal is sent to the brain, so the stimulus isn't seen.

What Do Rods and Cones Do? Early anatomists established that the photoreceptors in vertebrate eyes come in two distinct types: small rod-shaped cells and cone-shaped cells, called **rods** and **cones,** respectively (**FIGURE 47.12a**). When technical advances allowed changes in the membrane potentials of these cells to be recorded, it became clear that rods and cones differ in function as well as structure.

(a) Ommatidia are the functional units of insect eyes.

(b) Each ommatidium contains receptor cells that send axons to the CNS.

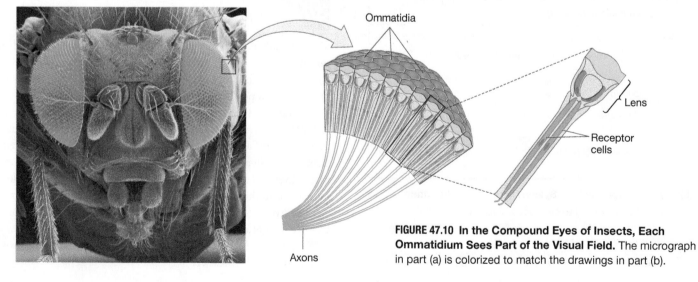

Ommatidia

Lens

Receptor cells

Axons

FIGURE 47.10 In the Compound Eyes of Insects, Each Ommatidium Sees Part of the Visual Field. The micrograph in part (a) is colorized to match the drawings in part (b).

(a) The structure of the vertebrate eye

(b) In the retina, cells are arranged in layers.

FIGURE 47.11 Simple Eyes Have a Single Lens That Focuses Incoming Light on Receptor Cells. (a) Light passes through the pupil of the eye and is focused onto the retina by the cornea and lens. **(b)** The photoreceptor cells that respond to light are in the "outermost" layer of the retina, furthest from the light source.

(a) Rods and cones contain stacks of membranes.

(b) Rhodopsin is a transmembrane protein complex.

(c) The retinal molecule inside rhodopsin changes shape when retinal absorbs light.

FIGURE 47.12 Rods and Cones Are Packed with Transmembrane Proteins That Contain the Pigment Retinal.
(a) Rods and cones have membranous disks containing thousands of opsin molecules (called rhodopsin in rods).
(b) Each opsin holds one retinal molecule. **(c)** Retinal changes conformation when it absorbs light. In response, opsin also changes shape.

✔**QUESTION** Explain why the change from *cis* to *trans* shown in part (c) would affect opsin's function.

Rods are sensitive to dim light but not to color. Cones, in contrast, are much less sensitive to faint light but are stimulated by different wavelengths (i.e., by colors). These discoveries explained why night vision is largely black and white—at night, the rods do most of the work.

Rods dominate most of the retina, but one small spot in the center of the retina has only cones. This is the **fovea** (see Figure 47.11a). When people focus on an object, their eyes move so that the image falls on the fovea of each eye. Based on these observations, biologists concluded that the high density of cones in the fovea maximizes the resolution of the image.

How Do Rods and Cones Detect Light?
As Figure 47.12a shows, rods and cones have segments that are packed with membrane-rich disks. The membranes contain large quantities of a transmembrane protein called **opsin.** Each opsin molecule is associated with a molecule of the pigment **retinal.** In rod cells, the two-molecule complex is called **rhodopsin (FIGURE 47.12b).**

Experiments with isolated retinal, opsin, and rhodopsin molecules confirmed that retinal changes shape when it absorbs light. Specifically, the number-11 carbon in the retinal molecule changes from the *cis* conformation to the *trans* conformation (**FIGURE 47.12c**). Retinal is a light switch.

The shape change that occurs in retinal triggers a series of events that culminate in a different stream of action potentials being sent to the brain. The sequence of events is unusual, though, because the receipt of a light stimulus does not open ion channels or trigger the release of a neurotransmitter to a sensory neuron.

In vertebrates, the molecular basis of vision is a shape change in retinal that closes ion channels and decreases the amount of neurotransmitter being released to the sensory neuron. In rod cells, electrical activity across the membrane, as well as neurotransmitter release, are maximized in the dark. When retinal has not been activated by light, sodium channels in the rod's plasma membrane are open (**FIGURE 47.13a**), and entry of sodium continually depolarizes the rod cells. Exposure to light transmits information by *inhibiting* both processes.

FIGURE 47.13b shows how shutting down happens:

1. Rhodopsin is activated when light causes retinal to change shape from the *cis* to *trans* conformation.

2. Rhodopsin activation causes a membrane-bound molecule called transducin to activate the enzyme phosphodiesterase (PDE).

3. PDE breaks down a nucleotide called cyclic guanosine monophosphate (cGMP) to guanosine monophosphate (GMP).

4. As cGMP levels decline, cGMP-gated sodium channels in the plasma membrane of the rod cell close.

5. When sodium channels close, Na^+ entry decreases and the membrane hyperpolarizes.

✔ If you understand this sequence of events, you should be able to explain why cGMP acts as both a second messenger and a ligand in this system, and how transducin compares to G

(a) Before stimulation (no light)

Rod

Disk membranes

cis

GDP

Rhodopsin (inactive)

Transducin (inactive)

PDE (inactive)

Na^+

cGMP

cGMP-gated sodium channel **(open)**

Plasma membrane of rod

Na^+ current depolarizes membrane

(b) After stimulation (light)

Light

trans

Rhodopsin (activated)

GTP

Transducin (activated)

PDE (activated)

cGMP   **GMP**

cGMP-gated sodium channel **(closed)**

Lack of Na^+ current hyperpolarizes membrane

FIGURE 47.13 A Signal-Transduction Pathway Connects Light Absorption to Changes in Membrane Potential. (a) An unstimulated photoreceptor. Notice that sodium ions flow into the cell when light is *not* being received. **(b)** A stimulated photoreceptor. Activation of rhodopsin leads to a reduction in cGMP concentration. With less cGMP available, cGMP-gated sodium channels close, and the membrane hyperpolarizes.

✔QUESTION The inflow of sodium ions into a photoreceptor cell is called "the dark current." Why?

proteins (introduced in Chapter 11). You should also be able to explain why shutting Na^+ channels results in hyperpolarization.

In response to the ensuing change in membrane potential, smaller quantities of the neurotransmitter glutamate are discharged at the synapse. The decrease in neurotransmitter indicates to the postsynaptic bipolar cell that the rod has absorbed

No color deficiency

Red–green color deficiency

FIGURE 47.14 People with Red–Green Color Blindness Cannot Distinguish Red from Green. These images show what a person with red–green color blindness would see, compared to a person with normal color vision.

light. As a result, a new pattern of action potentials is sent to the brain, via neurons called ganglion cells. Axons from ganglion cells are bundled into the optic nerve.

This system is exquisitely sensitive: Biologists have recorded a measurable change in the membrane potentials of rod cells in response to a single photon of light. But how do humans and other animals perceive color?

Color Vision: The Puzzle of Dalton's Eye   To answer this question, consider the research program initiated by John Dalton[1] in the late eighteenth century. At the age of 26, Dalton realized that he and his brother saw colors differently than other people did. To them, red sealing wax and green laurel leaves appeared to be the same color, and a rainbow exhibited only two hues. Dalton and his brother could not differentiate the colors red and green. This condition is called red–green color blindness (**FIGURE 47.14**).

In a lecture delivered in 1794, Dalton explained his perceptions by hypothesizing that red wavelengths failed to reach his retinas. Further, he hypothesized that because a normal eyeball is filled with clear fluid, and because blue fluids absorb red light, his defective vision resulted from the presence of bluish fluid rather than clear fluid in his eyes.

---

[1]Dalton was an accomplished physicist. He was the first proponent of the atomic theory of matter and formulated Dalton's law on the partial pressures of gases (introduced in Chapter 45). Red–green color blindness is sometimes called daltonism in his honor.

To test this hypothesis, Dalton left instructions that his eyes should be removed after his death and examined to see if the fluid inside was blue. When he died 50 years later, an assistant dutifully removed the eyes from Dalton's corpse and examined them. The fluid inside the eye was not blue at all, however, but slightly yellow—the normal color for an older person. Further, when the back was cut off one eye and colored objects were viewed through the lens, the objects looked perfectly normal. Dalton's hypothesis was incorrect.

Color Vision: Multiple Opsins   What, then, caused Dalton's color blindness? The key to answering this question was the discovery that the human retina contains three types of color-sensitive photoreceptors: blue, green, and red cones, named for the colors that they best perceive.

To follow up on this result, biologists analyzed opsin molecules from the three cell types and found that each had a distinct amino acid sequence. The three proteins are now called the blue, green, and red opsins (or S, M, and L, for short, medium, and long wavelengths, respectively). Although retinal is the light-absorbing molecule in all photoreceptor cells, the different opsin molecules cause each type to respond to a different range of wavelengths of light.

Based on these results, biologists hypothesized that the brain distinguishes colors by combining signals initiated by the three classes of opsins. **FIGURE 47.15**, for example, graphs how much light is absorbed across a range of wavelengths by the S, M, and

FIGURE 47.15 Color Vision Is Possible because Different Opsins Absorb Different Wavelengths of Light. Each human cone cell contains one of three different types of opsin. Each opsin absorbs a different range of wavelengths.

✔QUESTION The retinal molecules in S, M, and L opsins are identical. What is the likely reason that they respond to different wavelengths of light?

L opsins of humans. Notice that a wavelength of 560 nanometers (nm) stimulates L cones strongly, M cones to an intermediate degree, and S cones not at all. In response to the corresponding signals from these cells, the brain perceives the color yellow.

Does this hypothesis explain Dalton's color blindness? According to the data in Figure 47.15, wavelengths from green to red do not stimulate S opsin at all. It is thus unlikely that S opsin is involved in red–green color blindness. Did Dalton fail to distinguish red and green because his M or L cones were defective? Research has shown that red–green color-blind people lack either functional M or L cones, or both. Was the same true of Dalton?

This question was answered in the 1990s, when the genes for the M and L opsins were sequenced. Remarkably, Dalton's eyes had been preserved. Researchers managed to extract DNA from the 150-year-old tissue and analyze his M opsin genes and L opsin genes. They found that Dalton had a normal *L* allele but lacked a functional *M* allele. As a result, he did not have green-sensitive cones. The puzzle of Dalton's color vision was solved.

Red–green color blindness, or the inability to distinguish red and green due to an absence of either M or L cones, is estimated to affect about 5 to 10 percent of men and less than 0.5 percent of women. It is more prevalent in men than in women because it is an X-linked trait (see Chapter 14). In contrast, the inability to distinguish blue from other colors, which would result from absence of the S cone, is an extremely rare, autosomal dominant trait.

Do Other Animals See Color? What about other animals—do they see color the way humans do? The answer is, probably not. Animals that are active at night have relatively few cone cells and many rods, giving them high sensitivity to light but poor color vision. On the other hand, many vertebrate and invertebrate species have four or more types of opsins and probably perceive a world of colors that is much richer than ours.

In general, the types of opsins found in a species correlate with the environment it inhabits and its mode of life. For example,

- A marine fish called the coelacanth (pronounced *SEE-luh-kanth*), which lives in water 200 m deep, has two opsins that respond to the blue region of the spectrum (wavelengths of 478 nm and 485 nm). As a result, coelacanths perceive several distinct hues of blue that we would perceive as a single color. These opsins offered an adaptive advantage to coelacanths because wavelengths in the yellow and red parts of the spectrum do not penetrate well into deep water—only blue light exists in the coelacanth's habitat.

- In humans and other primates that eat fruit, two of the three opsins are sensitive to wavelengths around 550 nm. The presence of these opsins allows individuals to distinguish between the greens, yellows, and reds of unripe and ripe fruits.

- Many animals (for example, some insects and birds) can see ultraviolet light, which has shorter wavelengths than humans can see. This ability can be vital. Certain flowers have ultraviolet patterns that serve as signals for insect pollinators. Also, many birds have strong ultraviolet patterns in their plumage

that are invisible to us but are important criteria used by females of some species for selecting mates.

If you understand that . . .

- In vertebrates, light detection begins when retinal changes shape after absorbing light.
- Stimulation of rhodopsin or the photopsins by light causes conformation changes in retinal, which triggers a series of events resulting in the closing of sodium channels in the membrane of a photoreceptor cell. The resulting change in membrane voltage triggers a change in neurotransmitter release from the receptor cell.

✔ **You should be able to . . .**

1. Explain why retinal can be thought of as a "light switch."

2. Predict the type of vision loss that results from each of the following: detachment of the retina following a car accident, a mutation that knocks out the gene for S opsin, and an age-related clouding of the lens.

Answers are available in Appendix A.

47.4 Chemoreception: Sensing Chemicals

Chemoreception occurs when chemicals bind to chemoreceptors to initiate action potentials in sensory neurons. The sense of taste, called **gustation,** and the sense of smell, called **olfaction,** originate in chemoreceptors. Chemoreceptors detect the presence of particular molecules by undergoing a change in membrane potential when a specific compound is present. In this way, information about the presence of a particular chemical is transduced to an electrical signal in the body.

Until recently, taste and smell were poorly understood in comparison with vision and hearing. It is easy to see why. Eyes and ears respond to relatively simple stimuli—light and sound waves. The tongue and nose, in contrast, respond to thousands of different chemicals. Consequently, taste and smell were difficult to study until techniques became available for identifying how particular molecules bind to certain receptors. But in the past 20 years, research on the chemosenses has exploded.

Taste: Detecting Molecules in the Mouth

The chemoreceptor cells that sense taste are clustered in structures known as **taste buds.** Although humans have taste buds scattered around the mouth and throat, most taste buds are located on the tongue (**FIGURE 47.16**). A taste bud contains about 100 spindle-shaped taste cells, which contain taste receptors and synapse onto sensory neurons.

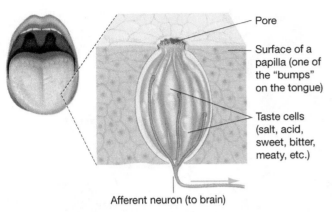

FIGURE 47.16 Taste Buds Contain Many Types of Chemoreceptors.

Labels in figure:
- Pore
- Surface of a papilla (one of the "bumps" on the tongue)
- Taste cells (salt, acid, sweet, bitter, meaty, etc.)
- Afferent neuron (to brain)

How do these receptors work on a molecular level, and how do they produce the sensation of taste? Early taste research focused on the hypothesis that four "basic tastes" existed: salty, sour, bitter, and sweet.

Salt and Sour Researchers who analyzed the membrane proteins in taste cells found strong evidence that salt and sour sensations result from the activity of ion channels.

- The sensation of saltiness is due primarily to sodium ions (Na^+) dissolved in food. These ions flow into certain taste cells through open Na^+ channels and depolarize the cells' membranes.

- Sourness is due in part to the presence of protons (H^+), which flow directly into certain taste cells through H^+ channels and depolarize the membrane.

The sour taste of grapefruit and other citrus fruits, for example, results from the release of protons by citric acid. In general, the lower the pH of a food, the more it depolarizes a taste cell's plasma membrane, and the more sour the food tastes.

Compared to salt and sour, the molecular mechanisms responsible for the sensations of bitterness and sweetness have been much more difficult to identify. Researchers have only recently been able to document that certain food molecules actually bind to specific receptors on taste cells to cause bitter and sweet tastes.

Why Do Many Different Foods Taste Bitter? Bitterness has been difficult to understand because molecules with very different structures, such as orange peel or unsweetened cocoa, are all perceived as bitter. How is this possible?

An answer began to emerge after researchers confirmed that some humans genetically lack the ability to taste certain bitter substances. In 1931, Arthur Fox was synthesizing phenylthiocarbamide (PTC) and accidentally blew some of it into the air. A nearby colleague complained of a bitter taste in his mouth, but Fox could not taste anything. Follow-up research confirmed that the ability to taste PTC is inherited and polymorphic. About 25 percent of Americans cannot sense the molecule.

To find the gene responsible for this trait, biologists compared the distribution of genetic markers observed in "tasters" and "nontasters." The mapping effort recently narrowed down this gene's location to several candidate chromosomal segments. (Chapter 20 introduced this type of gene hunt.) In one of the regions, researchers found a family of 40 to 80 genes that encode transmembrane receptor proteins.

Follow-up work has documented that each protein in the family binds to a different type of bitter molecule. A taste cell, however, can have many different receptor proteins from this family. As a result, many different molecules can depolarize the same cell and cause the sensation of bitterness.

Why are so many genes devoted to detecting bitterness? Many of the molecules that bind to these receptors are found in toxic plants; most animals react to bitter foods by spitting them out and avoiding them in the future. In essence, bitterness indicates, "This food is dangerous; don't swallow it."

The proliferation of genes responsible for detecting bitterness hints that the trait is extremely adaptive—meaning that individuals with the capacity to detect a wide array of bitter compounds produce more offspring than do individuals that lack this ability.

What Is the Molecular Basis of Sweetness and Other Tastes? Inspired by progress on bitter receptors, research teams used a similar approach—analyzing mutant mice that could *not* sense sweetness—to understand how the sweet receptor works.

In humans and mice, three closely related membrane receptors are responsible for detecting sweetness as well as glutamate and other amino acids. Glutamate triggers the sensation called **umami,** which is the meaty taste of the molecule monosodium glutamate (MSG). Glutamate is sensed by one particular pair of the three receptors. Sweetness is sensed by a different pair of receptors.

Recent work has solved a long-standing question about sweet sensation—why so many different types of sugars trigger the same sensation. As it turns out, a single receptor has binding sites for multiple types of sweet compounds, meaning that a variety of molecules can stimulate action potentials from the same receptor.

Currently, several teams are studying how the different taste sensations are conveyed to the brain and interpreted. Although taste is beginning to reveal its secrets, the complete story will probably not be known for many years.

Olfaction: Detecting Molecules in the Air

Taste allows animals to assess the quality of their food before swallowing it. Olfaction, in contrast, allows animals to monitor airborne molecules that convey information. Wolves and domestic dogs, for example, can distinguish millions of different airborne molecules at vanishingly small concentrations. The molecules that constitute odor contain information about the movements and activities of prey and other members of their own species.

Odorants Provide Information about the Environment Airborne molecules that convey information about food or the environment are called **odorants**. When they reach the nose, they diffuse

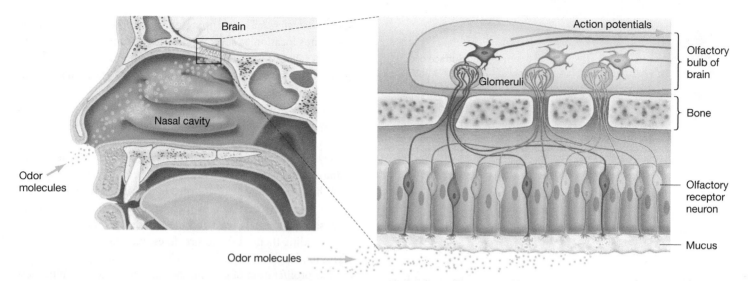

FIGURE 47.17 In Mammals, Chemoreceptor Cells in the Nose Respond to Specific Odorants. Each of the chemosensory neurons in the nose has one type of odor receptor protein on its dendrites. Sensory neurons with the same receptor project to the same glomerulus, or section within the olfactory bulb of the brain.

into a mucus layer in the roof of the nose (**FIGURE 47.17**). There, they activate olfactory neurons by binding to membrane-bound receptor proteins. Axons from these neurons project up to the **olfactory bulb,** the part of the brain where olfactory signals are processed and interpreted.

Understanding the anatomy of the odor-recognition system was a relatively simple task. Understanding how receptor neurons distinguish one molecule from another was much more difficult. Initially, investigators hypothesized that receptors respond to a small set of "basic odors," such as musky, floral, minty, and so on. The idea was that each basic odor would be detected by its own type of receptor, much like the way gustation works.

In 1991, Linda Buck and Richard Axel discovered a gene family in mice that comprises hundreds of distinct coding regions and encodes receptor proteins on the surface of olfactory receptor neurons. Follow-up experiments confirmed that each receptor protein binds to a small set of molecules.

Further work established that most, if not all, vertebrates possess this family of genes. Mammals typically have 800 to 1500 of these genes, meaning mammals could produce at least 800 to 1500 different olfactory receptor proteins. In humans, however, about half of these genes have mutations that probably render them nonfunctional. This observation may explain why the sense of smell is so poor in humans compared with that of most other mammals. But the sense of smell is even worse in whales and dolphins—70% to 80% of their olfactory receptor genes are nonfunctional. These animals rely far more on hearing and vision than on smell to sense their environment.

Buck and Axel's announcement inspired a series of questions. How many different receptors occur in the membrane of each neuron involved in odor reception? How does the brain make sense of the input from so many different receptors?

Recently, scientists determined that each olfactory neuron has only one type of receptor, and neurons with the same type of receptor are linked to distinct regions in the olfactory bulb of the brain. These regions are called **glomeruli** (meaning little balls).

Follow-up work from Axel's lab indicated that each smell recognized by mice is associated with the activation of a different subset of the 2000 glomeruli in the brain. For example, the activation of clumps 130, 256, and 1502 might be perceived as the smell "cinnamon." In essence, then, the sensing of odorants is similar to the visual system's use of three cones to perceive many colors; but odor reception works on a much larger scale.

Interestingly, odorant receptors recently have been identified in locations other than the typical olfactory epithelium—for example, in tissues of the heart and pancreas. Receptors have also been identified in the cell membranes of sperm cells, where they appear to play an important role in guiding the sperm toward the egg. Research on this complex and impressive sense continues at a furious pace.

Pheromones Provide Information about Members of the Same Species Recall from the introduction to this chapter that male silk moths are able to locate female moths from miles away. Males have much larger antennae than females, suggesting that they use these antennae in a sex-specific manner.

These factors led scientists to hypothesize that female moths release a chemical into the environment that binds to chemoreceptors on the males' antennae and acts as an attractant. In 1959, scientists identified this chemical and named it bombykol, after the scientific name of the silk moth (*Bombyx mori*). Bombykol was the first chemically characterized **pheromone,** meaning a chemical that is secreted into the environment and that affects the behavior or physiology of animals of the same species.

Thousands of pheromones have since been identified in invertebrates and vertebrates alike, performing such roles as alerting other members of a beehive to an intruder or signaling a male rodent that a female is ovulating. In insects, pheromones typically bind to receptors on the antennae. In tetrapod vertebrates, pheromone receptors are often localized in the **vomeronasal organ,** a sensory organ in the nasal region but distinct from the olfactory bulb. The vomeronasal organ and the olfactory bulb send signals to different parts of the brain, although some animals also sense odorants with the vomeronasal organ. For example, a snake may follow a scent trail of prey using its vomeronasal organ.

Do humans release pheromones? This is a hotly debated question. In 1971, Martha McClintock demonstrated that the menstrual cycles of women living in close contact with one another become synchronized as a result of a secretion from the women's armpits, but the responsible chemical has yet to be identified. Other studies have suggested that secretions from men's armpits can alter hormone levels in women, but both the amount of the chemical secreted and the reaction to it vary dramatically among men and women. Until scientists more fully understand the mechanisms by which these human "pheromones" achieve their effects, the jury is out.

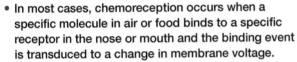

check your understanding

C Y U
If you understand that . . .

- In most cases, chemoreception occurs when a specific molecule in air or food binds to a specific receptor in the nose or mouth and the binding event is transduced to a change in membrane voltage.
- Odorants bind to chemoreceptors to initiate action potentials to the brain, where action potentials from specific receptors are interpreted as particular smells.
- Pheromones bind to chemoreceptors that send action potentials to brain regions that then affect the behavior or physiology of the animal.

✔ **You should be able to . . .**

1. Discuss why a loss in chemosensory ability occurs when you burn your tongue with extremely hot food.

2. Develop a hypothesis to explain why the vomeronasal organ appears to be vestigial in most primates.

Answers are available in Appendix A.

47.5 Other Sensory Systems

The stimuli and senses discussed so far are the ones you are likely most familiar with. But animals can sense much more than pressure waves, light, and chemicals. All animals can sense temperature and painful stimuli, and some can even perceive electric or magnetic fields. Let's start with temperature and examine each of these other stimuli and senses in turn.

Thermoreception: Sensing Temperature

Recall that many animals thermoregulate to maintain body temperature within an acceptable range (see Chapter 42). Virtually every physiological process, from digestion to metabolic rate, is temperature-dependent, so the ability to sense temperature changes in the environment and respond accordingly is crucial.

Thermoreception Helps Animals Thermoregulate Animals detect heat energy by **thermoreception** and adjust their behaviors or physiological processes, such as shivering and sweating, in response.

Some thermoreceptors are located in the central nervous system. In mammals, the hypothalamus is the brain region that senses departures from homeostatic body temperature and sends signals to effectors to restore homeostasis (see Figure 42.16).

Thermoreceptors also are commonly found on skin and other outer surfaces of animals, so that changes in the temperature of the environment can be sensed. As an example, several types of thermoreceptors have been identified in mammals. Some receptors depolarize in response to cooling, and others depolarize in response to heating. Picking up a cold object stimulates "cold receptors" in your skin, resulting in an increase in the rate of action potentials in sensory neurons that inform your brain that the object is cold.

Interestingly, extreme temperatures are sensed by a different type of receptor, called a nociceptor, that also senses other painful stimuli such as those produced by certain chemicals, excessive pressure, and tissue damage. If you touch a hot stove burner, the pain you feel arises primarily via stimulation of nociceptors.

Pit Vipers Have Extremely Sensitive Thermoreceptors The pit vipers are a group of snakes named after the two temperature-sensitive pits just beneath their nostrils (**FIGURE 47.18a**, see page 968). Inside each pit is a membrane lined with exquisitely sensitive thermoreceptors—a rattlesnake's thermoreceptors can sense changes in temperature as little as 0.003°C.

Pit vipers use these thermoreceptors to sense the heat energy given off by prey, predators, and even possible burrows in which to hide. The brains of these snakes may combine visual and thermal stimuli into a "thermal image" that might look something like an image from an infrared camera (**FIGURE 47.18b**). Even in complete darkness, rattlesnakes can strike prey with deadly precision.

✔ If you understand this, you should be able to predict whether a rattlesnake could strike effectively at its prey with (1) its eyes covered but pits exposed, (2) its eyes and pits covered with cotton cloth, and (3) its eyes and pits covered with an opaque heat-blocking material.

Electroreception: Sensing Electric Fields

All animals give off weak electrical impulses that arise from the activity of their nerves and muscles. Since water is a good conductor of these electrical impulses, many kinds of fishes use **electroreception,** or sensation of electric fields, to locate prey, detect predators, and navigate.

Sharks Use Electroreception to Hunt and Navigate In sharks, some parts of the lateral line system are specialized to detect electrical fields rather than pressure. Tiny pores scattered across a shark's head contain structures called **ampullae of Lorenzini**

(a) Pit vipers have temperature-sensitive pits.

(b) Warm animals emit infrared radiation.

FIGURE 47.18 Pit Vipers Use Thermoreception to Detect Prey. (a) Pits are lined with extremely sensitive thermoreceptors. **(b)** Pit vipers can detect infrared radiation given off by rodents and other prey.

(**FIGURE 47.19**). These ampullae are lined with hair cells that depolarize in response to a change in electrical polarity and send signals to the shark's brain via sensory neurons.

The sensitivity of these ampullae is remarkable—sharks can detect electrical currents as weak as a nanovolt (a billionth of a volt). This ability allows sharks to detect prey that are far away or even buried in the sand at the ocean floor. Combined with their exquisite sense of olfaction—sharks can detect a single drop of blood in a million drops of ocean water—electroreception makes them finely tuned predators.

Although data are scarce, scientists have some evidence to suggest that sharks also use electroreception to navigate. Ocean currents moving through the Earth's magnetic field generate weak electrical currents. Sharks have been observed in the ocean orienting themselves to these fields as well as to artificially created fields in the laboratory.

Electrogenic Fishes Generate Electric Fields Electrogenic fishes have specialized organs near their tails that generate

FIGURE 47.19 Ampullae of Lorenzini on Sharks' Heads Detect Electrical Fields to Help Locate Prey and Navigate.

electric fields stronger than those of regular nerves or muscles. The currents generated by the electric organ move in an arc through the water (**FIGURE 47.20**). Any item located within that arc will disrupt the currents, allowing the fish's electroreceptors to detect it. In this way, electrogenic fishes use their electric organs to locate prey, sense predators, navigate through murky water, and even communicate with other fishes.

Some electrogenic fishes have the ability to produce extremely strong currents that stun or kill their prey. The electric organs of electric eels take up over 80 percent of their body mass and can generate a 500-volt change in electrical potential and 1 ampere of current in the water around them. This amount of current is enough to kill a person swimming in water with an electric eel.

Magnetoreception: Sensing Magnetic Fields

The Earth produces a magnetic field as it rotates on its axis. Just as a compass uses this magnetic field to determine direction, animals may home in on magnetic fields while navigating. **Magnetoreception** has been described in many groups of organisms, including bacteria, fungi, invertebrates, and all vertebrate classes.

In general, studies of the mechanisms by which animals sense the Earth's magnetic fields are in their infancy. However, scientists are confident that at least several distinct mechanisms have evolved. Recall that sharks can sense magnetic fields indirectly via electric fields produced by ocean currents. Terrestrial organisms, in contrast, have the ability to sense magnetic fields directly.

In 1968, German scientists noticed that European robins being kept in the laboratory with no visual cues to the outside began to sit at one end of their cages at the beginning of the migratory season. This behavior led the scientists to hypothesize that birds use magnetoreception to determine direction as they migrate.

Since then, support for this hypothesis has accumulated. Application of artificial magnetic fields changes the robins' orientation. For example, when placed into circular chambers with artificial magnetic fields, birds always orient themselves in relation to the artificial field. Furthermore, disruption of the

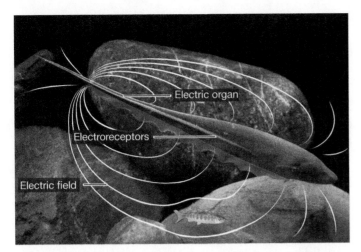

FIGURE 47.20 Electrogenic Fishes Can Create Strong Electric Fields. The fish emits an electric current with its electric organ. When a prey object disrupts the electric field, the fish detects its presence using electroreceptors.

magnetic field prevents birds from navigating properly. When scientists fitted homing pigeons with little caps that reversed the polarity of the magnetic field, the pigeons flew in the direction opposite to the one they were trained to fly in.

What enables birds to sense magnetic fields? It is likely a combination of factors. One hypothesis is that deposits of iron inside sensory neurons in the beak play a role in stimulating a response to changes in the magnetic field. In support of this hypothesis, cutting these neurons prevents birds from responding to artificial changes in the magnetic field in the lab. To complicate matters, however, magnetoreception in birds is apparently also dependent on vision. Covering the right eye—but not the left eye—of migrating birds interferes with their ability to navigate using magnetic cues.

This complication highlights a key feature of sensory perception: Animals do not use individual senses in isolation, but rather combine sensations of many types when locating prey, evading predation, or communicating with other animals. Whether it is a homing pigeon using magnetoreception and photoreception to find its way home, or a rattlesnake using a visual and thermal image to strike at a mouse, all animals depend on sensory systems that work together to provide the animal with the information it requires to survive.

check your understanding

If you understand that . . .

- Thermoreceptors, nociceptors, electroreceptors, and magnetoreceptors convey information about temperature, painful stimuli, electric fields, and magnetic fields, respectively.
- Animals combine input from multiple senses to locate prey, escape from predators, and communicate with other animals.

✔ **You should be able to . . .**

1. Identify one or more habitat types in which electrogenic fishes may have a survival advantage.

2. Propose a mechanism by which a female sea turtle migrates to lay eggs on the beach she was born on decades earlier.

Answers are available in Appendix A.

CHAPTER 47 REVIEW

For media, go to MasteringBiology

If you understand . . .

47.1 How Do Sensory Organs Convey Information to the Brain?

- If the membrane potential of a sensory cell is altered substantially enough in response to a stimulus, the pattern of action potentials that it sends to the brain changes. In this way, sensory stimuli as different as sound and light are transduced to electrical signals.

- The brain is able to distinguish different types of stimuli because axons from different types of sensory neurons project to different regions of the brain.

✔ You should be able to suggest a hypothesis to explain why people who have had limbs amputated experience "phantom pain"—the perception that their missing tissue hurts.

47.2 Mechanoreception: Sensing Pressure Changes

- Pressure receptors detect direct physical stimulation, including stimulation from sound.

- Hair cells, the major sensory detectors in the vertebrate ear, undergo a change in membrane potential in response to bending of their stereocilia.

- Sound waves of a certain frequency cause a certain part of the cochlea's basilar membrane to vibrate. Hair cells at this location stimulate action potentials in sensory neurons in response to the vibration.

- Hair cells in the lateral line systems of fishes and aquatic amphibians are stimulated by pressure changes in the water.

✔ You should be able to identify the key feature common to gravity sensing, hearing, and the lateral line system.

47.3 Photoreception: Sensing Light

- In the vertebrate eye, photoreceptors are located in rods and cones. Although these two cell types differ in structure and function, both contain molecules that consist of retinal paired with opsin (called rhodopsin in rod cells).

- The rhodopsin found in rods is stimulated by even the faintest light.

- Color vision is possible because cones contain opsins that respond to specific wavelengths of light absorbed by retinal.

- Humans distinguish colors based on the pattern of stimulation of three types of opsins found in cones. People who lack one of the functional cone opsins are color blind, meaning they cannot distinguish as many colors as people with all three opsins can.

✔ You should be able to explain how different animal species are able to see different colors.

47.4 Chemoreception: Sensing Chemicals

- Chemoreceptors detect the presence of certain foodborne or airborne molecules.

- Taste buds contain taste cells with membrane proteins that respond to toxins, salt, acid, and other types of molecules in food. Sodium ions and protons enter taste cells via channels and depolarize the membrane directly, producing the sensations of saltiness and acidity. Sugars and some toxic compounds bind to membrane receptors and trigger action potentials that are interpreted by the brain as sweet and bitter flavors, respectively.

- Smell, or olfaction, is used to scan molecules from the outside environment. Airborne chemicals are detected by hundreds of different odor-receptor proteins located in the membranes of receptor cells in the nose.

✔ You should be able to explain why people with nasal congestion complain that food tastes bland, and why the brain can perceive so many different flavors based on inputs from just four or five basic types of taste receptors.

47.5 Other Sensory Systems

- Thermoreceptors respond to changes in temperature.

- Nociceptors respond to painful stimuli, including extreme temperatures, certain chemicals, excessive pressure, and tissue damage.

- Electroreceptors contain modified hair cells that respond to electric fields.

- Magnetoreceptors respond to magnetic fields and are often used in navigation and orientation.

✔ You should be able to explain why the chemical capsaicin, found in chili peppers, feels hot on the tongue.

(MB) MasteringBiology

1. MasteringBiology Assignments

Tutorials and Activities Structure and Function of the Eye; Vertebrate Eye

Questions Reading Quizzes, Blue-Thread Questions, Test Bank

2. eText Read your book online, search, take notes, highlight text, and more.

3. The Study Area Practice Test, Cumulative Test, BioFlix® 3-D Animations, Videos, Activities, Audio Glossary, Word Study Tools, Art

You should be able to . . .

✔ **TEST YOUR KNOWLEDGE**   *Answers are available in Appendix A*

1. What is echolocation?
   a. use of echoes from high-frequency vocalizations to detect objects
   b. use of extremely low-frequency vocalizations to communicate over long distances
   c. vision based on input from many independent lenses, functioning like pixels on a computer screen
   d. variation in the structure of opsin proteins, which allows animals to see different colors

2. In the human ear, how do different hair cells respond to different frequencies of sound?
   a. Waves of pressure move through the fluid in the cochlea.
   b. Hair cells are "sandwiched" between membranes.
   c. Receptors in the stereocilia of each hair cell are different; each receptor protein responds to a certain range of frequencies.
   d. Because the basilar membrane varies in stiffness, it vibrates in certain places in response to certain frequencies.

3. Which of the following comparisons of rods and cones is *false*?
   a. Most human eyes have one type of rod and three types of cones.
   b. Rods are more sensitive to dim light than cones are.
   c. Nocturnal animals have fewer rods than diurnal animals.
   d. Both rods and cones use retinal and opsins to detect light.

4. Which of the following statements about taste is *true*?
   a. Sweetness is a measure of the concentration of hydrogen ions in food.
   b. Sodium ions from foods can directly depolarize certain taste cells.
   c. All bitter-tasting compounds have a similar chemical structure.
   d. Membrane receptors are involved in detecting acids.

5. True or false: A rattlesnake's facial pits detect the presence of prey using photoreception.

6. What type of sensory system do migrating birds use to detect direction?

7. Considering that sounds and odors both trigger changes in the patterns of action potentials from sensory cells, how does the brain perceive which sense is which when the action potentials reach the brain?
   a. The action potentials stimulated by sounds are different in size and shape from those stimulated by an odor.
   b. The axons from different sensory neurons go to different areas of the brain.
   c. Mechanoreception is not consciously perceived by the brain, whereas chemoreception is.
   d. Chemoreception is not consciously perceived by the brain, whereas mechanoreception is.

8. Give three examples of how the sensory abilities of an animal correlate with its habitat or method of finding food and mates.

9. QUANTITATIVE Scientists collected data on the date of onset of the menstrual cycles of a group of women who moved into a college dormitory together in the fall. The *y*-axis of the graph shows the average difference (in days) between the onset of a woman's cycle and the average onset date of the rest of the women. Describe how these data provide evidence for the existence of a human pheromone.

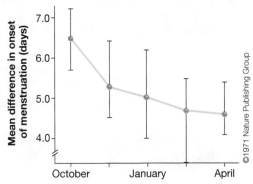

DATA: McClintock, M. K. 1971. *Nature* 229: 244–245, Figure 1.

10. Explain how the bending of stereocilia in a hair cell and binding by a bitter-tasting molecule can both result in an ion channel opening.

11. Scientists generally think that a "good hypothesis" is one that is reasonable and testable and inspires further research into the phenomenon. Using these criteria, was Dalton's hypothesis about color vision a good hypothesis? Was it correct? Explain your answer.

12. Design experiments to test the hypothesis that electric eels are both electrogenic and electroreceptive.

13. Which of the following animals is *least* likely to have a well-developed vomeronasal organ?
    a. fish
    b. salamander
    c. rodent
    d. snake

14. Compare and contrast the lateral line system of fishes with electroreception in sharks.

15. A company is marketing perfumes that it claims are attractive to the opposite sex because they contain "pure reaction-grade human pheromones." Explain why this is unlikely to be truth in advertising.

16. Houseflies have about 800 ommatidia in each of their compound eyes. Dragonflies, in contrast, have up to 10,000 ommatidia per eye. Houseflies feed by lapping up watery material from piles of excrement or rotting carcasses, which they locate by scent. Dragonflies are aerial predators and hunt by sight. How would you test the hypothesis that the large numbers of ommatidia in dragonflies make them more efficient hunters?

48 Animal Movement

In this chapter you will learn that

Muscle movement was a key innovation in the evolution of animals

↓ starting with

| Nervous system Ch. 46 | → initiates | Muscle contraction 48.1 |

↓ in

| Muscle tissues 48.2 | → can be | Smooth, cardiac, skeletal |

↓ combine with

| Skeletal systems 48.3 | → can be | Hydrostat, exoskeleton, endoskeleton |

↓ enable

| Locomotion 48.4 | on/in | Land, water, air |

Basilisk lizards are able to run on water—literally. This impressive escape strategy demonstrates the extent to which muscle-generated movements have diversified among animals. For most animals, complex muscle movements make the difference between life and death.

This chapter is part of the Big Picture. See how on pages 840–841.

You may have discovered while studying biology that plants and animals are more similar to each other than they initially appear. For example, plants and animals both require water and nutrients in specific quantities, have highly specialized tissues and complex reproductive structures, and launch defenses against parasites and predators. You can see a comparison of these and other traits in the Big Picture of Plant and Animal Form and Function on pages 840–841.

However, animals possess a quality that clearly distinguishes them from plants and other organisms: movement by virtue of muscle contractions. Muscle-generated movements were a key innovation in animal evolution. Rapid movements, along with sophisticated sensory structures

✔ When you see this checkmark, stop and test yourself. Answers are available in Appendix A.

(Chapter 47) and complex information processing systems (Chapter 46), were vital to animal diversification (Chapter 33), and these attributes made animals efficient eating machines in diverse ecosystems.

Muscles generate movement by exerting force and causing shape changes. Movement falls into two general categories:

1. *Movement of the entire organism relative to its environment.* **Locomotion,** movement of an animal under its own power, enables animals to seek food, water, mates, and shelter as well as avoid predators. Modes of locomotion include undulating, jetting, swimming, walking, running, jumping, gliding, and flying (introduced in Chapter 33).

2. *Movement of one part of the animal relative to other parts (not involved in locomotion).* This type of movement also has important functions—for example, to ventilate gills for gas exchange and to grasp prey. Even sessile organisms like sea anemones and barnacles use complex, muscle-generated movements to survive and reproduce.

How do animals accomplish their spectacular movements? This chapter starts by probing into the mechanism of muscle contraction that serves as the "engine" for most animal movements and then considers how muscle and skeletal systems work together to produce locomotion. The latter discussion introduces a research field called **biomechanics,** in which the principles of physics and engineering are applied to questions about the mechanical structure and function of organisms. Let's jump in.

48.1 How Do Muscles Contract?

The mechanism responsible for the contraction of muscle has fascinated and perplexed scientists for many centuries. Before the advent of microscopes and modern research techniques, scientists could only speculate about what makes muscles contract and relax.

Early Muscle Experiments

In the second century C.E., the Roman physician and philosopher Galen proposed that spirits flowed from nerves into muscles, inflating them and increasing their diameter. This "inflation" hypothesis persisted into the seventeenth century, when the French philosopher René Descartes suggested that nerves carry fluid from the pineal gland—which is in the brain and was considered to be the seat of the soul—to the muscles, making them shorten and swell.

Later that century, Dutch anatomist Jan Swammerdam tested Descartes' inflation hypothesis with a simple yet elegant experiment. He placed a piece of frog muscle into an airtight syringe with the nerve protruding through a small hole in the side and a small drop of water in the tip. He then stimulated the nerve, causing the muscle to contract. If the muscle's volume changed during contraction, the drop of water in the tip of the syringe would move. But it did not. The volume of the muscle remained constant.

Swammerdam's experiment demonstrated an important point: The contraction mechanism is inherent to the muscle itself—muscle is not like a balloon, and the nerve is not like a water hose filling a balloon. This insight was confirmed by Italian scientist Luigi Galvani in the 1790s, when he severed the nerve to a frog's leg muscle and then connected the two sides of the cut with a metal conductor. Contraction occurred. He concluded that the nerve and muscle possess "animal electricity" that can induce contraction.

If the shape of muscle does not change by inflation, what is the mechanism of muscle contraction?

The Sliding-Filament Model

Early microscopists established that the muscle tissue in vertebrate limbs and hearts is composed of slender fibers. A **muscle fiber** is a long, thin muscle cell. Within each of the muscle cells are many threadlike, contractile filaments called **myofibrils.** The myofibrils inside muscle fibers often look striped or striated due to the alternating light–dark units called **sarcomeres,** which repeat along the length of a myofibril (**FIGURE 48.1**).

FIGURE 48.1 Muscle Cells Contain Many Myofibrils, Which Contain Many Sarcomeres. Skeletal muscle cells (fibers) have a striped appearance due to repeating sarcomeres, which are units of alternating light–dark bands.

The microscopists observed that sarcomeres shorten as myofibrils contract. Sarcomeres then lengthen when the cell relaxes and an external force stretches the muscle. Based on these observations, it became clear that the question of how muscles contract simplifies to the question of how sarcomeres shorten.

Biologists knew that the answer must involve the two types of protein that had been found in sarcomeres: **actin** and **myosin**. But they did not know the shapes of these molecules or how they were arranged within the sarcomere. Did both types of molecules span the entire length of the sarcomere? Or were the filaments restricted to certain bands within the sarcomere?

In 1952, biologist Hugh Huxley produced electron micrographs of sarcomeres in cross section. He observed that there were two types of filaments, **thin filaments** and **thick filaments**, and that these filaments overlapped in the dark bands but not in the light bands. Huxley and his collaborator Jean Hanson also observed that sarcomeres stripped of their myosin had no dark bands. They concluded that the thick filaments must be composed of myosin, and the thin filaments must be composed of actin.

How did myosin and actin interact to shorten the sarcomere? In 1954, Huxley and Hanson hit on the key insight when they observed how the light and dark bands in sarcomeres changed when the muscles were relaxed versus contracted. Overall, the width of the dark bands did not change during a contraction, but the light bands became narrower.

To explain these observations, Huxley and Hanson proposed the **sliding-filament model** illustrated in **FIGURE 48.2**. The hypothesis was that the filaments slide past one another during a contraction. That is, the sarcomere can shorten with no change in lengths of the thin and thick filaments themselves:

- The distance from point A to point C does not change, and the distance from point B to point D does not change.

- Points A and B move closer to each other during contraction, as do points C and D.

Another pair of researchers, Andrew F. Huxley (no relation to Hugh) and Rolf Niedergerke, published the same result at the same time. Follow-up research has shown that the Huxley–Hanson model is correct in almost every detail.

Structurally, thin filaments are composed of two coiled chains of actin, a common component of the cytoskeleton of eukaryotic cells (see Chapter 7). One end of each thin filament is anchored to a structure called the **Z disc,** which forms the end wall of the sarcomere. The other end of a thin filament is free to interact with thick filaments. Thick filaments are composed of multiple strands of myosin. They span the center of the sarcomere and are free at both ends to interact with thin filaments.

To appreciate how the sliding-filament model works, consider the following analogy: Two large trucks are parked 50 m apart, facing each other. Each has a long rope attached to the front bumper. Six burly weightlifters stand in a line in front of each truck, grab onto the rope, and pull, hand over hand, so that the two trucks roll toward one another. ✔ If you understand the model, you should be able to explain which elements in the analogy represent the Z discs, which are the thin filaments, and which elements are the thick filaments.

How Do Actin and Myosin Interact?

How does this sliding action occur at the molecular level? Early work on the three-dimensional structure of myosin revealed that each myosin molecule is made up of a pair of subunits with "tails" coiled around one another and "heads" bent to the side. Each myosin head can bind to actin, and the head region can catalyze the hydrolysis of ATP into adenosine diphosphate (ADP) and a phosphate ion. These results suggested that myosin—not actin—was the site of active movement.

In addition, electron microscopy revealed that myosin and actin are locked together shortly after an animal dies, and its muscles enter the stiff state known as rigor mortis. Because ATP is unavailable in dead tissue, the data suggested that ATP is involved in getting myosin to release from actin once the two molecules have bound to each other.

Later, Ivan Rayment and colleagues solved the detailed three-dimensional structure of the myosin head (**FIGURE 48.3**). Using X-ray crystallographic techniques (see **BioSkills 11** in Appendix B), Rayment's group determined the location of the actin-binding site and examined how myosin's structure

Relaxed sarcomere

Thin filament (actin)   Thick filament (myosin)   Z disk

A   B   C   D

Contracted sarcomere

A   B   C   D

FIGURE 48.2 The Sliding-Filament Model Explains Important Aspects of Sarcomere Contraction. When a sarcomere contracts, the lengths of the thin filaments (distance from A to C) and thick filaments (distance from B to D) do not change. Rather, the filaments slide past one another.

✔**QUESTION** According to the model shown here, why is the dark band in a sarcomere (see Figure 48.1) dark and the light band light?

Myosin head

ATP binding site

Colors indicate different protein subunits of myosin head

Actin binding site

FIGURE 48.3 Myosin's "Head" Binds ATP and Actin. Each myosin molecule consists of two subunits with their tails coiled together and their two heads exposed. The heads contain binding sites for ATP and actin; when one of these subunits binds, its myosin head changes shape.

changed when ATP or ADP bound to it. As predicted, the protein's conformation changed significantly when bound to ATP versus ADP.

Based on these data, Rayment and co-workers proposed a four-step model for actin–myosin interaction (**FIGURE 48.4**):

Step 1 ATP binds to the myosin head, causing a conformational change that releases the head from the actin in the thin filament.

Step 2 When ATP is hydrolyzed to ADP and inorganic phosphate, the neck of the myosin straightens and the head pivots. The myosin head then binds to a new actin subunit farther down the thin filament. In this position, the myosin head is "cocked" in its high-energy state, ready for the power stroke.

Step 3 When inorganic phosphate is released, the neck bends back to its original position. This bending, called the power stroke, moves the entire thin filament relative to the thick filament.

Step 4 After ADP is released, the myosin head is ready to bind to another molecule of ATP.

As ATP binding, hydrolysis, and release continue, the two ends of the sarcomere are pulled closer together. (The transition from step 4 to step 1 cannot occur after death. Rigor mortis sets in as ATP supplies run out.)

The same basic ratcheting mechanism between actin and myosin is responsible for the amoeboid movement observed in amoebae and slime molds (see Chapter 30) as well as the streaming of cytoplasm observed in algae and land plants. Actin and myosin have played a critical role in the diversification of eukaryotes because they make movement possible in the absence of cilia and flagella.

PROCESS: MODEL FOR ACTIN–MYOSIN INTERACTION

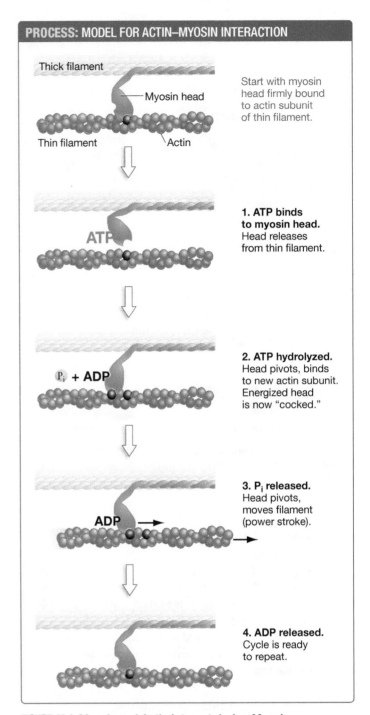

Thick filament

Myosin head

Thin filament — Actin

Start with myosin head firmly bound to actin subunit of thin filament.

ATP

1. ATP binds to myosin head. Head releases from thin filament.

P_i + ADP

2. ATP hydrolyzed. Head pivots, binds to new actin subunit. Energized head is now "cocked."

ADP →

3. P_i released. Head pivots, moves filament (power stroke).

4. ADP released. Cycle is ready to repeat.

FIGURE 48.4 Myosin and Actin Interact during Muscle Contraction. Summary of the current model of how myosin and actin interact as a sarcomere contracts. The four steps repeat rapidly. (Only one head of the myosin molecule is shown, for simplicity.)

Considering that ATP is almost always available in living muscles, how do muscles ever stop contracting and relax? Besides actin, thin filaments contain two key proteins called **tropomyosin** and **troponin.** Tropomyosin and troponin work together to block the myosin binding sites on actin. The myosin–actin interaction cannot then occur, and thick and thin

(a) Muscle relaxed: Tropomyosin and troponin work together to block the myosin binding sites on actin.

P_i + ADP

Troponin
Tropomyosin
Actin

Myosin head
Myosin binding sites (blocked)

(b) Contraction begins: When a calcium ion binds to troponin, the troponin–tropomyosin complex moves, exposing myosin binding sites.

P_i + ADP

Calcium ion
Troponin–tropomyosin complex, moved

FIGURE 48.5 Troponin and Tropomyosin Regulate Muscle Activity. Note that the myosin head is in its energized state when a muscle is relaxed.

filaments cannot slide past each other. As a result, the muscle relaxes (**FIGURE 48.5a**).

But when calcium ions bind to troponin, the troponin–tropomyosin complex moves in a way that exposes the myosin binding sites on actin. As **FIGURE 48.5b** shows, myosin then binds and contraction can begin.

How are calcium ions released so that contraction can begin? The process begins with the arrival of an action potential—an electrical signal from a motor neuron.

How Do Neurons Initiate Contraction?

You are probably sitting as you read. If so, contract your calf muscles to point your toes. Your nervous system just played a critical role in controlling the timing of your muscle contractions. First, your central nervous system—your brain—received input from an array of sensory cells in your peripheral nervous system, such as the ones in the retina of your eyes as you read this paragraph. Then your brain integrated this information and triggered action potentials in the motor neurons of the peripheral nervous system of your legs. (See Chapter 46 to review the structure and function of the nervous system.)

FIGURE 48.6 summarizes what happens when an action potential from a motor neuron arrives at a muscle cell:

PROCESS: HOW DO ACTION POTENTIALS TRIGGER MUSCLE CONTRACTION?

Motor neuron
Muscle cell

Motor neuron
Action potential
ACh
ACh receptor
Action potentials
T tubule
Sarcoplasmic reticulum
Thick filaments (myosin) Thin filaments (actin) Ca^{2+} ions

1. Action potential arrives; acetylcholine (ACh) is released from motor neuron.

2. ACh binds to ACh receptors on the muscle cell, triggering depolarization that leads to action potential.

3. Action potentials propagate across muscle cell's plasma membrane and into interior of cell via T tubules.

4. Ca^{2+} channels open in sarcoplasmic reticulum in response to action potentials in T tubules.

5. Ca^{2+} is released from sarcoplasmic reticulum. Sarcomeres shorten when troponin and tropomyosin move in response to Ca^{2+} and expose myosin binding sites in the thin filaments (see Figure 48.5).

FIGURE 48.6 Action Potentials Trigger Ca^{2+} Release. Action potentials at the neuromuscular junction trigger the release of Ca^{2+}, which binds to troponin–tropomyosin and allows myosin to form a cross-bridge with actin.

1. Action potentials trigger the release of the neurotransmitter **acetylcholine** from the motor neuron into the synaptic cleft between the motor neuron and the muscle cell. (See Figure 46.12 on page 939 for a review of the synaptic cleft.)

2. Acetylcholine diffuses across the synaptic cleft and binds to acetylcholine receptors on the plasma membrane of the muscle cell. By recording voltage changes in muscle cells, biologists showed that a membrane depolarization occurs in response to the binding of acetylcholine. If enough acetylcholine is applied to a muscle cell, depolarization triggers action potentials in the fiber itself.

3. The action potentials propagate along the length of the muscle fiber and spread into the interior of the fiber via invaginations of the muscle cell membrane called **T tubules.** (The T stands for transverse, meaning extending across.)

4. T tubules intersect with extensive sheets of smooth endoplasmic reticulum called the **sarcoplasmic reticulum.** When an action potential passes down a T tubule and reaches one of these intersections, a protein in the T-tubule membrane changes conformation and opens calcium channels in the sarcoplasmic reticulum.

5. Calcium causes the myosin binding sites on the actin filaments to be exposed, enabling the contraction to begin.

By this series of events, the interaction of the nervous system and muscle tissue at the neuromuscular junction precisely regulates muscle contractions—and thus complex movement.

check your understanding

If you understand that . . .
- Muscles shorten when thick filaments of myosin slide past thin filaments of actin in a series of binding events mediated by the hydrolysis of ATP.
- Muscle cells shorten in response to action potentials, which trigger the release of calcium ions that enable actin and myosin to interact.

✔ **You should be able to . . .**
1. Describe the sliding-filament model.
2. Predict the effect on muscle function of drugs that have the following actions: increase acetylcholine release at the neuromuscular junction, prevent conformational changes in troponin, and block uptake of calcium ions into the sarcoplasmic reticulum.

Answers are available in Appendix A.

48.2 Muscle Tissues

How do muscle cells and muscle tissues vary? After years of careful anatomical study, biologists concluded that animals have three classes of muscle tissue: (1) smooth muscle, (2) cardiac muscle, and (3) skeletal muscle. You, and all other vertebrates, have all three.

Smooth, cardiac, and skeletal muscle share several properties. They all contract as described by the sliding-filament model, and they all contract in response to electrical stimulation. However, the three classes of muscle also differ in important ways, summarized in **TABLE 48.1** (see page 978):

- *Voluntary versus involuntary* **Voluntary muscles** can contract in response to conscious thought (and also by unconscious reflexes) and are stimulated by neurons in the somatic division of the peripheral nervous system. **Involuntary muscles** contract only in response to unconscious electrical activity and are stimulated and inhibited by neurons in the autonomic division of the peripheral nervous system. (See Chapter 46 for a review of the somatic and autonomic divisions of the nervous system.)

- *Multinucleate versus uninucleate* Muscle cells may have one or many nuclei depending on the size of the cells.

- *Striated versus unstriated* In some muscle cells, the actin and myosin filaments are aligned in rows forming sarcomeres, giving the cells and tissues a banded appearance; for this reason, it is often called **striated muscle.** Other muscle cells are unstriated.

Let's apply these characteristics to each class of muscle tissue in more detail.

Smooth Muscle

Smooth muscle cells are unbranched, tapered at each end, and often organized into thin sheets. They lack the sarcomeres that are found in skeletal and cardiac muscle; hence, they are unstriated and appear smooth. Smooth muscle cells are relatively small and have a single nucleus.

Smooth muscle is essential to the function of the lungs, blood vessels, digestive system, urinary bladder, and reproductive system. Bronchioles in the lungs have a layer of smooth muscle that controls the size of airways; similarly, smooth muscles in the blood vessels can contract or relax to alter blood-flow patterns and blood pressure (Chapter 45). Layers of smooth muscle in the gastrointestinal tract help mix and move food (Chapter 44), and uterine smooth muscle is responsible for expelling the fetus during birth (see Chapter 50).

Smooth muscle is innervated by autonomic motor neurons, and it is thus involuntary. Acetylcholine from parasympathetic ("rest-and-digest") neurons aids digestion by stimulating contraction of smooth muscle in the stomach and intestine (Chapter 46).

In contrast, sympathetic ("fight-or-flight") neurons release the neurotransmitter **norepinephrine;** and the adrenal glands adjacent to the kidneys release the hormone **epinephrine,** also called **adrenaline** (see Chapter 49). Norepinephrine and epinephrine have the opposite effect to that of acetylcholine: they inhibit contraction of muscle in the stomach and intestine. Different smooth muscles are stimulated or inhibited by autonomic motor neurons depending on the type of neurotransmitter and neuron receptors they have.

	Smooth Muscle	Cardiac Muscle	Skeletal Muscle
	25 μm	25 μm	25 μm
Location	Intestines, arteries, other	Heart	Attached to the skeleton
Function	Move food, help regulate blood pressure, etc.	Pump blood	Move skeleton
Cell characteristics		Intercalated discs	Nuclei
	Single nucleus	1 or 2 nuclei	Multinucleate
	Unstriated	Striated	Striated
	Unbranched	Branched; intercalated discs form direct cytoplasmic connection end to end	Unbranched
	No sarcomeres	Contains sarcomeres	Contains sarcomeres
	Activity is "involuntary," meaning that signal from motor neuron is not required	Activity is "involuntary," meaning that signal from motor neuron is not required	Activity is "voluntary," meaning that signal from somatic motor neuron is required

Cardiac Muscle

Cardiac muscle makes up the walls of the heart and is responsible for pumping blood throughout the body. Unlike smooth muscle, cardiac muscle cells contain sarcomeres and are striated. Further, cardiac muscle cells have a unique branched structure, and they are directly connected end-to-end via specialized regions called intercalated discs. These discs are critical to the flow of electrical signals from cell to cell and thus to the coordination of the heartbeat (Chapter 45).

Like smooth muscle, cardiac muscle is involuntary—it contracts following spontaneous depolarizations. During rest, parasympathetic neurons release acetylcholine onto the heart. This neurotransmitter slows down the rate of depolarization of cardiac cells. The result is a slower heart rate.

During exercise, or when an animal is frightened, stressed, or otherwise stimulated, sympathetic neurons release norepinephrine onto the heart, and the adrenal gland releases epinephrine. Norepinephrine and epinephrine have an opposite effect to that of acetylcholine: they increase heart rate and strengthen the force of cardiac muscle contraction. The result is that more blood is pumped from the heart—an essential component of the fight-or-flight response of the sympathetic nervous system (Chapter 46).

Skeletal Muscle

Skeletal muscle consists of exceptionally long, unbranched muscle fibers. For example, a muscle fiber of a cat may be 0.4 mm wide and 40 mm long—enormous compared to most cells. These large cells result from the fusion of many smaller embryonic cells during development, accounting for the multiple nuclei spread out along the cell. Each muscle fiber is packed with myofibrils, each of which may contain thousands of sarcomeres, giving skeletal muscle its striated appearance.

Skeletal muscle is so named because it usually attaches to the skeleton. When skeletal muscle contracts, it exerts a pulling force on the skeleton, causing it to move—powering the sprint of cheetahs, the flight of hummingbirds, and the pinch of crab claws. In addition, skeletal muscle encircles the openings of the digestive and urinary tracts and controls swallowing, defecation, and urination.

A significant fraction of the body of many animals is composed of skeletal muscle. For example, 63 percent of the body weight of trout is skeletal muscle, and mammals—including humans—of all sizes are 40 to 45 percent muscle. Clearly, skeletal muscle plays an important role in animal biology. Some of the major skeletal muscles in the human body are shown in **FIGURE 48.7**.

FIGURE 48.7 Major Skeletal Muscles Make Up a Large Portion of the Human Body.

Labels (left figure, posterior view, top to bottom):
Trapezius
Deltoid
Triceps brachii
Latissimus dorsi
Gluteus maximus
Biceps femoris (hamstrings)
Gastrocnemius
Soleus
Achilles tendon

Labels (right figure, anterior view, top to bottom):
Pectoralis major
Biceps brachii
Rectus abdominus
Sartorius
Rectus femoris (quadriceps)

Skeletal-Fiber Types

Slow fiber	Fast fiber	Intermediate fiber
Red	White	Pink or red
High myoglobin concentration	Low myoglobin concentration	High myoglobin concentration
Derive most ATP via aerobic respiration (slow oxidative)	Derive most ATP via glycolysis (fast glycolytic)	Derive ATP from glycolysis and aerobic respiration (slow glycolytic)
Many mitochondria	Few mitochondria	Many mitochondria
Slow twitch	Fast twitch	Intermediate twitch
Fatigues slowly	Fatigues quickly	Intermediate fatigue

The twitch graphs show Tension (y-axis) vs. Time (msec) (x-axis) for each fiber type.

Skeletal muscle is distinguished from cardiac and smooth muscle in being voluntary. Skeletal muscle must be stimulated by somatic motor neurons to contract (Chapter 46). If these motor neurons are damaged, as can occur with a spinal cord injury, skeletal muscle cannot contract and becomes paralyzed.

Although all muscles contract as described by the sliding-filament model, not all skeletal-muscle fibers have the same contractile properties. The force output of skeletal muscles depends on (1) the relative proportion of different fiber types, (2) the organization of fibers within the muscle, and (3) how the muscle is used. Let's take a closer look at these sources of variation in muscle performance.

Skeletal-Fiber Types Skeletal muscle fibers can be divided into general types based on their structural and functional characteristics, summarized in **TABLE 48.2**:

- **Slow muscle fibers** (slow oxidative fibers) appear red because they contain a high concentration of myoglobin, an iron-bearing pigment that carries oxygen (similar to but distinct from hemoglobin in the blood). Slow fibers contract slowly because the myosin hydrolyzes ATP at a slow rate. They also fatigue slowly because they have many mitochondria and can generate steady quantities of ATP using oxidative phosphorylation—that is, aerobic respiration (Chapter 9)—thanks to the plentiful supply of oxygen delivered by myoglobin.

- **Fast muscle fibers** (fast glycolytic fibers) appear white because they have a low myoglobin concentration. They contract rapidly because the myosin hydrolyzes ATP at a rapid rate, but they also fatigue rapidly because their primary source of ATP is glycolysis rather than aerobic respiration.

- **Intermediate muscle fibers** (fast oxidative fibers) appear pink or red. Their contractile properties vary but are intermediate between slow and fast fibers because they derive ATP from both glycolysis and aerobic respiration.

The different fiber types are present in all skeletal muscles, but their relative abundances differ from muscle to muscle. Slow fibers are abundant in muscles specialized for endurance—such as the leg muscles of birds that excel at swimming or walking (the "dark meat" of chicken legs). In humans, the soleus muscle in the back of the calf is an example of a muscle with a high proportion of slow fibers—it helps to keep you upright when you stand.

Fast fibers contract and relax up to three times faster than slow fibers, making them well suited for bursts of activity. The "white meat" of chicken breasts, specialized only for quick bursts of flight to escape predators, are made primarily of fast fibers. The muscles that control your eye movements are another example.

✔ If you understand muscle fiber types, you should be able to predict whether the breast meat of pigeons (which are capable of prolonged flights) is composed of dark meat or white meat and explain why there is a color difference between these muscle types.

Can humans change their fiber types through training? Experiments have shown that endurance training can increase the density of blood vessels, mitochondria, and myoglobin in muscle fibers, enabling athletes to improve their muscle performance. However, training does not change slow fibers to fast fibers, nor the reverse. The ratio of muscle-fiber types in muscles is heritable—that is, genetically determined.

Skeletal-Fiber Organization The force of a muscle is proportional to the cross-sectional area of the muscle—the number of sarcomeres lined up side by side exerting a pull in synchrony. By contrast, the length change of the muscle is determined by the length of the muscle fibers—how many sarcomeres are lined up in a row in each fiber. Thus, the arrangement of muscle fibers within a muscle influences the contractile properties of the muscle.

For a given muscle volume, some muscles are organized to maximize length change, because the fibers are parallel to each other in long bands (**FIGURE 48.8**, left)—the longer the chain of

FIGURE 48.8 Muscle-Fiber Patterns Affect Contractile Properties. Most muscles have fibers arranged in either a parallel or pennate pattern. The orientation of muscle fibers affects the contractile properties of the muscle.

sarcomeres in series in the myofibril, the greater the length change. The sartorius muscle in the human thigh has parallel fibers.

Other muscles are organized to maximize force, because the fibers are in a diagonal, or pennate, pattern (penna means feather; **FIGURE 48.8**, right)—the greater the number of sarcomeres pulling in parallel, the greater the force produced. The gastrocnemius muscle in the human calf is a pennate muscle.

Context of Muscle Contraction The relative abundances of fiber types and organization of muscle fibers are not sufficient to account for the diversity of muscle contraction properties of skeletal muscles. Muscle tension also varies according to how extended the muscle is when it contracts and how rapidly it is allowed to shorten—if at all. Muscles like the quadriceps in your thigh can even exert a force while they are lengthening, such as when you ease your weight down a step.

These circumstances depend on the interaction between the muscle and the skeleton. Let's take a closer look at how the muscle and skeletal systems interact to produce movement.

48.3 Skeletal Systems

All a muscle can do is pull. How can complex movements be accomplished using an engine that can only pull? Also, muscles are limited in how much they can shorten. How, then, can they cause dramatic shape changes in animal bodies?

Muscle forces and shape changes are transmitted to other parts of the body and to the environment via the skeleton. Skeletal systems perform four main functions:

1. *Protection* from physical and biological assaults.

2. *Maintenance of body posture* despite the downward pull of gravity and the vagaries of wind and waves.

3. *Re-extension of shortened muscles* If no mechanism of re-extension existed, muscles would shorten only once.

4. *Transfer of muscle forces* to other parts of the body and to the environment, enabling a much greater range of force production and shape change than can be accomplished by muscle alone.

The relative importance of these roles varies among animals according to their lifestyles and environments. For example, natural selection has favored turtles with robust shells in some environments—a skeletal adaptation for protection. In other environments, natural selection has favored highly reduced shells—an adaptation for rapid locomotion. There are trade-offs between protection and mobility (see Chapter 25).

Despite the stunning diversity in animal bodies, virtually all animals can be considered to have one (or more) of three types of skeletal systems:

1. Hydrostatic skeletons use the hydrostatic pressure of enclosed body fluids or soft tissues to support the body.

2. Endoskeletons have rigid structures inside the body.

3. Exoskeletons have rigid structures on the outside of the body.

Let's consider how each skeletal system transmits muscle forces and shape changes.

Hydrostatic Skeletons

Despite their squishy appearance, soft-bodied animals do have skeletons—hydrostatic skeletons. First let's look at how they are built, and then consider how they function.

Structure **Hydrostatic skeletons** ("still-water skeletons"), or hydrostats, are constructed of an extensible body wall in tension surrounding a fluid or deformable tissue under compression. When fluid is under compression, its pressure increases. The pressurized internal fluid, rather than a rigid structure, enables soft-bodied animals to maintain posture, re-extend muscles, and transfer muscle forces to the environment.

Hydrostatic skeletons occur in diverse animals, from sea anemones and jellyfish to mollusks and many types of worms (Chapter 34). Hydrostats also support *parts* of animals, such as the tongues and penises of humans and the tube feet of echinoderms (see Chapter 35).

The structures of hydrostatic skeletons are diverse as well. The body wall of hydrostats may include different numbers and orientations of muscle layers and fiber-reinforced cuticles or connective tissues. The interior may include seawater, coelomic fluid, blood, or soft organs such as intestines. Hydrostatic skeletons composed mostly of muscle, such as tongues and tentacles, are called muscular hydrostats (Chapter 34).

Function How do animals with hydrostatic skeletons move? Consider an earthworm. Its body wall consists of a cuticle reinforced with collagen fibers as well as two layers of muscle—longitudinal muscles, oriented along the length of the animal, and circumferential muscles, oriented in bands around each segment. When the circumferential muscles contract, they make the segments narrower and squeeze the internal coelomic fluid and tissue, thus increasing internal pressure. The pressure pushes outward in all directions, extending the relaxed longitudinal muscles, lengthening the segment (**FIGURE 48.9**).

When the longitudinal muscles contract and the circumferential muscles relax, the reverse occurs—the segments become wider and shorter, pushing sideways against the soil. Alternating contractions of longitudinal and circumferential muscles pass down the earthworm in waves, called **peristalsis.** In this way, earthworms move forward (or backward) within their underground burrows.

Longitudinal and circumferential muscles in earthworms make up an **antagonistic muscle group,** a group of two or more muscles that re-extend one another via the skeleton.

✔ If you understand how a hydrostatic skeleton works, you should be able to explain how earthworms can push laterally against the walls of their burrows despite having muscles that can only pull.

Endoskeletons

Even though parts of you, like your tongue, are supported by a hydrostatic skeleton, your endoskeleton is what keeps you standing up and on the move.

Structure **Endoskeletons** ("inside skeletons") are rigid structures that occur within the body. Even the most ancient of animal lineages, the sponges, secrete spicules—stiff spikes of silica or calcium carbonate ($CaCO_3$)—that provide structural support for the body. In echinoderms, the endoskeleton consists of calcium carbonate plates just beneath the skin—fused into a rigid case in sea urchins, but suspended in a flexible matrix that enables bending of the arms in sea stars.

The vertebrate endoskeleton differs from those of most sponges and echinoderms—and from hydrostatic skeletons—in that it is composed of rigid levers (the bones) separated by joints. Vertebrates change the shapes of their bodies largely by changing the *joint angles* between bones in the limbs and between the limbs and the body, rather than changing the shapes of body segments themselves.

Vertebrate skeletons are composed of four main elements:

1. **Bones** are made up of cells in a hard extracellular matrix of calcium phosphate ($CaPO_4$) with small amounts of calcium carbonate and protein fibers. The adult human body contains 206 bones (**FIGURE 48.10a**, see page 982). The meeting places where adjacent bones interact are called **articulations,** or **joints.** Bones articulate in ways that limit the range of motion, for example enabling a swivel in the shoulder joint but a hinge in the elbow joint (**FIGURE 48.10b**).

2. **Cartilage** is made up of cells scattered in a gelatinous matrix of polysaccharides and protein fibers. Cartilage can be quite rigid, such as in the clam-crushing jaws of some stingrays, or more rubbery, such as in the pads that cushion the joints in your knees and back.

3. **Tendons** are bands of fibrous connective tissue, primarily collagen, that connect skeletal muscles to bones. The ropelike structure of tendons transmits muscle forces to precise locations on the bones, sometimes quite a distance away from the muscle itself.

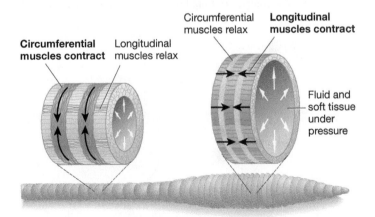

Circumferential muscles contract Longitudinal muscles relax

Circumferential muscles relax **Longitudinal muscles contract**

Fluid and soft tissue under pressure

FIGURE 48.9 Antagonistic Muscle Groups Cause Shape Changes in Hydrostatic Skeletons. The pressure of the internal fluid or tissue transmits forces between muscle groups and between muscles and the environment.

(a) Bones of the human endoskeleton

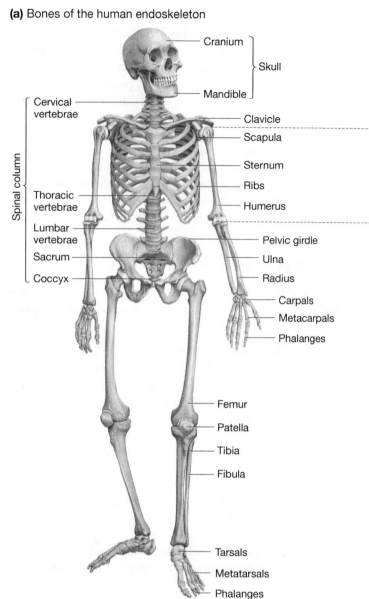

FIGURE 48.10 **Bones of the Human Endoskeleton.** Since bones are rigid and cannot change shape themselves, they articulate at joints that make specific types of movement possible, such as the swiveling and hinging shown here.

(b) Joints enable movement

flexors, muscles that pull bones closer together, decreasing the joint angle between them. They swing your lower leg back toward your thigh, reducing the angle of your knee joint. The quadriceps muscles in the front of your thigh are **extensors,** muscles that increase the angle of a joint. They straighten your leg at the knee joint (**FIGURE 48.11**).

The hamstrings and quadriceps muscles accomplish the large swing of the lower leg by inserting into different locations on the tibia (the shin bone). The articulation of the tibia with the femur (the thigh bone) serves as the pivot point for this lever. Along with enabling a shape change, the bone transmits the forces exerted within the thigh muscles to the foot—such as when the extension of your leg enables you to kick a ball.

The role of bones is primarily mechanical, but they also serve several physiological functions. Chief among these is storage of calcium and other minerals. When blood calcium falls to low levels, the bones release calcium to maintain blood–calcium homeostasis. The interior of long bones, called bone marrow, is also

4. **Ligaments** are bands of fibrous connective tissue, primarily collagen, that bind bones to other bones. Ligaments stabilize the joints.

Function Bones and cartilages are structures that do a good job of resisting compression (pushing) and bending, whereas tendons and ligaments do a good job of resisting tension (pulling). These structures combine with muscle in ways that enable the efficient transmission of muscle forces and shape changes.

Vertebrate skeletons move by means of changes in joint angles controlled by antagonistic muscle groups. For example, consider how your thigh muscles flex and extend your knee joint. The hamstring muscles in the back of your thigh are

FIGURE 48.11 **Endoskeletons Move by Contraction and Relaxation of Flexor and Extensor Muscles.**

the source of red blood cells needed to carry oxygen in the blood (Chapter 45) and white blood cells needed for the immune system (Chapter 51).

Exoskeletons

The mechanical function of rigid cuticle in exoskeletons is similar in many ways to that of endoskeletons. Exoskeletons occur primarily in arthropod animals, including insects, crustaceans, and arachnids (spiders, ticks, scorpions).

Structure An **exoskeleton** ("outside skeleton") is an exterior skeleton that encloses and protects an animal's body. The origin of the exoskeleton was a key innovation that preceded the spectacular diversification of arthropods. Arthropods, especially insects, are the most diverse and abundant animals on Earth (Chapter 34).

The material composition of exoskeletons varies. Insect exoskeletons consist of a cuticle formed from a composite of proteins and the polysaccharide chitin (Chapter 43). Chitinous ingrowths of the skeleton form **apodemes,** where muscles attach. Crustaceans such as crabs and lobsters have a cuticle that is mineralized with calcium carbonate, making their shells relatively thick and hard—and heavy. Most crustaceans are marine, so their buoyancy in water helps support their weight.

Function Like vertebrates, arthropods have paired flexor–extensor muscles that operate their jointed skeletons, causing movements that are based on changes in joint angles rather than changes in the dimensions of the segments themselves (**FIGURE 48.12**). Unlike vertebrates, however the muscles of arthropods must be packed *within* the skeletal tubes.

One solution to the problem of the interior placement of muscles is that many arthropod muscles have the pennate, or feather-like arrangement of muscle fibers illustrated in Figure 48.8. This arrangement boosts force output by effectively increasing the muscle cross-sectional area but not the muscle width during a contraction.

The disadvantage of pennate muscles is that their length change is small, so they have limited range of motion. Arthropods compensate for this constraint in part by the placement of their apodemes, which can transduce a small shortening of a muscle into a large change in joint angle.

The rigid levers of vertebrate skeletons can grow continuously as the rest of the body grows. But since the rigid exoskeletons of arthropods encase the growing soft tissue like a suit of armor, they must be shed—molted—periodically and replaced with a bigger one. Arthropods are vulnerable to predation during molts because their skeleton is soft and dysfunctional at that time (Chapter 34).

Hydrostatic skeletons, endoskeletons, and exoskeletons all transmit muscle forces and shape changes to other parts of the body and to the environment. While hydrostatic skeletons are the most widespread in terms of the number of animal phyla that contain species with skeletons of this type (virtually all—even arthropods, echinoderms, and vertebrates), the lever-based, segmented, jointed skeletons of arthropods and vertebrates win the prize for overall functional diversity (Chapters 34 and 35).

check your understanding

If you understand that . . .

- Hydrostatic skeletons, endoskeletons, and exoskeletons transmit muscle forces and shape changes to other parts of the animal and to the environment.
- Antagonistic muscles re-extend one another via the skeleton.
- In contrast to hydrostatic skeletons, shape changes in endoskeletons and exoskeletons involve changes in angles between rigid segments rather than in the segments themselves.

✔ **You should be able to . . .**

1. Compare and contrast the structure and function of hydrostatic skeletons, endoskeletons, and exoskeletons.

2. Predict what would happen if neurons simultaneously stimulated the biceps and triceps muscles to contract.

Answers are available in Appendix A.

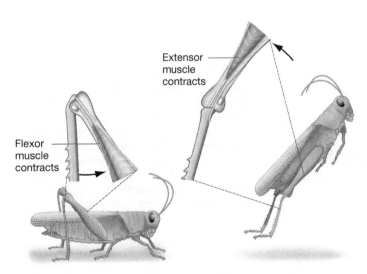

FIGURE 48.12 Exoskeletons Move by Contraction and Relaxation of Flexor and Extensor Muscles.

Extensor muscle contracts

Flexor muscle contracts

48.4 Locomotion

The most spectacular capability conferred on animals by the combination of muscle contractions and skeletal systems is efficient locomotion. Animals locomote to seek food, water, mates, and shelter, and to avoid predators. In the process, some animals migrate thousands of miles per year, and others perform astonishing feats of acrobatics. Locomotion has been shaped by natural selection and has been central to complex ecological relationships since the radiation of animals during the Cambrian Explosion more than 500 million years ago (Chapter 33).

Many of the diverse modes of locomotion are already familiar to you. Here are some examples.

- *On land* Crawling, walking, running, climbing, hopping, jumping, burrowing

- *In water* Undulating, jetting, swimming, rowing

- *In air* Flying, gliding

What variables do biologists analyze to unlock the secrets of animal locomotion?

How Do Biologists Study Locomotion?

Experimental studies on locomotion have increased exponentially in recent years, partly due to the conceptual breakthrough offered by the field of biomechanics—applying the principles of engineering to quantify the mechanics of organisms. Biomechanics studies the physical act of locomotion at different levels, including

- material properties
- structures
- motions
- forces
- energetics
- ecology and evolution

Let's examine each of these levels, starting with materials.

How Are the Material Properties of Tissues Important to Locomotion?
Because the active contractile properties of skeletal muscle are central to understanding the generation of forces in locomotion, a great deal of research has been devoted to analyzing the contractile properties of muscle, introduced in Section 48.1. However, the passive material properties of the skeletal elements are also essential to understanding the transmission of forces.

Consider how you move. A pioneer of biomechanics, R. McNeill Alexander, observed that the movement and energy exchange of the center of mass of a person walking is mechanically similar to that of an upside-down pendulum, whereas running is mechanically similar to a bouncing ball. Based on this insight, Alexander hypothesized that the large tendons of the lower legs of terrestrial animals—for example, the Achilles tendon at the back of your ankle (see Figure 48.7)—work as springs when animals run.

To test this hypothesis, Alexander measured the elastic properties of tendon by clamping a piece of tendon in an engineering device that measures the pulling force of the tissue as it is stretched to different lengths and released. The tendon returned 93 percent of the energy invested, losing only 7 percent to heat. This high rate of elastic-energy storage explains how tendons add a spring to the steps of runners, reducing the amount of muscle-generated power that must be generated from step to step.

The importance of the material properties of muscles and skeletal tissues is most obvious when they fail, whether in subtle or catastrophic ways. You may have firsthand experience with the debilitating consequences of broken bones or strained tendons.

How Is Musculoskeletal Structure Adapted for Locomotion?
Many biologists begin their study of locomotion by examining the size and shape of the skeletal elements. Careful measurements of skeleton geometries can reveal a great deal about the posture of the organism, range of motion of joints, and skeletal function in general. For example, a team of biologists recently used structural analysis to test the hypothesis that an early tetrapod amphibian, called *Ichthyostega*, could walk on all four limbs like today's salamanders (see **FIGURE 48.13**).

Hind limbs may have had limited contact with the ground due to the limited range of rotation of the hip joint

10 cm

FIGURE 48.13 The Relationship between Structure and Function Can Be Studied Using Computer Models. This computer image of an early tetrapod, *Ichthyostega*, was built using high-resolution scans of fossil bones. The 3D model measures the range of rotation of the major limb joints based on skeleton geometry, determining the overall range of motion of the limbs. The model found that the range of rotation of the shoulder and hip joints was more limited than previously thought, rejecting the hypothesis that these animals could walk on all fours.

FIGURE 48.14 The Motions of Locomotion Can Be Captured on Film. Eadweard Muybridge shot many photo series of animals during locomotion to enable precise analysis of limb and body motions over time.

Similar mechanical principles can be applied to relate the shapes of wings to flying ability, and the shapes of aquatic animals to their ability to swim through water. Sometimes, however, you just have to watch the action itself to understand how an animal uses its body to locomote.

What Does Locomotion Look Like in Living Animals?

All the motions of different parts of the skeleton, such as the angular rotation of limbs and the pattern of footfall on the ground, together produce locomotion.

Photographer Eadweard Muybridge is famous for his pioneering photo sequences of locomotion. His work with animals reportedly began when he was commissioned by a racehorse owner to settle a wager on whether horses are ever completely airborne during a gallop. His results are shown in the top sequence of **FIGURE 48.14**. High-speed video and digital images of many other animals have since been recorded, providing insights into many forms of locomotion, such as

- The gait of human sprinters (Figure 48.14, bottom)
- The complex wing-beat patterns of hovering bees and hummingbirds
- The upright, bipedal (two-footed) gait of basilisk lizards running across the surface of water (see chapter opening image)
- The footfall pattern that prevents centipedes from tripping on their own legs
- The aerial undulating of snakes that glide down from treetops
- The limb-like use of fins in lungfish "walking"
- Peristalsis in the muscular feet of crawling snails

For most animals, the pattern of movement during locomotion varies with speed. For example, horses walk at slow speeds, trot at intermediate speeds, and gallop at fast speeds—each gait has a distinct pattern of leg motions.

Computers facilitate the analysis of the many images captured in motion studies. The results are themselves insightful, but they also serve as an important stepping-stone to understanding the forces involved in locomotion.

What Forces Are Involved in Locomotion?

If an animal wants to move forward, it must push something backward, as predicted by Newton's third law of motion. Otherwise, the animal could move but not get anywhere, like a person with smooth shoes on slick ice. The types of forces that are important to locomotion vary according to whether the animal is locomoting on land, in water, in air, or some combination of the three.

On land, gravitational forces and inertial forces dominate. The gravitational force experienced by an animal is its weight, which is the product of its mass and the acceleration due to gravity (9.8 m/sec² on Earth). Weight is important on land because most terrestrial animals must hold themselves up to move forward. Inertial forces are proportional to mass and velocity, and they represent resistance of bodies and limbs to acceleration and deceleration. Note in the horse and human photo sequences in Figure 48.14 that the arms and legs must swing back and forth dramatically—an energy-intensive process of acceleration and deceleration.

In water, gravitational forces are less important than on land, due to the counteracting buoyant forces supporting the animal's weight in water. However, aquatic animals must overcome drag, the force that resists forward motion through fluids. Convergent evolution of torpedo-shaped bodies has occurred in diverse aquatic animals, from tuna to dolphins and ichthyosaurs, due to the strong selection for bodies that minimize drag during rapid locomotion (Chapter 28). Aquatic animals that move more slowly face less drag and thus are morphologically more diverse.

Water and air are both fluids, but air is a thousand times less dense than water. Buoyant forces are therefore negligible in air, making gravitational forces very important to fliers—most animals that locomote by flying or gliding have adaptations that make them lightweight. They must produce a force called lift to counteract gravity, and they must also minimize drag. As a result, fast fliers tend to have very streamlined shapes.

FIGURE 48.15 Visualization of Airflow Is Used to Analyze the Forces Involved in the Hovering of a Bat. The lift and drag forces acting on flying and swimming animals can be measured by observing fluid flow around the animal. The arrows represent the velocity of tiny water droplets illuminated by a laser in front of a high-speed camera.

How are forces measured? Scientists can measure ground force—the force with which a terrestrial animal strikes the substratum during a step—by coaxing an animal to walk, run, or hop upon an instrument called a force plate. Quantifying forces in fluids is more nuanced. It often requires indirect measurement by visualizing the airflow or waterflow around the animal, such as the flow of air around the wings of a hovering bat shown in **FIGURE 48.15**.

What Is the Cost of Locomotion? Animals must spend energy to find food and mates and escape from predators. However, the more energy an animal spends on locomotion, the less it can spend on producing offspring. There is strong selection pressure to minimize the cost of locomotion. How do biologists measure this cost?

To get a sense of variables that determine the cost of locomotion, consider the classic studies by Alexander on the gait transition from walking to running in humans. Alexander discovered that walking is an efficient mode of transport because you exchange potential energy at the top of your stride with kinetic energy midstride. However, the resulting pendulum-like motion of your center of mass is not efficient at higher speeds, when it becomes more cost effective to run using your spring-like tendons and other skeletal tissues to store energy between strides.

Alexander hypothesized that animals locomote using the most energy-efficient gait at each speed. To test this hypothesis in horses, physiologists Dan Hoyt and Richard Taylor trained horses to walk, trot, and gallop at a range of unnatural speeds—for example, trotting at a speed where the horse would normally have preferred to gallop. Hoyt and Taylor fitted each horse with an oxygen mask and ran it on a treadmill, so that they could measure oxygen consumption and speed simultaneously. The rate of oxygen consumption is a measure of energy use. (Oxygen consumption is proportional to ATP production during aerobic respiration; see Chapter 9.) The researchers then plotted energy use as a function of speed for each gait.

Hoyt and Taylor also filmed the horses moving freely around their paddock and measured the speeds at which the horses used different gaits when given free choice. The researchers then compared the lab data to the gait preference in the paddock.

The graph in **FIGURE 48.16** shows Hoyt and Taylor's results. The three curves at the top represent the energy used per distance traveled at the three gaits—the dips in the curves indicate the speeds at which the gaits were most efficient. The bars at the bottom of the graph show which gaits and speeds the horses chose when they were able to locomote freely in the paddock. The data support the hypothesis that the horses use the most energy-efficient gaits at different speeds and avoid intermediate speeds where the cost of locomotion is higher.

Similar studies have been conducted for diverse animals running, swimming, and flying, with similar results. Natural selection favors animals that locomote efficiently because they have more energy available for other vital activities.

Evolution and Ecology: What Is the Context? Some biologists study animal locomotion with a strictly mechanistic focus, investigating proximate causes via "how" questions. Alexander's study of the elastic energy storage of the Achilles tendon falls in this category. However, biologists also examine the ultimate causation of locomotion by asking "why" questions, which require an analysis of evolutionary history and ecological context.

For example, why do dolphins and ichthyosaurs have such similar torpedo-like shapes? Both taxa are fast-swimming predators in their marine environments, yet phylogenetic analysis shows that they are not closely related (see Figure 28.3). Rather, the torpedo-like shape evolved independently in both lineages because it substantially reduces drag during rapid locomotion in water, improving the efficiency of locomotion and thus increasing fitness (the number of viable, fertile offspring produced).

But locomotion cannot be "perfected" over evolutionary time. Natural environments are spatially complex and change over time, so the context for locomotion is not static. Further, there are numerous constraints and fitness trade-offs to different aspects of locomotion. Some trade-offs occur within locomotory systems (such as the trade-off between stability and maneuverability) while others occur between locomotion and other aspects of behavior (such as using appendages both for locomotion and feeding). Overall, a thorough study of locomotion requires an integrative approach with evolutionary and ecological context.

Size Matters

Animals that use muscles to power locomotion span a vast range of sizes—an astonishing 10 orders of magnitude (10,000,000,000), from tiny ants to giant whales. Many of these animals also grow over a large size range during their development. As is true for many other aspects of animal structure and function, size matters.

QUESTION: Do horses minimize the cost of locomotion?

HYPOTHESIS: Horses choose gaits that minimize energy use at different speeds.

NULL HYPOTHESIS: Horses do not choose gaits based on cost of locomotion.

EXPERIMENTAL SETUP:

1. Measure oxygen consumption of horses trained to walk, trot, and gallop at a range of speeds on a treadmill. Calculate energy used per distance travelled at different speeds.

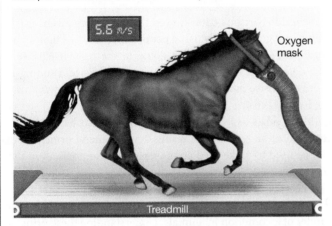

2. Videotape the same horses locomoting freely in the paddock, and measure the gaits and speeds they choose to use naturally.

PREDICTION: For each gait, there is a range of speeds where energy use is minimized. Horses will favor these gaits and speeds.

PREDICTION OF NULL HYPOTHESIS: There will be no correlation between chosen gaits and energy consumption.

RESULTS:

CONCLUSION: Horses choose gaits that minimize energy use at different speeds and avoid speeds with high energy consumption.

FIGURE 48.16 Horses Minimize the Cost of Locomotion by Choosing Appropriate Gaits.

SOURCE: Hoyt, D. F., and C. R. Taylor. 1981. Gait and the energetics of locomotion in horses. *Nature* 292: 239–240.

✔**QUANTITATIVE** Use the graph to estimate the relative energy expense of galloping rather than trotting at 3.5 meters/second (m/s).

FIGURE 48.17 Size Influences Skeleton Geometry. When scaled to the same size, you can see that the elephant has thicker bones and a more upright stance than a cat does, due to the disproportionate burden of gravity on larger animals. The femurs are highlighted for comparison.

Two organisms may be geometrically similar—that is, they may have exactly the same proportions—but if they are different sizes, they are more different than they seem. To start, the ratio of surface area to volume decreases as the organism gets larger, because surface area is proportional to length squared, while volume is proportional to length cubed (see Chapter 42). This concept has far-reaching implications for physiology. It also has important mechanical implications.

The weight of an animal is proportional to its volume, and the ability of leg bones to support the weight is proportional to their cross-sectional area. Thus, large terrestrial organisms have disproportionately hefty skeletal elements to avoid breaking their legs—something that Galileo observed 400 years ago when he compared skeletons of small and large organisms (**FIGURE 48.17**).

So, in the late 1980s, it was surprising when Andrew Biewener discovered that animals of different sizes maintain similar stresses in their skeletal tissues. How do they do that?

Biewener observed that posture and behavior are important variables. Small animals locomote in a more crouched posture and make great leaps, while large animals like elephants are more straight-legged, so that their skeletons, rather than their muscles, can support their body weight. Larger animals also locomote more gently. If a house cat were enlarged to be the size of an elephant, it would break its own bones when attempting a pounce.

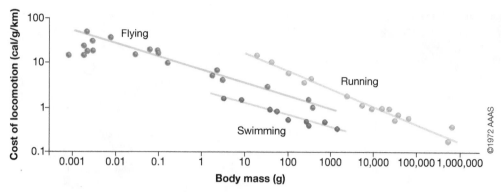

FIGURE 48.18 Cost of Locomotion for Swimming, Running, and Flying Animals Decreases with Mass.

DATA: Schmidt-Nielsen, K. 1972. Locomotion: Energy cost of swimming, flying, and running. *Science* 177: 222–228.

✔**QUANTITATIVE** About how much more costly is it to run than to swim for animals with a body mass of 100 g?

©1972 AAAS

Size is also of paramount importance in determining how organisms locomote through fluids. Fluids have an inherent viscosity, or stickiness, that is of minor concern to animals that are very large and/or fast (thus have high inertia), like dolphins, which can glide through water with little effort. But viscosity is of enormous concern to animals that are very small and/or slow (thus have low inertia), like plankton. If you were the size of a grain of rice and went for a swim in a lake, the water would feel like corn syrup.

Size also affects the cost of locomotion. Physiologist Knut Schmidt-Nielsen, in the early 1970s, was the first to make an energy comparison between different modes of locomotion. To account for widely different body sizes and distances traveled, he normalized all the data to the energy cost in calories per gram of body per kilometer traveled and then plotted the data on a log–log plot (see **BioSkills 6** in Appendix B).

Schmidt-Nielson's data showed that larger organisms use less energy per gram of body tissue per distance traveled than smaller organisms—all the lines slope down to the right (**FIGURE 48.18**). The data also show that for any given body size, running is a more expensive mode of transport than flying, and swimming is the least expensive.

Many other aspects of locomotion can also be plotted as a function of body size. The results reveal general principles that can be used to make predictions for diverse animals, even those that are extinct.

The take-home message? The laws of physics establish very definite constraints on the realm of possible sizes, shapes, and modes of locomotion in animals. Understanding these principles helps illuminate themes and variations among the diversity of animals.

CHAPTER 48 REVIEW

For media, go to MasteringBiology

If you understand that . . .

48.1 How Do Muscles Contract?

- The muscles of vertebrate limbs are composed of muscle cells, called muscle fibers, that contain threadlike contractile elements called myofibrils, each divided into contractile units called sarcomeres.

- Sarcomeres appear striated, or banded, due to the aligned arrangement of thick filaments (myosin) and thin filaments (actin).

- Sarcomeres shorten when thick filaments of myosin slide past thin filaments of actin in a series of binding events mediated by the hydrolysis of ATP.

- Calcium ions play an essential role in muscle contraction by making the actin in thin filaments available for binding by myosin.

- Acetylcholine released from somatic motor neurons is the neurotransmitter that stimulates contraction of skeletal muscle.

✔ You should be able to predict the primary symptom of botulism, which occurs when a toxin prevents release of acetylcholine from the neuromuscular junction. Explain your answer.

48.2 Muscle Tissues

- Smooth muscle lines bronchioles, blood vessels, the gastrointestinal tract, and certain reproductive organs. Smooth muscle cells are small and unstriated and have a single nucleus. Contractions are involuntary.

- Cardiac muscle occurs in the heart and forces blood through the circulatory system. Cardiac muscle cells are striated, branched, and connected to one another by intercalated discs. Contractions are involuntary.

- Most skeletal muscles are attached to the skeleton and are responsible for voluntary movement of the body. Skeletal muscle cells are long, striated, and multinucleate.

- Skeletal muscle fibers are specialized to contract slowly or quickly and to have a high or low endurance. These properties depend on the concentration of myoglobin present and the use of aerobic respiration and/or glycolysis for the production of ATP.

- Skeletal muscle fibers are organized in a parallel arrangement, which maximizes shortening, or a pennate pattern, which maximizes force production.

✔ You should be able to describe the major structural and functional differences among the three classes of muscle tissue.

48.3 Skeletal Systems

- Hydrostatic skeletons are composed of a body wall in tension surrounding a fluid or soft tissue under compression.

- Endoskeletons are internal skeletons, surrounded by soft tissue. In vertebrates, the jointed endoskeleton is composed of bones, cartilages, tendons, and ligaments.

- Exoskeletons are external skeletons, enclosing soft tissue. In arthropods, the muscles occur within the rigid, jointed cuticle composed of chitin, proteins, and sometimes minerals. Exoskeletons must be shed to enable growth.

- Movement is based on antagonistic muscle groups that act on a skeleton. In vertebrates and arthropods, these muscles are called flexors and extensors. They change the joint angle between rigid skeletal segments—especially in limbs.

✔ You should be able to compare and contrast the types of shape changes that are caused by muscle contractions in the different types of skeletons.

48.4 Locomotion

- Locomotion is movement relative to the environment and requires the transmission of muscle forces to the land, water, or air surrounding the animal.

- Locomotion can be studied at different levels: material properties, structures, motions, forces, and energetics, and within the larger context of evolution and ecology.

- Locomotion on land is usually dominated by gravitational and inertial forces. Swimmers must overcome drag. Fliers must overcome drag and must generate enough lift to overcome gravitational forces.

- Body size is important to the mechanics of locomotion.

✔ You should be able to give general examples of how physical constraints affect the locomotion of animals on land, in water, and in air.

(MB) **MasteringBiology**

1. MasteringBiology Assignments

Tutorials and Activities Muscle Contraction (1 of 2): Muscle Cells and Action Potentials; Muscle Contraction (2 of 2): The Sarcomere and the Sliding-Filament Model; Human Skeleton; Skeletal Muscle Structure; Structure and Contraction of Muscle Fibers

Questions Reading Quizzes, Blue-Thread Questions, Test Bank

2. eText Read your book online, search, take notes, highlight text, and more.

3. The Study Area Practice Test, Cumulative Test, BioFlix® 3-D Animations, Videos, Activities, Audio Glossary, Word Study Tools, Art

You should be able to . . .

✔ TEST YOUR KNOWLEDGE

Answers are available in Appendix A

1. Which of the following classes of muscle is/are voluntary?
   a. skeletal muscle
   b. cardiac muscle
   c. smooth muscle
   d. all of the above

2. Which of the following is a neurotransmitter that stimulates contraction of skeletal muscle?
   a. norepinephrine
   b. adrenaline
   c. acetylcholine
   d. calcium

3. In muscle cells, myosin molecules continue moving along actin molecules as long as
   a. ATP is present and troponin is not bound to Ca^{2+}.
   b. ADP is present and tropomyosin is released from intracellular stores.

   c. ADP is present and intracellular acetylcholine is high.
   d. ATP is present and intracellular Ca^{2+} is high.

4. True or false: The postural muscles of your legs are composed mostly of slow muscle fibers.

5. Which of the following is critical to the function of most exoskeletons, endoskeletons, and hydrostatic skeletons?
   a. Muscles interact with the skeleton in antagonistic groups.
   b. Muscles attach to each of these types of skeleton via tendons.
   c. Muscles extend joints by pushing skeletal elements.
   d. Segments of the body or limbs are extended when paired muscles relax in unison.

6. True or false: A large animal will experience twice the gravitational forces of a small animal half its size if their geometries are the same.

7. How did data on sarcomere structure inspire the sliding-filament model of muscle contraction? Explain why the observation that muscle cells contain many mitochondria and extensive smooth endoplasmic reticulum turned out to be logical once the molecular mechanism of muscular contraction was understood.

8. If a sprinter began an endurance running program, which of the following would occur?
   a. Some slow fibers would become fast fibers.
   b. Some fast fibers would become slow fibers.
   c. Some fibers would develop a higher mitochondrial density.
   d. Some fibers would develop a lower mitochondrial density.

9. Acetylcholine has very different effects on cardiac muscle and skeletal muscle. Explain.

10. Rigor mortis is the stiffening of a body after death that occurs when myosin binds to actin but cannot unbind. Why does myosin bind, and what prevents it from unbinding?

11. R. McNeill Alexander discovered that the arch of the human foot operates like a spring during running. Predict how a runner's oxygen consumption would change if a runner wore shoes that prevented the arches from changing shape. Explain your reasoning.

12. Explain how the physical constraints of locomotion on land, in water, and in air influence convergent evolution.

13. Predict the effect of ingestion of the toxin atropine on heart rate. Atropine is a naturally occurring compound in many poisonous nightshade plants. It blocks acetylcholine receptors in the heart. Explain your logic.

14. The force exerted by the shortening of a sarcomere depends on the length of the sarcomere when the stimulus is received. Based on your understanding of the sliding-filament model, predict how force would differ if the sarcomere were in a greatly stretched state at the beginning of the contraction versus in an average resting position. Explain your reasoning.

15. **QUANTITATIVE** The speed at which you switch from a walk to a run can be predicted using what is called the Froude number, based on the relative importance of gravitational and inertial forces of your pendulum-like walking gait.

$$\text{Froude number} = \frac{(\text{speed of locomotion})^2}{\text{gravitational acceleration} \times \text{leg length}}$$

Most mammals change from a walk to a run at a Froude number of 0.5. If gravitational acceleration is 9.8 m/s^2, and your leg length is 0.9 m, at what speed are you likely to switch from a walk to a run?
   a. 2.0 m
   b. 1.9 m/s
   c. 2.1 m/s
   d. 2.0 m/s^2

16. If you have seen the film *Jurassic Park*, you can imagine the nightmarish sight of *Tyrannosaurus rex* in high-speed pursuit of you as you flee in a jeep. Evaluate whether the data collected on the biomechanics of locomotion in living species are sufficient to determine how fast a large dinosaur like *T. rex* could run.

49 Chemical Signals in Animals

In this chapter you will learn that

Animals use hormones to regulate body functions and respond to their environment

by first examining

Cell-to-cell signaling **49.1**

and looking closer at

Chemical signaling diversity

Hormones and how they are studied

then asking

What do hormones do? **49.2**

How are hormones regulated? **49.3**

How do hormones act on target cells? **49.4**

The spectacular transformation that occurs during insect metamorphosis is triggered by chemical signals called hormones.

n response to sights, sounds, and other sensory stimuli, an animal's nervous system sends rapid messages, in the form of action potentials, to precise locations in the body. In many cases, these messages result in immediate, temporary responses such as muscle contractions and movement.

In response to changes in external or internal conditions, cells in the central nervous system (CNS) or the endocrine system release certain molecules. These molecules produce longer-term responses in a broad range of tissues and organs.

The **endocrine system** is a collection of organs and cells that secrete chemical signals into the bloodstream. A chemical signal that circulates through body fluids and affects distant target cells is called a **hormone.**

This chapter is part of the Big Picture. See how on pages 840–841.

✔ When you see this checkmark, stop and test yourself. Answers are available in Appendix A.

As you read this, a large suite of hormones is coursing through your circulatory system. These molecules are regulating the maturation of sperm or eggs by your reproductive system, changing the composition of the urine forming in your kidneys, and controlling the release of digestive enzymes in your gastrointestinal tract. Earlier in your life, changes in hormone concentrations led to the dramatic physical changes associated with puberty.

The goal of this chapter is to explore how hormones and other types of internal chemical signals work in animals. Together, animal nervous systems and endocrine systems process information about the internal and external environment—a function that is one of the five key attributes of life (see Chapter 1).

Let's begin with an overview of chemical signaling systems and then plunge into analyzing how hormones regulate the activity of target cells.

49.1 Cell-to-Cell Signaling: An Overview

Animal chemical signals are present in extremely low concentrations but can have enormous effects on their target cells. Unlike action potentials, which are electrical impulses that have a short-term effect on a single cell or on a small population of adjacent cells, the messages that chemical signals carry can have a relatively long-lasting effect.

In combination, electrical and chemical signals allow animals to coordinate the activities of cells throughout the body. They are the mechanism responsible for maintaining trillions of cells as an integrated unit called an individual.

Major Categories of Chemical Signals

The chemical signals found in animals have diverse structures and functions. **TABLE 49.1** summarizes how biologists go about organizing the diversity of chemical signals, based on where the molecules originate and where they act.

Notice that the names for most of the categories use the Greek word root *crin*, meaning separated. Its use captures something essential about how chemical signals act: They are released from cells and thus are separated from them.

It's important to note that these five classes of chemical messenger do not coincide with five structurally distinct classes of molecules. For example, the endocrine signals found in a particular organism routinely belong to several families of chemical compounds, ranging from amino acid derivatives to lipids. And a particular family of molecules—say, peptides or the lipids called steroids—may function as endocrine, autocrine, and paracrine signals in the same individual.

Autocrine Signals Act on the Same Cell That Secretes Them
Translated literally, autocrine means same-separated. The name is appropriate because **autocrine** signals affect the same cell that releases them.

Perhaps the best-studied autocrine signals are **cytokines** ("cell-movers"). Most cytokines amplify the response of a cell

Five Categories of Chemical Signals in Animals

Type of Chemical Signal	Source and Target
Autocrine signals act on the same cell that secretes them.	
Paracrine signals diffuse locally and act on nearby cells.	
Endocrine signals are hormones carried between cells by blood or other body fluids.	
Neural signals diffuse a short distance between neurons.	
Neuroendocrine signals (neurohormones) are hormones released from neurons.	

to a stimulus. An example is interleukin 2, which in the course of fighting an infection is synthesized and released by a type of white blood cell called a T cell. Interleukin 2 activates T cells to help eliminate the infection. It also causes the cells to divide repeatedly, producing more activated T cells for host defense.

Paracrine Signals Act on Neighboring Cells
Translated literally, **paracrine** means beside-separated. Paracrine signals diffuse locally and act on target cells near the source cell. Cytokines, for example, may act as paracrine signals as well as autocrine signals, because they can trigger responses by other nearby cells of the immune system (see Chapter 51).

It is common, in fact, to observe that a single chemical messenger can be assigned to more than one category of signal, based on its mode of action. Like some cytokines, the cell–cell signals named **insulin, glucagon,** and **somatostatin** cross categories. These molecules are produced by three distinct populations of cells within a region of the **pancreas** called the islets of Langerhans. The molecules act on nearby pancreatic cells as paracrine signals and ensure a smooth, steady response to

(a) Endocrine pathway

(b) Neuroendocrine pathway

(c) Neuroendocrine-to-endocrine pathway

FIGURE 49.1 Hormones Act via Three Pathways and Are Regulated by Negative Feedback.

changing blood-glucose levels. But they also act as hormones, in that they are released into the blood or other body fluids and affect distant cells—in this case, controlling the concentration of glucose in the blood.

Endocrine Signals Are Hormones   **Endocrine** ("inside-separated") signals are carried to distant cells by blood or other body fluids. The cells that produce endocrine signals may be organized into discrete organs called **glands** or may be interspersed among the cells of other organs—as are the islets of Langerhans in the pancreas.

Because hormones are well studied and particularly important to understanding how animals work, they serve as the focus of this chapter. Many or most of the principles discussed are relevant to the other categories of animal cell–cell signals as well, however.

Neural Signals Are Neurotransmitters   You might recall that when an action potential arrives at a synapse, it triggers the release of neurotransmitters that bind to receptors on the postsynaptic cell and induce a change in membrane potential—altering the tendency for the postsynaptic cell to fire action potentials (see Chapter 46).

Neural signaling can be very fast, because action potentials propagate rapidly and neurotransmitters have to diffuse only a short distance—across the tiny gap between two neurons, called the synaptic cleft. Neural signals also tend to be short-lived, because the signaling molecules are broken down or taken back up by the presynaptic cell.

Neuroendocrine Signals Act at a Distance   Even though they are released from neurons, **neuroendocrine** ("nerve-inside-separated") signals are considered hormones and are therefore

often called **neurohormones.** They share a key attribute with endocrine signals: They act on distant cells. Unlike neural signals, they do not act at the adjacent synapse.

Antidiuretic hormone (ADH; also called vasopressin) is a particularly well studied neuroendocrine signal. ADH is produced by neurons that originate in the hypothalamus of the brain. But instead of acting as a neural signal, ADH acts on cells in the collecting duct of the kidney to help regulate water excretion (see Chapter 43).

Hormone Signaling Pathways

In plants, sensory cells perceive changes in the environment and broadcast a hormonal signal that triggers an appropriate response from effector cells (see Chapter 40). Some animal hormones are also sent directly from endocrine cells to effector cells, in response to a stimulus (**FIGURE 49.1a**). But frequently, hormonal signaling in animals involves additional steps.

In many cases, information about external or internal conditions is gathered by sensory receptors and then integrated by neurons in the central nervous system (CNS) before the production of a hormonal signal. Neurons in the CNS respond by releasing neuroendocrine signals that act on effector cells directly (**FIGURE 49.1b**) or stimulate cells in the endocrine system, which respond by producing a hormone (**FIGURE 49.1c**).

All three types of signaling pathway—direct from an endocrine cell, direct from the CNS, or CNS-to-endocrine-system—are regulated by **negative feedback,** or **feedback inhibition.** In feedback inhibition, the product of a process inhibits its production. Negative feedback is key to homeostasis (see Chapter 42).

In the endocrine pathway, the hormone produced by the effector cells feeds back on endocrine cells, lowering production of the hormone and down-regulating the response (Figure 49.1a). An effector's response also feeds back to cells that initiate the neuroendocrine and neuroendocrine-to-endocrine pathways. A change in input from these cells then lowers production of the signal and reduces the response (see Figures 49.1b and 49.1c). Neuroendocrine-to-endocrine signaling pathways have an additional layer of regulation, because the endocrine signal usually inhibits production of the neuroendocrine signal (Figure 49.1c).

The take-home message? The nervous system and endocrine system are tightly integrated. Endocrine signals are released in response to electrical signals; in turn, endocrine signals modulate the electrical signals transmitted by the nervous system.

Feedback inhibition in the endocrine system is analogous to temperature control by a heat-sensitive thermostat. If the temperature is too high, the thermostat sends a signal that turns the furnace off; if the temperature is too low, the thermostat sends a signal that turns the furnace on. The result is a constant air temperature. In animal cell–cell signaling, feedback inhibition reduces production or secretion of the hormone, or both.

✔ If you understand this concept, you should be able to explain which parts of the thermostat analogy correspond to the sensory input, CNS, cell-to-cell signal, and effector in a hormone signaling pathway. You should also be able to predict what happens to hormone concentrations when feedback inhibition fails in a hormone signaling pathway.

Hypothalamus

Growth-hormone-releasing hormone: stimulates release of GH from pituitary gland

Corticotropin-releasing hormone (CRH): stimulates release of ACTH from pituitary gland

Thyrotropin-releasing hormone: stimulates release of TSH from thyroid gland

Gonadotropin-releasing hormone (GnRH): stimulates release of FSH and LH from pituitary gland

Antidiuretic hormone (ADH): promotes reabsorption of H_2O by kidneys

Oxytocin: induces labor and milk release from mammary glands in females

Pineal gland

Melatonin: regulates sleep-wake cycles and seasonal reproduction

Anterior pituitary gland

Growth hormone (GH): stimulates growth factors

Adrenocorticotropic hormone (ACTH): stimulates adrenal glands to secrete glucocorticoids such as cortisol

Thyroid-stimulating hormone (TSH): stimulates thyroid gland to secrete thyroid hormones

Follicle-stimulating hormone (FSH) and **luteinizing hormone (LH):** stimulate production of gametes and sex steroid hormones

Prolactin (PRL): stimulates mammary gland growth and milk production in females

■ Polypeptides
■ Amino acid derivatives
■ Steroids

Parathyroid glands (on dorsal side of thyroid gland)

Parathyroid hormone (PTH): increases blood Ca^{2+}

Thyroid gland

Thyroid hormones, thyroxine (T_4) and triiodothyronine (T_3): increase metabolic rate and heart rate; promote growth

Adrenal glands

Epinephrine: produces many effects related to short-term stress response

Cortisol: produces many effects related to short-term and long-term stress responses

Aldosterone: increases reabsorption of Na^+ by kidneys

Kidneys

Erythropoietin (EPO): stimulates synthesis of red blood cells

Pancreas (islets of Langerhans)

Insulin: decreases blood glucose

Glucagon: increases blood glucose

Ovaries (in females)

Estradiol: regulates development and maintenance of secondary sex characteristics in females; other effects

Progesterone: prepares uterus for pregnancy

Testes (in males)

Testosterone: regulates development and maintenance of secondary sex characteristics in males; other effects

FIGURE 49.2 Humans Possess a Diverse Array of Endocrine Glands and Hormones. This list is only partial. The heart, gastrointestinal tract, adipose (fat) tissue, and many other organs and cells also produce hormones.

What Makes Up the Endocrine System?

The endocrine system is the collection of cells, tissues, and organs responsible for hormone production and secretion. Organs that secrete a hormone into the bloodstream are called **endocrine glands.**

The tissues and organs that make up the endocrine system vary widely among animals. For example, neurons that manufacture and secrete hormones are particularly important in insects, where they regulate molting, metamorphosis, and other processes. Salmon have an unusual gland that secretes a hormone responsible for regulating calcium ion concentration.

Even within one species, the diversity of endocrine system components can be impressive. For example, **FIGURE 49.2** shows major human glands with endocrine functions. As mentioned earlier, hormone-secreting cells are not always organized into discrete glands. In many cases, they are located in other kinds of organs. (The islets of Langerhans were mentioned earlier as an example.)

It's important to note that not all glands in the body are part of the endocrine system. **Exocrine glands,** in contrast to endocrine glands, deliver their secretions through outlets called ducts into a space other than the circulatory system. Most of the digestive glands are either exocrine glands—an example is the salivary glands—or mixed endocrine and exocrine glands, such as the pancreas (see Chapter 44). The exocrine cells of the pancreas secrete digestive enzymes through ducts into the intestine. The endocrine portion of the pancreas consists of cells that secrete insulin and glucagon directly into the bloodstream.

At first glance, the diversity of hormones, glands, and their effects can seem overwhelming, especially considering that

Figure 49.2 represents just a partial catalog for a single species. But the picture can be simplified somewhat by recognizing that most animal hormones belong to one of just three major structural families: polypeptides, amino acid derivatives, or steroids.

Chemical Characteristics of Hormones

FIGURE 49.3 illustrates the three major classes of chemicals that can act as hormones in animals:

1. Polypeptides, which are chains of amino acids linked by peptide bonds (see Chapter 4)

2. Amino acid derivatives

3. Steroids, which are a family of lipids distinguished by a four-ring structure (see Chapter 6)

Secretin, a hormone produced in the small intestine to stimulate the exocrine portion of the pancreas, is a polypeptide; the hormone epinephrine is synthesized in the medulla of the **adrenal glands** from the amino acid tyrosine; and the hormone cortisol is synthesized in the cortex of the adrenal glands from the steroid cholesterol.

Hormone Concentrations Are Low, but Their Effects Are Large Hormones vary widely in structure but share a common characteristic: They have profound effects on individuals, even though they are present at vanishingly small concentrations. As an example, consider research that led to the discovery of **growth hormone (GH).**

Several researchers noted that rats and other laboratory animals stopped growing when their pituitary glands were removed. Based on this observation, it was widely suspected that

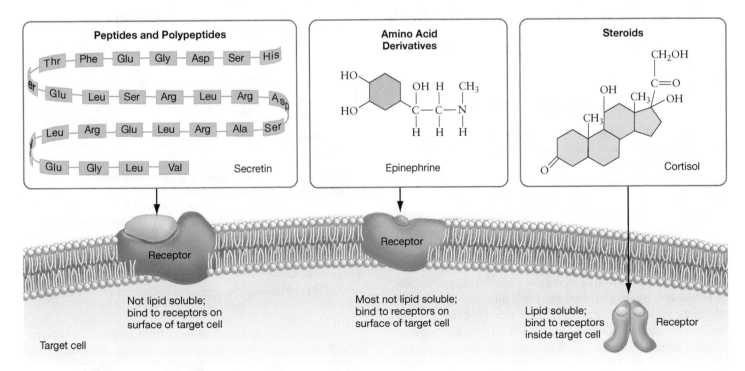

FIGURE 49.3 Most Animal Hormones Belong to One of Three Chemical Families.

the pituitary produces a chemical signal that promotes cell division and other aspects of growth.

To test this hypothesis, a research group purified a polypeptide from cow pituitary glands, injected the polypeptide into lab rats, and documented rapidly accelerated growth. When the researchers injected 0.01 mg of the molecule each day for nine days into rats that lacked pituitary glands, the width of the growth plates in the rats' leg bones increased by 50 percent. The individuals also gained an average of 10 g compared with rats that lacked a pituitary and did not receive the hormone treatment. Stated another way, an additional 0.09 mg of hormone led to a weight gain of 10,000 mg.

Further, 1 kg of cow pituitary tissue yielded a mere 0.04 g of growth hormone. By mass, the hormone makes up just 4 one-thousandths of 1 percent of the cow pituitary.

Only Some Hormones Can Cross Cell Membranes Given that small amounts of polypeptide, amino-acid-derived, and steroid hormones have large effects on the activity of cells, organs, and systems, how do the three types of hormones differ? The major difference is that steroids are lipid soluble, but polypeptides and most amino acid derivatives are not (see Figure 49.3).

Important exceptions to this rule are the **thyroid hormones, triiodothyronine** (also known as T_3) and **thyroxine** (also known as T_4), that are produced by the **thyroid gland.** The thyroid hormones are derived from the amino acid tyrosine but are lipid soluble.

Differences in solubility are important because steroids and thyroid hormones cross plasma membranes much more readily than do other types of hormones. To affect a target cell, all polypeptides and most amino acid derivatives bind to a receptor on the cell surface. Lipid-soluble hormones, in contrast, can diffuse through the plasma membrane and bind to receptors inside the cell.

Before exploring the consequences of this difference in more detail, let's consider how researchers discovered the array of hormones introduced here.

How Do Researchers Identify a Hormone?

Research on animal hormones began in earnest in the early 1900s, when researchers found that adding dilute hydrochloric acid (HCl) to the small intestine, to mimic the arrival of acidic material from the stomach, stimulated the pancreas to secrete compounds that neutralized the acid (see Chapter 44).

How could stimulating the small intestine lead to a response by the pancreas? The working hypothesis was that a chemical traveled from the small intestine to the pancreas by way of the bloodstream, to signal the arrival of acid.

To test this idea, researchers did the experiment summarized in **FIGURE 49.4**. An extract from the small intestine was injected into blood vessels in a dog's neck. A short time later, the pancreas secreted an alkaline solution. This result was strong evidence that the extract from the small intestine contained a cell–cell signaling molecule. The hormone was later purified and named **secretin.**

RESEARCH

QUESTION: How could stimulating the small intestine with acid cause the pancreas to secrete buffers that neutralize the acid?

HYPOTHESIS: A chemical messenger from the small intestine reaches the pancreas via the bloodstream.

NULL HYPOTHESIS: There is no blood-borne chemical messenger from the small intestine that acts on the pancreas.

EXPERIMENTAL SETUP:

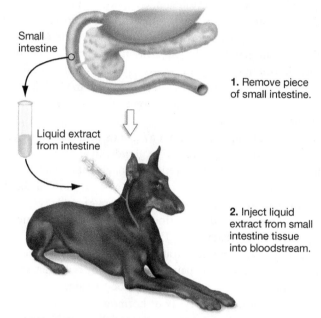

Small intestine

Liquid extract from intestine

1. Remove piece of small intestine.

2. Inject liquid extract from small intestine tissue into bloodstream.

PREDICTION: Liquid extract in bloodstream will signal pancreas to secrete buffer.

PREDICTION OF NULL HYPOTHESIS: Liquid extract in bloodstream will not signal pancreas to secrete buffer.

RESULTS:

Blood vessels

A short time after extract is injected into bloodstream, pancreas secretes buffer into intestine.

Buffers

CONCLUSION: A hormone from the small intestine (secretin) signals cells in the pancreas via the bloodstream.

FIGURE 49.4 Researchers Studied Digestive Processes to Find Experimental Evidence of a Hormone. Follow-up work succeeded in isolating the molecule secretin.

SOURCE: Bayliss, W. M., and E. H. Starling. 1902. The mechanism of pancreatic secretion. *Journal of Physiology* 28: 325–353.

✔ **QUESTION** What would be an appropriate control in this experiment?

In essence, using liquid extracts from suspected endocrine glands is a way to deliver a suspected hormone and evaluate the effect. The opposite strategy has also been productive: observing the consequences of removing specific organs or tissues. For example, removing the adrenal glands of an animal rapidly leads to death due to low blood sodium, low blood sugar, and low blood pressure. But injecting adrenal extracts into the blood of animals lacking adrenal glands corrects these abnormalities. Results like these provided strong evidence that hormones secreted by the adrenal glands help regulate blood-sugar concentrations and blood pressure.

A Breakthrough in Measuring Hormone Levels

Documenting an association between a particular gland or hormone and an effect in the body is just a first step. To understand hormone action, researchers have to figure out how these signals help animals stay alive and produce offspring.

The key to this goal is the ability to quantify the levels of hormones circulating in the bloodstream. Given that hormones are present in such small amounts, this ability eluded researchers for decades.

The breakthrough came in the 1950s when Rosalind Yalow developed the **radioimmunoassay,** a feat for which she later received the Nobel Prize in Medicine. In a radioimmunoassay, the quantity of hormone in a blood sample can be estimated by adding radioactively labeled hormones that compete with the unknown amount of hormone to bind to an antibody (see **BioSkills 9** in Appendix B). This technique revolutionized the field of endocrinology and the treatment of endocrine diseases, allowing the precise concentrations of hormones to be measured from blood samples drawn from animals or patients. Currently, radioimmunoassays are used every day in endocrine research and in medical laboratories. Much of what researchers know about hormone function was learned in studies employing the radioimmunoassay.

check your understanding

If you understand that . . .

- Cell-to-cell signaling occurs through a variety of pathways, and distant signaling is accomplished via the nervous system or endocrine system.
- Three main chemical families of hormones affect cells in different ways.

✔ **You should be able to . . .**

1. Discuss why it is often difficult to differentiate between the functions of the nervous and endocrine systems.
2. Predict the location of thyroid hormone receptors.

Answers are available in Appendix A.

49.2 What Do Hormones Do?

At the beginning of this chapter, you read that hormones are chemical messengers. If so, what do hormones "say"?

A first step in answering this question is to recognize that a single hormone can exert a variety of effects. For example, the thyroid hormones, T_3 and T_4, stimulate metabolism in humans and thus oxygen consumption throughout the body. But they also promote growth, increase heart rate, and stimulate the synthesis of many important macromolecules.

A second step in grasping what hormones do is to recognize that several different hormones may affect the same aspect of physiology. Insulin, glucagon, epinephrine, and cortisol all influence glucose levels in the blood.

Some hormones have extremely diverse effects; in other hormones, the functions appear to overlap. These observations begin to make sense when hormone action is viewed in the context of the whole organism. Hormones coordinate the activities of cells in three arenas: (1) development, growth, and reproduction; (2) response to environmental challenges; and (3) maintenance of homeostasis.

Let's analyze each of these functions in turn.

How Do Hormones Direct Developmental Processes?

In animals, as in plants, hormones play a key role in regulating growth and development. Growth hormones and sex hormones play crucial roles in promoting cell division, increasing overall body size, and promoting sexual differentiation as an individual matures; certain hormones direct the development of particular cells and tissues at critical junctures in an individual's life.

Let's explore two of the most dramatic examples of hormonal control—metamorphosis in amphibians and in insects—and then survey other developmental processes that are affected by hormone action.

T_3's Role in Amphibian Metamorphosis Frogs, toads, and salamanders are called amphibians ("double-lives") because in most species, juveniles live in water while adults live on land. The process of changing from an immature, aquatic larva to a sexually mature, terrestrial frog, toad, or salamander is an example of **metamorphosis** ("change-form") (**FIGURE 49.5**; see page 998).

Two sets of complementary experiments, published in 1912 and 1916, established that frog metamorphosis depends on thyroid hormones. Researchers could induce frog tadpoles to undergo metamorphosis by feeding them ground-up thyroid glands from horses; they could prevent metamorphosis by surgically removing the tadpoles' thyroid glands.

Follow-up work showed that the thyroid hormone triiodothyronine, or T_3, is responsible for many of the changes observed in metamorphosis. In response to a signal from the brain, the pituitary gland secretes thyroid-stimulating hormone, which stimulates the thyroid gland to produce thyroxine (T_4). The T_4 is then converted to T_3 at the target tissues.

In juvenile amphibians, cells respond to increased levels of T_3 by growing and forming new structures, such as legs. Other structures—such as a tadpole's tail—disintegrate or are absorbed. Still other tissues change structure and function. For example,

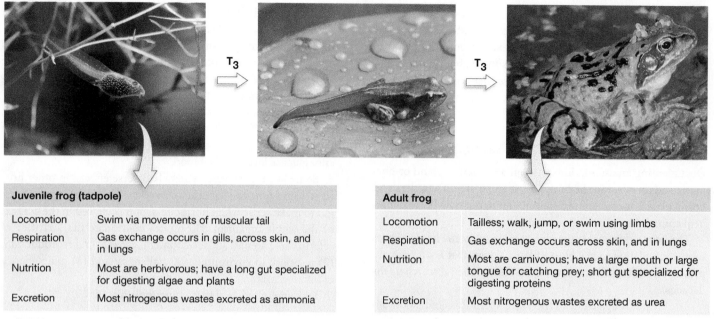

Juvenile frog (tadpole)	
Locomotion	Swim via movements of muscular tail
Respiration	Gas exchange occurs in gills, across skin, and in lungs
Nutrition	Most are herbivorous; have a long gut specialized for digesting algae and plants
Excretion	Most nitrogenous wastes excreted as ammonia

Adult frog	
Locomotion	Tailless; walk, jump, or swim using limbs
Respiration	Gas exchange occurs across skin, and in lungs
Nutrition	Most are carnivorous; have a large mouth or large tongue for catching prey; short gut specialized for digesting proteins
Excretion	Most nitrogenous wastes excreted as urea

FIGURE 49.5 Amphibian Metamorphosis Is a Continuous Process. When metamorphosis begins in a frog, toad, or salamander, the individual stays active and feeding. The continuous and gradual transition from juvenile to adult is mediated by T_3.

changes in existing cells are responsible for the switch from a tadpole's long intestine, specialized for digesting plant material, to an adult's short intestine, specialized for digesting insects and other prey. In the liver, cells respond to T_3 by manufacturing the enzymes required to excrete urea—the nitrogenous waste product released by adults—instead of the ammonia produced by tadpoles.

✔ If you understand the basic principles of hormone action, you should be able to suggest a hypothesis explaining why different frog cells can respond to T_3 in such different ways.

Hormone Interactions Regulate Insect Metamorphosis Some insects show a remarkable type of juvenile-to-adult transition, called holometabolous metamorphosis (see Chapter 33). In species that undergo this process, juveniles are called larvae. As in amphibians, insect larvae look completely different from adults, live in different habitats, and eat different food.

As larvae grow, they undergo a series of molts in which they shed their old exoskeleton, expand their bodies, and produce a new exoskeleton. After a specific number of these juvenile molts, however, they secrete a tough case called a pupal case. Inside the pupal case, specific populations of larval cells give rise to a completely new adult body. The rest of the larval body disintegrates (**FIGURE 49.6**).

In insects, metamorphosis depends on interactions between two hormones. If **juvenile hormone (JH)** is present at a high concentration in the larva, surges of the hormone **ecdysone** induce the growth of a juvenile insect via molting. But if JH levels are low, ecdysone triggers a complete remodeling of the body—metamorphosis—and the transition to adulthood and sexual maturity.

Sexual Development and Activity in Vertebrates In mammals and other vertebrates, long-distance cell-to-cell signals play key roles as embryos develop. Hormones also direct anatomical and physiological changes that occur later in life. Some of the most important of these changes involve the reproductive organs.

FIGURE 49.6 Insect Metamorphosis Occurs during a Resting Stage. When metamorphosis begins in a holometabolous insect, the individual enters a resting stage called the pupa.

Events early in development dictate whether the sex organs, or **gonads,** of a vertebrate embryo become male (**testes**) or female (**ovaries**). This process is called primary sex determination. In mammals, primary sex determination depends on genes on sex chromosomes.

Once testes or ovaries develop, they begin producing different hormones. In human males, the early testes produce two hormones:

1. A steroid hormone called **testosterone** induces early development of the male reproductive tract.

2. A polypeptide hormone called **Müllerian inhibitory substance** inhibits development of the female reproductive tract.

In females, the ovaries produce the steroid hormone **estradiol,** which is in the family of molecules called **estrogens.** Estradiol is required for further development of the female reproductive tract.

Sex hormones also play a key role in the juvenile-to-adult transition. When humans reach early adolescence, for example, surges of sex hormones lead to the physical and emotional changes associated with **puberty.** These developmental changes create the adult phenotype and the ability to produce offspring.

In boys, surges of sex hormones lead to changes that include enlargement of the penis and testes and growth of facial and body hair. In girls, an increased concentration of estradiol leads to the enlargement of breasts, the onset of menstruation, and other changes. In both sexes, a growth surge begins at puberty. This growth is stimulated by growth hormone (GH) produced in the pituitary gland (see Figure 49.2). GH regulates growth factors, which are signals that control the cell cycle (see Chapter 12). Growth originates in the epiphyseal plate, a small piece of cartilage separating the shaft from the end at both extremities of long bones.

Puberty is associated with a growth spurt because the effect of GH on the human skeleton is enhanced by the action of sex hormones, which surge during adolescence. Although GH and sex hormones continue to be produced long after puberty, growth in humans stops when the epiphyseal plates are replaced with bone, and no further growth is possible.

The sex hormones continue to play a key role in adults. In humans, sex hormones are instrumental in regulating sperm production and the menstrual cycle (see Chapter 50). The result of this steady release of sex hormones is that humans can mate year-round. In many animals, reproductive behavior is instead confined to specific times of the year. In these species, environmental cues such as increasing day length, warmth, or the onset of seasonal rains trigger the release of sex hormones (see Chapter 53).

How Photoperiod Affects Sex Hormone Release   The increase in day length—or increasing **photoperiod**—during spring is particularly important in stimulating the release of sex hormones in seasonally reproducing mammals, lizards, and birds. The lengthening spring photoperiod is sensed by photoreceptors, which are sensory receptors that respond to light (see Chapter 47). These photoreceptors send signals to the hypothalamus, a brain region that initiates a series of signals directing production of sex hormones.

The location of the photoreceptors depends on the animal. In mammals, photoreceptors in the retinas of the eyes sense light and send signals to the **pineal gland** via a pathway leading through the brain and spinal cord. The pineal gland secretes the hormone **melatonin,** which relays photoperiodic information to the hypothalamus. It also regulates sleep–wake cycles.

Maximal melatonin secretion occurs in the dark, so stimulation of the pineal gland by photoreceptors *reduces* melatonin secretion. Animals therefore experience a daily rhythm in melatonin levels, which are highest at night. As photoperiod increases in spring, there are fewer hours of dark, resulting in lower *overall* levels of melatonin (**FIGURE 49.7**). This decline in melatonin "informs" the hypothalamus to stimulate the testes and ovaries to make sex hormones.

Many diurnal lizard species do not rely solely on retinal photoreceptors. These lizards have a small hole in the top of their skull, covered only by a thin layer of skin; this hole is often called the pineal eye or parietal eye (**FIGURE 49.8**, see page 1000). The pineal gland lies just below this opening and responds to direct light. Interestingly, in birds, neither retinal nor pineal photoreceptors are responsible for relaying photoperiodic information to the hypothalamus. It appears that their photoreceptors are located diffusely throughout the brain.

The common thread among mammals, lizards, birds, and other vertebrates is that information about photoperiod is transduced from a signal detected by photoreceptors into a hormonal signal originating at the hypothalamus. The role of the hypothalamus is discussed in more detail in Section 49.3.

Some Chemicals Can Disrupt Hormone Signaling   In 1962, Rachel Carson stunned the world with her book *Silent Spring,* in which she described how commercially produced synthetic chemicals—even at low concentrations—can adversely affect humans and

FIGURE 49.7 Melatonin Levels Are Affected by Time of Day and Photoperiod in Seasonal Breeders. Melatonin levels are highest at night, and they are higher overall when photoperiod is short (8 h light, 16 h dark) than when it is long (16 h light, 8 h dark).

DATA: Vanecek, J. 1998. *Physiological Reviews* 78: 687–721.

Parietal eye

FIGURE 49.8 Some Diurnal Reptiles Have a Parietal Eye. This tiny hole in the skull allows sunlight to directly stimulate photoreceptors in the pineal gland.

wildlife. Since then, evidence has continued to accumulate that many pesticides, industrial chemicals, and other pollutants interfere with normal endocrine function in many animal species.

In the early 1990s, scientists coined the term **endocrine disruptor** to describe chemicals that interfere with normal hormonal signaling. This interference can happen in a number of ways, including binding of the chemical to hormone receptors and altering the metabolism of hormones, thereby affecting physiological processes and behavior.

Many endocrine disruptors are **xenoestrogens**—foreign chemicals that bind to estrogen receptors and induce estrogen-like effects. For example, exposure to the widely used herbicide atrazine causes feminization and even sex change of male frogs, induces reproductive-tissue abnormalities and reduced spawning in fishes, and is associated with low sperm counts and increased risk of birth defects in humans.

Another well-known xenoestrogen is bisphenol A (BPA), an industrial chemical used in many plastic or metal products, including containers for food and water. Laboratory animals exposed to BPA show abnormal development of reproductive and brain tissues, along with increased risk of some types of cancers.

As research continues into the possible health effects of xenoestrogens and other endocrine disruptors, many states and countries are limiting or banning their use. For example, atrazine is currently banned in the European Union, and BPA is banned in some countries for use in products such as baby bottles.

How Do Hormones Coordinate Responses to Stressors?

When an animal is thrust into a dangerous or unpredictable situation, hormones play a part in both the short-term and long-term responses. Let's explore each of these response categories in turn.

Short-Term Responses to Stress The short-term reaction, called the **fight-or-flight response,** is triggered by the sympathetic nervous system (see Chapter 46). If you were being chased by a grizzly bear, action potentials from your sympathetic nerves would stimulate your adrenal medulla and lead to the release of the hormone **epinephrine,** also known as **adrenaline.** (The Greek word roots *epi* and *nephron* mean "top-kidney"; the Latin word roots *ad* and *renal* also mean "top-kidney.")

To determine how epinephrine affects the body, researchers injected human volunteers with a saline solution—as a control—or epinephrine. Each point graphed in the "Results" section of **FIGURE 49.9** represents data from one of the volunteers; the data in the table are average values from the seven study participants.

The data indicate dramatic increases in an array of physiological processes: concentrations of free fatty acids and glucose in the blood, pulse rate, blood pressure, and oxygen consumption by the brain. In addition, the volunteers reported strong subjective feelings of anxiety and excitement.

Other experiments showed that epinephrine redirects blood away from the skin and digestive system and toward the heart, brain, and muscles. Epinephrine relaxes smooth muscle in blood vessels leading to these tissues—increasing blood delivery.

Taken together, the responses to epinephrine lead to a state of heightened alertness and increased energy use that prepares the body for rapid, intense action such as fighting or fleeing. Epinephrine coordinates the activities of cells in many organs and systems throughout the body to prepare an individual to cope with a life-threatening situation.

Long-Term Responses to Stress If you have ever experienced the fight-or-flight response, you may recall that the state is short lived. Once an epinephrine "rush" wears off, most people feel exhausted and want to rest and eat.

What happens if the stress continues and turns into a long-term condition? In the course of a lifetime, it is not unusual for a person to experience periods of starvation or fasting, prolonged emotional distress, or chronic illness. How do hormones help humans and other animals cope with extended stress?

Early studies of long-term stress in human subjects suggested a role for the hormone **cortisol,** which is produced in the adrenal cortex. Increased levels of cortisol were found in airplane pilots and crew members during long flights, athletes who were training for intense contests, parents of children undergoing treatment for cancer, and college students who were preparing for final exams. Why?

What Does Cortisol Do? In humans, cortisol's primary role is to ensure the continuing availability of glucose for use by the brain during long-term stress. Cortisol manages three main processes that maintain glucose production:

1. Cortisol induces the synthesis of liver enzymes that make glucose from amino acids and other chemical precursors.

2. Cortisol makes adipose tissue—fat tissue—and resting muscles resistant to insulin. Insulin normally stimulates **adipocytes**

RESEARCH

QUESTION: How does epinephrine affect the body?

HYPOTHESIS: Epinephrine causes changes involved in the fight-or-flight response.

NULL HYPOTHESIS: Epinephrine is not involved in the fight-or-flight response.

EXPERIMENTAL SETUP:

Control   Epinephrine

1. Inject human volunteers with saline solution or epinephrine.

Saline solution

2. Document changes in fatty-acid and glucose concentrations in blood, pulse rate, blood pressure, and oxygen consumption in brain.

PREDICTION: Epinephrine increases fatty-acid and glucose concentrations in blood, pulse rate, blood pressure, and brain oxygen consumption relative to controls.

PREDICTION OF NULL HYPOTHESIS: No differences in physiological state of individuals based on molecule injected.

RESULTS:

	Control	Epinephrine
Pulse rate (beats/min)	78.3	89.6
Blood pressure (average, mm Hg)	90.9	108.7
O_2 consumption in brain (cc O_2/100 g/min)	3.41	4.16

CONCLUSION: Epinephrine causes an array of changes associated with the fight-or-flight response.

FIGURE 49.9 Epinephrine Prepares the Body for Action.

SOURCES: King, B. D., L. Sokoloff, and R. L. Wechsler. 1952. The effects of *l*-epinephrine and *l*-nor-epinephrine upon cerebral circulation and metabolism in man. *Journal of Clinical Investigation* 31: 273–279. Mueller, P. S., and D. Horwitz. 1962. Plasma free fatty acid and blood glucose responses to analogues of norepinephrine in man. *Journal of Lipid Research* 3: 251–255.

✔ **QUESTION** Why did researchers bother to inject volunteers in the control group with saline? Why inject them with anything?

and resting muscle cells to remove glucose from the bloodstream. But when cortisol makes these cells resistant to insulin, glucose is reserved for use by the brain and exercising muscles.

3. Cortisol promotes the release of fatty acids—the body's major fuel molecules—from adipose tissue, for use by the heart and muscles.

Because of its importance in regulating blood glucose, cortisol is referred to as a **glucocorticoid.**

The long-term stress response comes at a high price, however, as any victim of a serious injury or illness knows. Glucocorticoids make amino acids available for glucose synthesis by promoting the degradation of contractile proteins in muscle. The resulting loss of muscle mass may cause severe weakness. Also, glucocorticoids impair wound healing and suppress immune and inflammatory responses. These processes are costly in terms of energy use, but reducing them makes the body more susceptible to infection.

The overall concept here is that the long-term stress response is a compromise—a fitness trade-off (see Chapter 42). The fuel requirements of the brain are met at the expense of other tissues and organs.

✔ If you understand the effects of chronically elevated cortisol in stressed individuals, you should be able to predict its effect on their ability to heal wounds and fight pathogens.

How Are Hormones Involved in Homeostasis?

Recall that homeostasis is the maintenance of relatively constant physical and chemical conditions inside the body. Homeostatic systems depend on three components (see Figure 42.14):

1. A sensory receptor that monitors conditions relative to a normal value, or set point

2. An integrator that processes information from the sensor

3. Effector cells that return conditions to the set point

In homeostatic systems, messages often travel from integrators to effectors in the form of hormones.

Leptin and Energy Reserves Healthy animals keep energy in reserve for use during periods of decreased food availability. This energy reserve typically takes the form of the lipid triglyceride (see Chapter 6). Triglyceride is an effective energy-storage molecule because large amounts of ATP can be generated when its three fatty-acid subunits are oxidized.

Although some triglyceride is present in muscle cells, most is stored in adipocytes. These cells make up the fat bodies found in insects and other species and the adipose tissue of mammals. Even a lean 70-kg human male stores enough energy in adipose tissue to survive for over 30 days without eating.

Is there a homeostatic system that regulates triglyceride stores? An answer started to emerge in the 1970s, when biologists began studying mutations in mouse genes called *obese* and *diabetic.* Homozygous *obese* (*ob/ob*) and *diabetic* (*db/db*) mice eat large quantities of food and move much less than heterozygous

ob/ob

ob/+

FIGURE 49.10 In Mice, Mutations in the *obese* and *diabetic* Genes Can Cause Obesity. These mice are siblings from the same litter. At the *ob* gene, the lean mouse is heterozygous and the obese mouse is homozygous.

or wild-type siblings do. The homozygotes also become extremely obese (**FIGURE 49.10**).

To test the hypothesis that the *obese* and *diabetic* gene products are involved in cell–cell signaling, researchers turned to an experimental technique called **parabiosis,** which works as follows:

1. Two closely related animals are surgically united by suturing the pelvis, shoulders, and abdominal walls together.

2. The newly created twin animal is allowed to recover from the procedure.

3. Within a short time, capillaries form between the two parabiotic partners. As a result, the two individuals develop a shared circulatory system.

4. The united circulatory system permits the passage of certain hormones and other long-lived molecules between the two partners. The system does not allow passage of molecules that are rapidly metabolized, such as glucose and fatty acids.

5. No new nerves grow between the two animals. As a result, they can influence one another through endocrine signals but not through electrical signals.

When researchers performed parabiosis between *db/db*, *ob/ob*, and lean mice, the results were striking (**FIGURE 49.11**):

• When a *db/db* animal was joined to either a lean animal or an *ob/ob* animal, the *db/db* mouse continued to eat and grow normally; but its partner stopped eating, lost weight, and eventually died of apparent starvation.

• When *ob/ob* animals were joined to lean animals, the *ob/ob* mice ate less food and gained weight less rapidly than did *ob/ob* mice joined to other *ob/ob* mice.

In addition, when two *db/db* animals were joined, both partners ate and grew as expected—both became obese. There was no difference between the two in survival.

To interpret these results, biologists hypothesized that mice produce a satiation, or "stop-eating," hormone—a negative feedback signal in homeostasis with respect to fat stores. The explanation

RESEARCH

QUESTION: Are the *diabetic* (*db*) and *obese* (*ob*) gene products involved in endocrine function?

HYPOTHESIS: The gene products are involved in hormonal signaling that affects appetite.

NULL HYPOTHESIS: The gene products are not involved in hormonal signaling.

EXPERIMENTAL SETUP:
Perform parabiotic surgery on closely related mice with different genotypes, so blood-borne products will pass between the mice in each pair.

db/db Lean   db/db ob/ob   ob/ob Lean

PREDICTION: *db/db* mice parabiosed to either *ob/ob* or lean mice will gain weight while their partners lose weight; *ob/ob* mice parabiosed to a lean mouse will lose weight.

PREDICTION OF NULL HYPOTHESIS: No changes in phenotypes of parabiotic mice will occur.

RESULTS:

db/db Lean   db/db ob/ob   ob/ob Lean

db/db
↑Body weight
↑Adipose tissue mass

Lean
↓ Food intake
↓ Body weight
↓ Adipose tissue mass
 (death by starvation)

db/db
↑Body weight
↑Adipose tissue mass

ob/ob
↓ Food intake
↓ Body weight
↓ Adipose tissue mass
 (death by starvation)

ob/ob
↓ Food intake
↓ Body weight
↓ Adipose tissue mass

Lean
No change

CONCLUSION: Gene products of *db* and *ob* are involved in hormonal signaling that affects appetite.

FIGURE 49.11 Parabiosis of Genetically Obese and Lean Mice Provides Evidence of a "Satiation Hormone."

SOURCE: Coleman, D. L. 1973. Effects of parabiosis of obese with diabetes and normal mice. *Diabetologia* 9: 294–298.

✔ **QUESTION** Suppose a mouse were doubly homozygous *(ob/ob db/db)*. Would its phenotype be the same as or different from singly homozygous individuals? Explain your reasoning.

was that *db/db* mice lack the receptor required for the hormone to affect target cells.

As *db/db* mice got fatter and fatter, they would produce more and more hormone—to no avail. This model explained why the signal from *db/db* mice greatly reduced food intake in their parabiotic partners but had no effect on the *db/db* mouse.

The satiation hormone itself was postulated to be encoded by the *obese* gene. As a result, *ob/ob* mice do not produce the hormone. They respond to the signal if it is available from a normal partner, however.

A key idea here is that the two genotypes produce the same phenotype because they disable different parts of the same hormone-signaling system. For a system to function, the hormone must be present and the target tissues must have receptors for that specific hormone.

This model was confirmed in 1994, when the *obese* gene product, called **leptin,** was shown to be a polypeptide hormone. Leptin is secreted into the blood by adipocytes and interacts with a specific receptor located in many tissues, including areas of the brain known to control feeding behavior.

In mice, leptin injections correct the obesity of *ob/ob* (leptin-deficient) mice but not of *db/db* (leptin-receptor-deficient) individuals. Unfortunately, leptin injections are not helpful for the vast majority of obese humans.

Follow-up work has shown that leptin levels in the blood vary in proportion to total adipose tissue mass. When adipose mass falls below a set point, the leptin level in the blood also falls. The brain senses the decrease in leptin level and generates both an increase in appetite and a decrease in energy expenditure. These responses promote eating and restore energy balance.

When sufficient food intake has occurred to restore triglyceride stores, leptin levels rise. The result is diminished appetite, increased energy expenditure, and the stabilization of adipose tissue mass. This is a striking example of homeostasis achieved by feedback inhibition.

ADH, Aldosterone, and Water and Electrolyte Balance Recall that when an individual is dehydrated, **antidiuretic hormone (ADH)** is released from the pituitary gland (see Chapter 43). ADH increases the permeability of the kidney's collecting ducts to water, causing water to be reabsorbed from urine and saved.

ADH is instrumental in achieving homeostasis with respect to water balance. For example, the ethanol in alcoholic beverages inhibits the release of ADH from the pituitary. People who imbibe large quantities of these beverages produce large quantities of dilute urine. The resulting water loss can lead to dehydration and nausea—symptoms associated with an alcoholic hangover.

When sodium concentration in body fluids is low, **aldosterone** is released from the adrenal cortex (see Chapter 43). Because aldosterone increases reabsorption of sodium ions in the distal tubules of the kidney, it plays a key role in homeostasis with respect to electrolyte concentrations and overall volume of body fluids. Adrenal hormones with this effect are called **mineralocorticoids.**

ADH saves water; aldosterone saves sodium. Together, they are key players in maintaining water and electrolyte balance.

EPO and Oxygen Availability Erythropoietin (EPO) is a crucial element in the homeostatic system for blood oxygen levels. When blood oxygen levels fall, the kidneys and other tissues release EPO, which stimulates the production of red blood cells. The more red blood cells, the higher the oxygen-carrying capacity of blood is. If you moved to a place at high elevation and experienced chronic oxygen deficit, your body would respond by releasing EPO and increasing red blood cell concentration. This explains why many athletes train at high altitudes.

Unfortunately, some endurance athletes have turned to EPO injections as a way to increase the oxygen-carrying capacity of their blood and give themselves a competitive edge. The practice is dangerous, as well as illegal. EPO abuse is thought to be responsible for the collapse and death of several cyclists during races in the mid-1990s.

The increased viscosity of blood in EPO abusers is accentuated during exercise, when blood plasma volume drops due to dehydration. The combination of high blood viscosity and low blood volume can impair blood flow through capillaries, increasing the risk of tissue damage and blood clotting. If clots form in blood vessels that lead to the heart or brain, heart attack or stroke may occur.

check your understanding

If you understand that . . .
- Hormones usually function in directing development and sexual maturation, preparing an individual for environmental challenges, or achieving homeostasis.
- Endocrine disruptors and stressors can disrupt hormone signaling and homeostasis.

✔ **You should be able to . . .**
1. Compare and contrast the roles of the hormones involved in metamorphosis of amphibians and insects.
2. Explain how the various changes induced by elevated cortisol result in a response to long-term stress.

Answers are available in Appendix A.

49.3 How Is the Production of Hormones Regulated?

Most hormones are released in response to an environmental cue or a message from an integrator in a homeostatic system. Often, the nervous system is closely involved. For example, environmental cues that signal the onset of the breeding season or the presence of a predator are received by sensory receptors and interpreted by the brain. Similarly, integration in most homeostatic systems is done by neurons in the central nervous system (CNS)—the brain and spinal cord (see Chapter 46).

Based on these observations, the short answer to the question posed in the title of this section is simple: In many cases, hormone production is directly or indirectly controlled by the nervous system.

The Hypothalamus and Pituitary Gland

The **pituitary gland,** located at the base of the brain, is directly connected to the brain region called the **hypothalamus.** This physical link between the hypothalamus and pituitary is the basis of the connection between the CNS and the endocrine system.

The pituitary has two distinct segments: the **anterior pituitary** and the **posterior pituitary.** In 1930, a biologist documented the consequences of removing the entire pituitary in laboratory rats: The animals stopped growing, could not maintain a normal body temperature, and suffered atrophy (shrinkage) of their genitals, thyroid glands, and adrenal cortexes. Not surprisingly, their life span shortened dramatically.

These experiments suggested that, in addition to secreting growth hormone, the pituitary secretes hormones that regulate the production of a wide variety of other hormones. As a case study, let's look at the pituitary hormone that acts on the adrenal glands.

Controlling the Release of Glucocorticoids

Early work on rats suggested that a molecule from the pituitary gland affects the adrenal gland. This molecule soon came to be called **adrenocorticotropic hormone,** or **ACTH** (*adreno* refers to the adrenal glands; *cortico* refers to the outer portion, or cortex, of the gland; and *tropic* means affecting the activity of).

ACTH was purified and characterized in 1943. When human volunteers were injected with ACTH, cortisol levels in their blood rose (**FIGURE 49.12a**). This result provided evidence that ACTH is a regulatory hormone. The adrenal cortex secretes glucocorticoids in response to ACTH released from the pituitary.

What regulates ACTH? Biologists from two laboratories independently showed that ACTH is released in response to a molecule produced by the hypothalamus. After years of effort, a different team of researchers succeeded in purifying a peptide—just 41 amino acids long—called **corticotropin-releasing hormone (CRH).** When the hypothalamus releases CRH, it stimulates cells in the anterior pituitary to secrete ACTH into the bloodstream.

Feedback Inhibition by Glucocorticoids

What *stops* glucocorticoid secretion? The key is to recognize that glucocorticoids themselves suppress ACTH production by the pituitary gland. Glucocorticoids accomplish feedback inhibition—they suppress their own production.

When human volunteers were injected with cortisol, ACTH levels in their bloodstream dropped dramatically (**FIGURE 49.12b**). Cortisol also inhibits release of CRH from the hypothalamus. Thus, if glucocorticoid levels become too high, ACTH levels fall. But if glucocorticoid levels become too low, ACTH levels rise and drive a compensatory increase in glucocorticoid production. **FIGURE 49.12c** summarizes the relationships among CRH, ACTH, and the glucocorticoid cortisol.

What happens when feedback inhibition fails? Certain pituitary tumors diminish the ability of cortisol to suppress ACTH production, leading to persistently high blood ACTH and

(a) Results of injecting ACTH into human volunteers

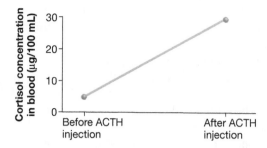

(b) Results of injecting cortisol into human volunteers

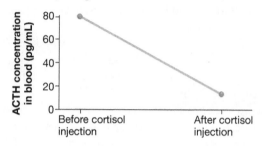

(c) Feedback inhibition by cortisol on ACTH

FIGURE 49.12 Cortisol Exerts Feedback Inhibition on ACTH Release. (a) ACTH stimulates release of cortisol. **(b)** Cortisol inhibits ACTH release. **(c)** The interaction between cortisol, ACTH, and CRH is an example of feedback inhibition.

DATA: Upton, G. V., A. Corbin, C. C. Mabry, et al. 1973. *Acta Endocrinologica* 73: 437–443.

✔ **EXERCISE** How would you use the data in part (a) to devise a test for adrenal failure in humans?

cortisol levels. The result is **Cushing's disease,** an unrelenting stress response that depletes the body's protein reserves. It is fatal if not treated.

Patterns in Glucocorticoid Release

Under ordinary circumstances, the production of CRH by the hypothalamus displays a daily rhythm and reaches its highest level in the early morning hours. This pattern drives a corresponding daily rhythm in ACTH production and blood cortisol level.

Ordinarily, the morning peak in blood cortisol levels coincides with arousal and initiation of the day's activities—and the effect is saving glucose for use by the brain. As an aside, the unpleasant symptoms of jet lag are due in part to the daily cortisol rhythm being out of synchrony with local time for several days after you arrive in a new time zone.

When the brain processes stimuli that produce pain or anxiety, it initiates a long-term stress response and a sustained increase in CRH production. Increased CRH production causes blood ACTH and cortisol levels to remain much higher throughout the day than they are in the unstressed state.

The Hypothalamic–Pituitary Axis—An Overview The CRH-ACTH-glucocorticoid relationship is just one of many hormone systems based on interactions among the hypothalamus, pituitary, and target glands or cells. The **hypothalamic–pituitary axis** actually forms two anatomically distinct systems (**FIGURE 49.13**). The

anterior pituitary develops from cells in an embryo's mouth and throat lining; the posterior pituitary is an extension of the brain.

Two distinct populations of neurons in the hypothalamus influence the posterior versus anterior sections of the pituitary gland. Both types of hypothalamic neurons synthesize and release neurohormones and are called **neurosecretory cells.** The neurosecretory cells release hormones under the control of brain regions responsible for integrating information about the external or internal environment. For example, information about an upcoming exam or athletic contest might trigger action potentials that lead to the release of CRH.

The Posterior Pituitary Even though both parts of the gland contain neurosecretory cells, the anterior pituitary and posterior pituitary function in different ways. As Figure 49.13a indicates, the posterior portion of the pituitary is an extension of the hypothalamus itself.

Neurosecretory cells that project from the hypothalamus produce the hormones ADH and oxytocin, which are then stored

(a) The posterior pituitary stores neurohormones.

(b) The anterior pituitary secretes regulatory hormones.

Hormone	ADH	Oxytocin
Target	Kidney nephrons	Uterine muscles, mammary glands
Response	Aquaporins activated; H_2O reabsorbed	Contractions during labor; ejection of milk during nursing

Hormone	ACTH	Follicle-stimulating hormone (FSH) and luteinizing hormone (LH)	Growth hormone (GH)	Prolactin (PRL)	Thyroid-stimulating hormone (TSH)
Target	Adrenal cortex	Testes or ovaries	Many tissues	Mammary glands	Thyroid
Response	Glucocorticoids production	Gamete and sex steroid hormone production	Cell division; growth	Mammary-gland growth; milk production	Thyroid hormone production

FIGURE 49.13 The Hypothalamus and Pituitary Interact Closely. (a) Developmentally and anatomically, the posterior pituitary is an extension of the hypothalamus. Neurosecretory cells in the hypothalamus extend directly into the posterior pituitary and secrete ADH (vasopressin) and oxytocin. **(b)** The hypothalamus and the anterior pituitary communicate indirectly, via blood vessels. Hormones produced by other populations of neurosecretory cells in the hypothalamus travel in the blood to the anterior pituitary, where they control the release of pituitary hormones.

in the posterior pituitary. From there, ADH and oxytocin are released into the bloodstream. This is an example of the neuro-endocrine pathway of hormone action. Recall that ADH aids in the reabsorption of water by the kidneys. **Oxytocin** helps induce labor and milk release in female mammals.

The Anterior Pituitary   Unlike the situation in the posterior pituitary, the hypothalamus and anterior pituitary are connected indirectly. Neurosecretory cells from the hypothalamus secrete stimulatory or inhibitory hormones into tiny blood vessels to the anterior pituitary. In response, the anterior pituitary alters its secretion of hormones that enter the general bloodstream and act on target tissues or glands. This is an example of the neuroendocrine-to-endocrine pathway of hormone action.

Many of the hormones produced by the anterior pituitary stimulate the production of other hormones. The anterior pituitary hormones include ACTH; **follicle-stimulating hormone (FSH)** and **luteinizing hormone (LH),** which are involved in stimulating the gonads to produce sex hormones and gametes; GH; **prolactin,** which stimulates mammary gland growth and milk production in mammals; and **thyroid-stimulating hormone (TSH),** which triggers the production of thyroid hormones.

Control of Epinephrine by Sympathetic Nerves

When biologists analyze how the nervous system and endocrine system interact to control the release of epinephrine, the distinction between the two systems begins to blur. Section 49.2 introduced how epinephrine acts as an endocrine signal. During the fight-or-flight response, sympathetic nerves trigger the release of epinephrine from the adrenal medulla into the bloodstream. But in addition, some sympathetic nerves release the related molecule **norepinephrine** directly onto target cells.

In effect, the endocrine system broadcasts a messenger by secreting it into the bloodstream whereas the nervous system delivers a chemical messenger directly to particular cells. Epinephrine and norepinephrine, which differ from one another only by the presence of an additional methyl group on epinephrine, are members of the family of molecules called **catecholamines.**

Catecholamines function as neurotransmitters as well as hormones. Similarities between hormones and neurotransmitters do not end there, however. In some cases, their mode of action is similar.

Some neurotransmitters initiate changes in gene expression in neurons (see Chapter 46). You might recall that in the sea slug *Aplysia*, repeated application of serotonin led to gene activation and changes in the behavior of the synapse. By altering synapses in this way, neurotransmitters play a central role in learning and memory.

Similarly, many hormones exert their effects by activating particular genes in target cells. Understanding how gene activation occurs in response to neurotransmitters and hormones is the subject of intense research at laboratories around the world. It is also the subject of Section 49.4.

If you understand that . . .

- Hormone concentrations are tightly regulated—in some cases by negative feedback, in other cases by stimulatory or inhibitory signals from the hypothalamus.

✔ **You should be able to . . .**

1. Explain how feedback inhibition occurs in production of ACTH.
2. Discuss the relationship between processing centers in the brain, neurosecretory cells in the hypothalamus, and hormone-secreting cells in the anterior pituitary.

Answers are available in Appendix A.

49.4 How Do Hormones Act on Target Cells?

The key to understanding how hormones act on target cells is to recognize that only some hormones are lipid soluble and cross plasma membranes readily (see Figure 49.3). More specifically, steroid hormones are small lipids that enter cells without difficulty. But the peptide and polypeptide hormones—as well as most amino acid derivatives, with the exception of thyroid hormones—do not cross plasma membranes easily because of their large size and electrical charge.

Differences in the lipid solubility of hormones are important because they relate to the location of hormone receptors in a target cell. Steroid and thyroid hormones bind to intracellular receptors, while most amino acid derivatives and all polypeptides bind to receptors on the surface of the plasma membrane.

To compare these two distinct paths of hormone action, let's consider how estradiol and epinephrine affect target cells. As a steroid and a nonsteroid, they serve as model systems for target cell responses to hormonal signals.

Steroid Hormones Bind to Intracellular Receptors

Estrogens are steroids that direct the development of female secondary sex characteristics in many animal species. In humans and other mammals, the most important estrogen is the molecule estradiol (formally, 17 β-estradiol).

Because of its importance in reproduction by humans and domesticated animals, estradiol's mode of action has been the topic of intense investigation for over 50 years. How do target cells receive the signal carried by estradiol?

Identifying the Estradiol Receptor   In 1966, biologists succeeded in isolating the estradiol receptor in laboratory rats. **FIGURE 49.14** indicates the experimental approach. In essence, the research team introduced labeled hormone molecules into females and

PROCESS: ISOLATING THE ESTRADIOL RECEPTOR

Radioactive estradiol

1. Inject 5 female rats with labeled hormone.

Uterus tissue

2. Remove uteri and homogenize tissue; remove particulates (cellular debris).

Uterus extract

Sucrose density gradient

3. Add soluble extract to centrifuge tube above sucrose density gradient. Centrifuge.

Lowest density

Highest density

4. Molecules in extract separate by density. Use instrument to locate radioactive estradiol.

Top of centrifuge tube

Labeled hormone + receptor form band here; can be isolated and analyzed

Bottom of centrifuge tube

20   60   100   140   180

Counts per minute

FIGURE 49.14 Labeled Hormones Can Be Used to Find Hormone Receptors. If radioactive estradiol binds to its receptor in the uterus, the hormone–receptor complex should form a distinct band of radioactivity when molecules from uterine cells are separated by centrifugation.

DATA: Toft, D., and J. Gorski. 1966. *Proceedings of the National Academy of Sciences, USA* 55: 1574–1581.

then used density-gradient centrifugation—a technique introduced in **BioSkills 10** in Appendix B—to separate molecules in target cells by size.

When centrifugation was complete, the biologists found that the labeled estradiol was concentrated in a narrow band

comprised of the labeled hormone bound to its receptor. After purifying the receptor molecule, they found that proteinase enzymes could destroy it. Based on this result, they inferred that the estradiol receptor was a protein.

Follow-up experiments established that the estradiol receptor is primarily located in the nucleus but is not associated with the nuclear envelope. Further, the receptor is found only in estradiol target tissues, including the uterus, hypothalamus, and mammary glands. The latter finding was particularly exciting, because it clarified how hormones act in a tissue-specific way.

This is a crucial point: Hormones are broadcast throughout the body via the bloodstream, but they act only on cells that express the appropriate receptor. Target cells respond to a particular hormone because they contain a receptor for that hormone.

Later, biologists found that the gene for the estradiol receptor is similar to the genes that encode receptors for the glucocorticoids, testosterone, and other steroid hormones. This result suggested that all steroid receptors are descended from an ancestral receptor molecule, and that the binding of any steroid hormone to its receptor affects the target cell via a similar mechanism.

What happens once the hormone–receptor complex is present inside the nucleus of a target cell?

Documenting Changes in Gene Expression During the 1970s and 1980s, work in several laboratories suggested that estradiol and other steroid hormones affect gene transcription after they bind to their receptors. For example, researchers injected laboratory animals with estradiol or other steroid hormones and documented changes in the mRNAs and proteins produced in target cells. These data showed that steroid hormones can cause dramatic changes in the amount or timing of mRNA production by a large number of genes.

How? The estradiol receptor, like other members of the steroid-hormone receptor family, has two copies of a distinctive DNA-binding region called a zinc finger. DNA-binding domains are sections of a protein that make physical contact with DNA. The presence of zinc fingers in the estradiol receptor suggested that once estradiol had bound to it, the hormone–receptor complex might affect gene expression by binding directly to DNA.

Follow-up work confirmed that steroid hormone–receptor complexes bind to specific sites in DNA called **hormone-response elements**. Hormone-response elements are located just "upstream" (in the 5′ direction) from the start of target genes. Gene expression changes when a regulatory molecule such as a steroid hormone–receptor complex binds to the hormone-response element for that gene.

FIGURE 49.15 (see page 1008) summarizes the current model of how steroid hormones affect target cells. Because each hormone–receptor complex leads to the production of many copies of the gene product, the signal from the hormone is amplified. In this way, a small number of hormone molecules produces a large change in the activity of target cells and tissues.

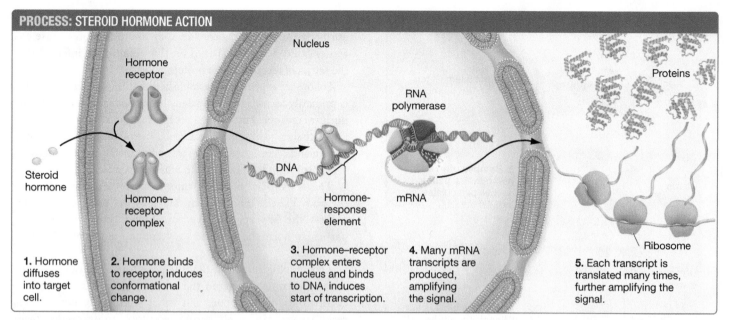

1. Hormone diffuses into target cell.

2. Hormone binds to receptor, induces conformational change.

3. Hormone–receptor complex enters nucleus and binds to DNA, induces start of transcription.

4. Many mRNA transcripts are produced, amplifying the signal.

5. Each transcript is translated many times, further amplifying the signal.

FIGURE 49.15 Steroid Hormones Bind to Receptors Inside Target Cells and Affect Gene Expression.

Hormones That Bind to Cell-Surface Receptors

Most polypeptide hormones and amino-acid-derived hormones are not lipid soluble. For these molecules to affect a cell, they must bind to receptors on the cell surface. Because the messenger never enters the target cell, its message must be transduced—changed into a form that is active inside the cell. This phenomenon is known as **signal transduction** (Chapter 11).

To explore how signal transduction occurs, let's first examine hormone receptors that reside in the plasma membrane and then explore the molecules that process the message inside the cell. In both cases, epinephrine will be used as a model system.

Identifying the Epinephrine Receptor

In 1948 a biologist published an exhaustive set of studies on how epinephrine affects dogs, cats, rats, and rabbits. The responses fell into two distinct categories, depending on the tissue being considered. To explain this observation, the researcher suggested that epinephrine binds two distinct types of receptor. He called these hypothetical proteins the alpha receptor and the beta receptor.

Follow-up work with molecules that block epinephrine receptors documented that there are actually two types of alpha receptors and two types of beta receptors. Thus, there are four distinct epinephrine receptors. Each is found in a distinct tissue type, and each induces a different response from the cell.

The discovery of four epinephrine receptors reinforces the concept of tissue specificity observed in experiments on the estradiol receptor. Hormones are transmitted throughout the body, not unlike a cell phone signal that is broadcast through the atmosphere. But their message is received only by cells with the appropriate receptor—just as a cell phone signal is received only by equipment with the appropriate antenna. Since there are four distinct epinephrine receptors, the same hormone can trigger different effects in different cells. What happens once epinephrine binds to one of these receptors?

What Acts as the Second Messenger?

Signal transduction occurs when a chemical message binds to a cell-surface receptor and triggers a response inside the cell.

How does a signal from epinephrine increase glucose levels in the blood? To answer this question, biologists focused on the enzyme **phosphorylase,** which catalyzes a reaction that cleaves glucose molecules off glycogen (**FIGURE 49.16a**). Phosphorylase exists in active and inactive forms; the enzyme switches between

(a) Phosphorylase catalyzes the production of glucose from glycogen.

Glycogen   Glycogen   Glucose

(b) Phosphorylase is activated in response to epinephrine.

FIGURE 49.16 Epinephrine Activates the Enzyme That Catalyzes the Formation of Glucose from Glycogen. (a) Phosphorylase is activated when an enzyme adds a phosphate group to it. **(b)** When epinephrine is added to cell-free extracts from liver tissue, the amount of activated phosphorylase increases dramatically.

DATA: Rall, T. W., E. W. Sutherland, and J. Berthet. 1957. *Journal of Biological Chemistry* 224: 463–475.

these states when it is phosphorylated or dephosphorylated by another enzyme.

Phosphorylase is present in liver cells—the primary source of blood glucose during the fight-or-flight response. As predicted, when researchers added epinephrine to extracts from homogenized (ground up) liver cells, much larger amounts of phosphorylase were activated relative to cell extracts that did not receive epinephrine (**FIGURE 49.16b**).

This observation suggested there was something in the homogenized cells that activated phosphorylase when epinephrine was present. By purifying components from the liver cell extracts and testing them one by one, researchers eventually found the ingredient that activated phosphorylase: a molecule called cyclic adenosine monophosphate, or **cyclic AMP (cAMP).**

The role of cAMP in epinephrine signaling was confirmed when researchers studied epinephrine's effects on the rat heart. During the fight-or-flight response, heart rate and the contractile force of the heart rise dramatically—increasing cardiac output. Cardiac output is a measure of how much blood the heart pumps per unit time.

The graphs in **FIGURE 49.17** show how cardiac muscle cells respond to epinephrine. In each graph, the x-axis plots time, in seconds, after epinephrine is applied.

- *Top* Almost immediately, there is a striking increase in cAMP levels inside the cells.
- *Middle* A few seconds later, the contractile force of the cells increases, peaking about 18 seconds after the hormone's arrival.
- *Bottom* Phosphorylase activity also rises, but peaks later—about 40 seconds after the signal.

To capture the importance of cAMP in triggering these effects, biologists refer to it as a second messenger. Recall that a **second messenger** is a nonprotein signaling molecule that increases in concentration inside a cell in response to a received signal—a molecule that binds at the surface (see Chapter 11).

A Phosphorylation Cascade How does cAMP transfer the hormonal signal on the cell surface to phosphorylase inside the cell? Follow-up work revealed the mechanism for epinephrine action on liver cells (**FIGURE 49.18**).

When epinephrine binds to its receptor, it activates a G protein (see Chapter 11), which then activates the enzyme adenylyl cyclase. Adenylyl cyclase catalyzes a reaction that converts ATP to cAMP. Next, cAMP initiates a chain of events called a **signal transduction cascade,** by binding to an enzyme called

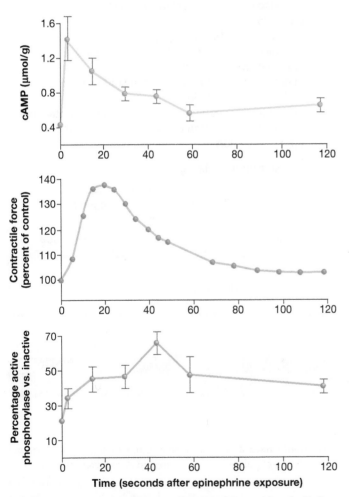

FIGURE 49.17 The Chemistry and Activity of Heart Muscle Cells Change in Response to Epinephrine.

DATA: Robison, G. A., R. W. Butcher, I. Oye, et al. 1965. *Molecular Pharmacology* 1: 168–177.

FIGURE 49.18 Epinephrine Triggers a Signal Transduction Cascade. Epinephrine's signal is amplified at each of steps 3–6.

cAMP-dependent protein kinase A. This enzyme responds by phosphorylating the enzyme phosphorylase kinase, which then phosphorylates phosphorylase.

In studying this model, it is crucial to recognize two points:

1. cAMP transmits the signal from the cell surface to the signaling cascade.

2. Together, cAMP production and the subsequent phosphorylation events amplify the original signal from epinephrine.

To appreciate the second point, consider that, in response to stimulation by the hormone–receptor complex, adenylyl cyclase is thought to catalyze the formation of at least 100 molecules of cAMP. In turn, each of these cAMP molecules activates many molecules of cAMP-dependent protein kinase A. Subsequently, each protein kinase molecule activates many molecules of phosphorylase kinase, and so on.

In this way, the binding of just a single molecule of epinephrine may trigger the release of millions or even billions of glucose molecules. Amplification through a signal transduction cascade explains why tiny amounts of hormones can have such huge effects on an individual.

The model in Figure 49.18 was inspired by experiments on the epinephrine receptor called the beta-1 receptor. But other researchers showed that when epinephrine binds to an alpha-1 receptor, a completely different signal transduction event occurs. In this and many other receptor systems, calcium ions (Ca^{2+}) serve as the second messenger in conjunction with another molecule called IP_3. Diacylglycerol (DAG) and 3′, 5′-cyclic GMP (cGMP) are also common second messengers in hormone response systems.

Why Do Different Target Cells Respond in Different Ways?

Researchers are increasingly impressed with the diversity and complexity of signal transduction cascades. For example, target cells that have the same receptor protein may have different second messengers or different enzyme systems that are available for activation. As a result, the same hormone and receptor can give rise to different responses in different target cells.

This finding helps explain one of the most fundamental observations about hormones: The same chemical messenger can trigger different responses in cells from different organs or in cells at different developmental stages. The reason is that the cells contain different receptors, second messengers, amplification steps, protein kinases, enzymes, or transcriptionally active genes.

To summarize this section, steroid and thyroid hormones tend to exert their effects through changes in gene expression. In contrast, polypeptide and most amino-acid-derived hormones activate a specific protein or set of proteins, usually by phosphorylation. Steroid and thyroid hormones activate transcription factors that lead to the production of new proteins; nonsteroid hormones trigger signal transduction cascades that activate existing proteins.

If you understand that . . .

- Hormones act on target cells by binding to receptors.
- The response to a hormone is tissue specific, because only certain cells contain receptors for particular hormones.
- In response to binding by lipid-soluble hormones, hormone–receptor complexes bind to DNA and induce changes in gene expression.
- In response to binding by lipid-insoluble hormones, activated cell-surface receptors induce the production of second messengers and the activation of signal transduction cascades that result in the phosphorylation of existing proteins.

✔ **You should be able to . . .**

1. Predict what would happen to the estradiol response if an individual had mutations that changed the DNA sequence of its hormone-response element.

2. Suppose a steroid hormone binds to a cell-surface receptor. Explain how its mode of action would compare to a polypeptide hormone.

Answers are available in Appendix A.

CHAPTER 49 **REVIEW** *For media, go to MasteringBiology*

If you understand . . .

49.1 **Cell-to-Cell Signaling: An Overview**

- Hormones are chemical messengers that are released from neurons or cells of the endocrine system, circulate in the blood or other body fluids, and trigger a response in target cells containing an appropriate receptor.

- Hormones have a variety of chemical structures. Most animal hormones are polypeptides, amino acid derivatives, or steroids.

✔ You should be able to explain the relationship between electrical signals from the nervous system and chemical signals from the nervous system and endocrine system, as they work in combination to coordinate the body's response to environmental change.

49.2 What Do Hormones Do?

- Together with the nervous system, hormones coordinate the activities of diverse cells and tissues. A single hormone may affect a wide array of cells and tissues and induce a variety of responses.

- Estradiol is an example of a hormone that regulates development and sexual maturation. Estradiol stimulates the formation of female sex characteristics in human embryos and the maturation of these tissues in adolescence.

- Melatonin is a hormone that regulates reproductive physiology in response to seasonal changes in day length.

- Epinephrine and cortisol are examples of hormones that help individuals cope with environmental changes. These hormones activate the short-term and long-term responses to stressors, triggering the fight-or-flight response or inducing changes that conserve glucose for use by the brain.

- Hormones are involved in a wide array of homeostatic interactions. For example, hormones are involved in directing cells that modify the concentrations of water, sodium ions, and other components of the blood and interstitial fluid. Homeostasis with respect to triglyceride stores is also under endocrine control.

✔ You should be able to explain why hormones—instead of electrical signals from the nervous system—are primarily responsible for regulating responses to environmental change, embryonic and sexual development, and homeostasis.

49.3 How Is the Production of Hormones Regulated?

- In many cases, the release of a hormone is regulated by chemical messengers from the anterior pituitary.

- Hormone-secreting cells in the anterior pituitary are regulated by hormones released by the hypothalamus region of the brain.

- The long-term stress response is a well-studied example of hormone regulation. The brain responds to long-term stress by triggering the release of the hypothalamic hormone CRH. CRH activates the release of ACTH by the pituitary gland, which stimulates the production of cortisol by cells in the adrenal cortex. Because cortisol inhibits the production of ACTH and CRH, the chain of events is regulated by feedback inhibition.

✔ You should be able to predict the consequences if negative feedback fails to occur after the release of CRH.

49.4 How Do Hormones Act on Target Cells?

- Animal hormones have two basic modes of action. Steroid and thyroid hormones are lipid soluble, cross plasma membranes readily, and often bind to receptors inside cells. Most polypeptide and amino-acid-derived hormones are not lipid soluble; they bind to receptors located in the membranes of target cells. In both cases, the response to a hormone is tissue specific because only certain cells express certain receptors.

- Most steroid and thyroid hormones act by inducing a change in gene expression.

- Polypeptide and most amino-acid-derived hormones trigger signal transduction cascades that activate one or more target proteins by phosphorylation.

- Although they are produced in tiny concentrations, hormones have large effects because they trigger gene expression or because their message is amplified through a signal transduction cascade.

✔ You should be able to predict whether a cell can respond to more than one hormone at a time, and explain why or why not.

(MB) MasteringBiology

1. MasteringBiology Assignments

Tutorials and Activities Coordination of the Endocrine and Nervous Systems; Endocrine System Anatomy; Hormone Actions on Target Cells; Human Endocrine Glands and Hormones; Peptide Hormone Action; Steroid Hormone Action; What Role Do Genes Play in Appetite Regulation?

Questions Reading Quizzes, Blue-Thread Questions, Test Bank

2. eText Read your book online, search, take notes, highlight text, and more.

3. The Study Area Practice Test, Cumulative Test, BioFlix® 3-D Animations, Videos, Activities, Audio Glossary, Word Study Tools, Art

You should be able to . . .

✔ **TEST YOUR KNOWLEDGE**   *Answers are available in Appendix A*

1. Both epinephrine and cortisol are involved in the body's response to stress. How do the two molecules differ?
   a. Cortisol is an amino acid derivative; epinephrine is a steroid.
   b. Cortisol binds to receptors on the plasma membranes of target cells; epinephrine binds to receptors in the interior of target cells.
   c. Epinephrine mediates the short-term response; cortisol mediates the short- and long-term responses.
   d. Cortisol controls the release of epinephrine from the adrenal glands.

2. How do steroid hormones differ from polypeptide hormones and most amino-acid-derived hormones?
   a. Steroids are lipid soluble and cross plasma membranes readily.
   b. Polypeptide and amino-acid-derived hormones are longer lived in the bloodstream and thus exert greater signal amplification.
   c. Polypeptide hormones are the most structurally complex and induce permanent changes in target cells.
   d. Only polypeptide and steroid hormones bind to receptors in the plasma membrane.

3. Which of the following developmental processes is *not* controlled by hormones?
   a. the initial development of male and female gonads, soon after fertilization
   b. overall growth
   c. molting in insects and other invertebrate animals
   d. metamorphosis in insects and other invertebrate animals

4. What is a hormone-response element?
   a. a receptor for a steroid hormone
   b. a receptor for a polypeptide hormone
   c. a segment of DNA where a hormone–receptor complex binds
   d. an enzyme that is activated in response to hormone binding and produces a second messenger

5. Where are the photoreceptors that sense day length located in mammals? In lizards? In birds?

6. True or False: In hormone systems, feedback inhibition occurs when the presence of a hormone inhibits release of the hormone.

✔ **TEST YOUR UNDERSTANDING**   *Answers are available in Appendix A*

7. Hormones are present in tiny concentrations, yet have large effects on target cells and on the individual as a whole. This is possible because a single hormone molecule
   a. can bind to many receptors at once.
   b. can activate many intracellular enzymes.
   c. can initiate transcription of DNA by binding to cell-surface receptors.
   d. can bind irreversibly to its receptor.

8. Why is the observation that one hormone may bind to more than one type of receptor important?

9. Compare and contrast the structure and function of the anterior and posterior pituitary glands.

10. Compare and contrast the modes of action of steroid and nonsteroid hormones.

11. Design a study to test the hypothesis that the symptoms of jet lag are caused by disruption of normal daily cortisol rhythms.

12. You see an advertisement on television for a leptin supplement that claims to result in dramatic weight loss. Explain why this supplement is unlikely to be effective.

✔ **TEST YOUR PROBLEM-SOLVING SKILLS**   *Answers are available in Appendix A*

13. You are a physician supervising a patient's recovery from the surgical removal of the posterior pituitary. Which of the following symptoms would you expect to have to treat?
    a. amenorrhea (cessation of the menstrual cycle)
    b. cessation of growth
    c. dehydration
    d. high blood-glucose levels

14. Suppose that during a detailed anatomical study of a marine invertebrate, you found a small, previously undescribed structure. How would you test the hypothesis that the structure is a gland that releases one or more hormones?

15. Suppose that a researcher announces the discovery of a hormone that affects the metabolism of fats in lab rats. Preliminary data indicate that the hormone is a polypeptide, about 50 amino acids long. The researcher suggests that the hormone activates transcription of a gene for a protein that stimulates fat breakdown. Explain why this researcher's hypothesis is reasonable or unreasonable.

16. QUANTITATIVE Scientists set out to test the hypothesis that atrazine is an endocrine disruptor that feminizes male amphibians. They treated male amphibians with atrazine, then later compared their circulating testosterone concentrations with those of control males and females. The results are shown here. Bars with different letters are significantly different from one another. Was the hypothesis supported? Why or why not?

DATA: Hayes, T. B., A. Collins, M. Lee, et al. 2002. Proceedings of the National Academy of Sciences, USA 99: 5476–5480.

50 Animal Reproduction

In this chapter you will learn that

Animals have an amazing diversity of reproductive strategies

by first comparing →

Asexual and sexual reproduction **50.1**

looking into →

Mechanisms of asexual reproduction

(some can switch modes)

Sexual reproduction and gametogenesis

leads to ↓

Fertilization and egg development **50.2**

finally exploring →

Reproductive structures and their functions **50.3**

modified by ↓

Sex hormones in mammals **50.4**

Pregnancy and birth **50.5**

The swollen, red rump of this female Hamadryas baboon indicates that she is about to produce an egg. She will probably mate with several males before the egg is fertilized. This chapter discusses how hormones regulate the female reproductive cycle and the consequences of multiple mating.

All the cells, tissues, organs, and systems introduced in Unit 8 exist for one reason: They allow animals to survive long enough and gather enough resources to reproduce. Stated another way, producing offspring is the reason that adaptations exist. Reproduction is the underlying purpose of virtually everything that an animal does. Replication is a fundamental attribute of life (see Chapter 1).

But if evolution by natural selection explains *why* animals reproduce, the goal of this chapter is to explore *how* reproduction occurs. Section 50.1 introduces research on animals that cycle between asexual and sexual modes of reproduction and explains how both modes occur at the cellular level. Section 50.2 surveys the diverse ways in which animals accomplish fertilization once gametes have formed. The final three sections focus on mammalian reproduction, using humans as the primary model organism.

This chapter is part of the Big Picture. See how on pages 840–841.

✔ When you see this checkmark, stop and test yourself. Answers are available in Appendix A.

The chapter includes topics of urgent practical interest, such as human birth control methods and the recent and dramatic declines in the death rate of human mothers during childbirth. Understanding and manipulating animal reproductive systems is an important issue for physicians, veterinarians, farmers, zookeepers, conservation biologists, and many others in biology-related professions.

50.1 Asexual and Sexual Reproduction

Several earlier chapters have explored how asexual and sexual reproduction differ. When reproduction is asexual, it occurs without fusion of gametes. **Asexual reproduction** is usually based on mitosis and results in offspring that are genetically identical to their parent. **Sexual reproduction,** in contrast, is based on meiosis and fusion of gametes. Due to genetic recombination during meiosis and the fusion of haploid gametes—usually from different parents—during fertilization, sexual reproduction results in offspring that are genetically different from each other and from their parents.

Let's first consider how animals engage in asexual reproduction to make genetically identical copies of themselves, then review the mechanisms responsible for sexual reproduction.

How Does Asexual Reproduction Occur?

In thousands of animal species, individuals can **clone** themselves—that is, produce large numbers of identical copies of themselves asexually.

There are three main mechanisms of asexual reproduction:

1. **Budding** An offspring begins to form within or on a parent (**FIGURE 50.1a**). The process is complete when the offspring—a miniature version of the parent—breaks free and begins to grow on its own.

2. **Fission** An individual simply splits into two or more descendants (**FIGURE 50.1b**).

3. **Parthenogenesis** (literally, "virgin-origin") Females produce offspring without fertilization by a male. Usually, parthenogenetic eggs are produced by mitosis, or by meiosis that occurs after chromosome number has doubled, but without crossing over and recombination. In both of these cases, offspring are genetically identical to the mother. Parthenogenesis occurs in a wide diversity of lineages, including certain invertebrates, fishes, lizards, snakes, and birds (**FIGURE 50.1c**).

Many animal species regularly switch between reproducing asexually and reproducing sexually. Why?

Switching Reproductive Modes: A Case History

Daphnia are crustaceans that live in freshwater habitats throughout the world. In a typical year, *Daphnia* produce only diploid, female offspring throughout the spring and summer, via parthenogenesis. The eggs produced by parthenogenesis develop in a structure called a brood pouch (**FIGURE 50.2**) and are released when the female molts her exoskeleton.

(a) Budding in hydra

(b) Fission in anemones

(c) Parthenogenesis in lizards

Unfertilized eggs develop into offspring

FIGURE 50.1 Mechanisms of Asexual Reproduction in Animals Are Diverse.

In late summer or early fall, however, many *Daphnia* females begin producing unfertilized, asexually produced eggs, some of which develop into males. Once the males have matured, sexual reproduction ensues: Haploid **sperm,** the male gametes, produced by meiosis in the males, fertilize haploid **eggs,** the female gametes, that females produce via meiosis.

Fertilization is the fusion of sperm and egg. In *Daphnia*, fertilized eggs are released into a durable case that falls to the bottom

FIGURE 50.2 Female *Daphnia* Can Produce Eggs by Parthenogenesis.

of the pond or lake for the winter. In spring, the sexually produced offspring hatch and the females begin reproducing asexually.

Biologists try to explain observations like this at two levels (see Chapter 53):

1. **Proximate causation** addresses *how* a trait is produced. When researchers identify the genetic, developmental, hormonal, or neural mechanisms responsible for a phenotype, they are working at the proximate level.

2. **Ultimate causation** addresses *why* a trait occurs, in terms of its effect on fitness. Researchers who work at the ultimate level try to understand the evolutionary history of traits.

Let's consider work on proximate aspects of the asexual–sexual switch in *Daphnia* and then consider ultimate causation.

What Environmental Cues Trigger the Switch? For decades, most researchers contended that day length triggered the asexual–sexual switch. The idea was that the shortening days of late summer or fall affected sensors in the brain (see Chapter 49), and these receptors produced electrical or hormonal signals that induced the production of males and haploid eggs.

In 1965, however, biologists showed that high population densities are also a factor. Researchers who brought *Daphnia pulex* populations into the lab and kept day length constant found a strong, positive correlation between population density and the percentage of females reproducing sexually. When you read the graph in **FIGURE 50.3a**, note that the *x*-axis is logarithmic (see **BioSkills 6** in Appendix B). The highest proportion of sexually reproducing offspring is found at high density.

Another group of investigators built on this result by pinpointing the specific aspects of crowding that affected the animals. These biologists brought a closely related species—*D. magna*—into the laboratory and altered day length, the amount of food available to individuals, and the quality of the water the *Daphnia* occupied. To vary water quality, the investigators used either clean water or "crowded" water taken from tanks where *D. magna* were being maintained at high density.

As **FIGURE 50.3b** shows, individuals in the study population switched to sexual reproduction only if they were exposed to poor-quality, "crowded" water, low food availability, *and* short day lengths. In short, *D. magna* needs three different cues from the environment to switch to sexual reproduction. Two of these cues were associated with high population density; the third is associated with the onset of winter.

(a) Sexual reproduction is more common in crowded populations of *Daphnia pulex* than in sparse populations.

(b) In *Daphnia magna*, which environmental cues trigger the switch to sexual reproduction?

Water quality	Food concentration	Day length	Sexual broods (%)
Clean	Low	Short	0
Crowded	Low	Short	44
Clean	Low	Long	0
Crowded	Low	Long	0
Clean	High	Short	0
Crowded	High	Short	0
Clean	High	Long	0
Crowded	High	Long	0

FIGURE 50.3 In *Daphnia*, Environmental Cues Signal the Switch from Asexual to Sexual Reproduction.
(a) There is a strong, positive correlation between the percentage of females that reproduce sexually and the density of the population (plotted on a log scale here). **(b)** Environmental conditions were varied experimentally for *Daphnia*. "Crowded" water was taken from tanks containing dense populations.

DATA: Kleiven, O. T., P. Larsson, and A. Hoboek. 1992. *Oikos* 65: 197–206.

✔ **QUESTION** How would you go about determining which molecule or molecules in "crowded" water serve as a signal that triggers sexual reproduction?

Why Do *Daphnia* Switch between Asexual and Sexual Reproduction? *Daphnia* appear to start sexual reproduction when conditions worsen. Why?

The leading hypothesis in answer to this question points out that sexually produced offspring are genetically diverse. When the environment changes, genetically diverse offspring are likely to include individuals that will survive better and reproduce more than will offspring that are identical to their parents. Parasites evolve rapidly, changing the environment for their hosts. According to the hypothesis, snails with greater genetic variability are likely to adapt better to those changes.

Genetically variable offspring have higher fitness in environments with rapidly evolving parasites, deteriorating physical conditions, or other types of rapid environmental change. Genetic variability increases the likelihood that at least some of the offspring will be able to survive and reproduce in such an environment. ✔ If you understand this concept, you should be able to predict the conditions under which asexual reproduction would be found as the only method of producing offspring.

To date, however, the variable-environment hypothesis has yet to be tested rigorously in *Daphnia*. Research on the adaptive significance of sex in this species and other organisms continues.

Mechanisms of Sexual Reproduction: Gametogenesis

The mitotic cell divisions, meiotic cell divisions, and developmental events that produce male and female gametes, or sperm and eggs, are collectively called **gametogenesis** (see Chapter 23). **Spermatogenesis** is the formation of sperm; **oogenesis** is the formation of eggs (**FIGURE 50.4**).

In the vast majority of animals, gametogenesis occurs in a sex organ, or **gonad.** Male gonads are called **testes;** female gonads are called **ovaries.** Early in development, reproductive cells known as germ cells enter the testes and ovaries and give rise to diploid cells that will undergo gametogenesis.

Spermatogenesis in Mammals   Figure 50.4a summarizes the events that take place during spermatogenesis. Note that in the male gonad, diploid cells called **spermatogonia** (singular: **spermatogonium**) divide by mitosis. Some of the resulting cells continue to function as spermatogonia; others change to form specialized cells that are committed to developing into sperm.

The specialized cells produced by spermatogonia are called **primary spermatocytes.** They undergo meiosis I and produce two **secondary spermatocytes,** which then undergo meiosis II. The result is four haploid cells called **spermatids.**

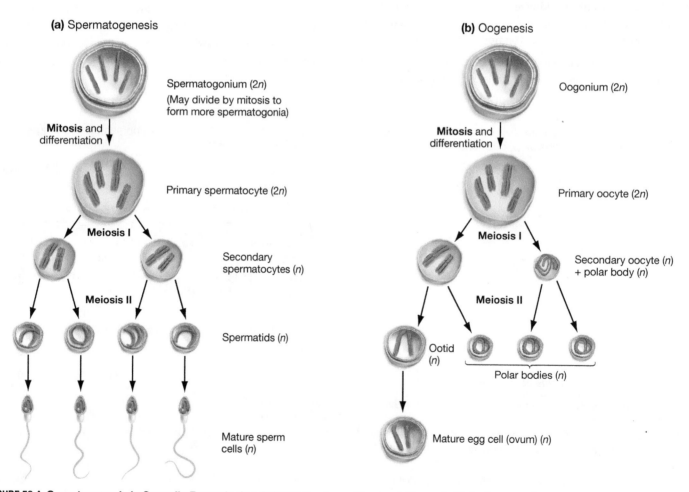

(a) Spermatogenesis

Spermatogonium (2*n*)
(May divide by mitosis to form more spermatogonia)

Mitosis and differentiation

Primary spermatocyte (2*n*)

Meiosis I

Secondary spermatocytes (*n*)

Meiosis II

Spermatids (*n*)

Mature sperm cells (*n*)

(b) Oogenesis

Oogonium (2*n*)

Mitosis and differentiation

Primary oocyte (2*n*)

Meiosis I

Secondary oocyte (*n*) + polar body (*n*)

Meiosis II

Ootid (*n*)

Polar bodies (*n*)

Mature egg cell (ovum) (*n*)

FIGURE 50.4 Gametogenesis in Sexually Reproducing Animals Produces Sperm or Eggs via Meiosis.

Each haploid spermatid matures into a sperm—a cell that is specialized for carrying a haploid genome from the male through the female reproductive tract and fertilizing an egg. The production of spermatogonia, primary spermatocytes, and sperm occurs continuously throughout a male's adult life.

Structure and Function of Sperm As a mammalian sperm cell matures, it acquires the four main compartments shown in **FIGURE 50.5**: the head, neck, midpiece, and tail.

- The head region contains the nucleus and an enzyme-filled structure called the **acrosome.** The enzymes stored in the acrosome allow the sperm to penetrate the barriers surrounding the egg.

- The neck encloses a centriole that will combine with a centriole contributed by the egg to form a centrosome (see Chapter 7). The centrosome is required for spindle formation during mitosis (see Chapter 12).

- The midpiece is packed with mitochondria (see Chapter 9), which produce the ATP required to power movement.

- The tail region consists of a **flagellum**—a long structure, composed of microtubules and surrounded by plasma membrane, that whips back and forth to make swimming possible (see Chapter 7).

Sperm are race cars—stripped down, streamlined, souped-up cells that are specialized for racing other sperm to the egg. Eggs, in comparison, are like semitrailers—bulky, far less mobile storage containers that are packed with valuable merchandise and securely locked.

Head
Neck
Midpiece
Tail

Acrosome
Nucleus
Centriole
Plasma membrane
Vacuole (not present in all sperm)
Mitochondria
Flagellum

0.5 μm

FIGURE 50.5 Mammalian Sperm Are Specialized for Motility and Fusing with an Egg Cell. The morphology of human sperm is typical of many mammal species.

Oogenesis in Mammals The top of Figure 50.4b highlights an important similarity between spermatogenesis and oogenesis: In the female gonad, diploid cells called **oogonia** (singular: **oogonium**) divide by mitosis. Some of the resulting cells continue to function as oogonia; others change to form specialized cells that are committed to producing an egg. In this respect, oogonia and spermatogonia are similar.

However, subsequent steps in gametogenesis are markedly different in females. The specialized cells produced by an oogonium are called **primary oocytes.** When these cells undergo meiosis, only one of the haploid products, known as a **secondary oocyte,** can eventually mature into an egg. The secondary oocyte is arrested in the final stages of meiosis II until it is fertilized by a sperm; it then completes meiosis to become an ootid, which matures into an **ovum,** or a mature egg cell.

The other cells produced by meiosis in females have a tiny amount of cytoplasm and do not mature into eggs. Because the distribution of cytoplasm is so unequal during each meiotic division in females, the smaller cells are called **polar bodies.** (Recall that polar refers to inequality or opposites.) Polar bodies degenerate shortly after their formation.

In addition, the production of primary oocytes stops early in development in many mammals; in humans, it stops before a female fetus is born. The primary oocytes enter prophase of meiosis I during fetal development but then stop developing for months, years, or decades.

Structure and Function of Eggs Animal gametes are easy to tell apart: Sperm are small and motile; eggs are large and nonmotile. Eggs are large mainly because they contain the nutrients required for the embryo's early development. The quantity of nutrients present in the egg varies widely among species, however.

- In mammals, embryos start to obtain nutrition through a maternal organ called the **placenta** within a week or two after fertilization. Thus, the egg only has to supply nutrients for early development and is relatively small.

- In species where females lay eggs into the environment, the stores in the egg are the *only* source of nutrients until organs have formed and a larva or juvenile hatches and begins to feed. In these species, the nutrients required for early development are provided by **yolk**—a fat- and protein-rich cytoplasm that is loaded into egg cells as they mature. Yolk may be present as one large mass or as many small granules.

Just outside the plasma membrane of eggs, a fibrous, mat-like sheet of glycoproteins called the **vitelline envelope** forms and surrounds the egg. In many aquatic animals, such as the sea urchin, a large gelatinous matrix known as a jelly layer surrounds the vitelline envelope to further enclose and protect the egg (see **FIGURE 50.6a**, on page 1018).

In the eggs of humans and other mammals, the vitelline envelope structure is unusually thick and is called the **zona pellucida**. This structure is surrounded by a layer of cells, called the corona radiata, that a sperm must penetrate before it can fertilize the oocyte (**FIGURE 50.6b**).

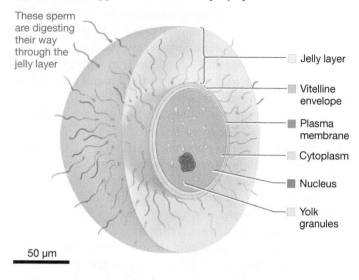

(a) Sea urchin eggs are surrounded by a jelly coat.

These sperm are digesting their way through the jelly layer

- Jelly layer
- Vitelline envelope
- Plasma membrane
- Cytoplasm
- Nucleus
- Yolk granules

50 μm

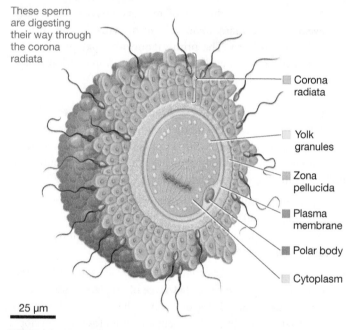

(b) Human oocytes are surrounded by a protective layer called the corona radiata.

These sperm are digesting their way through the corona radiata

- Corona radiata
- Yolk granules
- Zona pellucida
- Plasma membrane
- Polar body
- Cytoplasm

25 μm

FIGURE 50.6 Eggs Are Surrounded by Protective Structures.

50.2 Fertilization and Egg Development

Fertilization is the joining of a sperm and an egg to form a diploid **zygote.** Recall that during this process, a sperm makes contact with an egg and penetrates the egg's membrane. The molecular mechanisms of sperm–egg binding have been well characterized using the sea urchin as a model organism (see Chapter 23).

The following discussion focuses on the range of fertilization mechanisms observed among animals and the question of what happens to fertilized eggs. In many species, individuals release their gametes into their environment synchronously (at the same time), and external fertilization occurs. In other animals, sperm is deposited into the reproductive tracts of females, and internal fertilization occurs.

External Fertilization

Most animals that rely on external fertilization live in aquatic environments. The correlation between external fertilization and aquatic environments is logical, because gametes and embryos must be protected from drying. If external fertilization occurred in a terrestrial environment, either the gametes or the resulting zygote would likely die of desiccation.

Another observation is that species with external fertilization tend to produce exceptionally large numbers of gametes. For example, a female sea star *Asterias amurensis* typically releases 100,000,000 eggs into the surrounding seawater during spawning. Males release many times that number of sperm. The leading hypothesis to explain this pattern is that the probability of a sperm and egg meeting in an ocean or lake is small unless large numbers of gametes are present. In addition, production of a large number of embryos is beneficial because so many of them fall prey to predators.

Given that sperm and eggs from different individuals must be released into the environment synchronously for external fertilization to work, how is gamete release coordinated? The answer has two parts.

1. Gametogenesis usually occurs in response to environmental cues, such as lengthening days and warmer water temperatures, that indicate a favorable season for breeding.

2. Gametes are released in response to specific cues from individuals of the same species.

In fishes and other aquatic animals with well-developed eyes, external fertilization by spawning is often the culmination of an elaborate courtship ritual between a male and female. In contrast, courtship behavior appears to be much less important—or even absent—in species such as clams, sea urchins, and sea cucumbers. How do animals that cannot see their mates time the release of their gametes?

Recent research indicates that the chemical messengers called **pheromones** (see Chapter 47) might be involved in synchronizing gamete release. For example, biologists maintained two groups of sea cucumbers under natural conditions of light and temperature for 15 months. In one treatment, individuals were isolated in separate tanks; in another, they were kept in tanks containing other sea cucumbers. The researchers found that all the individuals that shared their tanks released gametes during the normal spawning period. In contrast, only about 10 percent of the individuals maintained in isolation released gametes.

Although these data suggest that pheromones might be involved in coordinating external fertilization in sea cucumbers, they are not definitive. Only recently have biologists begun to identify the actual pheromones involved in mating of marine invertebrates. Accumulation of data on pheromones in multiple species will help solidify the case that spawning is coordinated by pheromones.

Internal Fertilization

Internal fertilization occurs in the vast majority of terrestrial animals as well as in a significant number of aquatic animals. Internal fertilization occurs in one of two ways:

1. After copulation, in which males deposit sperm directly into the female reproductive tract with the aid of a copulatory organ, usually called a **penis.**

2. After males package their sperm into a structure called a **spermatophore,** which is then picked up and placed into the female's reproductive tract by the male or female.

In some salamander species, for example, the male places the spermatophore on the ground within its territory. Later the female picks it up with her **cloaca,** a chamber that both the reproductive and excretory systems flow into and that opens to the environment. In this case, the result is internal fertilization without any direct physical contact between the sexes.

What Is Sperm Competition? In terms of understanding animal behavior, perhaps the most important insight about internal fertilization originated with Geoff Parker. In 1970 Parker published experiments, on dung flies, that confirmed the existence of **sperm competition**—competition between sperm from different males to fertilize the eggs of the same female.

Parker's experiments consisted of a series of matings of one female with two males. In each experiment, the two males were selected in such a way that Parker could distinguish between their offspring. The proportion of offspring fathered by each male was not 50:50. Instead, whichever male was last to copulate fathered an average of 85 percent of the offspring produced.

In addition to suggesting that sperm were competing to fertilize eggs, these data indicated that, in this experiment, the second male won. Follow-up research has confirmed that **second-male advantage** is widespread, although not universal, in insects and some other animal groups. How does it occur?

Sperm Competition and Second-Male Advantage To explore why second-male advantage occurs, biologists turned to the fruit fly *Drosophila melanogaster*. A research group introduced a gene into male fruit flies that produced sperm with green fluorescent tails (**FIGURE 50.7**). When a mating by a green-spermed male was followed with a second mating—by a male having normal-colored sperm—many fewer green-tailed sperm were found in the female's sperm-storage area compared with the number observed when no second mating took place.

To interpret this finding, the biologists suggested that the sperm of the second male physically dislodged the first male's gametes from the female sperm-storage area and inserted themselves in their place. The researchers also showed that the fluid that accompanies sperm during fertilization is able to displace stored sperm from competing males. More recent research has shown that in addition to these factors, females are able to eject

---

RESEARCH

QUESTION: How does the "second-male advantage" occur in sperm competition?

HYPOTHESIS: In sperm-storage areas, sperm from the second male displaces sperm from the first male.

NULL HYPOTHESIS: The mechanism does not involve sperm displacement from sperm-storage areas.

EXPERIMENTAL SETUP:

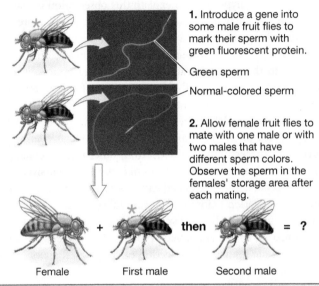

1. Introduce a gene into some male fruit flies to mark their sperm with green fluorescent protein.

Green sperm

Normal-colored sperm

2. Allow female fruit flies to mate with one male or with two males that have different sperm colors. Observe the sperm in the females' storage area after each mating.

Female + First male then Second male = ?

PREDICTION: When females mate twice, little sperm from the first male remains in storage.

PREDICTION OF NULL HYPOTHESIS: When females mate twice, most or all of the sperm deposited by the first male is still present.

RESULTS:

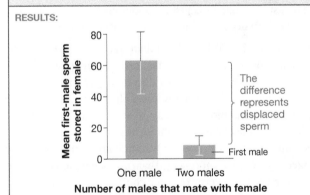

The difference represents displaced sperm

First male

CONCLUSION: If a female mates a second time, most sperm from the first male she mated with disappears.

FIGURE 50.7 Experimental Evidence Supports Second-Male Advantage in *Drosophila*. The graph shows the average number of green sperm stored in females when males with green sperm were the only male to mate or the first of two males to mate.

Price, C. S. C., K. A. Dyer, and J. A. Coyne. 1999. Sperm competition between *Drosophila* males involves both displacement and incapacitation. *Nature* 400: 449–452.

✔**QUESTION** Based on these data, is the claim that sperm from the second male physically displaces sperm from the first male valid?

sperm from their reproductive tracts. Together these mechanisms resulted in the second male's sperm fertilizing most of the eggs laid.

Why Is Testes Size Variable among Species? Research on sperm competition has recently contributed another major finding. In species where females routinely mate with multiple males before laying eggs or giving birth, males have extraordinarily large testes for their size and produce proportionately larger numbers of sperm.

The leading hypothesis to explain this observation is that fertilization is similar to a lottery in which each sperm represents a ticket. The more tickets a male enters in the competition, the higher his chance of "winning" fertilizations and passing his alleles on to the next generation. Males with exceptionally large testes produce exceptionally large numbers of sperm, and they are more likely to win the lottery.

The lottery model has been challenged, however, by evidence that females often store sperm and exert control over which sperm are successful in fertilization. In other words, females do not always accept the results of sperm competition passively. Females of some species actively choose which male performs the last copulation before fertilization takes place. In other species, females physically eject sperm from undesirable males. This phenomenon has been dubbed cryptic female choice. The name is appropriate because the selection of sperm by females is hidden from males.

Unusual Mating Strategies

Studies on animal sexual reproduction have documented some of the most remarkable behaviors and structures observed in nature. The list that follows is just a sampling, offered simply to illustrate the diversity of mating arrangements in animals.

- *Femmes fatales* When Australian redback spiders mate, the male does a somersault after inserting his penis-like organ. The somersault places his dorsal surface in front of the female's mouthparts. In many cases the female responds by eating the male. (Biologists refer to females who cannibalize males as femmes fatales—a phrase used to describe murderous human females in movies.) Cannibalized males copulate longer and fertilize more eggs than non-cannibalized males, probably because sperm transfer continues until the meal is over.

- *Giant sperm* In the fruit fly *Drosophila bifurca*, males average 1.5 mm in total body length. Their sperm, however, are each 6 cm long. The coiled, bulky sperm fill the female's sperm-storage area and make it impossible for another male's sperm to enter.

- *Infidelity* When researchers assess paternity in bird species that appear to be monogamous, they find that up to 60 percent of nests contain at least one offspring fathered by a male that is not mated to the resident female. In most cases, females actively solicit copulations from males holding nearby territories.

FIGURE 50.8 Snails Stab with Love Darts. Stabbing a mate with a mucus-covered love dart increases reproductive success.

- *Love darts* Many snails and slugs are **hermaphroditic,** meaning an individual has both male and female gonads. During mating, two individuals simultaneously receive and deposit sperm to fertilize each other's eggs. In some species, individuals fire mucus-covered "love darts" from their genitalia into the mating partner (**FIGURE 50.8**). A substance in the mucus increases the chance that sperm will successfully fertilize eggs.

- *Hypodermic insemination* Some bedbugs have hypodermic penises, meaning that they function like a hypodermic needle. Males force the organ through the female's abdominal wall and deposit sperm directly into her body cavity.

Why Do Some Females Lay Eggs While Others Give Birth?

For females, many aspects of fertilization and egg laying vary among species: for example, a female's size or age at first breeding; the number and size of eggs she produces; the number of times she reproduces over the course of a lifetime. Here let's consider a different aspect of variation in animal reproduction: whether eggs are laid outside the body or retained inside the body.

In **oviparous** ("egg-bearing") animals, the embryo completes most of its development inside an egg that is laid into the environment. In some oviparous species, such as sea stars, sea urchins, and most insects, no further care is provided by the parents; the eggs and embryos are left to fend for themselves. Birds, however, incubate their eggs and feed the young after hatching; fish may guard their eggs from predators and fan the clutches to oxygenate them.

In **viviparous** ("live-bearing") species, embryonic development takes place entirely within the mother's body. The embryo attaches to the reproductive tract of the mother and receives nutrition directly from her—via diffusion from her circulatory system. In **ovoviviparous** species, offspring develop inside the mother's body but are nourished by nutrient-rich yolk stored in the egg.

Why does oviparity exist in some groups and viviparity or ovoviviparity in others? Biologists tackled this question by studying the lizard genus *Sceloporus*. Some *Sceloporus* populations are oviparous; others are ovoviviparous.

To understand how oviparity and ovoviviparity evolved, the biologists analyzed a phylogenetic tree—based on molecular and morphological data—of many *Sceloporus* species (FIGURE 50.9). Two conclusions should make sense to you (see **BioSkills 7** in Appendix B for help with interpreting phylogenetic trees):

1. Because the basal branches of the tree represent oviparous species, egg laying probably represents the original or ancestral condition.

2. As the red branches on the tree show, ovoviviparity evolved independently in two groups.

Using a similar research strategy, biologists have found that ovoviviparous populations of sea stars have evolved from oviparous populations on several occasions.

Why did natural selection favor these changes between egg laying and live birth? Ovoviviparity or viviparity should evolve when it leads to higher numbers of surviving young.

Researchers have long hypothesized that natural selection favoring live birth should be especially strong in cold habitats. Low temperatures slow the development of embryos. Thus, in cold habitats, it might be advantageous for females to retain eggs inside their bodies so that they can thermoregulate to maintain the offspring at a more favorable temperature for development. This hypothesis is supported by the fact that ovoviviparous *Sceloporus* species live in the highlands of the southwestern United States and central Mexico, where temperatures can be cool.

Sceloporus lizards have been a productive group in which to study this question because the trait has evolved multiple times in extremely closely related species—making it easier to find correlations between live birth and other characteristics, such as living in cold habitats. The issue is much more difficult to study in mammals, where viviparity evolved just once. Monotremes (duck-billed platypus and echidna) are oviparous, while marsupials and placental mammals are viviparous (see Chapter 35). It

is still not clear why, and how, viviparity evolved in mammals before the marsupial–placental split. Is the cold-habitats hypothesis relevant? To date, nobody knows.

50.3 Reproductive Structures and Their Functions

The first two sections of this chapter considered (1) the broad contrast between asexual and sexual reproduction and (2) general patterns in fertilization and egg care. Now let's explore the mechanics of sexual reproduction in more detail. The first task is to understand the anatomy of the male and female reproductive systems—focusing primarily on mammals.

The Male Reproductive System

In humans, the external anatomy of the male reproductive system consists of the scrotum and the penis. The saclike **scrotum** holds the testes; the penis functions as the organ of copulation necessary for internal fertilization.

Biologists who have compared these structures among animal species have been struck by their variability and complexity. For example, a scrotum occurs only in certain mammal species. Whales, elephants, hedgehogs, moles, and many other mammal groups, along with all other vertebrates, lack the structure entirely and instead have testes that are located well within the abdominal cavity. Among species that do have a scrotum, the structure's size and shape vary from tiny to prominent. In many primate species, the scrotum is brightly colored and appears to function in courtship or other sexual display behaviors.

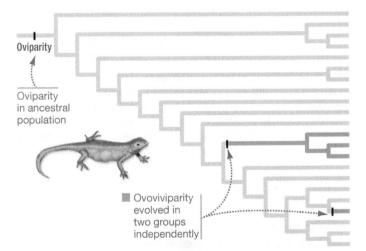

FIGURE 50.9 Ovoviviparity Has Evolved More than Once in *Sceloporus* Lizards. Each twig on this phylogenetic tree represents a *Sceloporus* species or population from central Mexico.

How Does External Anatomy Affect Sperm Competition? Diversity in scrotal morphology pales in comparison with variation in the structure of the penis or other types of male **genitalia,** meaning external copulatory organs. In many groups of insects and spiders, for example, closely related species are morphologically identical except for their distinctive genitalia. Why are these organs so diverse?

The leading hypothesis on this question is that certain genitalia shapes give males an advantage in sperm competition. To test this idea, biologists measured the size of the spines found on the tips of male genitalia in a species of seed beetle (**FIGURE 50.10a**) and then documented which males were most successful at fertilizing eggs when females mated with two males.

The *x*-axis in **FIGURE 50.10b** plots variation in the length of genital spines; the *y*-axis plots the proportion of eggs fertilized by the second male to copulate with a female, during experimental matings. The data points on the graph indicate genital spine length in each second-to-mate male tested. Note that second-to-mate males with the longest genital spines fathered a higher percentage of offspring than did others. The mechanism appears to be the ability of long spines to stick in the female's reproductive tract and prolong copulation time, even if the female is ready for copulation to end.

The upshot? Size matters, in seed beetles. Natural selection that occurs during sperm competition may explain why genitalia are so diverse among insect and spider species.

Internal Anatomy of Human Male Reproductive Organs Although it consists of many structures, the reproductive system in human males has just three basic functional components. **FIGURE 50.11** shows the relevant structures in side view and front view.

1. *Spermatogenesis and sperm storage* Sperm are produced in the testes and stored nearby in the **epididymis.**

2. *Production of accessory fluids* Complex solutions form in the **seminal vesicles, prostate gland,** and **bulbourethral gland.** These accessory fluids are added to sperm before **ejaculation,** or expulsion from the body. **TABLE 50.1** lists some components of these accessory fluids, which combine with sperm to form **semen.**

3. *Transport and delivery* A **vas deferens** is a muscular tube that transports sperm from the epididymis to the short **ejaculatory duct,** where it is mixed with accessory fluids to form semen. The semen then enters the **urethra,** a longer tube that passes through the penis and services both the reproductive and urinary systems in males. The semen is expelled during ejaculation.

The composition of the accessory fluids varies widely among animals. In many insects, spiders, and vertebrates, molecules in the accessory fluids congeal after they arrive in the female reproductive tract and plug it. Experiments have shown that these copulatory plugs can serve as an effective deterrent to future matings. In some species, though, females or second males actively remove them.

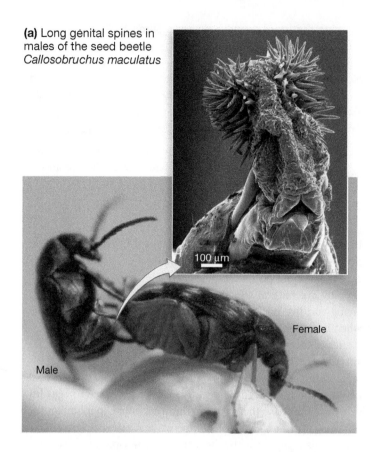

(a) Long genital spines in males of the seed beetle *Callosobruchus maculatus*

100 µm

Female

Male

(b) During sperm competition, males with longer genital spines father more offspring.

©2009 Elsevier

FIGURE 50.10 In Some Insects and Spiders, Variation of Male Genitalia among Individuals May Affect Reproductive Success. **(a)** An elaborate male reproductive structure in a seed beetle, used to transfer sperm to the female. The spiky structures at the top of the inset photo are called genital spines. **(b)** Data from experiments on reproductive success during sperm competition.

DATA: Hotzy, C., and G. Arnqvist. 2009. *Current Biology* 19: 404–407.

Another diverse aspect of male internal anatomy is a bone inside the penis called the **baculum.** Some mammal species, including humans, lack this feature. But in rodents, the shape of the baculum is so variable that it can be used to distinguish among species. And in seals, baculum size correlates with mating system—the mating practices observed in a species. For example,

(a) Side view

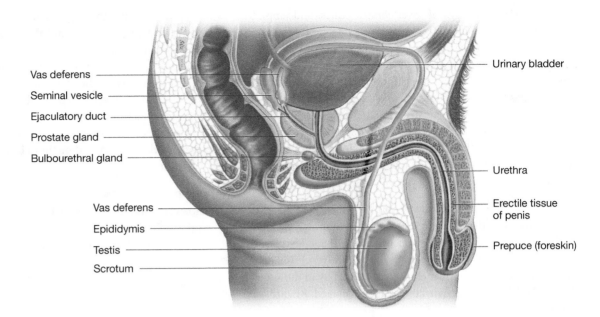

- Vas deferens
- Seminal vesicle
- Ejaculatory duct
- Prostate gland
- Bulbourethral gland

- Vas deferens
- Epididymis
- Testis
- Scrotum

- Urinary bladder
- Urethra
- Erectile tissue of penis
- Prepuce (foreskin)

(b) Front view

- Urinary bladder
- Seminal vesicle
- Prostate gland
- Bulbourethral gland
- Urethra
- Erectile tissue of penis
- Vas deferens
- Epididymis
- Testis

FIGURE 50.11 The Reproductive Tract in a Human Male Produces, Stores, and Transports Sperm.

SUMMARY TABLE 50.1 **Accessory Fluids in Human Semen**

Source	Content	Function
Seminal vesicles	Fructose (a sugar)	Source of chemical energy for sperm movement
	Prostaglandins	Stimulate smooth-muscle contractions in uterus
Prostate gland	Antibiotic compound	Prevent urinary tract infections in males
	Citric acid	Nutrient used by sperm
Bulbourethral gland	Alkaline mucus	Lubricate tip of penis; neutralize acids in urethra

in seal species in which females routinely mate with several males before becoming pregnant, males have not only large testes for their size but also a large baculum. The testes and baculum are much smaller in species in which females mate with a single male. In all species where it appears, the baculum helps stiffen the penis during copulation.

The Female Reproductive System

In animals, the most important part of the female reproductive system is the ovary—where meiosis occurs and mature egg cells, or ova, are produced. In the vast majority of species, the mature egg cell is a membrane-bound structure consisting of a haploid nucleus, a full complement of other organelles, and a large supply of nutrients in the form of yolk.

In the gelatinous eggs of invertebrates, fishes, and amphibians, the extensive outer layers and plasma membranes play a key role in binding sperm from the same species and initiating fertilization. Other animals produce membrane-rich amniotic eggs, which evolved as an innovation that allowed tetrapods to lay large eggs that do not desiccate on land (see Chapter 35).

To probe variation in egg structure and female reproductive systems further, let's consider two highly specialized examples, the reproductive systems of female birds and mammals. Birds lay an amniotic egg protected by a hard shell; most mammals are viviparous.

The Reproductive Tract of Female Birds FIGURE 50.12 diagrams the bird reproductive system. Starting with the ovary, the labels on the drawing indicate the sequence of events that takes place as the egg moves down the reproductive tract. The result is a hard-shelled egg—like the familiar chicken eggs you buy in the store—that can be laid into the environment and incubated. All birds are oviparous.

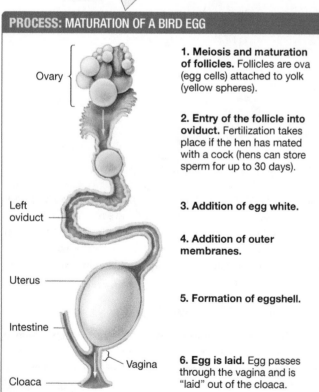

PROCESS: MATURATION OF A BIRD EGG

Ovary

Left oviduct

Uterus

Intestine

Vagina

Cloaca

1. **Meiosis and maturation of follicles.** Follicles are ova (egg cells) attached to yolk (yellow spheres).

2. **Entry of the follicle into oviduct.** Fertilization takes place if the hen has mated with a cock (hens can store sperm for up to 30 days).

3. **Addition of egg white.**

4. **Addition of outer membranes.**

5. **Formation of eggshell.**

6. **Egg is laid.** Egg passes through the vagina and is "laid" out of the cloaca.

FIGURE 50.12 All Birds Are Oviparous.

From the time the egg is released from the ovary until the zygote undergoes mitosis and the embryo begins to develop, a bird egg is a single cell. The ostrich egg, which can be over 15 cm (6 inches) in diameter, contains one of the largest single cells known in animals. The structure stores enough nutrients and water to sustain development until hatching.

Note that although male birds have two testes, females of most bird species have just one functional ovary. The presence of a single working ovary is thought to be an adaptation that reduces weight and makes flight more efficient.

External and Internal Anatomy of Human Females FIGURE 50.13 shows side and front views of the human female reproductive system. The external anatomy features the **labia minora** (singular: **labium minus**) and the **labia majora** (singular: **labium majus**), the opening of the urethra, and the opening of the vagina.

- The labia are folds of skin that cover the urethral and vaginal openings.
- The **clitoris** is an organ that develops from the same population of embryonic cells that gives rise to the penis in males. It becomes erect during sexual stimulation and is covered with a protective sheath called the *prepuce*, homologous with the prepuce (foreskin) that covers the end of the penis.
- The urethral opening, where urine is expelled, is separate from the reproductive structures.
- The **vagina,** or birth canal, is the chamber where semen is deposited during sexual intercourse and where the baby is delivered during childbirth.

The internal anatomy of the female reproductive system in humans and other mammals is dominated by structures serving the system's two functions:

1. *Production and transport of eggs* Eggs are produced in the paired ovaries. During **ovulation,** a secondary oocyte is expelled from the ovary and enters the **oviduct,** also known as the **fallopian tube,** where fertilization may take place. Fertilized eggs are then transported through the oviduct to the muscular sac called the **uterus.**
2. *Development of offspring* The uterus is where embryonic development takes place. During childbirth, the developed embryo (now called a fetus) passes through an opening in the **cervix**—the bottom part of the uterus—and into the vagina.

Passage of the fetus through the cervix and vagina is the norm in mammals. However, the spotted hyena has evolved a fascinating alternative. In female hyenas, the birth canal is enclosed within an inch-wide, 7-inch-long clitoris that resembles a male's penis (**FIGURE 50.14**). Mating and birthing occur through the clitoris, which often tears during delivery of the Chihuahua-sized cubs.

Although the adaptive significance of the hyena clitoris is unknown, scientists have determined that it results from very high levels of male-typical sex hormones in females. These hormones also cause females to be larger and more muscular than males and socially dominant over them.

(a) Side view

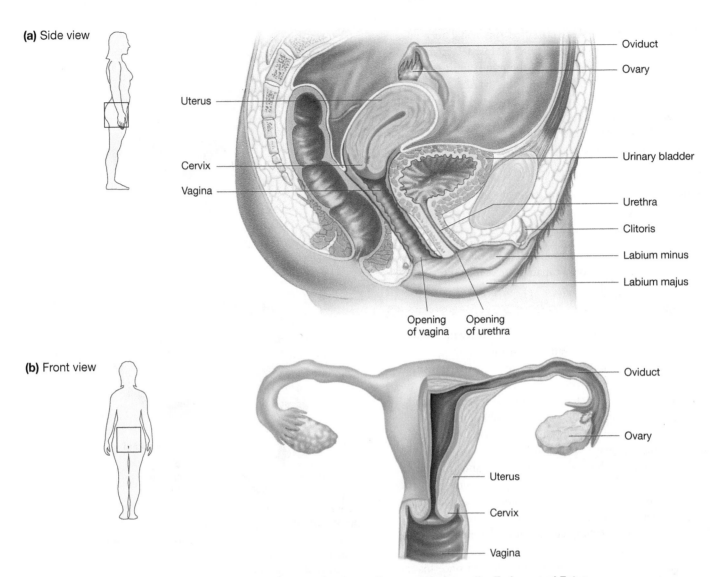

Oviduct

Ovary

Uterus

Cervix

Vagina

Urinary bladder

Urethra

Clitoris

Labium minus

Labium majus

Opening of vagina   Opening of urethra

(b) Front view

Oviduct

Ovary

Uterus

Cervix

Vagina

FIGURE 50.13 The Reproductive Tract in a Human Female Produces Eggs and Nurtures the Embryo and Fetus.

✔**QUESTION** Generate a hypothesis to explain why male and female gonads are paired in mammals. How would you test your hypothesis?

FIGURE 50.14 Female Spotted Hyenas Have Enlarged Clitori. High levels of testosterone are responsible for the male-like anatomy of female hyenas. They copulate and give birth through the long, narrow clitoris.

Let's take a further look at how sex hormones control reproduction, focusing on a mammal—specifically, humans—as an illustrative example. But it is important to note that the same or very similar hormones control reproduction in all vertebrates, from fishes to birds.

50.4 The Role of Sex Hormones in Mammalian Reproduction

Recall that the sex hormones **testosterone** and **estradiol** are steroids; the latter belongs to a class of hormones known as estrogens (see Chapter 49). Testosterone and estradiol bind to receptors within the cytoplasm or nucleus of target cells. The resulting hormone–receptor complexes bind to DNA and trigger changes in gene expression.

Testosterone and estradiol are classified as gonadal hormones because they are produced in the gonads. Most testosterone is synthesized in specialized cells inside the testes; most estradiol and other estrogens are synthesized in the ovaries. More specifically, the female sex hormones are produced by cells that surround each developing egg. These surrounding cells form a structure called a **follicle** around the egg.

The sex hormones play a key role in three events:

1. Development of the reproductive tract in embryos

2. Maturation of the reproductive tract during the transition from childhood to adulthood

3. Regulation of spermatogenesis and oogenesis in adults

To explore the action of sex hormones, let's take a closer look at their role in the transition from human juvenile to adult.

Which Hormones Control Puberty?

Puberty is the process that leads to sexual maturity in humans. In amphibians, the juvenile-to-adult transition is triggered by the hormone T_3 (triiodothyronine); in insects, the transition occurs in response to ecdysone (see Chapter 49). But in humans, the transition is directed by increased levels of gonadal hormones—testosterone in boys and estradiol in girls.

A group of physicians recently offered dramatic evidence for testosterone's role: a 2-year-old boy whose symptoms included an enlarged penis, the development of pubic hair, and facial acne. Through careful interviewing, the doctors discovered that the child's father was a bodybuilder who was smearing a testosterone cream on his own arms in an attempt to build muscle mass. The child was apparently absorbing enough testosterone from being carried by his father to trigger puberty-like symptoms. All his symptoms except for penile enlargement subsided once his father discontinued the testosterone cream.

What Regulates the Gonadal Hormones? Researchers suggested that sex hormone production is regulated by the hypothalamic–pituitary axis (see Chapter 49). Recall that chemical signals from the hypothalamus lead to the release of regulatory hormones from the pituitary gland, which then cause the release of hormones from other glands.

Two advances made it possible to test this hypothesis rigorously:

1. Researchers isolated a hormone called **gonadotropin-releasing hormone (GnRH)** from the hypothalamus.

2. Investigators noted that boys and girls who were entering puberty experienced pulses in the concentration of two pituitary hormones, **luteinizing hormone (LH)** and **follicle-stimulating hormone (FSH)**.

These observations inspired the hypothesis that GnRH directs LH and FSH pulses, which then trigger increases in testosterone and estradiol (**FIGURE 50.15**).

To test this idea, researchers administered pulses of GnRH to boys and girls who had deficits in their hypothalamus and were experiencing delays in the onset of puberty. As predicted, the

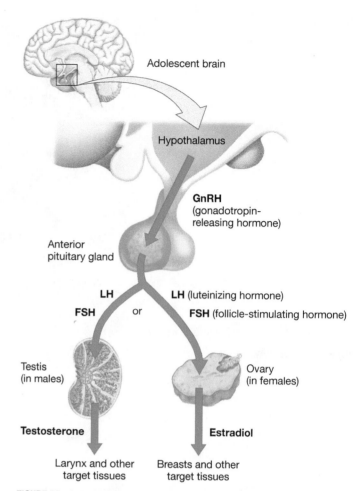

FIGURE 50.15 In Humans, Puberty Is Triggered by Hormones from the Hypothalamus and Pituitary.

✓ QUESTION How does control of testosterone and estradiol production compare with control of cortisol release by the adrenal gland? (See Figure 49.12c.)

GnRH treatment induced surges in LH and FSH, followed by puberty onset.

What Regulates the Hypothalamic and Pituitary Hormones? The model in Figure 50.15 raises a question: What triggers GnRH increases at the appropriate age? Although this question remains unanswered, there is some evidence that nutritional state is involved. For example:

• The current average age for the onset of menstruation in females in the United States is slightly over 12 years. This is much earlier than the average age of 17 years during the eighteenth and nineteenth centuries, when the general nutritional state of the population was poorer.

• Among girls living today, individuals with large fat stores tend to enter puberty earlier than do girls who are thin.

• The hormone leptin is produced by fat cells (see Chapter 49), and leptin levels are higher in people with more fat. In mice, leptin accelerates the onset of puberty, suggesting that leptin may be involved in triggering the GnRH surge.

Research continues on how age, nutritional condition, and perhaps other factors interact to promote the release of GnRH.

If you recall the discussion of how the adrenal hormone cortisol is controlled, however, you might suspect that the model of sex-hormone regulation in Figure 50.15 is simplified (see Chapter 49). Many hormones participate in negative feedback—also called feedback inhibition—meaning that the presence of the hormone inhibits the factor that triggers its release.

Do sex hormones participate in negative feedback? The short answer to this question is yes. To appreciate the details, let's investigate hormonal control of the human menstrual cycle.

Which Hormones Control the Menstrual Cycle in Mammals?

FIGURE 50.16 illustrates the sequence of events in the human ovary during the **menstrual cycle,** a monthly reproductive cycle. Although the cycle's length varies among women, 28 days is about average.

In conjunction with changes in the ovary illustrated in the figure, the lining of the uterus undergoes a dramatic thickening and regression. Ultimately, if fertilization and implantation do not occur, part of the uterine lining sloughs off and is expelled through the vagina.

Day 0 in the menstrual cycle is marked by the beginning of **menstruation**—the expulsion of the uterine lining. The remainder of the cycle has two distinct phases:

1. *Follicular phase*   A follicle matures during the **follicular phase,** which lasts an average of 14 days. Primary oocytes complete meiosis I during this phase. Ovulation occurs when the follicle is mature and releases its secondary oocyte into the oviduct.

2. *Luteal phase*   The **luteal phase** begins with ovulation and averages 14 days in length. Its name was inspired by the formation

and subsequent degeneration of a structure called the **corpus luteum** ("yellowish body") from the ruptured follicle.

The regular occurrence of ovulation throughout the year makes human females extremely unusual among mammals. Although some mammals ovulate multiple times during the year, most ovulate only during a single prescribed breeding season—often in response to environmental cues such as changing photoperiod; less often in response to cues from males.

In addition, only humans and other great apes menstruate. In the vast majority of mammals, the lining of the uterus is reabsorbed if pregnancy does not occur. These females are described as having an **estrous cycle** and are sexually receptive only during estrus—also known as being "in heat."

Whether an estrous or menstrual cycle occurs, the basic sequence of events, with a follicular phase preceding ovulation and a luteal phase following ovulation, is shared among mammals. Hormonal control of the estrous and menstrual cycles is also similar.

How Do Pituitary and Ovarian Hormones Change during a Menstrual Cycle?   By monitoring hormone concentrations in the blood or urine of a large number of women over the course of the menstrual cycle, researchers were able to document dramatic changes in the concentrations of estradiol and several other hormones. During each cycle,

- LH and FSH are produced in the anterior pituitary gland in response to GnRH;

- the steroid hormone **progesterone** is produced along with estrogens, including estradiol, in the ovaries.

Several observations jump out of the data in **FIGURE 50.17** (see page 1028):

1. LH levels are fairly constant except for a spike that begins just before ovulation, suggesting that LH might be the trigger for this event.

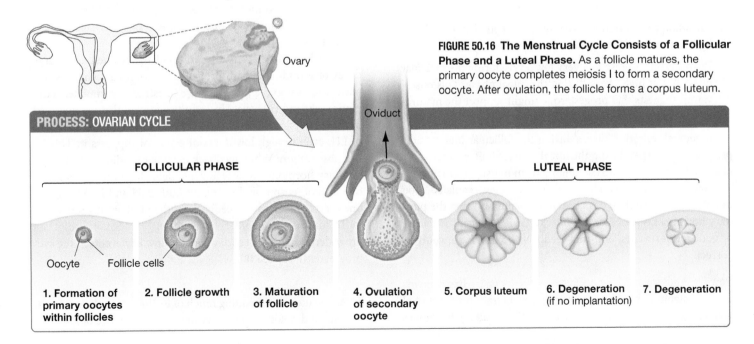

Ovary

Oviduct

FIGURE 50.16 The Menstrual Cycle Consists of a Follicular Phase and a Luteal Phase. As a follicle matures, the primary oocyte completes meiosis I to form a secondary oocyte. After ovulation, the follicle forms a corpus luteum.

PROCESS: OVARIAN CYCLE

FOLLICULAR PHASE

LUTEAL PHASE

Oocyte   Follicle cells

1. Formation of primary oocytes within follicles

2. Follicle growth

3. Maturation of follicle

4. Ovulation of secondary oocyte

5. Corpus luteum

6. Degeneration (if no implantation)

7. Degeneration

FOLLICULAR PHASE
Follicle matures

Ovulation

LUTEAL PHASE
Corpus luteum degenerates

FIGURE 50.17 Hormones Regulate Events in the Human Menstrual Cycle.

Ovarian cycle

Ovarian hormone cycle

Progesterone

Estradiol

Pituitary hormone cycle

LH

FSH

Menstrual (uterine) cycle

Menstruation

Day

2. FSH concentrations are relatively high during the follicular phase and low during the luteal phase, though they also make a small spike before ovulation.

3. Estradiol concentrations change in a complex way, highlighted by a peak late in the follicular phase.

4. Progesterone is present at very low levels during the follicular phase but at high levels during the luteal phase. This observation suggests that progesterone might support the maturation of the thickened uterine lining.

In general, estradiol surges during the follicular phase, while progesterone surges during the luteal phase. Similar patterns are observed in other mammals, ranging from marsupials to mice.

Now the question is: How are these changes regulated? You might hypothesize (based on reading Chapter 49) that the pituitary hormones released in response to GnRH trigger the release of gonadal hormones, and that feedback occurs. If so, you'd be correct.

How Do Pituitary and Ovarian Hormones Interact? Experiments helped establish that changes in the concentration of estradiol and progesterone affect the release of the pituitary hormones

LH and FSH. Researchers worked with three volunteers whose ovaries had been removed because of cancerous growths or other problems. The women were receiving low, maintenance-level doses of estradiol, which appeared to exert negative feedback on LH and FSH.

But when the investigators injected the women with larger doses of estradiol or with progesterone, dramatic changes took place. For example, a large increase in estradiol stimulated a dramatic spike in LH levels. This result suggested that positive feedback was occurring. High levels of estradiol increase the release of LH, even though low doses of estradiol suppress it. This is a key observation: When feedback occurs, estradiol's effect on the pituitary hormones is dosage dependent. In contrast, progesterone injections appeared to inhibit both FSH and LH. Progesterone exerts only negative feedback on the pituitary hormones.

To summarize the interplay between LH, FSH, estradiol, and progesterone, let's start at day 0 and follow key events as the cycle progresses (**FIGURE 50.18**).

Day 0–7

- As the uterus is shedding much of its lining, a follicle is beginning to develop in one ovary under the influence of FSH.

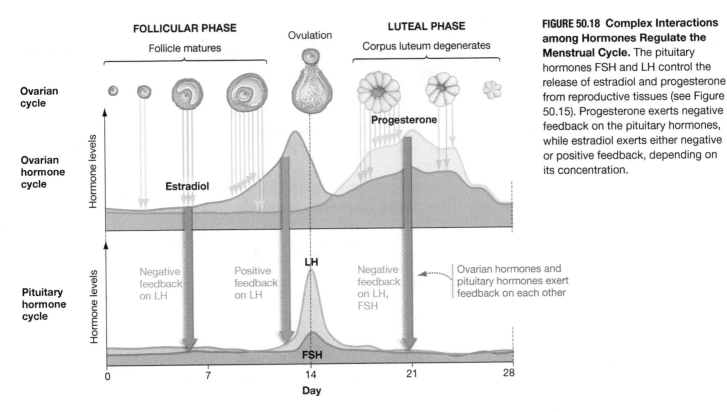

FIGURE 50.18 Complex Interactions among Hormones Regulate the Menstrual Cycle. The pituitary hormones FSH and LH control the release of estradiol and progesterone from reproductive tissues (see Figure 50.15). Progesterone exerts negative feedback on the pituitary hormones, while estradiol exerts either negative or positive feedback, depending on its concentration.

- The follicle produces estradiol and a small amount of progesterone.
- While its levels are still relatively low, estradiol suppresses LH secretion through negative feedback inhibition.

Day 8–14

- As the follicle grows, its production of estradiol gradually increases. The increase in estradiol stimulates mitosis and an increase in cell number in the uterine lining.
- The enlarged follicle produces large quantities of estradiol, which begin to exert positive feedback on LH secretion.
- Positive feedback results in a spike in LH levels, just after estradiol concentrations peak.
- The LH surge triggers ovulation and ends the follicular phase.

✔ If you understand that estradiol's effect on LH changes from negative regulation to positive regulation, you should be able to predict the effect of (1) injecting a female volunteer with high estradiol concentrations early in a cycle, and (2) using a drug to keep estradiol production low throughout the cycle.

Day 15–21

- As the corpus luteum develops from the remains of the ruptured follicle, it secretes large amounts of progesterone and small quantities of estradiol, in response to LH.
- The rise in progesterone lowers production of LH and FSH and activates the thickened uterine lining, creating a spongy tissue with a well-developed blood supply. In this way, progesterone fosters an environment that supports embryonic development if fertilization occurs.

Day 22–28

- If fertilization does not occur, the corpus luteum degenerates.
- Progesterone levels fall as the corpus luteum shrinks.
- When progesterone declines, the thickened lining of the uterus degenerates. This causes the menstrual bleeding that marks the first day of the next cycle.
- GnRH, LH, and FSH are released from the inhibitory control that progesterone exerts.
- LH and FSH levels rise, and a new menstrual cycle begins.

The interplay between ovarian and pituitary hormones is similar in other mammal species that have been studied to date.

Manipulating Hormone Levels to Prevent Pregnancy Data on hormonal control of the menstrual cycle opened new avenues in birth control research. Specifically, researchers have found that manipulating levels of progesterone and estradiol can prevent ovulation and serve as a safe and effective method of **contraception,** or preventing unwanted pregnancies.

Hormone-based contraceptive methods deliver synthetic versions of progesterone or of progesterone and estradiol. These hormones suppress the release of GnRH, FSH, and LH through negative feedback inhibition. Because an LH spike does not occur during the follicular phase of the cycle, the follicle does not mature and ovulation does not occur.

In the United States, birth control pills are the most widely used contraceptive method. Hormone-containing pills are taken for three weeks and then stopped for one week to allow menstruation to occur. Other popular hormonal contraceptives include injections and hormone-secreting patches and implants.

Type	Name	Mode of Action	Percent Effectiveness*
Hormone-based methods	The Pill, the Patch, the Ring, the Shot, the Implant	Provides continuous or cyclical delivery of progesterone, or progesterone plus estradiol.	92 to 99.9[†]
Barrier methods	Condom	Covers penis and prevents sperm from entering uterus.	85
	Female condom	Covers labia, vagina, and cervix and prevents sperm from entering uterus.	79
	Diaphragm	Covers cervix and prevents sperm from entering uterus.	84
	Sponge	Covers cervix and prevents sperm from entering uterus; also contains a molecule that immobilizes sperm.	84
	Spermicide	Foam or jelly covers cervix and prevents sperm from entering uterus; contains a molecule that immobilizes sperm.	71
Behavioral methods	Rhythm method	Couple refrains from vaginal intercourse around time of ovulation.	80
	Withdrawal	Man withdraws penis before ejaculation.	73
Fertilization prevention	Intrauterine device (IUD)	Small T-shaped structure inserted into uterus; induces uterus to produce substances hostile to sperm and eggs.	99
	Emergency contraception	Delivers progesterone or progesterone and estradiol after unprotected vaginal intercourse.	92
Pregnancy termination	Mifepristone	Blocks progesterone receptors so menstruation occurs even after fertilization and implantation.	92

*"Percent Effectiveness" indicates the average percentage of women who do not become pregnant during one year of typical use.

[†]Depends on delivery system used.

But as **TABLE 50.2** indicates, hormone-based methods are just one of several approaches to preventing pregnancy. Other methods work by preventing sperm from contacting the oocyte or by interfering with implantation of the embryo.

check your understanding

If you understand that . . .

- During the follicular phase of the menstrual cycle, FSH and LH stimulate the follicle to develop and secrete estradiol, which prepares the uterine lining for implantation of an embryo.
- During the luteal phase, progesterone protects the uterine lining. If fertilization does not occur, progesterone drops and the lining is shed.

✔ **You should be able to . . .**

1. Describe the function of FSH, and explain how FSH levels increase at the end of one cycle and the start of the next cycle.

2. Predict the consequences of a drug that inhibits the release of FSH.

Answers are available in Appendix A.

The "Percent Effectiveness" column on the far right of the table indicates the average percentage of women who do not become pregnant during one year of typical use of that method. Percent effectiveness usually increases dramatically if couples are able to use a method "perfectly"—meaning, exactly as specified for optimal effectiveness, during every episode of sexual intercourse.

When sperm and egg do unite successfully, the menstrual cycle is interrupted. The corpus luteum does not degenerate, and progesterone and estradiol levels stay high. Menstruation does not occur. Instead, the mother is now pregnant.

50.5 Pregnancy and Birth in Mammals

Viviparity allows the mother to provide a warm, protected environment for offspring during early development. Oviparous species that guard or incubate their eggs also provide warm, safe surroundings for their young.

Any investment that parents make in an offspring comes at a cost, however: The more a mother invests in each offspring, the lower the number of offspring she can produce.

Pregnancy and **lactation**—providing milk that nourishes offspring after birth—represent some of the most extreme forms of parental care known in animals. And in some mammal species, parental care continues long after lactation ends. Humans, for example, are largely or completely dependent on their parents for protection and nutrition until puberty or young adulthood.

Let's examine how mammals make this investment, starting with marsupials and then turning to the eutherian mammals.

Gestation and Early Development in Marsupials

You might recall that mammals comprise three major monophyletic groups: monotremes, eutherians, and marsupials (see Chapter 35). Although all mammals are endothermic, have fur, and lactate, the three lineages are distinguished by their mode of reproduction:

- Monotremes lay eggs and incubate them until hatching.

- In eutherians, mothers carry offspring internally for relatively long periods of development and nourish them via a placenta.

- In marsupials, the corpus luteum is not maintained as it is in eutherians, and no placenta forms.

In marsupials, young are ejected from the mother's body at the end of the estrous cycle. As a result, they are far less developed than eutherian mammal offspring, which undergo a lengthier **gestation**—the developmental period that takes place inside the mother.

The jaws, gut, lungs, and forelimbs of a newly born marsupial are relatively well developed at birth. As a result, the offspring is able to climb from its mother's vagina to a nipple, which is usually enclosed in a pouch created by a flap of skin (**FIGURE 50.19a**). The offspring clamps onto the nipple and continues to develop, fed by the mother's milk (**FIGURES 50.19b** and **50.19c**).

Even after growing large enough to leave the pouch and begin moving and feeding on its own, offspring will return to the pouch for protection. Marsupial mothers invest a great deal in their offspring, even though a relatively short period of development takes place inside their bodies.

Major Events during Human Pregnancy

Marsupials and eutherians differ sharply in terms of how long the developing embryo is retained inside the mother's body. Let's consider humans as a model organism in eutherian reproduction.

When a secondary oocyte is released from the human ovary, the cell is viable for less than 24 hours. Human sperm, in contrast, remain capable of fertilizing an egg for up to five days. Therefore, sexual intercourse in humans has to occur less than five days before ovulation or immediately after ovulation for pregnancy to result.

Although an ejaculation may contain hundreds of millions of sperm, most die as they travel through the uterus. Only 100 to 300 actually succeed in reaching the oviduct, where fertilization takes place.

You might recall that when sperm and oocyte meet, enzymes released from the head of the sperm create a path through the material surrounding the oocyte membrane. Once the membranes of the oocyte and sperm have fused, the oocyte nucleus completes meiosis II. The two nuclei then unite to form a diploid zygote.

Implantation Smooth-muscle contractions in the oviduct gradually move the zygote toward the uterus. As it travels, the cell begins to divide by mitosis. By the time it reaches the lining of the uterus, the embryo consists of a hollow ball of cells. It then undergoes **implantation**—meaning that it becomes embedded in the thickened, vascularized wall of the uterus. It will stay in the uterus for approximately 270 days (9 months).

Once the embryo is implanted in the uterine lining, its cells begin synthesizing and secreting the hormone human chorionic gonadotropin; this hormone is later produced in larger quantities by the placenta. **Human chorionic gonadotropin (hCG) is a chemical messenger that prevents the corpus luteum from degenerating.** When hCG is present, the ovary continues secreting progesterone, and the menstrual cycle is arrested. Since hCG is excreted in the mother's urine, it is the chemical used to detect pregnancy in pregnancy tests.

The First Trimester Human gestation is divided into 3-month stages called trimesters. Not long after implantation is complete, mass movements of cells result in the formation of the three major embryonic tissues, called ectoderm, endoderm, and mesoderm (see Chapter 23). By 8 weeks into gestation, these tissues have differentiated into the various organs and systems of

(a) Brushtail possum shortly after birth

5 mm

(b) 1.5 months after birth

20 mm

(c) 3.5 months after birth

20 mm

FIGURE 50.19 Marsupials Trade a Long Gestation Period for a Long Lactation Period. Relative to eutherians, marsupial offspring—like this brushtail possum—spend a short time developing in the uterus and a relatively long time being fed milk after birth and being protected in the mother's pouch.

the body. Also by this time, the heart has begun pumping blood through a circulatory system. The embryo at this stage is called a **fetus** (FIGURE 50.20a).

Early in development, the embryonic ectoderm contributes to several important membranes. One of these membranes, the **amnion,** completely surrounds the embryo. The amnion eventually fills with amniotic fluid, which provides the embryo with a protective cushion.

The other key event in the first trimester is the formation of the placenta. This organ, which starts to form on the uterine wall a few weeks after implantation, is composed of tissues from both mother and embryo (see Chapter 35). Because the placenta contains a dense supply of blood vessels from the mother, it provides nutrition for the growing fetus. Arteries transport blood from the circulatory system of the fetus, through the **umbilical cord,** to an extensive capillary bed in the placenta. This capillary bed provides a large surface area for the exchange of gases, nutrients, and wastes between maternal and fetal blood, even though the maternal and fetal blood do not commingle.

The placenta secretes a variety of hormones, including large amounts of progesterone and estrogens. Because these hormones suppress the release of GnRH, LH, and FSH through negative feedback, they prevent the maturation and ovulation of additional follicles. By the end of the first trimester, the placenta is producing more than enough progesterone to replace the amount that had been produced by the corpus luteum, which has degenerated by this time. In essence, the placenta takes over from the corpus luteum in secreting the hormones required to maintain the pregnancy.

✔ If you understand how hormones influence pregnancy, you should be able to explain (1) why women who produce low levels of progesterone from the corpus luteum are prone to miscarriage, and (2) why a pregnancy test administered to a woman in her third trimester would register negative.

The Second and Third Trimesters After the fetal organs and placenta form during the first trimester, the rest of development consists mainly of growth (**FIGURES 50.20b** and **50.20c**). During the last weeks of pregnancy, the brain and lungs undergo particularly dramatic growth and development. If a baby is born prematurely, intervention may be required to keep the baby alive until the lungs can complete their development.

The machinery and level of hospital care required by premature infants emphasize just how superbly adapted mothers are for nourishing a growing fetus in the uterus. It costs hundreds of thousands of dollars for health care providers to do what mothers do naturally in the last trimester. Let's take a closer look at this critical aspect of pregnancy.

How Does the Mother Nourish the Fetus?

In oviparous and ovoviviparous species, mothers produce relatively large eggs that contain all of the nutrients and fluids that the embryo needs for development until hatching. But in some viviparous species, eggs are relatively small and contain almost no nutrients.

In species such as humans, the developing embryo depends on the mother's body for oxygen, chemical energy in the form of sugars, amino acids and other raw materials for growth, and waste removal. What physiological changes occur in human mothers to accommodate these demands?

Oxygen Exchange between Mother and Fetus During pregnancy, a mother's respiratory and circulatory systems change in ways that increase the efficiency of nutrient transfer and gas exchange with the fetus. For example:

- A woman's total blood volume expands by as much as 50 percent during pregnancy.

- To accommodate the increase in blood volume, maternal blood vessels dilate (widen) and blood pressure drops.

- The mother's heart enlarges and beats faster, increasing her total cardiac output by almost 50 percent.

- The mother's breathing rate and breathing volume increase to accommodate the fetus's demand for oxygen and its production of carbon dioxide.

In addition, important adaptations heighten the efficiency of gas exchange between the mother and the embryo, in the placenta. In many species, such as sheep, maternal and fetal blood

(a) 1st trimester

(b) 2nd trimester

(c) 3rd trimester

FIGURE 50.20 Development of the Human Fetus Can Be Divided into Three Trimesters.

in the placenta flow in a countercurrent fashion (**FIGURE 50.21a**). Countercurrent flows maintain a concentration gradient that increases the efficiency of diffusion or other types of exchange (see Chapter 42).

Countercurrent flow does not occur in the human placenta. Instead, oxygen exchange between mother and fetus is efficient because of another mechanism. Maternal arteries in humans empty into a space at the junction of the maternal and fetal portions of the placenta. This space is packed with small projections called villi, which contain the fetal blood vessels. Thus, a large surface area from the fetus is bathed with highly oxygenated maternal blood. The fetal villi are analogous to the alveoli of the lungs (see Chapter 45), which provide a large surface area for gas exchange.

FIGURE 50.21b illustrates another adaptation that increases the rate of oxygen delivery to a human fetus. Notice that the axes on this graph give the partial pressure of oxygen in blood versus the percentage of hemoglobin that holds oxygen at that pressure. Recall that this type of graph is an **oxygen–hemoglobin equilibrium curve** (see Chapter 45).

The key message of this graph is that the data for fetal hemoglobin are shifted to the left of the data for adult hemoglobin. This means that the fetus's blood always has a higher affinity for

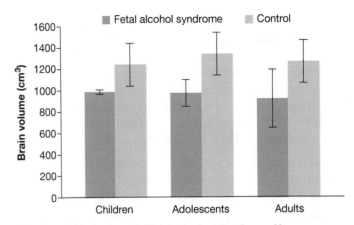

FIGURE 50.22 Patients with Fetal Alcohol Syndrome Have Reduced Brain Size. The effect is most pronounced in young people.

DATA: Swayze, V. W., II, Johnson, V. P., Hanson, J. W., et al. 1997. *Pediatrics* 99: 232–240.

oxygen than does the mother's blood. Even if the partial pressure of oxygen is the same in the mother and fetus, oxygen will move from the mother's blood to the fetus's blood because it is held more tightly by the fetal hemoglobin.

Biologists interpret this pattern as an adaptation. The high oxygen affinity of fetal hemoglobin ensures that the fetus is always able to acquire oxygen from the mother.

Toxic Chemicals Can Be Transferred from Mother to Fetus

Mothers and embryos exchange more than nutrients and wastes—they can also exchange dangerous molecules. As an example, consider the thalidomide tragedy. During the 1950s, hundreds of children were affected by their mothers' consumption of the tranquilizer thalidomide, which was prescribed to treat morning sickness. The molecule diffused into the fetal bloodstream and caused birth defects—often a dramatic shortening of the arms.

Although thalidomide is now banned for use by pregnant women, alcohol use continues to affect newborns. Compared to children of mothers who do not drink alcohol, children of mothers who imbibe ethanol are at high risk for hyperactivity, severe learning disabilities, and depression. Collectively, these symptoms are termed **fetal alcohol syndrome (FAS).**

To investigate why FAS occurs, researchers used a technique called magnetic resonance imaging (MRI) to compare the total brain volumes of patients suffering from fetal alcohol syndrome with those of healthy controls. They found that the brains of children with FAS were over 20 percent smaller than those of healthy children (**FIGURE 50.22**). In adolescents and adults with FAS, brain volume was 28 percent smaller than in healthy people of the same age, suggesting that prenatal exposure to alcohol stunts brain development throughout life.

The MRIs also revealed numerous structural abnormalities in the brain. These abnormalities, along with the reduced brain volume, are thought to be responsible for dramatic reductions in IQ and other measures of intelligence and learning ability. On the basis of results like these, public health officials strongly advise pregnant women against drinking *any* alcohol.

(a) Countercurrent blood flow in the sheep placenta

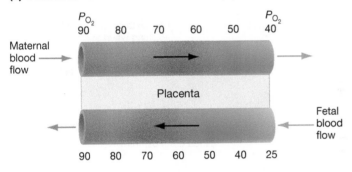

(b) Oxygen–hemoglobin saturation curves in the human fetus and mother

FIGURE 50.21 Mammals Have Adaptations That Increase Delivery of Oxygen to the Fetus.

Birth

Although the mechanisms responsible for initially triggering the birthing process are not completely understood, the posterior pituitary hormone **oxytocin** is important in stimulating smooth-muscle cells in the uterine wall to begin contractions. The contractions that expel the fetus from the uterus constitute **labor.**

When the birthing process starts, the uterus contracts at relatively low frequency. Then, as **FIGURE 50.23** shows,

1. The cervical canal at the base of the uterus begins to open, or dilate. Once the cervix is fully dilated, uterine contractions become more forceful, longer, and more frequent.

2. The fetus is expelled through the cervix and into the vagina.

PROCESS: BIRTHING SEQUENCE

— Placenta
— Umbilical cord
— Uterus
— Cervix

1. Dilation of cervix

2. Expulsion: delivery of the infant

3. Delivery of placenta

— Placenta (detaching from uterus)

— Umbilical cord

FIGURE 50.23 Human Birth Occurs in Three Stages.

✔**QUESTION** Human babies are normally born headfirst and face down, as shown. Predict the consequences of a baby positioned face up or feet first in the birth canal.

3. After the baby is delivered, the placenta remains attached to the uterine wall. At this point, caregivers clamp and cut the umbilical cord that connects the child and the placenta. When the mother delivers the placenta and accompanying membranes, birth is complete.

Although this description sounds straightforward, in reality a large number of complications are possible. For example, **FIGURE 50.24** shows the number of Swedish mothers who died, per 100,000 live births, in five-year intervals between 1760 and 1980. In 1760, approximately 1.4 mothers died for every 100 infants successfully delivered. In most cases, the cause of death was blood loss or infection following delivery. Note that the steepest drops in mortality rate occurred with the introduction of hand-washing in the late 1870s.

As a result of sterile techniques, antibiotics, and blood transfusion technology, Sweden's mortality rate has now declined to less than 0.007 percent. Improved nutrition, sanitation, and medical care have also reduced infant mortality rates in many countries.

The huge decline in the rate of death associated with childbirth qualifies as one of the great triumphs of modern medicine. Unfortunately, because many developing nations lack sterile facilities and antibiotics, the mortality of mothers and infants remains high in those countries.

First data on efficacy of hand-washing when assisting birth

First antibiotics (sulfa drugs)

Penicillin; first blood transfusions

90% of mothers receive prenatal care

FIGURE 50.24 Maternal Mortality in Childbirth Has Decreased Dramatically. These data are from Sweden; the scale on the vertical axis is logarithmic. *Note:* Data showing that hand-washing dramatically reduced maternal mortality during childbirth were published in 1861, but doctors resisted the idea until well after Pasteur had published the germ theory of disease in 1878.

DATA: Högberg, U., and I. Joelsson. 1985. *Acta Obstetricia et Gynecologica Scandinavica* 64: 583–592.

✔**QUANTITATIVE** If a woman living in Sweden in 1760 had 10 children over the course of her lifetime, what was her overall chance of dying in childbirth?

If you understand . . .

50.1 Asexual and Sexual Reproduction

- Asexual reproduction produces offspring with genetic material from the mother only.

- Sexually produced offspring are genetically unique because of recombination during meiosis and the fusion of haploid gametes from different parents during fertilization.

- Asexual reproduction is favored in constant environments, whereas sexual reproduction is favored in changing environments, or when food availability is low and availability of mates is high.

✔ You should be able to explain why sexually produced offspring would have higher fitness than asexually produced offspring in a population of animals that is highly parasitized.

50.2 Fertilization and Egg Development

- In human males, spermatogenesis is continuous throughout adult life, but in human females, all primary oocytes are formed early in development. Meiosis of oocytes is arrested for long periods of time, and cell division after meiosis is so unequal in females that just one egg—not four—is produced from each primary oocyte.

- Fertilization is external in many aquatic animals but internal in almost all terrestrial species.

- When sperm competition occurs, males have large testes relative to their body size, and the last male to mate usually fathers a disproportionately large number of offspring.

- Once eggs are fertilized, females, depending on their species, may lay the eggs or may retain them and give birth to live offspring. In some groups at least, viviparity may be an adaptation that increases the survival of young in cold habitats.

✔ You should be able to explain why fish that care for their young are expected to produce fewer eggs per year than fish that do not care for their young, and also whether sperm competition is common in fish.

50.3 Reproductive Structures and Their Functions

- In humans, the male reproductive system includes structures specialized for producing and storing sperm, synthesizing other components of semen, or transporting and delivering semen.

- The female reproductive system includes structures specialized for producing eggs, receiving sperm, and nourishing offspring during early development.

✔ You should be able to suggest ways to surgically sterilize human males and females for birth control.

50.4 The Role of Sex Hormones in Mammalian Reproduction

- In mammals, GnRH from the hypothalamus triggers the release of FSH and LH from the pituitary gland. The pituitary hormones regulate the production of the gonadal hormones, testosterone and estradiol, in the testes and ovaries, respectively.

- During the human menstrual cycle, estradiol and progesterone exert feedback (estradiol, positive and negative feedback; progesterone, negative only) on the production of FSH and LH. Interactions between the pituitary and ovarian hormones are responsible for regulating cyclical changes in the ovaries and uterus.

✔ You should be able to predict whether FSH and LH concentrations would increase or decrease following ovariectomy (surgical removal of ovaries).

50.5 Pregnancy and Birth in Mammals

- If fertilization occurs, the developing embryo and placenta secrete the hormone hCG, which arrests the menstrual cycle and allows pregnancy to continue.

- During the first trimester, the embryo becomes implanted in the thickened uterine wall, organs develop, and the nutritive organ called the placenta forms.

- To make rapid growth possible during the second and third trimesters, the mother's heart rate and blood volume increase. Nutrients and gases are exchanged efficiently in the placenta.

✔ You should be able to explain why the placenta is highly vascularized (consists mainly of blood vessels).

MB **MasteringBiology**

1. MasteringBiology Assignments

Tutorials and Activities Human Gametogenesis; Human Reproduction; Reproductive System of the Human Female; Reproductive System of the Human Male; Sex Hormones and Mammalian Reproduction

Questions Reading Quizzes, Blue-Thread Questions, Test Bank

2. eText Read your book online, search, take notes, highlight text, and more.

3. The Study Area Practice Test, Cumulative Test, BioFlix® 3-D Animations, Videos, Activities, Audio Glossary, Word Study Tools, Art

You should be able to . . .

1. What term describes the mode of asexual reproduction in which offspring develop from unfertilized eggs?
   a. parthenogenesis; **b.** budding; **c.** regeneration; **d.** fission

2. In sperm competition, what is "second-male advantage"?
   a. the observation that when females mate with two males, each male fertilizes the same number of eggs
   b. the observation that when females mate with two males, the second male fertilizes most of the eggs
   c. the observation that females routinely mate with at least two males before laying eggs or becoming pregnant
   d. the observation that accessory fluids prevent matings by second males—for example, by forming copulatory plugs

3. How are the human penis and clitoris similar?
   a. They develop from the same population of embryonic cells.
   b. Both develop during the earliest stages of puberty.

   c. Both contain the urethra.
   d. Both produce accessory fluids required during sexual intercourse.

4. Which of the following hormones would you expect to be more elevated in female hyenas compared to females of other species of mammals?
   a. follicle-stimulating hormone
   b. testosterone
   c. progesterone
   d. oxytocin

5. What two pituitary hormones are involved in regulating the human menstrual cycle?

6. True or false: The corpus luteum is retained upon implantation due to the presence of the hormone human chorionic gonadotropin (hCG).

7. What is the fundamental difference between sexual and asexual reproduction, in terms of the characteristics of offspring?

8. Summarize the experimental evidence that *Daphnia* require three cues to trigger sexual reproduction. Discuss what these cues indicate about the environment. Generate a hypothesis for why sexual reproduction is adaptive for these animals.

9. Contrast spermatogenesis with oogenesis in humans. How do these processes differ with respect to numbers of cells produced, gamete size, and timing of the second meiotic division?

10. Give examples of negative feedback and positive feedback in hormonal control of the human menstrual cycle. Why can high estradiol levels be considered a "readiness" signal from a follicle?

11. In hormone production, how do the follicle and corpus luteum compare?
    a. Both produce primarily estradiol.
    b. Both produce primarily progesterone.
    c. The follicle produces more estradiol than progesterone; the corpus luteum produces more progesterone than estradiol.
    d. The follicle produces mostly progesterone; the corpus luteum produces estradiol and progesterone.

12. Suppose that females of a certain insect species routinely mate with two males and choose their mates based on courtship displays or other traits. Predict whether females would be choosier about the first male or the second male that they mate with, or equally choosy about both males. Explain the logic behind your prediction.

13. Researchers have recently developed methods for cloning mammals. In effect, biologists can now induce asexual reproduction in species that do not normally reproduce asexually. Suppose this practice becomes so widespread in the future that most of the sheep in the world become genetically identical. Discuss some possible consequences of this development.

14. Suppose that you've been given a chance to study several populations of lizards—some of them viviparous, some of them oviparous. Which populations would you expect to produce especially large eggs, relative to their overall body size?

15. When marsupial eggs are developing, a shell membrane appears for a short time and then disintegrates. State a hypothesis to explain this observation.

16. **QUANTITATIVE** Consider the data in the table below, showing the average body mass index (BMI, where higher values represent heavier girls for a given height; see Chapter 44) of pre-pubertal

and post-pubertal girls. Which conclusion can you make from the data?

Age	BMI	
	Pre-pubertal	Post-pubertal
11	−0.22	0.75
12	−0.28	0.52
13	−0.56	0.34

DATA: Anderson, S. E., G. E. Dallal, and A. Must. 2003. *Pediatrics* 111: 844–850.

a. At a given age, there are more girls with low BMI than with high BMI.
b. At a given age, girls with high BMI are more likely to hit puberty than girls with low BMI.
c. 11-, 12-, and 13-year-olds are equally likely to hit puberty.
d. There is no relationship between BMI and age at puberty.

51 The Immune System in Animals

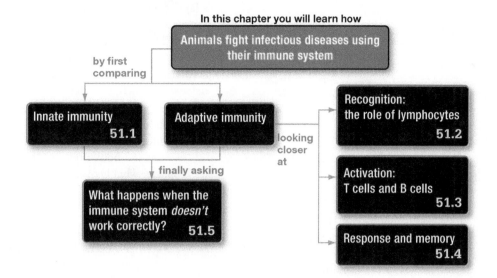

In this chapter you will learn how

Animals fight infectious diseases using their immune system

by first comparing

Innate immunity
51.1

Adaptive immunity

finally asking

What happens when the immune system *doesn't* work correctly? **51.5**

looking closer at

Recognition: the role of lymphocytes
51.2

Activation: T cells and B cells
51.3

Response and memory
51.4

Colorized scanning electron micrograph of human immune system cells attacking *Wuchereria bancrofti*—a parasitic nematode that clogs lymph vessels and causes the dramatic swelling called elephantiasis. This chapter explores how immune system cells are able to recognize parasitic worms, bacteria, and viruses as foreign and eliminate them.

This chapter is part of the Big Picture. See how on pages 840–841.

nfectious disease threatens the survival of every living organism. For example, animals alone may be infected by thousands of different disease-causing bacteria, viruses, parasitic worms, fungi, and protists—collectively referred to as **pathogens.** Given the ability of these pathogens to cause illness and death, it is remarkable that so many animals stay healthy for most of their lives.

If an animal contracts a bacterial or viral illness, however, it may eventually recover, even without medical intervention. Frequently, after recovering from an infection, the animal will be immune to (literally, "exempt" from) the disease—that is, it will not contract the same disease in the future. **Immunity** is a resistance to or protection against a disease-causing pathogen.

Thus, the **immune system** is responsible for defending animals against pathogens. The successful function of the immune system involves three key processes. It must **(1)** prevent the entry of

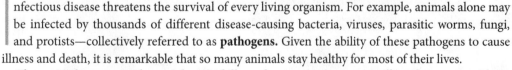

✔ When you see this checkmark, stop and test yourself. Answers are available in Appendix A.

Innate Immune System	Adaptive Immune System
Occurs in all animals	Occurs only in vertebrates
Has both cell-mediated and secreted components	Has both cell-mediated and secreted components
Rapid response	Slow response
Broadly specific response against types of pathogens	Specific response against pathogen strains
Response does not vary when infections recur (no memory)	Response is more rapid and efficient when infections recur (memory)

51.1 Innate Immunity

Innate immunity is so named because it is inherent in all animals and is ready to go from the moment of birth. In contrast, adaptive immunity occurs in only 1 percent of animals—the vertebrates—and is not fully developed until 6 months after birth. Clearly, the innate immune system on its own has succeeded in protecting a spectacular abundance and diversity of animals, both invertebrates and vertebrates. It is the first line of defense and includes exterior anatomical structures that protect animals from invading pathogens as well as interior detection and response systems.

To launch an investigation into innate immunity, let's first focus on how the body prevents entry by foreign invaders. Then let's consider what happens if some do get in.

Barriers to Entry

The most effective way for animals to avoid getting an infection is to prevent pathogens from entering their bodies in the first place. In humans and many other animals, the most important deterrent to infection is the exterior surface. For example,

- The armored bodies of insects are difficult for pathogens to enter because they are covered with the tough layer called cuticle, along with a layer of wax (see Chapter 43).

- Soft-bodied organisms like slugs, snails, and earthworms are covered with a protective layer of **mucus** (the adjective is mucous), a slimy mix of glycoproteins and water that traps pathogens and sloughs off.

- Human skin has an outer layer of dead cells that are reinforced with tough fibers of the protein keratin.

Besides providing a tough physical barrier, these outer surfaces also ensure a restrictive chemical environment. The oil secreted by skin cells is converted to fatty acids by bacteria that live

potential pathogens; (**2**) detect the presence of a pathogen by distinguishing it from the animal's own body; and (**3**) eliminate the pathogen.

When biologists began analyzing the immune system, they found that certain immune system cells are ready to respond to foreign invaders at all times, while others must be activated first. Cells that are always ready confer **innate immunity.** Cells that must first be selectively activated to tailor their response against a specific pathogen confer **adaptive immunity.**

In combination, the innate and adaptive immune responses form a powerful system for protecting individuals against a formidable and ever-changing array of pathogens. **TABLE 51.1** illustrates the different roles of the innate and adaptive immune systems. This chapter explores the different roles of these two systems and how their cells and secreted cellular products collaborate to battle infections. It then considers the problems that arise when the immune system does not function properly.

Airways (lining of trachea)
Most pathogens are trapped in mucus before they can reach the lungs. Beating cilia sweep pathogens up and out of the airway.

Mucus-secreting cells   Ciliated cells

Eyes
Blinking wipes tears across the eye. Tears contain the antibacterial enzyme lysozyme.

Ears
Hairs and earwax trap pathogens in the passageway of the external ear.

Nose
The nasal passages are lined with mucous secretions and hairs that trap pathogens.

Digestive tract
Pathogens are trapped in saliva and mucus, then swallowed. Most are destroyed by the low pH of the stomach.

FIGURE 51.1 The Body Protects Openings. The scanning electron micrograph of mucus-secreting and ciliated cells has been colorized.

harmlessly on the surface. The fatty acids lower the pH of the surface to about 5, creating a dry, acidic environment that prevents the growth of most pathogens.

Unless this protective exterior surface is broken by an injury, the most vulnerable places in animal bodies are the gaps in the surface where the digestive tract, reproductive tract, gas-exchange surfaces, and sensory organs make contact with the environment.

How Are Openings in the Body Protected? As **FIGURE 51.1** shows, gaps in the outer surface of animal bodies are protected by a complex array of chemical substances that discourage pathogen entry.

The mucus that protects the surface of soft-bodied invertebrates is equally important in protecting the surface openings in vertebrates. For example, many of the pathogens that you breathe in or ingest while eating or drinking stick to the mucus that lines your airways and digestive tract. Pathogens that are stuck in mucus cannot come in contact with the plasma membranes of epithelial tissues. In mammals, many of these pathogens are either coughed out or swallowed and killed in the acidic environment of the stomach. In the respiratory tract, they are swept out of harm's way by the beating of cilia (see Figure 51.1).

Gaps in the body that are not covered with mucous layers — the eyes, for example—are often protected by other types of secretions. Your ears are protected by waxy secretions, and your eyes by tears that contain the enzyme **lysozyme.** Lysozyme acts as an **antibiotic** by digesting bacterial cell walls.

How Do Pathogens Gain Entry? When preventive measures fail, as they sometimes do, pathogens gain entry to tissues beneath the skin. Flu viruses, for example, have an enzyme on their surface that disrupts the mucous lining of the respiratory tract. When the outer surface of the virus makes contact with a host cell beneath the mucous layer, the virus is able to enter the cell and begin an infection.

Falls, wounds, and other types of physical trauma provide another important mode of entry. When the skin is broken, bacteria and other pathogens gain direct access to the tissues inside.

To viruses, bacteria, and fungi, your body is a tropical paradise, brimming with resources. Given that a single bacterium could give rise to a population of 100 trillion in a day, something must be done, and fast, when the outer defenses of the body are penetrated. What happens then?

The Innate Immune Response

If foreign invaders penetrate the body's protective barrier, particular cells initiate the **innate immune response**—the body's first response to pathogens. As a group, these cells are called white blood cells to distinguish them from red blood cells. More formally, they are known as **leukocytes** ("white-cells").

The leukocytes involved in innate immunity provide an immediate, *generic* response that is directed against the general type of pathogen encountered. The response is considered generic since it is directed against broad groups of pathogens instead of

specific organisms. For example, the innate response is able to distinguish between fungi and bacteria but cannot identify a specific strain within either group.

How Are Pathogens Recognized by the Innate Immune System?
The answer to this question emerged from a breakthrough in the late 1990s, when results from unrelated studies came together. Earlier, German researchers had identified a protein that was important in the early development of fruit flies. When flies were mutated to no longer express this protein, their larvae developed abnormally. The strange appearance of the larvae inspired the researchers to name it the Toll protein (in German, "toll" means amazing).

Jules Hoffmann and his colleagues found that Toll protein also served an important immune function: Adult fruit flies lacking Toll could not defend themselves from fungal infection (see **FIGURE 51.2**). Just a year later, two separate labs identified Toll-like

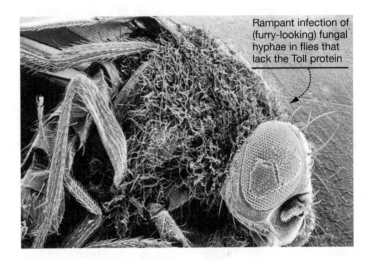

Rampant infection of (furry-looking) fungal hyphae in flies that lack the Toll protein

FIGURE 51.2 Toll Protein Is Critical to Immune Response.
Researchers learned this by comparing the survival rate of wild-type flies and mutants unable to express Toll protein after exposure to a fungal pathogen. The error bars represent the confidence interval for each point with $p \leq 0.05$. This means that there is 95 percent confidence that if the experiment were repeated, the value for each time point would fall within the bracketed range.

DATA: Lemaitre, B., et al. 1996. *Cell* 86: 973–983.

proteins in humans, demonstrating their universality. This was an exciting discovery, but the function of these Toll-like proteins was still not understood.

The riddle of Toll-like proteins was solved by Bruce Beutler and his colleagues, who studied how mutant mice reacted to antigens. An **antigen** is any foreign molecule that can initiate an immune system response. Earlier researchers observed that a molecule called lipopolysaccharide (LPS), harvested from the surface of Gram-negative bacteria (see Chapter 29), reliably caused a powerful immune response in animals. However, Beutler had a mutant mouse that did not respond to this antigen—and the mouse had defective Toll-like proteins.

Beutler had discovered that the Toll-like protein functions as a receptor that receives the signal that a pathogen is present, in this case a Gram-negative bacterium. This breakthrough was key to understanding how the innate immune system could become engaged.

Toll-like receptors (TLRs) are a subset of a larger group of proteins called **pattern-recognition receptors,** which serve as sentinels to signal the presence of molecules associated with pathogens. The TLRs have been observed in fungi and plants, suggesting that they arose in a common ancestor of eukaryotes. Eleven TLRs have been identified in humans, each one responding to a different kind of antigen. For example,

- TLR2 binds to zymosan, a common molecule in fungi.

- TLR4 binds to lipopolysaccharide (LPS) from Gram-negative bacteria (the one in Beutler's study).

- TLR5 binds to flagellin from the flagella of motile bacteria.

- TLR7 binds to single-stranded viral RNA.

What all these antigens have in common is that they are ubiquitous within the broad group in which they have been identified (e.g., *all* of the countless Gram-negative bacteria produce LPS), yet they do not occur in the host animal. Thus, these antigens serve as reliable signals of attack for the innate immune system. A mere 11 TLRs in humans is enough to detect virtually any type of invasion.

Pattern-Recognition Receptors Transduce Signals   When pattern-recognition receptors, such as TLRs, on the surface of a leukocyte receive the signal that an invader is present, they trigger a signal cascade within the cell that will have different consequences depending on which TLRs were activated (see Chapter 11 for an introduction to signal transduction). This step is vital to determining and orchestrating the innate immune responses to follow. For example,

- When the Toll protein in healthy flies detects a fungal infection, a signal cascade activates the expression and secretion of antimicrobial peptides, which function as antibiotics to destroy the fungal pathogens.

- When TLR4 in humans is activated by LPS, a signal cascade leads to the expression and secretion of **cytokines** ("cell movers"). Cytokines are a class of diverse molecules that signal other parts of the immune system in various ways, such as

attracting other immune cells to the site of infection or stimulating other immune cells into action.

- When TLR7 in humans detects the single-stranded RNA of a virus, the cell may produce and secrete a specific type of cytokine called interferon, which stimulates neighboring cells to improve their resistance to the imminent viral infection.

✓ QUANTITATIVE If you understand the role of pattern-recognition receptors, you should be able to return to Figure 51.2 and predict how the same two groups of flies would respond to a Gram-negative bacterial pathogen. Draw a line on the graph to represent flies lacking the Toll protein after treatment, and explain your reasoning. Assume the survival rate of wild-type flies would be similar to what is shown in the figure.

Once the general signal of an invasion is received by TLRs, the first response is sent out and a cascade of further actions occurs that result in a fully engaged immune response. For example, consider what would happen if you were to trip on the sidewalk while running across campus and scrape your elbow. If this wound resulted in an infection, you would observe redness, swelling, pain, and heat in the infected area. What is responsible for all of these changes?

The Inflammatory Response in Humans   The physical signs of an infected injury are a direct consequence of the innate immune response being called into action to contain and eliminate the infection. **FIGURE 51.3** summarizes the major steps in this **inflammatory** ("in-flames") **response,** a multistep, innate immune response observed in an array of animals.

Step 1 A break in the skin allows bacteria to enter the body. If capillaries and other small blood vessels are broken, the wound bleeds.

Step 2 Immediately after the injury, blood components called **platelets** release proteins that form clots and lessen bleeding. Other clotting proteins in the blood form cross-linked structures that help wall off the wound and reduce blood loss.

Step 3 Wounded tissues and leukocytes called **macrophages** secrete **chemokines** ("chemical-movers"), which are signaling molecules that recruit other cells to the site of infection. The localized production of chemokines is important because it forms a gradient that marks a path to the wound site.

Step 4 **Mast cells** release chemical messengers such as **histamine** that constrict blood vessels at the site of the wound—reducing blood flow and thus blood loss. Mast cells also release other signaling molecules that induce blood vessels slightly farther from the wound to dilate and become more permeable.

Step 5 The combination of dilated blood vessels and a chemokine gradient is like a 911 call that provides specific directions to the scene of a fire. Leukocytes called neutrophils move out of dilated blood vessels and migrate to the site of the infection. **Neutrophils** are major players in the innate response. They destroy invading cells by phagocytosis ("cell eating")—meaning that they engulf them. Once inside neutrophils, the invading cells are killed with a complex array of antimicrobial

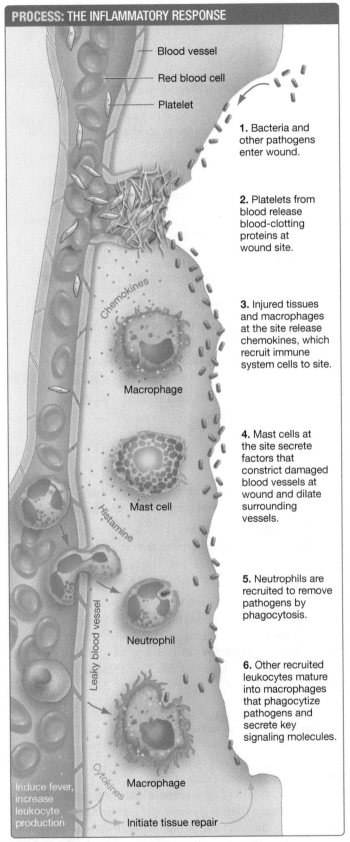

PROCESS: THE INFLAMMATORY RESPONSE

- Blood vessel
- Red blood cell
- Platelet

1. Bacteria and other pathogens enter wound.

2. Platelets from blood release blood-clotting proteins at wound site.

Chemokines

3. Injured tissues and macrophages at the site release chemokines, which recruit immune system cells to site.

Macrophage

4. Mast cells at the site secrete factors that constrict damaged blood vessels at wound and dilate surrounding vessels.

Histamine

Mast cell

5. Neutrophils are recruited to remove pathogens by phagocytosis.

Leaky blood vessel

Neutrophil

6. Other recruited leukocytes mature into macrophages that phagocytize pathogens and secrete key signaling molecules.

Cytokines

Macrophage

Induce fever, increase leukocyte production

Initiate tissue repair

FIGURE 51.3 The Inflammatory Response of Innate Immunity Has Many Elements.

✔QUESTION How do these events lead to the classical signs of an infection, which include rubor (reddening), calor (heat), and tumor (swelling), among others?

compounds—including lysozyme, which degrades bacterial cell walls. Cells that perform this activity are collectively referred to as phagocytes.

Step 6 Cells that will mature into macrophages arrive. Besides secreting chemokines, these macrophages produce additional cytokines that have an array of effects. Some of these cytokines stimulate bone marrow to make and release additional leukocytes, induce fever—an elevated body temperature that aids in healing—and activate cells involved in tissue repair and wound healing. Like neutrophils, macrophages are also phagocytes that help clear invaders from the area.

This overview actually simplifies the situation—many other cell types and cell–cell signals are involved in responding to pathogens at a site of infection.

The inflammatory response continues until all foreign material is eliminated and the wound is repaired. **TABLE 51.2** on page 1042 summarizes key cells and molecules involved in the response.

But what happens when the innate immune system of a vertebrate fails to contain and eliminate invading pathogens?

check your understanding

C Y U

If you understand that . . .
- The innate immune response occurs when macrophages and mast cells that reside in tissues, and neutrophils that circulate in the blood, react in a generic way to signals from invading pathogens.

✔ **You should be able to . . .**

Describe how the following steps that first-aid workers use when initially treating a wound mimic events in the innate immune response:

1. Applying direct pressure to close blood vessels
2. Cleaning the wound to remove dirt and debris
3. Applying bandages containing compounds that halt blood flow from the wound

Answers are available in Appendix A.

51.2 Adaptive Immunity: Recognition

In addition to the innate immune response, vertebrates have evolved the ability to recognize specific antigens, and can differentiate between different species and even different strains of pathogens. The leukocytes involved can customize their response to the particular invader, so this arm of the immune system is often referred to as the **adaptive immune response.**

Given the array of pathogens that exist, an animal is almost certain to be exposed to an enormous variety of antigens in the course of its lifetime. Is there a limit to how many different antigens its adaptive immune system responds to?

(a) Key Cells

Name	Primary Function
Mast cells	Release signals that constrict blood flow from wound and increase blood flow to wound area
Neutrophils	Kill invading cells via phagocytosis
Macrophages	Release cytokines that recruit other cells to wound site and stimulate a variety of activities; kill invading cells via phagocytosis

(b) Key Signaling Molecules

Name	Produced By	Received By	Message/Function
Histamine	Mast cells	Blood vessels	High concentration constricts blood vessels at wound site, among other activities
Chemokines*	Injured tissues and macrophages in tissues	Neutrophils and macrophages	Mark path to wound; promote dilation and increased permeability of blood vessels
Cytokines other than chemokines	Macrophages	Leukocytes	Mark path to wound
		Bone marrow	Increase production of macrophages and neutrophils
		CNS	Induce fever by raising set point for control of body temperature
		Local tissues	Stimulate cells involved in wound repair

*Note that chemokines are a subset of cytokines.

Research conducted in the early 1920s answered this question. Researchers synthesized organic compounds that do not exist in nature, injected the novel molecules into rabbits, and observed whether they activated the animals' adaptive immune system. To the amazement of the scientists, the rabbits were able to recognize and respond to every novel antigen tested. More specifically, each of the animals produced proteins in their blood called antibodies that specifically bound to the particular antigen that was injected. **Antibodies** are secreted proteins that bind to a specific part of a specific antigen.

The take-home message is that the adaptive immune system can specifically respond to an almost limitless array of antigens. This observation, and subsequent research on antibody production, led to the identification of four key characteristics of the adaptive immune response:

1. *Specificity* Antibodies and other components of the adaptive immune system bind only to specific sites on specific antigens.

2. *Diversity* The adaptive immune response recognizes an almost limitless array of antigens.

3. *Memory* The adaptive immune response can be reactivated quickly if it recognizes antigens from a previous infection.

4. *Self–nonself recognition* Molecules that are produced by the individual do not act as antigens, meaning that the adaptive

immune system can distinguish between self and nonself. Nonself molecules are antigens; self molecules are not.

The next question is, What are the cells and organs that are responsible for this adaptive immune response?

An Introduction to Lymphocytes

The leukocytes that carry out the major features of the adaptive immune response are called **lymphocytes.** In contrast to the diversity of leukocytes involved in the innate immune response, lymphocytes are primarily divided into two distinct cell types that differ in their role in the adaptive immune response and their site of maturation.

The Discovery of B Cells and T Cells In 1956 a group of researchers provided an important insight into the adaptive immune system, by accident. The biologists were investigating the immune system's response to the bacterium *Salmonella typhimurium*, a common cause of food poisoning in humans. To do this work, they needed to produce and isolate antibodies to a particular antigen from *S. typhimurium*—an antigen that is toxic. Their plan was to inject a large number of chickens with the antigen and collect the antibodies that the treated animals produced in response.

In addition to injecting many normal chickens, though, the biologists happened to include some chickens that had undergone experimental removal of an organ called the bursa. Six of the nine chickens that lacked a bursa and that were injected with the antigen died. The other three birds without a bursa survived, but failed to produce antibodies to the antigen. In contrast, all of the chickens with the bursa intact survived and produced large quantities of antibodies. To make sense of these results, the researchers proposed that the bursa is critical for antibody production and that antibodies are important in neutralizing the toxic antigens.

Not long after this observation was published, three groups of scientists independently conducted a related experiment. To explore the function of the thymus in mammals, these groups removed the organ from newborn mice. Mice lacking a thymus developed pronounced defects in their immune systems. For example, when pieces of skin from other mice were grafted onto the experimental individuals, their immune systems did not recognize the tissue as foreign. In contrast, individuals with an intact thymus quickly mounted an immune response that killed the foreign skin cells.

The results of these and follow-up experiments showed that lymphocytes from the bursa and thymus have different functions. The two types of lymphocytes became known as bursa-dependent and thymus-dependent lymphocytes, or B cells and T cells, respectively.

- **B cells** produce antibodies. Later work showed that in humans and other species where the bursa organ does not exist, B cells mature in bone marrow.

- **T cells** are involved in graft rejection along with an array of other immune functions, including recognizing and killing host cells that are infected with a virus. T cells mature in the thymus.

The requirement for maturation in a particular organ distinguishes the lymphocytes from other leukocytes that were introduced in Section 51.1. What is the origin of these lymphocytes, and where do they go after they mature?

Where Are Lymphocytes Found? The colored structures in **FIGURE 51.4** mark the major sites in the human body that play a key role in the life of a lymphocyte.

- *Lymphocyte origin* All lymphocytes are produced in **bone marrow**—a tissue that fills the internal cavities in bones.

- *Lymphocyte maturation* B cells mature in the bone marrow in humans and many other animals. T cells mature in the **thymus**—an organ located in the upper part of the chest of vertebrates (just behind the breastbone in humans).

- *Lymphocyte activation* Lymphocytes recognize antigens and become activated in the spleen and lymph nodes. The **spleen** is an organ located near the stomach in the abdominal cavity and is also involved in destroying old blood cells. **Lymph nodes** are small, oval organs that are located all around the body. Lymph nodes filter the lymph passing through them. Recall that **lymph** is a mixture of fluid and lymphocytes (see Chapter 45).

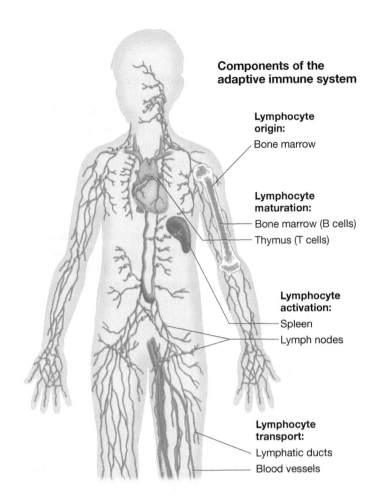

Components of the adaptive immune system

Lymphocyte origin:
Bone marrow

Lymphocyte maturation:
Bone marrow (B cells)
Thymus (T cells)

Lymphocyte activation:
Spleen
Lymph nodes

Lymphocyte transport:
Lymphatic ducts
Blood vessels

FIGURE 51.4 Lymphocytes Are Formed, Activated, and Transported in the Immune System.

The liquid portion of lymph originates in fluid that is forced out of capillaries by blood pressure.

- *Lymphocyte transport* Lymphocytes circulate through the blood and the secondary organs of the immune system—lymph nodes, spleen, and lymphatic ducts. Lymphatic ducts are thin-walled, branching tubules that transport lymph throughout the body and, along with lymph nodes, make up the **lymphatic system.**

Besides being found in bone marrow, the thymus, spleen, lymph nodes, lymphatic ducts, and blood vessels, large numbers of lymphocytes, along with other leukocytes, are associated with skin cells and with epithelial tissues that secrete mucus—primarily in the digestive tract and respiratory tract. Collectively, the immune system cells found in the gut and respiratory organs are called **mucosal-associated lymphoid tissue (MALT).** Leukocytes in the skin and MALT are important because they guard points of entry against pathogens.

Now let's turn to one of the most fundamental questions in immunology: How are B cells and T cells able to recognize so many different antigens? Let's begin with a look at the B-cell and T-cell receptors—the molecules that start the adaptive immune response.

Lymphocytes Recognize a Diverse Array of Antigens

By the 1960s, biologists understood that B cells can produce antibodies to a seemingly limitless number of antigens and that each antibody is specific to a particular antigen. They hypothesized that each B cell formed in the bone marrow has thousands of copies of a receptor on its surface that recognizes only one antigen. What's more, each B cell expresses a unique receptor relative to the other B cells. What would such a receptor look like?

The Discovery of B-Cell Receptors To test the hypothesis that B cells have antigen-specific receptors on their surfaces, researchers injected experimental animals with radioactively labeled antigens. This strategy was similar to the experiments with labeled estradiol that allowed biologists to isolate that hormone's receptor (see Chapter 49).

As predicted, the labeled antigens bound to a protein on the surface of only those B cells that produced antibody to the antigen. Chemical analysis of this **B-cell receptor (BCR)** protein

showed that it has the same overall structure as the antibodies produced by the B cells and secreted into the blood.

The BCR protein consists of two distinct polypeptides (**FIGURE 51.5a**). The smaller polypeptide is called the **light chain.** The larger polypeptide is roughly twice the size of the light chain and is called the **heavy chain.** Each BCR has two copies of the light chain and two copies of the heavy chain that are all held together by disulfide bonds. The heavy chain includes a transmembrane domain that anchors the protein in the plasma membrane of the B cell.

The antibodies produced by a B cell are identical in structure to its BCR, except that they lack the transmembrane domains. Instead of being inserted into the plasma membrane, antibodies are secreted from the cell and circulate throughout the body.

Both the BCR and the antibodies produced by B cells belong to a family of proteins called the **immunoglobulins (Ig).** Immunoglobulins are crucial to the adaptive immune response.

TABLE 51.3 shows the five classes of immunoglobulin proteins that act as antibodies. The five types are symbolized IgG, IgD, IgE, IgA, and IgM. Each class is distinguished by unique amino acid sequences in the heavy-chain region, and each has a distinct

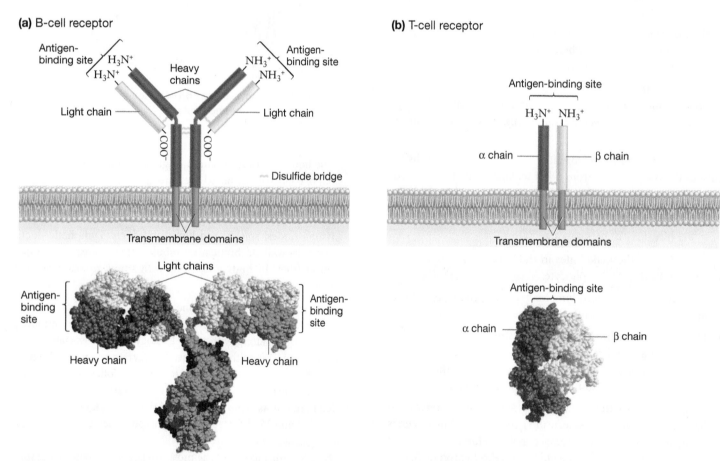

(a) B-cell receptor

Antigen-binding site
H_3N^+
H_3N^+
Heavy chains
NH_3^+
NH_3^+
Antigen-binding site
Light chain
Light chain
COO⁻
COO⁻
Disulfide bridge
Transmembrane domains

Light chains
Antigen-binding site
Antigen-binding site
Heavy chain
Heavy chain

(b) T-cell receptor

Antigen-binding site
H_3N^+ NH_3^+
α chain
β chain
Transmembrane domains

Antigen-binding site
α chain
β chain

FIGURE 51.5 B-Cell Receptors and T-Cell Receptors Have Transmembrane Domains and Antigen-Binding Sites. (a) Schematic and space-filling models of the B-cell receptor, which is shaped like a Y. **(b)** The shape of the T-cell receptor resembles one "arm" of the Y-shaped B-cell receptor. The space-filling models for each receptor do not show the transmembrane domains that anchor them in the plasma membrane.

TABLE 51.3   Five Classes of Immunoglobulins

Name	Structure (Secreted Form)	Function
IgG	Monomer	The most abundant type of secreted antibody. Circulates in blood and interstitial fluid. Protects against bacteria, viruses, and toxins.
IgD	Monomer	Present on membranes of immature B cells; rarely secreted. Serves as BCR.
IgE	Monomer	Secreted in minute amounts. Involved in response to parasitic worms. Also responsible for hypersensitive reaction that produces allergies.
IgA	Dimer	Most common antibody in breast milk, tears, saliva, and the mucus lining the respiratory and digestive tracts. Prevents bacteria and viruses from attaching to mucous membranes; helps immunize breastfed newborns.
IgM	Pentamer	First type of secreted antibody to appear during an infection. Binds many antigens at once; effective at clumping viruses and bacteria so that they can be killed. Monomeric form also serves as BCR.

FIGURE 51.6 Most Antigens Have Multiple Epitopes for B-Cell Receptors and T-Cell Receptors. The envelope of the influenza virus includes the protein hemagglutinin. This version of hemagglutinin has several distinct epitopes for BCRs and TCRs. One example of each type is highlighted as a space-filling model.

function in the immune response. Besides acting as antibodies, IgD and monomeric forms of IgM also serve as BCRs.

The Discovery of T-Cell Receptors   It took much longer for researchers to isolate and characterize the **T-cell receptor (TCR).** The technique used to identify BCRs was not useful for the receptors on T cells, suggesting that they cannot bind antigens on their own. It turns out that the T cells require other cells to (**1**) process the antigens and (**2**) present them to the TCRs. This means that for a TCR to recognize an antigen, the foreign molecule must first undergo a complex process called **antigen presentation.**

This is a fundamentally important distinction. B cells bind to antigens directly; T cells bind only to antigens that are displayed by other cells.

Other data showed that the TCR is composed of two protein chains: an alpha (α) chain and a beta (β) chain (**FIGURE 51.5b**). The overall shape of the TCR is similar to one of the two "arms" of an antibody or BCR molecule.

Antibodies and Receptors Bind to Epitopes   Antibodies, BCRs, and TCRs do not bind to the entire antigen. Instead, they bind to a selected region of the antigen called an **epitope.**

To understand the relationship between an antigen and an epitope, consider that every bacterium, virus, fungus, and protist is made up of a large number of different molecules. Many of these molecules are antigens, because they would be recognized

as being foreign to your cells. In turn, each antigen may have many different epitopes, where binding by antibodies and lymphocyte receptors actually takes place.

FIGURE 51.6 illustrates a protein called hemagglutinin, which is found on the surface of the influenza virus. This antigen has several distinct epitopes, two of which are identified in the figure. Often, the epitopes for BCRs are different from those recognized by TCRs because of how epitopes are presented to the different receptors: BCRs bind to epitopes in the context of the intact antigen, while TCRs bind to epitopes that have been processed and presented by other cells. Each epitope is recognized by a particular antibody, BCR, or TCR. It is not unusual for an antigen to have between 10 and 100 different epitopes.

How do the immunoglobulins recognize specific epitopes? The answer to this question emerged through detailed studies of the heavy and light chains.

What Is the Molecular Basis of Antibody Specificity and Diversity? In the 1950s, biologists developed an important model system for studying immunoglobulin structure. The model cells were B-cell tumors, or myelomas, that could be grown in laboratory culture.

Each type of myeloma arises from a single cell that produces a single type of antibody. When researchers compared the immunoglobulins produced by different myelomas, they found that the sequence of amino acids in the light chains consisted of two regions: one that is virtually identical among the myelomas, and another that is unique. These light-chain segments have come to be known as the **constant (C) regions** and **variable (V) regions,** respectively. Heavy chains also have a C region and a V region.

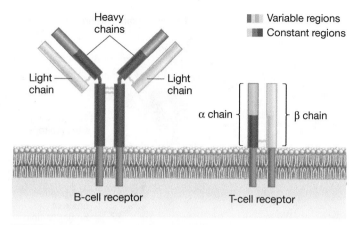

FIGURE 51.7 The Variable Regions of B-Cell Receptors and T-Cell Receptors Face Away from the Plasma Membrane.

FIGURE 51.7 makes two important points:

1. The *V* regions of heavy and light chains in a BCR are adjacent to one another and face away from the plasma membrane.

2. The *V* and *C* regions of a TCR are arranged in a similar manner as the BCR.

The presence of unique amino acid sequences in the *V* regions of every antibody, BCR, and TCR explains why each of these proteins binds to a unique epitope. Your body can respond to an almost limitless number of antigens because the number of different BCRs, antibodies, and TCRs is virtually limitless. How does all this variation come to be?

The Discovery of Gene Recombination In 1965, W. J. Dryer and J. Claude Bennett proposed a fantastic-sounding explanation for how immunoglobulin genes code for so many different variable regions and thus so many different proteins. They hypothesized that as a lymphocyte is maturing, a segment from a variable-region gene is cut and combined with a segment from the constant-region gene. Further, they proposed that this cutting and pasting is done in a different way in each lymphocyte. The result is that each lymphocyte has a novel "*V* + *C* gene" for the light-chain protein.

At the time, this type of genetic recombination in cells had not been observed, and most researchers considered the hypothesis wildly implausible.

Eleven long years passed before the first experimental evidence was obtained to support their hypothesis. In 1976, researchers showed that the amount of DNA in the *V* + *C* region of mature lymphocytes is shorter than it is in immature lymphocytes. This is exactly what the gene-recombination hypothesis predicts.

The flurry of studies inspired by this result showed that the genes for light chains have dozens of different *V* segments, several different joining (*J*) segments, and a single constant (*C*) segment. The heavy-chain gene includes diversity (*D*) segments along with a set of *V* and *J* segments similar to those in the light-chain

gene. The genes that encode the α and β chains of the TCR have a similar arrangement of distinct segments, each with multiple versions.

As a lymphocyte matures, the various gene regions are mixed and matched to produce unique receptors. **FIGURE 51.8** illustrates the steps involved in the production of the BCR light chain. A similar process occurs in the DNA encoding the heavy chain gene. Consider how these recombination events lead to BCR diversity:

1. In the light chain, one of the 40 *V* segments recombines with one of the 5 *J* segments. This step can produce $40 \times 5 = 200$ different light chains.

2. In the heavy chain, any one of the 51 *V* segments, 27 *D* segments, and 6 *J* segments can recombine.

3. The light-chain and heavy-chain rearrangements occur independently, but the final BCR includes the polypeptides from each gene to form the antigen-specific variable region.

✔ **QUANTITATIVE** If you understand how unique genes for the light and heavy chains of the BCR are made, you should be able to calculate the number of different antigen-specific BCRs that could made by this process in (1) a human and (2) a single human B cell.

In addition, gene segments do not always join precisely during DNA recombination. Some variation occurs where the *V* and *D* segments join and where the *D* and *J* segments join. As a result, an estimated 10^{10} to 10^{14} different BCRs can form in a single individual. TCR production is just as diverse.

Gene recombination is the molecular mechanism responsible for the specificity and diversity of the adaptive immune system. This process allows each lymphocyte to produce a unique BCR, antibody, or TCR—enabling it to recognize a unique epitope. The charges and geometry at each surface make the way the receptor and epitope fit together extremely specific.

How Does the Immune System Distinguish Self from Nonself?

If the adaptive immune system has the remarkable ability to generate specific, targeted responses against virtually any substance, what keeps it from reacting against itself? If a receptor responded to a **self molecule**—that is, a molecule belonging to the host—the receptor would trigger an immune response. Because this type of response is extremely rare compared to responses against foreign antigens, biologists hypothesized that there must be some mechanism for eliminating self-reactive B cells and T cells.

To test this hypothesis, researchers injected B cells and T cells possessing anti-self receptors into mice and found that the injected lymphocytes were eliminated. Follow-up work showed that if B cells and T cells maturing in the bone marrow and thymus have anti-self receptors, the cells are likely to be destroyed or permanently inactivated before they leave these organs. The conclusion? Some form of "self-education" occurs during the maturation phase of lymphocyte development.

PROCESS: V AND J SEGMENTS ARE RANDOMLY SELECTED AND JOINED DURING DNA RECOMBINATION

Light-chain DNA: $V_1\,V_2\,V_3\,...$ V_{40} $J_1\,J_2\,J_3\,...$ J_5 C

Recombined DNA: V_{12} J_3 C

Processed mRNA: V_{12} J_3 C

Light chain

B cell 2   B cell 1   B cell 3

1. Start with DNA encoding the light chain, containing full complement of V, J, and C segments.

2. DNA segments recombine. One V segment joins with one J segment and the C segment, removing the DNA in between them.

3. Transcription and RNA processing occur—removal of introns, addition of poly(A) tail, etc.

4. Translation results in a protein with a unique amino acid sequence. Depending on which V and J segments join during recombination, many different amino acid sequences are possible.

FIGURE 51.8 As B Cells Mature, Immunoglobulin-Gene Segments Recombine to Form a Single Gene. In the mature B cell 1, the final light-chain gene consists of the V_{12}, J_3, and C segments spliced together; the final heavy-chain gene might consist of the V_{48}, D_{22}, J_1, and C segments spliced together (not shown).

Researchers have also identified another type of T cell that acts as a regulatory cell. Regulatory T cells suppress certain parts of the immune system to limit the intensity of normal responses, and may help inhibit any self-reactive cells that slip through the self-education system. Defective or insufficient regulatory T cells may be partly responsible for the development of immune disorder diseases (see Section 51.5).

Even after the self-reactive lymphocytes are inactivated or removed, the structural diversity still present in the receptors of mature B and T cells is sufficient for recognizing virtually any foreign antigen. But this diversity alone is not enough to mount an effective immune response. The enormous repertoire of possible BCRs and TCRs means that only a few cells will express any particular epitope-specific receptor. To engage the invading pathogen successfully, these few lymphocytes must reproduce. How does this selective expansion take place?

check your understanding

If you understand that . . .

- The adaptive immune response is performed by lymphocytes that circulate in the lymph, blood, and organs of the immune system.
- Lymphocytes can respond to a wide array of antigens because gene segments recombine to produce receptors that bind to unique epitopes on antigens.
- Lymphocytes that express self-reactive receptors are eliminated or inactivated to prevent immune responses against the body's own cells.

✔ You should be able to . . .

1. Cite two key events that must occur during lymphocyte maturation to ensure a safe and effective adaptive immune response.
2. Explain the benefit of assembling the variable region from discrete segments rather than having multiple intact variable regions.

Answers are available in Appendix A.

51.3 Adaptive Immunity: Activation

Lymphocytes are normally in a resting, or inactive, state. Over the course of a day, an inactive lymphocyte may hang out in the skin or MALT, enter the lymphatic vessels, migrate through a lymph node, cross over into the blood, pass through the spleen, return to the blood, and so on.

If an inactive lymphocyte does not encounter the epitope that it is programmed to respond to, the cell eventually dies. As it turns out, this is the fate of most of the lymphocytes that originate from the bone marrow. During an infection, however, some of these inactive lymphocytes will go on to mount a massive attack that is targeted against the specific invader.

The Clonal Selection Theory

The diversity of the adaptive response is useful only if it can be directed against an infection. In the 1950s, Frank Bernet and colleagues developed the **clonal selection theory** to explain how only the most useful lymphocytes are activated during an infection. Their theory made several key claims about how the adaptive immune system works:

- *Antigens are recognized by receptors on B cells and T cells.* Each lymphocyte formed in the bone marrow or thymus expresses a unique receptor on its surface that binds to a unique epitope in an antigen.

- *Lymphocytes require receptor-epitope binding to become activated.* When the receptor on a lymphocyte binds to an epitope, a switch is thrown in the lymphocyte that may ultimately push the cell from a quiescent to an activated state.

- *Activated lymphocytes are cloned.* An activated lymphocyte divides and makes many identical copies of itself. In this way, specific cells are selected and cloned in response to an infection.

- *Activated lymphocytes endure.* Some of the cloned cells descended from an activated lymphocyte persist long after the pathogen is eliminated. As a result, the cloned cells are able to respond quickly and effectively if the infection recurs.

FIGURE 51.9 summarizes the assertions made by the clonal selection theory, using a B cell model. ✔ If you understand the clonal selection theory, you should be able to explain what "clonal" and "selection" refer to.

The activation of the relevant B cells and T cells, however, is a carefully controlled, stepwise process. The mechanism is reminiscent of the precautions that nations with powerful missiles take to avoid accidental deployment. For the most dangerous weapons, the signal to launch is checked and cross-checked, using a series of codes and signals. In the immune system, the checking and cross-checking occur through protein–protein interactions on the surfaces of cells, and the release and receipt of cytokines and other signaling molecules.

Let's first take a look at how T cells are activated and then examine how they play a role in the full activation of B cells.

T-Cell Activation

As the innate immune response is battling the invaders at the site of infection, leukocytes known as **dendritic** ("tree-like") **cells** are gobbling up antigens and debris via phagocytosis. These dendritic cells are like messengers that collect antigens from the battle scene and then report to the lymph nodes, where they present them to lymphocytes. Antigen presentation is a key event that links the innate and adaptive arms of the immune system; without this transfer of information, the adaptive immune response would not be activated to respond to the infection.

To understand how the activation system works, let's explore how the antigens taken up by dendritic cells are processed and presented to T lymphocytes. T cells are classified as CD4+ or

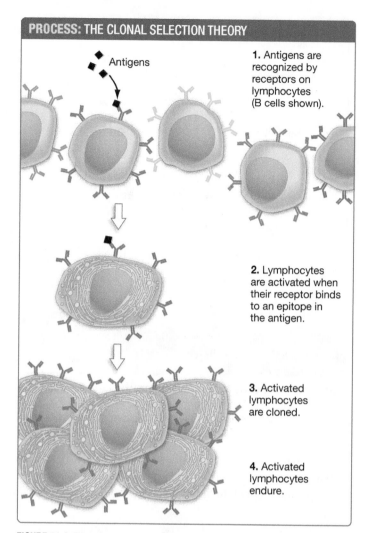

Antigens

1. Antigens are recognized by receptors on lymphocytes (B cells shown).

2. Lymphocytes are activated when their receptor binds to an epitope in the antigen.

3. Activated lymphocytes are cloned.

4. Activated lymphocytes endure.

FIGURE 51.9 The Clonal Selection Theory. The clonal selection theory maintains that certain lymphocytes are "selected" by binding to an antigen and that the selected cells then multiply.

CD8+, based on the presence of key proteins called **CD4** or **CD8** on their plasma membranes. CD4+ T cells and CD8+ T cells have distinct functions in the adaptive immune response.

Antigen Presentation via MHC Proteins Recall that the receptors on T cells can bind only to epitopes that have been processed and presented on the surface of other cells. The surface proteins responsible for presenting these epitopes are called **major histocompatibility (MHC) proteins.** MHC proteins have a groove that binds to small peptide fragments that are typically 8 to 20 amino acids in length (see **FIGURE 51.10**).

MHC proteins come in two types, called **class I** and **class II MHC proteins.** Dendritic cells present peptides in both classes of MHC proteins, but the origins of the peptides differ between the two: class I MHC proteins bind antigens inside the endoplasmic reticulum (ER); class II MHC proteins bind antigens inside endosomes. The proteins that are processed and loaded onto the class I MHC are derived from the cell's cytosol, while those loaded onto the class II MHC are obtained from the external environment.

Antigen-binding site

Antigen fragment

MHC protein

MHC transmembrane domains

FIGURE 51.10 MHC Structure Promotes Peptide Binding. A peptide processed from the hemagglutinin antigen fits onto the epitope-binding groove in this class II MHC protein like a hot dog in a bun. The class I MHC has a similar epitope-binding groove.

FIGURE 51.11 shows how dendritic cells process antigens and load the peptides onto the class II MHC molecules (a similar process takes place for class I molecules in the ER).

Step 1 Dendritic cells ingest antigens at a site of infection.

Step 2 The antigen is moved from the phagosome to an endosome, where an enzyme complex breaks the protein into small peptide fragments (see Chapter 7).

PROCESS: MHC ANTIGEN PRESENTATION

Antigen

1. **Dendritic cell ingests antigen** via phagocytosis.

Antigen fragment

2. **Enzymes break antigen into peptide fragments.**

Major histocompatibility (MHC) protein

3. **Peptides are loaded onto MHC** in endosomes (class II).

4. **MHC–peptide complex is transported** to cell surface.

Dendritic cell

MHC

Piece of antigen

5. **MHC protein presents peptide on surface** of cell membrane.

FIGURE 51.11 Dendritic Cells Present Antigens. Dendritic cells take in antigens, break them into pieces, and present the fragments in the groove of an MHC protein.

Step 3 Some of the peptides are bound to the groove of the class II MHC molecules, which were made in the rough ER and transported to the endosome.

Step 4 The MHC–peptide complexes are exported from the endosomes into vesicles for transport to the cell surface.

Step 5 The vesicles fuse with the plasma membrane, displaying the MHC–peptide complexes on the cell surface.

It's important to note that humans have several genes encoding class I and class II MHC proteins. As a result, you can produce several distinct proteins of each class that vary in the type of peptide that is presented. In addition, the MHC genes are among the most polymorphic of any genes known—meaning that many different alleles exist in the population (see Chapter 14). Because so many distinct alleles exist, a wide array of peptides can be bound and presented—allowing these cells to activate a response to many different pathogens.

How Are T Cells Activated by Antigen-Presenting Cells? **FIGURE 51.12** illustrates what happens when T cells recognize the MHC–peptide complexes on a dendritic cell. CD4+ T cells interact with class II MHC–bound epitopes on dendritic cells; CD8+ T cells interact with class I MHC–bound epitopes.

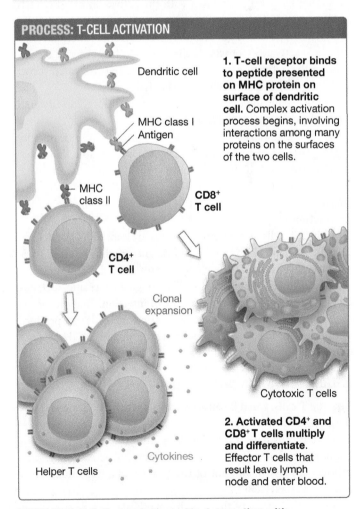

PROCESS: T-CELL ACTIVATION

Dendritic cell

MHC class I Antigen

MHC class II

CD8+ T cell

CD4+ T cell

1. **T-cell receptor binds to peptide presented on MHC protein on surface of dendritic cell.** Complex activation process begins, involving interactions among many proteins on the surfaces of the two cells.

Clonal expansion

Cytotoxic T cells

Helper T cells

Cytokines

2. **Activated CD4+ and CD8+ T cells multiply and differentiate.** Effector T cells that result leave lymph node and enter blood.

FIGURE 51.12 T Cells Are Activated by Interacting with MHC–Peptide Complexes.

To the T cell, the antigen-presenting cell carries the message "I've found antigen—response required." If the TCR binds, the activation process starts in response to interactions with proteins on the dendritic cell. In most cases, full activation of CD8$^+$ T cells also requires interactions with cytokines produced by activated CD4$^+$ T cells.

An activated T cell divides to produce a series of genetically identical daughter cells. This event, called clonal expansion, is a crucial step in the adaptive immune response. It leads to a large population of lymphocytes capable of responding specifically to the antigen that has entered the body. During this clonal expansion, many morphological changes also occur that prepare the cells for their specific functional roles in the immune response.

Cytotoxic T Cells and Helper T Cells

When activated CD8$^+$ T cells undergo clonal expansion, the daughter cells develop into **cytotoxic** ("cell-poison") **T cells,** also known as cytotoxic T lymphocytes (CTLs) or killer T cells. The adjectives cytotoxic and killer are appropriate: CD8$^+$ T cells kill cells that are infected with an intracellular pathogen.

In contrast, the daughter cells of activated CD4$^+$ lymphocytes differentiate into **helper T cells.** The adjective helper is also appropriate: Helper T cells assist with the activation of other cells in the immune response. There are two types of helper T cells, designated T_H1 and T_H2, and they have distinct functions: T_H1 cells help activate cytotoxic T cells; T_H2 cells help activate B cells. During the activation phase, the dendritic cell will often direct which type of helper T cell the CD4$^+$ lymphocyte becomes. The outcome usually depends on which types of TLRs were activated in the dendritic cell at the site of infection.

✔ If you understand the role of antigen presentation by MHC proteins, you should be able to create a table that summarizes the roles of the class I and class II MHC proteins with regard to the origin of the peptides, type of T cells that bind to them, and the activity that is stimulated in these T cells via the interaction.

Helper T cells and cytotoxic T cells are capable of participating in the adaptive immune response and are thus referred to as **effector** T cells. Following clonal expansion and maturation, effector T cells leave the lymph node through a lymphatic duct, enter the blood, and migrate to the site of infection. One role of the T_H2 cells is to identify and activate particular B cells that produce antibodies directed against the invading pathogens. How does this work?

B-Cell Activation and Antibody Secretion

Like the T cells, when B cells are activated, they replicate and undergo significant morphological changes. As **FIGURE 51.13a** shows, inactive B cells have a large nucleus, little cytoplasm, few mitochondria, and a ruffled membrane. Upon full activation, B cells produce a massive amount of rough ER and a large number of mitochondria (**FIGURE 51.13b**).

Recall that many of the proteins synthesized in a cell's rough ER are inserted into the plasma membrane or secreted from the

(a) Inactive lymphocyte

(b) Activated lymphocyte

FIGURE 51.13 Lymphocytes Exist in Two States: Inactive and Activated. (a) This inactive B cell has a small amount of cytoplasm and few organelles. **(b)** This activated B cell, called a plasma cell, has extensive rough endoplasmic reticulum (ER) and many mitochondria—suggesting that a great deal of protein synthesis is taking place.

cell (see Chapter 7). The increased rough ER in this activated B cell is required for manufacturing and secreting antibodies.

The activation process that leads to these changes involves several steps (**FIGURE 51.14**):

Step 1 The BCRs on B cells interact directly with antigens that are floating free in lymph or blood. The BCR–antigen interaction results in the first part of B-cell activation. The bound antigen is internalized and digested into fragments, which are then loaded onto class II MHC proteins. As a result, a B cell that encounters its antigen displays the antigen's epitopes on its own surface, and the peptides are cradled in the groove of a class II MHC protein.

Step 2 When an *activated* CD4$^+$ T_H2 cell with a complementary receptor arrives, it binds to the MHC–peptide complex on the B cell. The interaction between a B cell and a helper T cell supplies activation signals that stimulate the helper T cell.

Step 3 The helper T cell responds by releasing cytokines that provide a second signal to the B cell that fully activates it.

Step 4 The now fully activated B cell replicates, and some of the daughters differentiate into effector B lymphocytes called **plasma cells.** Plasma cells produce and secrete large quantities of antibodies. Recall that antibodies are identical to B-cell receptors, except that they lack a transmembrane domain and are secreted instead of being found in the plasma membrane.

Foreign protein

B-cell receptors

B cell

MHC class II protein

1. B cell recognizes invader. The B cell encounters and binds to an epitope on the antigen. The antigen is internalized, processed, and presented on the surface via class II MHC proteins.

B cell

2. B cell stimulates active T cell. The MHC–peptide complex interacts with complementary receptors on a helper T cell, stimulating it.

Activation

Cytokines

Helper T cell

3. B cell is activated by helper T cell. Cytokines from the activated helper T cell activate the B cell.

4. B cell gets to work. The activated B cell replicates, and some differentiate into plasma cells, which produce large quantities of antibodies.

Plasma cells

Antibodies

Antibodies will bind to antigens and mark them for destruction

FIGURE 51.14 B Cells Are Activated by Binding to Antigens and Interacting with Helper T Cells.

When B cells are first activated to replicate by binding to the free antigen, they migrate to a specialized area in the lymph node called the germinal center. There, the DNA sequences that code for the variable region of the immunoglobulin gene undergo rapid mutations that modify the variable region of the receptors. B cells with receptors that bind best to the free antigen live and produce daughter cells; those that bind to the antigen less effectively die.

This process of **somatic hypermutation** fine-tunes the immune response. The BCRs that result from somatic hypermutation bind to the antigen more tightly than those formed during maturation in the bone marrow.

Once the effector B cells and T cells have been activated, the adaptive immune response is in full swing. Cytotoxic and helper T cells move into the site of infection, and antibodies specific to the invading pathogen begin to circulate in the blood and lymph. The adaptive immune system has recognized the invaders and initiated its response.

Now let's look at what happens once the secreted antibodies and activated cells are recruited to the site of infection. For pathogens, the results are usually devastating.

check your understanding

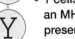

If you understand that . . .

- T cells are activated when their receptor binds to an MHC–peptide complex displayed on an antigen-presenting cell.
- B cells are stimulated when their receptor binds to an antigen; then they are fully activated via interactions with a T_H cell activated by the same antigen.
- Activated lymphocytes undergo dramatic changes in morphology and divide rapidly.

✔ **You should be able to . . .**

Generate a hypothesis to explain the observation that humans who are heterozygous for the genes encoding MHC proteins tend to be healthier than individuals who are homozygous for these genes.

Answers are available in Appendix A.

51.4 Adaptive Immunity: Response and Memory

In combination with the leukocytes involved in the innate immune system, the cells of the adaptive immune system are almost always successful in eliminating threats from bacteria, parasites, fungi, and viruses.

The array of mechanisms used by the adaptive immune system to dispose of foreign invaders are broadly grouped into two responses:

1. The **cell-mediated response** is promoted by T_H1 cells and involves cytotoxic T lymphocytes (activated $CD8^+$ cells) among others. This response primarily takes place via cell–cell contact.

2. The **humoral response** is promoted by T_H2 cells and involves the production of antibodies and other proteins secreted into the blood and lymph. (The Latin root *humor* means fluid.)

To understand how the adaptive immune system actually eliminates pathogens, let's examine how these two responses deal with invaders that remain extracellular versus those that reside within cells.

How Are Extracellular Pathogens Eliminated?

During the innate immune response to an infection, macrophages and dendritic cells at the site phagocytize some of the invaders. In addition to killing the foreign cells, these leukocytes process and present antigens via the MHC class II proteins. Macrophages display the processed epitopes on their surfaces at the site of infection, while dendritic cells move to the lymph nodes to present epitopes of the antigens. If the epitopes are recognized by helper T cells, the now activated T cells move into the site of infection.

If an activated T_H1 cell binds to the class II MHC proteins presented by these antigen-laden macrophages, two things happen. First, the phagocytic activity of the macrophages is enhanced. Second, the T_H1 cells secrete cytokines that recruit additional phagocytic cells to the site—increasing the inflammatory response.

In addition, antibodies from plasma cells begin attaching to extracellular bacteria, fungi, viruses, and other foreign material. These bound antibodies interfere with the infection in four ways (**FIGURE 51.15**):

1. *Opsonization ("preparation for eating")* Antibodies from plasma cells begin coating pathogens at the infection site. Pathogens that are tagged with antibodies are readily destroyed by phagocytes.

2. *Neutralization* Coated pathogens are blocked from interacting with—and thus infecting—host cells. Their participation in the infection is neutralized.

3. *Agglutination ("gluing together")* In many cases, antibodies cause the clumping of antigens, including cells and viruses, via a process called **agglutination.** Each antibody has at least two binding sites (pentameric IgM antibodies have 10), so a single antibody can bind epitopes on foreign cells and cross-link them. Clumped cells and viruses are not

able to infect the cells of the body and are easy targets for phagocytes.

4. *Co-stimulation of complement proteins* Antibodies that are bound to pathogens also stimulate a lethal group of proteins called the **complement system.** Complement proteins circulate in the bloodstream and assemble at antigen–antibody complexes. When complement proteins activate, they punch deadly holes in the plasma membranes of pathogens.

Within a few days, this combination of killing mechanisms—armies of phagocytic cells and complement proteins that home in on antibody-tagged material—usually eliminates all of the extracellular pathogens. But what about those pathogens that reside within the cells of the body?

How Are Intracellular Pathogens Eliminated?

Virtually all nucleated cells in the body express class I MHC proteins on their cell surface. Recall that these MHC proteins display peptides that were processed from cytosolic proteins. This means that if a cell were infected, peptides from the foreign proteins expressed inside the cell would be loaded into some of the MHC proteins and presented to circulating T cells.

Cells that display the antigens of intracellular pathogens are effectively waving a flag that says, "I'm infected. If you destroy me, you'll destroy the infection." As activated CD8$^+$ cells migrate into the area, those that recognize the class I MHC–peptide complex are stimulated to respond to the signal.

After cytotoxic T cells have recognized their target, they form a tight attachment with the cell to direct the secretion of molecules from the T cell to the target cell's surface without disturbing adjacent cells. Some of these secreted products assemble into pores in the target cell's plasma membrane that allow other molecules from the T cell to pass through. As shown in **FIGURE 51.16**,

PROCESS: HUMORAL RESPONSE

Complement protein
Pathogen
Antibody

Neutrophil

Uninfected host cell

Opsonization   Neutralization   Agglutination   Co-stimulation of complement proteins

FIGURE 51.15 The Humoral Response Eliminates Extracellular Pathogens.

these molecules pass directly into the cytoplasm of the target cell and activate a signaling cascade that causes the cell to self-destruct via **apoptosis** (see Chapter 22). The result of apoptosis is the death and fragmentation of a cell into smaller vesicles, called apoptotic bodies, which are ingested by phagocytes like macrophages.

Once the cytotoxic T cell has delivered this signal, often referred to as the "kiss of death," the T cell releases the dying cell and seeks out another infected cell to kill. Over time, all of the infected cells are eliminated.

If the pathogen is able to replicate only inside host cells, as is the case with viruses, this cell-mediated response limits the spread of the infection by preventing the production of new generations of pathogens.

Why Does the Immune System Reject Foreign Tissues and Organs?

The innate and adaptive immune responses have devastating effects on invading pathogens. Unfortunately, they are equally deadly in response to tissues or organs that are introduced into a patient to heal a wound or cure a disease.

Consider the problems that can arise with blood transfusions. You might recall that certain individuals have red blood cells with membrane glycoproteins called A and B (see Chapter 14). These molecules act as antigens if they are introduced into a person whose own blood cells lack those glycoproteins.

For example, if you have type A blood, it means that your blood cells have the A glycoprotein. If your blood is transfused into a person who lacks the A glycoprotein—meaning someone who has type B or type O blood—the recipient's immune system will recognize the A glycoprotein as an antigen and mount a devastating response against it. For a blood transfusion to be successful, the recipient may have to receive blood that lacks the A

and B glycoproteins entirely (type O) or that contains the same "antigens" found in his or her own blood.

Similar problems arise in organ transplants, except that the molecules directing the rejection of these tissues are the class I MHC proteins. The adaptive system mounts a response against the foreign MHC proteins, and the innate response is activated based on the *absence* of "self MHC" signals. If transplanted cells do not display the same class I MHC proteins as the host, components of the innate immune system kill them.

To prevent strong immune reactions against transplanted organs or tissues, physicians seek donors who have MHC proteins that are extremely similar to those of the recipient, as is often seen in siblings. Even with close relatives, however, molecular differences will exist between the donor and recipient. Thus, physicians must also treat the recipient with drugs that suppress the immune response.

Thanks to steady improvements in drug development and in systems for matching MHC types between donors and recipients, the success rate for organ transplants has improved dramatically in recent years.

As the blood transfusion and organ transplant examples show, the immune system rejects foreign cells because they either contain nonself molecules or lack molecules that serve as indicators of self. To your immune system, a mismatched blood transfusion or an organ transplant is indistinguishable from a massive influx of bacteria, viruses, or other foreign invaders.

Responding to Future Infections: Immunological Memory

In addition to producing the cells that implement the humoral and cell-mediated responses, activated B cells and T cells produce specialized daughter cells called memory cells. **Memory cells** do not participate in the initial adaptive response, or **primary immune response.** Instead, they provide a surveillance

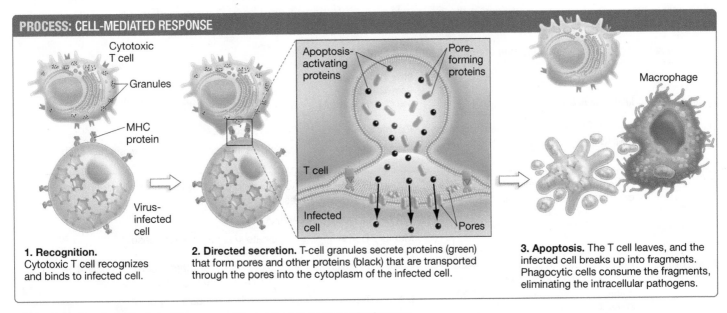

PROCESS: CELL-MEDIATED RESPONSE

Cytotoxic T cell

Granules

MHC protein

Virus-infected cell

Apoptosis-activating proteins

Pore-forming proteins

T cell

Infected cell

Pores

Macrophage

1. Recognition. Cytotoxic T cell recognizes and binds to infected cell.

2. Directed secretion. T-cell granules secrete proteins (green) that form pores and other proteins (black) that are transported through the pores into the cytoplasm of the infected cell.

3. Apoptosis. The T cell leaves, and the infected cell breaks up into fragments. Phagocytic cells consume the fragments, eliminating the intracellular pathogens.

FIGURE 51.16 The Cell-Mediated Response Eliminates Intracellular Pathogens.

service after the original infection has been cleared. Memory cells remain in the spleen and lymph nodes for years or decades, ready to provide a rapid response should an infection with the same antigen recur.

The production of memory cells is a hallmark of the vertebrate immune response. It occurs only to a limited degree, if at all, in invertebrates. How do memory cells protect an individual from future infections?

The Secondary Response Is Strong and Fast If the same antigen enters the body a second time, memory cells recognize certain epitopes of the antigen and trigger a second adaptive response, or **secondary immune response**. FIGURE 51.17 compares the rate of antibody production during the first and second exposures to a virus. The launching of a secondary immune response by means of memory cells is known as **immunological memory.**

The secondary immune response is faster and more efficient than the primary response. It is faster because the presence of an expanded pool of memory T and B cells increases the likelihood that lymphocytes with the correct antigen-specific receptors will find the antigen and activate quickly. It is more efficient because some of the memory B cells pass through another round of somatic hypermutation, the same process that occurred at the start of the primary immune response (see Section 51.3). Memory B cells with receptors that bind best to the antigen's epitope live and produce daughter cells.

To review immunological memory, starting with how B cells and T cells are activated, expanded, and differentiated into the various effector cells, study **TABLE 51.4.**

Vaccination Leads to Immunological Memory The production and effectiveness of memory cells explain phenomena that have been observed throughout recorded history. For example, when plague struck Athens in 430 B.C., the historian Thucydides noted that only people who had recovered from the illness could nurse the sick, because they did not become ill a second time. In the middle ages, Chinese and Turkish practitioners protected people

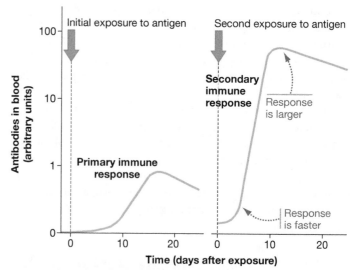

FIGURE 51.17 The Secondary Immune Response Is Faster and Stronger than the Primary Response. The graphs represent results that are commonly observed when biologists inject the same antigen into a mouse at two different times and measure the concentration of antibody in the blood (plotted on logarithmic scale). The change in antibody concentration over time indicates the speed and strength of the response to the antigen.

✔**EXERCISE** Imagine you injected a new antigen at the same time you began the second exposure to the original antigen. Draw a line on the graph to show what you predict would be the response to this new antigen in terms of antibody concentration.

from smallpox by intentionally exposing them to the dried crusts of smallpox pustules taken from infected individuals. These are examples of **immunization**—the conferring of immunity to a particular disease.

In the late 1700s Edward Jenner refined this immunization technique. In Jenner's day, milkmaids were considered pretty because their faces were not pockmarked with scars from smallpox infections. Jenner knew that cows suffered from a smallpox-like

SUMMARY TABLE 51.4 Activation and Function of Adaptive Immune System Cells

Type of Lymphocyte	Method of Activation	Cells That Result from Activation and Clonal Expansion	Function of Resulting Cells
B cell	Receptor binds to free antigen, then class II MHC–peptide complex interacts with TCR of T_H2 cell	Plasma cells	Secrete antibodies
		Memory B cells	Participate in secondary response (secrete antibodies)
CD4⁺ T cell	Receptor binds to class II MHC–peptide complex on dendritic cell or other antigen-presenting cell	T_H1 helper T cells	Activate cytotoxic T cells; regulate inflammatory response
		T_H2 helper T cells	Activate B cells
		Memory T cells	Participate in secondary response
CD8⁺ T cell	Receptor binds to class I MHC–peptide complex on dendritic cell or other antigen-presenting cell; T cell is then stimulated by cytokines of T_H1 cells	Cytotoxic T cells	Kill infected host cells that present complementary class I MHC–peptide complex
		Memory T cells	Participate in secondary response

disease called cowpox. He reasoned that milkmaids became immune to smallpox because they had been exposed to cowpox while milking cows.

To test this hypothesis, Jenner inoculated a boy with fluid from a cowpox pustule. Later he inoculated the same child with fluid from a smallpox pustule. That is, he first inoculated the boy with cowpox pathogens and then with smallpox pathogens.

As predicted, the boy did not contract smallpox. Jenner's technique was named **vaccination.** (The Latin root *vacca* means cow.) Vaccination is the introduction of a **vaccine,** a preparation containing antigens from a weakened or altered pathogen, to prime the body's immune system so it fights later infections effectively.

The data in Figure 51.17 explain why vaccination is an effective defense against certain pathogens—it speeds up the body's response to an infection. The antigens used in a vaccine are usually components of a pathogen's exterior, because antibodies can readily attach to them.

Consider the three general types of vaccines against viruses:

1. *Subunit vaccines* consist of isolated viral proteins: familiar examples include vaccines against hepatitis B and influenza.

2. *Inactivated viruses* have been damaged by chemical treatments—often exposure to formaldehyde—or exposure to ultraviolet light. They do not cause infections, but are antigenic. If you have been vaccinated for hepatitis A or polio, you may have received an inactivated virus.

3. *Attenuated viruses* are also called "live" virus vaccines because they consist of complete virus particles. Researchers make these viruses harmless by culturing them on cells from species other than the normal host. In adapting to growth on the atypical cells, the viruses usually lose the ability to grow

rapidly in their normal host cells. The smallpox and measles vaccines consist of attenuated viruses.

After vaccination, the body mounts a primary immune response that produces memory cells. If a second infection occurs later, these memory cells respond quickly and eliminate the threat before illness appears. Vaccinations function like fire drills or earthquake preparedness exercises—they prepare the immune system for a specific threat.

Unfortunately, viruses such as the human immunodeficiency virus (HIV) mutate so rapidly that traditional vaccines have been ineffective for immunization. Memory cells generated from HIV vaccination are unlikely to recognize the changed epitopes in the rapidly evolving strains. Currently, the only "cure" for HIV infection is prevention.

51.5 What Happens When the Immune System *Doesn't* Work Correctly?

The vertebrate immune system is a marvel of adaptation. A healthy immune system is able to defeat the vast majority of infections without medical intervention and with little impact on the body. The immune system is a formidable threat to pathogens, but if its response is dysfunctional, it can also be a liability to the health and even survival of the animal.

To appreciate this point, let's first examine what happens when the immune system activates inappropriate responses and then look at the consequences when it fails to respond.

Allergies

For some people, walking through a field of ragweed or petting a cat is a prescription for developing a runny nose, itchy eyes, and labored breathing. For other people, being stung by a bee or eating a peanut is a life-threatening experience. These are examples of allergies—immune responses that are overreacting to foreign substances.

An **allergy,** or allergic reaction, is an abnormal response to an antigen. For reasons that are still not clear, certain people produce the IgE class of antibodies in response to specific molecules found in cat dander, nuts, plant pollen, or other products. Molecules that trigger this response are called **allergens** instead of antigens.

Allergic reactions are considered inappropriate responses because the normal role of IgE antibodies is to defend against infections by parasitic worms. Recent research has also shown that in areas where infections with intestinal worms are common, allergies are almost unknown. These data suggest that allergies don't occur if IgEs are expressed normally—that is, in response to worm infections.

The presence of IgE antibodies in an allergic response is important because it triggers a series of events known as the **hypersensitive reaction.** When a susceptible person is first exposed to an allergen, the IgE antibodies that are produced bind to receptors on mast cells and certain other leukocytes. This binding

check your understanding

If you understand that . . .

- Extracellular pathogens are eliminated by the humoral response, which includes antibodies, complement proteins, and other secreted products that opsonize, neutralize, agglutinate, or directly lyse the invaders.
- Intracellular pathogens are eliminated by activated cytotoxic T cells that recognize epitopes displayed on infected cells and kill them before the pathogens inside can replicate.
- Immunological memory provides a rapid, efficient secondary response against future infections with the same pathogen.

✔ You should be able to . . .

Compare and contrast inactivated and attenuated vaccines, and predict which would result in a more effective cell-mediated immune response.

Answers are available in Appendix A.

occurs at the constant region of the IgE heavy chains—the "tail" of the antibody. Once this binding event occurs, the cells (and the person) are said to be sensitized.

If the person is later exposed to the same allergen, the variable regions of these IgE antibodies bind to the allergen molecules and send a signal to the sensitized cells to rapidly secrete large quantities of histamine, cytokines, and other compounds. In response to these molecules, blood vessels dilate and become more permeable, smooth-muscle cells contract, and other cells secrete mucus. Common manifestations of these hypersensitive reactions are hay fever, hives, and asthma.

In more severe responses, blood vessels can dilate to the point where blood pressure plummets, oxygen delivery to the brain is reduced dramatically, and the person loses consciousness. In addition, severe smooth-muscle contractions in the digestive system and respiratory system can induce vomiting, diarrhea, and complete constriction of the airway passages. This combination of events, known as anaphylactic shock, is lethal unless the person receives immediate medical attention.

What can be done about allergies? The number one approach is to avoid the allergen. If exposure does occur, drugs called antihistamines can sometimes reduce symptoms by blocking histamine receptors. In some cases, corticosteroids applied to the skin can reduce swelling. Longer-term solutions are still in the experimental stage, however. Developing effective treatments for allergies is an important research frontier in biomedicine.

Although allergic reactions are considered abnormal, they are still directed against substances that are foreign to the body. What happens when the immune system turns against the body's own tissues?

Autoimmune Diseases

An immune response directed against molecules or cells that normally exist in the host is known as **autoimmunity.** Autoimmune reactions often result in disease due to the destruction of the body's own cells and structures.

- Multiple sclerosis (MS) results from the production of cytotoxic T cells that attack the myelin sheath of nerve fibers (see Chapter 46). Because damage to myelin reduces the efficiency of nerve signaling, coordination problems result.

- Rheumatoid arthritis develops when self-reactive antibodies alter the lining of joints, causing painful inflammation.

- Type 1 diabetes mellitus occurs when cytotoxic T cells attack and kill insulin-secreting cells in the pancreas, resulting in a lack of insulin and inability to regulate blood glucose levels (see Chapter 44).

The cause and treatment of immune-mediated diseases are active areas of research. The mechanisms behind these diseases are still poorly understood, but an interesting connection is emerging. Beginning in the late 1960s, several studies have shown that the risk of developing allergies and autoimmune diseases increases in individuals who live in homes with a high level of sanitation.

Results from these studies have led to the development of the **hygiene hypothesis,** which claims that autoimmune and allergic responses arise in individuals who, because of hygienic practices, have experienced less exposure to pathogens and parasites. Recall that allergies are virtually absent in areas where intestinal worms are common. This effect is also seen for autoimmune diseases, which do not involve the IgE-based hypersensitivity reaction.

This correlation between hygiene and immune disorders points to the close ties that have been forged during the coevolution of the immune system and the invaders it defends us against. Low infant mortality and other benefits of sanitary lifestyles appear to come at the cost of inappropriate immune responses.

But, what happens when the immune system fails to respond at all?

Immunodeficiency Diseases

Children who are born with a genetic disorder called severe combined immunodeficiency (SCID) lack a normal immune system and are unable to fight off infections. The results are devastating. Without a functional immune system, the number of potential infectious agents grows to include "opportunistic" pathogens that would normally be incapable of causing disease.

You may recall that one form of SCID, designated SCID-X1, is caused by mutations in a gene on the X chromosome (see Chapter 20). The gene codes for a receptor protein necessary for the development of T cells. In another type of SCID, called Omenn syndrome, there is a genetic defect in one of the enzymes responsible for DNA recombination in maturing lymphocytes. As a result, cells are unable to generate normal T-cell and B-cell receptors.

In both cases, the adaptive immune response is badly impaired. Afflicted individuals suffer debilitating illness from infections that other children fight off easily. If not given a bone-marrow transplant or kept in a completely sterile environment, children with SCID typically will die before they are 2 years old.

Similarly, people who are infected with the **human immunodeficiency virus (HIV)** suffer from a progressive failure of the immune system. HIV infects and kills $CD4^+$ T cells and macrophages (see Chapter 36). Recall that these cells are required for both the humoral and cell-mediated immune responses. As the infection continues, populations of $CD4^+$ T cells gradually decline to a point where the immune system can no longer mount an effective response to infection.

Eventually, HIV-infected people develop **acquired immune deficiency syndrome (AIDS).** They succumb to illnesses that physicians almost never see in people with healthy immune systems.

Over the past three decades, the incidence of HIV infections has skyrocketed throughout the world, devastating the immune systems of millions of people. This, coupled with the marked increase in allergies and autoimmune diseases in the developed world, illustrates the importance of continued research into the many nuances of the immune system.

If you understand . . .

51.1 Innate Immunity

- Animals protect themselves from infection by establishing barriers that prevent the entry of pathogens.

- The innate immune system provides a rapid, generic response against broad classes of pathogens based on the recognition of pathogen-associated molecules by pattern-recognition receptors.

- During the innate immune system's inflammatory response, leukocytes respond to an infection by phagocytizing foreign material and cells and releasing cytokines that stimulate neighboring cells to respond to the infection and recruit more leukocytes into the area.

✓ You should be able to explain the role of Toll-like receptors in the innate immune response.

51.2 Adaptive Immunity: Recognition

- The vertebrate immune system possesses T cells and B cells that recognize specific epitopes on antigens via unique receptors found on the T- and B-cell surfaces.

- The receptor proteins expressed on the surfaces of T cells and B cells are generated through DNA recombination—a rearrangement of gene segments.

- Because every receptor that results from DNA recombination is slightly different, the immune system is able to recognize and respond to an almost limitless array of antigens.

✓ You should be able to predict how the BCR locus in the genomic DNA of immature B cells is different from that in mature B cells.

51.3 Adaptive Immunity: Activation

- The clonal selection theory explains how the most appropriate cells for controlling an infection are selected and expanded to mount an effective immune response.

- T cells are activated when their receptors recognize epitopes displayed in MHC proteins on antigen-presenting cells. Full activation and differentiation of T cells often requires secondary signals from cytokines.

- B cells are activated when their receptors bind to free antigens. B cells differentiate into plasma cells and memory cells only when fully activated by helper T cells.

✓ You should be able to explain how antigen-presenting cells of the innate immune response are required for the differentiation of B cells into plasma cells.

51.4 Adaptive Immunity: Response and Memory

- Extracellular pathogens are eliminated via phagocytosis and soluble components secreted by the humoral response, which include antibodies. Antibodies contribute to the response by opsonizing, neutralizing, or agglutinating pathogens, or by co-stimulating the lytic activity of the complement proteins.

- Intracellular pathogens are eliminated via the cell-mediated response, in which infected host cells are induced to self-destruct by cytotoxic T cells.

- The immune system rejects blood and tissue transplants based on the presence of nonself molecules, often in the form of carbohydrates or MHC proteins, or on a lack of self recognition.

- The production of memory lymphocytes, via a primary infection or vaccination, allows the immune system to respond rapidly and effectively to future infections by the same pathogen.

✓ You should be able to explain how viruses are eliminated by both humoral and cell-mediated immune responses.

51.5 What Happens When the Immune System *Doesn't* Work Correctly?

- Allergies are abnormal, hypersensitive responses that are generated against allergens, which are antigens that are not associated with infection.

- Autoimmune diseases occur when self-reactive lymphocytes elicit an immune response against the body's own tissues.

- Immunodeficiency diseases arise from genetic mutations or viral infections that disrupt the function of key components in the immune system.

✓ You should be able to predict the consequences of producing IgE against a self molecule.

MB MasteringBiology

1. **MasteringBiology Assignments**

   Tutorials and Activities Acquired Immunity; Adaptive Immune Response; Immune Responses; Inflammatory Response

   Questions Reading Quizzes, Blue-Thread Questions, Test Bank

2. **eText** Read your book online, search, take notes, highlight text, and more.

3. **The Study Area** Practice Test, Cumulative Test, BioFlix® 3-D Animations, Videos, Activities, Audio Glossary, Word Study Tools, Art

You should be able to . . .

1. What is the primary difference between the innate and adaptive responses?
   a. The innate response does not distinguish between pathogens, while the adaptive response does.
   b. Only the innate response is triggered by antigens.
   c. The adaptive response generates immunological memory and is more specific than the innate response.
   d. The innate response does not kill cells; the adaptive response does.

2. The overall role of the inflammatory response is to
   a. contain and eliminate foreign cells and material at the site of infection.
   b. increase heat at the site of infection to activate enzymes used in the immune response.
   c. produce antibodies that bind to invading cells and eliminate them.
   d. increase blood flow at the site of the wound to flush out invading pathogen.

3. What is the difference between an epitope and an antigen?
   a. An epitope is any foreign substance; an antigen is a foreign protein.
   b. An epitope is the part of an antigen where an antibody or lymphocyte receptor binds.
   c. An antigen is the part of an epitope where an antibody or lymphocyte receptor binds.
   d. Antigens are recognized by B cells and antibodies; epitopes are recognized by T cells.

4. How do B-cell receptors and antibodies differ?

5. What is one of the differences between $CD4^+$ and $CD8^+$ T cells?
   a. $CD4^+$ cells are immature and $CD8^+$ cells are mature.
   b. $CD4^+$ cells are activated and $CD8^+$ cells are not.
   c. $CD4^+$ cells interact with class II MHC proteins, and $CD8^+$ cells interact with class I MHC proteins.
   d. $CD4^+$ cells activate cell-mediated responses, and $CD8^+$ cells activate humoral responses.

6. What is the hygiene hypothesis?

7. Explain how B cells are fully activated to differentiate into plasma cells.

8. Compare and contrast how B-cell receptors and T-cell receptors recognize antigens.

9. What do vaccines need to contain in order to be effective? Why don't we have vaccines for HIV?

10. Why is clonal selection necessary for the adaptive immune response but not the innate immune response?
    a. The adaptive response uses receptors to recognize pathogens, and the innate response does not.
    b. There is more diversity in the receptors in the adaptive response compared to the innate response.
    c. Cells in the innate response do not require activation, and those in the adaptive response do.
    d. This process is used for targeting pathogens, and the innate response is used only to stop blood flow from the wound.

11. Compare and contrast the interaction between pathogens and the pattern-recognition receptors on leukocytes versus that between antigens and BCRs or TCRs.

12. Explain how DNA recombination leads to the production of almost limitless numbers of different B-cell receptors, T-cell receptors, and antibodies.

13. Which of the following outcomes would be expected if there were a mutation that inhibited somatic hypermutation?
    a. The diversity of pattern-recognition receptors would be significantly lowered.
    b. B and T lymphocytes would not be able to produce receptors that recognize antigens.
    c. The adaptive immune system would not be able to respond to pathogens.
    d. The secondary immune response to a repeat infection would produce the same antibodies as those made in the primary immune response.

14. Predict how self-reactive B cells are identified and eliminated during maturation.

15. It seems astonishing that the immune system can produce antibodies to compounds that were recently synthesized for the first time in the lab. Given that viruses and other pathogens evolve rapidly, why would natural selection favor individuals having genes that made extensive antibody diversity possible?

16. During World War II, physicians discovered that if they removed undamaged skin from a patient and grafted it onto the site of a burn on the same patient, the tissue healed well. But if the grafted tissue came from a different individual, the skin graft was rejected. What is responsible for this difference?

CHAPTER 1

IN-TEXT QUESTIONS AND EXERCISES

p. 4 Fig. 1.2 *analyze* If Pasteur had done any of the things listed, he would have had more than one variable in his experiment. This would allow critics to claim that he got different results because of the differences in broth types, heating, or flask types—not the difference in exposure to preexisting cells. The results would not be definitive.

p. 6 *apply* The average kernel protein content would decline, from 11 percent to a much lower value over time.

p. 6 CYU *apply* The data points would all be about 11 percent, indicating no change in average kernel protein content over time.

p. 7 Fig. 1.4 *apply* Molds and other fungi are more closely related to green algae because they differ from plants at two positions (5 and 8 from left) but differ from green algae at only one position (8).

p. 8 Fig. 1.6 *apply* The eukaryotic cell is roughly 10 times the size of the prokaryotic cell.

p. 9 CYU *evaluate* From the sequence data provided, species A and B differ only in one ribonucleotide of the rRNA sequence (position 10 from left). Species C differs from species A and B in four ribonucleotides (positions 1, 2, 9, and 10). A correctly drawn phylogenetic tree would indicate that species A and B appear to be closely related, and species C is more distantly related. See **FIGURE A1.1**.

p. 13 (1) *analyze* You could conclude that the ants weren't navigating normally, because they had been caught and released and transferred to a new channel. **(2)** *analyze* You could conclude that the ants can't navigate normally on their manipulated legs.

p. 13 CYU The key here is to test predation rates during the hottest part of the day (when desert ants actually feed) versus other parts of the day. The experiment would best be done in the field, where natural predators are present. One approach would be to capture a large number of ants, divide the group in two, and measure predation rates (number of ants killed per hour) when they are placed in normal habitat during the hottest part of the day versus an hour before (or after). **(1)** *analyze* The control group here is the normal condition—ants out during the hottest part of the day. If you didn't include a control, a critic could argue that predation did or did not occur because of your experimental setup or manipulation, not because of differences in temperature. **(2)** *analyze* You would need to make sure that there is no difference in body size, walking speed, how they were captured and maintained, or other traits that might make the ants in the two groups more or less susceptible to predators. They should also be put out in the same habitat, so the presence of predators is the same in the two treatments.

p. 12 Fig. 1.9 *analyze* The interpretation of the experiment would not likely change, but your confidence in the conclusions drawn would be reduced if you used just one ant.

IF YOU UNDERSTAND . . .

1.1. *understand* Dead cells cannot regulate the passage of materials between exterior and interior spaces, replicate, use energy, or process information. **1.2** *understand* Observations on thousands of diverse species supported the claim that all organisms consist of cells. The hypothesis that all cells come from preexisting cells was supported when Pasteur showed that new cells do not arise and grow in a boiled liquid unless they are introduced from the air. **1.3** *understand* If seeds with higher protein content leave the most

offspring, then individuals with low protein in their seeds will become rare over time. **1.4** *understand* A newly discovered species can be classified as a member of the Bacteria if the sequence of its rRNA contains some features found only in Bacteria. The same logic applies to classifying a new species in the Archaea or Eukarya. **1.5** *understand* (1) A hypothesis is an explanation of how the world works; a prediction is an outcome you should observe if the hypothesis is correct. (2) Experiments are convincing because they measure predictions from two opposing hypotheses. Both predicted actions cannot occur, so one hypothesis will be supported while the other will not.

YOU SHOULD BE ABLE TO . . .

✔ Test Your Knowledge

1. *remember* d **2.** *understand* d **3.** *remember* populations **4.** *understand* b **5.** *understand* An individual's ability to survive and reproduce **6.** *understand* c

✔ Test Your Understanding

7. *evaluate* That the entity they discovered replicates, processes information, acquires and uses energy, is cellular, and that its populations evolve. **8.** *understand* a **9.** *understand* Over time, traits that increased the fitness of individuals in this habitat became increasingly frequent in the population. **10.** *understand* Individuals with certain traits are selected, in the sense that they produce the most offspring. **11.** *analyze* Yes. If evolution is defined as "change in the characteristics of a population over time," then those organisms that are most closely related should have experienced less change over time. On a phylogenetic tree, species with substantially similar rRNA sequences would be diagrammed with a closer common ancestor—one that had the sequences they inherited—than the ancestors shared between species with dissimilar rRNA sequences. **12.** *understand* A null hypothesis specifies what a researcher should observe when the hypothesis being tested isn't correct.

✔ Test Your Problem-Solving Skills

13. *analyze* A scientific theory is not a guess—it is an idea whose validity can be tested with data. Both the cell theory and the theory of evolution have been validated by large bodies of observational and experimental data. **14.** *apply* If all eukaryotes living today have a nucleus, then it is logical to conclude that the nucleus arose in a common ancestor of all eukaryotes, indicated by the arrow you should have added to the figure. See **FIGURE A1.2**. If it had arisen in a common ancestor of Bacteria or Archaea, then species in those groups would have had to lose the trait—an unlikely event. **15.** *evaluate* The data set was so large and diverse that it was no longer reasonable to argue that noncellular life-forms had been discovered. **16.** *apply* b

🄱🄸🄶 PICTURE Doing Biology

p. 16 CYU (1) *understand* Biologists design and carry out a study, either observational or experimental, to test their ideas. As part of this process, they state their ideas as a hypothesis and null hypothesis and make predictions. They analyze and interpret the data they have gathered, and determine whether the data support their ideas. If not, they revisit their ideas and come up with an alternative hypothesis and design another study to test these new predictions. **(2)** *understand* There are many

possible examples. Consider, for example, the experiment on navigation in foraging desert ants (Chapter 1). In addition to testing how the ants use information on stride length and number to calculate how far they are from the nest (multicellular organism and population levels), researchers also could test how the "pedometer" works at the level of cells and molecules. **(3)** *analyze* A hypothesis is a testable statement to explain a specific phenomenon or a set of observations. The word theory refers to proposed explanations for very broad patterns in nature that are supported by a wide body of evidence. A theory serves as a framework for the development of new hypotheses. **(4)** *analyze* The next step is to relate your findings to existing theories and the current scientific literature, and then to communicate your findings to colleagues through informal conversations, presentations at scientific meetings, and eventually publication in peer-reviewed journals.

CHAPTER 2

IN-TEXT QUESTIONS AND EXERCISES

p. 20 Fig. 2.3 *apply* There are 15 electrons in phosphorus, so there must be 15 protons, which is the atomic number. Since the mass number is 31, then the number of neutrons is 16.

p. 23 *understand* *Water:* arrows pointing from hydrogens to oxygen atom; *ammonia:* arrows pointing from hydrogens to nitrogen atom; *methane:* double arrows between carbons and hydrogens; *carbon dioxide:* arrows pointing from carbon to oxygens; *molecular nitrogen:* double arrows between nitrogens.

p. 23 Fig. 2.7 *understand* Oxygen and nitrogen have high electronegativities. They hold shared electrons more tightly than C, H, and many other atoms, resulting in polar bonds.

p. 24 CYU *evaluate* See **FIGURE A2.1**.

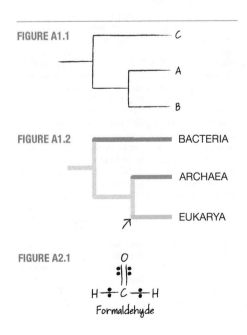

FIGURE A1.1

FIGURE A1.2

FIGURE A2.1
Formaldehyde

p. 25 (1) `evaluate` $^{\delta+}$H–O$^{\delta-}$–H$^{\delta+}$ **(2)** `apply` If water were linear, the partial negative charge on oxygen would have partial positive charges on either side. Compared to the actual, bent molecule, the partial negative charge would be much less exposed and less able to participate in hydrogen bonding.

p. 26 Fig. 2.14 `understand` Oils are nonpolar. They have long chains of carbon atoms bonded to hydrogen atoms, which share electrons evenly because their electronegativities are similar. When an oil and water are mixed, the polar water molecules interact with each other via hydrogen bonding rather than with the nonpolar oil molecules, which interact with themselves instead.

p. 28 Table 2.2 `understand` "Cause" (Row 1): electrostatic attractions between partial charges on water and opposite charges on ions; hydrogen bonds; water and other polar molecules. "Biological Consequences" (Row 2): ice to float; freezing solid. "Cause" (Row 4): lots of heat energy; break hydrogen bonds and change water to a gas.

p. 29 `apply` The proton concentration would be 3.2×10^{-9} M.

p. 29 Fig. 2.17 `apply` The concentration of protons would decrease because milk is more basic (pH 6.5) than black coffee (pH 5).

p. 30 `apply` The bicarbonate concentration would increase. The protons (H$^+$) released from carbonic acid would react with the hydroxide ions (OH$^-$) dissociated from NaOH to form H$_2$O, leaving fewer protons free to react with bicarbonate to reform carbonic acid.

p. 32 CYU (1) `apply` The reaction would be spontaneous based on the change in potential energy; the reactants have higher chemical energy than the products. The entropy, however, is not increased based on the number of molecules, although heat given off from the reaction still results in increased entropy. **(2)** `understand` The electrons are shifted farther from the nuclei of the carbon and hydrogen atoms and closer to the nuclei of the more electronegative oxygen atoms.

p. 32 Fig. 2.19 `remember` See **FIGURE A2.2**.

p. 33 Fig. 2.21 `analyze` The water-filled flask is the ocean; the gas-filled flask is the atmosphere; the condensed water droplets are rain; the electrical sparks are lightning.

p. 37 Table 2.3 `understand` All the functional groups in Table 2.3, except the sulfhydryl group (–SH), are highly polar. The sulfhydryl group is only very slightly polar.

IF YOU UNDERSTAND . . .

2.1. `understand` The bonds in methane and ammonia are all covalent, but differ in polarity: Methane has nonpolar covalent bonds while ammonia has polar covalent bonds. Sodium chloride does not have covalent bonds; instead, ionic bonds hold the ionized sodium and chloride together. **2.2** `understand` Assuming neutral pH, amino and hydroxyl groups would interact with the partial negative charge on water's oxygen, because they both carry a positive charge (partial for the hydrogen in the hydroxyl). The carboxyl group would interact with the partial positive charges on water's hydrogens, since it would carry a negative charge after losing the proton from its hydroxyl. **2.3** `understand` Like solar radiation, the energy in electricity generates free radicals that would promote the reaction. **2.4** `understand` "Top-down" approach: The reaction responsible for synthesizing acetic acid is observed in cells and can serve as an intermediate for the formation of a more complex molecule (acetyl CoA) that is used by cells throughout the tree of life. "Bottom-up" approach: This reaction can also occur under conditions that mimic the early Earth environment in deep-sea vents. **2.5** `apply` The hydroxyls would increase the solubility of octane by introducing polar covalent bonds, which would make the molecule more hydrophilic. The high electronegativity of oxygen would decrease the potential energy of the modified molecule.

YOU SHOULD BE ABLE TO . . .

✔ Test Your Knowledge

1. `understand` b **2.** `remember` a **3.** `remember` c **4.** `remember` d
5. `remember` Potential energy and entropy. **6.** `remember` The prebiotic soup model and the surface metabolism model.

✔ Test Your Understanding

7. `apply` c. Acetic acid has more highly electronegative oxygen atoms than the other molecules. When bonded to carbon or hydrogen, each oxygen will result in a polar covalent bond. **8.** `apply` Relative electronegativities would be F > O > H > Na. One bond would form with sodium, and it would be ionic. **9.** `understand` When oxygen is covalently bonded to hydrogen, the difference in electronegativities between the atoms causes the electrons to spend more time near the oxygen. In contrast, the atoms in H$_2$ and O$_2$ have the same electronegativities, so they equally share electrons in their covalent bonds. **10.** `apply` See **FIGURE A2.3**. **11.** `apply` The dissociation reaction of carbonic acid lowers the pH of the solution by releasing extra H$^+$ into the solution. If additional CO$_2$ is added, the sequence of reactions would be driven to the right, which would make the ocean more acidic. **12.** `understand` The carbon framework determines the overall shape of an organic molecule. The functional groups attached to the carbons determine the molecule's chemical behavior, because these groups are likely to interact with other molecules.

✔ Test Your Problem-Solving Skills

13. `analyze` b. **14.** `analyze` No, they don't conflict. Shells that are farther from the protons (positive charges) in the nucleus house electrons that have greater potential energy than shells closer to the nucleus. **15.** `analyze` One possible concept map relating the structure of water to its properties is shown below (see **FIGURE A2.4**). **16.** `create` In hot weather, water absorbs large amounts of heat due to its high specific heat and high heat of vaporization. In cold weather, water releases the large amount of heat that it has absorbed.

FIGURE A2.2

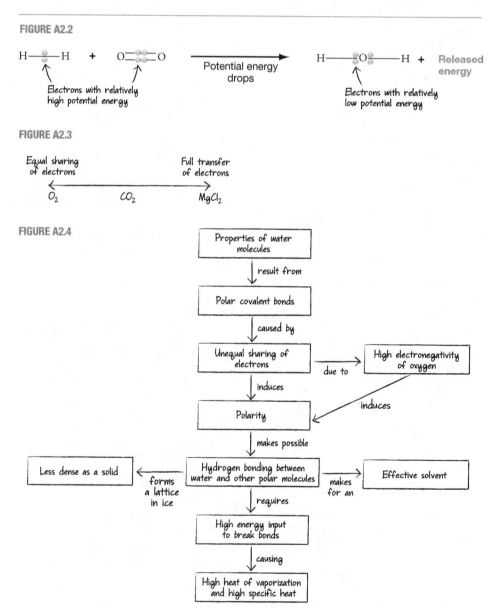

FIGURE A2.3

FIGURE A2.4

CHAPTER 3

IN-TEXT QUESTIONS AND EXERCISES

p. 43 Fig. 3.2 (understand) The green R-groups contain mostly C and H, which have roughly equal electronegativities. Electrons are evenly shared in C–H bonds and C–S bonds, so the groups are nonpolar. Cysteine has a sulfur that is slightly more electronegative than hydrogen, so it will be less nonpolar than the other green groups. All of the pink R-groups have a highly electronegative oxygen atom with a partial negative charge, making them polar.
p. 44 (apply) From most hydrophilic to most hydrophobic: (1) aspartate, (2) asparagine, (3) tyrosine, (4) valine. The most hydrophilic amino acids will have side chains with full charges (ionized), like aspartate, followed by those with the largest number of highly electronegative atoms, like oxygen or nitrogen. Highly electronegative atoms produce polar covalent bonds with carbon or hydrogen. The most hydrophobic will not have oxygen or nitrogen in their side chains, but instead will have the largest number of C–H bonds, which are nonpolar covalent.
p. 47 CYU (understand) See **FIGURE A3.1**.
p. 52 CYU (1) (understand) Secondary, tertiary, and quaternary structure all depend on bonds and other interactions between amino acids that are linked in a chain in a specific order (primary structure). **(2)** (apply) There would be 20^5 different peptides, or 3.2×10^6 different primary sequences.
p. 55 CYU (apply) Amino acid changes would be expected to be in the active site or in regions that affect the folded structure of this site. Either of these changes could result in a different active site that either binds a new substrate or catalyzes a different reaction.

IF YOU UNDERSTAND . . .

3.1. (understand) Look at the R-group of the amino acid. If there is a positive charge, then it is basic. If there is a negative charge, then it is acidic. If there is not a charge, but there is an oxygen atom, then it is polar uncharged. If there is no charge or oxygen, then it is nonpolar.
3.2 (apply) Nonpolar amino acid residues would be found in the interior of a globular protein, grouped with other nonpolar residues due to hydrophobic interactions.
3.3 (analyze) Both calmodulin and infectious prions require some form of induction to achieve their active conformations. Calcium ions are required for calmodulin to fold into its functional structure while prions are induced to change their shape by other, improperly folded prion proteins. **3.4** (analyze) *Catalysis:* Proteins are made of amino acids, which have many reactive functional groups, and can fold into different shapes that allow the formation of active sites. *Defense:* Similar to catalysis, the chemical properties and capacity for different shapes allows proteins to be made that can attach to virtually any type of invading virus or cell. *Signaling:* The flexibility in protein structure allows protein activities to be quickly turned on or off based on binding to signal molecules or ions.

YOU SHOULD BE ABLE TO . . .

✔ Test Your Knowledge

1. (remember) d **2.** (remember) The atoms and functional groups found in the side chains. **3.** (remember) b **4.** (remember) b
5. (understand) The order and type of amino acids (i.e., the primary structure) contains the information that directs folding. **6.** (understand) a

✔ Test Your Understanding

7. (understand) Because the nonpolar amino acid residues are not able to interact with the water solvent, they are crowded together in the interior of a protein and surrounded by a network of hydrogen-bonded water molecules. This crowding leads to the development of van der Waals interactions that help glue the nonpolar

side chains together. **8.** (understand) No, polymerization is a nonspontaneous reaction because the product molecules have lower entropy than the free form of the reactants and there would be nothing to prevent hydrolysis from reversing the reaction. **9.** (understand) Many possible correct answers, including (1) the presence of an active site in an enzyme that is precisely shaped to fit a substrate or substrates in the correct orientation for a reaction to occur; (2) the doughnut shape of porin that allows certain substances to pass through it; (3) the cable shape of collagen to provide structural support for cells and tissues. **10.** (create) Proteins are highly variable in overall shape and chemical properties due to variation in the composition of R-groups and the array of secondary through quaternary structures that are possible. This variation allows them to fulfill many different roles in the cell. Diversity in the shape and reactivity of active sites also makes them effective catalysts. **11.** (understand) c **12.** (create) In many proteins, especially those involved in cell signaling, their structure is affected by binding to other molecules or ions. Since the shape of the protein is directly involved in its function, the protein's activity is regulated by controlling how it is folded. If proteins were inflexible, this type of control could not occur.

✔ Test Your Problem-Solving Skills

13. (analyze) The side chain of proline is a cyclic structure that is covalently bonded to the nitrogen in the core amino group. This restricts the movement of the side chain relative to the core nitrogen, which further restricts the backbone when the nitrogen participates in a peptide bond with a neighboring amino acid. **14.** (create) See **FIGURE A3.2 15.** (apply) b. Phosphates have a negative charge, so they are most likely to form ionic bonds with the positively-charged side chains of basic amino acid residues. **16.** (apply) The inherited forms likely have some alteration in the primary structure such that the infectious form is spontaneously generated at a higher rate than normal. The amino acid sequence in these prions would likely differ from those transmitted between animals.

CHAPTER 4

IN-TEXT QUESTIONS AND EXERCISES

p. 59 (understand) See **FIGURE A4.1**.

FIGURE A3.1

FIGURE A3.2

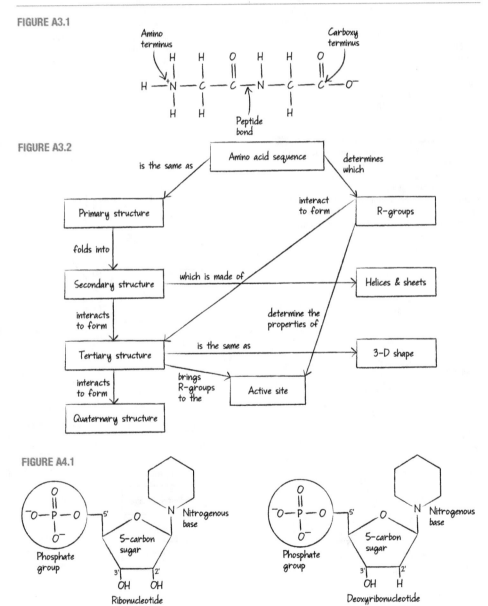

FIGURE A4.1

p. 60 Fig. 4.3 (understand) 5'-UAGC-'3
p. 61 CYU (understand) See FIGURE A4.2.
p. 63 (understand) If the two strands were parallel, the G-C pairing would align N—H groups together and C=O groups together, which would not allow hydrogen bonds to form.
p. 64 Fig. 4.8 (analyze) It is not spontaneous—energy must be added (as heat) for the reaction to occur.
p. 65 CYU (remember) See FIGURE A4.3.

IF YOU UNDERSTAND . . .

4.1. (understand) Cells activate nucleotides by linking additional phosphates to an existing 5' phosphate. Activation increases the chemical energy in the nucleotides enough to offset the decrease in entropy that will result from the polymerization reaction. **4.2** (understand) C-G pairs involve three hydrogen bonds, so they are more stable than A-T pairs with just two hydrogen bonds. **4.3** (analyze) A single-stranded RNA molecule has unpaired bases that can pair with other bases on the same RNA strand, thereby folding the molecule into stem-and-loop configurations. These secondary structures can further fold on themselves, giving the molecule a tertiary structure. Because DNA molecules are double stranded, with no unpaired bases, further internal folding is not possible. **4.4** (understand) Examples would include (1) the production of nucleotides, and (2) polymerization of RNA. It is thought that nucleotides were scarce during chemical evolution, so their catalyzed synthesis by a ribozyme would have been advantageous. Catalysis by an RNA replicase would have dramatically increased the reproductive rate of RNA molecules.

YOU SHOULD BE ABLE TO . . .

✔ Test Your Knowledge

1. (remember) c **2.** (remember) c **3.** (remember) a **4.** (remember) d **5.** (remember) One end has a free phosphate group on the 5' carbon; the other end has a free hydroxyl group bonded to the 3' carbon. **6.** (understand) DNA is a more stable molecule than RNA because it lacks a hydroxyl group on the 2' carbon and is therefore more resistant to cleavage, and because the two sugar-phosphate backbones are held together by many hydrogen bonds between nitrogenous bases.

✔ Test Your Understanding

7. (apply) In DNA, the secondary structure requires that every guanine pairs with a cytosine and every thymine pairs with an adenine, resulting in consistent ratios between the nucleotides. Chargaff's rules do not apply to RNA, since it is single-stranded and the pairing is not consistent throughout the molecules. **8.** (apply) a; if 30 percent is adenine, then 30 percent would be thymine, since they are base-paired together. This means that 40 percent consists of G-C base pairs, which would be equally divided between the two bases. **9.** (apply) The DNA sequence of the new strand would be 5'-ATCGATATC-3'. The RNA sequence would be the same, except each T would be replaced by a U. **10.** (understand) DNA has limited catalytic ability because it (1) lacks functional groups that can participate in catalysis and (2) has a regular structure that is not conducive to forming shapes required for catalysis. RNA molecules can catalyze some reactions because they (1) have exposed hydroxyl functional groups and (2) can fold into shapes that that can function in catalysis. **11.** (apply) No. Catalytic activity in ribozymes depends on the tertiary structure generated from single-stranded molecules. Double-stranded nucleic acids do not form tertiary structures. **12.** (understand) An RNA replicase would undergo replication and be able to evolve. It would process information in the sense of copying itself, and it would use energy to drive polymerization reactions. It would not be bound by a membrane and considered a cell, however, and it would not be able to acquire energy. It would best be considered as an intermediate step between nonlife and true life (as outlined in Chapter 1).

✔ Test Your Problem-Solving Skills

13. (create) See FIGURE A4.4 **14.** (apply) Yes—if the complementary bases lined up over the entire length of the two strands, they would twist into a double helix analogous to a DNA molecule. The same types of hydrogen bonds and hydrophobic interactions would occur as observed in the "stem" portion of hairpins in single-stranded RNA. **15.** (apply) In a triple helix, the bases are unlikely to align properly for hydrogen bonding to occur, so hydrophobic interactions would probably be more important. **16.** (apply) b; the high temperature would make it more likely that the secondary and tertiary structures would be denatured in the ribozymes. To overcome this effect, you would expect the hairpins to possess more G-C pairs, since they consist of three hydrogen bonds compared to the two found in A-T pairs.

CHAPTER 5

IN-TEXT QUESTIONS AND EXERCISES

p. 73 Fig. 5.2 (understand) See the structure of mannose in FIGURE A5.1.
p. 75 CYU (apply) See FIGURE A5.2.
p. 78 CYU (remember) They could differ in (1) location of linkages (e.g., 1,4 or 1,6); (2) types of linkages (e.g., α or β); (3) the sequence of the monomers (e.g., two galactose and then two glucose, versus alternating galactose and glucose); and/or (4) whether the four monomers are linked in a line or whether they branch.
p. 79 Fig. 5.6 (apply) The percentage of inhibition would not change for the intact glycoprotein bar. The purified

FIGURE A4.3

FIGURE A5.1

FIGURE A4.2

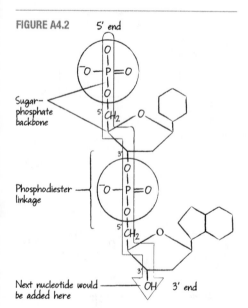

FIGURE A4.4

carbohydrate bar would be at zero inhibition, and the glycoprotein with digested carbohydrate bar would be similar to the intact glycoprotein bar.

p. 80 Fig. 5.7 `understand` All of the C—C and C—H bonds should be circled.

p. 81 CYU (1) `understand` *Aspect 1:* The β-1,4-glycosidic linkages in these molecules result in insoluble fibers that are difficult to degrade. *Aspect 2:* When individual molecules of these carbohydrates align, bonds form between them and produce fibers or sheets that resist pulling and pushing forces. **(2)** `apply` Most are probably being broken down into glucose, some of which in turn is being broken down in reactions that lead to the synthesis of ATP. Some will be resistant to digestion, such as the insoluble cellulose that makes up dietary fiber. This will help retain water and support the digestion and passage of fecal material.

IF YOU UNDERSTAND . . .

5.1. `understand` Molecules have to interact in an extremely specific orientation in order for a reaction to occur. Changing the location of a functional group by even one carbon can mean that the molecule will undergo completely different types of reactions. **5.2** `analyze` Glycosidic linkages can vary more in location and geometry than linkages between amino acids and nucleotides do. This variability increases the structural diversity possible in carbohydrates compared to proteins and nucleic acids. **5.3** `understand` (1) Polysaccharides used for energy storage are formed entirely from glucose monomers joined by α-glycosidic linkages; structural polysaccharides are made up of glucose or other sugars joined by β-glycosidic linkages. (2) The monomers in energy-storage polysaccharides are linked in a helical arrangement; the monomers in structural polysaccharides are linked in a linear arrangement. (3) Energy-storage polysaccharides may branch; structural polysaccharides do not. (4) Individual chains of energy-storage polysaccharides do not associate with each other; adjacent chains of structural polysaccharides are linked by hydrogen bonds or covalent bonds.

YOU SHOULD BE ABLE TO . . .

✔ **Test Your Knowledge**

1. `remember` d **2.** `remember` Monosaccharides can differ from one another in three ways: (1) the location of their carbonyl group; (2) the number of carbon atoms they contain; and (3) the orientations of their hydroxyl groups. **3.** `remember` a **4.** `remember` c **5.** `remember` a **6.** `understand` The electrons in the C═O bonds of carbon dioxide molecules are held tightly by the highly electronegative oxygen atoms, so they have low potential energy. The electrons in the C—C and C—H bonds of carbohydrates are shared equally, so they have much higher potential energy.

✔ **Test Your Understanding**

7. `understand` c. **8.** `apply` a; lactose is a disaccharide formed from a β-1,4-glycosidic linkage, so if two glucose molecules were linked with this bond, they would resemble units of cellulose and not be digested by human infants or adults. **9.** `understand` Carbohydrates are ideal for displaying the identity of the cell because they are so diverse structurally. This diversity enables them to serve as very specific identity tags for cells. **10.** `understand` When you compare the glucose monomers in an α-1,4-glycosidic linkage versus in a β-1,4-glycosidic linkage, the linkages are located on opposite sides of the plane of the glucose rings, and the glucose monomers are linked in the same orientation versus having every other glucose flipped in orientation. β-1,4-glycosidic linkages are much more likely to form linear fibers and sheets, so they resist degradation. **11.** `remember` Because (1) no mechanism is

known for the prebiotic polymerization of sugars; (2) no catalytic carbohydrates have been discovered that can perform polymerization reactions; and (3) sugar residues in a polysaccharide are not capable of complementary base pairing. **12.** `understand` Starch and glycogen both consist of glucose monomers joined by α-1,4-glycosidic linkages, and both function as storage carbohydrates. Starch is a mixture of unbranched and branched polysaccharides—called amylose and amylopectin, respectively. All glycogen polysaccharides are branched.

✔ **Test Your Problem-Solving Skills**

13. `analyze` Carbohydrates are energy-storage molecules, so minimizing their consumption may reduce total energy intake. Lack of available carbohydrate also forces the body to use fats for energy, reducing the amount of fat that is stored. **14.** `apply` d; lactose is a disaccharide of glucose and galactose, which can be cleaved by enzymes expressed in the human gut to release galactose. **15.** `analyze` Amylase breaks down the starch in the cracker into glucose monomers, which stimulate the sweet receptors in your tongue. **16.** `apply` When bacteria contact lysozyme, the peptidoglycan in their cell walls begins to degrade, leading to the death of the bacteria. Lysozyme therefore helps protect humans against bacterial infections.

CHAPTER 6

IN-TEXT QUESTIONS AND EXERCISES

p. 87 `understand` Fatty acids are amphipathic because their hydrocarbon tails are hydrophobic but their carboxyl functional groups are hydrophilic.

p. 87 Fig. 6.4 `apply` At the polar hydroxyl group in cholesterol and the polar head group in phospholipids.

p. 88 CYU (1) `analyze` Fats consist of three fatty acids linked to glycerol; steroids have a distinctive four-ring structure with variable side groups attached; phospholipids have a hydrophilic, phosphate-containing "head" region and a hydrocarbon tail. **(2)** `understand` In cholesterol, the hydrocarbon steroid rings and isoprenoid chain are hydrophobic; the hydroxyl group is hydrophilic. In phospholipids, the phosphate-containing head group is hydrophilic; the hydrocarbon chains are hydrophobic.

p. 89 `apply` Amino acids have amino and carboxyl groups that are ionized in water and nucleotides have negatively charged phosphates. Due to their charge and larger size, both would be placed below the small ions at the bottom of the scale (permeability$<10^{-12}$ cm/sec).

p. 90 CYU `create` See **TABLE A6.1**.

p. 91 Fig. 6.10 `apply` Increasing the number of phospholipids with polyunsaturated tails would increase permeability of the liposomes. Starting from the left, the first line (no cholesterol) would represent liposomes with 50% polyunsaturated phospholipids, the second line would be 20% polyunsaturated phospholipids, and the third line would contain only saturated phospholipids.

p. 92 `apply` If there is a difference in temperature, then there would be a difference in thermal motion. The solute concentration on the side with a higher temperature would decrease because the solute particles would be moving faster and hence be more likely to move to the cooler side of the membrane, where they would slow down.

p. 93 Fig. 6.13 `apply` Higher, because less water would have to move to the right side to achieve equilibrium.

FIGURE A5.2

Start with a monosaccharide. This one is a 3-carbon aldose (carbonyl group at end).

Variation 1: 3-carbon ketose (carbonyl group in middle)

Variation 2: 4-carbon aldose

Variation 3: 3-carbon aldose with different arrangement of hydroxyl group

TABLE A6.1

Factor	Effect on permeability	Reason
Temperature	Decreases as temperature decreases.	Lower temperature slows movement of hydrocarbon tails, allowing more interactions (membrane is more dense).
Cholesterol	Decreases as cholesterol content increases.	Cholesterol molecules fill in the spaces between the hydrocarbon tails, making the membrane more tightly packed.
Length of hydrocarbon tails	Decreases as length of hydrocarbon tails increases.	Longer hydrocarbon tails have more interactions (membrane is more dense).
Saturation of hydrocarbon tails	Decreases as degree of saturation increases.	Saturated fatty acids have straight hydrocarbon tails that pack together tightly, leaving few gaps.

p. 94 CYU create See **FIGURE A6.1**.

p. 96 Fig. 6.18 create Repeat the procedure using a lipid bilayer that is free of membrane proteins, such as synthetic liposomes constructed from only phospholipids. If proteins were responsible for the pits and mounds, then this control would not show these structures.

p. 97 apply Your arrow should point out of the cell. There is no concentration gradient for chloride, but the outside has a net positive charge, which favors outward movement of negative ions.

p. 97 Fig. 6.21 analyze No—the 10 replicates where no current was recorded probably represent instances where the CFTR protein was damaged and not functioning properly. (In general, no experimental method works "perfectly.")

p. 101 CYU understand Passive transport does not require an input of energy—it happens as a result of energy already present in existing concentration or electrical gradients. Active transport is active in the sense of requiring an input of energy from, for example, ATP. In cotransport, a second ion or molecule is transported against its concentration gradient along with (i.e., "co") an ion that is transported along its concentration gradient.

p. 101 Fig. 6.26 understand *Diffusion:* description as given; no proteins involved. *Facilitated diffusion:* Passive movement of ions or molecules that cannot cross a phospholipid bilayer readily along a concentration gradient; facilitated by channel or carrier proteins. *Active transport:* Active movement of ions or molecules that will build a gradient; facilitated by pump proteins powered by an energy source such as ATP.

IF YOU UNDERSTAND . . .

6.1. analyze Adding H_2 increases the saturation of the oil by converting C=C bonds into C—C bonds with added hydrogens. Lipids with more C—H bonds tend to be solid at room temperature. **6.2** understand Highly permeable and fluid bilayers possess short, unsaturated hydrocarbon tails while those that are highly impermeable and less fluid contain long, saturated hydrocarbon tails. **6.3** apply (1) The solute will diffuse until both sides are at equal concentrations. (2) Water will diffuse toward the side with the higher solute concentration. **6.4** apply See **FIGURE A6.2**.

YOU SHOULD BE ABLE TO . . .

✓ Test Your Knowledge

1. understand c **2.** remember a **3.** understand b **4.** understand d
5. understand For osmosis to occur, a concentration gradient and membrane that allows water to pass, but not the solute, must be present. **6.** analyze Channel proteins form pores in the membrane and carrier proteins undergo conformational changes to shuttle molecules or ions across the membrane.

✓ Test Your Understanding

7. apply b **8.** analyze No, because they have no polar end to interact with water. Instead, these lipids would float on the surface of water, or collect in droplets suspended in water, reducing their interaction with water to a minimum. **9.** understand Hydrophilic, phosphate-containing head groups interact with water; hydrophobic hydrocarbon tails associate with each other. A bilayer is more stable than are independent phospholipids in solution. **10.** apply Ethanol's polar hydroxyl group reduces the speed at which it can cross a membrane, but its small size and lack of charge would allow it to slowly cross membranes—between the rates of water and glucose transport. **11.** understand Only nonpolar, hydrophobic amino acid residues would be found in the portion of the protein that crosses the membrane. In the interior of the bilayer, these residues would be hidden from the water solvent and interact with the nonpolar lipid tails. **12.** apply Chloride ions from sodium chloride will move from the left side to the right through the CFTR. Water

will initially move from the right side to the left by osmosis, but as chloride ions move to the right, water will follow. Na^+ and K^+ ions will not move across the membrane.

✓ Test Your Problem-Solving Skills

13. apply c **14.** create Flip-flops should be rare, because they require a polar head group to pass through the hydrophobic portion of the lipid bilayer. To test this prediction, you could monitor the number of dyed phospholipids that transfer from one side of the membrane to the other in a given period of time. **15.** apply Organisms that live in very cold environments are likely to have highly unsaturated hydrocarbon tails. The kinks in unsaturated hydrocarbon tails keep membranes fluid and permeable, even at low temperature. Organisms that live in very hot environments would likely have phospholipids with saturated tails, to prevent membranes from becoming too fluid and permeable. **16.** analyze Adding a methyl group makes a drug more hydrophobic and thus more likely to pass through a lipid bilayer. Adding a charged group makes it hydrophilic and reduces its ability to pass through the lipid bilayer. These modifications would help target the drug to either the inside or outside of cells, respectively.

BIG PICTURE Chemistry of Life

p. 104 CYU (1) understand Oxygen is much more electronegative than hydrogen, so within water, the electrons are unequally shared in the O—H covalent bonds. The resulting partial negative charge around the oxygen and partial positive charges around the hydrogen atoms allow for hydrogen bonds to form among water molecules. **(2)** analyze Unlike other macromolecules, nucleic acids can serve as templates for their own replication. RNA is generally single-stranded and can adopt many different

three-dimensional structures. The flexibility in structure, combined with the presence of reactive hydroxyl groups, contribute to the formation of active sites that catalyze chemical reactions. One or more of these catalytic RNA molecules may have evolved the ability to self-replicate. DNA is not likely to have catalyzed its own replication, as it is most often double-stranded, with no clear tertiary structure, and it lacks the reactive hydroxyl groups. **(3)** remember In the amino acid, the nitrogen in the amino (NH_3^+) group and the carbon in the carboxyl (COO^-) group should be circled. In the nucleotide, the oxygen in the hydroxyl (OH) group and the phosphorus in the phosphate (PO_4^{2-}) group on the nucleotide should be circled. **(4)** understand A line representing a protein should be drawn such that it completely crosses the lipid bilayer at least once. The protein could be involved in a variety of different roles, including transport of substances across the membrane in the form of a channel, carrier, or pump.

CHAPTER 7

IN-TEXT QUESTIONS AND EXERCISES

p. 110 CYU remember (1) The nucleoid compacts the chromosome to fit inside the cell via supercoiling while still keeping it accessible for replication and transmission of information. (2) Photosynthetic membranes increase food production by providing a large surface area to hold the pigments and enzymes required for photosynthesis. (3) Flagella propel cells through liquid, often toward a food source. (4) The layer of thick, strong material stiffens the cell wall and provides protection from mechanical damage.

p. 114 Fig. 7.12 create Storing the toxins in vacuoles prevents the toxins from damaging the plant's own organelles and cells.

FIGURE A6.1

FIGURE A6.2

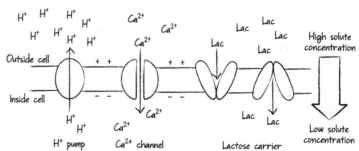

p. 116 CYU (1) (understand) Both organelles contain specific sets of enzymes. Lysosomal enzymes digest macromolecules in the acidic lumen of this organelle, releasing monomers that can be recycled into new macromolecules. Peroxisomes contain catalase and other enzymes that process fatty acids and toxins via oxidation reactions. **(2)** (understand) From top to bottom: administrative/information hub, protein factory, large molecule manufacturing and shipping (protein synthesis and folding center, lipid factory, protein finishing and shipping line, waste processing and recycling center), warehouse, fatty-acid processing and detox center, power station, food-manufacturing facility, support beams, perimeter fencing with secured gates, and leave blank.
p. 118 Fig. 7.16 (remember) See **FIGURE A7.1**.
p. 121 (analyze) (1) Nucleotides are small enough that they would diffuse through the nuclear pore complex along their gradients—a passive process that would not require energy. (2) Large proteins must be escorted through the nuclear pore complex, which is directional and requires energy, since the protein is concentrated inside the nucleus.
p. 121 Fig. 7.18 (apply) "Prediction": The labeled tail region fragments or the labeled core region fragments of the nucleoplasmin protein will be found in the cell nucleus. "Prediction of null hypothesis": Either both the fragments (no required signal) or neither of them (whole protein signal) will be found in the nucleus of the cell. "Conclusion": The send-to-nucleus signal is in the tail region of the nucleoplasmin protein.
p. 123 (apply) During the chase period, proteins appear to have first entered the Golgi after 7 minutes and then started to move into secretory granules after 37 minutes. This means that in this experiment, it took approximately 30 minutes for the fastest-moving proteins to pass through the Golgi.

p. 126 (apply) In receptor-mediated endocytosis, the conversion of a late endosome to a lysosome is dependent on receiving acid hydrolases from the Golgi. If this receptor is not present, then the enzymes will not be sent and the late endosome will not mature into a lysosome to digest the endocytosed products.
p. 126 CYU (1) (apply) Proteins that enter the nucleus are fully synthesized and have an NLS that interacts with another protein to get it into the organelle. The NLS is not removed. Proteins that enter the ER have a signal sequence that interacts with the SRP during translation. The ribosome is moved to the ER and synthesis continues, moving the protein into the ER. The signal is removed once it enters the organelle. **(2)** (apply) The protein would be in the lysosome. The ER signal would direct the protein into the ER before it is completely synthesized. The M-6-P tag will direct the protein from the Golgi to the late endosome to the lysosome. Thus the complete protein is never free in the cytosol, where the NLS could direct it into the nucleus.
p. 132 CYU (analyze) Actin filaments are made up of two strands of actin monomers, microtubules are made up of tubulin protein dimers that form a tube, and intermediate filaments are made up of a number of different protein subunits. Actin filaments and microtubules exhibit polarity (or directionality), and new subunits are constantly being added or subtracted at either end (but added faster to the plus end). All three elements provide structural support, but only actin filaments and microtubules serve as tracks for motors involved in movement and cell division.
p. 132 Fig. 7.30 (apply) The microtubule doublets of the axoneme would slide past each other completely, but the axoneme would not bend.

IF YOU UNDERSTAND . . .

7.1 (understand) (1) Cells will be unable to synthesize new proteins and will die. (2) In many environments, cells will be unable to resist the osmotic pressure of water entering the cytoplasm and will burst. (3) The cell shape will be different, and cells will not be able to divide.
7.2 (understand) (1) The cell will be unable to produce a sufficient amount of ATP and will die. (2) Reactive molecules, like hydrogen peroxide, will damage the cell, and it will likely die. (3) Nothing will happen, since plants do not have centrioles. **7.3** (analyze) The liver cell would be expected to have more peroxisomes and less rough endoplasmic reticulum than the salivary cells would. **7.4** (create) The addition or removal of phosphates would change the folded structure of the protein, exposing the NLS for nuclear transport. **7.5** (create) As a group, proteins have complex and highly diverse shapes and chemical properties that allow them to recognize a great number of different zip codes in a very specific manner. **7.6** (apply) The Golgi is positioned near the microtubule organizing center, which has microtubules running from the minus end out to the plus end (near the plasma membrane). Kinesin would be used to move these vesicles as it walks toward the plus end.

YOU SHOULD BE ABLE TO . . .

✓ **Test Your Knowledge**

1. (understand) b **2.** (remember) They have their own small, circular chromosomes; they produce their own ribosomes; and they divide in a manner that is similar to bacterial fission, independent of cellular division. **3.** (remember) c **4.** (remember) b **5.** (remember) a **6.** (understand) The phosphate links to the motor protein and causes it to change shape, which results in the protein moving along the filament.

✓ **Test Your Understanding**

7. (analyze) All cells are bound by a plasma membrane, are filled with cytoplasm, carry their genetic information (DNA) in chromosomes, and contain ribosomes

FIGURE A7.1

(a) Animal pancreatic cell: Exports digestive enzymes.

Rough ER — Secretory vesicles

(b) Animal testis cell: Exports lipid-soluble signals.

Smooth ER

(c) Plant leaf cell: Manufactures ATP and sugar.

Chloroplasts — Nucleus
Vacuole

(d) Brown fat cells: Burn fat to generate heat in lieu of ATP.

Mitochondria

(the sites of protein synthesis). Some prokaryotes have organelles not found in plants or animals, such as a magnetite-containing structure. Plant cells have chloroplasts, vacuoles, and a cell wall. Animal cells contain lysosomes and lack a cell wall. **8.** analyze a; the endoplasmic reticulum is responsible for synthesizing the membrane proteins required for the transport of solutes across the plasma membrane. **9.** create The NLS will be used to actively import the protein into the nucleus, leaving very little of the protein in the cytoplasm. Diffusion alone would not drive all the protein into the nucleus. **10.** create Ribosome in cytoplasm (signal is synthesized) → Ribosome at rough ER (protein is completed, folded, and glycosylated) → Transport vesicle → Golgi apparatus (protein is processed; has molecular zip code indicating destination) → Transport vesicle → Plasma membrane → Extracellular space. **11.** create This occurs in microfilaments and microtubules because they have ends that differ structurally and functionally—they have different filament growth rates. Intermediate filaments have identical ends, so there is no difference in the rate of assembly between the two ends. **12.** understand Polarized cytoskeletal filaments (microtubules or microfilaments) are present between the organelles. End-directed motor proteins use ATP to move these transport vesicles between them.

✔ Test Your Problem-Solving Skills

13. understand b; fimbriae are involved in bacterial attachment to surfaces and other cells, which would be important in the ability to grow on teeth. **14.** create The proteins must receive a molecular zip code that binds to a receptor on the surface of peroxisomes. They could diffuse randomly to peroxisomes or be transported in a directed way by motor proteins. **15.** create The tails cleaved from nucleoplasmin could be attached to the gold particles that were excluded from crossing the pore complex owing to their size. If these modified particles entered the nucleus, then the tail is not limited to the nucleoplasmin transport alone. **16.** apply The proteins would likely be found in the cytoplasm (e.g., actin and myosin) or imported into the mitochondria. Since there is a high energy demand, you would predict that there are many active mitochondria.

CHAPTER 8

IN-TEXT QUESTIONS AND ANSWERS

p. 139 apply (1) If ΔS is positive (products have more disorder than reactants), then according to the free energy equation, ΔG is more likely to be negative as temperature (T) increases even if ΔH is positive. The increased temperature represents added heat energy that may be used to drive an endothermic reaction to completion, making the reaction spontaneous. (2) Exothermic reactions may be nonspontaneous if they result in a decrease in entropy—meaning that the products are more ordered than the reactants (ΔS is negative).
p. 139 CYU (1) understand Gibbs equation: $\Delta G = \Delta H - T \Delta S$. ΔG symbolizes the change in the Gibbs free energy. ΔH represents the difference in enthalpy (heat, pressure, and volume) between the products and the reactants. T represents the temperature (in degrees Kelvin) at which the reaction is taking place. ΔS symbolizes the change in entropy (amount of disorder). **(2)** understand When ΔH is negative—meaning that the reactants have lower enthalpy than the products—and when ΔS is positive, meaning that the products have higher entropy (are more disordered) than the reactants.
p. 140 Fig. 8.4 understand Each point represents the data from a single test, not an average of many experiments, so it is not possible to calculate the standard error of the average.
p. 144 analyze Redox reactions transfer energy between molecules or atoms via electrons. When oxidized molecules are reduced, their potential energy increases. ATP

hydrolysis is often coupled with the phosphorylation of another molecule. This phosphorylation increases the potential energy of the molecule.
p. 144 CYU (1) understand Electrons in C—H bonds are not held as tightly as electrons in C—O bonds, so they have higher potential energy. **(2)** understand In part, because its three phosphate groups have four negative charges in close proximity. The electrons repulse each other, raising their potential energy.
p. 144 Fig. 8.9 understand The ΔG in the uncoupled reaction would be positive (>0), and each of the steps in the coupled reaction would have a negative (<0) ΔG.
p. 146 Fig. 8.12 understand No—a catalyst affects only the activation energy, not the overall change in free energy.
p. 147 remember (1) binding substrates, (2) transition state, (3) R-groups, (4) structure
p. 147 Fig. 8.14 analyze See **FIGURE A8.1**.
p. 149 CYU (1) create The rate of the reaction is based primarily on the activity of the enzyme. Once the temperature reaches a level that causes unfolding and inactivation of the enzyme, the rate decreases to the uncatalyzed rate. **(2)** apply The shape change would most likely alter the shape of the active site. If phosphorylation activates catalytic activity, the change to the active site would allow substrates to bind and be brought to their transition state. If phosphorylation inhibits catalytic activity, the shape change to the active site would likely prevent substrates from binding or no longer orient them correctly for the reaction to occur.
p. 150 apply The concentration of A and B would be higher than in the fully functional pathway since they are not depleted to produce C. If D is not being depleted by other reactions, then equilibrium would be established between C and D, resulting in lower concentrations of both.

IF YOU UNDERSTAND . . .

8.1 understand Reactions are spontaneous when the free energy in the products is lower than that of the reactants (ΔG is negative). Enthalpy and entropy are measures used to determine free-energy changes. Enthalpy measures the potential energy of the molecules, and entropy measures the disorder. For exergonic, spontaneous reactions, disorder normally increases and the potential energy stored in the products normally decreases relative to the reactants. **8.2** understand Energetic coupling transfers free energy released from exergonic reactions to drive endergonic reactions. Since endergonic reactions are required for sustaining life, without energetic coupling, life would not exist. **8.3** understand Amino acid R-groups lining the active site interact with the substrates, orienting them in a way that stabilizes the transition state, thereby lowering the activation energy needed for the reaction to proceed. **8.4** analyze Allosteric regulation and phosphorylation cause changes in the conformation of the enzyme that affects its catalytic function. Allosteric regulation involves non-covalent bonding, while phosphorylation is a covalent modification of the enzyme's primary structure. **8.5** apply In the first step of the pathway, the rate would increase as the intermediate, which is the product of the first reaction, is removed. In the last step, the rate would decrease due to the loss of the intermediate, which serves as the substrate for the last reaction.

YOU SHOULD BE ABLE TO . . .

✔ Test Your Knowledge

1. remember c **2.** remember a **3.** remember a **4.** remember The enzyme changes shape, but the change is not permanent. The enzyme shape will return to its original conformation after releasing the products. **5.** remember d **6.** remember When the product of a pathway feeds back to interact with an enzyme early in the same pathway to inhibit its function.

✔ Test Your Understanding

7. understand The shape of reactant molecules (the key) fits into the active site of an enzyme (the lock). Fischer's

original model assumed that enzymes were rigid; in fact, enzymes are flexible and dynamic. **8.** understand d. Energy, such as the thermal energy in fire, must be provided to overcome the activation energy barrier before the reaction can proceed. **9.** understand The phosphorylation reaction is exergonic because the electrons in ADP and the phosphate added to the substrate experience less electrical repulsion, and thus have less potential energy, than they did in ATP. A phosphorylated reactant (i.e., an activated intermediate) gains enough potential energy to shift the free energy change for the reaction from endergonic to exergonic. **10.** apply For the coupled reaction, step 1 has a ΔG of about -3 kcal/mol and step 2 has a ΔG of about -3 kcal/mol. The uncoupled reaction has a ΔG of about $+1.3$ kcal/mol. **11.** analyze Both are mechanisms that regulate enzymes; the difference is whether the regulatory molecule binds at the active site (competitive inhibition) or away from the active site (allosteric regulation). **12.** apply Catabolic reactions will often have a negative ΔG based on a decrease in enthalpy and increase in entropy. Anabolic reactions are the opposite—a positive ΔG that is based on an increase in enthalpy and decrease in entropy.

✔ Test Your Problem-Solving Skills

13. create See **FIGURE A8.2**. **14.** apply Without the coenzyme, the free-radical-containing transition state would not be stabilized and the reaction rate would drop dramatically. **15.** analyze The data suggest that the enzyme and substrate form a transition state that requires a change in the shape of the active site, and that each movement corresponds to one reaction. **16.** apply b. The sugar likely functions as an allosteric regulator to activate the enzyme.

CHAPTER 9

IN-TEXT QUESTIONS AND EXERCISES

p. 156 Fig. 9.2 remember Glycolysis: "What goes in" = glucose, NAD^+, ADP, inorganic phosphate; "What comes out" = pyruvate, NADH, ATP. Pyruvate processing: "What goes in" = pyruvate, NAD^+; "What comes out" = NADH, CO_2, acetyl CoA. Citric acid cycle: "What goes in" = acetyl CoA, NAD^+, FAD, GDP or ADP, inorganic phosphate; "What comes out" = NADH, $FADH_2$, ATP or GTP, CO_2. Electron transport and oxidative phosphorylation: "What goes in" = NADH, $FADH_2$, O_2, ADP, inorganic phosphate; "What comes out" = ATP, H_2O, NAD^+, FAD.

FIGURE A8.1

FIGURE A8.2

p. 161 [apply] If the regulatory site had a higher affinity for ATP than the active site, then ATP would always be bound at the regulatory site, and glycolysis would always proceed at a very slow rate.

p. 161 Fig. 9.9 [remember] "Positive control": AMP, NAD$^+$, CoA (reaction substrates). "Negative control by feedback inhibition": acetyl CoA, NADH, ATP (reaction products).

p. 165 CYU [remember] (1) and (2) are combined with the answer to p. 171 CYU (3) Start with 12 triangles on glucose. (These triangles represent the 12 pairs of electrons that will be moved to electron carriers during redox reactions throughout glycolysis and the citric acid cycle.) Move two triangles to the NADH circle generated by glycolysis and the other 10 triangles to the pyruvate circle. Then move these 10 triangles through the pyruvate dehydrogenase square, placing two of them in the NADH circle next to pyruvate dehydrogenase. Add the remaining eight triangles in the acetyl CoA circle. Next move the eight triangles in the acetyl CoA circle through the citric acid cycle, placing six of them in the NADH circle and two in the FADH$_2$ circle generated during the citric acid cycle. (4) These boxes are marked with stars in the diagram.

p. 165 Fig. 9.13 [apply] NADH would be expected to have the highest amount of chemical energy since its production is correlated with the largest drop in free energy in the graph.

p. 167 Fig. 9.15 [understand] The proton gradient arrow should start above in the inner membrane space and point down across the membrane into the mitochondrial matrix. *Complex I:* "What goes in" = NADH; "What comes out" = NAD$^+$, e$^-$, transported H$^+$. *Complex II:* "What goes in" = FADH$_2$; "What comes out" = FAD, e$^-$, H$^+$. *Complex III:* "What goes in" = e$^-$, H$^+$; "What comes out" = e$^-$, transported H$^+$. *Complex IV:* "What goes in" = e$^-$, H$^+$, O$_2$; "What comes out" = H$_2$O, transported H$^+$.

p. 169 [explain] "Indirect" is accurate because most of the energy released during glucose oxidation is not used to produce ATP directly. Instead, this energy is stored in reduced electron carriers that are used by the ETC to generate a proton gradient across a membrane. These protons then diffuse down their concentration gradient across the inner membrane through ATP synthase, which drives ATP synthesis.

p. 169 Fig. 9.17 [create] They could have placed the vesicles in an acidic solution that has a pH below that of the solution in the vesicle. This would set up a proton gradient across the membrane to test for ATP synthesis.

p. 171 CYU [understand] See **FIGURE A9.1**. To illustrate the chemiosmotic mechanism, take the triangles (electrons) piled on the NADH and FADH$_2$ circles and move them through the ETC. While moving these triangles, also move dimes from the mitochondrial matrix to the intermembrane space. As the triangles exit the ETC, add them to the oxygen to water circle. Once all the dimes have been pumped by the ETC into the intermembrane space, move them through ATP synthase back into the mitochondrial matrix to fuel the formation of ATP.

p. 173 CYU [understand] Electron acceptors such as oxygen have a much higher electronegativity than pyruvate. Donating an electron to O$_2$ causes a greater drop in potential energy, making it possible to generate much more ATP per molecule of glucose.

IF YOU UNDERSTAND . . .

9.1 [understand] The radioactive carbons in glucose can be fully oxidized by the central pathways to generate CO$_2$, which would be radiolabeled. Other molecules, like lipids and amino acids, would also be expected to be radiolabeled since they are made using intermediates from the central pathways in other anabolic pathways. **9.2** [apply] See **FIGURE A9.2**. **9.3** [understand] Pyruvate dehydrogenase accomplishes three different tasks that would be expected to require multiple enzymes and active sites: CO$_2$ release, NADH production, and linking of an acetyl group to CoA. **9.4** [apply] NADH would decrease if a drug poisoned the acetyl CoA and oxaloacetate-to-citrate enzyme, since the citric acid cycle would no longer be able to produce NADH in the steps following this reaction in the pathway. **9.5** [apply] The ATP synthase allows protons to reenter the mitochondrial matrix after they have been pumped out by the ETC. By blocking ATP synthase, you would expect the pH of the matrix to increase (decreased proton concentration). **9.6** [understand] Organisms that produce ATP by fermentation would be expected to grow more slowly than those that produce ATP via cellular respiration simply because fermentation produces fewer ATP molecules per glucose molecule than cellular respiration does.

YOU SHOULD BE ABLE TO . . .

✓ Test Your Knowledge

1. [understand] Glycolysis → Pyruvate processing → citric acid cycle → ETC and chemiosmosis. The first three steps are responsible for glucose oxidation; the final step produces the most ATP. **2.** [remember] b **3.** [understand] d **4.** [understand] Most of the energy is stored in the form of NADH. **5.** [remember] c **6.** [remember] a

✓ Test Your Understanding

7. [understand] Stored carbohydrates can be broken down into glucose that enters the glycolytic pathway. If carbohydrates are absent, products from fat and protein catabolism can be used to fuel cellular respiration or fermentation. If ATP is plentiful, anabolic reactions use

FIGURE A9.1

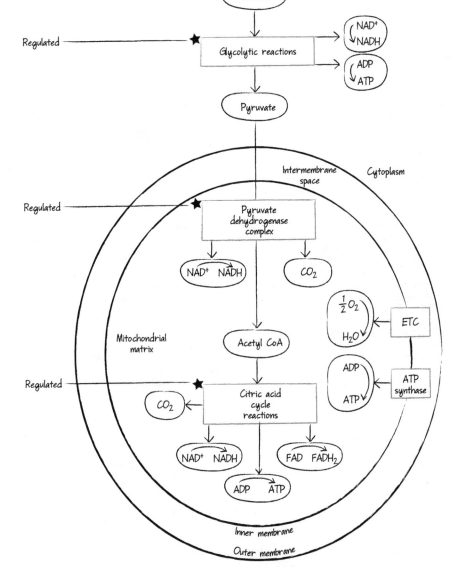

Eukaryotic mitochondrion

FIGURE A9.2

intermediates of the glycolytic pathway and the citric acid cycle to synthesize carbohydrates, fats, and proteins. **8.** analyze Both processes produce ATP from ADP and P_i, but substrate-level phosphorylation occurs when enzymes remove a "high-energy" phosphate from a substrate and directly transfer it to ADP, while oxidative phosphorylation occurs when electrons move through an ETC and produce a proton-motive force that drives ATP synthase. **9.** understand Aerobic respiration is much more productive because oxygen has extremely high electronegativity compared with other electron acceptors, resulting in a greater release of energy during electron transport and more proton pumping. **10.** apply b **11.** analyze Both phosphofructokinase and isocitrate dehydrogenase are regulated by feedback inhibition, where the product of the reaction or series of reactions inhibits the enzyme activity. They differ in that phosphofructokinase is regulated by allosteric inhibition while isocitrate dehydrogenase is controlled by competitive inhibition. **12.** understand Oxidative phosphorylation is possible via a proton gradient that is established by redox reactions in the ETC. ATP synthase consists of a membrane-associated F_o unit and a F_1 unit joined by a rotor shaft. When protons flow through the F_o unit, it spins the rotor shaft within the fixed F_1 unit. This spinning shaft causes structural changes in the F_1 that drives the synthesis of ATP from ADP and P_i.

✔ Test Your Problem-Solving Skills

13. create When complex IV is blocked, electrons can no longer be transferred to oxygen, the final acceptor, and cellular respiration stops. Fermentation could keep glycolysis going, but it is inefficient and unlikely to fuel a cell's energy needs over the long term. Cells that lack the enzymes required for fermentation would die first. **14.** apply Because mitochondria with few cristae would have fewer electron transport chains and ATP synthase molecules, they would produce much less ATP than mitochondria with numerous cristae. **15.** apply For each glucose molecule, two ATP are produced in glycolysis and two ATP are produced in the citric acid cycle via substrate-level phosphorylation. A total of 10 NADH and 2 $FADH_2$ molecules are produced from glycolysis, pyruvate oxidation, and the citric acid cycle. If each NADH were to yield 3 ATP, and each $FADH_2$ were to yield 2 ATP, then a total of 34 ATP would be produced via oxidative phosphorylation. Adding these totals would result in 38 ATP. A cell will not produce this much ATP, because the proton-motive force is used in other transport steps and because of other issues that may reduce the overall efficiency. **16.** apply b

CHAPTER 10

IN-TEXT QUESTIONS AND EXERCISES

p. 177 understand See FIGURE A10.1. This reaction is endergonic because there are more high-energy chemical bonds in the products compared with the reactants, and there is a decrease in entropy.
p. 177 Fig. 10.1 apply See FIGURE A10.1.
p. 180 Fig. 10.6 apply See FIGURE A10.2.

p. 182 Fig. 10.9 apply The energy state corresponding to a photon of green light would be located between the energy states corresponding to red and blue photons.
p. 183 CYU apply The outer pigments would be more likely to absorb blue photons (short wavelength, high energy), and interior pigments would absorb red photons (long wavelength, low energy). This establishes a pathway to direct photon energy toward the reaction center since resonance energy is transferred from higher to lower energy levels.
p. 184 Fig. 10.11 understand Yes—otherwise, changes in the production of oxygen could be due to differences in the number of chloroplasts, not differences in the rate of photosynthesis.
p. 186 analyze Light → Antenna complex → Reaction center → Pheophytin → ETC → Proton gradient → ATP synthase. Electrons from water are donated to the reaction center to replace those that were transferred to pheophytin.
p. 188 remember (1) Plastocyanin transfers electrons that move through the cytochrome complex in the ETC to the reaction center of photosystem (PS) I. (2) After they are excited by a photon and donated to the initial electron acceptor.
p. 190 CYU analyze In mitochondria, high-energy electrons are donated by NADH or $FADH_2$ (primary donors) and passed through an ETC to generate a proton-motive force. The low-energy electrons at the end of the chain are accepted by O_2 (terminal acceptor) to form water. In chloroplasts, low-energy electrons are donated by H_2O (primary donor), energized by photons or resonance energy, and passed through an ETC to generate a proton-motive force. These electrons are then excited a second time by photons or resonance energy, and the high-energy electrons are accepted by $NADP^+$ (terminal acceptor) to form NADPH.
p. 190 Fig. 10.18 understand The researchers didn't have any basis on which to predict these intermediates. They needed to perform the experiment to identify them.
p. 192 apply Each complete cycle requires 3 ATP and 2 NADPH molecules. To complete 6 runs through the cycle, a total of 18 ATP and 12 NADPH molecules are needed. By following the number of carbons, it is apparent that only three RuBP molecules are required, since they are fully regenerated every 3 cycles: 3 RuBP (15 carbons) fix and reduce 3 CO_2 to generate 6 G3P (18 carbons), yielding 1 G3P (3 carbons); the other 5 are used to regenerate 3 RuBP (15 carbons). The regeneration of RuBP means that only three must be required for continued runs through the Calvin cycle.
p. 194 Fig. 10.24 apply The morning would have the highest concentration of organic acids in the vacuoles of CAM plants, since these acids are made during the night and used up during the day.
p. 195 CYU (1) understand (a) C_4 plants use PEP carboxylase to fix CO_2 into organic acids in mesophyll cells. These organic acids are then transported into bundle-sheath cells, where they release carbon dioxide to rubisco. (b) CAM plants take in CO_2 at night and have enzymes that fix it into organic acids stored in the central vacuoles of photosynthesizing cells. During the day, the organic acids are processed to release CO_2 to rubisco.

(c) By diffusion through a plant's stomata when they are open. **(2)** apply The concentration of starch would be highest at the end of the day and lowest at the start of the day. Starch is made and stored in the chloroplasts of leaves during periods of high photosynthetic activity during the day. At night, it is broken down to make sucrose, which is transported throughout the plant to drive cellular respiration. (Cellular respiration also occurs during the day, but the impact is minimized due to the photosynthetic production of sugar.)

IF YOU UNDERSTAND . . .

10.1 understand The Calvin cycle depends on the ATP and NADPH produced by the light-capturing reactions, so it is not independent of light. **10.2** understand Most of the energy captured by pigments in chloroplasts is converted into chemical energy by reducing electron acceptors in ETCs. When pigments are extracted, the antenna complexes, reaction centers, and ETCs have been disassembled, so the energy is given off as fluorescence and heat. **10.3** understand Oxygen is produced by a critical step in photosynthesis: splitting water to provide electrons to PS II. If oxygen production increases, it means that more electrons are moving through the photosystems. **10.4** apply Each CO_2 that is fixed and reduced by the Calvin cycle requires 2 NADPH, which means that 12 NADPH molecules are required for a 6-carbon glucose. Each NADPH is made when two high-energy electrons reduce $NADP^+$. Each of these high-energy electrons originates from H_2O only after being excited by 2 photons (one in PS II and one in PS I). This means that 48 photons are required to produce 24 high-energy electrons to reduce 12 $NADP^+$ molecules for the fixation and reduction of 6 CO_2 to make glucose. Photorespiration would increase the number of photons required, since some of the CO_2 that is fixed would be released.

YOU SHOULD BE ABLE TO . . .

✔ Test Your Knowledge

1. remember d **2.** understand c **3.** remember c **4.** understand b **5.** remember The conversion of light energy to chemical energy occurs when electrons are transferred from excited pigments to an electron carrier in the photosystems.
6. remember The electron transport chain that accepts electrons from PS II. Plastocyanin is the molecule that transfers electrons from this chain to the PS I reaction center.

✔ Test Your Understanding

7. understand The electrons taken from water in PS II are excited twice by either photons or resonance energy. When excited in PS II, the electrons are transferred to PQ and used to build a proton-motive force that makes ATP. After reaching PS I, they are excited a second time and will either be used to reduce $NADP^+$ to make NADPH (noncyclic) or be transported back to PQ to produce more ATP (cyclic). **8.** analyze c **9.** understand The fixation phase is when CO_2 is fixed to RuBP by rubisco to form 3-phosphoglycerate. The reduction phase uses ATP to phosphorylate the carbons and NADPH to

---

FIGURE A10.1

Electrons are pulled _closer to C_ ; C is _reduced_

Electrons are pulled _farther from O_ ; O is _oxidized_

| 6 CO_2 (carbon dioxide) | 6 H_2O (water) | Input of energy | Potential energy increases | Glucose $(CH_2O)_4H$ | 6 O_2 (oxygen) |

O₂–seeking bacteria

Alga

reduce them with high-energy electrons to form G3P. The regeneration phase uses more ATP to convert some of the G3P to RuBP to continue the cycle. **10.** $\boxed{\text{understand}}$ Photorespiration occurs when levels of CO_2 are low and O_2 are high. Less sugar is produced because (1) CO_2 doesn't participate in the initial reaction catalyzed by rubisco and (2) when rubisco catalyzes the reaction with O_2 instead, one of the products is eventually broken down to CO_2 in a process that uses ATP. **11.** $\boxed{\text{understand}}$ In both C_4 and CAM plants, atmospheric CO_2 is brought in through stomata and first captured by fixing it to a 3-carbon molecule by PEP carboxylase. The C_4 pathway and CAM differ in the timing of this first fixation step—it occurs during the day in C_4 plants and during the night in CAM plants. They also differ in the location of the Calvin cycle with respect to this first fixation step. In C_4 plants, the two processes occur in different cells, while in CAM plants they occur in the same cell, but at different times (Calvin cycle during the day). **12.** $\boxed{\text{analyze}}$ Photosynthesis in chloroplasts produces sugar, which is used as a source of carbon for building organic molecules and energy for cellular respiration. Mitochondria harvest the energy stored in sugar to produce ATP, which is used to drive many cellular activities.

✔ Test Your Problem-Solving Skills

13. $\boxed{\text{apply}}$ (1) O_2, ATP, and NADPH would be formed by noncyclic electron flow. (2) No O_2 or NADPH would be formed, but ATP may be made by cyclic electron flow. (3) Initially, O_2 and NADPH would be formed by noncyclic electron flow, but no ATP would be made. Without ATP, the Calvin cycle would halt and, once all the NADP⁺ is used up, noncyclic electron flow would switch to cyclic electron flow. **14.** $\boxed{\text{evaluate}}$ Because rubisco evolved in a high CO_2, low O_2 environment, which would minimize the impact of photorespiration, the hypothesis is credible. But once O_2 levels increased, any change in rubisco that minimized photorespiration would give individuals a huge advantage over organisms with "old" forms of rubisco. There has been plenty of time for such changes to occur, making the "holdover" hypothesis less credible. **15.** $\boxed{\text{analyze}}$ b; the wavelength of light could excite PS I, but not PS II, resulting in cyclic electron flow since no electrons could be harvested from water by PS II. **16.** $\boxed{\text{create}}$ No—they are unlikely to have the same complement of photosynthetic pigments. Different wavelengths of light are available in various layers of a forest and water depths. It is logical to predict that plants and algae have pigments that absorb the available wavelengths efficiently. One way to test this hypothesis would be to isolate pigments from species in different locations and test the absorbance spectra of each.

BIG PICTURE Energy

p. 198 CYU (1) $\boxed{\text{understand}}$ Photosynthesis uses H_2O as a substrate and releases O_2 as a by-product; cellular respiration uses O_2 as a substrate and releases H_2O as a by-product. **(2)** $\boxed{\text{understand}}$ Photosynthesis uses CO_2 as a substrate; cellular respiration releases CO_2 as a by-product. **(3)** $\boxed{\text{analyze}}$ CO_2 fixation would essentially stop; CO_2 would continue to be released by cellular respiration. CO_2

levels in the atmosphere would increase rapidly, and production of new plant tissue would cease—meaning that most animals would quickly starve to death. **(4)** $\boxed{\text{analyze}}$ ATP "is used by" the Calvin cycle; photosystem I "yields" NADPH.

CHAPTER 11

IN-TEXT QUESTIONS AND EXERCISES

p. 204 CYU $\boxed{\text{analyze}}$ Plant cell walls and animal ECMs are both fiber composites. In plant cell walls the fiber component consists of cross-linked cellulose fibers, and the ground substance is pectin. In animal ECMs the fiber component consists of collagen fibrils, and the ground substance is proteoglycan.
p. 207 $\boxed{\text{apply}}$ Developing muscle cells could not adhere normally, and muscle tissue would not form properly. The embryo would die.
p. 207 Fig. 11.9 $\boxed{\text{apply}}$ "Prediction": Cells treated with an antibody that blocks membrane proteins involved in adhesion will not adhere. "Prediction of null hypothesis": All cells will adhere normally.
p. 209 CYU (1) $\boxed{\text{analyze}}$ The three structures differ in composition, but their function is similar. The middle lamella in plants is composed of pectins that glue adjacent cells together. Tight junctions are made up of membrane proteins that line up and "stitch" adjacent cells together. Desmosomes are "rivet-like" structures composed of proteins that link the cytoskeletons of adjacent cells. **(2)** $\boxed{\text{understand}}$ The plasma membranes of adjacent plant cells are continuous at plasmodesmata and share portions of the smooth endoplasmic reticulum. Gap junctions connect adjacent animal cells by forming protein-lined pores. Both structures result in openings between the cells that allow cytosol, including ions and small molecules, to be shared.
p. 211 Fig. 11.12 $\boxed{\text{create}}$ The steroid hormone likely changes the structure of the receptor such that it now exposes a nuclear localization signal, which is required for the protein to be transported into the nucleus.
p. 213 $\boxed{\text{analyze}}$ The spy is the signaling molecule that arrives at the cell surface (the castle gate). The guard is the G-protein-coupled receptor in the plasma membrane, and the queen is the G protein. The commander of the guard is the enzyme that is activated by the G protein to produce second messengers (the soldiers).
p. 214 $\boxed{\text{apply}}$ **(1)** The red dominos (RTK components) would be the first two dominos in the chain, followed by the black domino (Ras). This black domino would start two or more new branches, each one represented by a single domino of one color (e.g., green) to represent the activation of one type of kinase. Each of these green dominos would then again branch out, knocking down two or more branches consisting of single dominos of a different color (blue). The same branching pattern would result from each blue domino knocking down two or more yellow dominos. **(2)** Each single black domino (Ras) would require 10 green dominos, 100 blue dominos, and 1000 yellow dominos.
p. 215 CYU (1) $\boxed{\text{understand}}$ Each cell–cell signaling molecule binds to a specific receptor protein. A cell can respond to a signal only if it has the appropriate receptor. Only certain cell types will have the appropriate receptor for a given signaling molecule. **(2)** $\boxed{\text{understand}}$ Signals are amplified if one or more steps in a signal transduction pathway, involving either second messengers or a phosphorylation cascade, result in the activation of multiple downstream molecules.
p. 215 Fig. 11.16 $\boxed{\text{apply}}$ **(1)** cell responses A and C **(2)** cell responses A, B, and C **(3)** cell responses B and C.

IF YOU UNDERSTAND . . .

11.1 $\boxed{\text{apply}}$ **(1)** Cells without functional integrin molecules would likely die as a result of not being able to send the

appropriate anchorage-dependent survival signals. **(2)** Cells would be more sensitive to pulling or shearing forces; both cells and tissues would be weaker and more susceptible to damage. **11.2** $\boxed{\text{apply}}$ Cells would not be coordinated in their activity, so the heart tissue would not contract in unison and the heart would not beat. **11.3** $\boxed{\text{analyze}}$ Adrenalin binds to both heart and liver cells, but the activated receptors trigger different signal transduction pathways and lead to different cell responses. **11.4** $\boxed{\text{analyze}}$ The signal transduction pathways are similarly organized in both unicellular and multicellular organisms—consisting of signaling molecules, receptors, and second messengers. There is more variety in the means of transmitting the signal between cells in multicellular organisms compared to unicellular organisms. For example, there are no direct connections such as gap junctions or plasmodesmata in unicellular organisms.

YOU SHOULD BE ABLE TO . . .

✔ Test Your Knowledge

1. $\boxed{\text{remember}}$ Fiber composites consist of cross-linked fiber components that withstand tension and a ground substance that withstands compression. The cellulose microfibrils in plants and collagen fibrils in animals functionally resemble the steel rods in reinforced concrete. The pectin in plants and proteoglycan in animals functionally resemble the concrete ground substance. **2.** $\boxed{\text{remember}}$ b **3.** $\boxed{\text{remember}}$ a **4.** $\boxed{\text{understand}}$ b **5.** $\boxed{\text{remember}}$ d **6.** $\boxed{\text{remember}}$ Responses that affect which proteins are produced and those that affect the activity of existing proteins.

✔ Test Your Understanding

7. $\boxed{\text{understand}}$ b **8.** $\boxed{\text{understand}}$ If each enzyme in the cascade phosphorylates many copies of the enzyme in the next step of the cascade, the initial signal will be amplified many times over. **9.** $\boxed{\text{analyze}}$ All three are made up of membrane-spanning proteins that directly interact between adjacent cells. Tight junctions seal adjacent animal cells together; gap junctions allow a flow of material from the cytosol of one to the other. Desmosomes firmly secure adjacent cells to one another but do not affect the movement of substances between cells or into the cells. **10.** $\boxed{\text{understand}}$ When dissociated cells from two sponge species were mixed, the cells sorted themselves into distinct aggregates that contained only cells of the same species. By blocking membrane proteins with antibodies and isolating cells that would not adhere, researchers found that specialized groups of proteins, including cadherins, are responsible for selective adhesion. **11.** $\boxed{\text{create}}$ Signaling molecule crosses plasma membrane and binds to intracellular receptor (reception) → Receptor changes conformation, and the activated receptor complex moves to target site (processing) → Activated receptor complex binds to a target molecule (e.g., a gene or membrane pump), which changes its activity (response) → Signaling molecule falls off receptor or is destroyed; receptor changes to inactive conformation (deactivation). **12.** $\boxed{\text{understand}}$ Information from different signals may conflict or be reinforcing. "Crosstalk" between signaling pathways allows cells to integrate information from many signals at the same time instead of responding to each signal in isolation.

✔ Test Your Problem-Solving Skills

13. $\boxed{\text{apply}}$ d **14.** $\boxed{\text{analyze}}$ (a) The response would have to be extremely local—the activated receptor complex would have to affect nearby proteins. (b) No amplification could occur, because the number of signaling molecules dictates the amount of the response. (c) The only way to regulate the response would be to block the receptor or make it more responsive to the signaling molecule. **15.** $\boxed{\text{create}}$ Chitin forms chains that can cross-link with one another. This fungal cell wall would likely need to be either relaxed or destroyed, and new cell wall synthesis would be coordinated with the directional growth.

16. create Antibody binding to the two parts of the receptor may be causing them to dimerize. Since dimerization normally results after the signaling molecule binds to the receptor, the antibodies could be activating the receptor by mimicking this interaction. The result would be signal transduction even in the absence of the signaling molecule.

CHAPTER 12

IN-TEXT QUESTIONS AND EXERCISES

p. 223 understand **(1)** Genes are segments of chromosomes that code for RNAs and proteins. **(2)** Chromosomes are made of chromatin. **(3)** Sister chromatids are identical copies of the same chromosome, joined together. **p. 224 Fig. 12.5** apply **(1)** prophase cells would have 4 chromosomes with $2x$ DNA. **(2)** Anaphase cells would

have 8 chromosomes and $2x$ DNA. **(3)** Each daughter cell will have 4 chromosomes with x DNA. **p. 226** apply See **TABLE A12.1**. **p. 227 Fig. 12.6** apply The chromosome and black bar would move at the same rate toward the spindle pole. **p. 229 CYU (1)** apply See **FIGURE A12.1**. **(2)** apply Loss of the motors would result in two problems: (1) It would reduce the ability of chromosomes to attach to microtubules via their kinetochores; (2) cytokinesis would be inhibited since the Golgi-derived vesicles would not be moved to the center of the spindle to build the cell plate. **p. 230 Fig. 12.10** create Inject cytoplasm from an M-phase frog egg into a somatic cell that is in interphase. If the somatic cell starts mitosis, then the meiotic factor is not limited to gametes. **p. 230 Fig. 12.11** analyze In effect, MPF turns itself off after it is activated. If this didn't happen, the cell might undergo mitosis again before the cell has replicated its DNA.

p. 231 understand MPF activates proteins that get mitosis under way. MPF consists of a cyclin and a Cdk, and it is turned on by phosphorylation at the activating site and dephosphorylation at the inhibitory site. Enzymes that degrade cyclin reduce MPF levels. **p. 232 CYU (1)** remember $G_1 \rightarrow S \rightarrow G_2 \rightarrow M$. Checkpoints occur at the end of G_1 and G_2 and during M phase. **(2)** understand Cdk levels are fairly constant throughout the cycle, but cyclin increases during interphase and peaks in M phase. This accumulation of cyclin is a prerequisite for MPF activity, which turns on at the end of G_2, initiating M phase, and declines at the end of M phase.

IF YOU UNDERSTAND . . .

12.1 understand The G_1 and G_2 phases give the cell time to replicate organelles and grow before division as well as perform the normal functions required to stay alive. Chromosomes replicate during S phase and are separated from one another during M phase. Cytokinesis also occurs during M phase, when the parent cell divides into two daughter cells. **12.2** apply In cells that do not dissolve the nuclear envelope, the spindle must be constructed inside the nucleus to attach to the chromosomes and separate them. **12.3** apply Cells would prematurely enter M phase, shortening the length of G_2 and resulting in the daughter cells' being smaller than normal. **12.4** analyze The absence of growth factors in normal cells would cause them to arrest in the G_1 phase—eventually all the cells in the culture would be in G_1. The cancerous cells are not likely to be dependent on these growth factors, so the cells would not arrest and would continue through the cell cycle.

YOU SHOULD BE ABLE TO . . .

✓ **Test Your Knowledge**

1. remember b **2.** remember d **3.** remember d **4.** remember c
5. understand Daughter chromosomes were observed to move toward the pole faster than do the marked regions of fluorescently labeled kinetochore microtubules.
6. remember The cycle would arrest in M phase, and cytokinesis would not occur.

✓ **Test Your Understanding**

7. apply For daughter cells to have identical complements of chromosomes, all the chromosomes must be replicated during the S phase, the spindle apparatus must connect with the kinetochores of each sister chromatid in prometaphase, and the sister chromatids of each replicated chromosome must be partitioned in anaphase and fully separated into daughter cells by cytokinesis. **8.** create One possible concept map is shown in **FIGURE A12.2**. **9.** understand Microinjection experiments suggested that something in the cytoplasm of M-phase cells activated the transition from interphase to M phase. The control for this experiment was to inject cytoplasm from a G_2-arrested oocyte into another G_2-arrested oocyte. **10.** apply Protein kinases phosphorylate proteins. Phosphorylation changes a protein's shape, altering its function (activating or inactivating it). As a result, protein kinases regulate the function of proteins. **11.** understand Cyclin concentrations change during the cell cycle. At high concentration, cyclins bind to a specific cyclin-dependent kinase (or Cdk), forming a dimer. This dimer becomes active MPF by changing its shape through the phosphorylation (activating site) and dephosphorylation (inhibitory site) of Cdk. **12.** analyze b.

✓ **Test Your Problem-Solving Skills**

13. analyze a; adding up each phase allows you to determine that the cell cycle is 8.5 hours long. After 9 hours, the radiolabeled cells would have passed through a full cycle and be in either S phase or G_2—none would have entered M phase. **14.** apply The embryo passes through multiple rounds of the cell cycle, but cytokinesis does

FIGURE A12.1

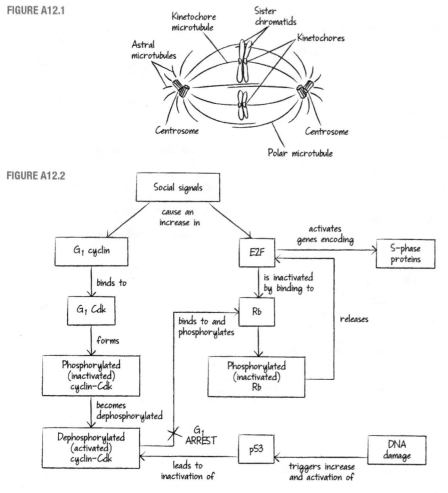

FIGURE A12.2

TABLE A12.1

	Prophase	Prometaphase	Metaphase	Anaphase	Telophase
Spindle apparatus	Starts to form	Contacts and moves chromosomes	Anchors poles to membrane and produces tension at kinetochores	Pulls chromatids apart	Defines site of cytokinesis
Nuclear envelope	Present	Disintegrates	Nonexistent	Nonexistent	Re-forms
Chromosomes	Condense	Attach to microtubules	Held at metaphase plate	Sister chromatids separate into daughter chromosomes	Collect at opposite poles

not occur during M phases. **15.** `apply` Early detection of cancers leads to a greater likelihood of survival. The widespread implementation of breast and prostate exams allows for the identification and removal of benign tumors before they become malignant. **16.** `analyze` Cancer requires many defects. Older cells have had more time to accumulate defects. Individuals with a genetic predisposition to cancer start out with some cancer-related defects, but this does not mean that the additional defects required for cancer to occur will develop.

CHAPTER 13

IN-TEXT QUESTIONS AND EXERCISES

p. 239 `apply` $n = 4$; the organism is diploid with $2n = 8$.

p. 240 `remember` See **FIGURE A13.1**. Because the two sister chromatids are identical and attached, it is sensible to consider them as parts of a single chromosome.

p. 242 `apply` There will be four DNA molecules in each gamete because the 8 replicated chromosomes in a diploid cell are reduced to 4 replicated chromosomes per cell at the end of meiosis I. In meiosis II, the sister chromatids of each replicated chromosome are separated. Each cell now contains 4 unreplicated chromosomes, each with a single molecule of DNA.

p. 245 `apply` Crossing over would not occur and the daughter cells produced by meiosis would be diploid, not haploid. There would be no reduction division.

p. 246 CYU (1) `understand` Use four long and four short pipe cleaners (or pieces of cooked spaghetti) to represent the chromatids of two replicated homologous chromosomes (four total chromosomes). Mark two long and two short ones with a colored marker pen to distinguish maternal and paternal copies of these chromosomes. Twist identical pipe cleaners (e.g., the two long colored ones) together to simulate replicated chromosomes. Arrange the pipe cleaners to depict the different phases of meiosis I as follows: *Early prophase I:* Align sister chromatids of each homologous pair to form two tetrads. *Late prophase I:* Form one or more crossovers between non-sister chromatids in each tetrad. (This is hard to simulate with pipe cleaners—you'll have to imagine that each chromatid now contains both maternal and paternal segments.) *Metaphase I:* Line up homologous pairs (the two pairs of short pipe cleaners and the two pairs of long pipe cleaners) at the metaphase plate. *Anaphase I:* Separate homologs. Each homolog still consists of sister chromatids joined at the centromere. *Telophase I and cytokinesis:* Move homologs apart to depict formation of two haploid cells, each containing a single replicated copy of two different chromosomes. **(2)** `understand` During anaphase I, homologs (not sister chromatids, as in mitosis) are separated, making the cell products of meiosis I haploid. **(3)** `understand` The pairing of homologs in metaphase I and their separation in anaphase I so that one goes to one daughter cell and the other to the other daughter cell means that each daughter cell obtains precisely one copy of each type of chromosome.

p. 248 `apply` Each gamete would inherit either all maternal or all paternal chromosomes. This would limit genetic variation in the offspring by precluding the many possible gametes containing various combinations of maternal and paternal chromosomes.

p. 248 CYU (1) `apply` See **FIGURE A13.2**. Maternal chromosomes are white and paternal chromosomes are black. Daughter cells with other possible combinations of chromosomes than shown could result from meiosis of this parent cell. **(2)** `understand` Crossing over would increase the genetic diversity of these gametes by creating many different combinations of maternal and paternal alleles along each of the chromosomes. **(3)** `analyze` Asexual reproduction generates no appreciable genetic diversity. Self-fertilization is preceded by meiosis so it generates gametes, through crossing over and independent

assortment, that have combinations of alleles not present in the parent. Outcrossing generates the most genetic diversity among offspring because it produces new combinations of alleles from two different individuals.

p. 249 Fig. 13.11 `apply` See **FIGURE A13.3**.

p. 251 Fig. 13.14 `apply` Asexual: 64 (16 individuals from generation three produce 4 offspring per individual). Sexually: 16 (8 individuals from generation three form 4 couples; each couple produces 4 offspring).

p. 252 Fig. 13.15 `apply` The rate of outcrossing is predicted to rise initially, as the pathogen selects for resistant roundworms, and then to fall as the roundworms in the population gain resistance and take advantage of the increased numbers of offspring offered by having more hermaphrodites that can reproduce by self-fertilization.

IF YOU UNDERSTAND . . .

13.1 `remember` See the right panel of Figure 13.7 and note how the cells transition from diploid to haploid in meiosis I. Also note that each chromosome contains two sister chromatids before and after meiosis I.

13.2 `understand` (a) See Figure 13.10 and note how the two different ways of aligning two homologous pairs of chromosomes at metaphase I of meiosis can create four different combinations of maternal and paternal chromosomes in daughter cells. (b) See Figure 13.11. Note that for each homologous pair with one crossover, two chromatids are recombinant and two are unaltered. Since each chromatid will produce a chromosome at the end of meiosis II, your drawing should show that two out of four of these chromosomes are recombinant.

13.3 `apply` (a) Your model should show events similar to those of Figure 13.12. (b) Your model should show two cells with one of each type of chromosome (n), one cell with an extra copy of one chromosome ($n + 1$), and one gamete without any copies of one chromosome ($n - 1$).

13.4 `remember` Sexual reproduction will likely occur during times when conditions are changing rapidly, because genetically diverse offspring may have an advantage in the new conditions.

YOU SHOULD BE ABLE TO . . .

✓ Test Your Knowledge

1. `remember` b **2.** `remember` a **3.** `remember` b **4.** `understand` b
5. `remember` 1/2 **6.** `remember` mitosis.

✓ Test Your Understanding

7. `remember` Homologous chromosomes are similar in size, shape, and gene content, and originate from different parents. Sister chromatids are exact copies of a chromosome that are generated when chromosomes are replicated (S phase of the cell cycle). **8.** `understand` Refer to Figure 13.7 as a guide for this exercise. The four pens represent the chromatids in one replicated homologous pair; the four pencils, the chromatids in a different homologous pair. To simulate meiosis II, make two "haploid cells"—each with a pair of pens and a pair of pencils representing two replicated chromosomes (one of each type in this species). Line them up in the middle of the cell; then separate the two pens and the two pencils in each cell such that one pen and one pencil go to each of four daughter cells. **9.** `understand` Meiosis I is a reduction division because homologs separate—daughter cells have just one of each type of chromosome instead of two. Meiosis II is not a reduction division because sister chromatids separate— daughter cells have unreplicated chromosomes instead of replicated chromosomes, but still just one of each type. **10.** `apply` b. **11.** `apply` Tetraploids produce diploid gametes, which combine with a haploid gamete from a diploid individual to form a triploid offspring. Mitosis proceeds normally in triploid cells because mitosis doesn't require forming pairs of chromosomes. But during meiosis in a triploid, homologous chromosomes can't pair up correctly. The third set of chromosomes does not have a homologous partner to pair with. **12.** `understand` Asexually produced individuals are genetically identical, so if one is susceptible to a new disease, all are. Sexually produced individuals are genetically unique, so if a new disease strain evolves, at least some plants are likely to be resistant.

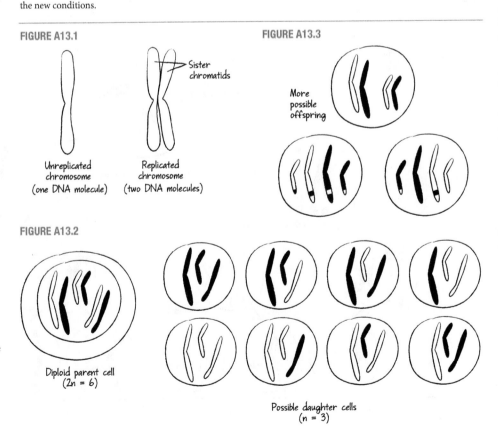

FIGURE A13.1

Unreplicated chromosome (one DNA molecule)

Replicated chromosome (two DNA molecules)

Sister chromatids

FIGURE A13.2

Diploid parent cell ($2n = 6$)

Possible daughter cells ($n = 3$)

FIGURE A13.3

More possible offspring

13. apply The gibbon would have 22 chromosomes in each gamete, and the siamang would have 25. Each somatic cell of the offspring would have 47 chromosomes. The offspring should be sterile because it has some chromosomes that would not form homologous pairs at prophase I of meiosis. **14.** apply c **15.** apply Aneuploidy is the major cause of spontaneous abortion. If spontaneous abortion is rare in older women, it would result in a higher incidence of aneuploid conditions such as Down syndrome in older women, as recorded in the Figure 13.13. **16. (a)** create Such a study might be done in the laboratory, controlling conditions in identical dishes. A population of rotifers infected with fungus could be established in each dish. One dish of rotifers would be kept moist; the other dishes of rotifers would be allowed to dry out. After various periods of time, water would be added to each dish and then the rotifers would be observed to see if fungal infections reappeared. **(b)** create Wind disperses the rotifer to new and often pathogen-free environments. In this case, the ticket to a sex-free existence is not genetic diversity but the evolution of an alternative means of evading pathogens made possible by fungus-infected rotifers ridding themselves of the pathogen when they dry.

CHAPTER 14

IN-TEXT QUESTIONS AND EXERCISES

p. 260 Fig. 14.3 evaluate An experiment is a failure if you didn't learn anything from it. That is not the case here.
p. 261 Table 14.2 analyze *Row 6: 3.14 : 1; Row 7: 266.*
p. 262 apply Filling in the top and side of a Punnett square requires writing out the types and ratios of gametes. For a cross involving one gene (monohybrid cross), this amounts to applying the principle of segregation as one allele is segregated from another. The phenotype ratios are 1 : 1 round : wrinkled; the genotype ratios are 1 : 1 *Rr : rr*.
p. 262 Fig. 14.4 understand No—the outcome (the expected offspring genotypes that the Punnett square generates) will be the same.

p. 263 CYU (1) apply See answer to Problem 13 in Test Your Problem-Solving Skills, below. **(2)** apply See answer to Problem 15 in Test Your Problem-Solving Skills.
p. 265 apply *AABb → AB* and *Ab. PpRr → PR, Pr, pR,* and *pr. AaPpRr → APR, APr, ApR, Apr, aPR, apR, apr,* and *aPr.*
p. 265 CYU (1) apply See answer to Problem 14 in Test Your Problem-Solving Skills. **(2)** apply See answer to Problem 16 in Test Your Problem-Solving Skills.
p. 269 CYU (1) understand See **FIGURE A14.1**. Segregation of alleles occurs when homologs that carry those alleles are separated during anaphase I. One allele ends up in each daughter cell. **(2)** understand See **FIGURE A14.2**. Independent assortment occurs because homologous pairs line up randomly at the metaphase plate during metaphase I. The figure shows two alternative arrangements of homologs in metaphase I. As a result, it is equally possible for a gamete to receive the following four combinations of alleles: *YR, Yr, yR, yr.*
p. 270 Fig. 14.12 apply *XWY, XWy, XwY, Xwy.*

p. 271 Fig. 14.13 analyze Random chance (or perhaps red-eyed, gray-bodied males don't survive well).
p. 273 Fig. 14.16 analyze The gene colored orange is *ruby*; the gene colored blue is *miniature wings*.
p. 275 Fig. 14.19 understand In this case of gene-by-gene interaction, the 9 : 3 : 3 : 1 ratio comes from four different forms of one trait (comb shape), whereas in a standard dihybrid cross, the four different phenotypes come from two different phenotypes for each of two genes.
p. 276 CYU (1) apply The comb phenotype results from interactions between alleles at two different genes, not a single gene. Matings between rose- and pea-comb chickens produce F$_2$ offspring that may have a new combination of alleles and thus new phenotypes. **(2)** apply Kernel color in wheat is influenced by alleles at many different genes, not a single gene. F$_2$ offspring have a normal distribution of phenotypes, not a 3 : 1 ratio.
p. 276 Fig. 14.21 understand Because there are many different genotypes that can produce intermediate coloration and fewer that can produce the extremes of coloration.

FIGURE A14.1

FIGURE A14.2

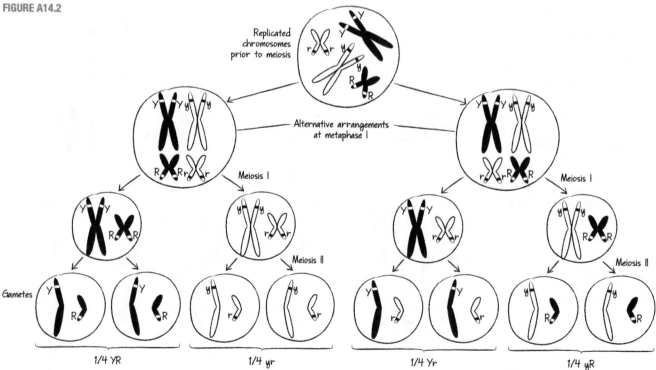

A heterozygous female and a color-blind male. 1 color-blind male : 1 color-blind female.

IF YOU UNDERSTAND . . .

14.1 understand Because crosses within a pure line never produce individuals with a different phenotype, this indicates that there must be only one allele in pure-line individuals. **14.2** apply B and b. **14.3** apply BR, Br, bR, and br, in equal proportions. The B and b alleles are located on different but homologous chromosomes, which separate into different daughter cells during meiosis I. The $BbRr$ notation indicates that the B and R genes are on different chromosomes. As a result, the chromosomes line up independently of each other in metaphase of meiosis I. The B allele is equally likely to go to a daughter cell with R as with r; likewise, the b allele is equally likely to go to a daughter cell with R as with r. **14.4** apply There are many ways to show this. If eye color in *Drosophila* is chosen as an example, then a pair of Punnett squares like those in Figure 14.11 illustrate how reciprocal crosses involving an X-linked recessive gene give different results. **14.5** understand Linkage refers to the physical connection of two alleles on the same chromosome. Crossing over breaks this linkage between particular alleles as segments of maternal and paternal homologs are exchanged. **14.6** apply See **FIGURE A14.3** for the pedigree. Note that the birth order of daughters and sons is arbitrary; other birth orders are equally valid. A Punnett square will show that all the sons are predicted to be X$^+$Y and all the daughters X$^+$X^c, where X$^+$ shows the dominant, X-linked allele for normal color vision and X^c shows the X-linked recessive color blindness allele.

YOU SHOULD BE ABLE TO . . .

✔ Test Your Understanding

1. understand d **2.** understand b **3.** understand a **4.** understand d
5. understand a **6.** understand a **7.** evaluate d **8.** remember b **9.** apply b
10. understand d

✔ Test Your Problem-Solving Skills

11. apply **Example Solution** Here you're given offspring phenotypes and you're asked to infer parental genotypes. To do this you have to propose hypothetical parental genotypes to test, make a Punnett square to predict the offspring genotypes, and then see if the predicted offspring phenotypes match the data. In this case, coming up with a hypothesis for the parental genotypes is relatively straightforward because the problem states that the trait is due to one gene and two alleles. No information on sex is given, so assume the gene is autosomal (the simplest case). Now look at the second entry in the chart. It shows a 3 : 1 ratio of offspring from a winged individual that self-fertilizes. This result is consistent with the hypothesis that W^+ is dominant and W^- recessive and that this parent's genotype is W^+W^-. Now let's look at the first cross in the chart. If W^+ is dominant, then a wingless parent must be W^-W^-. When you do the Punnett square to predict offspring genotypes from selfing, you find that all the offspring will produce wingless fruits, consistent with the data. In the third cross, all the offspring make winged fruits even though one of the parents produces wingless

FIGURE A14.3

fruits and thus is W^-W^-. This would happen only if the winged parent is W^+W^+. (If this reasoning isn't immediately clear to you, work the Punnett square.) In the fourth cross, you could get offspring that all make winged fruits if the parents were W^+W^+ and W^+W^+, or if the parents were W^+W^+ and W^+W^-. Either answer is correct. Again, you can write out the Punnett squares to see that this statement is correct.

12. apply **Example Solution** Here you are given parental and offspring phenotypes and are asked to infer the parental genotypes. As a starting point, assume that the coat colors are due to the simplest genetic system possible: one autosomal gene with two alleles, where one allele is dominant and the other recessive. Because female II produces only black offspring, it's logical to suppose that black is dominant to brown. Let's use B for black and b for brown. Then the male parent is bb. To produce offspring with a 1 : 1 ratio of black : brown coats, female I must be Bb. But to produce all black offspring, female II must be BB. This model explains the data, so you can accept it as correct.

13. apply 3/4; 1/256 (see BioSkills 5 in Appendix B); 1/2 (the probabilities of transmitting the alleles or having sons does not change over time). **14.** apply Your answer to the first three parts should conform to the F_1 and F_2 crosses diagrammed in Figure 14.5b, except that different alleles and traits are being analyzed. The recombinant gametes would be Yi and yI. Yes—there would be some individuals with yellow seeds and constricted pods and with green seeds and inflated pods. **15.** apply Cross 1: non-crested (Cc) × non-crested (Cc) = 22 non-crested ($C\_$); 7 crested (cc). Cross 2: crested (cc) × crested (cc) = 20 crested (cc). Cross 3: non-crested (Cc) × crested (cc) = 7 non-crested (Cc); 6 crested (cc). Non-crested (C) is the dominant allele. **16.** apply This is a dihybrid cross that yields progeny phenotypes in a 9 : 3 : 3 : 1 ratio. Let O stand for the allele for orange petals and o the allele for yellow petals; let S stand for the allele for spotted petals and s the allele for unspotted petals. Start with the hypothesis that O is dominant to o, that S is dominant to s, that the two genes are found on different chromosomes so they assort independently, and that the parent individual's genotype is $OoSs$. If you do a Punnett square for the $OoSs \times OoSs$ mating, you'll find that progeny phenotypes should be in the observed 9 : 3 : 3 : 1 proportions. **17.** analyze Let D stand for the normal allele and d for the allele responsible for Duchenne-type muscular dystrophy. The woman's family has no history of the disease, so her genotype is almost certainly DD. The man is not afflicted, so he must be DY. (The trait is X-linked, so he has only one allele; the "Y" stands for the Y chromosome.) Their children are not at risk. The man's sister could be a carrier, however—meaning she has the genotype Dd. If so, then half of the second couple's male children are likely to be affected. **18.** understand Your stages of meiosis should look like a simplified version of Figure 13.7, except with $2n = 4$ instead of $2n = 6$. The A and a alleles could be on the red and blue versions of the longest chromosome, and the B and b alleles could be on the red and blue versions of the smallest chromosome, similar to the way the hair- and eye-color genes are shown in Figure 13.10. The places you draw them are the locations of the A and B genes, but each chromosome has only one allele. Each pair of red and blue chromosomes is a homologous pair. Sister chromatids bear the same allele (e.g., both sister chromatids of the long blue chromosomes might bear the a allele). Chromatids from the longest and shortest chromosomes are not homologous. To identify the events that result in the principles of segregation and independent assortment, see Figures 14.7 and 14.8 and substitute A, a, and B, b for R, r and Y, y. **19.** apply Half their offspring should have the genotype iI^A and the type A blood phenotype. The other half of their offspring should have the genotype iI^B and the type B blood

phenotype. Second case: the genotype and phenotype ratios would be 1 : 1 : 1 : 1 I^AI^B (type AB) : I^Ai (type A) : I^Bi (type B) : ii (type O). **20.** apply Because the children of Tukan and Valco had no eyes and smooth skin, you can conclude that the allele for eyelessness is dominant to eyes and the allele for smooth skin is dominant to hooked skin. E = eyeless, e = two eye sets, S = smooth skin, s = hooked skin. Tukan is $eeSS$; Valco is $EEss$. The children are all $EeSs$. Grandchildren with eyes and smooth skin are eeS-. Assuming that the genes are on different chromosomes, one-fourth of the children's gametes are ee and three-fourths are S-. So 1/4 ee × 3/4 S- × 32 = 6 children would be expected to have two sets of eyes and smooth skin. **21.** create Although the mothers were treated as children by a reduction of dietary phenylalanine, they would have accumulated phenylalanine and its derivatives once they went off the low-phenylalanine diet as young adults. Children born of such mothers were therefore exposed to high levels of phenylalanine during pregnancy. For this reason, a low-phenylalanine diet is recommended for such mothers throughout the pregnancy. **22.** apply According to Mendel's model, palomino individuals should be heterozygous at the locus for coat color. If you mated palomino individuals, you would expect to see a combination of chestnut, palomino, and cremello offspring. If blending inheritance occurred, however, all the offspring should be palomino. **23.** apply Because this is an X-linked trait, the father who has hemophilia could not have passed the trait on to his son. Thus, the mother in couple 1 must be a carrier and must have passed the recessive allele on to her son, who is XY and affected. To educate a jury about the situation, you should draw what happens to the X and Y during meiosis and then make a drawing showing the chromosomes in couple 1 and couple 2, with a Punnett square showing how these chromosomes are passed to the affected and unaffected children. **24.** apply The curved-wing allele is autosomal recessive; the lozenge-eyed allele is sex-linked (specifically, X-linked) recessive. Let L be the allele for long wings and l be the allele for curved wings; let X^R be the allele for red eyes and X^r the allele for lozenge eyes. The female parent is LlX^RX^r; the male parent is LlX^RY. **25.** apply Albinism indicates the absence of pigment, so let b stand for an allele that gives the absence of blue and y for an allele that gives the absence of yellow pigment. If blue and yellow pigment blend to give green, then both green parents are $BbYy$. The green phenotype is found in $BBYY$, $BBYy$, $BbYY$, and $BbYy$ offspring. The blue phenotype is found in $BByy$ or $Bbyy$ offspring. The yellow phenotype is observed in $bbYY$ or $bbYy$ offspring. Albino offspring are $bbyy$. The phenotypes of the offspring should be in the ratio 9 : 3 : 3 : 1 as green : blue : yellow : albino. Two types of crosses yield $BbYy$ F_1 offspring: $BByy \times bbYY$ (blue × yellow) and $BBYY \times bbyy$ (green × albino). **26.** apply The chance that their first child will have hemophilia is 1/2. This is because all sons will have the disease and there is a 1/2 chance of having a firstborn son. The chance of having a carrier as their first child is also 1/2. This is because all daughters and none of the sons will be carriers and there is a 1/2 chance of having a firstborn daughter. (Recall that males cannot carry an X-linked recessive trait—with only one X chromosome, males either have the trait or not.) **27.** apply Autosomal dominant.

CHAPTER 15

IN-TEXT QUESTIONS AND EXERCISES

p. 286 Fig. 15.2 analyze The lack of radioactive protein in the pellet (after centrifugation) is strong evidence; they could also make micrographs of infected bacterial cells before and after agitation.
p. 287 Fig. 15.3 apply 5'-TAG-3'.
p. 288 Fig. 15.5 apply The same two bands should appear, but the upper band (DNA containing only ^{14}N) should get bigger and darker and the lower band (hybrid DNA)

should get smaller and lighter in color since each succeeding generation has relatively less heavy DNA.

p. 290 apply See FIGURE A15.1. The new strands grow in opposite directions, each in the 5′ → 3′ direction.

p. 292 apply Helicase, topoisomerase, single-strand DNA-binding proteins, primase, and DNA polymerase are all required for leading-strand synthesis. If any one of these proteins is nonfunctional, DNA replication will not occur.

p. 293 apply See FIGURE A15.2. If DNA ligase were defective, then the leading strand would be continuous, and the lagging strand would have gaps in it where the Okazaki fragments had not been joined.

p. 294 CYU (1) understand DNA polymerase adds nucleotides only to the free 3′ −OH on a strand. Primase synthesizes a short RNA sequence that provides the free 3′ end necessary for DNA polymerase to start working. **(2)** understand The need to begin DNA synthesis many times on the lagging strand requires many new primers. Since DNA polymerase I is needed to remove primers, it is required predominantly on the primer-rich lagging strand.

p. 296 Fig. 15.13 apply As long as the RNA template could bind to the "overhanging" section of single-stranded DNA, any sequence could produce a longer strand. For example, 5′-CCCAUUCCC-3′ would work just as well.

p. 297 CYU (1) understand This is because telomerase is needed only to replicate one end of a linear DNA and bacterial DNAs lack ends because they are circular. **(2)** understand Since telomerase works by extending one strand of DNA without any external template and because DNA synthesis requires a template, telomerase must contain an internal template to allow it to extend a DNA chain.

p. 299 Fig. 15.16 analyze They are lower in energy and not absorbed effectively by the DNA bases.

p. 300 CYU (1) apply The mutation rate would be predicted to rise because differences in base-pair stability and shape make it possible for DNA polymerase to distinguish correct from incorrect base pairs during DNA replication. **(2)** apply The mutation rate should increase because without a way to distinguish which strand to use as a template for repair, about half of mismatches on average would be repaired using the incorrect strand as a template. **(3)** remember The enzyme that removes the dimer and surrounding DNA is specific to nucleotide excision repair. DNA polymerase and DNA ligase work in both nucleotide excision repair and normal DNA synthesis.

p. 300 Fig. 15.18 apply Exposure to UV radiation can cause formation of thymine dimers. If thymine dimers are not repaired, they represent mutations. If such mutations occur in genes controlling the cell cycle, cells can grow abnormally, resulting in cancers.

IF YOU UNDERSTAND . . .

15.1 apply These results would not allow distinguishing whether DNA or protein was the genetic material.
15.2 understand The bases added during DNA replication are shown in red type.

Original DNA: **CAATTACGGA**
GTTAATGCCT
Replicated DNA: **CAATTACGGA**
GTTAATGCCT
CAATTACGGA
GTTAATGCCT

15.3 understand See FIGURE A15.3. **15.4** understand Because cancer cells divide nearly without limit, it's important for these cells to have active telomerase so that chromosomes don't shorten to the point where cell division becomes impossible. **15.5** understand If errors in DNA aren't corrected, they represent mutations. When DNA repair systems fail, the mutation rate increases. As the mutation rate increases, the chance that one or more cell-cycle genes will be mutated increases. Mutations in these genes often result in uncontrolled cell division, ultimately leading to cancer.

YOU SHOULD BE ABLE TO . . .

✓ Test Your Knowledge

1. remember d **2.** understand a **3.** remember topoisomerase
4. remember b **5.** remember c **6.** remember telomerase

✓ Test Your Understanding

7. remember Labeling DNA or labeling proteins. **8.** understand DNA is constantly damaged, and many pathways have evolved to repair this onslaught of damage. If a DNA repair pathway is inactivated by mutation, damage is inefficiently repaired. Consequently mutation rates increase, and the increased number of mutations increases the probability that cancer-causing mutations will occur. **9.** understand On the lagging strand, DNA polymerase moves away from the replication fork. When helicase unwinds a new section of DNA, primase must build a new primer on the template for the lagging strand (closer to the fork) and another polymerase molecule must begin synthesis at this point. This makes the lagging-strand synthesis discontinuous. On the leading strand, DNA polymerase moves in the same direction as helicase, so synthesis can continue, without interruption, from a single primer (at the origin of replication). **10.** understand Telomerase binds to the overhang at the end of a chromosome. Once bound, it begins catalyzing the addition of deoxyribonucleotides to the overhang in the

5′ → 3′ direction, lengthening the overhang. This allows primase, DNA polymerase, and ligase to catalyze the addition of deoxyribonucleotides to the lagging strand in the 5′ → 3′ direction, restoring the lagging strand to its original length. **11.** apply a (Because the ability to distinguish which strand contains the incorrect base would be lost.) **12.** analyze d (The regularity of DNA's structure allows enzymes to recognize any type of damage that distorts this regular structure.)

✓ Test Your Problem-Solving Skills

13. analyze c (If DNA polymerase could synthesize DNA 3′ → 5′ as well as the normal 5′ → 3′, then both newly synthesized DNA strands could be extended to follow the replication fork.) **14.** apply (a) In FIGURE A15.4, the gray lines represent DNA strands containing radioactivity. (b) After one round of replication in radioactive solution, one double-stranded DNA would be radioactive in both strands and the other would not be radioactive in either strand. After another round of DNA synthesis, this time in nonradioactive solution, one of the four DNA molecules would be radioactive in both strands and the other three DNA molecules would contain no radioactivity in any strand. **15.** analyze (a) The double mutant of both *uvrA* and *recA* is most sensitive to UV light; the single mutants are in between; and the wild type is

FIGURE A15.1

3′ ——————————————————— 5′
5′ —————————————————→ 3′
Direction of new strand synthesis

Direction of new strand synthesis
3′ ←————————————————— 5′
5′ ——————————————————— 3′

FIGURE A15.2

3′
5′ ————————→ 3′

5′
Okazaki fragments
3′ ▭ ▭ ▭ ▭ 5′
5′
3′

FIGURE A15.3

Lagging   Leading
3′ 5′
5′   3′
3′   5′
5′   3′
5′ 3′
Leading   Lagging

FIGURE A15.4

Chromosome → Replication in radioactive solution → Sister chromatids → Mitosis → Replication in nonradioactive solution →

least sensitive. (b) The *recA* gene contributes more to UV repair through most of the UV dose levels. But at very high UV doses, the *uvrA* gene is somewhat more important than the *recA* gene. **16.** apply About 4600 seconds or 77 minutes. This answer comes from knowing that replication proceeds bidirectionally, so replication from each fork is predicted to replicate half the chromosome. This is 4.6 million base pairs/2 = 2.3 million base pairs. At 500 base pairs per second, this requires 2.3 million base pairs/500 base pairs per second = 4600 seconds. To obtain the time in minutes, divide 4600 seconds by 60 seconds per minute.

CHAPTER 16

IN-TEXT QUESTIONS AND EXERCISES

p. 305 Fig. 16.1 apply No, it could not make citrulline from ornithine without enzyme 2. Yes, it would no longer need enzyme 2 to make citrulline.

p. 306 Fig. 16.2 create Many possibilities: strain of fungi used, exact method for creating mutants and harvesting spores to grow, exact growing conditions (temperature, light, recipe for growth medium—including concentrations of supplements), objective criteria for determining growth or no growth.

p. 309 CYU remember Change in DNA sequence, change in sequence of transcribed mRNA, potential change in amino acid sequence of protein, likely altered protein function (if amino acid sequence was altered), likely change in phenotype.

p. 312 analyze (1) The codons in Figure 16.4 are translated correctly. (2) See **FIGURE A16.1**. (3) There are many possibilities (just pick alternative codons for one or more of the amino acids); one is an mRNA sequence (running 5′ → 3′): 5′ GCG-AAC-GAU-UUC-CAG 3′. To get the corresponding DNA sequence, write this sequence but substitute Ts for Us: 5′ GCG-AAC-GAT-TTC-CAG 3′. Now write the complementary bases, which will be in the 3′ → 5′ direction: 3′ CGC-TTG-CTA-AAG-GTC 5′. When this second strand is transcribed by RNA polymerase, it will produce the mRNA given with the proper 5′ → 3′ orientation.

p. 312 CYU (1) apply Note the 3′ → 5′ polarity of the DNA sequences in the accompanying table, and in the subsequent answer. This means that the complementary mRNA codon will read 5′ → 3′. U (rather than T) is the base transcribed from A.

DNA	mRNA Codon	Amino Acid
ATA	UAU	Tyrosine
ATG	UAC	Tyrosine
ATT	UAA	Stop
GCA	UGC	Cysteine

(2) understand The ATA → ATG mutation would have no effect on the protein. The ATA → ATT mutation introduces a stop codon, so the resulting polypeptide would be shortened. This would result in synthesis of a mutant protein much shorter than the original protein. The ATA → GCA mutation might have a profound effect on the protein's conformation because cysteine's structure is different from tyrosine's.

p. 312 Fig. 16.7 apply See **FIGURE A16.1**.

p. 314 Fig. 16.9 analyze Chromosomes 2, 3, 6, 10, 13, 14, 15, 18, 19, 21, 22, and the X chromosome show aneuploidy. Virtually every chromosome has structural rearrangements, and translocations are the most obvious. These are seen when two or more different colors occur on the same chromosome.

IF YOU UNDERSTAND . . .

16.1 understand Ornithine, citrulline or arginine could be added to allow growth. As Figure 16.1 shows, these compounds are made after the steps catalyzed by the enzymes 1, 2, and 3 that are needed to produce arginine. **16.2** understand An inhibitor of RNA synthesis will eventually prevent the synthesis of new proteins because newly synthesized mRNA is needed for translation. **16.3** apply AUG UGG AAA/AAG CAA/CAG **16.4** understand Since redundancy is having more than one codon specify a particular amino acid, redundancy makes it possible for there to be a point mutation without altering the amino acid. This is a silent mutation. A silent mutation is likely to be neutral because there is no change in amino acid sequence.

YOU SHOULD BE ABLE TO . . .

✔ **Test Your Knowledge**

1. remember d **2.** understand a **3.** understand d **4.** remember Because there is no chemical complementarity between nucleotides and amino acids; and because in eukaryotes, DNA is in the nucleus but translation occurs in the cytoplasm. **5.** understand d **6.** remember A codon that signals the end of translation.

✔ **Test Your Understanding**

7. understand Because the Morse code and genetic code both use simple elements (dots and dashes; 4 different bases) in different orders to encode complex information (words; amino acid sequences).

8.

Substrate 1 $\xrightarrow{A}$ Substrate 2 $\xrightarrow{B}$ Substrate 3 $\xrightarrow{C}$

Substrate 4 $\xrightarrow{D}$ Substrate 5 $\xrightarrow{E}$ Biological Sciazine

apply Substrate 3 would accumulate. Hypothesis: The individuals have a mutation in the gene for enzyme D.

9. understand They supported an important prediction of the hypothesis: Losing a gene (via mutation) resulted in loss of an enzyme. **10.** understand c **11.** understand In a triplet code, addition or deletion of 1–2 bases disrupts the reading frame "downstream" of the mutation site(s), resulting in a dysfunctional protein. But addition or deletion of 3 bases restores the reading frame—the normal sequence is disrupted only between the first and third mutation. The resulting protein is altered but may still be able to function normally. Only a triplet code would show these patterns. **12.** understand A point mutation changes the nucleotide sequence of an existing allele, creating a new one, so it always changes the genotype. But because the genetic code is redundant, and because point mutations can occur in DNA sequences that do not code for amino acids, these point mutations do not change the protein product and therefore do not change the phenotype.

✔ **Test Your Problem-Solving Skills**

13. apply See **FIGURE A16.2**. **14.** analyze Every copying error would result in a mutation that would change the amino acid sequence of the protein and would likely affect its function. **15.** analyze Before the central dogma was understood, DNA was known to be the hereditary material, but no one knew how particular sequences of bases resulted in the production of RNA and protein products. The central dogma clarified how genotypes produce phenotypes. **16.** analyze c

CHAPTER 17

IN-TEXT QUESTIONS AND EXERCISES

p. 318 Fig. 17.1 remember RNA is synthesized in the 5′ → 3′ direction; the DNA template is "read" 3′ → 5′.

p. 321 CYU (1) apply Transcription would be reduced or absent because the missing nucleotides are in the −10 region, one of the two critical parts of the promoter. **(2)** understand NTPs are required because the three phosphate groups raise the monomer's potential energy enough to make the polymerization reaction exergonic.

p. 322 Fig. 17.5 apply There would be no loops—the molecules would match up exactly.

p. 323 CYU (1) understand The subunits contain both RNA (the *ribonucleo–* in the name) and proteins. **(2)** understand The cap and tail protect mRNAs from degradation and facilitate translation.

p. 326 Fig. 17.11 analyze If the amino acids stayed attached to the tRNAs, the gray line in the graph would stay high and the green line low. If the amino acids were transferred to some other cell component, the gray line would decline but the green line would be low.

p. 327 understand (1) The amino acid attaches on the top right of the L-shaped structure. (2) The anticodon is antiparallel in orientation to the mRNA codon, and it contains the complementary bases.

p. 332 CYU understand E is for exit—the site where uncharged tRNAs are ejected; P is for peptidyl (or peptide bond)—the site where peptide bond formation takes place; A is for aminoacyl—the site where aminoacyl tRNAs enter.

IF YOU UNDERSTAND . . .

17.1 apply Transcription would continue past the normal point because the insertion of nucleotides would disrupt the structure of the RNA hairpin that functions as a terminator. **17.2** apply The protein-coding segment of the gene is predicted to be longer in eukaryotes because of the presence of introns. **17.3** understand The tRNAs act as adaptors because they couple the information contained in the nucleotides of mRNA to that contained in the amino acid sequence of proteins. **17.4** apply An incorrect amino acid would appear often in proteins. This is because the altered synthetase would sometimes add the correct amino acid for a particular tRNA and at other

---

FIGURE A16.1

mRNA sequence:
5′ AUG-CUG-GAG-GGG-GUU-AGA-CAU 3′

Amino acid sequence:
Met-Leu-Glu-Gly-Val-Arg-His

FIGURE A16.2

Bottom DNA strand:
3′ AACTT-TAC(start)-GGG-CAA-ACC-TCT-AGC-CCA-ATG-TCG-ATC(stop)-AGTTTC 5′

mRNA sequence:
5′ AUG-CCC-GUU-UGG-AGA-UCG-GGU-UAC-AGC-UAG 3′

Amino acid sequence:
Met-Pro-Val-Trp-Arg-Ser-Gly-Tyr-Ser

times add the incorrect amino acid. **17.5** create One possible concept map is shown in **FIGURE A17.1**.

✓ YOU SHOULD BE ABLE TO . . .

✓ Test Your Knowledge

1. understand c **2.** remember c **3.** understand To speed the correct folding of newly synthesized proteins **4.** remember d **5.** remember At the 3′ end **6.** apply b

✓ Test Your Understanding

7. understand Basal transcription factors bind to promoter sequences in eukaryotic DNA and facilitate the binding of RNA polymerase. As part of the RNA polymerase holoenzyme, sigma binds to a promoter sequence in bacterial DNA and to allow RNA polymerase to initiate at the start of genes. **8.** analyze The wobble rules allow a single tRNA to pair with more than one type of mRNA codon. This is distinct from redundancy, in which more than one codon can specify a single amino acid. If the wobble rules did not exist, there would need to be one tRNA for each amino-acid-specifying codon in the redundant genetic code. **9.** apply b **10.** apply After a peptide bond forms between the polypeptide and the amino acid held by the tRNA in the A site, the ribosome moves down the mRNA. As it does, an uncharged tRNA leaves the E site. The now-uncharged tRNA that was in the P site enters the E site; the tRNA holding the polypeptide chain moves from the A site to the P site, and a new aminoacyl tRNA enters the A site. **11.** apply The ribosome's active site is made up of RNA, not protein. **12.** understand The separation allows the aminoacyl tRNA to place the amino acid into the ribosome's active site while reaching to the distant codon on the mRNA.

✓ Test Your Problem-Solving Skills

13. create Ribonucleases degrade mRNAs that are no longer needed by the cell. If an mRNA for a hormone that increased heart rate were never degraded, the hormone would be produced continuously and heart rate would stay elevated—a dangerous situation. **14.** apply d **15.** analyze The regions most crucial to the ribosome's function should be the most highly conserved: the active site, the E, P, and A sites, and the site where mRNAs initially bind. **16.** analyze The most likely locations are one of the grooves or channels where RNA, DNA, and ribonucleotides move through the enzyme—plugging one of them would prevent transcription.

CHAPTER 18

IN-TEXT QUESTIONS AND EXERCISES

p. 338 understand See **TABLE A18.1**.
p. 338 Fig. 18.1 analyze Write "Slowest response, most efficient resource use" next to the transcriptional control label. Write "Fastest response, least efficient resource use" next to the post-translational control label.
p. 339 Fig. 18.2 apply Plates from all three treatments must be identical and contain identical growth medium, except for the presence of the sugars labeled in the figure. Also, all plates must be grown under the same physical conditions (temperature, light) for the same time.
p. 340 Fig. 18.3 apply Use a medium with all 20 amino acids when producing a master plate of mutagenized *E. coli* colonies; then use a replica plate that contains all the amino acids except tryptophan. Choose cells from the master plate that did *not* grow on the replica plate.
p. 341 understand *lacZ* codes for the β-galactosidase enzyme, which breaks the disaccharide lactose into glucose and galactose. *lacY* codes for the galactoside permease enzyme, which transports lactose into the bacterial cell. *lacI* codes for a protein that shuts down production of the other *lac* products. When lactose is absent, the *lacI* product prevents transcription. This is logical because there is no reason for the cell to make β-galactosidase

and galactoside permease if there is no lactose to metabolize. But when lactose is present, it interacts with *lacI* in some way so that *lacZ* and *lacY* are induced (their transcription can occur). When lactose is present, the enzymes that metabolize it are expressed.
p. 343 apply A mutation that prevents lactose binding to repressor is predicted to prevent transcription of the *lac* operon under any condition. This is because the repressor would never come off the operator. A mutation in the operator that prevents repressor binding is predicted to lead to constitutive expression of the *lac* operon.
p. 343 Fig. 18.8 understand Put the "Repressor protein" on the operator. No transcription will take place. Then put the "RNA polymerase" on the promoter. No transcription will take place. Finally, put "lactose" on the repressor protein and then remove the resulting lactose–repressor complex from the operon. Transcription will begin.
p. 344 CYU (1) understand It is logical that the genes for metabolizing lactose should be expressed only when lactose is available. **(2)** understand See **FIGURE A18.1**.
p. 344 apply This mutation is predicted to lower rates of transcription initiation because AraC's binding to

RNA polymerase is essential for AraC to work as an activator.

IF YOU UNDERSTAND . . .

18.1 understand Production of β-galactosidase and galactosidase permease are under transcriptional control—transcription depends on the action of regulatory proteins. The activity of the repressor is under post-translational control. **18.2** apply Treat *E. coli* cells with a mutagen, and create a master plate that is grown at 33°C. Replica-plate this master plate and grow the replica plate at 42°C. Look for colonies that are on the master plate at 33°C but not on the replica plate at 42°C. **18.3** understand The operator is the parking brake; the repressor locks it in place, and the inducer releases it. **18.4** apply *ara* initiator mutations are likely to affect positive and negative control because AraC must bind to the *ara* initiator sequence for both forms of control. **18.5** create Mutations that create operators for the SOS regulon repressor protein would put new genes under control of the repressor and incorporate them into the regulon.

FIGURE A17.1

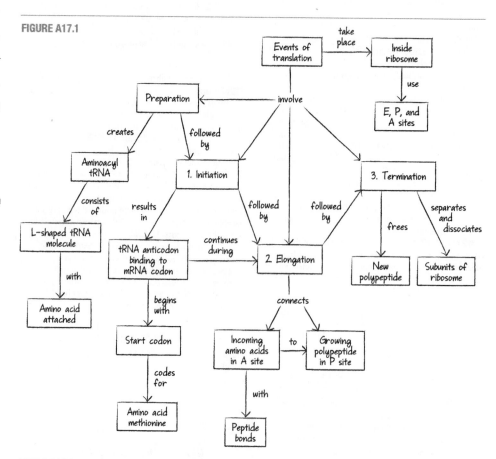

TABLE A18.1

Name	Function
lacZ	Gene for β-galactosidase
lacY	Gene for galactoside permease
Operator	Binding site for repressor
Promoter	Binding site for RNA polymerase
Repressor	Shuts down transcription
Lactose	Binds to repressor and stimulates transcription (removes negative control)
Glucose	At low concentration, increases transcription

✔ **Test Your Knowledge**

1. remember b **2.** understand d **3.** apply b **4.** remember DNA; RNA;
5. understand When glucose and another sugar are present in the environment, inducer exclusion prevents the use of the other sugar and allows only use of glucose.
6. remember a

✔ **Test Your Understanding**

7. understand The glycolytic enzymes are always needed in the cell because they are required to produce ATP, and ATP is always needed. **8.** understand Positive control means that a regulatory protein, when present, causes transcription to increase. Negative control means that a regulatory protein, when present, prevents transcription. **9.** apply b. **10.** apply Regulation of the *lac* operon should be normal. The location of the *lacI* gene isn't important, because the gene produces a protein that diffuses within the cell to the operator. **11.** analyze b (since the activator needs to bind to regulatory sequences to activate gene expression, preventing DNA binding would cripple the regulon and prevent cholera).**12.** analyze The *lac* operon would be strongly induced. Once inside the cell, the IPTG will bind to the repressor, causing it to release from DNA. IPTG cannot be broken down, so its concentration will remain high. Finally, since glucose is absent, there will be no inducer exclusion to inhibit IPTG transport through the galactoside permease transporter.

✔ **Test Your Problem-Solving Skills**

13. create Set up cultures with individuals that all come from the same colony of toluene-tolerating bacteria. Half the cultures should have toluene as the only source of carbon; half should have glucose or another common source of carbon. The glucose-containing medium serves as a control to ensure that cells can be grown in the lab. Cells will grow in both cultures if they are able to use toluene as a source of carbon; they will grow only in glucose-containing medium if toluene cannot be used as a carbon source. **14.** apply At a rate of 1×10^{-4} mutants per cell, you would on average find one mutant in every 10,000 (1×10^4) cells. Therefore, you should screen a bit more (~ 2–3 times more) than 10,000 cells to be reasonably sure of finding at least one mutant. **15.** apply b **16.** analyze Cells with functioning β-galactosidase will produce blue colonies; cells with *lacZ* mutations or *lacY* mutations will not produce β-galactosidase and will produce white colonies. The *lac* promoter could be mutated so that RNA polymerase cannot bind.

CHAPTER 19

IN-TEXT QUESTIONS AND EXERCISES

p. 352 Fig. 19.4 analyze Acetylation of histones decondenses chromatin and allows transcription to begin, so HATs are elements in positive control. Deacetylation condenses chromatin and inactivates transcription, so HDACs are elements in negative control.
p. 353 CYU (1) apply Many more genes than normal are predicted to be expressed because the inability to methylate DNA would lead to more decondensed chromatin. **(2)** understand Addition of acetyl groups to histones or methyl groups to DNA can cause chromatin to decondense or condense, respectively. Different patterns of acetylation or methylation will determine which genes in muscle cells versus brain cells can be transcribed and which are not available for transcription.
p. 353 Fig. 19.5 create They could do something to change histone modifications to see how this affects gene transcription instead of just making the observation that certain histone modifications and low rates of transcription go together.
p. 355 Fig. 19.6 analyze A typical eukaryotic gene usually contains introns and is regulated by multiple enhancers. Bacterial operons lack introns and enhancers. The

promoter-proximal element found in some eukaryotic genes is comparable to the *araC* binding site in the *ara* operon of bacteria. Bacterial operons have a single promoter but code for more than one protein; eukaryotic genes code for a single product.
p. 356 analyze The basal transcription factors found in muscle and nerve cells are similar or identical; the regulatory transcription factors found in each cell type are different.
p. 356 understand DNA forms loops when distant regulatory regions, such as silencers and enhancers, are brought close to the promoter through binding of regulatory transcription factors to mediator.
p. 356 CYU (1) analyze Bacterial regulatory sequences are found close to the promoter; eukaryotic regulatory sequences can be close to the promoter or far from it. Bacterial regulatory proteins interact directly with RNA polymerase to initiate or prevent transcription; eukaryotic regulatory proteins influence transcription by altering chromatin structure or binding to the basal transcription complex through mediator proteins.
(2) understand Certain regulatory proteins decondense chromatin at muscle- or brain-specific genes and then activate or repress the transcription of cell-type-specific genes. Muscle-specific genes are expressed only if muscle-specific regulatory proteins are produced and activated.
p. 358 understand Alternative splicing does not occur in bacteria because bacterial genes do not contain introns—in bacteria, each gene codes for a single product. Alternative splicing is part of step 3 in Figure 19.1.

p. 359 remember Step 4 and step 5. RNA interference either **(1)** decreases the life span of mRNAs or **(2)** inhibits translation.
p. 360 CYU (1) understand It became clear that a single gene can code for multiple products instead of a single one.
(2) understand The miRNAs interfere with mRNAs by targeting them for destruction or preventing them from being translated.
p. 362 CYU (1) understand Many different types of mutations can disrupt control of the cell cycle and initiate cancer. These mutations can affect any of the six levels of control over gene regulation outlined in Figure 19.1.
(2) understand The p53 protein is responsible for shutting down the cell cycle in cells with damaged DNA. If the protein does not function, then cells with damaged DNA—and thus many mutations—continue to divide. If these cells have mutations in genes that regulate the cell cycle, then they may continue to divide in an uncontrolled fashion.

IF YOU UNDERSTAND . . .

19.1 understand Because eukaryotic RNAs are not translated as soon as transcription occurs, it is possible for RNA processing to occur, which creates variation in the mRNAs produced from a primary RNA transcript and in their life spans. **19.2** understand The default state of eukaryotic genes is "off," because the highly condensed state of the chromatin makes DNA unavailable to RNA polymerase. **19.3** understand See **FIGURE A19.1**. **19.4** understand Alternative splicing makes it possible for a single gene to

FIGURE A18.1

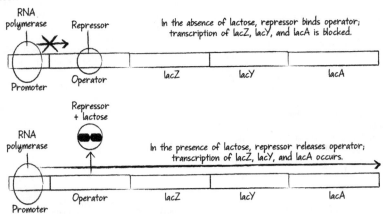

In the absence of lactose, repressor binds operator; transcription of lacZ, lacY, and lacA is blocked.

In the presence of lactose, repressor releases operator; transcription of lacZ, lacY, and lacA occurs.

FIGURE A19.1

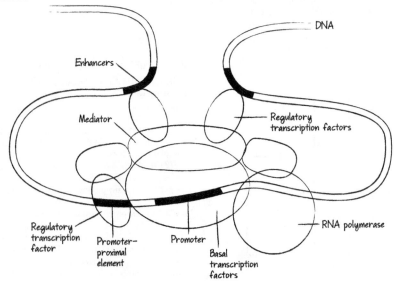

code for multiple products. **19.5** analyze One difference is that, in bacteria, genes that need to be turned on together are often clustered together in operons. Eukaryotes do not use operons. A similarity is in the way eukaryotes turn on many genes together and the strategy of bacterial regulons. Here, genes that are in many locations share a DNA regulatory sequence and are activated when a regulatory transcription factor binds to the regulatory sequence. **19.6** understand In mutant cells that lack a form of p53 that can bind to enhancers, DNA damage cannot arrest the cell cycle, the cell cannot kill itself, and damaged DNA is replicated. This leads to mutations that can move the cell farther down the road to cancer.

YOU SHOULD BE ABLE TO . . .

✓ Test Your Knowledge

1. remember c **2.** remember A tumor suppressor is a gene or protein that holds cell division in check unless conditions are right for the cell to divide. **3.** understand d **4.** remember The set of regulatory transcription factors present in a particular cell, not differences in DNA sequence, are largely responsible for which genes are expressed. **5.** remember b **6.** remember d

✓ Test Your Understanding

7. analyze (a) Enhancers and the *araC* site are similar because both are sites in DNA where regulatory proteins bind. They are different because enhancers generally are located at great distances from the promoter, whereas the *araC* site is located nearer the promoter. (b) Promoter-proximal elements and the *lac* operon operator are both regulatory sites in DNA located close to the promoter. (c) Basal transcription factors and sigma are proteins that must bind to the promoter before RNA polymerase can initiate transcription. They differ because sigma is part of the RNA polymerase holoenzyme, while the basal transcription complex recruits RNA polymerase to the promoter. **8.** understand If changes in the environment cause changes in how spliceosomes function, then the RNAs and proteins produced from a particular gene could change in a way that helps the cell cope with the new environmental conditions. **9.** analyze (a) Enhancers and silencers are both regulatory sequences located at a distance from the promoter. Enhancers bind regulatory transcription factors that activate transcription; silencers bind regulatory proteins that shut down transcription. (b) Promoter-proximal elements and enhancers are both regulatory sequences that bind positive regulatory transcription factors. Promoter-proximal elements are located close to the promoter; enhancers are far from the promoter. (c) Transcription factors bind to regulatory sites in DNA; mediator does not bind to DNA but instead forms a bridge between regulatory transcription factors and

basal transcription factors. **10.** apply Inhibition of a histone deacetylase is predicted to leave acetyl groups on histones longer than normal. This is predicted to keep gene transcription going longer than normal, leading to higher levels of particular proteins and a change in the phenotype of the cell. **11.** analyze c; this is because there are many more modifications in the histone code relative to the genetic code. **12.** apply The cell is predicted to arrest in the cell cycle and most likely will commit suicide through activation of apoptosis genes because of the continually active *p53*.

✓ Test Your Problem-Solving Skills

13. apply c; This is because promoters and enhancers are brought into close physical proximity when transcription begins (see Figure 19.8). Since rats of malnourished mothers initiate *Hnf4a* gene transcription infrequently, the promoter and enhancer will be together less often in these animals compared with rats born to well-nourished mothers. **14.** understand See **FIGURE A19.2**. The value for the normal diet should be shown as 1.0 and the value for the low-protein diet should be shown as 0.64 (0.64 comes from the ratio of the cpm of the low-protein diet divided by the cpm of the normal diet, or 7368/11,478). **15.** create You could treat a culture of DNA-damaged cells with a drug that stops transcription and then compare them with untreated DNA-damaged cells. If transcriptional control regulates p53 levels, then the p53 level would be lower in the treated cells versus control cells. If control of p53 levels is post-translational, then in both cultures the p53 level would be the same. *Other approaches:* Add labeled NTPs to damaged cells and see if they are incorporated into mRNAs for p53; or add labeled amino acids to damaged cells and see if they are incorporated into completed p53 proteins; or add labeled phosphate groups and see if they are added to p53 proteins. **16.** create The double-stranded RNA could be cut by the same enzyme that creates double-stranded miRNAs from miRNA precursors. This double-stranded RNA may be incorporated into RISC and converted into a single-stranded RNA. The single-stranded RNA held by RISC would work in RNA interference to trigger the destruction of the complementary viral mRNA.

Big Picture Genetic Information

p. 366 CYU (1) remember Star = DNA, mRNA, proteins. **(2)** understand RNA "is reverse transcribed by" reverse transcriptase "to form" DNA. **(3)** analyze E = splicing, etc.; E = meiosis and sexual reproduction (along with their links). **(4)** analyze Chromatin "makes up" chromosomes; independent assortment and recombination "contribute to" high genetic diversity.

CHAPTER 20

IN-TEXT QUESTIONS AND EXERCISES

p. 372 understand Isolate the DNA and cut it into small fragments with EcoRI, which leaves sticky ends. Cut copies of a plasmid or other vector with EcoRI. Mix the fragments and plasmids under conditions that promote complementary base pairing by sticky ends of fragments and plasmids. Use DNA ligase to catalyze formation of phosphodiester bonds and seal the sequences. **p. 372 Fig. 20.4** apply No—many times, because many copies of each type of mRNA were present in the pituitary cells, and many pituitary cells were used to prepare the library. **p. 373** apply The probe must be single stranded so that it will bind by complementary base pairing to the target DNA, and it must be labeled so that it can be detected. The probe will base-pair only with fragments that include a sequence complementary to the probe's sequence. A probe with the sequence 5′ AATCG 3′ will

bind to the region of the target DNA that has the sequence 5′ CGATT 3′ as shown here:

5′ AATCG 3′
3′ TCCGGTTAGCATTACCATTTT 5′

p. 374 CYU (1) understand When the endonuclease makes a staggered cut in a palindromic sequence and the strands separate, the single-stranded bases that are left will bind ("stick") to the single-stranded bases left where the endonuclease cut the same palindrome at a different location. **(2)** understand The word probe means to examine thoroughly. A DNA probe "examines" a large set of sequences thoroughly and binds to one—the one that has a complementary base sequence. **p. 375 Fig. 20.7** analyze The polymerase will begin at the 3′ end of each primer. On the top strand in part (b), it will move to the left; on the bottom strand, it will move to the right. As always, synthesis is in the 5′-to-3′ direction. **p. 376 CYU (1)** understand Denaturation makes DNA single stranded so the primer can bind to the sequence during the annealing step. Once the primer is in place, *Taq* polymerase can synthesize the rest of the strand during the extension step. It is a "chain reaction" because the products of each reaction cycle are used in the next reaction cycle—this is why the number of copies doubles in each cycle. **(2)** apply One of many possible answers is shown in **FIGURE A20.1**. **p. 378 CYU** understand If ddNTPs were present at high concentration, they would almost always be incorporated—meaning that only fragments from the first complementary base in the sequence would be produced. **p. 379** understand Using a map with many markers makes it more likely that there will be one marker that is very tightly linked to the gene of interest—meaning that a form of the marker will almost always be associated with the phenotype you are tracking. **p. 382 CYU** create Start with a genetic map with as many polymorphic markers as possible. Determine the genotype at these markers for a large number of individuals who have the same type of alcoholism—one that is thought to have a genetic component—as well as a large number of unaffected individuals. Look for particular versions of a marker that is almost always found in affected individuals. Genes that contribute to a predisposition to alcoholism will be near that marker. **p. 383 Fig. 20.11** apply The insertion will probably disrupt the gene and have serious consequences for the cell and potentially the individual.

IF YOU UNDERSTAND . . .

20.1 understand Special features of DNA are needed to allow replication in a cell. It is unlikely that any DNA fragment generated by cutting the DNA of one species with a restriction enzyme would replicate when inserted into a bacterial cell. By placing DNA fragments within plasmids that normally replicate in a bacterial cell, the inserted DNA can be replicated along with the plasmid. **20.2** analyze Advantages of cloning in cells: No knowledge of the sequence is required. Disadvantages of cloning in cells: It is slower and technically more difficult than PCR. PCR advantages: It is fast and easy, and it can amplify a DNA sequence that is rare in the sample. PCR disadvantage: It requires knowledge of sequences on either side of the target gene, so primers can be designed. **20.3** understand The length of each fragment is dictated by where a ddNTP was incorporated into the growing strand, and each ddNTP corresponds to a base on the template strand. Thus the sequence of fragment sizes corresponds to the sequence on the template DNA. **20.4** understand In this case, both affected and unaffected individuals would have the same marker, so there would be no way to associate a particular form of the marker with the gene and phenotype you are interested in. **20.5** understand An advantage of using retrovirus vectors is that any foreign genes they carry are integrated into

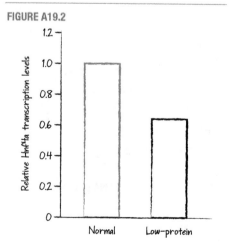

FIGURE A19.2

human chromosomes, so delivered genes become a permanent feature of the cell's genome. The concern is that they may integrate at sites that alter gene function in ways that harm cell function, or even worse, lead to cancer. **20.6** (understand) Genes can be inserted into the Ti plasmids carried by *Agrobacterium*, and the recombinant plasmids transferred to plant cells infected by the bacterium.

YOU SHOULD BE ABLE TO . . .

✔ Test Your Knowledge

1. (understand) They cut DNA at specific sites, known as recognition sites, to produce DNA fragments useful for cloning. **2.** (remember) b **3.** (understand) ddNTPs lack the −OH (hydroxyl) group on the 3′ carbon of deoxyribose sugar that is required to extend the DNA chain during synthesis. **4.** (understand) a **5.** (remember) b **6.** (understand) d

✔ Test Your Understanding

7. (understand) When a restriction endonuclease cuts a "foreign gene" sequence and a plasmid, the same sticky ends are created on the excised foreign gene and the cut plasmid. After the sticky ends on the foreign gene and the plasmid anneal, DNA ligase catalyzes closure of the DNA backbone, sealing the foreign gene into the plasmid DNA (see **FIGURE A20.2**). **8.** (apply) If the DNA at each cycle steadily doubles, it is predicted to yield a 33.6 million-fold increase (a 2^{25}-fold increase). **9.** (understand) A cDNA library is a collection of complementary DNAs made from all the mRNAs present in a certain group of cells. A cDNA library from a human nerve cell would be different from one made from a human muscle cell, because nerve cells and muscle cells express many different genes that are specific to their cell type. **10.** (remember) Genetic markers are genes or other loci

that have known locations in the genome. When these locations are diagrammed, they represent the physical relationships between landmarks—in other words, they form a map. **11.** (understand) d; these regulatory sequences promote expression of introduced genes in the endosperm—the rice grain eaten as food. **12.** (analyze) PCR and cellular DNA synthesis are similar in the sense of producing copies of a template DNA. Both rely on primers and DNA polymerase. The major difference between the two is that PCR copies only a specific target sequence, but the entire genome is copied during cellular DNA synthesis.

✔ Test Your Problem-Solving Skills

13. (create) You could use a computer program to identify likely promoter sequences in the sequence data and then look for sequences just downstream that have an AUG start codon and codons that could be part of the protein-coding exons of a potential protein about 500 amino acids (500 codons or 1500 bases) long. **14.** (analyze) False. Since somatic-cell modifications cannot be passed on to future generations, but germ-line modifications may be passed on to offspring, germ-line modification opens a new set of ethical questions. **15.** (analyze) In both techniques, researchers use an indicator to identify either a gene of interest or a colony of bacteria with a particular trait. The problem is the same—picking one particular thing (a certain gene or a cell with a particular mutation) out of a large collection. **16.** (analyze) (a) Primer 1b binds to the top right strand and would allow DNA polymerase to synthesize the top strand across the target gene. Primer 1a, however, binds to the upper left strand and would allow DNA polymerase to synthesize the upper strand *away* from the target gene. Primer 2a binds to the bottom left strand and would allow DNA polymerase to synthesize the bottom strand across the target gene. Primer 2b, however, binds

to the bottom right strand and would allow DNA polymerase to synthesize away from the target gene. (b) She could use primer 1b with primer 2a.

CHAPTER 21

IN-TEXT QUESTIONS AND EXERCISES

p. 391 (understand) If no overlap occurred, there would be no way of ordering the fragments correctly. You would be able to put fragments only in random order—not the correct order.
p. 391 Fig. 21.2 (understand) Shotgun sequencing is based on fragmenting the genome into many small pieces.
p. 393 Fig. 21.3 (understand) Because the mRNA codons that match up with each strand are oriented in the 5′-to-3′ direction.
p. 395 CYU (1) (understand) Parasites don't need genes that code for enzymes required to synthesize molecules they acquire from their hosts. **(2)** (understand) If two closely related species inherited the same gene from a common ancestor, the genes should be similar. But if one species acquired the same gene from a distantly related species via lateral gene transfer, then the genes should be much less similar.
p. 399 Fig. 21.9 (understand)
Chromosome 1:
ψβ2-ε-Gγ-Aγ-ψβ1-δ-ψβ2-ε-Gγ-Aγ-ψβ1-δ-β
Chromosome 2: β
p. 400 CYU (1) (understand) Eukaryotic genomes have vastly different numbers of transposable elements and other repeated sequences that don't directly contribute to phenotype. So genome size can vary widely without changing the number of coding genes or the complexity of the organism. **(2)** (apply) Because most of the genome is transcribed, this number is very close to the percentage of the genome that codes for exons. In humans, the ratio would be roughly 2%.
p. 401 (understand) Start with a microarray containing exons from a large number of human genes. Isolate mRNAs from brain tissue and liver tissue, and make labeled cDNAs from each. Probe the microarray with cDNA made from each type of tissue, and record where binding occurs. Binding events identify exons that are transcribed in each type of tissue. Compare the results to identify genes that are expressed in brain but not liver, or in liver but not brain.
p. 402 (understand) Cell replication is an emergent property because it is due to the interaction of proteins working in a network, yet is a property that could not be predicted from the analysis of any one of these proteins.

IF YOU UNDERSTAND . . .

21.1 (understand) If a search of human gene sequence databases revealed a gene that was similar in base sequence, and if follow-up work confirmed that the mouse and human genes were similar in their pattern of exons and introns and regulatory sequences, then the researchers could claim that they are homologous. **21.2** (understand) The genome of a parasite is predicted to be smaller and to have fewer genes, particularly for metabolism, than the genome of a nonparasite. **21.3** (understand) These features include variable numbers of transposable elements; noncoding repeated sequences; and noncoding, nonrepetitive DNA. These add to genome size without adding genes that directly influence phenotype. Additionally, mechanisms such as alternative splicing and the possibility of many noncoding regulatory RNAs can create situations in which few protein-coding genes but many different proteins are expressed in intricately regulated ways—features that would increase morphological complexity without increasing gene number. **21.4** (understand) Color-coded clusters shown in Figure 21.13 represent networks of interacting proteins that work together to carry out a particular cellular function. But it's clear from the figure that smaller, clusters are also connected into larger networks.

FIGURE A20.1

5′ CATGACTATTACGTATCGGGTACTATGCTATCGATCTAGCTACGCTAGCT 3′
3′ GTACTGATAATGCATAGCCCATGATACGATAGCTAGATCGATGCGATCGA 5′

Primer #1, which will anneal to the 3′ end of the top strand:

5′ AGCTAGCGTAGCTAGATCGAT 3′

Primer #2, which will anneal to the 3′ end of the bottom strand:

5′ CATGACTATTACGTATCGGGT 3′

The primers will bind to the separated strands of the parent DNA sequence as follows:

5′ CATGACTATTACGTATCGGGTACTATGCTATCGATCTAGCTACGCTAGCT 3′
                                 3′ TAGCTAGATCGATGCGATCGA 5′

5′ CATGACTATTACGTATCGGGT 3′
3′ GTACTGATAATGCATAGCCCATGATACGATAGCTAGATCGATGCGATCGA 5′

FIGURE A20.2

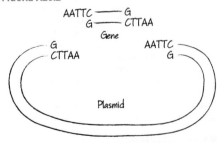

Sticky ends on cut plasmid and gene to be inserted are complementary and can base pair

DNA ligase forms four phosphodiester bonds between gene and plasmid DNA at the points indicated by arrows

YOU SHOULD BE ABLE TO . . .

✓ Test Your Knowledge

1. [remember] c **2.** [understand] a **3.** [remember] d **4.** [remember] b **5.** [understand] Finding two or more genes of similar sequence in the genome. **6.** [remember] Having a sequence that is clearly related to a functional gene but with a crippling mutation such as a stop codon or deletion.

✓ Test Your Understanding

7. [analyze] Computer programs are used to scan sequences in both directions to find ATG start codons, a gene-sized logical sequence with recognizable codons, and then a stop codon. One can also look for characteristic promoter, operator, and other regulatory sites. It is more difficult to identify open reading frames in eukaryotes because their genomes are so much larger and because of the presence of introns and repeated sequences. **8.** [evaluate] c **9.** [apply] Water flea gene density is about 31,000 genes/200 Mbp = 155 genes/Mbp. In humans, gene density is about 21,000 genes/3000 Mbp = 7 genes/Mbp. The relative gene density of water flea/human = 155/7 = 22. **10.** [analyze] Because with many different alleles, the chance that a match is coincidental is low relative to the chance that they match by identity (they come from the same person). **11.** [understand] A DNA microarray experiment identifies which genes are being expressed in a particular cell at a particular time. If a series of experiments shows that different genes are expressed in cells at different times or under different conditions, it implies that expression was turned on or off in response to changes in age or changes in conditions. **12.** [understand] Homology is a similarity among different species that is due to their inheritance from a common ancestor. If a newly sequenced gene is found to be homologous with a known gene of a different species, it is assumed that the gene products have similar function.

✓ Test Your Problem-Solving Skills

13. [analyze] If "gene A" is not necessary for existence, it can be lost by an event like unequal crossing over (on the chromosome with deleted segments) with no ill effects on the organism. In fact, individuals who have lost unnecessary genes are probably at a competitive advantage, because they no longer have to spend time and energy copying and repairing unused genes. **14.** [apply] If the grave were authentic, it might include two very different parental patterns along with five children whose patterns each represented a mix between the two parents. The other unrelated individuals would have patterns not shared by anyone else in the grave. **15.** [apply] d; mutation of such a central gene is likely to influence the phenotypes associated with all the genes it interacts with. If these interacting genes are involved in many different functions, the effects of mutation of a central gene are likely to be widespread, or pleiotropic. **16.** [apply] You would expect that the livers and blood of chimps and humans would function similarly, but that strong differences occur in brain function. The microarray data support this prediction and suggest chimp and human brains are different because certain genes are turned on or off at different times and expressed in different amounts.

CHAPTER 22

IN-TEXT QUESTIONS AND EXERCISES

p. 410 CYU [understand] Biologists have been able to grow entire plants from a single differentiated cell taken from an adult. They have also succeeded in producing animals by transferring the nucleus of a fully differentiated cell to an egg whose nucleus has been removed.
p. 411 Fig. 22.4 [apply] Anterior structures are predicted to be shifted farther back in the embryo, since higher-than-normal levels of the anterior-specifying Bicoid protein would be present in posterior regions.
p. 412 Fig. 22.6 [apply] Since Bicoid is located in the anterior cytoplasm and works to turn on genes that produce anterior structures, a prediction is that this manipulation would create a concentration gradient of Bicoid with its highest point in the middle of the mutant embryo. This would lead to the most anterior-like structures in the center of the embryo and more posterior structures forming on either side.
p. 414 Fig. 22.10 [create] Genes are usually considered homologous if they have similar DNA sequence and exon–intron structure.
p. 415 CYU (1) [understand] Bicoid is a transcription factor, and cells that experience a high versus medium versus low concentrations of Bicoid express Bicoid-regulated genes at different levels. **(2)** [understand] Bicoid proteins control the expression of gap genes, which—in effect—tell cells which third of the embryo they are in along the anterior–posterior axis. Gap genes control the expression of pair-rule genes, which organize the embryo into individual segments. Pair-rule genes control expression of segment polarity genes, which establish an anterior–posterior polarity to each individual segment. Segment polarity genes turn on homeotic genes, which then trigger genes for producing segment-specific structures like wings or legs.

IF YOU UNDERSTAND . . .

22.1 [apply] The seedling would be misshapen because controlling the plane of cleavage is essential for formation of normal structures in plants. **22.2.** [understand] Cells in the stem contain a complete set of genes, so if they can be de-differentiated and receive appropriate signals from other cells, they could differentiate into any cell type and cooperate to form any structure of the plant, including roots. **22.3** [analyze] Because segmentation genes work in a cascade, mutation of a pair-rule gene is expected to influence the expression of genes in the next level, segment polarity genes, but not affect the expression of earlier acting genes, such as gap genes. **22.4** [apply] This procedure is predicted to lead to the loss of limbs because the regulatory sequence replacement will cause the regions of expression of *Hoxc8* and *Hoxc6* to overlap.

YOU SHOULD BE ABLE TO . . .

✓ Test Your Knowledge

1. [remember] b **2.** [remember] stem cells **3.** [understand] a **4.** [remember] True **5.** [remember] a **6.** [remember] a conserved gene that can be expressed at different times and places during development to produce different types of structures in different organisms.

✓ Test Your Understanding

7. [understand] They occur in both plants and animals, and are responsible for the changes that occur as an embryo develops. **8.** [understand] If an adult cell used for cloning had some permanent change or loss of genetic information during development, it should not be able to direct the development of a viable adult. But if the adult cell is genetically equivalent to a fertilized egg, then it should be capable of directing the development of a new individual. **9.** [understand] The researchers exposed adult flies to treatments that induce mutations and then looked for embryos with defects in the anterior–posterior body axis or body segmentation. The embryos had mutations in genes required for body axis formation and segmentation. **10.** [understand] Development—specifically differentiation—depends on changes in gene expression. Changes in gene expression depend on differences in regulatory transcription factors. **11.** [create] Differentiation is triggered by the production of transcription factors, which induce other transcription factors, and so on—a sequence that constitutes a regulatory cascade—as development progresses. At each step in the cascade, a new subset of genes is activated—resulting in a step-by-step progression from undifferentiated to fully differentiated cells. **12.** [analyze] c. This result shows that a mouse tool-kit gene is similar enough (conserved) to a fly gene to be able to take over its function.

✓ Test Your Problem-Solving Skills

13. [apply] There would be more Bicoid protein farther toward the posterior and less in the anterior—meaning that the anterior segments would be "less anterior" in their characteristics and the posterior segments would be "more anterior." **14.** [analyze] The result means that the human gene was expressed in worms and that its protein product has performed the same function in worms as in humans. This is strong evidence that the genes are homologous and that their function has been evolutionarily conserved. **15.** [apply] c. Since the morphogen concentration is halved every 100 μm away from the posterior pole, dropping to 1/16 the highest concentration would require 4 "halvings" of the initial concentration ($1/2 \times 1/2 \times 1/2 \times 1/2 = 1/16$), and this would take place in 4 steps of 100 μm each—or 400 μm from the morphogen source. This is where the leg is predicted to be formed. **16.** [create] Tool-kit genes—such as *Hox* genes—are responsible for triggering the production of structures like wings or legs in particular segments. One hypothesis to explain the variation is that changes in gene expression of tool-kit genes led to different numbers of segments that express leg-forming *Hox* genes.

CHAPTER 23

IN-TEXT QUESTIONS AND EXERCISES

p. 421 [understand] If the bindin-like protein were blocked, it would not be able to bind to the egg-cell receptor for sperm. Fertilization would not take place.
p. 422 CYU (1) [apply] This change is predicted to allow the sperm of species A to bind to—and likely fertilize—the eggs of species B. **(2)** [understand] The entry of a sperm triggers a wave of calcium ion release around the egg-cell membrane. Thus, calcium ions act as a signal that indicates "A sperm has entered the egg." In response to the increase in calcium ion concentration, cortical granules fuse with the egg-cell membrane and release their contents to the exterior, causing the destruction of the sperm receptor and, in sea urchins, the formation of a fertilization envelope.
p. 424 CYU [understand] If cytoplasmic determinants could not be localized, they would be distributed to all blastomeres. This would specify the same fate for all blastomeres and their descendants instead of different fates.
p. 426 CYU (1) [understand] Ectoderm and endoderm, because these are the inside and outside layers. **(2)** [apply] The future anterior–posterior and dorsal–ventral region of the embryo could first be identified as the blastopore forms and cells begin migrating into the embryo. The blastopore marks the future posterior of the embryo, and the region opposite the blastopore is where the anterior will form. Cells that first migrate into the blastopore come from the dorsal side, so this is the future dorsal region; the ventral region is on the opposite side.
p. 427 Fig. 23.9 [understand] Because neural tube formation depends on rearrangements of the cytoskeleton, a drug that locked the cytoskeleton in place is predicted to prevent formation of the neural tube.
p. 428 Fig. 23.12 [analyze] It made gene expression possible in non-muscle cells. A "general purpose" promoter can lead to transcription of an inserted cDNA in any type of cell, including fibroblasts.

IF YOU UNDERSTAND . . .

23.1 [understand] Bindin and the egg-cell receptor for sperm bind to each other. For this to happen, their tertiary structures must fit together somewhat like the structures of a lock and key. **23.2** [understand] If the cytoplasmic determinant is localized to a certain part of the egg cytoplasm, blastomeres will either contain or not contain the determinant, because the egg's cytoplasm becomes divided between blastomeres. **23.3** [analyze] Both cells would contain cell–cell signals and transcription factors specific to mesodermal cells, but the anterior cell would contain anterior-specific signals and regulatory transcription factors while the posterior cell would contain posterior-specific signals and regulators. **23.4** [understand] Cell–cell signals from the notochord direct the formation of the neural tube; subsequently, cell–cell signals from the notochord, neural tube, and ectoderm direct the differentiation of somite cells. Cells expand during neural tube formation. Somite cells move to new positions, proliferate, and differentiate.

✓ YOU SHOULD BE ABLE TO . . .

✓ Test Your Knowledge

1. [remember] Bindin is a protein on the sperm-cell head that binds to a receptor on the egg plasma membrane to allow species-specific attachment of sperm and egg. **2.** [understand] c **3.** [understand] c **4.** [understand] d **5.** [remember] ectoderm; **6.** [remember] a

✓ Test Your Understanding

7. [evaluate] As mammals, mice form a placenta that allows the embryo to obtain nutrients from the mother. There's no need for lots of nutrient reserves to support development. Frogs develop outside the mother's body, and the embryo is on its own. There is a need for nutrient reserves, and the large egg provides this. Eggs contain stores of nutrients. **8.** [apply] This treatment is predicted to promote polyspermy, because there would be no release of Ca^{2+} inside the cell to trigger a block to polyspermy. **9.** [understand] They contain different types and/or concentrations of cytoplasmic determinants. **10.** [analyze] a; this is because ectodermal cells have to constrict on their dorsal margins and expand on their ventral margin to fold ectoderm into a neural tube. **11.** [understand] When transplanted early in development, somite cells become the cell type associated with their new location. But when transplanted later in development, somite cells become the cell type associated with the original location. These observations indicate that the same cell is not committed to a particular fate until later in development. **12.** [apply] More than 50,000 fibroblasts would need to be screened. This is because if 2×10^{-5} of all mRNAs and cDNAs coded for MyoD, then 50,000 cells would need to be screened on average to find one with a MyoD cDNA (50,000 comes from the quotient of $1/2 \times 10^{-5}$ MyoD cDNAs per cDNA). Since this is an average, in some groups of 50,000 cells, there may not be a MyoD cDNA. Therefore, to be reasonably sure of finding at least one MyoD cDNA, more than 50,000 cells should be screened.

✓ Test Your Problem-Solving Skills

13. [analyze] The interactions that allow proteins to bind to each other are extremely specific—certain amino acids have to line up in certain positions with specific charges and/or chemical groups exposed. Sperm–egg binding is species specific because each form of bindin binds to a different egg-cell receptor for sperm. **14.** [evaluate] Translation is not needed for cleavage to occur normally—meaning that all the proteins needed for early cleavage are present in the egg. **15.** [create] The yellow pigment may be a cytoplasmic determinant, or it may be only a molecule that is found in the same cells as the cytoplasmic determinant. Without isolating the pigment and adding it to cells to see if it acts as a cytoplasmic determinant, these two possibilities cannot be distinguished. **16.** [apply] a (See Figure 23.10 and note how the notochord is closest

to the somite region that forms bone. If an additional notochord were transplanted near the top of the neural tube, signals from the notochord would be expected to convert the nearby somite into bone-producing somite, adding to the amount of bone.)

CHAPTER 24

IN-TEXT QUESTIONS AND EXERCISES

p. 434 Fig. 24.2 [remember] See **FIGURE A24.1**.
p. 436 CYU (1) [understand] Cells located on the outside of the embryo become epidermal tissue; cells in the interior become vascular tissue; cells in between become ground tissue. **(2)** [apply] The *MONOPTEROS* gene would be over-expressed, producing higher levels of the MONOPTEROUS transcription factor. The embryonic cells would get information indicating that they are in the shoot apex, where MONOPTEROS levels are high. As a result, the root would not develop normally.
p. 437 Fig. 24.6 [analyze] Many animals have three body axes analogous to the three axes of leaves: anterior–posterior (like the leaf's proximal–distal), dorsal–ventral (like the leaf's adaxial–abaxial), and a left–right axis (like the leaf's mediolateral axis). The main plant body, in contrast, has only two axes: apical–basal and radial.
p. 438 CYU (1) [create] Simply planting a cut stem will often produce a new root meristem. This occurs whenever the cutting produces a new plant. **(2)** [create] If the plant could be genetically engineered to express higher levels of the PHAN transcription factor in leaves, the likely result would be a shift in leaf type from simple to compound.
p. 440 [apply] Mutual inhibition between the *A* and *C* genes helps create separate regions of gene expression. If *C* gene expression is lost, then the inhibition of *A* gene expression is also lost, and *A* genes are expressed in their normal location and where they are normally inhibited by *C* gene expression.
p. 441 CYU (1) [analyze] Plants generate many reproductive structures—flowers—many times and in many places during the plant's life. Plant reproductive cells are created from vegetative cells, not cells that are set aside early in development as reproductive cells. In contrast, animals form reproductive structures (ovaries and testes) in set locations, produce reproductive structures only once during development and set aside specialized reproductive cells (the germ line) early in their development. **(2)** [analyze] Like *Hox* genes, *MADS-box* genes code for DNA-binding proteins that function as transcriptional regulators. Both Hox proteins and MADS-domain proteins are involved in regulatory transcription cascades that lead to development of specific structures in specific locations. Although they perform similar functions, *Hox* genes and *MADS-box* genes evolved independently and differ in their base sequences. Thus, these genes and their protein products are not homologous.

IF YOU UNDERSTAND . . .

24.1 [apply] If there were only one master regulatory gene for patterning, the predicted result of loss of that gene's function by mutation would be disorganization of the entire apical–basal axis, not just removal of parts of it as seen in the actual mutants. **24.2** [create] Many things could be imagined, but all should point in the direction of many repeated body parts (for example, 21 wings in a variety of shapes and positions on the body) and a very large overall size (if the fly is old) due to continuous growth. **24.3** [apply] The prediction is that sepals would develop from every whorl. This is because when the *C* gene product is absent, *A* gene expression spreads across the flower. Since there is no *B* gene expression, only sepals would form—the structure associated with only *A* gene expression.

✓ YOU SHOULD BE ABLE TO . . .

✓ Test Your Knowledge

1. [understand] This is because mechanisms for development evolved independently in plants and animals. **2.** [understand] b **3.** [remember] b **4.** [remember] d **5.** [remember] c **6.** [understand] a.

✓ Test Your Understanding

7. [understand] c. **8.** [understand] Animal cells that lead to sperm or egg formation are "sequestered" early in development, so that they are involved only in reproduction and undergo few rounds of mitosis. Because mutations occur in every round of mitosis, plant eggs and sperm should contain many more mutations than animal eggs and sperm. **9.** [understand] Meristem cells are undifferentiated and divide to produce a cell that remains in the meristem and a cell that can commit to a particular pathway of differentiation. These are properties of stem cells. **10.** [understand] Just like the three tissue layers in plant embryos, the tissues produced in the SAMs and RAMs of a 300-year-old oak tree can differentiate into all of the specialized cell types found in a mature plant. **11.** [evaluate] The ABC model predicted that the *A*, *B*, and *C* genes would each be expressed in specific whorls in the developing flower. The experimental data showed that the prediction was correct. **12.** [understand] Cell proliferation is responsible for expansion of meristems and growth in size. Asymmetric cell division along with cell expansion is responsible for giving the adult plant a particular shape. Differentiation is responsible for generating functional tissues and organs in the adult plant. Cell–cell signals are responsible for setting up the apical–basal axis of the embryo and for guiding the pollen tube to the egg.

✓ Test Your Problem-Solving Skills

13. [create] Four. This is because there are four states of gene activity (in any cell, *D* can be on or off and *E* can be on or off) and therefore four structures that could be specified using this code. Sets of on-off switches are known as Boolean systems and are the basis for information storage in computers and (sometimes) in cells. **14.** [analyze] Because vegetative growth is continuous and occurs in all directions (unrestricted), it could be considered indeterminate. In contrast, reproductive growth produces mature reproductive organs and stops when those organs are complete. Because its duration is limited, it could be considered determinate. **15.** [apply] a; see Figure 24.10 to understand the reasoning. *A* and *B* together create petals, and *B* and *C* together create stamens. **16.** [analyze] One hundred times more likely. This is because 100 times more cell divisions were required to create the oak egg, and the rate of mutation per cell division is constant. (You don't need to consider the absolute rate of 1×10^{-6}/division, since the question asks about relative chances only).

FIGURE A24.1

Apical

Radial ←

Basal

CHAPTER 25

IN-TEXT QUESTIONS AND ANSWERS

p. 448 Fig. 25.4 `analyze` Under special creation, fossils with transitional features would be explained as separate types, unrelated to other organisms and just coincidentally having intermediate features.

p. 448 Fig. 25.5 `analyze` If vestigial traits result from inheritance of acquired characteristics, some individuals must have lost the traits during their own lifetimes and passed the reduced traits on to their offspring. For example, a certain monkey's long tail might have been bitten off by a predator. The new traits would then somehow have passed to the individual's eggs or sperm, resulting in shorter-tailed offspring, until humans with a coccyx resulted.

p. 453 CYU (1) `evaluate` If birds evolved from dinosaurs you would expect to find transitional fossil dinosaurs with feathers. Such fossils have been found. **(2)** `analyze` The DNA sequences of chimpanzees and humans are so similar because we share a recent common ancestor.

p. 455 (1) `understand` Relapse occurred because the few bacteria remaining after drug therapy were not eliminated by the patient's weakened immune system and began to reproduce quickly. **(2)** `understand` No—almost all of the cells present at the start of step 3 would have been resistant to the drug.

p. 456 `understand` When antibiotics are overused, some bacteria will be killed, including the bacteria that are harmless or beneficial to us. However, bacteria that are resistant to these antibiotics will flourish and multiply, reducing the likelihood that antibiotic treatment will be effective in the future.

p. 457 Fig. 25.16 `analyze` "Prediction": Beak measurements were different before and after the drought. "Prediction of null hypothesis": No difference was found in beak measurements before and after the drought.

p. 459 CYU (1) `understand` *Postulate 1:* Traits vary within a population. *Postulate 2:* Some of the trait variation is heritable. *Postulate 3:* There is variation in reproductive success (some individuals produce more offspring than others). *Postulate 4:* Individuals with certain heritable traits produce the most offspring. The first two postulates describe heritable variation; the second two describe differential reproductive success. **(2)** `understand` Beak size and shape and body size vary among individual finches, in part because of differences in their genotypes (some alleles lead to deeper or shallower beaks, for example). When a drought hit, individuals with deep beaks survived better and produced more offspring than individuals with shallow beaks.

p. 460 `analyze` In biology, an adaptation is any heritable trait that increases an individual's ability to produce offspring in a particular environment. In everyday English, adaptation is often used to refer to an individual's nonheritable adjustment to meet an environmental challenge, a phenomenon that biologists call acclimatization. The phenotypic changes resulting from acclimatization are not passed on to offspring.

IF YOU UNDERSTAND . . .

25.1 `analyze` Typological thinking is based on the idea that species are unchanging types and that variations within species are unimportant or even misleading. Population thinking recognizes the variation that occurs within a species as critically important. **25.2** `remember` Many answers are possible. For example, the evolution of an antibiotic-resistant TB bacterium demonstrates that species change through time. The homologous sequences in human and fruit fly genes demonstrate that these species are related by common ancestry. **25.3** `analyze` In biology, fitness is the ability of an individual to produce viable, fertile offspring, relative to that ability in other individuals in the population. In everyday English, fitness is a physical attribute that is acquired as a result of practice or exercise. **25.4** `apply` Under special creation, changes in *Mycobacterium tuberculosis* populations would be explained as individual creative events governed by an intelligent creator. Under the theory of evolution by inheritance of acquired characters, changes in *M. tuberculosis* populations would be explained by the cells trying to transcribe genes in the presence of the drug, and their *rpoB* gene becoming altered as a result. **25.5** `understand` Brain size in *H. sapiens* might be constrained by the need for babies to pass through the mother's birth canal, by the energy required to maintain a large brain as an adult, or by lack of genetic variation for even larger brain size. Flying speed in falcons might be constrained by loss of maneuverability (and thus less success in hunting), the energy demands of extremely rapid flight, or lack of genetic variation for even faster flight.

YOU SHOULD BE ABLE TO . . .

✓ Test Your Knowledge

1. `remember` b **2.** `remember` False **3.** `remember` a **4.** `understand` d
5. `remember` homologous **6.** `remember` c

✓ Test Your Understanding

1. `understand` Mutation produces new genetic variations, at random, without any forethought as to which variations might prove adaptive in the future. Individuals with mutations that are disadvantageous won't produce many offspring, but individuals with beneficial mutations will produce many offspring. The beneficial mutations will thus increase in frequency through selection. **2.** `analyze` Artificial selection is determined by human choice and is goal directed, whereas natural selection is the unplanned differential reproductive success of individuals that vary in their heritable traits. **3.** `analyze` b (In some environments, being big and strong lowers fitness.) **4.** `analyze` There were 751 finches before the drought and 90 finches afterward, so only 90/751 = 0.12, or 12 percent, survived. There were more finches with deep beaks before the drought only because the population was much larger, but the *average* beak depth increased after the drought. **5.** `analyze` The evidence for within-patient evolution is that DNA sequences at the start and end of treatment were identical except for a single nucleotide change in the *rpoB* gene. If rifampin were banned, *rpoB* mutant strains likely would have had lower fitness in the drug-free environment and would not continue to increase in frequency in *M. tuberculosis* populations. **6.** `apply` (1) Traits vary within rabbit populations, such as fur color and hearing ability. (2) Some of this variation is heritable. (3) More rabbits are produced than can survive. (4) Rabbits with certain heritable traits, such as the ability to camouflage in their environment or hear the approach of foxes, produce more offspring.

✓ Test Your Problem-Solving Skills

1. `analyze` The theory of evolution by natural selection predicts that white and brown deer mice are descendants of an ancestral population that varied in color. The white mice had higher fitness in the beach environment, where they were more likely to escape the notice of predators. Likewise, the brown mice had a higher fitness in the shady forest environment. Over time, the average color of each population changed. Special creation claims that the two colors of mice do not change and were created by divine intervention. Evolution by inheritance of acquired characteristic predicts that the mice in the different environments needed to change color, so they did so and then passed their traits onto their offspring. **2.** `analyze` b (No, because changes in height due to nutrition and reduced incidence of disease are not heritable.) **3.** `evaluate` Evidence to support a common ancestor of humans and chimpanzees about 6–7 million years ago includes the following: fossils from Africa that date from 6–7 million years ago with primate characteristics that are common to chimps and humans, and DNA sequence comparisons showing homologies between chimps and humans (the rate of change of some genes can be used to estimate time of divergence). **4.** `evaluate` The theory of evolution fits the six criteria as follows: (1) and (2): It provides a common underlying mechanism responsible for puzzling observations such as homology, geographic proximity of similar species, the law of succession in the fossil record, vestigial traits, and extinctions. (3) and (4): It suggests new lines of research to test predictions about the outcome of changing environmental conditions in populations, about the presence of transitional forms in the fossil record, and so on. (5): It is a simple idea that explains the tremendous diversity of living and fossil organisms and why species continue to change today. (6): The realization that all organisms are related by common descent and that none are higher or lower than others was a surprise.

CHAPTER 26

IN-TEXT QUESTIONS AND EXERCISES

p. 467 (1) `apply` The frequencies of the three genotypes are shown by $p^2 + 2pq + q^2 = 1$.

$$\text{Freq}(A_1A_1) = p^2 = (0.6)^2 = 0.36.$$
$$\text{Freq}(A_1A_2) = 2pq = 2(0.6)(0.4) = 0.48.$$
$$\text{Freq}(A_2A_2) = q^2 = (0.4)^2 = 0.16.$$

(2) `apply` Allele frequencies in the offspring gene pool:

$$\text{Freq}(A_1) = 0.36 + 1/2(0.48) = 0.60.$$
$$\text{Freq}(A_2) = 1/2(0.48) + 0.16 = 0.40.$$

(3) `apply` Evolution has not occurred, because the allele frequencies have not changed. **p. 469 Table 26.1** `apply` The observed allele frequencies, calculated from the observed genotype frequencies, are 0.43 for *M* and 0.57 for *N*. The expected genotypes, calculated from the observed allele frequencies under the Hardy–Weinberg principle, are 0.185 for *MM*, 0.49 for *MN*, 0.325 for *NN*. **p. 470 CYU** `apply` Given the observed genotype frequencies, the observed allele frequencies are

$$\text{Freq}(A_1) = 0.574 + 1/2(0.339) = 0.744.$$
$$\text{Freq}(A_2) = 1/2(0.339) + 0.087 = 0.256.$$

Given these allele frequencies, the genotype frequencies expected under the Hardy–Weinberg principle are A_1A_1: $0.744^2 = 0.554$; A_1A_2: $2(0.744 \times 0.256) = 0.381$; A_2A_2: $0.256^2 = 0.066$. There are 4 percent too few heterozygotes observed, relative to the expected proportion. One of the assumptions of the Hardy–Weinberg principle is not met at this gene in this population, at this time. **p. 470** `apply` The proportions of homozygotes should increase, and the proportions of heterozygotes should decrease, but allele frequencies will not change if no selection is occurring at this locus. **p. 474** `apply` There will be an excess of observed genotypes containing the favored allele compared to the proportion expected under Hardy–Weinberg proportions. **p. 478 (1)** `understand` Sperm are small and hence relatively cheap to produce, whereas eggs are large and require a large investment of resources to produce. **(2)** `understand` Sperm are inexpensive to produce, so reproductive success for males depends on their ability to find mates—not on their ability to find resources to produce sperm. The opposite pattern holds for females. Sexual selection is based on variation in ability to find mates, so is more intense in males—leading to more exaggerated traits in males. **p. 479** `apply` If allele frequencies are changing due to drift, the populations in Table 26.1 would behave like the simulated populations in Figure 26.12—frequencies would drift up and down over time, and diverge. **p. 479 Fig. 26.12** `analyze` See FIGURE A26.1. **p. 480 Fig. 26.13** `analyze` The smaller the population, the greater the chance that genetic drift will cause alleles to be fixed or lost in the population in a short time. Starting with a smaller population makes the experiment easier.

p. 481 Fig. 26.14 (analyze) Original population: freq(A_1) = $(9 + 9 + 11)/54 = 0.54$; New population: freq(A_1) = $(2 + 2 + 1)/6 = 0.83$. The frequency of A_1 has increased dramatically.

p. 482 CYU (1) (understand) When allele frequencies fluctuate randomly up and down, sooner or later the frequency of an allele will hit 0. That allele thus is lost from the population, and the other allele at that locus is fixed. **(2)** (understand) In small populations, sampling error is large. For example, the accidental death of a few individuals would have a large impact on allele frequencies.

IF YOU UNDERSTAND . . .

26.1 (apply) The frequencies of two alleles are shown by $p + q = 1$. The frequencies of the three genotypes are shown by $p^2 + 2pq + q^2 = 1$.

If freq(A_1) = 0.2, then freq(A_2) = 0.8.
Freq(A_1A_1) = $p^2 = (0.2)^2 = 0.04$.
Freq(A_1A_2) = $2pq = 2(0.2)(0.8) = 0.32$.
Freq(A_2A_2) = $q^2 = (0.8)^2 = 0.64$.

26.2 (apply) It increased homozygosity of recessive alleles associated with diseases such as hemophilia. As a result, the royal families were plagued by genetic diseases. **26.3** (understand) The Hardy–Weinberg principle predicts that allele frequencies will stay the same over time. Natural selection favors some alleles over others, causing the frequency of those alleles to increase. **26.4** (apply) Genetic drift will rapidly reduce genetic variation in small populations. Thus, it is a high priority to keep the populations of endangered animals as large as possible. **26.5** (apply) In some cases, gene flow could be used to increase genetic variation and reduce inbreeding depression. However, gene flow can also be deleterious if the new alleles reduce the fitness of the endangered population. **26.6** (apply) Mutation rates themselves are too slow to affect an endangered population in a short time—unless a deleterious or beneficial mutation exposes the population to selection. (In instances where chemicals cause mutations in organisms, such as pesticides causing abnormal development in frogs, actions could be taken to remove the source of chemicals.)

YOU SHOULD BE ABLE TO . . .

✓ Test Your Knowledge

1. (understand) It defines what genotype and allele frequencies should be expected if evolutionary processes and nonrandom mating are *not* occurring. **2.** (remember) b **3.** (understand) a **4.** (remember) b **5.** (remember) True **6.** (understand) c

✓ Test Your Understanding

1. (apply) d; freq(A_1) = 0.9, so freq(A_2) = 0.1. Freq(A_2A_2) = $q^2 = (0.1)^2 = 0.01$. The number of babies with cystic

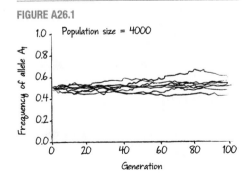

FIGURE A26.1

fibrosis = $2500(0.01) = 25$. **2.** (apply) Inbreeding causes an increase in the number of homozygotes. Individuals who are homozygous for recessive deleterious alleles are exposed to—and eliminated by—directional natural selection. **3.** (evaluate) This statement is false. Mutations do not occur because an organism wants or needs them. They just happen by accident and can be beneficial, neutral, or deleterious. **4.** (understand) Sexual selection is most intense on the sex that makes the least investment in the offspring, so that sex tends to have exaggerated traits that make individuals successful in competition for mates. In this way, sexual dimorphism evolves. **5.** (analyze) An allele in males, because a successful male can have as many as 100 offspring—10 times more than a successful female. **6.** (create) Your concept map should have linking verbs that relate the following information: Selection may decrease genetic variation or maintain it (e.g., if heterozygotes are favored). Genetic drift reduces it. Gene flow may increase or decrease it (depending on whether immigrants bring new alleles or emigrants remove alleles). Mutation increases it.

✓ Test Your Problem-Solving Skills

1. (apply) If we let p stand for the frequency of the loss-of-function allele, we know that $p^2 = 0.0001$; therefore $p = \sqrt{0.0001} = 0.01$. By subtraction, the frequency of normal alleles is 0.99. Under the Hardy–Weinberg principle, the frequency of heterozygotes is $2pq$, or $2 \times 0.01 \times 0.99$, which is 0.0198. **2.** (analyze) c **3.** (analyze) If males never lift a finger to help females raise children, the fundamental asymmetry of sex is pronounced and sexual dimorphism should be high. If males invest a great deal in raising offspring, then the fundamental asymmetry of sex is small and sexual dimorphism should be low. **4.** (apply) Captive-bred young, transferred adults, and habitat corridors could be introduced to counteract the damaging effects of drift and inbreeding in the small, isolated populations. It is important, though, to introduce individuals only from similar habitats and connect patches of similar habitat with corridors—to avoid introducing alleles that lower fitness.

CHAPTER 27

IN-TEXT QUESTIONS AND EXERCISES

p. 493 CYU (1) (understand) Biological species concept: Reproductive isolation means that no gene flow is occurring. Morphological species concept: If populations are evolving independently, they may have evolved morphological differences. Phylogenetic species concept: If populations are evolving independently, they should have synapomorphies that identify them as independent twigs on the tree of life. **(2)** (understand) Biological species concept: It cannot be used to evaluate fossils, species that reproduce asexually, or species that do not occur in the same area and therefore never have the opportunity to mate. Morphospecies concept: It cannot identify species that differ in traits other than morphology and is subjective—it can lead to differences of opinion that cannot be resolved by data. Phylogenetic species concept: Reliable phylogenetic information exists only for a small number of organisms.
p. 494 Fig. 27.4 (analyze) The disruption of the salt marsh was a vicariance event that split the range of seaside sparrows into two areas (Gulf and Atlantic coasts), limiting gene flow and eventually resulting in two monophyletic groups.
p. 496 Fig. 27.7 (analyze) It creates genetic isolation—the precondition for divergence—by separating apple maggot flies from hawthorn flies, which mate on hawthorn fruits.
p. 497 (create) The difference in timing of ripening of apples and hawthorns causes divergence by separating apple flies and hawthorn flies not only in space (apple fruits versus hawthorn fruits) but also in time. The

earlier ripening of apples means that apple maggots develop in relatively warm conditions in the fall compared to hawthorn maggots. Natural selection may favor traits in apple maggots that make them adapted to warmer temperatures and traits in hawthorn maggots that make them adapted to cooler temperatures.
p. 498 (apply) A diploid grape plant experiences a defect in meiosis resulting in the formation of diploid gametes. The individual self-fertilizes, producing tetraploid offspring. The tetraploid grape plants self-fertilize or mate with other tetraploid individuals, producing a tetraploid population.
p. 499 (apply) A tetraploid ($4n$) species (in this case, wheat) gives rise to diploid ($2n$) gametes, and a diploid wheat species gives rise to haploid (n) gametes. When a haploid gamete (n) fertilizes a diploid gamete ($2n$), a sterile triploid offspring results. If an error in mitosis occurs and chromosome number doubles before meiosis, then hexaploid wheat is formed.
p. 499 CYU (understand) Use one of the cases given in the chapter to illustrate dispersal (Galápagos finches), vicariance (snapping shrimp), or habitat specialization (apple maggot flies). In each case, drift will cause allele frequencies to change randomly. Differences in the habitats occupied by the isolated populations will cause allele frequencies to change under natural selection. Any mutation that occurs in one population will not occur in the other due to the lack, or low level, of gene flow. Or use polyploidization in maidenhair fern or *Tragopogon* to illustrate how a massive mutation can itself create immediate genetic isolation and divergence.
p. 501 Fig. 27.12 (evaluate) If the same processes were at work in the past that are at work in an experiment with living organisms, then the outcome of the experiment should be a valid replication of the outcome that occurred in the past.
p. 502 (analyze) Gene regions in the experimental hybrid would not align with those of *H. annuus* and *H. petiolaris*, meaning that the gray bar representing the DNA of the experimental hybrid could have had no orange or red regions corresponding to *H. annuus* and *H. petiolaris*.

IF YOU UNDERSTAND THAT . . .

27.1 (apply) Human populations would not be considered separate species under the biological concept, because human populations can successfully interbreed. They would not be considered separate species under the morphological species concept, because all human populations have the same basic morphology. Although human races differ in minor superficial attributes such as skin color and hair texture, they have virtually identical anatomy and physiology in all other regards. Nor would human populations be separate species under the phylogenetic species concept, because they all arose from a very recent common ancestor. (DNA comparisons have revealed that human races are remarkably similar genetically and do not differ enough genetically to qualify even for subspecies status.) **27.2** (create) (Many possibilities) Collect a group of fruit flies that are living in your kitchen, separate the individuals into different cages, and expose the two new populations to dramatically different environmental conditions—heat, lighting, food sources. In essence: Create genetic isolation and then conditions for divergence due to drift, mutation, and selection. **27.3** (evaluate) The sample experiment described above is an example of vicariance, because the experimenter fragmented the habitat into isolated cages. (A different mechanism of speciation may have been involved in your example.) **27.4** (apply) Hybrid individuals will increase in frequency, and natural selection will favor parents that hybridize. The hybrid zone will increase. The two parent populations may go extinct if hybrids live in the same environment.

✓ Test Your Knowledge

1. `remember` a 2. `remember` b 3. `understand` a 4. `understand` a
5. `understand` d 6. `remember` False

✓ Test Your Understanding

1. `analyze` Direct observation: Galápagos finch colonization, apple maggot flies, *Tragopogon* allopolyploids, maidenhair ferns. Indirect studies: snapping shrimp, sunflowers. 2. `understand` a 3. `understand` Based on biological and morphospecies criteria (breeding range and morphological traits, respectively), one species with six subspecies of seaside sparrows were recognized. However, based on phylogenetic criteria (comparison of gene sequences), biologists concluded that seaside sparrows represent only two monophyletic groups. 4. `evaluate` Some flies breed on apple fruits, others breed on hawthorn fruits. This reduces gene flow. Because apple fruits and hawthorn fruits have different scents and other characteristics, selection is causing the populations to diverge. 5. `apply` Decreasing. Gene flow tends to equalize allele frequencies among populations. 6. `apply` c

✓ Test Your Problem-Solving Skills

1. `analyze` b 2. `create` Sexual selection can promote divergence because it affects gene flow directly. For example, if blond women began mating only with blond men, and dark-haired women began mating only with dark-haired men, gene flow would decrease between these two populations even though they live in the same geographic area. Given enough time, these two populations would diverge due to mutation, genetic drift, and natural selection. 3. `evaluate` Two things: (1) Some flies happen to have alleles that allow them to respond to apple scents; others happen to have alleles that allow them to respond to hawthorn scents. There is no "need" involved. (2) The alleles for scent response exist; they are not acquired by spending time on the fruit. 4. `analyze` If the isolated populations and habitat fragments are small enough, the species is likely to dwindle to extinction due to inbreeding and loss of genetic variation or catastrophes like a severe storm or a disease outbreak. If the isolated populations survive, they are likely to diverge into new species because they are genetically isolated and because the habitats may differ.

CHAPTER 28

IN-TEXT QUESTIONS AND EXERCISES

p. 507 Fig. 28.1 `apply` Many trees are possible. See **FIGURE A28.1**.

p. 510 `understand` Because whales and hippos share SINEs 4–7 and no other species has these four, they are synapomorphies that define them as a monophyletic group. The similarity is unlikely to be due to homoplasy because the chance of four SINEs inserting in exactly the same place in two different species, independently, is very small.

p. 510 CYU `analyze` Hair and limb structures in humans and whales are examples of homology because they are traits that can be traced to a common ancestor. All mammals have hair and similar limb bone structure. However, extensive hair loss and advanced social behavior in whales and humans are examples of homoplasy. These traits are not common to all mammalian species and likely arose independently during the evolution of specific mammalian lineages.

p. 511 Fig. 28.4 `apply` See **FIGURE A28.2**. SINEs group (19, 20) identifies peccaries and pigs as a monophyletic group.

p. 513 `apply` Since some organisms fossilize better than others—such as mollusks with hard shells—a paleontologist would be mistaken to conclude that mollusks were more abundant than other organisms at a certain time just because mollusks are more abundant in the fossil

record. Similarly, a paleontologist would be mistaken to conclude that no organisms occupied an ancient desert based on the lack of fossil evidence; organisms in deserts do not fossilize well. Several other examples are possible.

p. 517 `create` In habitats on the mainland, complex communities have been established for a long time, so tarweeds in California and *Anolis* lizards on the mainland experience greater competition for resources and are limited to specific niches.

p. 520 CYU `understand` In terms of ecological opportunity, many resources were available because no other animals (or other types of organisms) existed to exploit them. Also, the evolution of new species in new niches made new niches available for predators. Morphological innovations like limbs and complex mouthparts were important because they made it possible for animals to live in habitats other than the benthic area, and to consume new types of food.

p. 520 Fig. 28.14 `analyze` The dinosaurs went extinct during the end-Cretaceous. About 15 percent of families went extinct.

IF YOU UNDERSTAND . . .

28.1 `apply` Humans are the only mammal that walks upright on two legs, so this character is a synapomorphy that defines the hominins. **28.2** `apply` Snail shells are hard structures, and snails often live in marine environments that fossilize readily, so snail fossils are relatively abundant. Abundant fossils are more helpful than rare fossils for estimating changes in relative abundance over time. **28.3** `analyze` They could eat food that was available off the substrate. And once animals lived off (and away from) the bottom, it created an opportunity for other animals who could eat them to evolve. **28.4** `evaluate` This statement is false. Human-induced extinctions today are not due to poor adaptation to the normal environment. Humans are changing habitats rapidly and in unusual ways (such as clear-cutting forests), so most species cannot adapt to these changes rapidly enough to avoid going extinct. In this way, human-induced extinctions are more similar to mass extinctions than to background extinctions.

YOU SHOULD BE ABLE TO . . .

✓ Test Your Knowledge

1. `remember` c 2. `understand` a 3. `understand` b 4. `understand` False
5. `understand` d 6. `remember` True

✓ Test Your Understanding

1. `evaluate` The fossil record is biased because recent, abundant organisms with hard parts that live underground or in environments where sediments are being deposited are most likely to fossilize. Even so, it is the only data available on what organisms that lived in the past looked like, and where they lived. 2. `create` An enormous amount of animal diversity appeared in a time frame that was short compared to the sweep of earlier Earth history. The most common analogy is to an explosion, referring to the sudden and spectacular radiation of lineages. Another analogy is your hand, where one line (your arm) comes into a node (your palm) and several lineages (your fingers) come out. Many other answers are possible. 3. `understand` Homoplasy is rare relative to homology, so phylogenetic trees that minimize the total change required are usually more accurate. In the case of the artiodactyl astragalus, parsimony was misleading because the astragalus was lost when whales evolved limblessness—creating two changes (a loss following a gain) instead of one (a gain). 4. `create` (Many answers are possible.) Adaptive radiation of *Anolis* lizards after they colonized new islands in the Caribbean. *Hypothesis:* After lizards arrived on each new island, where there were no predators or competitors, they rapidly diversified to occupy four distinct types of habitats on each island. Adaptive radiation during the Cambrian period

following a morphological innovation. *Hypothesis:* Additional *Hox* genes made it possible to organize a large, complex body; the evolution of complex mouthparts and limbs made it possible for animals to move and find food in new ways. 5. `apply` About 90 percent of species went extinct during the end-Permian, so about 10 percent survived, written as 0.1 in decimals. $1880 \times 0.1 = 188$ would make it to graduation. 6. `apply` d

✓ Test Your Problem-Solving Skills

1. `understand` b 2. `evaluate` Many trees are possible. See **FIGURE A28.3**. Sponges are the outgroup. The most parsimonious tree has the fewest number of changes (five in this case). Adding more characters to the matrix can improve the reliability of the results. 3. `understand` It would be helpful to have (1) a large crater or other physical evidence from a major impact dated at 251 mya, (2) shocked quartz and microtektites in rock layers dating to the end-Permian era, and (3) high levels of iridium or other elements common in space rocks and rare on Earth, dated to the time of the extinctions. 4. `create` Many concept map topologies are possible. Maps should follow this logic: Oxygen is an effective final electron acceptor in cellular respiration because of its high electronegativity. Organisms that use it as a final electron acceptor can produce more usable energy than organisms that do not use oxygen, but only if it is available. With more available energy, aerobic organisms can grow larger and move faster.

B I G PICTURE Evolution

p. 526 CYU (1) `understand` Circle = inbreeding, sexual selection, nonrandom mating, natural selection, genetic drift,

---

FIGURE A28.1

FIGURE A28.2

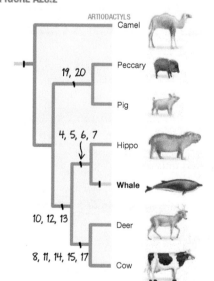

mutation, and gene flow. **(2)** analyze Adaptation "increases" fitness; synapomorphies "identify branches on" the tree of life. **(3)** analyze Several answers possible, e.g., adaptive radiations and mass extinctions → can be studied using → the fossil record. **(4)** analyze Genetic drift, mutation, and gene flow "are random with respect to" fitness.

CHAPTER 29

IN-TEXT QUESTIONS AND EXERCISES

p. 529 Fig. 29.1 analyze Prokaryotic—it would require just one evolutionary change, the origin of the nuclear envelope in Eukarya. If it were eukaryotic, it would require that the nuclear envelope was lost in both Bacteria and Archaea—two changes are less parsimonious (see Chapter 28).

p. 529 Table 29.1 apply See FIGURE A29.2.

p. 534 Fig. 29.5 analyze Weak—different culture conditions may have revealed different species.

p. 535 CYU (1) create Conditions should mimic a spill—sand or stones with a layer of crude oil, or seawater with oil floating on top. Add samples, from sites contaminated with oil, that might contain cells capable of using molecules in oil as electron donors or electron acceptors. Other conditions (temperature, pH, etc.) should be realistic. **(2)** create Use metagenomic analysis. After isolating DNA from a soil sample, fragment and sequence the DNA. Compare the sequences to those from known organisms and use the data to place the species on the tree of life.

p. 538 apply cyanobacteria—photoautotrophs; *Clostridium aceticum*—chemoorganoautotroph; *Nitrosopumilus* sp.—chemolithoautotrophs; helicobacteria— photoheterotrophs; *Escherichia coli*—chemoorganoheterotroph; *Beggitoa*—chemolithotrophic heterotrophs.

p. 539 Fig. 29.10 apply Table 29.5 contains the answers. For example, for organisms called sulfate reducers you would have H_2 as the electron donor, SO_4^{2-} as the electron acceptor, and H_2S as the reduced by-product. For humans the electron donor is glucose ($C_6H_{12}O_6$), the electron acceptor is O_2, and the reduced by-product is water (H_2O).

p. 541 understand Different species of bacteria that have different forms of bacterial chlorophyll will not absorb the same wavelengths of light and will therefore not compete.

p. 542 Fig. 29.12 apply Aerobic respiration. More free energy is released when oxygen is the final electron acceptor than when any other molecule is used, so more ATP can be produced and used for growth.

p. 543 apply If bacteria and archaea did not exist, then (1) the atmosphere would have little or no oxygen, and (2) almost all nitrogen would exist in molecular form (the gas N_2).

p. 543 Fig. 29.13 apply Add a label for "animals" in the upper right quarter of the circle, and draw arrows leading from "Organic compounds with amino groups" to "animals" and from "animals" to "NH_3."

p. 544 CYU evaluate Eukaryotes can only (1) fix carbon via the Calvin–Benson pathway, (2) use aerobic respiration with organic compounds as electron donors, and (3) perform oxygenic photosynthesis. Among bacteria and archaea, there is much more diversity in pathways for carbon fixation, respiration, and photosynthesis, along with many more fermentation pathways.

p. 545a, p. 545b, p. 546a, p. 546b, p. 547 apply See FIGURE A29.1.

p. 548 Fig. 29.21 apply About 600,000 *N. maritimus* cells would fit end to end on a meter stick.

IF YOU UNDERSTAND . . .

29.1 analyze Eukaryotes have a nuclear envelope that encloses their chromosomes; bacteria and archaea do not. Bacteria have cell walls that contain peptidoglycan, and archaea have phospholipids containing isoprene subunits in their plasma membranes. Thus, the exteriors of a bacterium and archaeon are radically different. Archaea and eukaryotes also have similar machinery for processing genetic information. **29.2** apply After extracting and sequencing DNA from bacteria that live in an insect gut, analyze the resulting sequences for genes that are similar to known genes involved in nitrogen fixation. **29.3** understand They are compatible, and thrive in each others' presence—one species' waste product is the other species' food.

YOU SHOULD BE ABLE TO . . .

✔ Test Your Knowledge

1. understand c **2.** remember b **3.** remember d **4.** remember a **5.** remember False—the enzyme nitrogenase is poisoned by oxygen. **6.** remember peptidoglycan

✔ Test Your Understanding

1. understand c **2.** understand An electron donor provides the potential energy required to produce ATP. **3.** evaluate Yes. The array of substances that bacteria and archaea can use as electron donors, electron acceptors, and fermentation substrates, along with the diversity of ways that they can fix carbon and perform photosynthesis, allows them to live just about anywhere. **4.** apply They should get energy from reduced organic compounds "stolen" from their hosts. **5.** understand Large amounts of potential energy are released and ATP produced when oxygen is the electron acceptor, because oxygen is so electronegative.

FIGURE A28.3

Five changes: most parsimonious

Six changes

Eleven changes

FIGURE A29.1

FIGURE A29.2

Large body size and high growth rates are not possible without large amounts of ATP. **6.** [understand] They are paraphyletic because the prokaryotes include some (bacteria and archaea) but not all (eukaryotes) groups derived from the common ancestor of all organisms living today.

✓ Test Your Problem-Solving Skills

1. [understand] c (It is pathogenic on a wide variety of organisms). **2.** [analyze] This result supports their hypothesis because the drug poisons the enzymes of the electron transport chain and prevents electron transfer to Fe^{3-}, which is required to drive magnetite synthesis. If magnetite had still formed, another explanation would have been needed. **3.** [create] Hypothesis: A high rate of tooth cavities in Western children is due to an excess of sucrose in the diet, which is absent from the diets of East African children. To test this hypothesis, switch a group of East African children to a diet that contains sucrose, switch a group of Western children to a diet lacking sucrose, and monitor the presence of *S. mutans* in both groups of children. **4.** [create] Look in waters or soil polluted with benzene-containing compounds. Put samples from these environments in culture tubes where benzene is the only source of carbon. Monitor the cultures and study the cells that grow efficiently.

CHAPTER 30

IN-TEXT QUESTIONS AND EXERCISES

p. 529 CYU (1) [understand] Opisthokonts have a flagellum at the base or back of the cell; alveolate cells contain unique support structures called alveoli; stramenopiles—meaning "straw hairs"—have straw-like hairs on their flagella. **(2)** [understand] In direct sequencing, DNA is isolated directly from the environment and analyzed to place species on the tree of life. It is not necessary to actually see the species being studied.
p. 557 CYU (1) [understand] *Plasmodium* species are transmitted to humans by mosquitoes. If mosquitoes can be prevented from biting people, they cannot spread the disease. **(2)** [apply] Iron added → primary producers (photosynthetic protists and bacteria) bloom → more carbon dioxide taken up from atmosphere during photosynthesis → consumers bloom, eat primary producers → bodies of primary producers and consumers fall to bottom of ocean → large deposits of carbon-containing compounds form on ocean floor.
p. 560 [apply] See FIGURE A30.1.
p. 560 Fig. 30.8 [analyze] Two—one derived from the original bacterium and one derived from the eukaryotic cell that engulfed the bacterium.
p. 561 [apply] A photosynthetic bacterium (e.g., a cyanobacterium) could have been engulfed by a larger eukaryotic cell. If it was not digested, it could continue to photosynthesize and supply sugars to the host cell.
p. 562 [apply] Membrane infoldings observed in bacterial species today support the hypothesis's plausibility—they confirm that the initial steps actually could have happened. The continuity of the nuclear envelope and ER are consistent with the hypothesis, which predicts that the two structures are derived from the same source (infolded membranes).
p. 564 [apply] The chloroplast genes label should come off of the branch that leads to cyanobacteria. (If you had a phylogeny of just the cyanobacteria, the chloroplast branch would be located somewhere inside.)
p. 565 [analyze] The acquisition of the mitochondrion and the chloroplast represent the transfer of entire genomes, and not just single genes, to a new organism.
p. 566 [apply] Yes—when food is scarce or population density is high, the environment is changing rapidly (deteriorating). Offspring that are genetically unlike their parents may be better able to cope with the new and challenging environment.

p. 569a [understand] (1) Alternation of generations refers to a life cycle in which there are multicellular haploid phases and multicellular diploid phases. A gametophyte is the multicellular haploid phase that produces gametes; the sporophyte is the multicellular diploid phase that produces spores. A spore is a cell that grows into a multicellular individual but is not produced by fusion of two cells. A zygote is a cell that grows into a multicellular individual but *is* produced by fusion of two cells (gametes). Gametes are haploid cells that fuse to form a zygote. (2) See FIGURE A30.2.
p. 569b [apply] Most likely, the two types of amoebae evolved independently. The alternative hypothesis is that the common ancestor of alveolates, stramenopiles, rhizarians, plants, opisthokonts, and amoebozoa were amoeboid and that this growth form was lost many times. See FIGURE A30.1.
p. 569c, p. 569d, p. 569e, p. 570a, p. 570b, p. 570c [apply] See FIGURE A30.1.
p. 571 [analyze] If euglenids could take in food via phagocytosis (ingestive feeding), then that would have provided a mechanism by which a smaller photosynthetic protist could have been engulfed and incorporated into the cell via secondary endosymbiosis.
p. 572a, p. 572b, p. 573, p 574. [apply] See FIGURE A30.1.

IF YOU UNDERSTAND . . .

30.1 [understand] Photosynthetic protists use CO_2 and light to produce sugars and other organic compounds, so they furnish the first or primary source of organic material in an ecosystem. **30.2** [remember] Eukaryotic cells contain many organelles, including a nucleus, endomembrane system, and an extensive cytoskeleton. Most prokaryotic cells contain few or no organelles, and no nucleus or endomembrane system and lack a complex cytoskeleton. **30.3** [understand] (1) Outside membrane was from host eukaryote; inside from engulfed cyanobacterium. (2) From the outside in, the four membranes are derived from the eukaryote that engulfed a chloroplast-containing eukaryote, the plasma membrane of the eukaryote that was engulfed, the outer membrane of the engulfed cell's chloroplast, and the inner membrane of its chloroplast.

YOU SHOULD BE ABLE TO . . .

✓ Test Your Knowledge

1. [understand] b **2.** [remember] b **3.** [remember] a **4.** [remember] b **5.** [remember] False. **6.** [remember] diatoms.

✓ Test Your Understanding

1. [understand] a **2.** [understand] Because all eukaryotes living today have cells with a nuclear envelope, it is valid to infer that their common ancestor also had a nuclear envelope. Because bacteria and archaea do not have a nuclear envelope, it is valid to infer that the trait arose in the common ancestor of eukaryotes. **3.** [understand] The host cell provided a protected environment and carbon compounds for the endosymbiont; the endosymbiont provided increased ATP from the carbon compounds **4.** [understand] It confirmed a fundamental prediction made by the hypothesis and could not be explained by any alternative hypothesis. **5.** [understand] All alveolates have alveoli, which are unique structures that function in supporting the cell. Among alveolates there are species that are (1) ingestive feeders, photosynthetic, or parasitic, and that (2) move using cilia, flagella, or a type of amoeboid movement.

✓ Test Your Problem-Solving Skills

1. [analyze] d **2.** [analyze] If the apicoplast that is found in *Plasmodium* (the organism that causes malaria) is genetically similar to chloroplasts, and if glyphosate poisons chloroplasts, it is reasonable to hypothesize that glyphosate will poison the apicoplast and potentially kill the *Plasmodium*. This would be a good treatment strategy

for malaria because humans have no chloroplasts, provided that the glyphosate produces no other effects that would be detrimental to humans. **3.** [evaluate] Primary producers usually grow faster when CO_2 concentration increases, but to date they have not grown fast enough to make CO_2 levels drop—CO_2 levels have been increasing steadily over decades. **4.** [analyze] Given that lateral gene transfer can occur at different points in a phylogenetic history, specific genes can become part of a lineage by a different route from that taken by other genes of the organism. In the case of chlorophyll *a*, its history traces back to a bacterium being engulfed by a protist and forming a chloroplast.

FIGURE A30.1

FIGURE A30.2

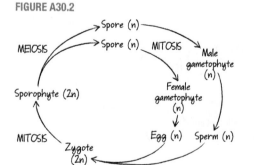

CHAPTER 31

IN-TEXT QUESTIONS AND EXERCISES

p. 582 CYU (understand) (1) Green algae and land plants share an array of morphological traits that are synapomorphies, including the chlorophylls they contain; (2) green algae appear before land plants in the fossil record; and (3) on phylogenetic trees estimated from DNA sequence data, green algae and land plants share a most recent common ancestor—green algae are the initial groups to diverge, and land plants diverge subsequently.

p. 583 Fig. 31.6 (remember) If you find the common ancestor of all the green algae (at the base of the Ulvophyceae), the lineages that are collectively called green algae don't include that common ancestor and all of its descendants—only some of its descendants. The same is true for non-vascular plants (the common ancestor here is at the base of the Hepaticophyta) and seedless vascular plants (the common ancestor here is at the base of the Lycophyta).

p. 586 Fig. 31.8 (understand) Water flows more easily through a short, wide pipe than through a long, skinny one because there is less resistance from the walls of the pipe.

p. 587 Fig. 31.9 (understand) To map an innovation on a phylogenetic tree, biologists determine the location(s) on the tree that is (are) consistent with all descendants from that point having the innovation (unless, in rare cases, the innovation was lost in a few descendants).

p. 589 (understand) Alternation of generations occurs when there are multicellular haploid individuals and multicellular diploid individuals in a life cycle. One hypothesis is that this life cycle evolved because the plants with a large sporophyte were more successful at using the wind to disperse many spores, while a small gametophyte produces gametes that can swim.

p. 590 (apply) In the hornwort photo, the sporophyte is the spike-like green and yellow structure; the gametophyte is the leafy-looking structure underneath. The horsetail gametophyte is the microscopic individual on the left; the sporophyte is the much larger individual on the right.

p. 590 Fig. 31.14 (understand) There are multicellular haploid stages and multicellular diploid stages in these plants.

p. 592 Fig. 31.18 (analyze) Gymnosperm gametophytes are microscopic, so they are even smaller than fern gametophytes. The gymnosperm gametophyte is completely dependent on the sporophyte for nutrition, while fern gametophytes are not. Fern gametophytes are photosynthetic and even supply nutrition to the young sporophytes.

p. 593 Fig. 31.19 (analyze) Consistent—the fossil data suggest that gymnosperms evolved earlier, and gymnosperms have larger gametophytes than angiosperms.

p. 595 Fig. 31.21 (analyze) To eliminate all other variables except for the one being tested, such as the possibility of pollination by other insects in the field. Scales differ because the two insect species visited flowers at very different frequencies.

p. 596 CYU (1) (understand) Cuticle prevents water loss from the plant; UV-absorbing compounds allow plants to be exposed to high light intensities without damage to their DNA; vascular tissue moves water up from the soil and moves photosynthetic products down to the roots.
(2) (apply) See FIGURE A31.1.
p. 600a, p. 600b, p. 601, p. 603a, p. 603b, p. 604, p. 605, p. 606 (apply) See FIGURE A31.1.
p. 596 Fig. 31.23 (apply) See FIGURE A31.1.

IF YOU UNDERSTAND . . .

31.1 (apply) Rates of soil formation will decline; rates of soil loss will increase. **31.2** (understand) All green algae and land plants contain chlorophyll *a* and *b*, have similar arrangements of thylakoid membranes, and store starch.
31.3 (analyze) Both spores and seeds have a tough, protective coat, so they can survive while being dispersed to a new location. Seeds have the advantage of carrying a store of nutrients with them—when a spore germinates, it must make its own food via photosynthesis right away.

YOU SHOULD BE ABLE TO . . .

✔ Test Your Knowledge

1. (remember) c **2.** (remember) c **3.** (remember) a **4.** (remember) c **5.** (remember) (a) Gametangia are found in all land plant groups except angiosperms; (b) pollen is found in gymnosperms and angiosperms; (c) seeds are found in gymnosperms and angiosperms; (d) fruit is found in angiosperms.
6. (understand) In a gametophyte-dominant life cycle, the gametophyte is larger and longer lived than the sporophyte and produces most of the nutrition. In a sporophyte-dominant life cycle, the sporophyte generation is the larger, longer-lived, and photosynthetic phase of the life cycle.

✔ Test Your Understanding

1. (remember) c **2.** (understand) Plants build and hold soils required for human agriculture and forestry, and they increase water supplies that humans can use for drinking, irrigation, or industrial use. Plants release oxygen that we breathe. **3.** (analyze) Cuticle prevents water loss from leaves but also prevents entry of CO_2 required for photosynthesis. Stomata allow CO_2 to diffuse but can close to minimize water loss. Liverwort pores allow gas exchange but cannot be closed if conditions become dry. **4.** (understand) They provided the support needed for plants to grow upright and not fall over in response to wind or gravity. Erect growth allowed plants to compete for light. **5.** (apply) Homosporous plants produce a single type of spore that produces both egg and sperm. Heterosporous plants produce two different types of spores that develop into two different gametophytes that produce either egg or sperm. In a tulip, the microsporangium is found within the stamen, and the megasporangium is found within the ovule. Microspores divide by mitosis to form male gametophytes (pollen grains); megaspores divide by mitosis to form the female gametophyte. **6.** (understand) Animal pollinators increase the efficiency of pollination so plants can save energy by making less pollen.

✔ Test Your Problem-Solving Skills

1. (understand) d **2.** (analyze) A "reversion" to wind pollination might be favored by natural selection because it is costly to produce a flower that can attract animal pollinators. Because wind-pollinated species grow in dense clusters, they can maximize the chance that the wind will carry pollen from one individual to another (less likely if the individuals are far apart). Wind-pollinated deciduous trees flower in early spring before their developing leaves begin to block the wind. **3.** (create) The combination represents a compromise between efficiency and safety. Tracheids can still transport water if vessels become blocked by air bubbles. **4.** (create) Alter one characteristic of a flower, and present the flower to the normal pollinator. As a control, present the normal (unaltered) flower to the normal pollinator. Record the amount of time the pollinator spends in the flower, the amount of pollen removed, or some other measure of pollination success. Repeat for other altered characteristics. Analyze the data to determine which altered characteristic affects pollination success the most.

CHAPTER 32

IN-TEXT QUESTIONS AND EXERCISES

p. 619 CYU (1) (understand) Because they are made up of a network of thin, branching hyphae, mycelia have a large surface area, which makes absorption efficient.
(2) (remember) Swimming spores and gametes, zygosporangia, basidia, asci.

FIGURE A31.1

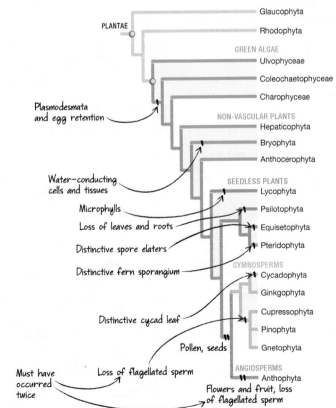

p. 622 Fig. 32.10 *analyze* Labeled nutrient experiments identify which nutrients are exchanged and in which direction. They explain *why* plants do better in the presence of mycorrhizae, and why the fungus also benefits. **p. 625** *analyze* Human sperm and egg undergo plasmogamy followed by karyogamy during fertilization, but heterokaryotic cells do not occur in humans or other eukaryotes besides fungi. **p. 625 Fig. 32.12** *apply* The haploid mycelium. **p. 626 Fig. 32.13** *analyze* Mycelia produced by asexual reproduction are genetically identical to their parent; mycelia produced by sexual reproduction are genetically different from both parents (each spore has a unique genotype). **p. 628 CYU (1)** *understand* When birch tree seedlings are grown in the presence and absence of EMF, individuals denied their normal EMF are not able to acquire sufficient nitrogen and phosphorus. **(2)** *analyze* Each type of fungus involves meiosis and the production of haploid spores. In a zygosporangium, meiosis occurs in a multinucleate cell; in basidia and asci, this cell has a single diploid nucleus. In asci, meiosis is followed by one round of mitosis. **p. 628, p. 630a, p. 630b, p. 630c, p. 631, p. 632a, p. 632b** *apply* See **FIGURE A32.1**.

IF YOU UNDERSTAND . . .

32.1 *apply* Loss of mycorrhizal fungi would decrease nutrient delivery to plants and reduce their growth inside the experimental plots, compared to control plots with intact fungi. Also, lack of fungi would slow decay of dead plant material, causing a dramatic buildup of dead organic material. **32.2** *understand* Like humans, fungi are composed of eukaryotic cells. Bacteria are prokaryotic and thus have many more unique targets for antibiotics. **32.3** *understand* Haploid hyphae fuse, forming a heterokaryotic mycelium. When karyogamy occurs, a diploid nucleus forms—just as when gametes fuse. **32.4** *understand* See the diversity boxes.

YOU SHOULD BE ABLE TO . . .

✓ Test Your Knowledge
1. *remember* a **2.** *remember* d **3.** *remember* b **4.** *remember* b **5.** *understand* Pores in septa allow nutrients to move from regions of acquisition to regions of mycelial growth; **6.** *remember* commensal.

✓ Test Your Understanding
7. *understand* c **8.** *evaluate* Along with a few bacteria, fungi are the only organisms that can digest wood completely. If the wood is not digested, carbon remains trapped in wood. Without fungi, CO_2 would be tied up and unavailable for photosynthesis, and the presence of undecayed organic matter would reduce the space available for plants to grow. **9.** *analyze* Fungi produce enzymes that degrade cellulose and lignin. **10.** *analyze* Plant roots have much smaller surface area than EMF or AMF. Hyphae are much smaller than the smallest portions of plant roots so can penetrate dead material more efficiently. Extracellular digestion—which plant roots cannot do—allows fungi to break large molecules into small compounds that can be absorbed. **11.** *apply* DNA and RNA contain phosphorus. Mycorrhizal fungi effectively increase the volume of soil mined by root systems and can make adequate amounts of phosphorus available for supporting rapid cell division. **12.** *analyze* Both compounds are processed via extracellular digestion. Different enzymes are involved, however. The degradation of lignin is uncontrolled and does not yield useful products; digestion of cellulose is controlled and produces useful glucose molecules.

✓ Test Your Problem-Solving Skills
13. *apply* c **14.** *create* (1) Confirm that the chytrid fungus is found only in sick frogs and not healthy frogs.

(2) Isolate the chytrid fungus and grow it in a pure culture. (3) Expose healthy frogs to the cultured fungus and see if they become sick. (4) Isolate the fungus from the experimental frogs, grow it in culture, and test whether it is the same as the original fungus. **15.** *create* Each of the different cellulase enzymes attacks cellulose in a different way, so producing all the enzymes together increases the efficiency of the fungus in breaking down cellulose completely. It is likely that lignin peroxidase is produced along with cellulases, so they can act in concert to degrade wood. To test this idea, you could harvest enzymes secreted from a mycelium before and after it contacts wood in a culture dish, and see if the cellulases and lignin peroxidase appear together once the mycelium begins growing on the wood. **16.** *create* Collect a large array of colorful mushrooms that are poisonous; also capture mushroom-eating animals, such as squirrels. Present a hungry squirrel with a choice of mushrooms that have been dyed or painted a drab color versus others treated with a solution that is identical to the dye or paint used but uncolored. Record which mushrooms the squirrel eats. Repeat the test with many squirrels and many mushrooms.

CHAPTER 33

IN-TEXT QUESTIONS AND EXERCISES

p. 639 Fig. 33.2 *apply* The label and bar for "Segmentation?" should go on the branch to the left of the clam shell and "Mollusca" label. The label and bar for "Toolkit genes for segmentation?" should go on the same branch as cephalization, CNS, and coelom. **p. 640** *apply* The genetic tool kit required for multicellularity includes genes that regulate the cell cycle, and thus cell division. When something goes wrong in this process, uncontrolled growth called cancer can result. Thus, cancer may be as ancient as multicellularity itself. **p. 642 Fig. 33.7** *analyze* Stained *Hox* and *dpp* gene products would either not be found in *Nematostella* at all or would not occur in the same anterior–posterior and dorsal–ventral pattern as observed in bilaterians. **p. 646 CYU (1)** *apply* Yes, your body has a tube-within-a-tube design. Humans are vertebrates, vertebrates are chordates, chordates are bilaterians, and all bilaterians have a tube-within-a-tube design, consisting of a bilaterally symmetrical, slender body with an inner tube (gut), an outer tube (mostly skin), muscles and organs derived from mesoderm in between, and one end (head) cephalized with sensory organs and brain. **(2)** *understand* When an animal moves through an environment in one direction, its ability to acquire food and perceive and respond to threats is greater if its feeding, sensing, and information-processing structures are at the leading end. **p. 651 Fig. 33.14** *evaluate* The larva and metamorphosis arrow should be circled in the life cycle. It might be adaptive to skip the feeding larval stage if more food resources are available for the mother (who must expend energy to produce the yolk) than for the larva, if predation pressure on larvae is high, or if other environmental factors for larvae are unfavorable, resulting in high mortality of the larvae. **p. 652 CYU (1)** *understand* The mouthparts of deposit feeders are relatively simple because these animals simply gulp relatively soft material. The mouthparts of mass feeders are more complex because they have to tear off and process chunks of relatively hard material. **(2)** *create* Gametes that are shed into aquatic environments can float or swim. This cannot happen on land, so internal fertilization is more common there. Also, gametes are more at risk of drying out on land, and internal fertilization eliminates that risk. **p. 653** *apply* The origin of cilia-powered swimming should be marked on the Ctenophora branch.

p. 654 *apply* The origin of cnidocytes should be marked on the Cnidaria branch.

IF YOU UNDERSTAND . . .

33.1 *apply* It should increase dramatically—animals would no longer be consuming plant material. **33.2** *understand* Your drawing should show the tube-within-a-tube design of a worm, with one end labeled head and containing the mouth and brain. From the brain, one or two major nerve tracts should run the length of the body. The gut should go from the mouth to the other end, ending in the anus. In between the gut and outer body wall, there should be blocks or layers of muscle. **33.3** *apply* There are many possible answers. For example, within the phylum Arthropoda: The sensory systems have diversified (e.g., eyes are important for sight in flies, antennae are important for smelling in moths); feeding strategies have diversified (e.g., butterflies are fluid feeders but caterpillars are mass feeders); ecological roles have diversified (e.g., butterflies are herbivores but spiders are carnivores); limbs have diversified (e.g., butterflies have legs and wings but caterpillars have lobe-like limbs); and some insects, like grasshoppers, have direct development whereas others, like mosquitoes, have indirect development. **33.4** *remember* If you see a

FIGURE A32.1

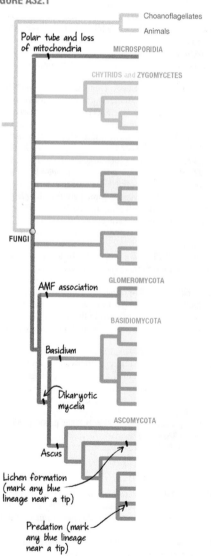

sessile, multicellular animal with no symmetry or radial symmetry, with outcurrent holes, and choanocytes, it is a sponge. If you see a radially symmetric, transparent, planktonic organism moving via rows of cilia, it is a ctenophore. If you see a diploblastic, outwardly radially symmetric animal that is either a single polyp or colony of multiple polyps, or it is a medusa, then the animal is a cnidarian.

YOU SHOULD BE ABLE TO . . .

✓ Test Your Knowledge

1. `remember` a **2.** `remember` True **3.** `remember` protostomes, deuterostomes **4.** `understand` a **5.** `remember` d **6.** `remember` d

✓ Test Your Understanding

1. `understand` Diploblasts have two types of embryonic tissue (ectoderm, endoderm); triploblasts have three (ectoderm, mesoderm, endoderm). Mesoderm made it possible for an enclosed, mesoderm-lined cavity to develop, creating a coelom. **2.** `analyze` Many unicellular organisms are heterotrophic, but they can consume only small bits of food. Animals are multicellular, so they are larger and can consume larger amounts of food—making them important consumers in food webs. **3.** `apply` There are about 18.4 named arthropods for every named species of chordate (1,160,000/63,000), about 1.3 named mollusks for every named species of chordate (85,000/63,000), and about 0.4 named species of nematode worm for every named species of chordate (25,000/63,000). These numbers are likely to be underestimates because there are probably many more arthropods, mollusks, and nematode worms than have been described, yet relatively few of the larger, more familiar chordates (including vertebrates) left to be described. **4.** `understand` In oviparous species, the mother adds nutrient-rich yolk to the egg that nourishes the developing embryo. In viviparous species, the mother transfers nutrients directly from her body to the growing embryo. **5.** `evaluate` You could stain the gene products of the segmentation genes in chordates (such as a mouse embryo) and observe whether the expression pattern correlates with the expression pattern observed in arthropods (such as a fly embryo). **6.** `apply` c

✓ Test Your Problem-Solving Skills

1. `analyze` Yes—if the same gene is found in nematodes and humans, it was likely found in the common ancestor of protostomes and deuterostomes. If so, then fruit flies should also have this gene. **2.** `apply` Asexual reproduction is likely to be favored when the sea anemones have plenty of resources such as space and food. Sexual reproduction is likely to be favored when the sea anemones are stressed due to lack of resources, predation, infection, or other factors. In this case, it is adaptive to reproduce in a way that enables dispersal to a new site and/or genetically diverse offspring, some of which will be more likely to survive selection than a clone of identical animals would. **3.** `apply` Since echinoderms evolved from bilaterally symmetric ancestors with nerve cords, they should have nerve cords too. However, the organization of the nerve cords should be more diffuse, because echinoderms need to take in and process information from multiple directions—not just one. **4.** `apply` d

CHAPTER 34

IN-TEXT QUESTIONS AND EXERCISES

p. 659 Fig. 34.3 `apply` We know that aquatic living was the ancestral trait because the outgroups (e.g., Cnidaria, Porifera) were aquatic. We also have fossil evidence that the aquatic lifestyle was ancestral.
p. 660, p. 662, p. 665, p. 666, p. 667, p. 671, p. 674 `apply` See FIGURE A34.1.

p. 662 Fig. 34.6 `evaluate` Different phyla of worms have different feeding strategies and mouthparts, and they eat different foods, so they are not all in direct competition for food.
p. 664 CYU (1) `apply` Snails on land would be expected to have gills or lungs inside the body or otherwise protected from drying out, thinner shells so that they are not as heavy, and desiccation-resistant eggs. **(2)** `evaluate` The tentacles of the octopus could be probed during development for *Hox* genes known to be important in establishing the foot of snails. If the expression pattern is similar, homology would be supported.
p. 664 Fig. 34.10 `apply` You should have drawn a circle around all the tentacles.

IF YOU UNDERSTAND . . .

34.1 `remember` Ecological opportunity was important in the diversification of protostomes because many lineages made the transition from water to land, radiating into new niches. Genetic opportunity was important because changes in the expression of developmental tool-kit genes enabled diverse modifications of body plans based on existing genes. **34.2** `understand` One source of diversification is the loss of ancestral traits; just because a species does not possess a trait does not mean that its ancestor didn't. **34.3** `analyze` Both arthropods and vertebrates have segmented bodies, which are especially amenable to modularity since a change in the expression pattern of developmental tool-kit genes can change the number, size, and type of segments, exposing morphological diversity to the adaptive forces of natural selection.

YOU SHOULD BE ABLE TO . . .

✓ Test Your Knowledge

1. `remember` False **2.** `remember` a **3.** `remember` c **4.** `understand` c **5.** `remember` Both groups are bilaterally symmetric triploblasts with the protostome pattern of development. **6.** `remember` b

✓ Test Your Understanding

1. `remember` d **2.** `apply` Both annelids and arthropods have segmented bodies, unlike other protostomes such as flatworms and nematodes. **3.** `evaluate` Clams are relatively sessile as adults. Because larvae can swim, they provide a mechanism to disperse to new locations. To test this idea, you could measure the distance traveled by clam larvae and measure their success in recruiting to new habitats compared to larvae that are artificially limited to stay near their parents. **4.** `apply` Land plants would require cuticles or other mechanisms to keep their tissues from drying out, a mechanism of keeping gametes from

FIGURE A34.1

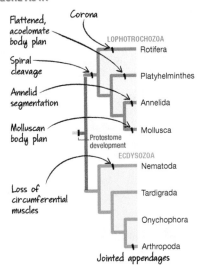

drying out, and structural support to retain body shape despite the effects of gravity. **5.** `evaluate` The ability to fly allowed insects to disperse to new habitats and find new food sources efficiently. **6.** `analyze` It might be adaptive for juveniles of hemimetabolous individuals to use the same food source as the adults when the food source is plentiful.

✓ Test Your Problem-Solving Skills

1. `analyze` See FIGURE A34.2. **2.** `analyze` If the ancestors of brachiopods and mollusks lived in similar habitats and experienced natural selection that favored similar traits, then they would have evolved to have similar forms and habitats. This is called convergent evolution (see Chapter 28). **3.** `evaluate` This hypothesis is reasonable because simple characteristics often occur before more elaborate ones; this could be true for the radula and foot in aplacophorans. An alternate hypothesis would be that these structures' simple forms are derived characteristics that are adaptations for parasitism, but the evidence suggests that aplacophorans are carnivores and detritivores, not parasites. The position of the aplacophorans in the phylogeny could be tested by using DNA sequence data rather than morphological characters in a phylogenetic analysis. **4.** `apply` b

CHAPTER 35

IN-TEXT QUESTIONS AND EXERCISES

p. 682 Fig. 35.1 `apply` Invertebrates are a paraphyletic group. The group does not include all the descendants of the common ancestor because vertebrates are excluded.
p. 684 CYU `understand` If it's an echinoderm, it should have five-part radial symmetry, a calcium carbonate endoskeleton just underneath the skin, and a water vascular system (e.g., visible tube feet).
p. 684, p. 685a, p. 685b `apply` See FIGURE A35.1.

FIGURE A34.2

FIGURE A35.1

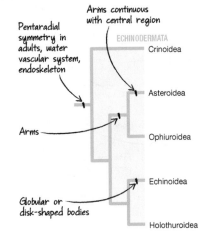

p. 690 Fig. 35.12 [apply] Mammals and birds are equally related to amphibians, because birds and mammals share a common ancestor, and this ancestor shares a common ancestor with amphibians.

p. 691 [apply] See FIGURE A35.2

p. 693 Fig. 35.16 [analyze] If this phylogeny were estimated on the basis of limb traits, it would be based on the *assumption* that the limb evolved from fish fins in a series of steps. But because the phylogeny was estimated from other types of data, it is legitimate to use it to analyze the evolution of the limb without making a circular argument.

p. 694 Fig. 35.18 [analyze] The yolk sac is smaller here—their function in an amniotic egg has been taken over by the placenta.

p. 695 CYU (1) [understand] Jaws allow animals to capture food efficiently and process it by crushing or tearing. The ability to process food more efficiently increased the importance of jawed vertebrates as herbivores and carnivores. The tetrapod limb enables vertebrates to locomote on land. The amniotic egg enables vertebrates to lay eggs on land; amniotic eggs are resistant to drying out. The placenta protects the growing fetus and enables the mother to be mobile. Parental care reduces the number of offspring that can be raised, but increases their survivorship. Flight enabled pterosaurs, bats, and birds to expand into new habitats. **(2)** [analyze] Protostomes and deuterostomes

living on land both have mechanisms for preventing their eggs from drying out (e.g., thick membranes in snail and insect eggs; amniotic egg in tetrapods), preventing their skin from drying out (e.g., waxy cuticle in insects, scaly skin in reptiles), and keeping their respiratory surfaces from drying out (e.g., lungs in land snails and tracheal system in insects; lungs in tetrapods).

p. 696 [apply] See FIGURE A35.3.

p. 698a, p. 698b, p. 699, p. 700 [apply] See FIGURE A35.4.

p. 701a, p. 701b, p. 702a, p. 702b, p. 703a, p. 703b [apply] See FIGURE A35.3.

p. 705 Table 35.1 [analyze] 44 percent, 26 percent; relative to body size, *H. floresiensis* has a smaller brain than *H. sapiens.*

p. 706 Fig. 35.37 [apply] Five hominin species existed 1.8 mya, and three existed 100,000 years ago.

p. 706 Fig. 35.38 [analyze] The forehead became much larger, and the face became "flatter"; the brow ridges are less prominent in later skulls than in earlier skulls.

IF YOU UNDERSTAND . . .

35.1 [apply] (1) Mussels and clams will increase dramatically; (2) kelp density will increase.

35.2 [understand] Your cartoon should have a dorsal hollow nerve chord along the top, dorsal edge; another line representing the notochord below the nerve cord; a tail

labeled at one end; and a series of lines or bumps representing pharyngeal gill slits near the other end.

35.3 [understand] Like today's lungfish, the earliest tetrapods could have used their limbs for pulling themselves along the substrate in shallow-water habitats. **35.4** [understand] The earliest fossils of *H. sapiens* are found in Africa, and phylogenetic analyses of living human populations indicate that the most basal groups are all African.

YOU SHOULD BE ABLE TO . . .

✔ Test Your Knowledge

1. [remember] True **2.** [remember] a **3.** [understand] d **4.** [remember] mammals **5.** [understand] d **6.** [remember] b

✔ Test Your Understanding

1. [understand] c **2.** [understand] Pharyngeal gill slits function in suspension feeding. The notochord furnishes a simple endoskeleton that stiffens the body; electrical signals that coordinate movement are carried by the dorsal hollow nerve cord to the muscles in the post-anal tail, which beats back and forth to make swimming possible. Cephalochordates and Urochordates are chordates, but they are not vertebrates. **3.** [analyze] If jaws are derived forms of gill arches, then the same genes and the same cells should be involved in the development of the jaw and the gill arches. **4.** [understand] Increased parental care allows

---

FIGURE A35.2

FIGURE A35.3

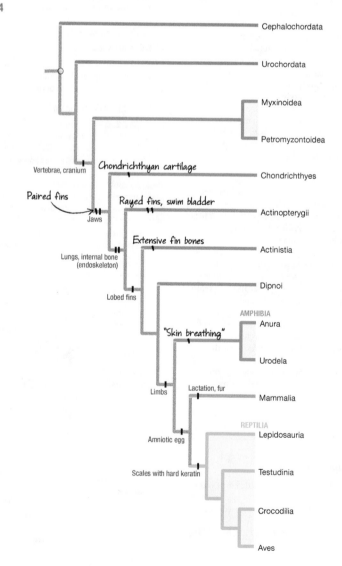

FIGURE A35.4

offspring to be better developed, and thus have increased chances of survival, before they have to live on their own. **5.** `evaluate` Yes: Over a short time interval, many species that occupy an array of foods and habitats evolved. Changes in tooth and jaw structure, tool use, and body size suggest that different hominin species exploited different types of food. **6.** `evaluate` A phylogeny of modern humans would reject the out-of-Africa hypothesis if the most ancestral lineages in the phylogeny were from any other region but Africa, such as Asia or Europe.

✔ Test Your Problem-Solving Skills

1. `apply` b **2.** `evaluate` Xenoturbellidans either retain traits that were present in the common ancestor of all deuterostomes, or they have lost many complex morphological characteristics. **3.** `evaluate` Vertebrate limbs are diverse in form and function (they can serve as arms, flippers, wings, etc.) but have homologous bones within. Molecular probes could be used to visualize the expression patterns of tool-kit genes like *Hox* in the embryonic limbs of different vertebrate species to show how changes in the expression of homologous genes can cause changes to the structure and function of limbs. **4.** `analyze` See **FIGURE A35.5**. It is important to use indigenous people for this study to minimize the amount of gene flow (transfer of alleles from one population to another) that occurred during the history of the study populations due to immigration and interbreeding.

CHAPTER 36

IN-TEXT QUESTIONS AND EXERCISES

p. 717 Fig. 36.7 `apply` Starting with 1 virion, approximately 80 virions were produced after 40 minutes. If all of these virions infected new cells, you expect each of the 80 infected cells to produce 80 virions by 90 minutes, or a total of 6400 virions. The cells appear to replicate once every 30 minutes, which means that they will have replicated 3 times during this period. The total number of cells at 90 minutes would be approximately 2^3, or 8 cells. There would be 800 times as many virions as cells.
p. 717 Fig. 36.8 `analyze` No—it only shows that CD4 is required. (Subsequent work showed that other proteins are involved as well.)
p. 721 `analyze` In T4, the viral genome is injected into the host while HIV inserts its genome via membrane fusion. HIV must first convert its genome into double-stranded DNA before replication, which is not necessary for T4, which has a DNA genome. Both viruses use double-stranded DNA to make mRNA that is translated into viral proteins. T4 virions are made in the cytosol of the cell and are released when the cell bursts. HIV virions bud from the cell surface, which does not require cell death.
p. 721 Fig. 36.12 `create` Budding may disrupt the integrity of the host-cell plasma membrane enough to kill the cell.
p. 722 CYU `analyze` Like cars in the assembly line of an automobile plant, new virions are assembled from premanufactured parts, and large numbers of progeny are produced per generation. Cells, on the other hand, reproduce by the division of a single integrated unit, resulting in only two progeny per generation.
p. 724 Fig. 36.15 `apply` You should have the following bars and labels: on branch to HIV-2 (sooty mangabey to human); on branch to HIV-1 strain O (chimp to human); on branch to HIV-1 strain N (chimp to human); on branch to HIV-1 strain M (chimp to human).

p. 725 CYU (1) `analyze` The escaped-genes and degenerate-cell hypotheses state that viruses originated from cells, while the RNA-world hypothesis suggests that viruses originated in parallel with, maybe even influenced, the origin of cells. The escaped-gene and RNA-world hypotheses state that the viruses originated from parasitic molecules. The degenerate-cell hypothesis states that viruses originated from parasitic organisms. **(2)** `apply` A mutation that made transmission between humans more efficient would make it more dangerous. Such mutation could occur via genomic reassortment in pigs, where an avian virus and a human virus could co-infect cells and produce recombinants.

IF YOU UNDERSTAND . . .

36.1 `understand` HIV infects helper T cells. These cells play a central role in the human immune response. As latently infected cells are activated by the immune response, HIV replication causes cell death. Over several years, steady decline in helper T cells leads to a compromised immune response that allows other infections to eventually kill the infected person. **36.2** `understand` (1) Fusion inhibitors block viral envelope proteins or host cell receptors. (2) Protease inhibitors prevent processing/assembly of viral proteins. (3) Reverse transcriptase inhibitors block reverse transcriptase, preventing replication of the genome. **36.3** `apply` There is heritable variation among virions, due to random changes that occur as their genomes are copied. There is also differential reproductive success among virions in their ability to successfully infect host cells. This differential success is due to the presence of certain heritable traits. **36.4** `analyze` Both classes have positive-sense single-stranded RNA genomes. Class IV viruses use an RNA replicase to make a negative-sense copy of the genome, which is then used as a template for producing mRNA and genomic RNAs. Class VI viruses use reverse transcriptase to convert their RNA genome into dsDNA, which is then integrated into the host genome. There it serves as a template to for transcribing viral mRNA and genomic RNAs by host-cell RNA polymerase.

YOU SHOULD BE ABLE TO . . .

✔ Test Your Knowledge

1. `remember` d **2.** `remember` b **3.** `understand` In class VI and VII viruses, reverse transcriptase converts positive-sense single-stranded RNA to double-stranded DNA.
4. `remember` b **5.** `remember` c **6.** `remember` The unique type of genetic material included in the virion (classes I–V) in addition to the manner in which the genome is replicated (compare classes VI and VII with classes IV and I, respectively).

✔ Test Your Understanding

7. `analyze` A virus with an envelope exits host cells by budding. A virus that lacks an envelope exits host cells by lysis (or other mechanisms that don't involve budding). **8.** `analyze` (1) Rate of viral genome replication is much higher via the lytic cycle. In lysogeny, the viral genome can replicate only when the host cell replicates. (2) Only the lytic cycle produces virions. (3) The lytic

cycle results in host cell death, while the host cell continues to survive during lysogeny. **9.** `create` Viruses rely on host-cell enzymes to replicate, whereas bacteria do not. Therefore, many drugs designed to disrupt the virus life cycle cannot be used, because they would kill host cells as well. Only viral-specific proteins are good targets for drug design. **10.** `evaluate` Each major hypothesis to explain the origin of viruses is associated with a different type of genome. Escaped genes: single- and double-stranded DNA or possibly single-stranded RNA. Degenerate cells: double-stranded DNA. RNA world: single- or double-stranded RNA. **11.** `understand` The phylogenies of SIVs and HIVs show that the two shared common ancestors, but that SIVs are ancestral to the HIVs. Also, there are plausible mechanisms for SIVs to be transmitted to humans through butchering or contact with pets, but fewer or no plausible mechanisms for HIVs to be transmitted to monkeys or chimps. **12.** `apply` a; the single-stranded positive-sense RNA genome of this virus can serve as an mRNA to produce viral proteins, while all of the other possibilities require viral enzymes to transcribe the genome into mRNA.

✔ Test Your Problem-Solving Skills

13. `create` Culture the *Staphylococcus* strain outside the human host and then add the virus to determine whether the virus kills the bacterium efficiently. Then test the virus on cultured human cells to determine whether the virus harms human cells. Then test the virus on monkeys or other animals to determine if it is safe. Finally, you could test the virus on human volunteers. **14.** `evaluate` Prevention is currently the most cost-effective program, but it does not help people who are already infected. Treatment with effective drugs not only prolongs lives but also reduces virus loads in infected people, so that they have less chance of infecting others. **15.** `apply` b; by prolonging the infection in this manner, these viruses are more likely to be transmitted to a new host. **16.** `evaluate` Viruses cannot be considered to be alive by the definition given in (1), because viruses are not capable of replicating by themselves. By the definition given in (2), it can be argued that viruses are alive because they store, maintain, replicate, and use genetic information—although they cannot perform all these tasks on their own.

CHAPTER 37

IN-TEXT QUESTIONS AND EXERCISES

p. 732 Fig. 37.1 `apply` New branches and leaves should develop to the right (the plant will also lean that way); new lateral roots will develop to the left.
p. 734 Fig. 37.3 `analyze` Lawn grasses are shallow rooted, so when the top part of soil dries out, the plants cannot grow.
p. 735 Table 37.1 `understand` For aerobic respiration, which generates ATP needed to keep cells alive.
p. 736 Fig. 37.5 `evaluate` The individuals are genetically identical—thus, any differences in size or shape in the different habitats are due to phenotypic plasticity and not genetic differences.
p. 738 Fig. 37.6 `analyze` Large surface area provides more area to capture photons, but also more area from which to lose water and be exposed to potentially damaging tearing forces from wind.

FIGURE A35.5

p. 739 CYU (1) remember See **FIGURE A37.1 (2)** understand The generalized body of a plant has a taproot with many lateral roots and broad leaves. Examples of deviations from this generalized body structure follow. Modified roots: Unlike taproots, *fibrous roots* do not have one central root, and *adventitious roots* arise from stems. Modified stems: *Stolons* grow along the soil and grow roots and leaves at each node, and *rhizomes* grow horizontally underground. Both of these modified stems function in asexual reproduction. Modified leaves: The *needle-shaped leaves* of cacti don't lose water to transpiration compared with typical broad leaves, and they protect the plant from herbivory. *Tendrils* on climbing plants are modified leaves that do not photosynthesize, but wrap around trees or other substrates to facilitate climbing.

p. 745 Fig. 37.16 evaluate Many possible answers; for example, in leaf cells, genes encoding proteins involved in photosynthesis would be expressed. These genes wouldn't be expressed in the root cells.

p. 747 CYU (1) understand The apical meristem gives rise to the three primary meristems. The apical meristem is a single mass of cells localized at the tip of a root or shoot; the primary meristems are localized in distinctive locations behind the apical meristem. **(2)** understand Epidermal cells are flattened and lack chloroplasts; they secrete the cuticle (in shoots) or extend water and nutrient-absorbing root hairs (in roots) and protect the plant. Parenchyma cells are "metabolic workhorse cells" found throughout the plant body; they perform photosynthesis and synthesize, store, and/or transport materials. Tracheids are long, slim cells with pits in their secondary cell walls; they are found in xylem and are dead when mature, and they conduct water and solutes up the plant. Sieve-tube elements are long, thin cells found in phloem; they are alive when mature and conduct sugars and other solutes up and down the plant.

p. 750 understand Secondary growth adds cells only to the inside and outside of the vascular cambium, so the nail remains the same distance above the ground as the tree grows.

p. 750 Fig. 37.24 remember You should have labeled a light band as early wood, and the dark band to its right as late wood; both bands together comprise one growth ring.

p. 751 CYU (1) apply The rings are small on the shaded side and large on the sunny side. **(2)** remember See **FIGURE A37.2.**

IF YOU UNDERSTAND . . .

37.1 analyze Phenotypic plasticity is more important in (1) environments where conditions vary because it gives individuals the ability to change the growth pattern of their roots, shoots, and stems to access sunlight, water, and other nutrients as the environment changes; and (2) in long-lived species because it gives individuals a mechanism to change their growth pattern as the environment changes throughout their lifetime.

37.2 analyze The zone of cellular maturation is where root hairs emerge. Root hairs are the site where most of the water and ions are absorbed in the root system. **37.3** apply (1) In both cases, the structure would stop growing; (2) the shoot would lack epidermal cells and would die.

37.4 apply (1) The plant would not produce secondary xylem and phloem, so its girth and transport ability would be reduced, though it would still produce bark. (2) The trunk would have much wider tree rings on one side than the other side.

YOU SHOULD BE ABLE TO . . .

✔ Test Your Knowledge

1. remember b **2.** understand a **3.** remember c **4.** remember a **5.** remember Cactus spines are modified leaves; thorns are modified stems. **6.** analyze Collenchyma and sclerenchyma cells both have thickened cell walls and function in providing structural support. Unlike collenchyma cells, sclerenchyma cells have lignified cell walls and are dead at maturity.

✔ Test Your Understanding

7. remember b **8.** evaluate The general function of both systems is to acquire resources: The shoot system captures light and carbon dioxide; the root system absorbs water and nutrients. Vascular tissue is continuous throughout both the shoot and root systems. Diversity in roots and shoots enables plants of different species to live together in the same environment without directly competing for resources. **9.** analyze Continuous growth enhances phenotypic plasticity because it allows plants to grow and respond to changes or challenges in their environment (such as changes in light and water availability). **10.** apply Cuticle reduces water loss; stomata facilitate gas exchange. Plants from wet habitats should have a relatively large number of stomata and thin cuticle. Plants living in dry habitats should have relatively few stomata and thick cuticle. **11.** analyze Parenchyma cells in the ground tissue perform photosynthesis and/or synthesize and store materials. In vascular tissue, parenchyma cells in rays conduct water and solutes across the stem; other parenchyma cells differentiate into the sieve-tube elements and companion cells in phloem. **12.** understand Cells produced to the inside of the vascular cambium differentiate into secondary xylem; cells produced to the outside of the vascular cambium differentiate into secondary phloem.

✔ Test Your Problem-Solving Skills

13. apply c **14.** remember Asparagus—stem; Brussels sprouts—lateral buds; celery—petiole; spinach—leaf (petiole and blade); carrots—taproot; potato—modified stem. **15.** create They grow continuously, so do not have alternating groups of large and small cells that form rings. **16.** analyze Girdling disrupts transport of solutes in secondary phloem. The tree's root system starves.

CHAPTER 38

IN-TEXT QUESTIONS AND EXERCISES

p. 756 apply (1) Add a solute (e.g., salt or sugar) to the left side of the U-tube, so that solute concentration is higher than on the right side. (2) Water would move from the right side of the U-tube into the left side.

p. 757 apply Pull the plunger up.

p. 757 Fig. 38.2 apply It increases (becomes less negative), which reduces the solute potential gradient between the two sides.

p. 758 CYU (1) apply When a cell is in equilibrium with pure water, the negative solute potential of the cell will be balanced by the positive pressure potential (turgor pressure). **(2)** analyze Irrigation water contains low concentrations of salts so the solute potential will be near zero. As water evaporates, salts remain in the soil, lowering the solute potential.

p. 760 Fig. 38.7 understand See **FIGURE A38.1.**

p. 761 apply The individual would not be able to exclude toxic solutes from the xylem. If these solutes were transported throughout the plant in high enough concentrations, they could damage tissues. The mutant could also not maintain ions in the xylem and so could not generate root pressure.

p. 764 Fig. 38.12 analyze The line on the graph would go up, and large jumps would occur each time the light level was turned up, because increased transpiration rates would increase negative pressure (tension) at the leaf surface and thus increase the water-potential gradient.

p. 765 CYU (1) understand At the air–water interface under a stoma, water molecules cannot hydrogen-bond in all directions—they can bond only with water molecules below them. This creates tension at the surface. When transpiration occurs, the curvature of the meniscus increases—increasing tension. **(2)** apply When a nearby stoma closes, the humidity inside the air space increases, transpiration decreases, and surface tension on menisci decreases. The same changes occur when a rain shower starts. When air outside the leaf dries, though, the opposite occurs: The humidity inside the air space decreases, transpiration increases, and surface tension on menisci increases.

p. 768 Fig. 38.17 understand See **FIGURE A38.2.**

p. 771 CYU (1) understand In early spring, growing leaves are sinks and phloem sap moves toward them. In midsummer, leaves are sources and phloem sap moves away from them. **(2)** create The movement of water and solutes in sieve elements is by bulk flow, so stationary organelles would provide resistance to the flow.

p. 771 Fig. 38.22 create With time the plant will metabolize the radioactive sucrose, and levels of radioactivity will decline.

FIGURE A37.1

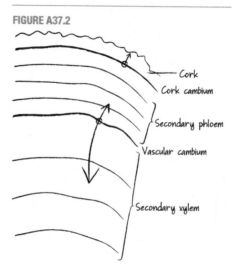

Apical bud
Internode
Node
Leaf
Node
Axillary bud
Stem
Shoot system
Lateral root
Root system
Tap root

FIGURE A37.2

Cork
Cork cambium
Secondary phloem
Vascular cambium
Secondary xylem

IF YOU UNDERSTAND . . .

38.1 analyze Because xylem cells are dead at maturity, there are no plasma membranes for water to cross and thus no solute potential. The only significant force acting on water in xylem is pressure. **38.2** evaluate It drops, because transpiration rates from stomata decrease, reducing the water potential gradient between roots and leaves. **38.3** understand CAM plants open their stomata only at night, and thus they transpire when temperatures are cooler. **38.4** analyze To create high pressure in phloem near sources and low pressure near sinks, water has to move from xylem to phloem near sources and from phloem back to xylem near sinks. If xylem were not close to phloem, this water movement could not occur, and the pressure gradient in phloem wouldn't exist.

YOU SHOULD BE ABLE TO . . .

✔ Test Your Knowledge

1. understand d **2.** apply b **3.** understand a **4.** remember c **5.** remember Endodermal cells. **6.** remember Companion cells.

✔ Test Your Understanding

7. understand c **8.** apply See **FIGURE A38.3**. **9.** analyze The force responsible for root pressure is the high solute potential of root cells at night, when they continue to accumulate ions from soil and water follows by osmosis. The mechanism is active, in the sense that ions are imported against their concentration gradient. Capillarity is driven by adhesion of water molecules to the sides of xylem cells that creates a pull upward, and by cohesion with water molecules below. Cohesion-tension, generated by transpiration, is driven by the loss of water molecules from menisci in leaves, which creates a large negative pressure (tension). Both capillarity and cohesion-tension are passive processes. **10.** understand Water moves up xylem because of transpiration-induced tension created at the air–water interface under stomata, which is communicated down to the roots by the cohesion of water molecules. Sap flows in phloem because of a pressure gradient that exists between source and sink cells, driven by differences in sucrose concentration and flows of water into or out of nearby xylem. **11.** understand By pumping protons out, companion cells create a strong electrochemical gradient favoring entry of protons. A cotransport protein (a symporter) uses this proton gradient to import sucrose molecules *against* their concentration gradient. **12.** apply When it germinates—sucrose stored inside the seed is released to the growing embryo, which cannot yet make enough sucrose to feed itself.

✔ Test Your Problem-Solving Skills

13. apply b **14.** create Plants would not grow as quickly or as tall, because the taller they grew, the more energy they would need to use to transport water and thus the less they would have available for growth. **15.** analyze Because they cannot readily replace water that is lost to transpiration, they have to close stomata. During a heat wave, transpiration cannot cool the plants' tissues. The trees bake to death. **16.** create Closing aquaporins will slow or stop movement of water from cells into the xylem and out of the plant via transpiration. Because water does not leave the cells, they are able to maintain turgor (normal solute potentials).

CHAPTER 39

IN-TEXT QUESTIONS AND EXERCISES

p. 776 Fig. 39.1 evaluate To test the hypothesis that most of a plant's mass comes from CO_2 in the atmosphere, design an experiment where plants are grown in the presence or absence of CO_2 and compare growth, and/or grow plants in the presence of labeled CO_2 and document the presence of labeled carbon in plant tissues.

p. 777 Fig. 39.3 evaluate Because soil is so complex, it would be difficult to defend the assumption that the soils with and without added copper are identical except for the difference in copper concentrations. Also, it would be better to compare treatments that had no copper versus normal amounts of copper, instead of comparing with and without added copper.

p. 783 Fig. 39.10 analyze The proton gradient arrow in (b) should begin above the membrane and cross the

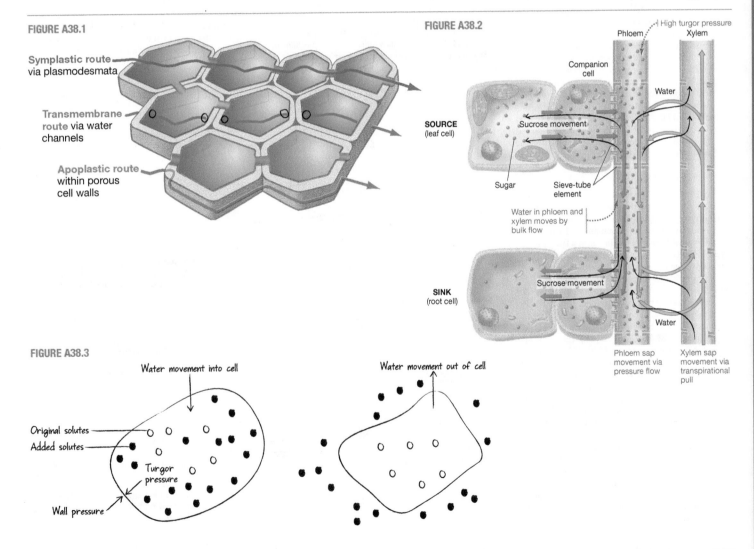

FIGURE A38.1

Symplastic route via plasmodesmata

Transmembrane route via water channels

Apoplastic route within porous cell walls

FIGURE A38.2

SOURCE (leaf cell)

Companion cell

Phloem Xylem

High turgor pressure

Water

Sucrose movement

Sugar Sieve-tube element

Water in phloem and xylem moves by bulk flow

SINK (root cell)

Sucrose movement

Water

Phloem sap movement via pressure flow

Xylem sap movement via transpirational pull

FIGURE A38.3

Water movement into cell

Water movement out of cell

Original solutes

Added solutes

Turgor pressure

Wall pressure

membrane pointing down. The electrical gradient arrow in (c) should also begin above the membrane (marked positive) and cross the membrane pointing down (toward negative).

p. 784a (apply) (1) If there is no voltage across the root-hair membrane, there is no electrical gradient favoring entry of cations, and absorption stops. (2) There is no route for cations to cross the root hair membrane, so absorption stops.

p. 784b (apply) (1) If there is no proton gradient across the root-hair membrane, there is no gradient favoring entry of protons, so symport of anions stops. (2) There is no route for anions to cross the root-hair membrane, so absorption stops.

p. 786 CYU (1) (understand) Proton pumps establish an electrical gradient across root-hair membranes, and the inside of the membrane is much more negative than the outside. Cations follow this electrical gradient into the cell. Anions are transported with protons via symporters.
(2) (create) See **FIGURE A39.1**.

p. 788 CYU (1) (analyze) Many possible answers, including the following: If N-containing ions are abundant in the soil, the energetic cost of maintaining nitrogen-fixing bacteria may outweigh the benefits in terms of increased nitrogen availability. If mycorrhizal fungi provide a plant with nitrogen-containing ions, the energetic cost of maintaining nitrogen-fixing bacteria may outweigh the benefits in terms of increased nitrogen availability. Plants that grow near species with N-fixing bacteria might gain nitrogen by absorbing nitrogen-containing ions after root nodules die, or by "stealing it" (roots of different species will often grow together). There could be a genetic constraint (see Chapter 25): Plant species may simply lack the alleles required to manage the relationship. Note: To be considered correct, your hypothesis should be plausible (based on the underlying biology) and testable. **(2)** (analyze) Many possible answers, including: Because the bacterial cell enters root cells, it is important for the plant to have reliable signals that it is not a parasitic bacterium. It is advantageous for the plant to be able to reject nitrogen-fixing bacteria if usable nitrogen is already abundant in the soil.

p. 790 Fig. 39.17 (analyze) Mistletoe is parasitic (it extracts water and certain nutrients from host trees); bromeliads are not (they are epiphytes and don't affect the fitness of their host plants).

IF YOU UNDERSTAND . . .

39.1 (evaluate) Add nitrogen to an experimental plot and compare growth within the plot to growth of a similar plot nearby. Do many replicates of the comparison, to convince yourself and others that the results are not due to unusual circumstances in one or a few plots.
39.2 (understand) Loam is a mixture of sand, clay, and organic matter so it can retain water and nutrients, and provide aeration, all properties that benefit plants.
39.3 (apply) A higher membrane voltage would allow cations in soil to cross root-hair membranes more readily through membrane channels and would allow anions to enter more readily via symporters. Nutrient absorption should increase. **39.4** (evaluate) Grow pea plants in the presence of rhizobia, exposed to air with (1) N_2 containing the heavy isotope of nitrogen and (2) radioactive carbon dioxide. Allow the plant to grow and then analyze the rhizobia and plant tissues. If the rhizobia–pea plant interaction is mutualistic, then ^{15}N-containing compounds and radioactive-carbon-containing compounds should be observed in both rhizobia and plant tissues. As a control, grow pea plants in the presence of the labeled compounds but without rhizobia. If the mutualism hypothesis is correct, the plant should contain labeled carbon but no heavy nitrogen. **39.5** (analyze) Despite the abundance of rain in tropical rain forests, many epiphytes there have limited access to water because their roots don't reach the soil. Like desert plants, these epiphytes are adapted to conserve water.

✓ Test Your Knowledge

1. (remember) b **2.** (understand) d **3.** (remember) d **4.** (understand) c **5.** (remember) Mycorrhizal fungi obtain most of their nitrogen from proteins in soil organic matter. Rhizobia obtain nitrogen from N_2 gas. **6.** (remember) Plant

✓ Test Your Understanding

7. (remember) b **8.** (evaluate) (1) Grow a large sample of corn plants with a solution containing all essential nutrients needed for normal growth and reproduction. Grow another set of genetically identical (or similar) corn plants in a solution containing all essential nutrients except iron. Compare growth, color, and seed production (and/or other attributes) in the two treatments. (2) Grow corn plants in solutions containing all essential nutrients in normal amounts, but with varying concentrations of iron. Determine the lowest iron concentration at which plants exhibit normal growth rates. **9.** (understand) If nutrients in soil are scarce, there is intense natural selection favoring alternative ways of obtaining nutrients—for example, by digesting insects or stealing nutrients.
10. (apply) Higher in soils with both sand and clay. Clay fills spaces between sand grains and slows (1) passage of water through soil and away from roots, and (2) leaching of anions in soil water. Clay particles also provide negative charges that retain cations so they are available to plants. The presence of sand keeps soil loose enough to allow roots to penetrate. **11.** (understand) Some metal ions are poisonous to plants, and high levels of other ions are toxic. Passive mechanisms of ion exclusion—such as a lack of ion channels that allow passage into root cells—do not require an expenditure of ATP. Active mechanisms of exclusion—such as production of metallothioneins—require an expenditure of ATP.
12. (analyze) (1) The amounts required and the mechanisms of absorption are different. Much larger amounts of C, H, and O are needed than mineral nutrients. CO_2 enters

plants via stomata, and water is absorbed passively along a water potential gradient; mineral nutrients enter via active uptake in root hairs or via mycorrhizae. (2) Macronutrients are required in relatively large quantities and usually function as components of macromolecules; micronutrients are required in relatively small quantities and usually function as enzyme cofactors.

✓ Test Your Problem-Solving Skills

13. (apply) The soil lost 60 g while the tree gained 74,000 g. Dividing 60 by 74,000 and converting to a percentage equals 0.08%, far less than 4%. The water that van Helmont used may have contained many of the macro- and micronutrients that were incorporated into plant mass. To test this hypothesis, conduct an experiment of the same design but have one treatment with pure water and one treatment with water containing solutes ("hard water"). Compare the percentage of soil mass incorporated into the willow tree in both treatments. Another explanation: Willow may be unusual. To test this hypothesis, repeat the experiment with willow and several other species under identical conditions, and compare the percentage of soil mass incorporated into the plants. A third hypothesis: Van Helmont's measurements were inaccurate. Repeat the experiment. **14.** (analyze) Acid rain inundates the soil with protons, which bind with the negatively charged clay particles—displacing the cations that are normally bound there. Cations may then be leached away. **15.** (evaluate) If the hypothesis is correct, vanadate should inhibit phosphorus uptake. To test this hypothesis, grow plants in solution with radioactively labeled phosphorus without vanadate and in solution with vanadate. Measure the amount of labeled phosphorus in root cells with and without vanadate. **16.** (apply) Alder should decompose much faster, because it provides more nitrogen to the fungi and bacteria that are responsible for decomposition—meaning that they can grow faster.

FIGURE A39.1

1. Passive exclusion:
Root hairs have no membrane proteins to transport Cd^{2+}

2. Active exclusion:
Metallothionein proteins bind to Cd^{2+} that enter cells, making them harmless

3. Active exclusion:
Proteins in tonoplast move Cd^{2+} into vacuole for safe storage

4. Passive exclusion:
Endodermal cells have no membrane proteins to transport Cd^{2+} past the Casparian strip

CHAPTER 40

IN-TEXT QUESTIONS AND EXERCISES

p. 796 CYU (1) `understand` The only cells that can respond to a specific environmental signal or hormone are cells that have an appropriate receptor for that signal or hormone. **(2)** `remember` Once a receptor is activated, it triggers production of many second messengers or the activation of many proteins in a phosphorylation cascade.

p. 797 Fig. 40.4 `evaluate` Wild-type plants already have a normal *PHOT1* gene and can respond to directional blue light. Inserting a defective gene into a wild-type plant would have no effect.

p. 798 Fig. 40.5 `evaluate` For "Tip removed": Cut the tip and replace it, to control for the hypothesis that the wound itself causes bending (not the loss of the tip). For "Tip covered": Cover the tip with a transparent cover instead of opaque one, to control for the hypothesis that the cover itself affects bending (not excluding light).

p. 799 CYU (1) `apply` Using a syringe or other device, apply auxin to the west side of each stem, just below the tip. **(2)** `apply` Stems would become very long and would not respond to directional light.

p. 801 Table 40.1 `analyze` The average germination rate of lettuce seeds last exposed to red light is about 99 percent. The average for seeds last exposed to far-red light is about 50 percent. Both of these values are much higher than the germination rate of buried seeds that receive no light at all; their germination rate is only 9 percent.

p. 803 CYU (1) `understand` Lettuce grows best in full sunlight. If a seed receives red light, it indicates that the seed is in full sunlight—meaning that conditions for growth are good. But if a seed receives far-red light, it indicates that the seed is in shade—meaning that conditions for growth are poor. **(2)** `understand` Night length changes very predictably throughout the year, so it is a reliable signal for plants to perceive.

p. 803 Fig. 40.10 `apply` An appropriate control treatment would be to graft a leaf from a plant that had never been exposed to a short-night photoperiod onto a new plant that also had never been exposed to the correct conditions for flowering. If the hypothesis is correct, the grafted leaf should not induce flowering.

p. 804 Fig. 40.12 `evaluate` (One possibility) Direct pressure from a statolith changes the shape of the receptor protein. The shape change triggers a signal transduction cascade, leading to a gravitropic cell response.

p. 805 CYU (1) `analyze` In both roots and shoots, auxin is redistributed asymmetrically in response to a signal. In phototropism, auxin is redistributed to the shaded side of the shoot tip in response to blue light. In gravitropism, auxin is redistributed to the lower part of the root or shoot in response to gravity. **(2)** `understand` In root cells, auxin concentrations indicate the direction of gravity. The gravitropic response is triggered by changes in the distribution of auxin in root-tip cells.

p. 809 Fig. 40.19 `analyze` The dwarfed individuals can put more of their available energy into reproduction because they are using less energy for growth than taller individuals.

p. 810 Fig. 40.21 `analyze` Without these data, a critic could argue that the stomata closed in response to water potentials in the leaf, not a signal from the roots.

p. 813 CYU (1) `analyze` ABA: Stomata should close and photosynthesis should stop compared to individuals that do not receive extra ABA. GA: Their stems should elongate rapidly, compared to plants that do not receive extra GA. **(2)** `understand` The rotten apple produces ethylene, which, being a gas, stimulates all of the other apples in the bushel basket to ripen too fast.

p. 816 `analyze` Sacrificing a small number of cells due to the hypersensitive response is far less costly for the plant than death.

p. 819 CYU (1) `analyze` Because specific receptor proteins recognize and match specific proteins produced by pathogens, having a wide array of these proteins allows individuals to recognize and respond to a wide array of pathogens. **(2)** `understand` If herbivores are not present, synthesizing large quantities of proteinase inhibitors wastes resources (ATP and substrates) that could be used for growth and reproduction.

IF YOU UNDERSTAND . . .

40.1 `analyze` An external signal arrives at a receptor cell/protein (a person sees a barn on fire); the person calls (the receptor transduces the signal); the dispatcher rings a siren (a second messenger is produced or phosphorylation cascade occurs); the fire department members get to the pumper truck (gene expression or some other cell activity changes). **40.2** `analyze` Sun-adapted (high-light) plants are likely to have a significant positive phototropic response (i.e., growth toward light), whereas shade-adapted (low-light) plants are less likely to respond phototropically. **40.3** `evaluate` They do not perform photosynthesis and therefore do not need to switch behavior based on being located in full sunlight versus shade. **40.4** `understand` Starch is denser than water, so it falls quickly to the bottom of cells. **40.5** `understand` A reduction in plant height in a windy environment might keep them upright. **40.6** `evaluate` Those hormones need to be transported throughout the plant, because cells throughout the body need to respond to the signal—not in a single direction. **40.7** `understand` In the hypersensitive response, a few cells are sacrificed to prevent a pathogen from spreading throughout the plant.

YOU SHOULD BE ABLE TO . . .

✔ Test Your Knowledge

1. `remember` d **2.** `remember` b **3.** `remember` d **4.** `understand` b **5.** `remember` hypersensitive response **6.** `remember` plasma membrane

✔ Test Your Understanding

7. `remember` a **8.** `remember` The P_{fr}–P_r switch in phytochromes lets plants know whether they are in shade or sunlight, and triggers responses that allow sun-adapted plants to avoid shade. Phototropins are activated by blue light, which indicates full sunlight, and trigger responses that allow plants to photosynthesize at full capacity. **9.** `analyze` Transduce means to convert energy from one form to another. There are many examples: A touch may be converted to an electrochemical potential, light energy or pressure into a shape change in a protein, and so on. **10.** `understand` (Many possibilities) Cells or tissue involved: Auxin directs gravitropism in roots but phototropism in shoots. Developmental stage or age: GA breaks dormancy in seeds but extends stems in older individuals. Concentration: Large concentrations of auxin in stems promote cell elongation; small concentrations do not. Other hormones: The concentration of GA relative to ABA influences whether germination proceeds. **11.** `apply` Ethylene promotes senescence; increased ethylene sensitivity in leaves (combined with reduced auxin concentrations) causes abscission; fruits exposed to ethylene ripen faster than fruits that are not exposed to ethylene. Produce dealers can maximize profits by delaying ripening during transit of fruit by removing ethylene and then adding ethylene just before fruits are displayed. **12.** `apply` When one leaf on the mutant plant is chewed on by a herbivore, the damaged leaf would not be able to signal to the rest of the plant, so it would not be able to fend off future herbivore attacks.

✔ Test Your Problem-Solving Skills

13. `apply` a **14.** `evaluate` Small-seeded plants need to perform photosynthesis early in seedling development or they will starve. Therefore, they need to germinate in direct sunlight. The food reserves in large-seeded plants can support seedling growth for a relatively long time without photosynthesis. Therefore, they do not need to germinate in direct sunlight. **15.** `evaluate` In these seeds, (a) little or no ABA is present, or (b) ABA is easily leached from the seeds, so its inhibitory effects are eliminated. Test hypothesis (a) by determining whether ABA is present in the seeds. Test hypothesis (b) by comparing the amount of ABA present before and after running water—enough to mimic the amount of rain that falls in a tropical rain forest over a few days or weeks—over the seeds. **16.** `apply` New types of plant resistance provide a selective pressure for pathogens to evolve new mechanisms to evade the resistance.

CHAPTER 41

IN-TEXT QUESTIONS AND EXERCISES

p. 825 `understand` (1) The microsporocyte divides by meiosis to generate male spores; the megasporocyte divides by meiosis to generate female spores. (2) In angiosperms, the female gametophyte stays within the ovary at the base of the flower even after it matures and produces an egg cell. Fertilization and seed development take place in the same location.

p. 829 CYU (1) `analyze` Both are diploid cells that divide by meiosis to produce spores. The megasporocyte produces a megaspore (female spore); a microsporocyte produces a microspore (male spore). **(2)** `analyze` Both are multicellular individuals that produce gametes by mitosis. The female gametophyte is larger than the male gametophyte and produces an egg; the male gametophyte produces sperm.

p. 829 Fig. 41.8 `understand` A gametophyte is the multicellular individual that produces gametes by mitosis. The pollen grain is a gametophyte because it is multicellular and produces sperm by mitosis.

p. 832 CYU (1) `understand` Insects feed on nectar and/or pollen in flowers. Flowers provide food rewards to insects to encourage visitation. The individuals that attract the most pollinators produce the most offspring. **(2)** `understand` One product of double fertilization is the diploid zygote, which will eventually grow by mitosis into a mature sporophyte. The other product is the triploid endosperm nucleus, which will grow by mitosis to form a source of nutrients for the embryo.

p. 833 `remember` The endosperm nucleus in the central cell is triploid; the zygote is diploid; the synergid and other cells remaining from the female gametophyte are haploid.

p. 834 `analyze` A cotyledon and a radicle develop into organs with distinct functions, but both are composed of the same tissues: Epidermis is derived from protoderm, ground tissue is derived from ground meristem, and vascular tissue is derived from the procambium.

p. 835 Fig. 41.14 `evaluate` Hypothesis: Fruit changes color when it ripens as a signal to fruit eaters. The color change is advantageous because seeds are not mature in unripe fruit and are unlikely to survive. Fruit eaters disperse mature seeds in ripe fruit, which are more likely to survive.

p. 836 Fig. 41.16 `evaluate` Hackberries served as a control to demonstrate that animals that avoided chilies were hungry,

p. 837 Fig. 41.17 `understand` Beans—their cotyledons are aboveground.

IF YOU UNDERSTAND . . .

41.1 `remember` The sporophyte is diploid, the gametophyte is haploid, spores and gametes are haploid, and zygotes are diploid. **41.2** `apply` There would be four embryo sacs in each ovule, and thus four eggs. **41.3** `apply` Endosperm tissue is generated when two haploid, female nuclei fuse with one haploid sperm nucleus, forming a cell with three sets of chromosomes. In contrast, a zygote is formed by the fusion of one haploid sperm nucleus and one haploid egg, forming a cell with two sets of chromosomes. **41.4** *analyze* A cold requirement for germination ensures that seeds from plants native to habitats that experience cold winters will not germinate in the fall and fail to become established before winter.

✔ Test Your Knowledge

1. *remember* b **2.** *understand* c **3.** *remember* a **4.** *understand* d
5. *remember* Megasporocytes (female) and microsporocytes (male) undergo meiosis to generate megaspores and microspores, respectively. These divide mitotically to give rise to female and male gametophytes—the embryo sac and pollen grain, respectively. **6.** *remember* Pollination is the transfer of pollen from an anther to a stigma. Fertilization is the fusion of haploid gametes.

✔ Test Your Understanding

7. *understand* c **8.** *analyze* Asexual reproduction is a quick, efficient way for a plant to produce large numbers of offspring. The disadvantage is that offspring are genetically identical to the parent and thus are vulnerable to the same diseases and able to thrive only in habitats similar to those inhabited by the parent. **9.** *analyze* Sepals are the outermost structures in a flower and usually protect the structures within; petals are located just inside the sepals and usually function to advertise the flower to pollinators. In wind- and bee-pollinated flowers, sepals probably do not vary much in overall function or general structure. Petals are often lacking in wind-pollinated species; they tend to be broad and flat (to provide a landing site) and colored purple, blue, or yellow in bee-pollinated species (often with ultraviolet markings).
10. *analyze* Outcrossing increases genetic diversity among offspring, making them more likely to thrive if environmental conditions change from the parental generation. However, outcrossing requires cross-pollination to be successful. Self-fertilization results in relatively low genetic diversity among offspring but ensures that pollination succeeds. **11.** *remember* The female gametophyte develops inside the ovule; the ovary develops into the pericarp of the fruit after fertilization. A mature ovule contains both gametophyte and sporophyte tissue; the ovary and carpel are all sporophyte tissue. **12.** *analyze* The endosperm of a corn seed and the cotyledons of a bean seed both contain nutrients needed for the seed to germinate. The bean cotyledons have absorbed the nutrients of the endosperm.

✔ Test Your Problem-Solving Skills

13. *remember* c **14.** *apply* Subtracting the number of used pollen grains per plant (600) from the number of pollen grains generated per plant (10 million) divided by 10 million reveals that 99.994% of the pollen is wasted. **15.** *evaluate* In response to "cheater" plants, insects should be under intense selection pressure to detect and avoid species that do not offer a food reward. Individuals would have to be able to distinguish scents, colors, or other traits that identify cheaters. In response to cheater insects, plants possessing unusually thick or toxic petals or sepals that prevented insects from bypassing the anthers on their way to the nectaries would be more successful at producing seed in the next generation. **16.** *evaluate* Acorns have a large, edible mass and are usually animal dispersed (e.g., by squirrels that store them and forget some). Cherries have an edible fruit and are animal dispersed. Burrs stick to animals and are dispersed as the animals move around. Dandelion seeds float in wind. To estimate the distance that each type of seed is dispersed from the parent, (1) set up "seed traps" to capture seeds at various distances from the parent plant; (2) sample locations at various distances from the parent and analyze young individuals—using techniques introduced in Chapter 21 to determine if they are offspring from the parent being studied; (3) mark seeds, if possible, and find them again after dispersal.

BIG PICTURE Plant and Animal Form and Function

p. 840 CYU (1) *understand* Plants and animals may accumulate compounds in the epidermis and skin (such as waxes and other lipids) that reduce water loss. Both plants and animals evolved closable openings in this barrier to allow regulated gas exchange. Plants and animals may also change the amount of the body exposed to the sun (in plants by moving leaves, in animals by changing posture or taking shelter) to reduce evaporative water loss.
(2) *understand* Many plants and animals keep their eggs inside their bodies and depend on sperm being deposited near them rather than releasing both to the environment where they could dry out. **(3)** *understand* Very large animals and plants need to build strong supporting structures such as thick bones or trunks to withstand the force of gravity. They also need elaborate internal transport systems to bring nutrients to each cell and carry wastes away.
(4) *analyze* Cellular respiration fits into one of the key functions for survival—Nutrition. Both plants and animals depend on sugar (produced by plants via photosynthesis) as their energy source to make ATP via cellular respiration.

CHAPTER 42

IN-TEXT QUESTIONS AND EXERCISES

p. 844 Fig. 42.2 *create* There would be no difference between experimental and control males.
p. 847 Fig. 42.4 *analyze* Based on the projections called axons that go to specific cells and tissues, neurons transmit signals to specific tissues rather than broadly throughout the body.
p. 850 CYU *analyze* *Connective tissues* consist of cells embedded in a matrix. The density of the matrix determines the rigidity of the connective tissue and its function in padding/protection, structural support, or transport. *Nervous tissue* is composed of neurons and support cells. Neurons have long projections that function as "biological wires" for transmitting electrical signals. *Muscle tissue* has cells that can contract and functions in movement. *Epithelial tissue* has tightly packed layers of cells with distinct apical and basal surfaces. They protect underlying tissues and regulate the entry and exit of materials into these tissues.
p. 850 *apply* A newborn.
p. 851 Fig. 42.10 *apply* The puppy has to eat relatively more because it has a higher mass-specific metabolic rate than an adult dog.
p. 852 Fig. 42.11 *apply* There should be no oxygen uptake on either side of the chamber.
p. 853 CYU (1) *understand* As three-dimensional size increases, volume increases more quickly than does surface area, so the surface area/volume ratio decreases.
(2) *analyze* The squirrel has a higher mass-specific metabolic rate because it is a smaller animal. It therefore needs more energy per unit body mass than a bear.
p. 858 CYU (1) *analyze* Endotherms can remain active during the winter and at night and sustain high levels of aerobic activities such as running or flying, but they require large amounts of food energy. Ectotherms need much less food and can devote a larger proportion of their food intake to reproduction, but they have a hard time maintaining high activity levels at night or in cold weather. **(2)** *apply* The whale's tongue would become colder because not as much heat would be transferred from warm arterial blood to cool venous blood with concurrent flow as with countercurrent flow, and as a result more heat from the arterial blood would be lost to the environment.

IF YOU UNDERSTAND . . .

42.1 *understand* Some adaptations to cold environments in mammals include the evolution of thick coats of hair,

---

countercurrent heat exchangers in the extremities, and dark coloration to absorb radiation. Acclimatization may include further thickening and darkening of hair upon exposure to cold. **42.2** *apply* Loosening tight junctions would compromise the barrier between the external and internal environments. It would allow water and other substances (including toxins) to pass more readily across the epithelial boundary. **42.3** *create* King Kong is endothermic, and his huge mass would generate a great deal of heat. His relatively small surface area, especially as compared to a normal-sized gorilla, would not be able to dissipate the heat—especially if it were covered with fur. **42.4** *create* Individuals could acclimatize by normal homeostatic mechanisms (increased sweating, seeking shade, etc.). Individuals with alleles that allowed them to function better at higher temperatures would produce more offspring than individuals without those alleles. Over time, this would lead to adaptation via natural selection. **42.5** *create* *Apatosaurus* was likely a homeotherm because of its huge size. Because of its low surface area/volume ratio, metabolically produced heat would be trapped inside the dinosaur's body. Even if exposed to cold or warm temperatures, the dinosaur's temperature would not change much because it takes a long time for it to lose or gain heat across such a relatively small surface area.

✔ Test Your Knowledge

1. *understand* True **2.** *remember* Epithelial **3.** *understand* b
4. *understand* d **5.** *understand* a **6.** *analyze* a

✔ Test Your Understanding

7. *understand* It is where environmental changes are sensed first. As an effector, it has a large surface area for losing or gaining heat. **8.** *analyze* A warm frog can move, breathe, and digest much faster than a cold frog, because enzymes work faster (rates of chemical reactions increase) at 35°C than at 5°C. A warm frog needs much more food, though, to support this high metabolic rate. **9.** *understand* *Absorptive sections* have many folds and projections that increase their surface area for absorption. *Capillaries* have a high surface area because they are thin and highly branched, making exchange of fluids and gases efficient. *Beaks of Galápagos finches* have sizes and shapes that correlate with the type of food each species eats. Large beaks are used to crack large seeds; long, thin beaks are used to pick insects off surfaces; etc. *Fish gills* are thin, flattened structures with a large surface area, which facilitates the efficient exchange of gases and wastes. **10.** *analyze* A larger sphere has a relatively smaller surface area for its interior volume than does a smaller sphere, because surface area increases with size at a lower rate than volume does. Diffusion will be more efficient in the smaller sphere. **11.** *apply* See **FIGURE A42.1**. The graph shows that as size (length) increases, the volume increases more rapidly than the surface area. **12.** *understand* c. This is an example of negative feedback because insulin is released in response to an increase in a parameter (blood sugar concentration)

FIGURE A42.1

and acts to reduce that parameter so that homeostasis is maintained. In the other examples, homeostasis is not maintained.

✓ Test Your Problem-Solving Skills

13. create Large individuals require more food, so are more susceptible to starvation if food sources can't be defended (e.g., when seeds are scattered around). They may also be slower and/or more visible to predators.
14. create You would have to show changes in the frequencies of alleles whose products are affected by temperature—for example, increases in the frequencies of alleles for enzymes that operate best at higher temperature. **15.** create The heat-dissipating system should have a very high surface area/volume ratio. Based on biological structures, it should be flattened (thin) and highly folded and branched, and/or contain tubelike projections. **16.** understand a. Turtles are ectothermic, so they do not expend much energy to produce body heat. Mice are endothermic, and expend energy to regulate body temperature. Mice will therefore have a higher basal metabolic rate and consume more food.

CHAPTER 43

IN-TEXT QUESTIONS AND EXERCISES

p. 863 Fig. 43.1 understand The black and white molecules are moving down their concentration gradients, and the concentration of red molecules does not affect the concentration of black or white molecules.
p. 867 Fig. 43.6 understand The sodium–potassium ATPase performs primary active transport; the sodium–chloride–potassium cotransporter performs secondary active transport; the chloride and potassium channels are involved in passive transport.
p. 869 CYU apply The osmolarity of the fish's tissue would decrease because the pumps would no longer be able to import sodium.
p. 869 Fig. 43.8 analyze The underside of the abdomen is shaded by the grasshopper's body. This location is relatively cool and should reduce the loss of water during respiration.
p. 871 CYU (1) understand Uric acid is insoluble in water and can be excreted without much water loss. **(2)** understand When electrolytes are reabsorbed in the hindgut epithelia, water follows along an osmotic gradient.
p. 873 understand Blood contains cells and large molecules as well as ions, nutrients, and wastes; the filtrate contains only ions, nutrients, and wastes.
p. 876 apply (1) Less water reabsorption will occur because the osmotic gradient is not as steep. (2) Filtrate osmolarity will be lower because less water has been reabsorbed. (3) Salt reabsorption will be reduced because the concentration of NaCl in the filtrate is lower.
p. 878 apply Ethanol consumption inhibits water reabsorption, leading to a larger volume of less concentrated urine. Nicotine consumption increases water reabsorption, leading to a smaller volume of more concentrated urine.
p. 878 CYU apply Water intake increases blood pressure and filtration rate in the renal corpuscle. It leads to lowered electrolyte concentration in the blood and filtrate and the production of large volumes of dilute urine. Eating large amounts of salt results in concentrated, hyperosmotic urine. Water deprivation triggers ADH release and the production of concentrated, hyperosmotic urine.
p. 879 Fig. 43.19 create Carrying a large volume of water in the bladder may be energetically costly to the animal. The animal also might not be able to run as quickly and may have an increased predation risk.

IF YOU UNDERSTAND . . .

43.1 understand A sodium gradient establishes an osmotic gradient that moves water. It also sets up an electrochemical gradient that can move ions against their electrochemical gradient by secondary active transport through a cotransporter, or with their electrochemical gradient by passive diffusion through a channel.
43.2 analyze Cartilaginous fishes such as sharks maintain a high concentration of urea in their blood, which raises their osmolarity to about the same as seawater. Therefore, they lose very little water to the environment. Bony fishes drink seawater and maintain their osmolarity lower than that of seawater by actively transporting excess ions out of the gills. Sharks actively excrete sodium chloride through a rectal gland. **43.3** understand When comparing chloride cells in gills of salmon in the ocean versus freshwater, the cells are in different locations on the gills, different forms of Na^+/K^+-ATPase exist, and $Na^+K^+Cl^-$ transporters are located on the basolateral side of the membrane in ocean salmon and on the apical side of the membrane in freshwater salmon. **43.4** apply Aquatic insects produce more urea, and especially more ammonia, than terrestrial insects because these nitrogenous wastes are energetically less expensive to produce than uric acid. Ammonia and urea require more water to be excreted, but aquatic insects have access to plenty of water. **43.5** create The longer the loop of Henle, the steeper the osmotic gradient in the interstitial fluid in the medulla. Steep osmotic gradients allow a great deal of water to be reabsorbed as urine passes through the collecting duct, resulting in highly concentrated urine.

YOU SHOULD BE ABLE TO . . .

✓ Test Your Knowledge

1. remember True **2.** understand c **3.** apply a **4.** remember d
5. understand c **6.** remember bladder

✓ Test Your Understanding

7. create b. Fishes excrete nitrogenous wastes in the form of ammonia. Although the ammonia can easily diffuse across the fish's gills into the water, freshwater ponds are often small and ammonia can build up to potentially toxic levels. **8.** analyze Ammonia is toxic and must be diluted with large amounts of water to be excreted safely. Urea and uric acid are safer and do not have to be excreted with large amounts of water but are more expensive to produce in terms of energy expenditure. Fish excrete ammonia; mammals excrete urea; insects excrete uric acid. You would expect the embryos inside terrestrial eggs to excrete uric acid, since it is the least toxic and is insoluble in water. **9.** understand Mitochondria are needed to produce ATP. A key component of salt regulation in fish is the Na^+/K^+-ATPase, which requires ATP to function. Because ATP fuels establishment of the ion gradients necessary for the cotransport mechanisms used by chloride cells as well as other epithelial cells, an abundance of mitochondria would be expected in a chloride cell. **10.** create The wax layer should be thicker in insects inhabiting dry habitats, because the animals there are under more osmotic stress (more evaporation due to low humidity; less available drinking water). Insects that live in extremely humid habitats may lack the ability to close their spiracles; animals that live in humid habitats may excrete primarily ammonia or urea and have relatively short loops of Henle. In each case, you could test these predictions by comparing individuals of closely related species that live in dry versus humid habitats.
11. create Without a loop of Henle, there would be no concentration gradient in the medullary interstitial fluid, so water would not be absorbed from the pre-urine in the collecting duct; concentrated urine could not be formed. **12.** create Freshwater has such a low osmolarity that if invertebrates osmoconformed to it, they would lack sufficient quantities of ions to conduct electrical currents in their excitable tissues.

✓ Test Your Problem-Solving Skills

13. apply c. Without aquaporins in the collecting duct, water cannot be reabsorbed, which would result in increased urine volume and decreased urine osmolarity.
14. create The mammal's urine volume would increase dramatically. Ouabain would poison Na^+/K^+-ATPase pumps in the nephron, which would prevent reabsorption of ions and nutrients. Without the osmotic gradient established by reabsorption of ions and nutrients, the water would remain in the nephron and exit as urine.
15. analyze The observation that the Na^+ concentration decreased by 30 percent, even as volume also decreased, indicates that Na^+ ions were reabsorbed to a greater degree than water. The observation that the urea concentration increased by 50 percent indicates that urea was not reabsorbed. **16.** analyze The hypothesis that the mussels are osmoconformers is supported because their hemolymph is very close in osmolarity to that of the wide range of water osmolarities into which they are placed.

CHAPTER 44

IN-TEXT QUESTIONS AND EXERCISES

p. 891 understand Nutrient absorption occurs in the small intestine, but not in the esophagus or stomach, and the rate of absorption increases with available surface area.
p. 894 Fig. 44.14 understand Frog eggs don't normally make the sodium-glucose cotransporter protein, so the researchers could be confident that if the protein appeared, it was from the injected RNA. They would not be confident of this if the RNAs were injected into rabbit epithelial cells, where the protein was probably already present.
p. 896 apply (1) Decreased energy yield from foods due to reduced glucose absorption, (2) increased fecal glucose content due to the passing of unabsorbed glucose into the colon, (3) watery feces due to decreased water reabsorption in the large intestine (lower osmotic gradient). (An additional effect would be increased flatulence due to the metabolism of unabsorbed glucose by bacteria that produce gases as waste products.)
p. 896 CYU (1) understand *Mouth:* Food is taken in; teeth physically break down food into smaller particles; salivary amylase begins to break down carbohydrates; lingual lipase initiates the digestion of fats. *Esophagus:* Food is moved to the stomach via peristaltic contractions. *Stomach:* HCl denatures proteins; pepsin begins to digest them. *Small intestine:* Pancreatic enzymes complete the digestion of carbohydrates, proteins, lipids, and nucleic acids. Most of the water and all of the nutrients are absorbed here. *Large intestine:* Water is reabsorbed and feces are formed. *Anus:* Feces accumulate in the rectum and are expelled out the anus. **(2)** apply If the release of bile salts is inhibited, fats would not be digested and absorbed efficiently, and they would pass into the large intestine. The individual would likely produce fatty feces and lose weight over time. Trypsin inactivation would inhibit protein digestion and reduce absorption of amino acids, likely leading to weight loss and muscle atrophy due to protein deprivation. If the Na^+-glucose cotransporter is blocked, then glucose would not be absorbed into the bloodstream. The individual would likely become sluggish from lack of energy.
p. 897 Fig. 44.16 understand In type 2 diabetes mellitus, the top, black arrow from glucose (in the blood) to glycogen (inside liver and muscle cells) is disrupted. In type 1 diabetes mellitus, the green arrow on the top left, indicating insulin produced by the pancreas, is disrupted.
p. 898 understand Individuals with type 1 diabetes mellitus have normal insulin receptors on their liver cells but do not produce and release insulin from the pancreas. Individuals with type 2 diabetes mellitus produce and release insulin, but they have fewer insulin receptors and/or these receptors do not respond as well to insulin. In both types of diseased individuals, glucose concentrations in the blood remain high. In individuals without disease, both insulin production and insulin receptors

are normal, so liver cells respond to insulin by taking up glucose from the blood and storing it.

p. 898 Fig. 44.17 *analyze* The data do not prove that obesity causes type 2 diabetes mellitus. The study provides correlational data and cannot directly assign causation. However, the study provides strong evidence of an association between obesity and type 2 diabetes.

p. 899 CYU (1) *understand* Type 1 diabetes mellitus is an autoimmune disease, meaning that the body's immune system mistakenly kills the insulin-producing cells of the pancreas. Type 2 diabetes mellitus is caused by a combination of factors, including genetic predisposition, obesity, and poor diet. **(2)** *apply* When the blood sugar of individuals with type 1 diabetes mellitus gets too high, they should inject themselves with insulin, which will trigger the absorption and storage of glucose by their cells. When their blood glucose gets too low, they should eat something with a high concentration of sugar (such as orange juice or a candy bar) to increase their blood glucose levels quickly. (If glucose is dangerously low, a glucagon shot may be administered.)

IF YOU UNDERSTAND . . .

44.1 *analyze* One serving of skim milk has approximately 80 kcal, while whole milk has about 152 kcal. **44.2** *apply* Its mouth or tongue would be modified to make probing of flowers and sucking nectar efficient. It would not require teeth, and its digestive tract would be relatively simple. For example, the stomach would not have to pulverize food or digest large amounts of protein, and the large intestine would not have to process and store large amounts of bulky waste. **44.3** *analyze* Amylase will work best at a neutral or slightly alkaline pH (typical of mouth and small intestine), whereas pepsin will work best at a low pH (typical of the stomach). **44.4** *understand* Type 1 diabetes mellitus leads to high glucose because insulin is not produced, and therefore glucose cannot be moved from blood into cells. Type 2 diabetes mellitus leads to high blood glucose because insulin does not bind in sufficient amounts to its receptor, so glucose cannot be transported from blood into cells.

YOU SHOULD BE ABLE TO . . .

✓ Test Your Knowledge

1. *remember* a **2.** *understand* d **3.** *remember* d **4.** *remember* b **5.** *remember* insulin, glucagon **6.** *remember* False

✓ Test Your Understanding

7. *understand* The bird crop is an enlarged sac that can hold quickly ingested food; in leaf-eating species it is filled with symbiotic organisms and functions as a fermentation vessel. The cow rumen is an enlarged portion of the stomach; the elephant cecum is enlarged relative to other species. Both structures are filled with symbiotic organisms and function as fermentation vessels. **8.** *analyze* Digestive enzymes break down macromolecules (proteins, carbohydrates, nucleic acids, lipids). If they weren't produced in an inactive form, they would destroy the cells that produce and secrete them. **9.** *understand* b **10.** *understand* Pima living in the United States are genetically similar to Pima living in Mexico, but they have higher rates of type 2 diabetes. This suggests that the difference in diabetes rates is due to environmental factors such as diet and exercise. **11.** *understand* Humans have sharp canines and incisors, which may help to bite and tear meat. Humans also have flat molars, which help in grinding grains and other plant material. **12.** *apply* In a type 1 diabetic who took too much insulin, the blood glucose would become very low. The person would become sluggish and disoriented, and might slip into a diabetic coma.

✓ Test Your Problem-Solving Skills

13. *create* A proton pump inhibitor reduces activity of the proton pumps in parietal cells, reducing the amount of HCl produced. The contents of the stomach would

therefore become less acidic, helping to prevent acid reflux. **14.** *analyze* b **15.** *evaluate* The result is still valid because the injection caused secretion—if it is also true that lack of injection led to no secretion. The criticism is also somewhat valid, however, because the researchers couldn't rule out the hypothesis that signaling from nerves plays some sort of role, too. **16.** *create* Terrestrial animals are exposed to increased risk of water loss, and the large intestine is where water reabsorption occurs. In most cases, fish do not need to reabsorb large amounts of water from their feces.

CHAPTER 45

IN-TEXT QUESTIONS AND EXERCISES

p. 905 CYU (1) *analyze* Oxygen partial pressure is high in mountain streams because the water is cold, mixes constantly, and has a high surface area (due to white water). Oxygen partial pressure is low at the ocean bottom because that is far from the surface, where gas exchange takes place, and there is relatively little mixing. **(2)** *create* *Warm-water species:* Large amount of air, because the oxygen-carrying capacity of warm water is low. *Vigorous algal growth:* Small amount of air, because algae contribute oxygen to the water through photosynthesis. *Sedentary animals:* Small amount of air, because sedentary animals require relatively little oxygen.

p. 906 Fig. 45.4 *analyze* External gills are ventilated passively and are efficient because they are in direct contact with water. They are exposed to predators and mechanical damage, however. Internal gills are protected but have to be ventilated by some type of active mechanism for water flow.

p. 907 Fig. 45.6 *apply* If "concurrent flow" occurred, oxygen transfer would stop, because the partial pressure gradient driving diffusion would fall to zero partway along the length of the capillary.

p. 909 Fig. 45.7 *analyze* Using many animals increases confidence that the results are true for most or all members of the population and are not due to one or a few unusual individuals or circumstances.

p. 912 CYU (1) *analyze* Common features include large surface area, short diffusion distance (a thin gas exchange membrane), and a mechanism that keeps fresh air or water moving over the gas exchange surface. Only fish gills use a countercurrent exchange mechanism; only tracheae deliver oxygen directly to cells without using a circulatory system; only lungs (mammalian) contain "dead space"—areas that are not involved in gas exchange. **(2)** *apply* The P_{O_2} would go down as oxygen is used up, the P_{CO_2} would go up as CO_2 diffuses into the blood but cannot be exhaled, and pH would drop as the CO_2 dissolves in blood to form bicarbonate and H^+ ions.

p. 913 *create* There would be an even larger change in the oxygen saturation of hemoglobin in response to an even smaller change in the partial pressure of oxygen.

p. 914 Fig. 45.16 *analyze* According to the data in the figure, when the tissue P_{O_2} is 30 mm Hg, the oxygen saturation of hemoglobin is about 15 percent for blood at pH 7.2 and about 25 percent at pH 7.4. Therefore, about 85 percent of the oxygen is released from hemoglobin at pH 7.2, but only about 75 percent of the oxygen is released at pH 7.4.

p. 915 Fig. 45.18 *analyze* In the lungs, a strong partial pressure gradient favors diffusion of dissolved CO_2 from blood into the alveoli. As the partial pressure of CO_2 in the blood declines, hydrogen ions leave hemoglobin and react with bicarbonate to form more CO_2, which then diffuses into the alveoli and is exhaled from the lungs.

p. 916 CYU *analyze* The curve for Tibetans should be shifted to the left relative to people adapted to sea level—meaning that their hemoglobin has a higher affinity for oxygen at all partial pressures.

p. 921 Fig. 45.24 *analyze* Air from the alveoli mixes with air in the "dead space" in the bronchi and trachea on its way

out of the body. This dead space air is from the previous inhalation ($P_{O_2} = 160$ mm Hg; $P_{CO_2} = 0.3$), so when the alveolar air mixes with the dead space air, the partial pressures in the exhaled air achieve levels intermediate between that of inhaled and alveolar air.

p. 922 *apply* Without the delay at the AV node, the ventricles would not have the chance to fully fill with blood from the atria. As a consequence, the volume of blood ejected from the ventricles would decline.

p. 924 CYU (1) *understand* Blood plasma that leaks out of the capillaries and is not reabsorbed into capillaries enters lymphatic vessels that eventually merge with blood vessels. **(2)** *understand* See **FIGURE A45.1**.

IF YOU UNDERSTAND . . .

45.1 *understand* Cellular respiration. **45.2** *create* It is harder to extract oxygen from water than it is to extract it from air. This limits the metabolic rate of water breathers. **45.3** *analyze* Large animals have a relatively small body surface area relative to their volume. If they had to rely solely on gas exchange across their skin, the skin surface area would not be large enough to exchange the volume of oxygen needed to meet their metabolic needs. **45.4** *create* Since O_2 cannot compete as well as CO for binding sites in hemoglobin, O_2 transport decreases. As oxygen levels in the blood drop, tissues (most crucially in the brain) become deprived of oxygen, and suffocation occurs. **45.5** *apply* Their circulatory systems should have relatively high pressures—achieved by independent systemic and pulmonary circulations powered by a four-chambered heart—to maximize the delivery of oxygenated blood to metabolically active tissues.

YOU SHOULD BE ABLE TO . . .

✓ Test Your Knowledge

1. *understand* b **2.** *remember* c **3.** *understand* d **4.** *understand* d **5.** *remember* c **6.** *understand* An open circulatory system is less efficient than a closed circulatory system because it is not capable of directing the hemolymph toward specific organs, and because the pressure is lower, resulting in slower flow of hemolymph.

✓ Test Your Understanding

7. *understand* During exercise, P_{O_2} decreases in the tissues and P_{CO_2} increases. The increase in P_{CO_2} lowers tissue pH. The drop in P_{O_2} and pH causes more oxygen to be released from hemoglobin. **8.** *understand* In blood, CO_2 is converted to carbonic acid, which dissociates into a proton and a bicarbonate ion. Because the protons bind to deoxygenated hemoglobin, they do not cause a dramatic drop in blood pH (hemoglobin acts as a buffer). Also, small decreases in blood pH trigger a homeostatic response that increases breathing rate and expulsion of CO_2. **9.** *understand* Because airflow through bird lungs is unidirectional, the gas exchange surfaces are

FIGURE A45.1

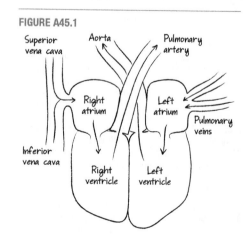

continuously ventilated with fresh, oxygenated air. The trachea and bronchi in a mammalian lung do not have a gas exchange surface, and bidirectional airflow in these species means that "stale" air has to be expelled before "fresh" air can be inhaled. As a result, the alveoli are not ventilated continuously. **10.** [understand] Lungs increase the temperature of the air and are moist to allow greater solubility of gases (increasing k); alveoli present a large surface area (large A); the epithelium of alveoli is thin (small D); and constant delivery of deoxygenated blood to alveoli maintains a steep partial pressure gradient, favoring diffusion of oxygen into the body ($P_2 - P_1$ is high). **11.** [analyze] Yes—the trait compensates for the small P_{O_2} gradient between stagnant water and the blood of the carp. **12.** [analyze] c

✔ Test Your Problem-Solving Skills

13. [create] Cold water carries more oxygen than warm water does, so icefish blood can carry enough oxygen to supply the tissues with oxygen even in the absence of hemoglobin. The oxygen and carbon dioxide are simply dissolved in the blood. **14.** [analyze] The insect tracheal system delivers O_2 directly to respiring cells; in humans, O_2 is first taken up by the blood of the circulatory system, which then delivers it to the cells. The human respiratory system also has specialized muscles devoted to ventilating the lungs. The open circulatory system of the insect is a low-pressure system, which makes use of pumps (hearts) and body movements to circulate hemolymph. The closed circulatory system of humans is a high-pressure system, which can respond to rapid changes in O_2 demand by tissues. **15.** [understand] If the pulmonary circulation were under pressure as high as that found in the systemic circulation, large amounts of fluid would be forced out of capillaries in the lungs. There is a conflict between the thin surface required for efficient gas exchange and the thickness needed in blood vessels to withstand high pressure. **16.** [analyze] b

CHAPTER 46

IN-TEXT QUESTIONS AND EXERCISES

p. 931 Fig. 46.3 [apply] No—as K^+ leaves the cell along its concentration gradient, the interior of the cell becomes more negative. As a result, an electrical gradient favoring movement of K^+ into the cell begins to counteract the concentration gradient favoring movement of K^+ out of the cell. Eventually, the two opposing forces balance out, and there is no net movement of K^+.

p. 932a [apply] $E_{Ca2+} = 58$ mV $\times$ log 1 mM/0.0001 mM)/ (+2) = 116 mV.

p. 932b [apply] The membrane potential would be near zero because the Na^+ and K^+ would diffuse across the membrane along their concentration gradients.

p. 934 CYU (1) [apply] The resting potential would fall (be much less negative) because K^+ could no longer leak out of the cell. **(2)** [understand] The size of an action potential from a particular neuron does not vary, so it cannot contain information. Only the frequency of action potentials from the same neuron varies.

p. 934 [understand] See FIGURE A46.1.

p. 935 (1) [understand] Na^+ channels open when the cell membrane depolarizes, and as more channels open, more Na^+ flows into the cell, further depolarizing the cell, causing more voltage-gated Na^+ channels to open. **(2)** [apply] Sodium would flow inward as potassium flowed outward, and their charges would cancel each other out. The membrane would not depolarize.

p. 938 CYU (1) [understand] Once the threshold level of depolarization is attained, the probability that the voltage-gated sodium channels will open approaches 100 percent. But below threshold, the massive opening of Na^+ channels does not occur. This is why the action potential is all or none. **(2)** [apply] The membrane would depolarize and stay depolarized.

p. 938 Fig. 46.10 [understand] Get a solution taken from the synapse between the heart muscle and the vagus nerve *without* the nerve being stimulated. Expose a second heart to this solution. There should be no change in heart rate.

p. 940 [analyze] The inside of the cell becomes more positive (depolarized). This shifts the membrane potential closer to the threshold, making it more likely that the postsynaptic cell will fire an action potential.

p. 942 CYU [apply] The EPSP would not be as strong if an enzyme that broke down excitatory neurotransmitter increased in concentration.

p. 943 Fig. 46.16 [analyze] In the rest-and-digest mode, pupils take in less light stimulation, the heartbeat slows to conserve energy, liver conserves glucose, and digestion is promoted by release of gallbladder products. In the fight-or-flight mode, the pupils open to take in more light, the heartbeat increases to support muscle activity, and the liver releases glucose into the blood to fuel the fight or the flight (for example, from a predator).

p. 944 Fig. 46.18 [understand] Point to your forehead and top of your head for the frontal lobe, the top and top rear of your head for the parietal lobe, the back of your head for the occipital lobe, and the sides of your head (just above your ear openings) for the temporal lobes.

p. 946 Fig. 46.21 [analyze] No—for example, part (b) indicates that the size of the brain area devoted to the trunk is no bigger than the size of the brain area devoted to the thumb.

p. 948 CYU (1) [understand] One method is to study individuals with known brain lesions and correlate the location of the defect with a deficit in mental or physical function. Another is to directly stimulate brain areas in conscious patients during brain surgery and record the response. **(2)** [remember] Learning and memory involve changes in the number, sensitivity, and placement of synapses and neurons.

IF YOU UNDERSTAND . . .

46.1 [understand] See FIGURE A46.2. **46.2** [understand] Because the behavior of voltage-gated Na^+ and K^+ channels does not change. Every time a membrane depolarizes past the threshold and all of the voltage-gated Na^+ channels are open, the same amount of Na^+ will enter the cell, and the cell will depolarize to the same extent. In addition, K^+ channels open at the same time and allow the same amount of K^+ to leave the cell, so that it repolarizes to the same extent. **46.3** [understand] Summation means that EPSPs and IPSPs produced at various sites combine to determine the response or lack of response in a postsynaptic cell. Thus the behavior of a postsynaptic cell depends on the integration of input from many synapses. **46.4** [apply] The animal would likely die because its basic bodily functions (circulation, breathing, digestion, etc.) would be disrupted.

FIGURE A46.1

FIGURE A46.2

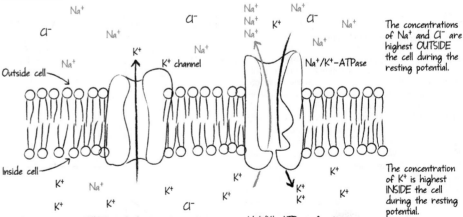

The concentrations of Na^+ and Cl^- are highest OUTSIDE the cell during the resting potential.

The concentration of K^+ is highest INSIDE the cell during the resting potential.

K^+ channels function to increase the negative charge on the inside of a resting neuron until the membrane voltage reaches the equilibrium potential.

Na^+/K^+-ATPases function to establish the Na^+ and K^+ gradients across the membrane.

✓ **Test Your Knowledge**

1. `remember` b 2. `remember` d 3. `understand` c 4. `remember` a 5. `remember` b 6. `understand` See **FIGURE A46.3**.

✓ **Test Your Understanding**

7. `understand` The Na⁺/K⁺-ATPase pumps 3 Na⁺ out of the cell for every 2 K⁺ it brings in. Since more positive charges leave the cell than enter it, there is a difference in charge on the two sides of the membrane and thus a voltage. 8. `analyze` The ligand-gated channel opens or closes in response to binding by a small molecule (for example, a neurotransmitter); the voltage-gated channel opens or closes in response to changes in membrane potential. 9. `understand` EPSPs and IPSPs are produced by flow of ions through channels in the membrane, which change the membrane potential in the postsynaptic cell. If a flow of ions at one point depolarizes the cell but a nearby flow of ions hyperpolarizes the cell, then in combination the flows of ions cancel each other out. But if two adjacent flows of ions depolarize the cell, then their effects are additive and sum to produce a total that is more likely to reach threshold. 10. `analyze` The somatic system responds to external stimuli and controls voluntary skeletal muscle activity, such as movement of arms and legs. The autonomic system controls internal involuntary activities, such as digestion, heart rate, and gland activities. 11. `apply` A drug that activates the sympathetic nervous system would increase heart rate. 12. `understand` a

✓ **Test Your Problem-Solving Skills**

13. `apply` The neurotransmitters will stay in the synaptic cleft longer, so the amount of binding to ligand-gated channels will increase. Ion flows into the postsynaptic cell will increase dramatically, affecting its membrane potential and likelihood of firing action potentials. 14. `analyze` The benefit of these approaches is that they allow the specific role of the brain region to be assessed in a living person. However, diseased or damaged brains may not respond like healthy and undamaged brains. Also, the extent of lesions and the exact location of electrical stimulation may be difficult to determine, making correlations between the regions affected and the response imprecise. 15. `analyze` c (Because the current is reduced from normal levels, but not eliminated completely, there must be more than one type of potassium channel present. The venom probably knocks out just one specific type of channel.) 16. `create` A reasonable hypothesis is that the new neurons arise from stem cells in the brain.

CHAPTER 47

IN-TEXT QUESTIONS AND EXERCISES

p. 959 CYU (1) `apply` A punctured eardrum wouldn't vibrate correctly and would result in hearing loss at all frequencies in the affected ear. If the stereocilia are too short to come into contact with the tectorial membrane, vibration of the basilar membrane will not cause them to bend, and sound will not be detected. A loss in basilar membrane flexibility would result in the inability to hear lower-pitched sounds, such as human speech. **(2)** `understand` The lateral line system senses pressure waves in the water, allowing fishes to detect the presence of such objects as prey, predators, or potential mates.
p. 959 Fig. 47.9 `create` Vision could potentially be used to detect prey in diurnal catfish. To control for this, catfish could be blinded or "blindfolded" with an opaque material over their eyes to permit study of the lateral line system.
p. 961 Fig. 47.12 `create` The change in retinal's shape would induce a change in opsin's shape. A protein's

function correlates with its shape, so opsin's function is likely to change as well.
p. 962 `analyze` cGMP is a second messenger in the sense that it signals the sodium channel that transducing is inactive, and it is also a ligand that opens sodium channels. Transducin is like a G protein because it switches from "off" to "on" in response to a receptor protein, activating a key protein as a result. Closing Na⁺ channels stops the entry of positive charges into the cell, making the inside of the cell more negative relative to the outside.
p. 962 Fig. 47.13 `understand` This ion flow occurs when the photoreceptor cell is not receiving light—when the cell is in the dark.
p. 963 Fig. 47.15 `create` The S, M, and L opsin proteins are each different in structure. These structural differences affect the ability of retinal to absorb specific frequencies of light and change shape.
p. 964 CYU (1) `understand` Retinal acts like an on–off switch that indicates whether light has fallen on a rod cell. When retinal absorbs light, it changes shape. The shape change triggers events that result in a change in action potentials, signaling that light has been absorbed. **(2)** `apply` Retina detachment from a jarring car accident would separate the retina from the optic nerve, resulting in blindness. The mutation would produce blue–purple color blindness, because the S opsin responds to wavelengths in that region of the spectrum. A clouded lens would reduce the amount of light that reaches the retina, reducing visual sensitivity.
p. 967 CYU (1) `understand` Extremely hot food damages taste receptors, so those proteins would no longer be able to respond to their chemical triggers. **(2)** `create` One hypothesis to explain why the vomeronasal organ is vestigial in most primates is that primates have evolved other complex methods of communication, including gestures, facial expressions, and language, that render sensation with the vomeronasal organ less important.
p. 967 `apply` (1) A rattlesnake with its eyes covered will strike effectively because its pits will sense the prey. (2) A rattlesnake with its eyes and pits covered with cotton cloth will strike effectively. The cloth will block vision but the heat from the animal will penetrate the cloth to stimulate the thermoreceptors in the pits. (3) A rattlesnake with its eyes and pits covered with an opaque heat-blocking material will not be able to strike effectively, because the material blocks vision and heat energy.

p. 969 CYU (1) `create` Electrogenic fishes may have a survival advantage over other fishes in murky water, where vision would be compromised. **(2)** `create` Female sea turtles likely use magnetic cues to locate their natal beaches.

IF YOU UNDERSTAND . . .

47.1 `create` The sensory neurons that used to serve that limb are cut, but may still fire action potentials that stimulate brain areas associated with awareness of pain in the missing limb. **47.2** `analyze` Each of these senses involves ion channels that open in response to application of pressure to cell membranes. **47.3** `understand` Because different animal species have different sets of opsin molecules that respond to different wavelengths of light absorbed by retinal they see different colors. **47.4** `create` Smells are part of the sensation of flavor, and people with nasal congestion cannot smell. In addition to smell being important, the small number of taste receptors can send many different frequencies of action potentials to the brain, and the taste receptors can be stimulated in many different combinations. **47.5** `understand` Capsaicin binds to thermoreceptors that are also activated at high temperatures, leading to the perception of heat. It can also bind to some nociceptors.

YOU SHOULD BE ABLE TO . . .

✓ **Test Your Knowledge**

1. `understand` a 2. `remember` d 3. `understand` c 4. `remember` b 5. `remember` false 6. `remember` magnetoreception

✓ **Test Your Understanding**

7. `understand` b 8. `understand` The many possible examples include several blue opsins allowing coelacanths to distinguish the blue wavelengths present in their deep-sea habitat, infrasound hearing allowing elephants to hear over long distances, ultrasonic hearing allowing bats to hunt by echolocation and moths to avoid bat predators, red and yellow opsins allowing fruit-eating mammals to distinguish ripe from unripe fruit. 9. `analyze` The data provide evidence that a pheromone acted to synchronize the women's menstrual cycles because the average time between onset of a woman's cycle and that of the rest of the women declined over time following move-in to the dorms in fall. It declined most rapidly after they first moved in together. The leveling of the graph over time

FIGURE A46.3

DEPOLARIZATION: Na⁺ channels open, allowing influx of Na⁺ into cell.

REPOLARIZATION: K⁺ channels are open, allowing K⁺ to exit the cell; Na⁺ channels are inactivated.

HYPERPOLARIZATION: K⁺ channels have remained open and are closing slowly, temporarily hyperpolarizing the cell.

Resting potential

Membrane potential (mV)

40

0

−65

Resting potential: K⁺, Na⁺, and Cl⁻ are at equilibrium with respect to their electrochemical gradients.

Resting potential is reestablished.

0   2   4

Time (msec)

shows that the onset dates became more synchronized. **10.** (understand) Ion channels in hair cells are thought to open in response to physical distortion caused by the bending of stereocilia. Ion channels in taste receptors, in contrast, change shape and open when certain bitter-tasting molecules bind to them. **11.** (evaluate) Dalton's hypothesis was reasonable because blue fluid absorbs red light and would prevent it from reaching the retina, and this could be tested by dissecting his eyes. Although the hypothesis was rejected, it inspired a rigorous test and required researchers to think of alternative explanations. **12.** (create) To determine whether electric eels are electrogenic, place an eel in a tank and measure the electricity of water to determine if the eel is emitting an electric field. To determine whether they are electroreceptive, place an eel in a tank in the dark and add an object the size and shape of a prey item that emits no scent or taste. Block the lateral line system with a chemical as in Figure 47.9. If the eel locates the object, it must do so using electroreception.

✓ Test Your Problem-Solving Skills

13. (understand) a; vomeronasal organs evolved in tetrapods, presumably as an adaptation for communication in terrestrial habitats. **14.** (analyze) Both the lateral line system and electroreceptors use hair cells to sense changes in the water. Hair cells in the lateral line system contain mechanoreceptors that are depolarized in response to changes in water pressure. In contrast, electroreceptors cause hair cells to depolarize in response to changes in the electric field. **15.** (evaluate) It is unlikely that a perfume contains human pheromones because they have never actually been characterized, although their presence is suspected. **16.** (create) Block all but a few hundred ommatidia in the center of each compound eye of many dragonflies with an opaque material, and observe any differences that might occur in their ability to detect and pursue prey compared to individuals whose ommatidia had been covered by a transparent material.

CHAPTER 48

IN-TEXT QUESTIONS AND EXERCISES

p. 974 (understand) The trucks are the Z discs, the ropes are the thin filaments, and the burly weightlifters are the thick filaments.
p. 974 Fig. 48.2 (analyze) The dark band includes thin filaments and a dense concentration of bulbous structures extending from the thick filament; the light band consists of thin filaments only.
p. 977 CYU (1) (understand) In a sarcomere, thick myosin filaments are sandwiched between thin actin filaments. When the heads on myosin contact actin and change conformation, they pull the actin filaments toward one another, shortening the whole sarcomere. **(2)** (apply) Increased acetylcholine release would result in an increased rate of muscle-cell contraction. Preventing conformational changes in troponin would prevent muscle contraction. Blocking the uptake of calcium ions into the sarcoplasmic reticulum would lead to sustained muscle contraction.
p. 980 (apply) The breast meat of pigeons is composed of dark meat because these muscles are specialized for endurance. Their dark color is due to a high concentration of myoglobin.
p. 981 (create) When earthworms shorten their longitudinal muscles, the segment containing these muscles shortens and squeezes the internal fluid, which becomes pressurized and pushes in all directions, expanding the circumference of the segment and pushing laterally against the burrow.
p. 983 CYU (1) (analyze) Exoskeletons are made of chitin, proteins, and other substances like calcium carbonate. Endoskeletons are made of calcium phosphate, calcium

carbonate, and proteins. Exoskeletons occur on the outsides of animals, whereas endoskeletons occur on the insides. Both types of skeletons are composed of rigid levers separated by joints such that motions are a result of changes in joint angles. Both types of skeletons attach to skeletal muscle and serve to transmit muscle forces. Hydrostatic skeletons also transmit muscle forces but are made of soft tissues that vary from animal to animal. Shape changes in hydrostatic skeletons occur not from changes in joint angles, but from shape changes of the bodies themselves. **(2)** (apply) No movement of the arm would occur, because for these antagonistic muscles to produce movement, one of them must be contracted while the other is relaxed.
p. 986 Fig. 48.16 (apply) Galloping at 3.5 m/s would cost about 75 percent more energy.
p. 987 Fig. 48.18 (apply) About 10 times more costly.

IF YOU UNDERSTAND . . .

48.1 (apply) Paralysis, because acetylcholine has to bind to its receptors on the membrane of postsynaptic muscle fibers for action potentials to propagate in the muscle and cause contraction. **48.2** (analyze) Skeletal muscle is multinucleate, unbranched, striated, and voluntary. Cardiac muscle is branched, striated, and involuntary. Smooth muscle is smooth and involuntary. **48.3** (analyze) When muscles contract in hydrostatic skeletons, the body deforms by becoming longer and thinner, or shorter and narrower, or by bending side to side. In vertebrates and arthropods, muscle contractions cause changes in the joint angles between rigid segments rather than shape changes in the segments themselves. **48.4** (understand) On land, an animal is constrained by its weight; larger animals cannot move as freely and are in greater danger of breaking than are smaller animals. In water, animals are constrained by their shapes. Some shapes would cause high drag, which would inhibit locomotion. In air, animals are constrained by their weight. Larger animals must create more lift to overcome gravity.

YOU SHOULD BE ABLE TO . . .

✓ Test Your Knowledge

1. (remember) a **2.** (remember) c **3.** (understand) d **4.** (remember) True
5. (understand) a **6.** (understand) False.

✓ Test Your Understanding

7. (create) The key observation was that the banding pattern of sarcomeres changed during contraction. Even though the entire unit became shorter, only some portions moved relative to each other. This observation suggested that some portions of the structure slid past other portions. Muscle fibers have to have many mitochondria because large amounts of ATP are needed to power myosin heads to move along actin filaments; large amounts of calcium stored in smooth ER are needed to initiate contraction by binding to troponin.
8. (apply) c **9.** (analyze) Acetylcholine reduces the rate and force of contraction of cardiac muscle, whereas it stimulates skeletal muscle to contract. **10.** (create) ATP is no longer produced after death, so calcium cannot be actively transported from the cytosol into the sarcoplasmic reticulum to enable the binding of myosin and actin. The proteins do not unbind, because this process requires ATP. **11.** (apply) The oxygen consumption of the runner would increase because the arches of his or her feet would no longer store as much elastic energy, which normally reduces the energetic cost of running.
12. (create) Because diverse animals experience the same physical constraints where they live, natural selection favors the same kind of adaptations to those constraints, resulting in convergent evolution. For example,

dolphins and ichthyosaurs are distantly related but have a similar body shape due to their similar ability to swim rapidly through water.

✓ Test Your Problem-Solving Skills

13. (apply) In cardiac muscle, the binding of acetylcholine to its receptors causes the heart rate to slow. Ingestion of nightshade would increase heart rate because it blocks acetylcholine receptors. **14.** (apply) If the sarcomere is in a stretched state before stimulation, the initial force production would be reduced because less overlap would occur between actin and myosin; thus, fewer myosin heads could engage in the pull. **15.** (apply) c:

$$0.5 = v^2/(9.8 \text{ m/s}^2 \times 0.9 \text{ m})$$
$$0.5 = v^2/8.8 \text{ m}^2/\text{s}^2$$
$$v^2 = 0.5 \times 8.8 \text{ m}^2/\text{s}^2$$
$$v = \text{square root of } 4.4 \text{ m}^2/\text{s}^2$$
$$v = 2.1 \text{ m/s}$$

16. (evaluate) The bones of *T. rex* are available and can be analyzed to determine their structural properties, such as the mechanical advantage of the leg joints and the ability of the bones to withstand the different forces that the skeleton would experience at different speeds based on data from living animals. A very close estimate could be made, but the exact speed would be hard to determine without observing the dinosaur in action, since behavior affects speed.

CHAPTER 49

IN-TEXT QUESTIONS AND EXERCISES

p. 994 (apply) The current temperature is analogous to the sensory input; the thermostat, to the CNS; the thermostat signal, to the cell-to-cell signal; and the furnace, to the effector. If feedback inhibition fails, hormone production will continually increase.
p. 996 Fig. 49.4 (understand) Injecting the dog with a liquid extract from a different part of the body, or with a solution that is similar in chemical composition to the extract from the pancreas (e.g., in pH or ions present) but lacking molecules produced by the pancreas.
p. 997 CYU (1) (understand) It is difficult to differentiate between the nervous and endocrine systems because some neurons secrete hormones and some endocrine glands respond to neural signals. **(2)** (apply) Thyroid hormone receptors are intracellular because thyroid hormones are lipid soluble.
p. 998 (create) The cells could differ in receptors, signal transduction systems, or response systems (genes or proteins that can be activated).
p. 1001 (understand) Stressed individuals would be less able to heal or to fight pathogens.
p. 1001 Fig. 49.9 (understand) The saline injection controlled for any stress induced by the injection procedure and for introducing additional fluids.
p. 1002 Fig. 49.11 (apply) Same—a defect in both signal and receptor has the same effect—no response—as a defective signal alone or a defective receptor alone.
p. 1003 CYU (1) (analyze) In amphibians, an increase in thyroid hormones stimulates metamorphosis from a tadpole to an adult without a resting stage. In insects, a decrease in juvenile hormone and increase in ecdysone stimulate metamorphosis during a resting (pupal) stage. **(2)** (understand) By reducing immune system function, promoting release of fatty acids from storage cells and use of amino acids from muscles for energy production, and preventing release of glucose in response to signals from insulin, cortisol conserves glucose supplies for use by the brain.
p. 1004 Fig. 49.12 (create) Inject ACTH and monitor cortisol levels. If cortisol does not increase, adrenal failure is likely.

p. 1006 CYU (1) `understand` ACTH triggers the release of cortisol, but cortisol inhibits ACTH release by blocking the release of CRH from the hypothalamus and suppressing ACTH production. High levels of cortisol tend to lower cortisol levels in the future. **(2)** `understand` Processing centers in the brain are responsible for synthesizing a wide array of sensory input. To start a response to this sensory input, they stimulate neurosecretory cells in the hypothalamus. Neurohormones from these cells travel to the anterior pituitary, where they trigger the production and release of other hormones.

p. 1010 CYU (1) `apply` The steroid-hormone–receptor complex would probably fail to bind to the hormone-response element. Then gene expression would not change in response to the hormone—the arrival of the hormone would have little or no effect on the target cell. **(2)** `analyze` The mode of action would be similar to that of a peptide—hormone binding would have to trigger a signal transduction event that resulted in production of a second messenger or a phosphorylation cascade.

IF YOU UNDERSTAND . . .

49.1 `understand` The production and release of hormones are often controlled, directly or indirectly, by electrical or chemical signals from the nervous system. A good example is the hypothalamic–pituitary axis, where neuroendocrine signals regulate endocrine signals. **49.2** `analyze` Electrical signals are short lived and localized. All of the processes listed in the question are long-term changes in the organism that require responses by tissues and organs throughout the body. **49.3** `apply` The hypothalamus would continue to release CRH, stimulating the pituitary to release ACTH, which in turn would stimulate continued release of cortisol from the adrenal cortex. Sustained, high circulating levels of cortisol have negative effects. **49.4** `apply` A cell can respond to more than one hormone at a time, because cells produce different receptors for each hormone.

YOU SHOULD BE ABLE TO . . .

✔ Test Your Knowledge

1. `understand` c **2.** `understand` a **3.** `remember` a **4.** `remember` c **5.** `remember` Mammals—eye and pineal gland; lizards—pineal gland; birds—diffuse in brain **6.** `remember` True

✔ Test Your Understanding

7. `understand` b **8.** `create` This is one reason that the same hormone can trigger different effects in different tissues. For example, epinephrine binds to four different types of receptors in different tissues—eliciting a different response from each. **9.** `analyze` The posterior pituitary is an extension of the hypothalamus; the anterior pituitary is independent—it communicates with the hypothalamus via chemical signals in blood vessels. The posterior pituitary is a storage area for hypothalamic hormones; the anterior pituitary synthesizes and releases an array of hormones in response to releasing hormones from the hypothalamus. **10.** `analyze` Steroid hormones act directly, and alter gene expression. They usually bind to receptors inside the cell, forming a complex that binds to DNA and activates transcription. Nonsteroid hormones act indirectly, and activate proteins. They bind to receptors on the cell surface and trigger production of a second messenger or a phosphorylation cascade, ending in activation of proteins already present in the cell. **11.** `create` Subject volunteers to travel and jet lag, measure ACTH and cortisol levels, and correlate these results with their perceived level of jet lag symptoms. **12.** `create` Studies have shown that treatment with leptin is largely ineffective in reducing food intake in obese humans.

✔ Test Your Problem-Solving Skills

13. `apply` c (dehydration would result, because ADH would not be present) **14.** `create` There are two basic strategies: (1) remove the structure from some individuals and compare their behavior and condition to untreated individuals in the same environment, or (2) make a liquid extract from the structure, inject it into some individuals, and compare their behavior and condition to individuals in the same environment that were injected with water or a saline solution. **15.** `evaluate` The researcher's hypothesis initially sounds unreasonable, because peptide hormones cannot enter cells to initiate gene transcription. However, a second messenger activated by the hormone could possibly lead to changes in gene transcription. **16.** `evaluate` The graph shows that atrazine treatment reduces the testosterone concentrations in male frogs to near the levels of females. This result supports the hypothesis that atrazine is an endocrine disruptor that feminizes male amphibians.

CHAPTER 50

IN-TEXT QUESTIONS AND EXERCISES

p. 1015 Fig. 50.3 `create` Isolate and identify the molecules found in "crowded" water. Test each molecule by adding it, at the same concentration found in crowded water, to clean water occupied by a single *Daphnia* and recording whether the female produces a male-containing brood. Repeat with many test females, and for each molecule identified, record the number of male-containing broods produced in clean water.

p. 1016 `apply` Asexual reproduction would be expected in environments where conditions change little over time.

p. 1019 Fig. 50.7 `evaluate` No—the data are consistent with the displacement hypothesis, but there is no direct evidence for it. There are other plausible explanations for the data.

p. 1021 CYU (1) `analyze` Oviparity usually requires less energy input from the mother after egg laying, and mothers do not have to carry eggs around as long—meaning that they can lay more eggs and be more mobile. However, before egg laying, mothers have to produce all the nutrition the embryo will require in the egg, and eggs may not be well protected after laying. Viviparity usually increases the likelihood that the developing offspring will survive until birth, but it limits the number of young that can be produced to the space available in the mother's reproductive tract. If viviparous young can be nourished longer than oviparous young, then they may be larger and more capable of fending for themselves. **(2)** `create` Divide a population of sperm into two groups. Subject one group of sperm to spermkillerene at concentrations observed in the female reproductive tract (the experimental group) but not the other group (the control). Document the number of sperm that are still alive over time in both groups.

p. 1025 Fig. 50.13 `create` (Note: There is more than one possible answer—this is an example.) Having two gonads is an "insurance policy" against loss or damage. To test this idea, surgically remove one gonad from a large number of male and female rats. Do a similar operation on a large number of similar male and female rats, but do not remove either gonad. Once the animals have recovered, place them in a barn or other "natural" setting and let them breed. Compare reproductive success of individuals with paired versus unpaired gonads.

p. 1026 Fig. 50.15 `analyze` It is similar. In both cases, the hypothalamus produces a releasing factor (GnRH or CRH) that acts on the pituitary. The releasing factor stimulates release of regulatory hormones from the anterior pituitary (LH and FSH or ACTH). These hormones travel via the bloodstream and act on the gonads or adrenals to induce the release of hormones from these glands. In both cases, the hormones are involved in negative feedback control of the regulatory hormones from the pituitary.

p. 1029 `apply` (1) There should be an LH spike early in the cycle, followed by early ovulation. (This might not actually occur, though, if the follicle is not yet ready to rupture.) (2) LH levels should remain low—no mid-cycle spike and no ovulation.

p. 1030 CYU (1) `understand` FSH triggers maturation of an ovarian follicle. Its level rises at the end of a cycle because it is no longer inhibited by progesterone, which stops being produced at high levels when the corpus luteum degenerates. **(2)** `apply` The drug would keep FSH levels low, meaning that follicles would not mature and would not begin producing estradiol and progesterone. The uterine lining would not thicken.

p. 1032 `analyze` (1) The uterine lining may not be maintained adequately—if it degenerates, a miscarriage is likely. (2) Pregnancy tests detect hCG in urine. By the third trimester, hCG is no longer being produced by the placenta.

p. 1034 Fig. 50.23 `apply` The birth will be more difficult, as limbs or other body parts can get caught.

p. 1034 Fig. 50.24 `apply` A mother's chance of dying in childbirth in 1760 was a little over 1000 in 100,000 live births, or close to 1.0 percent. If she gave birth 10 times, she would have a 10 percent chance of dying in childbirth.

IF YOU UNDERSTAND . . .

50.1 `create` Parasites could wipe out a population of asexually produced animals because the animals do not vary genetically. However, a population of sexually produced animals would have enough genetic variability that some individuals could present better resistance to the parasites. **50.2** `create` Parental care demands resources (time, nutrients) that cannot be used to produce more eggs. Sperm competition is not likely when external fertilization occurs, and most fishes use external fertilization. **50.3** `create` The most common methods are surgical ligation (tying off) of the fallopian tubes in a woman to stop the delivery of the egg to the uterus and cutting the vas deferens (vasectomy) in a man to stop the delivery of sperm into the semen. **50.4** `apply` FSH and LH levels would increase following ovariectomy, because sex steroids would no longer be present to exert negative feedback upon the anterior pituitary. **50.5** `understand` The placenta is highly vascularized to provide a large surface area for transfer of oxygen and nutrients from the mother's circulation to the fetus's, and of wastes from the fetus's circulation to the mother's.

YOU SHOULD BE ABLE TO . . .

✔ Test Your Knowledge

1. `remember` a **2.** `understand` b **3.** `remember` a **4.** `remember` b **5.** `remember` Luteinizing hormone (LH) and follicle-stimulating hormone (FSH) **6.** `remember` True

✔ Test Your Understanding

7. `remember` Every offspring that is produced sexually is genetically unique; offspring that are produced asexually have DNA from their mother only and often are genetically identical to her. **8.** `create` *Daphnia* females produced males only when they were exposed to short day lengths *and* water from crowded populations *and* low food levels. These conditions are likely to occur in the fall. Sexual reproduction could be adaptive in fall if genetically variable offspring are better able to thrive in conditions that occur the following spring. **9.** `analyze` Spermatogenesis generates four haploid sperm cells from each primary spermatocyte; oogenesis produces only one haploid egg cell from each primary oocyte. Egg cells are much larger than sperm cells because they contain more cytoplasm. In males, the second meiotic division occurs right after the first meiotic division, but in females it is delayed until fertilization. **10.** `understand` LH triggers release of estradiol, but at low levels estradiol inhibits further release of LH. LH and FSH trigger release of progesterone, but progesterone inhibits further release of LH and FSH. High levels of estradiol trigger release of more LH. The follicle can produce high levels of estradiol only if it has grown and matured—meaning that it is ready for ovulation to occur. **11.** (analyze) c **12.** (apply) Females should be choosier about their second mate because the sperm of the second male has an advantage in sperm competition.

✔ Test Your Problem-Solving Skills

13. (create) If all sheep were genetically identical, it is less likely that any individuals could survive a major adverse event (e.g., a disease outbreak or other environmental challenge) than if the population were composed of genetically diverse individuals, which are likely to vary in their ability to cope with the new conditions. **14.** (apply) Oviparous populations should produce larger eggs than viviparous populations. Because oviparous species deposit eggs in the environment, the eggs must contain all the nutrients and water required for the entire period of embryonic development. But because embryos in viviparous species receive nourishment directly from the mother, it is likely that their eggs are smaller. **15.** (create) One hypothesis might be that the shell membrane is a vestigial trait (see Chapter 25)—specifically, an "evolutionary holdover" from an ancestor of today's marsupials that laid eggs. **16.** (analyze) b

CHAPTER 51

IN-TEXT QUESTIONS AND EXERCISES

p. 1040 (apply) See **FIGURE A51.1**. Each type of Toll protein is a pattern-recognition receptor that responds to a different group of pathogens. Since bacteria and fungi are very different pathogens, you would not expect a receptor that recognizes fungi to be required for a response against bacteria, so flies lacking Toll would survive as well as the wild-type control.
p. 1041 CYU (analyze) (1) Applying direct pressure constricts blood vessels, mimicking the effect of histamine released from mast cells. (2) Cleaning the wound to remove dirt and debris mimics the phagocytic activity of macrophages and neutrophils to eliminate foreign cells and material. (3) Applying bandages impregnated with platelet-recruiting compounds mimics the effect of chemokines released from injured tissues and macrophages, which attract circulating leukocytes and platelets to facilitate blood clotting and wound repair.
p. 1041 Fig. 51.3 (understand) Rubor (reddening) is due to increased blood flow to the infected or injured area. Mast cells trigger dilation of vessels by releasing signaling molecules. Calor (heat) is the result of fever, which is activated by the release of cytokines from macrophages that are in the area of infection. Tumor (swelling) occurs

due to dilation of blood vessels triggered by mast cells and macrophages.
p. 1046 (apply) (1) In a human, 200 different light chains are possible (see text). The recombination of the V, D, and J segments results in $51 \times 27 \times 6 = 8262$ possible heavy chains. Since these chains combine to produce BCRs, a total of $200 \times 8262 = 1,652,400$ are possible within a human. (2) Only one type of antibody can be produced from a single human B cell.
p. 1047 CYU (1) (understand) The production of BCRs and TCRs via DNA recombination is required for an effective response. Selection against self-reactive lymphocytes is required for a safe response. **(2)** (understand) The cell is able to make a larger number of differing variable regions per unit of DNA when the variable region is segmented. For example, 3 segments of V and 3 segments of J can randomly assemble into 9 different variable light chains, whereas only 6 light chains would be possible with the same amount of DNA if the V regions were preassembled.
p. 1048 (understand) "Clonal" refers to the cloning—producing many exact copies—of cells that are "selected" by the binding of their receptor to a specific antigen (or MHC–peptide complex, which is introduced later).
p. 1050 (understand) See **TABLE A51.1**.
p. 1051 CYU (create) Individuals who are heterozygous for MHC genes have greater variability in their MHC proteins than do individuals who are homozygous for these genes. The increased variability in MHC proteins allows a greater variety of peptides to be presented to T cells, which thus would be able to recognize, attack, and eliminate a wider range of pathogens compared with T cells in homozygotes. As a result, heterozygous individuals would likely suffer fewer infections than homozygous individuals.
p. 1054 Fig. 51.17 (apply) See **FIGURE A51.2**. The response to the new antigen would resemble the primary response that occurred after the first injection of the original antigen.

p. 1055 CYU (analyze) Attenuated viruses are able to infect the cells of a vaccinated host, but not well enough to cause disease. Inactivated viruses are not able to infect cells at all, and instead remain as extracellular particles. Since attenuated viruses can infect cells, their antigens will be processed and presented by class I MHC complexes, which will promote a stronger cell-mediated response than the inactivated viruses.

IF YOU UNDERSTAND . . .

51.1 (understand) Toll-like receptors recognize molecular patterns of broad groups of pathogens, such as viruses, fungi, and bacteria. Upon binding to these molecules, the receptors cause signals to be transduced that activate cellular responses directed at eliminating these types of pathogens. **51.2** (analyze) The BCR locus would be smaller in mature B cells compared to immature cells, since recombination between the gene segments resulted in removing sections of DNA. **51.3** (understand) Antigen-presenting cells activate the CD4$^+$ T cells that are required for the full activation of B cells to differentiate into plasma cells. **51.4** (understand) Extracellular viruses are agglutinated and inactivated by antibodies produced in the humoral response. Intracellular viruses are eliminated by the destruction of their host via cytotoxic T cells in the cell-mediated response. **51.5** (apply) The IgE antibodies generate hypersensitive responses. When these are severe, they can lead to anaphylactic shock. If IgE antibodies were generated against self molecules, you would expect a strong response that will lead to either anaphylactic shock or chronic inflammation.

YOU SHOULD BE ABLE TO . . .

✔ Test Your Knowledge

1. (remember) c **2.** (understand) a **3.** (remember) b **4.** (understand) They are identical except that a B-cell receptor has a transmembrane domain that allows it to be located in the B-cell

---

TABLE A51.1

MHC type	Origin of Peptide	Type of T cell that binds	Activity stimulated
Class I MHC	Host cell cytosol	CD8+	Activates T cell to secrete molecules that kill infected cell
Class II MHC	Extracellular environment	CD4+	Activates T cell to secrete cytokines that support the immune response of other cells

---

FIGURE A51.2

FIGURE A51.1

plasma membrane. **5.** `remember` c **6.** `understand` The hygiene hypothesis claims that immune disorders such as allergies and autoimmunity have increased in frequency due to the implementation of hygienic practices that decrease common bacterial, viral, and parasitic infections.

✓ Test Your Understanding

7. `understand` B cells are first activated to divide via interactions between the BCRs and their complementary antigens. They are then fully activated to differentiate and continue to divide when their class II MCH–peptide complex interacts with an activated CD4$^+$ T cell that has a complementary TCR. **8.** `analyze` Both the BCR and TCR interact with epitopes of antigens via binding sites located in the variable regions of their polypeptide chains. A TCR is like one "arm" of a BCR. The BCR can interact with an antigen directly, while the TCR interacts with its epitope in the context of an MHC protein presented by another cell. **9.** `analyze` Vaccines have to contain an antigen that can stimulate an appropriate primary immune response. Vaccines have not worked for HIV, because the antigens on this virus are constantly changed through mutation, rendering it unrecognizable to memory cells generated following vaccination. **10.** `analyze` c **11.** `analyze` Pattern-recognition receptors on leukocytes bind to surface molecules that are present on many pathogens. BCRs and TCRs bind to particular epitopes of antigens that are pathogen specific. Pattern-recognition-receptor binding directly activates the cell, while BCRs and TCRs often require additional signals to become fully activated. **12.** `understand` The mixing and matching of different combinations of gene segments from the variable and joining regions of the light-chain immunoglobulin gene, along with diversity regions of the heavy-chain gene, gives lymphocytes a unique sequence in both chains of the BCR. The variable regions in the alpha and beta chains of the TCR are assembled from segments in a similar manner.

✓ Test Your Problem-Solving Skills

13. `apply` d **14.** `create` If B cells have receptors that bind to self molecules during maturation, they would become activated during the maturation process. Instead of the signal causing the cells to divide, these immature B cells might respond to the signal by becoming permanently inactivated or might be induced to undergo apoptosis. **15.** `analyze` Natural selection favors individuals that can create a large array of antibodies, because the high mutation rates and rapid evolution observed in pathogens means that they will constantly present the immune system with new antigens. If these antigens were not recognized by the immune system, the pathogens would multiply freely and kill the individual. **16.** `analyze` Tissue from the patient's own body is marked with the major histocompatibility (MHC) proteins that the patient's immune system recognizes as self. This results in the preservation, rather than the destruction, of the transplanted tissue. Tissue from a different person is marked with MHC proteins that are unique to that individual. The body of the transplant recipient will recognize the MHC proteins (and other molecules) of the donor as foreign, resulting in a full immune response and rejection of the grafted tissue.

CHAPTER 52

IN-TEXT QUESTIONS AND EXERCISES

p. 1062 Fig. 52.1 `apply` It would increase, because cattle would no longer succumb to the disease carried by tsetse flies. (In Africa, cattle are limited more by biotic conditions than abiotic conditions.)
p. 1064 Fig. 52.4 `apply` Since there were 28 plots and four treatments, there were probably about seven replicate plots per treatment. This is important because the abiotic and biotic characteristics of individual plots are likely to vary in natural landscapes.
p. 1066 `apply` See **FIGURE A52.1**.
p. 1067 CYU (1) `apply` There is little rain at the North Pole because dense, dry air descends and absorbs moisture.
(2) `apply` The highest risk of skin cancer is in the tropics where the Sun is often directly overhead, resulting in a large amount of solar radiation per unit area.
p. 1069 Fig. 52.10 `analyze` Most of the remaining wildlands occur in Canada, at the top of the image.
p. 1070a `understand` Vines and epiphytes increase productivity because they are photosynthetic organisms that fill space between small trees and large trees—they capture light and use nutrients that might not be used in a forest that lacked vines and epiphytes.
p. 1070b `understand` Most leaves have a large surface area to capture light. Light is abundant in deserts, however, and plants with a large surface area would be susceptible to high water loss and/or overheating.
p. 1071a `understand` Crop grains like wheat and corn are grasses, so they thrive in the grassland biome.
p. 1071b `apply` There is a continuous grass cover and scattered trees. (A biome like this is called a savanna.)
p. 1072a `understand` Boreal forests should move north.
p. 1072b `understand` High elevations present an abiotic environment (precipitation and temperature) that is similar to the conditions in arctic tundra. As a result, the plant species present will have similar adaptations to cope with these physical conditions. These similarities are the result of convergent evolution.
p. 1073 Fig. 52.17 `analyze` Visible wavelengths enter the chamber from the top and through the glass, warming the plants and ground inside the chamber. Some of this heat energy is retained within the mini-greenhouse.
p. 1074 CYU (1) `analyze` Tropical dry forests are probably less productive than tropical wet forests because they have less water available to support photosynthesis during some periods of the year. **(2)** `apply` Your answer will depend on where you live. For example, if global warming continues at the current projected rates, several effects can be expected. If you live in a coastal area, you can expect water levels to rise. In general, plant communities will change because average temperature and moisture—and variation in temperature and moisture—will change. Also, development is likely to increase in most areas, transforming the landscape.
p. 1077a `apply` The littoral zone, because light is abundant and nutrients are available from the substrate.
p. 1077b `apply` Bogs are nitrogen poor, so plants that are able to capture and digest insects have a large advantage. This advantage does not exist in marshes and swamps, where nutrients are more readily available.
p. 1078a `apply` Cold water contains more oxygen than warm water, so it can support a higher rate of cellular respiration.
p. 1078b `apply` Species that live in estuaries must be able to tolerate variable salinity; freshwater marsh species do not. The abiotic environment (salt concentration) is so different that few species grow well in both habitats.
p. 1079 `apply` No—the aphotic zone is lightless, so natural selection favors individuals that do not invest energy in developing and maintaining eyes.

IF YOU UNDERSTAND . . .

52.1 `understand` Populations are made up of individual organisms; communities are made up of populations of different species; ecosystems are made up of groups of communities along with the abiotic environment; the biome is made up of all the world's ecosystems. **52.2** `apply` The Argentine ants would likely increase their range into the moist agricultural fields, displacing native ants. **52.3** `apply` All latitudes would get equal amounts of sunlight all year round. There would be no seasons and no changes in climate with latitude. **52.4** `apply` If precipitation does not increase, then increased transpiration rates will put plants under water stress and reduce productivity. **52.5** `apply` Like deserts, open oceans have plenty of sunlight but very low productivity. Unlike deserts, which are water limited, open oceans are nutrient limited.

YOU SHOULD BE ABLE TO . . .

✓ Test Your Knowledge

1. `remember` Organismal, population, community, ecosystem, and biosphere/global **2.** `remember` c **3.** `remember` b **4.** `remember` d **5.** `remember` c **6.** `remember` a

✓ Test Your Understanding

7. `create` Examples of possible questions: *Organismal:* How do humans cope with extremely hot (or cold) weather conditions? *Population:* How large will the total human population be in 50 years? *Community:* How are humans affecting ocean species such as cod and tuna? *Ecosystem:* How does nitrogen runoff from agricultural fields affect the availability of oxygen in the Gulf of Mexico? *Global:* How will human-induced changes in global temperature affect sea level? **8.** `apply` *Biotic:* These ants compete with native ants. *Abiotic:* Argentine ants require moist conditions, not too cold, not too hot. These factors interact; Argentine ants outcompete native ants in moist environments, but competition slows down their dispersal. **9.** `apply` b **10.** `analyze` Productivity in intertidal and neritic zones is high because sunlight is readily available and because nutrients are available from estuaries and deep-ocean currents. Productivity in the oceanic zone is extremely low—even though light is available at the surface—because nutrients are scarce. The deepest part of the oceanic zone may have nutrients available from the substrate but lacks light to support photosynthesis. **11.** `analyze` As elevation increases, biomes often change in a similar way as increasing in latitude (e.g., you might go from temperate forest to a boreal-type forest to tundra). **12.** `analyze` The natural biome of Eastern North America is primarily temperate forest. However, most of this area is now occupied by human-influenced biomes such as cities, suburbs, and farmland.

FIGURE A52.1

13. apply a **14.** create A single population of lizards was split when South America split from Africa due to continental drift. Fossils are too old to represent animals dispersed by human travel. **15.** evaluate Yes. As average global temperatures increase, some organisms may be able to adapt by moving their range. As mountaintops warm, however, organisms adapted to cold conditions will have nowhere to go. Ecologists have good reason to be concerned about these possible extinctions. **16.** create Temperature is relatively constant in the tropics all year, whereas temperature fluctuates dramatically at higher latitudes. Organisms that are physiologically adapted to constant temperatures may not have as much genetic variation for traits that enable adaptation to higher temperatures, compared to organisms at higher latitudes, which are adapted to tolerate different temperatures at different times of the year.

CHAPTER 53

IN-TEXT QUESTIONS AND EXERCISES

p. 1086 Fig. 53.4 analyze Even though half of each plot serves as a control, the natural landscape varies within plots, so there is a chance that foraging in the two subplots could be different due to factors not measured in the experiment. By measuring foraging activity before adding owls and seeds, the researchers established an additional control and a baseline.

p. 1087 CYU apply If owls were allowed to hunt within the treatment subplots, the threat of predation would increase. A larger amount of supplementary seed would be required to compensate for the increased risk of foraging. Gerbils do not consciously calculate costs and benefits. One possible proximate explanation is that the presence of hunting owls would increase the stress level in the gerbils (e.g., the hormone cortisol), which could suppress the hunger of the gerbils or reduce their drive to find food.

p. 1088 Fig. 53.6 analyze To form a better control, bring females into the lab and give them the same food and housing conditions as the treatment groups, but expose them to artificial lighting that simulates the short daylight conditions of winter.

p. 1089 CYU (1) analyze As in lizards, a rise in sex hormones in humans is a proximate cause of sexual readiness. **(2)** analyze Humans, like lizards, tend to choose a mate who will increase their fitness. For example, women often prefer men who will be able to help provide for offspring.

p. 1090 Fig. 53.7 apply The data would either be distributed randomly around the circles, showing no preference for direction, or the turtles would swim the same direction in both cases, suggesting that they are using compass orientation rather than map orientation.

p. 1093 Fig. 53.10 apply See FIGURE A53.1.

p. 1095 CYU (1) analyze Auditory communication allows individuals to communicate over long distances but can be heard by predators. Olfactory communication is effective in the dark, and scents can continue to carry information long after the signaler has left, but scents do not carry long distances. Visual communication is effective during the day but can be seen by predators. **(2)** apply The ability to detect deceit protects the individual from fitness costs (e.g., being eaten, having a mate's eggs fertilized by someone else); avoiding or punishing "liars" should also lower the frequency of deceit (because it becomes less successful). Alleles associated with detecting and avoiding or punishing liars should be favored by natural selection and increase in frequency.

p. 1096 Fig. 53.14 apply Between first cousins, $r = 1/2n \times 1/2n \times 1/2n = 1/8$.

p. 1097 apply Humans often live near kin, are good at recognizing kin, and in many cases can give kin resources for protection that increase fitness.

p. 1097 Fig. 53.15 create The control would be to drag an object of similar size through the colony, at similar times of day. This would test the hypothesis that prairie dogs are reacting to the presence of the experimenter and the disturbance—not to a predator.

IF YOU UNDERSTAND . . .

53.1 apply At the proximate level of causation, human teens experience an increase in the sex hormones testosterone and estrogen, which increases sex drive. At the ultimate level of causation, teens with a higher sex drive will, on average, have more offspring, which will increase their fitness and pass along their alleles at a higher rate than individuals with a lower sex drive. **53.2** apply You would have to consider variables that provide costs and benefits to the web-building spiders. For example, spinning silk and building webs uses up energy, so the size of the web, the amount of material used to build the web, and the frequency of building webs should be considered, as well as the effect of web-building on vulnerability to predation. But large, well-built webs can increase prey capture, so the rate of capture of prey of various types of webs also should be considered. **53.3** create Only a bowerbird with plenty of resources could afford the time and energy to build a highly decorated bower. Thus, the decoration behavior is likely an honest signal of male fitness. **53.4** create Migration is condition-dependent in some birds. If enough food is available locally, the costs of migration outweigh the benefits, so the birds' fitness will be higher if they stay put. **53.5** apply Deceitful communication should decrease. If large males are gone, few nests are available and more female-mimic males would compete at them. Most female-mimic males would have no (or fewer) eggs to fertilize and no male to take care of the eggs they did fertilize. There would be less natural selection pressure favoring deceitful communication. **53.6** understand (1) Long-lived, so that extensive interactions with kin and nonrelatives are possible; (2) good memory, to record reciprocal interactions; (3) kin nearby, to make inclusive fitness gains possible; (4) ability to share fitness benefits (e.g., food) between individuals.

YOU SHOULD BE ABLE TO . . .

✓ Test Your Knowledge

1. understand c **2.** understand Adaptive significance of behavior; how behavior affects fitness **3.** remember d **4.** understand c **5.** understand d **6.** understand c

✓ Test Your Understanding

7. understand The fitness of honeybees depends on how much energy they can bring into the hive by foraging. Hives with honeybees that successfully communicate locations of food resources will be favored by natural

---

FIGURE A53.1

The straight run of the upward waggle dance is twice as long as in Fig. 53.10

---

selection such that the frequency of alleles that determine the communication behavior will increase in the population. **8.** evaluate Evolution favors organisms that forage optimally—maximizing benefits and minimizing costs—increasing the frequency of optimal foraging in populations over time. However, individuals in populations vary and environmental circumstances vary. Some individuals will forage suboptimally and suffer reduced fitness. **9.** apply Only males that are in good shape (well nourished and free of disease) have the resources required to perform their displays. **10.** understand Longer day lengths indicate the arrival of spring, when renewed plant growth and insect activity make more food available. Courtship displays from males indicate the presence of males that can fertilize eggs. **11.** analyze This statement is true; sea turtles need both a map and a compass. A map provides information about the spatial relationships of places on a landscape; a compass allows sea turtles to orient the map—so that it aligns with the actual landscape. **12.** apply c

✓ Test Your Problem-Solving Skills

13. create One possibility would be to set up a testing arena with food supplies at various distances from the food source where adult flies are originally released. Identify rover and sitter adults by genotype and test them individually; then record how far each travels during a set time interval. If the rover and sitter alleles affect foraging movements in adults, then rovers should be more likely to move (and to move farther) than sitters. **14.** analyze a, b, c, d; the howler monkeys make condition-dependent decisions about what to eat and are likely to prefer high-energy foods, when available, or foods with nutrients or compounds that would increase their fitness. Hamilton's rule could be considered if some monkeys give the highest-quality foods to close relatives rather than consuming it themselves. **15.** analyze Individuals have an r of 1/2 with full siblings and an r of 1/8 with first cousins. If the biologist will lose his life, he needs to save two siblings, or eight first cousins, to keep the "lost copy" of his altruism alleles in the population. **16.** apply People would be expected to donate blood under two conditions: (1) They either received blood before or expect to have a transfusion in the future, and/or (2) they receive some other benefit in return, such as a good reputation among people who might be able to help them in other ways.

CHAPTER 54

IN-TEXT QUESTIONS AND EXERCISES

p. 1105 Fig. 54.2 apply See FIGURE A54.1 (see A:48)

p. 1106 (1) apply Compared to northern populations, fecundity should be high and survivorship low in southern populations. **(2)** apply Fecundity can be much higher if females lay eggs instead of retaining them in their bodies and giving birth to live young.

p. 1109 Fig. 54.6 create At high population density, competition for food limits the amount of energy available to female song sparrows for egg production.

p. 1112 CYU create One possibility is competition for food. To test this idea, compare carrying capacity in identical fenced-in areas that differ only in the amount of food added. (You could also test a hypothesis of space limitation by doubling the size of the enclosure but keeping the amount of food the same.)

p. 1114 Fig. 54.10 apply When the population of hares was low, the rate of increase was not density dependent on food. That is, plenty of food was available for the few rabbits, so competition for this resource was low.

p. 1117 CYU (1) understand Developed nations have roughly equal numbers of individuals in each age class because the fertility rate has been constant and survivorship high for many years. Developing countries have many more children and young people than older people, because the fertility rate and survivorship have been increasing. **(2)** `understand` Because survivorship is high in most human populations, changes in overall growth rate depend almost entirely on fertility rates.

p. 1118 Fig. 54.14 `apply` The total population size of first-generation females in the third year = 1000; in the fourth year = 240. The total population size of first- and second-generation females in the third year = 2308; in the fourth year = 1230. The number of 1-year-olds created by third-generation females (blue) is $981 \times 0.33 = 324$. The number of newborns (green) created by these 1-year-olds is $324 \times 4.0 = 1296$. Thus, the total population size at the end of the fourth year is expected to be $200 + 792 + 1296 + 324 + 198 + 40 = 2850$.

p. 1119 `apply` The population appears to be increasing over time: From the original 1000 females, there are 2308 females in the third year.

IF YOU UNDERSTAND . . .

54.1 `create` First you would need to observe snowshoe hares to determine how many pellets one individual can produce in one day, on average. Then you could use a quadrat to measure the number of fresh pellets produced in a known area of habitat in one day and use your pellets/hare number to convert this value to hares/area. **54.2** `analyze` In ancient Rome, survivorship was probably low and fecundity high—women started reproducing at a young age and did not live long, on average. In Rome today, the population has high survivorship and low fecundity. **54.3** `apply` See **FIGURE A54.2 54.4** `apply` The population density of mosquitoes would not be constant over time. Ignoring the number of predators that might be present, you would expect high densities of these short-lived insects after rains, followed by low densities during extended dry periods. **54.5** `understand` Fewer children are being born per female, but there are many more females of reproductive age due to high fecundity rates in the previous generation. **54.6** `apply` Small, isolated

populations are likely to be wiped out by bad weather, a disease outbreak, or changes in the habitat. But in a metapopulation, migration between the individual small populations helps reestablish populations and maintain the overall size of the metapopulation.

YOU SHOULD BE ABLE TO . . .

✓ Test Your Knowledge

1. `remember` b **2.** `remember` births, deaths, immigration, emigration **3.** `remember` a **4.** `remember` True **5.** `understand` d **6.** `understand` c

✓ Test Your Understanding

7. `remember` c **8.** `apply` See **TABLE A54.1 9.** `create` The population has undergone near-exponential growth recently because advances in nutrition, sanitation, and medicine have allowed humans to live at high density without suffering from decreased survivorship and fecundity. Eventually, however, growth rates must slow as density-dependent effects such as disease and famine increase death rates and lower birth rates. **10.** `understand` (a) Corridors allow individuals to move between populations, increasing gene flow and making it possible to recolonize habitats where populations have been lost. (b) Maintaining unoccupied habitat makes it possible for the habitat to be recolonized. **11.** `apply` *Lacerta vivipara* is most vulnerable at the southern edge of its ranges, such as in the mountains of Spain. These lizards are cold adapted and may become locally extinct in areas that become too warm **12.** `analyze` r_{max} gives the population growth rate in the absence of density-dependent limitation; r is the actual growth rate, which is usually affected by density-dependent factors.

✓ Test Your Problem-Solving Skills

13. `apply` Fewer older individuals will be left in the population; there will be relatively more young individuals. If too many older individuals are taken, population growth rate may decline sharply as reproduction stops or slows. (But if relatively few older individuals are taken, more

resources are available to younger individuals and their survivorship and fecundity, and the population's overall growth rate, may increase.) **14.** `create` The sunflowers and beetles are both metapopulations. To preserve them, you must preserve as many of the sunflower patches as possible (or plant more) and maintain corridors (that may be smaller sunflower patches) along which the beetles can migrate between the patches. **15.** `apply` As a sexually transmitted disease, AIDS will reduce the number of sexually active adults. If the epidemic continues unabated, the numbers of both reproductive-age adults and children will decline, causing a top-heavy age distribution dominated by older adults and the elderly. See **FIGURE A54.3. 16.** `analyze` c

CHAPTER 55

IN-TEXT QUESTIONS AND EXERCISES

p. 1126 `apply` Your graph should look like Figure 55.4, with *Semibalanus* on the left, *Chthamalus* is on the right, and the *x*-axis indicating height in the intertidal zone. The realized niche of *Chthamalus* is in the upper intertidal zone.

p. 1127 Fig. 55.5 `apply` If Connell had done the treatments on different rocks, a critic could argue that differences in survival were due to differences in the nature of the rocks—not differences in competition.

p. 1128 `apply` Your graph should look like the bottom graph in Figure 55.6, showing that the niches of *Semibalanus* and *Chthamalus* are nonoverlapping with respect to height in the intertidal zone.

p. 1131 Fig. 55.10 `apply` This experimental design tested the hypothesis that mussels can sense the presence of crabs. If the crabs had been fed mussels, a critic could argue that the mussels detect the presence of damaged mussels—not crabs. (As it turns out, the mussels can do both. The experimenters did the same experiment with broken mussel shells in the chamber instead of a fish-fed crab.)

FIGURE A54.1

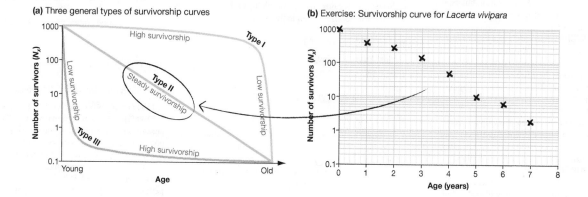

(a) Three general types of survivorship curves

(b) Exercise: Survivorship curve for *Lacerta vivipara*

TABLE A54.1

Trait	Life–History Continuum		
	Left	Middle	Right
	High fecundity Low survivorship	Intermediate	Low fecundity High survivorship
Growth habit	Herbaceous	Shrub	Tree
Disease- and predator-fighting ability	Low	Medium	High
Seed size	Small	Medium	Large
Seed number	Many	Moderate	Few
Body size	Small	Medium	Large

FIGURE A54.2

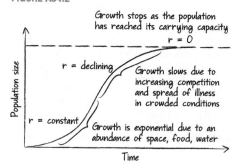

Growth stops as the population has reached its carrying capacity $r = 0$

r = declining

Growth slows due to increasing competition and spread of illness in crowded conditions

r = constant

Growth is exponential due to an abundance of space, food, water

p. 1133 Fig. 55.12 create Put equal numbers of infected and uninfected ants in a pen that includes a bird predator. Count how many of each type are eaten over time.

p. 1134 Fig. 55.13 apply Many possible answers. For example, acacia trees may expend energy in producing the large bulbs the ants live in and the food they eat. The *Crematogaster* ants may expend energy and risk injury or death in repelling herbivores. The cleaner shrimp may occasionally get injured or eaten by the fish they are cleaning, or their diets may be somewhat restricted by the association. The host fish may miss meals or be at greater risk of predation when they are undergoing cleaning by the shrimp. In both examples of mutualism, however, the overall associations are positive for both parties.

p. 1135 CYU (1) understand The individuals do not choose or try to have traits that reduce competition—they simply have those traits (or not). Resource partitioning just happens, because individuals with traits that allow them to exploit different resources produce more offspring, which also have those traits. (Recall from Chapter 25 that natural selection occurs on individuals, but adaptive responses such as resource partitioning are properties of populations.) **(2)** understand When species interact via consumption, a trait that gives one species an advantage will exert natural selection on individuals of the other species who have traits that reduce that advantage. This reciprocal adaptation will continue indefinitely. An example is the interaction of *Plasmodium* with the human immune system: The human immune system has evolved the ability to detect proteins from the *Plasmodium* and kill infected cells; in response, *Plasmodium* has evolved different proteins that the immune system does not detect.

p. 1135 Fig. 55.14 apply These steps controlled for the alternative hypothesis that differences in treehopper survival were due to differences in the plants they occupied—not the presence or absence of ants.

p. 1137 Fig. 55.15 analyze Predictable, at least to a degree.

p. 1142 CYU (1) understand The shade provided by early successional species increases humidity, and decomposition of their tissues adds nutrients and organic material to the soil. These conditions favor growth by later successional species, which can outcompete the early successional species. **(2)** understand The presence or absence of a plant species where nitrogen fixation occurs would dramatically alter nutrient conditions, and thus the speed of succession and the types of species that could become established. For example, species that require high nitrogen would be favored on sites where alder grew, and species that can tolerate low nitrogen would thrive on sites where alder is absent.

p. 1142 Fig. 55.22 apply The effect would be to make the island more remote, which would lower the rate of immigration and move the whole immigration curve downward. The rate of extinction would increase, shifting the curve upward. The overall effect would be to decrease the number of species, because the island effectively would have become more remote.

p. 1143 analyze The calculations for community 1 are as follows. (Note that your final answer is sensitive to the number of digits you use in your calculations; see BioSkills 1 in Appendix B).

Species A	Species B
$p_A = 10/18 = 0.555$	$p_B = 1/18 = 0.0555$
$\ln 0.555 = -0.589$	$\ln 0.0555 = -2.891$
$-0.589 \times 0.555 = -0.327$	$-2.891 \times 0.0555 = -0.160$

Species C	Species D
$p_C = 1/18 = 0.0555$	$p_D = 3/18 = 0.167$
$\ln 0.0555 = -2.891$	$\ln 0.167 = -1.790$
$-2.891 \times 0.0555 = -0.160$	$-1.790 \times 0.167 = -0.299$

Species E	Species F
$p_E = 2/18 = 0.111$	$p_F = 1/18 = 0.0555$
$\ln 0.111 = -2.198$	$\ln 0.0555 = -2.891$
$-2.198 \times 0.111 = -0.244$	$-2.891 \times 0.0555 = -0.160$

Summing the values of $p \times \ln p$ for each species and multiplying by -1 gives the Shannon index of species diversity for community 1:

$$(-1)[(-0.327) + (-0.160) + (-0.160) + (-0.299) + (-0.244) + (-0.160)] = 1.350$$

Similar calculations would give species diversity values of 1.794 for community 2 and 1.610 for community 3.

IF YOU UNDERSTAND . . .

55.1 understand A mutualistic relationship becomes a parasitic one if one of the species stops receiving a benefit. The treehopper–ant mutualism becomes parasitic in years when spiders are rare, because the treehoppers no longer derive a benefit but pay a fitness cost (producing honeydew that the ants eat). **55.2** apply An invasive species could have a large impact on community structure by replacing the dominant plant species that creates physical structure, which in turn affects other members of the community; by replacing a keystone species that has an indirect effect on many other species; or by outcompeting many species, causing a direct reduction in species richness. **55.3** apply All species at Glacier Bay must be able to survive in a cold climate with the local amount of precipitation. The species earliest in the succession must also be able to grow on rock exposed as the glacier melts. But chance, historical differences in seed sources, and presence or absence of alder (and nitrogen fixation) created differences in the species present. **55.4** apply Many national parks are surrounded by altered habitats, so they are functionally similar to islands. Species richness in the park is more likely to be preserved if it is large and located nearby other wilderness areas.

YOU SHOULD BE ABLE TO . . .

✔ Test Your Knowledge

1. remember True **2.** remember a **3.** remember d **4.** understand d **5.** remember fecundity **6.** remember c

✔ Test Your Understanding

7. evaluate Yes—the treehopper–ant mutualism is parasitic or mutualistic, depending on conditions; competition can evolve into no competition over time if niche differentiation occurs; arms races mean that the outcome of host–parasite interactions can change over time, and so forth (many other examples). **8.** understand b **9.** evaluate (a) If community composition is predictable, then the species present should not change over time. But if composition is not predictable, then that species should undergo significant changes over time. (b) If community composition is predictable, then two sites with identical abiotic factors should develop identical communities. If community composition is not predictable, then sites with identical abiotic factors should develop variable communities. In most tests, the data best match the predictions of the "not predictable" hypothesis, though communities show elements of both. **10.** analyze Disturbance is any short-lived event that changes the distribution of resources. Compared to low-frequency fires, high-frequency fires would tend to be less severe (less fuel builds up) and would tend to exert more intense natural selection for adaptations to resist the effects of fire. Compared to low-severity fires, high-severity fires would open up more space for pioneering species and would tend to exert more intense natural selection for adaptations to resist the effects of fire. **11.** apply Early successional species are adapted to disperse to new environments (small seeds) and grow and reproduce quickly (reproduce at an early age, grow quickly). They can tolerate severe abiotic conditions (high temperature, low humidity, low nutrient availability) but have little competitive ability. These species are able to enter a new environment (with no competitors) and thrive. These attributes are considered adaptations because they increase the fitness of these species. **12.** analyze The idea is that high productivity will lead to high population density of consumers, leading to competition and intense natural selection favoring niche differentiation that leads to speciation.

✔ Test Your Problem-Solving Skills

13. analyze Natural selection will favor orchid individuals that have traits that resist bee attack: thicker flower walls, nectar storage in a different position, a toxin in the flower walls, and so on. Individuals could also be favored if their anthers were in a position that accomplished pollination even if bees eat through the walls of the nectar-storage structure. **14.** apply b **15.** apply The exact answer will depend on the location of the campus. The first species to appear must possess good dispersal ability, rapid growth, quick reproductive periods, and tolerance for very harsh and severe conditions. The two-acre plot is likely to be colonized first by pioneer species that have very "weedy" characteristics. But once colonization is under way, the course of succession will depend more on how the various species interact with each other. The presence of one species can inhibit or facilitate the arrival and establishment of another. For example, an early-arriving species might provide the shade and nutrients required by a late-arriving species. The site's history and nearby ecosystems may influence which species appear at each stage; for instance, an undisturbed ecosystem nearby could be a source for native species. The pattern and rate of this succession is also influenced by the overall environmental conditions affecting it. Only species with traits appropriate to the local climate are likely to colonize the site. **16.** create One reasonable experiment would involve constructing artificial ponds and introducing different numbers of plankton species

FIGURE A54.3

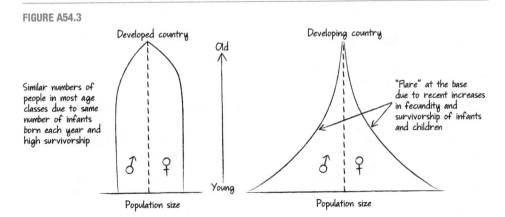

Developed country

Similar numbers of people in most age classes due to same number of infants born each year and high survivorship

♂ ♀

Population size

Old

Young

Developing country

"Flare" at the base due to recent increases in fecundity and survivorship of infants and children

♂ ♀

Population size

to different ponds, but the same total number of individuals. (Any natural immigration to the ponds would have to be prevented.) After a period of time, remove all of the plankton and measure the biomass present. Make a graph with number of species on the *x*-axis and total biomass on the *y*-axis. If the hypothesis is correct, the line of best fit through the data should have a positive slope.

CHAPTER 56

IN-TEXT QUESTIONS AND EXERCISES

p. 1152 *apply* To grow a kilogram of beef, first you have to grow 10 kg of grain or grass and feed it to the cow. Only 10 percent of this 10 kg will be used for growth and reproduction—the other 9 kg is used for cellular respiration or lost as heat.

p. 1152 *apply* Crustaceans and fish are ectothermic, so they are much more efficient at converting primary production into the biomass in their bodies than are endothermic birds and mammals.

p. 1153 *apply* Tuna are top predators in marine food webs. Mercury occurs in low concentrations in the water, but accumulates in higher concentrations at each trophic level. A person who eats a lot of tuna will accumulate mercury over time, potentially causing mercury poisoning. Sardines are primary consumers, a low trophic level, and therefore are not in danger of biomagnification of mercury.

p. 1155 Fig. 56.10 *apply* Average NPP = 125 g/m^2/year. Area = 65%. Although NPP in the ocean is very small, the ocean is so vast that it ends up being the largest contributor to total NPP.

p. 1156 CYU (1) *understand* At each trophic level, most of the energy that is consumed is lost to cellular respiration and ultimately heat, metabolism, or other maintenance activities, which leaves only a small percentage of energy for biomass production (growth and reproduction).
(2) *apply* Since NPP depends on temperature, it is logical to predict that NPP will increase with global warming. (However, Section 56.3 explains that global warming changes precipitation patterns, ocean stratification, and other factors that can limit NPP.)

p. 1157 *understand* When the trees of tropical rain forests are hauled away and the remnants of the forest are burned, the nutrients are exported from the site, leaving only poor soil to nourish the crops.

p. 1158 Fig. 56.13 *create* One logical hypothesis is that the total amount exported increases as tree roots and other belowground organic material decay and begin to wash into the stream; the amount exported should begin to decline as nitrate reserves become exhausted—eventually, there is no more nitrate to wash away.

p. 1159 Fig. 56.14 *apply* Much more water should evaporate. One possibility is that more water vapor will be blown over land and increase precipitation over land.

p. 1162 CYU *apply* **Water:** perhaps the most direct impacts concern rates of groundwater replenishment. Converting biomes into farms or suburbs decreases groundwater recharge and increases runoff. Irrigation pumps groundwater to the surface, where much of it runs off into streams. **Nitrogen:** Humans introduce huge quantities of nitrogen into ecosystems as fertilizer. **Carbon:** Humans are releasing huge quantities of stored carbon, in the form of fossil fuels, into the atmosphere as CO_2. Whereas humans are depleting supplies of freshwater below natural levels, we are increasing quantities of N and C above natural levels.

p. 1164 Fig. 56.22 *analyze* Photosynthesis increases in summer, resulting in removal of CO_2 from the atmosphere.

p. 1165 Fig. 56.23 *apply* $2 \times 7000 = 14,000$ million metric tons of CO_2, double the current total.

p. 1168 *apply* The pH of the water will fall (acidity will increase) because CO_2 from your breath reacts with water to form carbonic acid.

IF YOU UNDERSTAND . . .

56.1 *apply* Tofu is a plant, a primary producer at the base of the productivity pyramid. Cows are primary consumers, one level up. It would be about 10 times more efficient to eat tofu, because 10 times the quantity of plants would be necessary to provide one pound of beef.
56.2 *apply* Decomposition is slowest in cold, wet temperatures on land and in anoxic conditions in the ocean (or in freshwater). Stagnant water often becomes anoxic as decomposers use up available oxygen that is not replenished by diffusion from the atmosphere. **56.3** *apply* Many possible answers, including (1) reducing carbon dioxide emissions by finding alternatives to fossil fuels should decrease the amount of carbon dioxide released into the atmosphere, (2) recycling programs decrease carbon dioxide emissions from manufacturing plants and power plants, and (3) reforestation can tie up carbon dioxide in wood.

YOU SHOULD BE ABLE TO . . .

✓ Test Your Knowledge

1. *remember* True **2.** *understand* d **3.** *understand* a **4.** *remember* c
5. *remember* c **6.** *remember* b

✓ Test Your Understanding

7. *apply* a **8.** *analyze* Sharks normally eat fish that eat the herbivorous fish that keep the growth of algae on corals in check. When sharks are absent, more of the intermediate fish survive, which decreases the number of herbivorous fish, causing algae to thrive on the coral reef. **9.** *analyze* Warmer, wetter climates speed decomposition; cool temperatures slow it. Decomposition regulates nutrient availability because it releases nutrients from detritus and allows them to reenter the food web.
10. *evaluate* Yes, corals would be a good choice. Corals are important ecologically because they have a high NPP, providing energy to diverse animals in reef food webs. Corals are also vulnerable to climate change from the increase in water temperature, which can cause bleaching (loss of photosynthetic algae), as well as from the effects of ocean acidification, which slows the rate of coral growth. **11.** *analyze* Forest fires are an example of positive feedback; drier, hotter summers increase the risk of fires, and fires release CO_2 to the atmosphere, increasing global warming and the risk of fires. Forest growth is an example of negative feedback; as CO_2 increases in the atmosphere, the rate of forest growth in some areas may increase (due to increased availability of CO_2, increased temperature, and/or increased precipitation), reducing the concentration of CO_2 in the atmosphere. **12.** *analyze* The open ocean has almost no nutrient input from the land and has little upwelling to supply nutrients from the deep ocean. In contrast, intertidal and coastal areas receive large inputs of nutrients from rivers as well as from upwellings from ocean depths.

✓ Test Your Problem-Solving Skills

13. *create* One possibility is to add radioactive nitrogen to the water of the ponds in natural concentrations (control ponds) and elevated nitrogen concentrations (treatment ponds) and then follow this isotope through the primary producers, primary consumers, and secondary consumers in the system by measuring the amount of radioisotope in the tissues of organisms, and quantity of biomass, at each trophic level in the treatment versus control ponds **14.** *evaluate* c **15.** *apply* Without herbivores, there is no link in the nitrogen cycle between primary producers and secondary consumers. All of the plant nitrogen would go to the primary decomposers and back

into the soil. If decomposition is rapid enough, nitrogen would cycle quickly between primary producers, decomposers, the soil, and back to primary producers.
16. *apply* Atmospheric oxygen would increase due to extensive photosynthesis, but carbon dioxide levels would decrease because little decomposition was occurring. The temperature would drop because fewer greenhouse gases would be trapping heat reflected from the Earth's surface.

CHAPTER 57

IN-TEXT QUESTIONS AND EXERCISES

p. 1177 CYU *create* Do an all-taxon survey: organize experts and volunteers to collect, examine, and identify all species present. This could include direct sequencing studies (see Chapter 29) to document the bacteria and archaea present in different habitats on campus.

p. 1178 Fig. 57.7 *apply* Nearly all endangered species are threatened by more than one factor and therefore appear on the graph more than once, leading to species totals greater than 100 percent.

p. 1180 Fig. 57.9 *analyze* Because the treatments and study areas were assigned randomly, there is no bias involved in picking certain areas that are unusual in terms of their biomass or species diversity. They should represent a random sample of biomass and species diversity in fragments of various sizes versus intact forest.

p. 1183 Using the equation $S = (18.9)A^{0.15}$, plug in 5000 km^2 for A and solve for S such that $S = 18.9 \times 5000^{0.15} = 68$ species.

p. 1184 CYU (1) *understand* Fragmentation reduces habitat quality by creating edges that are susceptible to invasion and loss of species, due to changed abiotic conditions. Genetic problems occur inside fragments as species become inbred and/or lose genetic diversity via genetic drift. **(2)** *create* Habitat destruction and fragmentation in Borneo are obvious from the clear-cut areas. Overexploitation could be assessed by interviewing local citizens and law enforcement officers to the history of poaching orangutans for the pet trade or for other uses. Effects from invasive species could be assessed by published or local accounts of community structure over time. Pollution could be assessed by direct measurement of soil, water, air, and possibly tissue samples (in dead orangutans, feces, or blood samples). The effects of climate change could be assessed based on local weather records.

p. 1185 Fig. 57.12 *analyze* *Null hypothesis:* NPP is not affected by species richness or functional diversity of species. *Prediction:* NPP will be greater in plots with more species and more functional groups. *Prediction of null:* There will be no difference in NPP based on species richness or number of functional groups.

p. 1188 CYU *create* Establish several study plots in an area of Borneo where some of the forest has been converted to palm plantation but the adjacent forest (of similar elevation and topography) has rain forest still intact—half of the plots in the plantation and half in the rain forest. Monitor soil depth, soil nutrient levels, precipitation, and stream sediment levels regularly, and compare after several years.

p. 1191 Fig. 57.16 *apply* See **FIGURE A57.1**.

IF YOU UNDERSTAND . . .

57.1 *apply* At the genetic level, you could measure the genetic diversity of the orangutan population by comparing DNA sequences obtained from blood or feces samples. This information could help you to determine if a genetic bottleneck had occurred during the clear-cutting period. At the species level, you could monitor the increase in species richness of trees, birds, mammals, and other groups as the restoration progressed. At the ecosystem level, you could measure the ability of the growing forest to retain nutrients in the soil, prevent erosion, trap carbon, increase the prevalence of rain, and other functions **57.2** (apply) Small, geographically isolated populations of orangutans have low genetic diversity due to lack of gene flow and genetic drift and can suffer from inbreeding depression. Low genetic diversity makes them vulnerable to changes in their habitat, and small population size makes them susceptible to catastrophic events such as storms, disease outbreaks, or fires. **57.3** (apply) If many species are present, and each uses the available resources in a unique way, then a larger proportion of all available resources should be used—leading to higher biomass production. **57.4** (evaluate) Twelve percent of land area is already protected, but a mass extinction event is still going on. This is because 12 percent is not sufficient to save most species, and because many or most protected areas are not located in biodiversity hotspots.

YOU SHOULD BE ABLE TO . . .

✔ Test Your Knowledge

1. (remember) d **2.** (remember) a **3.** (remember) b **4.** (remember) d
5. (understand) c **6.** (understand) overexploitation

✔ Test Your Understanding

7. (analyze) b **8.** (analyze) By comparing the number of species estimated to reside in the park before and after the survey, biologists will have an idea of how much actual species diversity is being underestimated in other parts of the world—where research was at the level of Great Smoky Mountains National Park before the survey. The limitation is that it may be difficult to extrapolate from the data—no one knows if the situation at this park is typical of other habitats. **9.** (evaluate) No one correct answer; there are good arguments for both. In a case like the orangutans of Borneo, it is not possible to save the orangutans without saving their ecosystem. On the other hand, it is much easier to attract money and volunteers to help save a charismatic primate like an orangutan than to save a patch of forest. Both types of efforts are needed, their relative importance depends on the circumstance. **10.** (evaluate) In experiments with species native to North American grasslands, study plots that have more species produce more biomass, change less during a disturbance (drought), and recover from a disturbance faster than study plots with fewer species. **11.** (understand) Wildlife corridors facilitate the movement of individuals. Corridors allow areas to be recolonized if a species is lost and enable the introduction of new alleles that can counteract genetic drift and inbreeding in small, isolated

populations. **12.** (apply) Ecosystem services are beneficial effects that ecosystems have on humans. It is rare to own or pay for these services, so no one has a vested interest in maintaining them. This is a primary reason that ecosystems are destroyed, even though the services they offer are valuable. The exception would be landowners who are paid *not* to log their forests, to maintain ecosystem functions like carbon storage.

✔ Test Your Problem-Solving Skills

13. (apply) d **14.** (create) Catalog existing biodiversity at the genetic, species, and ecosystem levels. Using these data, find areas that have the highest concentration of biodiversity at each level. Protect as many of these areas as possible, and connect them with corridors of habitat. **15.** (evaluate) There is no "correct" answer. One argument is that conservationists could lobby officials in Brazil and Indonesia to learn from the mistakes made in developed nations, and preserve enough forests to maintain biodiversity and keep ecosystem services intact—avoiding the expenditures that developed countries had to make to clean up pollution and restore ecosystems and endangered species. **16.** (analyze) Species with specialized food or habitat requirements, large size (and thus large requirements for land area), small population size, and low reproductive rate are vulnerable to extinction. A good example is the koala bear. Species that are particularly resistant to pressure from humans possess the opposite of many of these traits. A good example of a resistant species is the Norway rat.

BIG PICTURE Ecology

p. 1196 CYU (1) (analyze) Species interactions "are agents of" natural selection. **(2)** (analyze) As the human population climbs, more resources are needed to support the population. The harvesting of resources affects other species both directly (e.g. overfishing) and indirectly (e.g. climate change), resulting in decreased biodiversity. **(3)** (analyze) Both energy and nutrients flow through trophic levels in food webs. However, energy flows in one direction, ultimately dissipating as heat, whereas nutrients continue to cycle through ecosystems. **(4)** (apply) Human behavior "increases emissions of" CO_2 "causes" greenhouse effect "causes" climate change.

Bioskills

BIOSKILLS 1; p. B:2 CYU (1) (apply) 3.1 miles **(2)** (apply) 37°C
(3) (apply) Multiply your weight in pounds by 1/2.2 (0.45).

(4) (apply) 4 **(5)** (apply) Two significant figures. When you multiply, the answer can have no more significant figures than the least accurate measurement—in this case, 1.6.

BIOSKILLS 2; p. B:3 CYU (1) (apply) "different-yoked-together" **(2)** (apply) "sugary-loosened" **(3)** (apply) "study-of-form" **(4)** (apply) "three-bodies"

BIOSKILLS 3; p. B:6 CYU (1) (apply) about 18% **(2)** (apply) a dramatic drop (almost 10%) **(3)** (understand) No—the order of presentation in a bar chart does not matter (though it's convenient to arrange the bars in a way that reinforces the overall message). **(4)** (apply) 11 **(5)** (apply) 68 inches

BIOSKILLS 4; p. B:7 CYU (apply) Test 2, the estimate based on the larger sample—the more replicates or observations you have, the more precise your estimate of the average should be.

BIOSKILLS 5; p. B:8 CYU (1) (apply) $1/2 \times 1/2 \times 1/2 \times 1/2 = 1/16$ **(2)** (apply) $1/6 + 1/6 + 1/6 = 1/2$

BIOSKILLS 6; p. B:9 CYU (1) (understand) exponential **(2)** (apply) $\ln N_t = \ln N_0 + rt$

BIOSKILLS 7; p. B:9 Fig. B7.1. (analyze) See **FIGURE BA.1**. **p. B:10 Fig. B7.2.** (analyze) See **FIGURE BA.2**. **p. B:11 Fig. B7.3.** (analyze) Figure B7.3d is different.

FIGURE BA.1

FIGURE BA.2

FIGURE A57.1

The bars would be very short and could be positive or negative each year—none of them significantly different (error bars would all overlap)

BIOSKILLS 8; p. B:12 Fig. B8.1. apply See **FIGURE BA.3**.

BIOSKILLS 9; p. B:13 Fig. B9.1 understand DNA and RNA are acids that tend to drop a proton in solution, giving them a negative charge.
p. B:15 CYU analyze The lane with no band comes from a sample where RNA X is not present. The same size RNA X is present in the next two lanes, but the faint band has very few copies while the dark band has many. In the fourth lane, the band is formed by a smaller version of RNA X, and relatively few copies are present.

BIOSKILLS 10; p. B:17 CYU (1) understand size and/or density. **(2)** apply Mitochondria, because they are larger in size compared with ribosomes.

BIOSKILLS 11; p. B:19 CYU (1) analyze No—it's just that no mitochondria happened to be present in this section

sliced through the cell. **(2)** explain Understanding a molecule's structure is often critical to understanding how it functions in cells.

BIOSKILLS 12; p. B:21 CYU (1) analyze It may not be clear that the results are relevant to noncancerous cells that are not growing in cell culture—that is, that the artificial conditions mimic natural conditions. **(2)** analyze It may not be clear that the results are relevant to individuals that developed normally, from an embryo—that is, that the artificial conditions mimic natural conditions.

BIOSKILLS 13; p. B:23 Fig. B13.1 analyze This is human body temperature—the natural habitat of *E. coli*.
p. B:24 CYU (1) analyze *Caenorhabditis elegans* would be a good possibility, because the cells that normally die have already been identified. You could find mutant individuals that lacked normal cell death; you could compare

the resulting embryos to normal embryos and be able to identify exactly which cells change as a result. **(2)** analyze Any of the multicellular organisms in the list would be a candidate, but *Dictyostelium discoideum* might be particularly interesting because cells stick to each other only during certain points in the life cycle. **(3)** analyze *Mus musculus*—as the only mammal in the list, it is the organism most likely to have a gene similar to the one you want to study.

BIOSKILLS 14; p. B:26 CYU synthesize Many examples are possible. See Figure 1.9 in Chapter 1, as an example of the format to use for your Research Box.

BIOSKILLS 15; p. B:27 CYU (1) analyze See **FIGURE BA.4**. **(2)** analyze See **FIGURE BA.4**.

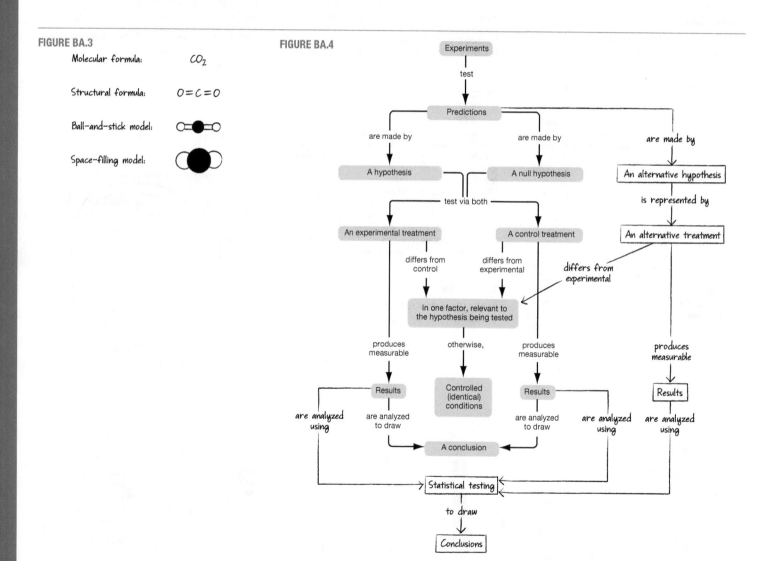

FIGURE BA.3

Molecular formula: CO_2

Structural formula: $O = C = O$

Ball-and-stick model:

Space-filling model:

FIGURE BA.4

| BIOSKILL 1 | the metric system and significant figures |

Scientists ask questions that can be answered by observing or measuring things—by collecting data. What units are used to make measurements? When measurements are reported, how can you tell how reliable the data are?

The Metric System

The metric system is the system of units of measure used in every country of the world but three (Liberia, Myanmar, and the United States). It is also the basis of the SI system—the International System of Units (abbreviated from the French, *Système international d'unités*)—used in scientific publications.

The popularity of the metric system is based on its consistency and ease of use. These attributes, in turn, arise from the

system's use of the base 10. For example, each unit of length in the system is related to all other measures of length in the system by a multiple of 10. There are 10 millimeters in a centimeter; 100 centimeters in a meter; 1000 meters in a kilometer.

Measures of length in the English system, in contrast, do not relate to each other in a regular way. Inches are routinely divided into 16ths; there are 12 inches in a foot; 3 feet in a yard; 5280 feet (or 1760 yards) in a mile.

If you have grown up in the United States and are accustomed to using the English system, it is extremely important to begin developing a working familiarity with metric units and values. **Tables B1.1** and **B1.2** (see B:2) should help you get started with this process.

As an example, consider the following question: An American football field is 120 yards long, while rugby fields are 144 meters

TABLE B1.1 **Metric System Units and Conversions**

Measurement	Unit of Measurement and Abbreviation	Metric System Equivalent	Converting Metric Units to English Units
Length	kilometer (km)	1 km = 1000 m = 10^3 m	1 km = 0.62 mile
	meter (m)	1 m = 100 cm	1 m = 1.09 yards = 3.28 feet = 39.37 inches
	centimeter (cm)	1 cm = 0.01 m = 10^{-2} m	1 cm = 0.3937 inch
	millimeter (mm)	1 mm = 0.001 m = 10^{-3} m	1 mm = 0.039 inch
	micrometer (μm)	1 μm = 10^{-6} m = 10^{-3} mm	
	nanometer (nm)	1 nm = 10^{-9} m = 10^{-3} μm	
Area	hectare (ha)	1 ha = 10,000 m^2	1 ha = 2.47 acres
	square meter (m^2)	1 m^2 = 10,000 cm^2	1 m^2 = 1.196 square yards
	square centimeter (cm^2)	1 cm^2 = 100 mm^2 = 10^{-4} m^2	1 cm^2 = 0.155 square inch
Volume	liter (L)	1 L = 1000 mL	1 L = 1.06 quarts
	milliliter (mL)	1 mL = 1000 μL = 10^{-3} L	1 mL = 0.034 fluid ounce
	microliter (μL)	1 μL = 10^{-6} L	
Mass	kilogram (kg)	1 kg = 1000 g	1 kg = 2.20 pounds
	gram (g)	1 g = 1000 mg	1 g = 0.035 ounce
	milligram (mg)	1 mg = 1000 μg = 10^{-3} g	
	microgram (μg)	1 μg = 10^{-6} g	
Temperature	Kelvin (K)*		K = °C + 273.15
	degrees Celsius (°C)		°C = $\frac{5}{9}$ (°F − 32)
	degrees Fahrenheit (°F)		°F = $\frac{9}{5}$°C + 32

*Absolute zero is −273.15 °C = 0 K.

Prefixes Used in the Metric System

Prefix	Abbreviation	Definition
nano–	n	$0.000\ 000\ 001 = 10^{-9}$
micro–	μ	$0.000\ 001 = 10^{-6}$
milli–	m	$0.001 = 10^{-3}$
centi–	c	$0.01 = 10^{-2}$
deci–	d	$0.1 = 10^{-1}$
–	–	$1 = 10^{0}$
kilo–	k	$1000 = 10^{3}$
mega–	M	$1\ 000\ 000 = 10^{6}$
giga–	G	$1\ 000\ 000\ 000 = 10^{9}$

long. In yards, how much longer is a rugby field than an American football field? To solve this problem, first convert meters to yards: 144 m × 1.09 yards/m = 157 yards (note that the unit "m" cancels out). The difference in yards is thus: 157 – 120 = 37 yards. If you did these calculations on a calculator, you might have come up with 36.96 yards. Why has the number of yards been rounded off? The answer lies in significant figures. Let's take a closer look.

Significant Figures

Significant figures or "sig figs"—the number of digits used to report the measurement—are critical when reporting scientific data. The number of significant figures in a measurement, such as 3.524, is the number of digits that are known with some degree of confidence (3, 5, and 2) plus the last digit (4), which is an estimate or approximation. How do scientists know how many digits to report?

Rules for Working with Significant Figures

The rules for counting significant figures are summarized here:

- All nonzero numbers are always significant.
- Leading zeros are never significant; these zeros do nothing but set the decimal point.
- Embedded zeros are always significant.

- Trailing zeros are significant *only* if the decimal point is specified (Hint: Change the number to scientific notation. It is easier to see the "trailing" zeros.)

Table B1.3 provides examples of how to apply these rules. The bottom line is that significant figures indicate the precision of measurements.

Precision versus Accuracy

If biologists count the number of bird eggs in a nest, they report the data as an exact number—say, 3 eggs. But if the same biologists are measuring the diameter of the eggs, the numbers will be inexact. Just how inexact they are depends on the equipment used to make the measurements. For example, if you measure the width of your textbook with a ruler several times, you'll get essentially the same measurement again and again. Precision refers to how closely individual measurements agree with each other. So, you have determined the length with precision, but how do you know if the ruler was accurate to begin with?

Accuracy refers to how closely a measured value agrees with the correct value. You don't know the accuracy of a measuring device unless you calibrate it, by comparing it against a ruler that is known to be accurate. As the sensitivity of equipment used to

TABLE B1.3 **Rules for Working with Significant Figures**

Example	Number of Significant Figures	Scientific Notation	Rule
35,200	5	3.52×10^{4}	All nonzero numbers are always significant
0.00352	3	3.52×10^{-3}	Leading zeros are not significant
1.035	4	$1.035\ (\times 10^{0})$	Imbedded zeros are always significant
200	1	2×10^{2}	Trailing zeros are significant only if the decimal point is specified
200.0	4	2.000×10^{2}	Trailing zeros are significant only if the decimal point is specified

make a measurement increases, the number of significant figures increases. For example, if you used a kitchen scale to weigh out some sodium chloride, it might be accurate to 3 ± 1 g (1 significant figure); but an analytical balance in the lab might be accurate to 3.524 ± 0.001 g (4 significant figures).

In science, only the numbers that have significance—that are obtained from measurement—are reported. It is important to follow the "sig fig rules" when reporting a measurement, so that data do not appear to be more accurate than the equipment allows.

Combining Measurements

How do you deal with combining measurements with different degrees of accuracy and precision? The simple rule to follow is that the accuracy of the final answer can be no greater than the least accurate measurement. So, when you multiply or divide measurements, the answer can have no more significant figures than the least accurate measurement. When you add or subtract measurements, the answer can have no more decimal places than the least accurate measurement.

As an example, consider that you are adding the following measurements: 5.9522, 2.065, and 1.06. If you plug these numbers into your calculator, the answer your calculator will give you is 9.0772. However, this is incorrect—you must round your answer off to the nearest value, 9.08, to the least number of decimal places in your data.

It is important to nail down the concept of significant figures and to practice working with metric units and values. The Check Your Understanding questions in this BioSkill should help you get started with this process.

BIOSKILL 2 some common Latin and Greek roots used in biology

Greek or Latin Root	English Translation	Example Term
a, an	not	anaerobic
aero	air	aerobic
allo	other	allopatric
amphi	on both sides	amphipathic
anti	against	antibody
auto	self	autotroph
bi	two	bilateral symmetry
bio	life, living	bioinformatics
blast	bud, sprout	blastula
co	with	cofactor
cyto	cell	cytoplasm
di	two	diploid
ecto	outer	ectoparasite
endo	inner, within	endoparasite
epi	outer, upon	epidermis
exo	outside	exothermic
glyco	sugary	glycolysis
hetero	different	heterozygous
homo	alike	homozygous
hydro	water	hydrolysis
hyper	over, more than	hypertonic
hypo	under, less than	hypotonic
inter	between	interspecific
intra	within	intraspecific
iso	same	isotonic
logo, logy	study of	morphology
lyse, lysis	loosen, burst	glycolysis
macro	large	macromolecule

Greek or Latin Root	English Translation	Example Term
meta	change, turning point	metamorphosis
micro	small	microfilament
morph	form	morphology
oligo	few	oligopeptide
para	beside	parathyroid gland
photo	light	photosynthesis
poly	many	polymer
soma	body	somatic cells
sym, syn	together	symbiotic, synapsis
trans	across	translation
tri	three	trisomy
zygo	yoked together	zygote

check your understanding

If you understand BioSkill 2

✓ **You should be able to . . .**

Provide literal translations of the following terms:

1. heterozygote
2. glycolysis
3. morphology
4. trisomy

Graphs are the most common way to report data, for a simple reason. Compared to reading raw numerical values in a table or list, a graph makes it much easier to understand what the data mean.

Learning how to read and interpret graphs is one of the most basic skills you'll need to acquire as a biology student. As when learning piano or soccer or anything else, you need to understand a few key ideas to get started and then have a chance to practice—a lot—with some guidance and feedback.

Getting Started

To start reading a graph, you need to do three things: read the axes, figure out what the data points represent—that is, where they came from—and think about the overall message of the data. Let's consider each step in turn.

What Do the Axes Represent?

Graphs have two axes: one horizontal and one vertical. The horizontal axis of a graph is also called the *x*-axis or the abscissa. The vertical axis of a graph is also called the *y*-axis or the ordinate. Each axis represents a variable that takes on a range of values. These values are indicated by the ticks and labels on the axis. Note that each axis should *always* be clearly labeled with the unit or treatment it represents.

FIGURE B3.1 shows a scatterplot—a type of graph where continuous data are graphed on each axis. Continuous data can take an array of values over a range. In contrast, discrete data can take only a restricted set of values. If you were graphing the average height of men and women in your class, height is a continuous variable, but gender is a discrete variable.

For the example in this figure, the *x*-axis represents time in units of generations of maize; the *y*-axis represents the average percentage of the dry weight of a maize kernel that is protein.

To create a graph, researchers plot the independent variable on the *x*-axis and the dependent variable on the *y*-axis (Figure B3.1a). The terms independent and dependent are used because the values on the *y*-axis depend on the *x*-axis values. In our example, the researchers wanted to show how the protein content of maize kernels in a study population changed over time. Thus, the protein concentration plotted on the *y*-axis depended on the year (generation) plotted on the *x*-axis. The value on the *y*-axis always depends on the value on the *x*-axis, but not vice versa.

In many graphs in biology, the independent variable is either time or the various treatments used in an experiment. In these cases, the *y*-axis records how some quantity changes as a function of time or as the outcome of the treatments applied to the experimental cells or organisms.

(a) Read the axes—what is being plotted?

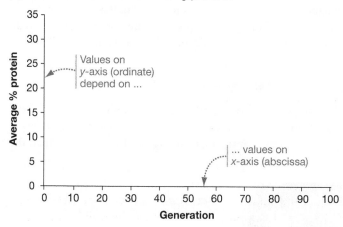

(b) Look at the bars or data points—what do they represent?

(c) What's the punchline?

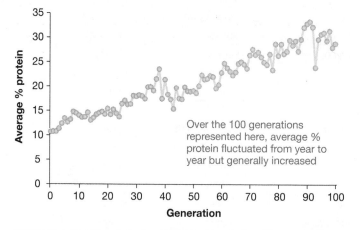

FIGURE B3.1 Scatterplots Are Used to Graph Continuous Data.

What Do the Data Points Represent?

Once you've read the axes, you need to figure out what each data point is. In our maize kernel example, the data point in Figure B3.1b represents the average percentage of protein found in a sample of kernels from a study population in a particular generation.

If it's difficult to figure out what the data points are, ask yourself where they came from—meaning, how the researchers got them. You can do this by understanding how the study was done and by understanding what is being plotted on each axis. The y-axis will tell you what they measured; the x-axis will usually tell you when they measured it or what group was measured. In some cases—for example, in a plot of average body size versus average brain size in primates—the x-axis will report a second variable that was measured.

In other cases, a data point on a graph may represent a relative or arbitrary unit of measurement. The data point shows the ratio of the amount of a substance, intensity, or other quantities, relative to a predetermined reference measurement. For example, the y-axis might show the percentage of relative activity of an enzyme—the rate of the enzyme-catalyzed reaction, scaled to the highest rate of activity observed (100 percent)—in experiments conducted under conditions that are identical except for one variable, such as pH or temperature (see Figure 8.15).

What Is the Overall Trend or Message?

Look at the data as a whole, and figure out what they mean. Figure B3.1c suggests an interpretation of the maize kernel example. If the graph shows how some quantity changes over time, ask yourself if that quantity is increasing, decreasing, fluctuating up and down, or staying the same. Then ask whether the pattern is the same over time or whether it changes over time.

When you're interpreting a graph, it's extremely important to limit your conclusions to the data presented. Don't extrapolate beyond the data, unless you are explicitly making a prediction based on the assumption that present trends will continue. For example, you can't say that the average percentage of protein content was increasing in the population before the experiment started, or that it will continue to increase in the future. You can say only what the data tell you.

Types of Graphs

Many of the graphs in this text are scatterplots like the one shown in Figure B3.1c, where individual data points are plotted. But you will also come across other types of graphs in this text.

Scatterplots, Lines, and Curves

Scatterplots sometimes have data points that are by themselves, but at other times data points will be connected by dot-to-dot lines to help make the overall trend clearer, as in Figure B3.1c, or may have a smooth line through them.

A *smooth line* through data points—sometimes straight, sometimes curved—is a mathematical "line of best fit." A line of best fit represents a mathematical function that summarizes the relationship between the x and y variables. It is "best" in the sense of fitting the data points most precisely. The line may pass through some of the points, none of the points, or all of the points.

Curved lines often take on characteristic shapes depending on the relationships between the x and y variable. For example, a bell-shaped curve depicts a normal distribution in which most data points are clumped near the middle, while a sigmoid or S-shaped curve exhibits small changes at first, which then accelerate and approach maximal value over time. Data from studies on population growth (see Chapter 54), enzyme kinetics (see Chapter 8), and oxygen–hemoglobin dissociation (see Chapter 45) typically fall on a curved line.

Bar Charts, Histograms, and Box-and-Whisker Plots

Scatterplots, or line-of-best-fit graphs, are the most appropriate type of graph when the data have a continuous range of values and you want to show individual data points. But other types of graphs are used to represent different types of distributions:

- *Bar charts* plot data that have discrete or categorical values instead of a continuous range of values. In many cases the bars might represent different treatment groups in an experiment, as in **FIGURE B3.2a** (see B:6). In this graph, the height of the bar indicates the average value. Statistical tests can be used to determine whether a difference between treatment groups is significant (see **BIOSKILLS 4**).

- *Histograms* illustrate frequency data and can be plotted as numbers or percentages. **FIGURE B3.2b** shows an example where height is plotted on the x-axis, and the number of students in a population is plotted on the y-axis. Each rectangle indicates the number of individuals in each interval of height, which reflects the relative frequency, in this population, of people whose heights are in that interval. The measurements could also be recalculated so that the y-axis would report the proportion of people in each interval. Then the sum of all the bars would equal 100 percent. Note that if you were to draw a smooth curve connecting the top of the bars on this histogram, the smooth curve would represent the shape of a bell.

- *Box-and-whisker plots* allow you to easily see where most of the data fall. Each box indicates where half of the data numbers are. The whiskers indicate the lower extreme and the upper extreme of the data. The vertical line inside each box indicates the median—meaning that half of the data are above this value and half are below (see Figure 1.9 for an example).

When you are looking at a bar chart that plots values from different treatments in an experiment, ask yourself if these values are the same or different. If the bar chart reports averages over discrete ranges of values, ask what trend is implied—as you would for a scatterplot.

(a) Bar chart

(b) Histogram

FIGURE B3.2 Bar Charts and Histograms. (a) Bar charts are used to graph data that are discontinuous or categorical. **(b)** Histograms show the distribution of frequencies or values in a population.

When you are looking at a histogram, ask whether there is a "hump" in the data—indicating a group of values that are more frequent than others. Is the hump in the center of the distribution of values, toward the left, or toward the right? If so, what does it mean?

Similarly, when you are looking at a box-and-whisker plot, ask yourself what information the graph gives you. What is the range of values for the data? Where are half the data points? Below what value is three quarters of the data?

Getting Practice

Working with this text will give you lots of practice with reading graphs—they appear in almost every chapter. In many cases we've inserted an arrow to represent your instructor's hand at the whiteboard, with a label that suggests an interpretation or draws your attention to an important point on the graph. In other cases, you should be able to figure out what the data mean on your own or with the help of other students or your instructor.

check your understanding

If you understand BioSkill 3

✔ **You should be able to . . .**

1. **QUANTITATIVE** Determine the total change in average percentage of protein in maize kernels, from the start of the experiment until the end.

2. **QUANTITATIVE** Determine the trend in average percentage of protein in maize kernels between generation 37 and generation 42.

3. Explain whether the conclusions from the bar chart in Figure B3.2a would be different if the data and label for Treatment 3 were put on the far left and the data and label for Treatment 1 on the far right.

4. **QUANTITATIVE** Determine approximately how many students in this class are 70 inches tall, by using Figure B3.2b.

5. **QUANTITATIVE** Determine the most common height in the class graphed in Figure B3.2b.

BIOSKILL 4 using statistical tests and interpreting standard error bars

When biologists do an experiment, they collect data on individuals in a treatment group and a control group, or several such comparison groups. Then they want to know whether the individuals in the two (or more) groups are different. For example, in one experiment student researchers measured how fast a product formed when they set up a reaction with three different concentrations of reactants (introduced in Chapter 8). Each treatment—meaning, each combination of reactant concentrations—was replicated many times.

FIGURE B4.1 graphs the average reaction rate for each of the three treatments in the experiment. Note that Treatments 1, 2, and 3 represent increasing concentrations of reactants. The thin

"I-beams" on each bar indicate the standard error of each average. The standard error is a quantity that indicates the uncertainty in the calculation of an average.

For example, if two trials with the same concentration of reactants had a reaction rate of 0.075 and two trials had a reaction rate of 0.025, then the average reaction rate would be 0.050. In this case, the standard error would be large. But if two trials had a reaction rate of 0.051 and two had a reaction rate of 0.049, the average would still be 0.050, but the standard error would be small.

In effect, the standard error quantifies how confident you are that the average you've calculated is the average you'd observe if

FIGURE B4.1 **Standard Error Bars Indicate the Uncertainty in an Average.**

you did the experiment under the same conditions an extremely large number of times. It is a measure of precision (see **BIOSKILLS 1**).

Once they had calculated these averages and standard errors, the students wanted to answer a question: Does reaction rate increase when reactant concentration increases?

After looking at the data, you might conclude that the answer is yes. But how could you come to a conclusion like this objectively, instead of subjectively?

The answer is to use a statistical test. This can be thought of as a three-step process.

1. Specify the null hypothesis, which is that reactant concentration has no effect on reaction rate.

2. Calculate a test statistic, which is a number that characterizes the size of the difference among the treatments. In this case, the test statistic compares the actual differences in reaction rates among treatments to the difference predicted by the null hypothesis. The null hypothesis predicts that there should be no difference.

3. The third step is to determine the probability of getting a test statistic at least as large as the one calculated just by chance. The answer comes from a reference distribution—a mathematical function that specifies the probability of getting various values of the test statistic if the null hypothesis is correct. (If you take a statistics course, you'll learn which test

statistics and reference distributions are relevant to different types of data.)

You are very likely to see small differences among treatment groups just by chance—even if no differences actually exist. If you flipped a coin 10 times, for example, you are unlikely to get exactly five heads and five tails, even if the coin is fair. A reference distribution tells you how likely you are to get each of the possible outcomes of the 10 flips if the coin is fair, just by chance.

In this case, the reference distribution indicated that if the null hypothesis of no actual difference in reaction rates is correct, you would see differences at least as large as those observed only 0.01 percent of the time just by chance. By convention, biologists consider a difference among treatment groups to be statistically significant if there is less than a 5 percent probability of observing it just by chance. Based on this convention, the student researchers were able to claim that the null hypothesis is not correct for reactant concentration. According to their data, the reaction they studied really does happen faster when reactant concentration increases.

You'll likely be doing actual statistical tests early in your undergraduate career. To use this text, though, you only need to know what statistical testing does. And you should take care to inspect the standard error bars on graphs in this book. As a *very* rough rule of thumb, averages often turn out to be significantly different, according to an appropriate statistical test, if there is no overlap between two times the standard errors.

check your understanding

If you understand BioSkill 4
✓ You should be able to . . .

QUANTITATIVE Determine which of the following tests used to estimate the average height of individuals in a class is likely to have the smallest standard error, and why.

- Measuring the height of two individuals chosen at random to estimate the average.

- Measuring the height of every student who showed up for class on a particular day to estimate the average.

In several cases in this text, you'll need to combine probabilities from different events in order to solve a problem. One of the most common applications is in genetics problems. For example, Punnett squares work because they are based on two fundamental rules of probability. Each rule pertains to a distinct situation.

The Both-And Rule

The both-and rule—also known as the product rule or multiplication rule—applies when you want to know the probability that two or more independent events occur together. Let's use the rolling of two dice as an example. What is the probability of rolling two sixes? These two events are independent, because the probability of rolling a six on one die has no effect on the probability of rolling a six on the other die. (In the same way, the probability of getting a gamete with allele R from one parent has no effect on the probability of getting a gamete with allele R from the other parent. Gametes fuse randomly.)

The probability of rolling a six on the first die is 1/6. The probability of rolling a six on the second die is also 1/6. The probability of rolling a six on *both* dice, then, is $1/6 \times 1/6 = 1/36$. In other words, if you rolled two dice 36 times, on average you would expect to roll two sixes once.

In the case of a cross between two parents heterozygous at the R gene, the probability of getting allele R from the father is 1/2 and the probability of getting R from the mother is 1/2. Thus, the probability of getting both alleles and creating an offspring with genotype RR is $1/2 \times 1/2 = 1/4$.

The Either-Or Rule

The either-or rule—also known as the sum rule or addition rule—applies when you want to know the probability of an event happening when there are several different ways for the same event or outcome to occur. In this case, the probability that the event will occur is the sum of the probabilities of each way that it can occur.

For example, suppose you wanted to know the probability of rolling either a one or a six when you toss a die. The probability of drawing each is 1/6, so the probability of getting one or the other is $1/6 + 1/6 = 1/3$. If you rolled a die three times, on average you'd expect to get a one or a six once.

In the case of a cross between two parents heterozygous at the R gene, the probability of getting an R allele from the father and an r allele from the mother is $1/2 \times 1/2 = 1/4$. Similarly, the probability of getting an r allele from the father and an R allele from the mother is $1/2 \times 1/2 = 1/4$. Thus, the combined probability of getting the Rr genotype in either of the two ways is $1/4 + 1/4 = 1/2$.

check your understanding

If you understand BioSkill 5

✔ **You should be able to . . .**

1. **QUANTITATIVE** Calculate the probability of getting four "tails" if four students each toss a coin.

2. **QUANTITATIVE** Calculate the probability of getting a two, a three, or a six after a single roll of a die.

You have probably been introduced to logarithms and logarithmic notation in algebra courses, and you will encounter logarithms at several points in this course. Logarithms are a way of working with powers—meaning, numbers that are multiplied by themselves one or more times.

Scientists use exponential notation to represent powers. For example,

$$a^x = y$$

means that if you multiply a by itself x times, you get y. In exponential notation, a is called the base and x is called the exponent. The entire expression is called an exponential function.

What if you know y and a, and you want to know x? This is where logarithms come in. You can solve for exponents using logarithms. For example,

$$x = \log_a y$$

This equation reads, x is equal to the logarithm of y to the base a. Logarithms are a way of working with exponential functions. They are important because so many processes in biology (and chemistry and physics, for that matter) are exponential. To understand what's going on, you have to describe the process with an exponential function and then use logarithms to work with that function.

Although a base can be any number, most scientists use just two bases when they employ logarithmic notation: 10 and e (sometimes called Euler's number after Swiss mathematician Leonhard Euler). What is e? It is a rate of exponential growth shared by many natural processes, where e is the limit of $(1 + \frac{1}{n})^n$ (as n tends to infinity). Mathematicians have shown that the base e is an irrational number (like π) that is approximately equal to 2.718. Like 10, e is just a number; $10^0 = 1$ and, likewise, $e^0 = 1$. But both 10 and e have qualities that make them convenient to use in biology (as well as chemistry and physics).

Logarithms to the base 10 are so common that they are usually symbolized in the form $\log y$ instead of $\log_{10} y$. A logarithm to the base e is called a natural logarithm and is symbolized ln (pronounced *EL-EN*) instead of log. You write "the natural logarithm of y" as ln y.

Most scientific calculators have keys that allow you to solve problems involving base 10 and base e. For example, if you know y, they'll tell you what log y or ln y are—meaning that they'll solve for x in our first example equation. They'll also allow you to find a number when you know its logarithm to base 10 or base e. Stated another way, they'll tell you what y is if you know x, and y is equal to e^x or 10^x. This is called taking an antilog. In most cases, you'll use the inverse or second function button on your calculator to find an antilog (above the log or ln key).

To get some practice with your calculator, consider this equation:

$$10^2 = 100$$

If you enter 100 in your calculator and then press the log key, the screen should say 2. The logarithm tells you what the exponent is. Now press the antilog key while 2 is on the screen. The calculator screen should return to 100. The antilog solves the exponential function, given the base and the exponent.

If your background in algebra isn't strong, you'll want to get more practice working with logarithms—you'll see them frequently during your undergraduate career. Remember that once you understand the basic notation, there's nothing mysterious about logarithms. They are simply a way of working with exponential functions, which describe what happens when something is multiplied by itself a number of times—like cells that divide and then divide again and then again.

Using logarithms will also come up when you are studying something that can have a large range of values, like the concentration of hydrogen ions in a solution or the intensity of sound that the human ear can detect. In cases like this, it's convenient to express the numbers involved as exponents. Using exponents makes a large range of numbers smaller and more manageable. For example, instead of saying that hydrogen ion concentration in a solution can range from 1 to 10^{-14}, the pH scale allows you to simply say that it ranges from 1 to 14. Instead of giving the actual value, you're expressing it as an exponent. It just simplifies things.

check your understanding

If you understand BioSkill 6

✔ **You should be able to . . .**

Use the equation $N_t = N_0 e^{rt}$ (see Chapter 54).

1. Explain what type of function this equation describes.

2. **QUANTITATIVE** Determine how you would write the equation, after taking the natural logarithm of both sides.

Phylogenetic trees show the evolutionary relationships among species, just as a genealogy shows the relationships among people in your family. They are unusual diagrams, however, and it can take practice to interpret them correctly.

To understand how evolutionary trees work, consider **FIGURE B7.1**. Notice that a phylogenetic tree consists of a root (the most ancestral branch in the tree), branches, nodes, and tips.

- Branches represent populations through time. In this text, branches are drawn as horizontal lines. In most cases the length of the branch is arbitrary and has no meaning, but in some cases branch lengths are proportional to time or the extent of genetic difference among populations (if so, there will be a scale at the bottom of the tree). The vertical lines on the tree represent splitting events, where one group broke into two independent groups. Their length is arbitrary—chosen simply to make the tree more readable.

- Nodes (also called forks) occur where an ancestral group splits into two or more descendant groups (see point A in Figure B7.1). Thus, each node represents the most recent common ancestor of the two or more descendant populations that emerge from it. If more than two descendant groups emerge from a node, the node is called a polytomy (see node C). A polytomy usually means that the populations split from one another so quickly that it is not possible to tell which split off earlier or later.

- Tips (also called terminal nodes) are the tree's endpoints, which represent groups living today or a dead end—a branch ending in extinction. The names at the tips can represent species or larger groups such as mammals or conifers.

Recall that a taxon (plural: taxa) is any named group of organisms (see Chapter 1). A taxon could be a single species, such as *Homo sapiens*, or a large group of species, such as Primates. Tips connected by a single node on a tree are called sister taxa.

The phylogenetic trees used in this text are all rooted. This means that the first, or most basal, node on the tree—the one on the far left in this book—is the most ancient. To determine where the root on a tree occurs, biologists include one or more outgroup species when they are collecting data to estimate a particular phylogeny. An outgroup is a taxonomic group that is known to have diverged before the rest of the taxa in the study. Outgroups are used to establish whether a trait is ancestral or derived. An ancestral trait is a characteristic that existed in an ancestor; a derived trait is a characteristic that is a modified form of the ancestral trait, found in a descendant.

In Figure B7.1, Taxon 1 is an outgroup to the monophyletic group consisting of taxa 2–6. A monophyletic group consists of an ancestral species and all of its descendants. The root of a tree is placed between the outgroup and the monophyletic group being studied. This position in Figure B7.1 is node A. Note that black hash marks are used to indicate a derived trait that is shared among the red branches, and another derived trait that is shared among the orange branches.

Understanding monophyletic groups is fundamental to reading and estimating phylogenetic trees. Monophyletic groups may also be called lineages or clades and can be identified using the "one-snip test": If you cut any branch on a phylogenetic tree, all of the branches and tips that fall off represent a monophyletic group. Using the one-snip test, you should be able to convince yourself that the monophyletic groups on a tree are nested. In Figure B7.1, for example, the monophyletic group comprising node A and taxa 1–6 contains a monophyletic group consisting of node B and taxa 2–6, which includes the monophyletic group represented by node C and taxa 4–6.

To put all these new terms and concepts to work, consider the phylogenetic tree in **FIGURE B7.2**, which shows the relationships between common chimpanzees and six human and humanlike species that lived over the past 5–6 million years. Chimps functioned as an outgroup in the analysis that led to this tree, so the root was placed at node A. The branches marked in red identify a monophyletic group called the hominins.

To practice how to read a tree, put your finger at the tree's root, at the far left, and work your way to the right. At node A, the ancestral population split into two descendant populations. One of these populations eventually evolved into today's chimps; the other gave rise to the six species of hominins pictured. Now

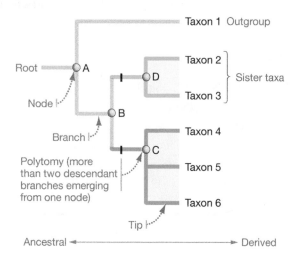

FIGURE B7.1 Phylogenetic Trees Have Roots, Branches, Nodes, and Tips.

✔**EXERCISE** Circle all four monophyletic groups present.

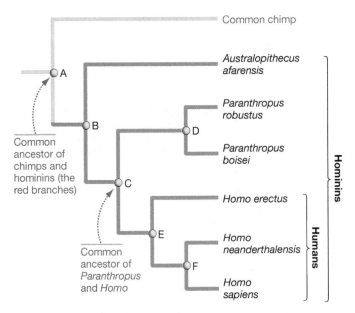

FIGURE B7.2 **An Example of a Phylogenetic Tree.** A phylogenetic tree showing the relationships of species in the monophyletic group called hominins.

✔**EXERCISE** All of the hominins walked on two legs—unlike chimps and all of the other primates. Add a mark on the phylogeny to show where upright posture evolved, and label it "origin of walking on two legs." Circle and label a pair of sister species. Label an outgroup to the monophyletic group called humans (species in the genus *Homo*).

continue moving your finger toward the tips of the tree until you hit node C. It should make sense to you that at this splitting event, one descendant population eventually gave rise to two *Paranthropus* species, while the other became the ancestor of humans—species in the genus *Homo*. As you study Figure B7.2, consider these two important points:

1. There are many equivalent ways of drawing this tree. For example, this version shows *Homo sapiens* on the bottom. But the tree would be identical if the two branches emerging from node E were rotated 180°, so that the species appeared in the order *Homo sapiens*, *Homo neanderthalensis*, *Homo erectus*. Trees are read from root to tips, not from top to bottom or bottom to top.

2. No species on any tree is any higher or lower than any other. Chimps and *Homo sapiens* have been evolving exactly the same amount of time since their divergence from a common ancestor—neither species is higher or lower than the other. It is legitimate to say that more ancient groups like *Australopithecus afarensis* have traits that are ancestral or more basal—meaning, that appeared earlier in evolution—compared to traits that appear in *Homo sapiens*, which are referred to as more derived.

FIGURE B7.3 presents a chance to test your tree-reading ability. Five of the six trees shown in this diagram are identical in terms of the evolutionary relationships they represent. One differs. The key to understanding the difference is to recognize that the ordering of tips does not matter in a tree—only the ordering of nodes (branch points) matters. You can think of a tree as being like a mobile: The tips can rotate without changing the underlying relationships.

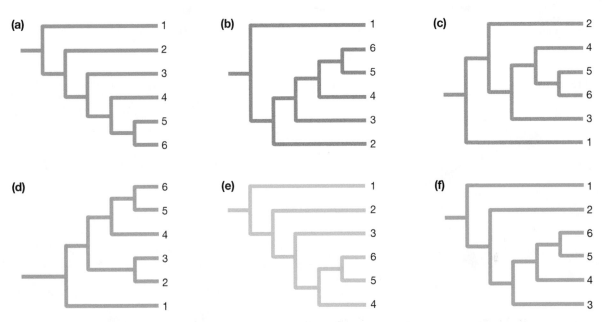

FIGURE B7.3 **Alternative Ways of Drawing the Same Tree.**

✔**QUESTION** Five of these six trees describe exactly the same relationships among taxa 1 through 6. Identify the tree that is different from the other five.

If you haven't had much chemistry yet, learning basic biological chemistry can be a challenge. One stumbling block is simply being able to read chemical structures efficiently and understand what they mean. This skill will come much easier once you have a little notation under your belt and you understand some basic symbols.

Atoms are the basic building blocks of everything in the universe, just as cells are the basic building blocks of your body. Every atom has a one- or two-letter symbol. **Table B8.1** shows the symbols for most of the atoms you'll encounter in this book. You should memorize these. The table also offers details on how the atoms form bonds as well as how they are represented in some visual models.

When atoms attach to each other by covalent bonding, a molecule forms. Biologists have a couple of different ways of representing molecules—you'll see each of these in the book and in class.

- Molecular formulas like those in **FIGURE B8.1a** simply list the atoms present in a molecule. Subscripts indicate how many of each atom are present. If the formula has no subscript, only one atom of that type is present. A methane (natural gas) molecule, for example, can be written as CH_4. It consists of one carbon atom and four hydrogen atoms.

TABLE B8.1 Some Attributes of Atoms Found in Organisms

Atom	Symbol	Number of Bonds It Can Form	Standard Color Code*
Hydrogen	H	1	white
Carbon	C	4	black
Nitrogen	N	3	blue
Oxygen	O	2	red
Sodium	Na	1	—
Magnesium	Mg	2	—
Phosphorus	P	5	orange or purple
Sulfur	S	2	yellow
Chlorine	Cl	1	—
Potassium	K	1	—
Calcium	Ca	2	—

*In ball-and-stick or space-filling models.

- Structural formulas like those in **FIGURE B8.1b** show which atoms in the molecule are bonded to each other. Each bond is indicated by a dash. The structural formula for methane in-

	Methane	Ammonia	Water	Oxygen
(a) Molecular formulas:	CH_4	NH_3	H_2O	O_2

(b) Structural formulas:

(c) Ball-and-stick models:

(d) Space-filling models:

FIGURE B8.1 Molecules Can Be Represented in Several Different Ways.

✔**EXERCISE** Carbon dioxide consists of a carbon atom that forms a double bond with each of two oxygen atoms, for a total of four bonds. It is a linear molecule. Write carbon dioxide's molecular formula and then draw its structural formula, a ball-and-stick model, and a space-filling model.

dicates that each of the four hydrogen atoms forms one covalent bond with carbon, and that carbon makes a total of four covalent bonds. Single covalent bonds are symbolized by a single dash; double bonds are indicated by two dashes.

Even simple molecules have distinctive shapes, because different atoms make covalent bonds at different angles. Ball-and-stick and space-filling models show the geometry of the bonds accurately.

- In a ball-and-stick model, a stick is used to represent each covalent bond (see **FIGURE B8.1c**).

- In space-filling models, the atoms are simply stuck onto each other in their proper places (see **FIGURE B8.1d**).

To learn more about a molecule when you look at a chemical structure, ask yourself three questions:

1. *Is the molecule polar—meaning that some parts are more negatively or positively charged than others?* Molecules that contain nitrogen or oxygen atoms are often polar, because these atoms have such high electronegativity (see Chapter 2). This trait is important because polar molecules dissolve in water.

2. *Does the structural formula show atoms that might participate in chemical reactions?* For example, are there charged atoms or amino or carboxyl ($-COOH$) groups that might act as a base or an acid?

3. *In ball-and-stick and especially space-filling models of large molecules, are there interesting aspects of overall shape?* For example, is there a groove where a protein might bind to DNA, or a cleft where a substrate might undergo a reaction in an enzyme?

BIOSKILL 9 separating and visualizing molecules

To study a molecule, you have to be able to isolate it. Isolating a molecule is a two-step process: the molecule has to be separated from other molecules in a mixture and then physically picked out or located in a purified form. **BIOSKILLS 9** focuses on the techniques that biologists use to separate nucleic acids and proteins and then find the particular one they are interested in.

Using Electrophoresis to Separate Molecules

In molecular biology, the standard technique for separating proteins and nucleic acids is called gel electrophoresis or, simply, electrophoresis (literally, "electricity-moving"). You may be using electrophoresis in a lab for this course, and you will certainly be analyzing data derived from electrophoresis in this text.

The principle behind electrophoresis is simple. Proteins (when denatured and coated with a special detergent) and nucleic acids carry a charge. As a result, these molecules move when placed in an electric field. Negatively charged molecules move toward the positive electrode; positively charged molecules move toward the negative electrode.

To separate a mixture of macromolecules so that each one can be isolated and analyzed, researchers place the sample in a gelatinous substance. More specifically, the sample is placed in a "well"—a slot in a sheet or slab of the gelatinous substance. The "gel" itself consists of long molecules that form a matrix of fibers. The gelatinous matrix has pores that act like a sieve through which the molecules can pass.

When an electrical field is applied across the gel, the molecules in the well move through the gel toward an electrode. Molecules that are smaller or more highly charged for their size move faster than do larger or less highly charged molecules. As they move, then, the molecules separate by size and by charge. Small and highly charged molecules end up at the bottom of the gel; large, less-charged molecules remain near the top.

An Example "Run"

FIGURE B9.1 (see B:14) shows the electrophoresis setup used in an experiment investigating how RNA molecules polymerize. In this case, the investigators wanted to document how long RNA molecules became over time, when ribonucleoside triphosphates were present in a particular type of solution.

Step 1 shows how they loaded samples of macromolecules, taken on different days during the experiment, into wells at the top of the gel slab. This is a general observation: Each well holds a different sample. In this and many other cases, the researchers also filled a well with a sample containing fragments of known size, called a size standard or "ladder."

In step 2, the researchers immersed the gel in a solution that conducts electricity and applied a voltage across the gel. The molecules in each well started to run down the gel, forming a lane. After several hours of allowing the molecules to move, the researchers removed the electric field (step 3). By then, molecules of different size and charge had separated from one another. In this case, small RNA molecules had reached the bottom of the gel. Above them were larger RNA molecules, which had run more slowly.

PROCESS: GEL ELECTROPHORESIS

Samples of macromolecules collected on different days

2 4 6 8 14

Fragments of known size

Wells —

Gel —

Power supply

− (top)

+ (bottom)

Molecules that are smaller and more highly charged run farther than molecules that are larger and less highly charged

1. Load cavities ("wells") in gel with samples.

2. Hook up power supply and run gel. Molecules separate over time as some migrate faster than others.

3. Remove gel after samples have run its length.

FIGURE B9.1 Macromolecules Can Be Separated via Gel Electrophoresis.

✔**QUESTION** DNA and RNA run toward the positive electrode. Why are these molecules negatively charged?

Why Do Separated Molecules Form Bands?

When researchers visualize a particular molecule on a gel, using techniques described in this section, the image that results consists of bands: shallow lines that are as wide as a lane in the gel. Why?

To understand the answer, study **FIGURE B9.2**. The left panel shows the original mixture of molecules. In this cartoon, the size of each dot represents the size of each molecule. The key is to realize that the original sample contains many copies of each specific molecule, and that these copies run down the length of the gel together—meaning, at the same rate—because they have the same size and charge.

It's that simple: Molecules that are alike form a band because they stay together.

Using Thin Layer Chromatography to Separate Molecules

Gel electrophoresis is not the only way to separate molecules. Researchers also use a method called thin layer chromatography. This method was developed in the early 1900s by botanists who were analyzing the different-colored pigments from leaves of a plant (see Chapter 10), hence the name chromatography from the Greek words khroma for "color" and graphein, "to write."

In this method, rather than loading the sample into the well of a gel, the samples are deposited or "spotted" near the bottom of a stiff support, either glass or plastic, that is coated with a thin layer of silica gel, cellulose, or a similar porous material. The coated support is placed in a solvent solution. As the solvent

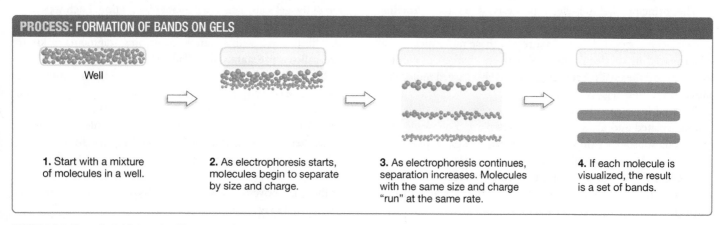

PROCESS: FORMATION OF BANDS ON GELS

Well

1. Start with a mixture of molecules in a well.

2. As electrophoresis starts, molecules begin to separate by size and charge.

3. As electrophoresis continues, separation increases. Molecules with the same size and charge "run" at the same rate.

4. If each molecule is visualized, the result is a set of bands.

FIGURE B9.2 On a Gel, Molecules That Are Alike Form Bands.

wicks upward through the coating by capillary action, it carries the molecules in the mixture with it. Molecules are carried at different rates, based on their size and solubility in the solvent.

Visualizing Molecules

Once molecules have been separated using electrophoresis or thin layer chromatography, they have to be detected. Unfortunately, although plant pigments are colored, proteins and nucleic acids are invisible unless they are tagged in some way. Let's first look at two of the most common tagging systems and then consider how researchers can tag and visualize specific molecules of interest and not others.

Using Radioactive Isotopes and Autoradiography

When molecular biology was getting under way, the first types of tags in common use were radioactive isotopes—forms of atoms that are unstable and release energy in the form of radiation.

In the polymerization experiment diagrammed in Figure B9.1, for example, the researchers had attached a radioactive phosphorus atom to the monomers—ribonucleoside triphosphates—used in the original reaction mix. Once polymers formed, they contained radioactive atoms. When electrophoresis was complete, the investigators visualized the polymers by laying X-ray film over the gel. Because radioactive emissions expose film, a black dot appears wherever a radioactive atom is located in the gel. So many black dots occur so close together that the collection forms a dark band.

This technique for visualizing macromolecules is called autoradiography. The autoradiograph that resulted from the polymerization experiment is shown in **FIGURE B9.3**. The samples, taken on days 2, 4, 6, 8, and 14 of the experiment, are noted along the bottom. The far right lane contains macromolecules of known size; this lane is used to estimate the size of the molecules in the experimental samples. The bands that appear in each sample lane represent the different polymers that had formed.

Reading a Gel

One of the keys to interpreting or "reading" a gel, or the corresponding autoradiograph, is to realize that darker bands contain more radioactive markers, indicating the presence of many radioactive molecules. Lighter bands contain fewer molecules.

To read a gel, then, you look for (1) the presence or absence of bands in some lanes—meaning, some experimental samples—versus others, and (2) contrasts in the darkness of the bands—meaning, differences in the number of molecules present.

For example, several conclusions can be drawn from the data in Figure B9.3. First, a variety of polymers formed at each stage. After the second day, for example, polymers from 12 to 18 monomers long had formed. Second, the overall length of polymers produced increased with time. At the end of the fourteenth day, most of the RNA molecules were between 20 and 40 monomers long.

FIGURE B9.3 Autoradiography Is a Technique for Visualizing Macromolecules. The molecules in a gel can be visualized in a number of ways. In this case, the RNA molecules in the gel exposed an X-ray film because they had radioactive atoms attached. When developed, the film is called an autoradiograph.

Starting in the late 1990s and early 2000s, it became much more common to tag nucleic acids with fluorescent tags. Once electrophoresis is complete, fluorescence can be detected by exposing the gel to an appropriate wavelength of light; the fluorescent tag fluoresces or glows in response (fluorescence is explained in Chapter 10).

Fluorescent tags have important advantages over radioactive isotopes: (1) They are safer to handle. (2) They are faster—you don't have to wait hours or days for the radioactive isotope to expose a film. (3) They come in multiple colors, so you can tag several different molecules in the same experiment and detect them independently.

Using Nucleic Acid Probes

In many cases, researchers want to find one specific molecule—a certain DNA sequence, for example—in the collection of molecules on a gel. How is this possible? The answer hinges on using a particular molecule as a probe.

You'll learn in more detail about how probes work in this text (Chapter 20). Here it's enough to get the general idea: A probe is a marked molecule that binds specifically to your molecule of interest. The "mark" is often a radioactive atom, a fluorescent tag, or an enzyme that catalyzes a color-forming or light-emitting reaction.

If you are looking for a particular DNA or RNA sequence on a gel, for example, you can expose the gel to a single-stranded probe that binds to the target sequence by complementary base pairing. Once it has bound, you can detect the band through autoradiography or fluorescence.

- **Southern blotting** is a technique for making DNA fragments that have been run out on a gel single stranded, transferring them from the gel to a nylon membrane, and then probing them to identify segments of interest. The technique was named after its inventor, Edwin Southern.

- **Northern blotting** is a technique for transferring RNA fragments from a gel to a nylon membrane and then probing them to detect target segments. The name is a lighthearted play on Southern blotting—the protocol from which it was derived.

Using Antibody Probes

How can researchers find a particular protein out of a large collection of different proteins? The answer is to use an antibody. An antibody is a protein that binds specifically to a section of a different protein (see Chapter 51 if you want more detail on the structure of antibodies and their function in the immune system).

To use an antibody as a probe, investigators attach a tag molecule—often an enzyme that catalyzes a color-forming reaction—to the antibody and allow it to react with proteins in a mixture. The antibody will stick to the specific protein that it binds to and then can be visualized thanks to the tag it carries.

If the proteins in question have been separated by gel electrophoresis and transferred to a membrane, the result is called a western blot. The name western is an extension of the Southern and northern patterns.

Using Radioimmunoassay and ELISA to Measure Amounts of Molecules

Another important method that makes use of antibodies is called a radioimmunoassay. This method is used when investigators want to measure tiny amounts of a molecule, such as a hormone in the blood. In this case, a known quantity of a hormone is labeled with a radioactive tag. This tagged hormone is then mixed with a known amount of antibody, and the two bind to one another. Next, a sample of blood, containing an unknown quantity of that same hormone, is added. The hormone from the blood and the radiolabeled hormone compete for antibody binding sites. As the concentration of unlabeled hormone increases, more of it binds to the antibody, displacing more of the radiolabeled hormone. The amount of unbound radiolabeled hormone is then measured. Using known standards as a reference, the amount of hormone in the blood can be determined.

Another commonly used technique based on similar principles is called ELISA (enzyme-linked immunosorbent assay). In this case, the amount of a particular molecule is measured using colorimetric signals instead of a radioactive signal.

check your understanding

If you understand BioSkill 9

✓ You should be able to . . .

Interpret a gel that has been stained for "RNA X." One lane contains no bands. Two lanes have a band in the same location, even though one of the bands is barely visible and the other is extremely dark. The fourth lane has a faint band located below the bands in the other lanes.

Biologists use a technique called differential centrifugation to isolate specific cell components. Differential centrifugation is based on breaking cells apart to create a complex mixture and then separating components in a centrifuge. A centrifuge accomplishes this task by spinning cells in a solution that allows molecules and other cell components to separate according to their density or size and shape. The individual parts of the cell can then be purified and studied in detail, in isolation from other parts of the cell.

The first step in preparing a cell sample for centrifugation is to release the cell components by breaking the cells apart. This can be done by putting them in a hypotonic solution, by exposing them to high-frequency vibration, by treating cells with a detergent, or by grinding them up. Each of these methods breaks apart plasma membranes and releases the contents of the cells.

The resulting pieces of plasma membrane quickly reseal to form small vesicles, often trapping cell components inside. The solution that results from the homogenization step is a mixture of these vesicles, free-floating macromolecules released from the cells, and organelles. A solution like this is called a cell extract or cell homogenate.

When a cell homogenate is placed in a centrifuge tube and spun at high speed, the components that are in solution tend to move outward, along the red arrow in **FIGURE B10.1a**. The effect is similar to a merry-go-round, which seems to push you outward in a straight line away from the spinning platform. In response to this outward-directed force, the cell homogenate exerts a centripetal (literally, "center-seeking") force that pushes the homogenate away from the bottom of the tube. Larger, denser molecules or particles resist this inward force more readily than do smaller, less dense ones and so reach the bottom of the centrifuge tube faster.

To separate the components of a cell extract, researchers often perform a series of centrifuge runs. Steps 1 and 2 of

(a) How a centrifuge works

When the centrifuge spins, the macromolecules tend to move toward the bottom of the centrifuge tube (red arrow)

The solution in the tube exerts a centripetal force, which resists movement of the molecules to the bottom of the tube (blue arrow)

Motor

Very large or dense molecules overcome the centripetal force more readily than smaller, less dense ones. As a result, larger, denser molecules move toward the bottom of the tube faster.

(b) PROCESS: DIFFERENTIAL CENTRIFUGATION

Low-speed centrifugation

Medium-speed centrifugation

High-speed centrifugation

Supernatant

Pellet

1. Start with uniform cell homogenate in centrifuge tube.

2. After low-speed spin, pellet contains large components. Transfer supernatant to new tube.

3. After medium-speed spin, pellet contains medium components. Transfer supernatant to new tube.

4. After high-speed spin, pellet contains small components.

(c) PROCESS: SUCROSE DENSITY–GRADIENT CENTRIFUGATION

Lower-density solution

Higher-density solution

Sample

1. Add sample to tube of variable-density solution.

2. Run centrifuge. Cell components are separated into distinct bands based on size and/or density.

3. To extract specific cell components for analysis, puncture bottom of tube with needle and collect drops from specific bands.

FIGURE B10.1 Cell Components Can Be Separated by Centrifugation. (a) The forces inside a centrifuge tube allow cell components to be separated. **(b)** Through a series of centrifuge runs made at increasingly higher speeds, an investigator can separate fractions of a cell homogenate by size via differential centrifugation. **(c)** A high-speed centrifuge run can achieve extremely fine separation among cell components by sucrose density–gradient centrifugation.

FIGURE B10.1b illustrate how an initial treatment at low speed causes larger, heavier parts of the homogenate to move below smaller, lighter parts. The material that collects at the bottom of the tube is called the pellet, and the solution and solutes left behind form the supernatant ("above-swimming"). The supernatant is placed in a fresh tube and centrifuged at increasingly higher speeds and longer durations. Each centrifuge run continues to separate cell components based on their size and density.

To accomplish separation of macromolecules or organelles, researchers frequently follow up with centrifugation at extremely high speeds. One strategy is based on filling the centrifuge tube with a series of sucrose solutions of increasing density (**FIGURE B10.1c**). The density gradient allows cell components to separate on the basis of small differences in size, shape, and density. When the centrifuge run is complete, each cell component occupies a distinct band of material in the tube, based on how quickly each component moves through the increasingly dense gradient of sucrose solution during the centrifuge run. A researcher can then collect the material in each band for further study.

If you understand BioSkill 10
✓ You should be able to . . .

1. Explain the physical basis for separating molecules or cell components via centrifugation given that electrophoresis separates molecules by charge and size.

2. Determine which cell component—ribosomes or mitochondria—you would expect to find closer to the bottom of the tube if you centrifuge a cell homogenate by sucrose density–gradient centrifugation.

BIOSKILL 11 — biological imaging: microscopy and x-ray crystallography

A lot of biology happens at levels that can't be detected with the naked eye. Biologists use an array of microscopes to study small multicellular organisms, individual cells, and the contents of cells. And to understand what individual macromolecules or macromolecular machines like ribosomes look like, researchers use data from a technique called X-ray crystallography.

You'll probably use dissecting microscopes and compound light microscopes to view specimens during your labs for this course, and throughout this text you'll be seeing images generated from other types of microscopy and from X-ray crystallographic data. Among the fundamental skills you'll be acquiring as an introductory student, then, is a basic understanding of how these techniques work. The key is to recognize that each approach for visualizing microscopic structures has strengths and weaknesses. As a result, each technique is appropriate for studying certain types or aspects of cells or molecules.

Light and Fluorescence Microscopy

If you use a dissecting microscope during labs, you'll recognize that it works by magnifying light that bounces off a whole specimen—often a live organism. You'll be able to view the specimen in three dimensions, which is why these instruments are sometimes called stereomicroscopes, but the maximum magnification possible is only about 20 to 40 times normal size (20X to 40X).

To view smaller objects, you'll probably use a compound microscope. Compound microscopes magnify light that is passed *through* a specimen. The instruments used in introductory labs are usually capable of 400X magnifications; the most sophisticated compound microscopes available can achieve magnifications of about 2000X. This is enough to view individual bacterial or eukaryotic cells and see large structures inside cells, like condensed chromosomes (see Chapter 12). To prepare a specimen for viewing under a compound light microscope, the tissues or cells are usually sliced to create a section thin enough for light to pass through efficiently. The section is then dyed to increase contrast and make structures visible. In many cases, different types of dyes are used to highlight different types of structures.

To visualize specific proteins, researchers use a technique called immunostaining. After preparing tissues or cells for viewing, the specimen is stained with fluorescently tagged antibodies. In this case, the cells are viewed under a fluorescence microscope. Ultraviolet or other wavelengths of light are passed through the specimen. The fluorescing tag emits visible light in response. The result? Beautiful cells that glow green, red, or blue.

Electron Microscopy

Until the 1950s, the compound microscope was the biologist's only tool for viewing cells directly. But the invention of the electron microscope provided a new way to view specimens. Two basic types of electron microscopy are now available: one that allows researchers to examine cross sections of cells at extremely high magnification, and one that offers a view of surfaces at somewhat lower magnification.

Transmission Electron Microscopy

The transmission electron microscope (TEM) is an extraordinarily effective tool for viewing cell structure at high magnification. TEM forms an image from electrons that pass through a specimen, just as a light microscope forms an image from light rays that pass through a specimen.

Biologists who want to view a cell under a transmission electron microscope begin by "fixing" the cell, meaning that they treat it with a chemical agent that stabilizes the cell's structure and contents while disturbing them as little as possible. Then the researcher permeates the cell with an epoxy plastic that stiffens the structure. Once this epoxy hardens, the cell can be cut into extremely thin sections with a glass or diamond knife. Finally, the sectioned specimens are impregnated with a metal—often lead. (The reason for this last step is explained shortly.)

FIGURE B11.1a outlines how the transmission electron microscope works. A beam of electrons is produced by a tungsten filament at the top of a column and directed downward. (All of the air is pumped out of the column, so that the electron beam isn't scattered by collisions with air molecules.) The electron beam passes through a series of lenses and through the specimen. The lenses are actually electromagnets, which alter the path of the beam much like a glass lens in a dissecting or compound microscope bends light. The electromagnet lenses magnify and focus the image on a screen at the bottom of the column. There the electrons strike a coating of fluorescent crystals, which emit visible light in response—just like a television screen. When the microscopist moves the screen out of the way and allows the electrons to expose a sheet of black-and-white film or to be detected by a digital camera, the result is a micrograph—a photograph of an image produced by microscopy.

The image itself is created by electrons that pass through the specimen. If no specimen were in place, all the electrons would pass through and the screen (and micrograph) would be uniformly bright. Unfortunately, cell materials by themselves would also appear fairly uniform and bright. This is because an atom's ability to deflect an electron depends on its mass. In turn, an atom's mass is a function of its atomic number. The hydrogen, carbon, oxygen, and nitrogen atoms that dominate biological molecules have low atomic numbers. This is why cell biologists must saturate cell sections with lead solutions. Lead has a high atomic number and scatters electrons effectively. Different macromolecules take up lead atoms in different amounts, so the metal acts as a "stain" that produces contrast. With TEM, areas of dense metal scatter the electron beam most, producing dark areas in micrographs.

The advantage of TEM is that it can magnify objects up to 250,000×—meaning that intracellular structures are clearly visible. The downsides are that researchers are restricted to observing dead, sectioned material, and they must take care that the preparation process does not distort the specimen.

Scanning Electron Microscopy

The scanning electron microscope (SEM) is the most useful tool biologists have for looking at the surfaces of structures. Materials are prepared for scanning electron microscopy by coating their surfaces with a layer of metal atoms. To create an image of this surface, the microscope scans the surface with a narrow beam of electrons. Electrons that are reflected back from the surface or that are emitted by the metal atoms in response to the beam then strike a detector. The signal from the detector controls a second electron beam, which scans a TV-like screen and forms an image magnified up to 50,000 times the object's size.

Because SEM records shadows and highlights, it provides images with a three-dimensional appearance (**FIGURE B11.1b**). It cannot magnify objects nearly as much as TEM can, however.

(a) Transmission electron microscopy: High magnification of cross sections

Tungsten filament
(source of electrons)

Condenser lens

Specimen

Objective lens

Projector lens

Image on
fluorescent screen

0.2 μm

Cross section of *E. coli* bacterium

(b) Scanning electron microscopy:
Lower magnification of surfaces

1 μm

Surface view of *E. coli* bacteria

FIGURE B11.1 There Are Two Basic Types of Electron Microscopy.

Studying Live Cells and Real-Time Processes

Until the 1960s, biologists were unable to get clear, high-magnification images of living cells. But a series of innovations over the past 50 years has made it possible to observe organelles and subcellular structures in action.

The development of video microscopy, where the image from a light microscope is captured by a video camera instead of by an eye or a film camera, proved revolutionary. It allowed specimens to be viewed at higher magnification, because video cameras are more sensitive to small differences in contrast than are the human eye or still cameras. It also made it easier to keep live specimens functioning normally, because the increased light sensitivity of video cameras allows them to be used with low illumination, so specimens don't overheat. And when it became possible to digitize video images, researchers began using computers to remove out-of-focus background material and increase image clarity.

A more recent innovation was the use of a fluorescent molecule called green fluorescent protein, or GFP, which allows researchers to tag specific molecules or structures and follow their movement in live cells over time. This was a major advance over immunostaining, in which cells have to be fixed. GFP is naturally synthesized in jellyfish that fluoresce, or emit light. By affixing GFP to another protein (using genetic engineering techniques described in Chapter 20) and then inserting it into a live cell, investigators can follow the protein's fate over time and even videotape its movement. For example, researchers have video-taped GFP-tagged proteins being transported from the rough ER through the Golgi apparatus and out to the plasma membrane. This is cell biology: the movie.

GFP's influence has been so profound that the researchers who developed its use in microscopy were awarded the 2008 Nobel Prize in Chemistry.

Visualizing Structures in 3-D

The world is three-dimensional. To understand how microscopic structures and macromolecules work, it is essential to understand their shape and spatial relationships. Consider three techniques currently being used to reconstruct the 3-D structure of cells, organelles, and macromolecules.

- *Confocal microscopy* is carried out by mounting cells that have been treated with one or more fluorescing tags on a microscope slide and then focusing a beam of ultraviolet or other wavelengths of light at a specific depth within the specimen. The fluorescing tag emits visible light in response. A detector for this light is then set up at exactly the position where the emitted light comes into focus. The result is a sharp image of a precise plane in the cell being studied (**FIGURE B11.2a**). Note that if you viewed the same specimen under a conventional fluorescence microscope, the image would be blurry because it results from light emitted by the entire cell (**FIGURE B11.2b**). By altering the focal plane, a researcher can record images from

(a) Confocal fluorescence image of single cell

(b) Conventional fluorescence image of same cell

25 µm

FIGURE B11.2 Confocal Microscopy Provides Sharp Images of Living Cells. (a) The confocal image of this mouse intestinal cell is sharp, because it results from light emitted at a single plane inside the cell. **(b)** The conventional image of this same cell is blurred, because it results from light emitted by the entire cell.

an array of depths in the specimen; a computer can then be used to generate a 3-D image of the cell.

- *Electron tomography* uses a transmission electron microscope to generate a 3-D image of an organelle or other subcellular structure. The specimen is rotated around a single axis while the researcher takes many "snapshots." The individual images are then pieced together with a computer. This technique has provided a much more accurate view of mitochondrial structure than was possible using traditional TEM (see Chapter 7).

- *X-ray crystallography, or X-ray diffraction analysis*, is the most widely used technique for reconstructing the 3-D structure of molecules. As its name implies, the procedure is based on bombarding crystals of a molecule with X-rays. X-rays are scattered in precise ways when they interact with the electrons surrounding the atoms in a crystal, producing a diffraction pattern that can be recorded on X-ray film or other types of detectors (**FIGURE B11.3**). By varying the orientation of the X-ray beam as it strikes a crystal and documenting the

check your understanding

 If you understand BioSkill 11

✔ You should be able to . . .

 1. Interpret whether the absence of mitochondria in a transmission electron micrograph of a cancerous human liver means that the cell lacks mitochondria.

 2. Explain why the effort to understand the structure of biological molecules is worthwhile even though X-ray crystallography is time consuming and technically difficult. What's the payoff?

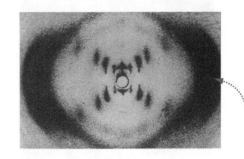

The patterns are determined by the structure of the molecules within the crystal

FIGURE B11.3 X-Ray Crystallography. When crystallized molecules are bombarded with X-rays, the radiation is scattered in distinctive patterns. The photograph at the right shows an X-ray film that recorded the pattern of scattered radiation from DNA molecules.

diffraction patterns that result, researchers can construct a map representing the density of electrons in the crystal. By relating these electron-density maps to information about the primary structure of the nucleic acid or protein, a 3-D model of the molecule can be built. Virtually all of the molecular models used in this book were built from X-ray crystallographic data.

BIOSKILL 12 cell and tissue culture methods

For researchers, there are important advantages to growing plant and animal cells and tissues outside the organism itself. Cell and tissue cultures provide large populations of a single type of cell or tissue and the opportunity to control experimental conditions precisely.

Animal Cell Culture

The first successful attempt to culture animal cells occurred in 1907, when a researcher cultivated amphibian nerve cells in a drop of fluid from the spinal cord. But it was not until the 1950s and 1960s that biologists could routinely culture plant and animal cells in the laboratory. The long lag time was due to the difficulty of re-creating conditions that exist in the intact organism precisely enough for cells to grow normally.

To grow in culture, animal cells must be provided with a liquid mixture containing the nutrients, vitamins, and hormones that stimulate growth. Initially, this mixture was serum, the liquid portion of blood; now, serum-free media are available for certain cell types. Serum-free media are preferred because they are much more precisely defined chemically than serum.

In addition, many types of animal cells will not grow in culture unless they are provided with a solid surface that mimics the types of surfaces that enable cells in the intact organisms to adhere. As a result, cells are typically cultured in flasks (**FIGURE B12.1a**, left; see B:22).

Even under optimal conditions, though, normal cells display a finite life span in culture. In contrast, many cultured cancerous cells grow indefinitely. This characteristic correlates with a key feature of cancerous cells in organisms: Their growth is continuous and uncontrolled.

Because of their immortality and relative ease of growth, cultured cancer cells are commonly used in research on basic aspects of cell structure and function. For example, the first human cell type to be grown in culture was isolated in 1951 from a malignant tumor of the uterine cervix. These cells are called HeLa cells in honor of their donor, Henrietta Lacks, who died soon thereafter from her cancer. HeLa cells continue to grow in laboratories around the world (Figure B12.1a, right).

Plant Tissue Culture

Certain cells found in plants are totipotent—meaning that they retain the ability to divide and differentiate into a complete, mature plant, including new types of tissue. These cells, called parenchyma cells, are important in wound healing and asexual reproduction. But they also allow researchers to grow complete adult plants in the laboratory, starting with a small number of parenchyma cells.

Biologists who grow plants in tissue culture begin by placing parenchyma cells in a liquid or solid medium containing all the

(a) Animal cell culture: immortal HeLa cancer cells

(b) Plant tissue culture: tobacco callus

Shoots

Callus

FIGURE B12.1 **Animal and Plant Cells Can Be Grown in the Lab.**

nutrients required for cell maintenance and growth. In the early days of plant tissue culture, investigators found not only that specific growth signals called hormones were required for successful growth and differentiation but also that the relative abundance of hormones present was critical to success.

The earliest experiments on hormone interactions in tissue cultures were done with tobacco cells in the 1950s by Folke Skoog and co-workers. These researchers found that when the hormone called auxin was added to the culture by itself, the cells enlarged but did not divide. But if the team added roughly equal amounts of auxin and another growth signal called cytokinin to the cells, the cells began to divide and eventually formed a callus, or an undifferentiated mass of parenchyma cells.

By varying the proportion of auxin to cytokinins in different parts of the callus and through time, the team could stimulate the growth and differentiation of root and shoot systems and produce whole new plants (**FIGURE B12.1b**). A high ratio of auxin to cytokinin led to the differentiation of a root system, while a high ratio of cytokinin to auxin led to the development of a shoot system. Eventually Skoog's team was able to produce a complete plant from just one parenchyma cell.

The ability to grow whole new plants in tissue culture from just one cell has been instrumental in the development of genetic engineering (see Chapter 20). Researchers insert recombinant genes into target cells, test the cells to identify those that successfully express the recombinant genes, and then use tissue culture techniques to grow those cells into adult individuals with a novel genotype and phenotype.

check your understanding

If you understand BioSkill 12

✔ **You should be able to . . .**

1. Identify a limitation of how experiments on HeLa cells are interpreted.

2. State a disadvantage of doing experiments on plants that have been propagated from single cells growing in tissue culture.

Research in biological science starts with a question. In most cases, the question is inspired by an observation about a cell or an organism. To answer it, biologists have to study a particular species. Study organisms are often called model organisms, because investigators hope that they serve as a model for what is going on in a wide array of species.

Model organisms are chosen because they are convenient to study and because they have attributes that make them appropriate for the particular research proposed. They tend to have some common characteristics:

- *Short generation time and rapid reproduction* This trait is important because it makes it possible to produce offspring quickly and perform many experiments in a short amount of time—you don't have to wait long for individuals to grow.

- *Large numbers of offspring* This trait is particularly important in genetics, where many offspring phenotypes and genotypes need to be assessed to get a large sample size.

- *Small size, simple feeding and habitat requirements* These attributes make it relatively cheap and easy to maintain individuals in the lab.

The following notes highlight just a few model organisms supporting current work in biological science.

Escherichia coli

Of all model organisms in biology, perhaps none has been more important than the bacterium *Escherichia coli*—a common inhabitant of the human gut. The strain that is most commonly worked on today, called K-12 (**FIGURE B13.1a**; see B:24), was originally isolated from a hospital patient in 1922.

During the last half of the twentieth century, key results in molecular biology originated in studies of *E. coli*. These results include the discovery of enzymes such as DNA polymerase, RNA polymerase, DNA repair enzymes, and restriction endonucleases; the elucidation of ribosome structure and function; and the initial characterization of promoters, regulatory transcription factors, regulatory sites in DNA, and operons. In many cases, initial discoveries made in *E. coli* allowed researchers to confirm that homologous enzymes and processes existed in an array of organisms, often ranging from other bacteria to yeast, mice, and humans.

The success of *E. coli* as a model for other species inspired Jacques Monod's claim that "Once we understand the biology of *Escherichia coli*, we will understand the biology of an elephant." The genome of *E. coli* K-12 was sequenced in 1997, and the strain continues to be a workhorse in studies of gene function, biochemistry, and particularly biotechnology. Much remains to be learned, however. Despite over 60 years of intensive study, the function of about a third of the *E. coli* genome is still unknown.

In the lab, *E. coli* is usually grown in suspension culture, where cells are introduced to a liquid nutrient medium, or on plates containing agar—a gelatinous mix of polysaccharides. Under optimal growing conditions—meaning before cells begin to get crowded and compete for space and nutrients—a cell takes just 30 minutes on average to grow and divide. At this rate, a single cell can produce a population of over a million descendants in just 10 hours. Except for new mutations, all of the descendant cells are genetically identical.

Dictyostelium discoideum

The cellular slime mold *Dictyostelium discoideum* is not always slimy, and it is not a mold—meaning a type of fungus. Instead, it is an amoeba. Amoeba is a general term that biologists use to characterize a unicellular eukaryote that lacks a cell wall and is extremely flexible in shape. *Dictyostelium* has long fascinated biologists because it is a social organism. Independent cells sometimes aggregate to form a multicellular structure.

Under most conditions, *Dictyostelium* cells are haploid (n) and move about in decaying vegetation on forest floors or other habitats. They feed on bacteria by engulfing them whole. When these cells reproduce, they can do so sexually by fusing with another cell then undergoing meiosis, or asexually by mitosis, which is more common. If food begins to run out, the cells begin to aggregate. In many cases, tens of thousands of cells cohere to form a 2-mm-long mass called a slug (**FIGURE B13.1b**). (This is not the slug that is related to snails.)

After migrating to a sunlit location, the slug stops and individual cells differentiate according to their position in the slug. Some form a stalk; others form a mass of spores at the tip of the stalk. (A spore is a single cell that develops into an adult organism, but it is not formed from gamete fusion like a zygote is.) The entire structure, stalk plus mass of spores, is called a fruiting body. Cells that form spores secrete a tough coat and represent a durable resting stage. The fruiting body eventually dries out, and the wind disperses the spores to new locations, where more food might be available.

Dictyostelium has been an important model organism for investigating questions about eukaryotes:

- Cells in a slug are initially identical in morphology but then differentiate into distinctive stalk cells and spores. Studying this process helped biologists better understand how cells in plant and animal embryos differentiate into distinct cell types.

(a) Bacterium *Escherichia coli* (strain K-12)

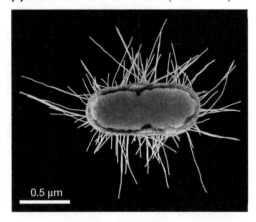

0.5 μm

(b) Slime mold *Dictyostelium discoideum*

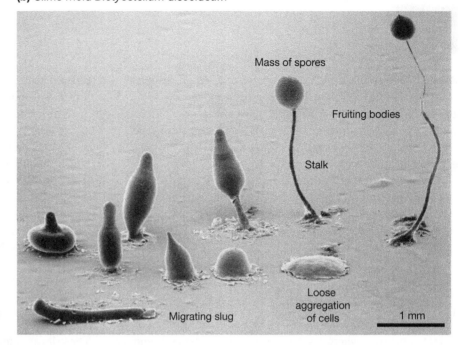

Mass of spores

Fruiting bodies

Stalk

Migrating slug

Loose aggregation of cells

1 mm

(c) Thale cress *Arabidopsis thaliana*

5 cm

(e) Fruit fly *Drosophila melanogaster*

0.5 mm

(f) Roundworm *Caenorhabditis elegans*

0.1 mm

(d) Yeast *Saccharomyces cerevisiae*

5 μm

(g) Mouse *Mus musculus*

1 cm

FIGURE B13.1 Model Organisms.

✔ **QUESTION** *E. coli* is grown at a temperature of 37°C. Why?

- The process of slug formation has helped biologists study how animal cells move and how they aggregate as they form specific types of tissues.

- When *Dictyostelium* cells aggregate to form a slug, they stick to each other. The discovery of membrane proteins responsible for cell–cell adhesion helped biologists understand some of the general principles of multicellular life (highlighted in Chapter 11).

Arabidopsis thaliana

In the early days of biology, the best-studied plants were agricultural varieties such as maize (corn), rice, and garden peas. When biologists began to unravel the mechanisms responsible for oxygenic photosynthesis in the early to mid-1900s, they relied on green algae that were relatively easy to grow and manipulate in the lab—often the unicellular species *Chlamydomonas reinhardii*—as an experimental subject.

Although crop plants and green algae continue to be the subject of considerable research, a new model organism emerged in the 1980s and now serves as the preeminent experimental subject in plant biology. That organism is *Arabidopsis thaliana*, commonly known as thale cress or wall cress (**FIGURE B13.1c**).

Arabidopsis is a member of the mustard family, or Brassicaceae, so it is closely related to radishes and broccoli. In nature it is a weed—meaning a species that is adapted to thrive in habitats where soils have been disturbed.

One of the most attractive aspects of working with *Arabidopsis* is that individuals can grow from a seed into a mature, seed-producing plant in just four to six weeks. Several other attributes make it an effective subject for study: It has just five chromosomes, has a relatively small genome with limited numbers of repetitive sequences, can self-fertilize as well as undergo cross-fertilization, can be grown in a relatively small amount of space and with a minimum of care in the greenhouse, and produces up to 10,000 seeds per individual per generation.

Arabidopsis has been instrumental in a variety of studies in plant molecular genetics and development, and it is increasingly popular in ecological and evolutionary studies. In addition, the entire genome of the species has now been sequenced, and studies have benefited from the development of an international "*Arabidopsis* community"—a combination of informal and formal associations of investigators who work on *Arabidopsis* and use regular meetings, e-mail, and the Internet to share data, techniques, and seed stocks.

Saccharomyces cerevisiae

When biologists want to answer basic questions about how eukaryotic cells work, they often turn to the yeast *Saccharomyces cerevisiae*.

S. cerevisiae is unicellular and relatively easy to culture and manipulate in the lab (**FIGURE B13.1d**). In good conditions, yeast cells grow and divide almost as rapidly as bacteria. As a result,

the species has become the organism of choice for experiments on control of the cell cycle and regulation of gene expression in eukaryotes. For example, research has confirmed that several of the genes controlling cell division and DNA repair in yeast have homologs in humans; and when mutated, these genes contribute to cancer. Strains of yeast that carry these mutations are now being used to test drugs that might be effective against cancer.

S. cerevisiae has become even more important in efforts to interpret the genomes of organisms like rice, mice, zebrafish, and humans. It is much easier to investigate the function of particular genes in *S. cerevisiae* by creating mutants or transferring specific alleles among individuals than it is to do the same experiments in mice or zebrafish. Once the function of a gene has been established in yeast, biologists can look for the homologous gene in other eukaryotes. If such a gene exists, they can usually infer that it has a function similar to its role in *S. cerevisiae*. It was also the first eukaryote with a completely sequenced genome.

Drosophila melanogaster

If you walk into a biology building on any university campus around the world, you are almost certain to find at least one lab where the fruit fly *Drosophila melanogaster* is being studied (**FIGURE B13.1e**).

Drosophila has been a key experimental subject in genetics since the early 1900s. It was initially chosen as a focus for study by T. H. Morgan, because it can be reared in the laboratory easily and inexpensively, matings can be arranged, the life cycle is completed in less than two weeks, and females lay a large number of eggs. These traits made fruit flies valuable subjects for breeding experiments designed to test hypotheses about how traits are transmitted from parents to offspring (see Chapter 14).

More recently, *Drosophila* has also become a key model organism in the field of developmental biology. The use of flies in developmental studies was inspired largely by the work of Christianne Nüsslein-Volhard and Eric Wieschaus, who in the 1980s isolated flies with genetic defects in early embryonic development. By investigating the nature of these defects, researchers have gained valuable insights into how various gene products influence the development of eukaryotes (see Chapter 22). The complete genome sequence of *Drosophila* has been available to investigators since the year 2000.

Caenorhabditis elegans

The roundworm *Caenorhabditis elegans* emerged as a model organism in developmental biology in the 1970s, due largely to work by Sydney Brenner and colleagues. (*Caenorhabditis* is pronounced *see-no-rab-DIE-tiss*.)

C. elegans was chosen for three reasons: (**1**) Its cuticle (soft outer layer) is transparent, making individual cells relatively easy to observe (**FIGURE B13.1f**); (**2**) adults have exactly 959 nonreproductive cells; and, most important, (**3**) the fate of each cell in an embryo can be predicted because cell fates are invariant among individuals. For example, when researchers examine a 33-cell *C. elegans* embryo, they know exactly which of the 959 cells in the adult will be derived from each of those 33 embryonic cells.

In addition, *C. elegans* are small (less than 1 mm long), are able to self-fertilize or cross-fertilize, and undergo early development in just 16 hours. The entire genome of *C. elegans* has now been sequenced.

Mus musculus

The house mouse *Mus musculus* is the most important model organism among mammals. For this reason, it is especially prominent in biomedical research, where researchers need to work on individuals with strong genetic and developmental similarities to humans.

The house mouse was an intelligent choice of model organism in mammals because it is small and thus relatively inexpensive to maintain in captivity, and because it breeds rapidly. A litter can contain 10 offspring, and generation time is only 12 weeks—meaning that several generations can be produced in a year. Descendants of wild house mice have been selected for docility and other traits that make them easy to handle and rear; these populations are referred to as laboratory mice (**FIGURE B13.1g**).

Some of the most valuable laboratory mice are strains with distinctive, well-characterized genotypes. Inbred strains are virtually homogenous genetically (see Chapter 26) and are useful in experiments where gene-by-gene or gene-by-environment interactions have to be controlled. Other populations carry mutations that knock out genes and cause diseases similar to those observed in humans. These individuals are useful for identifying the cause of genetic diseases and testing drugs or other types of therapies.

check your understanding

C
Y
U

If you understand BioSkill 13

✔ **You should be able to . . .**

Determine which model organisms described here would be the best choice for the following studies. In each case, explain your reasoning.

1. A study of why specific cells in an embryo die at certain points in normal development. One goal is to understand the consequences for the individual when programmed cell death does not occur when it should.

2. A study of proteins that are required for cell–cell adhesion.

3. Research on a gene suspected to be involved in the formation of breast cancer in humans.

BIOSKILL 14 primary literature and peer review

As part of the process of doing science, biologists communicate their results to the scientific community through publications in scientific journals that report on their original research discoveries (see Chapter 1). These published reports are referred to, interchangeably, as the primary literature, research papers, or primary research articles.

What Is the Primary Literature?

Scientists publish "peer-reviewed" papers. This means that several experts in the field have carefully read the paper and considered its strengths and weaknesses. Reviewers write a critique of the paper and make a recommendation to the journal editor as to whether the paper should be published. Often reviewers will suggest additional experiments that need to be completed before a paper is considered acceptable for publication. The peer review process means that research discoveries are carefully vetted before they go to press.

A primary research paper can be distinguished from secondary sources—such as review articles, textbooks, and magazine articles—by looking for key characteristics. A primary research paper includes a detailed description of methods and results, written by the researchers who did the work. A typical paper contains a Title, Abstract, Introduction, Materials and Methods (or Experimental Design), Results and Discussion (**Table B14.1**), although the order and name of the sections varies among journals.

Getting Started

At first, trying to read the primary literature may seem like a daunting task. A paper may be peppered with unfamiliar terms and acronyms. If you tried to read a research paper from start to finish, like you might read a chapter in this textbook, it would be a frustrating experience. But, with practice, the scientific literature becomes approachable, and it is well worth the effort. The primary literature is the cutting edge, the place to read firsthand about the process of doing science. Becoming skilled at reading and evaluating scientific reports is a powerful way to learn how to think critically—to think like a biologist.

To get started, try breaking down reading the primary research article into a series of steps:

1. Read the authors' names. Where are they from? Are they working as a team or alone? After delving into the literature,

Studying Live Cells and Real-Time Processes

Until the 1960s, biologists were unable to get clear, high-magnification images of living cells. But a series of innovations over the past 50 years has made it possible to observe organelles and subcellular structures in action.

The development of video microscopy, where the image from a light microscope is captured by a video camera instead of by an eye or a film camera, proved revolutionary. It allowed specimens to be viewed at higher magnification, because video cameras are more sensitive to small differences in contrast than are the human eye or still cameras. It also made it easier to keep live specimens functioning normally, because the increased light sensitivity of video cameras allows them to be used with low illumination, so specimens don't overheat. And when it became possible to digitize video images, researchers began using computers to remove out-of-focus background material and increase image clarity.

A more recent innovation was the use of a fluorescent molecule called green fluorescent protein, or GFP, which allows researchers to tag specific molecules or structures and follow their movement in live cells over time. This was a major advance over immunostaining, in which cells have to be fixed. GFP is naturally synthesized in jellyfish that fluoresce, or emit light. By affixing GFP to another protein (using genetic engineering techniques described in Chapter 20) and then inserting it into a live cell, investigators can follow the protein's fate over time and even videotape its movement. For example, researchers have videotaped GFP-tagged proteins being transported from the rough ER through the Golgi apparatus and out to the plasma membrane. This is cell biology: the movie.

GFP's influence has been so profound that the researchers who developed its use in microscopy were awarded the 2008 Nobel Prize in Chemistry.

Visualizing Structures in 3-D

The world is three-dimensional. To understand how microscopic structures and macromolecules work, it is essential to understand their shape and spatial relationships. Consider three techniques currently being used to reconstruct the 3-D structure of cells, organelles, and macromolecules.

- *Confocal microscopy* is carried out by mounting cells that have been treated with one or more fluorescing tags on a microscope slide and then focusing a beam of ultraviolet or other wavelengths of light at a specific depth within the specimen. The fluorescing tag emits visible light in response. A detector for this light is then set up at exactly the position where the emitted light comes into focus. The result is a sharp image of a precise plane in the cell being studied (**FIGURE B11.2a**). Note that if you viewed the same specimen under a conventional fluorescence microscope, the image would be blurry because it results from light emitted by the entire cell (**FIGURE B11.2b**). By altering the focal plane, a researcher can record images from

(a) Confocal fluorescence image of single cell   **(b)** Conventional fluorescence image of same cell

25 μm

FIGURE B11.2 Confocal Microscopy Provides Sharp Images of Living Cells. (a) The confocal image of this mouse intestinal cell is sharp, because it results from light emitted at a single plane inside the cell. **(b)** The conventional image of this same cell is blurred, because it results from light emitted by the entire cell.

an array of depths in the specimen; a computer can then be used to generate a 3-D image of the cell.

- *Electron tomography* uses a transmission electron microscope to generate a 3-D image of an organelle or other subcellular structure. The specimen is rotated around a single axis while the researcher takes many "snapshots." The individual images are then pieced together with a computer. This technique has provided a much more accurate view of mitochondrial structure than was possible using traditional TEM (see Chapter 7).

- *X-ray crystallography, or X-ray diffraction analysis*, is the most widely used technique for reconstructing the 3-D structure of molecules. As its name implies, the procedure is based on bombarding crystals of a molecule with X-rays. X-rays are scattered in precise ways when they interact with the electrons surrounding the atoms in a crystal, producing a diffraction pattern that can be recorded on X-ray film or other types of detectors (**FIGURE B11.3**). By varying the orientation of the X-ray beam as it strikes a crystal and documenting the magnification. TEM forms an image from electrons that pass through a specimen, just as a light microscope forms an image from light rays that pass through a specimen.

Biologists who want to view a cell under a transmission electron microscope begin by "fixing" the cell, meaning that they treat it with a chemical agent that stabilizes the cell's structure and contents while disturbing them as little as possible. Then the researcher permeates the cell with an epoxy plastic that stiffens the structure. Once this epoxy hardens, the cell can be cut into extremely thin sections with a glass or diamond knife. Finally, the sectioned specimens are impregnated with a metal—often lead. (The reason for this last step is explained shortly.)

FIGURE B11.1a outlines how the transmission electron microscope works. A beam of electrons is produced by a tungsten filament at the top of a column and directed downward. (All of the air is pumped out of the column, so that the electron beam isn't scattered by collisions with air molecules.) The electron beam passes through a series of lenses and through the specimen. The lenses are actually electromagnets, which alter the path of the beam much like a glass lens in a dissecting or compound microscope bends light. The electromagnet lenses magnify and focus the image on a screen at the bottom of the column. There the electrons strike a coating of fluorescent crystals, which emit visible light in response—just like a television screen. When the microscopist moves the screen out of the way and allows the electrons to expose a sheet of black-and-white film or to be detected by a digital camera, the result is a micrograph—a photograph of an image produced by microscopy.

The image itself is created by electrons that pass through the specimen. If no specimen were in place, all the electrons would pass through and the screen (and micrograph) would be uniformly bright. Unfortunately, cell materials by themselves would

also appear fairly uniform and bright. This is because an atom's ability to deflect an electron depends on its mass. In turn, an atom's mass is a function of its atomic number. The hydrogen, carbon, oxygen, and nitrogen atoms that dominate biological molecules have low atomic numbers. This is why cell biologists must saturate cell sections with lead solutions. Lead has a high atomic number and scatters electrons effectively. Different macromolecules take up lead atoms in different amounts, so the metal acts as a "stain" that produces contrast. With TEM, areas of dense metal scatter the electron beam most, producing dark areas in micrographs.

The advantage of TEM is that it can magnify objects up to 250,000×—meaning that intracellular structures are clearly visible. The downsides are that researchers are restricted to observing dead, sectioned material, and they must take care that the preparation process does not distort the specimen.

Scanning Electron Microscopy

The scanning electron microscope (SEM) is the most useful tool biologists have for looking at the surfaces of structures. Materials are prepared for scanning electron microscopy by coating their surfaces with a layer of metal atoms. To create an image of this surface, the microscope scans the surface with a narrow beam of electrons. Electrons that are reflected back from the surface or that are emitted by the metal atoms in response to the beam then strike a detector. The signal from the detector controls a second electron beam, which scans a TV-like screen and forms an image magnified up to 50,000 times the object's size.

Because SEM records shadows and highlights, it provides images with a three-dimensional appearance (**FIGURE B11.1b**). It cannot magnify objects nearly as much as TEM can, however.

(a) Transmission electron microscopy: High magnification of cross sections

Tungsten filament (source of electrons)

Condenser lens

Specimen

Objective lens

Projector lens

Image on fluorescent screen

Cross section of *E. coli* bacterium

0.2 μm

(b) Scanning electron microscopy: Lower magnification of surfaces

Surface view of *E. coli* bacteria

1 μm

FIGURE B11.1 There Are Two Basic Types of Electron Microscopy.

If you haven't had much chemistry yet, learning basic biological chemistry can be a challenge. One stumbling block is simply being able to read chemical structures efficiently and understand what they mean. This skill will come much easier once you have a little notation under your belt and you understand some basic symbols.

Atoms are the basic building blocks of everything in the universe, just as cells are the basic building blocks of your body. Every atom has a one- or two-letter symbol. **Table B8.1** shows the symbols for most of the atoms you'll encounter in this book. You should memorize these. The table also offers details on how the atoms form bonds as well as how they are represented in some visual models.

When atoms attach to each other by covalent bonding, a molecule forms. Biologists have a couple of different ways of representing molecules—you'll see each of these in the book and in class.

- Molecular formulas like those in **FIGURE B8.1a** simply list the atoms present in a molecule. Subscripts indicate how many of each atom are present. If the formula has no subscript, only one atom of that type is present. A methane (natural gas) molecule, for example, can be written as CH_4. It consists of one carbon atom and four hydrogen atoms.

TABLE B8.1 Some Attributes of Atoms Found in Organisms

Atom	Symbol	Number of Bonds It Can Form	Standard Color Code*
Hydrogen	H	1	white
Carbon	C	4	black
Nitrogen	N	3	blue
Oxygen	O	2	red
Sodium	Na	1	—
Magnesium	Mg	2	—
Phosphorus	P	5	orange or purple
Sulfur	S	2	yellow
Chlorine	Cl	1	—
Potassium	K	1	—
Calcium	Ca	2	—

*In ball-and-stick or space-filling models.

- Structural formulas like those in **FIGURE B8.1b** show which atoms in the molecule are bonded to each other. Each bond is indicated by a dash. The structural formula for methane in-

	Methane	Ammonia	Water	Oxygen
(a) Molecular formulas:	CH_4	NH_3	H_2O	O_2

FIGURE B8.1 Molecules Can Be Represented in Several Different Ways.

✓**EXERCISE** Carbon dioxide consists of a carbon atom that forms a double bond with each of two oxygen atoms, for a total of four bonds. It is a linear molecule. Write carbon dioxide's molecular formula and then draw its structural formula, a ball-and-stick model, and a space-filling model.

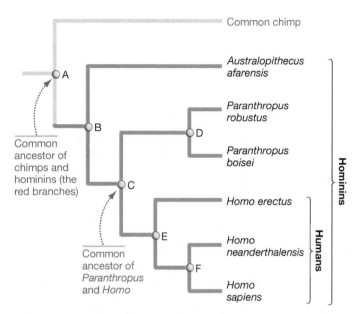

FIGURE B7.2 **An Example of a Phylogenetic Tree.** A phylogenetic tree showing the relationships of species in the monophyletic group called hominins.

✔ **EXERCISE** All of the hominins walked on two legs—unlike chimps and all of the other primates. Add a mark on the phylogeny to show where upright posture evolved, and label it "origin of walking on two legs." Circle and label a pair of sister species. Label an outgroup to the monophyletic group called humans (species in the genus *Homo*).

continue moving your finger toward the tips of the tree until you hit node C. It should make sense to you that at this splitting event, one descendant population eventually gave rise to two *Paranthropus* species, while the other became the ancestor of humans—species in the genus *Homo*. As you study Figure B7.2, consider these two important points:

1. There are many equivalent ways of drawing this tree. For example, this version shows *Homo sapiens* on the bottom. But the tree would be identical if the two branches emerging from node E were rotated 180°, so that the species appeared in the order *Homo sapiens*, *Homo neanderthalensis*, *Homo erectus*. Trees are read from root to tips, not from top to bottom or bottom to top.

2. No species on any tree is any higher or lower than any other. Chimps and *Homo sapiens* have been evolving exactly the same amount of time since their divergence from a common ancestor—neither species is higher or lower than the other. It is legitimate to say that more ancient groups like *Australopithecus afarensis* have traits that are ancestral or more basal—meaning, that appeared earlier in evolution—compared to traits that appear in *Homo sapiens*, which are referred to as more derived.

FIGURE B7.3 presents a chance to test your tree-reading ability. Five of the six trees shown in this diagram are identical in terms of the evolutionary relationships they represent. One differs. The key to understanding the difference is to recognize that the ordering of tips does not matter in a tree—only the ordering of nodes (branch points) matters. You can think of a tree as being like a mobile: The tips can rotate without changing the underlying relationships.

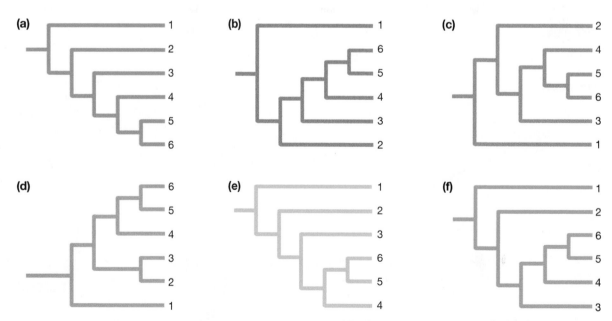

FIGURE B7.3 **Alternative Ways of Drawing the Same Tree.**

✔ **QUESTION** Five of these six trees describe exactly the same relationships among taxa 1 through 6. Identify the tree that is different from the other five.

TABLE B14.1 Sections of a Primary Research Paper

Section	Characteristics
Title	Short, succinct, eye-catching
Abstract	Summary of Methods, Results, Discussion. Explains why the research was done and why the results are significant.
Introduction	Background information (what past work was done, why the work was important). States the objectives and hypotheses of the study and explains why the study is important.
Materials and Methods	Explains how the work was done and where it was done.
Results	Explains what the data show.
Discussion	Explains why the data show what they show, how the analysis relates to the objectives from the Introduction, and the significance of findings and how they advance the field.

certain familiar names will crop up again and again. You'll begin to recognize the experts in a particular field.

2. Read the title. It should summarize the key finding of the paper and tell you what you can expect to learn from the paper.

3. Read the abstract. The abstract summarizes the entire paper in a short paragraph. At this point, it might be tempting to stop reading. But sometimes authors understate or overstate the significance and conclusions of their work. You should never cite an article as a reference after having read only the abstract.

4. Read the Introduction. The first couple of paragraphs should make it clear what the objectives or hypotheses of the paper are; the remaining paragraphs will give you the background information you need to understand the point of the paper.

5. Flip through the article and look at the figures, illustrations, and tables, including reading the legends.

6. Read the Results section carefully. Ask yourself these questions: Do the results accurately describe the data presented

in the paper? Were all the appropriate controls carried out in an experiment? Are there additional experiments that you think should have been performed? Are the figures and tables clearly labeled?

7. Consult the Materials and Methods section to help understand the research design and the techniques used.

8. Read the Discussion. The first and last paragraphs usually summarize the key findings and state their significance. The Discussion is the part of the paper where the results are explained in the context of the scientific literature. The authors should explain what their results mean.

Getting Practice

The best way to get practice is to start reading the scientific literature as often as possible. You could begin by reading some of the references cited in this textbook. You can get an electronic copy of most articles through online databases such as PubMed, Science-Direct, or Google Scholar, or through your institution's library.

After reading a primary research paper, you should be able to paraphrase the significance of the paper in a few sentences, free of technical jargon. You should also be able to both praise and criticize several points of the paper. As you become more familiar with reading the scientific literature, you're likely to start thinking about what questions remain to be answered. And, you may even come up with "the next experiment."

check your understanding

If you understand BioSkill 14

✔ **You should be able to . . .**

After choosing a primary research paper on a topic in biology that you would like to know more about, select one figure in the Results section that reports on the experiment and construct a Research box (like the ones in this textbook) that depicts this experiment.

A concept map is a graphical device for organizing and expressing what you know about a topic. It has two main elements: **(1)** concepts that are identified by words or short phrases and placed in a box or circle, and **(2)** labeled arrows that physically link two concepts and explain the relationship between them. The concepts are arranged hierarchically on a page with the most general concepts at the top and the most specific ideas at the bottom.

The combination of a concept, a linking word, and a second concept is called a proposition. Good concept maps also have cross-links—meaning, labeled arrows that connect different elements in the hierarchy as you read down the page.

Concept maps, initially developed by Joseph Novak in the early 1970s, have proven to be an effective studying and learning tool. They can be particularly valuable if constructed by a group, or when different individuals exchange and critique concept maps they have created independently. Although concept maps vary widely in quality and can be graded using objective criteria, there are many equally valid ways of making a high-quality concept map on a particular topic.

When you are asked to make a concept map in this text, you will usually be given at least a partial list of concepts to use. As an example, suppose you were asked to create a concept map on experimental design and were given the following concepts: results, predictions, control treatment, experimental treatment, controlled (identical) conditions, conclusions, experiment, hypothesis to be tested, null hypothesis. One possible concept map is shown in **FIGURE B15.1**.

Good concept maps have four qualities:

1. They exhibit an organized hierarchy, indicating how each concept on the map relates to larger and smaller concepts.

2. The concept words are specific—not vague.

3. The propositions are accurate.

4. There is cross-linking between different elements in the hierarchy of concepts.

As you practice making concept maps, go through these criteria and use them to evaluate your own work as well as the work of fellow students.

check your understanding

If you understand BioSkill 15

✔ **You should be able to . . .**

1. Add an "Alternative hypothesis" concept to the map in Figure B15.1, along with other concepts and labeled linking arrows needed to indicate its relationship to other information on the map. (Hint: Recall that investigators often contrast a hypothesis being tested with an alternative hypothesis that does not qualify as a null hypothesis.)

2. Add a box for the concept "Statistical testing" (see **BIOSKILLS 4**) along with appropriately labeled linking arrows.

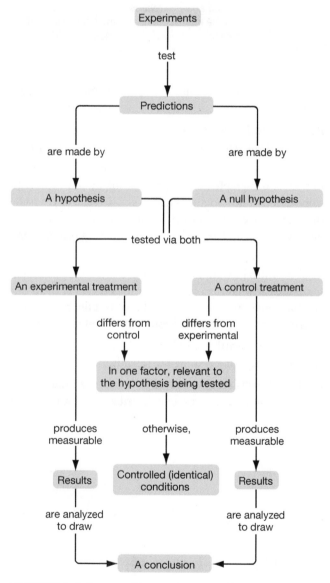

FIGURE B15.1 A Concept Map on Principles of Experimental Design.

Most students have at one time or another wondered why a particular question on an exam seemed so hard, while others seemed easy. The explanation lies in the type of cognitive skills required to answer the question. Let's take a closer look.

Categories of Human Cognition

Bloom's Taxonomy is a classification system that instructors use to identify the cognitive skill levels at which they are asking students to work, particularly on practice problems and exams. Bloom's Taxonomy is also a very useful tool for students to know—it can help you to figure out the appropriate level at which you should be studying to succeed in a course.

Bloom's Taxonomy distinguishes six different categories of human thinking: remember, understand, apply, analyze, evaluate, and create. One of the most useful distinctions lies not in the differences among the six categories, but rather in the difference between high-order cognitive (HOC) and low-order cognitive (LOC) skills. **FIGURE B16.1** shows how the different levels of the taxonomy can be broken into HOC and LOC skills.

Skills that hallmark LOCs include recall, explanation, or application of knowledge in the exact way that you have before (remember, understand, apply), while skills that typify HOCs include the application of knowledge in a new way, as well as the breakdown, critique, or creation of information (analyze, evaluate, and create). Most college instructors will assume students are proficient at solving LOC questions and will expect you to frequently work at the HOC levels. HOC problems usually require use of basic vocabulary and applying knowledge—working at this level helps students to master the LOC levels.

Six Study Steps to Success

Bloom's Taxonomy provides a useful guide for preparing for an exam, using the following six steps:

1. *Answer in-chapter questions while reading the chapter.* All questions in this book have been assigned Bloom's levels, so you can review the question answers and the Bloom's level while you study.

2. *Identify the Bloom's level(s) of the questions that you are having greatest difficulty answering.* While working through

FIGURE B16.1 Bloom's Taxonomy.

the text, take note of the content and Bloom's level(s) that you find the most challenging.

3. *Use the Bloom's Taxonomy Study Guide* (**Table B16.1;** see B:30) *to focus your study efforts at the appropriate Bloom's level.* Table B16.1 lists specific study methods that can help you practice your understanding of the material at both the LOC and HOC levels, whether you are studying alone or with a study group.

4. *Complete the end-of-chapter questions as if you're taking an exam, without looking for the answers.* If you look at the chapter text or jump to the answers, then you really aren't testing your ability to work with the content and have reduced the questions to the lowest Bloom's level of remember.

5. *Grade your answers and note the Bloom's level of the questions you got wrong.* At what level of Bloom's Taxonomy were the questions you missed?

6. *Use the Bloom's Taxonomy Study Guide to focus your study efforts at the appropriate Bloom's level.* If you missed a lot of questions, then spend more time studying the material and find other resources for quizzing yourself.

By following these steps and studying at the HOC levels, you should succeed in answering questions on in-class exams.

	Individual Study Activities	Group Study Activities
Create (HOC) Generate something new	• Generate a hypothesis or design an experiment based on information you are studying • Create a model based on a given data set • Create summary sheets that show how facts and concepts relate to each other • Create questions at each level of Bloom's Taxonomy as a practice test and then take the test	• Each student puts forward a hypothesis about biological process and designs an experiment to test it. Peers critique the hypotheses and experiments • Create a new model/summary sheet/concept map that integrates each group member's ideas
Evaluate (HOC) Defend or judge a concept or idea	• Provide a written assessment of the strengths and weaknesses of your peers' work or understanding of a given concept based on previously determined criteria	• Provide a verbal assessment of the strengths and weaknesses of your peers' work or understanding of a given concept based on previously described criteria, and have your peers critique your assessment
Analyze (HOC) Distinguish parts and make inferences	• Analyze and interpret data in primary literature or a textbook without reading the author's interpretation and then compare the authors' interpretation with your own • Analyze a situation and then identify the assumptions and principles of the argument • Compare and contrast two ideas or concepts • Construct a map of the main concepts by defining the relationships of the concepts using one- or two-way arrows	• Work together to analyze and interpret data in primary literature or a textbook without reading the author's interpretation, and defend your analysis to your peers • Work together to identify all of the concepts in a paper or textbook chapter, construct individual maps linking the concepts together with arrows and words that relate the concepts, and then grade each other's concept maps
Apply (HOC or LOC) Use information or concepts in new ways (HOC) or in the same ways (LOC)	• Review each process you have learned and then ask yourself: What would happen if you increase or decrease a component in the system, or what would happen if you alter the activity of a component in the system? • If possible, graph a biological process and create scenarios that change the shape or slope of the graph	• Practice writing out answers to old exam questions on the board, and have your peers check to make sure you don't have too much or too little information in your answer • Take turns teaching your peers a biological process while the group critiques the content
Understand (LOC) Explain information or concepts	• Describe a biological process in your own words without copying it from a book or another source • Provide examples of a process • Write a sentence using the word • Give examples of a process	• Discuss content with peers • Take turns quizzing each other about definitions, and have your peers check your answer
Remember (LOC) Recall information	• Practice labeling diagrams • List characteristics • Identify biological objects or components from flash cards • Quiz yourself with flash cards • Take a self-made quiz on vocabulary • Draw, classify, select, or match items • Write out the textbook definitions	• Check a drawing that another student labeled • Create lists of concepts and processes that your peers can match • Place flash cards in a bag and take turns selecting one for which you must define a term • Do the preceding activities, and have peers check your answers

Periodic Table of Elements

Legend:

1 — Atomic number
H — Atomic symbol
Hydrogen — Name
1.008 — Atomic weight*

- Most common elements in living things
- Other major elements (minerals) found in living things
- Important trace elements (minerals) found in living things
- Elements mostly found in non-living things
- Elements not found in nature (synthesized by scientists)

Periodic Table

1	2	3	4	5	6	7	8	9	10	11	12	13	14	15	16	17	18
1 **H** Hydrogen 1.008																	2 **He** Helium 4.003
3 **Li** Lithium 6.941	4 **Be** Beryllium 9.012											5 **B** Boron 10.81	6 **C** Carbon 12.01	7 **N** Nitrogen 14.01	8 **O** Oxygen 16.00	9 **F** Fluorine 19.00	10 **Ne** Neon 20.18
11 **Na** Sodium 22.99	12 **Mg** Magnesium 24.31											13 **Al** Aluminum 26.98	14 **Si** Silicon 28.09	15 **P** Phosphorus 30.97	16 **S** Sulfur 32.07	17 **Cl** Chlorine 35.45	18 **Ar** Argon 39.95
19 **K** Potassium 39.10	20 **Ca** Calcium 40.08	21 **Sc** Scandium 44.96	22 **Ti** Titanium 47.87	23 **V** Vanadium 50.94	24 **Cr** Chromium 52.00	25 **Mn** Manganese 54.94	26 **Fe** Iron 55.85	27 **Co** Cobalt 58.93	28 **Ni** Nickel 58.69	29 **Cu** Copper 63.55	30 **Zn** Zinc 65.38	31 **Ga** Gallium 69.72	32 **Ge** Germanium 72.64	33 **As** Arsenic 74.92	34 **Se** Selenium 78.96	35 **Br** Bromine 79.90	36 **Kr** Krypton 83.80
37 **Rb** Rubidium 85.47	38 **Sr** Strontium 87.61	39 **Y** Yttrium 88.91	40 **Zr** Zirconium 91.22	41 **Nb** Niobium 92.91	42 **Mo** Molybdenum 95.96	43 **Tc** Technetium [98]†	44 **Ru** Ruthenium 101.1	45 **Rh** Rhodium 102.9	46 **Pd** Palladium 106.4	47 **Ag** Silver 107.9	48 **Cd** Cadmium 112.4	49 **In** Indium 114.8	50 **Sn** Tin 118.7	51 **Sb** Antimony 121.8	52 **Te** Tellurium 127.6	53 **I** Iodine 126.9	54 **Xe** Xenon 131.3
55 **Cs** Cesium 132.9	56 **Ba** Barium 137.3	*	72 **Hf** Hafnium 178.5	73 **Ta** Tantalum 180.9	74 **W** Tungsten 183.8	75 **Re** Rhenium 186.2	76 **Os** Osmium 190.2	77 **Ir** Iridium 192.2	78 **Pt** Platinum 195.1	79 **Au** Gold 197.0	80 **Hg** Mercury 200.6	81 **Tl** Thallium 204.4	82 **Pb** Lead 207.2	83 **Bi** Bismuth 209.0	84 **Po** Polonium [209]	85 **At** Astatine [210]	86 **Rn** Radon [222]
87 **Fr** Francium [223]	88 **Ra** Radium [226]	**	104 **Rf** Rutherfordium [265]	105 **Db** Dubnium [268]	106 **Sg** Seaborgium [271]	107 **Bh** Bohrium [270]	108 **Hs** Hassium [277]	109 **Mt** Meitnerium [276]	110 **Ds** Damstadtium [281]	111 **Rg** Roentgenium [280]	112 **Cn** Copernicium [285]	113 **Uut** Ununtrium [284]	114 **Fl** Flerovium [289]	115 **Uup** Ununpentium [288]	116 **Lv** Livermorium [293]	117 **UUs** Ununseptium [294]	118 **UUo** Ununoctium [294]

*Lanthanides

57 **La** Lanthanum 138.9	58 **Ce** Cerium 140.1	59 **Pr** Praseodymium 140.9	60 **Nd** Neodymium 144.2	61 **Pm** Promethium [145]	62 **Sm** Samarium 150.4	63 **Eu** Europium 152.0	64 **Gd** Gadolinium 157.3	65 **Tb** Terbium 158.9	66 **Dy** Dysprosium 162.5	67 **Ho** Holmium 164.9	68 **Er** Erbium 167.3	69 **Tm** Thulium 168.9	70 **Yb** Ytterbium 173.1	71 **Lu** Lutetium 175.0

**Actinides

89 **Ac** Actinium [227]	90 **Th** Thorium 232.0	91 **Pa** Protactinium 231.0	92 **U** Uranium 238.0	93 **Np** Neptunium [237]	94 **Pu** Plutonium [244]	95 **Am** Americium [243]	96 **Cm** Curium [247]	97 **Bk** Berkelium [247]	98 **Cf** Californium [251]	99 **Es** Einsteinium [252]	100 **Fm** Fermium [257]	101 **Md** Mendelevium [258]	102 **No** Nobelium [259]	103 **Lr** Lawrencium [262]

DATA: Wieser, M. E., and M. Berglund. 2009. *Pure and Applied Chemistry* 81: 2131–2156.

*Atomic weights are reported to four significant figures.

†For elements with a variable number of protons and/or neutrons, the mass number of the longest-lived isotope of the element is reported in brackets.

Glossary

5′ cap A modified guanine (G) nucleotide (7-methylguanylate) added to the 5′ end of eukaryotic mRNAs. Helps protect the mRNA from being degraded and promotes efficient initiation of translation.

abdomen A region of the body; in insects, one of the three prominent body regions called tagmata.

abiotic Not alive (e.g., air, water, and some components of soil). Compare with **biotic**.

aboveground biomass The total mass of living plants in an area, excluding roots.

abscisic acid (ABA) A plant hormone that inhibits growth; it stimulates stomatal closure and triggers dormancy.

abscission In plants, the normal (often seasonal) shedding of leaves, fruits, or flowers.

abscission zone The region at the base of a petiole where cell wall degradation occurs; results in the dropping of leaves.

absorption In animals, the uptake of ions and small molecules, derived from food, across the lining of the intestine and into the bloodstream.

absorption spectrum The amount of light of different wavelengths absorbed by a pigment. Usually depicted as a graph of light absorbed versus wavelength. Compare with **action spectrum**.

acclimation A change in a study organism's phenotype that occurs in response to laboratory conditions.

acclimatization A change in an individual's phenotype that occurs in response to a change in natural environmental conditions.

acetyl CoA A molecule produced by oxidation of pyruvate (the final product of glycolysis) in a reaction catalyzed by pyruvate dehydrogenase. Can enter the citric acid cycle and is used as a carbon source in the synthesis of fatty acids, steroids, and other compounds.

acetylation Addition of an acetyl group ($-COCH_3$) to a molecule. Acetylation of histone proteins is important in controlling chromatin condensation.

acetylcholine (Ach) A neurotransmitter, released by nerve cells at neuromuscular junctions, that triggers contraction of skeletal muscle cells but slows the rate of contraction in cardiac muscle cells. Also used as a neurotransmitter between neurons.

acid Any compound that gives up protons or accepts electrons during a chemical reaction or that releases hydrogen ions when dissolved in water.

acid-growth hypothesis The hypothesis that auxin triggers elongation of plant cells by increasing the activity of proton pumps, making the cell wall more acidic and leading to expansion of the cell wall and an influx of water.

acoelomate A bilaterian animal that lacks an internal body cavity (coelom). Compare with **coelomate** and **pseudocoelomate**.

acquired immune deficiency syndrome (AIDS) A human disease characterized by death of immune system cells (in particular helper T cells) and subsequent vulnerability to other infections. Caused by the human immunodeficiency virus (HIV).

acrosome A caplike structure, located on the head of a sperm cell, that contains enzymes capable of dissolving the outer coverings of an egg.

ACTH See **adrenocorticotropic hormone**.

actin A globular protein that can be polymerized to form filaments. Actin filaments are part of the cytoskeleton and constitute the thin filaments in skeletal muscle cells.

actin filament A long fiber, about 7 nm in diameter, composed of two intertwined strands of polymerized actin protein; one of the three types of cytoskeletal fibers. Involved in cell movement. Also called a *microfilament*. Compare with **intermediate filament** and **microtubule**.

action potential A rapid, temporary change in electrical potential across a membrane, from negative to positive and back to negative. Occurs in cells, such as neurons and muscle cells, that have an excitable membrane.

action spectrum The relative effectiveness of different wavelengths of light in driving a light-dependent process such as photosynthesis. Usually depicted as a graph of some measure of the process versus wavelength. Compare with **absorption spectrum**.

activation energy The amount of energy required to initiate a chemical reaction; specifically, the energy required to reach the transition state.

activator A protein that binds to a DNA regulatory sequence to increase the frequency of transcription initiation by RNA polymerase.

active site The location in an enzyme molecule where substrates (reactant molecules) bind and react.

active transport The movement of ions or molecules across a membrane against an electrochemical gradient. Requires energy (e.g., from hydrolysis of ATP) and assistance of a transport protein (e.g., pump).

adaptation Any heritable trait that increases the fitness of an individual with that trait, compared with individuals without that trait, in a particular environment.

adaptive immune response See **adaptive immunity**.

adaptive immunity Immunity to a particular pathogen or other antigen conferred by activated B and T cells in vertebrates. Characterized by specificity, diversity, memory, and self–nonself recognition. Also called *adaptive immune response*. Compare with **innate immunity**.

adaptive radiation Rapid evolutionary diversification within one lineage, producing many descendant species with a wide range of adaptive forms.

adenosine triphosphate (ATP) A molecule consisting of an adenine base, a sugar, and three phosphate groups that can be hydrolyzed to release energy. Universally used by cells to store and transfer energy.

adhesion The tendency of certain dissimilar molecules to cling together due to attractive forces. Compare with **cohesion**.

adipocyte A fat cell.

adrenal gland Either of two small endocrine glands, one above each kidney. The outer portion (cortex) secretes several steroid hormones; the inner portion (medulla) secretes epinephrine and norepinephrine.

adrenaline See **epinephrine**.

adrenocorticotropic hormone (ACTH) A peptide hormone, produced and secreted by the anterior pituitary, that stimulates release of steroid hormones (e.g., cortisol, aldosterone) from the adrenal cortex.

adult A sexually mature individual.

adventitious root A root that develops from a plant's shoot system instead of from the plant's root system.

aerobic Referring to any metabolic process, cell, or organism that uses oxygen as an electron acceptor. Compare with **anaerobic**.

afferent division The part of the nervous system that transmits information about the internal and external environment to the central nervous system. Consists mainly of sensory neurons. Compare with **efferent division**.

age class All the individuals of a specific age in a population.

age-specific fecundity The average number of female offspring produced by a female in a certain age class.

age structure The proportion of individuals in a population that are of each possible age.

agglutination Clumping together of cells or viruses by antibodies or other cross-linking molecules.

aggregate fruit A fruit (e.g., raspberry) that develops from a single flower that has many separate carpels. Compare with **multiple** and **simple fruit**.

AIDS See **acquired immune deficiency syndrome**.

albumen A solution of water and protein (particularly albumins), found in amniotic eggs, that nourishes the growing embryo. Also called *egg white*.

alcohol fermentation Catabolic pathway in which pyruvate produced by glycolysis is converted to ethanol in the absence of oxygen.

aldosterone A hormone that stimulates the kidney to conserve salt and water and promotes retention of sodium; produced in the adrenal cortex.

allele A particular version of a gene.

allergen Any molecule (antigen) that triggers an allergic response (an allergy).

allergy An IgE-mediated abnormal response to an antigen, usually characterized by dilation of blood vessels, contraction of smooth muscle cells, and increased activity of mucus-secreting cells.

allopatric speciation Speciation that occurs when populations of the same species become geographically isolated, often due to dispersal or vicariance. Compare with **sympatric speciation**.

allopatry Condition in which two or more populations live in different geographic areas. Compare with **sympatry**.

allopolyploidy (adjective: allopolyploid) The state of having more than two full sets of chromosomes (polyploidy) due to hybridization between different species. Compare with **autopolyploidy**.

allosteric regulation Regulation of a protein's function by binding of a regulatory molecule, usually to a specific site distinct from the active site, that causes a change in the protein's shape.

α-amylase See **amylase**.

α-helix (alpha-helix) A protein secondary structure in which the polypeptide backbone coils into a spiral shape stabilized by hydrogen bonding.

alternation of generations A life cycle involving alternation of a multicellular haploid stage (gametophyte) with a multicellular diploid stage (sporophyte). Occurs in most plants and some protists.

alternative splicing In eukaryotes, the splicing of primary RNA transcripts from a single gene in different ways to produce different mature mRNAs and thus different polypeptides.

altruism Any behavior that has a fitness cost to the individual (lowered survival and/or reproduction) and a fitness benefit to the recipient. See **reciprocal altruism**.

alveolus (plural: alveoli) Any of the tiny air-filled sacs of a mammalian lung.

ambisense RNA virus A ssRNA virus whose genome consists of at least one strand that contains both positive-sense and negative-sense regions.

aminoacyl tRNA A transfer RNA molecule that is covalently bound to an amino acid.

aminoacyl-tRNA synthetase An enzyme that catalyzes the addition of a particular amino acid to its corresponding tRNA molecule.

ammonia (NH₃) A small molecule, produced by the breakdown of proteins and nucleic acids, that is very toxic to cells. Is a strong base that gains a proton to form the ammonium ion (NH_4^+).

amnion The innermost of the membranes surrounding the embryo in an amniotic egg.

amniotes A major lineage of vertebrates (Amniota) that reproduce with amniotic eggs. Includes all reptiles (including birds) and mammals—all tetrapods except amphibians.

amniotic egg An egg that has a watertight shell or case enclosing a membrane-bound water supply (the amnion), food supply (yolk sac), and waste sac (allantois).

amoeboid motion See **cell crawling**.

amphibians A lineage of vertebrates, many of which breathe through their skin and feed on land but lay

their eggs in water. Represent the earliest tetrapods; include frogs, salamanders, and caecilians.

amphipathic Containing hydrophilic and hydrophobic regions.

ampullae of Lorenzini Structures on the heads of sharks that contain cells with electroreceptors.

amylase Any enzyme that can break down starch by catalyzing hydrolysis of the glycosidic linkages between the glucose residues.

amyloplasts Starch-storing organelles (plastids) in plants. In root cap cells, they settle to the bottom of the cell and may be used as gravity detectors.

anabolic pathway Any set of chemical reactions that synthesizes large molecules from smaller ones. Generally requires an input of energy. Compare with **catabolic pathway**.

anadromous Having a life cycle in which adults live in the ocean (or large lakes) but migrate up freshwater streams to breed and lay eggs.

anaerobic Referring to any metabolic process, cell, or organism that uses an electron acceptor other than oxygen, including fermentation or anaerobic respiration. Compare with **aerobic**.

anaphase A stage in mitosis or meiosis during which chromosomes are moved to opposite poles of the spindle apparatus.

anatomy The study of the physical structure of organisms.

ancestral trait A trait found in the ancestors of a particular group.

aneuploidy (adjective: aneuploid) The state of having an abnormal number of copies of a certain chromosome.

angiosperm A flowering vascular plant that produces seeds within mature ovaries (fruits). The angiosperms form a single lineage. Compare with **gymnosperm**.

animal A member of a major lineage of eukaryotes (Animalia) whose members typically have a complex, large, multicellular body; eat other organisms; and are mobile.

animal model Any disease that occurs in a non-human animal and has parallels to a similar disease of humans. Studied by medical researchers in hopes that findings may apply to human disease.

anion A negatively charged ion.

annelids Members of the phylum Annelida (segmented worms). Distinguished by a segmented body and a coelom that functions as a hydrostatic skeleton. Annelids belong to the lophotrochozoan branch of the protostomes.

annual Referring to a plant whose life cycle normally lasts only one growing season—less than one year. Compare with **perennial**.

anoxygenic Referring to any process or reaction that does not produce oxygen. Photosynthesis in purple sulfur and purple nonsulfur bacteria, which does not involve photosystem II, is anoxygenic. Compare with **oxygenic**.

antagonistic muscle group A set of two or more muscles that reextend one another by transmitting their forces via the skeleton.

antenna (plural: antennae) A long appendage of the head that is used to touch or smell.

antenna complex Part of a photosystem, containing an array of chlorophyll molecules and accessory pigments, that receives energy from light and directs the energy to a central reaction center during photosynthesis.

anterior Toward an animal's head and away from its tail. The opposite of posterior.

anterior pituitary The part of the pituitary gland containing endocrine cells that produce and release a variety of peptide hormones in response to other hormones from the hypothalamus. Compare with **posterior pituitary**.

anther The pollen-producing structure at the end of a stamen in flowering plants (angiosperms).

antheridium (plural: antheridia) The sperm-producing structure in most land plants except angiosperms.

anthropoids One of the two major lineages of primates, including apes, humans, and all monkeys. Compare with **prosimians**.

antibiotic Any substance, such as penicillin, that can kill or inhibit the growth of bacteria.

antibody A protein produced by B cells that can bind to a specific part of an antigen, tagging it for removal by the immune system. All monomeric forms of antibodies consist of two light chains and two heavy chains, which vary between different antibodies. Also called *immunoglobulin*.

anticodon The sequence of three bases (a triplet) in a transfer RNA molecule that can bind to an mRNA codon with a complementary sequence.

antidiuretic hormone (ADH) A peptide hormone, secreted from the posterior pituitary gland, that stimulates water retention by the kidney. Also called *vasopressin*.

antigen Any foreign molecule, often a protein, that can stimulate an innate or adaptive response by the immune system.

antigen presentation Process by which small peptides, derived from ingested particulate antigens (e.g., bacteria) or intracellular antigens (e.g., viruses in infected cell) are complexed with MHC proteins and transported to the cell surface, where they are displayed and can be recognized by T cells.

antiparallel Describing the opposite orientation of nucleic acid strands that are hydrogen bonded to one another, with one strand running in the $5' \rightarrow 3'$ direction and the other in the $3' \rightarrow 5'$ direction.

antiporter A carrier protein that allows an ion to diffuse down an electrochemical gradient, using the energy of that process to transport a different substance in the opposite direction *against* its concentration gradient. Compare with **symporter**.

antiviral Any drug or other agent that can interfere with the transmission or replication of viruses.

aorta In terrestrial vertebrates, the major artery carrying oxygenated blood away from the heart.

aphotic zone Deep water receiving no sunlight. Compare with **photic zone**.

apical Toward the top. In plants, at the tip of a branch. In animals, on the side of an epithelial layer that faces the environment and not other body tissues. Compare with **basal**.

apical–basal axis The shoot-to-root axis of a plant.

apical bud A bud at the tip of a stem or branch, where growth occurs to lengthen the stem or branch.

apical dominance Inhibition of lateral bud growth by the apical meristem at the tip of a plant branch.

apical meristem A group of undifferentiated plant cells, at the tip of a shoot or root, that is responsible for primary growth. Compare with **cambium**.

apodeme Any of the chitinous ingrowths of the exoskeleton to which muscles attach.

apomixis The formation of mature seeds without fertilization occurring; a type of asexual reproduction.

apoplast In plants, the region outside plasma membranes consisting of the porous cell walls and the intervening extracellular air space. Compare with **symplast**.

apoptosis Series of tightly controlled changes that lead to the self-destruction of a cell. Occurs frequently during embryological development and as part of the immune response to remove infected or cancerous cells. Also called *programmed cell death*.

appendix A blind sac (having only one opening) that extends from the cecum in some mammals.

aquaporin A type of channel protein through which water can move by osmosis across a plasma membrane.

aquifer An underground layer of porous rock, sand, or gravel that is saturated with water.

ara operon A set of three genes in *E. coli* that are transcribed into a single mRNA and required for metabolism of the sugar arabinose. Transcription of the *ara* operon is controlled by the AraC regulatory protein.

araC The regulatory gene (written as *araC*) or regulatory protein (when written as AraC) of the *E. coli ara* operon.

arbuscular mycorrhizal fungi (AMF) Fungi from the Glomeromycota lineage whose hyphae enter the root cells of their host plants. Also called *endomycorrhizal fungi*.

Archaea One of the three taxonomic domains of life, consisting of unicellular prokaryotes distinguished by cell walls made of certain polysaccharides not found in bacterial or eukaryotic cell walls, plasma membranes composed of unique isoprene-containing phospholipids, and ribosomes and RNA polymerase similar to those of eukaryotes. Compare with **Bacteria** and **Eukarya**.

archegonium (plural: archegonia) The egg-producing structure in most land plants except angiosperms.

arteriole Any of the many tiny vessels that carry blood from arteries to capillaries.

arteriosclerosis Hardening and loss of elasticity of arteries.

artery Any thick-walled blood vessel that carries blood (oxygenated or not) under relatively high pressure away from the heart to organs throughout the body. Compare with **vein**.

arthropods Members of the phylum Arthropoda. Distinguished by a segmented body; a hard, jointed exoskeleton; paired, jointed appendages; and an extensive body cavity called a hemocoel. Arthropods belong to the ecdysozoan branch of the protostomes.

articulation A movable point of contact between two rigid components of a skeleton, such as between bones of a vertebrate endoskeleton or between segments of cuticle in an arthropod exoskeleton. See **joint**.

artificial selection Deliberate manipulation by humans, as in animal and plant breeding, of the genetic composition of a population by allowing only individuals with desirable traits to reproduce.

ascus (plural: asci) Specialized spore-producing cell found at the ends of hyphae in "sac fungi" (Ascomycota).

asexual reproduction Any form of reproduction where offspring inherit DNA from only one parent. Includes binary fission, budding, and parthenogenesis. Compare with **sexual reproduction**.

astral microtubules Mitotic and meiotic microtubules that have arisen from the two spindle poles and interact with proteins on the plasma membrane.

asymmetric competition Ecological competition between two species in which one species suffers a much greater fitness decline than the other. Compare with **symmetric competition**.

atomic number The number of protons in the nucleus of an atom, giving the atom its identity as a particular chemical element.

atomic weight The average mass of an element that is based on the relative proportions of all the naturally occurring isotopes.

ATP synthase A large membrane-bound protein complex that uses the energy of protons flowing through it to synthesize ATP.

ATP See **adenosine triphosphate**.

atrioventricular (AV) node A region of the heart between the right atrium and right ventricle where electrical signals from the atrium are slowed briefly before spreading to the ventricle. This delay allows the ventricle to fill with blood before contracting. Compare with **sinoatrial (SA) node**.

atrium (plural: atria) A thin-walled chamber of the heart that receives blood from veins and pumps it to a neighboring chamber (the ventricle).

autocrine Relating to a chemical signal that affects the same cell that produced and released it.

autoimmunity A pathological condition in which the immune system attacks self cells or tissues of an individual's own body.

autonomic nervous system The part of the peripheral nervous system that controls internal organs and involuntary processes, such as stomach contraction, hormone release, and heart rate. Includes parasympathetic and sympathetic nerves. Compare with **somatic nervous system**.

autophagy The process by which damaged organelles are surrounded by a membrane and delivered to a lysosome to be recycled.

autopolyploidy (adjective: autopolyploid) The state of having more than two full sets of chromosomes (polyploidy) due to a mutation that doubled the chromosome number. All the chromosomes come from the same species. Compare with **allopolyploidy**.

autosomal inheritance The inheritance patterns that occur when genes are located on autosomes rather than on sex chromosomes.

autosome Any chromosome other than a sex chromosome (i.e,. any chromosome other than the X or Y in mammals).

autotroph Any organism that can synthesize reduced organic compounds from simple inorganic sources such as CO_2 or CH_4. Most plants and some bacteria and archaea are autotrophs. Also called *primary producer*. Compare with **heterotroph**.

auxin Indoleacetic acid (IAA), a plant hormone that stimulates phototropism and cell elongation.

axillary bud A bud that forms at a node and may develop into a lateral (side) branch. Also called *lateral bud*.

axon A long projection of a neuron that can propagate an action potential.

axon hillock The site in a neuron where an axon joins the cell body and where action potentials are first triggered.

axoneme A structure found in eukaryotic cilia and flagella and responsible for their motion; composed of two central microtubules surrounded by nine doublet microtubules (9 + 2 arrangement).

B cell A type of lymphocyte that matures in the bone marrow and, with T cells, is responsible for adaptive immunity. Produces antibodies and also functions in antigen presentation. Also called *B lymphocyte*.

B-cell receptor (BCR) An immunoglobulin protein embedded in the plasma membrane of mature B cells and to which antigens bind. Apart from the transmembrane domain, it is identical in structure to antibodies.

BAC library A collection of all the sequences found in the genome of a species, inserted into **bacterial artificial chromosomes (BACs)**.

background extinction The average rate of low-level extinction that has occurred continuously throughout much of evolutionary history. Compare with **mass extinction**.

Bacteria One of the three taxonomic domains of life, consisting of unicellular prokaryotes distinguished by cell walls composed largely of peptidoglycan, plasma membranes similar to those of eukaryotic cells, and ribosomes and RNA polymerase that differ from those in archaeans or eukaryotes. Compare with **Archaea** and **Eukarya**.

bacterial artificial chromosome (BAC) An artificial version of a bacterial chromosome that can be used as a cloning vector to produce many copies of large DNA fragments.

bacteriophage Any virus that infects bacteria.

baculum A bone inside the penis; usually present in mammals with a penis that lacks erectile tissue.

balancing selection A mode of natural selection in which no single allele is favored in all populations of a species at all times. Instead, there is a balance among alleles in terms of fitness and frequency.

ball-and-stick model A representation of a molecule where atoms are shown as balls—colored and scaled to indicate the atom's identity—and covalent bonds are shown as rods or sticks connecting the balls in the correct geometry.

bar coding The use of well-characterized gene sequences to identify species.

bark The protective outer layer of woody plants, composed of cork cells, cork cambium, and secondary phloem.

baroreceptors Specialized nerve cells in the walls of the heart and certain major arteries that detect changes in blood pressure and trigger appropriate responses by the brain.

basal Toward the base. In plants, toward the root or at the base of a branch where it joins the stem. In animals, on the side of an epithelial layer that abuts underlying body tissues. Compare with **apical**.

basal body The microtubule organizing center for cilia and flagella in eukaryotic cells. Consists of nine triplets of microtubules arranged in a circle and establishes the structure of axonemes. Structurally identical with a centriole.

basal lamina A thick, collagen-rich extracellular matrix that underlies most epithelial tissues in animals and connects it to connective tissue.

basal metabolic rate (BMR) The total energy consumption by an organism at rest in a comfortable environment. For aerobes, often measured as the amount of oxygen consumed per hour.

basal transcription factor Proteins, present in all eukaryotic cells, that bind to promoters and help initiate transcription. Compare with **regulatory transcription factor**.

base Any compound that acquires protons or gives up electrons during a chemical reaction or accepts hydrogen ions when dissolved in water.

basidium (plural: basidia) Specialized spore-producing cell at the ends of hyphae in club fungi, members of the Basidiomycota.

basilar membrane The membrane on which the bottom portion of hair cells sits in the vertebrate cochlea.

basolateral Toward the bottom and sides. In animals, the side of an epithelial layer that faces other body tissues and not the environment.

Batesian mimicry A type of mimicry in which a harmless or palatable species resembles a dangerous or poisonous species. Compare with **Müllerian mimicry**.

beak A structure that exerts biting forces and is associated with the mouth; found in birds, cephalopods, and some insects.

behavior Any action by an organism, often in response to a stimulus.

behavioral ecology The study of how organisms respond to particular abiotic and biotic stimuli from their environment.

beneficial In genetics, referring to any mutation, allele, or trait that increases an individual's fitness.

benign tumor A mass of abnormal tissue that appears due to unregulated growth but does not spread to other organs. Benign tumors are not cancers. Compare with **malignant tumor**.

benthic Living at the bottom of an aquatic environment.

benthic zone The area along the bottom of an aquatic environment.

β-pleated sheet (beta-pleated sheet) A protein secondary structure in which the polypeptide backbone folds into a sheetlike shape stabilized by hydrogen bonding.

bilateral symmetry An animal body pattern in which one plane of symmetry divides the body into a left side and a right side. Typically, the body is long and narrow, with a distinct head end and tail end. Compare with **radial symmetry**.

bilaterian A member of a major lineage of animals (Bilateria) that are bilaterally symmetrical at some point in their life cycle, have three embryonic germ layers, and have a coelom. All protostomes and deuterostomes are bilaterians.

bile A complex solution produced by the liver, stored in the gallbladder, and secreted into the intestine. Contains steroid derivatives called bile salts that are responsible for emulsification of fats during digestion.

binary fission The process of cell division used for asexual reproduction of many prokaryotic cells. The genetic material is replicated and partitioned to opposite sides of a growing cell, which is then divided in half to create two genetically identical cells.

biodiversity The diversity of life considered at three levels: genetic diversity (variety of alleles and/or genes in a population, species, or group of species); species diversity (variety and relative abundance of species present in a certain area); and ecosystem diversity (variety of communities and abiotic components in a region).

biodiversity hotspot A region that is extraordinarily rich in species.

biofilm A complex community of bacteria enmeshed in a polysaccharide-rich, extracellular matrix that allows them to attach to a surface.

biogeochemical cycle The pattern of circulation of an element or molecule among living organisms and the environment.

biogeography The study of how species and populations are distributed geographically.

bioinformatics The field of study concerned with managing, analyzing, and interpreting biological information, particularly DNA sequences.

biological species concept The definition of a species as a population or group of populations that are reproductively isolated from other groups. Members of a species have the potential to interbreed in nature to produce viable, fertile offspring but cannot interbreed successfully with members of other species. Compare with **morphospecies** and **phylogenetic species concepts**.

bioluminescence The emission of light by a living organism via an enzyme-catalyzed reaction.

biomagnification In animal tissues, an increase in the concentration of particular molecules that may occur as those molecules are passed up a food chain.

biomass The total mass of all organisms in a given population or geographical area; usually expressed as total dry weight.

biome A large terrestrial or marine region characterized by distinct abiotic characteristics and dominant types of vegetation.

biomechanics A field of biology that applies the principles of physics and engineering to analyze the mechanical structure and function of organisms.

bioprospecting The effort to find commercially useful compounds by studying organisms—especially species that are poorly studied to date.

bioremediation The use of living organisms, usually bacteria or archaea, to degrade environmental pollutants.

biosphere The thin zone surrounding the Earth where all life exists; the sum of all terrestrial and aquatic ecosystems.

biotechnology The application of biological techniques and discoveries to medicine, industry, and agriculture.

biotic Living, or produced by a living organism. Compare with **abiotic**.

bipedal Walking primarily on two legs; characteristic of hominins.

bipolar cell A type of cell in the vertebrate retina that receives information from one or more photoreceptors and passes it to other bipolar cells or ganglion cells.

bivalent The structure formed by synapsed homologous chromosomes during prophase of meiosis I. Also known as a *tetrad*.

bivalves A lineage of mollusks that have shells made of two parts, or valves, such as clams and mussels.

bladder A mammalian organ that holds urine until it can be excreted.

blade The wide, flat part of a plant leaf.

blastocoel Fluid-filled cavity in the blastula of many animal species.

blastocyst The mammalian blastula. A roughly spherical structure composed of trophoblast cells on the exterior and a cluster of cells (the inner cell mass) that fills part of the interior space.

blastomere A cell created by cleavage divisions in early animal embryos.

blastopore An opening (pore) in the surface of some early embryos, through which cells move during gastrulation.

blastula In vertebrate development, a ball of cells (blastomere cells) typically surrounding a fluid-filled cavity (the blastocoel). The blastula is formed by cleavage of a zygote and undergoes gastrulation. See **blastocyst**.

blood A type of connective tissue consisting of red blood cells and leukocytes suspended in a fluid portion, an extracellular matrix called plasma. Transports materials throughout the vertebrate body.

body mass index (BMI) A mathematical relationship of weight and height used to assess obesity in humans. Calculated as weight (kg) divided by the square of height (m^2).

body plan The basic architecture of an animal's body, including the number and arrangement of limbs, body segments, and major tissue layers.

bog A freshwater wetland that has no or almost no water flow, resulting in very low oxygen levels and acidic conditions.

Bohr shift The rightward shift of the oxygen–hemoglobin equilibrium curve that occurs with decreasing pH. It results in hemoglobin being more likely to release oxygen in the acidic environment of exercising muscle.

bone A type of vertebrate connective tissue consisting of living cells and blood vessels within a hard extracellular matrix composed of calcium phosphate ($CaPO_4$) and small amounts of calcium carbonate ($CaCO_3$) and protein fibers.

bone marrow The soft tissue filling the inside of large bones; contains stem cells that develop into red blood cells and leukocytes throughout life.

Bowman's capsule The hollow, double-walled, cup-shaped portion of a nephron that surrounds a glomerulus in the vertebrate kidney.

brain A large mass of neurons, located in the head region of an animal, that is involved in information processing; may also be called the cerebral ganglion.

brain stem The most posterior portion of the vertebrate brain, connecting to the spinal cord and responsible for autonomic body functions such as heart rate, respiration, and digestion.

braincase See **cranium**.

branch (1) A part of a phylogenetic tree that represents populations through time. (2) Any extension of a plant's shoot system.

brassinosteroids A family of steroid hormones found in plants; stimulate growth.

bronchiole Any of the small tubes in mammalian lungs that carry air from the bronchi to the alveoli.

bronchus (plural: bronchi) In mammals, one of a pair of large tubes that lead from the trachea to each lung.

bryophyte See **non-vascular plants**.

budding Asexual reproduction in which an outgrowth from the parent breaks free as an independent individual; occurs in yeasts and some invertebrates.

buffer A substance that, in solution, acts to minimize changes in the pH of that solution when acid or base is added.

bulbourethral gland In male mammals, either of a small pair of glands at the base of the urethra that secrete an alkaline mucus (part of semen), which lubricates the tip of the penis and neutralizes acids in the urethra during copulation. In humans, also called *Cowper's glands*.

bulk flow The directional mass movement of a fluid due to pressure differences, such as movement of water through plant xylem and phloem, and movement of blood in animals.

bulk-phase endocytosis Nonspecific uptake of extracellular fluid by pinching off the plasma membrane to form small membrane-bound vesicles; considered to be a means of retrieving membrane from the surface following exocytosis. Compare with **receptor-mediated endocytosis**.

bundle-sheath cell A type of cell found around the vascular tissue (veins) of plant leaves.

C_3 pathway The most common form of photosynthesis in which atmospheric CO_2 is fixed by rubisco to form 3-phosphoglycerate, a three-carbon molecule. Used in first phase of the Calvin cycle.

C_4 pathway A variant type of photosynthesis in which atmospheric CO_2 is first fixed by PEP carboxylase into four-carbon acids, rather than the three-carbon molecules of the classic C_3 pathway. Used to concentrate CO_2 to reduce photorespiration in the Calvin cycle while stomata are closed.

C_4 photosynthesis A variant type of photosynthesis in which atmospheric CO_2 is first fixed into four-carbon sugars, rather than the three-carbon sugars of classic C_3 photosynthesis. Enhances photosynthetic efficiency in hot, dry environments by reducing loss of oxygen due to photorespiration.

cadherin Any of a class of cell-surface proteins involved in selective cell–cell adhesion. Important for coordinating movements of cells and the establishment of tissues during embryological development.

callus In plants, a mass of undifferentiated cells that can generate roots and other tissues necessary to create a mature plant.

Calvin cycle In photosynthesis, the set of reactions that use NADPH and ATP formed in the light-dependent reactions to drive the fixation of CO_2, reduction of the fixed carbon to produce sugar, and the regeneration of the substrate used to fix CO_2. Also called *light-independent reactions*.

calyx All of the sepals of a flower.

CAM See **crassulacean acid metabolism**.

cambium (plural: cambia) In woody plants, tissue that consists of two types of cylindrical meristems that increase the width of roots and shoots through the process of secondary growth. See **vascular cambium** and **cork cambium**.

Cambrian explosion The rapid diversification of animal body types and lineages that occured between the species present in the Doushantuo faunas (around 570 mya), Ediacaran faunas (565–542 mya), and the Early Cambrian faunas (525–515 mya).

cancer General term for any tumor whose cells grow in an uncontrolled fashion, invade nearby tissues, and spread to other sites in the body.

capillarity The tendency of water to move up a narrow tube due to adhesion, cohesion, and surface tension (also called capillary action).

capillary Any of the many small, thin-walled blood vessels that permeate all tissues and organs, and allow exchange of gases and other molecules between blood and body cells.

capillary bed A thick network of capillaries.

capsid A shell of protein enclosing the genome of a virus particle.

carapace In crustaceans, a large platelike section of the exoskeleton that covers and protects the cephalothorax (e.g., a crab's "shell").

carbohydrate Any of a class of molecules that contain a carbonyl group, several hydroxyl groups, and several to many carbon-hydrogen bonds. See **monosaccharide** and **polysaccharide**.

carbon cycle, global The worldwide movement of carbon among terrestrial ecosystems, the oceans, and the atmosphere.

carbon fixation The process of converting gaseous carbon dioxide into an organic molecule, often associated with photosynthesis. See also **PEP carboxylase** and **rubisco**.

carbonic anhydrase An enzyme that catalyzes the formation of carbonic acid (H_2CO_3) from carbon dioxide and water.

carboxylic acids Organic acids with the form R-COOH (a carboxyl group).

cardiac cycle One complete heartbeat cycle, including systole and diastole.

cardiac muscle The muscle tissue of the vertebrate heart; responsible for pumping blood. Consists of long, branched fibers that are electrically connected and that initiate their own contractions; not under voluntary control. Compare with **skeletal** and **smooth muscle**.

cardiovascular disease A group of diseases of the heart and blood vessels caused by poor diet, obesity, inactivity, genetics, tobacco use, age, and other factors.

carnivore (adjective: carnivorous) An animal whose diet consists predominantly of meat, or other animals. Most members of the mammalian taxon Carnivora are carnivores. Some plants are carnivorous, trapping and killing small animals and then absorbing nutrients from the prey's body. Compare with **herbivore** and **omnivore**.

carotenoid Any of a class of accessory pigments, found in chloroplasts, that absorb wavelengths of light not absorbed by chlorophyll; typically appear yellow, orange, or red. Includes carotenes and xanthophylls.

carpel The female reproductive organ in a flower. Consists of the stigma, to which pollen grains adhere; the style, through which the pollen tube grows; and the ovary, which houses the ovule. Compare with **stamen**.

carrier protein A membrane protein that facilitates diffusion of a small molecule (e.g., glucose) across a membrane by a process involving a reversible change in the shape of the protein. Also called *carrier* or *transporter*. Compare with **channel protein**.

carrier A heterozygous individual carrying a normal allele and a recessive allele for an inherited trait; does not display the phenotype of the recessive trait but can pass the recessive allele to offspring.

carrying capacity (K) The maximum population size of a certain species that a given habitat can support.

cartilage A type of vertebrate connective tissue that consists of relatively few cells scattered in a stiff matrix of polysaccharides and protein fibers. Provides structural support.

Casparian strip In plant roots, a waxy layer containing suberin, a water-repellent substance that prevents movement of water through the walls of endodermal cells, thus blocking the apoplastic pathway of water and ion movement into the vascular tissue.

cast A type of fossil, formed when the decay of a body part leaves a void that is then filled with minerals that later harden.

catabolic pathway Any set of chemical reactions that breaks down large, complex molecules into smaller ones, releasing energy in the process. Compare with **anabolic pathway**.

catalysis (verb: catalyze) Acceleration of the rate of a chemical reaction due to a decrease in the free energy of the transition state, called the activation energy.

catalyst Any substance that increases the rate of a chemical reaction without itself undergoing any permanent chemical change.

catecholamines A class of small compounds, derived from the amino acid tyrosine, that are used as hormones or neurotransmitters. Include epinephrine, norepinephrine, and dopamine.

cation A positively charged ion.

cation exchange The release (displacement) of cations, such as magnesium and calcium from soil particles, by protons in acidic soil water. The released cations are available for uptake by plants.

CD4 A membrane protein on the surface of some T cells in humans. CD4$^+$ T cells can give rise to helper T cells.

CD8 A membrane protein on the surface of some T cells in humans. CD8$^+$ T cells can give rise to cytotoxic T cells.

Cdk See **cyclin-dependent kinase**.

cDNA See **complementary DNA**.

cDNA library A set of cDNAs from a particular cell type or stage of development. Each cDNA is carried by a plasmid or other cloning vector and can be separated from other cDNAs. Compare with **genomic library**.

cecum A blind sac between the small intestine and the colon. Is enlarged in some species (e.g., rabbits) that use it as a fermentation vat for digestion of cellulose.

cell A highly organized compartment bounded by a thin, flexible structure (plasma membrane) and containing concentrated chemicals in an aqueous (watery) solution. The basic structural and functional unit of all organisms.

cell body The part of a neuron that contains the nucleus; where incoming signals are integrated. Also called the *soma*.

cell crawling A form of cellular movement involving actin filaments in which the cell produces bulges (pseudopodia) that stick to the substrate and pull the cell forward. Also called *amoeboid motion*.

cell cycle Ordered sequence of events in which a eukaryotic cell replicates its chromosomes, evenly partitions the chromosomes to two daughter cells, and then undergoes division of the cytoplasm.

cell-cycle checkpoint Any of several points in the cell cycle at which progression of a cell through the cycle can be regulated.

cell division Creation of new cells by division of preexisting cells.

cell-mediated (immune) response The type of immune response that involves generation of cytotoxic

T cells from CD8$^+$ T cells. Defends against pathogen-infected cells, cancer cells, and transplanted cells. Compare with **humoral (immune) response**.

cell membrane See **plasma membrane**.

cell plate A flattened sac-like structure formed in the middle of a dividing plant cell from Golgi-derived vesicles containing cell wall material; ultimately divides the cytoplasm into two separate cells.

cell sap An aqueous solution found in the vacuoles of plant cells.

cell theory The theory that all organisms are made of cells and that all cells come from preexisting cells.

cell wall A fibrous layer found outside the plasma membrane of most bacteria and archaea and many eukaryotes.

cellular respiration A common pathway for production of ATP, involving transfer of electrons from compounds with high potential energy through an electron transport chain and ultimately to an electron acceptor (often oxygen).

cellulase An enzyme that can break down cellulose by catalyzing hydrolysis of the glycosidic linkages between the glucose residues.

cellulose A structural polysaccharide composed of glucose monomers joined by β-1,4-glycosidic linkages. Found in the cell wall of algae, plants, and some bacteria and fungi.

Cenozoic era The most recent interval of geologic time, beginning 65.5 million years ago, during which mammals became the dominant vertebrates and angiosperms became the dominant plants.

central dogma The scheme for information flow in the cell: DNA → RNA → protein.

central nervous system (CNS) Large numbers of neurons aggregated into clusters called ganglia in bilaterian animals. In vertebrates, the central nervous system consists of the brain and spinal cord. Compare with **nerve net** and **peripheral nervous system (PNS)**.

centriole One of two small cylindrical structures found together within the centrosome near the nucleus of a eukaryotic cell (not found in plants). Consists of microtubule triplets and is structurally identical with a basal body.

centromere Constricted region of a replicated chromosome where the two sister chromatids are joined and the kinetochore is located.

centrosome Structure in animal and fungal cells, containing two centrioles, that serves as a microtubule organizing center for the cell's cytoskeleton and for the spindle apparatus during cell division.

cephalization The formation in animals of a distinct anterior region (the head) where sense organs and a mouth are clustered.

cephalochordates One of the three major chordate lineages (Cephalochordata), comprising small, mobile organisms that live in marine sands and suspension feed; also called *lancelets* or *amphioxus*. Compare with **urochordates** and **vertebrates**.

cephalopods A lineage of mollusks including the squid, octopuses, and nautiluses. Distinguished by large brains, excellent vision, tentacles, and a reduced or absent shell.

cerebellum Posterior section of the vertebrate brain; involved in coordination of complex muscle movements, such as those required for locomotion and maintaining balance.

cerebrum The most anterior section of the vertebrate brain. Divided into left and right hemispheres and four lobes: frontal lobe, involved in complex decision making (in humans); occipital lobe, receives and interprets visual information; parietal lobe, involved in integrating sensory and motor functions; and temporal lobe, functions in memory, speech (in humans), and interpreting auditory information.

cervix The bottom portion of the uterus, containing a canal that leads to the vagina.

chaetae (singular: chaeta) Bristle-like extensions found in some annelids.

channel protein A protein that forms a pore in a cell membrane. The structure of most channels allows them to admit just one or a few types of ions or molecules. Compare with **carrier protein**.

character Any genetic, morphological, physiological, or behavioral characteristic of an organism to be studied.

character displacement The evolutionary tendency for the traits of similar species that occupy overlapping ranges to change in a way that reduces interspecific competition.

chelicerae A pair of clawlike appendages found around the mouth of certain arthropods called chelicerates (spiders, mites, and allies).

chemical bond An attractive force binding two atoms together. Covalent bonds, ionic bonds, and hydrogen bonds are types of chemical bonds.

chemical energy The potential energy stored in covalent bonds between atoms.

chemical equilibrium A dynamic but stable state of a reversible chemical reaction in which the forward reaction and reverse reactions proceed at the same rate, so that the concentrations of reactants and products remain constant.

chemical evolution A The theory that simple chemical compounds in the early atmosphere and ocean combined via chemical reactions to form larger, more complex substances, eventually leading to the origin of life and the start of biological evolution.

chemical reaction Any process in which one compound or element is combined with others or is broken down; involves the making and/or breaking of chemical bonds.

chemiosmosis An energetic coupling mechanism whereby energy stored in an electrochemical proton gradient is used to drive an energy-requiring process such as production of ATP.

chemokine Any of a subset of cytokines that acts as a chemical signal attracting leukocytes to a site of tissue injury or infection.

chemolithotroph An organism (bacteria or archaea) that produces ATP by oxidizing inorganic molecules with high potential energy such as ammonia (NH_3) or methane (CH_4). Also called *lithotroph*. Compare with **chemoorganotroph**.

chemoorganotroph An organism that produces ATP by oxidizing organic molecules with high potential energy such as sugars. Also called *organotroph*. Compare with **chemolithotroph**.

chemoreception A sensory system in which receptors are activated in response to the binding of chemicals.

chemoreceptor A sensory cell or organ specialized for detection of specific molecules or classes of molecules.

chiasma (plural: chiasmata) The X-shaped structure formed during meiosis by crossing over between non-sister chromatids in a pair of homologous chromosomes.

chitin A structural polysaccharide composed of *N*-acetyl-glucosamine (NAG) monomers joined end to end by β-1,4-glycosidic linkages. Found in cell walls of fungi and many algae, and in external skeletons of insects and crustaceans.

chitons A lineage of marine mollusks that have a protective shell formed of eight calcium carbonate plates.

chlorophyll Any of several closely related green pigments, found in chloroplasts, that absorb light during photosynthesis.

chloroplast A chlorophyll-containing organelle, bounded by a double membrane, in which photosynthesis occurs; found in plants and photosynthetic protists. Also the location of starch, amino acid, fatty acid, purine, and pyrimidine synthesis.

choanocyte A specialized, flagellated feeding cell found in choanoflagellates (protists that are the closest living relatives of animals) and sponges (the most ancient animal phylum).

cholecystokinin A peptide hormone secreted by cells in the lining of the small intestine. Stimulates the secretion of digestive enzymes from the pancreas and of bile from the liver and gallbladder.

chordate Any member of the phylum Chordata. Chordates are deuterostomes distinguished by a dorsal hollow nerve cord, pharyngeal gill slits, a notochord, and a post-anal tail. Includes vertebrates, cephalochordata, and urochordata.

chromatid One of the two identical double-stranded DNAs composing a replicated chromosome that is connected at the centromere to the other strand.

chromatin The complex of DNA and proteins, mainly histones, that compose eukaryotic chromosomes. Can be highly compact (heterochromatin) or loosely coiled (euchromatin).

chromatin remodeling The process by which the DNA in chromatin is unwound from its associated proteins to allow transcription or replication. May involve chemical modification of histone proteins or reshaping of the chromatin by large multiprotein complexes in an ATP-requiring process.

chromatin remodeling complexes Sets of enzymes that use energy from ATP hydrolysis shift nucleosomes on DNA to expose regulatory sequences to transcription factors.

chromosome theory of inheritance The principle that genes are located on chromosomes and that patterns of inheritance are determined by the behavior of chromosomes during meiosis.

chromosome Gene-carrying structure consisting of a single long molecule of double-stranded DNA and associated proteins (e.g., histones). Most prokaryotic cells contain a single, circular chromosome; eukaryotic cells contain multiple noncircular (linear) chromosomes located in the nucleus.

cilium (plural: cilia) One of many short, filamentous projections of some eukaryotic cells, containing a core of microtubules. Used to move the cell as well as to circulate fluid or particles around the surface of a stationary cell. See **axoneme**.

circadian clock An internal mechanism found in most organisms that regulates many body processes (sleep–wake cycles, hormonal patterns, etc.) in a roughly 24-hour cycle.

circulatory system The system responsible for moving oxygen, carbon dioxide, and other materials (hormones, nutrients, wastes) within the animal body.

cisternae (singular: cisterna) Flattened, membrane-bound compartments that make up the Golgi apparatus.

cisternal maturation The process of cargo movement through the Golgi apparatus by residing in cisternae that mature from *cis* to *trans* via the import and export of different Golgi enzymes.

citric acid cycle A series of eight chemical reactions that start with citrate (deprotonated citric acid) and ends with oxaloacetate, which reacts with acetyl CoA to form citrate—forming a cycle that is part of the pathway that oxidizes glucose to CO_2. Also known as the *Krebs cycle* or *tricarboxylic acid* (*TCA*) *cycle*.

clade See **monophyletic group**.

cladistic approach A method for constructing a phylogenetic tree that is based on identifying the unique traits (shared, derived characters, called synapomorphies) of each monophyletic group.

Class I MHC protein A type of MHC protein that is present on the plasma membrane of virtually all nucleated cells and functions in presenting antigen to $CD8^+$ T cells.

Class II MHC protein A type of MHC protein that is present only on the plasma membrane of certain cells in the immune response, such as dendritic cells, macrophages, and B cells. It functions in presenting epitopes of antigens to $CD4^+$ T cells.

cleavage In animal development, the series of rapid mitotic cell divisions, with little cell growth, that produces successively smaller cells (blastomeres) and transforms a zygote into a multicellular blastula.

cleavage furrow A pinching in of the plasma membrane that occurs as cytokinesis begins in animal cells and deepens until the cytoplasm is divided into two daughter cells.

climate The prevailing, long-term weather conditions in a particular region.

climax community The stable, final community that develops from ecological succession.

clitoris A rod of erectile tissue in the external genitalia of female mammals. Is formed from the same embryonic tissue as the male penis and has a similar function in sexual arousal.

cloaca In a few mammals and many nonmammalian vertebrates, a body cavity opening to the outside and used by both the excretory and reproductive systems.

clonal selection theory The dominant explanation of the generation of an adaptive immune response. According to the theory, the immune system retains a vast pool of inactive lymphocytes, each with a unique receptor for a unique epitope. Lymphocytes that encounter their complementary epitopes are stimulated to divide (selected and cloned), producing daughter cells that combat infection and confer immunity.

clone (1) An individual that is genetically identical to another individual. (2) A lineage of genetically identical individuals or cells. (3) As a verb, to make one or more genetic replicas of a cell or individual.

cloning vector A plasmid or other agent used to transfer recombinant genes into cultured host cells. Also called simply *vector*.

closed circulatory system A circulatory system in which the circulating fluid (blood) is confined to blood vessels and flows in a continuous circuit. Compare with **open circulatory system**.

cnidocyte A specialized stinging cell found in cnidarians (e.g., jellyfish, corals, and anemones) and used in capturing prey.

co-receptor Any membrane protein that acts with some other membrane protein in a cell interaction or cell response.

cochlea The organ of hearing in the inner ear of mammals, birds, and crocodilians. A coiled, fluid-filled tube containing specialized pressure-sensing neurons (hair cells) that detect sounds of different pitches.

coding strand See **non-template strand**.

codominance An inheritance pattern in which heterozygotes exhibit both of the traits seen in each type of homozygous individual.

codon A sequence of three nucleotides in DNA or RNA that codes for an amino acid or a start or stop signal for protein synthesis.

coefficient of relatedness (*r*) A measure of how closely two individuals are related. Calculated as the probability that an allele in two individuals is inherited from the same ancestor.

coelom An internal, usually fluid-filled body cavity that is completely or partially lined with mesoderm.

coelomate An animal that has a true coelom, completely lined with mesoderm. Compare with **acoelomate** and **pseudocoelomate**.

coenocytic Containing many nuclei and a continuous cytoplasm through a filamentous body, without the body being divided into distinct cells. Some fungi are coenocytic.

coenzyme A small organic molecule that is a required cofactor for an enzyme-catalyzed reaction. Often donates or receives electrons or functional groups during the reaction.

coenzyme A (CoA) A molecule that is required for many cellular reactions and that is often transiently linked to other molecules, such as acetyl groups (see **acetyl CoA**).

coenzyme Q A nonprotein molecule that shuttles electrons between membrane-bound complexes in the mitochondrial electron transport chain. Also called **ubiquinone** or *Q*.

coevolution A pattern of evolution in which two interacting species reciprocally influence each other's adaptations over time.

coevolutionary arms race A series of adaptations and counter-adaptations observed in species that interact closely over time and affect each other's fitness.

cofactor An inorganic ion, such as a metal ion, that is required for an enzyme to function normally. May be bound tightly to an enzyme or associate with it transiently during catalysis.

cohesion The tendency of certain like molecules (e.g., water molecules) to cling together due to attractive forces. Compare with **adhesion**.

cohesion-tension theory The theory that water movement upward through plant vascular tissues is due to loss of water from leaves (transpiration), which pulls a cohesive column of water upward.

cohort A group of individuals that are the same age and can be followed through time.

coleoptile A modified leaf that covers and protects the stems and leaves of grass seedlings.

collagen A fibrous, pliable, cable-like glycoprotein that is a major component of the extracellular matrix of animal cells. Various subtypes differ in their tissue distribution, some of which are assembled into large fibrils in the extracellular space.

collecting duct In the vertebrate kidney, a large straight tube that receives filtrate from the distal tubules of several nephrons. Involved in the regulated reabsorption of water.

collenchyma cell In plants, an elongated cell with cell walls thickened at the corners that provides support to growing plant parts; usually found in strands along leaf veins and stalks. Compare with **parenchyma cell** and **sclerenchyma cell**.

colon The portion of the large intestine where feces are formed by compaction of wastes and reabsorption of water.

colony An assemblage of individuals. May refer to an assemblage of semi-independent cells or to a breeding population of multicellular organisms.

commensalism (adjective: commensal) A symbiotic relationship in which one organism (the commensal) benefits and the other (the host) is not harmed. Compare with **mutualism** and **parasitism**.

communication In ecology, any process in which a signal from one individual modifies the behavior of another individual.

community All of the species that interact with each other in a certain area.

companion cell In plants, a cell in the phloem that is connected via many plasmodesmata to adjacent sieve-tube elements. Companion cells provide materials to maintain sieve-tube elements and function in the loading and unloading of sugars into sieve-tube elements.

compass orientation A type of navigation in which movement occurs in a specific direction.

competition In ecology, the interaction of two species or two individuals trying to use the same limited resource (e.g., water, food, living space). May occur between individuals of the same species

(intraspecific competition) or different species (interspecific competition).

competitive exclusion principle The principle that two species cannot coexist in the same ecological niche in the same area because one species will outcompete the other.

competitive inhibition Inhibition of an enzyme's ability to catalyze a chemical reaction via binding of a nonreactant molecule that competes with the substrate(s) for access to the active site.

complement system A set of proteins that circulate in the bloodstream and can destroy bacteria by forming holes in the bacterial plasma membrane.

complementary base pairing The association between specific nitrogenous bases of nucleic acids stabilized by hydrogen bonding. Adenine pairs only with thymine (in DNA) or uracil (in RNA), and guanine pairs only with cytosine.

complementary DNA (cDNA) DNA produced in the laboratory using an RNA transcript as a template and reverse transcriptase; corresponds to a gene but lacks introns. Also produced naturally by retroviruses.

complementary strand A newly synthesized strand of RNA or DNA that has a base sequence complementary to that of the template strand.

complete digestive tract A digestive tract with two openings, usually called a mouth and an anus.

complete metamorphosis See **holometabolous metamorphosis**.

compound eye An eye formed of many independent light-sensing columns (ommatidia); occurs in arthropods. Compare with **simple eye**.

concentration gradient Difference across space (e.g., across a membrane) in the concentration of a dissolved substance.

condensation reaction A chemical reaction in which two molecules are joined covalently with the removal of an −OH from one and an −H from another to form water. Also called a *dehydration reaction*. Compare with **hydrolysis**.

conduction (1) Direct transfer of heat between two objects that are in physical contact. Compare with **convection**. (2) Transmission of an electrical impulse along the axon of a nerve cell.

cone cell A photoreceptor cell with a cone-shaped outer portion that is particularly sensitive to bright light of a certain color. Also called simply *cone*. Compare with **rod cell**.

connective tissue An animal tissue consisting of scattered cells in a liquid, jellylike, or solid extracellular matrix. Includes bone, cartilage, tendons, ligaments, and blood.

conservation biology The effort to study, preserve, and restore threatened genetic diversity, populations, communities, and ecosystems.

constant (C) region The invariant amino acid sequence in polypeptides that are used to make antibodies, B-cell receptors, and T-cell receptors. Apart from antibody class types (IgG, IgM, etc.), this region remains constant within an individual. Compare with **variable (V) region**.

constitutive Always occurring; always present. Commonly used to describe enzymes and other proteins

that are synthesized continuously or mutants in which one or more genetic loci are constantly expressed due to defects in gene control.

constitutive defense A defensive trait that is always manifested even in the absence of a predator or pathogen. Also called *standing defense*. Compare with **inducible defenses**.

constitutive mutant An abnormal (mutated) strain that produces a product at all times, instead of under certain conditions only.

consumer See **heterotroph**.

consumption Predation or herbivory.

continental shelf The portion of a geologic plate that extends from a continent under seawater.

continuous strand See **leading strand**.

contraception Any of several methods to prevent pregnancy.

control In a scientific experiment, a group of organisms or samples that do not receive the experimental treatment but are otherwise identical to the group that does.

convection Transfer of heat by movement of large volumes of a gas or liquid. Compare with **conduction**.

convergent evolution The independent evolution of similar traits in distantly related organisms due to adaptation to similar environments and a similar way of life.

cooperative binding The tendency of the protein subunits of hemoglobin to affect each other's oxygen binding such that each bound oxygen molecule increases the likelihood of further oxygen binding.

coprophagy The eating of feces.

coral reef A large assemblage of colonial marine corals that usually serves as shallow-water, sunlit habitat for many other species as well.

co-receptor Any membrane protein that acts with some other membrane protein in a cell interaction or cell response.

core enzyme The enzyme responsible for catalysis in a multipart holoenzyme.

cork cambium One of two types of cylindrical meristem, consisting of a ring of undifferentiated plant cells found just under the cork layer of woody stems and roots; produces new cork cells on its outer side. Compare with **vascular cambium**.

cork cell A cell in the protective outermost layer of a woody stem and root that produces and accumulates waxes that make the cell less permeable to water and gases.

corm A rounded, thick underground stem that can produce new plants via asexual reproduction.

cornea The transparent sheet of connective tissue at the very front of the eye in vertebrates and some other animals. Protects the eye and helps focus light.

corolla All of the petals of a flower.

corona The cluster of cilia at the anterior end of a rotifer; in many species it facilitates suspension feeding.

corpus callosum A thick band of neurons that connects the two hemispheres of the cerebrum in the mammalian brain.

corpus luteum A yellowish structure that secretes progesterone in an ovary. Is formed from a follicle that has recently ovulated.

cortex (1) In animals, the outermost region of an organ, such as the kidney or adrenal gland. (2) In plants, a layer of ground tissue found outside the vascular bundles of roots and outside the pith of a stem.

corticotropin-releasing hormone (CRH) A peptide hormone, produced and secreted by the hypothalamus, that stimulates the anterior pituitary to release ACTH.

cortisol A steroid hormone, produced and secreted by the adrenal cortex, that increases blood glucose and prepares the body for stress. The major glucocorticoid hormone in some mammals. Also called *hydrocortisone*.

cost–benefit analysis Decisions or analyses that weigh the fitness costs and benefits of a particular action.

cotransporter A transmembrane protein that facilitates diffusion of an ion down its previously established electrochemical gradient and uses the energy of that process to transport some other substance, in the same or opposite direction, *against* its concentration gradient. Also called *secondary active transporter*. See **antiporter** and **symporter**.

cotyledon The first leaf, or seed leaf, of a plant embryo. Used for storing and digesting nutrients and/or for early photosynthesis.

countercurrent exchanger In animals, any anatomical arrangement that allows the maximum transfer of heat or a soluble substance from one fluid to another. The two fluids must be flowing in opposite directions and have a heat or concentration gradient between them.

covalent bond A type of chemical bond in which two atoms share one or more pairs of electrons. Compare with **hydrogen bond** and **ionic bond**.

cranium A bony, cartilaginous, or fibrous case that encloses and protects the brain of vertebrates. Forms part of the skull. Also called *braincase*.

crassulacean acid metabolism (CAM) A variant type of photosynthesis in which CO_2 is fixed and stored in organic acids at night when stomata are open and then released to feed the Calvin cycle during the day when stomata are closed. Helps reduce water loss and CO_2 loss by photorespiration.

cristae (singular: crista) Sac-like invaginations of the inner membrane of a mitochondrion. Location of the electron transport chain and ATP synthase.

Cro-Magnon A prehistoric European population of modern humans (*Homo sapiens*) known from fossils, paintings, sculptures, and other artifacts.

crop A storage organ in the digestive system of certain vertebrates.

cross A mating between two individuals that is used for genetic analysis.

cross-pollination Pollination of a flower by pollen from another individual, rather than by self-fertilization. Also called *crossing*.

cross-talk Interactions among signaling pathways, triggered by different signals, that modify a cellular response.

crossing over The exchange of segments of non-sister chromatids between a pair of homologous chromosomes that occurs during meiosis I.

crosstalk Interactions among signaling pathways that modify a cellular response.

crustaceans A lineage of arthropods that includes shrimp, lobster, and crabs. Many have a carapace (a platelike portion of the exoskeleton covering the cephalothorax) and mandibles for biting or chewing.

cryptic species A species that cannot be distinguished from similar species by easily identifiable morphological traits.

culture In cell biology, a collection of cells or a tissue growing under controlled conditions, usually in suspension or on the surface of a dish of solid growth medium.

Cushing's disease A human endocrine disorder caused by loss of feedback inhibition of cortisol on ACTH secretion. Characterized by high ACTH and cortisol levels and wasting of body protein reserves.

cuticle A protective coating secreted by the outermost layer of cells of an animal or a plant; often functioning to reduce evaporative water loss.

cyanobacteria A lineage of photosynthetic bacteria formerly known as blue-green algae. Likely the first life-forms to carry out oxygenic photosynthesis.

cyclic AMP (cAMP) Cyclic adenosine monophosphate; a small molecule, derived from ATP, that is widely used by cells in signal transduction and transcriptional control.

cyclic electron flow Path of electrons in which excited electrons of photosystem I are transferred back to plastoquinone (PQ), the start of the electron transport chain normally associated with photosystem II. Instead of reducing $NADP^+$ to make NADPH, the electron energy is used to make ATP via photophosphorylation. Compare with **noncyclic electron flow**.

cyclin One of several regulatory proteins whose concentrations fluctuate cyclically throughout the cell cycle. Involved in the control of the cell cycle via cyclin-dependent kinases.

cyclin-dependent kinase (Cdk) Any of several related protein kinases that are functional only when bound to a cyclin and are activated by other modifications. Involved in control of the cell cycle.

cytochrome *c* (cyt *c*) A soluble protein that shuttles electrons between membrane-bound complexes in the mitochondrial electron transport chain.

cytokine Any of a diverse group of signaling proteins, secreted largely by cells of the immune system, whose effects include stimulating leukocyte production, recruiting cells to the site of infection, tissue repair, and fever. Generally function to regulate the type, intensity, and duration of an immune response.

cytokinesis Division of the cytoplasm to form two daughter cells. Typically occurs immediately after division of the nucleus by mitosis or meiosis.

cytokinins A class of plant hormones that stimulate cell division and retard aging.

cytoplasm All the contents of a cell, excluding the nucleus, bounded by the plasma membrane.

cytoplasmic determinant A regulatory transcription factor or signaling molecule that is distributed unevenly in the cytoplasm of the egg and that directs early pattern formation in an embryo.

cytoplasmic streaming The directed flow of cytosol and organelles that facilitates distribution of materials within some large plant and fungal cells. Occurs along actin filaments and is powered by myosin.

cytoskeleton In eukaryotic cells, a network of protein fibers in the cytoplasm that are involved in cell shape, support, locomotion, and transport of materials within the cell. Prokaryotic cells have a similar but much less extensive network of fibers.

cytosol The fluid portion of the cytoplasm, excluding the contents of membrane-enclosed organelles.

cytotoxic T cell A type of $CD8^+$ effector T cell that induces apoptosis in infected and cancerous cells. Recognizes target cells via interactions with complementary class I MHC–peptide complexes. Also called *cytotoxic T lymphocyte (CTL)* and *killer T cell*. Compare with **helper T cell**.

dalton (Da) A unit of mass equal to 1/12 the mass of one carbon-12 atom; about the mass of 1 proton or 1 neutron.

daughter strand The strand of DNA that is newly replicated from an existing template strand of DNA.

day-neutral plant A plant whose flowering time is not affected by the relative length of day and night (the photoperiod). Compare with **long-day** and **short-day plant**.

dead space Air passages that are not involved in gas exchange with the blood; examples are the trachea and bronchi.

deciduous Describing a plant that sheds leaves or other structures at regular intervals (e.g., each autumn).

decomposer See **detritivore**.

decomposer food chain An ecological network of detritus, decomposers that eat detritus, and predators and parasites of the decomposers.

deep sequencing A method to learn the types of mRNAs or DNA sequences present in cells, and their relative amounts, involving the preparation and sequencing of cDNA libraries.

definitive host The host species in which a parasite reproduces sexually. Compare with **intermediate host**.

dehydration reaction See **condensation reaction**.

deleterious In genetics, referring to any mutation, allele, or trait that reduces an individual's fitness.

deletion In genetics, refers to the loss of part of a chromosome.

demography The study of factors that determine the size and structure of populations through time.

dendrite A short extension from a neuron's cell body that receives signals from other neurons.

dendritic cell A type of leukocyte that ingests and digests foreign antigens, moves to a lymph node, and presents the antigens' epitopes, in the context of MHC proteins on its membrane, to $CD4^+$ and $CD8^+$ T cells.

dense connective tissue A type of connective tissue, distinguished by having an extracellular matrix dominated by collagen fibers. Found in tendons and ligaments.

density dependent In population ecology, referring to any characteristic that varies depending on population density.

density independent In population ecology, referring to any characteristic that does not vary with population density.

deoxyribonucleic acid (DNA) A nucleic acid composed of deoxyribonucleotides that carries the genetic information of a cell. Generally occurs as two intertwined strands, but these can be separated. See also **double helix**.

deoxyribonucleoside triphosphate (dNTP) A monomer used by DNA polymerase to polymerize DNA. Consists of the sugar deoxyribose, a base (A, T, G, or C), and three phosphate groups.

deoxyribonucleotide See **nucleotide**.

depolarization A change in membrane potential from its resting negative state to a less negative or a positive state; a normal phase in an action potential. Compare with **hyperpolarization**.

depolarization Change in membrane potential from its resting negative state to a less negative or to a positive state; a normal phase in an action potential. Compare with **hyperpolarization**.

deposit feeder An animal that eats its way through a food-containing substrate.

derived trait A trait that is clearly homologous with a trait found in an ancestor of a particular group, but that has a new form.

dermal tissue system The tissue forming the outer layer of a plant; also called *epidermis*.

descent with modification The phrase used by Darwin to describe his hypothesis of evolution by natural selection.

desmosome A type of cell–cell attachment structure, consisting of cadherin proteins, that is anchored to intermediate filaments. Serves to link the cytoskeletons of adjacent animal cells and form strong cell–cell attachments throughout a tissue. Compare with **gap junction** and **tight junction**.

detergent A type of small amphipathic molecule used to solubilize hydrophobic molecules in aqueous solution.

determination In development, the commitment of a cell to a particular differentiated fate. Once a cell is fully determined, it can differentiate only into a particular cell type (e.g., liver cell, brain cell).

detritivore An organism whose diet consists mainly of dead organic matter (detritus). Various bacteria, fungi, protists, and animals are detritivores. Also called *decomposer*.

detritus A layer of dead organic matter that accumulates at ground level or on seafloors and lake bottoms.

deuterostomes A major lineage of bilaterian animals that share a pattern of embryological development, including formation of the anus earlier than the mouth, and formation of the coelom by pinching off of layers of mesoderm from the gut. Includes echinoderms and chordates. Compare with **protostomes**.

developmental homology A similarity in embryonic form, or in the fate of embryonic tissues, that is due to inheritance from a common ancestor.

diabetes mellitus A disease caused by defects in insulin production (type 1) or in the response of cells to insulin (type 2). Characterized by abnormally high blood glucose levels and huge amounts of glucose-containing urine.

diaphragm An elastic, sheetlike structure. In mammals, the muscular sheet of tissue that separates the chest and abdominal cavities. It contracts and moves downward during inhalation, expanding the chest cavity.

diastole The portion of the cardiac cycle during which the atria or ventricles of the heart are relaxed. Compare with **systole**.

diastolic blood pressure The force exerted by blood against artery walls during relaxation of the heart's left ventricle. Compare with **systolic blood pressure**.

dicot Any flowering plant (angiosperm) that has two cotyledons (embryonic leaves) upon germination. The dicots do not form a monophyletic group. Also called *dicotyledonous plant*. Compare with **eudicot** and **monocot**.

dideoxy sequencing A laboratory technique for determining the exact nucleotide sequence of DNA. Relies on the use of dideoxynucleoside triphosphates (ddNTPs), which terminate DNA replication.

diencephalon The part of the mammalian brain that relays sensory information to the cerebellum and functions in maintaining homeostasis.

differential centrifugation Procedure for separating cellular components according to their size and density by spinning a cell homogenate in a series of centrifuge runs. After each run, the supernatant is removed from the deposited material (pellet) and spun again at progressively higher speeds.

differential gene expression Expression of different sets of genes in cells with the same genome. Responsible for creating different cell types.

differentiation The process by which any unspecialized cell becomes a distinct specialized cell type (e.g., liver cell, brain cell), usually by changes in gene expression. Also called *cell differentiation*.

diffusion Spontaneous movement of a substance from one region to another, often with a net movement from a region of high concentration to one of low concentration (i.e., down a concentration gradient).

digestion The physical and chemical breakdown of food into molecules that can be absorbed into the body of an animal.

digestive tract The long tube that begins at the mouth and ends at the anus. Also called *alimentary canal* or *gastrointestinal (GI) tract*.

dihybrid cross A mating between two parents that are heterozygous for two different genes.

dikaryotic Describing a cell or fungal mycelium having two haploid nuclei that are genetically distinct.

dimer An association of two molecules that may be identical (homodimer) or different (heterodimer).

dioecious Describing an angiosperm species that has male and female reproductive structures on separate plants. Compare with **monoecious**.

diploblast (adjective: diploblastic) An animal whose body develops from two basic embryonic cell layers

or tissues—ectoderm and endoderm. Compare with **triploblast**.

diploid (1) Having two sets of chromosomes (2*n*). (2) A cell or an individual organism with two sets of chromosomes, one set inherited from the mother and one set from the father. Compare with **haploid**.

direct sequencing A technique for identifying and studying microorganisms that cannot be grown in culture. Involves detecting and amplifying copies of certain specific genes in the microorganisms' DNA, sequencing these genes, and then comparing the sequences with the known sequences from other organisms.

directional selection A mode of natural selection that favors one extreme phenotype with the result that the average phenotype of a population changes in one direction. Generally reduces overall genetic variation in a population. Compare with **disruptive selection** and **stabilizing selection**.

disaccharide A carbohydrate consisting of two monosaccharides (sugar residues) linked together.

discontinuous strand See **lagging strand**.

discrete trait An inherited trait that exhibits distinct phenotypes rather than the continuous variation characteristic of a quantitative trait such as body height.

dispersal The movement of individuals from their place of origin (birth, hatching) to a new location.

disruptive selection A mode of natural selection that favors extreme phenotypes at both ends of the range of phenotypic variation. Maintains overall genetic variation in a population. Compare with **stabilizing selection** and **directional selection**.

distal tubule In the vertebrate kidney, the convoluted portion of a nephron into which filtrate moves from the loop of Henle. Involved in the regulated reabsorption of sodium and water. Compare with **proximal tubule**.

disturbance In ecology, any strong, short-lived disruption to a community that changes the distribution of living and/or nonliving resources.

disturbance regime The characteristic disturbances that affect a given ecological community.

disulfide bond A covalent bond between two sulfur atoms, typically in the side chains of certain amino acids (e.g., cysteine). Often contributes to tertiary and quaternary levels of protein structure.

DNA See **deoxyribonucleic acid**.

DNA cloning Any of several techniques for producing many identical copies of a particular gene or other DNA sequence.

DNA fingerprinting Any of several methods for identifying individuals by unique features of their genomes. Commonly involves using PCR to produce many copies of certain short tandem repeats (microsatellites) and then analyzing their lengths.

DNA helicase An enzyme that breaks hydrogen bonds between nucleotides of DNA, "unzipping" a double-stranded DNA molecule.

DNA library See **cDNA library** and **genomic library**.

DNA ligase An enzyme that joins pieces of DNA by catalyzing the formation of a phosphodiester bond between the pieces.

DNA methylation The addition of a methyl group ($-CH_3$) to a DNA molecule.

DNA methyltransferase A class of eukaryotic enzymes that add a methyl group to cytosines in DNA. Methylation of DNA leads to chromatin condensation and is an important means of regulating gene expression in eukaryotes.

DNA microarray A set of single-stranded DNA fragments, representing thousands of different genes that are permanently fixed to a small glass slide. Can be used to determine which genes are expressed in different cell types, under different conditions, or at different developmental stages.

DNA polymerase Any enzyme that catalyzes synthesis of DNA from deoxyribonucleotide triphosphates (dNTPs).

domain (1) A taxonomic category, based on similarities in basic cellular biochemistry, above the kingdom level. The three recognized domains are Bacteria, Archaea, and Eukarya. (2) A section of a protein that has a distinctive tertiary structure and function.

dominant Referring to an allele that determines the same phenotype when it is present in homozygous or heterozygous form.. Compare with **recessive**.

dormancy A temporary state of greatly reduced metabolic activity and growth in plants or plant parts (e.g., seeds, spores, bulbs, and buds).

dorsal Toward an animal's back and away from its belly. The opposite of ventral.

dorsal hollow nerve chord See **nerve chord**.

double fertilization An unusual form of reproduction seen in flowering plants, in which one sperm cell fuses with an egg to form a zygote and the other sperm cell fuses with two polar nuclei to form the triploid endosperm.

double helix The secondary structure of DNA, consisting of two antiparallel DNA strands wound around each other.

Down syndrome A human developmental disorder caused by trisomy of chromosome 21.

downstream In genetics, the direction in which RNA polymerase moves along a DNA strand. Compare with **upstream**.

duplication In genetics, refers to an additional copy of part of a chromosome.

dynein A class of motor proteins that uses the chemical energy of ATP to "walk" toward the minus end of a microtubule. Dyneins are responsible for bending of cilia and flagella, play a role in chromosome movement during mitosis, and can transport vesicles and organelles.

early endosome A small transient organelle that is formed by the accumulation of vesicles from receptor-mediated endocytosis and is an early stage in the formation of a lysosome.

ecdysone An insect hormone that triggers either molting (to a larger larval form) or metamorphosis (to the adult form), depending on the level of juvenile hormone.

ecdysozoans A major lineage of protostomes (Ecdysozoa) that grow by shedding their external skeletons (molting) and expanding their bodies. Includes arthropods, nematodes, and other groups. Compare with **lophotrochozoans**.

echinoderms A major lineage of deuterostomes (Echinodermata) distinguished by adult bodies with five-sided radial symmetry, a water vascular system, and tube feet. Includes sea urchins, sand dollars, and sea stars.

echolocation The use of echoes from vocalizations to obtain information about locations of objects in the environment.

ecological selection Also known as environmental selection. A type of natural selection that favors individuals with heritable traits that enhance their ability to survive and reproduce in a certain physical and/or biological environment, excluding their ability to obtain a mate. Compare with **sexual selection**.

ecology The study of how organisms interact with each other and with their surrounding environment.

ecosystem All the organisms that live in a geographic area, together with the nonliving (abiotic) components that affect or exchange materials with the organisms; a community and its physical environment.

ecosystem diversity The variety of biotic components in a region along with abiotic components, such as soil, water, and nutrients.

ecosystem function The sum of biological and chemical processes that are characteristic of a given ecosystem—such as primary production, nitrogen cycling, and carbon storage.

ecosystem services All of the benefits that humans derive, directly or indirectly, from ecosystem functions.

ecotourism Tourism that is based on observing wildlife or experiencing other aspects of natural areas.

ectoderm The outermost of the three basic cell layers (germ layers) in most animal embryos; gives rise to the outer covering and nervous system. Compare with **endoderm** and **mesoderm**.

ectomycorrhizal fungi (EMF) Fungi whose hyphae form a dense network that covers their host plant's roots but do not enter the root cells.

ectoparasite A parasite that lives on the outer surface of the host's body.

ectotherm An animal that gains most of its body heat from external sources as opposed to metabolic processes. Compare with **endotherm**.

effector Any cell, organ, or structure with which an animal can respond to external or internal stimuli. Usually functions, along with a sensor and integrator, as part of a homeostatic system.

efferent division The part of the nervous system that carries commands from the central nervous system to the body. Consists primarily of motor neurons.

egg A mature female gamete and any associated external layers (such as a shell). Larger and less mobile than the male gamete. In animals, also called *ovum*.

ejaculation The release of semen from the copulatory organ of a male animal.

ejaculatory duct A short duct through which sperm move during ejaculation; connects the vas deferens to the urethra.

electric current A flow of electrical charge past a point. Also called *current*.

electrical potential Potential energy created by a separation of electrical charges between two points. Also called *voltage*.

electrocardiogram (EKG) A recording of the electrical activity of the heart, as measured through electrodes on the skin.

electrochemical gradient The combined effect of an ion's concentration gradient and electrical (charge) gradient across a membrane that affects the diffusion of ions across the membrane.

electrogenic fish Any of various kinds of fishes having specialized electric organs that emit a current into the water to detect objects.

electrolyte Any compound that dissociates into ions when dissolved in water. In nutrition, any of the major ions necessary for normal cell function.

electromagnetic spectrum The entire range of wavelengths of radiation extending from short wavelengths (high energy) to long wavelengths (low energy). Includes gamma rays, X-rays, ultraviolet, visible light, infrared, microwaves, and radio waves (from short to long wavelengths).

electron acceptor A reactant that gains an electron and is reduced in a reduction–oxidation reaction.

electron carrier Any molecule that readily accepts electrons from and donates electrons to other molecules. Protons may be transferred with the electrons in the form of hydrogen atoms.

electron donor A reactant that loses an electron and is oxidized in a reduction–oxidation reaction.

electron shell A group of orbitals of electrons with similar energies. Electron shells are arranged in roughly concentric layers around the nucleus of an atom, and electrons in outer shells have more energy than those in inner shells. Electrons in the outermost shell, the valence shell, often are involved in chemical bonding.

electron transport chain (ETC) Any set of membrane-bound protein complexes and mobile electron carriers involved in a coordinated series of redox reactions in which the potential energy of electrons is successively decreased and used to pump protons from one side of a membrane to the other.

electronegativity A measure of the ability of an atom to attract electrons toward itself from an atom to which it is bonded.

electroreception A sensory system in which receptors are activated by electric fields.

electroreceptor A sensory cell or organ specialized to detect electric fields.

element A substance, consisting of atoms with a specific number of protons. Elements preserve their identity in chemical reactions.

elongation (1) The process by which RNA lengthens during transcription. (2) The process by which a polypeptide chain lengthens during translation.

elongation factors Proteins involved in the elongation phase of translation, assisting ribosomes in the synthesis of the growing peptide chain.

embryo A young, developing organism; the stage after fertilization and zygote formation.

embryo sac The female gametophyte in flowering plants.

embryogenesis The production of an embryo from a zygote. Embryogenesis is an early event in development of animals and plants.

Embryophyta An increasingly popular name for the lineage called land plants; reflects their retention of a fertilized egg.

embryophyte A plant that nourishes its embryos inside its own body. All land plants are embryophytes.

emergent property A property that stems from the interaction of simpler elements and that is impossible to predict from the study of individual elements.

emergent vegetation Any plants in an aquatic habitat that extend above the surface of the water.

emerging disease Any infectious disease, often a viral disease, that suddenly afflicts significant numbers of humans for the first time; often due to changes in the host species for a pathogen or to radical changes in the genetic material of the pathogen.

emigration The migration of individuals away from one population to other populations. Compare with **immigration**.

emulsification (verb: emulsify) The dispersion of fat into an aqueous solution. Usually requires the aid of an amphipathic substance such as a detergent or bile salts, which can break large fat globules into microscopic fat droplets.

endangered species A species whose numbers have decreased so much that it is in danger of extinction throughout all or part of its range.

endemic species A species that lives in one geographic area and nowhere else.

endergonic Referring to a chemical reaction that requires an input of energy to occur and for which the Gibbs free-energy change (ΔG) is greater than zero. Compare with **exergonic**.

endocrine Relating to a chemical signal (hormone) that is released into the bloodstream by a producing cell and acts on a distant target cell.

endocrine disruptor An exogenous chemical that interferes with normal hormonal signaling.

endocrine gland A gland that secretes hormones directly into the bloodstream or interstitial fluid instead of into ducts. Compare with **exocrine gland**.

endocrine system All of the glands and tissues that produce and secrete hormones into the bloodstream.

endocytosis General term for any pinching off of the plasma membrane that results in the uptake of material from outside the cell. Includes phagocytosis, pinocytosis, and receptor-mediated endocytosis. Compare with **exocytosis**.

endoderm The innermost of the three basic cell layers (germ layers) in most animal embryos; gives rise to the digestive tract and organs that connect to it (liver, lungs, etc.). Compare with **ectoderm** and **mesoderm**.

endodermis In plant roots, a cylindrical layer of cells that separates the cortex from the vascular tissue and location of the Casparian strip.

endomembrane system A system of organelles in eukaryotic cells that synthesizes, processes, transports, and recycles proteins and lipids. Includes the endoplasmic reticulum (ER), Golgi apparatus, and lysosomes.

endomycorrhizal fungi See **arbuscular mycorrhizal fungi (AMF)**.

endoparasite A parasite that lives inside the host's body.

endophyte (adjective: endophytic) A fungus that lives inside the aboveground parts of a plant in a symbiotic relationship. Compare with **epiphyte**.

endoplasmic reticulum (ER) A network of interconnected membranous sacs and tubules found inside eukaryotic cells. See **rough** and **smooth endoplasmic reticulum**.

endoskeleton Bony and/or cartilaginous structures within the body that provide support. Examples are the spicules of sponges, the plates in echinoderms, and the bony skeleton of vertebrates. Compare with **exoskeleton**.

endosperm A triploid ($3n$) tissue in the seed of a flowering plant (angiosperm) that serves as food for the plant embryo. Functionally analogous to the yolk in some animal eggs.

endosymbiont An organism that lives in a symbiotic relationship inside the body of its host.

endosymbiosis An association between organisms of two different species in which one lives inside the cell or cells of the other.

endosymbiosis theory The theory that mitochondria and chloroplasts evolved from prokaryotes that were engulfed by host cells and took up a symbiotic existence within those cells, a process termed primary endosymbiosis. In some eukaryotes, chloroplasts may have originated by secondary endosymbiosis; that is, when a cell engulfed a chloroplast-containing protist and retained its chloroplasts.

endotherm An animal whose primary source of body heat is internally generated. Compare with **ectotherm**.

endothermic Referring to a chemical reaction that absorbs heat. Compare with **exothermic**.

energetic coupling In cellular metabolism, the mechanism by which energy released from an exergonic reaction (commonly, hydrolysis of ATP) is used to drive an endergonic reaction.

energy The capacity to do work or to supply heat. May be stored (potential energy) or available in the form of motion (kinetic energy).

enhancer A regulatory sequence in eukaryotic DNA that may be located far from the gene it controls or within introns of the gene. Binding of specific proteins to an enhancer enhances the transcription of certain genes.

enrichment culture A method of detecting and obtaining cells with specific characteristics by placing a sample, containing many types of cells, under a specific set of conditions (e.g., temperature, salt concentration, available nutrients) and isolating those cells that grow rapidly in response.

enthalpy (*H*) A quantitative measure of the amount of potential energy, or heat content, of a system plus the pressure and volume it exerts on its surroundings.

entropy (*S*) A quantitative measure of the amount of disorder of any system, such as a group of molecules.

envelope (viral) A membrane that encloses the capsids of some viruses. Normally includes specialized proteins that attach to host-cell surfaces.

environmental sequencing See **metagenomics**.

enzyme A protein catalyst used by living organisms to speed up and control biological reactions.

epicotyl In some embryonic plants, a portion of the embryonic stem that extends above the cotyledons.

epidemic The spread of an infectious disease throughout a population in a short time period. Compare with **pandemic**.

epidermis The outermost layer of cells of any multicellular organism.

epididymis A coiled tube wrapped around each testis in reptiles, birds, and mammals. The site of the final stages of sperm maturation.

epigenetic inheritance Pattern of inheritance involving differences in phenotype that are not due to differences in the nucleotide sequence of genes.

epinephrine A catecholamine hormone, produced and secreted by the adrenal medulla, that triggers rapid responses related to the fight-or-flight response. Also called *adrenaline*.

epiphyte (adjective: epiphytic) A nonparasitic plant that grows on the trunks or branches of other plants and is not rooted in soil.

epithelial tissues See **epithelium**.

epithelium (plural: epithelia) An animal tissue consisting of sheetlike layers of tightly packed cells that line an organ, a gland, a duct, or a body surface. Also called *epithelial tissue*.

epitope A small region of a particular antigen to which an antibody, B-cell receptor, or T-cell receptor binds.

equilibrium potential The membrane potential at which there is no net movement of a particular ion into or out of a cell.

ER signal sequence A short amino acid sequence that marks a polypeptide for transport to the endoplasmic reticulum, where synthesis of the polypeptide chain is completed and the signal sequence removed. See **signal recognition particle**.

erythropoietin (EPO) A peptide hormone, released by the kidney in response to low blood-oxygen levels, that stimulates the bone marrow to produce more red blood cells.

esophagus The muscular tube that connects the mouth to the stomach.

essential amino acid Any amino acid that an animal cannot synthesize and must obtain from the diet. May refer specifically to one of the eight essential amino acids of adult humans: isoleucine, leucine, lysine, methionine, phenylalanine, threonine, tryptophan, and valine.

essential nutrient Any chemical element, ion, or compound that is required for normal growth, reproduction, and maintenance of a living organism and that cannot be synthesized by the organism.

EST See **expressed sequence tag**.

ester linkage The covalent bond formed by a condensation reaction between a carboxyl group and a hydroxyl group. Ester linkages join fatty acids to glycerol to form a fat or phospholipid.

estradiol The major estrogen produced by the ovaries of female mammals and many other vertebrates. Stimulates development of the female reproductive tract, growth of ovarian follicles, and growth of breast tissue in mammals.

estrogens A class of steroid hormones, including estradiol, estrone, and estriol, that generally promote female-like traits. Secreted by the gonads, fat tissue, and some other organs.

estrous cycle A female reproductive cycle in which the uterine lining is reabsorbed rather than shed in the absence of pregnancy and the female is sexually receptive only briefly during mid-cycle (estrus). It is seen in all mammals except Old World monkeys and apes (including humans). Compare with **menstrual cycle**.

ethylene A gaseous plant hormone associated with senescence that induces fruits to ripen and flowers to fade.

eudicot A member of a monophyletic group (lineage) of angiosperms that includes complex flowering plants and trees (e.g., roses, daisies, maples). All eudicots have two cotyledons, but not all dicots are members of this lineage. Compare with **dicot** and **monocot**.

Eukarya One of the three taxonomic domains of life, consisting of unicellular organisms (most protists, yeasts) and multicellular organisms (fungi, plants, animals) distinguished by a membrane-bound cell nucleus, numerous organelles, and an extensive cytoskeleton. Compare with **Archaea** and **Bacteria**.

eukaryote A member of the domain Eukarya; an organism whose cells contain a nucleus, numerous membrane-bound organelles, and an extensive cytoskeleton. May be unicellular or multicellular. Compare with **prokaryote**.

eusociality A complex social structure in which workers sacrifice most or all of their direct reproduction to help rear the queen's offspring. Common in insects such as ants, bees, wasps, and termites.

eutherians A lineage of mammals (Eutheria) whose young develop in the uterus and are not housed in an abdominal pouch. Also called *placental mammals*.

evaporation The energy-absorbing phase change from a liquid state to a gaseous state. Many organisms evaporate water as a means of heat loss.

evo-devo Popular term for evolutionary developmental biology, a research field focused on how changes in developmentally important genes have led to the evolution of new phenotypes.

evolution (1) The theory that all organisms on Earth are related by common ancestry and that they have changed over time, and continue to change, via natural selection and other processes. (2) Any change in the genetic characteristics of a population over time, especially a change in allele frequencies.

ex situ conservation Preserving species outside of natural areas (e.g., in zoos, aquaria, or botanical gardens).

excitable membrane A plasma membrane that is capable of generating an action potential. Neurons, muscle cells, and some other cells have excitable membranes.

excitatory postsynaptic potential (EPSP) A change in membrane potential, usually depolarization, at a neuron dendrite that makes an action potential more likely.

exergonic Referring to a chemical reaction that can occur spontaneously, releasing heat and/or increasing entropy, and for which the Gibbs free-energy change (ΔG) is less than zero. Compare with **endergonic**.

exocrine gland A gland that secretes some substance through a duct into a space other than the circulatory system, such as the digestive tract or the skin surface. Compare with **endocrine gland**.

exocytosis Secretion of intracellular molecules (e.g., hormones, collagen), contained within membrane-bound vesicles, to the outside of the cell by fusion of vesicles to the plasma membrane. Compare with **endocytosis**.

exon A transcribed region of a eukaryotic gene or region of a primary transcript that is retained in the mature RNA. Except for 5′ and 3′ UTRs, mRNA exons code for amino acids. Compare with **intron**.

exoskeleton A hard covering secreted on the outside of the body, used for body support, protection, and muscle attachment. Prominent in arthropods. Compare with **endoskeleton**.

exothermic Referring to a chemical reaction that releases heat. Compare with **endothermic**.

exotic species A nonnative species that is introduced into a new area. Exotic species often are competitors, pathogens, or predators of native species.

expansins A class of plant proteins that break hydrogen bonds between components in the primary cell wall to allow it to expand for cell growth.

exponential population growth The accelerating increase in the size of a population that occurs when the growth rate is constant and density independent. Compare with **logistic population growth**.

expressed sequence tag (EST) A portion of a transcribed gene (synthesized from an mRNA in a cell), used to find the physical location of that gene in the genome.

extant species A species that is living today.

extensor A muscle that pulls two bones farther apart from each other, increasing the angle of the joint, as in the extension of a limb or the spine. Compare with **flexor**.

extinct species A species that has died out.

extracellular digestion Digestion that takes place outside of an organism, as occurs in many fungi that make and secrete digestive enzymes.

extracellular matrix (ECM) A complex meshwork in which animal cells are embedded, consisting of proteins (e.g., collagen, proteoglycan, laminin) and polysaccharides produced by the cells.

extremophile A bacterium or archaean that thrives in an "extreme" environment (e.g., high-salt, high-temperature, low-temperature, or low-pressure).

F$_1$ generation First filial generation. The first generation of offspring produced from a mating (i.e., the offspring of the parental generation).

facilitated diffusion Passive movement of a substance across a membrane with the assistance of transmembrane carrier proteins or channel proteins.

facilitation In ecological succession, the phenomenon in which early-arriving species make conditions more favorable for later-arriving species. Compare with **inhibition** and **tolerance**.

facultative anaerobe Any organism that can survive and reproduce by performing aerobic respiration when oxygen is available or fermentation when it is not.

FAD/FADH$_2$ Oxidized and reduced forms, respectively, of flavin adenine dinucleotide. A nonprotein electron carrier that functions in the citric acid cycle and oxidative phosphorylation.

fallopian tube A narrow tube connecting the uterus to the ovary in humans, through which the egg travels after ovulation. Site of fertilization and cleavage. In nonhuman animals, called *oviduct*.

fast muscle fiber Type of skeletal muscle fiber that is white in color, generates ATP by glycolysis, and contracts rapidly but fatigues easily. Also called *fast glycolytic*, or *Type IIb, fiber*.

fat A lipid consisting of three fatty acid molecules joined by ester linkages to a glycerol molecule. Also called *triacylglycerol* or *triglyceride*.

fatty acid A lipid consisting of a hydrocarbon chain bonded at one end to a carboxyl group. Used by many organisms to store chemical energy; a major component of animal and plant fats and phospholipids.

fauna All the animal species characteristic of a particular region, period, or environment.

feather A specialized skin outgrowth, composed of β-keratin, present in all birds as well as in some non-avian dinosaurs. Used for flight, insulation, display, and other purposes.

feces The waste products of digestion.

fecundity The average number of female offspring produced by a single female in the course of her lifetime.

feedback inhibition A type of control in which high concentrations of the product of a metabolic pathway inhibit one of the enzymes early in the pathway. A form of negative feedback.

fermentation Any of several metabolic pathways that regenerate oxidizing agents, such as NAD$^+$, by transferring electrons to a final electron acceptor in the absence of an electron transport chain. Allows pathways such as glycolysis to continue to make ATP.

ferredoxin In photosynthetic organisms, an iron- and sulfur-containing protein in the electron transport chain of photosystem I. Can transfer electrons to the enzyme NADP$^+$ reductase, which catalyzes formation of NADPH.

fertility The average number of surviving children that each woman has during her lifetime.

fertilization Fusion of the nuclei of two haploid gametes to form a zygote with a diploid nucleus.

fertilization envelope A physical barrier that forms around a fertilized egg in many animals. The fertilization envelope prevents fertilization by more than one sperm (polyspermy).

fetal alcohol syndrome (FAS) A condition marked by hyperactivity, severe learning disabilities, and depression. Thought to be caused by exposure of an individual to high blood alcohol concentrations during embryonic development.

fetus In live-bearing animals, the unborn offspring after the embryonic stage. It usually is developed sufficiently to be recognizable as belonging to a certain species.

fiber In plants, a type of elongated sclerenchyma cell that provides support to vascular tissue. Compare with **sclereid**.

Fick's law of diffusion A mathematical relationship that describes the rates of diffusion of gases.

fight-or-flight response Rapid physiological changes that prepare the body for emergencies. Includes increased heart rate, increased blood pressure, and decreased digestion.

filament Any thin, threadlike structure, particularly (1) the threadlike extensions of a fish's gills or (2) part of a stamen: the slender stalk that bears the anthers in a flower.

filter feeder See **suspension feeder**.

filtrate Any fluid produced by filtration, in particular the fluid ("pre-urine") in the Malpighian tubules of insects and the nephrons of vertebrate kidneys.

filtration A process of removing large components from a fluid by forcing it through a filter. Occurs in a renal corpuscle of the vertebrate kidney, allowing water and small solutes to pass from the blood into the nephron.

fimbria (plural: fimbriae) A long, needlelike projection from the cell membrane of prokaryotes that is involved in attachment to nonliving surfaces or other cells.

finite rate of increase (λ) The rate of increase of a population over a given period of time. Calculated as the ending population size divided by the starting population size. Compare with **intrinsic rate of increase**.

first law of thermodynamics The principle of physics that energy is conserved in any process. Energy can be transferred and converted into different forms, but it cannot be created or destroyed.

fission (1) A form of asexual reproduction in which a prokaryotic cell divides to produce two genetically similar daughter cells by a process similar to mitosis of eukaryotic cells. Also called *binary fission*. (2) A form of asexual reproduction in which an animal splits into two or more individuals of approximately equal size; common among invertebrates.

fitness The ability of an individual to produce viable offspring relative to others of the same species.

fitness trade-off See **trade-off**.

fixed action pattern (FAP) Highly stereotyped behavior pattern that occurs in a certain invariant way in a certain species. A form of innate behavior.

flaccid Limp as a result of low internal (turgor) pressure (e.g., a wilted plant leaf). Compare with **turgid**.

flagellum (plural: flagella) A long, cellular projection that undulates (in eukaryotes) or rotates (in prokaryotes) to move the cell through an aqueous environment. See **axoneme**.

flatworms Members of the phylum Platyhelminthes. Distinguished by a broad, flat, unsegmented body that lacks a coelom. Flatworms belong to the lophotrochozoan branch of the protostomes.

flavin adenine dinucleotide See **FAD/FADH₂**.

flexor A muscle that pulls two bones closer together, decreasing the joint angle, as in the flexing of a limb or the spine. Compare with **extensor**.

floral meristem A group of undifferentiated plant stem cells that can give rise to the four organs making up a flower.

florigen In plants, a protein hormone that is synthesized in leaves and transported to the shoot apical meristem, where it stimulates flowering.

flower In angiosperms, the part of a plant that contains reproductive structures. Typically includes a calyx, a corolla, and one or more stamens and/or carpels. See **perfect** and **imperfect flower**.

fluid connective tissue A type of connective tissue, distinguished by having a liquid extracellular matrix; includes blood.

fluid feeder Any animal that feeds by sucking or mopping up liquids such as nectar, plant sap, or blood.

fluid-mosaic model The widely accepted hypothesis that the plasma membrane and organelle membranes consist of proteins embedded in a fluid phospholipid bilayer.

fluorescence The spontaneous emission of light from an excited electron falling back to its normal (ground) state.

follicle In a mammalian ovary, a sac of supportive cells containing an egg cell.

follicle-stimulating hormone (FSH) A peptide hormone, produced and secreted by the anterior pituitary; it stimulates (in females) growth of eggs and follicles in the ovaries or (in males) sperm production in the testes.

follicular phase In a menstrual cycle, the first major phase, during which follicles grow and estradiol levels increase; ends with ovulation.

food Any nutrient-containing material that can be consumed and digested by animals.

food chain A relatively simple pathway of energy flow through a few species, each at a different trophic level, in an ecosystem. Might include, for example, a primary producer, a primary consumer, a secondary consumer, and a decomposer. A subset of a **food web**.

food web The complex network of interactions among species in an ecosystem formed by the transfer of energy and nutrients among trophic levels. Consists of many food chains.

foot One of the three main parts of the mollusk body; a muscular appendage, used for movements such as crawling and/or burrowing into sediment.

foraging Searching for food.

forebrain One of the three main regions of the vertebrate brain; includes the cerebrum, thalamus, and hypothalamus. Compare with **hindbrain** and **midbrain**.

fossil Any physical trace of an organism that existed in the past. Includes tracks, burrows, fossilized bones, casts, and so on.

fossil record All of the fossils that have been found anywhere on Earth and that have been formally described in the scientific literature.

founder effect A change in allele frequencies that often occurs when a new population is established from a small group of individuals (founder event) due to sampling error (i.e., the small group is not a representative sample of the source population).

fovea In the vertebrate eye, a portion of the retina where incoming light is focused; contains a high proportion of cone cells.

frameshift mutation The addition or deletion of a nucleotide in a coding sequence that shifts the reading frame of the mRNA.

free energy The energy of a system that can be converted into work. It may be measured only through the change in free energy in a reaction. See **Gibbs free-energy change**.

free radicals Any substance containing one or more atoms with an unpaired electron. Unstable and highly reactive.

frequency The number of wave crests per second traveling past a stationary point. Determines the pitch of sound and the color of light.

frequency-dependent selection A pattern of selection in which certain alleles are favored only when they are rare; a form of balancing selection.

fronds The large leaves of ferns.

frontal lobe In the vertebrate brain, one of the four major areas in the cerebrum.

fruit In flowering plants (angiosperms), a mature, ripened plant ovary (or group of ovaries), along with the seeds it contains and any adjacent fused parts; often functions in seed dispersal. See **aggregate**, **multiple**, and **simple fruit**.

fruiting body A structure formed in some prokaryotes, fungi, and protists for spore dispersal; usually consists of a base, a stalk, and a mass of spores at the top.

functional genomics The study of how a genome works; that is, when and where specific genes are expressed and how their products interact to produce a functional organism.

functional group A small group of atoms bonded together in a precise configuration and exhibiting particular chemical properties that it imparts to any organic molecule in which it occurs.

fundamental niche The total theoretical range of environmental conditions that a species can tolerate. Compare with **realized niche**.

fungi A lineage of eukaryotes that typically have a filamentous body (mycelium) and obtain nutrients by absorption.

fungicide Any substance that can kill fungi or slow their growth.

G protein Any of various proteins that are activated by binding to guanosine triphosphate (GTP) and inactivated when GTP is hydrolyzed to GDP. In G-protein-coupled receptors, signal binding directly triggers the activation of a G protein, leading to production of a second messenger or initiation of a phosphorylation cascade.

G₁ phase The phase of the cell cycle that constitutes the first part of interphase before DNA synthesis (S phase).

G₂ phase The phase of the cell cycle between synthesis of DNA (S phase) and mitosis (M phase); the last part of interphase.

gallbladder A small pouch that stores bile from the liver and releases it as needed into the small intestine during digestion of fats.

gametangium (plural: gametangia) (1) The gamete-forming structure found in all land plants except angiosperms. Contains a sperm-producing antheridium and an egg-producing archegonium. (2) The gamete-forming structure of some chytrid fungi.

gamete A haploid reproductive cell that can fuse with another haploid cell to form a zygote. Most multicellular eukaryotes have two distinct forms of gametes: egg cells (ova) and sperm cells.

gametogenesis The production of gametes (eggs or sperm).

gametophyte In organisms undergoing alternation of generations, the multicellular haploid form that arises from a single haploid spore and produces gametes. Compare with **sporophyte**.

ganglion (plural: ganglia) A mass of neurons in a centralized nervous system.

ganglion cell In the retina, a type of neuron whose axons form the optic nerves.

gap junction A type of cell–cell attachment structure that directly connects the cytosolic components of adjacent animal cells, allowing passage of water, ions, and small molecules between the cells. Compare with **desmosome** and **tight junction**.

gastrin A hormone produced by cells in the stomach lining in response to the arrival of food or to a neural signal from the brain. Stimulates other stomach cells to release hydrochloric acid.

gastropods A lineage of mollusks distinguished by a large, muscular foot and a unique feeding structure, the radula. Include slugs and snails.

gastrulation The process of coordinated cell movements, including the moving of some cells from the outer surface of the embryo to the interior, that results in the formation of three germ layers (endoderm, mesoderm, and ectoderm) and the axes of the embryo.

gated channel A channel protein that opens and closes in response to a specific stimulus, such as the binding of a particular molecule or a change in voltage across the membrane.

gemma (plural: gemmae) A small reproductive structure that is produced asexually in some plants during the gametophyte phase and can grow into a mature gametophyte; most common in non-vascular plants, particularly liverworts and club mosses, and in ferns

gene A section of DNA (or RNA, for some viruses) that encodes information for building one or more related polypeptides or functional RNA molecules along with the regulatory sequences required for its transcription.

gene duplication The formation of an additional copy of a gene, typically by misalignment of chromosomes during crossing over. Thought to be an important evolutionary process in creating new genes.

gene expression The set of processes, including transcription and translation, that convert information in DNA into aproduct of a gene, most commonly a protein.

gene family A set of genetic loci whose DNA sequences are extremely similar. Thought to have arisen by duplication of a single ancestral gene and subsequent mutations in the duplicated sequences.

gene flow The movement of alleles between populations; occurs when individuals leave one population, join another, and breed.

gene pool All the alleles of all the genes in a certain population.

gene therapy The treatment of an inherited disease by introducing a normal form of the gene.

generation The average time between a mother's first offspring and her daughter's first offspring.

genetic bottleneck A reduction in allelic diversity resulting from a sudden reduction in the size of a large population (population bottleneck) due to a random event.

genetic code The set of all codons and their meaning.

genetic correlation A type of evolutionary constraint in which selection on one trait causes a change in another trait as well; may occur when the same gene(s) affect both traits.

genetic diversity The diversity of alleles or genes in a population, species, or group of species.

genetic drift Any change in allele frequencies due to random events. Causes allele frequencies to drift up and down randomly over time, and eventually can lead to the fixation or loss of alleles.

genetic equivalence Having all different cell types of a multicellular individual possess the same genome.

genetic homology Similarity in DNA nucleotide sequences, RNA nucleotide sequences, or amino acid sequences due to inheritance from a common ancestor.

genetic map A list of genes on a chromosome that indicates their position and relative distances from one another. Also called a *linkage map*. Compare with **physical map**.

genetic marker A genetic locus that can be identified and traced in populations by laboratory techniques or by a distinctive visible phenotype.

genetic recombination A change in the combination of alleles on a given chromosome or in a given individual. Also called *recombination*.

genetic screen Any technique that identifies individuals with a particular type of mutation.

genetic variation (1) The number and relative frequency of alleles present in a particular population. (2) The proportion of phenotypic variation in a trait that is due to genetic rather than environmental influences in a certain population in a certain environment.

genetics The study of the inheritance of traits.

genital (plural: genitalia) Any external copulatory organ.

genome All the hereditary information in an organism, including not only genes but also stretches of DNA that do not contain genes.

genome annotation The process of analyzing a genome sequence to identify key features such as genes, regulatory sequences, and splice sites.

genomic library A set of DNA segments representing the entire genome of a particular organism. Each segment is carried by a plasmid or other cloning vector and can be separated from other segments. Compare with **cDNA library**.

genomics The field of study concerned with sequencing, interpreting, and comparing whole genomes from different organisms.

genotype All the alleles of every gene present in a given individual. Often specified only for the alleles of a particular set of genes under study. Compare with **phenotype**.

genus (plural: genera) In Linnaeus' system, a taxonomic category of closely related species. Always italicized and capitalized to indicate that it is a recognized scientific genus.

geologic time scale The sequence of eons, eras, and periods used to describe the geologic history of Earth.

germ cell In animals, any cell that can potentially give rise to gametes. Also called *germ-line cells*.

germ layer In animals, one of the three basic types of tissue formed during gastrulation; gives rise to all other tissues. See **endoderm**, **mesoderm**, and **ectoderm**.

germ line In animals, any of the cells that are capable of giving rise to reproductive cells (sperm or egg). Compare with **germ cell**.

germ theory of disease The theory that infectious diseases are caused by bacteria, viruses, and other microorganisms.

germination The process by which a seed becomes a young plant.

gestation The period of development inside the mother, from implantation to birth, in those species that have live birth.

gibberellins A class of hormones, found in plants and fungi, that stimulate growth. Gibberellic acid (GA) is one of the major gibberellins.

Gibbs free-energy change (ΔG) A measure of the change in enthalpy and entropy that occurs in a given chemical reaction. ΔG is less than 0 for spontaneous reactions and greater than 0 for nonspontaneous reactions.

gill Any organ in aquatic animals that exchanges gases and other dissolved substances between the blood and the surrounding water. Typically, a filamentous outgrowth of a body surface.

gill arch In aquatic vertebrates, a curved region of tissue between the gills. Gills are suspended from the gill arches.

gill filament In fish, any of the many long, thin structures that extend from gill arches into the water and across which gas exchange occurs.

gill lamella (plural: gill lamellae) Any of hundreds to thousands of sheetlike structures, each containing a capillary bed, that make up a gill filament.

gland An organ whose primary function is to secrete some substance, either into the blood (endocrine gland) or into some other space such as the gut or skin (exocrine gland).

glia Collective term for several types of cells in nervous tissue that are not neurons and do not conduct electrical signals but perform other functions, such as providing support, nourishment, or electrical insulation. Also called *glial cells*.

global carbon cycle See **carbon cycle, global**.

global climate change The global sum of all the local changes in temperature and precipitation patterns that accompany global warming (or in some past events, global cooling).

global gene regulation The regulation of multiple bacterial genes that are not part of one operon.

global nitrogen cycle See **nitrogen cycle, global**.

global warming A sustained increase in Earth's average surface temperature.

global water cycle See **water cycle, global**.

glomalin A glycoprotein that is abundant in the hyphae of arbuscular mycorrhizal fungi; when hyphae decay, it is an important component of soil.

glomerulus (plural: glomeruli) (1) In the vertebrate kidney, a ball-like cluster of capillaries, surrounded by Bowman's capsule, at the beginning of a nephron. (2) In the brain, a ball-shaped cluster of neurons in the olfactory bulb.

glucagon A peptide hormone produced by the pancreas in response to low blood glucose. Raises blood glucose by triggering breakdown of glycogen and stimulating gluconeogenesis. Compare with **insulin**.

glucocorticoids A class of steroid hormones, produced and secreted by the adrenal cortex, that increase blood glucose and prepare the body for stress. Include cortisol and corticosterone. Compare with **mineralocorticoids**.

gluconeogenesis Synthesis of glucose, often from non-carbohydrate sources (e.g., proteins and fatty acids). In plants, used to produce glucose from products of the Calvin cycle. In animals, occurs in the liver in response to low insulin levels and high glucagon levels.

glucose Six-carbon monosaccharide whose oxidation in cellular respiration is the major source of ATP in animal cells.

glyceraldehyde-3-phosphate (G3P) The phosphorylated three-carbon compound formed as the result of carbon fixation in the first step of the Calvin cycle.

glycerol A three-carbon molecule that forms the "backbone" of phospholipids and most fats.

glycogen A highly branched storage polysaccharide composed of glucose monomers joined by α-1,4- and α-1,6-glycosidic linkages. The major form of stored carbohydrate in animals.

glycolipid Any lipid molecule that is covalently bonded to a carbohydrate group.

glycolysis A series of 10 chemical reactions that oxidize glucose to produce pyruvate, NADH, and ATP.

Used by organisms as part of fermentation or cellular respiration.

glycoprotein Any protein with one or more covalently bonded carbohydrates, typically oligosaccharides.

glycosidic linkage The covalent linkage formed by a condensation reaction between two sugar monomers; joins the residues of a polysaccharide.

glycosylation Addition of a carbohydrate group to a molecule.

glyoxysome Specialized type of peroxisome found in plant cells and packed with enzymes for processing the products of photosynthesis.

gnathostomes Animals with jaws. Most vertebrates are gnathostomes.

Golgi apparatus A eukaryotic organelle, consisting of stacks of flattened membranous sacs (cisternae), that functions in processing and sorting proteins and lipids destined to be secreted or directed to other organelles. Also called *Golgi complex*.

gonad An organ, such as a testis or an ovary, that produces reproductive cells.

gonadotropin-releasing hormone (GnRH) A peptide hormone, produced and secreted by the hypothalamus, that stimulates release of follicle-stimulating hormone (FSH) and luteinizing hormone (LH) from the anterior pituitary.

grade In taxonomy, a group of species that share some, but not all, of the descendants of a common ancestor. Also called a *paraphyletic group*.

Gram-negative Describing bacteria that look pink when treated with a Gram stain. These bacteria have a cell wall composed of a thin layer of peptidoglycan and an outer phospholipid layer. Compare with **Gram-positive**.

Gram-positive Describing bacteria that look purple when treated with a Gram stain. These bacteria have cell walls composed of a thick layer of peptidoglycan and no outer phospholipid later. Compare with **Gram-negative**.

Gram stain A dye that distinguishes the two general types of cell walls found in bacteria. Used to routinely classify bacteria as Gram-negative or Gram-positive.

granum (plural: grana) In chloroplasts, a stack of flattened, membrane-bound thylakoid discs where the light reactions of photosynthesis occur.

gravitropism The growth or movement of a plant in a particular direction in response to gravity.

grazing food chain The ecological network of herbivores and the predators and parasites that consume them.

great apes See **hominids**.

green algae A paraphyletic group of photosynthetic organisms that contain chloroplasts similar to those in green plants. Often classified as protists, green algae are the closest living relatives of land plants and form a monophyletic group with them.

greenhouse gas An atmospheric gas that absorbs and reflects infrared radiation, so that heat radiated from Earth is retained in the atmosphere instead of being lost to space.

gross primary productivity In an ecosystem, the total amount of carbon fixed by photosynthesis (or more

rarely, chemosynthesis), including that used for cellular respiration, over a given time period. Compare with **net primary productivity**.

ground meristem The middle layer of a young plant embryo. Gives rise to the ground tissue system.

ground tissue An embryonic tissue layer that gives rise to parenchyma, collenchyma, and sclerenchyma—tissues other than the epidermis and vascular tissue. Also called *ground tissue system*.

groundwater Any water below the land surface.

growth factor Any of a large number of signaling molecules that are secreted by certain cells and that stimulate other cells to grow, divide, or differentiate.

growth hormone (GH) A peptide hormone, produced and secreted by the mammalian anterior pituitary, that promotes lengthening of the long bones in children and muscle growth, tissue repair, and lactation in adults. Also called *somatotropin*.

guanosine triphosphate (GTP) A nucleotide consisting of guanine, a ribose sugar, and three phosphate groups. Can be hydrolyzed to release free energy. Commonly used in RNA synthesis and also functions in signal transduction in association with G proteins.

guard cell One of two specialized, crescent-shaped cells forming the border of a plant stoma. Guard cells can change shape to open or close the stoma. See also **stoma**.

gustation The perception of taste.

guttation Excretion of water droplets from plant leaves; visible in the early morning. Caused by root pressure.

gymnosperm A vascular plant that makes seeds but does not produce flowers. The gymnosperms include five lineages of green plants (cycads, ginkgoes, conifers, redwoods, and gnetophytes). Compare with **angiosperm**.

H⁺-ATPase See **proton pump**.

habitat degredation The reduction of the quality of a habitat.

habitat destruction Human-caused destruction of a natural habitat, replaced by an urban, suburban, or agricultural landscape.

habitat fragmentation The breakup of a large region of a habitat into many smaller regions, separated from others by a different type of habitat.

Hadley cell An atmospheric cycle of large-scale air movement in which warm equatorial air rises, moves north or south, and then descends at approximately 30° N or 30° S latitude.

hair cell A pressure-detecting sensory cell that has tiny "hairs" (stereocilia) jutting from its surface. Found in the inner ear, lateral line system, and ampullae of Lorenzini.

hairpin A secondary structure in RNA consisting of a stable loop formed by hydrogen bonding between purine and pyrimidine bases on the same strand.

Hamilton's rule The proposition that an allele for altruistic behavior will be favored by natural selection only if $Br C$, where B = the fitness benefit to the recipient, C = the fitness cost to the actor, and r = the coefficient of relatedness between recipient and actor.

haploid (1) Having one set of chromosomes (1*n* or *n* for short). (2) A cell or an individual organism with one set of chromosomes. Compare with **diploid**.

haploid number The number of different types of chromosomes in a cell. Symbolized as *n*.

Hardy–Weinberg principle A principle of population genetics stating that genotype frequencies in a large population do not change from generation to generation in the absence of evolutionary processes (e.g., mutation, gene flow, genetic drift, and selection), and nonrandom mating.

haustorium (plural: haustoria) Highly modified stem or root of a parasitic plant. The haustorium penetrates the tissues of a host and absorbs nutrients and water.

hearing The sensation of the wavelike changes in air pressure called sound.

heart A muscular pump that circulates blood throughout the body.

heart murmur A distinctive sound caused by backflow of blood through a defective heart valve.

heartwood The older xylem in the center of an older stem or root, containing protective compounds and no longer functioning in water transport.

heat Thermal energy that is transferred from an object at higher temperature to one at lower temperature.

heat of vaporization The energy required to vaporize 1 gram of a liquid into a gas.

heat-shock proteins Proteins that facilitate refolding of proteins that have been denatured by heat or other agents.

heavy chain The larger of the two types of polypeptide chains in an antibody or B-cell receptor molecule; composed of a variable (*V*) region, which contributes to the antigen-binding site, and a constant (*C*) region. Differences in heavy-chain constant regions determine the different classes of immunoglobulins (IgA, IgE, etc.). Compare with **light chain**.

helper T cell A CD4$^+$ effector T cell that secretes cytokines and in other ways promotes the activation of other lymphocytes. Activated by interacting with complementary class II MHC–peptide complexes on the surface of antigen-presenting cells such as dendritic cells.

heme A small molecule that binds to each of the four polypeptides that combine to form hemoglobin; contains an iron ion that can bind oxygen.

hemimetabolous metamorphosis A type of metamorphosis in which the animal increases in size from one stage to the next, but does not dramatically change its body form. Also called *incomplete metamorphosis*.

hemocoel A body cavity, present in arthropods and some mollusks, containing a pool of circulatory fluid (hemolymph) bathing the internal organs. Unlike a coelom, a hemocoel is not lined in mesoderm.

hemoglobin An oxygen-binding protein consisting of four polypeptide subunits, each containing an oxygen-binding heme group. The major oxygen carrier in mammalian blood.

hemolymph The circulatory fluid of animals with open circulatory systems (e.g., insects) in which the fluid is not confined to blood vessels.

herbaceous Referring to a plant that is not woody.

herbivore (adjective: herbivorous) An animal that eats primarily plants and rarely or never eats meat. Compare with **carnivore** and **omnivore**.

herbivory The practice of eating plant tissues.

heredity The transmission of traits from parents to offspring via genetic information.

heritable Referring to traits that can be transmitted from one generation to the next.

hermaphrodite An organism that produces both male and female gametes.

heterokaryotic Describing a cell or fungal mycelium containing two or more haploid nuclei that are genetically distinct.

heterospory (adjective: heterosporous) In seed plants, the production of two distinct types of spores: microspores, which become the male gametophyte, and megaspores, which become the female gametophyte. Compare with **homospory**.

heterotherm An animal whose body temperature varies markedly. Compare with **homeotherm**.

heterotroph Any organism that cannot synthesize reduced organic compounds from inorganic sources and that must obtain them from other organisms. Some bacteria, some archaea, and virtually all fungi and animals are heterotrophs. Also called *consumer*. Compare with **autotroph**.

heterozygote advantage A pattern of natural selection that favors heterozygous individuals compared with homozygotes. Tends to maintain genetic variation in a population, thus is a form of balancing selection.

heterozygous Having two different alleles of a gene.

hexose A monosaccharide (simple sugar) containing six carbon atoms.

hibernation An energy-conserving physiological state, marked by a decrease in metabolic rate, body temperature, and activity, that lasts for a prolonged period (weeks to months). Occurs in some animals in response to winter cold and scarcity of food. Compare with **torpor**.

hindbrain One of the three main regions of the vertebrate brain, responsible for balance and sometimes hearing; includes the cerebellum and medulla oblongata. Compare with **forebrain** and **midbrain**.

histamine A molecule released from mast cells during an inflammatory response that, at high concentrations, causes blood vessels to constrict to reduce blood loss from tissue damage.

histone One of several positively charged (basic) proteins associated with DNA in the chromatin of eukaryotic cells.

histone acetyltransferases (HATs) A class of eukaryotic enzymes that loosen chromatin structure by adding acetyl groups to histone proteins.

histone code The hypothesis that specific combinations of chemical modifications of histone proteins contain information that influences chromatin condensation and gene expression.

histone deacetylases (HDACs) A class of eukaryotic enzymes that condense chromatin by removing acetyl groups from histone proteins.

holoenzyme A multipart enzyme consisting of a core enzyme (containing the active site for catalysis) along with other required proteins.

holometabolous metamorphosis A type of metamorphosis in which the animal completely changes its form; includes a distinct larval stage. Also called *complete metamorphosis*.

homeobox A DNA sequence of about 180 base pairs that codes for a DNA binding motif called the homeodomain in the resulting protein. Genes containing a homeobox usually play a role in controlling development of organisms from fruit flies to humans.

homeostasis (adjective: homeostatic) The array of relatively stable chemical and physical conditions in an animal's cells, tissues, and organs. May be achieved by the body's passively matching the conditions of a stable external environment (conformational homeostasis) or by active physiological processes (regulatory homeostasis) triggered by variations in the external or internal environment.

homeotherm An animal that has a constant or relatively constant body temperature. Compare with **heterotherm**.

homeotic mutation A mutation that causes one body part to be substituted for another.

hominids Members of the family Hominidae, which includes humans and extinct related forms, chimpanzees, gorillas, and orangutans. Distinguished by large body size, no tail, and an exceptionally large brain. Also called *great apes*.

hominins Any extinct or living species of bipedal apes, such as *Australopithecus africanus*, *Homo erectus*, and *Homo sapiens*.

homologous See **homology**.

homologous chromosomes In a diploid organism, chromosomes that are similar in size, shape, and gene content. Also called *homologs*.

homology (adjective: homologous) Similarity among organisms of different species due to their inheritance from a common ancestor. Features that exhibit such similarity (e.g., DNA sequences, proteins, body parts) are said to be homologous. Compare with **homoplasy**.

homoplasy (adjective: homoplastic) Similarity among organisms of different species due to reasons other than common ancestry, such as convergent evolution. Features that exhibit such similarity (e.g., the wings of birds and bats) are said to be homoplastic, or convergent. Compare with **homology**.

homospory (adjective: homosporous) In seedless vascular plants, the production of just one type of spore. Compare with **heterospory**.

homozygous Having two identical alleles of a gene.

hormone Any of many different signaling molecules that circulate throughout the plant or animal body and can trigger characteristic responses in distant target cells at very low concentrations.

hormone-response element A specific sequence in DNA to which a steroid hormone–receptor complex can bind and affect gene transcription.

host An individual that has been invaded by an organism such as a parasite or a virus, or that provides habitat or resources to a commensal organism.

host cell A cell that has been invaded by a parasitic organism or a virus and provides an environment that is conducive to the pathogen's growth and reproduction.

***Hox* genes** A class of genes found in several animal phyla, including vertebrates, that are expressed in a distinctive pattern along the anterior–posterior axis in early embryos and control formation of specific structures. *Hox* genes code for transcription factors with a DNA-binding sequence called a homeobox.

human Any member of the genus *Homo,* which includes modern humans (*Homo sapiens*) and several extinct species.

human chorionic gonadotropin (hCG) A glycoprotein hormone produced by a human embryo and placenta from about week 3 to week 14 of pregnancy. Maintains the corpus luteum, which produces hormones that preserve the uterine lining.

Human Genome Project The multinational research project that sequenced the human genome.

human immunodeficiency virus (HIV) A retrovirus that causes acquired immune deficiency syndrome (AIDS) in humans.

humoral (immune) response The type of immune response that is mediated through the production and secretion of antibodies, complement proteins, and other soluble factors that eliminate extracellular pathogens. Compare with **cell-mediated (immune) response**.

humus The decayed organic matter in soils.

hybrid The offspring of parents from two different strains, populations, or species.

hybrid zone A geographic area where interbreeding occurs between two species, sometimes producing fertile hybrid offspring.

hydrocarbon An organic molecule that contains only hydrogen and carbon atoms.

hydrogen bond A weak interaction between two molecules or different parts of the same molecule resulting from the attraction between a hydrogen atom with a partial positive charge and another atom (usually O or N) with a partial negative charge. Compare with **covalent bond** and **ionic bond**.

hydrogen ion (H⁺) A single proton with a charge of 1+; typically, one that is dissolved in solution or that is being transferred from one atom to another in a chemical reaction.

hydrolysis A chemical reaction in which a molecule is split into smaller molecules by reacting with water. In biology, most hydrolysis reactions involve the splitting of polymers into monomers. Compare with **condensation reaction**.

hydrophilic Interacting readily with water. Hydrophilic compounds are typically polar compounds containing partially or fully charged atoms. Compare with **hydrophobic**.

hydrophobic Not readily interacting with water. Hydrophobic compounds are typically nonpolar compounds that lack partially or fully charged atoms. Compare with **hydrophilic**.

hydrophobic interactions Very weak interactions between nonpolar molecules, or nonpolar regions of the same molecule, when exposed to an aqueous solvent. The surrounding water molecules support these interactions by interacting with one another and encapsulating the nonpolar molecules.

hydroponic growth Growth of plants in liquid cultures instead of soil.

hydrostatic skeleton A system of body support involving a body wall in tension surrounding a fluid or soft tissue under compression.

hydroxide ion (OH⁻) An oxygen atom and a hydrogen atom joined by a single covalent bond and carrying a negative charge; formed by dissociation of water.

hygiene hypothesis The claim that immune disorders arise in individuals less likely to have been exposed to pathogens and parasites, especially in early childhood. Provides an explanation for the increased risk of allergies and autoimmune disease in countries with high levels of sanitation.

hyperosmotic Comparative term designating a solution that has a greater solute concentration, and therefore a lower water concentration, than another solution. Compare with **hyposmotic** and **isosmotic**.

hyperpolarization A change in membrane potential from its resting negative state to an even more negative state; a normal phase in an action potential. Compare with **depolarization**.

hypersensitive reaction An intense allergic response by cells that have been sensitized by previous exposure to an allergen.

hypersensitive response In plants, the rapid death of a cell that has been infected by a pathogen, thereby reducing the potential for infection to spread throughout a plant. Compare with **systemic acquired resistance**.

hypertension Abnormally high blood pressure.

hypertonic Comparative term designating a solution that, if inside a cell or vesicle, results in the uptake of water and swelling or even bursting of the membrane-bound structure. This solution has a greater solute concentration than the solution on the other side of the membrane. Used when the solute is unable to pass through the membrane. Compare with **hypotonic** and **isotonic**.

hypha (plural: hyphae) One of the long, branching strands of a fungal mycelium (the mesh-like body of a fungus). Also found in some protists.

hypocotyl The stem of a very young plant; the region between the cotyledon (embryonic leaf) and the radicle (embryonic root).

hyposmotic Comparative term designating a solution that has a lower solute concentration, and therefore a higher water concentration, than another solution. Compare with **hyperosmotic** and **isosmotic**.

hypothalamic–pituitary axis The functional interaction of the hypothalamus and the anterior pituitary gland, which are anatomically distinct but work together to regulate most of the other endocrine glands in the body.

hypothalamus A part of the brain that functions in maintaining the body's internal physiological state by regulating the autonomic nervous system, endocrine system, body temperature, water balance, and appetite.

hypothesis A testable statement that explains a phenomenon or a set of observations.

hypotonic Comparative term designating a solution that, if inside a cell or vesicle, results in the loss of water and shrinkage of the membrane-bound structure. This solution has a lower solute concentration than the solution on the other side of the membrane. Used when the solute is unable to pass through the membrane. Compare with **hypertonic** and **isotonic**.

immigration The migration of individuals into a particular population from other populations. Compare with **emigration**.

immune system The system whose primary function is to defend the host organism against pathogens. Includes several types of cells (e.g., leukocytes). In vertebrates, several organs are also involved where specialized cells develop or reside (e.g., lymph nodes and thymus).

immunity (adjective: immune) State of being protected against infection by disease-causing pathogens.

immunization The conferring of immunity to a particular disease by artificial means.

immunoglobulin (Ig) Any of the class of proteins that are structurally related to antibodies.

immunological memory The ability of the immune system to "remember" an antigen and mount a rapid, effective adaptive immune response to a pathogen encountered years or decades earlier. Based on the formation of memory lymphocytes.

impact hypothesis The hypothesis that a collision between the Earth and an asteroid caused the mass extinction at the K–P boundary, 65 million years ago.

imperfect flower A flower that contains male parts (stamens) *or* female parts (carpels) but not both. Compare with **perfect flower**.

implantation The process by which an embryo buries itself in the uterine or oviductal wall and forms a placenta. Occurs in mammals and some other viviparous vertebrates.

in situ hybridization A technique for detecting specific DNAs and mRNAs in cells and tissues by use of labeled complementary probes. Can be used to determine where and when particular genes are expressed in embryos.

inbreeding Mating between closely related individuals. Increases homozygosity of a population and often leads to a decline in the average fitness via selection (inbreeding depression).

inbreeding depression In inbred offspring, fitness declines due to deleterious recessive alleles that are homozygous, thus exposed to selection.

inclusive fitness The combination of (1) direct production of offspring (direct fitness) and (2) extra production of offspring by relatives in response to help provided by the individual in question (indirect fitness).

incomplete digestive tract A digestive tract that has just one opening.

incomplete dominance An inheritance pattern in which the heterozygote phenotype is in between the homozygote phenotypes.

incomplete metamorphosis See **hemimetabolous metamorphosis**.

independent assortment, principle of The concept that each pair of hereditary elements (alleles of the same gene) segregates (separates) independently of alleles of other genes during meiosis. One of Mendel's two principles of genetics.

indeterminate growth A pattern of growth in which an individual continues to increase its overall body size throughout its life.

induced fit Change in the shape of the active site of an enzyme, as the result of initial weak binding of a substrate, so that it binds substrate more tightly.

inducer A small molecule that triggers transcription of a specific gene, often by binding to and inactivating a repressor protein.

inducible defense A defensive trait that is manifested only in response to the presence of a consumer (predator or herbivore) or pathogen. Compare with **constitutive defense**.

infection thread An invagination of the plasma membrane of a root hair through which beneficial nitrogen-fixing bacteria enter the roots of their host plants (legumes).

inflammatory response An aspect of the innate immune response, seen in most cases of infection or tissue injury, in which the affected tissue becomes swollen, red, warm, and painful.

ingestion The act of bringing food into the digestive tract.

inhibition In ecological succession, the phenomenon in which early-arriving species make conditions less favorable for the establishment of certain later-arriving species. Compare with **facilitation** and **tolerance**.

inhibitory postsynaptic potential (IPSP) A change in membrane potential, usually hyperpolarization, at a neuron dendrite that makes an action potential less likely.

initiation (1) In an enzyme-catalyzed reaction, the stage during which enzymes orient reactants precisely as they bind at specific locations within the enzyme's active site. (2) In DNA transcription, the stage during which RNA polymerase and other proteins assemble at the promoter sequence and open the strands of DNA to start transcription. (3) In translation, the stage during which a complex consisting of initiation factor proteins, a ribosome, an mRNA, and an aminoacyl tRNA corresponding to the start codon is formed.

initiation factors A class of proteins that assist ribosomes in binding to a messenger RNA molecule to begin translation.

innate behavior Behavior that is inherited genetically, does not have to be learned, and is typical of a species.

innate immune response See **innate immunity**.

innate immunity A set of barriers to infection and generic defenses against broad types of pathogens. Produces an immediate response that involves many different leukocytes, which often activate an inflammatory response. Compare with **acquired immunity**.

inner cell mass (ICM) A cluster of cells, in the interior of a mammalian blastocyst, that eventually develop into the embryo. Contrast with **trophoblast**.

inner ear The innermost portion of the mammalian ear, consisting of a fluid-filled system of tubes that includes the cochlea (which receives sound vibrations from the middle ear) and the semicircular canals (which function in balance).

insects A terrestrial lineage of arthropods distinguished by three tagmata (head, thorax, abdomen), a single pair of antennae, and unbranched appendages.

insulin A peptide hormone produced by the pancreas in response to high levels of glucose (or amino acids) in blood. Enables cells to absorb glucose and coordinates synthesis of fats, proteins, and glycogen. Compare with **glucagon**.

integral membrane protein Any membrane protein that spans the entire lipid bilayer. Also called *transmembrane protein*. Compare with **peripheral membrane protein**.

integrated pest management In agriculture or forestry, systems for managing insects or other pests that include carefully controlled applications of toxins, introduction of species that prey on pests, planting schemes that reduce the chance of a severe pest outbreak, and other techniques.

integrator A component of an animal's nervous system that functions as part of a homeostatic system by evaluating sensory information and triggering appropriate responses. See **effector** and **sensor**.

integrin Any of a class of cell-surface proteins that bind to laminins and other proteins in the extracellular matrix, thus holding cells in place.

intercalated disc A type of specialized connection between adjacent heart muscle cells that contains gap junctions, allowing electrical signals to pass between the cells.

intermediate filament A long fiber, about 10 nm in diameter, composed of one of various proteins (e.g., keratins, lamins); one of the three types of cytoskeletal fibers. Used to form networks that help maintain cell shape and hold the nucleus in place. Compare with **actin filament** and **microtubule**.

intermediate host The host species in which a parasite reproduces asexually. Compare with **definitive host**.

intermediate muscle fiber Type of skeletal muscle fiber that is pink in color, generates ATP by both glycolysis and aerobic respiration, and has contractile properties that are intermediate between slow fibers and fast fibers. Also called fast oxidative fiber.

interneuron A neuron that passes signals from one neuron to another. Compare with **motor neuron** and **sensory neuron**.

internode The section of a plant stem between two nodes (sites where leaves attach).

interphase The portion of the cell cycle between one M phase and the next. Includes the G_1 phase, S phase, and G_2 phase.

intersexual selection The sexual selection of an individual of one gender for mating by an individual of the other gender (usually by female choice).

interspecific competition Competition between members of different species for the same limited resource. Compare with **intraspecific competition**.

interstitial fluid The plasma-like fluid found in the region (interstitial space) between cells.

intertidal zone The region between the low-tide and high-tide marks on a seashore.

intrasexual selection Competition among members of one gender for an opportunity to mate (usually male–male competition).

intraspecific competition Competition between members of the same species for the same limited resource. Compare with **interspecific competition**.

intrinsic rate of increase (r_{max}) The rate at which a population will grow under optimal conditions (i.e., when birthrates are as high as possible and death rates are as low as possible). Compare with **finite rate of increase**.

intron A region of a eukaryotic gene that is transcribed into RNA but is later removed. Compare with **exon**.

invasive species An exotic (nonnative) species that, upon introduction to a new area, spreads rapidly and competes successfully with native species.

inversion A mutation in which a segment of a chromosome breaks from the rest of the chromosome, flips, and rejoins in reversed orientation.

invertebrates A paraphyletic group composed of animals without a backbone; includes about 95 percent of all animal species. Compare with **vertebrates**.

involuntary muscle Muscle that contracts in response to stimulation by involuntary (parasympathetic or sympathetic), but not voluntary (somatic), neural stimulation.

ion An atom or a molecule that has lost or gained electrons and thus carries an electric charge, either positive (cation) or negative (anion), respectively.

ion channel A type of channel protein that allows certain ions to diffuse across a plasma membrane down an electrochemical gradient.

ionic bond A chemical bond that is formed when an electron is completely transferred from one atom to another so that the atoms remain associated due to their opposite electric charges. Compare with **covalent bond** and **hydrogen bond**.

iris A ring of pigmented muscle just behind the cornea in the vertebrate eye that contracts or expands to control the amount of light entering the eye through the pupil.

isosmotic Comparative term designating a solution that has the same solute concentration and water concentration as another solution. Compare with **hyperosmotic** and **hyposmotic**.

isotonic Comparative term designating a solution that, if inside a cell or vesicle, results in no net uptake or loss of water and thus no effect on the volume of the membrane-bound structure. This solution has the same solute concentration as the solution on the other side of the membrane. Compare with **hypertonic** and **hypotonic**.

isotope Any of several forms of an element that have the same number of protons but differ in the number of neutrons.

joint A place where two components (bones, cartilages, etc.) of a skeleton meet. May be movable (an articulated joint) or immovable (e.g., skull sutures).

juvenile An individual that has adult-like morphology but is not sexually mature.

juvenile hormone An insect hormone that prevents larvae from metamorphosing into adults.

karyogamy Fusion of two haploid nuclei to form a diploid nucleus. Occurs in many fungi, and in animals and plants during fertilization of gametes.

karyotype The distinctive appearance of all of the chromosomes in an individual, including the number of chromosomes and their length and banding patterns (after staining with dyes).

keystone species A species that has an exceptionally great impact on the other species in its ecosystem relative to its abundance.

kidney In terrestrial vertebrates, one of a paired organ situated at the back of the abdominal cavity that filters the blood, produces urine, and secretes several hormones.

kilocalorie (kcal) A unit of energy often used to measure the energy content of food. A kcal of energy raises 1 kg of water 1°C.

kin selection A form of natural selection that favors traits that increase survival or reproduction of an individual's kin at the expense of the individual.

kinesin A class of motor proteins that uses the chemical energy of ATP to "walk" toward the plus end of a microtubule. Used to transport vesicles, particles, organelles and chromosomes.

kinetic energy The energy of motion. Compare with **potential energy**.

kinetochore A protein complex at the centromere where microtubules attach to the chromosome. Contains motor proteins and microtubule-binding proteins that are involved in chromosome segregation during M phase.

kinetochore microtubules Microtubules in the spindle formed during mitosis and meiosis that are attached to the kinetochore on a chromosome.

kinocilium (plural: kinocilia) A single cilium that juts from the surface of many hair cells and functions in detection of sound or pressure.

knock-out allele A mutant allele that does not produce a functional product. Also called a *null allele* or *loss-of-function allele*.

Koch's postulates Four criteria used to determine whether a suspected infectious agent causes a particular disease.

labium majus (plural: labia majora) One of two outer folds of skin that surround the labia minora, clitoris, and vaginal opening of female mammals.

labium minus (plural: labia minora) One of two folds of skin inside the labia majora and surrounding the opening of the urethra and vagina.

labor The strong muscular contractions of the uterus that expel the fetus during birth.

***lac* operon** A set of three genes in *E. coli* that are transcribed into a single mRNA and required for lactose metabolism. Studies of the *lac* operon revealed many insights about gene regulation.

lactation (verb: lactate) Production of milk to feed offspring, from mammary glands of mammals.

lacteal A small lymphatic vessel extending into the center of a villus in the small intestine. Receives chylomicrons containing fat absorbed from food.

lactic acid fermentation Catabolic pathway in which pyruvate produced by glycolysis is converted to lactic acid in the absence of oxygen.

lagging strand In DNA replication, the strand of new DNA that is synthesized discontinuously in a series of short pieces that are later joined. Also called *discontinuous strand*. Compare with **leading strand**.

laminins An abundant protein in the extracellular matrix that binds to other ECM components and to integrins in plasma membranes; helps anchor cells in place. Predominantly found in the basal lamina; many subtypes function in different tissues.

large intestine The distal portion of the digestive tract, consisting of the cecum, colon, and rectum. Its primary function is to compact the wastes delivered from the small intestine and absorb enough water to form feces.

larva (plural: larvae) An immature stage of an animal species in which the immature and adult stages have different body forms.

late endosome A membrane-bound vesicle that arises from an early endosome, accepts lysosomal enzymes from the Golgi, and matures into a lysosome.

latency In viruses that infect animals, the ability to coexist with the host cell in a dormant state without producing new virions. The viral genetic material is replicated as the host cell replicates. Genetic material may or may not be integrated in the host genome, depending on the virus.

lateral bud A bud that forms at the nodes of a stem and may develop into a lateral (side) branch. Also called *axillary bud*.

lateral gene transfer Transfer of DNA between two different species.

lateral line system A pressure-sensitive sensory organ found in many aquatic vertebrates.

lateral root A plant root that extends horizontally from another root.

leaching Loss of nutrients from soil via percolating water.

leading strand In DNA replication, the strand of new DNA that is synthesized in one continuous piece. Also called *continuous strand*. Compare with **lagging strand**.

leaf The main photosynthetic organ of vascular plants.

leak channel Ion channel that allows ions to leak across the membrane of a neuron in its resting state.

learning An enduring change in an individual's behavior that results from specific experience(s).

leghemoglobin An iron-containing protein similar to hemoglobin. Found in infected cells of legume root nodules where it binds oxygen, preventing it from poisoning a bacterial enzyme needed for nitrogen fixation.

legumes Members of the pea plant family. Many form symbiotic associations with nitrogen-fixing bacteria in their roots.

lens A transparent structure that focuses incoming light onto a retina or other light-sensing apparatus of an eye.

lenticel Spongy segment in bark that allows gas exchange between cells in a woody stem and the atmosphere.

leptin A hormone produced and secreted by fat cells (adipocytes) that acts to stabilize fat-tissue mass partly by inhibiting appetite and increasing energy expenditure.

leukocyte Any of several types of blood cells, including neutrophils, macrophages, and lymphocytes, that reside in tissues and circulate in blood and lymph. Functions in tissue repair and defense against pathogens. Also called *white blood cell*.

lichen A symbiotic association of a fungus, often in the Ascomycota lineage, and a photosynthetic alga or cyanobacterium.

life cycle The sequence of developmental events and phases that occurs during the life span of an organism, from fertilization to offspring production.

life history The sequence of events in an individual's life from birth to reproduction to death, including how an individual allocates resources to growth, reproduction, and activities or structures that are related to survival.

life table A data set that summarizes the probability that an individual in a certain population will survive and reproduce in any given year over the course of its lifetime.

ligament Connective tissue that joins bones of an endoskeleton together.

ligand Any molecule that binds to a specific site on a receptor molecule.

ligand-gated channel An ion channel that opens or closes in response to binding by a certain molecule. Compare with **voltage-gated channel**.

light chain The smaller of the two types of polypeptide chains in an antibody or B-cell receptor molecule; composed of a variable (V) region, which contributes to the antigen-binding site, and a constant (C) region. Compare with **heavy chain**.

lignin A substance, found in the secondary cell walls of some plants, that is exceptionally stiff and strong; a complex polymer built from six-carbon rings. Most abundant in woody plant parts.

limiting nutrient Any essential nutrient whose scarcity in the environment significantly reduces growth and reproduction of organisms.

limnetic zone Open water (not near shore) that receives enough sunlight to support photosynthesis.

lineage See **monophyletic group**.

LINEs (long interspersed nuclear elements) The most abundant class of transposable elements in human genomes; can create copies of itself and insert them elsewhere in the genome. Compare with **SINEs**.

lingual lipase An enzyme produced by glands in the tongue. It breaks down fat molecules into fatty acids and monoglycerides.

linkage In genetics, a physical association between two genes because they are on the same chromosome; the inheritance patterns resulting from this association.

linkage map See **genetic map**.

lipid Any organic substance that does not dissolve in water, but dissolves well in nonpolar organic solvents. Lipids include fatty acids, fats, oils, waxes, steroids, and phospholipids.

lipid bilayer The basic structural element of all cellular membranes; consists of a two-layer sheet of phospholipid molecules with their hydrophobic tails oriented toward the inside and their hydrophilic heads toward the outside. Also called *phospholipid bilayer*.

littoral zone Shallow water near shore that receives enough sunlight to support photosynthesis. May be marine or freshwater; often flowering plants are present.

liver A large, complex organ of vertebrates that performs many functions, including storage of glycogen, processing and conversion of food and wastes, and production of bile.

lobe-finned fish Fish with fins supported by bony elements that extend down the length of the structure.

locomotion Movement of an organism under its own power.

locus (plural: loci) A gene's physical location on a chromosome.

logistic population growth The density-dependent decrease in growth rate as population size approaches the carrying capacity. Compare with **exponential population growth**.

long interspersed nuclear elements See **LINEs**.

long-day plant A plant that blooms in response to short nights (usually in late spring or early summer in the Northern Hemisphere). Compare with **day-neutral** and **short-day plant**.

loop of Henle In the vertebrate kidney, a long U-shaped loop in a nephron that extends into the medulla. Functions as a countercurrent exchanger to set up an osmotic gradient that allows reabsorption of water from the collecting duct.

loose connective tissue A type of connective tissue consisting of fibrous proteins in a soft matrix. Often functions as padding for organs.

lophophore A specialized feeding structure found in some lophotrochozoans and used in suspension (filter) feeding.

lophotrochozoans A major lineage of protostomes (Lophotrochozoa) that grow by extending the size of their skeletons rather than by molting. Many phyla have a specialized feeding structure (lophophore) and/or ciliated larvae (trochophore). Includes rotifers, flatworms, segmented worms, and mollusks. Compare with **ecdysozoans**.

loss-of-function allele See **knock-out allele**.

LUCA The *l*ast *u*niversal *c*ommon *a*ncestor of cells. This theoretical entity is proposed to be the product of chemical evolution and provided characteristics of life that are shared by all living organisms on Earth today.

lumen The interior space of any hollow structure (e.g., the rough ER) or organ (e.g., the stomach).

lung Any respiratory organ used for gas exchange between blood and air.

luteal phase The second major phase of a menstrual cycle, after ovulation, when the progesterone levels are high and the body is preparing for a possible pregnancy.

luteinizing hormone (LH) A peptide hormone, produced and secreted by the anterior pituitary, that stimulates estrogen production, ovulation, and formation of the corpus luteum in females and testosterone production in males.

lymph The mixture of fluid and white blood cells that circulates through the ducts and lymph nodes of the lymphatic system in vertebrates.

lymph node Any of many small, oval structures that lymph moves through in the lymphatic system. Filters the lymph and screens it for pathogens and other antigens. Major sites of lymphocyte activation.

lymphatic system In vertebrates, a body-wide network of thin-walled ducts (or vessels) and lymph nodes, separate from the circulatory system. Collects excess fluid from body tissues and returns it to the blood; also functions as part of the immune system.

lymphocyte A cell that circulates through the bloodstream and lymphatic system and is responsible for the development of adaptive immunity. In most cases belongs to one type of leukocyte—either B cells or T cells.

lysogeny In viruses that infect bacteria (bacteriophages), the ability to coexist with the host cell in a dormant state without producing new virions. The viral genetic material is integrated in the host chromosome and is replicated as the host cell replicates. Compare with **lytic cycle**.

lysosome A small, acidified organelle in an animal cell containing enzymes that catalyze hydrolysis reactions and can digest large molecules. Compare with **vacuole**.

lysozyme An enzyme that functions in innate immunity by digesting bacterial cell walls. Occurs in lysosomes of phagocytes and is secreted in saliva, tears, mucus, and egg white.

lytic cycle A type of viral replicative growth in which the production and release of virions kills the host cell. Compare with **lysogeny**.

M phase The phase of the cell cycle during which cell division occurs. Includes mitosis or meiosis and often cytokinesis.

M-phase-promoting factor (MPF) A complex of a cyclin and cyclin-dependent kinase that, when activated, phosphorylates a number of specific proteins needed to initiate mitosis in eukaryotic cells.

macromolecular machine A group of proteins that assemble to carry out a particular function.

macromolecule Any very large organic molecule, usually made up of smaller molecules (monomers) joined together into a polymer. The main biological macromolecules are proteins, nucleic acids, and polysaccharides.

macronutrient Any element (e.g., nitrogen) that is required in large quantities for normal growth, reproduction, and maintenance of a living organism. Compare with **micronutrient**.

macrophage A type of leukocyte in the innate immune system that participates in the inflammatory response by secreting cytokines and phagocytizing invading pathogens and apoptotic cells. Also serves as an antigen-presenting cell to activate lymphocytes.

MADS box A DNA sequence that codes for a DNA-binding motif in proteins; present in floral organ identity genes in plants. Functionally similar sequences are found in some fungal and animal genes.

magnetoreception A sensory system in which receptors are activated in response to magnetic fields.

magnetoreceptor A sensory cell or organ specialized for detecting magnetic fields.

major histocompatibility protein See **MHC protein**.

maladaptive Describing a trait that lowers fitness.

malaria A human disease caused by five species of the protist *Plasmodium* and passed to humans by mosquitoes.

malignant tumor A tumor that is actively growing and disrupting local tissues or is spreading to other organs. Cancer consists of one or more malignant tumors. Compare with **benign tumor**.

Malpighian tubules A major excretory organ of insects, consisting of blind-ended tubes that extend from the gut into the hemocoel. Filter hemolymph to form "pre-urine" and then send it to the hindgut for further processing.

mammals One of the two lineages of amniotes (vertebrates that produce amniotic eggs) distinguished by hair (or fur) and mammary glands. Includes the monotremes (platypuses), marsupials, and eutherians (placental mammals).

mammary glands Specialized exocrine glands that produce and secrete milk for nursing offspring. A diagnostic feature of mammals.

mandibles Any mouthpart used in chewing. In vertebrates, the lower jaw. In insects, crustaceans, and myriapods, the first pair of mouthparts.

mantle One of the three main parts of the mollusk body; the thick outer tissue that protects the visceral mass and may secrete a calcium carbonate shell.

marsh A wetland dominated by grasses and other nonwoody plants.

marsupials A lineage of mammals (Marsupiala) that nourish their young in an abdominal pouch after a very short period of development in the uterus.

mass extinction The extinction of a large number of diverse evolutionary groups during a relatively short period of geologic time (about 1 million years). May occur due to sudden and extraordinary environmental changes. Compare with **background extinction**.

mass feeder An animal that ingests chunks of food.

mass number The total number of protons and neutrons in an atom.

mast cell A type of leukocyte that is stationary (embedded in tissue) and helps trigger the inflammatory response, including secretion of histamine, to infection or injury. Particularly important in allergic responses and defense against parasites.

maternal chromosome A chromosome inherited from the mother.

mechanical advantage The ratio of force exerted on a load to the muscle force of the effort. A measure of the force efficiency of a mechanical system.

mechanoreception A sensory system in which receptors are activated in response to changes in pressure.

mechanoreceptor A sensory cell or organ specialized for detecting distortions caused by touch or pressure. One example is hair cells in the cochlea.

mediator Regulatory proteins in eukaryotes that form a physical link between regulatory transcription factors that are bound to DNA, the basal transcription complex, and RNA polymerase.

medium A liquid or solid that supports the growth of cells.

medulla The innermost part of an organ (e.g., kidney or adrenal gland).

medulla oblongata In vertebrates, a region of the brain stem that along with the cerebellum forms the hindbrain.

medusa (plural: medusae) The free-floating stage in the life cycle of some cnidarians (e.g., jellyfish). Compare with **polyp**.

megapascal (MPa) A unit of pressure (force per unit area) equivalent to 1 million pascals (Pa).

megasporangium (plural: megasporangia) In heterosporous species of plants, a spore-producing structure that produces megaspores, which go on to develop into female gametophytes.

megaspore In seed plants, a haploid (n) spore that is produced in a megasporangium by meiosis of a diploid ($2n$) megasporocyte; develops into a female gametophyte. Compare with **microspore**.

meiosis In sexually reproducing organisms, a special two-stage type of cell division in which one diploid ($2n$) parent cell produces haploid (n) cells (gametes); results in halving of the chromosome number. Also called *reduction division*.

meiosis I The first cell division of meiosis, in which synapsis and crossing over occur and homologous chromosomes are separated from each other, producing daughter cells with half as many chromosomes (each composed of two sister chromatids) as the parent cell.

meiosis II The second cell division of meiosis, in which sister chromatids are separated from each other. Similar to mitosis.

melatonin A hormone, produced by the pineal gland, that regulates sleep–wake cycles and seasonal reproduction in vertebrates.

membrane potential A difference in electric charge across a cell membrane; a form of potential energy. Also called *membrane voltage*.

memory Retention of learned information.

memory cells A type of lymphocyte responsible for maintaining immunity for years or decades after an infection. Descended from a B cell or T cell activated during a previous infection or vaccination.

meniscus (plural: menisci) The concave boundary layer formed at most air–water interfaces due to adhesion and surface tension.

menstrual cycle A female reproductive cycle seen in Old World monkeys and apes (including humans) in which the uterine lining is shed (menstruation) if no pregnancy occurs. Compare with **estrous cycle**.

menstruation The periodic shedding of the uterine lining through the vagina that occurs in females of Old World monkeys and apes, including humans.

meristem (adjective: meristematic) In plants, a group of undifferentiated cells, including stem cells, which can divide and develop into various adult tissues throughout the life of a plant. See also **apical meristem** and **ground meristem**.

mesoderm The middle of the three basic cell layers (germ layers) in most animal embryos; gives rise to muscles, bones, blood, and some internal organs (kidney, spleen, etc.). Compare with **ectoderm** and **endoderm**.

mesoglea A gelatinous material, containing scattered ectodermal cells, that is located between the ectoderm and endoderm of cnidarians (e.g., jellyfish, corals, and anemones).

mesophyll cell A type of cell, found near the surfaces of plant leaves, that is specialized for the light-dependent reactions of photosynthesis.

Mesozoic era The interval of geologic time, from 251 million to 65.5 million years ago, during which gymnosperms were the dominant plants and dinosaurs the dominant vertebrates. Ended with extinction of the dinosaurs (except birds).

messenger RNA (mRNA) An RNA molecule transcribed from DNA that carries information (in codons) that specifies the amino acid sequence of a polypeptide.

meta-analysis A comparative analysis of the results of many smaller, previously published studies.

metabolic pathway A linked series of biochemical reactions that build up or break down a particular molecule; the product of one reaction is the substrate of the next reaction.

metabolic rate The total energy use by all the cells of an individual. For aerobic organisms, often measured as the amount of oxygen consumed per hour.

metabolic water The water that is produced as a by-product of cellular respiration.

metagenomics The inventory of all the genes in a community or ecosystem by sequencing, analyzing, and comparing the genomes of the component organisms. Also called *environmental sequencing*.

metagenomics The inventory of all the genes in a community or ecosystem by sequencing, analyzing, and comparing the genomes of the component organisms. Sequencing of all or most of the genes present in an environment directly (also called *environmental sequencing*).

metallothioneins Small plant proteins that bind to and prevent excess metal ions from acting as toxins.

metamorphosis Transition from one developmental stage to another, such as from the larval to the adult form of an animal.

metaphase A stage in mitosis or meiosis during which chromosomes line up in the middle of the cell.

metaphase plate The plane along which chromosomes line up in the middle of the spindle during metaphase of mitosis or meiosis; not an actual structure.

metapopulation A population made up of many small, physically isolated populations connected by migration.

metastasis The spread of cancerous cells from their site of origin to distant sites in the body where they may establish additional tumors.

methanogen A prokaryote that produces methane (CH_4) as a by-product of cellular respiration.

methanotroph An organism (bacteria or archaea) that uses methane (CH_4) as its primary electron donor and source of carbon.

methyl salicylate (MeSA) A molecule that is hypothesized to function as a signal, transported among tissues, that triggers systematic acquired resistance in plants—a response to pathogen attack.

MHC protein Any of a large set of mammalian cell-surface glycoproteins involved in marking cells as self and in antigen presentation to T cells. Also called *MHC molecule*. Compare with **Class I** and **Class II MHC protein**.

microbe Any microscopic organism, including bacteria, archaea, and various tiny eukaryotes.

microbiology The field of study concerned with microscopic organisms.

microfibril Bundled strands of cellulose that serve as the fibrous component in plant cell walls.

microfilament See **actin filament**.

micronutrient Any element (e.g., iron, molybdenum, magnesium) that is required in very small quantities for normal growth, reproduction, and maintenance of a living organism. Compare with **macronutrient**.

micropyle The tiny pore in a plant ovule through which the pollen tube reaches the embryo sac.

microRNA (miRNA) A small, single-stranded RNA associated with proteins in an RNA-induced silencing complex (RISC). Processed from a longer premiRNA gene transcript. Can bind to complementary sequences in mRNA molecules, allowing the associated proteins of RISC to degrade the bound mRNA or inhibit its translation. See **RNA interference**.

microsatellite A noncoding stretch of eukaryotic DNA consisting of a repeating sequence 2 to 6 base pairs long. Also called *short tandem repeat* or *simple sequence repeat*.

microsporangium (plural: microsporangia) In heterosporous species of plants, a spore-producing structure that produces microspores, which go on to develop into male gametophytes.

microspore In seed plants, a haploid (n) spore that is produced in a microsporangium by meiosis of a diploid ($2n$) microsporocyte; develops into a male gametophyte. Compare with **megaspore**.

microtubule A long, tubular fiber, about 25 nm in diameter, formed by polymerization of tubulin protein dimers; one of the three types of cytoskeletal fibers. Involved in cell movement and transport of materials within the cell. Compare with **actin filament** and **intermediate filament**.

microtubule organizing center (MTOC) General term for any structure (e.g., centrosome and basal body) that organizes microtubules in cells.

microvilli (singular: microvillus) Tiny protrusions from the surface of an epithelial cell that increase the surface area for absorption of substances.

midbrain One of the three main regions of the vertebrate brain; includes sensory integrating and relay centers. Compare with **forebrain** and **hindbrain**.

middle ear The air-filled middle portion of the mammalian ear, which contains three small bones (ossicles) that transmit and amplify sound from the tympanic membrane to the inner ear. Is connected to the throat via the eustachian tube.

migration (1) In ecology, a seasonal movement of large numbers of organisms from one geographic location or habitat to another. (2) In population genetics, movement of individuals from one population to another.

millivolt (mV) A unit of voltage equal to 1/1000 of a volt.

mimicry A phenomenon in which one species has evolved (or learns) to look or sound like another species. See **Batesian mimicry** and **Müllerian mimicry**.

mineral One of various inorganic substances that are important components of enzyme cofactors or of structural materials in an organism.

mineralocorticoids A class of steroid hormones, produced and secreted by the adrenal cortex, that regulate electrolyte levels and the overall volume of body fluids. Aldosterone is the principal one in humans. Compare with **glucocorticoids**.

minisatellite A noncoding stretch of eukaryotic DNA consisting of a repeating sequence that is 6 to 100 base pairs long. Also called *variable number tandem repeat (VNTR)*.

mismatch repair The process by which mismatched base pairs in DNA are fixed.

missense mutation A point mutation (change in a single base pair) that changes one amino acid for another within the sequence of a protein.

mitochondrial matrix Central compartment of a mitochondrion, which is lined by the inner membrane; contains mitochondrial DNA, ribosomes, and the enzymes for pyruvate processing and the citric acid cycle.

mitochondrion (plural: mitochondria) A eukaryotic organelle that is bounded by a double membrane and is the site of aerobic respiration and ATP synthesis.

mitogen-activated protein kinases (MAPK) Enzymes that are involved in signal transduction pathways that often lead to the induction of cell replication. Different types are organized in a series, where one kinase activates another via phosphorylation. See also **phosphorylation cascade**.

mitosis In eukaryotic cells, the process of nuclear division that results in two daughter nuclei genetically identical with the parent nucleus. Subsequent cytokinesis (division of the cytoplasm) yields two daughter cells.

mode of transmission The type of inheritance observed as a trait is passed from parent to offspring. Some common types are autosomal recessive, autosomal dominant, and X-linked recessive.

model organism An organism selected for intensive scientific study based on features that make it easy to work with (e.g., body size, life span), in the hope that findings will apply to other species.

molarity A common unit of solute concentration equal to the number of moles of a dissolved solute in 1 liter of solution.

mole The amount of a substance that contains 6.022×10^{23} of its elemental entities (e.g., atoms, ions, or molecules). This number of molecules of a compound will have a mass equal to the molecular weight of that compound expressed in grams.

molecular chaperone A protein that facilitates the folding of newly synthesized proteins into their correct three-dimensional shape. Usually works by an ATP-dependent mechanism.

molecular formula A notation that indicates only the numbers and types of atoms in a molecule, such as H_2O for the water molecule. Compare with **structural formula**.

molecular weight The sum of the atomic weights of all of the atoms in a molecule; roughly, the total number of protons and neutrons in the molecule.

molecule A combination of two or more atoms held together by covalent bonds.

molting A method of body growth, used by ecdysozoans, that involves the shedding of an external protective cuticle or skeleton, expansion of the soft body, and growth of a new external layer.

monocot Any flowering plant (angiosperm) that has a single cotyledon (embryonic leaf) upon germination. Monocots form a monophyletic group. Also called a monocotyledonous plant. Compare with **dicot**.

monoecious Describing an angiosperm species that has both male and female reproductive structures on each plant. Compare with **dioecious**.

monohybrid cross A mating between two parents that are both heterozygous for one given gene.

monomer A small molecule that can covalently bind to other similar molecules to form a larger macromolecule. Compare with **polymer**.

monophyletic group An evolutionary unit that includes an ancestral population and all of its descendants but no others. Also called a *clade* or *lineage*. Compare with **paraphyletic group** and **polyphyletic group**.

monosaccharide A molecule that has the molecular formula $(CH_2O)_n$ and cannot be hydrolyzed to form any smaller carbohydrates. Also called *simple sugar*. Compare with **oligosaccharide** and **polysaccharide**.

monosomy The state of having only one copy of a particular type of chromosome in an otherwise diploid cell.

monotremes A lineage of mammals (Monotremata) that lay eggs and then nourish the young with milk. Includes just five living species: the platypus and four species of echidna, all with leathery beaks or bills.

morphogen A molecule that exists in a concentration gradient and provides spatial information to embryonic cells.

morphology The shape and appearance of an organism's body and its component parts.

morphospecies concept The definition of a species as a population or group of populations that have measurably different anatomical features from other groups. Also called *morphological species concept*. Compare with **biological species concept** and **phylogenetic species concept**.

motor neuron A nerve cell that carries signals from the central nervous system (brain and spinal cord) to an effector, such as a muscle or gland. Compare with **interneuron** and **sensory neuron**.

motor protein A class of proteins whose major function is to convert the chemical energy of ATP into motion. Includes dynein, kinesin, and myosin.

MPF See **M-phase-promoting factor**.

mRNA See **messenger RNA**.

mucosal-associated lymphoid tissue (MALT) Collective term for lymphocytes and other leukocytes associated with skin cells and mucus-secreting epithelial tissues in the gut and respiratory tract. Plays an important role in preventing entry of pathogens into the body.

mucous cell A type of cell found in the epithelial layer of the stomach; responsible for secreting mucus into the stomach.

mucus (adjective: mucous) A slimy mixture of glycoproteins (called mucins) and water that is secreted in many animal organs for lubrication. Serves as a barrier to protect surfaces from infection.

Müllerian inhibitory substance A peptide hormone, secreted by the embryonic testis, that causes regression (withering away) of the female reproductive ducts.

Müllerian mimicry A type of mimicry in which two (or more) harmful species resemble each other. Compare with **Batesian mimicry**.

multicellularity The state of being composed of many cells that adhere to each other and do not all express the same genes, with the result that some cells have specialized functions.

multiple allelism The existence of more than two alleles of the same gene.

multiple fruit A fruit (e.g., pineapple) that develops from many separate flowers and thus many carpels. Compare with **aggregate** and **simple fruit**.

multiple sclerosis (MS) A human autoimmune disease in which the immune system attacks the myelin sheaths that insulate axons of neurons.

muscle fiber A single muscle cell.

muscle tissue An animal tissue consisting of bundles of long, thin, contractile cells (muscle fibers). Functions primarily in movement.

mutagen Any physical or chemical agent that increases the rate of mutation.

mutant An individual that carries a mutation, particularly a new or rare mutation.

mutation Any change in the hereditary material of an organism (DNA in most organisms, RNA in some viruses). The only source of new alleles in populations.

mutualism (adjective: mutualistic) A symbiotic relationship between two organisms (mutualists) that benefits both. Compare with **commensalism** and **parasitism**.

mutualist Organism that is a participant and partner in a mutualistic relationship. See **mutualism**.

mycelium (plural: mycelia) A mass of underground filaments (hyphae) that form the body of a fungus. Also found in some protists and bacteria.

mycorrhiza (plural: mycorrhizae) A mutualistic association between certain fungi and the roots of most vascular plants, sometimes visible as nodules or nets in or around plant roots.

myelin sheath Multiple layers of myelin, derived from the cell membranes of certain glial cells, wrapped around the axon of a neuron and providing electrical insulation.

myocardial infarction Death of cardiac muscle cells when deprived of oxygen.

myoD A transcription factor that is a master regulator of muscle cell differentiation (short for "*myoblast determination*").

myofibril Long, slender structure composed of contractile proteins organized into repeating units (sarcomeres) in vertebrate heart muscle and skeletal muscle.

myosin Any one of a class of motor proteins that use the chemical energy of ATP to move along actin filaments in muscle contraction, cytokinesis, and vesicle transport.

myriapods A lineage of arthropods with long segmented trunks, each segment bearing one or two pairs of legs. Includes millipedes and centipedes.

NAD$^+$/NADH Oxidized and reduced forms, respectively, of nicotinamide adenine dinucleotide. A nonprotein electron carrier that functions in many of the redox reactions of metabolism.

NADP$^+$/NADPH Oxidized and reduced forms, respectively, of nicotinamide adenine dinucleotide phosphate. A nonprotein electron carrier that is reduced during the light-dependent reactions in photosynthesis and extensively used in biosynthetic reactions.

natural experiment A situation in which a natural change in conditions enables comparisons of groups, rather than a manipulation of conditions by researchers.

natural selection The process by which individuals with certain heritable traits tend to produce more surviving offspring than do individuals without those traits, often leading to a change in the genetic makeup of the population. A major mechanism of evolution.

nauplius A distinct planktonic larval stage seen in many crustaceans.

Neanderthal A recently extinct European species of hominin, *Homo neanderthalensis*, closely related to but distinct from modern humans.

nectar The sugary fluid produced by flowers to attract and reward pollinating animals.

nectary A nectar-producing structure in a flower.

negative control Of genes, when a regulatory protein shuts down expression by binding to DNA on or near the gene.

negative feedback A self-limiting, corrective response in which a deviation in some variable (e.g., concentration of some compound) triggers responses aimed at returning the variable to normal. Represents a means of maintaining homeostasis. Compare with **positive feedback**.

negative pressure ventilation Ventilation of the lungs by expanding the rib cage so as to "pull" air into the lungs. Compare with **positive pressure ventilation**.

negative-sense RNA virus An ssRNA virus whose genome contains sequences complementary to those in the mRNA required to produce viral proteins. Compare with **ambisense virus** and **positive-sense virus**.

nematodes See **roundworms**.

nephron One of many tiny tubules inside the kidney that function in the formation of urine.

neritic zone Shallow marine waters beyond the intertidal zone, extending down to about 200 meters, where the continental shelf ends.

nerve A long, tough strand of nervous tissue, typically containing thousands of axons, wrapped in connective tissue; carries impulses between the central nervous system and some other part of the body.

nerve cord In chordate animals, a hollow bundle of nerves extending from the brain along the dorsal (back) side of the animal, with cerebrospinal fluid inside a central channel. One of the defining features of chordates.

nerve net A nervous system in which neurons are diffuse instead of being clustered into large ganglia or tracts; found in cnidarians and ctenophores.

nervous tissue An animal tissue consisting of nerve cells (neurons) and various supporting cells.

net primary productivity (NPP) In an ecosystem, the total amount of carbon fixed by photosynthesis over a given time period minus the amount oxidized during cellular respiration. Compare with **gross primary productivity**.

net reproductive rate (R_0) The growth rate of a population per generation; equivalent to the average number of female offspring that each female produces over her lifetime.

neural tube A folded tube of ectoderm that forms along the dorsal side of a young vertebrate embryo; gives rise to the brain and spinal cord.

neuroendocrine Referring to nerve cells (neurons) that release hormones into the blood or to such hormones themselves.

neurogenesis The birth of new neurons from central nervous system stem cells.

neurohormones Hormones produced by neurons.

neuron A cell that is specialized for the transmission of nerve impulses. Typically has dendrites, a cell body, and a long axon that forms synapses with other neurons. Also called *nerve cell*.

neurosecretory cell A nerve cell (neuron) that produces and secretes hormones into the bloodstream. Principally found in the hypothalamus. Also called *neuroendocrine cell*.

neurotoxin Any substance that specifically destroys or blocks the normal functioning of neurons.

neurotransmitter A molecule that transmits signals from one neuron to another or from a neuron to a muscle or gland. Examples are acetylcholine, dopamine, serotonin, and norepinephrine.

neutral In genetics, referring to any mutation or mutant allele that has no effect on an individual's fitness.

neutrophil A type of leukocyte, capable of moving through body tissues, that engulfs and digests pathogens and other foreign particles; also secretes various compounds that attack bacteria and fungi.

niche The range of resources that a species can use and the range of conditions that it can tolerate. More broadly, the role that species plays in its ecosystem.

niche differentiation The evolutionary change in resource use by competing species that occurs as the result of character displacement.

nicotinamide adenine dinucleotide See **NAD$^+$/NADH**.

nitrogen cycle, global The movement of nitrogen among terrestrial ecosystems, the oceans, and the atmosphere.

nitrogen fixation The incorporation of atmospheric nitrogen (N_2) into ammonia (NH_3), which can be used to make many organic compounds. Occurs in only a few lineages of bacteria and archaea.

nociceptor A sensory cell or organ specialized to detect tissue damage, usually producing the sensation of pain.

Nod factors Molecules produced by nitrogen-fixing bacteria that help them recognize and bind to roots of legumes.

node (1) In animals, any small thickening (e.g., a lymph node). (2) In plants, the part of a stem where leaves or leaf buds are attached. (3) In a phylogenetic tree, the point where two branches diverge, representing the point in time when an ancestral group split into two or more descendant groups. Also called *fork*.

node of Ranvier One of the unmyelinated sections that occurs periodically along a neuron's axon and serves as a site where an action potential can be regenerated.

nodule Globular structure on roots of legume plants that contain symbiotic nitrogen-fixing bacteria.

noncyclic electron flow Path of electron flow in which electrons pass from photosystem II, through an electron transport chain, to photosystem I, and ultimately to NADP$^+$ during the light-dependent reactions of photosynthesis. See also **Z scheme**.

nondisjunction An error that can occur during meiosis or mitosis, in which one daughter cell receives two copies of a particular chromosome and the other daughter cell receives none.

nonpolar covalent bond A covalent bond in which electrons are equally shared between two atoms of the same or similar electronegativity. Compare with **polar covalent bond**.

nonsense mutation A point mutation (change in a single base pair) that converts an amino-acid-specifying codon into a stop codon.

non-sister chromatids The chromatids of a particular type of chromosome (after replication) with respect to the chromatids of its homologous chromosome. Crossing over occurs between non-sister chromatids. Compare with **sister chromatids**.

non-template strand The strand of DNA that is not transcribed during synthesis of RNA. Its sequence corresponds to that of the mRNA produced from the other strand. Also called *coding strand*.

non-vascular plants A paraphyletic group of land plants that lack vascular tissue and reproduce using spores. The non-vascular plants include three lineages of green plants (liverworts, mosses, and hornworts). These lineages are sometimes called *bryophytes*.

norepinephrine A catecholamine used as a neurotransmitter in the sympathetic nervous system. Also produced by the adrenal medulla and functions as a hormone that triggers rapid responses relating to the fight-or-flight response.

notochord A supportive but flexible rod that occurs in the back of a chordate embryo, ventral to the developing spinal cord. Replaced by vertebrae in most adult vertebrates. A defining feature of chordates.

nuclear envelope The double-layered membrane enclosing the nucleus of a eukaryotic cell.

nuclear lamina A lattice-like sheet of fibrous nuclear lamins, which are one type of intermediate filament. Lines the inner membrane of the nuclear envelope, stiffening the envelope and helping to organize the chromosomes.

nuclear lamins Intermediate filaments that make up the nuclear lamina layer—a lattice-like layer inside the nuclear envelope that stiffens the structure.

nuclear localization signal (NLS) A short amino acid sequence that marks a protein for delivery to the nucleus.

nuclear pore An opening in the nuclear envelope that connects the inside of the nucleus with the cytoplasm and through which molecules such as mRNA and some proteins can pass.

nuclear pore complex A large complex of dozens of proteins lining a nuclear pore, defining its shape and regulating transport through the pore.

nuclease Any enzyme that can break down RNA or DNA molecules.

nucleic acid A macromolecule composed of nucleotide monomers. Generally used by cells to store or transmit hereditary information. Includes ribonucleic acid and deoxyribonucleic acid.

nucleoid In prokaryotic cells, a dense, centrally located region that contains DNA but is not surrounded by a membrane.

nucleolus In eukaryotic cells, a specialized structure in the nucleus where ribosomal RNA processing occurs and ribosomal subunits are assembled.

nucleosome A repeating, bead-like unit of eukaryotic chromatin, consisting of about 200 nucleotides of DNA wrapped twice around eight histone proteins.

nucleotide excision repair The process of removing a damaged region in one strand of DNA and replacing it with the correct sequence using the undamaged strand as a template.

nucleotide A molecule consisting of a five-carbon sugar (ribose or deoxyribose), a phosphate group, and one of several nitrogen-containing bases. DNA and RNA are polymers of nucleotides containing deoxyribose (deoxyribonucleotides) and ribose (ribonucleotides), respectively. Equivalent to a nucleoside plus one phosphate group.

nucleus (1) The center of an atom, containing protons and neutrons. (2) In eukaryotic cells, the large organelle containing the chromosomes and surrounded by a double membrane. (3) A discrete clump of neuron cell bodies in the brain, usually sharing a distinct function.

null allele See **knock-out allele**.

null hypothesis A hypothesis that specifies what the results of an experiment will be if the main hypothesis being tested is wrong. Often states that there will be no difference between experimental groups.

nutrient Any substance that an organism requires for normal growth, maintenance, or reproduction.

occipital lobe In the vertebrate brain, one of the four major areas in the cerebrum.

oceanic zone The waters of the open ocean beyond the continental shelf.

odorant Any volatile molecule that conveys information about food or the environment.

oil An unsaturated fat that is liquid at room temperature.

Okazaki fragment Short segment of DNA produced during replication of the lagging strand template. Many Okazaki fragments make up the lagging strand in newly synthesized DNA.

olfaction The perception of odors.

olfactory bulb A bulb-shaped projection of the brain just above the nose. Receives and interprets odor information from the nose.

oligodendrocyte A type of glial cell that wraps around axons of some neurons in the central nervous system, forming a myelin sheath that provides electrical insulation. Compare with **Schwann cell**.

oligopeptide A chain composed of fewer than 50 amino acids linked together by peptide bonds. Often referred to simply as *peptide*.

oligosaccharide A linear or branched polymer consisting of less than 50 monosaccharides joined by glycosidic linkages. Compare with **monosaccharide** and **polysaccharide**.

ommatidium (plural: ommatidia) A light-sensing column in an arthropod's compound eye.

omnivore (adjective: omnivorous) An animal whose diet regularly includes both meat and plants. Compare with **carnivore** and **herbivore**.

oncogene Any gene whose protein product stimulates cell division at all times and thus promotes cancer development. Often is a mutated form of a gene involved in regulating the cell cycle. See **proto-oncogene**.

one-gene, one-enzyme hypothesis The hypothesis that each gene is responsible for making one enzyme. This hypothesis has expanded to include genes that produce proteins other than enzymes or that produce RNAs as final products.

oogenesis The production of egg cells (ova).

oogonium (plural: oogonia) In an ovary, any of the diploid cells that can divide by mitosis to create primary oocytes (which can undergo meiosis) and more oogonia.

open circulatory system A circulatory system in which the circulating fluid (hemolymph) is not confined to blood vessels. Compare with **closed circulatory system**.

open reading frame (ORF) Any DNA sequence, ranging in length from several hundred to thousands of base pairs long, that is flanked by a start codon and a stop codon. ORFs identified by computer analysis of DNA may be functional genes, especially if they have other features characteristic of genes (e.g., promoter sequence).

operator In prokaryotic DNA, a binding site for a repressor protein; located near the start of an operon.

operculum The stiff flap of tissue that covers the gills of teleost fishes.

operon A region of prokaryotic DNA that codes for a series of functionally related genes and is transcribed from a single promoter into one mRNA.

opsin A transmembrane protein that is covalently linked to retinal, the light-detecting pigment in rod and cone cells.

optic nerve A bundle of neurons that runs from the eye to the brain.

optimal foraging The concept that animals forage in a way that maximizes the amount of usable energy they take in, given the costs of finding and ingesting their food and the risk of being eaten while they're at it.

orbital The region of space around an atomic nucleus in which an electron is present most of the time.

organ A group of tissues organized into a functional and structural unit.

organ system Groups of tissues and organs that work together to perform a function.

organelle Any discrete, membrane-bound structure within a cell (e.g., mitochondrion) that has a characteristic structure and function.

organic For a compound, containing carbon and hydrogen and usually containing carbon–carbon bonds. Organic compounds are widely used by living organisms.

organism Any living entity that contains one or more cells.

organogenesis A stage of embryonic development that follows gastrulation and that creates organs from the three germ layers.

origin of replication The site on a chromosome at which DNA replication begins.

osmoconformer An animal that does not actively regulate the osmolarity of its tissues but conforms to the osmolarity of the surrounding environment.

osmolarity The concentration of dissolved substances in a solution, measured in osmoles per liter.

osmoregulation The process by which a living organism controls the concentration of water and salts in its body.

osmoregulator An animal that actively regulates the osmolarity of its tissues.

osmosis Diffusion of water across a selectively permeable membrane from a region of low solute concentration (high water concentration) to a region of high solute concentration (low water concentration).

ossicles, ear In mammals, three bones found in the middle ear that function in transferring and amplifying sound from the outer ear to the inner ear.

ouabain A plant toxin that poisons the sodium–potassium pumps of animals.

out-of-Africa hypothesis The hypothesis that modern humans (*Homo sapiens*) evolved in Africa and spread to other continents, replacing other *Homo* species without interbreeding with them.

outcrossing Reproduction by fusion of the gametes of different individuals, rather than by self-fertilization.

outer ear The outermost portion of the mammalian ear, consisting of the pinna (ear flap) and the ear canal. Funnels sound to the tympanic membrane.

outgroup A taxon that is closely related to a particular monophyletic group but is not part of it.

oval window A membrane separating the fluid-filled cochlea from the air-filled middle ear; sound vibrations pass through it from the middle ear to the inner ear in mammals.

ovary The egg-producing organ of a female animal, or the fruit- and seed-producing structure in the female part of a flower.

overexploitation Unsustainable removal of wildlife from the natural environment for use by humans.

oviduct See **fallopian tube**.

oviparous In animals, producing eggs that are laid outside the body where they develop and hatch. Compare with **ovoviviparous** and **viviparous**.

ovoviviparous In animals, producing eggs that are retained inside the body until they are ready to hatch. Compare with **oviparous** and **viviparous**.

ovulation The release of an egg from an ovary of a female vertebrate. In humans, an ovarian follicle releases an egg at the end of the follicular phase of the menstrual cycle.

ovule In flowering plants, the structure inside an ovary that contains the female gametophyte and eventually (if fertilized) becomes a seed.

ovum (plural: ova) See **egg**.

oxidation The loss of electrons from an atom or molecule during a redox reaction, either by donation of an electron to another atom or molecule, or by the shared electrons in covalent bonds moving farther from the atomic nucleus.

oxidative phosphorylation Production of ATP molecules by ATP synthase using the proton gradient established via redox reactions of an electron transport chain.

oxygen–hemoglobin equilibrium curve The graphed depiction of the percentage of hemoglobin in the blood that is bound to oxygen at various partial pressures of oxygen.

oxygenic Referring to any process or reaction that produces oxygen. Photosynthesis in plants, algae, and cyanobacteria, which involves photosystem II, is oxygenic. Compare with **anoxygenic**.

oxytocin A peptide hormone, secreted by the posterior pituitary, that triggers labor and milk production in females and that stimulates pair bonding, parental care, and affiliative behavior in both sexes.

p53 A tumor-suppressor protein (molecular weight of 53 kilodaltons) that responds to DNA damage by stopping the cell cycle, turning on DNA repair machinery, and, if necessary, triggering apoptosis. Encoded by the *p53* gene.

pacemaker cell Any of a group of specialized cardiac muscle cells in the sinoatrial (SA) node of the vertebrate heart that have an inherent rhythm and can generate an electrical impulse that spreads to other heart cells.

paleontologists Scientists who study the fossil record and the history of life.

Paleozoic era The interval of geologic time, from 542 million to 251 million years ago, during which fungi, land plants, and animals first appeared and diversified. Began with the Cambrian explosion and ended with the extinction of many invertebrates and vertebrates at the end of the Permian period.

pancreas A large gland in vertebrates that has both exocrine and endocrine functions. Secretes digestive enzymes into a duct connected to the intestine and secretes several hormones (notably, insulin and glucagon) into the bloodstream.

pancreatic amylase An enzyme produced by the pancreas that breaks down glucose chains by catalyzing hydrolysis of the glycosidic linkages between the glucose residues.

pancreatic lipase An enzyme that is produced in the pancreas and acts in the small intestine to break bonds in complex fats, releasing small lipids.

pandemic The spread of an infectious disease in a short time period over a wide geographic area and affecting a very high proportion of the population. Compare with **epidemic**.

parabiosis An experimental technique for determining whether a certain physiological phenomenon is regulated by a hormone; consists of surgically uniting two individuals so that hormones can pass between them.

paracrine Relating to a chemical signal that is released by one cell and affects neighboring cells.

paraphyletic group A group that includes an ancestral population and *some* but not all of its descendants. Compare with **monophyletic group**.

parapodia (singular: parapodium) Appendages found in some annelids from which bristle-like structures (chaetae) extend.

parasite An organism that lives on a host species (ectoparasite) or in a host species (endoparasite) and that damages its host.

parasitism (adjective: parasitic) A symbiotic relationship between two organisms that is beneficial to one organism (the parasite) but detrimental to the other (the host). Compare with **commensalism** and **mutualism**.

parasitoid An organism that has a parasitic larval stage and a free-living adult stage. Most parasitoids are insects that lay eggs in the bodies of other insects.

parasympathetic nervous system The part of the autonomic nervous system that stimulates responses for conserving or restoring energy, such as reduced heart rate and increased digestion. Compare with **sympathetic nervous system**.

parenchyma cell In plants, a general type of cell with a relatively thin primary cell wall. These cells, found in leaves, the centers of stems and roots, and fruits, are involved in photosynthesis, storage, and transport. Compare with **collenchyma cell** and **sclerenchyma cell**.

parental care Any action by which an animal expends energy or assumes risks to benefit its offspring (e.g., nest building, feeding of young, defense).

parental generation The adults used in the first experimental cross of a breeding experiment.

parental strand A strand of DNA that is used as a template during DNA synthesis.

parietal cell A cell in the stomach lining that secretes hydrochloric acid.

parietal lobe In the vertebrate brain, one of the four major areas in the cerebrum.

parsimony The logical principle that the most likely explanation of a phenomenon is the most economical or simplest. When applied to comparison of alternative phylogenetic trees, it suggests that the one requiring the fewest evolutionary changes is most likely to be correct.

parthenogenesis Development of offspring from unfertilized eggs; a type of asexual reproduction.

partial pressure The pressure of one particular gas in a mixture of gases; the contribution of that gas to the overall pressure.

particulate inheritance The observation that genes from two parents do not blend together in offspring, but instead remain separate or particle-like.

pascal (Pa) A unit of pressure (force per unit area).

passive transport Diffusion of a substance across a membrane. When this event occurs with the assistance of membrane proteins, it is called *facilitated diffusion*.

patch clamping A technique for studying the electrical currents that flow through individual ion channels by sucking a tiny patch of membrane to the hollow tip of a microelectrode.

paternal chromosome A chromosome inherited from the father.

pathogen (adjective: pathogenic) Any entity capable of causing disease, such as a microbe, virus, or prion.

pattern formation The series of events that determines the spatial organization of an entire embryo or parts of an embryo, for example, setting the major body axes early in development.

pattern-recognition receptor On leukocytes, a class of membrane proteins that bind to molecules commonly associated with foreign cells and viruses and signal responses against broad types of pathogens. Part of the innate immune response.

peat Semi-decayed organic matter that accumulates in moist, low-oxygen environments such as *Sphagnum* (moss) bogs.

pectin A gelatinous polysaccharide found in the primary cell wall of plant cells. Attracts and holds water, forming a gel that resists compression forces and helps keep the cell wall moist.

pedigree A family tree of parents and offspring, showing inheritance of particular traits of interest.

penis The copulatory organ of male mammals, used to insert sperm into a female.

pentose A monosaccharide (simple sugar) containing five carbon atoms.

PEP carboxylase An enzyme that catalyzes addition of CO_2 to phosphoenolpyruvate, a three-carbon compound, forming a four-carbon organic acid. See also **C_4 pathway** and **crassulacean acid metabolism (CAM)**.

pepsin A protein-digesting enzyme present in the stomach.

peptide See **oligopeptide**.

peptide bond The covalent bond formed by a condensation reaction between two amino acids; links the residues in peptides and proteins.

peptidoglycan A complex structural polysaccharide found in bacterial cell walls.

perennial Describing a plant whose life cycle normally lasts for more than one year. Compare with **annual**.

perfect flower A flower that contains both male parts (stamens) and female parts (carpels). Compare with **imperfect flower**.

perforation In plants, a small hole in the primary and secondary cell walls of vessel elements that allows passage of water.

pericarp The part of a fruit, formed from the ovary wall, that surrounds the seeds and protects them. Corresponds to the flesh of most edible fruits and the hard shells of most nuts.

pericycle In plant roots, a layer of cells just inside the endodermis that give rise to lateral roots.

peripheral membrane protein Any membrane protein that does not span the entire lipid bilayer and associates with only one side of the bilayer. Compare with **integral membrane protein**.

peripheral nervous system (PNS) All the components of the nervous system that are outside the central nervous system (the brain and spinal cord). Includes the somatic nervous system and the autonomic nervous system.

peristalsis Rhythmic waves of muscular contraction. In the digestive tract, pushes food along. In animals with hydrostatic skeletons, enables crawling.

permafrost A permanently frozen layer of icy soil found in most tundra and some taiga.

permeability The tendency of a structure, such as a membrane, to allow a given substance to diffuse across it.

peroxisome An organelle found in most eukaryotic cells that contains enzymes for oxidizing fatty acids and other compounds, including many toxins, rendering them harmless. See **glyoxysome**.

petal Any of the leaflike organs arranged around the reproductive organs of a flower. Often colored and scented to attract pollinators.

petiole The stalk of a leaf.

pH A measure of the concentration of protons in a solution and thus of acidity or alkalinity. Defined as the negative of the base-10 logarithm of the proton concentration: $pH = -\log[H^+]$.

phagocytosis Uptake by a cell of small particles or cells by invagination and pinching off of the plasma membrane to form small, membrane-bound vesicles; one type of **endocytosis**.

pharyngeal gill slits A set of parallel openings from the throat to the outside that function in both feeding and gas exchange. A diagnostic trait of chordates.

pharyngeal jaw A secondary jaw in the back of the throat; found in some fishes, it aids in food processing. Derived from modified gill arches.

phenology The timing of events during the year, in environments where seasonal changes occur.

phenotype The detectable traits of an individual. Compare with **genotype**.

phenotypic plasticity Within-species variation in phenotype that is due to differences in environmental conditions. Occurs more commonly in plants than animals.

pheophytin The molecule in photosystem II that accepts excited electrons from the reaction center chlorophyll and passes them to an electron transport chain.

pheromone A chemical signal, released by an individual into the external environment, that can trigger changes in the behavior or physiology or both of another member of the same species.

phloem A plant vascular tissue that conducts sugars between roots and shoots; contains sieve-tube elements and companion cells. Primary phloem develops from the procambium of apical meristems; secondary phloem, from the vascular cambium. Compare with **xylem**.

phosphatase An enzyme that removes phosphate groups from proteins or other molecules. Phosphatases are often used in the inactivation of signaling pathways that involve the phosphorylation and activation of proteins.

phosphodiester linkage Chemical linkage between adjacent nucleotide residues in DNA and RNA. Forms when the phosphate group of one nucleotide condenses with the hydroxyl group on the sugar of another nucleotide. Also known as *phosphodiester bond*.

phosphofructokinase The enzyme that catalyzes synthesis of fructose-1,6-bisphosphate from fructose-6-phosphate, a key reaction in glycolysis (step 3). Also called *6-phosphofructokinase*.

phospholipid A class of lipid having a hydrophilic head (including a phosphate group) and a hydrophobic tail (consisting of two hydrocarbon chains). Major components of the plasma membrane and organelle membranes.

phosphorylase An enzyme that breaks down glycogen by catalyzing hydrolysis of the α-glycosidic linkages between the glucose residues.

phosphorylation (verb: phosphorylate) The addition of a phosphate group to a molecule.

phosphorylation cascade A series of enzyme-catalyzed phosphorylation reactions commonly used in signal transduction pathways to amplify and convey a signal inward from the plasma membrane.

photic zone In an aquatic habitat, water that is shallow enough to receive some sunlight (whether or not it is enough to support photosynthesis). Compare with **aphotic zone**.

photon A discrete packet of light energy; a particle of light.

photoperiod The amount of time per day (usually in hours) that an organism is exposed to light.

photoperiodism Any response by an organism to the relative lengths of day and night (i.e., photoperiod).

photophosphorylation Production of ATP molecules by ATP synthase using the proton-motive force generated as light-excited electrons flow through an electron transport chain during photosynthesis.

photoreception A sensory system in which receptors are activated by light.

photoreceptor A molecule, a cell, or an organ that is specialized to detect light.

photorespiration A series of light-driven chemical reactions that consumes oxygen and releases carbon dioxide, basically undoing photosynthesis. Usually occurs when there are high O_2 and low CO_2 concentrations inside plant cells; often occurs when stomata must be kept closed to prevent dehydration.

photoreversibility A change in conformation that occurs in certain plant pigments when they are exposed to the particular wavelengths of light that they absorb; triggers responses by the plant.

photosynthesis The complex biological process that converts the energy of light into chemical energy stored in glucose and other organic molecules. Occurs in most plants, algae, and some bacteria.

photosystem One of two types of units, consisting of a central reaction center surrounded by antenna complexes, that is responsible for the light-dependent reactions of photosynthesis.

photosystem I A photosystem that contains a pair of P700 chlorophyll molecules and uses absorbed light energy to reduce $NADP^+$ to NADPH.

photosystem II A photosystem that contains a pair of P680 chlorophyll molecules and uses absorbed light energy to produce a proton-motive force for the synthesis of ATP. Oxygen is produced as a by-product when water is split to obtain electrons.

phototroph An organism (most plants, algae, and some bacteria) that produces ATP through photosynthesis.

phototropins A class of plant photoreceptors that detect blue light and initiate various responses.

phototropism Growth or movement of an organism in a particular direction in response to light.

phylogenetic species concept The definition of a species as the smallest monophyletic group in a phylogenetic tree. Compare with **biological species concept** and **morphospecies concept**.

phylogenetic tree A branching diagram that depicts the evolutionary relationships among species or other taxa.

phylogeny The evolutionary history of a group of organisms.

phylum (plural: phyla) In Linnaeus' system, a taxonomic category above the class level and below the kingdom level. In plants, sometimes called a *division*.

physical map A map of a chromosome that shows the number of base pairs between various genetic markers. Compare with **genetic map**.

physiology The study of how an organism's body functions.

phytochrome A specialized plant photoreceptor that exists in two shapes depending on the ratio of red to far-red light and is involved in the timing of certain physiological processes, such as flowering, stem elongation, and germination.

pigment Any molecule that absorbs certain wavelengths of visible light and reflects or transmits other wavelengths.

piloting A type of navigation in which animals use familiar landmarks to find their way.

pineal gland An endocrine gland, located in the brain, that secretes the hormone melatonin.

pioneering species Those species that appear first in recently disturbed areas.

pit In plants, a small hole in the secondary cell walls of tracheids and vessel elements that allows passage of water.

pitch The sensation produced by a particular frequency of sound. Low frequencies are perceived as low pitches; high frequencies, as high pitches.

pith In the shoot systems of plants, ground tissue located to the inside of the vascular bundles.

pituitary gland A small gland located directly under the brain and physically and functionally connected to the hypothalamus. Produces and secretes an array of hormones that affect many other glands and organs.

placenta A structure that forms in the pregnant uterus from maternal and fetal tissues. Delivers oxygen to the fetus, exchanges nutrients and wastes between mother and fetus, anchors the fetus to the uterine wall, and produces some hormones. Occurs in most mammals and in a few other vertebrates.

placental mammals See **eutherians**.

plankton Drifting organisms (animals, plants, archaea, or bacteria) in aquatic environments.

Plantae The monophyletic group that includes red, green, and glaucophyte algae, and land plants.

plasma The non-cellular portion of blood.

plasma cell A B cell that produces large quantities of antibodies after being activated by interacting with antigen and a CD4$^+$ T cell via peptide presentation. Also called an *effector B cell*.

plasma membrane A membrane that surrounds a cell, separating it from the external environment and selectively regulating passage of molecules and ions into and out of the cell. Also called *cell membrane*.

plasmid A small, usually circular, supercoiled DNA molecule independent of the cell's main chromosome(s) in prokaryotes and some eukaryotes.

plasmodesmata (singular: plasmodesma) Physical connections between two plant cells, consisting of membrane-lined gaps in the cell walls through which the two cells' plasma membranes, cytoplasm, and smooth ER can connect directly. Functionally similar to gap junctions in animal cells.

plasmogamy Fusion of the cytoplasm of two individuals. Occurs in many fungi.

plastocyanin A small protein that shuttles electrons originating from photosystem II to the reaction center of photosystem I during photosynthesis.

plastoquinone (PQ) A nonprotein electron carrier in the chloroplast electron transport chain. Receives excited electrons from photosystem II (noncyclic) or photosystem I (cyclic) and passes them to more electronegative molecules in the chain. Also transports protons from the stroma to the thylakoid lumen, generating a proton-motive force.

platelet A small membrane-bound cell fragment in vertebrate blood that functions in blood clotting. Derived from large cells in the bone marrow.

pleiotropy (adjective: pleiotropic) The ability of a single gene to affect more than one trait.

ploidy The number of complete chromosome sets present. *Haploid* refers to a ploidy of 1; *diploid*, a ploidy of 2; *triploid*, a ploidy of 3; and *tetraploid*, a ploidy of 4.

point mutation A mutation that results in a change in a single base pair in DNA.

polar (1) Asymmetrical or unidirectional. (2) Carrying a partial positive charge on one side of a molecule and a partial negative charge on the other. Polar molecules are generally hydrophilic.

polar body Any of the tiny, nonfunctional cells that are made as a by-product during meiosis of a primary oocyte, due to most of the cytoplasm going to the ovum.

polar covalent bond A covalent bond in which electrons are shared unequally between atoms differing in electronegativity, resulting in the more electronegative atom having a partial negative charge and the other atom, a partial positive charge. Compare with **nonpolar covalent bond**.

polar microtubules Mitotic and meiotic microtubules that have arisen from the two spindle poles and overlap with each other in the middle of the spindle apparatus.

polar nuclei In flowering plants, the nuclei in the female gametophyte that fuse with one sperm nucleus to produce the endosperm. Most species have two.

pollen grain In seed plants, a male gametophyte enclosed within a protective coat of sporopollenin.

pollen tube In flowering plants, a structure that grows out of a pollen grain after it reaches the stigma, extends down the style, and through which two sperm cells are delivered to the ovule.

pollination The process by which pollen reaches the carpel of a flower (in flowering plants), transferred from anther to stigma, or reaches the ovule directly (in conifers and their relatives).

pollination syndrome Suites of flower characters that are associated with certain types of pollinators and that have evolved through natural selection imposed by the interaction between flowers and pollinators.

poly(A) signal In eukaryotes, a short sequence of nucleotides near the 3′ end of pre-mRNAs that signals cleavage of the RNA and addition of the poly(A) tail.

poly(A) tail In eukaryotes, a sequence of about 100–250 adenine nucleotides added to the 3′ end of newly transcribed messenger RNA molecules.

polygenic inheritance Having many genes influence one trait.

polymer Any long molecule composed of small repeating units (monomers) bonded together. The main biological polymers are proteins, nucleic acids, and polysaccharides.

polymerase chain reaction (PCR) A laboratory technique for rapidly generating millions of identical copies of a specific stretch of DNA. Works by incubating the original DNA sequence of interest with primers, nucleotides, and DNA polymerase.

polymerization (verb: polymerize) The process by which many identical or similar small molecules (monomers) are covalently bonded to form a large molecule (polymer).

polymorphic species A species that has two or more distinct phenotypes in the same interbreeding population at the same time.

polymorphism (adjective: polymorphic) (1) The occurrence of more than one allele at a genetic locus in a population. (2) The occurrence of more than two distinct phenotypes of a trait in a population.

polyp The immotile (sessile) stage in the life cycle of some cnidarians (e.g., jellyfish). Compare with **medusa**.

polypeptide A chain of 50 or more amino acids linked together by peptide bonds. Compare with **oligopeptide** and **protein**.

polyphyletic group An unnatural group based on convergent (homoplastic) characteristics that are not present in a common ancestor. Compare with **monophyletic group**.

polyploidy (adjective: polyploid) The state of having more than two full sets of chromosomes, either from the same species (autopolyploidy) or from different species (allopolyploidy).

polyribosome A messenger RNA molecule along with more than one attached ribosome and their growing peptide strands.

polysaccharide A linear or branched polymer consisting of many monosaccharides joined by glycosidic linkages. Compare with **monosaccharide** and **oligosaccharide**.

polytomy A node in a phylogenetic tree that depicts an ancestral branch dividing into three or more descendant branches; usually indicates that insufficient data were available to resolve which taxa are more closely related.

population A group of individuals of the same species living in the same geographic area at the same time.

population density The number of individuals of a population per unit area.

population dynamics Changes in the size and other characteristics of populations through time and space.

population ecology The study of how and why the number of individuals in a population changes over time and space.

population thinking The ability to analyze trait frequencies, event probabilities, and other attributes of populations of molecules, cells, or organisms.

pore In land plants, an opening in the epithelium that allows gas exchange. See also **stoma**.

positive control Of genes, when a regulatory protein triggers expression by binding to DNA on or near the gene.

positive feedback A physiological mechanism in which a change in some variable stimulates a response that increases the change. Relatively rare in organisms but is important in generation of the action potential. Compare with **negative feedback**.

positive pressure ventilation Ventilation of the lungs by using positive pressure in the mouth to "push" air into the lungs. Compare with **negative pressure ventilation**.

positive-sense RNA virus An ssRNA virus whose genome contains the same sequences as the mRNA required to produce viral proteins. Compare with **ambisense virus** and **negative-sense virus**.

posterior Toward an animal's tail and away from its head. The opposite of anterior.

posterior pituitary The part of the pituitary gland that contains the ends of hypothalamic neurosecretory cells and from which oxytocin and antidiuretic hormone are secreted. Compare with **anterior pituitary**.

postsynaptic neuron A neuron that receives signals, usually via neurotransmitters, from another neuron at a synapse. Compare with **presynaptic neuron**.

post-translational control Regulation of gene expression by modification of proteins (e.g., addition of a phosphate group or sugar residues) after translation.

postzygotic isolation Reproductive isolation resulting from mechanisms that operate after mating of individuals of two different species occurs. The most common mechanisms are the death of hybrid embryos or reduced fitness of hybrids.

potential energy Energy stored in matter as a result of its position or molecular arrangement. Compare with **kinetic energy**.

prebiotic soup model Hypothetical explanation for chemical evolution whereby small molecules reacted with one another in a mixture of organic molecules condensed into a body of water, typically in reference to the early oceans.

Precambrian The interval between the formation of the Earth, about 4.6 billion years ago, and the appearance of most animal groups about 542 million years ago. Unicellular organisms were dominant for most of this era, and oxygen was virtually absent for the first 2 billion years.

pre-mRNA In eukaryotes, the primary transcript of protein-coding genes. Pre-mRNA is processed to form mRNA.

predation The killing and eating of one organism (the prey) by another (the predator).

predator Any organism that kills other organisms for food.

prediction A measurable or observable result of an experiment based on a particular hypothesis. A correct prediction provides support for the hypothesis being tested.

pressure-flow hypothesis The hypothesis that sugar movement through phloem tissue is due to differences in the turgor pressure of phloem sap.

pressure potential (ψ_P) A component of the potential energy of water caused by physical pressures on a solution. It can be positive or negative. Compare with **solute potential (ψ_S)**.

presynaptic neuron A neuron that transmits signals, usually by releasing neurotransmitters, to another neuron or to an effector cell at a synapse.

prezygotic isolation Reproductive isolation resulting from any one of several mechanisms that prevent individuals of two different species from mating.

primary active transport A form of active transport in which a source of energy like ATP is directly used to move ions against their electrochemical gradients.

primary cell wall The outermost layer of a plant cell wall, made of cellulose fibers and gelatinous polysaccharides, that defines the shape of the cell and withstands the turgor pressure of the plasma membrane.

primary consumer An herbivore; an organism that eats plants, algae, or other primary producers. Compare with **secondary consumer**.

primary decomposer A decomposer (detritivore) that consumes detritus from plants.

primary growth In plants, an increase in the length of stems and roots due to the activity of apical meristems. Compare with **secondary growth**.

primary immune response An adaptive immune response to a pathogen that the immune system has not encountered before. Compare with **secondary immune response**.

primary meristem In plants, three types of partially differentiated cells that are produced by apical meristems, including protoderm, ground meristem, and procambium. Compare with **apical meristem** and **cambium**.

primary oocyte Any of the large diploid cells in an ovarian follicle that can initiate meiosis to produce a haploid secondary oocyte and a polar body.

primary producer Any organism that creates its own food by photosynthesis or from reduced inorganic compounds and that is a food source for other species in its ecosystem. Also called *autotroph*.

primary spermatocyte Any of the diploid cells in the testis that can initiate meiosis I to produce two secondary spermatocytes.

primary structure The sequence of amino acid residues in a peptide or protein; also the sequence of nucleotides in a nucleic acid. Compare with **secondary**, **tertiary**, and **quaternary structure**.

primary succession The gradual colonization of a habitat of bare rock or gravel, usually after an environmental disturbance that removes all soil and previous organisms. Compare with **secondary succession**.

primary transcript In eukaryotes, a newly transcribed RNA molecule that has not yet been processed to a mature RNA. Called *pre-mRNA* when the final product is a protein.

primase An enzyme that synthesizes a short stretch of RNA to use as a primer during DNA replication.

primates The lineage of mammals that includes prosimians (lemurs, lorises, etc.), monkeys, and great apes (including humans).

primer A short, single-stranded RNA molecule that base-pairs with a DNA template strand and is elongated by DNA polymerase during DNA replication.

probe A radioactively or chemically labeled single-stranded fragment of a known DNA or RNA sequence that can bind to and detect its complementary sequence in a sample containing many different sequences.

proboscis A tubular, often extensible feeding appendage with which food can be obtained.

procambium A primary meristem tissue that gives rise to the vascular tissue.

product Any of the final materials formed in a chemical reaction.

progesterone A steroid hormone produced and secreted by the corpus luteum in the ovaries after ovulation and by the placenta during gestation; protects the uterine lining.

programmed cell death Regulated cell death that is used in development, tissue maintainance, and destruction of infected cells. Can occur in different ways; apoptosis is the best-known mechanism.

prokaryote A member of the domain Bacteria or Archaea; a unicellular organism lacking a nucleus and containing relatively few organelles or cytoskeletal components. Compare with **eukaryote**.

prolactin A peptide hormone, produced and secreted by the anterior pituitary, that promotes milk production in female mammals and has a variety of effects on parental behavior and seasonal reproduction in other vertebrates.

prometaphase A stage in mitosis or meiosis during which the nuclear envelope breaks down and microtubules attach to kinetochores.

promoter A short nucleotide sequence in DNA that binds a sigma factor (in bacteria) or basal transcription factors (in eukaryotes) to enable RNA polymerase to begin transcription. In bacteria, several contiguous genes are often transcribed from a single promoter. In eukaryotes, each gene generally has its own promoter.

promoter-proximal element In eukaryotes, regulatory sequences in DNA that are close to a promoter and that can bind regulatory transcription factors.

proofreading The process by which a DNA polymerase recognizes and removes a wrong base added during DNA replication and then continues synthesis.

prophase The first stage in mitosis or meiosis during which chromosomes become visible and the spindle apparatus forms. Synapsis and crossing over occur during prophase of meiosis I.

prosimians One of the two major lineages of primates, a paraphyletic group including lemurs, pottos, and lorises. Compare with **anthropoids**.

prostate gland A gland in male mammals that surrounds the base of the urethra and secretes a fluid that is a component of semen.

prosthetic group A non-amino acid atom or molecule that is permanently attached to an enzyme or other protein and is required for its function.

protease An enzyme that can break up proteins by cleaving the peptide bonds between amino acid residues.

proteasome A macromolecular machine that destroys proteins that have been marked by the addition of ubiquitin.

protein A macromolecule consisting of one or more polypeptide chains composed of 50 or more amino acids linked together. Each protein has a unique sequence of amino acids and generally possesses a characteristic three-dimensional shape.

protein kinase An enzyme that catalyzes the addition of a phosphate group to another protein, typically activating or inactivating the substrate protein.

proteinase inhibitors Defense compounds, produced by plants, that induce illness in herbivores by inhibiting digestive enzymes.

proteoglycan A type of highly glycosylated protein found in the extracellular matrix of animal cells that attracts and holds water, forming a gel that resists compression forces.

proteome The complete set of proteins produced by a particular cell type.

proteomics The systematic study of the interactions, localization, functions, regulation, and other features of the full protein set (proteome) in a particular cell type.

protist Any eukaryote that is not a green plant, animal, or fungus. Protists are a diverse paraphyletic group. Most are unicellular, but some are multicellular or form aggregations called colonies.

protocell A hypothetical pre-cell structure consisting of a membrane compartment that encloses replicating macromolecules, such as ribozymes.

protoderm The exterior layer of a young plant embryo that gives rise to the epidermis.

proto-oncogene Any gene that encourages cell division in a regulated manner, typically by triggering specific phases in the cell cycle. Mutation may convert it into an oncogene. See **oncogene**.

proton pump A membrane protein that can hydrolyze ATP to power active transport of protons (H^+ ions) across a membrane against an electrochemical gradient. Also called H^+-ATPase.

proton-motive force The combined effect of a proton gradient and an electric potential gradient across a membrane, which can drive protons across the membrane. Used by mitochondria and chloroplasts to power ATP synthesis via the mechanism of chemiosmosis.

protostomes A major lineage of animals that share a pattern of embryological development, including formation of the mouth earlier than the anus, and formation of the coelom by splitting of a block of mesoderm. Includes arthropods, mollusks, and annelids. Compare with **deuterostomes**.

proximal tubule In the vertebrate kidney, the convoluted section of a nephron into which filtrate moves from Bowman's capsule. Involved in the largely unregulated reabsorption of electrolytes, nutrients, and water. Compare with **distal tubule**.

proximate causation In biology, the immediate, mechanistic cause of a phenomenon (how it happens), as opposed to why it evolved. Also called *proximate explanation*. Compare with **ultimate causation**.

pseudocoelomate An animal that has a coelom that is only partially lined with mesoderm. Compare with **acoelomate** and **coelomate**.

pseudogene A DNA sequence that closely resembles a functional gene but is not transcribed. Thought to have arisen by duplication of the functional gene followed by inactivation due to a mutation.

pseudopodium (plural: pseudopodia) A temporary bulge-like extension of certain protist cells used in cell crawling and ingestion of food.

puberty The various physical and emotional changes that an immature human undergoes in reaching reproductive maturity. Also the period when such changes occur.

pulmonary artery A short, thick-walled artery that carries oxygen-poor blood from the heart to the lungs.

pulmonary circulation The part of the circulatory system that sends oxygen-poor blood to the lungs. It is separate from the rest of the circulatory system (the systemic circulation) in mammals and birds.

pulmonary vein A short, thin-walled vein that carries oxygen-rich blood from the lungs to the heart. Humans have four such veins.

pulse–chase experiment A type of experiment in which a population of cells or molecules at a particular moment in time is marked by means of a labeled molecule (pulse) and then their fate is followed over time (chase).

pump Any membrane protein that can hydrolyze ATP and change shape to power active transport of a specific ion or small molecule across a plasma membrane against its electrochemical gradient. See **proton pump**.

pupa (plural: pupae) A metamorphosing insect that is enclosed in a protective case.

pupil The hole in the center of the iris through which light enters a vertebrate or cephalopod eye.

pure line In animal or plant breeding, a strain that produces offspring identical with themselves when self-fertilized or crossed to another member of the same population. Pure lines are homozygous for most, if not all, genetic loci.

purifying selection Selection that lowers the frequency of or even eliminates deleterious alleles.

purines A class of small, nitrogen-containing, double-ringed bases (guanine, adenine) found in nucleotides. Compare with **pyrimidines**.

pyrimidines A class of small, nitrogen-containing, single-ringed bases (cytosine, uracil, thymine) found in nucleotides. Compare with **purines**.

pyruvate dehydrogenase A large enzyme complex, located in the mitochondrial matix, that is responsible for converting pyruvate to acetyl CoA during cellular respiration.

quantitative trait A trait that exhibits continuous phenotypic variation (e.g., human height), rather than the distinct forms characteristic of discrete traits.

quaternary structure In proteins, the overall three-dimensional shape formed from two or more polypeptide chains (subunits); determined by the number, relative positions, and interactions of the subunits. In single stranded nucleic acids, the hydrogen bonding between two or more distinct strands will form this level of structure through hydrophobic interactions between complementary bases. Compare with **primary**, **secondary**, and **tertiary structures**.

quorum sensing Cell–cell signaling in unicellular organisms, in which cells of the same species communicate via chemical signals. It is often observed that cell activity changes dramatically when the population reaches a threshold size, or quorum.

radial symmetry An animal body pattern that has at least two planes of symmetry. Typically, the body is in the form of a cylinder or disk, and the body parts radiate from a central hub. Compare with **bilateral symmetry**.

radiation Transfer of heat between two bodies that are not in direct physical contact. More generally, the emission of electromagnetic energy of any wavelength.

radicle The root of a plant embryo.

radioactive isotope A version of an element that has an unstable nucleus, which will release radiation energy as it decays to a more stable form. Decay often results in the radioisotope becoming a different element.

radioimmunoassay A competitive binding assay in which the quantity of hormone in a sample can be estimated. Uses radioactively labeled hormones that compete with the unknown hormone to bind with an antibody.

radula A rasping feeding appendage in mollusks such as gastropods (snails, slugs).

rain shadow The dry region on the side of a mountain range away from the prevailing wind.

range The geographic distribution of a species.

Ras protein A type of G protein that is activated by enzyme-linked cell-surface receptors, including receptor tyrosine kinases. Activated Ras then initiates a phosphorylation cascade, culminating in a cell response.

ray-finned fishes Members of the Actinopterygii, a diverse group of fishes with fins supported by bony rods arranged in a ray pattern.

rays In plant shoot systems with secondary growth, a lateral row of parenchyma cells produced by vascular cambium. Transport water and nutrients laterally across the stem.

Rb protein A tumor suppressor protein that helps regulate progression of a cell from the G_1 phase to the S phase of the cell cycle. Defects in Rb protein are found in many types of cancer.

reactant Any of the starting materials in a chemical reaction.

reaction center Centrally located component of a photosystem containing proteins and a pair of specialized chlorophyll molecules. It is surrounded by antenna complexes that transmit resonance energy to excite the reaction center pigments.

reading frame A series of non-overlapping, three-base-long sequences (potential codons) in DNA or RNA. The reading frame for a polypeptide is set by the start codon.

realized niche The portion of the fundamental niche that a species actually occupies given limiting factors such as competition with other species. Compare with **fundamental niche**.

receptor-mediated endocytosis Uptake by a cell of certain extracellular macromolecules, bound to specific receptors in the plasma membrane, by pinching off the membrane to form small membrane-bound vesicles.

receptor tyrosine kinase (RTK) Any of a class of enzyme-linked cell-surface signal receptors that undergo phosphorylation after binding a signaling molecule. The activated, phosphorylated receptor then triggers a signal transduction pathway inside the cell.

recessive Referring to an allele whose phenotypic effect is observed only in homozygous individuals. Compare with **dominant**.

reciprocal altruism Altruistic behavior that is exchanged between a pair of individuals at different times (i.e., sometimes individual A helps individual B, and sometimes B helps A).

reciprocal cross A cross in which the mother's and father's phenotypes are the reverse of that examined in a previous cross.

recombinant Possessing a new combination of alleles. May refer to a single chromosome or DNA molecule, or to an entire organism.

recombinant DNA technology A variety of techniques for isolating specific DNA fragments and introducing them into different regions of DNA or a different host organism.

rectal gland A salt-excreting gland in the digestive system of sharks, skates, and rays.

rectum The last portion of the digestive tract. It is where feces are held until they are expelled.

red blood cell A hemoglobin-containing cell that circulates in the blood and delivers oxygen from the lungs to the tissues.

redox reaction Any chemical reaction that involves either the complete transfer of one or more electrons from one reactant to another, or a reciprocal shift in the position of shared electrons within one or more of the covalent bonds of two reactants. Also called *reduction–oxidation reaction.*

reduction The gain of electrons by an atom or molecule during a redox reaction, either by acceptance of an electron from another atom or molecule, or by the shared electrons in covalent bonds moving closer to the atomic nucleus.

reduction–oxidation reaction See **redox reaction**.

reflex An involuntary response to environmental stimulation. May involve the brain (conditioned reflex) or not (spinal reflex).

refractory No longer responding to stimuli that previously elicited a response. An example is the tendency of voltage-gated sodium channels to remain closed immediately after an action potential.

regulatory sequence Any segment of DNA that is involved in controlling transcription of a specific gene by binding a regulatory transcription factor protein.

regulatory transcription factor General term for proteins that bind to DNA regulatory sequences (eukaryotic enhancers, silencers, and promoter-proximal elements), but not to the promoter itself, leading to an increase or decrease in transcription of specific genes. Compare with **basal transcription factor**.

regulon A large set of genes in bacteria that are controlled by a single type of regulatory molecule. Regulon genes are transcribed in response to environmental cues and allow cells to respond to changing environments.

reinforcement In evolutionary biology, the natural selection for traits that prevent interbreeding between recently diverged species.

release factors Proteins that trigger termination of translation when a ribosome reaches a stop codon.

renal corpuscle In the vertebrate kidney, the ball-like structure at the beginning of a nephron, consisting of a glomerulus and the surrounding Bowman's capsule. Acts as a filtration device.

replacement rate The number of offspring each female must produce over her entire life to "replace" herself and her mate, resulting in zero population growth. The actual number is slightly more than 2 because some offspring die before reproducing.

replica plating A method of identifying bacterial colonies that have certain mutations by transferring cells from each colony on a master plate to a second (replica) plate and observing their growth when exposed to different conditions.

replication fork The Y-shaped site at which a double-stranded molecule of DNA is separated into two single strands for replication.

replicative growth The process by which viruses produce new virions.

replisome The macromolecular machine that copies DNA; includes DNA polymerase, helicase, primase, and other enzymes.

repolarization Return to a resting potential after a membrane potential has changed; a normal phase in an action potential.

repressor (1) In bacteria, a protein that binds to an operator sequence in DNA to prevent transcription when an inducer is not present and that comes off DNA to allow transcription when an inducer binds to the repressor protein. (2) In eukaryotes, a protein that binds to a silencer sequence in DNA to prevent or reduce gene transcription.

reproductive development The phase of plant development that involves development of the flower and reproductive cells. Follows vegetative development and occurs when a shoot apical meristem (SAM) transitions to a flower-producing meristem.

reptiles One of the two lineages of amniotes (vertebrates that produce amniotic eggs) distinguished by adaptations for life and reproduction on land. Living reptiles include turtles, snakes and lizards, crocodiles and alligators, and birds. Except for birds, all are ectotherms.

resilience, community A measure of how quickly a community recovers following a disturbance.

resistance, community A measure of how much a community is affected by a disturbance.

respiratory system The collection of cells, tissues, and organs responsible for gas exchange between an animal and its environment.

resting potential The membrane potential of a cell in its resting, or normal, state.

restriction endonucleases Bacterial enzymes that cut DNA at a specific base-pair sequence (restriction site). Also called *restriction enzymes*.

retina A thin layer of light-sensitive cells (rods and cones) and neurons at the back of a simple eye, such as that of cephalopods and vertebrates.

retinal A light-absorbing pigment that is linked to the protein opsin in rods and cones of the vertebrate eye.

retrovirus A virus with an RNA genome that reverse-transcribes its RNA into a double-stranded DNA sequence, which is then inserted into the host's genome as part of its replicative cycle.

reverse transcriptase An enzyme that can synthesize double-stranded DNA from a single-stranded RNA template.

rhizobia (singular: rhizobium) Members of the bacterial genus *Rhizobium*; nitrogen-fixing bacteria that live in root nodules of members of the pea family (legumes).

rhizoid The hairlike structure that anchors a nonvascular plant to the substrate.

rhizome A modified stem that runs horizontally underground and produces new plants at the nodes (a form of asexual reproduction). Compare with **stolon**.

rhodopsin A transmembrane complex that is instrumental in detection of light by rods of the vertebrate eye. Is composed of the transmembrane protein opsin covalently linked to retinal, a light-absorbing pigment.

ribonucleic acid (RNA) A nucleic acid composed of ribonucleotides that usually is single stranded. Functions include structural components of ribosomes (rRNA), transporters of amino acids (tRNA), and messages of the DNA code required for protein synthesis (mRNA), among others.

ribonucleotide See **nucleotide**.

ribosomal RNA (rRNA) An RNA molecule that forms part of the ribosome.

ribosome A large macromolecular machine that synthesizes proteins by using the genetic information encoded in messenger RNA. Consists of two subunits, each composed of ribosomal RNA and proteins.

ribosome binding site In a bacterial mRNA molecule, the sequence just upstream of the start codon to which a ribosome binds to initiate translation. Also called the *Shine–Dalgarno sequence*.

ribozyme Any RNA molecule that can act as a catalyst, that is, speed up a chemical reaction.

ribulose bisphosphate (RuBP) A five-carbon compound that combines with CO_2 in the first step of the Calvin cycle during photosynthesis.

RNA See **ribonucleic acid**.

RNA interference (RNAi) Degradation of an RNA molecule or inhibition of its translation following its binding by a short RNA (microRNA) whose sequence is complementary to a portion of the mRNA.

RNA polymerase An enzyme that catalyzes the synthesis of RNA from ribonucleotides using a DNA template.

RNA processing In eukaryotes, the changes that a primary RNA transcript undergoes to become a mature RNA molecule. For pre-mRNA it includes the addition of a 5′ cap and poly(A) tail and splicing to remove introns.

RNA replicase A viral enzyme that can synthesize RNA from an RNA template. Also called an *RNA-dependent RNA polymerase*.

RNA world hypothesis Proposal that chemical evolution produced RNAs that could catalyze key reactions involved in their own replication and basic metabolism, which led to the evolution of proteins and the first life-form.

rod cell A photoreceptor cell with a rod-shaped outer portion that is particularly sensitive to dim light but not used to distinguish colors. Also called simply *rod*. Compare with **cone cell**.

root (1) An underground part of a plant that anchors the plant and absorbs water and nutrients. (2) The most ancestral branch in a phylogenetic tree.

root apical meristem (RAM) A group of undifferentiated plant stem cells at the tip of a plant root that can differentiate into mature root tissue.

root cap A small group of cells that covers and protects the root apical meristem. Senses gravity and determines the direction of root growth.

root hair A long, thin outgrowth of the epidermal cells of plant roots, providing increased surface area for absorption of water and nutrients.

root pressure Positive pressure of xylem sap in the vascular tissue of roots. Generated during the night as a result of the accumulation of ions from the soil and subsequent osmotic movement of water into the xylem.

root system The belowground part of a plant.

rotifer Member of the phylum Rotifera. Distinguished by a cluster of cilia, called a corona, used in suspension feeding in marine and freshwater environments. Rotifers belong to the lophotrochozoan branch of the protostomes.

rough endoplasmic reticulum (rough ER) The portion of the endoplasmic reticulum that is dotted with ribosomes. Involved in synthesis of plasma membrane proteins, secreted proteins, and proteins localized to the ER, Golgi apparatus, and lysosomes. Compare with **smooth endoplasmic reticulum**.

roundworms Members of the phylum Nematoda. Distinguished by an unsegmented body with a pseudocoelom and no appendages. Roundworms belong to the ecdysozoan branch of the protostomes. Also called *nematodes*.

rubisco The enzyme that catalyzes the first step of the Calvin cycle during photosynthesis: the addition of a molecule of CO_2 to ribulose bisphosphate. See also **carbon fixation**.

ruminant Member of a group of hoofed mammals (e.g., cattle, sheep, deer) that have a four-chambered stomach specialized for digestion of plant cellulose. Ruminants regurgitate cud, a mixture of partially digested food and cellulose-digesting bacteria, from the largest chamber (the rumen) for further chewing.

salinity The proportion of solutes dissolved in water in natural environments, often designated in grams of solute per kilogram of water (cited as parts per thousand).

salivary amylase An enzyme that is produced by the salivary glands and that can break down starch by catalyzing hydrolysis of the glycosidic linkages between the glucose residues.

salivary gland A type of vertebrate gland that secretes saliva (a mixture of water, mucus-forming glycoproteins, and digestive enzymes) into the mouth.

sampling error The selection of a nonrepresentative sample from some larger population, due to chance.

saprophyte An organism that feeds primarily on dead plant material.

sapwood The younger xylem in the outer layer of wood of a stem or root, functioning primarily in water transport.

sarcomere The repeating contractile unit of a skeletal muscle cell; the portion of a myofibril located between adjacent Z disks.

sarcoplasmic reticulum Sheets of smooth endoplasmic reticulum in a muscle cell. Contains high concentrations of calcium, which can be released into the cytoplasm to trigger contraction.

saturated Referring to lipids in which all the carbon-carbon bonds are single bonds. Such compounds have relatively high melting points. Compare with **unsaturated**.

scanning electron microscope (SEM) A microscope that produces images of the surfaces of objects by reflecting electrons from a specimen coated with a layer of metal atoms. Compare with **transmission electron microscope**.

scarify To scrape, rasp, cut, or otherwise damage the coat of a seed. Necessary in some species to trigger germination.

Schwann cell A type of glial cell that wraps around axons of some neurons outside the brain and spinal cord, forming a myelin sheath that provides electrical insulation. Compare with **oligodendrocyte**.

scientific name The unique, two-part name given to each species, with a genus name followed by a species name—as in *Homo sapiens*. Scientific names are always italicized, and are also known as Latin names.

sclereid In plants, a relatively short type of sclerenchyma cell that usually functions in protection, such as in seed coats and nutshells. Compare with **fiber**.

sclerenchyma cell In plants, a cell that has a thick secondary cell wall and provides support; typically contains the tough structural polymer lignin and usually is dead at maturity. Includes fibers and sclereids. Compare with **collenchyma cell** and **parenchyma cell**.

scrotum A sac of skin containing the testes and suspended just outside the abdominal body cavity of many male mammals.

second law of thermodynamics The principle of physics that the entropy of the universe or any closed system always increases.

second-male advantage The reproductive advantage, in some species, of a male who mates with a female last, after other males have mated with her.

second messenger A nonprotein signaling molecule produced or activated inside a cell in response to stimulation at the cell surface. Commonly used to relay the message of a hormone or other extracellular signaling molecule.

secondary active transport Transport of an ion or molecule in a defined direction that is often against its electrochemical gradient, in company with an ion or molecule being transported along its electrochemical gradient. Also called *cotransport*.

secondary cell wall The thickened inner layer of a plant cell wall formed by certain cells as they mature and have stopped growing; in water-conducting cells, contains lignin. Provides support or protection.

secondary consumer A carnivore; an organism that eats herbivores. Compare with **primary consumer**.

secondary growth In plants, an increase in the width of stems and roots due to the activity of cambium. Compare with **primary growth**.

secondary immune response The adaptive immune response to a pathogen that the immune system has encountered before. Normally much faster and more efficient than the primary response, due to immunological memory. Compare with **primary immune response**.

secondary metabolites Molecules that are closely related to compounds in key synthetic pathways and that often function in defense.

secondary oocyte A cell produced by meiosis I of a primary oocyte in the ovary. If fertilized, will complete meiosis II to produce an ootid (which develops into an ovum) and a polar body.

secondary spermatocyte A cell produced by meiosis I of a primary spermatocyte in the testis. Can undergo meiosis II to produce spermatids.

secondary structure In proteins, localized folding of a polypeptide chain into regular structures (i.e., alpha-helix and beta-pleated sheet) stabilized by hydrogen bonding between atoms of the peptide backbone. In nucleic acids, elements of structure (e.g., helices and hairpins) stabilized by hydrogen bonding and hydrophobic interactions between complementary bases. Compare with **primary**, **tertiary**, and **quaternary structures**.

secondary succession Gradual colonization of a habitat after an environmental disturbance (e.g., fire, windstorm, logging) that removes some or all previous organisms but leaves the soil intact. Compare with **primary succession**.

secretin A peptide hormone produced by cells in the small intestine in response to the arrival of food from the stomach. Stimulates secretion of bicarbonate (HCO_3^-) from the pancreas.

sedimentary rock A type of rock formed by gradual accumulation of sediment, particularly sand and mud, as in riverbeds and on the ocean floor. Most fossils are found in sedimentary rocks.

seed A plant reproductive structure consisting of an embryo, associated nutritive tissue (endosperm), and an outer protective layer (seed coat). In angiosperms, develops from the fertilized ovule of a flower.

seed bank A repository where seeds, representing many different varieties of domestic crops or other species, are preserved.

seed coat A protective layer around a seed that encases both the embryo and the endosperm.

segment A well-defined, repeated region of the body along the anterior–posterior body axis, containing structures similar to other nearby segments.

segmentation Division of the body or a part of it into a series of similar structures; exemplified by the body segments of insects and worms and by the somites of vertebrates.

segmentation genes A group of genes that control the formation and patterning of body segmentation in embryonic development. Includes maternal genes, gap genes, pair-rule genes, and segment polarity genes.

segregation, principle of The concept that each pair of hereditary elements (alleles of the same gene) separate from each other during meiosis. One of Mendel's two principles of genetics.

selective adhesion The tendency of cells of one tissue type to adhere to other cells of the same type.

selective permeability The property of a membrane that allows some substances to diffuse across it much more readily than other substances.

selectively permeable membrane Any membrane across which some solutes can move more readily than others.

self-fertilization The fusion of two gametes from the same individual to form offspring. Also called *selfing*.

self molecule A molecule that is synthesized by an organism and is a normal part of its cells and/or body; as opposed to nonself, or foreign, molecules.

semen The combination of sperm and accessory fluids that is released by male mammals and reptiles during ejaculation.

semiconservative replication The way DNA replicates, in which each strand of an existing DNA molecule serves as a template to create a new complementary DNA strand. It is called semiconservative because each newly replicated DNA molecule conserves one of the parental strands and contains another, newly replicated strand.

seminal vesicle In male mammals, either of a pair of reproductive glands that secrete a sugar-containing fluid into semen, which provides energy for sperm movement.

senescence The genetically programmed, active process of aging.

sensor Any cell, organ, or structure with which an animal can sense some aspect of the external or internal environment. Usually functions, along with an integrator and effector, as part of a homeostatic system.

sensory neuron A nerve cell that carries signals from sensory receptors to the central nervous system. Compare with **interneuron** and **motor neuron**.

sepal One of the protective leaflike organs enclosing a flower bud and later part of the outermost portion of the flower.

septum (plural: septa) Any wall-like structure. In fungi, septa divide the filaments (hyphae) of mycelia into cell-like compartments.

serotonin A neurotransmitter involved in many brain functions, including sleep, pleasure, and mood.

serum The liquid that remains when cells and clot material are removed from clotted blood. Contains water, dissolved gases, growth factors, nutrients, and other soluble substances. Compare with **plasma**.

sessile Permanently attached to a substrate; not capable of moving to another location.

set point A normal or target value for a regulated internal variable, such as body heat or blood pH.

sex chromosome Chromosomes that differ in shape or in number in males and females. For example, the X and Y chromosomes of many animals. Compare with **autosome**.

sex-linked inheritance Inheritance patterns observed in genes carried on sex chromosomes. In this case, females and males have different numbers of alleles of a gene. Often creates situations in which a trait appears more often in one sex. Also called *sex-linkage*.

sexual dimorphism Any trait that differs between males and females.

sexual reproduction Any form of reproduction in which genes from two parents are combined via fusion of gametes, producing offspring that are genetically distinct from both parents. Compare with **asexual reproduction**.

sexual selection A type of natural selection that favors individuals with traits that increase their ability to obtain mates. Acts more strongly on males than females. (Compare with **ecological selection**.)

shell A hard, protective outer structure.

Shine–Dalgarno sequence See **ribosome binding sequence**.

shoot In a plant embryo, the combination of hypocotyl and cotyledons, which will become the aboveground portions of the plant.

shoot apical meristem (SAM) A group of undifferentiated plant stem cells at the tip of a plant stem that can differentiate into mature shoot tissues.

shoot system The aboveground part of a plant comprising stems, leaves, and flowers (in angiosperms).

short-day plant A plant that blooms in response to long nights (usually in late summer or fall in the Northern Hemisphere). Compare with **day-neutral** and **long-day plant**.

short tandem repeats (STRs) Relatively short DNA sequences that are repeated, one after another, down the length of a chromosome. See **microsatellite**.

shotgun sequencing A method of sequencing genomes that is based on breaking the genome into small pieces, sequencing each piece separately, and then figuring out how the pieces are connected.

sieve plate In plants, a pore-containing structure at each end of a sieve-tube element in phloem.

sieve-tube element In plants, an elongated sugar-conducting cell in phloem that lacks nuclei and has sieve plates at both ends, allowing sap to flow to adjacent cells.

sigma A bacterial protein that associates with the core RNA polymerase to allow recognition of promoters.

signal In behavioral ecology, any information-containing behavior or characteristic.

signal receptor Any cellular protein that binds to a particular signaling molecule (e.g., a hormone or neurotransmitter) and triggers a response by the cell. Receptors for lipid-insoluble signals are transmembrane proteins in the plasma membrane; those for many lipid-soluble signals (e.g., steroid hormones) are located inside the cell.

signal recognition particle (SRP) An RNA–protein complex that binds to the ER signal sequence in a polypeptide as it emerges from a ribosome and transports the ribosome–polypeptide complex to the ER membrane, where synthesis of the polypeptide is completed.

signal transduction The process by which a stimulus (e.g., a hormone, a neurotransmitter, or sensory information) outside a cell is converted into an intracellular signal required for a cellular response. Usually involves a specific sequence of molecular events, or signal transduction pathway, that may lead to amplification of the signal.

signal transduction cascade See **phosphorylation cascade**.

silencer A regulatory sequence in eukaryotic DNA to which repressor proteins can bind, inhibiting gene transcription.

silent mutation A point mutation that changes the sequence of a codon without changing the amino acid that is specified.

simple eye An eye with only one light-collecting apparatus (e.g., one lens), as in vertebrates and cephalopods. Compare with **compound eye**.

simple fruit A fruit (e.g., apricot) that develops from a single flower that has a single carpel or several fused carpels. Compare with **aggregate** and **multiple fruit**.

simple sequence repeat See **microsatellite**.

SINEs (short interspersed nuclear elements) The second most abundant class of transposable elements in human genomes; can create copies of itself and insert them elsewhere in the genome. Compare with **LINEs**.

single nucleotide polymorphism (SNP) A site on a chromosome where individuals in a population have different nucleotides. Can be used as a genetic marker to help track the inheritance of nearby genes.

single-strand DNA-binding proteins (SSBP) A protein that attaches to separated strands of DNA during replication or transcription, preventing them from re-forming a double helix.

sink Any tissue, site, or location where an element or a molecule is consumed or taken out of circulation (e.g., in plants, a tissue where sugar exits the phloem). Compare with **source**.

sinoatrial (SA) node In the right atrium of the vertebrate heart, a cluster of cardiac muscle cells that initiates the heartbeat and determines the heart rate. Compare with **atrioventricular (AV) node**.

siphon A tubelike appendage of many mollusks, often used for feeding or propulsion.

sister chromatids The paired strands of a recently replicated chromosome, which are connected at the centromere and eventually separate during anaphase of mitosis and meiosis II. Compare with **non-sister chromatids**.

sister species Closely related species that occupy adjacent branches in a phylogenetic tree.

skeletal muscle The muscle tissue attached to the bones of the vertebrate skeleton. Consists of long, unbranched muscle fibers with a characteristic striped (striated) appearance; controlled voluntarily. Compare with **cardiac** and **smooth muscle**.

sliding-filament model The hypothesis that thin (actin) filaments and thick (myosin) filaments slide past each other, thereby shortening the sarcomere. Shortening of all the sarcomeres in a myofibril results in contraction of the entire myofibril.

slow muscle fiber Type of skeletal muscle fiber that is red in color due to the abundance of myoglobin, generates ATP by oxidative phosphorylation, and contracts slowly but does not fatigue easily. Also called *slow oxidative*, or *Type I, fiber*.

small intestine The portion of the digestive tract between the stomach and the large intestine. The site of the final stages of digestion and of most nutrient absorption.

small nuclear ribonucleoproteins See **snRNPs**.

smooth endoplasmic reticulum (smooth ER) The portion of the endoplasmic reticulum that does not have ribosomes attached to it. Involved in synthesis and secretion of lipids. Compare with **rough endoplasmic reticulum**.

smooth muscle The unstriated muscle tissue that lines the intestine, blood vessels, and some other organs. Consists of tapered, unbranched cells that can sustain long contractions. Not voluntarily controlled. Compare with **cardiac** and **skeletal muscle**.

snRNPs (small nuclear ribonucleoproteins) Complexes of proteins and small RNA molecules that function as components of spliceosomes during splicing (removal of introns from pre-mRNAs).

sodium–potassium pump A transmembrane protein that uses the energy of ATP to move sodium ions out of the cell and potassium ions in. Also called Na^+/K^+-ATPase.

soil organic matter Organic (carbon-containing) compounds found in soil.

solute Any substance that is dissolved in a liquid.

solute potential (ψ_S) A component of the potential energy of water caused by a difference in solute concentrations at two locations. Can be zero (pure water) or negative. Compare with **pressure potential (ψ_P)**.

solution A liquid containing one or more dissolved solids or gases in a homogeneous mixture.

solvent Any liquid in which one or more solids or gases can dissolve.

soma See **cell body**.

somatic cell Any type of cell in a multicellular organism except eggs, sperm, and their precursor cells. Also called *body cells*.

somatic hypermutation Mutation that occurs in the variable regions of immunoglobulin genes when B cells are first activated and in memory cells, resulting in novel variation in the receptors that bind to antigens.

somatic nervous system The part of the peripheral nervous system (outside the brain and spinal cord) that controls skeletal muscles and is under voluntary control. Compare with **autonomic nervous system**.

somatostatin A hormone secreted by the pancreas and hypothalamus that inhibits the release of several other hormones.

somite A block of mesoderm that occurs in pairs along both sides of the developing neural tube in a vertebrate embryo. Gives rise to muscle, vertebrae, ribs, and the dermis of the skin.

sori In ferns, a cluster of spore-producing structures (sporangia) usually found on the underside of fronds.

source Any tissue, site, or location where a substance is produced or enters circulation (e.g., in plants, the tissue where sugar enters the phloem). Compare with **sink**.

space-filling model A representation of a molecule where atoms are shown as balls—color-coded and scaled to indicate the atom's identify—attached to each other in the correct geometry.

speciation The evolution of two or more distinct species from a single ancestral species.

species An evolutionarily independent population or group of populations. Generally distinct from other species in appearance, behavior, habitat, ecology, genetic characteristics, and so on.

species–area relationship The mathematical relationship between the area of a certain habitat and the number of species that it can support.

species diversity The variety and relative abundance of the species present in a given ecological community.

species richness The number of species present in a given ecological community.

specific heat The amount of energy required to raise the temperature of 1 gram of a substance by 1°C; a measure of the capacity of a substance to absorb energy.

sperm A mature male gamete; smaller and more mobile than the female gamete.

sperm competition Competition between the sperm of different males to fertilize eggs inside the same female.

spermatid An immature sperm cell.

spermatogenesis The production of sperm. Occurs continuously in a testis.

spermatogonium (plural: spermatogonia) Any of the diploid cells in a testis that can give rise to primary spermatocytes.

spermatophore A gelatinous package containing sperm cells that is produced by males of species that have internal fertilization without copulation.

sphincter A muscular valve that can close off a tube, as in a blood vessel or a part of the digestive tract.

spicule Stiff spike of silica or calcium carbonate that provides structural support in the body of many sponges.

spindle apparatus The array of microtubules responsible for moving chromosomes during mitosis and meiosis; includes kinetochore microtubules, polar microtubules, and astral microtubules.

spines In plants, modified leaves that are stiff and sharp and that function in defense.

spiracle In insects, a small opening that connects air-filled tracheae to the external environment, allowing for gas exchange.

spleen A dark red organ, found near the stomach of most vertebrates, that filters blood, stores extra red blood cells in case of emergency, and plays a role in immunity.

spliceosome In eukaryotes, a large, complex assembly of snRNPs (small nuclear ribonucleoproteins) that catalyzes removal of introns from primary RNA transcripts.

splicing The process by which introns are removed from primary RNA transcripts and the remaining exons are connected together.

sporangium (plural: sporangia) A spore-producing structure found in seed plants, some protists, and some fungi (e.g., chytrids).

spore (1) In bacteria, a dormant form that generally is resistant to extreme conditions. (2) In eukaryotes, a single haploid cell produced by mitosis or meiosis (not by fusion of gametes) that is capable of developing into an adult organism.

sporophyte In organisms undergoing alternation of generations, the multicellular diploid form that arises from two fused gametes and produces haploid spores. Compare with **gametophyte**.

sporopollenin A watertight material that encases spores and pollen of modern land plants.

stabilizing selection A mode of natural selection that favors phenotypes near the middle of the range of phenotypic variation. Reduces overall genetic variation in a population. Compare with **disruptive selection** and **directional selection**.

stamen The male reproductive structure of a flower. Consists of an anther, in which pollen grains are produced, and a filament, which supports the anther. Compare with **carpel**.

standing defense See **constitutive defense**.

stapes The last of three small bones (ossicles) in the middle ear of vertebrates. Receives vibrations from the tympanic membrane and by vibrating against the oval window passes them to the cochlea.

starch A mixture of two storage polysaccharides, amylose and amylopectin, both formed from α-glucose monomers. Amylopectin is branched, and amylose is unbranched. The major form of stored carbohydrate in plants.

start codon The AUG triplet in mRNA at which protein synthesis begins; codes for the amino acid methionine.

statocyst A sensory organ of many arthropods that detects the animal's orientation in space (e.g., whether the animal is flipped upside down).

statolith A tiny stone or dense particle found in specialized gravity-sensing organs in some animals such as lobsters, and in gravity-sensing tissues of plants.

statolith hypothesis The hypothesis that amyloplasts (dense, starch-storing plant organelles) serve as statoliths in gravity detection by plants.

stem cell Any relatively undifferentiated cell that can divide to produce a daughter cell that remains a stem cell and a daughter cell that can differentiate into specific cell types.

stems Vertical, aboveground structures that make up the shoot system of plants.

stereocilium (plural: stereocilia) One of many stiff outgrowths from the surface of a hair cell that are involved in detection of sound by terrestrial vertebrates or of waterborne vibrations by fishes.

steroid A class of lipid with a characteristic four-ring hydrocarbon structure.

sticky ends The short, single-stranded ends of a DNA molecule cut by a restriction endonuclease. Tend to form hydrogen bonds with other sticky ends that have complementary sequences.

stigma The sticky tip at the end of a flower carpel to which pollen grains adhere.

stolon A modified stem that runs horizontally over the soil surface and produces new plants at the nodes (a form of asexual reproduction). Compare with **rhizome**.

stoma (plural: stomata) Generally, a pore or opening. In plants, a microscopic pore, surrounded by specialized cells that open the pore, on the surface of a leaf or stem through which gas exchange occurs. See also **guard cells**.

stomach A tough, muscular pouch in the vertebrate digestive tract between the esophagus and small intestine. Physically breaks up food and begins digestion of proteins.

stop codon Any of three mRNA triplets (UAG, UGA, or UAA) that cause termination of protein synthesis. Also called a *termination codon*.

strain The lowest, most specific level of taxonomy that refers to a population of individuals that are genetically very similar or identical.

striated muscle Muscle tissue containing protein filaments organized into repeating structures that give the cells and tissues a banded appearance.

stroma The fluid matrix of a chloroplast in which the thylakoids are embedded. Site where the Calvin cycle reactions occur.

structural formula A two-dimensional notation in which the chemical symbols for the constituent atoms are joined by straight lines representing single (–), double (=), or triple (≡) covalent bonds. Compare with **molecular formula**.

structural homology Similarities in adult organismal structures (e.g., limbs, shells, flowers) that are due to inheritance from a common ancestor.

style The slender stalk of a flower carpel connecting the stigma and the ovary.

suberin Waxy substance found in the cell walls of cork tissue and in the Casparian strip of endodermal cells.

subspecies A population that has distinctive traits and some genetic differences relative to other populations of the same species but that is not distinct enough to be classified as a separate species.

substrate (1) A reactant that interacts with a catalyst, such as an enzyme or ribozyme, in a chemical reaction. (2) A surface on which a cell or organism sits.

substrate-level phosphorylation Production of ATP or GTP by the transfer of a phosphate group from an intermediate substrate directly to ADP or GDP. Occurs in glycolysis and in the citric acid cycle.

succession In ecology, the gradual colonization of a habitat after an environmental disturbance (e.g., fire, flood), usually by a series of species. See **primary** and **secondary succession**.

sucrose A disaccharide formed from glucose and fructose. One of the two main products of photosynthesis.

sugar Synonymous with carbohydrate, though usually used in an informal sense to refer to small carbohydrates (monosaccharides and disaccharides).

summation The additive effect of different postsynaptic potentials on a nerve or muscle cell, such that several subthreshold stimulations can cause or inhibit an action potential.

supporting connective tissue A type of connective tissue distinguished by having a firm extracellular matrix. Includes bone and cartilage.

surface metabolism model Hypothetical explanation for chemical evolution whereby small molecules reacted with one another through catalytic activity associated with a surface, such as the mineral deposits found in deep-sea hydrothermal vents.

surface tension The cohesive force that causes molecules at the surface of a liquid to stick together, thereby resisting deformation of the liquid's surface and minimizing its surface area.

survivorship On average, the proportion of offspring that survive to a particular age.

survivorship curve A graph depicting the percentage of a population that survives to different ages.

suspension feeder Any organism that obtains food by filtering small particles or small organisms out of water or air. Also called *filter feeder*.

sustainability The planned use of environmental resources at a rate no faster than the rate at which they are naturally replaced.

sustainable agriculture Agricultural techniques that are designed to maintain long-term soil quality and productivity.

swamp A wetland that has a steady rate of water flow and is dominated by trees and shrubs.

swim bladder A gas-filled organ of many ray-finned fishes that regulates buoyancy.

symbiosis (adjective: symbiotic) Any close and prolonged physical relationship between individuals of two different species. See **commensalism**, **mutualism**, and **parasitism**.

symmetric competition Ecological competition between two species in which both suffer similar declines in fitness. Compare with **asymmetric competition**.

sympathetic nervous system The part of the autonomic nervous system that stimulates fight-or-flight responses, such as increased heart rate, increased blood pressure, and decreased digestion. Compare with **parasympathetic nervous system**.

sympatric speciation The divergence of populations living within the same geographic area into different species as the result of their genetic (not physical) isolation. Compare with **allopatric speciation**.

sympatry Condition in which two or more populations live in the same geographic area, or close enough to permit interbreeding. Compare with **allopatry**.

symplast In plants, the space inside the plasma membranes. The symplast of adjacent cells is often connected through plasmodesmata. Compare with **apoplast**.

symporter A cotransport protein that allows an ion to diffuse down an electrochemical gradient, using the energy of that process to transport a different substance in the same direction *against* its concentration gradient. Compare with **antiporter**.

synapomorphy A shared, derived trait found in two or more taxa that is present in their most recent common ancestor but is missing in more distant ancestors. Useful for inferring evolutionary relationships.

synapse The interface between two neurons or between a neuron and an effector cell.

synapsis The physical pairing of two homologous chromosomes during prophase I of meiosis. Crossing over is observed during synapsis.

synaptic cleft The space between two communicating nerve cells (or between a neuron and effector cell) at a synapse, across which neurotransmitters diffuse.

synaptic plasticity Long-term changes in the responsiveness or physical structure of a synapse that can occur after particular stimulation patterns. Thought to be the basis of learning and memory.

synaptic vesicle A small neurotransmitter-containing vesicle at the end of an axon that releases neurotransmitter into the synaptic cleft by exocytosis.

synaptonemal complex A network of proteins that holds non-sister chromatids together during synapsis in meiosis I.

synthesis (S) phase The phase of the cell cycle during which DNA is synthesized and chromosomes are replicated.

system A defined set of interacting chemical components under observation.

systemic acquired resistance (SAR) A slow, widespread response of plants to a localized infection that protects healthy tissue from invasion by pathogens. Compare with **hypersensitive response**.

systemic circulation The part of the circulatory system that sends oxygen-rich blood from the lungs out to the rest of the body. It is separate from the pulmonary circulation in mammals and birds.

systemin A peptide hormone, produced by plant cells damaged by herbivores, that initiates a protective response in undamaged cells.

systems biology The study of the structure of networks and how interactions between individual network components such as genes or proteins can lead to emergent biological properties.

systole The portion of the cardiac cycle during which the heart muscles are contracting. Compare with **diastole**.

systolic blood pressure The force exerted by blood against artery walls during contraction of the heart's left ventricle. Compare with **diastolic blood pressure**.

T cell A type of lymphocyte that matures in the thymus and, with B cells, is responsible for adaptive immunity. Involved in activation of B cells (CD4+ helper T cells) and destruction of infected cells (CD8+ cytotoxic T cells). Also called *T lymphocytes*.

T-cell receptor (TCR) A type of transmembrane protein found on T cells that can bind to antigens displayed on the surfaces of other cells. Composed of two polypeptides, called the alpha chain and beta chain, that consist of variable and constant regions. See **antigen presentation**.

T tubule Any of the membranous tubes that extend into the interior of muscle cells, propagating action potentials throughout the cell and triggering release of calcium from the sarcoplasmic reticulum.

tagmata (singular: tagma) Prominent body regions in arthropods, such as the head, thorax, and abdomen in insects.

taiga A vast forest biome throughout subarctic regions, consisting primarily of short coniferous trees. Characterized by intensely cold winters, short summers, and high annual variation in temperature.

taproot A large, vertical main root of a plant's root system.

taste bud A sensory structure, found chiefly in the mammalian tongue, containing spindle-shaped cells that respond to chemical stimuli.

TATA-binding protein (TBP) A protein that binds to the TATA box in eukaryotic promoters and is a component of the basal transcription complex.

TATA box A short DNA sequence in many eukaryotic promoters about 30 base pairs upstream from the transcription start site.

taxon (plural: taxa) Any named group of organisms at any level of a classification system.

taxonomy The branch of biology concerned with the classification and naming of organisms.

tectorial membrane A membrane, located in the vertebrate cochlea, that takes part in the transduction of sound by bending the stereocilia of hair cells in response to sonic vibrations.

telomerase An enzyme that adds DNA to the ends of chromosomes (telomeres) by catalyzing DNA synthesis from an RNA template that is part of the enzyme.

telomere The end of a linear chromosome that contains a repeated sequence of DNA.

telophase The final stage in mitosis or meiosis during which daughter chromosomes (homologous chromosomes in meiosis I) have separated and new nuclear envelopes begin to form around each set of chromosomes.

temperate Having a climate with pronounced annual fluctuations in temperature (i.e., warm summers and cold winters) but typically neither as hot as the tropics nor as cold as the poles.

temperature A measurement of thermal energy present in an object or substance, reflecting how much the constituent molecules are moving.

template strand An original nucleic acid strand used to make a new, complementary copy based on hydrogen bonding between nitrogeneous bases.

temporal lobe In the vertebrate brain, one of the four major areas in the cerebrum.

tendon A band of tough, fibrous connective tissue that connects a muscle to a bone.

tentacle A long, thin, muscular appendage typically used for feeling and feeding. Occurs in different forms in diverse animals such as cephalopod mollusks and sea anemones.

termination (1) In enzyme-catalyzed reactions, the final stage in which the enzyme returns to its original conformation and products are released. (2) In transcription, the dissociation of RNA polymerase from DNA. (3) In translation, the dissociation of a ribosome from mRNA when it reaches a stop codon.

territory An area that is actively defended by an animal from others of its species and that provides exclusive or semi-exclusive use of its resources by the owner.

tertiary consumers In a food chain or food web, organisms that feed on secondary consumers. Compare with **primary consumer** and **secondary consumer**.

tertiary structure The overall three-dimensional shape of a single polypeptide chain, resulting from multiple interactions among the amino acid side chains and the peptide backbone. In single-stranded nucleic acids, the three-dimensional shape is formed by hydrogen bonding and hydrophobic interactions between complementary bases. Compare with **primary**, **secondary**, and **quaternary structure**.

testcross The breeding of an individual that expresses a dominant phenotype but has an unknown genotype with an individual having only recessive alleles for the traits of interest. Used to order to infer the unknown genotype from observation of the phenotypes seen in offspring.

testis (plural: testes) The sperm-producing organ of a male animal.

testosterone A steroid hormone, produced and secreted by the testes, that stimulates sperm production and various male traits and reproductive behaviors.

tetrad The structure formed by synapsed homologous chromosomes during prophase of meiosis I. Also known as a *bivalent*.

tetrapod Any member of the lineage that includes all vertebrates with two pairs of limbs (amphibians, mammals, and reptiles, including birds).

texture A quality of soil, resulting from the relative abundance of different-sized particles.

theory An explanation for a broad class of phenomena that is supported by a wide body of evidence. A theory serves as a framework for the development of new hypotheses.

thermal energy The kinetic energy of molecular motion.

thermocline A steep gradient (cline) in environmental temperature, such as occurs in a thermally stratified lake or ocean.

thermophile A bacterium or archaean that thrives in very hot environments.

thermoreception A sensory system in which receptors are activated by changes in heat energy.

thermoreceptor A sensory cell or an organ specialized for detection of changes in temperature.

thermoregulation Regulation of body temperature.

thick filament A filament composed of bundles of the motor protein myosin; anchored to the center of the sarcomere. Compare with **thin filament**.

thigmotropism Growth or movement of an organism in response to contact with a solid object.

thin filament A filament composed of two coiled chains of actin and associated regulatory proteins; anchored at the Z disk of the sarcomere. Compare with **thick filament**.

thorax A region of the body; in insects, one of the three prominent body regions, along with the head and abdomen, called tagmata.

thorn A modified plant stem shaped as a sharp, protective structure. Helps protect a plant against feeding by herbivores.

threshold potential The membrane potential that will trigger an action potential in a neuron or other excitable cell. Also called simply *threshold*.

thylakoid A membrane-bound network of flattened sac-like structures inside a plant chloroplast that functions in converting light energy to chemical energy. A stack of thylakoid discs is a granum.

thymus An organ, located in the anterior chest or neck of vertebrates, in which immature T cells that originated in the bone marrow undergo maturation.

thyroid gland A gland in the neck that releases thyroid hormone (which increases metabolic rate) and calcitonin (which lowers blood calcium).

thyroid hormones Either of two hormones, triiodothyronine (T_3) or thyroxine (T_4), produced by the thyroid gland. See **triiodothyronine** and **thyroxine**.

thyroid-stimulating hormone (TSH) A peptide hormone, produced and secreted by the anterior pituitary, that stimulates release of thyroid hormones from the thyroid gland.

thyroxine (T_4) A lipid-soluble hormone, derived from the amino acid tyrosine, containing four iodine atoms and produced and secreted by the thyroid gland. Acts primarily to increase cellular metabolism. In mammals, T_4 is converted to the more active hormone triiodothyronine (T_3) in the liver.

Ti plasmid A plasmid carried by *Agrobacterium* (a bacterium that infects plants) that can integrate into a plant cell's chromosomes and induce formation of a gall.

tight junction A type of cell–cell attachment structure that links the plasma membranes of adjacent animal cells, forming a barrier that restricts movement of substances in the space between the cells. Most abundant in epithelia (e.g., the intestinal lining). Compare with **desmosome** and **gap junction**.

tip The end of a branch on a phylogenetic tree. Represents a specific species or larger taxon that has not (yet) produced descendants—either a group living today or a group that ended in extinction. Also called *terminal node*.

tissue A group of cells that function as a unit, such as muscle tissue in an animal or xylem tissue in a plant.

tolerance In ecological succession, the phenomenon in which early-arriving species do not affect the probability that subsequent species will become established. Compare with **facilitation** and **inhibition**.

tonoplast The membrane surrounding a plant vacuole.

tool-kit genes A set of key developmental genes that establishes the body plan of animals and plants; present at the origin of the multicellular lineages and elaborated upon over evolutionary time by a process of duplication and divergence. Includes *Hox* genes.

top-down control The hypothesis that population size is limited by predators or herbivores (consumers).

topoisomerase An enzyme that prevents the twisting of DNA ahead of the advancing replication fork by cutting the DNA, allowing it to unwind, and rejoining it.

torpor An energy-conserving physiological state, marked by a decrease in metabolic rate, body temperature, and activity, that lasts for a short period (overnight to a few days or weeks). Occurs in some small mammals when the ambient temperature drops significantly. Compare with **hibernation**.

totipotent Capable of dividing and developing to form a complete, mature organism.

toxin A poison produced by a living organism, such as a plant, animal, or microorganism.

trachea (plural: tracheae) (1) In insects, any of the small air-filled tubes that extend throughout the body and function in gas exchange. (2) In terrestrial vertebrates, the airway connecting the larynx to the bronchi. Also called *windpipe*.

tracheid In vascular plants, a long, thin, water-conducting cell that has pits where its lignin-containing secondary cell wall is absent, allowing water movement between adjacent cells. Compare with **vessel element**.

trade-off In evolutionary biology, an inescapable compromise between two traits that cannot be optimized simultaneously. Also called *fitness trade-off*.

trait Any observable characteristic of an individual.

transcription The process that uses a DNA template to produce a complementary RNA.

transcription factor General term for a protein that binds to a DNA regulatory sequence to influence transcription. It includes both regulatory and basal transcription factors.

transcriptional activator A eukaryotic regulatory transcription factor that binds to regulatory DNA sequences in enhancers or promoter-proximal elements to promote the initiation of transcription.

transcriptional control Regulation of gene expression by various mechanisms that change the rate at which genes are transcribed to form messenger RNA. In negative control, binding of a regulatory protein to DNA represses transcription; in positive control, binding of a regulatory protein to DNA promotes transcription.

transcriptome The complete set of genes transcribed in a particular cell.

transduction The conversion of information from one mode to another. For example, the process by which a stimulus outside a cell is converted into a response by the cell.

transfer RNA (tRNA) An RNA molecule that has an anticodon at one end and an amino acid attachment site at the other. Each tRNA carries a specific amino acid and binds to the corresponding codon in messenger RNA during translation.

transformation (1) Incorporation of external DNA into a cell. Occurs naturally in some bacteria; can be induced in the laboratory. (2) Conversion of a normal mammalian cell to one that divides uncontrollably.

transgenic A plant or animal whose genome contains DNA introduced from another individual, often from a different species.

transition state A high-energy intermediate state of the reactants during a chemical reaction that must be achieved for the reaction to proceed. Compare with **activation energy**.

transitional feature A trait that is intermediate between a condition observed in ancestral (older) species and the condition observed in derived (younger) species.

translation The process by which a polypeptide (a string of amino acids joined by peptide bonds) is synthesized from information in codons of messenger RNA.

translational control Regulation of gene expression by various mechanisms that alter the life span of messenger RNA or the efficiency of translation.

translocation (1) In plants, the movement of sugars and other organic nutrients through the phloem by bulk flow. (2) A type of mutation in which a piece of a chromosome moves to a nonhomologous chromosome. (3) The movement of a ribosome down a messenger RNA during translation.

transmembrane protein See **integral membrane protein**.

transmission The passage or transfer of (1) a disease from one individual to another or (2) electrical impulses from one neuron to another.

transpiration Loss of water vapor from aboveground plant parts. Occurs primarily through stomata.

transporter See **carrier protein**.

transposable elements Any of several kinds of DNA sequences that are capable of moving themselves, or copies of themselves, to other locations in the genome. Include LINEs and SINEs.

tree of life The phylogenetic tree that includes all organisms.

trichome A hairlike appendage that grows from epidermal cells in the shoot system of some plants. Trichomes exhibit a variety of shapes, sizes, and functions depending on species.

triiodothyronine (T₃) A lipid-soluble hormone, derived from the amino acid tyrosine, containing three iodine atoms and produced and secreted by the thyroid gland. Acts primarily to increase cellular metabolism. In mammals, T_3 has a stronger effect than does the related hormone thyroxine (T_4).

triose A monosaccharide (simple sugar) containing three carbon atoms.

triplet code A code in which a "word" of three letters encodes one piece of information. The genetic code is a triplet code because a codon is three nucleotides long and encodes one amino acid.

triploblast (adjective: triploblastic) An animal whose body develops from three basic embryonic cell layers or tissues: ectoderm, mesoderm, and endoderm. Compare with **diploblast**.

trisomy The state of having three copies of one particular type of chromosome in an otherwise diploid cell.

tRNA See **transfer RNA**.

trochophore A larva with a ring of cilia around its middle that is found in some lophotrochozoans.

trophic cascade A series of changes in the abundance of species in a food web, usually caused by the addition or removal of a key predator.

trophic level A feeding level in an ecosystem.

trophoblast The exterior of a blastocyst (the structure that results from cleavage in embryonic development of mammals).

tropomyosin A regulatory protein present in thin (actin) filaments that blocks the myosin-binding sites on these filaments in resting muscles, thereby preventing muscle contraction.

troponin A regulatory protein, present in thin (actin) filaments, that can move tropomyosin off the myosin-binding sites on these filaments, thereby triggering muscle contraction. Activated by high intracellular calcium.

true navigation The type of navigation by which an animal can reach a specific point on Earth's surface. Also called *map orientation*.

trypsin A protein-digesting enzyme present in the small intestine that activates several other protein-digesting enzymes.

tube foot One of the many small, mobile, fluid-filled extensions of the water vascular system of echinoderms; the part extending outside the body is called a podium, while the bulb within the body is the ampulla. Used in locomotion, feeding, and respiration.

tuber A modified plant rhizome that functions in storage of carbohydrates.

tuberculosis A disease of the lungs caused by infection with the bacterium *Mycobacterium tuberculosis*.

tumor A mass of cells formed by uncontrolled cell division. Can be benign or malignant.

tumor suppressor A protein (e.g., p53 or Rb) that prevents cell division, such as when the cell has DNA damage. Mutant genes that code for tumor suppressors are associated with cancer.

tundra The treeless biome in polar and alpine regions, characterized by short, slow-growing vegetation, permafrost, and a climate of long, intensely cold winters and very short summers.

turbidity Cloudiness of water caused by sediments and/or microscopic organisms.

turgid Swollen and firm as a result of high internal pressure (e.g., a plant cell containing enough water for the cytoplasm to press against the cell wall). Compare with **flaccid**.

turgor pressure The outward pressure exerted by the fluid contents of a living plant cell against its cell wall.

turnover In lake ecology, the complete mixing of upper and lower layers of water of different temperatures; occurs each spring and fall in temperate-zone lakes.

tympanic membrane The membrane separating the middle ear from the outer ear in terrestrial vertebrates, or similar structures in insects. Also called the *eardrum*.

ubiquinone See **coenzyme Q**.

ulcer A hole in an epithelial layer, exposing the underlying tissues to damage.

ultimate causation In biology, the reason that a trait or phenomenon is thought to have evolved; the adaptive advantage of that trait. Also called *ultimate explanation*. Compare with **proximate causation**.

umami The taste of glutamate, responsible for the "meaty" taste of most proteins and of monosodium glutamate.

umbilical cord The cord that connects a developing mammalian embryo or fetus to the placenta and through which the embryo or fetus receives oxygen and nutrients.

unequal crossover An error in crossing over during meiosis I in which the two non-sister chromatids match up at different sites. Results in gene duplication in one chromatid and gene loss in the other.

unsaturated Referring to lipids in which at least one carbon-carbon bond is a double bond. Double bonds produce kinks in hydrocarbon chains and decrease the compound's melting point. Compare with **saturated**.

upstream In genetics, opposite to the direction in which RNA polymerase moves along a DNA strand. Compare with **downstream**.

urea The major nitrogenous waste of mammals, adult amphibains, and cartilaginous fishes. Compare with **ammonia** and **uric acid**.

ureter In vertebrates, a tube that transports urine from one kidney to the bladder.

urethra The tube that drains urine from the bladder to the outside environment. In male vertebrates, also used for passage of semen during ejaculation.

uric acid A whitish excretory product of birds, reptiles, and terrestrial arthropods. Used to remove from the body excess nitrogen derived from the breakdown of amino acids. Compare with **ammonia** and **urea**.

urochordates One of the three major chordate lineages (Urochordata), comprising sessile or floating, filter-feeding animals that have a polysaccharide covering (tunic) and two siphons through which water enters and exits; include tunicates and salps. Compare with **cephalochordates** and **vertebrates**.

uterus The organ in which developing embryos are housed in mammals and some other viviparous vertebrates.

vaccination Artificially producing immunological memory against a pathogen by using isolated antigens or altered versions of the pathogen to stimulate an adaptive immune response in the absence of disease.

vaccine A preparation designed to stimulate an immune response against a particular pathogen without causing illness. Vaccines consist of inactivated (killed) pathogens, live but weakened (attenuated) pathogens, or parts of a pathogen (subunit vaccine).

vacuole A large organelle in plant and fungal cells that usually is used for bulk storage of water, pigments, oils, or other substances. Some vacuoles contain enzymes and have a digestive function similar to lysosomes in animal cells.

vagina The birth canal of female mammals; a muscular tube that extends from the uterus through the pelvis to the exterior.

valence The number of unpaired electrons in the outermost electron shell of an atom; often determines how many covalent bonds the atom can form.

valence electron An electron in the outermost electron shell, the valence shell, of an atom. Valence electrons tend to be involved in chemical bonding.

valence shell The outermost electron shell of an atom.

valve In circulatory systems, any of the flaps of tissue that prevent backward flow of blood, particularly in veins and in the heart.

van der Waals interactions A weak electrical attraction between two nonpolar molecules that have been brought together through hydrophobic interactions. Often contributes to tertiary and quaternary structures in proteins.

variable number tandem repeat See **minisatellite**.

variable (V) region The amino acid sequence that changes in polypeptides used to make antibodies, B-cell receptors, and T-cell receptors, This portion of the protein is highly variable within an individual and forms the epitope-binding site. Compare with **constant (C) region**.

vas deferens (plural: vasa deferentia) A muscular tube that stores and transports semen from the epididymis to the ejaculatory duct. Also called the *ductus deferens*.

vasa recta In the vertebrate kidney, a network of blood vessels that runs alongside the loop of Henle of a nephron. Functions in reabsorption of water and solutes from the filtrate.

vascular bundle In a plant stem, a cluster of xylem and phloem strands that run the length of the stem.

vascular cambium One of two types of cylindrical meristem, consisting of a ring of undifferentiated plant cells in the stem and root of woody plants; produces secondary xylem (wood) and secondary phloem. Compare with **cork cambium**.

vascular tissue In plants, tissue that transports water, nutrients, and sugars. Made up of the complex tissues xylem and phloem, each of which contains several cell types. Also called *vascular tissue system*.

vector A biting insect or other organism that transfers pathogens from one species to another. See also **cloning vector**.

vegetative development The phase of plant development that involves growth and the the production of all plant structures except the flower.

vein Any blood vessel that carries blood (oxygenated or not) under relatively low pressure from the tissues toward the heart. Compare with **artery**.

veliger A distinctive type of larva, found in mollusks.

vena cava (plural: venae cavae) Either of two large veins that return oxygen-poor blood to the heart.

ventral Toward an animal's belly and away from its back. The opposite of dorsal.

ventricle (1) A thick-walled chamber of the heart that receives blood from an atrium and pumps it to the body or to the lungs. (2) Any of several small fluid-filled chambers in the vertebrate brain.

venule Any of the body's many small veins (blood vessels that return blood to the heart).

vertebrae (singular: vertebra) The cartilaginous or bony elements that form the backbones of vertebrate animals.

vertebrates One of the three major chordate lineages (Vertebrata), comprising animals with a dorsal column of cartilaginous or bony structures (vertebrae) and a skull enclosing the brain. Includes fishes, amphibians, mammals, and reptiles (including birds). Compare with **cephalochordates** and **urochordates**.

vessel element In vascular plants, a short, wide, water-conducting cell that has gaps through both the primary and secondary cell walls, allowing unimpeded passage of water between adjacent cells. Compare with **tracheid**.

vestigial trait A reduced or incompletely developed structure that has no function or reduced function, but is clearly similar to functioning organs or structures in closely related species.

vicariance The physical splitting of a population into smaller, isolated populations by a geographic barrier.

villi (singular: villus) Small, fingerlike projections (1) of the lining of the small intestine or (2) of the fetal portion of the placenta adjacent to maternal arteries. Function to increase the surface area available for absorption of nutrients and gas exchange.

virion The infectious extracellular particle that is produced from a viral infection; used for transmitting the virus between hosts. It consists of a DNA or RNA genome enclosed within a protein shell (capsid) that may be further enveloped in a phospholipid bilayer. Compare with **virus**.

virulence (adjective: virulent) Referring to the ability of pathogens to cause severe disease in susceptible hosts.

virus An obligate, intracellular parasite that is acellular, but uses host-cell biosynthetic machinery to replicate. Compare with **virion**.

visceral mass One of the three main parts of the mollusk body; contains most of the internal organs and external gill.

visible light The range of wavelengths of electromagnetic radiation that humans can see, from about 400 to 700 nanometers.

vitamin Any of various organic micronutrients that usually function as coenzymes.

vitelline envelope A fibrous sheet of glycoproteins that surrounds mature egg cells in many vertebrates. Surrounded by a thick gelatinous matrix (the jelly layer) in some aquatic species. In mammals, called the *zona pellucida*.

viviparous Producing live young (instead of eggs) that develop within the body of the mother before birth. Compare with **oviparous** and **ovoviviparous**.

volt (V) A unit of electrical potential (voltage).

voltage Potential energy created by a separation of electric charges between two points. Also called *electrical potential*.

voltage clamping A technique for imposing a constant membrane potential on a cell. Widely used to investigate ion channels.

voltage-gated channel An ion channel that opens or closes in response to changes in membrane voltage. Compare with **ligand-gated channel**.

voluntary muscle Muscle that contracts in response to stimulation by voluntary (somatic), but not involuntary (parasympathetic or sympathetic), neural stimulation.

vomeronasal organ A paired sensory organ, located in the nasal region, containing chemoreceptors that bind odorants and pheromones.

wall pressure The inward pressure exerted by a cell wall against the fluid contents of a living plant cell.

Wallace line A line in the Indonesian region that demarcates two areas, each of which is characterized by a distinct set of animal species.

water potential (ψ) The potential energy of water in a certain environment compared with the potential energy of pure water at room temperature and atmospheric pressure. In living organisms, ψ equals the solute potential (ψ_S) plus the pressure potential (ψ_P).

water-potential gradient A difference in water potential in one region compared with that in another region. Determines the direction that water moves, always from regions of higher water potential to regions of lower water potential.

water table The upper limit of the underground layer of soil that is saturated with water.

water vascular system In echinoderms, a system of fluid-filled tubes and chambers that functions as a hydrostatic skeleton.

watershed The area drained by a single stream or river.

Watson–Crick pairing See **complementary base pairing**.

wavelength The distance between two successive crests in any regular wave, such as light waves, sound waves, or waves in water.

wax A class of lipid with extremely long, saturated hydrocarbon tails. Harder and less greasy than fats.

weather The specific short-term atmospheric conditions of temperature, moisture, sunlight, and wind in a certain area.

weathering The gradual wearing down of large rocks by rain, running water, temperature changes, and wind; one of the processes that transform rocks into soil.

weed Any plant that is adapted for growth in disturbed soils.

white blood cell Any of several types of blood cells, including neutrophils, macrophages, and lymphocytes, that circulate in blood and lymph and function in defense against pathogens.

wild type The most common phenotype seen in a wild population.

wildlife corridor Strips of wildlife habitat connecting populations that otherwise would be isolated by human-made development.

wilt To lose turgor pressure in a plant tissue.

wobble hypothesis The hypothesis that some tRNA molecules can pair with more than one mRNA codon by tolerating some non-standard base pairing in the third base, so long as the first and second bases are correctly matched.

wood Xylem resulting from secondary growth; forms strong supporting material. Also called *secondary xylem*.

worm An animal with a long, thin, tubelike body lacking limbs.

Woronin body A dense organelle in certain fungi that plugs pores in damaged septa to prevent leakage of cytoplasm.

X-linked inheritance Inheritance patterns for genes located on the mammalian X chromosome. Also called *X-linkage*.

X-ray crystallography A technique for determining the three-dimensional structure of large molecules, including proteins and nucleic acids, by analysis of the diffraction patterns produced by X-rays beamed at crystals of the molecule.

xenoestrogens Foreign chemicals that bind to estrogen receptors or otherwise induce estrogen-like effects.

xeroderma pigmentosum (XP) A human disease characterized by extreme sensitivity to ultraviolet light. Caused by an autosomal recessive allele that inactivates the nucleotide excision DNA repair system.

xylem A plant vascular tissue that conducts water and ions; contains tracheids and/or vessel elements.

Primary xylem develops from the procambium of apical meristems; secondary xylem, or wood, from the vascular cambium. Compare with **phloem**.

Y-linked inheritance Inheritance patterns for genes located on the mammalian Y chromosome. Also called *Y-linkage*.

yeast Any fungus growing as a single-celled form. Also, a specific lineage of Ascomycota.

yolk The nutrient-rich cytoplasm inside an egg cell; used as food for the growing embryo.

Z disk The structure that forms each end of a sarcomere. Contains a protein that binds tightly to actin, thereby anchoring thin filaments.

Z scheme Model for changes in the potential energy of electrons as they pass from photosystem II to photosystem I and ultimately to NADP$^+$ during the light-dependent reactions of photosynthesis. See also **noncyclic electron flow**.

zero population growth (ZPG) A state of stable population size due to fertility staying at the replacement rate for at least one generation.

zona pellucida The gelatinous layer around a mammalian egg cell. In other vertebrates, called the *vitelline envelope*.

zone of (cellular) division In plant roots, a group of apical meristematic cells just behind the root cap where cells are actively dividing.

zone of (cellular) elongation In plant roots, a group of young cells, derived from primary meristem tissues and located behind the apical meristem, that are increasing in length.

zone of (cellular) maturation In plant roots, a group of plant cells, located several millimeters behind the root cap, that are differentiating into mature tissues.

zygosporangium (plural: zygosporangia) The distinctive spore-producing structure in fungi that are members of the Zygomycota.

zygote The cell formed by the union of two gametes; a fertilized egg.

Credits

Photo Credits

46.23 Genevieve Anderson 46.24b Reproduced by permission of Elsevier Science from K.C. Martin, et al. "Synapse-specific, long-term facilitation of *Aplysia* sensory to motor synapses: A function for local protein synthesis in memory storage." *Cell* 91: 927–938 (1997), Fig. 2a. 46.25a Nigel J. Dennis/Photo Researchers, Inc.

Chapter 47 Opener Neil Hardwick/Alamy Images 47.3 Carole Hackney 47.7 Stephen Dalton/Nature Picture Library 47.8 Piet Spaans Viridiflavus/Used under the Creative Commons licence. http://creativecommons.org/licenses/by-sa/2.5/deed.en 47.9 Kirsten Pohlmann, Leibniz-Institut für Gewässerökologie und Binnenfischerei, and Thomas Breithaupt, University of Hull 47.10 Andrew Syred/Photo Researchers, Inc. 47.12 Dr. Don Fawcett/T. Kwwabara/Photo Researchers, Inc. 47.1 Dmitry Smirnov 47.14LR David Quillin 47.18a Corbis/Alamy Images 47.18b Ted Kinsman/Photo Researchers, Inc. 47.19 Dr. Steven Campana, Bedford Institute of Oceanography, Canada ©2003 Canadian Shark Research Lab. 47.20 Christopher Zarnick/Animal Graphics

Chapter 48 Opener Bence Mate/Nature Picture Library 48.1 Don W. Fawcett/Photo Researchers, Inc. 48.T1L Biophoto Associates/Photo Researchers, Inc. 48.T1M Manfred Kage/Peter Arnold/Photolibrary/Getty Images 48.T1R Nina Zanetti/Pearson Education 48.T2 National Library of Medicine 48.12TB Eadweard Muybridge Collection/Kingston Museum/Photo Researchers, Inc. 48.15 Anders Hedenström, Lund University, Sweden

Chapter 49 Opener Anthony Mercieca/Photo Researchers, Inc. 49.5L Stephen Dalton/Photo Researchers, Inc. 49.5M Duncan McEwan/Nature Picture Library 49.5R Juniors/SuperStock 49.6L Joël Héras/Biosphoto/Photo Researchers, Inc. 49.6M Andrew Darrington/Alamy 49.6R Perennou Nuridsany/Photo Researchers, Inc. 49.8 Kenneth M. Highfill/Photo Researchers, Inc. 49.10 The Jackson Laboratory, Bar Harbor, Maine

Chapter 50 Opener Jurgen & Christine Sohns/FLPA 50.1a LUTRA/NHPA/Photoshot 50.1b Andrew J. Martinez/Photo Researchers, Inc. 50.1c Charles J. Cole 50.2 Bruce J. Russell/Biomedia Associates 50.5 Privatdozent Dr. med. habil. Holger Jastrow 50.7TB Reprinted by permission from Catherine S.C. Price et al., "Sperm competition between *Drosophila* males involves both development and incapacitation." *Nature* 400: 449–452 (1999), Figs. 2 and 3. ©1999 Macmillan Magazines Limited. Image courtesy of Jerry A. Coyne, University of Chicago. 50.8 Premaphotos/Alamy 50.10a Johanna Liljestrand Rönn 50.10inset Fleur Champion de Crespigny 50.19ab John Cancalosi/Nature Picture Library 50.19c John Cancalosi/Alamy 50.20a Claude Edelmann/Petit Format/Photo Researchers, Inc. 50.20b Lennart Nilsson/Scanpix Sweden AB 50.20c Petit Format/Nestle/Science Source/Photo Researchers, Inc.

Chapter 51 Opener Eye of Science/Photo Researchers, Inc. 51.1 Susumu Nishinaga/Photo Researchers, Inc. 51.2 From "The Dorsoventral Regulatory Gene Cassette *spatzle/Toll/cactus* Controls the Potent Antifungal Response in *Drosophila* Adults." Lemaitre, B.; Nicolas, E.; Michaut, L., Reichhart, J-M.; Hoffmann, J. A. *Cell* 86(6): pp.973–983. 51.13a Steve Gschmeissner/Photo Researchers, Inc. 51.13b Steve Gschmeissner/SPL/Photo Researchers, Inc.

Chapter 52 Opener Associated Press/Gerald Herbert 52.T1-1 Paul Nicklen/National Geographic/Getty Images 52.T1-2 Sergey Gorshkov/Nature Picture Library 52.T1-3 Juniors Bildarchiv GmbH/Alamy Images 52.T1-4 Lynn M. Stone/Nature Picture Library 52.T1-5 Russ Heinl/All Canada Photos/SuperStock 52.2T Purestock/Getty Images 52.2B Comstock/Getty Images 52.3 Bruce Coleman/Alamy Images 52.UN1 stillfx/AGE fotostock 52.UN2 Anton Foltin/Shutterstock 52.UN3 David Hartnett 52.UN4 Inga Nielsen/Shutterstock 52.UN5 Bill Brooks/Alamy 52.UN6 Pixtal Images/Photolibrary 52.17 Fairfax Photos/Sandy Scheltema/Courtesy of the AGE/The AGE Photo Sales 52.18b Gerald Nowak/WaterFrame/AGE fotostock 52.UN7 imagebroker.net/SuperStock 52.UN8 Yuriy Kulyk/Shutterstock 52.UN9 John Warburton Lee/SuperStock

Chapter 53 Opener Tim Laman/National Geographic/Getty Images 53.1 Alex Wild Photography 53.5a walkingmoon/iStockphoto 53.11a Suzanne L. & Joseph T. Collins/Photo Researchers, Inc. 53.11b Robert L. Dunne/Photoshot 53.12 Dr. Bryan D. Neff 53.13 Tom McHugh/Photo Researchers, Inc.

Chapter 54 Opener Reinhard Dirscherl/AGE fotostock 54.1inset Hecker/Sauer/AGE fotostock 54.7 USDA/NRCS/Natural Resources Conservation Service 54.8a JC Schou/Biopix.dk (www.Biopix.dk) 54.12R Janine Wiedel Photolibrary/Alamy 54.15 Jeff Rotman/Oxford Scientific/Getty Images 54.16 Robert Thompson/Nature Picture Library

Chapter 55 Opener Scott Tuason/Image Quest Marine 55.1a PREMAPHOTOS/Nature Picture Library 55.1b David Tipling/Alamy Images 55.8 Henry Walter Bates 1862. "Contributions to an insect fauna of the Amazon Valley." *Lepidoptera: Heliconidae. Trans. Linn. Soc.* 23: 495–566. 55.T1-1 Francesco Tomasinelli/Photo Researchers, Inc. 55.T1-2 Stephen Dalton/Photo Researchers, Inc. 55.T1-3 Dirk Ercken/Shutterstock 55.T1-4 DJ Mattaar/Shutterstock 55.T1-5 NatureDiver/Shutterstock 55.12a–c Stephen P. Yanoviak 55.13a Raul Gonzalez Perez/Photo Researchers, Inc. 55.13b Reinhard Dirscherl/Alamy Images 55.16a Image Quest Marine 55.16b Richard Wong/Alamy 55.17 Neale Clark/Robert Harding World Imagery/Alamy Images 55.18 Tony C. Caprio 55.19(1) Elenamiv/Shutterstock 55.19(2–4) Michael P. Gadomski/Photo Researchers, Inc. 55.20L Glacier Bay National Park Photo 55.20M Christopher L. Fastie, Middlebury College 55.20R Glacier Bay National Park Photo

Chapter 56 Opener Eric Nathan/AGE fotostock 56.12a Richard Hartnup 56.12b Randall J. Schaetzl 56.13 John Campbell, USDA Forest Service, Northern Research 56.18 Associated Press 56.T1-1 avorym/Fotolia 56.T1-2 Environmental Stock Images by Streeter Photography/Alamy Images 56.T1-3 Robert Pickett/Papilio/Alamy 56.T1-4 Arterra Picture Library/Alamy Images 56.T1-5 Mark Conlin/Image Quest Marine

Chapter 57 Opener Art Wolfe/Newscom 57.1 Sergey Uryadnikov/AGE fotostock 57.2 DLILLC/Corbis 57.3 Craig McClain 57.8TB Planet Observer/Science Source/Photo Researchers, Inc. 57.T1-1 David Woodfall/Photo Researchers, Inc. 57.T1-2 Jeff Rotman/Photo Researchers, Inc. 57.T1-3 Theo Allofs/Getty Images 57.T1-4 Simon Booth/Photo Researchers, Inc. 57.T1-5 Peter Scoones/Photo Researchers, Inc. 57.13 From D. Tilman, 2005. "Causes, consequences and ethics of biodiversity." *Nature* 405: 208–211, Fig. 1. ©2000 Macmillan Magazines Ltd. 57.T2-1 Fedor Kondratenko/Shutterstock 57.T2-2 Brian Brake/Photo Researchers, Inc. 57.T2-3 Specta/Shutterstock 57.T2-4 koh sze kiat/Shutterstock 57.14 TPG/Associated Press 57.15T Mark Conlin/Alamy 57.16T Ellen Damschen 57.17 STERN/Black Star/Newscom 57.19 Brian Bailey/Taxi/Getty Images

Appendix A: Answers A7.1ab Don W. Fawcett/Photo Researchers, Inc. A7.1c Biophoto Associates/Photo Researchers, Inc. A7.1d From "Caveolin-1 expression is essential for proper nonshivering thermogenesis in brown adipose tissue." Cohen, A. W., Schubert, W., Brasaemle, D. L., Scherer, P. E., Lisanti, M. P. *Diabetes.* 2005 Mar; 54(3): 679–86, Fig. 6.

Appendix B: BioSkills B9.3 Reproduced by permission from J. P. Ferris, et al. "Synthesis of long prebiotic oligomers on mineral surfaces." *Nature* 381: 59–61 (1996), Fig. 2. ©1996 Macmillan Magazines Ltd. Image courtesy of James P. Ferris, Rensselaer Polytechnic Institute. B11.1a Biology Media/Photo Researchers, Inc. B11.1b Janice Carr/Centers for Disease Control and Prevention (CDC) B11.2ab Michael W. Davidson/Molecular Expressions B11.3 Rosalind Franklin/Photo Researchers, Inc. B12.1aL National Cancer Institute B12.1aR E.S. Anderson/Photo Researchers, Inc. B12.1b Sinclair Stammers/Photo Researchers, Inc. B13.1a Kwangshin Kim/Photo Researchers, Inc. B13.1b Richard L. (Larry) Blanton, Ph.D. B13.1c Holt Studios International/Photo Researchers, Inc. B13.1d Custom Medical Stock Photo, Inc. B13.1e Graphic Science/Alamy Images B13.1f Sinclair Stammers/Photo Researchers, Inc. B13.1g dra_schwartz/iStockphoto

Illustration and Text Credits

Chapter 1 1.3 Based on S. P. Moose, J. W. Dudley, and T. R. Rocheford. 2004. Maize selection passes the century mark: A unique resource for 21st century genomics. *Trends in Plant Science* 9 (7): 358–364, Fig. 1a; and the Illinois long-term selection experiment for oil and protein in corn (University of Illinois at Urbana–Champaign). 1.7a Based on T. P. Young and L. A. Isbell. 1991. Sex differences in giraffe feeding ecology: Energetic and social constraints. *Ethology* 87: 79–80, Figs. 5a, 6a. 1.9 Adapted by permission of AAAS and the author from M. Wittlinger, R. Wehner, and H. Wolf. 2006. The ant odometer: Stepping on stilts and stumps. *Science* 312: 1965–1967, Figs. 1, 2, 3. (http://www.sciencemag.org/content/312/5782/1965.short).

Chapter 2 T2-1 Data: D. R. Lide (editor). 2008. Standard thermodynamic properties of chemical substances. In *CRC Handbook of Physics and Chemistry*. 89th ed. Boca Raton, FL: CRC Press. 2.21 After S. L. Miller. 1953. A production of amino acids under possible primitive Earth conditions. *Science* 117 (3046): 528–529.

Chapter 3 Opener PDB ID: 2DN2. S. Y. Park, T. Yokoyama, N. Shibayama, et al. 2006. 1.25 Å resolution crystal structures of human hemoglobin in the oxy, deoxy and carbonmonoxy forms. *Journal of Molecular Biology* 360: 690–701. 3.8a PDB ID: 1TGH. Z. S. Juo, T. K. Chiu, P. M. Leiberman, et al. 1996. How proteins recognize the TATA box. *Journal of Molecular Biology* 261: 239–254. 3.8b PDB ID: 2X9K. F. Korkmaz-Ozkan, S. Koster, W. Kuhlbrandt, et al. 2010. Correlation between the OmpG secondary structure and its pH-dependent alterations monitored by FTIR. *Journal of Molecular Biology* 401: 56–67. 3.8c PDB ID: 2PTC. M. Marquart, J. Walter, J. Deisenhofer, et al. 1983. The geometry of the reactive site and of the peptide groups in trypsin, trypsinogen and its complexes with inhibitors. *Acta Crystallographica* Section B39: 480–490. 3.8d PDB ID: 1CLG. J. M. Chen, C. D. E. King, S. H. Feairheller, et al. 1991. An energetic evaluation of a "Smith" collagen microfibril model. *Journal of Protein Chemistry* 10: 535–552. 3.11b, left PDB ID: 2MHR. S. Sheriff, W. A. Hendrickson, and J. L. Smith. 1987. Structure of myohemerythrin in the azidomet state at 1.7/1.3 resolution. *Journal of Molecular Biology* 197: 273–296. 3.11b, middle PDB ID: 1FTP. N. H. Haunerland, B. L. Jacobson, G. Wesenberg, et al. 1994. Three-dimensional structure of the muscle fatty-acid-binding protein isolated from the desert locust *Schistocerca gregaria*. *Biochemistry* 33: 12378–12385. 3.11b, right PDB ID: 1IXA. M. Baron, D. G. Norman, T. S. Harvey, et al. 1992.

The three-dimensional structure of the first EGF-like module of human factor IX: Comparison with EGF and TGF-alpha. *Protein Science* 1: 81–90. **3.12a** PDB ID: 1D1L. P. B. Rupert, A. K. Mollah, M. C. Mossing, et al. 2000. The structural basis for enhanced stability and reduced DNA binding seen in engineered second-generation Cro monomers and dimers. *Journal of Molecular Biology* 296: 1079–1090. **3.12b** PDB ID: 2DN2. S.Y. Park, T. Yokoyama, N. Shibayama, et al. 2006. 1.25 Å resolution crystal structures of human hemoglobin in the oxy, deoxy and carbonmonoxy forms. *Journal of Molecular Biology* 360: 690–701. **3.14a** PDB ID: 1DMO. M. Zhang, T. Tanaka, and M. Ikura. 1995. Calcium-induced conformational transition revealed by the solution structure of apo calmodulin. *Nature Structural Biology* 2: 758–767. **3.14b** PDB ID: 3CLN. Y. S. Babu, C. E. Bugg, and W. J. Cook. 1988. Structure of calmodulin refined at 2.2 A resolution. *Journal of Molecular Biology* 204: 191–204. **3.16** PDB ID: 2PTC. M. Marquart, J. Walter, J. Deisenhofer, et al. 1983. The geometry of the reactive site and of the peptide groups in trypsin, trypsinogen and its complexes with inhibitors. *Acta Crystallographica* Section B39: 480–490.

Chapter 4 **T4-1** PDB ID: 1EHZ. H. Shi and P. B. Moore. 2000. The crystal structure of yeast phenylalanine tRNA at 1.93 Å resolution: A classic structure revisited. *RNA* 6: 1091–1105. **4.11** PDB ID: 1X8W. F. Guo, A. R. Gooding, and T. R. Cech. 2004. Structure of the *Tetrahymena* ribozyme: Base triple sandwich and metal ion at the active site. *Molecular Cell* 16: 351–362.

Chapter 5 **5.6** H. M. Florman, K. B. Bechtol, and P. M. Wassarman. 1984. Enzymatic dissection of the functions of the mouse egg's receptor for sperm. *Developmental Biology* 106: 243–255. Also H. M. Florman and P. M. Wassarman. 1985. O-linked oligosaccharides of mouse egg ZP3 account for its sperm receptor activity. *Cell* 41: 313–324; J. D. Bleil and P. M. Wassarman. 1988. Galactose at the nonreducing terminus of O-linked oligosaccharides of mouse egg zona pellucida glycoprotein ZP3 is essential for the glycoprotein's sperm receptor activity. *PNAS* 85: 6778–6782.

Chapter 6 **Opener** CHARMM-GUI Archive—Library of Pure Lipid Bilayer (www.charmm-gui.org/?doc=archive&lib=lipid_pure), POPE Bilayer Library (pope_n256.pdb). Reference: S. Jo, T. Kim, and W. Im. 2007. Automated builder and database of protein/membrane complexes for molecular dynamics simulations. *PLoS ONE* 2 (9): e880. **6.10** Data: J. de Gier, J. G. Mandersloot, and L. L. van Deenen. 1968. Lipid composition and permeability of liposomes. *Biochimica et Biophysica Acta* 150: 666–675. **6.21** Data: C. E. Bear, F. Duguay, A. L. Naismith, et al. 1992. Purification and functional reconstitution of the cystic fibrosis transmembrane conductance regulator (CFTR). *Cell* 68: 809–818. **6.22** PDB ID: 2ZZ9. K. Tani, T. Mitsuma, Y. Hiroaki, et al. 2009. Mechanism of aquaporin-4's fast and highly selective water conduction and proton exclusion. *Journal of Molecular Biology* 389: 694–706. **6.23** PDB ID: 1K4C. Y. Zhou, J. H. Morais-Cabral, A. Kaufman, et al. 2001. Chemistry of ion coordination and hydration revealed by a K^+ channel–Fab complex at 2.0 Å resolution. *Nature* 414: 43–48. **6.23** PDB ID: 3FB7. L. G. Cuello, V. Jogini, D. M. Cortes, et al. Open KcsA potassium channel in the presence of Rb^+ ion. (To be published.) **7.18** A. D. Mills, R. A. Laskey, P. Black, et al. 1980. An acidic protein which assembles nucleosomes in vitro is the most abundant protein in *Xenopus* oocyte nuclei. *Journal of Molecular Biology* 139: 561–568. Also C. Dingwall, S. V. Sharnick, and R. A. Laskey. 1982. A polypeptide domain that specifies migration of nucleoplasmin into the nucleus. *Cell* 30: 449–458.

Chapter 7 **7.5** Reprinted with kind permission from Springer Science+Business Media B.V. from David S. Goodsell, *The Machinery of Life*. 2nd ed., 2009.

Chapter 8 **8.8** PDB ID: 1Q18. V. V. Lunin, Y. Li, J. D. Schrag, et al. 2004. Crystal structures of *Escherichia coli* ATP-dependent glucokinase and its complex with glucose. *Journal of Bacteriology* 186: 6915–6927. **8.8** PDB ID: 2Q2R. A. T. Cordeiro, A. J. Caceres, D. Vertommen, et al. 2007. The crystal structure of *Trypanosoma cruzi* glucokinase reveals features determining oligomerization and anomer specificity of hexose-phosphorylating enzymes. *Journal of Molecular Biology* 372: 1215–1226. **8.14a&b** Data: N. N. Nawani, B. P. Kapadnis, A. D. Das, et al. 2002. Purification and characterization of a thermophilic and acidophilic chitinase from Microbispora sp. V2. *Journal of Applied Microbiology* 93: 865–975, Figs. 7, 8a. Also N. N. Nawani and B. P. Kapadnis. 2001. One-step purification of chitinase from *Serratia marcescens* NK1, a soil isolate. *Journal of Applied Microbiology* 90: 803–808, Figs. 3, 4. **8.15** PDB ID: 2ERK. B. J. Canagarajah, A. Khokhlatchev, M. H. Cobb, et al. 1997. Activation mechanism of the MAP kinase ERK2 by dual phosphorylation. *Cell* (Cambridge, MA) 90: 859–869. **8.15** PDB ID 3ERK. Z. Wang, B. J. Canagarajah, J. C. Boehm, et al. 1998. Structural basis of inhibitor selectivity in MAP kinases. *Structure* 6: 1117–1128.

Chapter 9 **9.7** PDB ID: 4PFK. P. R. Evans and P. J. Hudson. 1981. Phosphofructokinase: Structure and control. *Philosophical Transactions R. Soc. Lond. B: Biol. Sci.* 293: 53–62. **9.13** Data: X. Li, R. K. Dash, R. K. Pradhan, et al. 2010. A database of thermodynamic quantities for the reactions of glycolysis and the tricarboxylic acid cycle. *Journal of Physical Chemistry B* 114: 16068–16082, Table 4. **9.14** Data: D.F . Wilson, M. Erecinska, and P. L. Dutton. 1974. Thermodynamic relationships in mitochondrial oxidative phosphorylation. *Annual Review of Biophysics and Bioengineering* 3: 203–230, Tables 1, 3. Also V. D. Sled, N. I. Rudnitzky, Y. Hatefi, et al. 1994. Thermodynamic analysis of flavin in mitochondrial NADH: Ubiquinone oxidoreductase (complex I). *Biochemistry* 33: 10069–10075.

9.17 E. Racker and W. Stoeckenius. 1974. Reconstitution of purple membrane vesicles catalyzing light-driven proton uptake and adenosine triphosphate formation. *Journal of Biological Chemistry* 249: 662–663.

Chapter 10 **10.6** T. W. Engelmann. 1882. Oxygen excretion from plant cells in a microspectrum. *Botanische Zeitung* 40: 419–426. **10.7** Reprinted with kind permission from Springer Science+Business Media B.V. from The Photosynthetic Process. In *Concepts in Photobiology: Photosynthesis and Photomorphogenesis*, edited by G. S. Singhal, G. Renger, S. K. Sopory et al. Dordrecht: Kluwer Academic; co-published with Narosa Publishing House (New Delhi), pp. 11–51, Fig. 5. **10.11** R. Govindjee, Govindjee, and G. Hoch. 1964. Emerson enhancement effect in chloroplast reactions. *Plant Physiology* 39: 10–14. **10.18** A. A. Benson, J. A. Bassham, M. Calvin, et al. 1950. The path of carbon in photosynthesis. V. Paper chromatography and radioautography of the products. *Journal of the American Chemistry Society* 72: 1710–1718. **10.20a** PDB ID: 1RCX. T. C. Taylor and I. Andersson. 1997. The structure of the complex between rubisco and its natural substrate ribulose 1,5-bisphosphate. *Journal of Molecular Biology* 265: 432–444.

Chapter 11 **11.4** Reprinted with kind permission from Springer Science+Business Media B.V. from David S. Goodsell, *The Machinery of Life*. 2nd ed., 2009. **11.7b** B. Alberts, A. Johnson, J. Lewis, et al. 2002. *Molecular Biology of the Cell*. 4th ed., Fig. 19.5, p. 1069. **11.9** K. Hatta and M. Takeichi. 1986. Expression of N-cadherin adhesion molecules associated with early morphogenetic events in chick development. *Nature* 320: 447–449. Also M. Takeichi. 1988. The cadherins: Cell–cell adhesion molecules controlling animal morphogenesis. *Development* 102: 639–655.

Chapter 12 **12.6** G. J. Gorbsky, P. J. Sammack, and G. G. Borisey. 1987. Chromosomes move poleward during anaphase along stationary microtubules that coordinately dissemble from their kinetochore ends. *Journal of Cell Biology* 104: 9–18. **12.9** Based on J. L. Ptacin, S. F. Lee, E. C. Garner, et al. 2010. A spindle-like apparatus guides bacterial chromosome segregation. *Nature Cell Biology* 12: 791–798, Fig. 5. **12.10** Y. Masui and C. L. Markert. 1971. Cytoplasmic control of nuclear behavior during meiotic maturation of frog oocytes. *Journal of Experimental Zoology* 177: 129–145. **12.13** Data: the website of the National Cancer Institute (www.cancer.gov), Common Cancer Types, November 2010.

Chapter 13 **13.13** Data: National Down Syndrome Society, 2012. **13.15** L. T. Morran, O. G. Schmidt, I. A. Gelarden, et al. 2011. Running with the red queen: Host-parasite coevolution selects for biparental sex. *Science* 333: 216–218.

Chapter 14 **14.3** G. Mendel. 1866. Versuche über Pflanzen-hybriden. *Verhandlungen des naturforschenden Vereines in Brünn* 4: 3–47. English translation available from Electronic Scholarly Publishing (www.esp.org). **T14.2** Data: G. Mendel. 1866. Versuche über Pflanzen-hybriden. *Verhandlungen des naturforschenden Vereines in Brünn* 4: 3–47. **14.13** Data: T. H. Morgan. 1911. An attempt to analyze the constitution of the chromosomes on the basis of sex-limited inheritance in *Drosophila*. *Journal of Experimental Zoology* 11: 365–414. **14.20** Data: Pan American Health Organization/WHO. 2004. *Epidemiological Bulletin* 25: 9–13, Graph 1.

Chapter 15 **15.2** A. D. Hershey and M. Chase. 1952. Independent functions of viral protein and nucleic acid in growth of bacteriophage. *Journal of General Physiology* 36: 39–56. **15.5** Adapted by permission of Dr. Matthew Meselson after M. Meselson and F. W. Stahl. 1958. The replication of DNA in *Escherichia coli*. *PNAS* 44: 671–682, Fig. 6. **15.14** Data: R. C. Allsopp, H. Vaziri, C. Patterson, et al. 1992. Telomere length predicts replicative capacity of human fibroblasts. *PNAS* 89: 10114–10118. **15.18a** Data: J. E. Cleaver. 1970. DNA repair and radiation sensitivity in human (xeroderma pigmentosum) cells. *International Journal of Radiation Biology* 18: 557–565, Fig. 3. **15.18b** Data: J .E. Cleaver. 1972. Xeroderma pigmentosum: Variants with normal DNA repair and normal sensitivity to ultraviolet light. *Journal of Investigative Dermatology* 58: 124–128, Fig. 1. **15.UN2** Graph adapted by permission of the Radiation Research Society from P. Howard-Flanders and R. P. Boyce. 1966. DNA repair and genetic recombination: Studies on mutants of *Escherichia coli* defective in these processes. *Radiation Research Supplement* 6: 156–184, Fig. 8.

Chapter 16 **16.2** Data: A. M. Srb and N. H. Horowitz. 1944. The ornithine cycle in *Neurospora* and its genetic control. *Journal of Biological Chemistry* 154: 129–139.

Chapter 17 **17.2a** PDB ID: 3IYD. B. P. Hudson, J. Quispe, S. Lara-Gonzalez, et al. 2009. Three-dimensional EM structure of an intact activator-dependent transcription initiation complex. *PNAS* 106: 19830–19835. **17.11** Based on M. B. Hoagland, M. L. Stephenson, J. F. Scott, et al. 1958. A soluble ribonucleic acid intermediate in protein synthesis. *Journal of Biological Chemistry* 231: 241–257, Fig. 6. **17.13** PDB ID: 1ZJW: I. Gruic-Sovulj, N. Uter, T. Bullock, et al. 2005. tRNA-dependent aminoacyl-adenylate hydrolysis by a nonediting class I aminoacyl-tRNA synthetase. *Journal of Biological Chemistry* 280: 23978–23986. **17.14b** PDB IDs: 3FIK, 3FIH. E. Villa, J. Sengupta, L. G. Trabuco, et al. 2009. Ribosome-induced changes in elongation factor Tu conformation control GTP hydrolysis. *PNAS* 106: 1063–1068.

Chapter 18 **18.2** Data: A. B. Pardee, F. Jacob, and J. Monod. 1959. The genetic control and cytoplasmic expression of "inducibility" in the synthesis of ß-glactosidase by *E. coli*. *Journal of Molecular Biology* 1: 165–178.

Chapter 19 **Opener** PDB ID: 1ZBB. T. Schalch, S. Duda, D. F. Sargent, et al. 2005. X-ray structure of a tetranucleosome and its implications for the chromatin fibre. *Nature* 436: 138–141. **19.5** I. Sandovici, N. H. Smith, and M. D. Nitert. 2011. Maternal diet and aging alter the epigenetic control of a promoter–enhancer interaction at the *Hnf4a* gene in rat pancreatic islets. *PNAS* 108: 5449–5454. **19.7b** PDB ID: 1MDY. P. C. Ma, M. A. Rould, H. Weintraub, et al. 1994. Crystal structure of MyoD bHLH domain–DNA complex: Perspectives on DNA recognition and implications for transcriptional activation. *Cell* (Cambridge, MA) 77: 451–459.

Chapter 21 **21.1** Data: European Nucleotide Archive/EMBL–Bank Release Notes. Release 110, December 2011 (www.ebi.ac.uk/embl/). **21.4a&b** Data: (a) Y. Hou and S. Lin. 2009. *PLoS ONE* 4 (9): e6978, Supplemental Table S1. (b) KEGG: *Kyoto Encyclopedia of Genes and Genomes*, KEGG Organisms: Complete genomes (www.genome.jp/kegg/). **21.8** Reproduced by permission of GENDIA from www.paternity.be/information_EN.html#identitytest, Examples 1 and 2. **21.10** Based on G. T. Ryan. 2005. Synergy between sequence and size in large-scale genomics. *Nature Reviews Genetics* 6, Box 3, p. 702. **21.13** V. Pancaldi, O. S. Sarac, C. Rallis, et al. 2012. Predicting the fission yeast protein interaction network. *G3: Genes, Genomes, Genetics* 2 (4): 453–567, Fig. 5.

Chapter 22 **22.10a&b** Reproduced by permission from Macmillan Publishers Ltd after J. C. Pearson, D. Lemons, and W. McGinnis. 2005. Modulating *Hox* gene functions during animal body patterning. *Nature Reviews Genetics* 6: 893–904, Fig. 1.

Chapter 23 **23.12** H. Weintraub, S. J. Tapscott, R. L. Davis, et al. 1989. Activation of muscle-specific genes in pigment, nerve, fat, liver, and fibroblast cell lines by forced expression of MyoD. *PNAS* 86: 5434–5438.

Chapter 25 **25.4** Based on E. B. Daeschler et al. 2006. A Devonian tetrapod-like fish and the evolution of the tetrapod body plan. *Nature* 440: 757–763, Fig. 6; P. E. Ahlberg and J. A. Clack. 2006. A firm step from water to land. *Nature* 440: 747–749, Fig. 1; N. H. Shubin et al. 2006. The pectoral fin of *Tiktaalik roseae* and the origin of the tetrapod limb. *Nature* 440: 764–771, Fig. 4; M. Hildebrand and G. Goslow. 2001. *Analysis of Vertebrate Structures*, 5th ed. John Wiley and Sons, Inc. **25.11** *Indohyus* reproduced by permission from Macmillan Publishers Ltd after J. G. M. Thewissen, L. N. Cooper, M. T. Clementz, et al. 2007. Whales originated from aquatic artiodactyls in the Eocene epoch of India. *Nature* 450: 1190–1194, Fig. 5. *Rhodhocetus* reproduced by permission of AAAS after P. D. Gingerich, M. ul Haq, I. S. Zalmout, et al. 2001. Origin of whales from early Artiodactyls: Hands and feet of Eocene Protocetidae from Pakistan. *Science* 293: 2239–2242, Fig. 3. (http://www.sciencemag.org/content/293/5538/2239.abstract). *Dorudon* reproduced by permission after P. D. Gingerich, M. ul Haq, W. Von Koenigswald, et al. 2009. New protocetid whale from the middle Eocene of Pakistan: Birth on land, precocial development, and sexual dimorphism. *PLoS ONE* 4 (2): e4366, Fig. 1B. New data for Dorudon from M. D. Uhen. 2004. Form, function, and anatomy of *Dorudon atrox* (Mammalia, Cetacea): An Archaeocete from the Middle to Late Eocene of Egypt. University of Michigan Papesrs on Paleontology No. 34. *Delphinapterus* reproduced by permission of Skulls Unlimited International, Inc. (www.SkullsUnlimited.com). **25.14** Adapted by permission from D. P. Genereux and C. T. Bergstrom. 2005. Evolution in action: Understanding antibiotic resistance. In *Evolutionary Science and Society: Educating a New Generation*, edited by J. Cracraft and R. W. Bybee. Colorado Springs, CO: BSCS, 145–153, Fig 3. Data: Centers for Disease Control, 2004. **25.16** Data: P. T. Boag and P. R. Grant. 1981. Intense natural selection in a population of Darwin's finches (*Geospizinae*) in the Galápagos. *Science New Series* 214 (4516): 82–85, Table 1. **25.17** Body size and beak shape graph reproduced by permission of AAAS from P. R. Grant and B. R. Grant. 2002. *Science* 296:707–711, Fig. 1. Beak size graph reprinted by permission of AAAS from P. R. Grant and B. R. Grant. 2006. Evolution of character displacement in Darwin's finches. *Science* 313: 224–226, Fig. 2. (http://www.sciencemag.org/content/296/5568/707.abstract; http://www.sciencemag.org/content/313/5784/224.abstract).

Chapter 26 **T26.1** Data: W. C. Boyd. 1950. *Genetics and the Races of Man*. Boston: Little, Brown. Data are from Chapter 8, Blood Groups: Table 27, p. 238. **T26.2** Data: T. Markow, P. H. Hedrick, K. Zuerlein, et al. 1993. HLA polymorphism in the Havasupai: Evidence for balancing selection. *American Journal of Human Genetics* 53: 943–952, Table 3. **26.4** Data: W. E. Johnson, E. P. Onorato, M. E. Roelke, et al. 2010. Genetic restoration of the Florida panther. *Science* 329: 1641–1645. **26.5b** Data: C. R. Brown and M. B. Brown. 1998. Intense natural selection on body size and wing and tail asymmetry in cliff swallows during severe weather. *Evolution* 52: 1461–1475. **26.6b** Data: M. N. Karn, H. Lang-Brown, H. J. MacKenzie, et al. 1951. Birth weight, gestation time and survival in sibs. *Annals of Eugenics* 15: 306–322. **26.7b** Data: T. B. Smith. 1987. Bill size polymorphism and intraspecific niche utilization in an African finch. *Nature* 329: 717–719. **26.8b** Data: J. D. Blount, N. B. Metcalf, T. R. Birkhead, et al. 2003. Carotenoid modulation of immune function and sexual attractiveness in zebra finches. *Science* 300: 125–127, Fig. 1b. **26.10 b&c** Data: B. J. Le Boeuf and R. S. Peterson. 1969. Social status and mating activity in elephant seals. *Science* 163: 91–93. **26.12** Reproduced by permission of Pearson Education, Inc., from S. Freeman and J. Herron. 2004. *Evolutionary Analysis*. 3rd ed., Figs. 6.15a, 6.15c. ©2004. **26.13** W. E. Kerr and S. Wright. 1954. Experimental studies of the distribution of gene frequencies in very small populations of *Drosophilia melanogaster*: I. Forked. *Evolution* 8: 172–177. **26.16b** Data: H. Araki, B. Cooper, and M. S. Blouin. 2009. Carry-over effect of captive breeding reduces reproductive fitness of wild-born descendants in the wild. *Biology Letters* 5 (5): 621–624, Fig. 2b. **26.17** Data: S. F. Elena, V. S. Cooper, and R. E. Lenski. 1996. Punctuated evolution caused by selection of rare beneficial mutations. *Science New Series* 272: 1802–1804. **26.18b** N. Moran and T. Jarvik. 2010. Lateral transfer of genes from fungi underlies carotenoid production in aphids. *Science* 328: 624–627. **27.3a** Based on J. C. Avise and W. S. Nelson. 1989. Molecular genetic relationships of the extinct dusky seaside sparrow. *Science* 243 (4891): 646–648, Fig. 2.

Chapter 27 **27.3b** J. C. Avise and W. S. Nelson. 1989. Molecular genetic relationships of the extinct dusky seaside sparrow. *Science* 243 (4891): 646–648. **27.5b** Data: N. Knowlton, L. A. Weigt, L. A. Solorzano, et al. 1993. Divergence in proteins, mitochondrial DNA, and reproductive compatibility across the Isthmus of Panama. *Science* 260 (5114): 1629–1632, Fig. 1. **27.7** Adapted by permission of Wiley-Blackwell from H. R. Dambroski, C. Linn Jr., S. H. Berlocher, et al. 2005. The genetic basis for fruit odor discrimination in *Rhagoletis* flies and its significance for sympatric host shifts. *Evolution* 59 (9): 1953–1964, Figs. 1A, 1B. **27.11b** Data: S. E. Rohwer, E. Bermingham, and C. Wood. 2001. Plumage and mitochondrial DNA haplotype variation across a moving hybrid zone. *Evolution* 55: 405–422. **27.12** Data: L. H. Rieseberg, B. Sinervo, C. R. Linder, et al. 1996. Role of gene interactions in hybrid speciation: Evidence from ancient and experimental hybrids. *Science* 272 (5262): 741–745.

Chapter 28 **28.4c** Data adapted by permission of the publisher from M. A. Nikaido, P. Rooney, and N. Okada. 1999. Phylogenetic relationships among cetartiodactyls based on insertions of short and long interspersed elements: Hippopotamuses are the closest extant relatives of whales. *PNAS* 96: 10261–10266, Figs. 1, 7. ©1999 National Academy of Sciences, USA. **28.6, 28.7, 28.8** Data: International Commission on Stratigraphy, "International Stratigraphic Chart, 2009" (www.stratigraphy.org/column.php?id=Chart/TimeScale). The data on this site are modified from F. M. Gradstein and J. C. Ogg (eds.). 2004. *A Geologic Time Scale 2004*. Cambridge, UK: Cambridge University Press; and J. G. Ogg, G. Ogg, and F. M. Gradstein. 2008. *The Concise Geologic Time Scale*. Cambridge, UK: Cambridge University Press. **28.9** B. G. Baldwin, D. W. Kyhos, and J. Dvorak. 1990. Chloroplast DNA evolution and adaptive radiation in the Hawaiian Silversword Alliance (*Asteraceae–Madiinae*). *Annals of the Missouri Botanical Garden* 77 (1): 96–109, Fig. 2. **28.10c** J. B Losos, K. I. Warheit, and T. W. Schoener. 1997. Adaptive differentiation following experimental island colonization in Anolis lizards. *Nature* 387: 70–73. **28.13** Data: J. W. Valentine, D. H. Erwin, and D. Jablonski. 1996. Developmental evolution of metazoan body plans: The fossil evidence. *Developmental Biology* 173: 373–381, Fig. 3. Also R. de Rosa et al. 1999. *Hox* genes in brachiopods and priapulids and protostome evolution. *Nature* 399: 772–776; D. Chourrout, F. Delsuc, P. Chourrout, et al. 2006. Minimal *ProtoHox* cluster inferred from bilaterian and cnidarian *Hox* complements. *Nature* 442: 684–687. **28.14** Data: M. J. Benton. 1995. Diversification and extinction in the history of life. *Science* 268: 52–58.

Chapter 29 **29.12** Data: D. F. Wilson, M. Erecinska, and P. L. Dutton. 1974. Thermodynamic relationships in mitochondrial oxidative phosphorylation. *Annual Review of Biophysics and Bioengineering* 3: 203–230, Tables 1, 3. **30.1** S. M. Adl, A. G. Simpson, M. A. Farmer, et al. 2005. The new higher level classification of eukaryotes with emphasis on the taxonomy of protists. *Journal of Eukaryotic Microbiology* 52: 399–451. Also N. Arisue, M. Hasegawa, and T. Hashimoto. 2005. Root of the eukaryota tree as inferred from combined maximum likelihood analyses of multiple molecular sequence data. *Society for Molecular Biology and Evolution, Molecular Biology and Evolution* 22: 409–420; V. Hampl, L. Hug, J. W. Leigh, et al. 2009. Phylogenomic analyses support the monophyly of Excavata and resolve relationships among eukaryotic "supergroups." *PNAS* 106 (10): 3859–3864, Figs. 1, 2, 3; J. D. Hackett, H. S. Yoon, S. Li, et al. 2007. Phylogenomic analysis supports the monophyly of cryptophytes and haptophytes and the association of Rhizaria with chromalveolates. *Molecular Biology and Evolution* 24: 1702–1713, Fig. 1; P. Schaap, T. Winckler, M. Nelson, et al. 2006. Molecular phylogeny and evolution of morphology in the social amoebas. *Science* 314: 661–663.

Chapter 30 **30.1, 30.7** S. M. Adl, A. G. Simpson, M. A. Farmer, et al. 2005. The new higher level classification of eukaryotes with emphasis on the taxonomy of protists. *Journal of Eukaryotic Microbiology* 52: 399–451. Also N. Arisue, M. Hasegawa, and T. Hashimoto. 2005. Root of the eukaryota tree as inferred from combined maximum likelihood analyses of multiple molecular sequence data. *Society for Molecular Biology and Evolution, Molecular Biology and Evolution* 22: 409–420; V. Hampl, L. Hug, J. W. Leigh, et al. 2009. Phylogenomic analyses support the monophyly of Excavata and resolve relationships among eukaryotic "supergroups." *PNAS* 106 (10): 3859–3864, Figs. 1, 2, 3; J. D. Hackett, H. S. Yoon, S. Li, et al. 2007. Phylogenomic analysis supports the monophyly of cryptophytes and haptophytes and the association of Rhizaria with chromalveolates. *Molecular Biology and Evolution* 24: 1702–1713, Fig. 1; P. Schaap, T. Winckler, M. Nelson, et al. 2006. Molecular phylogeny and evolution of morphology in the social amoebas. *Science* 314: 661–663.

Chapter 31 **31.6, 31.9, 31.23** Y.-L. Qiu, L. Li, B. Wang, et al. 2006. The deepest divergences in land plants inferred from phylogenomic evidence. *PNAS* 103 (42): 15511–15516, Fig. 1. Also K. S. Renzaglia, S. Schuette, R. J. Duff, et al. 2007. Bryophyte phylogeny: Advancing the molecular and morphological frontiers. American Bryological and Lichenological Society, *The Bryologist* 110 (2): 179–213; J.-F. Pombert, C. Otis, C. Lemieux, et al. 2005. The chloroplast genome sequence of the green alga *Pseudendoclonium akinetum* (*Ulvophyceae*) reveals unusual structural features and new insights into the branching order of chlorophyte lineages. *Molecular Biology and Evolution* 22: 1903–1918. **31.21** Based on M. E. Hoballah, T. Gubitz, J. Stuurman, et al. 2007. Single gene-mediated shift in pollinator attraction in Petunia. *Plant Cell* 19: 779–790, Fig. 6g, 6h. **31.25** P. S. Soltis and D. E. Soltis. 2004. The origin and diversification of angiosperms. *American Journal of Botany* 91 (10): 1614–1626, Figs. 1, 2, 3.

Chapter 32 **32.7** S. M. Adl, A. G. Simpson, M. A. Farmer, et al. 2005. The new higher level classification of eukaryotes with emphasis on the taxonomy of protists. *Journal of Eukaryotic Microbiology* 52: 399–451. Also N. Arisue, M. Hasegawa, and T. Hashimoto. 2005. Root of the Eukaryota tree as inferred from combined maximum likelihood analyses of multiple molecular sequence data. *Molecular Biology and Evolution* 22: 409–420; V. Hampl, L. Hug, J. W. Leigh, et al. 2009. Phylogenomic analyses support the monophyly of Excavata and resolve relationships among eukaryotic "supergroups." *PNAS* 106: 3859–3864, Figs. 1, 2, 3; J. D. Hackett, H. S. Yoon, S. Li, et al. 2007. Phylogenomic analysis supports the monophyly of cryptophytes and haptophytes and the association of Rhizaria with chromalveolates. *Molecular Biology and Evolution* 24: 1702–1713, Fig 1; P. Schaap, T. Winckler, M. Nelson, et al. 2006. Molecular phylogeny and evolution of morphology in the social amoebas. *Science* 314: 661–663. **32.8** T. Y. James, F. Kauff, C. L. Schoch, et. al. 2006. Reconstructing the early evolution of fungi using a six-gene phylogeny. *Nature* 443: 818–822, Fig. 1. Also D. J. McLaughlin, D. S. Hibbett, F. Lutzoni, et al. 2009. The search for the fungal tree of life. *Trends in Microbiology* 17: 488–497; D. S. Hibbett, M. Binder, J. F. Bischoff, et al. 2007. A higher-level phylogenetic classification of the Fungi. *Mycological Research* 111: 509–547; H. Wang, X. Zhao, L. Gao, et al. 2009. A fungal phylogeny based on 82 complete genomes using the composition vector method. *BMC Evolutionary Biology* 9: 195. **32.10** After H. Bücking and W. Heyser. 2001. Microautoradiographic localization of phosphate and carbohydrates in mycorrhizal roots of *Populus tremula* x *Populus alba* and the implications for transfer processes in ectmycorrhizal associations. *Tree Physiology* 21: 101–107.

Chapter 33 **33.2** A. M. A. Aguinaldo, J. M. Turbeville, L. S. Linford, et al. 1997. Evidence for a clade of nematodes, arthropods, and other moulting animals. *Nature* 387: 489–493, Figs. 1, 2, 3. Also C. W. Dunn, A. Hejnol, D. Q. Matus, et al. 2008. Broad phylogenomic sampling improves resolution of the animal tree of life. *Nature* 452: 745–750, Figs. 1, 2. **33.7** Based on J. R. Finnerty, K. Pang, P. Burton, et al. 2004. Origins of bilateral symmetry: *Hox* and *Dpp* expression in a sea anemone. *Science* 304: 1335–1337.

Chapter 34 **34.1** A. D. Chapman. 2009. *Numbers of Living Species in Australia and the World*. 2nd ed. Canberra: Australian Biological Resources Study. **34.2** A. M. A. Aguinaldo, J. M. Turbeville, L. S. Linford, et al. 1997. Evidence for a clade of nematodes, arthropods, and other moulting animals. *Nature* 387: 489–493, Figs. 1, 2, 3. Also C. W. Dunn, A. Hejnol, D. Q. Matus, et al. 2008. Broad phylogenomic sampling improves resolution of the animal tree of life. *Nature* 452: 745–750, Figs. 1, 2. **34.28, 34.30** Based on data in J. Regier, J. W. Schultz, A. Zwick, et al. 2010. Arthropod relationships revealed by phylogenomic analysis of nuclear protein-coding sequences. *Nature* 463: 1079–1083. **34.UN1** Based on M. Averof and N. H. Patel. 1997. Crustacean appendage evolution associated with changes in *Hox* gene expression. *Nature* 388: 682–686, Fig. 4.

Chapter 35 **35.1** After C. W. Dunn, A. Hejnol. D. Q. Matus, et al. 2008. Broad phylogenomic sampling improves resolution of the animal tree of life. *Nature* 452: 745–750, Figs. 1, 2. Also F. Delsuc, H. Brinkmann, D. Chourrout, et al. 2006. Tunicates and not cephalochordates are the closest living relatives of vertebrates. *Nature* 439: 965–968, Fig. 1. **35.12** J. E. Blair and S. B. Hedges. 2005. Molecular phylogeny and divergence times of deuterostome animals. *Molecular Biology and Evolution* 22: 2275–2284, Figs. 1, 3, 4. **35.13** Data: American Museum of Natural History; BirdLife International (birdlife.org); FishBase (www.fishbase.org); IUCN 2010. *The IUCN Red List of Threatened Species* (http://www.iucnredlist.org); The Reptile Database (reptile-database.org). **35.16** Based on E. B. Daeschler, N. H. Shubin, F. A. Jenkins, et al. 2006. A Devonian tetrapod-like fish and the evolution of the tetrapod body plan. *Nature* 440: 757–763, Fig. 6. Also N. H. Shubin, E. B. Daeschler, F. A. Jenkins, et al. 2006. The pectoral fin of *Tiktaalik roseae* and the origin of the tetrapod limb. *Nature* 440:764–771, Fig. 4. **35.20** Color plate by Michael DiGiorgio. From Q. Li, K.-Q. Gao, J. Vinther, et al. 2010. Plumage color patterns of an extinct dinosaur. *Science* 327, Fig. 4, p. 1371. Reprinted by permission of the artist, Michael DiGiorgio. **35.22** J. E. Blair and S. B. Hedges. 2005. Molecular phylogeny and divergence times of deuterostome animals. *Molecular Biology and Evolution*, 22: 2275–2284, Figs. 1, 3, 4. Also F. R. Liu, M. M. Miyamoto, N. P. Freire, et al. 2001. Molecular and morphological supertrees for eutherian (placental) mammals. *Science*

291: 1786–1789; A. B. Prasad, M. W. Allard, E. D. Green, et al. 2008. Confirming the phylogeny of mammals by use of large comparative sequence data sets. Oxford University Press, *Molecular Biology and Evolution* 25: 1795–1808. **35.36** A. B. Prasad, M. W. Allard, E. D. Green, et al. 2008. Confirming the phylogeny of mammals by use of large comparative sequence data sets. Oxford University Press, *Molecular Biology and Evolution* 25 (9):1795–1808, Figs. 1, 2, 3. **35.37** Data: D. E. Lieberman. 2009. Palaeoanthropology: *Homo floresiensis* from head to toe. *Nature* 459: 41–42. **T35.1** S. L. Robson and B. Wood. 2008. Hominin life history: Reconstruction and evolution. *Journal of Anatomy* 212: 394–425. **35.39** J. Z. Li, D. M. Absher, H. Tang, et al. 2008. Worldwide human relationships inferred from genome-wide patterns of variation. *Science* 319: 1100–1104, Fig. 1. **35.40** After L. L. Cavalli-Sforza and M. W. Feldman. 2003. The application of molecular genetic approaches to the study of human evolution. *Nature Genetics Supplement* 33: 266–275, Fig. 3; Data: Genographic Project at National Geographic: https://genographic.nationalgeographic.com/genographic/lan/en/atlas.html; dates from August 2009 summary map: http://ngm.nationalgeographic.com/big-idea/02/queens-genes; M. Rasmussen, X. Guo, Y. Wang, et al. 2011. An Aboriginal Australian genome reveals separate human dispersals into Asia. *Science* 334: 94–98.

Chapter 36 **36.2a** Data: E. Arias. 2010. United States life tables, 2006. *National Vital Statistics Reports* 58 (21): 1–40, Table 10. Hyattsville, MD: National Center for Health Statistics. **36.3** Adapted with permission of Annual Reviews from G. Pantaleo and A. S. Fauci. 1996. Immunopathogensis of HIV infection. *Annual Review of Microbiology* 50: 825–854. Fig. 1a, p. 829. **36.7** Data courtesy of the Undergraduate Biotechnology Laboratory, California Polytechnic State University, San Luis Obispo. **36.8** A. G. Dalgleish, P. C. Beverley, P. R. Clapham, et al. 1984. The CD4 (T4) antigen is an essential component of the receptor for the AIDS retrovirus. *Nature* 312: 763–767. Also D. Klatzmann, E. Champagne, S. Chamaret, et al. 1984. T-lymphocyte T4 molecule behaves as the receptor for human retrovirus LAV. *Nature* 312: 767–768. **36.15** F. Gao, E. Bailes, D. L. Robertson, et al. 1999. Origin of HIV-1 in the chimpanzee *Pan troglodytes troglodytes*. *Nature* 397: 436–441, Fig. 2.

Chapter 37 **37.3** Illustration adapted by permission from "Root Systems of Prairie Plants," by Heidi Natura. ©1995 Conservation Research Institute. **37.5** After J. Clausen, D. D. Keck, and M. Hiesey. 1945. Experimental studies on the nature of species II: Plant evolution through amphiploidy and autoploidy, with examples from the *Madiinae*. Publication No. 564. Washington, DC: Carnegie Institute of Washington.

Chapter 38 **38.3** Data: R. G. Cline and W. S. Campbell. 1976. Seasonal and diurnal water relations of selected forest species. *Ecology* 57: 367–373, Fig. 1. **38.12** Adapted by permission of the American Society of Plant Biologists from C. M. Wei, M. T. Tyree, and E. Steudle. 1999. Direct measurement of xylem pressure in leaves of intact maize plants. A test of the cohesion-tension theory taking hydraulic architecture into consideration. *Plant Physiology* 121: 1191–1205, Fig. 6. ©1999 by American Society of Plant Biologists. **38.22** J. R. Gottwald, P. J. Krysan, J. C. Young, et al. 2000. Genetic evidence for the in planta role of phloem-specific plasma membrane sucrose transporters. *PNAS* 97: 13979–13984.

Chapter 39 **39.3** D. I. Arnon and P. R. Stout. 1939. The essentiality of certain elements in minute quantity for plants with special reference to copper. *Plant Physiology* 14: 371–375.

Chapter 40 **40.4** Based on S. Inoue, T. Kinoshita, M. Matsumoto, et al. 2008. Blue light-induced autophosphorylation of phototropin is a primary step for signaling. *PNAS* 105: 5626–5631. **40.5** After C. Darwin and F. Darwin. 1897. *The Power of Movement in Plants*. New York: D. Appleton. **40.10** After A. Lang, M. K. Chailakhyan, and I. A. Frolova. 1977. Promotion and inhibition of flower formation in a day-neutral plant in grafts with a short-day plant and a long-day plant. *PNAS* 74: 2412–2416. **T40.1** H. A. Borthwick, S. B. Hendricks, M. W. Parker, et al. 1952. A reversible photoreaction controlling seed germination. *PNAS* 38: 662–666, Table 1. **40.21** After P. G. Blackman and W. J. Davies. 1985. Root to shoot communication in maize plants of the effect of soil drying. *Journal of Experimental Botany* 36: 39–48.

Chapter 41 **41.10** Y.-L. Qiu, L. Li, B. Wang, et al. 2006. The deepest divergences in land plants inferred from phylogenomic evidence. *PNAS* 103 (42): 15511–15516, Fig. 1. Also K. S. Renzaglia, S. Schuette, R. J. Duff, et al. 2007. Bryophyte phylogeny: Advancing the molecular and morphological frontiers. American Bryological and Lichenological Society, *The Bryologist* 110 (2): 179–213; J.-F. Pombert, C. Otis, C. Lemieux, et al. 2005. The chloroplast genome sequence of the green alga *Pseudendoclonium akinetum* (*Ulvophyceae*) reveals unusual structural features and new insights into the branching order of chlorophyte lineages. *Molecular Biology and Evolution* 22: 1903–1918. **41.17** J. J. Tewksbury and G. P. Nabhan. 2001. Directed deterrence by capsaicin in chilies. *Nature* 412: 403–404.

Chapter 42 **42.2** A. M. Kerr, S. N. Gershman, and S. K. Sakaluk. 2010. Experimentally induced spermatophore production and immune responses reveal a trade-off in crickets. *Behavioral Ecology* 21: 647–654. **42.10** Data: K. Schmidt-Nielsen. 1984. *Scaling: Why Is Animal Size So Important?* Cambridge, UK: Cambridge University Press. **42.11** Adapted by permission of The Company of Biologists from

P. R. Wells and A. W. Pinder. 1996. The respiratory development of Atlantic salmon. II. Partitioning of oxygen uptake among gills, yolk sac and body surfaces. *Journal of Experimental Biology* 199: 2737–2744, Figs. 1A, 3.

Chapter 43 **43.13** Anterolateral view reproduced by permission from Macmillan Publishers Ltd after S. E. Pierce, J. A. Clack, and J. R. Hutchinson. 2012. Three-dimensional limb joint mobility in the early tetrapod *Ichthyostega*. *Nature* 486: Fig. 1, p. 523. **43.14** Adapted by permission of Elsevier from K. J. Ullrich, K. Kramer, and J. W. Boyer. 1961. Present knowledge of the counter-current system in the mammalian kidney. *Progress in Cardiovascular Diseases* 3: 395–431, Figs. 5, 7. ©1961. **43.19** J. R. Davis and D. F. DeNardo. 2007. The urinary bladder as a physiological reservoir that moderates dehydration in a large desert lizard, the Gila monster *Heloderma suspectum*. *Journal of Experimental Biology* 210 (8): 1472–1480. Reprinted with permission from The Company of Biologists.

Chapter 44 **44.14** After E. M. Wright. 1993. The intestinal Na$^+$/glucose cotransporter. *Annual Review of Physiology* 55: 575–589. Also K. Takata, T. Kasahara, M. Kasahar, et al. 1992. Immunohistochemical localization of Na$^+$-dependent glucose transporter in rat jejunum. *Cell and Tissue Research* 267: 3–9. **44.17a** Graph reproduced with permission from The American Diabetes Association from L. O. Schulz, P. H. Bennett, E. Ravussin, et al. 2006. Effects of traditional and Western environments on prevalence of type 2 diabetes in Pima Indians in Mexico and the U.S. *Diabetes Care* 29: 1866–1871, Fig. 1. ©2006 American Diabetes Association. **44.17b** Data: L. O. Schulz, P. H. Bennett, E. Ravussin, et al. 2006. Effects of traditional and Western environments on prevalence of type 2 diabetes in Pima Indians in Mexico and the U.S. *Diabetes Care* 29: 1866–1871, Table 2.

Chapter 45 **45.7** Based on Y. Komai. 1998. Augmented respiration in a flying insect. *Journal of Experimental Biology* 201: 2359–2366, Fig. 7. **45.9** R. A. Berner. 1999. Atmospheric oxygen over Phanerozoic time. *PNAS* 96: 10955–10957. Fig. 2. Copyright ©1999 National Academy of Sciences, USA. Reprinted with permission. **45.12** W. Bretz and K. Schmidt Nielsen. 1971. Bird respiration: Flow patterns in the duck lung. *Journal of Experimental Biology* 54: 103–118. **45.21** Data: E. H. Starling. 1896. On the absorption of fluids from the connective tissue spaces. *Journal of Physiology* 19: 312–326. **45.23** J. E. Blair and S. B. Hedges. 2005. Molecular phylogeny and divergence times of deuterostome animals. *Molecular Biology and Evolution* 22: 2275–2284, Figs. 1, 3, 4.

Chapter 46 **46.10** Data: O. Loewi. 1921. Über humorale Übertragbarkeit der Herznervenwirkung. *Pflügers Archiv European Journal of Physiology* 189: 239–242. **46.22** Data: R. Q. Quiroga, L. Reddy, G. Kreiman, et al. 2005. Invariant visual representation by single neurons in the human brain. *Nature* 435: 1102–1107, Fig. 1B. **46.24a&c** Based on K. C. Martin, A. Casadio, H. Zhu, et al. 1997. Synapse-specific, long-term facilitation of *Aplysia* sensory to motor synapses: A function for local protein synthesis in memory storage. *Cell* 91: 927–938.

Chapter 47 **47.2a&b** Data: J. E. Rose, J. E. Hind, D. J. Anderson, et al. 1971. Some effects of stimulus intensity on response of auditory nerve fibers in the squirrel monkey. *Journal of Neurophysiology* 34: 685–699. **47.9** K. Pohlmann, J. Atema, and T. Breithaupt. 2004. The importance of the lateral line in nocturnal predation of piscivorous catfish. *Journal of Experimental Biology* 207: 2971–2978. **47.15** Based on D. M. Hunt, K. S. Dulai, J. K. Bowmaker, et al. 1995. The chemistry of John Dalton's color blindness. *Science* 267: 984–988, Fig. 3. **47.UN1** Reprinted by permission from Macmillan Publishers Ltd. from M. K. McClintock. 1971. Menstrual synchrony and suppression. *Nature* 229: 244–245, Fig. 1.

Chapter 48 **48.3** PDB ID: 1KWO. D. M. Himmel, S. Gourinath, L. Reshetnikova, et al. 2002. Crystallographic findings on the internally uncoupled and near-rigor states of myosin: Further insights into the mechanics of the motor. *PNAS* 99: 12645–12650. **48.7** After Exercise Information—Interactive Muscle Map (www.askthetrainer.com/exercise-information/). **48.15** Reprinted by permission from Macmillan Publishers Ltd from D. F. Hoyt and C. R. Taylor. 1981. Gait and the energetics of locomotion in horses. *Nature* 292: 239–240, Fig. 2, p. 240. **48.16** S. Vogel. 2003. *Comparative Biomechanics: Life's Physical World*. 2nd ed. Princeton, NJ: Princeton University Press, Fig. 3.5, p. 60. Reprinted by permission of Princeton University Press. **48.17** Reprinted with permission from AAAS from K. Schmidt-Nielsen. 1972. Locomotion: Energy cost of swimming, flying, and running. *Science* 177: 222–228. (http://www.sciencemag.org/content/177/4045/222.extract).

Chapter 49 **49.4** W. M. Bayliss and E. H. Starling. 1902. The mechanism of pancreatic secretion. *Journal of Physiology* 28: 325–353. **49.7** Graph from J. Vanecek. 1998. Cellular mechanisms of melatonin action. *Psychological Reviews* 78: 687–721, Fig. 1, p. 688. Original Data: H. Illnerová, K. Hoffmann, J. Vanácek, et al. 1984. Adjustment of pineal melatonin and N-acetyltransferase rhythms to change from long to short photoperiod in the Djungarian hamster Phodopus sungorus. *Neuroendocrinology* 38: 226–231. **49.9** Data: B. D. King, L. Sokoloff, and R. L. Wechsler. 1952. The effects of l-epinephrine and l-nor-epinephrine upon cerebral circulation and metabolism in man. *Journal of Clinical Investigation* 31: 273–279. Also P. S. Mueller and D. Horwitz. 1962. Plasma free fatty acid and blood glucose

responses to analogues of norepinephrine in man. *Journal of Lipid Research* 3: 251–255. **49.11** Reprinted by permission from Macmillan Publishers Ltd. from D. L. Coleman. 2010. A historical perspective on Leptin. *Nature Medicine* 16: 1097–1099, Fig. 2. **49.12a&b** Data: G. V. Upton, A. Corbin, C. C. Mabry, et al. 1973. Evidence for the internal feedback phenomenon in human subjects: Effects of ACTH on plasma CRF. *Acta Endocrinologica* 73: 437–443. **48.13** Anterolateral view reproduced by permission from Macmillan Publishers Ltd after S. E. Pierce, J. A. Clack, and J. R. Hutchinson. 2012. Three-dimensional limb joint mobility in the early tetrapod *Ichthyostega*. *Nature* 486: Fig. 1, p. 523. **49.14** Data: D. Toft and J. Gorski. 1966. A receptor molecule for estrogens: Isolation from the rat uterus and preliminary characterization. *PNAS* 55: 1574–1581. **49.16b** Data: T. W. Rall, E. W. Sutherland, and J. Berthet, et al. 1957. The relationship of epinephrine and glucagon to liver phosphorylase. *Journal of Biological Chemistry* 224: 463–475. **49.17** Reproduced from E. W. Sutherland. 1971. Studies on the mechanism of hormone action. (A lecture delivered in Stockholm, December 11, 1971.) ©1971 Nobel Foundation; E. W. Sutherland. 1972. Studies on the mechanism of hormone action. *Science* 177: 401–408, Fig. 6. Also G. A. Robison, R. W. Butcher, I. Oye, et al. 1965. The effect of epinephrine on adenosine 3′,5′-phosphate levels in the isolated perfused rat heart. *Molecular Pharmacology* 1: 168–177. **49.UN1** Data: T. B. Hayes, A. Collins, M. Lee, et al. 2002. Hermaphroditic, demasculinized frogs after exposure to the herbicide atrazine at low ecologically relevant doses. *PNAS* 99: 5476–5480.

Chapter 50 **50.3a** Based on R. G. Stross and J. C. Hill. 1965. Diapause induction in *Daphnia* requires two stimuli. *Science* 150: 1462–1464, Fig. 3. **50.3b** Data: O. T. Kleiven, P. Larsson, and A. Hobæk. 1992. Sexual reproduction in *Daphnia magna* requires three stimuli. *Oikos* 65: 197–206, Table 4. **50.7** Based on C. S. C. Price, K. A. Dyer, and J. A. Coyne. 1999. Sperm competition between *Drosophila* males involves both displacement and incapacitation. *Nature* 400: 449–452, Fig. 3c. **50.9** R. Shine and M. S. Y. Lee. 1999. A reanalysis of the evolution of viviparity and egg-guarding in squamate reptiles. The Herpetologists' League, Inc., *Herpetologica* 55 (4): 538–549, Figs. 1, 2, 3. **50.10b** Reproduced by permission of Elsevier from C. Hotzy and G. Arnqvist. 2009. Sperm competition favors harmful males in seed beetles. *Current Biology* 19: 404–407, Fig. 2. ©2009. **50.17, 50.18** Data: R. Stricker, R. Eberhart, M.-C. Chevallier, et al. 2006. Establishment of detailed reference values for luteinizing hormone, follicle stimulating hormone, estradiol, and progesterone during different phases of the menstrual cycle on the Abbott ARCHITECT° analyzer. *Clinical Chemistry and Laboratory Medicine* 44: 883–887, Tables 1A, 1B. **50.22b** Data: V. W. Swayze II., V. P. Johnson, J. W. Hanson, et al. 1997. Magnetic resonance imaging of brain anomalies in fetal alcohol syndrome. *Pediatrics* 99: 232–240. **50.24** Adapted by permission of the publisher and author from U. Högberg and I. Joelsson. 1985. The decline in maternal mortality in Sweden, 1931–1980. *Acta Obstetricia et Gynecologica Scandinavica* 64: 583–592, Fig. 1. Taylor & Francis Group, http://www.informaworld.com.

Chapter 51 **51.2** Based on B. Lemaitre, E. Nicolas, L. Michaut, et al. 1996. The dorsoventral regulatory gene cassette *spätzle/Toll/cactus* controls the potent antifungal response in *Drosophila* adults. *Cell* 86: 973–983. **51.5a** PDB ID: 1IGT. L. J. Harris, S. B. Larson, K. W. Hasel, et al. 1997. Refined structure of an intact IgG2a monoclonal antibody. *Biochemistry* 36: 1581–1597. **51.5b** PDB ID: 1TCR. K. C. Garcia, M. Degano, R. L. Stanfield, et al. 1996. An αβ T cell receptor structure at 2.5 Å and its orientation in the TCR–MHC complex. *Science* 274: 209–219. **51.6** PDB ID: 1HMG. M. Knossow, R. S. Daniels, A. R. Douglas, et al. 1984. Three-dimensional structure of an antigenic mutant of the influenza virus haemagglutinin. *Nature* 311 (5987): 678–680. **51.10** PDB ID: 1DLH. L. J. Stern, J. H. Brown, T. S. Jardetzky, et al. 1994. Crystal structure of the human class II MHC protein HLA-DR1 complexed with an influenza virus peptide. *Nature* 368: 215–221. **51.17** Based on B. Alberts, A. Johnson, J. Lewis, et al. 2002. *Molecular Biology of the Cell*. 4th ed. New York: Garland Science, Taylor & Francis Group. Modified from Fig. 24-10. **52.1** Data: Food and Agricultural Organization of the United Nations.

Chapter 52 **52.4** S. B. Menke, R. N. Fisher, W. Jetz, et al. 2007. Biotic and abiotic controls of Argentine ant invasion success at local and landscape scales. *Ecology* 88 (12): 3164–3173. **52.9** D. M. Olson, E. Dinerstein, E. D. Wikramanayake, et al. 2001. Terrestrial ecoregions of the world: A new map of life on Earth. *BioScience* 51: 933–938. **52.11, 52.12, 52.13, 52.14, 52.15, 52.16** Data: E. Aguado and J. E. Burt. 1999. *Understanding Weather and Climate*. Upper Saddle River, NJ: Prentice Hall. **52.18a** Based on M. B. Saffo. 1987. New light on seaweeds. *BioScience* 37 (9): 654–664, Fig. 1. ©1987 American Institute of Biological Sciences.

Chapter 53 **53.3** After M. Sokolowski. 2001. *Drosophila*: Genetics meets behavior. *Nature Reviews Genetics* 2: 879–890, Fig. 2. **53.4** Z. Abramsky, M. L. Rosenzweig, and A. Subach. 2002. The costs of apprehensive foraging. *Ecology* 85: 1330–1340. **53.5b** Adapted by permission of the author from D. Crews. 1975. Psychobiology of reptilian reproduction. *Science* 189: 1059–1065, Fig. 1. **53.6** Adapted by permission of the author from D. Crews. 1975. Psychobiology of reptilian reproduction. *Science* 189: 1059–1065, Fig. 2. **53.7** Reprinted by permission from Macmillan Publishers Ltd from K. J. Lohmann, C. M. F. Lohmann, L. M. Ehrhart, et al. 2004. Geomagnetic map used in sea-turtle navigation. *Nature* 428: 909. **53.8** C. Egevang, et al. 2010. Tracking of

Arctic terns *Sterna paradisaea* reveals longest animal migration. *PNAS* 107(5): 2078–2081, Fig. 1B. ©2010 National Academy of Sciences, USA. Reprinted with permission. **53.15** Data: J. L. Hoogland. 1983. Nepotism and alarm calling in the black-tailed prairie dog (*Cynomys ludovicianus*). *Animal Behavior* 31: 472–479.

Chapter 54 **T54.1** Data reproduced from H. Strijbosch and R. C. M. Creemers. 1988. Comparative demography of sympatric populations of *Lacerta vivipara* and *Lacerta agilis*. *Oecologia* 76: 20–26. **54.5b** Data: G. F. Gause. 1934. *The Struggle for Existence.* Baltimore, MD: Williams & Wilkins. **54.6a** G. E. Forrester and M. A. Steele. 2000. Variation in the presence and cause of density-dependent mortality in three species of reef fishes. *Ecology* 81 (9): 2416–2427, Fig. 5. Copyright © 2000 by the Ecological Society of America. Reprinted with permission. **54.6b** Data: P. Arcese and J. N. M. Smith. 1988. Effects of population density and supplemental food on reproduction in song sparrows. *Journal of Animal Ecology* 57: 119–136, Fig. 2. **54.9** Data: D. A. MacLulich. 1937. Fluctuations in the numbers of the varying hare (*Lepus americanus*). *University of Toronto Studies Biological Series* 43. **54.10** Graphs reproduced by permission of AAAS from C. J. Krebs, S. Boutin, R. Boonstra, et al. 1995. Impact of food and predation on the snowshoe hare cycle. *Science* 269: 1112–1115, Fig. 3. (http://www.sciencemag.org/content/269/5227/1112). **54.11a** Data: United Nations, Department of Economic and Social Affairs, Population Division. 2011. *World Population Prospects: The 2010 Revision.* **54.11b** Data: United Nations, Department of Economic and Social Affairs, Population Division. 2011. *World Population Prospects: The 2010 Revision.* **54.13** United Nations, Department of Economic and Social Affairs, Population Division. 2011. *World Population Prospects: The 2010 Revision.*

Chapter 55 **55.2a&b, 55.3a&b** G. F. Gause. 1934. *The Struggle for Existence.* Baltimore. MD: Williams & Wilkins. **55.5** Data: J. H. Connell. 1961. The influence of interspecific competition and other factors on the distribution of the barnacle *Chthamalus stellatus*. *Ecology* 42: 710–723. **55.7** P. R. Grant. 1999. *Ecology and Evolution of Darwin's Finches.* ©1986 Princeton University Press. Fig. 15, p. 78. Reprinted by permission of Princeton University Press. **55.10** Adapted by permission of Ecological Society of America from G. H. Leonard, M. D. Bertness, and P. O. Yund. 1999. Crab predation, waterborne cues, and inducible defenses in the blue mussel, *Mytilus edulis*. *Ecology* 80: 1–14. **55.14** Based on J. H. Cushman and T. G. Whitham. 1989. Conditional mutualism in a membracid–ant association: temporal, age-specific, and density-dependent effects. *Ecology* 70 (4): 1040–1047. **55.15** Based on D. G. Jenkins and A. L. Buikema. 1998. Do similar communities develop in similar sites? A test with zooplankton structure and function. *Ecological Monographs* 68: 421–443. **55.16** Adapted with kind permission from Springer Science+Business Media B.V. from R. T. Paine. 1974. Intertidal community structure. *Oecologia* 15: 93–120. ©1974 Springer-Verlag. **55.18b** Data: T. W. Swetnam. 1993. Fire history and climate change in giant sequoia groves. *Science* 262: 885–889. **55.21a** Adapted by permission of Wiley-Blackwell from R. H. MacArthur and E. O. Wilson. 1963. An equilibrium theory of insular zoogeography. *Evolution* 17 (4): 373–387, Fig. 4. **55.21 b & c** Data: R. H. MacArthur and E. O. Wilson. 1963. An equilibrium theory of insular zoogeography. *Evolution* 17 (4): 373–387, Fig. 5. **55.23** Data: W. V. Reid and K. R. Miller. 1989. *Keeping Options Alive: The Scientific Basis for Conserving Biodiversity.* Washington, DC: World Resources Institute. **56.10** Data: H. Leith and R. H. Whittaker (eds.). 1975. *Primary Productivity in the Biosphere. Ecological Studies and Synthesis*, Vol. 14. New York: Springer-Verlag.

Chapter 56 **56.5** Based on D. E. Canfield, A. N. Glazer, and P. G. Falkowski. 2010. The evolution and future of Earth's nitrogen cycle. *Science* 330: 192–196. **56.13** Data: G. E. Likens, F. H. Bormann, N. M. Johnson, et al. 1970. Effects of forest cutting and herbicide treatment on nutrient budgets in the Hubbard Brook watershed-ecosystem. *Ecological Monographs* 40: 23–47. **56.14** T. Oki and S. Kanae. 2006. Global hydrological cycles and world water resources. *Science* 313: 1068–1072. **56.15** U.S Geological Survey: Changes in water levels and storage in the High Plains Aquifer, predevelopment to 2009. Fact Sheet, July 2011, Fig. 1. **56.16** N. Gruber and J. N. Galloway. 2008. An Earth-system perspective of the global nitrogen cycle. *Nature* 451: 293–296. **56.17** Based on D. E. Canfield, A. N. Glazer, and P. G. Falkowski. 2010. The evolution and future of Earth's nitrogen cycle. *Science* 330: 192–196.

56.19 B. Hönisch, A. Ridgwell, D. N. Schmidt, et al. 2012. The geological record of ocean acidification. *Science* 335: 1058–1063. **56.20** Carbon Dioxide Information Analysis Center (http://cdiac.ornl.gov/). **56.22, top** NOAA. **56.22, middle & bottom** T. R. Karl, J. M. Melillo, and T. C. Peterson (eds.). 2009. *Global Climate Change Impacts in the United States.* New York: Cambridge University Press, p. 6. **56.23** Carbon Dioxide Information Analysis Center (http://cdiac.ornl.gov/). **56.24** T. R. Karl, J. M. Melillo, and T. C. Peterson (eds). 2009. *Global Climate Change Impacts in the United States.* New York: Cambridge University Press, p. 34. **56.25** Reproduced by permission of AAAS from M. Zhao and S. W. Running. 2010. Drought-induced reduction in global terrestrial net primary production from 2000 through 2009. *Science* 329: 940–943, Fig. 2. (http://www.sciencemag.org/content/329/5994/940.abstract).

Chapter 57 **57.2** F.-R. Liu, M. M. Miyamoto, N. P. Freire, et al. 2001. Molecular and morphological supertrees for eutherian (placental) mammals. *Science* 291: 1786–1789, Fig. 1. Also A. B. Prasad, M. W. Mallard, E. D. Green, et al. 2008. Confirming the phylogeny of mammals by use of large comparative sequence data sets. Oxford University Press, *Molecular Biology and Evolution* 25 (9):1795–1808; U. Arnason, A. Gullberg, and A. Janke. 2004. Mitogenomic analyses provide new insights into cetacean origin and evolution. Elsevier, *Gene* 333: 27–34, Fig. 1. **57.4a&b** Adapted by permission of Macmillan Publishers Ltd after C. D. L. Orme, R. G. Davies, M. Burgess, et al. 2005. Global hotspots of species richness are not congruent with endemism or threat. *Nature* 436: 1016–1019, Figs. 1a, 1c. **57.5** D. M. Olson and E. Einerstein. 2002. The Global 200: Priority ecoregions for global conservation. *Annals of the Missouri Botanical Garden* 89: 199–224. Fig. 2, p. 201. Reprinted by permission of the Missouri Botanical Garden Press. **57.6** Data: IUCN 2012. *The IUCN Red List of Threatened Species. Version 2012.2.* (http://www.iucnredlist.org). **57.7** Reproduced by permission of the University of California Press Journals and the author from O. Venter et al. 2006. Threats to endangered species in Canada. *BioScience* 56 (11): 903–910, Fig 2. ©2006 American Institute of Biological Sciences. **57.9** Data: W. F. Laurance, S. G. Laurance, L. V. Ferreira, et al. 2000. Conservation: Rainforest fragmentation kills big trees. *Nature* 404: 836. **57.9** Based on W. F. Laurance, S. G. Laurance, L.V . Ferreira, et al. 1997. Biomass collapse in Amazonian forest fragments. *Science* 278: 1117–1118, Fig. 2. **57.10** Adapted by permission of Elsevier from J. M. Diamond. 1975. The island dilemma: Lessons of modern biogeographic studies for the design of natural reserves. *Biological Conservation* 7: 129–146. ©1975. **57.11** Reproduced by permission of AAAS from B. Sinervo, F. Mendez de la Cruz, D. B. Miles, et al. 2010. Erosion of lizard diversity by climate change and altered thermal niches. *Science* 328: 894–899, Fig. 3. (http://www.sciencemag.org/content/328/5980/894.abstract). **57.12** Graphs reproduced by permission of AAAS from D. Tilman, J. Knops, D. Wedin, et al. 1997. The influence of functional diversity and composition on ecosystem processes. *Science* 277: 1300–1302, Fig. 1. (http://www.sciencemag.org/content/277/5330/1300.short). **57.14** Data: W. E. Johnson, D. P. Onorato, M. E. Roelke, et al. 2010. Genetic restoration of the Florida panther. *Science* 329: 1641–1645. **57.15** Reproduced by permission of AAAS from E. I. Damschen, N. M. Haddad, J. L. Orrock, et al. 2006. Corridors increase plant species richness at large scales. *Science* 313: 1284–1286, Fig. 2B. (http://www.sciencemag.org/content/313/5791/1284.abstract). **57.17** Data: H. P. Jones and O. J. Schmitz. 2009. Rapid recovery of damaged ecosystems. *PLoS One* 4 (5): e5653, Fig. 4. **57.UN1, top** Data: A. P. Dobson. 1996. *Conservation and Biodiversity.* Scientific American Library. New York: Freeman. Original source of England maps: Reproduced by permission of the author from D. S. Wilcove, C. H. McLellan, and A. P. Dobson. 1986. Habitat fragmentation in the temperate zone. In *Conservation Biology: The Science of Scarcity and Diversity*, edited by M. E. Soule. Sunderland, MA: Sinauer Associates, pp. 237–256, Fig. 1. **57.UN1, bottom** Data: A. P. Dobson. 1996. *Conservation and Biodiversity.* Scientific American Library. New York: Freeman. Original source of U.S. Maps: W. B. Greeley, Chief, U.S. Forest Service. 1925. Relation of geography to timber supply. *Economic Geography* 1: 1–11.

Appendix B: BioSkills **TB16.1** A. Crowe, C. Dicks, and M. P. Wenderoth. 2008. Biology in bloom: Implementing Bloom's Taxonomy to enhance student learning in Biology. *CBE–Life Sciences Education* 7: 368–381, Table 3.

Index

AIDS (acquired immune deficiency syndrome), 309, **712**–14, **1056**. *See also* HIV (human immunodeficiency virus)

Air
  animal locomotion in, 984
  animal olfaction (smell) and, 965–66
  behavior of oxygen and carbon dioxide in, 903–4
  circulation patterns of, 1065–66
  lichens and quality of, 631 (*see also* Atmosphere)
  nitrogen fixation from, 787–89
  plant essential nutrients from, 780t
  water potential in, 758–59
Alarm calling, prairie dog, 1095–97
Albino sea turtle, 465f
Albumen, **693**
Alcohol fermentation, **172**–73f
Alcohol use, 1033
Alcohols, 37t
Aldehydes, 37t, 73
Aldoses, 73
Aldosterone, **877**, **1003**
Alexander, R. McNeill, 984, 986
Algae
  blooms of, 555–56, 1148f, 1160
  brown, 7, 552f, 557, 568–69, 574
  green (*see* Green algae)
  in lichens, 623, 631
  photosynthesis in, 180–81f, 184, 190–91
  protist lineages of, 569, 572, 574
  red, 572
  rRNA sequences for, 7
Alkaline mucus, 1023t
Alkaline soil, 781–82
Alkalinity, pH scale of, 29
Allantois membrane, 694
All-cells-from-cells hypothesis, 3–5
All-taxa surveys, 1175
Alleles, **238**, **261**
  in central dogma of molecular biology, 308–9
  codominance and incomplete dominance of, 272, 273f
  female choice for good, 475–76
  foraging, 1085
  frequencies of, in populations, 465–71f
  genes and, 238, 239f, 247 (*see also* Genes)
  genetic bottlenecks and, 481–82
  genetic drift and, 479
  in human inheritance, 277–79
  linkage, crossing over, and recombination of, 270–71, 274f
  in Mendelian genetics, 258t, 277t
  multiple alleleism and, 271–72
  mutations and, 305, 313, 483–84, 486
  in particulate inheritance, 261–63
  plant pathogen resistance, 815
  wild-type, 267–69
Allergens, **1055**–56
Allergies, **1055**–56
Alligators, 703
Allolactose, 341
Allopatric speciation, **494**–95
Allopatry, **494**–95, 500
Allopolyploids, **498**–99
Allosteric regulation, **149**, **343**
Aloe vera, 739f
α-amylase, **809**
α-glucose, 74, 75f, 76

α-glycosidic linkages, 75f–76
α-helix (alpha helix), **48**–49
α-tubulin, 129
Alpine skypilots, 831
ALS (Lou Gehrig's disease), 408
Alternate leaves, 738f
Alternation of generations, **567**, **588**, **823**
  fungal, 625–26f
  in land plants, 588–89, 823–25
  in protists, 567–69
Alternative splicing, RNA, 357–**58**, 360–61, **400**
Altman, Sidney, 67–68
Altruism, **1095**–98
Alvarez, Luis and Walter, 521
Alveolata, 558t, 570, 573
Alveoli, 570, **909**, 910f
Amazon rain forest, 1168, 1179, 1180. *See also* Tropical wet forests
Ambisense RNA viruses, **716**
American chestnut trees, 815
American elm trees, 613
Amines, 37t
Amino acids, **42**. *See also* Proteins
  amphipathic proteins and, 94, 95f
  as animal nutrients, 883
  carbohydrates and, 78
  in central dogma of molecular biology, 308
  functional groups and, 38
  genetic code for, 310–12
  hormones as derivatives of, 995–96
  major, found in organisms, 43t
  in metabolic pathways, 157
  as neurotransmitters, 941t
  in one-gene, one-enzyme hypothesis on gene expression, 305–7
  Oparin–Haldane theory on chemical evolution of, 41–42
  in peptidoglycan, 76
  in point mutations, 313–14
  polymerization of, to form proteins, 44–47
  ribosomal polypeptide synthesis from, in translation, 328–31
  sequences of, in genetic homologies, 450–51
  sequences of, in protein primary structure, 47–48
  side chains (R-groups) of, 42–44
  specification of, in translation, 325–26
  structure of, 42
  transfer of, to proteins by tRNAs, 326–28
Aminoacyl tRNAs, 326–**27**, 328–31
Aminoacyl-tRNA synthetases, **327**, 328f
Amino functional group, 37t–38, 42, 46, 48–49f
Amino-terminus, 46, 328
Ammonia, **865**
  in archaea metabolism, 539t
  molecular structure of, B:12f
  in nitrogen fixation, 542–44, 787
  as nitrogenous waste, 865, 866t
  in origin-of-life experiments, 33–34
  simple molecules of, 23f, 24f
Amnion, **1032**
Amniotes, **689**
Amniotic eggs, **689**–91, 693–94
Amniotic fluid, 1032
Amoeba, as model organism, B:23–B:25
Amoebic dysentery, 555t

Amoeboid motion, **565**
Amoebozoa, 558t, 569, 570
AMP (adenosine monophosphate), 162
Amphibians, **696**
  fungal parasites of, 629
  hormones in metamorphosis of, 997–98
  impacts of global climate change on, 1167
  lineages of, 696, 700
  mechanoreception by lateral line system of, 958–59
Amphioxus, 686, 687
Amphipathic compounds, **87**, 87, 94, 95f
Amplification
  cell–cell signal, 211–12
  chemical signal, 1010
  sensory signal, 953
Ampullae of Lorenzini, **967**–68
Amylases, **81**, 895t
Amylopectin, 76
Amyloplasts, **804**
Amylose, 76
Anabolic pathways, **151**–52, 157–58. *See also* Catabolic pathways
Anadromous organisms, **697**
Anaerobic respiration, 170–**71**, **542**. *See also* Aerobic respiration; Fermentation
Analyze (Bloom's taxonomy skill), B:29f–B:30t
Anaphase, **225**–27, 245f
Anaphase I, 242f, 243–44, 245f
Anaphase II, 243f, 244
Anaphylactic shock, 1056
Anatomy, **843**. *See also* Animal form and function
Anatomy, levels of organization, 849f
Ancestral groups, phylogenetic trees and, B:10–B:11
Ancestral traits, 491–92, **506**–10, B:10–B:11. *See also* Common ancestry
Anchor roots, 734–35f
Anemones, 653–54, 887f
Anfinson, Christian, 52
Angiosperms, **581**. *See also* Plant(s); Seed plants
  adaptive radiation of, 517, 595–97f
  Anthophyta lineage of, 608–9f
  in Cenozoic era, 515
  characteristics of, 581
  diversification of, 582
  flowers in, 592–93, 595f
  form and function of, 732 (*see also* Plant form and function)
  fruits in, 594–95
  life cycle of, 824f–25
  molecular phylogeny of, 583
  parasitic, 789, 790f
  pollination in, 593–94, 608–9f
  reproduction in, 823 (*see also* Plant reproduction)
Anguilla, 481
Anhydrite, 522
Animal(s), **637**. *See also* Animal cells; Animal development; Animal form and function; Animal movement; Animal nutrition; Animal reproduction
  appearance of, in Paleozoic era, 515
  biological methods for studying, 637–38
  Cambrian explosion and adaptive radiation of, 518–20, 636–37
  carbohydrate hydrolysis by digestive enzymes in, 80–81
  cell–cell attachments in, 204–7

chemical signals of, 992–93
chitin as structural polysaccharide in, 76, 78
cloning of, 409
cytokinesis in, 228
electrolytes of, 862
endocrine systems and hormones of, 991–92, 993, 994f–95 (*see also* Hormones, animal)
evolutionary innovations of, 637–46
food sources and feeding strategies of, 646–49
gap junctions in tissues of, 207–8
gas exchange and circulation in, 902–3 (*see also* Circulatory systems; Respiratory systems)
glycogen as storage polysaccharide in, 76
as heterotrophs, 176
immune systems of (*see* Immune systems)
impact of global warming on, 1166–68
key lineages of non-bilaterian, 652–54
life cycles of, 651–52
major phyla of, 638t
mass extinctions of, 520–23
as models of disease, 382
nervous systems of (*see* Nervous systems)
nitrogenous wastes of, 865–66
osmoregulation by (*see* Osmoregulation)
phylogenetic relationship of fungi to, 8, 618
phylogenies of, 637f, 639f
pollination by, 593–94, 608–9f, 830–32
populations of (*see* Population[s])
prions and diseases of, 53
protostome and deuterostome, 644–45 (*see also* Deuterostomes; Protostomes)
relative diversity of lineages of, 658f
seed dispersal by, 835–37
sensory systems of (*see* Sensory systems, animal)
shared traits of, 637
Animal cells. *See also* Eukaryotic cells
  anatomy and physiology at level of, 849f
  blood, 912, 982–83, **1039**–41, 1042t
  cultures of, B:21, B:22f (*see also* Cultures, cell and tissue)
  extracellular matrix in, 202–4
  lysosomes in, 114
  meiosis and mitosis in cell cycle of, 219–20 (*see also* Cell cycle)
  plant cells vs., 110f, 742–43 (*see also* Plant cells)
  sensory, 929
  shared development processes of, 406–9
  whole-cell dynamism of, 118–19
Animal development, 419–31. *See also* Development
  cleavage in, 423–24
  developmental principles in, 419–20
  early marsupial, 1031
  fertilization in, 420–23
  gastrulation in, 424–26
  germ cells in, 438
  germ line in, 433
  hormones in, 997–1000
  Hox genes in, 414–15
  human embryo image, 419f
  lophotrochozoan vs. ecdysozoan, 670
  ordered phases of, 420f
  organogenesis in, 426–29
  plant development vs., 432–33, 438 (*see also* Plant development)

Atlantic cod, 1153
Atlantic salmon, 483
Atmosphere. *See also* Air
 behavior of oxygen and carbon dioxide in, 903–4
 Cambrian explosion and oxygen in, 519–20
 chemical evolution conditions in, 30, 33–35
 end-Permian mass extinction and carbon dioxide in, 521
 greenhouse effect from carbon dioxide in, 1163
 plant nitrogen fixation of nitrogen from, 787–89
 prokaryotic oxygen revolution and oxygen in, 541–42
 water potential in, 758–59
Atomic level, anatomy and physiology at, 849*f*
Atomic number, **19**
Atomic weight, **20**, 29
Atoms
 chemical bonding of, in molecules, 21–24
 in chemical structures, B:12–B:13
 energy transformation in, 138*f*
 linking carbon, into organic molecules, 36
 radioactive decay of, 446–47
 structure of, 19–21
ATP (adenosine triphosphate), **60**, **143**
 actin filaments and, 128–29
 in active transport by pumps, 99–100
 in attachment of amino acids to tRNAs, 327
 in Calvin cycle, 191–92
 cell usage of, 119
 in cytokinesis, 228
 in endergonic reactions, 143–44
 as energy source for nucleic acid polymerization, 60
 energy transfer from glucose to, 81
 in enzyme catalysis, 145
 in glycolysis, 158*f*–61
 GTP vs., 162, 325
 microtubules, vesicle transport, and, 130–31
 in muscle contraction, 974–76, 979
 in photosynthesis, 178
 in photosystems, 185–89
 prokaryotic metabolic diversity in producing, 538–41
 synthesis of, by ATP synthase and chemiosmosis, 168–70
 synthesis of, from glucose oxidation in cellular respiration, 154–56
 synthesis of, in fermentation, 172–73
 synthesis of, in mitochondria, 115–16
 yield of, in cellular respiration, 170–72
ATP synthase, **168**–**69**, 170–71, 185–86, 189
Atria, **919**, 921
Atrioventricular (AV) node, **922**
Atrioventricular valves, 919
Attachments, intercellular. *See* Cell–cell attachments
Attenuated virus vaccines, 1055
Australia, 701
Australian redback spiders, 1020
Australopithecus genus, 705*t*, 706, B:11
Autocrine signals, **992**

Autoimmune diseases, 1056
Autoimmunity, **1056**
Autonomic nervous system, **942**–43, 977–78
Autophagy, **126**
Autopolyploids, **497**–98
Autoradiography, 948–49, B:15–B:16
Autosomal genes, 258*t*, 268
Autosomal inheritance, **269**, 277–78, 299
Autosomal traits
 Huntington's disease as, 378
 pituitary dwarfism as, 369
 recessive and dominant, 277–78
Autosomes, **238**, 240*t*
Autotrophs, **176**–77, **538**, 1149–50
Auxin, **798**
 in apical dominance, 806–7
 as blue-light phototropic hormone, 797–800
 in cell and tissue cultures, B:22
 as gravitropic signal, 804–5
 as morphogen in plant development, 412
 as plant growth regulator, 813*t*
 in setup of body axes in embryogenesis, 435–36
 structure and function of, 210*t*
Averages, standard error and, B:6–B:7
Averof, Michalis, 671
Aves, 703. *See also* Birds
Avian malaria, 1062
Axel, Richard, 966
Axes, body, 425, 435–36
Axes, graph, B:4–B:5
Axes, leaf, 436–38
Axillary buds, **735**
Axon hillock, **942**
Axons, **847**, **930**
 action potential propagation down, 936–38
 measuring membrane potentials of, 932–33
 in neuron anatomy, 847, 930
 squid, 130
Azoneme, **132**

B

B-cell receptors (BCRs), **1043**–47, 1048, 1050–51
B-cell tumors, 1045–46
B cells, **1043**
 activation and function of, 1054*t*
 activation of, 1050–51
 antigen receptors of, 1044–47
 as lymphocytes, 1042–43
 memory cells and, 1053–55
Baboons, 1013*f*
BAC libraries, **390**–91
Bacillus anthracis, 532
Bacillus thuringiensis, 385, 545
Backbones
 nucleic acid, 59–63, 65
 peptide, 46, 48–50
Background extinctions, 520–21. *See also* Extinctions
Bacteria, **529**. *See also* Bacteria domain; Prokaryotes
 aerobic vs. anaerobic respiration in, 171
 archaea vs., 528–29
 in bioremediation, 151

cell division by binary fission in, 228–29
 in cell theory experiment, 3–5
 citric acid cycle in, 162
 DNA of, 107–8
 DNA synthesis in, 291*f*
 evolution of antibiotic resistance in, 454–56
 flagella of, 131
 gene regulation in (*see* Gene regulation, bacterial)
 introducing recombinant plasmids into cells of, for transformation, 372
 isolating with enrichment cultures, 534*f*
 as model organism (*see Escherichia coli* [*E. coli*])
 nitrogen-fixing, 542–43, 787–89, 1133, 1158, 1160
 optimal temperature and pH for enzyme function in, 148
 origin of viruses in symbiotic, 723
 pathogenic, 531–33
 peptidoglycan as structural polysaccharide in, 76, 78, 529
 phospholipids in, 87
 photosynthesis in, 177, 184, 186–87, 515
 proteins as defense against, 54
 RNA polymerase in, 318
 transcription and translation in, 318–20*f*, 324*f*–25, 329–30
 in transgenic plant development, 385–86
Bacteria domain. *See also* Bacteria
 characteristics of, 529*t*
 key lineages of, 544, 545–47
 phylogeny of, 7*f*, 8, 536*f*
 as prokaryotic, 107, 528–29 (*see also* Prokaryotes)
Bacterial artificial chromosomes (BACs), **390**–91
Bacteriophages, 50, 715, **716**
Bacteriorhodopsin, 169, 541
Baculum, **1022**–23
Baker's yeast. *See Saccharomyces cerevisiae*
Balancing selection, **474**–75*t*
Ball-and-stick models, **24**, B:12*f*–B:13
Baltimore, David, 725
Bands, molecular, B:14–B:16
Banting, Frederick, 897
Baobab trees, 731*f*
Bar charts, B:5–B:7
Bar coding, species, **1174**
Bark, **750**, 750
Barnacles, 677, 1126, 1127*f*
Barometric pressure, animal sensing of, 646, 647*t*
Barorceptors, **924**
Barriers
 to conception, 1030*t*
 to dispersal, 1063
 to pathogens, 848–49, 1038–39
 plant cuticle as, 743, 815
 to polyspermy, 421–23
Bartel, David, 68–69
Basal body, **132**
Basal cells, **434**
Basal lamina, epithelium, **849**
Basal metabolic rate (BMR), **851**–52
Basal mutants, 435–36
Basal transcription factors, **320**, 325*t*, **356**
Basalts, 521
Base, exponential notation, B:9

Base(s), **28**–30, 38. *See also* Nitrogenous bases, nucleotide
Base *e*, exponential notation, B:9
Base 10
 exponential notation and, B:9
 metric system scientific notation and, B:1–B:2
Basic-odor hypothesis, 966
Basic side chains, 43*f*, 44
Basidia, **617**–18, 626–27
Basidiomycetes, 617–18, 626–27, 630
Basilar membrane, **956**–57
Basilisk lizards, 972*f*
Basket stars, 684
Basolateral side, epithelium, **849**
Bateman, A. J., 475
Bateman-Trivers theory, 475
Bates, Henry Walter, 1129–30
Batesian mimicry, 1129–**30**
Bateson, William, 273–75
Bats, 694, 952–53, 957–58
Bayliss, William, 893
Bdelloids, 650, 665
Bdellovibrios, 547
Beadle, George, 305
Beagle naval ship, 449–50
Beaks, cepalopod, **669**
Beaks, finch, 494–95, 1128
Beaumont, William, 890
Bedbugs, 1020
Bees, 594, 826–27, 856, 1092–93, 1097–98, 1133
Beeswax, 86*f*
Beetles, 138, 139*f*, 268, 1166*t*, 1167
Behavior, **1082**–85, 1129*f*. *See also* Behavioral ecology
Behavioral contraception methods, 1030*t*
Behavioral ecology, **1083**
 altruism in, 1095–98
 behaviors in, 1082–83, 1084–85
 communication in, 1091–95
 five questions of, 1085
 foraging in, 1085–87
 mating in, 1087–89
 migration in, 1089–91
 proximate and ultimate causation in, 1083–84
Behavioral prezygotic isolation, 491*t*
Bendall, Faye, 184
Beneficial alleles, **484**–85
Beneficial mutations, 313
Benign tumors, **233**
Benthic habitats, 519
Benthic organisms, 651, **652**
Benthic zone, lake and pond, **1077**
Benthic zone, ocean, **1079**
Benzene, 27*t*
Benzo[α]pyrene, 299
Beriberi, 147
Bermingham, Eldredge, 500–501
Bertness, Mark, 1130, 1131*f*
Beschta, Robert, 1153
Best, Charles, 897
Beta-blockers, 210
β-carotene, 180–81*f*, 385
β-galactosidase, 338–43
β-globin genes, 351, 399
β-glucose, 74, 75*f*, 76
β-glycosidic linkages, 75*f*–76
β-pleated sheet (beta-pleated sheet), **48**–49
β-tubulin, 129
Beutler, Bruce, 1040
Biewener, Andrew, 987–88
Bikonta, 558

Casein, 358
Casparian strip, **760**–61, 785
Casts, fossil, **512**
Catabolic pathways, **151**–52, 157. See also Anabolic pathways
Catalase, 115
Catalysis. See also Enzymes
  DNA as poor catalytic molecule, 64–65
  environmental factors affecting, 148
  enzyme helpers and, 147
  enzyme regulation and, 149–50
  enzymes and, 54
  eukaryotic organelles and, 110
  factors limiting rate of, 147
  of hydrolysis of carbohydrates to release glucose, 80–81
  lowering of activation energy in, 145–47
  nucleic acid polymerization and, 60–61
  polysaccharides and, 78
  RNA as catalytic molecule, 67–68
  smooth ER and, 114
  substrate positioning in, 144–45
Catalysts, 35–**36**, **144**. See also Catalysis; Enzymes
Catalyze, **54**
Catastrophic events, habitat fragmentation and, 1179
Catecholamines, **1006**
Caterpillars, 818–19
Cation exchange, **781**–82
Cations, 22f, 781–83
Causation, proximate and ultimate, 1083–84. See also Proximate causation; Ultimate causation
Cavali-Sforza, Luigi, 707
Cavities, body, 643–44
CD4 protein, **717**–78, **1048**
CD4$^+$ T cells, 1048–50, 1054t
CD8 protein, **1048**
CD8$^+$ T cells, 1048–50, 1052–53, 1054t
cDNA libraries, **372**–73, 381t
Cech, Thomas, 67–68
Cecum, **896**
Cedar Creek, 1150, 1185–86
Cell(s), **2**, 106–35. See also Animal cells; Plant cells
  animal anatomy and physiology at, 849f
  cancer and uncontrolled growth of, 233, 297, 361–63
  cell cycle of cell division of (see Cell cycle)
  cell theory and key concepts about, 2, 3–5, 106–7
  cultures of (see Cultures, cell and tissue)
  determination of, 427–29, 436–38
  developmental processes of, 406–8 (see also Development)
  eukaryotic vs. prokaryotic, 8, 107, 110t (see also Eukaryotic cells)
  functions of carbohydrates in, 78–80
  inserting genes into human, 383
  interactions between (see Cell–cell interactions)
  life cycles and diploid vs. haploid, 566–69
  lipid bilayers and first, 93–94
  living organisms and, 2–3
  microscopic images of, 3f
  microscopy in study of (see Microscopy)
  nucleus of (see Nucleus, cell)
  plasma membranes and intracellular enviroment of, 84, 100–101 (see also Plasma membranes)

protein functions in, 54
separating components of (see Centrifugation)
in top-down research approach to chemical evolution, 32
using microscopes to study structure of protist, 557–58
viral host (see Host cells)
vocabulary for describing chromosomal makeup of, 240t
water in living, 25
Cell adhesion proteins, 206–7
Cell body, **930**
Cell–cell attachments, 204–7, B:25
Cell–cell gaps, 207–9
Cell–cell interactions, 200–218. See also Chemical signaling
  connection and communication between adjacent cells, 204–9
  as developmental process, 406t, 407
  extracellular layers and, 201–4
  signaling between distant cells, 209–15
  signaling between unicellular organisms, 215–16
  for unicellular and multicellar organisms, 200–201
Cell–cell signaling
  carbohydrates in, 78–80
  chemical signals in, 992–93 (see also Chemical signaling)
  crosstalk and synthesizing input from many signals in, 215
  hormones and, in multicellular organisms, 209 (see also Hormones, animal)
  hormone structures and functions for, 210t
  in natural selection, 458–59f
  plant, 794–96
  proteins and, 54
  signal processing in, 210–14
  between unicellular organisms, 215–16
  in vegetative development, 436–38
Cell crawling, **129**
Cell cycle, **221**
  cancer as out-of-control cell division in, 232–34
  cell-cycle checkpoints and regulation of, 231–32
  cell division by meiosis and mitosis and cytokinesis in, 219–20
  cell replication in phases of, 220–23
  cytokinins and, 808
  discovery of cell–cell regulatory molecules, 229–31
  mitosis in M phase of, 223–29
  overview of, 223f
  regulation of, 229–32
Cell-cycle checkpoints, **231**–34
Cell-cycle control
  cancer and, 361–63
  cancer development from faulty, 233–34
  in cell division, 406
  checkpoints in, 231–32
  M phase–promoting factor in, 229–31
Cell death. See Apoptosis; Programmed cell death
Cell determination, 427–**29**, 436–38
Cell differentiation, 406t, **407**–10, 428f–29, 434–36, B:23
Cell division, **219**
  animal cytokinesis as, 128f–29 (see also Cytokinesis)
  bacterial, by binary fission, 228–29
  cancer and out-of-control, 232–34

cell cycle of, 219–20 (see also Cell cycle)
as developmental process, 406–7
plant, 433–34, 808
telomerase and somatic cell, 297
Cell-elongation responses, plant, 799–800
Cell expansion, 406t, 407
Cell extracts, B:17–B:18
Cell-fusion experiments, 229–30f
Cell homogenates, B:17–B:18
Cell identity, carbohydrates in, 78–80
Cell-mediated response, **1051**–53
Cell membranes. See Plasma membranes
Cell movement
  actin filaments and, 127–29
  as developmental process, 406t, 407
  flagella, cilia, and, 131–33
  gastrulation and, 424–26
  microtubules and vesicle transport, 129–31
  proteins and, 54
Cell plate, **228**
Cell proliferation, 406–8
Cell sap, **743**
Cell signaling. See Cell–cell signaling
Cell-surface hormone receptors, 1008–10
Cell theory, 2, **3**–6, 106
Cell walls, 76, **109**, **742**
  cellulose as structural polysaccharide in, 72f, 76, 78
  as characteristic of domains of life, 529t
  cohesion-tension theory and secondary, 762–63
  in cytokinesis, 228
  eukaryotic, 116, 117t
  ingestive feeding and, 563–64
  plant cell–cell attachments and, 204
  in plants, 76, 201–2, 580, 742, 745–46
  prokaryotic diversity in, 109, 529, 536–38
  protist, 562
Cellular respiration, **156**, **539**
  aerobic vs. anaerobic respiration in, 170–72
  ATP, ADP, glucose, and metabolism in, 154–55
  ATP synthesis in, 166–72
  catabolic pathways of, 152
  citric acid cycle in, 162–65
  in energy flow, 1149
  evolution of, in prokaryotes, 538
  fermentation as alternative pathway to, 155, 172–73
  free energy changes in, 165f
  gas exchange and, 903
  in global carbon cycle, 1161–62f
  glucose oxidation in four-step process of, 155–56, 164–65
  glycolysis as processing of glucose to pyruvate in, 158–61
  in metabolism, 156–58
  overview of, 156f, 164f
  photosynthesis and, 177, 195
  processing of pyruvate to acetyl CoA in, 161–62
  prokaryotic, 538–40
  summary of, 164f, 171f
Cellular slime molds, 216, 569, B:23–B:25
Cellulases, 624, **891**
Cellulose, **76**, **745**
  bird digestion of, 889
  in cell walls, 202, 745
  fungal digestion of, 612–15, 623, 624

as plant structural polysaccharide, 72f, 76, 78
ruminant digestion of, 891
structure of, 77t
Cenozoic era, **515**
Census of Marine Life, 1175
Centipedes, 675
Central dogma of molecular biology, 307–9, 317–18
Central mutants, 435
Central nervous system (CNS), **643**, **929**
  brain in, 942, 944–46 (see also Brains)
  hormones and, 991–94, 1003–6 (see also Hormones, animal)
  neurons in, 929
  origin of, in animal evolution, 643
  spinal cord in, 942, 943–44
  vertebrate, 688
Centrifugation
  in discovery of Okazaki fragments, 293
  in estrogen receptor research, 1007
  in Hershey–Chase experiment, 286
  in Meselson–Stahl experiment, 289
  separating cell components with, 118–19, B:17–B:18
Centrioles, **129**, **224**, 226t
Centromeres, **220**, 221f, 223–26, **241**
Centrosomes, **129**, **224**, 226t
Cephalization, **643**, **929**
Cephalochordates, **686**–87
Cephalopods, **662**, 669
Cerebellum, **688**, **944**
Cerebral ganglion, 929
Cerebrum, **688**, **944**–46
Cervix, **1024**–25f
Cestodes, 666
Cetaceans, evolution of, 452–53
CFTR (cystic fibrosis transmembrane conductance regulator), 97–98, 867–68
Chaetae, **667**
Chagas disease, 555t
Chambers, heart, 919–21
Channel proteins, **98**, **769**–72, **864**–65. See also Ion channels
Chaperone proteins, 360
Character displacement, **1127**–28
Characters, **506**. See also Phenotypes; Traits
Chargaff, Erwin, 61–62
Charophyceae (stoneworts), 582, 587, 600
Chase, Martha, 285–86
Checkpoints, cell-cycle, 231–34
Chelicerae, **678**
Chemical bonds, **21**
  carbon–carbon, 36
  covalent, 21–23
  ionic, 22–23f
Chemical digestion, **888**. See also Digestion, animal
Chemical energy, **31**
  carbohydrates in storage and release of, 80–81
  cellular respiration and, in fats and storage carbohydrates, 155 (see also Cellular respiration)
  in chemical reactions, 137–40f
  conversion of light energy to, in photosynthesis (see Photosynthesis)
  in energetic coupling of exergonic and endergonic reactions, 141–44
  as potential energy in chemical bonds, 31
  primary producers and, 1149–50
  secondary active transport and, 100

Detritus, **564**, **1077**, **1150**–51, 1156–57
Deuterostomes, **644**, **681**–710
  chordates, 686–88
  echinoderms, 682–86f
  embryonic development of, 681
  origin of, in animal evolution, 644–45
  phylogenetic tree of, 638t, 682f
  primates and hominins, 704–8
  protostomes vs., 658
  vertebrates, 688–703 (see also
    Vertebrates)
Devegetation, nutrient cycling and,
  1158–59
Development, 405–18. See also Animal
    development; Plant development
  challenges of embryonic, 405–6
  evolutionary change from changes in
    gene expression in, 416
  five essential processes of, 406t
  genetic equivalence and differential
    gene expression in, 408–10
  model organisms in, B:25–B:26
  shared processes of, 406–8
  triggering of differential gene
    expression by chemical signals in,
    410–15
Developmental biology, 405–6,
    B:25–B:26. See also Development
Developmental homologies, **450**
Devonian explosion, 520
Diabetes mellitus, **897**
  chromatin remodeling and, 352–53
  cytotoxic T cells and type 1, 1056
  insulin and types of, 369, 897–99
Diacylglycerol (DAG), 213t
Diaphragm, vertebrate, **910**
Diaphragms, contraceptive, 1030t
Diarrhea, 555t, 727, 896
Diastole, **923**
Diastolic blood pressure, **923**
Diatomaceous earth, 574
Diatoms, 556, 574
Dicots, **596**
Dictyostelium, 216, 569
Dictyostelium discoideum (cellular slime
    mold), B:23–B:25
Dideoxyribonucleoside triphosphates
    (ddNTPs), 377–78
Dideoxy sequencing, **377**
  in genetic engineering, 381t
  Huntington's disease gene search and,
    381–82
  value of, 376–78
Diencephalon, **944**
Diet, animal and human, 883–84, 885t.
    See also Animal nutrition
Dietary fiber, 78
Differential centrifugation, **118**–19,
    B:17–B:18. See also
    Centrifugation
Differential gene expression, **349**
  body axes and, 410
  in cell differentiation, 409–10
  eukaryotic gene regulation and, 349
  evolutionary conservation of chemical
    signals and regulatory genes in,
    414–15
  morphogens in setup of body axes in,
    410–12
  multiple uses of regulators in, 415
  regulatory genes and body positional
    information in, 412–14
  regulatory transcription factors in,
    354–56
  transcriptional control as, in
    development, 409–10

Differentiation, 406t, **407**–10,
    428f–29, 434–36. See also Cell
    differentiation
Diffusion, **92**, **862**. See also Facilitated
    diffusion; Osmosis
  across lipid bilayers, 91–92
  across prokaryotic cell walls, 109
  gas exchange by, 905–6, 916
  in osmoregulation, 862, 863f
  as passive transport, 101f
  skin-breathing and, 852
  surface area adaptions for, 852–53
Diffusion spectrum imaging, 928f
Digestion, animal, **886**–97. See also
    Animal nutrition
  carbohydrate hydrolysis by digestive
    enzymes in, 81
  complete human digestive tract in, 887f
  dietary fiber and human, 78
  digestive processes in, 888
  digestive tracts in, 886–87
  large intestine in, 896–97
  of lipids in small intestine, 895f
  lysosomes and, 114
  mammalian digestive enzymes in, 895t
  mouth and esophagus in, 888–89
  pancreatic enzymes and, 118
  plant proteinase inhibitors and, 817
  small intestine in, 892–96
  as step in animal nutrition, 883f
  stomachs in, 889–92
Digestion, fungal, 623–24
Digestive processes, animal, 888
Digestive system, human, 848f
Digestive tracts, **886**–87
Digits, significant figures and metric
    system, B:2–B:3
Dihybrid crosses, **263**–65
Dikaryotic fungi, **624**–25
Dimers, 50–51f, **129**
Dinoflagellates, 555–56, 573
Dinosaurs
  birds as, 694–95, 703
  flying, 517
  mass extinction of, 515, 521–23
Dioecious species, **827**, 828f
Diploblasts, **640**–41
Diploid cells, life cycles dominated by,
    566–69
Diploid number, 240t
Diploid organisms, **239**, 240t
Diplomonads, 571
Dipnoi, 699
Direct sequencing, **535**, **557**
  studying fungi with, 615
  studying prokaryotes with, 535
  studying protists with, 558–59
Directed-pollination hypothesis, 593–94
Directional selection, **472**–73, 475t
Disaccharides, **75**–76
Discontinuous replication hypothesis,
    292
Discontinuous strands, DNA. See
    Lagging strands, DNA
Discrete data, graphing, B:4–B:6
Discrete population growth, 1110f–11f
Discrete traits, **275**, 277t
Discussion, primary literature, B:27t
Diseases, animal
  fungal, 628–29
  in herbivore populations, 1131
  prions and, 53
  viral, 726–28
Diseases, plant
  fungal, 613, 632
  viral, 715–16, 727–28

Diseases and disorders, human
  achromatopsia from genetic
    bottleneck, 481–82
  allergies, 1055–56
  anthrax, 532
  autoimmune, 1056
  bacterial, 531t, 545–47
  beriberi, 147
  blindness from vitamin A deficiencies,
    385
  cancer (see Cancer)
  cardiovascular disease and
    arteriosclerosis, 924
  caused by protists, 554–56
  color blindness, 278–79
  Cushing's disease, 1004
  cystic fibrosis, 97–98, 867–68
  diabetes mellitus (see Diabetes
    mellitus)
  diarrhea, 896
  elephantiasis, 919, 1037f
  emerging viruses and emerging,
    723–25
  food poisoning, 532
  fungal, 613, 618, 628, 632
  from gas exchange and circulation
    failures, 903
  gene therapy for, 383–85
  genetic, 277–79
  genetic bottlenecks and, 481–82
  genetic testing for, 382
  germ theory of, 532
  hemophilia, 278
  HIV/AIDS [see AIDS [acquired
    immune deficiency syndrome];
    HIV [human immunodeficiency
    virus])
  Huntington's disease, 278, 378–82
  hypertension, 923
  hypophosphatemia (vitamin
    D-resistant rickets), 279
  immunodeficiency, 1056
  malaria, 573, 1132
  Marfan syndrome, 272–73
  multiple sclerosis (MS), 937–38, 1056
  muscular dystrophy, 200f
  myocardial infarctions, 925
  nematodes and, 674
  neurodegenerative, 408
  organs and systems parasitized by
    viral, 713f
  phenylketonuria, 273
  pituitary dwarfism, 369–70
  polymerase chain reaction and
    inherited, 376
  prions and spongiform
    encephalopathies, 53
  prokaryotes and, 531–33
  proteins in defense against, 54
  protostomes and, 658
  red-green color blindness, 963–64
  rheumatoid arthritis, 1056
  severe combined immunodeficiency
    (SCID), 383–84, 1056
  sickle-cell disease, 48, 278
  smallpox, 715, 726
  systems biology and, 402
  trypanosomiasis, 1062
  ulcers, 891
  vaccination, immunization, and,
    1054–55
  viral, 712–13, 726–28
  whole-genome sequencing
    and, 392
  xeroderma pigmentosum (XP),
    299–301

Dispersal, **1063**
  allopatric speciation by, 494–95
  in geographic distribution and
    abundance of species, 1062–63
  patterns of, for populations, 1102–3
  seed, 834–36
Dispersive replication, DNA, 288f, 289
Disruptive selection, **473**–74, 475t,
    496–97
Dissecting microscopes, B:18
Distal tubule, 872, **877**–78
Distortion, experimental, 13
Distribution, biodiversity, 1176–77
Distribution, population, 1102–3
Distribution, reference, B:7
Distribution of organisms. See
    Geographic distribution and
    abundance of organisms
Disturbance regimes, **1138**–39, 1145,
    1175, 1178, 1191–92
Disturbances, **1138**–39
Disulfide bonds, 38, **49**–50f
Dittmer, Howard, 733
Divergence, genetic. See Genetic
    divergence
Diversification
  in adaptive radiations, 516
  animal, 646–52
  of animal feeding strategies, 648–49
  in Cambrian explosion, 518–20
  fungal, 619–28
  land plant, 583–97f
  prokaryotic, 536–44
  protist, 559–69
  protostome, 657–58
  viral, 722–25
Diversity. See also Biodiversity; Species
    diversity
  adaptive immune response, 1042
  of animal large intestines, 896–97
  of animal mouthparts, 884–86
  antibody, 1045–46
  ecological, of prokaryotes, 541–44
  fungal, 615
  of hearts, 919, 920f
  in testes size, 1020
  of vertebrate lungs, 909
  viral, 712
Division, cell. See Cell division
Dixon, Henry, 762
Dll gene, 649–50f
DNA (deoxyribonucleic acid), 58.
    See also Nucleic acids; RNA
    (ribonucleic acid)
  amplification of, with polymerase
    chain reaction, 374–76
  cancer from damage to, 362–63
  cell-cycle checkpoints and damaged,
    231–32
  in central dogma of molecular biology,
    307–9
  in characteristics of domains of life,
    529t
  chromosomes, genes, and, 220 (see also
    Chromosomes; Genes)
  cloning of, 370, 374–76
  comparing *Homo sapiens* and *Homo
    neanderthalensis*, 376
  components of, 58–59
  in condensed chromatin as protected
    from DNase enzyme, 351
  damage. See Damaged DNA
  dedeoxy sequencing of, 376–78
  directionality of strands of, 59–60
  discovery of, as genetic material in
    genes, 284–86, 304

DNA (deoxyribonucleic acid)
(continued)
discovery of double helix secondary
structure of, 61–63, 286–87
electron micrograph of, 317f
eukaryotic, 112
expression of (see Gene expression)
as genetic information–containing
molecule, 64
metabolic pathways and, 157
methylation of, 351
mitochondrial, 115–16
model of condensed eukaryotic, 348f
mutations in, 483–84
next-generation technologies for
sequencing of, 378
in nucleus, 119–20
physical model of, 57f, 62f
polymerization of
deoxyribonucleotides to
form, 59–61
as poor catalytic molecule, 64–65
primary structure and secondary
structure of, 286f
prokaryotic, 107–8
as probe(s), 372–73, 381t, 401–2,
B:15–B:16
prokaryotic vs. eukaryotic, 110t
repair of damaged, 297–99
RNA structure vs. structure of, 66t
using restriction endonucleases and
DNA ligase to cut and paste,
370–72
visualizing, B:15–B:16
DNA cloning, 370, 374–76
DNA fingerprinting, 397–98f
DNA helicase, 290–91, 294t
DNA libraries, 372–73, 381t
DNA ligases, 293–94t, 370, 371–72, 380t
DNA methylation, 351
DNA methyl-transferases, 351
DNA microarrays, 400–402
DNA polymerases, 289
in DNA synthesis, 289–94t, 295
in genetic engineering, 380t
in polymerase chain reaction, 374–76
proofreading by, 297
RNA polymerases vs., 318
in telomere synthesis, 295–96
DNA sequences
cancer and regulatory, 363
in genetic code, 310–12
genetic homology and, 450–51
phylogenies based on, 510, 511f,
558–59
regulatory, in transcription initiation,
353–54
silent mutations in, as nonadaptive
traits, 461
transcription factors and specific,
355–56
DNA sequencing, 376–78. See also
Whole-genome sequencing
DNA synthesis, 284–303
comprehensive model of, 289–95
correcting mistakes in, 298–99
in dideoxy sequencing, 377–78
discovery of DNA as genetic material
in genes and, 284–86
discovery of secondary structure of
DNA and, 286–87
DNA polymerase in, 289–90

electron micrograph of, 284f
opening and stabiliizing of double
helix in, 290–91
process of, 64
proteins required for bacterial, 294t
repairing damaged DNA and
correcting mistakes in, 297–301
starting of replication in, 290
synthesis of lagging strand in, 292–95
synthesis of leading strand in, 291–92
telomere replication in, 295–97
testing of early hypotheses about DNA
replication and, 287–89
xeroderma pigmentosum (XP) case
study of damaged DNA, 299–301
DNA viruses, 715–16, 726, 726f
DNase enzyme, 351
Dogs, life cycle of, 241–42
Dolly (cloned sheep), 409
Dolphins, 452–53, 509–10, 985, 986
Domains, taxonomic, 7f, 8, 107, 528–29
Dominant alleles, 258t, 272, 273f, 277t
Dominant traits, 259. See also Recessive
traits
codominance and incomplete
dominance of, 272, 273f, 277t
human, 277–79
Huntington's disease as, 378
in Mendelian genetics, 259–60
Dopamine, 941t
Dormancy, plant, 751
Dormancy, seed, 808, 809–10, 836–37
Dormancy, viral, 721–22
Dorsal body axis, 410
Dorsal hollow nerve cord, 686
Dot-to-dot lines, scatterplot, B:5
Double bonds, 23
Double fertilization, plant, 593, 832–33
Double helix secondary structure, DNA,
62, 286
discovery of, 61–63, 286–87
electron micrograph of, 284f
opening and stabiliizing of, in DNA
synthesis, 290–91, 294t
transcription factors and, 355–56
Double-stranded DNA (dsDNA)
viruses, 726
Double-stranded RNA (dsRNA)
viruses, 727
Double-stranded viruses, 716, 726–27
Doubly compound leaves, 737–38f
Doughnut-shaped proteins, 47f
Doushantuo microfossils, 518
Downstream DNA, 319
Drag, animal locomotion and, 985
Drosophila bifurca. See Fruit flies
Drosophila melanogaster. See Fruit flies
Drought, natural selection in Galápagos
finches and, 456–59f
Drugs. See also Diseases and disorders,
human; Medicine
anti-AIDS, 309
anticancer, 607
anti-HIV, 719, 720
antiviral, 714, 719, 722
beta-blocker, 210
derived from green plants, 580
derived from land plants, 580t
ephedrine, 608
evolution of bacterial resistance to,
454–56
fungal infections and fungicides, 618

fungi in production of, 629, 632
homology and, 451
plants and, 580
resistance to, 532–33
Dryer, W. J., 1046
Dry habitats
carbon dioxide concentration
mechanisms for, 193–95
gas exchange in, 905
plant adaptations to, 754–55, 757–58,
765–66
Drying, seed maturation and, 834
Dscam gene, 358
Duero river hydroelectric dam, 154f
Duplications, chromosome, 314
Dusky seaside sparrow, 492–93
Dutch elm disease, 632, 1063
Dwarf plants, defective gibberellin in,
808–9
Dwarfism, pituitary. See Pituitary
dwarfism
Dyes, microscopy, B:18
Dynamite tree, 834–35
Dynein, 133
Dysentery, 555t
Dystrophin, 200f

E

E site, ribosome, 328–29f
E2F protein, 234
Eardrums, 956
Early endosome, 126
Early prophase I, 242, 244
Ear ossicles, 956
Ears, mammalian, 955–58
Earthworms, 658, 659
Ebola virus, 723, 728
Ecdysone, 670, 998
Ecdysozoans, 638t, 645, 658, 659f, 670–
78. See also Protostomes
Echidnas, 700
Echinoderms, 638t, 641, 682–86f
Echinoidea, 685–86f
Echolocation, 957–58
Ecological selection, 478
Ecology, 1059–81. See also Community
ecology; Conservation biology;
Ecosystem ecology; Global
ecology; Population ecology
aquatic biomes in, 1074–79
biodiversity scenarios in, 1182–84
climate patterns in, 1065–67
geographic distribution and
abundance of organisms in,
1061–65
levels of study in, 1060–61
organismal, 1060–61
pyramid of production in, 1152
summary of levels of study of, 1060t
terrestrial biomes in, 1068–74
Economic benefits
of biodiversity, 1187–88
of fungi, 613–15
Ecosystem diversity, 1174–75
Ecosystem ecology, 1148–59
in ecology, 1060t, 1061
ecosystem energy flow in, 1149–56
ecosystem nutrient cycling in, 1156–59
ecosystems and biosphere in, 1148–49
(see also Global ecology)
Ecosystem function, 1174
in biodiversity, 1174–75
conservation of, 1191–93
Ecosystem services, 578, 1187
benefits of, 1187–88

of green plants, 578–79
protostome, 658
quantifying, 1192–93
Ecosystems, 578, 1061, 1148
animal locomotion in, 986
animals in, 647–48
biosphere and, 1148–49 (see also
Global ecology)
ecosystem ecology and, 1060t, 1061
(see also Ecosystem ecology)
energy flow in (see Energy flow,
ecosystem)
fungi in, 613–15
human impacts on, 1149
impacts of global climate change on,
1168–69
most productive, 1154–56
nutrient cycling in, 1156–59
prokaryotes in, 541–44
protists in, 556–57
protostomes in, 658
restoration of, 1191–93
services of (see Ecosystem services)
Ecotourism, 1192
Ectoderm, 424f–25, 640–41
Ectomoss, 665
Ectomycorrhizal fungi (EMF), 620–21,
622f, 632
Ectoparasites, 648, 666, 667
Ectotherms, 697, 855–56, 1152
Edge effects, habitat fragmentation
and, 1180
Ediacaran fauna fossils, 518
Education, human population growth
and, 1117
Effector B cells, 1050–51
Effector cells, hormone, 993–94, 1001
Effector genes, 414f
Effector T cells, 1050–51
Effectors, 854
Efferent division, PNS, 942
Eggs, 237, 823, 1014
amniotic, 689–91, 693–94
bird, 1024f
cell-cycle regulation and frog, 229–30f
fertilization of, 5, 420–21, 433,
1018–20
gametogenesis of, 241
glycoproteins in fertilization of, 79–80
human, 237f
insect, 659
land plant, 587–88
mammalian, 1023–25
meiosis and, 219, 237–38
of monotreme mammals, 701
plant, 438, 823, 827–28
polyspermy prevention and, 421–23
reptile and bird, 696
in sexual reproduction, 1014–15
sexual selection and, 475
structure and function of, 1017–18f
variations in development of, 1020–21
Either-or rule, probability, B:8
Ejaculation, 1022
Ejaculatory duct, 1022–23t
Elastase, 895t
Electric current, 931
Electrical activation, heart, 921–22
Electrical energy, chemical evolution
and, 31, 33–34
Electrical fields
electroreception of, 646, 647t, 967–68
in gel electrophoresis, B:13–B:14
generation of, by electrogenic fishes,
968, 969f
Electrical potential, 930

Electrical signaling, animal, 928–42
  action potentials in, 933–38 (see also
    Action potentials)
  calculating equilibrium potentials with
    Nernst equation, 932f
  electrical activation of hearts, 921–22
  hormones and, 993–94
  membrane potentials in, 930–33
  nervous tissue and, 847
  neurons in, 928–30
  resting potential in, 931–32
  synapses in, 938–42 (see also
    Synapses)
  types of nervous systems and, 929 (see
    also Nervous systems)
Electrical signaling, plant, 806
Electrical stimulation of brain areas,
    945–46
Electrically charged side chains, 43f, 44
Electrocardiogram (EKG), 922
Electrochemical gradients, 97, 931
  active transport and, 99–100
  facilitated diffusion and, 96–99
  neurons and, 931
  in photosystems, 185–86
  in plant nutrient uptake, 783–84
  proton gradients as, 168–70
Electrogenic fishes, 968, 969f
Electrolytes, 862, 884. See also
    Osmoregulation
Electromagnetic spectrum, 179–83. See
    also Light-capturing reactions,
    photosynthetic
Electron acceptors, 142, 170–72
Electron carriers, 142
Electron donors, 142
Electron microscopy
  of bacterial translation, 324f
  in cell research, 119
  of DNA replication, 291f
  of DNA synthesis, 284f
  of fossilized plants, 585
  freeze-fracture, of membrane proteins,
    95–96f
  of introns and exons, 321–22
  of karyotypes, 304f, 314f
  of protist cells, 557–58
  in study of muscle contraction, 974
  of transport vesicles, 130f
  types of, B:18–B:19
Electron shells, 20–21
Electron tomography, B:20
Electron transport chains (ETCs), 166
  aerobic vs. anaerobic respiration and,
    170–72
  in cellular respiration, 156, 166–68
  chemiosis hypothesis on, 169–70
  discovery of ATP synthase complex
    and, 168–69
  organization of, 166–67
  in photosystem II, 185–86
  prokaryotic cellular respiration and,
    539–40
  reactions of, 168t
Electronegativity, 21–23f, 31–32, 141–42,
    166, B:13
Electrons
  in atomic structure, 19–21
  covalent bonds of, 21–23
  in energy transformations, 137–38
  excitation of, in photosynthetic
    pigments, 182–84
  ionic bonds and, 22–23f
  in microscopy (see Electron
    microscopy)
  in photosystems, 185–90

potential energy and, 31
prokaryotic donors and acceptors of,
    538–41
transfer of energy in redox reactions
    via, 141–43
Electrophoresis. See Gel electrophoresis
Electroreception, 967–68
Electroreceptors, 953
Elements, 19, B:12
Elephantiasis, 674, 919, 1037f
Elephant seals, 476–77
Elephants, 392, 491f–92, 851, 957
Elevational gradients
  shoot systems and, 736
  tree ranges and, 1137
Elimination, animal nutrition and, 883f,
    888, 896
ELISA (enzyme-linked immunosorbent
    assay), B:16
Ellis, Erle, 1069
Ellis, Hillary, 408
Elongation, stem, 799–802
Elongation factors, 331
Elongation phase, transcription, 320
Elongation phase, translation, 330–31
EMBL Nucleotide Sequence
    Database, 390f
Embryo sac, 828
Embryogenesis, plant, 433–36, 833–34
Embryonic development. See also
    Development
  animal modes of, 650t, 651
  apoptosis in, 408
  deuterostome, 681
  developmental biology and, 405–6
  embryonic tissue layers in, 640–41
  gastrulation in, 644–45
  lophotrochozoan, 660–61
  mammalian pregnancy, gestation, and
    birth in, 1030–34
Embryonic stem cells, 407
Embryonic tissues, animal, 424f–25,
    640–41, 1031–32
Embryonic tissues, plant, 434–35
Embryophytes, 587–88
Embryos, 219, 405
  cell cycle and, 219
  creation of, in cleavage, 423–24
  creation of plant seeds as, in
    embryogenesis, 433–36
  developmental homology in, 450
  gastrulation of, 424–26
  human, 419f
  inner cell mass and, 424
  land plant retention of, 587–88
  organogenesis in, 426–29
Emergent properties, 402
Emergent vegetation, 1077
Emerging diseases, 723–25
Emerging viruses, 723–25
Emerson, Robert, 184
Emigration, 482, 1103
Emulsification, 894
Endangered species, 1178. See also
    Threats, biodiversity
  biodiversity and, 1178
  conservation of, 492–93, 1189–90
  population ecology and conservation
    of, 1118–20
Endangered Species Act, 492–93
End-Cretaceous mass extinction,
    521–23
End-Permian mass extinction, 515, 521
Endemic species, 1176
Endergonic reactions, 139, 141–44
Endocrine disruptors, 999–1000

Endocrine glands, 993, 994f–95
Endocrine signals, 992t, 993. See also
    Hormones, animal
Endocrine systems, 991–92, 993, 994f–
    95. See also Hormones, animal
Endocytosis, 126, 718–19
Endoderm, 424f–25, 640–41
Endodermis, plant, 760, 785
Endomembrane system, 114, 117t, 121–27
Endomycorrhizal fungi, 621. See also
    Arbuscular mycorrhizal fungi
    (AMF)
Endoparasites, 648, 666
Endophytic fungi, 619, 620–23
Endoplasmic reticulum (ER), 112–13,
    117t, 121–24, 561–62
Endorphins, 941t
Endoskeletons, 682, 689, 980, 981–83
Endosperm, 593, 833
Endosymbionts, 547
Endosymbiosis, 560
  in green algae, 598
  in protists, 560–61, 564–65
Endosymbiosis theory, 116, 560–61
Endothermic processes, 30
Endothermic reactions, 138
Endotherms, 695–96, 703, 855–58
Endpoints, phylogenetic tree, B:10–B:11
Endurance training, human, 980
Energetic coupling, 141–44
Energy, 31. See also Chemical energy
  active transport and, 99–100
  in animal locomotion, 980, 985–86, 988
  animal metabolic rates of consumption
    of, 851–52
  capacity of water to absorb, 27–28
  carbohydrates in storage of, 72, 76–77t,
    80–81
  flow of (see Energy flow)
  in formation of micelles and lipid
    bilayers, 88
  hormones in homeostasis of animal,
    1000–1003
  life and forms of, 136–37
  living organisms and, 2
  metabolic diversity of prokaryotes and,
    538–41
  nucleic acid polymerization and, 60
  prokaryotic methods (feeding
    strategies) for obtaining, 538t
  renewable sources of, 1189
  RNA world hypothesis and, 69
  specific heats of some liquids, 27t
  transfer of, in energetic coupling,
    141–44
  transfer of, from glucose to ATP, 81
  transformations of, in chemical
    reactions, 137–40f (see also
    Chemical reactions)
  types and transformations of, 31
Energy flow
  ecosystem, 1149–53
  global, 1153–56
Energy hypothesis on species
    richness, 1145
Enforced altruism, 1097–98
Engelmann, T. W., 180–81f
English system, metric system and,
    B:1–B:3
Enhancement effect, photosystem,
    184, 188
Enhancers, 354, 355f, 356, 357f
Enkephalins, 941t
Enrichment cultures, 534, 540
Entamoeba histolytica, 555t
Enthalpy, 137–38, 139f

Entropy, 31–32, 52, 138
Envelope, fertilization, 422–23
Envelope, viral, 715
Enveloped viruses, 715, 720–21
Environment(s). See also Biomes;
    Ecosystems
  animal color vision and, 963–64
  animal sensing of changes in, 952–53
  animal sensory organs and, 646, 647t
  animals and extreme, 842–43
  background extinctions vs. mass
    extinctions and, 520–21
  balancing selection and, 474
  changes of, in time line of life, 515–16
  as context for succession, 1141, 1142
  cues from, in mating, 1087–88
  ecology and, 1059–60 (see also
    Ecology)
  effects of, on phenotypes, 273, 277t
  environmental sequencing
    (metagenomics) of, 395
  eukaryotic differential gene expression
    and internal, 349
  factors in, affecting enzyme
    function, 148
  gene expression and, 337
  natural selection and changes in,
    456–59f
  plant adaptations to dry and salty,
    754–55, 757–58, 765–66
  plant development in response to, 433
  plant sensory systems and, 793–94
  plant shoot systems and, 736
  plasma membranes and intracellular,
    100–101
  prokaryotic impacts on, 541–49
  seed germination and conditions in,
    836–37
  in triggering switch between
    reproductive modes in
    Daphnia, 1015
  whole-genome sequencing and study
    of, 392
Environmental selection, 478
Environmental sequencing
    (metagenomics), 395, 534–35,
    1173, 1175
Enzymatic combustion, fungal, 623–24
Enzyme kinetics, 147
Enzyme-linked immunosorbent assay
    (ELISA), B:16
Enzyme-linked receptors, 213–14
Enzymes, 54. See also Catalysis
  active transport and, 99–100
  animal digestive, 114, 118, 889,
    892–93, 895t
  in attachment of amino acids to
    tRNAs, 327, 328f
  catalysis of chemical reactions by, 54,
    144–47
  in cell–cell signal deactivation, 214
  in central dogma of molecular biology,
    308–9
  in chromatin remodeling, 351
  in citric acid cycle, 162–63f, 164t
  in DNA mismatch repair, 298–99
  in DNA synthesis, 289–95
  environmenal factors affecting
    function of, 148
  eukaryotic organelles and, 110
  folding and unfolding of, 52
  fungal, 614, 623–24
  in genetic engineering, 380t
  in glycolysis, 159–61
  helpers for, 147
  homeostasis and, 854

natural selection, 465, 472–78 (see also Natural selection)

nonrandom mating, 470–72

speciation and, 489–90 (see also Speciation)

summary of, 486t

Ex situ conservation, 1189–90

Exam preparation, Bloom's taxonomy and, B:20–B:30t

Excavata, 558t, 569, 571

Exceptions, central dogma, 308–9

Exceptions, Mendelian extensions and. See Extensions, Mendelian exceptions and

Excitable membranes, **934**

Excitatory postsynaptic potentials (EPSPs), **940**–42, 947–48

Excretory systems. See also Kidneys
insect, 870–71
nitrogenous wastes and, 864–65, 866t
shark, 866–68
waste elimination by, 883f, 888, 896

Exergonic reactions, **139**, 141–44. See also Spontaneous chemical reactions

Exhalation, 909–11f

Exit (E) site, ribosome, 328–29f

Exocrine glands, **995**

Exocytosis, **126**

Exons, **322**, 357–58, 381–82

Exoskeletons, 109, **670**–71, **689**, 980, **983**, 1038

Exothermic processes, **30**

Exothermic reactions, **138**

Exotic pets, 1181

Exotic species, **1063**, **1181**

Expansins, **202**, **800**

Expansion, cell, 406t, 407

Experiments
cell-fusion, 229–30f
design of, 11–14, B:28
global climate change, 1073
hypothesis testing with, 3–5, 9–11
natural, 457
pulse–chase (see Pulse–chase experiments)
single-trait, by Gregor Mendel, 259–63
in study of evolution, 484–85
two-trait, by Gregor Mendel, 263–66

Exponential notation, B:9

Exponential population growth, **1107**–8

Export, nutrient, 1157–59

Expressed sequence tags (ESTs), **393**

Extant species, **446**

Extension, polymerase chain reaction, 375–76

Extensions, Mendelian exceptions and, 269–77t
allele codominance and incomplete dominance of, 272
gene-by-environment interactions, 273
gene-by-gene interactions, 273–75
linkage, 270–71
multiple allelism, 271–72
overview of, 277t
pleiotropic genes, 272–73
quantitative traits in polygenic inheritance, 275–76

Extensors, **982**–83

External fertilization, animal, 650–51, 1018–19

Extinct species, **447**

Extinctions
animal, 637
current rates of, as biodiversity threat, 1178

global climate change and, 1166t, 1167

human population size and species, 1117

mass extinctions, 505f, 520–23, 1178

metapopulation migration vs., 1112f–13

predicting effects of biodiversity threats on rates of, 1182–84

species richness and rates of, 1143–44

Extracellular appendages, prokaryotic, 109

Extracellular digestion, fungal, **623**

Extracellular layers, 201–4

Extracellular matrix (ECM), **202**–4, **637**, 639, 847

Extraterrestrial life, 531

Extremophiles, **530**–31

Eye color, fruit fly, 268–69

eyeless gene, 451

Eyes
cancers of human, 234
insect, 675, 960
primate, 705
vertebrate, 960–64, 999, 1000f

F

F_1 generation, **259**

Facilitated diffusion, **99**, **769**, **864**
in osmoregulation, 864–65
as passive transport, 101f
in translocation, 769–72
via carrier proteins, 99
via channel proteins, 96–99

Facilitation, enzyme catalysis transition state, 146f–47

Facilitation, species richness and species, 1186

Facilitation, successional, **1141**

Facultative anaerobes, **173**

FAD (flavin adenine dinucleotide). See Flavin adenine dinucleotide (FAD)

$FADH_2$, 142, 147, 156, 162–72

Faith vs. science, 9

Fallopian tubes, **1024**–25f

Families, gene, 398–99

Fast (fast glycolytic) muscle fibers, **979**–80

Fat cells, 118

Fate, cell, 425, 427–29, 434–35. See also Cell determination

Fats, 86–87, 157

Fatty acids, **85**
cortisol and, 1001
fermentation and, 172
membranes and, 94
in metabolic pathways, 157
saturated and unsaturated, 85–86
in transgenic plants, 385

Fauna, **518**

FBN1 gene, 272–73

Feather stars, 684

Feathers, 517, 695, **703**

Features. See Traits

Feces, **888**
formation of, in large intestine, 888, 896
insect, 870–71

Fecundity, **1105**
in density-dependent logistic population growth, 1109–12
in human population growth, 1115–18
in life histories, 1105–7
in life tables, 1105, 1106f
in preservation of endangered species, 1118, 1119f

Feder, Jeffrey, 496–97

Feedback
in action potential depolarization, 935
in global climate change, 1164–66
in homeostasis, 854
in thermoregulation, 857

Feedback inhibition, **150**, **993**
in citric acid cycle, 162–63f
by glucocorticoids, 1004
in glycolysis regulation, 160–61
in hormone signaling pathways, 993–94
of metabolic pathways, 150, 151f
in pyruvate processing, 162

Feeding strategies, animal
diversification of, 648–49
ecdysozoan, 674–78
echinoderm, 683–84, 685–86f
fish, 692
invertebrate, 687–88
lophotrochozoan, 665–69
mouthpart adaptations for, 884
in trophic structure, 1150–51
vertebrate, 697–703
worm, 662

Feeding strategies, prokaryote, 538

Feeding strategies, protist, 563–65

Feet, primate, 705

Female choice, 475–76, 1020

Female gametes. See Eggs

Female reproductive systems
in birds, 1024
external and internal anatomy of human, 1024–25f
hormonal control of menstrual cycle in, 1027–30
mating strategies and, 1020
oogenesis in mammalian, 1017–18f
ovaries and eggs in, 1023–24
pregnancy and birth in mammalian, 1030–34

Females
in age structure of human population, 1115–16
childbirth and maternal mortality rates in human, 1034
chromosomes of, 238
eggs as reproductive cells of, 237
embryos and animal, 651
gametangia of, 587
human population growth and, 1115–18
male mimicking of, 1094–95
mate choice by, 1088–89
mating cues for, 1087–88
plant flower parts as, 827–28
sex chromosomes of, 268
sex hormones of, 999
sex-linked diseases of, 278–79
sexual selection and, 475–78

Femmes fatales, 1020

Fermentation, **172**, **540**
as alternative to cellular respiration, 155, 172–73
prokaryotic, 538, 539t, 540–41
in vitro study of, 158

Ferns, 522–23, 581, 604, 789

Ferredoxin, 186–87, 188

Fertility rates, global human population and, **1117**–18

Fertilization, **237**
cell cycle and, 5
fungal, 624–25
genetic variation and types of, 248–49
glycoproteins in cell identity for, 79–80
meiosis and, 237–38, 241–42

Fertilization, animal, **420**, **1014**
contraceptive methods to prevent human, 1030t
development and, 420–23
egg development after, 1020–21
external, 1018–19
human pregnancy and, 1031
image of human, 237f
internal, 1019–20
in reproduction, 650–51, 1014–15
unusual mating strategies and, 1020 (see also Mating)

Fertilization, plant, **823**
angiosperm pollination as, 593–94 (see also Pollination)
in development, 433
double, 593
model organism, B:25–B:26
pollen grains and, 590–91
in reproduction, 823, 832–33

Fertilization, protist, **566**

Fertilization envelope, **422**–23

Fertilizers, soil, 533, 787, 1155, 1158, 1160

Fetal alcohol syndrome (FAS), **1033**

Fetal hemoglobin, 915

Fetus, **1032**
gestation of, 1031–32
nourishment of, 1030–31, 1032–33

Fiber, dietary. See Dietary fiber

Fiber composites, extracellular layer, 201, 202f

Fibers, muscle. See Muscle fibers

Fibers, plant, 579–80, **745**–47

Fick, Adolf, 905–6

Fick's law of diffusion, 905–6

Fight-or-flight response, 921–22, 977–78, **1000**, 1001f, 1006, 1008–10

Filament, stamen, **827**

Filament proteins, 129

Filamentous algae, 180–81f

Filaments, cytoskeletal, 127–31

Filter feeders, **648**

Filtrate, **870**
insect, 870–71
mammal, 873–78

Filtration, renal corpuscle, **873**

Fimbriae, **109**

Finches, 449–50, 456–59f, 494–95, 845–46f, 1127

Finite rate of increase, 1110f–11f

Finnerty, John, 642

Fins, fish, 692–93

Fire
disturbance regimes and, 1138–39
seed germination and, 837
tree bark and, 750

Fireflies, 1094

Firmicutes, 544, 545

Firs, 607

First law of thermodynamics, **31**, 137–38

Fischer, Emil, 54, 144–45

Fishes
cichlid jaws, 884–86
electrogenic, 968, 969f
embryonic development of, 405f
evolution of, 689
female choice for paternal care in, 476
gene flow in, 482–83
gills of, 906–7
lineages of, 696, 698–99
mating and external fertilization in, 1018–19
mechanoreception by lateral line system of, 958–59
mutualisms with, 1133, 1134f

Fishes *(continued)*
  nitrogenous wastes of, 865
  osmoregulation by, 862–64, 866–69
  overfishing and, 1153, 1181
  populations of, 1101*f*
  in vertebrate phylogeny, 691
Fission, 654, 667, **1014**
Fitness, **6**, **454**, **1124**
  altruism and, 1095–98
  effects of evolutionary processes on, 465–66, 486*t*
  gene flow and, 482–83
  genetic drift and, 479
  hemoglobins and, 915
  heterozygous individuals and, 470
  inbreeding depression and, 471–72
  maladaptive traits and, 192
  mutations and, 484–85
  natural selection, adaptation, and biological, 6, 454
  point mutations and, 313
  sexual selection and, 475
  species interactions and, 1124
Fitness trade-offs, **462**, **1105**, **1126**
  in animal form and function, 843–45
  in competition, 1126–27
  as constraint on natural selection, 462
  endothermy vs. ectothermy, 856
  in life histories, 1105–6
  of mutualisms, 1134–35
  nitrogenous wastes and, 865
  parental care and, 694
  between skeletal protection and mobility, 980
Fixation phase, Calvin cycle, 191
Fixed action patterns (FAPs), **1084**
Flaccid cells, **756**
Flagella, **109**, **557**, **1017**
  bacterial vs. eukaryotic, 131
  cell movement by, 132–33
  in characteristics of domains of life, 529*t*
  fungal, 617, 618
  prokaryotic, 109
  protist, 557, 565–66
  of sperm, 1017
Flagellin, 131
Flattening adaptations, animal surface-area, 852–53
Flatworms, 638*t*, 666
Flavin adenine dinucleotide (FAD), **142**, 163*f*–65
Flavin-containing protein (FMN), 166
Flavins, 166
Flemming, Walther, 220
Flexibility, behavioral, 1084–85
Flexibility, polypeptide, 46
Flexors, **982**–83
Flies. *See* Fruit flies
Flight
  animal locomotion in, 984
  bird, 703
  evolution of vertebrate, 694–95
  insect, 671–72
Flightless birds, 703
Flocking, 1129*f*
Flood basalts, 521
Floods, succession after, 1139
Floral meristems, 438, **439**–41
Floral mimics, 739*f*
Florida panthers, 471–72, 1190
Florigen, **803**
Flour beetle, 869

FLOWERING LOCUS T (FT) gene, 803
Flowers, **592**, **823**
  angiosperm radiation and, 596
  buds and, 735
  fading of, 812
  in female gametophyte production, 827–28
  floral meristems and formation of, 438–41
  light reception and flowering of, 802–3
  in male gametophyte production, 828–29
  as morphological innovations, 517
  in plant reproduction, 592–93, 823
  pollination of, 593–94, 608–9*f*
  as reproductive structures, 826–29
Flu virus. *See* Influenza viruses
Fluid connective tissue, 846*t*, **847**
Fluid feeders, 648*t*, **649**, **884**
Fluid-mosaic model, **94**–**95**, 96*f*
Fluidity, plasma membrane, 90, 91*f*
Fluids, human seminal, 1023*t*
Flukes, 666
Fluorescence, **182**–84, 377, B:15–B:16
Fluorescence microscopy, B:18, B:20
Folate, 884*t*
Folding
  animal surface-area adaptations in, 852–53
  protein, 48–49, 51–53, 331–32, 359–60
  small intestine, 892
Follicle-stimulating hormone (FSH), 210*t*, **1006**, **1026**, 1027–30
Follicles, **1026**
Follicular phase, **1027**
Food, **883**
  angiosperms as human, 608
  animal communication and, 1092–93
  animal mouthparts and, 884–86
  animal nutrition and, 882–83
  animal sources of, 646–48
  disruptive selection and, 474
  as ecosystem provisioning service, 1187
  foraging behaviors and (*see* Foraging)
  fungi and human, 613–14, 630
  giraffe necks and competition for, 9–11
  green plants as human, 579
  migration and, 1091
  model organism, B:23
  plant pathogens and human, 815
  plant reproduction and human, 823
  in population cycles, 1114–15
  protostomes as human, 658
  transgenic crops as, 385–86
Food and Drug Administration (FDA), 374
Food and Nutrition Board, U.S. National Academy of Sciences, 883
Food chains, **556**, **1150**–51
Food poisoning, 532
Food webs, **1138**, **1150**–**51**, 1169
Foot, mollusk, **662**–63
Foraging, 11–14, **1085**–87
Foraminiferans, 572
Forces, animal locomotion, 985–86
Forebrain, **688**
Foreign cells, immune system and, 1046–47, 1053
Forensic biology, 376, 397–98*f*
Forestry
  green plants in, 579–80
  sustainable, 781

Forests
  boreal, 1072
  decomposition rates in, 1157
  fires in, 1165
  net primary production of, 1155, 1156
  temperate, 1071
  tropical wet, 1070
Forks, phylogenetic tree, B:10–B:11
Form-function correlations, 19, 118, 845–46, 884–85, 984–85
Formaldehyde, 34, 73, 74
Formulas, molecular and structural, 24, B:12–B:13
Forrester, Graham, 1109–11
Fossil fuels
  in biodiversity loss, 1189
  in global carbon cycle, 1161–62
  global climate change and, 1163
  in global nitrogen cycle, 1160
Fossil record, **446**, **511**
  animals in, 636, 637, 638
  bryozoans in, 665
  of Cambrian explosion, 518–19*f*
  cetaceans in, 452–53
  DNA in, 64–65
  evidence for evolutionary change in, 446–49
  feathered dinosaurs in, 694–95
  fish in, 692–93
  fossil formation and, 511–12
  fungi in, 616
  hominins and humans in, 705–8
  life's time line in, 513–16
  limitations of, 512–13
  mass extinctions in, 520–23
  pollen grains in, and ranges of species, 1137
  prokaryotes in, 530
  studying green plants using, 581–82
  studying history of life using, 511 (*see also* History of life)
  vertebrates in, 689–91
Fossilization, 511–12
Fossils, **446**, **511**
  amplification of DNA in, with polymerase chain reaction, 375–76
  formation of, 511–12
  ginkgoes as living, 606
  spores, 581–82
  trilobytes, 505*f*
  vascular tissue in, 585
Founder effect, **481**, 495
Four-o'-clocks, 272, 273*f*
Fovea, 961*f*–**62**
Fox, Arthur, 965
Fraenkel-Conrat, Heinz, 715–16
Fragmentation, 601, 603, 652, 654, 667
Frameshift mutations, **313**–14*t*
Franklin, Rosalind, 61–63
Free energy, **137**–39, 145–47, 165*f*
Free radicals, **34**, 182, 298
Free ribosomes, 123–24
Freeze-fracture electron microscopy, 95–96*f*
Frequencies, genetic maps and recombination, 274*f*
Frequency, sound, **955**–58
Frequency-dependent selection, **474**
Frequency distributions, histograms of, 275–76, B:5–B:6
Freshwater biomes
  characteristics of, 1076–78
  fish osmoregulation in, 863–64, 868
  global water cycle and, 1159–60
Frogs, 229–30*f*, 317*f*, 424*f*–25, 700, 909, 997–98

Fronds, **604**
Frontal lobe, **944**, 945
Fructose, 157, 195, 1023*t*
Fructose-1,6-bisphosphate, 158*f*–59, 160*f*
Fructose-6-phosphate, 158*f*–59, 160*f*
Fructose bisphosphate, 158
Fruit flies
  bicoid gene of, 411–12
  chromosomes of, 238
  foraging alleles in, 1085
  genetic drift in, 479–81
  genetic map of, 274*f*
  genome of, 392
  giant sperm of, 1020
  Hox genes of, 412–16, 508–9
  innate immunity in, 1039–40
  in linked gene experiments, 270–71
  as model organism, 675, B:24*f*, B:25
  mRNA sequences from one gene in, 358
  number of chromosomes in, 238
  as protostome, 658
  reinforcement of divergence in, 500
  second-male advantage in, 1019–20
  in testing of chromosome theory of inheritance, 267–69
Fruiting bodies, **547**, 563, B:23–B:25
Fruits, **594**, **823**
  angiosperm radiation and, 596
  animal color vision and, 964
  auxin and, 807
  development of, for seed dispersal, 833, 834–36
  ethylene and ripening of, 812–13*f*
  in plant reproduction, 594–95, 823
Fuel, green plants as, 579
Function-form correlations, 19, 118, 845–46, 884–85, 984–85
Functional genomics, **389**–90, 400–402. *See also* Genomics
Functional groups, **37**–38, 44
Functions, exponential, B:9
Fundamental niches, **1125**
Fungi, **612**–35
  adaptations of, as decomposers, 612–13, 623–24
  appearance of, in Paleozoic era, 515
  biological methods for studying, 615–19
  biological reasons for studying, 613–15
  in cell theory experiment, 3–5
  chitin as structural polysaccharide in, 76, 78
  diversification themes of, 582, 619–28
  economic and ecological impacts of, 613–14
  in ecosystems, 612–13
  key lineages of, 618–19, 628–32
  lichens as green algae and, 598
  life cycles of, 625–28
  molecular phylogenies of, 618–19
  morphological traits of, 615–18
  mycorrhizal, 614, 784, 1133
  phylogenetic relationship of animals to, 8, 618
  reproductive variation in, 624–25
  saprophytic, and carbon cycle on land, 614–15
  in symbiotic relationships, 613, 619–23
  vacuoles in cells of, 114–15
Fungicides, **618**
Fusion
  fungal, 624–25
  gamete, 241, 246–47, 420–21
  interbreeding and, 499, 502*t*

Boldface page numbers indicate a glossary entry; page numbers followed by an *f* indicate a figure; those followed by *t* indicate a table.

Genetic engineering, 368–88
  agricultural biotechnology and
    development of transgenic golden
    rice in, 385–86
  in agriculture, 579
  amplification of DNA with polymerase
    chain reaction in, 374–76
  common techniques used in, 381t
  common tools used in, 380t
  dideoxy DNA sequencing in, 376–78
  gene therapies in, 383–85
  genetic mapping in, 378–82
  of human growth hormone for pituitary
    dwarfism using recombinant DNA
    technology, 369–74
  molecular biology techniques and
    recombinant DNA technology in,
    368–69
  plant cell and tissue cultures in, B:22
  in video microscopy, B:20
Genetic equivalence, 408, 409
Genetic homologies, 450–51
Genetic isolation, 490, 494–99
Genetic libraries. See DNA libraries
Genetic maps, 271, 379
  benefits of finding disease genes
    with, 382
  crossing over and, in chromosome
    theory of inheritance, 271
  development of first, 274f
  ethical concerns over genetic testing
    and, 382
  finding Huntington's disease gene
    with, 378–82
  in genetic engineering, 381t
Genetic markers, 379, 479
  finding Huntington's disease gene
    with, 379
  in hybridization research, 502
  in studies of genetic drift, 479–80
Genetic recombination, 247–48,
    270–71, 274f
Genetic restoration, 1190
Genetic screens, 305–6, 339
Genetic testing, 382
Genetic variation, 472. See also Genetic
    diversity
  asexual vs. sexual reproduction
    and, 247
  from crossing over, 248
  drug resistance and, 455–56
  effects of evolutionary processes on,
    466, 486t
  effects of modes of natural selection
    on, 472, 475t
  genetic drift and, 479
  from independent assortment, 247–48
  lack of, as genetic constraint, 462
  mutations and, 485
  natural selection and, 454, 456
  sexual reproduction and, 566
  types of fertilization and, 248–49
Genetically modified food, 385–86
Genetics, 257. See also Genes
  behavior causation in (see Proximate
    causation)
  chromosome theory of inheritance
    and, 256–57 (see also Chromosome
    theory of inheritance)
  combining probabilities in, B:8
  developing tree of life from rRNA
    sequences in, 6–8

DNA in, 64 (see also DNA
    [deoxyribonucleic acid])
  genetic disorders (see Genetic
    disorders, human)
  Mendelian (see Mendelian genetics)
  model organisms in, B:23
  protostome, 660
  RNA in, 67 (see also RNA [ribonucleic
    acid])
  RNA world hypothesis and, 69
Genitalia, 1022–23
Gennett, J. Claude, 1046
Genome annotation, 392–93
Genome sequencing, 389f, 391, 725, 1173
Genomes, 289, 389
  emerging viruses from reassortment
    of, 723–24
  eukaryotic, 393, 395–400, 628 (see also
    Eukaryotic genomes)
  genetic homology and, 451
  genomics and, 389–90 (see also
    Genomics)
  identifying genes in, 392–93
  in Meselson–Stahl experiment, 289
  model organism, B:23–B:26
  origin of multicellularity in, 640
  prokaryotic, 392–95
  Saccharomyces cerevisiae, 618
  selecting organisms for sequencing
    of, 392
  technologies for sequencing, 376, 378
  viral, 715–16, 719–22
  whole-genome sequencing of, 390–93
Genomic libraries, 372, 381t
Genomic reassortment, viruses and,
    723–24
Genomics, 389–404
  bacterial and archaeal genomes in,
    393–95
  comparative, in studying evolutionary
    innovations of animals, 638
  development of, 389–90
  eukaryotic genomes in, 395–400
  functional genomics and, 400–402
  genome sequencers in, 389f, 391
  Human Genome Project in, 389–90,
    399–400
  metagenomics (environmental
    sequencing) in, 395, 534–35
  proteomics and, 402
  systems biology and, 402
  whole-genome sequencing in, 390–93
Genotypes, 261
  central dogma and linking phenotypes
    and, 308, 309f
  effect of inbreeding on frequencies of,
    470–71f
  exceptions and extensions to
    Mendelian rules on, 277t
  frequencies of, for human MN blood
    group, 468–69
  Hardy–Weinberg principle and
    frequencies of, 466–70
  mating, body odor, and, 470
  in Mendelian genetics, 258t
  mutations as changes in, 313–15 (see
    also Mutation[s])
  in particulate inheritance, 261–63
  phenotypes and, 261, 273–75 (see also
    Phenotypes)
  predicting, with Punnett square,
    262–63

Genus, taxonomic, 8
Geographic distribution and abundance
    of organisms
  abiotic factors in, 1062
  allopatric speciation and, 494–95
  biodiversity hotspots and, 1176–77
  biogeography and, 1079
  biotic factors in, 1062
  ecology and, 1059–65 (see also
    Ecology)
  estimating effects of global climate
    change on, 1183–84
  evidence for evolutionary change in,
    449–50
  global climate change and, 1073–74,
    1166t, 1167
  historical factors in, 1062–63
  interaction of biotic and abiotic factors
    in, 1063–65
  island biogeography and, 1144
  mapping current and past, 1137
  in population ecology, 1102–3
  protected areas and, 1190–91
  species interactions and, 1124
Geologic time scale, 446–47
Geothermal radiation, 541
Gerbils, 1086–87
Germ cells, 438
Germ layers, 424f–25, 639, 640–41
Germ line, 433
Germ theory of disease, 532
Germination, 433, 800–801, 809–10,
    832, 836–37
Gestation, 694, 1031–33
GH1 gene, pituitary dwarfism, 369, 373
Giant axon, squid, 130
Giant sequoias, 1138–39
Giant sperm, 1020
Giardia, 555t, 557, 571
Giardiasis, 555t, 571
Gibberellins, 808–10, 813t, 836–37
Gibbons, Ian, 132–33
Gibbs free-energy change, 138–39
Gila monsters, 878f, 879
Gilbert, Walter, 322
Gill arches, 692
Gill filaments, 906–7f
Gill lamellae, 852–53f, 906–7
Gill pouches, 450
Gills, 668, 687, 852, 906
  developmental changes in gas
    exchange with, 851–52
  mollusk, 662, 668
  in origin of insect wings, 671–72
  in osmoregulation, 862–64, 868
  protostome, 659
  structure and function of, 906–7
  vertebrate, 686–87
Ginkgoes, 606
Giraffe neck hypothesis testing, 9–11
Gizzards, avian, 891–92
Glabe, Charles, 421
Glacier Bay succession case history,
    1141–42
Glands, 848, 993, 994f–95
Glanville fritillaries (butterflies), 1112–
    13, 1118–20
Glaucophyte algae, 569
Gleason, Henry, 1136–37
Glia, 937
Global air circulation patterns, 1065–66
Global biogeochemical cycles, 1156,
    1159–62. See also Nutrient cycling
Global carbon cycle, 556, 1161–62
Global climate change, 1163–69
  causes of, 1163–64

devegetation and, 579
  effects of, on aquatic biomes, 1076
  effects of, on terrestrial biomes,
    1073–74
  estimating effects of, on species
    distributions, 1183–84
  global warming and, 1163
  impact of, on ecosystem net primary
    productivity, 1168–69
  impact of, on organisms, 1166–68
  life-history traits of endangered species
    and, 1118
  local consequences of, 1169
  positive and negative feedback in,
    1164–66
  protists and limitation of, 556–57
  Sphagnum moss and, 602
  summary of impacts on organisms
    of, 1166t
  temperature and precipitation changes
    from, 1164, 1166
  as threat to biodiversity, 1181t, 1182
Global ecology, 1159–69
  biosphere and ecosystems in, 1060t,
    1061, 1148–49 (see also Ecosystem
    ecology)
  global biogeochemical cycles in,
    1159–62
  global climate change in, 1163–69
  global net primary productivity
    patterns in, 1153–56
Global gene regulation, 344–45
Global human population, 1101–2,
    1117–18
Global net primary productivity
    patterns, 1153–56
Global nitrogen cycle, 1160–61
Global warming, global climate change
    and, 1163. See also Global climate
    change
Global water cycle, 1159–60, 1166
Globin genes, 399
Globular proteins, 47
Glomalin, 621
Glomeromycota, 618, 630
Glomeruli, 873, 878t, 966
Glucagon, 54, 897, 992
Glucocorticoids, 1000–1001, 1004–6
Glucokinase, 145
Gluconeogenesis, 195, 897
Glucose, 154
  animal absorption of, 893–94
  ATP synthesis from, 81
  ATP synthesis from oxidation of, in
    cellular respiration, 154–56, 170–72
  ATP synthesis from oxidation of, in
    fermentation, 155, 172–73
  carrier protein for, in human red blood
    cells of, 99
  configurations of, 73–74
  cortisol and availability of, 1000–1001
  in diabetes mellitus and nutritional
    homeostasis, 897–99
  enzyme catalysis of phosphorylation
    of, 145
  free energy changes in oxidation
    of, 165f
  glycogen and, 76
  glycolysis as processing of, to pyruvate,
    158–61
  hydrolysis of carbohydrates by
    enzymes to release, 80–81
  lac operon regulation by, 343
  lactose metabolism and preference for,
    338–39f
  in metabolic pathways, 157

as organic molecule, 36f
in photosynthesis, 177, 195
in redox reactions, 141–42
secondary active transport of, 100
spontaneous chemical reactions and, 32f
summary of oxidation of, 164f
Glucose-6-phosphate, 158f–59, 160f
GLUT-1 membrane carrier protein, 99
Glutamate, 48, 941t, 965
Glutamine, 328, 381–82
Glyceraldehyde-3-phosphate (G3P), **191**, 195
Glycerol, 27t, **86**, 157
Glycogen, **76**
as animal storage polysaccharide, 76
fungi and, 618
hydrolysis of, to release glucose, 80–81
in metabolic pathways, 157
structure of, 77t
Glycolipids, **109**
Glycolysis, **156**
in cellular respiration, 156
discovery of, 158–59
enzymes and reactions of, 160t
free energy changes in, 165f
in muscle contraction, 979
NADH and FADH$_2$ in, 164–65
overview of, 158f–59f
reactions of, 159
regulation of, 159–61
Glycophosphate, 385
Glycoproteins, 78, **79**–80, **124**, 1053
Glycosidic linkage, **75**
Glycosylation, **124**
Glyoxysomes, **115**
GMP (Guanosine monophosphate), 962
Gnathostomes, 638t, **688**, 691
Gnetophytes, 608
Golden rice, 385–86
Goldman, Steven, 948
Golgi, Camillo, 929–30
Golgi apparatus, **113**
in cytokinesis, 228
ECM components and, 202–3
post-translational protein modifications in, 332
protein modification, sorting, and transport inside, 124–26
structure and function of, 113–14, 117t
Gonadal hormones, 1025–26
Gonadotropin-releasing hormone (GnRH), **1026**, 1027–30
Gonads, **999**, **1016**–18f. *See also* Ovaries, animal; Testes
Google Scholar (online database), B:27
Gorillas, 705
Gracile austalopithecines, 706
Graded postsynaptic potentials, 940–41
Grades (paraphyletic lineages), **691**
Grafting, plant, 802–3
Gram-negative cells, **536**
Gram-positive cells, **536**
Gram stain technique, **536**
Grana, **116**, **178**–79f, 189
Grant, James, 1178
Grant, Peter and Rosemary, 456–58, 461, 494–95, 1127
Graphic concept maps, B:28
Graphs, B:4–B:6
Grasses, 1071
Grasslands, temperate, 1071
Gravitational forces, animal locomotion and, 985
Gravitropism, plant, **803**–5, 818t

Gravity
animal sensing of, 647t
hearts and, 919
Gray whales, 857–58
Grazing food chains, **1150**–51
Great apes, **705**
Great chain of being, 445
Great tits, 1166
Greek and Latin word roots, B:3
Green algae, **597**. *See also* Algae
key lineages of, 569, 597–98, 599–600
land plants and, as green plants, 577–78 (*see also* Green plants)
in lichens, 598, 623, 631
as model organism, B:25
photosystems and, 184
rRNA sequences for, 7
similarities between land plants and, 580
Green fluorescent protein (GFP), B:20
Greenhouse effect, 1163. *See also* Global climate change
Greenhouse gases, **1163**, 1192–93
Green iguanas, 481
Green light, 178, 180
Green opsin, 963–64
Green plants, 577–611
analysis of morphological traits of, 580–81
biological methods for studying, 580–83
ecosystem services of, 578–79
in fossil record, 581–82
green algae and land plants as, 577–78 (*see also* Green algae; Land plants)
green algae lineages, 597–98, 599–600
importance of, to humans, 579–80
key lineages of, 597–609f
molecular phylogenies of, 582–83
non-vascular plant lineages, 598, 601–2
seedless vascular plant lineages, 598–99, 603–5
seed plant lineages, 599, 606–9f
Green sea turtles, 1090–91
Greider, Carol, 296–97
Grooves, chemical structure, B:13
Gross primary productivity (GPP), **1149**
Groudine, Mark, 351
Ground finches, 494–95
Ground force, animal locomotion, 986
Ground meristem, **740**, 833
Ground tissue, **435**, 437f
Ground tissue systems, **740**, 742, 744–46, 748t, 833
Groundwater, **1159**–60
Groups, control, 13, B:6–B:7
Growing seasons, plant translocation and, 766
Growth. *See also* Development
cancer as out-of-control, 232–34
in energy flow, 1149
lophotrochozoan vs. ecdysozoan, 670
mammalian, 999
mitosis and, 220
plant, 436, 437f, 585, 805–6
population (*see* Population growth)
Growth factors, **233**–34
Growth hormone (GH), **995**–96, 999
Growth hormone therapy, 369–70. *See also* Human growth hormone (HGH)
Growth regulators, plant, 806–15
Growth rings, tree, 751, 1139
GTP (guanosine triphosphate), **162**
ATP vs., in citric acid cycle, 162
ATP vs., in translation, 326

in cell–cell signal deactivation, 214
in cell–cell signal processing, 211
Guanine, 58–59, 61–63, 65–66, 287, 326
Guanosine diphosphate (GDP), 211, 214
Guanosine monophosphate (GMP), 962
Guanosine triphosphate (GTP). *See* GTP (guanosine triphosphate)
Guard cells, **193**, **584**–85, **744**, 810, **811**–12
Gustation (taste), **964**–65
Guttation, **761**
Gymnosperms, **582**. *See also* Seed plants
diversification of, 582
lineages of, 599, 606–8
in Mesozoic era, 515
molecular phylogeny of, 583
seed evolution in, 592

H

H$^+$-ATPases. *See* Proton pumps
Haberl, Helmut, 1156
Habitat degradation, **1179**–80, 1181t
Habitat destruction, 1117–20, **1178**–79, 1181t
Habitat fragmentation, **1179**–80, 1181t
Habitats
animal communication modes and, 1093
carrying capacity of, 1108–9f, 1112
fossil record bias for, 512
metapopulation, 1112–13, 1118–20
model organism, B:23
nitrogenous wastes and, 865–66
prezygotic isolation and, 491t
prokaryotic diversity in, 530
species richness and area of, 1182–83
sympatric speciation and, 496–97
Hadley, George, 1065
Hadley cell, **1065**–66
Haeckel, Ernst, 1060
Haemophilus influenzae, 392
Hagfish, 691, 697
Hair cells, mechanoreception by, 954–58
Hairpin RNA secondary structure, **66**, 320
Haldane, J. B. S., 41–42
Hamadryas baboons, 1013f
Hamilton, William D., 1096–97
Hamilton's rule, **1096**–97
Hamstring muscles, 982
Hands, primate, 705
Hanski, Ilkka, 1113
Hanson, Jan, 974
Hantaan virus, 725
Haploid cells, life cycles dominated by, 566–69
Haploid number, **239**, 240t
Haploid organisms, **239**, 240t
Harbeespoort Dam, 1148f
Hardiness zones, plant, 1167
Hardy, G. H., 466–67
Hardy–Weinberg principle, **466**–70
Harrison, William, 369f
Hartwell, Leland, 231
Hatch, Hal, 194
Haustoria, **789**, 790f
Havasupai tribe, 469–70
Hawaiian silverswords, 516, 736
Hawkmoths, 908
Hawksworth, David, 615
Hawthorn maggot flies, 496–97
Hay fever, 1056
Head
arthropod, 670
cephalization and, 643

insect, 675
sperm, 422
Head-to-tail axes, 412, 413f
Health, human. *See* Diseases and disorders, human
Hearing, 646, 647t, **955**–58
Heart attacks, 925
Heart murmurs, **921**
Heart-wood, **751**
Hearts, **916**, 919–23
blood vessels and, 917–18
cardiac cycle of, 923
cardiac muscle in, 847f–48, 978
in closed circulatory systems, 917
electrical activation of, 921–22
evolution of multichambered, with multiple circulations, 919–21
human, 921
in open circulatory systems, 916–17
phylogenetic tree of, 920f
Heat, 31
animal thermoregulation and (*see* Thermoregulation)
in chemical reactions, 137–40f
denatured proteins and, 52–53
DNA replication and, 64
in endothermic and exothermic processes, 30
in nucleic acid polymerization, 60
in spontaneous chemical reactions, 31–32
sterilization by, 3–4
as thermal energy, 31
thermoreception of, 967, 968f
water and, 27–28
Heat exchange mechanisms, animal, 855, 857–58
Heat of vaporization, **28**
Heat-shock proteins, 52–53, **857**
Heavy chain polypeptides, **1044**–46
HeLa cells, B:21, B:22f
Helianthus, 501–2
Helicase. *See* DNA helicase
Helicobacter pylori, 891
Heliobacteria, 186, 539t
Helper T cells, 713–14, 717–18, **1050**–51, 1054t
Hemagglutinin, 1045
Heme, **912**
Hemichordates, 638t, 682
Hemimetabolous metamorphosis, **672**
Hemocoel, **663**, 671
Hemoglobin, **912**
acclimatization vs. adaptation for oxygen content of, 459–60
β-globin in, 351
in blood oxygen transport, 54, 912–15
in carbon dioxide transport and blood pH buffering, 914–16
globin genes and, 399
levels of structure of, 51t
sickle-cell disease and, 48
space-filling model of, 41f
Hemolymph, **870**, 916–17
Hemophilia, 278
Henle, Jacob, 874
Hepaticophyta (liverworts), 601
Hepatitis viruses, 727, 1055
Herbaceous plants, **734**
Herbicides, 182, 385, 1182
Herbivores, **647**, **737**, **1128**
animals as, 647–48
limitations on consumption by, 1131–32
in nutrient cycling, 1157–58

Boldface page numbers indicate a glossary entry; page numbers followed by an
f indicate a figure; those followed by *t* indicate a table.

I:20 INDEX

Boldface page numbers indicate a glossary entry; page numbers followed by an *f* indicate a figure; those followed by *t* indicate a table.

fungus, 625–27
land plant, 589–90, 823–25
plant, 433*f*
protist, 566–69
Life histories, 1105, **1106**–7, 1118. *See also* Life tables
Life spans, protein, 360
Life tables, **1104**–5, 1106*f*, 1118, 1119*f*. *See also* Life histories
Lifelong plant growth, 436
Ligaments, **982**
Ligand-gated channels, **940**
Ligands, **940**
Ligase. *See* DNA ligases
Light-capturing reactions, photosynthetic, 177–90. *See also* Photosynthesis
absorption of light energy by pigments, 179–82
in chloroplasts, 178–79*f*
discovery of, 177–78
electromagnetic spectrum and, 179
electron excitation in, 182–84
photosystem I in, 186–87
photosystem II in, 184–86
Z-scheme model for photosystem interactions in, 187–90
Light chain polypeptides, **1044**–46
Light energy
animal sight and, 647*t* (*see also* Photoreception, animal)
in aquatic biomes, 1074, 1075*f*
capture of, in photosynthesis (*see* Light-capturing reactions, photosynthetic)
carbohydrate storage of, as chemical energy, 80
chemical evolution and, 31, 34–35
in chemical reactions, 137
electromagnetic spectrum of, 179
land plant adaptations to, 584–85
in mating, 1087–88
in plant water transport, 765
primary producers and, 1149–50
prokaryotic photosynthesis and, 541
in terrestrial biomes, 1068–69
transformation of, into biomass, 1149–50
in tropical and polar climates, 1066
vegetative development and, 436
Light microscopy, 119, 241, 557–58, B:18
Lightning, chemical evolution and, 31, 33–34
Lignin, **202**, **585**, **745**
in decomposition rates, 1157
fungal digestion of, 612–15, 623–24
in secondary cell walls, 202, 585, 745, 763
Likens, Gene, 1149–50
Limbs
animal movement and, 649–50*f*
arthropod, 670–71
evolutionary loss of, 416, 644
as structural homology, 450
tetrapod, 448, 689, 692–93
Limiting factors, population size and growth, 1109–12
Limiting nutrients, **777**
Limnetic zone, **1077**
Lincoln, Abraham, 273
Lindeman, Raymond, 1150
Line of best fit, scatterplot, B:5
Lineages, **491**–93, **507**, B:10–B:11. *See also* Clades; Monophyletic groups
Lines, scatterplot, B:5
Lingual lipase, **889**, 895*t*

Linkage, **270**–71, 274*f*, 277*t*
Linkage maps, **379**. *See also* Genetic maps
Linked genes. *See* Linkage
Linnaeus, Carolus, 8–9
Lipid bilayers, **88**. *See also* Plasma membranes
artificial membranes as experimental, 88–89
in chemical evolution, 93–94
membrane proteins and, 94–101
permeability of, 89–92
space-filling model of, 84*f*
Lipid-insoluble hormones, 209, 210–11
Lipid-soluble hormones, 209, 210, 211*f*, 996, 1006
Lipids, **85**
animal digestion of, 888, 889, 894–96
bond saturation and hydrocarbon structure of, 85–86
in characteristics of domains of life, 529*t*
chemical evolution of, 88
nucleic acid polymerization and, 60–61
prokaryotic glycolipids and, 109
smooth ER and, 114
structure of membrane, 87–88, 90
types of, 86–87
Lipopolysaccharide (LPS), 845
Liposomes, 88–89, 90, 92–93
Liquids, specific heat of some, 27*t*
Literature, primary, B:26–B:27
Littoral zone, **1077**
Live virus vaccines, 1055
Liver, 115, 162, 848*f*, 887*f*, **894**
Liverworts, 601
Lizards
adaptive radiations of, 516
extinction rates of, 1183–84
impacts of global climate change on, 1167
life table and life history of, 1104–6
mating in, 1087–89
movement of, 972*f*
niche concept and interspecific competition of, 1125
pineal eye of, 999, 1000*f*
range of, 1102–3
as reptiles, 702
urine formation in, 878*f*–79
Ln notation, logarithms and, B:9
Loams, 778
Lobe-finned fishes, **699**
Lobes, brain, 944
Lobsters, 677
Lock-and-key model, enzyme catalysis, 54, 144–45
Locomotion, animal, **973**, 983–88. *See also* Animal movement
Locomotion, protist, 565–66
Locus (loci), gene, **266**
Loewi, Otto, 938–39
Log notation, logarithms and, B:9
Logarithms, B:9
Logistic population growth, **1108**–12
Lohmann, Kenneth, 1090–91
Long-day plants, **802**
Long interspersed nuclear elements (LINEs), **396**, 397*f*
Loop of Henle, 872, **874**–77*f*, 878*t*
Loose connective tissue, 846*t*, **847**
Lophophores, **660**–61*f*
Lophotrochozoans, 638*t*, **645**, 658, 659*f*, 660–70. *See also* Protostomes
Lorenz, Konrad, 1084

Lorises, 704
Losos, Jonathan, 516
Loss-of-function alleles, **305**
Loss-of-function mutations, 471, 481–82, 485
Lou Gehrig's disease (ALS), 408
Love darts, 1020
Lovejoy, Thomas, 1180
Low-order cognitive (LOC) skills, B:20–B:30*t*
LUCA (last universal common ancestor), **8**, **32**
Lumen, **112**, **178**
Luminescence, 377
Lungfish, 448, 692–93, 699
Lungs, 692–93, **909**–12, 916
LUREs, 832
Luteal phase, **1027**
Luteinizing hormone (LH), **1006**, **1026**, 1027–30
Lycophytes (club mosses), 603
Lyme disease, 546
Lymph, 919, **1043**
Lymph nodes, **1043**
Lymphatic system, 233*f*, 919, **1043**
Lymphocytes, **1042**–48. *See also* B cells; T cells
Lynx, 1113–15
Lysine, 311
Lysogeny, **721**–22
Lysosomes, **114**, 117*t*
Lysozyme, 716, **1039**
Lytic cycle, viral, **716**. *See also* Replicative growth, viral

M

M phase, **220**, 223–29, 232
M phase-promoting factor (MPF), **229**–31
MacArthur, Robert, 1143–44
MacArthur–Wilson model, 1143–45
McClintock, Barbara, 396
McClintock, Martha, 967
McGinnis, William, 415
Macromolecular machines, **50**
Macromolecules, **44**, 54, 157–58, B:13–B:14, B:20–B:21. *See also* Carbohydrates; Lipids; Molecules; Nucleic acids; Organic molecules; Proteins
Macronutrients, **777**, 780*t*
Macrophages, **1040**–41, 1042*t*
Mad cow disease, 53
MADS box, **441**
Maggot flies, 496–97
Magnesium, 777, 780*t*, 885*t*
Magnetic fields, 646, 647*t*, 968–69, 1090–91
Magnetite, 108, 534, 1090
Magnetoreception, 968–69
Magnetoreceptors, **953**
Magnifications, microscope, B:18
Maidenhair ferns, 498
Maize, 5–6, 579
Major histocompatibility (MHC) proteins, **1048**–51, 1053
Maladaptive traits, **192**
Malaria, 554–55, 573, 1132
Malaria, avian, 1062
Male gametes. *See* Sperm
Male reproductive systems, 1016–17, 1019–23
Males
in age structure of human population, 1115–16

chromosomes of, 238
female mate choice and, 475–76, 1088–89
female-mimicking behaviors of, 1094–95
gametangia of, 587
male-male competition of, 476–77
plant flower parts, 827
plant gametophytes, 828–29
sex chromosomes of, 268
sex hormones of, 999
sex-linked diseases of, 278–79
sexual selection and, 475–78
sperm as reproductive cells of, 237
visual cues from, in mating, 1088
Malignant tumors, **233**
Malpighian tubules, **870**–71
Malthus, Thomas Robert, 453, 1116
Mammals, **696**. *See also* Animal(s); Vertebrates
amniotic eggs in, 693–94
in Cenozoic era, 515
citric acid cycle in, 162
cleavage in, 423–24
cloning of, 409
digestive enzymes of, 895*t*
diversification of, 522–23
ecological efficiency of biomass production by, 1152
fungal mutualisms with, 629
gametogenesis in, 1016–18*f*
hormones in growth of, 999
house mouse as model organism for, B:26
lineages of, 696, 700–701
mouthparts of, 884, 885*f*
nitrogenous wastes of, 865
osmoregulation in, 871–79
ovarian cycle of, 1027*f*
overhunting of, 1181
parental care in, 694
pregnancy in, 1029–34
primates, hominids, and humans, 704–8 (*see also* Human[s])
reproduction in, 1021–30
seed dispersal by, 835
Mammary glands, **696**
Mandibles, arthropod, **677**
Manipulation, altruism and, 1097–98
Mantle, **662**–63
Map orientation (true navigation), migration and, 1089, 1090–91
Mapping
biodiversity hotspots and, 1177
communities, 1137
concept, B:28
genetic (*see* Genetic maps)
species richness and endemism, 1176
Marfan syndrome, 272–73
Margulis, Lynn, 560
Marine biomes
carbon cycle in, 556–57
characteristics of, 1075, 1078–79
impacts of global climate change on net primary productivity of, 1168–69
loss of biodiversity from overfishing of, 1181
net primary production of, 1154–56
osmoregulation in, 862–63, 866–68
Mark-recapture studies, 1103*f*, 1113, 1115
Markow, Therese, 469–70
Marshes, **1077**
Marsupials, 696, **701**, 1031
Martindale, Mark, 642, 663–64

Mass, metric units and conversions for, B:1t
Mass extinctions, 505f, **520**–23, 1178. *See also* Extinctions
Mass feeders, 648t, **649**, 662f, **884**
Mass number, **19**
Mast cells, **1040**–41, 1042t
Mate choice, female, 475–76, 1088–89
Materials and methods, primary literature, B:27t
Maternal chromosome, **239**
Maternal nurture, mammalian, 1030–33
Mathematical models, population ecology, 1061
Mating
　body odor and, 470
　deceitful communication in, 1094–95
　environmental cues in proximate causation of, 1087–88
　fertilization and, 1018–20
　fungal hyphae mating types in, 624
　giraffe necks and, 9–11
　Hardy–Weinberg principle and random vs. nonrandom, 468–70
　inbreeding as nonrandom, 470–72
　in Mendelian genetics, 257–58 (*see also* Crosses)
　penis size and, 1022–23
　pheromones and, 215–16, 966–67
　sexual selection and, 475–78, 1088–89
　speciation in interbreeding of isolated populations, 499–502
　unusual strategies for, 1020
Matthaei, Heinrich, 311–12
Maturation
　bird egg, 1024f
　lymphocyte, 1043
　seed, 834
Mayr, Ernst, 490, 495, 1083
Measles virus, 728, 1055
Measurements, combining, B:3
Mechanical energy, 137
Mechanical prezygotic isolation, 491t
Mechanistic causation. *See* Proximate causation
Mechanoreception, **954**–59
Mechanoreceptors, **953**, 954
Mediator, **356**
Medicine. *See also* Diseases and disorders, human; Drugs
　bacteria in, 531–33, 545–47
　ecosystem provisioning services and, 1187
　preventive, for HIV/AIDS, 722
Mediolateral leaf axis, 436–37f
Medium, growth, **339**
Medulla, kidney, **872**
Medulla oblongata, **688**
Medullary respiratory center, 911–12
Medusa, **653**–54
Megapascal (MPa), **756**
Megasporangia, **590**, 592, 828
Megaspores, **590**, 592, **828**
Meiosis, 219, **238**, 823
　asexual vs. sexual reproduction and, 247
　chromosome types and, 238
　discovery of, 237–38
　in eukaryotes, 553
　as explanation for Mendelian principles, 266–67
　genetic variation from, 247–49

heredity and chromosome changes in, 247
meiosis I and meiosis II in, 240–46
mistakes in, 249–51
mitosis and, in cell cycle, 219–20, 244t, 245f (*see also* Cell cycle; Mitosis)
overview of, 239–42
paradox of sexual reproduction and, 251–53
in plant reproduction, 823
ploidy concept and, 238–39
sex-linked inheritance and, 268–69
synapsis and crossing over in prophase I of, 246
theories on existence of sexual reproduction and, 251–53
types of mistakes in, 249–51
vocabulary for describing cell chromosomal makeup, 240t
Meiosis I, **240**–44, 245f
Meiosis II, **240**–41, 244–46
Melatonin, **999**
Melotto, Maeli, 816
Membrane channels, plasmodesmata and, 208–9
Membrane lipids, 87. *See also* Lipid bilayers; Lipids; Plasma membranes
Membrane potentials, **783**, **806**, **930**. *See also* Action potentials
　calculating, with Nernst equation, 932f
　as electrical potentials across plasma membranes, 930–33
　in photoreception, 962–63
　in plant nutrient uptake, 783–84
　in plant wind/touch responses, 806
　in sensory transduction and, 953–54
Membrane proteins, 94–101
　active transport by pumps, 99–100
　amphipathic proteins as, 94
　cell adhesion proteins and, 206–7
　cellular slime molds and, B:25
　development of fluid-mosaic model of plasma membranes and, 94–96f
　in extracellur matrix, 204
　facilitated diffusion via, 96–99
　functions of, 54
　isolating with detergents, 96
　in osmoregulation, 867–88
　plant cell–cell signals and, 795
　plasma membranes, intracellular environment, and, 100–101
　summary of passive and active transport by, 101f
　systems for studying, 96
　in translocation, 769–72
　in water and electolyte movement, 864–65
Membranes. *See also* Endomembrane system; Plasma membranes
　artificial, 88–89
　internal, 108, 110t
　mitochondrial, 115, 166–68
　thylakoid, 116
Memory, 945, **946**–49. *See also* Immunological memory; Learning
Memory cells, **1053**–55
Mendel, Gregor, 257–66. *See also* Mendelian genetics
Mendelian genetics, 256–79. *See also* Genetics
　applying rules of, to human inheritance, 277–79

chromosome theory of inheritance and, 256–57, 266–69 (*see also* Chromosome theory of inheritance)
dihybrid crosses in, 263–65
discrete traits in, 275
exceptions and extensions to rules of, 269–77t (*see also* Extensions, Mendelian exceptions and)
experimental system of Gregor Mendel in, 257–61
importance of, 256–57
meiosis and principles of, 266–67
particulate inheritance in, 261–63
rediscovery of, 266
single-trait experiments in, 259–63
testcrosses in, 265–66
two-trait experiments in, 263–66
vocabulary of, 258t
Meniscus, 26, **762**
Menke, Sean, 1064
Menstrual cycle, mammalian, **1027**–30
Menstruation, **1027**
Meristems, **407**, **434**, **740**
　cambrium, 748–50
　floral, in reproductive development, 438–41
　in plant development, 407, 412, 434
　production of primary plant bodies by, 740–42
　in vegetative development, 436, 437f
Meselson, Matthew, 288f, 289
Meselson–Stahl experiment, 288f, 289
Mesoderm, 424f–25, 426–29, **641**
Mesoglea, **653**
Mesophyll cells, **193**–94
Mesozoic era, **515**
Messenger RNAs (mRNAs), **119**, **307**. *See also* RNA (ribonucleic acid)
　alternative splicing of, 357–58
　alternative splicing of, and human genome size, 400
　bacterial vs. eukaryotic processing of, 361t
　in cDNA libraries, 372f
　eukaryotic processing of, 321–24, 349, 356–59
　in functional genomics, 400–402
　in gene expression, 307–8
　in genetic code, 310–12
　in morphogen research, 411–12
　RNA interference and, 358–59
　synthesis of, in transcription, 318–21f, 325t
　in transcription and translation, 308
　translation of, into polypeptides, 324–26, 328–31 (*see also* Translation)
　viral, 719
　visualizing, via in situ hybridization, 411f
Meta-analysis, **1131**
Metabolic diversity
　animal, 646–49
　fungal, 623–24
　prokaryotic, 538–41, 545–49
　protist, 563–65
Metabolic pathways, **150**–52, 156–58, 172–73f, **305**–7. *See also* Cellular respiration
Metabolic rate, **851**–52
Metabolic water, **843**, 864
Metabolomics, 390
Metagenomics (environmental sequencing), 395, 534–35, 1173, 1175
Metallothioneins, 785–**86**

Metamorphosis, **651**, **997**
　amphibian, 696, 997–98
　in animal life cycles, 651
　arthropod, 672
　echinoderm, 685
　insect, 675, 991f, **998**
Metaphase, 224f, **225**, 245f
Metaphase I, 242f, 243, 244, 245f
Metaphase II, 243f, 244
Metaphase plate, **225**, **243**
Metapopulations, **1103**
　dynamics of, 1112–13
　habitat fragmentation and, 1179–80
　in population ecology, 1103
　preserving endangered, 1118–20
　theory of island biogeography and, 1143–44
Metastasis, cancer, **233**
Meteorites, 33, 74
Methane, 23, 24f, 33–34, 541, 549, 1163, B:12–B:13
Methanotrophs, **541**
Methionine, 311, 324, 329–30
Methly salicylate (MeSA), **816**
Methods, biological. *See also* Bioskills; Process boxes
　enrichment cultures, 534, 540
　environmental sequencing (*see* Environmental sequencing [metagenomics])
　genetic engineering (*see* Genetic engineering)
　Gram stains, 536–37
　histograms, 457
　quantitative (*see* Quantitative Methods boxes)
　replica plating, 339–41
　for studying animals, 637–38
　for studying fungi, 615–18
　for studying green plants, 580–83
　for studying prokaryotes, 533–36
　for studying protists, 557–59
　for studying viruses, 714–22
Methods and materials, primary literature, B:27t
Metric system, B:1–B:3
Meyerowitz, Elliot, 439–41
MHC proteins. *See* Major histocompatibility (MHC) proteins
Mice, 308, 309f, 313, 414–15, 851, 1001–3, B:24f, B:25
Micelles, 88, 94
Microarrays, DNA, 400–402
Microbes, **530**. *See also* Archaea; Bacteria
Microbiology, **530**
Microelectrodes, measuring membrane potentials with, 932–33
Microfibrils, **202**
Microfilaments (actin filaments), 106f, 127–29, 228
Micrographs, B:19
Micronutrients, **777**, 780t
Micropyle, **828**
MicroRNAs (miRNAs), 358–**59**
Microsatellites, **396**–98f
Microscopy
　of muscle fibers, 973–74
　studying cell structure with, 2, 107, 118–19, 557–58
　types of, B:18–B:21
　video, 170, B:20
Microsporangia, **590**, 592, 828–29
Microspores, **590**, 592, **829**
Microsporidia, 618, 628
Microsporocytes, 828–29

Microtektites, 521–22f
Microtubule organizing centers (MTOCs), **129**, 224–25, 226t
Microtubules, **129**. See also Cytokinesis
image of, 106f
in meiosis, 241
in mitosis, 224–27
structure and function of, 128t, 129–31
Microvilli, **873–74**, **892**
Midbrain, **688**
Middle ear, **956**
Middle lamella, 204
Migration, **1089**
allopatric speciation by, 494–95
hypotheses on ultimate causation of, 1091
magnetoreception in bird, 968–69
metapopulation extinction vs., 1112f–13
navigation strategies in proximate causation of, 1089–91
Miller, Stanley, 33–34, 59
Millipedes, 675
Millivolt (mV), **930**
Milner, Brenda, 945
Mimicry, 1094–95, **1129**–30
Mineralocorticoids, **1003**
Minerals, **884**
as essential nutrients for animals and humans, 884, 885t
nucleic acid polymerization and, 60–61
as nutrients for plants, 776–79f, 780t (see also Nutrients, plant)
synthesis of monosaccharides from, 74
Minisatellites, **396**–98f
Miscarriages, 470
Misfolding, protein, 53
Mismatch repair, DNA, **298**–99
Missense mutations, **313**–14t
Mistakes, DNA synthesis, 297–301, 313–14
Mistletoe, 789, 790f
Mitchell, Peter, 169
Mitochondria, **115**
citric acid cycle in, 162–65
functions of, 115–16
origin of, in protists, 560–61
oxidation of NADH and $FADH_2$ in, 166–68
in parietal cells, 891
in proximal tubules, 874
pyruvate processing in, 161–62
replacement rate for, 119
structure and function of, 117t
Mitochondrial DNA (mtDNA), 500–501, 560–61
Mitochondrial matrix, **115**, **161**, 162–65
Mitogen-activated protein kinase (MAPK), 150, **213**–14
Mitosis, **219**–29
asexual reproduction vs. sexual reproduction and, 247
cell cycle and, 222–23
daughter cell production by cytokinesis in, 227–29
discovery of phases of, 220–22
events in, 223–26
meiosis and, in cell cycle, 219–20, 244t, 245f (see also Cell cycle; Meiosis)
movement of chromosomes during anaphase of, 226–27
Mitosis-promoting factor (MPF), 406
Mitotic spindle forces, 226
Mockingbirds, 449–50

Model organisms, **257**, B:23–B:26. See also Escherichia coli (E. coli); Fruit flies; Mustard plant; Roundworms; Saccharomyces cerevisiae
bread mold, 305, 629
cellular slime mold (Dictyostelium discoideum), B:23–B:25
characteristics of, B:23
in developmental biology, 408
genomes of, 392
liverworts, 601
peas as, in Mendelian genetics, 257–66
protostome, 658
Models
animal, of disease, 382
condensed eukaryotic DNA, 348f
DNA synthesis, 289–95
eukaryotic transcription initiation, 356, 357f
mathematical, of genetic drift, 480f
molecular, 24, B:12–B:13
Mode of transmission, **277**
Modified leaves, 739–40, 790
Modified roots, 734–35
Modified stems, 736–37
Moisture
in decomposition rates, 1157
terrestrial biomes and, 1068–69 (see also Precipitation; Water)
Molaison, Henry Gustav, 945
Molarity, **29**
Mold, bread, 305, 629
Mole, **29**
Molecular biology, 304, 305, 307–9, 368–69, B:13–B:16. See also Genetic engineering
Molecular chaperones, **52–53**, **331**–32
Molecular formulas, 24, B:12–B:13
Molecular level, anatomy and physiology at, 849f
Molecular machines, 108
Molecular phylogenies
of animals, 638, 640f
eukaryote and protist, 558, 559f
of fungi, 618–19
green plant, 582–83
human, 707–8
vertebrate, 691
Molecular weight, **29**
Molecules, **21**. See also Macromolecules
atomic structure and, 19–21
in bottom-up research approach to chemical evolution, 32–33
carbon and organic, 36–38 (see also Organic molecules)
cell-cycle regulatory, 229–31
in chemical reactions, 24
covalent bonding and, 21–23
development of phylogenetic tree of life from rRNA, 6–8
DNA (see DNA [deoxyribonucleic acid])
enzyme catalysis and substrate, 54
ionic bonding and, 22–23f
measuring amounts of, using radioimmunoassay, B:16
nuclear transport of, 119–21
reading chemical structures of, B:13–B:13
research on proteins as first self-replicating, 42, 55
RNA (see RNA [ribonucleic acid])
separating and visualizing, B:13–B:16
shapes and geometry of simple, 23–24
simple, from carbon, hydrogen, nitrogen, and oxygen, 23

solutes as (see Solutes)
visualizing, in 3-D, B:20–B:21
Mollusks, 638t, 657–58, 662–64, 668–69, 1175
Molting, **670**, 983
Molybdenum, as plant nutrient, 777, 780t
Momentum, population growth, 1116
Monarch butterflies, 1091
Monkeys, 704–5, 724–25, 1098
Monoamines, 941t
Monocots, **596**, 742f, 834, 837
Monod, Jacques, 307–8, 338–39f, 342–43, B:23
Monoecious species, **827**, 828f
Monogamy, 722
Monohybrid crosses, **259**–61
Monomers, **44**–46, 58–59, 72–75. See also Polymerization
Monophyletic groups, 491–93, **507**–8, **536**, 557, B:10–B:11. See also Clades; Lineages
Monosaccharides, **72**–75. See also Carbohydrates
Monosodium glutamate (MSG), 965
Monotremes, 696, **700**, 1031
Morgan, Thomas Hunt, 267–71, 274f, B:25
Morphogens, 410–14, 435–36
Morphology, **107**, **450**
adaptive radiations and, 517
of eukaryotic lineages, 558t
evolution of animal (see Evolutionary innovations, animal)
fungal, 615–18
of green plants, 580–81
lophotrochozoan, 660–61
morphospecies concept and, 490–91
phylogenies based on, 510, 511f
of prokaryotic and eukaryotic cells, 107
of protists, 557–58
structural homology in, 450–51
viral, 715, 725
Morphospecies concept, **490**–91, 492t
Mortality rates. See Death rates, animal
Mosquitos, 555, 672
Mosses, 176f, 581, 601–2, 824f–25
Moths, 952–53, 966–67
Motility. See Movement
Motor neurons, **929**, 976–77
Motor proteins, 54, **128**–31, 133, 228, 241
Mountain pine beetles, 1166t, 1167
Mountain ranges, climate and, 1066–67
Mouse, 392, B:24f, B:25. See also Mice
Mouths
digestion in, 888–89
human, 887f
lophotrochozoan, 662
mammalian, 701
mouthpart adaptations for, 884–86
protostome vs. deuterostome, 644–45, 658
taste as chemoreception by, 964–65
Movement
animal (see Animal movement)
cell (see Cell movement)
prokaryotic, 109, 536, 537f
protist, 565–66
wind/touch responses and plant, 806
Mucins, 889
Mucosal-associated lymphoid tissue (MALT), **1043**
Mucous cells, **890**
Mucus, **889**, **1038**, 1042

Mudslides, 779–81
Müller, Fritz, 1130
Müllerian inhibitory substance, **999**
Müllerian mimicry, **1130**
Multicellularity, **204**, **562**, 846
cell–cell interactions for, 200–201
cell–cell signaling and, 209–10t
in characteristics of domains of life, 529t
in eukaryotes, 8, 553
gene regulation and, 348–49
intercellular connections and, 204–9
origin of, in animal evolution, 638–40
in Paleozoic era, 515
protist, 562–63
Multichambered hearts, 919–21
Multidrug resistance, 456
Multinucleate muscle cells, 977
Multiple allelism, **271**–72, 277t
Multiple fertilization, prevention of, 421–23
Multiple fruits, **834**, 835f
Multiple sclerosis (MS), **937**–38, 1056
Multiplication rule, probability, B:8
Mumps virus, 728
Münch, Ernst, 768
Muscle cells, 162, 200f, 427–29, 637
Muscle contraction, 637, 973–77. See also Animal movement
actin-myosin interactions in, 974–76
animal locomotion and, 984–88
endoskeletons and, 982
in evolution of animal movement, 972–73
exoskeletons and, 983
initiation of, by neurons, 976–77
of skeletal muscles, 979–80
sliding-filament model of, 973–74
Muscle fibers, **973**–74, 979–80
Muscle tissues, **847**–48, 973–74, 977–80
Muscular dystrophy, 200f
Muscular hydrostats, 981
Musculoskeletal structure, 984–85. See also Skeletal systems
Mushrooms, 612f, 614, 617, 626–27, 630. See also Fungi
Mus musculus, 392, B:24f, B:26. See also Mice
Mussels, 662, 668, 1130, 1131f, 1137–38
Mustard plant
in blue-light phototropism research, 797
in cytokinin research, 808
embryogenesis in, 434f
in flowering hormone research, 803
genome of, 392
in gravitropic response research, 804
as model organism for plants, 433, B:24f, B:25
in phloem loading research, 770–71
in plant active ion exclusion research, 786–87
in population ecology, 1106–8
reproductive development of, 438–41
stages of development of, 433f
vegetative development of, 436–38
in wind/touch response research, 805
Mutagens, **339**
Mutants, **268**. See also Mutation(s)
creating experimental, 305, 339
histone, 351
homeotic, 413
in study of flower development, 439–41
in study of hormones and homeostasis, 1001–3

Boldface page numbers indicate a glossary entry; page numbers followed by an *f* indicate a figure; those followed by *t* indicate a table.

I:26   INDEX

in DNA synthesis mistakes, 298–99
in point mutations, 313–14
purines and pyrimidines, 58–59 (*see also* Purines; Pyrimidines)
in RNA synthesis, 308
sequences of, in genetic code, 310–12
Nitrogenous wastes, animal, 865–66, 870–71. *See also* Kidneys
Nitrous oxide, 1163
Nobel Prizes, 67, 997, B:20
Nociceptors, **953**, 967
Nocturnal animals, 964
Nod factors, **788**
Node of Ranvier, **937**
Nodes, phylogenetic tree, **506**, B:10–B:11
Nodes, stem, **735**
Nodules, 787–88
Nonadaptive traits, 461
Non-bilaterian animals, 638*t*, 652–54
Noncyclic electron flow, Z-scheme model, 188–89
Nondisjunction, 497*f*, 498
Nonenveloped viruses, 715, 720–21
Nonpolar covalent bonds, **21**–23*f*
Nonpolar side chains, 43*f*, 44
Nonrandom mating, 468–72. *See also* Sexual selection
Nonself recognition, 1042, 1046–47
Nonsense mutations, **313**–14*t*
Non-sister chromatids, 240*t*, **242**
Nonspontaneous chemical reactions. *See* Endergonic reactions
Nonvascular plants, **580**–83, 590, 591*f*, 598, 601–2
Norepinephrine, 941*t*, 977–78, **1006**
Northern blotting, B:16
Notation
exponential, B:9
scientific, B:2–B:3
Notochords, **426**–27, **686**
Nottebohm, Fernando, 948
Novak, Joseph, B:28
Nuclear envelope, **112**
in characteristics of domains of life, 529*t*
Eurkarya and, 553
in meiosis, 242–43, 244
origin of, in protists, 561–62
structure and function of, 119–20
Nuclear lamina, **112**, 119, 129
Nuclear lamins, **129**
Nuclear localization signal (NLS), **120**
Nuclear pore complex, **119**–20
Nuclear pores, **119**
Nuclear transfer (cloning), 409
Nucleases, 893, 895*t*
Nucleic acids, **58**–71. *See also* DNA (deoxyribonucleic acid); RNA (ribonucleic acid)
carbohydrates in, 78
in cells, 106
chemical evolution and nucleotide production for, 59
components of, 58–61
DNA structure and function, 61–65
polymerization of nucleotides to form, 59–61
as probes, B:15–B:16
in protocells, 94
RNA structure and function, 65–68
RNA world hypothesis on RNA as first self-replicating molecule, 57–58, 68–69
separating and visualizing molecules of, B:13–B:16
Nucleoids, **107**–8

Nucleolus, **112**, 119
Nucleoside triphosphates, 60, 120
Nucleosomes, **350**
Nucleotide excision repair, DNA, **299**–301
Nucleotides, **58**–61, 390
Nucleus, atomic, 19–21
Nucleus, cell, **110**
chromosomes in, 110, 112
division of, by meiosis and mitosis in cell cycle, 219–20 (*see also* Cell cycle; Meiosis; Mitosis)
eukaryotic cells vs. prokaryotic cells and, 8, 106, 110, 117*t*
in fungal fusion, 624–25
in muscle cells, 977
transport of molecules into and out of, 119–21
Nudibranchs, 668
Null alleles, **305**
Null hypothesis, **12**, 468–70, B:7
Nüsslein-Volhard, Christiane, 411–12, B:25
Nutrient availability, aquatic biome, 1075–79
Nutrient cycling, 1156–59. *See also* Biogeochemical cycles
Nutrient deficiency, plant, 777–78
Nutrient pollution, 1148*f*, 1182
Nutrients, **1148**
cell-cycle checkpoints and cell, 231
in ecosystems and biosphere, 1148–49 (*see also* Biogeochemical cycles; Nutrient cycling)
Nutrients, animal, **883**–84, 885*t*. *See also* Animal nutrition
Nutrients, plant, 732. *See also* Plant nutrition
availability of, in soil, 781–82
deficiencies in, 777–78
essential, 776–77, 780*t*
mycorrhizal fungi and (*see* Mycorrhizal fungi)
nitrogen fixation and, 787–89
uptake of, 782–87

O

Obesity, diabetes mellitus and, 898–99
O'Brien, Stephen, 1190
Observable traits, 258–59. *See also* Phenotypes; Traits
Observational studies, global climate change, 1073
Occipital lobe, **944**
Oceanic zone, **1079**
Oceans, **1079**
acidification of, 1166*t*, 1167–68, 1182
as aquatic biomes, 1079
changes of, in time line of life, 515–16
chemical evolution in early, 25–30, 74 (*see also* Water)
effects of, on climate, 1066–67
in end-Permian extinction, 521
estuaries and, 1078
global net primary productivity of, 1154–56
hydrothermal vents in (*see* Deep-sea hydrothermal vents)
impacts of global climate change on, 1166, 1168–69, 1182
protein polymerization in early, 44–45
protists in, 552*f*
upwelling of, 1075, 1155
Octopuses, 662, 669
Odorants, 496–97, **965**–66

Odum, Howard, 1150
Oil (petroleum), 556, 1059*f*
Oils, 86
Okazaki, Reiji, 292
Okazaki fragments, 292, **293**–95
Oldfield mice, 308, 309*f*, 313
Old World monkeys, 705
Oleanders, 765
Olfaction (smell), **964**, 965–67
Olfactory bulbs, **966**–67
Oligochaetes, 667
Oligodendrocytes, **937**
Oligopeptides, 46, 55
Oligosaccharides, **72**, 79. *See also* Carbohydrates
Omasum, 891
Ommatidia, **960**
Omnivores, 579, **647**–48
On the Origin of Species by Means of Natural Selection (book), 445, 446, 453, 489
Oncogenes, **361**
One-gene, one-enzyme hypothesis, **305**–7
One-snip test, phylogenetic tree, B:10
Onychophorans, 672–73
Oogenesis, **1016**, 1017–18*f*
Oogonia, **1017**
Oparin, Alexander I., 41–42
Oparin–Haldane chemical evolution theory, 41–42
Open circulatory systems, **916**–17
Open reading frames (ORFs), **392**–93
Operators, **343**
Operculum, **906**
Operon model, gene regulation, 342–44
Operons, **342**–44
Opisthokonts, 558*t*, 569
Opium, 815
Opposite leaves, 738*f*
Opsins, 961*f*–**62**, 963–64
Opsonization, pathogen, 1052
Optic nerve, **960**, 961*f*
Optimal foraging, **1085**–87
Oral rehydration therapy, 896
Orangutans, 705, 1172–73
Orbitals, electron, **20**–21
Ordinate, graph, B:4–B:5
Organ systems, animal, 713*f*, 848*f*, **849**–50
Organelles, **108**
benefits of eukaryotic, 110
in characteristics of domains of life, 529*t*
eukaryotic, 110–17*t*
plant cell, 129, 743
prokaryotic, 108
prokaryotic vs. eukaryotic, 110*t*
separating, from cells (*see* Centrifugation)
Organic matter, 778–79*f*, 782*t*, 1069. *See also* Biomass
Organic molecules, **36**. *See also* Carbohydrates; Lipids; Macromolecules; Molecules; Nucleic acids; Proteins
carbon–carbon bonds in, 36
functional groups and, 37–38
Oparin–Haldane theory on chemical evolution of, 41–42
prokaryote strategies for obtaining, 538–41
proteins (*see* Proteins)
spontaneous formation of, in origin-of-life experiments, 33–36
Organic soils, 781

Organismal ecology, 1060–61
Organisms, **2**
aerobic vs. anaerobic respiration in, 170–72
animal anatomy and physiology at level of, 849*f*
in aquatic biomes, 1077–79
atoms of, 19, 20*f*
bristlecone pine as oldest living, 432*f*
characteristics of living, 2–3 (*see also* Cell(s); Energy; Evolution; Genes)
distribution and abundance of (*see* Geographic distribution and abundance of organisms)
hybrid, 259
impact of global warming on, 1166–68
major amino acids found in, 43*f*
microbes as microscopic, 530
microscopy of living, B:20
misconceptions about, in evolution, 460–61
model (*see* Model organisms)
multicellular (*see* Multicellularity)
number of chromosomes of, 223
organismal ecology and, 1060–61
phylogenetic tree of all living, 6–8
recombinant, 270–71, 274*f*
selecting, for whole-genome sequencing, 392
taxonomy and scientific names for, 8–9
unicellular (*see* Unicellular organisms)
viruses vs. living, 711–12
water in cells of, 25
Organogenesis, **426**–29, 433, 1032
Organs, animal, **848**
anatomy and physiology at level of, 849*f*
female reproductive, 1023–25
gas exchange, 905–12, 916
hydrostats in, 981
immune system and transplants of, 1053
male reproductive, 1021–23
muscle tissue in, 977–78
organogenesis of, 426–29, 1032
organ systems and, 849–50
sensory, 646, 647*t*
tissues of, 848
viruses and human, 713*f*
Organs, plant, 433, 439
Origin-of-life research
earliest, 33–36
extremophiles and, 530–31
monosaccharides in, 74
nucleic acid formation, 60–61
nucleotide formation, 59
origin of viruses and, 723
polysaccharides in, 78
on proteins as first self-replicating molecules, 42, 44–45, 55
RNA as first self-replicating molecule, 57–58, 68–69
Origin of replication, **290**, 291*f*
Origin of Species, The (book), 5, 445, 446, 453, 489
Orr, Allen, 500
Oryx, 842–43, 855, 864
Oshima, Yasuji, 354
Osmoconformers, **862**, **866**
Osmolarity, **862**, 875
Osmoregulation, **862**
diffusion, osmosis, and, 862
excretion and, 862–66
in fishes, 866–68
hormones of, 1003
overview of, 871

Osmoregulation (continued)
  in terrestrial insects, 869–71
  in terrestrial vertebrates and
    mammals, 871–79
Osmoregulators, **862**–64, 866
Osmosis, **92**, **756**, **862**
  across lipid bilayers, 92–93
  across prokaryotic cell walls, 109
  in kidneys, 872
  in osmoregulation, 862, 863f, 871
  of water and solutes into roots, 765
Osmotic gradient, loop of Henle and,
    874–77f
Osmotic stress, 862–66, 869
Ouabain, **867**, 871
Out-of-Africa hypothesis, **707**–8
Outcross fertilization, **830**
Outcrossing, **248**–49
Outer ear, **956**
Outgroups, **506**
Outgroup species, phylogenetic tree,
    B:10–B:11
Oval window, ear, **956**
Ovarian cycle, mammalian, 1027f. See
    also Menstrual cycle, mammalian
Ovaries, animal, **999**, **1016**
  in cleavage, 423–24
  in female reproductive system,
    1023–24
  hormones of, 994f, 999, 1027–29
  oogenesis in mammalian, 1016,
    1017–18f
Ovaries, plant, **592**, **827**
  angiosperm, 592–95
  female gametophyte production in,
    827–28
  fruit and, 833
Overexploitation, 1178, **1180**–81
Overfishing, 1153, 1181
Overhunting, 1181
Overwatering, 778
Oviducts, **1024**–25f
Oviparous species, 650t, **651**, **694**,
    1020–21
Ovoviviparous species, 650t, **651**, **694**,
    1020–21
Ovulation, **1024**
Ovules, plant, **827**
  in embryogenesis, 433–36
  female gametophyte production in,
    827–28
  fruit and, 833
  in plant reproduction, 592–93
Ovum, **1017**
Oxidation, **141**. See also Redox
    (reduction-oxidation) reactions
  of acetyl CoA to carbon dioxide,
    162–65
  of glucose in cellular respiration,
    155–56, 164–65, 170–72
  of glucose in glycolysis, 158–61
  of NADH and FADH$_2$, 166–72
  peroxisomes and, 115
  of pyruvate, 161–62
  of water in photosystem II, 186
Oxidative phosphorylation, **156**, 170–71,
    979. See also Phosphorylation
Oxygen
  aerobic respiration and, 170–72
  in animal diversification, 646
  animal gas exchange and, 902–3
  behavior of, in air, 903–4

behavior of, in water, 904–5
Cambrian explosion and atmospheric,
    519–20
consumption of, in animal locomotion,
    986, 987f
cyanobacteria and, 545
in decomposition rates, 1157
discovery of, 177
electronegativity of, 21, 25
in electron transport chain, 166–67
exchange of, between mother and fetus
    during pregnancy, 1032–33
in fermentation, 172
hemoglobin and blood levels of, 48,
    54, 912–15
homeostasis for blood levels of, 1003
in living organisms, 19
molecular structure of, 24f, B:12f
passage of, through stomata of leaves,
    192–93
photosynthetic production of (see
    Photosynthesis)
Precambrian lack of, 515
prokaryotic oxygen revolution and,
    541–42
in protein secondary structure, 48–49
in redox reactions, 141–42
simple molecules of, 23
in soil, 778
in streams, 1078
Oxygen-hemoglobin equilibrium curve,
    913–14, **1033**
Oxygenic photosynthesis, **186**, 189, 539t,
    541–42, 545, 578
Oxytocin, 1005–6, **1034**
Ozone, photons and, 34

P

P site, ribosome, 328–29f
p53 protein, **231**–32, 233
p53 tumor-suppressor gene, 361–63
Pääbo, Svante, 376
Pace, Norman, 530
Pacemaker cells, **921**–22
Packaging, DNA, 360–61t
Paine, Robert, 1137–38, 1153
Palade, George, 121–23
Paleontologists, **513**
Paleozoic era, **515**, 909
Palindromes, 371
Palms, coconut, 1106–7
Pancreas, **892**, **992**
  digestive enzymes of, 892–93
  hormones of, 994f, 995, 996–97
  human, 848f, 887f
  paracrine signals and, 992–93
  pulse–chase experiment on cells of,
    121–23
Pancreatic amylase, **893**, 895t
Pancreatic lipase, **894**, 895t
Pandemics, **712**–13
Panthers, 471–72, 1190
Papermaking, 579–80, 607
Parabasalids, 571
Parabiosis, **1002**–3
Paracrine signals, **992**–93
Paralytic shellfish poisoning, 555–56
Paramecium species, 1108–9f, 1125–26
Paranthropus species, B:11
Paraphyletic groups, 508t, **553**, 582–83
Parapodia, 649t, **667**

Parasites, **564**, **648**, **789**, **1128**
  animals as, 648
  annelid, 667
  coevolution of, 1132
  endosymbiont bacterial, 547
  flatworms, 666
  fungal, 613, 628–29, 632
  human intestinal, 571
  manipulation of hosts by, 1132–33
  nematodes, 674
  parasitoid organisms as, 818
  plant, 789, 790f
  plants as, 789, 790f
  protist, 564, 569
  roundworms as, in elephantiasis, 1037f
  sexual reproduction and, 566
  viral, 285–86
Parasitic relationships, **620**. See also
    Parasites
Parasitism, **1128**
Parasitoid organisms, 485, **818**–19
Parasympathetic nervous system,
    942–43
Parathyroid glands, 994f
Parenchyma cells, **744**, 745f, 747, 748t,
    749, B:21–B:22
Parental care, **694**
  mammalian, 1030–33
  reptilian and bird, 703
  vertebrate, 694
Parental generation, **259**
Parental strands, DNA, 287–89
Parietal cells, **890**
Parietal lobe, **944**
Parker, Geoff, 1019–20
Parsimony, **508**
Parthenogenesis, **665**, **1014**
Partial pressure, **904**, 905
Particulate inheritance, 261–63
Pascal (Pa), **756**
Passive transport, **98**, **769**, **864**. See also
    Diffusion; Facilitated diffusion;
    Membrane proteins
  mechanisms of, 101f
  membrane proteins and, 98–99
  in osmoregulation, 864–65
  of water and solutes into roots, 759–61
Pasteur, Louis, 3–5, 532
Patch clamping, **935**
Paternal care, 476
Paternal chromosome, **239**
Paternity
  bird, 1020
  DNA fingerprinting in, 398f
Pathogenic bacteria, 531–33
Pathogens, **743**, **815**, **1037**
  animal immune system responses
    to, 1037–38 (see also Immune
    systems)
  plant defense responses to,
    815–17, 818t
Pattern component
  of chemical evolution theory, 18–19
  of germ theory of disease, 532
  of scientific theories, 3, 445
  of theory of evolution by natural
    selection, 5, 446
Pattern formation, body, **410**–15
Pattern-recognition receptors, **1040**
PCR. See Polymerase chain reaction
    (PCR)
Peas, Mendelian genetics and, 257–66.
    See also Mendelian genetics
Peat, **602**
Pectins, **202**
Pedigrees, 277–79

Pedometer hypothesis, ant navigation,
    11–14
Peer-reviewed primary literature,
    B:26–B:27
Pelc, Stephen, 221–22
Pellets, centrifugation, B:18
Penfield, Wilder, 945–46
Penicillin, 613, 632
Penis, **1019**–23
Pentoses, 59, **73**–74
PEP carboxylase, **193**–94
Pepsin, **890**, 895t
Peptide bonds, **45**
  formation of, in translation, 328,
    330–31
  glycosidic linkages vs., 76
  in protein polymerization, 45–46
  protein structure and, 48–49
Peptides, 45–**46**, 941t
Peptidoglycan, **76**, 77t, 78, 529, 536–37
Peptidyl (P) site, ribosome, 328–29f
Perennial plants, **599**, **734**
Perfect flowers, **827**
Perforations, **746**
Pericarp, **834**, 835f
Pericycle, **760**
Periodic table of elements, 19f
Periodontitis, 531
Peripheral membrane proteins, **95**
Peripheral nervous system (PNS), **929**,
    942–43
Peristalsis, **889**, **981**
Permafrost, **1072**
Permeability, plasma membrane, 88–94.
    See also Selective permeability,
    plasma membrane
Permian mass extinction, 521
Permineralized fossils, 512
Peroxisomes, **115**, 117t, 580
Persistent organic pollutants (POPs),
    1152–53, 1182
Pest control ecosystem services, 1188
Pesticides, 1182
Petals, **439**, **593**, **826**–27
Petiole, leaf, **737**
Petrified wood, 512
Petromyzontoidea, 697
pH, **29**
  acid-base reactions and, 29–30
  of blood, 912, 914–16
  optimal, for enzyme function, 148
  of soil, 781–82
  of wetlands, 1077
Phagocytosis, **126**, 563–64, 1040–41,
    1042t, 1052
Phanerozoic eon, 514f, 515
Pharmaceutical pollutants, 1182
Pharmaceuticals. See Drugs
Pharyngeal gill slits, **686**
Pharyngeal jaws, **692**, 884–86
Pharyngeal pouches, 686–87
Phenology, **1166**–67
Phenotypes, **258**
  behavior and, 1083
  central dogma and linking genotypes
    and, 308, 309f
  codominance and incomplete
    dominance in, 272, 273f
  effects of environment on, 273
  effects of gene interactions on,
    273–75
  effects of modes of natural selection
    on, 475t
  exceptions and extensions to
    Mendelian rules on, 277t
  finding disease genes and, 382

Plant reproduction (*continued*)
  flowers as reproductive structures of, 592–93, 826–29
  fruits in, 594–95
  heterospory in, 590, 591*f*, 592*f*
  importance of, 822–23
  plant life cycle and, 823–25
  pollen in, 590–91
  pollination in, 593–94, 608–9*f*, 830–32
  seeds in, 591–92, 833–37
  sexual, 586–87, 823
Plasma, **912**
Plasma cells, **1050**–51, 1054*t*
Plasma membranes, **84**–103
  artificial, 88–89
  carbohydrates in cell identity functions of, 78–80
  cell walls and, 116
  in characteristics of domains of life, 529*t*
  crossing of, by hormones, 996
  in cytokinesis, 228
  diffusion and osmosis across, 91–94
  electron transport chain and ATP synthase in, 171
  eukaryotic, 117*t*
  excitable, 934
  exocytosis and, 126
  extracellular layers and, 201
  fluidity of, 90, 91*f*
  importance of, to cells, 84–85, 100–101
  lipids and, 85–91, 529 (*see also* Lipid bilayers)
  lipid-soluble vs. lipid-insoluble hormones and, 209
  membrane proteins and fluid-mosaic model for, 94–101
  movement of electrolytes and water across, 864–65
  nuclear envelope and, 561–62
  plant passive exclusion at, 785
  in polyspermy prevention, 422–23
  prokaryotic, 109, 536–38
  selective permeability of (*see* Selective permeability, plasma membrane)
  smooth ER and, 114
  transport of solutes across, in plants, 769–71
Plasmids, **108**, **370**
  in DNA cloning, 370
  in genetic engineering, 380*t*
  inserting genes into, 370, 371*f*
  introducing recombinant, into bacterial cells for transformation, 372
  lateral gene transfer and, 395
  origin of viruses in, 723
  prokaryotic, 108
  prokaryotic vs. eukaryotic, 110*f*
  in transgenic plant development, 385–86
Plasmodesmata, **208**–9, **600**, **743**, **806**
Plasmodial slime molds, 570
Plasmodium, 554–55, 573, 1132
Plasmogamy, **624**–26*f*
Plastocyanin, **187**–88
Plastoquinone (PQ), **185**–88
Platelets, **912**, **1040**–41*f*
Platyhelminthes, 638*t*, 666
Platypuses, 392, 700
Pleiotropic genes, **272**–73, 277*t*, 461–62
Ploidy, 238–**39**, 240*t*, 241–42
Pneumatophore roots, 734–35*f*

Point mutations, 483
  types of, **313**–14
Poisons. *See* Toxins
Polar bears, 1166*t*, 1167
Polar bodies, **1017**
Polar climate, 1066, 1165
Polar covalent bonds, **21**–23*f*
Polar microtubules, **224**
Polar nuclei, **828**
Polar transport, auxin, **807**
Polarity, 21–23*f*, 25, 43*f*, 44, B:13
Polio virus, 727, 1055
Pollen grains, **591**, **829**
  fossilized, 512*f*
  germination of, 832
  in plant reproduction, 433*f*, 590–91
  seed plants and, 599
  stamens and production of, 827, 828–29
Pollen tube, **832**
Pollination, **593**, **830**
  of angiosperms, 593–94, 608–9*f*
  animal speciation and, 831–32
  as ecosystem supporting service, 1188
  evolution of, 830–31
  fertilization after, 832–33
  flower petals and, 826–27
  photoreception of UV light in, 964
  phylogenetic tree of, 831*f*
  in plant development, 433
  self-fertilization vs. cross-fertilization in, 258 (*see also* Crosses; Self-fertilization)
  selfing vs. outcrossing in, 830
  syndromes, 830
Pollination syndromes, **594**, **830**, 831*f*
Pollinators, 593–94, 608–9*f*, 831–32
Pollution
  biomagnification of pollutants in, 1152–53
  bioremediation of, using fungi, 632
  bioremediation of, using prokaryotes, 533
  ecology and, 1059–61
  nitrogen, 1160
  nutrient, 1148*f*
  prokaryotes and nitrate, 543–44
  soil, 785–87
  as threat to biodiversity, 1178, 1181*t*, 1182
Polychaetes, 667
Polygenic inheritance, 275–**76**, 277*t*, 458–59*f*
Polymerase chain reaction (PCR), **374**
  amplification of fossil DNA with, 376
  in DNA fingerprinting, 397–98*f*
  extremophiles and, 531
  in genetic engineering, 381*t*
  in next-generation sequencing technologies, 378
  requirements of, 374–76
  uses of, in genetic testing, 382
Polymerases. *See* DNA polymerases; RNA polymerase
Polymerization, **44**
  of amino acids to form proteins, 44–47
  chemical evolution and, 78
  of monosaccharides to form polysaccharides, 75–76
  of nucleotides to form nucleic acids, 59–61
  in origin-of-life experiments, 68–69

Polymers, **44**. *See also* Polymerization
  polysaccharides as, 72, 75–78
  proteins as, 44–47
Polymorphic genetic markers, **379**
Polymorphic species, **491**
Polymorphic traits, **272**, 277*t*, 485
Polypeptides, **46**. *See also* Proteins
  as hormones, 995–96, 999
  peptide bonds in formation of proteins and, 45–46
  synthesis of, by ribosomes, 119
  synthesis of, in translation, 324–26, 328–31 (*see also* Translation)
Polyphenylalanine, 311
Polyphyletic groups, 508*t*
Polyplacophorans, 669
Polyploid organisms, **239**, 240*t*
Polyploidy, 314, **497**–99
Polyps, **653**–54
Polyribosomes, 324*f*–**25**
Polysaccharides, **72**–73, 75–81, 117, 202–3
Poly(A) signal, **320**
Poly(A) tail, mRNA, **323**–24
Polyspermy prevention, 421–23
Polytomy, 506*t*, **618**, B:10
Polyunsaturated fats, 85, 86
Ponds, **1077**
Population(s), **5**, **446**, **466**, **1061**, **1102**
  analyzing changes in frequencies of alleles in, 466–70
  conservation of, 1189–91
  contemporary evolution of, 449
  contemporary speciation in, 451
  evolution of, by natural selection, 2, 5–6, 444, 446, 459 (*see also* Evolution; Natural selection)
  genetic diversity of, 1173–74
  genetic drift in, 478, 479, 480*f*, 481–82
  genetic variation in, 472, 475*t*
  growth of (*see* Population growth)
  Hardy–Weinberg principle and, 466
  human (*see* Human population)
  monophyletic, 491–92
  in phylogenetic trees, B:10–B:11
  population ecology and, 1060*t*, 1061, 1102 (*see also* Population ecology)
  population viability analysis for, 1182, 1190
  size and structure of (*see* Demography)
  speciation in, 494–502
Population cycles, 1113–15
Population density, 216, 1015, **1102**–3, 1108–12
Population dynamics, **1112**–15
Population ecology, **1102**. *See also* Population(s)
  in conservation of biodiversity, 1118–20
  demography in, 1103–7
  geographic distribution and abundance of species in, 1102–3
  human population growth in, 1115–18
  population dynamics in, 1112–15
  population growth in, 1107–12
  populations and, 1060*t*, 1061, 1101–2
Population growth, 1106*f*, 1107–12, 1118, 1119*f*
Population pyramids, 1115–16
Population thinking, **446**
Population viability analysis (PVA), 1182, 1190
Pore-like proteins, 98
Pores, **584**–85, **744**
Porifera, 638–41, 652
Positive control, **341**, 344, 345*f*

Positive feedback, **935**, 1164–66
Positive pressure ventilation, **909**
Positive-sense single-stranded RNA viruses, **716**, 727
Post-anal tails, 686
Posterior body axis, **410**
Posterior pituitary, **1004**, 1005–6
Postsynaptic neurons, **939**
Postsynaptic potentials, 940–42
Post-transcriptional control, bacterial, 360–61
Post-transcriptional control, eukaryotic, 356–61
Post-translational control, 337–38*f*
Post-translational modification, 360, 361*t*
Postzygotic isolation, **490**, 491*t*, 499–500
Potassium
  channels for, 98, 931, 934, 935
  electrolytes as, 862
  as human nutrient, 885*t*
  as plant nutrient, 777
  potassium leak channels, 931
Potatoes, 554, 555*t*, 574, 815
Potential energy, **31**, **137**
  aerobic respiration and, 171
  in ATP, 143–44, 154–55
  in chemical reactions, 31–32, 137–39
  phosphate functional group and, 60
  in storage polysaccharides, 80
  water potential and, 755
Prairie dogs, 1095–97
Prairies, 1071, 1191
Prebiotic soup, 57–61, 68–69, 93–94
Prebiotic soup model, chemical evolution, 33–35, 41–42, 44–45, 55
Precambrian, **513**–14
Precipitation
  global climate change and, 1073, 1163, 1166
  in global water cycle, 1159
  rain shadows and, 1066–67
  in terrestrial biomes, 1068–72
  in tropical wet forests, 1065–66
Precision
  accuracy vs., B:2–B:3
  standard error and, B:7
Predation, **1128**. *See also* Predators
  in animal diversification, 646
  by ascomycete fungi, 632
  deceitful communication and, 1094–95
  efficiency of animal, 1130–31
  foraging and, 1086–87
  by keystone species, 1137–38
  lateral line system in fishes and, 958–59
  in population cycles, 1114–15
  as threat to biodiversity, 1178
  top-down control in, and trophic cascades, 1153, 1154*f*
  in triggering of Cambrian explosion, 519–20
Predators, **647**. *See also* Predation
  animals as, 647–48
  deuterostomes as, 681*f*
  echinoderm, 685
  noxious compounds in vacuoles and, 115
  predator-removal programs and, 1130–31
  sharks as, 698
  top-down control by, and trophic cascades, 1153, 1154*f*
Predictability, community structure, 1135–37
Predictions, **3**
  biodiversity, 1182–84
  experimental design and, 13

Boldface page numbers indicate a glossary entry; page numbers followed by an *f* indicate a figure; those followed by *t* indicate a table.

Boldface page numbers indicate a glossary entry; page numbers followed by an *f* indicate a figure; those followed by *t* indicate a table.

Boldface page numbers indicate a glossary entry; page numbers followed by an *f* indicate a figure; those followed by *t* indicate a table.

I:34 INDEX

hemoglobin, 51*t*
  protein, 48–49
  RNA, 65–66
  tRNA, 327
Secondary succession, **1139**–40
Secondary xylem, 748–50
Secretin, **893**, 995, **996**–97
Secretions, barrier, 1038–39
Secretory pathway, endomembrane system, 121–26
Secretory vesicles, 123
Sedimentary rocks, **446**, 447*f*
Sediments, fossils and, 512
Seed banks, **1189**
Seed coat, **833**
Seed plants. *See also* Angiosperms; Gymnosperms
  characteristics of, 581
  evolution of, 831
  heterospory in, 590, 591*f*, 592*f*
  lineages of, 599, 606–9*f*
  molecular phylogeny of, 583
  pollen grains in, 590–91
  seeds in, 591–92
Seeding, bioremediation with, 533
Seedless vascular plants
  characteristics of, 581
  homospory in, 590, 591*f*
  lineages of, 598–99, 603–5
  molecular phylogeny of, 583
Seeds, **433**, 581, **823**, 833–37
  creation of, in embryogenesis, 433–36
  dormancy of, 836–37
  embryogenesis and, 833–34
  fruit development and dispersal of, 834–36
  germination of, 832, 836–37
  gibberellins and abscisic acid in dormancy and germination of, 809–10
  ground tissue, sclereids, and, 746
  in plant reproduction, 591–92, 823
  red/far-red light reception and germination of, 800–801
  role of drying in maturation of, 834
  seed coat and, 833
  seed plants and, 581, 599
  vacuoles in, 114
Segment identity genes. *See* Hox genes
Segment polarity genes, 412, 413*f*, 414*f*
Segmentation, **645**–46
Segmentation genes, **412**, 413*f*
Segmented bodies, arthropod, 670–71
Segmented worms, 638*t*, 662, 667
Segments, embryo, **412**
Segregation, principle of, **262**
  meiosis as explanation for, 266–67
  particulate inheritance and, 261–63
  Punnett squares and, 262–63
Selective adhesion, **206**–7
Selective breeding, 257. *See also* Crosses
Selective permeability, plasma membrane, **89**, **862**. *See also* Permeability, plasma membrane
  importance of, to cells, 84, 107
  intracellular environments and, 100–101
  ion channel proteins and, 98
  lipid bilayers and, 89
  osmoregulation and, 862
  osmosis and, 92–93
Self-fertilization, **248**–49, **258**, 470–71*f*, 497–98, **830**, B:25–B:26
Self molecule, **1046**–47
Self recognition, adaptive immune response and, 1042, 1046–47

Self-replicating molecules
  DNA as, 64–65
  lipid bilayers and, 93–94
  research on proteins as first, 42, 44–45, 55
  RNA world hypothesis on RNA as first, 57–58, 68–69
Self-sacrificing behavior, 461. *See also* Altruism
Selfish DNA sequences, 396, 397*f*
Selfish genes, 461
Semen, **1022**–23*t*
Semiconservative replication, DNA, **287**–89
Seminal vesicles, **1022**–23*f*
Senescence, plant, 812–13
Sensors, **854**
Sensory neurons, **929**
Sensory organs, animal, 646, 647*t*
Sensory receptors, homeostatic, 1001
Sensory systems, animal, 952–71
  chemoreception of chemicals in, 964–67
  electroreception of electric fields in, 967–68
  information transmission to brains in, 954
  magnetoreception of magnetic fields in, 968–69
  mechanoreception of pressure changes in, 954–59
  photoreception of light in, 959–64
  processes and receptors in, 953
  senses and sensory organs in, 646, 647*t*
  in sensing of environmental changes, 952–53
  sensory transduction in, 953–54
  thermoreception of temperature in, 967
Sensory systems, plant, 793–821
  defense responses to pathogens and herbivores, 815–19
  environments and, 793–94
  germination, stem elongation, and flowering responses to red and far-red light, 800–803
  gravitropic responses to gravity, 803–5
  growth regulators, 814*t*
  information processing and, 794–96
  phototropic responses to blue light, 796–800
  responses to wind and touch, 805–6
  selected, 818*t*
  youth, maturity, and aging as growth responses, 806–15
Sepals, **439**, 593, **826**
Separation
  of cell components, B:17–B:18
  of molecules, B:13–B:15
Septa, **616**
Sequences, amino acid, 46, 47–48
Sequences, DNA. *See* DNA sequences
Sequences, rRNA, 6–8
Sequencing technology, 389*f*, 391, 1173. *See also* Direct sequencing; DNA sequencing; Metagenomics (environmental sequencing); Whole-genome sequencing
Serotonin, 941*t*, **947**
Serum, blood, 233, B:21
Services, ecosystem. *See* Ecosystem services
Sessile organisms, **574**, **638**
Set point, **854**
Severe combined immunodeficiency (SCID), 383–84, 1056

Sex chromosomes, **238**, 240*t*, 268–69, 277*t*
Sex hormones
  binding of, to intracellular receptors, 1006–8*f*
  contraception and, 1029–30
  control of menstrual cycle by, 1027–30
  control of puberty by, 1026–27
  mammalian, 1025–30
  in mating, 1087–88
  testosterone and estradiol as steroids and, 1025–26
  in vertebrate sexual development and activity, 998–99
Sex-linked inheritance (sex-linkage), **268**
  in chromosome theory of inheritance, 268–69
  exceptions and extensions to Mendelian rules on, 277*t*
  human, 278–79
  linkage vs., 270
Sex pheromones, 215–16
Sexual abstinence, 722
Sexual activity. *See* Mating
Sexual competition, 9–11
Sexual dimorphism, **477**–78
Sexual reproduction, **247**, **823**, **1014**
  animal, 650–51, 998–99, 1016–18*f* (*see also* Animal reproduction)
  asexual reproduction vs., 247, 251–52, 566, 1014
  changing-environment hypothesis on, 253
  in eukaryotes, 553
  fertilization in (*see* Fertilization)
  fungal, 617–18, 624–27
  genetic variation from, 247–49
  hormones in, 998–99
  meiosis and, 219, 237–38 (*see also* Meiosis)
  paradox of, 251–52
  plant, 586–87, 823–25 (*see also* Plant reproduction)
  protist, 566
  purifying selection hypothesis on, 252–53
  switching between asexual and, by *Daphnia*, 1014–16
Sexual selection, **472**, 475–78, **1088**–89
Sexually transmitted diseases, 546, 547, 571. *See also* AIDS (acquired immune deficiency syndrome); HIV (human immunodeficiency virus)
Shade, far-red light as, 800–801
Shade leaves, 738
Shape
  active transport pump, 99–100
  animal locomotion and body, 985, 986
  enzyme, 54, 149–50
  flowers, 593–94
  genetic control of leaf, 436–38
  membrane carrier protein, 99
  organic molecule, 36*f*, 38
  plant receptor cell, 794–95
  prokaryotic diversity in, 536, 537*f*
  protein, 47–53, 331–32
  retinal change in, 962
  shoot system, 735–36
  signal receptor, 210
  simple molecule, 23–24, B:12–B:13
  transcription factors, 355–56
  viral, 715
Shared ancestry. *See* Common ancestry
Sharks, 512, 698, 866–68, 967–68
Sharp, Phillip, 321–22

Sheep, cloned, 409
Shells, **562**
  eggs, 693–94
  mollusk, 663, 668
  turtle, 702
Shine-Dagarno sequences, **329**–30
Shocked quartz, 521–22*f*
Shoot apical meristem (SAM), **434**
  in reproductive development, 438–39
  in vegetative development, 436, 437*f*
Shoot systems, **732**
  adventitious roots and, 734–35
  characteristics of, 735
  collenchyma cells and shoot support in, 744–45
  formation of, 434
  functions of, 732
  modified stems in, 736–37
  morphological diversity of, 735–36
  organization of primary, 742
  phenotypic plasticity in, 736
  water transport in (*see* Water transport, plant)
Shoots, **434**
Short-day plants, **802**
Short tandem repeats (STRs), **396**–98*f*
Shotgun sequencing, **390**–91
Shrimp, 495, 671, 677, 1133, 1134*f*
SI system, B:1
Sickle-cell disease, 48, 278
Side chains, amino acid
  amphipathic proteins and, 94
  in enzyme catalysis, 144–47
  function of, 42–44
  in major amino acids, 43*f*
  peptide bonds and, 45–46
  polarity of, and water solubility, 44
  protein structure and, 48–50
  of steroids, 87
Sieve plates, **747**, **767**
Sieve-tube elements, **747**, **767**, 770–71
Sig figs, metric system and, B:2–B:3
Sight, 481–82, 646, 647*t*
Sigma proteins, 319–20*f*, 325*t*
Sigmoid curves, scatterplot, B:5
Signal hypothesis, endomembrane system, 123–24
Signal processing, cell–cell, 210–14
Signal receptors, **210**. *See also* Enzyme-linked receptors; G-protein-coupled receptors
Signal recognition particle (SRP), **124**
Signal response, cell–cell, 214
Signal transduction, **210**, **794**, **1008**
  cross-talk in, 813–14
  via G-protein-coupled receptors, 211–13
  in hormone binding to cell-surface receptors, 1008–10
  of lipid-insoluble signals, 210–11
  in mechanoreception, 955
  in photoreception, 962–63
  in plant cells, 794–96
  via enzyme-linked receptors, 211, 213–14
Signal transduction cascades, **1009**–10
Signaling, cell–cell. *See* Cell–cell signaling
Signaling, chemical. *See* Chemical signaling
Signaling pathways, animal hormone, 993–94
Signals, communication, **1092**
Significant figures, metric system and, B:2–B:3
Silencers, **354**

Boldface page numbers indicate a glossary entry; page numbers followed by an *f* indicate a figure; those followed by *t* indicate a table.

Sugars, **72**
  as carbohydrates, 72–73 (*see also* Carbohydrates)
  chemical evolution and nucleotide production from, 59
  dry seeds and, 834
  glucagon and blood levels of, 54
  insulin and blood levels of, 47–48
  in metabolic pathways, 157
  in nucleotides, 58–60
  photosynthetic production of, 176–78 (*see also* Photosynthesis)
  plant transport of (*see* Translocation)
  in RNA (*see* Ribose)
Sulfate, 171
Sulfhydryl functional group, 37*t*–38
Sulfur, 38, 49, 177, 324
  as human nutrient, 885*t*
  as plant nutrient, 780*t*
  in proteins vs. in DNA, 285–86
Sum rule, probability, B:8
Summation, 941–**42**
Sun, compass orientation by, 1089–90
Sun leaves, 738
Sundews, 790
Sunfish, 1094–95
Sunflowers, 489*f*, 501–2, 1082
Sunlight. *See* Light energy
Supercoiled bacterial DNA, 107–8
Supernatants, centrifugation, B:18
Support structures, protist, 562
Supporting connective tissue, 846*t*, **847**
Supporting services, ecosystem, 1187*t*, 1188
Surface area
  animal adaptations that increase, 852–53
  animal gas exchange and, 905, 916
  fungal mycelia, 616–17
Surface/area relationships, plant, 732–33
Surface area/volume, animal body, 850–53
Surface metabolism model, chemical evolution, **33**, 35–36, 45, 60
Surface tension, **26**, **762**
Surgery, cancer, 233
Survival, evolution and, 5–6, 9–11
Survivorship, **1104**
  in density-dependent logistic population growth, 1109–12
  in human population growth, 1115–18
  in life histories, 1105–7
  in life tables, 1104–5, 1106*f*
  in preservation of endangered species, 1118, 1119*f*
Survivorship curves, 1104–**5**
Suspension feeders, 648, 662*f*, **884**
Suspensor, plant, 434
Sussman, Michael, 770
Sustainability, **1189**
Sustainable agriculture, **781**
Sustainable forestry, 781
Sutton, Walter, 257, 266–67
Swallowing, 889
Swammerdam, Jan, 973
Swamps, **1077**
Sweating, 864
Sweden
  age structure of human population in, 1115–16
  maternal mortality rates in, 1034
Sweet potato hawkmoth, 908

Sweetness taste receptors, 965
Swetnam, Thomas, 1139
Swim bladder, **698**
Swimming, 109, 132–33, 617, 984, 985, 988
Symbiosis, **560**, 619–23, 723, **891**. *See also* Mutualism(s)
Symbiotic relationships, **619**, **784**, 787–89. *See also* Mutualism(s)
Symbols, chemical structure, B:12–B:13
Symmetric competition, **1125**
Symmetry, body
  animal bilateral, 641–44
  echinoderm, 682–83*f*
Sympathetic nervous system, 924, **943**, 1006
Sympathetic neurons, 977
Sympatric speciation, **495–99**
Sympatry, **495**, 500
Symplast, **208–9***f*, **760**
Symplastic route, 760–61
Symporters, **770**, **865**
Synapomorphies, 491–**92**, 507–8, 553
Synapses, **939**. *See also* Electrical signaling, animal; Nervous systems
  memory, learning, and changes in, 946–48
  neurotransmitter functions and, 938–39, 940
  neurotransmitter release and structure of, 939–40
  postsynaptic potentials and, 940–42
Synapsis, 242
Synaptic cleft, **939**
Synaptic plasticity, 946, **947–48**
Synaptic vesicles, **939**
Synaptonemal complex, **246**
Synergids, 832
Synthesis (S) phase. *See* S (synthesis) phase
Syphilis, 546
Systemic acquired resistance (SAR), **816**
Systemic circulation, **919**–21
Systemin, 210*t*, **817**
Systems, **30**–32
Systems biology, 390, **402**. *See also* Genomics
Systole, **923**
Systolic blood pressure, **923**
Szilard, Leo, 342
Szostak, Jack, 94, 296

T

T-cell receptors (TCRs), 1044*f*, **1045**–47, 1048
T cells, **1043**
  activation and function of, 1054*t*
  activation of, 1048–50
  antigen receptors of, 1045–47
  discovery of, 1042–43
  in gene therapy for severe combined immunodeficiency (SCID), 384
  HIV and helper, 713–14, 717–18
  as lymphocytes, 1042–43
  memory cells and, 1053–55
T-DNA, 385–86
T (transverse) tubules, **977**
Tabin, Clifford, 458–59
Table salt. *See* Sodium chloride (table salt)
Tagging, 324, B:15–B:16, B:20

Tagmata, **670**
Taiga, **1072**
Tails, mRNA, 323–24
Tails, post-anal, 686
Tails, vestigial, 448
Talking trees hypothesis, 817–18
Tannins, 115, 815
Tapeworms, 666
Taproots, **733**, 734–35
Taq polymerase, 374–76, 380*t*, 531
Tardigrades, 638*t*, 672–73
Target crops, agricultural biotechnology and, 385
Tarsiers, 704
Taste (gustation), 646, 647*t*, 964–65
Taste buds, **964**–65
TATA box, **320**, **354**
TATA box-binding protein (TBP), 47, **354**
Tatum, Edward, 305
Taxol drug, 607
Taxon (taxa), **8**, B:10–B:11
Taxon-specific surveys, 1175
Taxonomic bias, fossil record, 512
Taxonomic diversity, 1174–75
Taxonomy, **8**
  fossil record and taxonomic bias, 511
  scientific names for organisms in, 8–9
Taxonomy, Bloom's, B:29–B:30*t*
Taylor, Richard, 986, 987*f*
Tectorial membrane, **956**–57
Teeth
  adaptations of, 884, 885*f*
  in fossil record, 512
  mammalian, 701
  reptilian, 703
Teleosts, 699
Telomerase, **296**–97
Telomeres, **295**–97
Telophase, 225*f*, **226**, 245*f*
Telophase I, 243–44, 245*f*
Telophase II, 243*f*, 244–45
Temperate forests, **1071**, 1158–59
Temperate grasslands, **1071**, 1184–87
Temperate zone, **1071**
Temperature, **31**
  as abiotic factor in geographic distribution and abundance of organisms, 1062
  animal regulation of body (*see* Thermoregulation)
  animal sensing of, 646, 647*t*
  in aquatic biomes, 1076
  blood flow and body, 924
  in climates, 1066–67
  in decomposition rates, 1157
  effect of, on hemoglobin, 914
  effects of, on reaction rates of chemical reactions, 139–40*f*
  global climate change and, 1073, 1163, 1164–66, 1182
  in mating, 1087–88
  as measure of thermal energy, 31
  membrane permeability and, 90–91
  metric units and conversions for, B:1*t*
  in net primary production of terrestrial biomes, 1155
  optimal, for enzyme function, 148
  spontaneous chemical reactions and, 138–39
  stratification of, in oceans, 1168
  in terrestrial biomes, 1068–72
  thermoreception of, 967, 968*f*
  tropical species richness and, 1145
  of water for gas solubility, 904
Template strands, DNA, **64**, **318**, 319–20*f*

Temporal avoidance mechanisms, plant self-fertilization, 830
Temporal bias, fossil record, 512–13
Temporal lobe, **944**, 945–46
Temporal prezygotic isolation, 491*t*
Tendon tissue, 846*t*
Tendons, **981**–82, 984, 986
Tendrils, 739
Tension-resisting elements, 201
Tentacles, 649*t*, 654, **669**
Terminal node, phylogenetic tree, 506*t*, B:10–B:11
Termination phase
  enzyme catalysis, 146*f*–47
  transcription, **320**, 321*f*
  translation, 331, 332*f*
Termites, 571
Terrestrial biomes, 1068–74
  animal locomotion in, 984, 985
  animal transition from aquatic environments to, 659–60
  arctic tundra, 1072
  boreal forests, 1072
  characteristics of natural, 1068–69
  effects of global climate change on, 1073–74
  gas exchange in, 907
  human impacts on, and anthropogenic, 1069
  impacts of global climate change on net primary productivity of, 1168
  insect osmoregulation in, 869–71
  internal fertilization in, 1019
  net primary production of, 1154–56
  nutrient cycling in, 1156–59
  osmoregulation in, 864
  plant adaptations to dry and salty, 754–55, 757–58, 765–66
  plant transition from aquatic environments to, 584–95 (*see also* Land plants)
  protostomes in, 659–60
  species richness of, 1176
  subtropical deserts, 1070
  temperate forests, 1071
  temperate grasslands, 1071
  tropical wet forests, 1070
  vertebrate osmoregulation in, 871–79 (*see also* Kidneys)
Territories, **477**, **1083**–84
Tertiary consumers, 556, **1150**
Tertiary structure, **49**
  DNA vs. RNA, 66*t*
  hemoglobin, 51*t*
  protein, 49–50
  RNA, 66
  tRNA, 327
Test preparation, Bloom's taxonomy and, B:29–B:30*t*
Testcrosses, 258*t*, 265–66
Testes, **999**, **1016**
  cells of, 118
  hormones of, 994*f*, 999
  male reproductive systems and, 1021–23
  spermatogenesis in mammalian, 1016–17
  variable size of, among species, 1020
Testosterone, **999**, **1025**
  in mammalian reproduction, 1025–26
  in mating, 1087–88
  puberty and, 1026–27
  testes and, 118
Tests, statistical, B:5, B:7
Testudinians, 702
Tetrads, 240*t*, **242**

Boldface page numbers indicate a glossary entry; page numbers followed by an *f* indicate a figure; those followed by *t* indicate a table.

Boldface page numbers indicate a glossary entry; page numbers followed by an *f* indicate a figure; those followed by *t* indicate a table.

Variable number tandem repeats (VNTRs), **396**–98*f*
Variable (V) regions, **1045**–46
Variables, experimental, 13
Vas deferens, **1022**–23*t*
Vasa recta, **876**, 877*f*
Vascular bundles, **742**
Vascular cambium, **748**
  auxin and, 807
  secondary growth functions of, 748–50
Vascular plants. *See* Seedless vascular plants; Seed plants
Vascular tissue, plant, **435**, **580**, **759**. *See also* Phloem tissue; Xylem tissue
  elaboration of, into tracheids and vessel elements, 585–86
  formation of, 435, 437*f*
  in land plants, 580–81, 585
  root and shoot systems and, 732
  translocation (sugar transport) and, 766–72
  water transport and, 759–65
Vascular tissue systems, plant, **740**. *See also* Phloem tissue; Xylem tissue
  embryogenesis and, 833
  phloem structure in, 747
  primary plant body and tissues of, 740–41, 742, 748*t*
  secondary growth of, 748–51
  xylem structure in, 746–47
Vectors, cloning, **370**. *See also* Plasmids
Vegetation, emergent, 1077
Vegetative development, **433**, 436–38
  flexibility of cell determination in, 438
  genetic control of leaf shape in, 436–38
  meristems in lifelong growth and, 436, 437*f*
  reproductive development vs., 433 (*see also* Reproductive development, plant)
Veins, 871, **917**–18, 921, 924
Veliger, **668**
Velvet worms, 672–73
Venae cavae, **921**
Venter, Oscar, 1178
Ventilation. *See also* Respiratory systems
  of fish gills, 906
  in gas exchange, 903
  homeostatic control of, 911–12
  of insect tracheae, 907
  of vertebrate lungs, 909–11*f*
Ventral body axis, **410**
Ventricles, **919**, 921
Venules, **917**–18
Venus flytraps, 790, 806
Vertebrae, **688**
Vertebral column, 688
Vertebrates, **645**, **682**, 688–703. *See also* Animal(s)
  adaptive immunity in, 1042 (*see also* Adaptive immune response; Immune systems)
  amphibians, 696
  body plan of, 686–87
  closed circulatory systems of, 917 (*see also* Circulatory systems)
  as deuterostome chordates, 682
  endoskeletons of, 981–83
  evolution of, 688–91
  fishes, 696
  in fossil record, 689–91
  hormones in sexual development and activity of, 998–99
  key lineages of, 696–703
  lungs of, 909–12 (*see also* Respiratory systems)

mammals, 696
molecular phylogenies of, 691
morphological innovations of, 692–93
nervous systems of, 942–49 (*see also* Nervous systems)
osmoregulation in terrestrial, 871–79 (*see also* Kidneys)
phylogenetic tree of, 690*f*
primates, hominins, and humans, 704–8 (*see also* Human[s])
relative abundance of species of, 691*f*
reproductive innovations of, 693–94
reptiles, 696–97
segmentation in, 645
sensory systems of (*see* Sensory systems, animal)
vertebrae, cranium, and brain structure of, 688
wings and flight in, 694–95
Vervet monkeys, 1098
Vesicles
  artificial membrane-bound, 88–89
  in chemical evolution, 94
  in cytokinesis, 228
  in electron transport chain, 168–69
  in endomembrane protein transport, 123–27
  microtubules and transport by, 130–31
  osmosis across membrane-bound, 92–93
  prokaryotic, 108
  rough ER in, 114
  secretory, 123
  synaptic, 939
Vessel elements, **586**, **746**–47
Vessels, angiosperm, 596
Vestigial traits, **448**
  of cetaceans, 453
  as evidence for evolutionary change, 448–49
  as nonadaptive, 461
Vicariance, **494**, 495
Video microscopy
  of ATP synthase, 170
  in biological imaging, B:20
  of transport vesicles, 130*f*
Villi, **852**–53*f*, **892**
Virchow, Rudolf, 3, 219
Virions, **714**–15. *See also* Replicative growth, viral
Virulent infections, 532, **712**
Virus vaccines, 1055
Viruses, **711**–30
  abundance and diversity of, 712
  analyzing coexistence of, with host cells, 721–22
  analyzing genetic material of, 715–16
  analyzing morphological traits of, 715
  analyzing phases of replicative growth of lytic cycle of, 716–21
  bacteriophage λ, 50
  Baltimore classification system for, 725–26
  biological methods of studying, 714–22
  biological reasons for studying, 712–14
  characteristics of living organisms vs. characteristics of, 712*t*
  diversification themes of, 722–25
  emerging diseases and emerging, 723–25
  eukaryotic translation and, 359–60
  in gene therapy, 383–85
  Hershey–Chase experiment on DNA in genes of, 285–86

human epidemics and pandemics from, 712–14
identification of emerging, 725
infection of host cells by, 711–12
key lineages of, 725–28
living organisms vs., 711–12
nanobiology in isolation of, 714
origins of, 723
photomicrograph of, 711*f*
proteins as defense against, 54
RNA in, 309
in splitting eukaryotic genes, 321–22
vaccines against, 1055
Visceral mass, **662**–63
Viscosity, animal locomotion and fluid, 988
Visible light, **179**
Vision, animal. *See* Photoreception, animal; Sight
Visual communication, 1093
Visual cues, mating and, 1088
Visualization, molecular, B:15–B:16
Vitamin A deficiency, 385
Vitamin B$_1$, 884*t*
Vitamin B$_6$, 305
Vitamin B$_{12}$, 884*t*
Vitamin C, 884*t*
Vitamin D, 884*t*
Vitamin D-resistant rickets (hypophasphatemia), 279
Vitamins, **883**
  coenzymes and, 147
  as essential nutrients for animals and humans, 883, 884*t*
  lipids as, 87
  transgenic crops and deficiencies of, 385
Vitelline envelope, **1017**–18*f*
Viviparous species, **651**, **694**, **1020**–21
Vivipary, 836
Volcanic gases, 24, 30, 34–35, 45
Volt, **930**
Voltage, **783**, **930**
Voltage clamping, **935**
Voltage-gated channels, 98–99, **934**
Volume, in chemical reactions, 138
Volume, metric units and conversions for, B:1*t*
Volume/surface area, animal body, 850–53
Volume/surface area relationships, plant, 732–33
Voluntary muscles, **977**, 979
Vomeronasal organ, **967**
Von Békésy, Georg, 956–57
Von Frisch, Karl, 1084, 1092–93
von Linné, Karl, 9

W

Waggle dance, honeybee, 1092–93
Waldeyer, Wilhelm, 220
Walking, 986
Wall cress. *See* Mustard plant
Wall pressure, **756**
Wallace, Alfred Russel, 5, 444, 446, 453, 1063
Wallace line, **1063**
Warblers, 1062
Wasps, 485, 818–19
Wasserman, Paul, 79–80
Wastes, 883*f*, 888, 896. *See also* Nitrogenous wastes, animal
Water
  absorption of, in large intestines, 896
  absorption of, in small intestines, 896

in acid-base reactions, 28–30
angiosperm vessels for conducting, 596
animal heat exchange by evaporation of, 855
animal locomotion in, 984, 985, 988
animal requirements for, 842–43
aquaporins and transport of, across plasma membranes, 98
behavior of oxygen and carbon dioxide in, 904–5
cohesion, adhesion, and surface tension of, 26
covalent bonds of, 21–23
in decomposition rates, 1157
density of, as liquid and as solid, 26–27
depth and flow of in aquatic biomes, 1074–75, 1077–79
efficiency of, as solvent, 25–26
energy absorbing capacity of, 27–28
green plant holding of, and moderation of climate, 578
insolubility of lipids in, 85, 86–87
land plant adaptations to prevent loss of, 584–85
large intestine absorption of, 888
lipid bilayers and, 89
molecular structure of, 24*f*, B:12*f*
in net primary production of terrestrial biomes, 1155
nitrate pollution of, 543–44
noncyclic electron flow between, and NADP$^+$ in Z-scheme model, 188–89
in nutrient cycling, 1158
osmosis as diffusion of, across lipid bilayers, 92–93
oxidation of, in photosystem II, 186
photosynthetic oxygen production from, 177–78
plant nutrients from, 780*t*
plant transport of (*see* Water transport, plant)
polarity of amino acid side chains and solubility in, 44
production of, by cellular respiration, 165
production of, by electron transport chain, 166*f*–67
properties of, 26–28
purification of, as ecosystem regulating service, 1188
in seed germination, 837
specific heat of, 27*t*
Water balance, animal. *See* Osmoregulation
Water bears, 672–73
Water cycle, global, 1159–60, 1166
Water depth, 1074, 1075*f*, 1077–79
Water flea, 1167
Water flow, 1074–75, 1077–79
Water loss, animal
  terrestrial adaptations to prevent, 662–63
  insect minimization of, 869–70
Water loss, plant
  biochemical pathway adaptations to, 766
  regulation of, 754–55, 765–66
  seed maturation and, 834
  stomata and regulation of, 743–44
  turgor pressure, wilting, and plant, 758
Water molds, 574

Boldface page numbers indicate a glossary entry; page numbers followed by an *f* indicate a figure; those followed by *t* indicate a table.